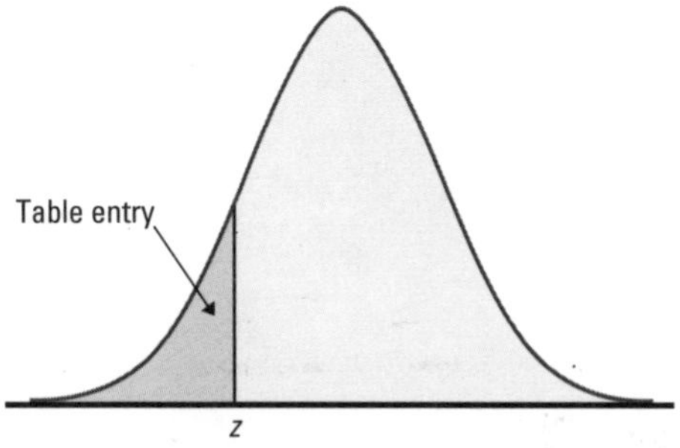

Table entry for *z* is the area under the standard normal curve to the left of *z*.

Table A *Standard Normal probabilities*

z	.00	.01	.02	.03	.04	.05	.06	.07	.08	.09
−3.4	.0003	.0003	.0003	.0003	.0003	.0003	.0003	.0003	.0003	.0002
−3.3	.0005	.0005	.0005	.0004	.0004	.0004	.0004	.0004	.0004	.0003
−3.2	.0007	.0007	.0006	.0006	.0006	.0006	.0006	.0005	.0005	.0005
−3.1	.0010	.0009	.0009	.0009	.0008	.0008	.0008	.0008	.0007	.0007
−3.0	.0013	.0013	.0013	.0012	.0012	.0011	.0011	.0011	.0010	.0010
−2.9	.0019	.0018	.0018	.0017	.0016	.0016	.0015	.0015	.0014	.0014
−2.8	.0026	.0025	.0024	.0023	.0023	.0022	.0021	.0021	.0020	.0019
−2.7	.0035	.0034	.0033	.0032	.0031	.0030	.0029	.0028	.0027	.0026
−2.6	.0047	.0045	.0044	.0043	.0041	.0040	.0039	.0038	.0037	.0036
−2.5	.0062	.0060	.0059	.0057	.0055	.0054	.0052	.0051	.0049	.0048
−2.4	.0082	.0080	.0078	.0075	.0073	.0071	.0069	.0068	.0066	.0064
−2.3	.0107	.0104	.0102	.0099	.0096	.0094	.0091	.0089	.0087	.0084
−2.2	.0139	.0136	.0132	.0129	.0125	.0122	.0119	.0116	.0113	.0110
−2.1	.0179	.0174	.0170	.0166	.0162	.0158	.0154	.0150	.0146	.0143
−2.0	.0228	.0222	.0217	.0212	.0207	.0202	.0197	.0192	.0188	.0183
−1.9	.0287	.0281	.0274	.0268	.0262	.0256	.0250	.0244	.0239	.0233
−1.8	.0359	.0351	.0344	.0336	.0329	.0322	.0314	.0307	.0301	.0294
−1.7	.0446	.0436	.0427	.0418	.0409	.0401	.0392	.0384	.0375	.0367
−1.6	.0548	.0537	.0526	.0516	.0505	.0495	.0485	.0475	.0465	.0455
−1.5	.0668	.0655	.0643	.0630	.0618	.0606	.0594	.0582	.0571	.0559
−1.4	.0808	.0793	.0778	.0764	.0749	.0735	.0721	.0708	.0694	.0681
−1.3	.0968	.0951	.0934	.0918	.0901	.0885	.0869	.0853	.0838	.0823
−1.2	.1151	.1131	.1112	.1093	.1075	.1056	.1038	.1020	.1003	.0985
−1.1	.1357	.1335	.1314	.1292	.1271	.1251	.1230	.1210	.1190	.1170
−1.0	.1587	.1562	.1539	.1515	.1492	.1469	.1446	.1423	.1401	.1379
−0.9	.1841	.1814	.1788	.1762	.1736	.1711	.1685	.1660	.1635	.1611
−0.8	.2119	.2090	.2061	.2033	.2005	.1977	.1949	.1922	.1894	.1867
−0.7	.2420	.2389	.2358	.2327	.2296	.2266	.2236	.2206	.2177	.2148
−0.6	.2743	.2709	.2676	.2643	.2611	.2578	.2546	.2514	.2483	.2451
−0.5	.3085	.3050	.3015	.2981	.2946	.2912	.2877	.2843	.2810	.2776
−0.4	.3446	.3409	.3372	.3336	.3300	.3264	.3228	.3192	.3156	.3121
−0.3	.3821	.3783	.3745	.3707	.3669	.3632	.3594	.3557	.3520	.3483
−0.2	.4207	.4168	.4129	.4090	.4052	.4013	.3974	.3936	.3897	.3859
−0.1	.4602	.4562	.4522	.4483	.4443	.4404	.4364	.4325	.4286	.4247
−0.0	.5000	.4960	.4920	.4880	.4840	.4801	.4761	.4721	.4681	.4641

(Continued)

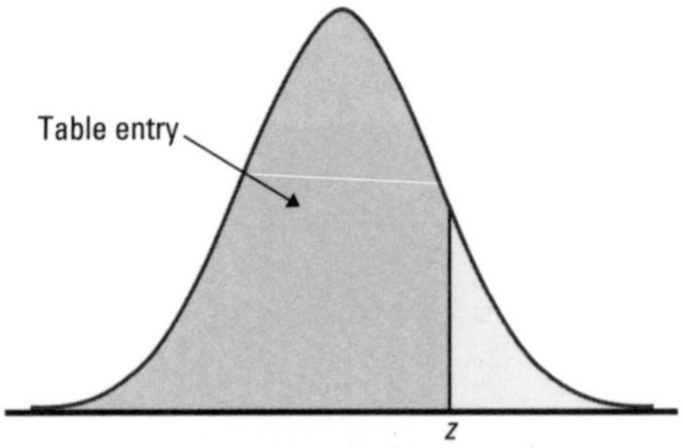

Table entry for z is the area under the standard normal curve to the left of z.

Table A Standard Normal probabilities (continued)

z	.00	.01	.02	.03	.04	.05	.06	.07	.08	.09
0.0	.5000	.5040	.5080	.5120	.5160	.5199	.5239	.5279	.5319	.5359
0.1	.5398	.5438	.5478	.5517	.5557	.5596	.5636	.5675	.5714	.5753
0.2	.5793	.5832	.5871	.5910	.5948	.5987	.6026	.6064	.6103	.6141
0.3	.6179	.6217	.6255	.6293	.6331	.6368	.6406	.6443	.6480	.6517
0.4	.6554	.6591	.6628	.6664	.6700	.6736	.6772	.6808	.6844	.6879
0.5	.6915	.6950	.6985	.7019	.7054	.7088	.7123	.7157	.7190	.7224
0.6	.7257	.7291	.7324	.7357	.7389	.7422	.7454	.7486	.7517	.7549
0.7	.7580	.7611	.7642	.7673	.7704	.7734	.7764	.7794	.7823	.7852
0.8	.7881	.7910	.7939	.7967	.7995	.8023	.8051	.8078	.8106	.8133
0.9	.8159	.8186	.8212	.8238	.8264	.8289	.8315	.8340	.8365	.8389
1.0	.8413	.8438	.8461	.8485	.8508	.8531	.8554	.8577	.8599	.8621
1.1	.8643	.8665	.8686	.8708	.8729	.8749	.8770	.8790	.8810	.8830
1.2	.8849	.8869	.8888	.8907	.8925	.8944	.8962	.8980	.8997	.9015
1.3	.9032	.9049	.9066	.9082	.9099	.9115	.9131	.9147	.9162	.9177
1.4	.9192	.9207	.9222	.9236	.9251	.9265	.9279	.9292	.9306	.9319
1.5	.9332	.9345	.9357	.9370	.9382	.9394	.9406	.9418	.9429	.9441
1.6	.9452	.9463	.9474	.9484	.9495	.9505	.9515	.9525	.9535	.9545
1.7	.9554	.9564	.9573	.9582	.9591	.9599	.9608	.9616	.9625	.9633
1.8	.9641	.9649	.9656	.9664	.9671	.9678	.9686	.9693	.9699	.9706
1.9	.9713	.9719	.9726	.9732	.9738	.9744	.9750	.9756	.9761	.9767
2.0	.9772	.9778	.9783	.9788	.9793	.9798	.9803	.9808	.9812	.9817
2.1	.9821	.9826	.9830	.9834	.9838	.9842	.9846	.9850	.9854	.9857
2.2	.9861	.9864	.9868	.9871	.9875	.9878	.9881	.9884	.9887	.9890
2.3	.9893	.9896	.9898	.9901	.9904	.9906	.9909	.9911	.9913	.9916
2.4	.9918	.9920	.9922	.9925	.9927	.9929	.9931	.9932	.9934	.9936
2.5	.9938	.9940	.9941	.9943	.9945	.9946	.9948	.9949	.9951	.9952
2.6	.9953	.9955	.9956	.9957	.9959	.9960	.9961	.9962	.9963	.9964
2.7	.9965	.9966	.9967	.9968	.9969	.9970	.9971	.9972	.9973	.9974
2.8	.9974	.9975	.9976	.9977	.9977	.9978	.9979	.9979	.9980	.9981
2.9	.9981	.9982	.9982	.9983	.9984	.9984	.9985	.9985	.9986	.9986
3.0	.9987	.9987	.9987	.9988	.9988	.9989	.9989	.9989	.9990	.9990
3.1	.9990	.9991	.9991	.9991	.9992	.9992	.9992	.9992	.9993	.9993
3.2	.9993	.9993	.9994	.9994	.9994	.9994	.9994	.9995	.9995	.9995
3.3	.9995	.9995	.9995	.9996	.9996	.9996	.9996	.9996	.9996	.9997
3.4	.9997	.9997	.9997	.9997	.9997	.9997	.9997	.9997	.9997	.9998

Teacher's Edition
0716777037
9780716777038

Publisher: Craig Bleyer
Executive Editor: Ruth Baruth
Associate Acquisitions Editor: Laura Hanrahan
Developmental Editor: Shona Burke
Senior Media Editor: Roland Cheyney
Marketing Manager: Cindi WeissGoldner
Associate Editor: Brendan Cady
Editorial Assistant: Laura Capuano
Photo Editor: Bianca Moscatelli
Photo Researcher: Brian Donnelly
Cover Designer: Vicki Tomaselli
Text Designer: Michael Stratton
Project Editors: Rebecca Dodson/Mary Louise Byrd
Cover and Interior Illustrations: Mark Chickinelli
Illustrations: Network Graphics
Illustration Coordinator: Bill Page
Production Manager: Julia DeRosa
Composition: Black Dot Group
Printing and Binding: Quebecor World Book Group

Library of Congress Control Number: 2006937045

ISBN-13: 978-0-7167-7309-2
ISBN-10: 0-7167-7309-0

Printed in the United States of America

Second Printing

W. H. Freeman and Company
41 Madison Avenue
New York, NY 10010
Houndmills, Basingstoke RG21 6XS, England
www.whfreeman.com

Third Edition

The Practice of Statistics

TI-83/84/89 Graphing Calculator Enhanced

Daniel S. Yates
Statistics Consultant

David S. Moore
Purdue University

Daren S. Starnes
The Lawrenceville School

W. H. Freeman and Company
New York

Contents

About the Authors

Daniel S. Yates has taught Advanced Placement Statistics in the Electronic Classroom (a distance learning facility) affiliated with Henrico County Public Schools in Richmond, Virginia. Prior to high school teaching, he was on the mathematics faculty at Virginia Tech and Randolph-Macon College. He has a Ph.D. in Mathematics Education from Florida State University. He has served as President of the Greater Richmond Council of Teachers of Mathematics and the Virginia Council of Teachers of Mathematics. Named a Tandy Technology Scholar in 1997, Dan is a 2000 recipient of the College Board/Siemens Foundation Advanced Placement Teaching Award. Although recently retired from classroom teaching, he stays in step with trends in teaching by frequently conducting College Board workshops for new and experienced AP Statistics teachers and by monitoring and participating in the AP Statistics electronic discussion group. Dan is coauthor of *Statistics Through Applications*, a popular textbook for non–AP Statistics classes.

David S. Moore is Shanti S. Gupta Distinguished Professor of Statistics at Purdue University and 1998 President of the American Statistical Association. David is an elected fellow of the American Statistical Association and of the Institute of Mathematical Statistics and an elected member of the International Statistical Institute. He has served as program director for statistics and probability at the National Science Foundation.

David has devoted his attention to the teaching of statistics. He was the content developer for the Annenberg/Corporation for Public Broadcasting college-level telecourse *Against All Odds: Inside Statistics* and for the series of video modules *Statistics: Decisions Through Data*, intended to aid the teaching of statistics in schools. He is the author of influential articles on statistics education and of several leading textbooks, including *Introduction to the Practice of Statistics* (written with George P. McCabe), *The Basic Practice of Statistics*, and most recently, *The Practice of Business Statistics*.

Daren S. Starnes is Master Teacher in Mathematics at the Lawrenceville School near Princeton, New Jersey. He earned his M.A. in Mathematics from the University of Michigan. In 1997, he received a GTE GIFT Grant to integrate AP Statistics and AP Environmental Science. He was named a Tandy Technology Scholar in 1999. Daren has led numerous one-day and weeklong AP Statistics institutes for new and experienced AP teachers, and he has been a reader, table leader, and question leader for the AP Statistics exam. In 2001–2002, he served as coeditor of the Technology Tips column in the NCTM journal *The Mathematics Teacher*. In January 2004, Daren was appointed to a three-year term as a member of the ASA/NCTM Joint Committee on the Curriculum in Statistics and Probability. More recently, he served as Chair of the Western Regional Council of the College Board. Daren is also coauthor of *Statistics Through Applications*, a new textbook for non–AP high school statistics.

Preface

The Practice of Statistics: TI-83/84/89 Graphing Calculator Enhanced (*TPS*), Third Edition, is an introductory text that focuses on data and statistical reasoning. It is intended for high school, college, and university students whose primary technological tool is the TI-83, TI-84, or TI-89 graphing calculator. This book is based on the successful college textbooks *The Basic Practice of Statistics* (*BPS*) by David Moore and *Introduction to the Practice of Statistics* (*IPS*) by David Moore and George McCabe. *The Practice of Statistics* was the first book written specifically for the College Board AP[1] Statistics course.

Statisticians have reached general consensus about the nature of a modern introductory statistics course. A joint committee of the American Statistical Association and the Mathematical Association of America summarized this consensus as follows:

- Emphasize statistical thinking
- Present more data and concepts with less theory and fewer formulas
- Foster active learning

More recently, the PreK–12 Report of the Guidelines for Assessment and Instruction in Statistics Education (GAISE) Project has emphasized exactly the same themes. The College Board developed its Advanced Placement Statistics syllabus around these guidelines. As a result, this text closely follows the AP syllabus and helps prepare students for the AP exam.

The Practice of Statistics takes advantage of the simulation, graphing, and computation capabilities of the TI-83/84/89 to promote active learning. By using the TI-83/84/89, students can focus on statistical concepts rather than on calculations. Although the book is elementary in the level of mathematics required and in the statistical procedures presented, it aims to give students an understanding of the main ideas of statistics and useful skills for working with data. Examples and exercises, though intended for beginners, use real data and give enough background to allow students to consider the meaning of their calculations.

The Third Edition

In revising *The Practice of Statistics,* we have attempted to build on the elements that made the first two editions successful:

- Clear explanations of statistical ideas and terminology.
- Lots of interesting examples and exercises. Highlighted titles emphasize the variety and abundance of real-life applications.

[1] © 1996 by College Entrance Examination Board and Education Testing Service. All rights reserved. College Board, Advanced Placement Program, AP, College Explorer, and the acorn logo are registered trademarks of the College Entrance Examination Board. The College Entrance Examination Board was not involved in the production of, and does not endorse, this product.

- Essential content for the AP exam.

- Activities and simulations to motivate statistical concepts.
- Technology Toolboxes show students how to use the statistical capabilities of the TI-83/84/89 and provide tips for interpreting standard computer output.

- Inference Toolbox introduced in Chapter 10 as a way for students to organize their inference procedures. Examples that use this structure are marked with a toolbox icon.

On the basis of classroom experiences, input from users of the second edition, and a careful review of recent AP Statistics exam questions, we have made some important content-related and organizational changes. Here is a summary:

- NEW **Preliminary Chapter** Gives students an overview of what statistics is all about.
- **Chapter 2** NEW Section 2.1 focuses on measures of relative standing within a distribution (percentiles and z-scores).
- **Chapter 3** Scatterplots and correlation combined in Section 3.1; linear regression content consolidated in Section 3.2; cautions about correlation and regression moved from Chapter 4 into new Section 3.3.
- **Chapter 4** Section 4.1 rewritten to emphasize transformations involving powers as well as logarithms for achieving linearity; material on monotonic functions deleted; relationships in categorical data shifted to Section 4.2; Section 4.3 devoted to the topic of establishing causation.
- **Chapter 5** Material on simulations moved to Chapter 6.
- **Chapter 6** NEW Section 6.1 uses simulations as a tool for estimating probabilities; Section 6.2 on Probability Models incorporates the idea of probability (old Section 6.1).
- **Chapter 7** Combining variances of dependent random variables deleted.

We have **reorganized the statistical inference chapters** (Chapters 10–15) to distinguish more clearly the roles of confidence intervals and significance tests. These chapters introduce the t distributions earlier and help students discriminate between problems involving means and those involving proportions.

- **Chapter 10** Focuses exclusively on constructing and interpreting confidence intervals, first (Section 10.1) for a population mean in the "ideal" case when σ is known, then (Section 10.2) in the more realistic setting when σ is unknown (using the t distributions), and finally (Section 10.3) for a population proportion. (This chapter combines parts of Sections 10.1, 11.1, and 12.1 from the second edition.)
- **Chapter 11** Addresses (Section 11.1) the underlying logic behind and (Section 11.2) the major steps involved in performing a significance test. Calculations are limited to the "ideal" case when σ is known. Section 11.3 addresses proper use and potential abuses of significance tests. Section 11.4,

Using Inference to Make Decisions, has been rewritten to emphasize interpretation of Type I and Type II errors in context, as well as the relationship between Type I error, Type II error, and the power of a significance test. (This chapter combines part of Section 10.1 with Sections 10.2 through 10.4 of the second edition.)

- **Chapter 12** Examines the details of carrying out significance tests for (Section 12.1) a population mean and (Section 12.2) a population proportion. (This chapter combines parts of Sections 11.1 and 12.1 from the second edition.)
- **Chapter 13** Shows how to compare (Section 13.1) two population means and (Section 13.2) two population proportions using confidence intervals and significance tests. (This chapter combines Sections 11.2 and 12.2 from the second edition.)
- **Chapter 14** Chi-Square Procedures was Chapter 13 in the previous edition.
- **Chapter 15** Inference for Regression (Chapter 14 in the second edition), has been condensed to a single section that includes conditions for inference about the slope of the regression model. Prediction intervals and confidence intervals for the mean response have been deleted.
- Analysis of Variance (Chapter 15 in the second edition) has been moved to the CD and Web site.

New Features

- An engaging **Case Study** begins every chapter, which students complete in the **Case Closed** at the end of the chapter.
- The **Data Analysis Toolbox** helps students organize their work for exercises involving data analysis, first introduced in Chapter 1.
- A **full-color design**, which includes color illustrations as well as **cartoons** and **photographs** that relate to specific examples or exercises, helps to stimulate student interest and enhances important concepts in the text.

- **Cautions** help students avoid common errors and misconceptions.
- **Statistical applet** icons mark examples, exercises, and activities that use applets from the book's Web site.
- **Statistical "nuggets"** in the margins briefly comment on topics of statistical interest in the world around us.
- Many new active-learning **Activities** have been added.
- Approximately one-third of the **Examples** and **Exercises**, which use current or updated data from many application areas, are new in this edition.
- **Part Review Exercises** with 10 free-response questions follow each of the four parts of the book. An additional 10 multiple-choice questions for each part have been added to the Platinum Resource Binder.

- **"Retro" exercises**—cumulative review exercises—are mixed in with other exercises and are identified in the Annotated Teacher's Edition with a special icon.
- **CrunchIt! statistical software** is available for a nominal fee through the book's companion Web site. Developed by Webster West of Texas A&M University, CrunchIt! is the easiest true statistical software for student use. Check out, for example, CrunchIt!'s flexible and straightforward process for entering data, often a real barrier to software use. CrunchIt! output is incorporated throughout the book.

Why Did You Do That?

There is no single best way to organize a presentation of statistics for beginners. That said, our choices reflect thinking about both content and pedagogy. Here are comments on several "frequently asked questions" about the order and selection of material in *The Practice of Statistics.*

Why does the distinction between population and sample not appear in Part I? This is a sign that there is more to statistics than inference. In fact, statistical inference is appropriate only in rather special circumstances. The chapters in Part I present tools and tactics for describing data—any data. These tools and tactics do not depend on the idea of inference from sample to population. Many data sets in these chapters (for example, the several sets of data about the 50 states) do not lend themselves to inference because they represent an entire population. John Tukey of Bell Labs and Princeton, the philosopher of modern data analysis, insisted that the population–sample distinction be avoided when it is not relevant. He used the word "batch" for data sets in general. We see no need for a special word, but we think Tukey is right.

Why not begin with data production? It is certainly reasonable to do so—the natural flow of a planned study is from design to data analysis to inference. But in their future employment most students will use statistics mainly in settings other than planned research studies. We place the design of data production (Chapter 5) after data analysis to emphasize that data-analytic techniques apply to any data. One of the primary purposes of statistical designs for producing data is to make inference possible, so the discussion in Chapter 5 motivates the subsequent study of probability.

Why do Normal distributions appear in Part I? Density curves such as the Normal curves are just another tool to describe the distribution of a quantitative variable, along with stemplots, histograms, and boxplots. Professional statistical software offers to make density curves from data just as it offers histograms. We prefer not to suggest that this material is essentially tied to probability, as the traditional order does. And we find it very helpful to break up the indigestible lump of probability that troubles students so much. Meeting Normal distributions early helps to do so and strengthens the "probability distributions are like data distributions" way of approaching probability.

Why not delay correlation and regression until late in the course, as is traditional? *The Practice of Statistics* begins by offering experience working with data and gives a conceptual structure for this nonmathematical but essential part of statistics. Students profit from more experience with data and from seeing the conceptual structure worked out in relations among variables as well as in describing single-variable data. Correlation and least-squares regression are very important descriptive tools and are often used in settings where there is no population–sample distinction, such as studies of all the employees of a firm. Perhaps most important, our approach asks students to think about what kind of relationship lies behind the data (confounding, lurking variables, association doesn't imply causation, and so on), without overwhelming them with the demands of formal inference methods. Inference in the correlation and regression setting is a bit complex, demands software, and often comes right at the end of the course. We find that delaying all mention of correlation and regression to that point means that students often don't master the basic uses and properties of these methods.

Why use the z procedures for a population mean to introduce the reasoning of inference? This is a pedagogical issue, not a question of statistics in practice. Sometime in the golden future we may start with resampling methods. We think that permutation tests make the reasoning of tests clearer than any traditional approach. For now the main choices are z for a mean and z for a proportion. We find z for means quite a bit more accessible to students.

Positively, we can say up front that we are going to explore the reasoning of inference in an overly simple setting. Remember, exactly Normal population and true simple random sample are as unrealistic as known σ. All the issues of practice—robustness against lack of Normality and application when the data aren't an SRS as well as the need to estimate σ—are put off until, with the reasoning in hand, we discuss the practically useful t procedures. This separation of initial reasoning from messier practice works well.

Negatively, starting with inference for p introduces many side issues: no exactly Normal sampling distribution, but a Normal approximation to a discrete distribution; use of $\hat{p}$ in both the numerator and the denominator of the test statistic to estimate both the parameter p and $\hat{p}$'s own standard deviation; loss of the direct link between test and confidence interval. Recent computational and theoretical work has demonstrated convincingly that the standard confidence intervals for proportions can be trusted only for very large sample sizes. The standard intervals often have a true confidence level much less than what was requested, and requiring larger samples encounters a maze of "lucky" and "unlucky" sample sizes until very large samples are reached. Fortunately, there is a simple cure: just add two successes and two failures to your data. We present these "plus four intervals" in the Platinum Resource Binder, along with guidelines for use.

For a detailed list of supplements available with *The Practice of Statistics*, Third Edition, go to www.whfreeman.com/tps3e.

Acknowledgments

Many people contributed significantly during the development and production of the third edition. First and foremost, we are indebted to the folks at W. H. Freeman for their ongoing support of *TPS*. We owe special thanks to Laura Hanrahan and Shona Burke for shepherding the project through to completion, keeping us on task and on deadline while maintaining focus on the overall appearance of the book. Their good judgment about many design issues—from color to photographs to font size—is evident in the finished product. Laura Capuano and Mary Louise Byrd did an outstanding job of keeping manuscript moving in the right direction at the right time during the production process. We thank Rebecca Dodson at Black Dot Group for her skillful oversight of the transmittal process. To our newest copy editor, Debbie Faust, we offer appreciation for her willingness to ask all the tough questions about our initial manuscript chapters. And to our longtime copy editor, Pam Bruton, we express sincere gratitude for another job done exceptionally well.

As we undertook major revisions in the third edition, we received invaluable statistical and authorial advice from Brad Hartlaub, who currently serves as Chief Reader for AP Statistics. Jackie Miller spent countless hours reviewing manuscript at various stages of the project. Mary Mortlock graciously agreed to write the odd-numbered answers that appear in the back of the book. *TPS* owes much of its success over these three editions to David Moore. He continues to provide us with fresh examples, data sets, and exercises, as well as contemporary thinking about the evolving discipline of statistics. His statistical and pedagogical influence shaped many of our decisions in this edition.

To our colleagues who reviewed chapters of the manuscript and offered many constructive suggestions, we give our thanks.

Christopher E. Barat, Villa Julie College, Stevenson, MD
Jason Bell, Canal Winchester High School, Canal Winchester, OH
Zack Bigner, Elkins High School, Missouri City, TX
Naomi Bjork, University High School, Irvine, CA
Robert Blaschke, Lynbrook High School, San Jose, CA
Alla Bogomolnaya, Orange High School, Pepper Pike, OH
Andrew Bowen, Grand Island Central School District, Grand Island, NY
Jacqueline Briant, Bishop Feehan High School, Attleboro, MA
Marlys Jean Brimmer, Ridgeview High School, Bakersfield, CA
Floyd E. Brown, Science Hill High School, Johnson City, TN
James Cannestra, Germantown High School, Germantown, WI
Joseph T. Champine, King High School, Corpus Christi, TX
Jared Derksen, Rancho Cucamonga High School, Rancho Cucamonga, CA
George J. DiMundo, Coast Union High School, Cambria, CA
Jeffrey S. Dinkelmann, Novi High School, Novi, MI
Ronald S. Dirkse, American School in Japan, Tokyo, Japan

Cynthia L. Dishburger, Whitewater High School, Fayetteville, GA
Michael Drake, Springfield High School, Erdenheim, PA
Mark A. Fahling, Gaffney High School, Gaffney, SC
David Ferris, Noblesville High School, Noblesville, IN
David Fong, University High School, Irvine, CA
Terry C. French, Lake Braddock Secondary School, Burke, VA
Glenn Gabanski, Oak Park and River Forest High School, Oak Park, IL
Jason Gould, Eaglecrest High School, Centennial, CO
Dr. Gene Vernon Hair, West Orange High School, Winter Garden, FL
Stephen Hansen, Napa High School, Napa, CA
Katherine Hawks, Meadowcreek High School, Norcross, GA
Gregory D. Henry, Hanford West High School, Hanford, CA
Duane C. Hinders, Foothill College, Los Altos Hills, CA
Beth Howard, Saint Edwards, Sebastian, FL
Michael Irvin, Legacy High School, Broomfield, CO
Beverly A. Johnson, Fort Worth Country Day School, Fort Worth, TX
Matthew L. Knupp, Danville High School, Danville, KY
Kenneth Kravetz, Westford Academy, Westford, MA
Lee E. Kucera, Capistrano Valley High School, Mission Viejo, CA
Christina Lepi, Farmington High School, Farmington, CT
Jean E. Lorenson, Stone Ridge School of the Sacred Heart, Bethesda, MD
Thedora R. Lund, Millard North High School, Omaha, NE
Philip Mallinson, Phillips Exeter Academy, Exeter, NH
Dru Martin, Ramstein American High School, Ramstein, Germany
Richard L. McClintock, Ticonderoga High School, Ticonderoga, NY
Louise McGuire, Pattonville High School, Maryland Heights, MO
Jennifer Michaelis, Collins Hill High School, Suwanee, GA
Dr. Jackie Miller, Ohio State University
Jason M. Molesky, Lakeville South High School, Lakeville, MN
Wayne Nirode, Troy High School, Troy, OH
Heather Pessy, Mt. Lebanon High School, Pittsburgh, PA
Sarah Peterson, University Preparatory Academy, Seattle, WA
Kathleen Petko, Palatine High School, Palatine, IL
German J. Pliego, University of St. Thomas
Stoney Pryor, A&M Consolidated High School, College Station, TX
Judy Quan, Alameda High School, Alameda, CA
Stephanie Ragucci, Andover High School, Andover, MA
James M. Reeder, University School, Hunting Valley, OH
Joseph Reiken, Bishop Garcia Diego High School, Santa Barbara, CA
Roger V. Rioux, Cape Elizabeth High School, Cape Elizabeth, ME
Tom Robinson, Kentridge Senior High School, Kent, WA
Albert Roos, Lexington High School, Lexington, MA
Linda C. Schrader, Cuyahoga Heights High School, Cuyahoga Heights, OH
Daniel R. Shuster, Royal High School, Simi Valley, CA
David Stein, Paint Branch High School, Burtonsville, MD
Vivian Annette Stephens, Dalton High School, Dalton, GA
Charles E. Taylor, Flowing Wells High School, Tucson, AZ
Reba Taylor, Blacksburg High School, Blacksburg, VA
Shelli Temple, Jenks High School, Jenks, OK
David Thiel, Math/Science Institute, Las Vegas, NV

William Thill, Harvard-Westlake School, North Hollywood, CA
Richard Van Gilst, Westminster Christian Academy, St. Louis, MO
Joseph Robert Vignolini, Glen Cove High School, Glen Cove, NY
Ira Wallin, Elmwood Park Memorial High School, Elmwood Park, NJ
Linda L. Wohlever, Hathaway Brown School, Shaker Heights, OH

I am extremely pleased that my good friend and colleague Daren Starnes partnered with me on the second and now the third edition of this textbook. Daren brings a flair for clear exposition that is rare in mathematics and statistics textbooks. His influence on every page helps make the subject accessible and understandable. Thanks, Daren!

I have also benefited from conversations with Floyd Bullard (North Carolina School of Science and Mathematics), particularly with regard to experimental designs. Eliot Scheuer (Nicolet High School in Milwaukee) has graciously given us permission to adapt one of my favorite activities (8A) that he originated. To the teachers who have attended my presentations and workshops, and to my students over the years who have taught me much, a resounding thank you.

Finally, thanks to my wife, Betty Jo, who typed, proofread, repaired, and compiled. Her constant support, both editorially and emotionally, is greatly appreciated.

Dan Yates

As a rookie author on the second edition of *TPS*, I benefited from the wise counsel of my respected friend and coauthor, Dan Yates. Our collaboration on that project was so successful that we teamed up to write *Statistics Through Applications (STA)*, which realized Dan's goal to produce a book for a non–AP level statistics course. During the development of the third edition, Dan gave me considerable latitude to explore several new ideas that I thought might enhance the book's organization. For his faith in my judgment and his ongoing support and encouragement, I offer my genuine appreciation.

To my students, especially the "varsity stats class" of 2005–2006, thank you for serving as willing guinea pigs during a hands-on trial run of the third edition. Your feedback was invaluable. Thanks also to the many teachers who have participated in my summer institutes for sharing your insights about statistics teaching.

My final thank you goes to my wife, Judy. Without her unwavering support these past two years, doing everything from entering data sets to organizing chapter files, there would be no third edition. She has an exceptional eye for detail and a limitless supply of patience. Thanks, love.

Daren Starnes

Third Edition

The Practice of Statistics

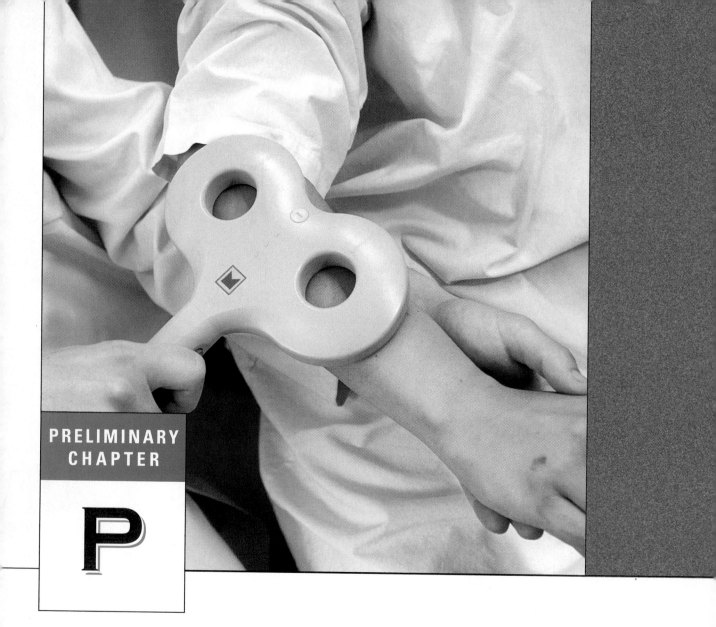

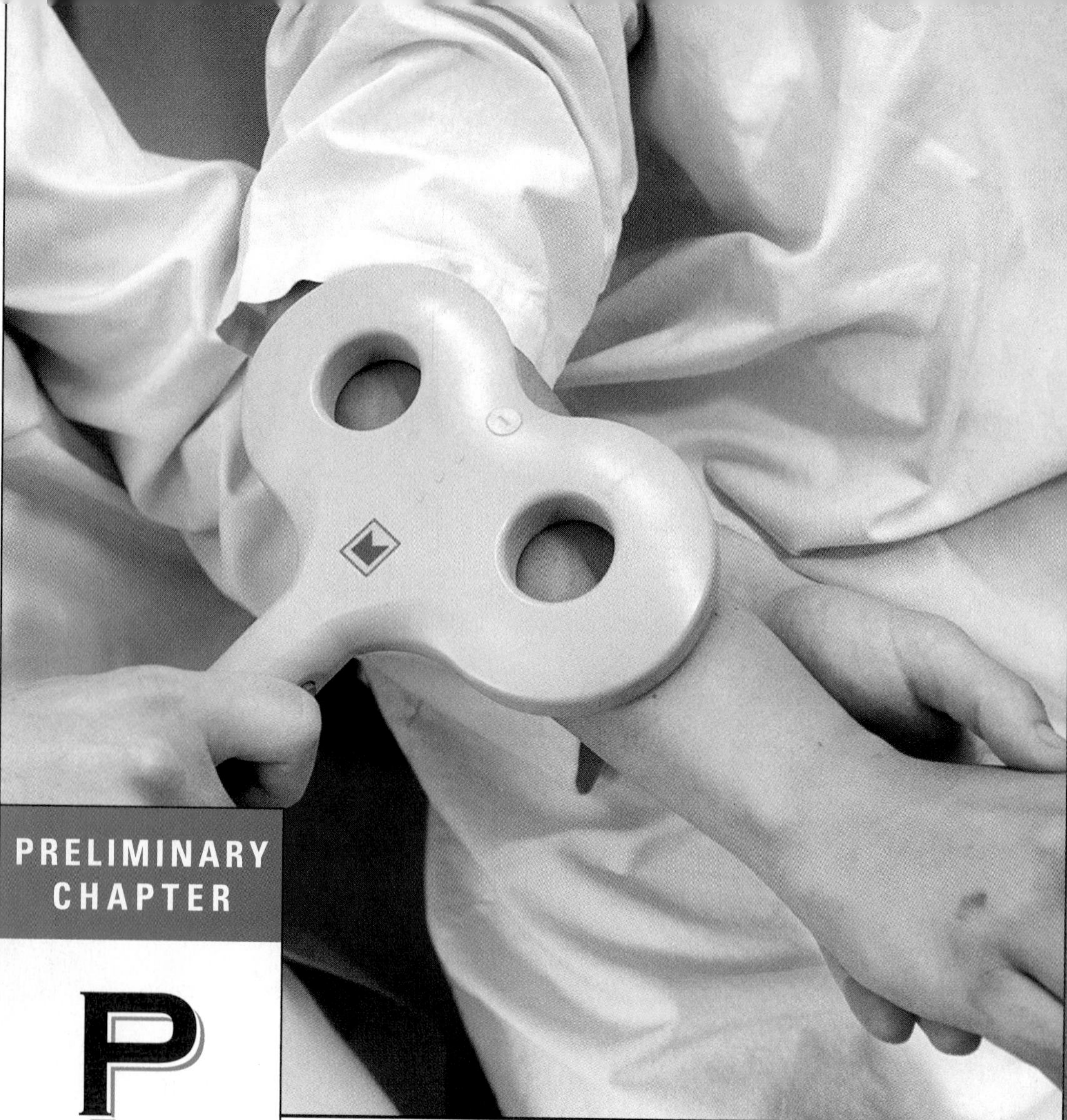

PRELIMINARY CHAPTER

P

What Is Statistics?

CASE STUDY

Can magnets help reduce pain?

Early research showed that magnetic fields affected living tissue in humans. Lately, doctors have begun to use magnets to treat patients' pain. Scientists wondered whether magnets would reduce the chronic pain felt by polio sufferers. They designed a study to find out.

Fifty patients with polio who reported steady pain were recruited for the study. A doctor identified a painful site on each patient and asked him or her to rate the pain on a scale from 0 (mild pain) to 10 (severe pain). Then, the doctor selected a sealed envelope containing a magnet from a box that contained both active and inactive magnets. That way, neither the doctor nor the patient knew which type of magnet was being used. The chosen magnet was applied to the site of the pain for 45 minutes. After "treatment," each patient was again asked to rate the level of pain from 0 to 10.

In all, 29 patients were given active magnets and 21 patients received inactive magnets. All but one of the patients rated their initial pain as an 8, 9, or 10. So scientists decided to focus on patients' final pain ratings. Here they are, grouped by the type of magnet used:[1]

Active: 0, 4, 7, 0, 4, 2, 5, 5, 3, 2, 2, 2, 3, 4, 6, 4, 3, 0, 2, 0, 4, 4, 5, 9, 10, 10, 10, 10, 7

Inactive: 4, 7, 5, 8, 8, 6, 8, 10, 10, 6, 10, 8, 10, 10, 10, 10, 9, 9, 10, 10, 9

What do the data tell us about whether the active magnets helped reduce pain? In this Preliminary Chapter, you will investigate the tools that statisticians use for data production, exploratory data analysis, probability, and drawing conclusions from data that are needed to help answer the scientists' question.

Activity

Water, water everywhere!

Materials: Three small paper cups per student; enough tap water for two cups per student and enough bottled water for one cup per student; one six-sided die and one index card per student

Bottled water is becoming an increasingly popular alternative to ordinary tap water. But can people really tell the difference if they aren't told which is which? Do you think *you* can tell the difference between bottled water and tap water? This Activity will give you a chance to discover answers to both of these questions.

1. Before class begins, your teacher will prepare numbered stations with cups of water. You will be given an index card with a station number on it.
2. Go to the corresponding station. Pick up three cups (labeled A, B, and C) and take them back to your seat.
3. Your task is to determine which one of the three cups contains the bottled water. Drink all of the water in Cup A first, then the water in Cup B, and finally the water in Cup C. Write down the letter of the cup that you think held the bottled water. Do not discuss your results with any of your classmates yet!
4. While you taste, your teacher will make a chart on the board like this one:

Station number	Bottled-water cup?	Truth

5. When you are instructed to do so, go to the board and record your station number and the letter of the cup you identified as containing bottled water.
6. Your teacher will now reveal the truth about the cups of drinking water. How many students in the class identified the bottled water correctly? What percent of the class is this?
7. Let's assume that no one can really distinguish tap water from bottled water. In that case, students would simply be guessing which cup of water tastes different. What percent of the class would you expect to guess correctly? How does this compare to the percent of correct identifications in your class?

8. Do you believe that the students in your class can actually tell the difference between tap water and bottled water? Before you answer, let's perform a brief *simulation*. We'll assume that you and your classmates are guessing at which cup holds the bottled water. Then each student would have a 1-in-3 chance of identifying the correct cup. Roll your die once for each student in your class. Let rolling a 1 or a 2 represent a correct guess. Let rolling a 3 through 6 represent an incorrect guess. (Note that this assignment of numbers gives each individual a 2-in-6 chance of being correct, which is the same as a 1-in-3 chance.) Record the number of times that you get a 1 or a 2. This result simulates the number of correct identifications made by the class.
9. On a number line drawn on the board by your teacher, mark an X above the number of correct identifications in your simulation. Based on the class's simulation results, how many correct identifications would make you doubt that students were just guessing? Why?
10. Look back to your class's actual tasting results in Step 6. What do you conclude about students' abilities to distinguish tap water from bottled water?

Introduction

Do cell phones cause brain cancer? How well do SAT scores predict college success? Should arthritis sufferers take Celebrex to ease their pain, or are the risks too great? What percent of U.S. children are overweight? How strong is the evidence for global warming? These are just a few of the questions that statistics can help answer. But what is statistics? And why should you study it?

statistics

Statistics is the science (and art) of learning from data. Data are usually numbers, but they are not "just numbers." *Data are numbers with a context.* The number 10.5, for example, carries no information by itself. But if we hear that a friend's new baby weighed 10.5 pounds at birth, we congratulate her on the healthy size of the child. The context engages our background knowledge and allows us to make judgments. We know that a baby weighing 10.5 pounds is quite large, and that a human baby is unlikely to weigh 10.5 ounces or 10.5 kilograms. The context makes the number informative.

You can find lots of data in newspapers and magazines and on the Internet. Such data are ripe for exploration. Part I of this book focuses on *exploratory data analysis.* In Chapters 1 through 4, you'll develop tools and strategies for organizing, describing, and analyzing data.

Sometimes data provide insights about questions we have asked. More often, researchers follow a careful plan for *producing data* to answer specific questions. In Part II of the book, you'll discover how they do it. Chapter 5 shows you how to design *surveys*, *experiments*, and *observational studies* correctly. Such well-produced data help us get the most reliable answers to difficult questions.

Probability is the study of chance behavior. When you flip coins, roll dice, deal cards, or play the lottery, the results are uncertain. But the laws of probability can tell you how likely (or unlikely) certain outcomes are. You'll learn how to calculate probabilities in Part III of the book, in Chapters 6 through 9.

population
sample

How can we draw conclusions about a large group (***population***) of *individuals*—people, animals, or things—based on information about a much smaller group (***sample***)? This is the challenge of *statistical inference.* Inference questions ask us to test claims about or provide estimates for unknown population values. Valid inference depends on appropriate data production, skillful data analysis, and careful use of probability. In Part IV of the book, we discuss the logic behind statistical inference and some of its methods. In Chapters 10 through 15, you'll learn some common ways of testing claims and computing estimates.

This Preliminary Chapter is intended to give you a snapshot of what statistics is all about. Where do data come from? What should you do with data once you have them? How can probability help you? What conclusions can you draw? Keep reading for some answers.

In your lifetime, you will be bombarded with data and statistical information. Opinion poll results, television ratings, gas prices, unemployment rates, medical study outcomes, and standardized test scores are discussed daily in the media. People make important decisions based on such data. Statistics will help you make sense of information like this. A solid understanding of statistics will enable you to make sound decisions based on data in your everyday life.

Data Production: Where Do You Get Good Data?

You want data on a question that interests you. Maybe you want to know what causes of death are most common among young adults, or whether the math performance of American schoolchildren is getting better.

It is tempting just to draw conclusions from our own experience, making no use of more representative data. You think (without really thinking) that the students at your school are typical. We hear a lot about AIDS, so we assume it must be a leading cause of death among young people. Or we recall an unusual incident that sticks in our memory exactly because it is unusual. When an airplane crash kills several hundred people, we fear that flying is unsafe, even though data on all flights show that flying is much safer than driving. Here's an example that shows why **data beat personal experiences.**

Example P.1 ***Power lines and cancer***

Got data?

Does living near power lines cause leukemia in children? The National Cancer Institute spent 5 years and $5 million gathering data on this question. The researchers compared 638 children who had leukemia with 620 who did not. They went into the homes and actually measured the magnetic fields in children's bedrooms, in other rooms, and at the front door. They recorded facts about power lines near the family home and also near the mother's residence when she was pregnant. Result: no connection between leukemia and exposure to magnetic fields of the kind produced by power lines was found. The editorial that accompanied the study report in the *New England Journal of Medicine* proclaimed, "It is time to stop wasting our research resources" on the question.[2]

Now consider a devastated mother whose child has leukemia and who happens to live near a power line. In the public mind, the striking story wins every time. A statistically literate person, however, knows that data are more reliable than personal experience because they systematically describe an overall picture rather than focus on a few incidents.

A better tactic is to head for the library or the Internet. There you will find plenty of data, not gathered specifically to answer your questions but available for your use. Recent data can be found online, but locating them can be challenging. Government statistical offices are the primary source for demographic, social, and economic data. Many nations have a single statistical office, like Statistics Canada (www.statcan.ca) or Mexico's INEGI (www.inegi.gob.mx). The United States does not have a national statistical office. More than 70 federal agencies collect data. Fortunately, you can reach most of them through the government's handy FedStats site (www.fedstats.gov).

Example P.2 ***Causes of death and math scores***

Finding data on the Internet

If you visit the National Center for Health Statistics Web site, www.cdc.gov/nchs, you will learn that accidents are the most common cause of death for U.S. citizens aged 20 to 24, accounting for over 40% of all deaths. Homicide is next, followed by suicide. AIDS ranks seventh, behind heart disease and cancer, at 1% of all deaths. The data also show that it is dangerous to be a young man: the overall death rate for men aged 20 to 24 is three times that for women, and the death rate from homicide is more than five times higher among men.

Figure P.1 *The Web sites of government statistical offices are prime sources of data. Here is the home page of the National Assessment of Educational Progress.*

If you go to the National Center for Education Statistics Web site, www.nces.ed.gov, you will find the latest National Assessment of Educational Progress (Figure P.1), which provides full details about the math skills of schoolchildren. Math scores have slowly but steadily increased since 1990. All racial/ethnic groups, both girls and boys, and students in most states are getting better in math.

The library and the Internet are sources of ***available data.***

Available Data

Available data are data that were produced in the past for some other purpose but that may help answer a present question.

Available data are the only data used in most student reports. Because producing new data is expensive, we all use available data whenever possible. However, the clearest answers to present questions often require data produced to answer those specific questions. The main statistical designs for producing data are *surveys*, *experiments*, and *observational studies*.

surveys

Surveys are popular ways to gauge public opinion. The idea of a survey is pretty simple:

- Select a *sample* of people to represent a larger *population*.

- Ask the individuals in the sample some questions and record their responses.
- Use sample results to draw some conclusions about the population.

In practice, however, getting valid survey results is not so easy. As the following example shows, **where the data come from is important.**

Example P.3 *Having kids or not?*
Good and bad survey results

The advice columnist Ann Landers once asked her readers, "If you had it to do over again, would you have children?" A few weeks later, her column was headlined "70% OF PARENTS SAY KIDS NOT WORTH IT." Indeed, 70% of the nearly 10,000 parents who wrote in said they would not have children if they could make the choice again. Do you believe that 70% of all parents regret having children?

You shouldn't. The people who took the trouble to write Ann Landers are not representative of all parents. Their letters showed that many of them were angry at their children. All we know from these data is that there are some unhappy parents out there. A statistically designed poll, unlike Ann Landers's appeal, targets specific people chosen in a way that gives all parents the same chance to be asked. Such a poll later showed that 91% of parents would have children again.

The lesson: if you are careless about how you get your data, you may announce 70% "No" when the truth is close to 90% "Yes."

census You may have wondered: why not survey everyone in the population (a ***census***) rather than a sample? Usually, it would take too long and cost too much. Our goal in choosing a sample is a picture of the population, disturbed as little as possible by the act of gathering information. Sample surveys are one kind of ***observational study.***

In other settings, we gather data from an ***experiment.*** In doing an experiment, we don't just observe individuals or ask them questions. We actually do something to people, animals, or objects to observe the response. Experiments can answer questions such as "Does aspirin reduce the chance of a heart attack?" and "Do more college students prefer Pepsi to Coke when they taste both without knowing which they are drinking?" Experiments, like samples, provide useful data only when properly designed. The distinction between experiments and observational studies is one of the most important ideas in statistics.

Observational Study versus Experiment

In an **observational study,** we observe individuals and measure variables of interest but do not attempt to influence the responses.

In an **experiment,** we deliberately do something to individuals in order to observe their responses.

The next example illustrates the difference between an observational study and an experiment.

Example P.4

Estrogen and heart attacks

Observational study versus experiment

Should women take hormones such as estrogen after menopause, when natural production of these hormones ends? In 1992, several major medical organizations said "Yes." Women who took hormones seemed to reduce their risk of a heart attack by 35% to 50%. The risks of taking hormones appeared small compared with the benefits.

The evidence in favor of hormone replacement came from a number of studies that simply compared women who were taking hormones with others who were not. But women who chose to take hormones were typically richer and better educated, and they saw doctors more often than women who did not take hormones. These women did many things to maintain their health. It isn't surprising that they had fewer heart attacks.

Experiments were needed to get convincing data on the link between hormone replacement and heart attacks. In the experiments, women did not decide what to do. A coin toss assigned each woman to one of two groups. One group took hormone replacement pills; the other took dummy pills that looked and tasted the same as the hormone pills. All kinds of women were equally likely to get either treatment. By 2002, several experiments with women of different ages showed that hormone replacement does *not* reduce the risk of heart attacks. The National Institutes of Health, after reviewing the evidence, concluded that the earlier observational studies were wrong. Taking hormones after menopause fell quickly out of favor.[3]

Observational studies are essential sources of data about topics from the opinions of voters to the behavior of animals in the wild. But an observational study, even one based on a statistical sample, is a poor way to gauge the effect of a change. To see the response to a change, we must actually impose the change. When our goal is to understand cause and effect, experiments are the best source of convincing data.

Exercises

P.1 Need a Jolt? Jamie is a hard-core computer programmer. She and all her friends prefer Jolt cola (caffeine equivalent to two cups of coffee) to either Coke or Pepsi (caffeine equivalent to less than one cup of coffee). Explain why Jamie's preference is not good evidence that most young people prefer Jolt to Coke or Pepsi.

P.2 Cell phones and brain cancer One study of cell phones and the risk of brain cancer looked at a group of 469 people who have brain cancer. The investigators matched each cancer patient with a person of the same age, sex, and race who did not have brain cancer, then asked about the use of cell phones.[4] Result: "Our data suggest that the use of hand-held cellular phones is not associated with risk of brain cancer."

(a) Is this an observational study or an experiment? Justify your answer.

(b) Based on this study, would you conclude that cell phones do not increase the risk of brain cancer? Why or why not?

P.3 Learning biology with computers An educational software company wants to compare the effectiveness of its computer animation for teaching biology with that of a textbook presentation. The company gives a biology pretest to each of a group of high school juniors, then divides them into two groups. One group uses the animation, and the other studies the text. The company retests all students and compares the increase in biology test scores in the two groups.

(a) Is this an experiment or an observational study? Justify your answer.

(b) If the group using the computer animation has a much higher average increase in test scores than the group using the textbook, what conclusions, if any, could the company draw?

P.4 Survey, experiment, or observational study? What is the best way to answer each of the questions below: a survey, an experiment, or an observational study that is not a survey? Explain your choices. For each question, write a few sentences about how such a study might be carried out.

(a) Are people generally satisfied with how things are going in the country right now?

(b) Do college students learn basic accounting better in a classroom or using an online course?

(c) How long do your teachers wait on the average after they ask the class a question?

P.5 I'll drink to that! In adults, moderate use of alcohol is associated with better health. Some studies suggest that drinking wine rather than beer or spirits yields added health benefits.

(a) Explain the difference between an observational study and an experiment to compare people who drink wine with people who drink beer.

(b) Suggest some characteristics of wine drinkers that might benefit their health. In an observational study, these characteristics are mixed up with the effects of drinking wine on people's health.

P.6 Get a job! Find some information on this question: what percent of college undergraduates work part-time or full-time while they are taking classes? Start with the National Center for Education Statistics Web site, www.nces.ed.gov. Keep a detailed written record of your search.

Data Analysis: Making Sense of Data

data analysis

The first step in understanding data is to hear what the data say, to "let the statistics speak for themselves." But numbers speak clearly only when we help them speak by organizing, displaying, summarizing, and asking questions. That's ***data analysis.***

Any set of data contains information about some group of ***individuals.*** The characteristics we measure on each individual are called ***variables.***

The importance of data integrity It has been accepted that global warming is a serious ecological problem. But Yale University researchers examined satellite and weather-balloon data collected since 1979 by NOAA (National Oceanic and Atmospheric Administration). They discovered that the satellites had drifted in orbit, throwing off the timing of temperature measurements. Nights looked as warm as days. Corrective action has shown that the pace of global warming over the past 30 years has actually been quite slow, a total increase of about 1 degree Fahrenheit. The lesson: always ask, "How were the data produced?"

Individuals and Variables

Individuals are the objects described by a set of data. Individuals may be people, but they may also be animals or things.

A **variable** is any characteristic of an individual. A variable can take different values for different individuals.

A college's student data base, for example, includes data about every currently enrolled student. The students are the *individuals* described by the data set. For each individual, the data contain the values of *variables* such as age, gender, choice of major, and grade point average. In practice, any set of data is accompanied by background information that helps us understand it.

When you meet a new set of data, ask yourself the following *key questions:*

1. **Who** are the individuals described by the data? How many individuals are there?
2. **What** are the variables? In what units is each variable recorded? Weights, for example, might be recorded in pounds, in thousands of pounds, or in kilograms.
3. **Why** were the data gathered? Do we hope to answer some specific questions? Do we want to draw conclusions about individuals other than the ones we actually have data for?
4. **When, where, how, and by whom** were the data produced? Where did the data come from? Are these available data or new data produced to answer current questions? Are the data from an experiment or an observational study? From a census or a sample? Who directed the data production? Can we trust the data?

Some variables, like gender and college major, simply place individuals into categories. Others, like age and grade point average (GPA), take numerical values for which we can do arithmetic. It makes sense to give an average GPA for a group of students, but it does not make sense to give an "average" gender. We can, however, count the numbers of female and male students and do arithmetic with these counts.

Categorical and Quantitative Variables

A **categorical variable** places an individual into one of several groups or categories.

A **quantitative variable** takes numerical values for which arithmetic operations such as adding and averaging make sense.

Example P.5 *Education in the United States*
Four key questions

Here is a small part of a data set that describes public education in the United States:

State	Region	Population (1000s)	SAT verbal	SAT math	Percent taking	Percent no HS	Teachers' pay ($1000)
CA	PAC	35,894	499	519	54	18.9	54.3
CO	MTN	4,601	551	553	27	11.3	40.7
CT	NE	3,504	512	514	84	12.5	53.6

Answer the four key questions about these data.

1. Who? The *individuals* described are the states. There are 51 of them, the 50 states and the District of Columbia, but we give data for only 3: California (CA), Colorado (CO), and Connecticut (CT). Each row in the table describes one individual.

2. What? The rest of the columns each contain the values of one variable for all the individuals. This is the usual arrangement in data tables. Seven *variables* are recorded for each state. The second column lists which region of the country the state is in. Region is a categorical variable. The Census Bureau divides the nation into nine regions. These three are Pacific (PAC), Mountain (MTN), and New England (NE). The third column contains state populations, in thousands of people. Population is a quantitative variable. Be sure to notice that the *units* are thousands of people. California's 35,894 stands for 35,894,000 people.

The remaining five variables are the average scores of the states' high school seniors on the SAT verbal and mathematics exams, the percent of seniors who take the SAT, the percent of students who did not complete high school, and average teachers' salaries in thousands of dollars. These are all quantitative variables. Each of these variables needs more explanation before we can fully understand the data.

Beginning in March 2005, the new SAT consisted of three tests: Critical Reading, Math, and Writing.

3. Why? Some people will use these data to evaluate the quality of individual states' educational programs. Others may compare states using one or more of the variables. Future teachers might want to know how much they can expect to earn.

4. When, where, how, and by whom? The population data come from the Current Population Survey, conducted by the federal government. They are fairly accurate as of July 1, 2004, but don't show later changes in population. State SAT averages came from the College Board's Web site, www.collegeboard.com, and were based on a census of all test takers that year. The percent of students who did not graduate in each state was determined by the 2003 Current Population Survey. Average teacher salaries were reported in the 2003 *Statistical Abstract of the United States,* using data provided by the National Education Association for 2002. These data are estimates based on samples of teachers from each state.

A variable generally takes values that vary (hence the name "variable"!). Categorical variables sometimes have similar counts in each category and sometimes don't. For example, if you recorded values of the variable "birth month" for the students at your school, you would expect about an equal number of students in each of the categories (January, February, March, . . .). If you measured the variable "favorite type of music," however, you might see very different counts in the categories classical, gospel, rock, rap, and so on. Quantitative variables may take values that are very close together or values that are quite spread out. We call the pattern of variation of a variable its ***distribution.***

Distribution

The **distribution** of a variable tells us what values the variable takes and how often it takes these values.

exploratory data analysis

Statistical tools and ideas can help you examine data in order to describe their main features. This examination is sometimes called ***exploratory data analysis.*** (We prefer data analysis.) Like an explorer crossing unknown lands, we first simply describe what we see. Each example we meet will have some background information to help us, but our emphasis is on examining the data. Here are two basic strategies that help us organize our exploration of a set of data:

- Begin by examining each variable by itself. Then move on to study relationships among the variables.
- Begin with a graph or graphs. Then add numerical summaries of specific aspects of the data.

We will organize our learning the same way. Chapters 1 and 2 examine single-variable data, and Chapters 3 and 4 look at relationships among variables. In both settings, we begin with graphs and then move on to numerical summaries.

Describing Categorical Variables

The values of a categorical variable are labels for the categories, such as "male" and "female." The distribution of a categorical variable lists the categories and gives either the *count* or the *percent* of individuals who fall in each category.

Example P.6

Do you wear your seat belt?
Describing categorical variables

Each year, the National Highway and Traffic Safety Administration (NHTSA) conducts an observational study on seat belt use. The table below shows the percent of front-seat passengers who were observed to be wearing their seat belts in each region of the United States in 1998 and 2003.[5]

Region	Percent wearing seat belts, 2003	Percent wearing seat belts, 1998
Northeast	74	66.4
Midwest	75	63.6
South	80	78.9
West	84	80.8

What do these data tell us about seat belt usage by front-seat passengers?

The *individuals* in this observational study are front-seat passengers. For each individual, the values of two *variables* are recorded: region (Northeast, Midwest, South, or West) and seat belt use (yes or no). Both of these variables are categorical.

bar graph

Figure P.2(a) shows a ***bar graph*** for the 2003 data. Notice that the vertical scale is measured in percents.

Figure P.2a *(a) A bar graph showing the percent of front-seat passengers who wore their seat belts in each of four U.S. regions in 2003.*

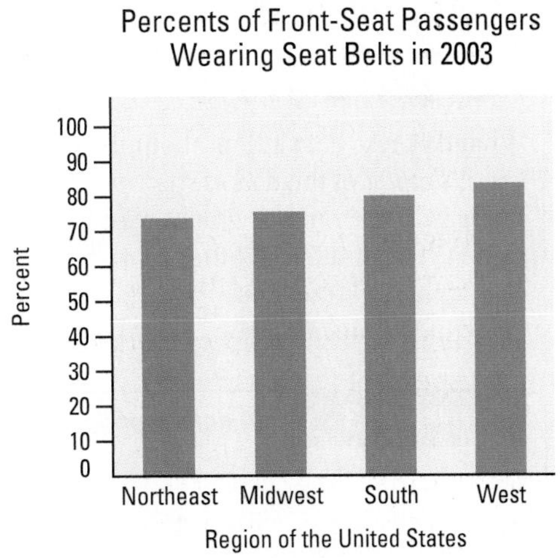

side-by-side bar graph

Front seat passengers in the South and West seem more concerned about wearing seat belts than those in the Northeast and Midwest. In all four regions, a high percent of front-seat passengers were wearing seat belts. Figure P.2(b) (on the next page) shows a ***side-by-side bar graph*** comparing seat belt usage in 1998 and 2003. Seat belt usage increased in all four regions over the five-year period.

Figure P.2b (b) *A side-by-side bar graph comparing the percent of front-seat passengers who wore their seat belts in the four U.S. regions in 1998 and 2003.*

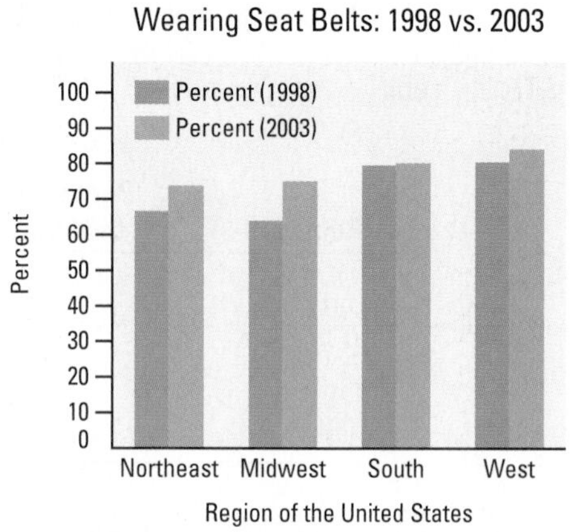

Describing Quantitative Variables

Several types of graphs can be used to display quantitative data. One of the simplest to construct is a ***dotplot.***

dotplot

Example P.7 *GOOOOAAAAALLLLLL!*
Describing quantitative variables

The number of goals scored by the U.S. women's soccer team in 34 games played during the 2004 season is shown below:[6]

3 0 2 7 8 2 4 3 5 1 1 4 5 3 1 1 3 3 3 2 1
2 2 2 4 3 5 6 1 5 5 1 1 5

What do these data tell us about the performance of the U.S. women's team in 2004?

A *dotplot* of the data is shown in Figure P.3. Each dot represents the goals scored in a single game. From this graph, we can see that the team scored between 0 and 8 goals per game. Most of the time, they scored between 1 and 5 goals. Their most frequent number of goals scored (the ***mode***) was 1. They averaged 3.059 goals per game. (Check our calculation of the ***mean*** on your calculator.)

Figure P.3 *A dotplot of goals scored by the U.S. women's soccer team in 2004.*

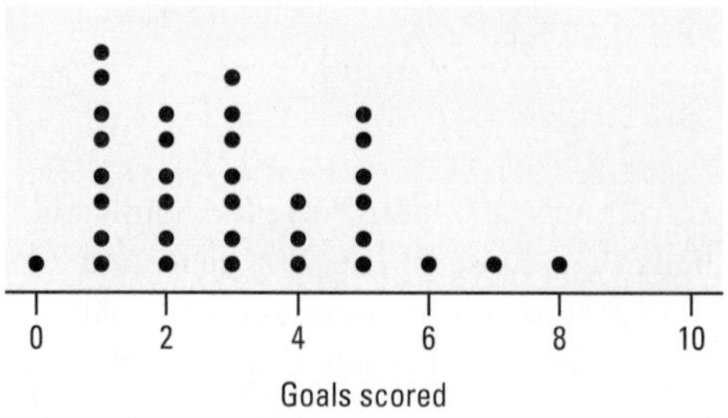

Making a statistical graph is not an end in itself. After all, a computer or graphing calculator can make graphs faster than we can. The purpose of a graph is to help us understand the data. After you (or your calculator) make a graph, always ask, "What do I see?"

Exploring Relationships between Variables

Quite often in statistics, we are interested in examining the relationship between two variables. For instance, we may want to know how the percent of students taking the SAT in U.S. states is related to those states' average SAT math scores, or perhaps how seat belt usage is related to region of the country. As the next example illustrates, **many relationships between two variables are influenced by other variables lurking in the background.**

Example P.8 *On-time flights*

Describing relationships between variables

Air travelers would like their flights to arrive on time. Airlines collect data about on-time arrivals and report them to the Department of Transportation. Here are one month's data for flights from several western cities for two airlines:

	On time	Delayed
Alaska Airlines	3274	501
America West	6438	787

You can see that the percents of late flights were

$$\text{Alaska Airlines} \quad \frac{501}{3775} = 13.3\%$$

$$\text{America West} \quad \frac{787}{7225} = 10.9\%$$

It appears that America West does better.

This isn't the whole story, however. For each flight (individual), we have data on two categorical variables: the airline and whether or not the flight was late. Let's add data on a third categorical variable, departure city.[7] The following table summarizes the results.

	Alaska Airlines		America West	
Departure city	**On time**	**Delayed**	**On time**	**Delayed**
Los Angeles	497	62	694	117
Phoenix	221	12	4840	415
San Diego	212	20	383	65
San Francisco	503	102	320	129
Seattle	1841	305	201	61
Total	**3274**	**501**	**6438**	**787**

The "Total" row shows that the new table describes the same flights as the earlier table. Look again at the percents of late flights, first for Los Angeles:

$$\text{Alaska Airlines} \quad \frac{62}{559} = 11.1\%$$

$$\text{America West} \quad \frac{117}{811} = 14.4\%$$

Alaska Airlines is better. The percents of late flights for Phoenix are

$$\text{Alaska Airlines} \quad \frac{12}{233} = 5.2\%$$

$$\text{America West} \quad \frac{415}{5255} = 7.9\%$$

Alaska Airlines is better again. In fact, as Figure P.4 shows, Alaska Airlines has a lower percent of late flights at every one of these cities.

Figure P.4 *Comparing the percents of delayed flights for two airlines at five airports.*

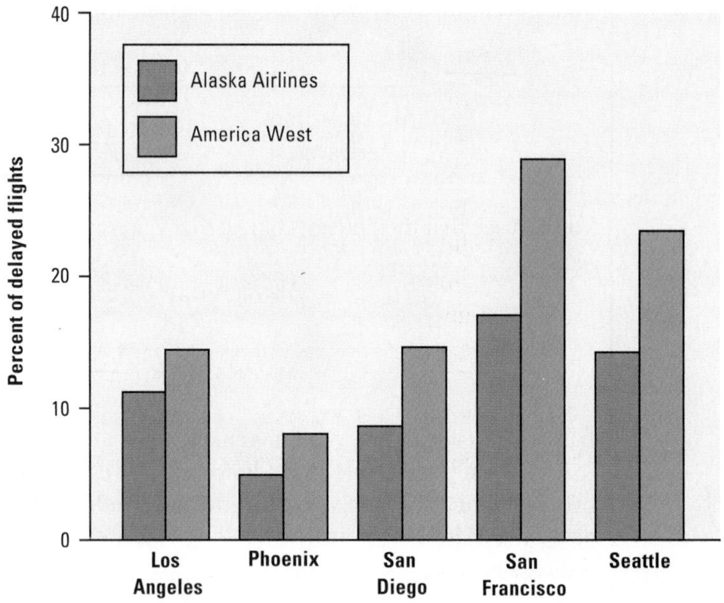

How can it happen that Alaska Airlines wins at every city but America West wins when we combine all the cities? Look at the data: America West flies most often from sunny Phoenix, where there are few delays. Alaska Airlines flies most often from Seattle, where fog and rain cause frequent delays. What city we fly from has a major influence on the chance of a delay, so including the city data reverses our conclusion. (We'll see other examples like this one in Chapter 4 when we examine *Simpson's paradox*.) The message is worth repeating: many relationships between two variables (like airline and whether the flight was late) are influenced by other variables lurking in the background (like departure city).

Exercises

P.7 Cool car colors Here are data on the most popular car colors for vehicles made in North America during the 2003 model year.[8]

Color	Percent of vehicles
Silver	20.1
White	18.4
Black	11.6
Medium/dark gray	11.5
Light brown	8.8
Medium/dark blue	8.5
Medium red	6.9

(a) Display these data in a bar graph. Be sure to label your axes and title your graph.

(b) Describe what you see in a few sentences. What percent of vehicles had other colors?

P.8 Comparing car colors Favorite vehicle colors may differ among types of vehicle. Here are data on the most popular colors in 2003 for luxury cars and SUVs, trucks, and vans. The entry "—" means "less than 1%."

Color	Luxury car percent	SUV/truck/van percent
Black	10.9	11.6
Light brown	—	6.3
Medium/dark blue	3.8	9.3
Medium/dark gray	23.3	8.8
Medium/dark green	—	7.0
Medium red	3.9	6.2
White	30.4	22.3
Silver	18.8	17.0

(a) Make a side-by-side bar graph to compare colors by vehicle type.

(b) Write a few sentences describing what you see.

P.9 U.S. women's soccer scores In Example P.7 (page 16), we examined the number of goals scored by the U.S. women's soccer team in games during the 2004 season. Here are data on the goal differential for those same games, computed as U.S. score minus opponent's score.

3 0 2 7 8 2 4 1 4 1 −2 3 4 3 0 1 2 2 3 2 0
1 1 1 1 3 5 6 1 4 5 0 −2 5

(a) Make a dotplot of these data.

(b) Describe what you see in a few sentences.

P.10 Olympic gold! Olympic athletes like Michael Phelps, Natalie Coughlin, Amanda Beard, and Paul Hamm captured public attention by winning gold medals in the 2004 (a)

Summer Olympic Games in Athens, Greece. Table P.1 displays the total number of gold medals won by a sample of countries in the 2004 Summer Olympics.

Table P.1 *Gold medals won by selected countries in the 2004 Summer Olympics*

Country	Gold medals	Country	Gold medals
Sri Lanka	0	Netherlands	4
Qatar	0	India	0
Vietnam	0	Georgia	2
Great Britain	9	Kyrgyzstan	0
Norway	5	Costa Rica	0
Romania	8	Brazil	4
Switzerland	1	Uzbekistan	2
Armenia	0	Thailand	3
Kuwait	0	Denmark	2
Bahamas	0	Latvia	0
Kenya	1	Czech Republic	1
Trinidad and Tobago	0	Hungary	8
Greece	6	Sweden	4
Mozambique	0	Uruguay	0
Kazakhstan	1	United States	35

Source: BBC Olympics Web site. news.bbc.co.uk/sport1/hi/olympics_2004.

Make a dotplot to display these data. Describe the distribution of number of gold medals won.

(b) Overall, 202 countries participated in the 2004 Summer Olympics, of which 57 won at least one gold medal. Do you believe that the sample of countries listed in the table is representative of this larger population? Why or why not?

P.11 A class survey Here is a small part of the data set that describes the students in an AP Statistics class. The data come from anonymous responses to a questionnaire on the first day of class.

	A	B	C	D	E	F
1	GENDER	HAND	HEIGHT	HOMEWORK TIME	MUSIC	COINS IN POCKET
2	F	L	65	200	RAP	50
3	M	L	72	30	COUNTRY	35
4	M	R	62	95	ROCK	35
5	F	L	64	120	R&B	0
6	M	R	63	220	CLASSICAL	0
7	F	R	58	60	ROCK	76
8	F	R	67	150	TOP 40	215
9						

Sheet1 Sheet2 Sheet3

Answer the key questions (who, what, why, when, where, how, and by whom) for these data. For each variable, tell whether it is categorical or quantitative. Try to identify the units of measurement for any quantitative variables.

P.12 Medical study variables Data from a medical study contain values of many variables for each of the people who were the subjects of the study. Which of the following variables are categorical and which are quantitative?

(a) Gender (female or male)
(b) Age (years)
(c) Race (Asian, black, white, or other)
(d) Smoker (yes or no)
(e) Systolic blood pressure (millimeters of mercury)
(f) Level of calcium in the blood (micrograms per milliliter)

Probability: What Are the Chances?

probability

Consider tossing a single coin. The result is a matter of chance. It can't be predicted in advance, because the result will vary if you toss the coin repeatedly. But there is still a regular pattern in the results, a pattern that becomes clear only after many tosses. This remarkable fact is the basis for the idea of ***probability***.

Example P.9 *Coin tossing*

Probability: what happens in the long run

When you toss a coin, there are only two possible outcomes, heads or tails. Figure P.5 shows the results of tossing a coin 1000 times. For each number of tosses from 1 to 1000, we have plotted the proportion of those tosses that gave a head. The first toss was a head, so the proportion of heads starts at 1. The second toss was a tail, reducing the proportion

Figure P.5 *The behavior of the proportion of coin tosses that give a head, from 1 to 1000 tosses of a coin. In the long run, the proportion of heads approaches 0.5, the probability of a head.*

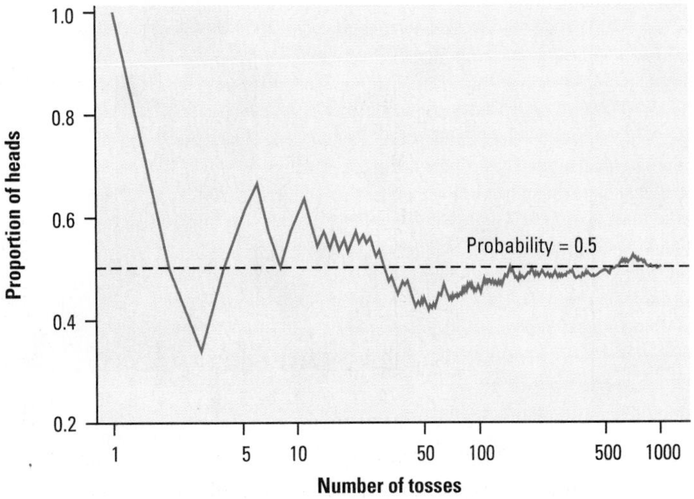

of heads to 0.5 after two tosses. The next three tosses gave a tail followed by two heads, so the proportion of heads after five tosses is 3/5, or 0.6.

The proportion of tosses that produce heads is quite variable at first, but it settles down as we make more and more tosses. Eventually this proportion gets close to 0.5 and stays there. We say that 0.5 is the *probability* of a head. The probability 0.5 appears as a horizontal line on the graph.

Example P.9 illustrates the big idea of probability: **chance behavior is unpredictable in the short run but has a regular and predictable pattern in the long run.** Casinos rely on this fact to make money every day of the year. We can use probability rules to analyze games of chance, like roulette, blackjack, and Texas hold 'em.

Probability plays an even more important role in the study of *variation*. If we toss a coin 30 times, will we get exactly 15 heads? Perhaps. Could we get as few as 11 heads? More than 24 heads? Probability tells us that there's about a 10% chance of getting 11 or fewer heads and less than a 1-in-1000 chance of getting more than 24 heads. If we toss our coin 30 times over and over and over again, the number of heads we obtain will vary. Probability quantifies the pattern of chance variation.

Example P.10 *Water, water everywhere*

Using probability to measure "how likely"

How can probability help us determine whether students can distinguish bottled water from tap water? Let's return to the Activity (page 4). Suppose that in Mr. Bullard's class, 13 out of 21 students made correct identifications. If we assume that the students in his

Figure P.6 *Graph showing the probability for each possible number of correct guesses in Mr. Bullard's class.*

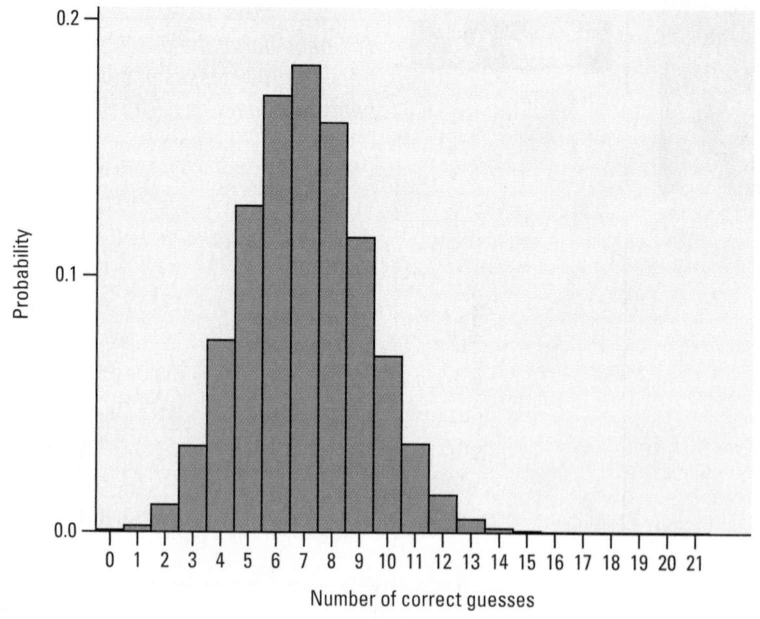

class *cannot* tell bottled water from tap water, then each one is basically guessing, with a 1-in-3 chance of being correct. So we'd expect about one-third of his 21 students, that is, about 7 students, to guess correctly. How likely is it that as many as 13 of his 21 students would guess correctly?

Figure P.6 is a graph of the probability values for the number of correct guesses in Mr. Bullard's class. As you can see from the graph, the chance of guessing 13 or more correctly is quite small. In fact, the actual probability of doing so is 0.0068.

So what do we conclude? Either Mr. Bullard's students are guessing, and they have incredibly good luck, or the students are not guessing. Since the students have less than a 1% chance of getting so many right "just by chance," we feel pretty sure that they are not guessing. It seems that they can detect the difference in taste between tap and bottled water.

As the previous example shows, probability allows us to decide whether an observed outcome is too unlikely to be due to chance variation. Too many students were able to identify which of their three cups contained a different type of water for us to believe that they were guessing. In effect, we tested the claim that the students were guessing. This is our first encounter with ***statistical inference.*** Notice the important role that probability played in leading us to a conclusion.

statistical inference

Statistical Inference: Drawing Conclusions from Data

How prevalent is cheating on tests? Representatives from the Gallup Organization were determined to find out. They conducted an Internet survey of 1200 students, aged 13 to 17, between January 23 and February 10, 2003. The question they posed was "Have you, yourself, ever cheated on a test or exam?" Forty-eight percent of those surveyed said "Yes." If Gallup had asked the same question of *all* 13- to 17-year-old students, would exactly 48% have answered "Yes"?

Gallup is trying to estimate the unknown percent of students in this age group who would say they have cheated on a test. (Notice that we didn't say the percent of students who actually *had* cheated on a test!) Their best estimate, given the survey results, would be 48%. But the folks at Gallup know that samples vary. If they had selected a different sample of 1200 students to respond to the survey, then they would probably have gotten a different estimate. *Variation is everywhere!*

Fortunately, probability provides a description of how the sample results will vary in relation to the true population percent. Based on the sampling method that Gallup used, we can say that their estimate of 48% is very likely to be within 3% of the true population percent. That is, we can be quite confident that between 45% and 51% of *all* teenage students would say that they have cheated on a test.

Statistical inference allows us to use the results of properly designed experiments, sample surveys, and other observational studies to draw conclusions that go beyond the data themselves. Whether we are testing a claim, as in the bottled versus tap water Activity, or computing an estimate, as in the Gallup survey, we rely on probability to help us answer research questions with a known degree of confidence. Unfortunately, we cannot be *certain* that our conclusions are correct. The following example shows you why.

Example P.11 ***Do mammograms help?***
Experiments and inference

Most women who reach middle age have regular mammograms to detect breast cancer. Do mammograms really reduce the risk of dying of breast cancer? To seek answers, doctors rely on "randomized clinical trials" that compare different ways of screening for breast cancer. We will see later that data from randomized comparative experiments are the gold standard. The conclusion from 13 such trials is that mammograms reduce the risk of death in women aged 50 to 64 years by 26%.[9]

On average, then, women who have regular mammograms are less likely to die of breast cancer. Of course, the results are different for different women. Some women who have mammograms every year die of breast cancer, and some who never have mammograms live to 100 and die when they crash their motorcycles. In spite of this individual variation, the results of the 13 clinical trials provide convincing evidence that women who have mammograms are less likely to die from breast cancer. That's because probability tells us that the large difference in death rates between women who had regular mammograms and those who didn't was unlikely to have occurred by chance. Can we be *sure* that mammograms reduce risk on the average? No, we can't be sure. **Because variation is everywhere, we cannot be certain about our conclusions.** However, statistics helps us better understand variation so that we can make reasonable conclusions.

Statistics gives us a language for talking about uncertainty that is used and understood by statistically literate people everywhere. In the case of mammograms, the doctors use that language to tell us that "mammography reduces the risk of dying of breast cancer by 26% (95% confidence interval, 17% to 34%)." According to Arthur Nielsen, head of the country's largest market research firm, that 26% is "a shorthand for a range that describes our actual knowledge of the underlying condition."[10] The range is 17% to 34%, and we are 95% confident that the true percent lies in that range. You will soon learn how to understand this language.

We can't escape variation and uncertainty. Learning statistics enables us to deal more effectively with these realities.

Statistical Thinking and You

The purpose of this book is to give you a working knowledge of the ideas and tools of practical statistics. Because data always come from a real-world context, doing statistics means more than just manipulating data. *The Practice of Statistics* is full of data, and each set of data has some brief background to help you understand what the data say. Examples and exercises usually express some brief understanding gained from the data. In practice, you would know much more about the background of the data you work with and about the questions you hope the data will answer. No textbook can be fully realistic. But it is important to form the habit of asking, "What do the data tell me?" rather than just concentrating on making graphs and doing calculations. This book tries to encourage good habits.

Still, statistics involves lots of calculating and graphing. The text presents the techniques you need, but you should use a calculator or computer software to automate calculations and graphs as much as possible.

Ideas and judgment can't (at least yet!) be automated. They guide you in telling the computer what to do and in interpreting its output. This book tries to explain the most important ideas of statistics, not just teach methods.

You learn statistics by doing statistical problems. This book offers four types of exercises, arranged to help you learn. Short problem sets appear after each major idea. These are straightforward exercises that help you solidify the main points before going on. The Section Exercises at the end of each numbered section help you combine all the ideas of the section. Chapter Review Exercises look back over the entire chapter. Finally, the Part Review Exercises provide challenging, cumulative problems like you might find on a final exam. At each step you are given less advance knowledge of exactly what statistical ideas and skills the problems will require, so each step requires more understanding.

Each chapter ends with a Chapter Review that includes a detailed list of specific things you should now be able to do. Go through that list, and be sure you can say "I can do that" to each item. Then try some chapter exercises.

The basic principle of learning is persistence. The main ideas of statistics, like the main ideas of any important subject, took a long time to discover and take some time to master. Once you put it all together—data analysis, data production, probability, and inference—statistics will help you answer important questions for yourself and for those around you.

Exercises

P.13 TV viewing habits You are preparing to study the television-viewing habits of high school students. Describe two categorical variables and two quantitative variables that you might record for each student. Give the units of measurement for the quantitative variables.

P.14 Roll the dice What is the probability of getting a "6" if you roll a fair six-sided die? Explain carefully what your answer means.

P.15 Tap water or bottled water, I Refer to Example P.10 (page 22). Which of the following results would provide more convincing evidence that Mr. Bullard's class could tell bottled water from tap water: 12 out of 21 correct identifications or 14 out of 21 correct identifications? Explain your answer.

P.16 Tap water or bottled water, II Refer to Example P.10 (page 22). Estimate the probability of getting 11 or more correct answers if the students were simply guessing. What would you conclude about whether Mr. Bullard's students could distinguish bottled water from tap water?

P.17 Spinning pennies Hold a penny upright on its edge under your forefinger on a hard surface, then snap it with your other forefinger so that it spins for some time before falling.

2, 3, 4, 5, 6, 7, 8, 9, 10, 11, 12
2 3 4 5 6 7 6 5 4 3 2

Is the coin equally likely to land heads or tails? Spin the coin a total of 20 times, recording whether it lands heads or tails each time.

(a) Make a graph like the one in Figure P.5 (page 21) that shows the proportion of heads after each toss.

(b) Based on your results, estimate the proportion of all spins of the coin that would be heads.

(c) What would you conclude about whether the coin lands heads half the time? Justify your answer.

(d) IN CLASS: Pool your results with those of your classmates. Would you change the conclusion you made in (c)? Why or why not?

P.18 Abstinence or not? An August 2004 Gallup Poll asked 439 teens aged 13 to 17 whether they thought young people should abstain from sex until marriage. 56% said "Yes."

(a) If Gallup had asked *all* teens aged 13 to 17 this question, would exactly 56% have said "Yes"? Explain.

(b) In this sample, 48% of the boys and 64% of the girls said "Yes." Are you convinced that a higher percent of girls than boys aged 13 to 17 feel this way? Why or why not?

CASE CLOSED!

Can magnets help reduce pain?

At the end of each chapter, you will be asked to use what you have learned to resolve the Case Study that opened the chapter. Just like in a court proceeding, you can exclaim "Case Closed!" when you have finished.

Start by reviewing the information in the magnets and pain relief Case Study (page 3). Then answer each of the following questions in complete sentences. Be sure to communicate clearly enough for any of your classmates to understand what you are saying.

1. Data analysis
   a. Answer the key questions: who, what, why, when, where, how, and by whom?
   b. Construct separate dotplots of the pain ratings for the individuals in the active- and inactive-magnet groups. Draw your plots one above the other using the same scale.
   c. Describe what you see in your graphs.
   d. Calculate the *mean* (average) pain rating for each group. Now calculate the difference between the two means.

2. Producing data
   a. Were these available data or new data produced to answer a current question?
   b. Is the design of the study an experiment, a survey, or an observational study that is not a survey? Justify your answer.
   c. Why did researchers let chance decide (by picking from sealed envelopes) whether each patient received an active or inactive magnet?
   d. Would it matter if the patients or doctors knew which type of magnet they had? Explain.

Suppose that the active magnets don't really reduce pain. Then each patient should report the same final pain level whether he or she is assigned to the active- or inactive-magnet group. If the active and inactive magnets are equally effective, we should not observe a very large difference in the mean pain ratings of the two groups. Is the difference you calculated in Question 1(d) large or small? Before you answer, take a look at Figure P.7.

Figure P.7 displays the results of a *simulation* performed using Fathom statistical software. The computer reassigned the patients into the active-

Figure P.7 *Graph from Fathom statistical software displaying the difference in average pain score for the two groups in the magnets and pain study for 10,000 trials of a computer simulation.*

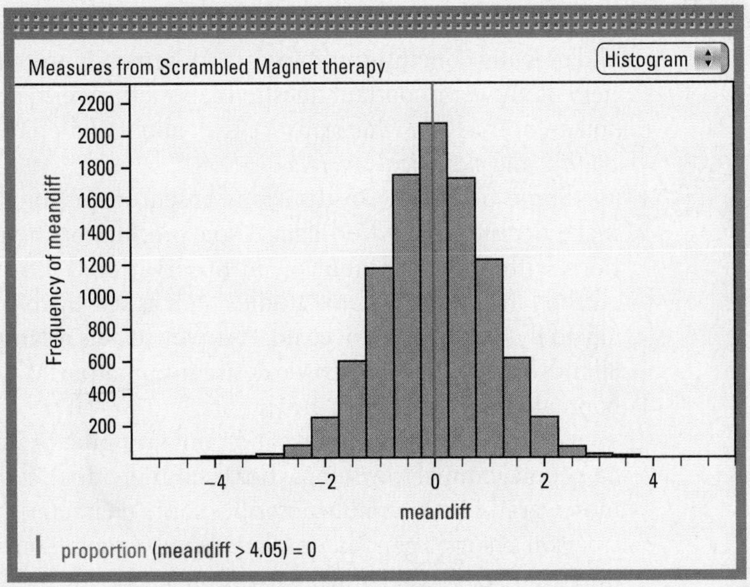

(continued)

and inactive-magnet groups 10,000 times, keeping each patient's final pain score the same as in the actual experiment. Each time, it computed the difference between the mean pain scores reported by the two groups. The graph displays the values of these 10,000 differences.

3. Probability
   a. Use the graph to estimate what percent of the time the difference in the groups' mean pain ratings is greater than 0. Explain your method.
   b. Based on the graph, how likely is it that the difference in mean pain ratings is greater than the one observed in this study (4.05) if the active magnets don't relieve pain?
4. Inference
   a. What would you estimate is the difference in mean pain relief when using active versus inactive magnets? Why?
   b. If you were testing the claim that the active magnets did not help reduce pain any better than the inactive magnets, what would you conclude? Explain.

Chapter Review

Summary

Statistics is the art and science of collecting, organizing, describing, analyzing, and drawing conclusions from data. When used properly, the tools of statistics can help us answer important questions about the world around us. This chapter gave you an overview of what statistics is all about: *data production, data analysis, probability*, and *statistical inference.*

Some people make decisions based on personal experiences. Statisticians make decisions based on data. **Data production** helps us answer specific questions with an **experiment** or an **observational study.** Experiments are distinguished from observational studies such as **surveys** by doing something intentionally to the individuals involved. A survey selects a **sample** from the **population** of all individuals about which we desire information. We base conclusions about the population on data about the sample.

A data set contains information on a number of **individuals.** Individuals may be people, animals, or things. For each individual, the data give values for one or more **variables.** A variable describes some characteristic of an individual, such as a person's height, gender, or salary. Some variables are **categorical** and others are **quantitative.** A categorical variable places each individual into a category, like male or female. A quantitative variable has numerical values that measure some characteristic of each individual, like height in centimeters or annual salary in dollars. Remember to ask the key questions—who, what, why, when, where, how, and by whom?—about any data set.

The **distribution** of a variable describes what values the variable takes and how often it takes these values. To describe a distribution, begin with a graph. You can use **bar graphs** to display categorical variables. A **dotplot** is a simple graph you can use to show the distributions of quantitative variables. When examining any graph, ask yourself "What do I see?"

Exploratory data analysis uses graphs and numerical summaries to describe the variables in a data set and the relations among them. The conclusions of an exploratory analysis may not generalize beyond the specific data studied.

Probability is the language of chance. Chance behavior is unpredictable in the short run but follows a predictable pattern over many repetitions. When we're dealing with chance behavior, the rules of probability help us determine the likelihood of particular outcomes.

Statistical inference produces answers to specific questions, along with a statement of how confident we can be that the answer is correct. The conclusions of statistical inference are usually intended to apply beyond the individuals actually studied. Successful statistical inference requires production of data intended to answer the specific questions posed.

What You Should Have Learned

Here is a review list of the most important skills you should have acquired from your study of this chapter.

A. Where Do Data Come From?

1. Explain why we should not draw conclusions based on personal experiences.
2. Recognize whether a study is an experiment, a survey, or an observational study that is not a survey.
3. Determine the best method for producing data to answer a specific question: experiment, survey, or other observational study.
4. Locate available data on the Internet to help you answer a question of interest.

B. Dealing with Data

1. Identify the individuals and variables in a set of data.
2. Classify each variable as categorical or quantitative. Identify the units in which each quantitative variable is measured.
3. Answer the key questions—who, what, why, when, where, how, and by whom?—about a given set of data.

C. Describing Distributions

1. Make a bar graph of the distribution of a categorical variable. Interpret bar graphs.

2. Make a dotplot of the distribution of a quantitative variable. Describe what you see.

3. Given a relationship between two variables, identify variables lurking in the background that might affect the relationship.

D. **Probability**

1. Interpret probability as what happens in the long run.

2. Use simulations to determine how likely an outcome is to occur.

E. **Statistical Inference**

1. Use the results of simulations and probability calculations to draw conclusions that go beyond the data.

2. Give reasons why conclusions cannot be certain in a given setting.

Web Links

These sites are excellent sources for available data:

U.S. Census Bureau Home Page www.census.gov

Data and Story Library lib.stat.cmu.edu/DASL/

Chapter Review Exercises

P.19 TV violence A typical hour of prime-time television shows three to five violent acts. Linking family interviews and police records shows a clear association between time spent watching TV as a child and later aggressive behavior.[11]

(a) Explain why this is an observational study rather than an experiment.

(b) Suggest several other variables describing a child's home life that may be related to how much TV he or she watches. Explain why these variables make it difficult to conclude that more TV *causes* aggressive behavior.

P.20 How safe are teen drivers? Find some information to help answer this question. Start with the National Highway and Traffic Safety Administration Web site, www.nhtsa.gov. Keep a detailed written record of your search.

P.21 Give it some gas! Here is a small part of a data set that describes the fuel economy (in miles per gallon) of 2004 model motor vehicles:

Make and Model	Vehicle type	Transmission type	Number of cylinders	City MPG	Highway MPG
Acura NSX	Two-seater	Automatic	6	17	24
BMW 330I	Compact	Manual	6	20	30
Cadillac Seville	Midsize	Automatic	8	18	26
Ford F150 2WD	Standard pickup truck	Automatic	6	16	19

Answer the key questions (who, what, why, when, where, how, and by whom?) for these data. Visit the government's fuel economy Web site www.fueleconomy.gov for more information about how these data were produced. For each variable, tell whether it is categorical or quantitative. Be sure to identify the units of measurement for any quantitative variables.

P.22 Wearing bicycle helmets According to the 2003 Youth Risk Behavior Survey, 85.9% of high school students reported rarely or never wearing bicycle helmets. The table below shows additional results from this survey, broken down by gender and grade in school.

	Rarely or never wore bicycle helmets		
Grade	**Female (%)**	**Male (%)**	**Total (%)**
9	80.3	86.4	83.9
10	85.9	88.1	87.1
11	86.8	87.6	87.3
12	86.1	87.5	86.9

(a) Make a bar graph to show the percent of students in each grade who said they rarely or never wore bicycle helmets. Write a few sentences describing what you see.

(b) Now make a side-by-side bar graph to compare the percents of males and females at each grade level who said they rarely or never wore bicycle helmets. Describe what you see in a few sentences.

P.23 Three of a kind You read in a book on poker that the probability of being dealt three of a kind in a five-card poker hand is 1/50. Explain in simple language what this means.

P.24 Baseball and steroids Late in 2004, baseball superstar Barry Bonds admitted using creams and ointments that contained steroids. Bonds said he didn't know that these substances contained steroids. A Gallup Poll asked a random sample of U.S. adults whether they thought Bonds was telling the truth: 42% said "probably not" and 33% said "definitely not."

(a) Why did Gallup survey a random sample of U.S. adults rather than a sample of people attending a Major League Baseball game?

(b) If Gallup had surveyed all U.S. adults instead of a sample, about what percent of the responses would be "probably not"? "Definitely not"? Explain.

(c) Can we conclude based on these results that Barry Bonds is lying? Why or why not?

P.25 Magnets and pain, I Refer to Case Closed! (page 26). Suppose the difference in the mean pain scores of the active and inactive groups had been 2.5 instead of 4.05. What conclusion would you draw about whether magnets help relieve pain in postpolio patients? Explain.

P.26 Magnets and pain, II Refer to the chapter-opening Case Study (page 3). The researchers decided to analyze the patients' final pain ratings. It also makes sense to examine the *difference* between patients' initial pain ratings and their final pain ratings in the active and inactive groups. Here are the data:

Active: 10, 6, 1, 10, 6, 8, 5, 5, 6, 8, 7, 8, 7, 6, 4, 4, 7, 10, 6, 10, 6, 5, 5, 1, 0, 0, 0, 0, 1
Inactive: 4, 3, 5, 2, 1, 4, 1, 0, 0, 1, 0, 0, 0, 0, 0, 0, 0, 1, 0, 0, 1

(a) Construct a dotplot for the active group's data. Describe what you see.

(b) Now make a dotplot for the inactive group's data immediately beneath using the same scale as the graph you made in (a). Write a few sentences comparing the changes in pain ratings for patients in the active and inactive groups.

(c) Calculate the mean (average) change in pain rating for each group.

(d) Figure P.8 shows the results of 10,000 repetitions of a computer simulation. As in Case Closed! (page 26), the computer redistributed the patients into the active- and inactive-magnet groups 10,000 times. Each time, it computed the difference between the mean "decrease in pain" scores reported by the two groups. The graph displays the values of these 10,000 differences. If you were testing the claim that the active magnets did not help reduce pain any better than the inactive magnets, what would you conclude? Explain.

Figure P.8 ***Graph from Fathom statistical software displaying the difference in average decrease in pain for the two groups in the magnets and pain study for 10,000 trials of a computer simulation.***

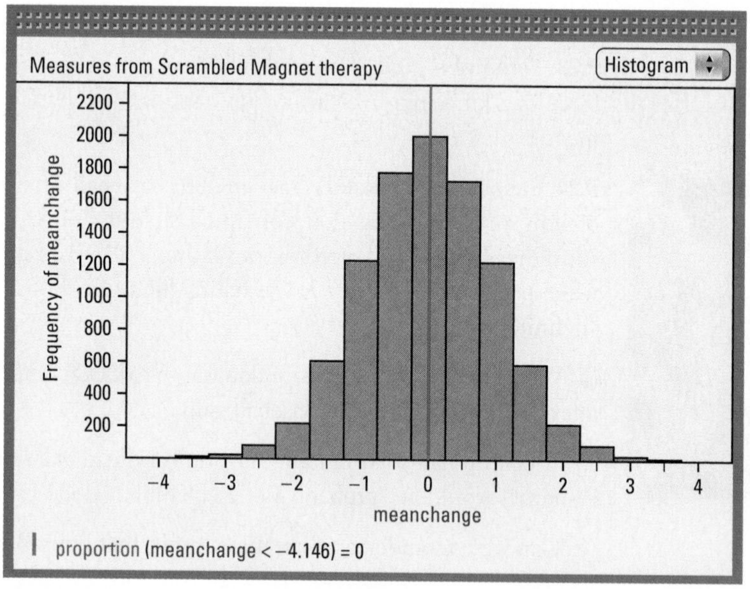

P.27 Are you driving a gas guzzler? Table P.2 displays the highway gas mileage for 30 model year 2004 midsize cars.

Table P.2 *Highway gas mileage for 2004 model midsize cars*

Model	MPG	Model	MPG
Acura 3.5RL	24	Jaguar XJR	24
Audi A6 Quattro	25	Lexus GS300	25
BMW 745I	26	Lexus LS430	25
Buick Regal	30	Lincoln-Mercury LS	24
Cadillac Deville	26	Lincoln-Mercury Sable	26
Cadillac Seville	26	Mercedes-Benz E320	27
Chevrolet Malibu	34	Mercedes-Benz E500	20
Chrysler Sebring	30	Mitsubishi Diamante	25
Dodge Stratus	28	Mitsubishi Galant	26
Honda Accord	34	Nissan Maxima	28
Hyundai Sonata	27	Saab 9-3	28
Infiniti G35	26	Saturn L300	28
Infiniti Q45	23	Toyota Camry	33
Jaguar S-Type 3.0	26	Volkswagen Passat	31
Jaguar Vanden Plas	28	Volvo S80	28

Source: U.S. Environmental Protection Agency, *Model Year 2004 Fuel Economy Guide*, found online at www.fueleconomy.gov.

Make a dotplot of these data. Describe what you see in a few sentences.

P.28 Mozart and test scores The Kalamazoo (Michigan) Symphony once advertised a "Mozart for Minors" program with this statement: "Question: Which students scored 51 points higher in verbal skills and 39 points higher in math? Answer: Students who had experience in music."[12]

(a) How do you think these data were obtained—from an experiment, a survey, or an observational study that wasn't a survey? Justify your answer.

(b) Can we conclude that the "Mozart for Minors" program *caused* an increase in students' test scores? Explain. (*Hint:* Think of a variable lurking in the background.)

(c) Describe an experiment to test whether "Mozart for Minors" really leads to higher test scores.

PART

I

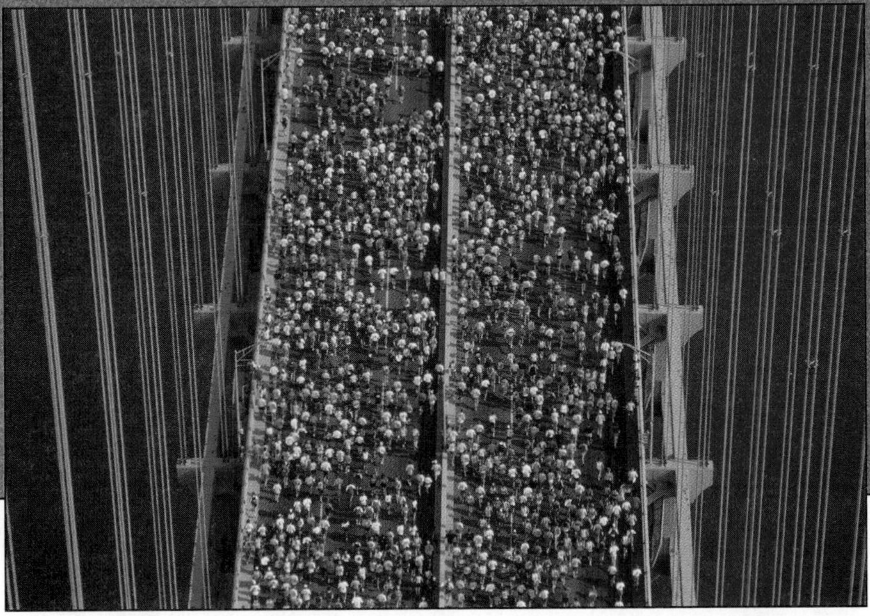

Analyzing Data: Looking for Patterns and Departures from Patterns

CHAPTER

1

Exploring Data

1.1 Displaying Distributions with Graphs

1.2 Describing Distributions with Numbers

Chapter Review

Nielsen ratings

What does it mean to say that a TV show was ranked number 1? The Nielsen Media Research company randomly samples about 5100 households and 13,000 individuals each week. The TV viewing habits of this sample are captured by metering equipment, and data are sent automatically in the middle of the night to Nielsen. Broadcasters and companies that want to air commercials on TV use the data on who is watching TV and what they are watching. The results of this data gathering appear as ratings on a weekly basis. For more information on the Nielsen TV ratings, go to www.nielsenmedia.com, and click on "Inside TV Ratings." Then under "Related," select "What Are TV Ratings?"

Here are the top prime-time shows for viewers aged 18 to 49 during the week of November 22–28, 2004.

Show	Network	Viewers (millions)
1. *Desperate Housewives*	ABC	16.2
2. *CSI*	CBS	10.9
3. *CSI: Miami*	CBS	10.5
4. *Extreme Makeover: Home Edition*	ABC	9.7
5. *Two and a Half Men*	CBS	8.8
6. *Without a Trace*	CBS	8.2
7. *Raymond*	CBS	8.0
8. *Law & Order: SVU*	NBC	7.8
9. *Monday Night Football*	ABC	7.8
10. *Survivor: Vanuatu*	CBS	7.8
11. *Seinfeld Story*	NBC	7.6
12. *Boston Legal*	ABC	7.4
13. *Apprentice*	NBC	7.1
14. *Fear Factor*	NBC	6.5
15. *Amazing Race*	CBS	6.1
16. *CSI: NY*	CBS	6.1
17. *NFL Monday Showcase*	ABC	5.7
18. *According to Jim*	ABC	5.5
19. *60 Minutes*	CBS	5.4
20. *Biggest Loser*	NBC	5.4

Source: *USA Today*, December 2, 2004.

Which network is winning the ratings battle? At the end of this chapter, you will be asked to use what you have learned to answer this question.

Activity 1A

How fast is your heart beating?

Materials: Clock or watch with second hand

A person's pulse rate provides information about the health of his or her heart. Would you expect to find a difference between male and female pulse rates? In this activity, you and your classmates will collect some data to try to answer this question.

1. To determine your pulse rate, hold the fingers of one hand on the artery in your neck or on the inside of your wrist. (The thumb should not be used, because there is a pulse in the thumb.) Count the number of pulse beats in one minute. As Jeremy notes in the cartoon above, you need sufficient data, so do this three times, and calculate your *average* individual pulse rate. Why is doing this three times better than doing it once?
2. Record the pulse rates for the class in a table, with one column for males and a second column for females. Are there any unusual pulse rates?
3. For now, simply calculate the average pulse rate for the males and the average pulse rate for the females, and compare.

Introduction

When you go to the movie theater, you see previews of upcoming movies. Similarly, our goal in the Preliminary Chapter was to give you a taste of what statistics is about and a sense of what lies ahead in this book. Now we are ready for the details. In this chapter, we will explore ways to describe data, both graphically and numerically. Initially, we will focus on ways to plot our data. We will learn how to construct histograms, stemplots, and boxplots, and how to decide which plot is best for different sets of circumstances. We'll talk about distributions, because distributions will come up again and again during our study of statistics. We'll also discuss ways to describe distributions numerically, including measures of center and spread. Our recurring theme throughout will be looking for patterns and departures from patterns.

1.1 Displaying Distributions with Graphs

Graphs for Categorical Variables

Recall our distinction between quantitative and categorical variables in the Preliminary Chapter. The values of a categorical variable are labels for the categories, such as "female" and "male." The distribution of a categorical variable lists the categories and gives either the count or the percent of individuals who fall in each category. The next two examples show you how to use graphical displays to examine the distribution of a categorical variable.

Example 1.1 *Radio station formats*
Bar graphs and pie charts

The radio audience rating service Arbitron places the country's 13,838 radio stations into categories that describe the kind of programs they broadcast. Here is the distribution of station formats:[1]

Format	Count of stations	Percent of stations
Adult contemporary	1,556	11.2
Adult standards	1,196	8.6
Contemporary hit	569	4.1
Country	2,066	14.9
News/Talk/Information	2,179	15.7
Oldies	1,060	7.7
Religious	2,014	14.6
Rock	869	6.3
Spanish language	750	5.4
Other formats	1,579	11.4
Total	**13,838**	**99.9**

roundoff error

It's a good idea to check data for consistency. The counts should add to 13,838, the total number of stations. They do. The percents should add to 100%. In fact, they add to 99.9%. What happened? Each percent is rounded to the nearest tenth. The exact percents would add to 100, but the rounded percents only come close. This is ***roundoff error***. Roundoff errors don't point to mistakes in our work, just to the effect of rounding off results.

Columns of numbers take time to read. You can use a pie chart or a bar graph to display the distribution of a categorical variable more vividly. Figure 1.1 illustrates both displays for the distribution of radio stations by format.

Figure 1.1 *(a) Pie chart of radio stations by format. (b) Bar graph of radio stations by format.*

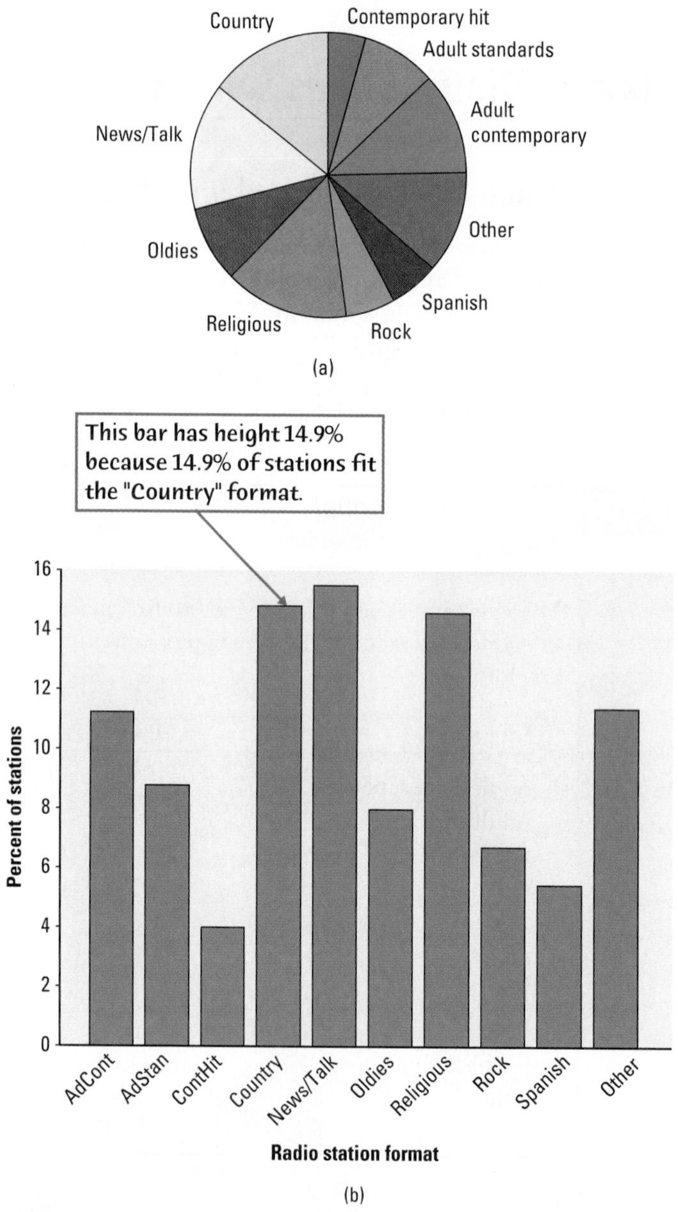

pie chart

CAUTION

Pie charts are awkward to make by hand, but software will do the job for you. **A pie chart must include all the categories that make up a whole. Use a pie chart only when you want to emphasize each category's relation to the whole.**

bar graph

We need the "Other formats" category in Example 1.1 to complete the whole (all radio stations) and allow us to make a pie chart. ***Bar graphs*** are easier to make and also easier to read, as Figure 1.1(b) illustrates. Bar graphs are more flexible than pie charts. Both graphs can display the distribution of a categorical variable, but a bar graph can also compare any set of quantities that are measured in the same units. Bar graphs and pie charts help an audience grasp data quickly.

Example 1.2 *Do you listen while you walk?*

When pie charts won't work

Portable MP3 music players, such as the Apple iPod, are popular—but not equally popular with people of all ages. Here are the percents of people in various age groups who own a portable MP3 player:[2]

Age group (years)	Percent owning an MP3 player
12–17	27
18–24	18
25–34	20
35–44	16
45–54	10
55–64	6
65+	2

It's clear that MP3 players are popular mainly among young people.

We can't make a pie chart to display these data. Each percent in the table refers to a different age group, not to parts of a single whole. The bar graph in Figure 1.2 compares the seven age groups. We see at a glance that MP3 player ownership generally declines with age.

Figure 1.2 *Bar graph comparing the percents of several age groups who own portable MP3 players.*

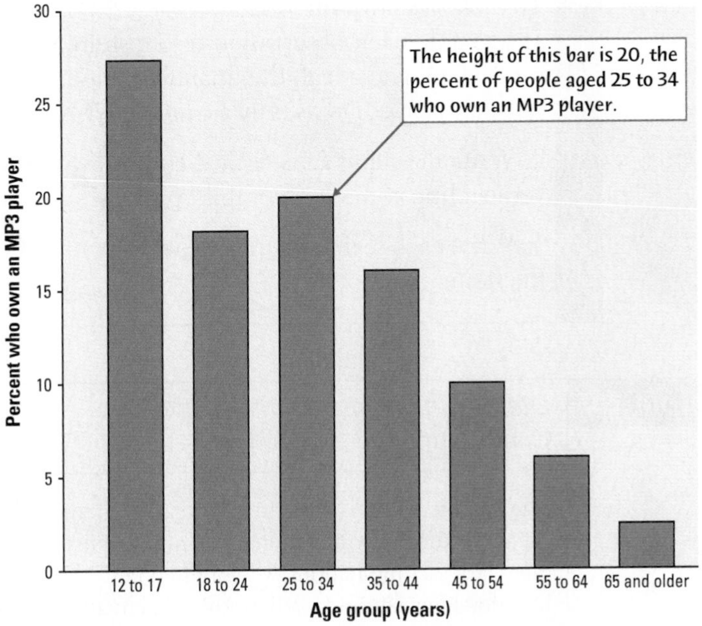

There is one question that you should always ask when you look at data, as the following example illustrates.

Example 1.3 *MP3 downloads*
Do the data tell you what you want to know?

Let's say that you plan to buy radio time to advertise your Web site for downloading MP3 music files. How helpful are the data in Example 1.1? Not very. You are interested, not in counting *stations*, but in counting *listeners*. For example, 14.6% of all stations are religious, but they have only a 5.5% share of the radio audience. In fact, you aren't even interested in the entire radio audience, because MP3 users are mostly young people. You really want to know what kinds of radio stations reach the largest numbers of young people. **Always think about whether the data you have help answer your questions.**

CAUTION

We now turn to the kinds of graphs that are used to describe the distribution of a quantitative variable. We will explain how to make the graphs by hand, because knowing this helps you understand what the graphs show. However, making graphs by hand is so tedious that software is almost essential for effective data analysis unless you have just a few observations.

Stemplots

A ***stemplot*** (also called a stem-and-leaf plot) gives a quick picture of the shape of a distribution while including the actual numerical values in the graph. Stemplots work best for small numbers of observations that are all greater than 0.

Stemplot

To make a **stemplot:**

1. Separate each observation into a **stem,** consisting of all but the final (rightmost) digit, and a **leaf,** the final digit. Stems may have as many digits as needed, but each leaf contains only a single digit.

2. Write the stems in a vertical column with the smallest at the top, and draw a vertical line at the right of this column.

3. Write each leaf in the row to the right of its stem, in increasing order out from the stem.

Example 1.4 *Literacy in Islamic countries*
Constructing and interpreting a stemplot

The Islamic world is attracting increased attention in Europe and North America. Table 1.1 shows the percent of men and women at least 15 years old who were literate in 2002 in the major Islamic nations. We omitted countries with populations of less than 3 million. Data for a few nations, such as Afghanistan and Iraq, are not available.[3]

Table 1.1 *Literacy rates in Islamic nations*

Country	Female percent	Male percent	Country	Female percent	Male percent
Algeria	60	78	Morocco	38	68
Bangladesh	31	50	Saudi Arabia	70	84
Egypt	46	68	Syria	63	89
Iran	71	85	Tajikistan	99	100
Jordan	86	96	Tunisia	63	83
Kazakhstan	99	100	Turkey	78	94
Lebanon	82	95	Uzbekistan	99	100
Libya	71	92	Yemen	29	70
Malaysia	85	92			

Source: United Nations data found at www.earthtrends.wri.org.

To make a stemplot of the percents of females who are literate, use the first digits as stems and the second digits as leaves. Algeria's 60% literacy rate, for example, appears as the leaf 0 on the stem 6. Figure 1.3 shows the steps in making the plot.

Figure 1.3 *Making a stemplot of the data in Example 1.4. (a) Write the stems. (b) Go through the data and write each leaf on the proper stem. For example, the values on the 8 stem are 86, 82, and 85 in the order of Table 1.1. (c) Arrange the leaves on each stem in order out from the stem. The 8 stem now has leaves 2, 5, and 6.*

```
2 |        2 | 9          2 | 9
3 |        3 | 1 8        3 | 1 8
4 |        4 | 6          4 | 6
5 |        5 |            5 |
6 |        6 | 0 3 3      6 | 0 3 3
7 |        7 | 1 1 0 8    7 | 0 1 1 8
8 |        8 | 6 2 5      8 | 2 5 6
9 |        9 | 9 9 9      9 | 9 9 9
  (a)        (b)            (c)
```

cluster

The overall pattern of the stemplot is irregular, as is often the case when there are only a few observations. There do appear to be two ***clusters*** of countries. The plot suggests that we might want to investigate the variation in literacy. For example, why do the three central Asian countries (Kazakhstan, Tajikistan, and Uzbekistan) have very high literacy rates?

back-to-back stemplot

When you wish to compare two related distributions, a ***back-to-back stemplot*** with common stems is useful. The leaves on each side are ordered out from the common stem. Here is a back-to-back stemplot comparing the distributions of female and male literacy rates in the countries of Table 1.1.

Female		Male
9	2	
8 1	3	
6	4	
	5	0
3 3 0	6	8 8
8 1 1 0	7	0 8
6 5 2	8	3 4 5 9
9 9 9	9	2 2 4 5 6
	10	0 0 0

The values on the left are the female percents, as in Figure 1.3, but ordered out from the stem, from right to left. The values on the right are the male percents. It is clear that literacy is generally higher among males than among females in these countries.

splitting stems

trimming

Stemplots do not work well for large data sets where each stem must hold a large number of leaves. Fortunately, there are two modifications of the basic stemplot that are helpful when plotting the distribution of a moderate number of observations. You can double the number of stems in a plot by ***splitting stems*** into two: one with leaves 0 to 4 and the other with leaves 5 through 9. When the observed values have many digits, ***trimming*** the numbers by removing the last digit or digits before making a stemplot is often best. You must use your judgment in deciding whether to split stems and whether to trim, though statistical software will often make these choices for you. Remember that the purpose of a stemplot is to display the shape of a distribution. Here is an example that makes use of both of these modifications.

Example 1.5 *Virginia college tuition*

Trimming and splitting stems

Tuition and fees for the 2005–2006 school year for 37 four-year colleges and universities in Virginia are shown in Table 1.2. In addition, there are 23 two-year community colleges that each charge \$2135. The data for these 60 schools were entered into a Minitab worksheet and the following stemplot was produced.

Let's see how the software has simplified and then plotted the data. The last leaf is the cost of tuition and fees for the University of Richmond (UR). Notice that the "Leaf Unit" is given as 1000. The actual figure for UR is \$34,850. In the stemplot, the stem is 3 and the leaf is 4. Since the leaf unit is given as 1000, the plotted value for UR is then \$34,000. Minitab has truncated, or chopped off, the last three digits, leaving 34 thousand, and this truncated number is plotted. This modification is called "trimming." Notice that the number is not rounded off.

Table 1.2 *Tuition and fees at Virginia colleges and universities for the 2005–2006 school year*

Virginia colleges	Tuition and fees ($)	Virginia colleges	Tuition and fees ($)
Averett	18,430	Patrick Henry	14,645
Bluefield	10,615	Randolph—Macon	22,625
Christendom	14,420	Randolph—Macon Women's	21,740
Christopher Newport	12,626	Richmond	34,850
DeVry	12,710	Roanoke	22,109
Eastern Mennonite	18,220	Saint Paul's	9,420
Emory and Henry	16,690	Shenandoah	19,240
Ferrum	16,870	Sweet Briar	21,080
George Mason	15,816	University of Virginia	22,831
Hampton	14,996	University of Virginia—Wise	14,152
Hampton—Sydney	22,944	Virginia Commonwealth	17,262
Hollins	21,675	Virginia Intermont	15,200
Liberty	13,150	Virginia Military Institute	19,991
Longwood	12,901	Virginia State	11,462
Lynchburg	22,885	Virginia Tech	16,530
Mary Baldwin	19,991	Virginia Union	12,260
Marymount	17,090	Washington and Lee	25,760
Norfolk State	14,837	William and Mary	21,796
Old Dominion	14,688		

Source: cgi.money.cnn.com/tools/collegecost/collegecost.html.

Figure 1.4 *Stemplot of tuitions and fees for 60 colleges and universities in Virginia, made by Minitab.*

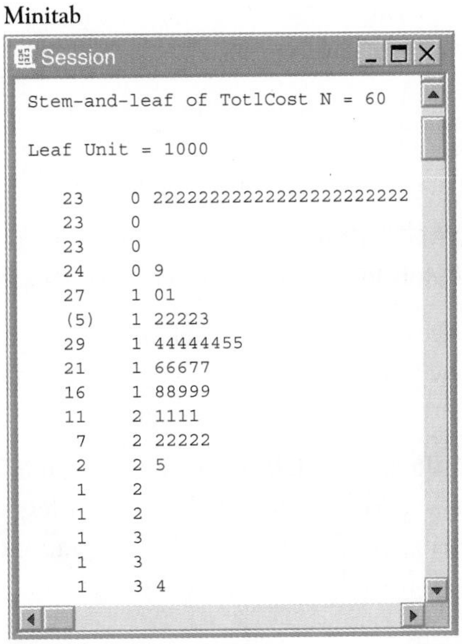

If each stem were a different number, then the plot would appear much more compact:

0 2222222222222222222222229
1 0122223444444556667788999
2 1111222225
3 4

Each stem now represents 10,000. The problem with this plot is that some important details of the distribution, such as the overall shape, gaps, and outliers, are obscured. So Minitab has split the stems so that the range of values for each stem is now 2000 rather than 10,000. This has the effect of stretching out the stemplot so that a more informative picture of the distribution can be seen. Other statistical software will handle trimming and splitting the stems in a similar manner.

Exercises

1.1 What's in the garbage? The formal name for garbage is "municipal solid waste." Here is a breakdown of the materials that made up American municipal solid waste in 2000:[4]

Material	Weight (million tons)	Percent of total
Food scraps	25.9	11.2
Glass	12.8	5.5
Metals	18.0	7.8
Paper, paperboard	86.7	37.4
Plastics	24.7	10.7
Rubber, leather, textiles	15.8	6.8
Wood	12.7	5.5
Yard trimmings	27.7	11.9
Other	7.5	3.2
Total	**231.9**	**100.0**

(a) Add the weights for the nine materials given, including "Other." Each entry, including the total, is separately rounded to the nearest tenth. So the sum and the total may differ slightly because of roundoff error.

(b) Make a bar graph of the percents. The graph gives a clearer picture of the main contributors to garbage if you order the bars from tallest to shortest.

(c) If you use software, also make a pie chart of the percents. Comparing the two graphs, notice that it is easier to see the small differences among "Food scraps," "Plastics," and "Yard trimmings" in the bar graph.

1.2 Do the data tell you what you want to know? To help you plan advertising for a Web site for downloading MP3 music files, you want to know what percent of owners of portable MP3 players are 18 to 24 years old. The data in Example 1.2 do *not* tell you what you want to know. Why not?

1.3 Cheese and chemistry As cheddar cheese matures, a variety of chemical processes take place. The taste of mature cheese is related to the concentration of several chemicals in the final product. In a study of cheddar cheese from the Latrobe Valley of Victoria, Australia, samples of cheese were analyzed for their chemical composition. The final concentrations of lactic acid in the 30 samples, as a multiple of their initial concentrations, are given below.[5]

0.86	1.53	1.57	1.81	0.99	1.09	1.29	1.78	1.29	1.58
1.68	1.90	1.06	1.30	1.52	1.74	1.16	1.49	1.63	1.99
1.15	1.33	1.44	2.01	1.31	1.46	1.72	1.25	1.08	1.25

A dotplot and a stemplot from the Minitab statistical software package are shown in Figure 1.5. Recall that dotplots were discussed in the Preliminary Chapter.

Figure 1.5 *Minitab dotplot and stemplot for cheese data, for Exercise 1.3.*

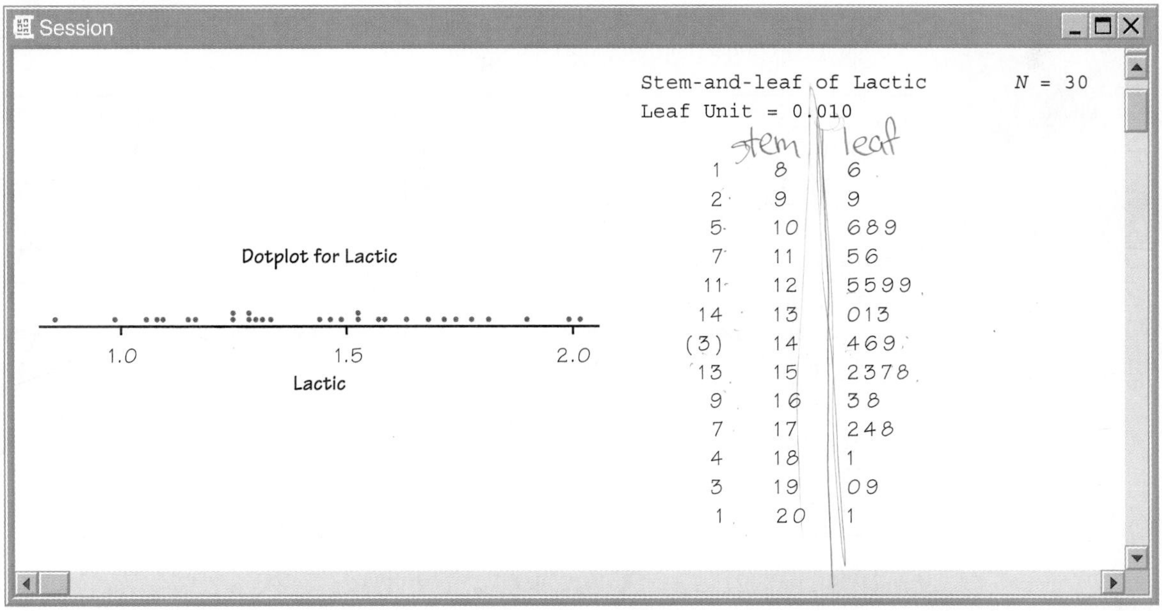

(a) Which plot does a better job of summarizing the data? Explain why.

(b) What do the numbers in the left column in the stemplot tell us? How does Minitab identify the row that contains the center of the distribution?

(c) The final concentration of lactic acid in one of the samples stayed approximately the same (as its initial concentration). Identify this sample in both plots.

1.4 Virginia college tuitions Tuitions and fees for the 2005–2006 school year for 60 two- and four-year colleges and universities in Virginia are shown in Table 1.2 (page 45) and the stemplot in Figure 1.4 (page 45) in Example 1.5. They ranged from a low of \$2135 for the two-year community colleges to a high of \$34,850 for the University of Richmond.

(a) Which stem and leaf represent Liberty University? Which stem and leaf represent Virginia State University? Identify the colleges represented in the row 2 1111.

(b) There is a string of 23 twos at the top of the stemplot. Which colleges do these numbers refer to? The stem for these entries is 0. What range of numbers would be plotted with this stem?

1.5 DRP test scores There are many ways to measure the reading ability of children. One frequently used test is the Degree of Reading Power (DRP). In a research study on third-grade students, the DRP was administered to 44 students.[6] Their scores were

40	26	39	14	42	18	25	43	46	27	19
47	19	26	35	34	15	44	40	38	31	46
52	25	35	35	33	29	34	41	49	28	52
47	35	48	22	33	41	51	27	14	54	45

Display these data graphically. Write a paragraph describing the distribution of DRP scores.

1.6 Shopping spree, I A marketing consultant observed 50 consecutive shoppers at a supermarket. One variable of interest was how much each shopper spent in the store. Here are the data (in dollars), arranged in increasing order:

3.11	8.88	9.26	10.81	12.69	13.78	15.23	15.62	17.00	17.39
18.36	18.43	19.27	19.50	19.54	20.16	20.59	22.22	23.04	24.47
24.58	25.13	26.24	26.26	27.65	28.06	28.08	28.38	32.03	34.98
36.37	38.64	39.16	41.02	42.97	44.08	44.67	45.40	46.69	48.65
50.39	52.75	54.80	59.07	61.22	70.32	82.70	85.76	86.37	93.34

(a) Round each amount to the nearest dollar. Then make a stemplot using tens of dollars as the stems and dollars as the leaves.

(b) Make another stemplot of the data by splitting stems. Which of the plots shows the shape of the distribution better?

(c) Write a few sentences describing the amount of money spent by shoppers at this supermarket.

Histograms

Stemplots display the actual values of the observations. This feature makes stemplots awkward for large data sets. Moreover, the picture presented by a stemplot divides the observations into groups (stems) determined by the number system rather than by judgment. Histograms do not have these limitations. A ***histogram*** breaks the range of values of a variable into *classes* and displays only the count or percent of the observations that fall into each class. You can choose any convenient number of classes, but you should *always choose classes of equal width*. Histograms are slower to construct by hand than stemplots and do not display the actual values observed. For these reasons we prefer stemplots for small data sets. The construction of a histogram is best shown by example. Any statistical software package will of course make a histogram for you, as will your calculator.

histogram

Example 1.6 *IQ scores*

Making a histogram

You have probably heard that the distribution of scores on IQ tests follows a bell-shaped pattern. Let's look at some actual IQ scores. Table 1.3 displays the IQ scores of 60 fifth-grade students chosen at random from one school.[7]

Table 1.3 *IQ test scores for 60 randomly chosen fifth-grade students*

145	139	126	122	125	130	96	110	118	118
101	142	134	124	112	109	134	113	81	113
123	94	100	136	109	131	117	110	127	124
106	124	115	133	116	102	127	117	109	137
117	90	103	114	139	101	122	105	97	89
102	108	110	128	114	112	114	102	82	101

Source: James T. Fleming, "The measurement of children's perception of difficulty in reading materials," *Research in the Teaching of English*, 1 (1967), pp. 136–156.

Step 1. Divide the range of the data into classes of equal width. The scores in Table 1.3 range from 81 to 145, so we choose as our classes

$$75 \leq \text{IQ score} < 85$$
$$85 \leq \text{IQ score} < 95$$
$$\vdots$$
$$145 \leq \text{IQ score} < 155$$

Be sure to specify the classes precisely so that each individual falls into exactly one class. A student with IQ 84 would fall into the first class, but IQ 85 falls into the second.

frequency
frequency table

Step 2. Count the number of individuals in each class. These counts are called ***frequencies,*** and a table of frequencies for all classes is a ***frequency table.***

Class	Count	Class	Count
75 to 84	2	115 to 124	13
85 to 94	3	125 to 134	10
95 to 104	10	135 to 144	5
105 to 114	16	145 to 154	1

Step 3. Draw the histogram and label your axes. First, on the horizontal axis mark the scale for the variable whose distribution you are displaying. That's IQ score. The scale runs from 75 to 155 because that is the span of the classes we chose. The vertical axis contains the scale of counts. Each bar represents a class. The base of the bar covers the class, and the bar height is the class count. There is no horizontal space between the bars unless a class is empty, so that its bar has height zero. Figure 1.6 is our histogram. It does look roughly "bell-shaped."

Figure 1.6 *Histogram of the IQ scores of 60 fifth-grade students, for Example 1.6.*

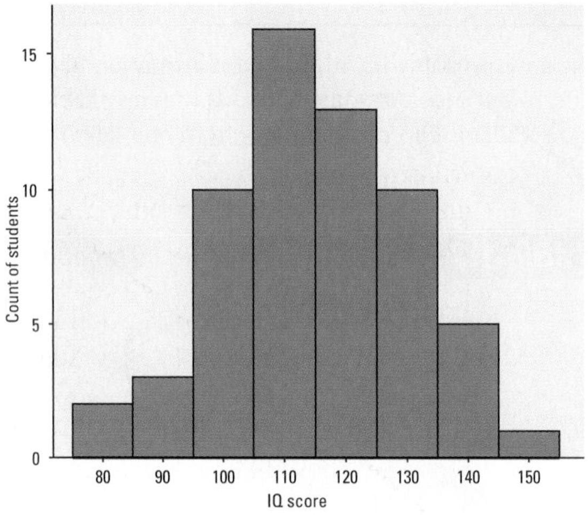

A good way to get a feel for how to optimize the presentation of a histogram is to play with an interactive histogram on the computer. The histogram function in the *One-Variable Statistical Calculator* applet on the student CD and Web site is particularly useful because you can change the number of classes by dragging with the mouse. So it's easy to see how the choice of classes affects the histogram.

APPLET

Activity 1B

The one-variable statistical calculator

The *One-Variable Statistical Calculator* applet on the book's Web site, www.whfreeman.com/tps3e, will make stemplots and histograms. It is intended mainly as a learning tool rather than a replacement for statistical software.

1. Go to the Web site and launch the *One-Variable Statistical Calculator* applet. Your screen should look like this:

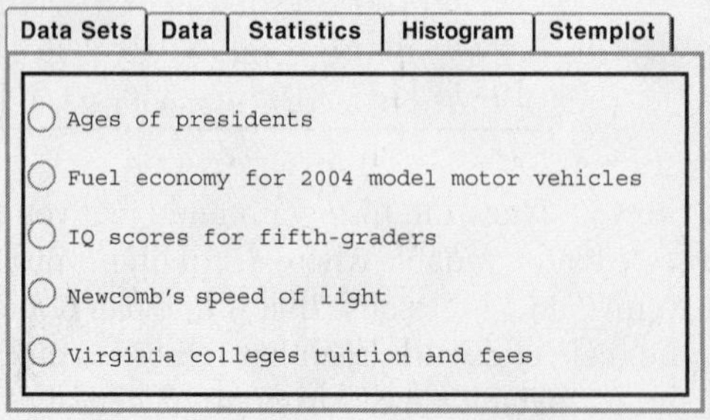

2. Choose the "Virginia colleges tuition and fees" data set, and then click on the "Histogram" tab.
   a. Sketch the default histogram that the applet first presents. If the default graph does not have eight classes, drag it to make a histogram with eight classes and sketch the result.
   b. Make a histogram with one class and also a histogram with the greatest number of classes that the applet allows. Sketch the results.
   c. Drag the graph until you find the histogram that you think best pictures the data. How many classes did you choose? Note that if you hold the mouse button down, you can see a popup box that tells you the number of classes. Sketch your final histogram.
   d. Click on STEMPLOT. Does the applet replicate the stemplot in Figure 1.4 (page 45) in your text? Why do you think the applet won't replicate the picture in your textbook exactly?
3. Select the data set "IQ scores of fifth-graders."
   a. See if you can replicate the histogram in Figure 1.6. Drag the scale until your histogram looks as much like the histogram in Figure 1.6 as you can make it. How many classes are there in your histogram? Describe the shape of your histogram.
   b. Click on STEMPLOT. Sketch the shape of the stemplot. Is the shape of the stemplot similar to the shape of your histogram?
   c. Select SPLIT STEMS. Does this improve the appearance of the stemplot? In what way?
4. This applet can create a histogram for almost any data set. Select a data set of your choice, click on DATA, and enter your data. Produce a histogram and drag the scale until you are happy with the result. Sketch the histogram. Click on STEMPLOT. Split the stems as needed.

Histogram Tips

Here are some important properties of histograms to keep in mind when you are constructing a histogram.

- Our eyes respond to the area of the bars in a histogram, so *be sure to choose classes that are all the same width.* Then area is determined by height and all classes are fairly represented.

- There is no one right choice of the classes in a histogram. Too few classes will give a "skyscraper" graph, with all values in a few classes with tall bars. Too many will produce a "pancake" graph, with most classes having one or no observations. Neither choice will give a good picture of the shape of the distribution. Five classes is a good minimum. Bottom line: *Use your judgment in choosing classes to display the shape.*

- Statistical software and graphing calculators will choose the classes for you. The default choice is often a good one, but you can change it if you want. *Beware of letting the device choose the classes.*
- *Use histograms of percents for comparing several distributions with different numbers of observations.* Large sets of data are often reported in the form of frequency tables when it is not practical to publish the individual observations. In addition to the frequency (count) for each class, we may be interested in the fraction or percent of the observations that fall in each class. A histogram of percents looks just like a frequency histogram such as Figure 1.6. Simply relabel the vertical scale to read in percents.

Histograms versus Bar Graphs

Although histograms resemble bar graphs, their details and uses are distinct. A histogram shows the distribution of counts or percents among the values of a single quantitative variable. A bar graph displays the distribution of a categorical variable. The horizontal axis of a bar graph identifies the values of the categorical variable. Draw bar graphs with blank space between the bars to separate the items being compared. Draw histograms with no space, to indicate that all values of the variable are covered.

The vital few? Skewed distributions can show us where to concentrate our efforts. Ten percent of the cars on the road account for half of all carbon dioxide emissions. A histogram of CO_2 emissions would show many cars with small or moderate values and a few with very high values. Cleaning up or replacing these cars would reduce pollution at a cost much lower than that of programs aimed at all cars. Statisticians who work at improving quality in industry make a principle of this: distinguish "the vital few" from "the trivial many."

Examining Distributions

Constructing a graph is only the first step. The next step is to interpret what you see. When you describe a distribution, you should pay attention to the following features.

Examining a Distribution

In any graph of data, look for the **overall pattern** and for striking **deviations** from that pattern.

You can describe the overall pattern of a distribution by its **shape, center,** and **spread.**

An important kind of deviation is an **outlier,** an individual value that falls outside the overall pattern.

We will learn more about describing center and spread numerically in Section 1.2. For now, we can describe the center of a distribution by its *midpoint,* the value with roughly half the observations taking smaller values and half taking larger values. We can describe the spread of a distribution by giving the *smallest and largest values.* Stemplots and histograms display the shape of a distribution in the same way. Just imagine a stemplot turned on its side so that the larger values lie to the right. Here are some things to look for in describing shape:

modes
unimodal

- Does the distribution have one or several major peaks, called ***modes?*** A distribution with one major peak is called ***unimodal***.

symmetric
skewed

- Is the distribution approximately symmetric or is it skewed in one direction? A distribution is ***symmetric*** if the values smaller and larger than its midpoint are mirror images of each other. It is ***skewed*** to the right if the *right tail* (larger values) is much longer than the *left tail* (smaller values).

Some variables commonly have distributions with predictable shapes. Many biological measurements on specimens from the same species and sex—lengths of bird bills, heights of young women—have symmetric distributions. Salaries, savings, and home prices, on the other hand, usually have right-skewed distributions. There are many moderately priced houses, for example, but the few very expensive mansions give the distribution of house prices a strong right skew.

Example 1.7 ***IQ scores and Virginia tuitions and fees***
Interpreting a histogram

What does the histogram of IQ scores in Figure 1.6 (page 50) tell us?

Shape: The distribution is *roughly symmetric* with a *single peak* in the center. We don't expect real data to be perfectly symmetric, so we are satisfied if the two sides of the histogram are roughly similar in shape and extent. **Center:** You can see from the histogram that the midpoint is not far from 110. Looking at the actual data shows that the midpoint is 114. **Spread:** The spread is from about 80 to about 150 (actually 81 to 145). There are no outliers or other strong deviations from the symmetric, unimodal pattern.

The distribution of Virginia's tuitions and fees in Figure 1.4 (page 45) can be summarized as follows:

Shape: The distribution of all two- and four-year colleges is right-skewed. However, if the 23 community college costs are deleted, that is, if we look only at four-year colleges, then the distribution is roughly bell-shaped. **Center:** There are two middle numbers ($n = 60$): 12,901 and 13,150. **Spread:** The data range from 2135 to 34,850. The University of Richmond is a striking deviation. One way to look at this is that 16 students could attend a Virginia community college for about the same cost as that for 1 student at the University of Richmond.

Dealing with Outliers

CAUTION

With small data sets, you can spot outliers by looking for observations that stand apart (either high or low) from the overall pattern of a histogram or stemplot. **Identifying outliers is a matter of judgment. Look for points that are clearly apart from the body of the data, not just the most extreme observations in a distribution.** You should search for an explanation for any outlier. Sometimes outliers point to errors made in recording the data. In other cases, the outlying observation may be caused by equipment failure or other unusual circumstances. In the next section we'll learn a rule of thumb that makes identifying outliers more precise.

Example 1.8 *Electronic components*

Outliers

Manufacturing an electronic component requires attaching very fine wires to a semiconductor wafer. If the strength of the bond is weak, the component may fail. Here are measurements on the breaking strength (in pounds) of 23 connections:[8]

0	0	550	750	950	950	1150	1150
1150	1150	1150	1250	1250	1350	1450	1450
1450	1550	1550	1550	1850	2050	3150	

Figure 1.7 is a histogram of these data.

Figure 1.7 *Histogram of a distribution with both low and high outliers, for Example 1.8.*

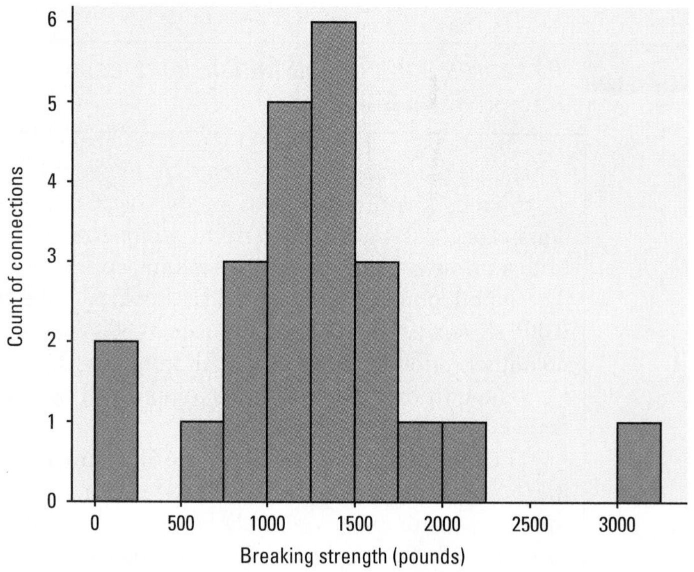

We expect the breaking strengths of supposedly identical connections to have a roughly symmetric overall pattern, showing chance variation among the connections. Figure 1.7 does show a symmetric pattern centered at about 1250 pounds—but it also shows three outliers that stand apart from this pattern, two low and one high.

The engineers were able to explain all three outliers. The two low outliers had strength 0 because the bonds between the wire and the wafer were not made. The high outlier at 3150 pounds was a measurement error. Further study of the data can simply omit the three outliers. **Note that, in general, it is not a good idea to just delete or ignore outliers.** One immediate finding is that the variation in breaking strength is too large—550 pounds to 2050 pounds when we ignore the outliers. The process of bonding wire to wafer must be improved to give more consistent results.

Exercises

1.7 Stock returns The total return on a stock is the change in its market price plus any dividend payments made. Total return is usually expressed as a percent of the beginning price. Figure 1.8 is a histogram of the distribution of total returns for all 1528 stocks listed on the New York Stock Exchange in one year.[9] Note that it is a histogram of the percents in each class rather than a histogram of counts.

Figure 1.8 *Histogram of the distribution of percent total return for all New York Stock Exchange common stocks in one year, for Exercise 1.7.*

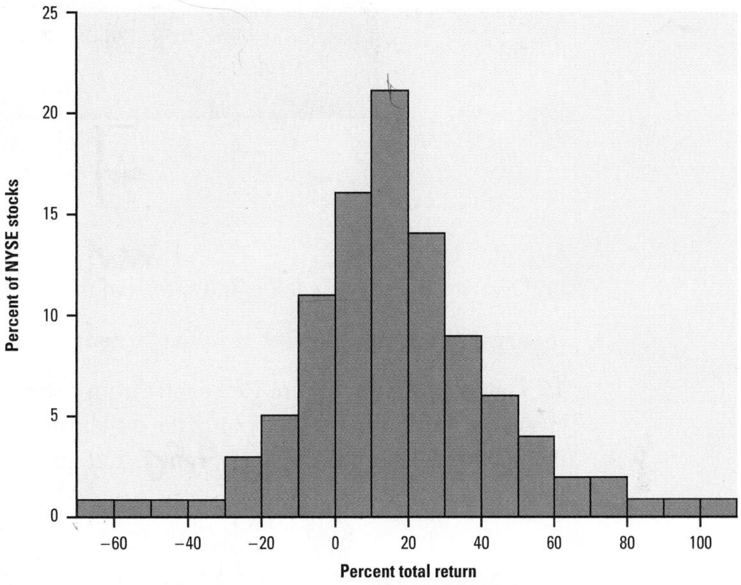

(a) Describe the overall shape of the distribution of total returns.

(b) What is the approximate center of this distribution? (For now, take the center to be the value with roughly half the stocks having lower returns and half having higher returns.)

(c) Approximately what were the smallest and largest total returns? (This describes the spread of the distribution.)

(d) A return less than zero means that an owner of the stock lost money. About what percent of all stocks lost money?

1.8 Freezing in Greenwich, England Figure 1.9 is a histogram of the number of days in the month of April on which the temperature fell below freezing at Greenwich, England.[10] The data cover a period of 65 years.

Figure 1.9 *The distribution of the number of frost days during April at Greenwich, England, over a 65-year period, for Exercise 1.8.*

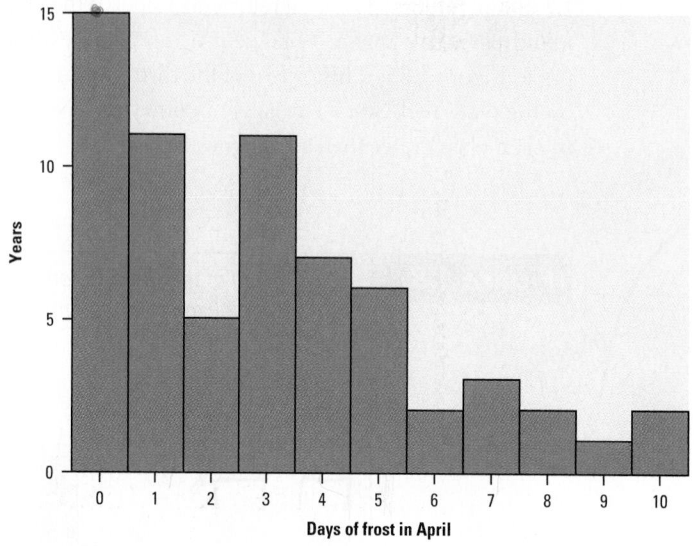

(a) Describe the shape, center, and spread of this distribution. Are there any outliers?

(b) In what percent of these 65 years did the temperature never fall below freezing in April?

1.9 Lightning storms Figure 1.10 comes from a study of lightning storms in Colorado. It shows the distribution of the hour of the day during which the first lightning flash for that day occurred. Describe the shape, center, and spread of this distribution.

Figure 1.10 *The distribution of the time of the first lightning flash each day at a site in Colorado, for Exercise 1.9.*

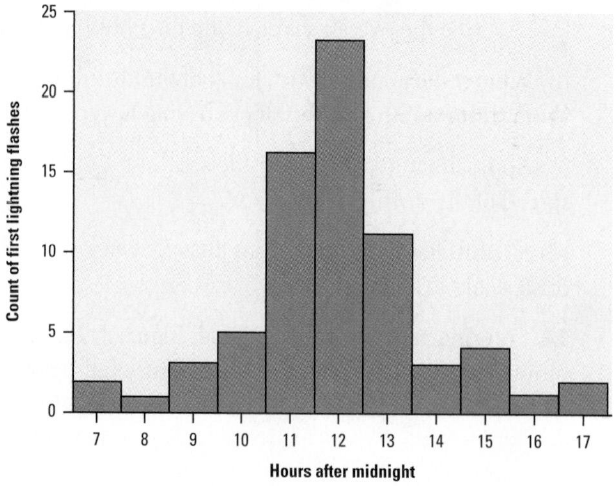

Time Plots

Whenever data are collected over time, it is a good idea to plot the observations in time order. **Displays of the distribution of a variable that ignore time order, such as stemplots and histograms, can be misleading when there is systematic change over time.**

CAUTION

Time Plot

A **time plot** of a variable plots each observation against the time at which it was measured. Always put time on the horizontal scale of your plot and the variable you are measuring on the vertical scale. Connecting the data points by lines helps emphasize any change over time.

Example 1.10 *Gas prices*

Time plot with seasonal variation

Figure 1.14 is a time plot of the average retail price of regular gasoline each month for the years 1996 to 2006.[13] You can see a drop in prices in 1998 when an economic crisis in Asia reduced demand for fuel. You can see rapid price increases in 2000 and 2003 due to instability in the Middle East and OPEC production limits. These deviations are so large that overall patterns are hard to see. Since 2002 there has been a generally upward trend in gas prices. In the second half of 2005, after Hurricane Katrina disrupted the production and flow of oil from the Gulf of Mexico, gas prices peaked near \$3.00 per gallon.

seasonal variation

There is nonetheless a clear *trend* of increasing price. Much of this trend just reflects inflation, the rise in the overall price level during these years. In addition, a close look at the plot shows ***seasonal variation***, a regular rise and fall that recurs each year. Americans drive more in the summer vacation season, so the price of gasoline rises each spring, then drops in the fall as demand goes down.

Figure 1.14 *Time plot of the average monthly price of regular gasoline from 1996 to 2006, for Example 1.10.*

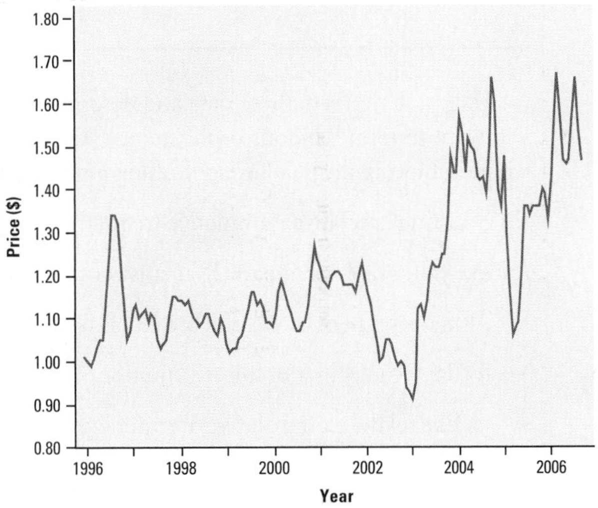

Exercises

1.13 Shopping spree, II Figure 1.15 is an ogive of the amount spent by grocery shoppers in Exercise 1.6 (page 48).

Figure 1.15 *Ogive of amount spent by grocery shoppers, for Exercise 1.13.*

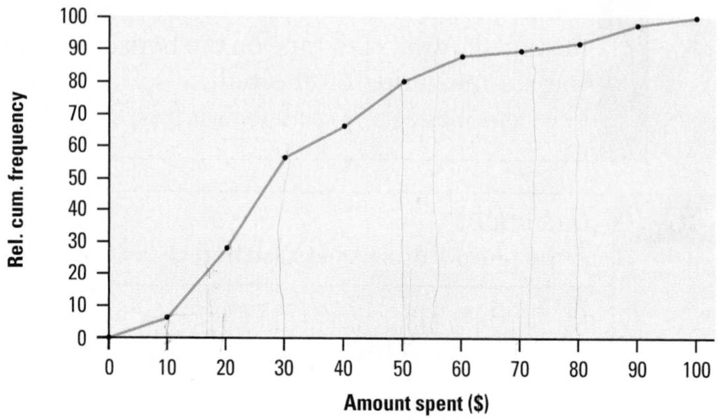

(a) Estimate the center of this distribution. Explain your method.

(b) What is the relative cumulative frequency for the shopper who spent $17.00?

(c) Draw the histogram that corresponds to the ogive.

1.14 Glucose levels People with diabetes must monitor and control their blood glucose level. The goal is to maintain "fasting plasma glucose" between about 90 and 130 milligrams per deciliter (mg/dl) of blood. Here are the fasting plasma glucose levels for 18 diabetics enrolled in a diabetes control class, five months after the end of the class:[14]

141	158	112	153	134	95	96	78	148
172	200	271	103	172	359	145	147	255

(a) Make a stemplot of these data and describe the main features of the distribution. (You will want to round and also split stems.) Are there outliers? How well is the group as a whole achieving the goal for controlling glucose levels?

(b) Construct a relative cumulative frequency graph (ogive) for these data.

(c) Use your graph from part (b) to answer the following questions:

- What percent of blood glucose levels were between 90 and 130?
- What is the center of this distribution?
- What relative cumulative frequency is associated with a blood glucose level of 130?

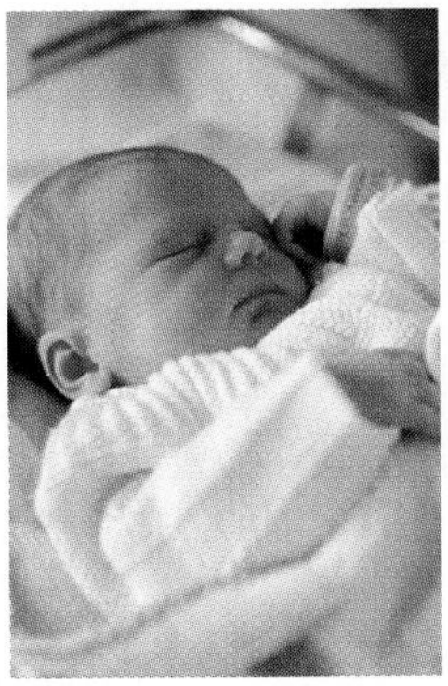

1.15 Birthrates The table below shows the number of births in the United States and the birthrates at 10-year intervals from 1960 to 2000. The birthrate is the number of births per 1000 population.[15]

Year	Total number	Rate
1960	4,257,850	23.7
1970	3,731,386	18.4
1980	3,612,258	15.9
1990	4,092,994	16.7
2000	4,058,814	14.4

(a) Construct a time plot for the birthrate, 1960 to 2000.

(b) Is there a trend in the birthrate? If so, describe the trend in a sentence or two.

(c) List some factors that you think might explain what you see in your birthrate time plot.

(d) Construct a time plot for the total number of births.

(e) Describe what is happening over time for the total number of births.

(f) Briefly explain how you can have such different plots for the two variables.

1.16 Life expectancy Most people are aware that life expectancy, the number of years a person can expect to live, is much longer now than it was, say, a century ago. Here are the numbers for women provided by the National Center for Health Statistics.

Year	Life expectancy (female)	Year	Life expectancy (female)
1900	48.3	1960	73.1
1910	51.8	1970	74.7
1920	54.6	1980	77.5
1930	61.6	1990	78.8
1940	65.2	2000	79.5
1950	71.1		

(a) Construct a time plot for these data.

(b) Describe what you see about the life expectancy of females over the last hundred years.

1.17 The speed of light Light travels fast, but it is not transmitted instantaneously. Light takes over a second to reach us from the moon and over 10 billion years to reach us from the most distant objects observed so far in the expanding universe. Because radio and radar also travel at the speed of light, an accurate value for that speed is important in communicating with astronauts and orbiting satellites. An accurate value for the speed of light is also important to computer designers because electrical signals travel at light speed. The first reasonably accurate measurements of the speed of light were made over a hundred years ago by A. A. Michelson and Simon Newcomb. Table 1.5 contains 66 measurements made by Newcomb between July and September 1882.

Table 1.5 *Newcomb's measurements of the passage time of light*

28	26	33	24	34	−44	27	16	40	−2	29	22	24	21
25	30	23	29	31	19	24	20	36	32	36	28	25	21
28	29	37	25	28	26	30	32	36	26	30	22	36	23
27	27	28	27	31	27	26	33	26	32	32	24	39	28
24	25	32	25	29	27	28	29	16	23				

Source: S. M. Stigler, "Do robust estimators work with real data?" *Annals of Statistics,* 5 (1977), pp. 1055–1078.

Newcomb measured the time in seconds that a light signal took to pass from his laboratory on the Potomac River to a mirror at the base of the Washington Monument and back, a total distance of about 7400 meters. Just as you can compute the speed of a car from the time required to drive a mile, Newcomb could compute the speed of light from the passage time. Newcomb's first measurement of the passage time of light was 0.000024828 second, or 24,828 nanoseconds. (There are 10^9 nanoseconds in a second.) The entries in Table 1.5 record only the deviation from 24,800 nanoseconds.

(a) Construct an appropriate graphical display for these data. Justify your choice of graph.

(b) Describe the distribution of Newcomb's speed of light measurements.

(c) Make a time plot of Newcomb's values. They are listed in order from left to right, starting with the top row.

(d) What does the time plot tell you that the graph you made in part (a) does not? *Lesson:* Sometimes you need to make more than one graphical display to uncover all of the important features of a distribution.

1.18 Civil unrest The years around 1970 brought unrest to many U.S. cities. Here are data on the number of civil disturbances in each three-month period during the years 1968 to 1972:

Period		Count	Period		Count
1968	Jan.–Mar.	6	1970	July–Sept.	20
	Apr.–June	46		Oct.–Dec.	6
	July–Sept.	25	1971	Jan.–Mar.	12
	Oct.–Dec.	3		Apr.–June	21
1969	Jan.–Mar.	5		July–Sept.	5
	Apr.–June	27		Oct.–Dec.	1
	July–Sept.	19	1972	Jan.–Mar.	3
	Oct.–Dec.	6		Apr.–June	8
1970	Jan.–Mar.	26		July–Sept.	5
	Apr.–Jun e	24		Oct.–Dec.	5

(a) Make a time plot of these counts. Connect the points in your plot by straight-line segments to make the pattern clearer.

(b) Describe the trend and the seasonal variation in this time series. Can you suggest an explanation for the seasonal variation in civil disorders?

Section 1.1 Summary

The **distribution** of a variable tells us what values it takes and how often it takes these values.

To describe a distribution, begin with a graph. **Bar graphs** and **pie charts** display the distributions of categorical variables. These graphs use the counts or percents of the categories. **Stemplots** and **histograms** display the distributions of quantitative variables. Stemplots separate each observation into a stem and a one-digit leaf. Histograms plot the **frequencies** (counts) or percents of equal-width classes of values.

When examining a distribution, look for **shape, center,** and **spread,** and for clear **deviations** from the overall shape. Some distributions have simple shapes, such as **symmetric** or **skewed.** The number of **modes** (major peaks) is another aspect of overall shape. Not all distributions have a simple overall shape, especially when there are few observations.

Outliers are observations that lie outside the overall pattern of a distribution. Always look for outliers and try to explain them.

A **relative cumulative frequency graph** (also called an **ogive**) is a good way to see the relative standing of an observation.

When observations on a variable are taken over time, make a **time plot** that graphs time horizontally and the values of the variable vertically. A time plot can reveal **trends** or other changes over time.

Section 1.1 Exercises

1.19 Ranking colleges Popular magazines rank colleges and universities on their "academic quality" in serving undergraduate students. Describe five variables that you would like to see measured for each college if you were choosing where to study. Give reasons for each of your choices.

1.20 Shopping spree, III Enter the amount of money spent in a supermarket from Exercise 1.6 (page 48) into your calculator. Then use the information in the Technology Toolbox (page 59) to construct a histogram. Use `ZoomStat/ZoomData` first to see what the calculator chooses for class widths. Then, in the calculator's WINDOW, choose new settings that are more sensible. Compare your histogram with the stemplots you made in Exercise 1.6. List at least one advantage that each plot has that the other plots don't have.

1.21 College costs The Department of Education estimates the "average unmet need" for undergraduate students—the cost of school minus estimated family contributions and financial aid. Here are the averages for full-time students at four types of institution in the 1999–2000 academic year:[16]

Public 2-year	Public 4-year	Private nonprofit 4-year	Private for profit
\$4495	\$4818	\$8257	\$8296

Make a bar graph of these data. Write a one-sentence conclusion about the unmet needs of students. Explain clearly why it is incorrect to make a pie chart.

1.22 New-vehicle survey The J. D. Power Initial Quality Study polls more than 50,000 buyers of new motor vehicles 90 days after their purchase. A two-page questionnaire asks about "things gone wrong." Here are data on problems per 100 vehicles for vehicles made by Toyota and by General Motors in recent years. Toyota has been the industry leader in quality. Make two time plots in the same graph to compare Toyota and GM. What are the most important conclusions you can draw from your graph?

	1998	1999	2000	2001	2002	2003	2004
GM	187	179	164	147	130	134	120
Toyota	156	134	116	115	107	115	101

1.23 Senior citizens, I The population of the United States is aging, though less rapidly than in other developed countries. Here is a stemplot of the percents of residents aged 65 and over in the 50 states, according to the 2000 census. The stems are whole percents and the leaves are tenths of a percent.

```
 5 | 7
 6 |
 7 |
 8 | 5
 9 | 6 7 9
10 | 6
11 | 0 2 2 3 3 6 7 7
12 | 0 0 1 1 1 1 3 4 4 5 7 8 9
13 | 0 0 0 1 2 2 3 3 3 4 5 5 6 8
14 | 0 3 4 5 7 9
15 | 3 6
16 |
17 | 6
```

(a) There are two outliers: Alaska has the lowest percent of older residents, and Florida has the highest. What are the percents for these two states?

(b) Ignoring Alaska and Florida, describe the shape, center, and spread of this distribution.

1.24 Senior citizens, II Make another stemplot of the percent of residents aged 65 and over in the states other than Alaska and Florida by splitting stems in the plot from the previous exercise. Which plot do you prefer? Why?

1.25 The statistics of writing style Numerical data can distinguish different types of writing, and sometimes even individual authors. Here are data on the percent of words of 1 to 15 letters used in articles in *Popular Science* magazine:[17]

Length:	1	2	3	4	5	6	7	8	9	10	11	12	13	14	15
Percent:	3.6	14.8	18.7	16.0	12.5	8.2	8.1	5.9	4.4	3.6	2.1	0.9	0.6	0.4	0.2

(a) Make a histogram of this distribution. Describe its shape, center, and spread.

(b) How does the distribution of lengths of words used in *Popular Science* compare with the similar distribution in Figure 1.11 (page 57) for Shakespeare's plays? Look in particular at short words (2, 3, and 4 letters) and very long words (more than 10 letters).

1.26 Student survey A survey of a large high school class asked the following questions:

1. Are you female or male? (In the data, male = 0, female = 1.)
2. Are you right-handed or left-handed? (In the data, right = 0, left = 1.)
3. What is your height in inches?
4. How many minutes do you study on a typical weeknight?

Figure 1.16 shows histograms of the student responses, in scrambled order and without scale markings. Which histogram goes with each variable? Explain your reasoning.

Figure 1.16 *Match each histogram with its variable, for Exercise 1.26.*

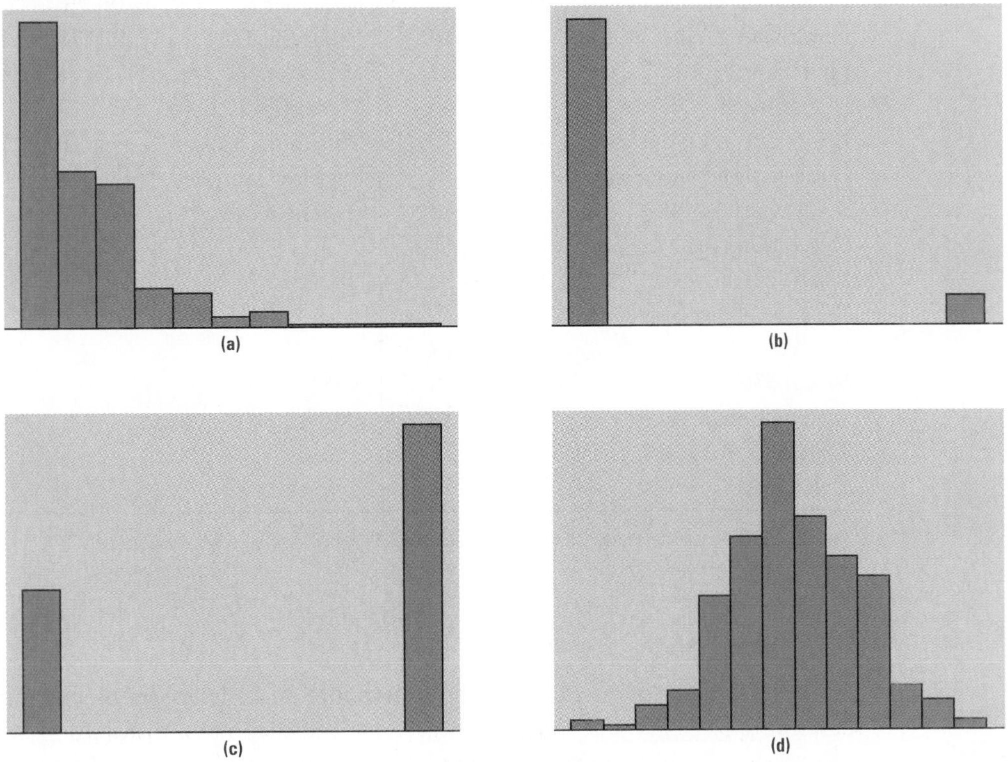

1.2 Describing Distributions with Numbers

Interested in a sporty car? Worried that it might use too much gas? The Environmental Protection Agency lists most such vehicles in its "two-seater" or "minicompact" categories. Table 1.6 on the next page gives the city and highway gas mileage for cars in these groups. (The mileages are for the basic engine and transmission combination for each car.) We want to compare two-seaters with minicompacts and city mileage with highway mileage. We can begin with graphs, but numerical summaries make the comparisons more specific.

Table 1.6 *Fuel economy (miles per gallon) for 2004 model motor vehicles*

Two-seater Cars			Minicompact Cars		
Model	**City**	**Highway**	**Model**	**City**	**Highway**
Acura NSX	17	24	Aston Martin Vanquish	12	19
Audi TT Roadster	20	28	Audi TT Coupe	21	29
BMW Z4 Roadster	20	28	BMW 325CI	19	27
Cadillac XLR	17	25	BMW 330CI	19	28
Chevrolet Corvette	18	25	BMW M3	16	23
Dodge Viper	12	20	Jaguar XK8	18	26
Ferrari 360 Modena	11	16	Jaguar XKR	16	23
Ferrari Maranello	10	16	Lexus SC 430	18	23
Ford Thunderbird	17	23	Mini Cooper	25	32
Honda Insight	60	66	Mitsubishi Eclipse	23	31
Lamborghini Gallardo	9	15	Mitsubishi Spyder	20	29
Lamborghini Murcielago	9	13	Porsche Cabriolet	18	26
Lotus Esprit	15	22	Porsche Turbo 911	14	22
Maserati Spyder	12	17			
Mazda Miata	22	28			
Mercedes-Benz SL500	16	23			
Mercedes-Benz SL600	13	19			
Nissan 350Z	20	26			
Porsche Boxster	20	29			
Porsche Carrera 911	15	23			
Toyota MR2	26	32			

Source: U.S. Environmental Protection Agency, "Model Year 2004 Fuel Economy Guide," found online at www.fueleconomy.gov.

A brief description of a distribution should include its *shape* and numbers describing its *center* and *spread*. We describe the shape of a distribution based on inspection of a histogram or a stemplot. Now you will learn specific ways to use numbers to measure the center and spread of a distribution. You can calculate these numerical measures for any quantitative variable. But to interpret measures of center and spread, and to choose among the several measures you will learn, you must think about the shape of the distribution and the meaning of the data. The numbers, like graphs, are aids to understanding, not "the answer" in themselves.

Measuring Center: The Mean

Numerical description of a distribution begins with a measure of its center or average. The two common measures of center are the ***mean*** and the ***median.*** The mean is the "average value," and the median is the "middle value." These are two different ideas for "center," and the two measures behave differently. We need precise rules for the mean and the median.

The Mean $\bar{x}$

To find the **mean** $\bar{x}$ of a set of observations, add their values and divide by the number of observations. If the n observations are $x_1, x_2, \ldots, x_n$, their mean is

$$\bar{x} = \frac{x_1 + x_2 \ldots + x_n}{n}$$

or, in more compact notation,

$$\bar{x} = \frac{\sum x_i}{n}$$

The Σ (capital Greek sigma) in the formula for the mean is short for "add them all up." The bar over the x indicates the mean of all the x-values. Pronounce the mean $\bar{x}$ as "x-bar." This notation is so common that writers who are discussing data use $\bar{x}$, $\bar{y}$, etc. without additional explanation. The subscripts on the observations x_i are just a way of keeping the n observations distinct. They do not necessarily indicate order or any other special facts about the data.

Example 1.11 *Mean highway mileage for two-seaters*
Calculating $\bar{x}$

The mean highway mileage for the 21 two-seaters in Table 1.6 is

$$\bar{x} = \frac{x_1 + x_2 + \ldots + x_n}{n}$$

$$= \frac{24 + 28 + 28 \ldots + 32}{21}$$

$$= \frac{518}{21} = 24.7 \text{ miles per gallon}$$

In practice, you can key the data into your calculator and request 1-Var Stats, or into software and request descriptive statistics.

The data for Example 1.11 contain an outlier: the Honda Insight is a hybrid gas-electric car that doesn't belong in the same category as the 20 gasoline-powered two-seater cars. If we exclude the Insight, the mean highway mileage drops to 22.6 mpg. The single outlier adds more than 2 mpg to the mean highway mileage. This illustrates an important weakness of the mean as a measure of center: **the mean is sensitive to the influence of a few extreme observations.** These may be outliers, but a skewed distribution that has no outliers will also pull the mean toward its long tail. Because the mean cannot resist the influence of extreme observations, we say that it is not a ***resistant measure*** of center. A measure that is resistant does more than limit the influence of outliers. Its value does not respond strongly to changes in a few observations, no matter how large those changes may be. The mean fails this requirement because we can make the mean as large as we wish by making a large enough increase in just one observation.

CAUTION

resistant measure

Measuring Center: The Median

We used the midpoint of a distribution as an informal measure of center in the previous section. The *median* is the formal version of the midpoint, with a specific rule for calculation.

The Median M

The **median** M is the midpoint of a distribution, the number such that half the observations are smaller and the other half are larger. To find the median of a distribution:

1. Arrange all observations in order of size, from smallest to largest.
2. If the number of observations n is odd, the median M is the center observation in the ordered list. Find the *location* of the median by counting $(n + 1)/2$ observations up from the bottom of the list.
3. If the number of observations n is even, the median M is the average of the two center observations in the ordered list. The location of the median is again $(n + 1)/2$ from the bottom of the list.

Note that the formula $(n + 1)/2$ does *not* give the median, just the location of the median in the ordered list. Medians require little arithmetic, so they are easy to find by hand for small sets of data. Arranging even a moderate number of observations in order is tedious, however, so that finding the median by hand for larger sets of data is unpleasant. Even simple calculators have an $\bar{x}$ button, but you will need computer software or a graphing calculator to automate finding the median.

Example 1.12 *Median highway mileage for two-seaters*
Finding the median by hand

To find the median highway mileage for 2004 model two-seater cars, arrange the data in increasing order:

13 15 16 16 17 19 20 22 23 23 **23** 24 25 25 26 28 28 28 29 32 66

Be sure to list *all* observations, even if they repeat the same value. The median is the bold 23, the 11th observation in the ordered list. You can find the median by eye—there are 10 observations to the left and 10 to the right. Or you can use the rule $(n + 1)/2 = 22/2 = 11$ to locate the median in the list.

What happens if we drop the Honda Insight? The remaining 20 cars have highway mileages

13 15 16 16 17 19 20 22 23 **23** **23** 24 25 25 26 28 28 28 29 32

Because the number of observations $n = 20$ is even, there is no center observation. There is a center pair—the bold pair of 23s have 9 observations to their left and 9 to their right. The median M is the average of the center pair, which is 23. The rule $(n + 1)/2 = 21/2 = 10.5$ for the position of the median in the list says that the median is at location "ten and one-half," that is, halfway between the 10th and 11th observations.

You see that the median is more resistant than the mean. Removing the Honda Insight did not change the median at all. Even if we mistakenly enter the Insight's mileage as 660 rather than 66, the median remains 23. The very high value is simply one observation to the right of center.

Activity 1C

The Mean and Median applet

The *Mean and Median* applet on the book's Web site, www.whfreeman.com/tps3e, allows you to place observations on a line and see their mean and median visually.

1. Go to the Web site and launch the *Mean and Median* applet. The applet consists of a horizontal line and a trash can.
2. Place two observations on the line by clicking below it. Why does only one arrow appear?
3. Place three observations on the line by clicking below it, two close together near the center of the line and one somewhat to the right of these two.
   a. Pull the single rightmost observation out to the right. (Place the cursor on the point, hold down the mouse button, and drag the point.) How does the mean behave? How does the median behave? Explain briefly why each measure acts as it does.
   b. Now drag the rightmost point to the left as far as you can. What happens to the mean? What happens to the median as you drag this point past the other two? (Watch carefully.)
4. Place five observations on the line by clicking below it.
   a. Add one additional observation *without changing the median*. Where is your new point?
   b. Use the applet to convince yourself that when you add yet another observation (there are now seven in all), the median does not change no matter where you put the seventh point. Explain why this must be true.

Mean versus Median

The median and mean are the most common measures of the center of a distribution. The mean and median of a symmetric distribution are close together. If the distribution is exactly symmetric, the mean and median are exactly the same. In a skewed distribution, the mean is farther out in the long tail than is the median. For example, the distribution of the sizes of the endowments of colleges and universities is strongly skewed to the right. Most institutions have modest endowments, but a few are very wealthy. The median endowment of colleges and universities in 2003 was $70 million—but the mean endowment was over $320 million. The few wealthy institutions pulled the mean up but did not affect the median. **Don't confuse the "average" value of a variable (the mean) with its "typical" value, which we might describe by the median.**

CAUTION

We can now give a better answer to the question of how to deal with outliers in data. First, look at the data to identify outliers and investigate their causes. You can then correct outliers if they are wrongly recorded, delete them for good reason, or otherwise give them individual attention. If you are interested only in Virginia four-year colleges, for example, then it makes sense to delete the 23 community colleges from consideration. Then among Virginia four-year colleges, the University of Richmond's $34,850 figure is now an outlier, as you will see shortly. It would be inappropriate to delete this high outlier. If you have no clear reason to drop outliers, you may want to use resistant methods, so that outliers have little influence over your conclusions. The choice is often a matter of judgment. The government's fuel economy guide lists the Honda Insight with the other two-seaters in Table 1.6. We might choose to report median rather than mean gas mileage for all two-seaters to avoid giving too much influence to one car model. In fact, we think that the Insight doesn't belong, so we will omit it from further analysis of these data.

Exercises

1.27 Quiz grades Joey's first 14 quiz grades in a marking period were

86 84 91 75 78 80 74 87 76 96 82 90 98 93

(a) Use the formula to calculate the mean. Check using "1-Var Stats" on your calculator.

(b) Suppose Joey has an unexcused absence for the 15th quiz, and he receives a score of zero. Determine his final quiz average. What property of the mean does this situation illustrate? Write a sentence about the effect of the zero on Joey's quiz average that mentions this property.

(c) What kind of plot would best show Joey's distribution of grades? Assume an eight-point grading scale (A: 93 to 100; B: 85 to 92; etc.). Make an appropriate plot, and be prepared to justify your choice.

1.28 SSHA scores, I The Survey of Study Habits and Attitudes (SSHA) is a psychological test that evaluates college students' motivation, study habits, and attitudes toward school. A private college gives the SSHA to a sample of 18 of its incoming first-year women students. Their scores are

154	109	137	115	152	140	154	178	101
103	126	126	137	165	165	129	200	148

(a) Make a stemplot of these data. The overall shape of the distribution is irregular, as often happens when only a few observations are available. Are there any potential outliers? About where is the median of the distribution (the score with half the scores above it and half below)? What is the spread of the scores (ignoring any outliers)?

(b) Find the mean score from the formula for the mean. Then enter the data into your calculator. You can find the mean from the home screen as follows:

TI-83/84	TI-89
• Press 2nd STAT (LIST) ▶▶ (MATH).	• Press CATALOG then 5 (M).
• Choose 3: mean(, enter list name, press ENTER.	• Choose mean(, type list name, press ENTER.

(c) Find the median of these scores. Which is larger: the median or the mean? Explain why.

1.29 Baseball player salaries Suppose a Major League Baseball team's mean yearly salary for a player is \$1.2 million, and that the team has 25 players on its active roster. What is the team's annual payroll for players? If you knew only the median salary, would you be able to answer the question? Why or why not?

1.30 Mean salary? Last year a small accounting firm paid each of its five clerks \$22,000, two junior accountants \$50,000 each, and the firm's owner \$270,000. What is the mean salary paid at this firm? How many of the employees earn less than the mean? What is the median salary? Write a sentence to describe how an unethical recruiter could use statistics to mislead prospective employees.

1.31 U.S. incomes The distribution of household incomes in the United States is strongly skewed to the right. In 2003, the mean and median household incomes in America were \$43,318 and \$59,067. Which of these numbers is the mean and which is the median? Explain your reasoning.

1.32 Home run records Who is baseball's greatest home run hitter? In the summer of 1998, Mark McGwire and Sammy Sosa captured the public's imagination with their pursuit of baseball's single-season home run record (held by Roger Maris). McGwire eventually set a new standard with 70 home runs. Barry Bonds broke McGwire's record when he hit 73 home runs in the 2001 season. How does this accomplishment fit Bonds's career? Here are Bonds's home run counts for the years 1986 (his rookie year) to 2004:

1986	1987	1988	1989	1990	1991	1992	1993	1994
16	25	24	19	33	25	34	46	37

1995	1996	1997	1998	1999	2000	2001	2002	2003	2004
33	42	40	37	34	49	73	46	45	45

(a) Calculate the mean and median of Bonds's home run data. What do these two numbers tell you about the distribution of the data?

(b) Make a stemplot of these data.

(c) What would you say is Bonds's home run production for a typical year (1986 to 2004)? Explain your reasoning in a sentence or two.

Measuring Spread: The Quartiles

A measure of center alone can be misleading. Two nations with the same median family income are very different if one has extremes of wealth and poverty and the other has little variation among families. A drug with the correct mean concentration of active ingredient is dangerous if some batches are much too high and others much too low. We are interested in the *spread* or *variability* of incomes and drug potencies as well as their centers. *The simplest useful numerical description of a distribution consists of both a measure of center and a measure of spread.*

range

One way to measure spread is to calculate the ***range***, which is the difference between the largest and smallest observations. For example, the number of home runs Barry Bonds has hit in a season has a *range* of 73 − 16 = 57. The range shows the full spread of the data. But it depends on only the smallest observation and the largest observation, which may be outliers.

pth percentile

We can describe the spread or variability of a distribution by giving several percentiles. The ***pth percentile*** of a distribution is the value such that p percent of the observations fall at or below it. The median is just the 50th percentile, so the use of percentiles to report spread is particularly appropriate when the median is our measure of center. The most commonly used percentiles other than the median are the ***quartiles***. The first quartile is the 25th percentile, and the third quartile is the 75th percentile. (The second quartile is the median itself.) To calculate a percentile, arrange the observations in increasing order and count up the required percent from the bottom of the list. Our definition of percentiles is a bit inexact, because there is not always a value with exactly p percent of the data at or below it. We will be content to take the nearest observation for most percentiles, but the quartiles are important enough to require an exact rule.

The Quartiles Q_1 and Q_3

To calculate the ***quartiles:***

1. Arrange the observations in increasing order and locate the median M in the ordered list of observations.
2. The **first quartile** Q_1 is the median of the observations whose position in the ordered list is to the left of the location of the overall median.
3. The **third quartile** Q_3 is the median of the observations whose position in the ordered list is to the right of the location of the overall median.

Here is an example that shows how the rules for the quartiles work for both odd and even numbers of observations.

Example 1.13 *Highway mileage*
Calculating quartiles

The highway mileages of the 20 gasoline-powered two-seater cars in Table 1.6 (page 70), arranged in increasing order, are

13 15 16 16 17 19 20 22 23 23 | 23 24 25 25 26 28 28 28 29 32

The median is midway between the center pair of observations. We have marked its position in the list by |. The first quartile is the median of the 10 observations to the left of the position of the median. Check that its value is $Q_1 = 18$. Similarly, the third quartile is the median of the 10 observations to the right of the |. Check that $Q_3 = 27$.

When there is an odd number of observations, the median is the unique center observation, and the rule for finding the quartiles excludes this center value. The highway mileages of the 13 minicompact cars in Table 1.6 are (in order)

19 22 23 23 23 26 **26** 27 28 29 29 31 32

The median is the bold 26. The first quartile is the median of the 6 observations falling to the left of this point in the list, $Q_1 = 23$. Similarly, $Q_3 = 29$.

We find other percentiles more informally if we are working without software. For example, we take the 90th percentile of the 13 minicompact mileages to be the 12th in the ordered list, because $0.90 \times 13 = 11.7$, which we round to 12. The 90th percentile is therefore 31 mpg.

Example 1.14 *CrunchIt! and Minitab*

Numerical summaries with computer software

Statistical software often provides several numerical measures in response to a single command. Figure 1.17 displays such output from the CrunchIt! and Minitab software for the highway mileages of two-seater cars (without the Honda Insight). Both tell us that there are

Figure 1.17 *Numerical descriptions of the highway gas mileage of two-seater cars from CrunchIt! and Minitab software.*

CrunchIt!

Summary statistics

Column	n	Mean	Variance	Std. Dev.	Std. Err.	Median	Range	Min	Max	Q1	Q3
Hwy mpg	20	22.6	27.936842	5.2855315	1.1818807	23	19	13	32	18	27

Minitab

Session

Descriptive Statistics: Hwy mpg

```
          Total
Variable  Count   Mean  StDev  Variance  Minimum     Q1  Median     Q3  Maximum
Hwy mpg      20  22.60   5.29     27.94    13.00  17.50   23.00  27.50    32.00


Variable  Range  IQR
Hwy mpg   19.00  10.00
```

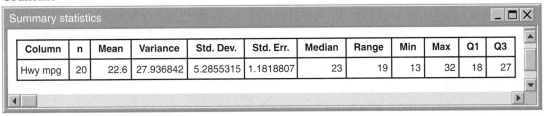

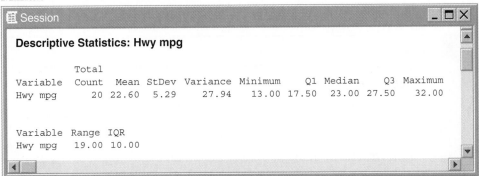

20 observations and give the mean, median, quartiles, and smallest and largest data values. Both also give other measures, some of which we will meet soon. CrunchIt! is basic online software that offers no choice of output. Minitab allows you to choose the descriptive measures you want from a long list.

The quartiles from CrunchIt! agree with our values from Example 1.13. But Minitab's quartiles are a bit different. For example, our rule for hand calculation gives first quartile $Q_1 = 18$. Minitab's value is $Q_1 = 17.5$. **There are several rules for calculating quartiles, which often give slightly different values. The differences are always small. For describing data, just report the values that your software gives.**

The Five-Number Summary and Boxplots

In Section 1.1, we used the smallest and largest observations to indicate the spread of a distribution. These single observations tell us little about the distribution as a whole, but they give information about the tails of the distribution that is missing if we know only Q_1, M, and Q_3. To get a quick summary of both center and spread, combine all five numbers.

The Five-Number Summary

The **five-number summary** of a set of observations consists of the smallest observation, the first quartile, the median, the third quartile, and the largest observation, written in order from smallest to largest. In symbols, the five-number summary is

$$\text{Minimum} \quad Q_1 \quad M \quad Q_3 \quad \text{Maximum}$$

These five numbers offer a reasonably complete description of center and spread. The five-number summaries for highway gas mileages are

$$13 \quad 18 \quad 23 \quad 27 \quad 32$$

for two-seaters and

$$19 \quad 23 \quad 26 \quad 29 \quad 32$$

for minicompacts. The median describes the center of the distribution; the quartiles show the spread of the center half of the data; the minimum and maximum show the full spread of the data. The five-number summary leads to another visual representation of a distribution, the ***boxplot***. Figure 1.18 shows boxplots for both city and highway gas mileages for our two groups of cars.

Boxplot

A **boxplot** is a graph of the five-number summary.

- A central box spans the quartiles Q_1 and Q_3.
- A line in the box marks the median M.
- Lines extend from the box out to the smallest and largest observations.

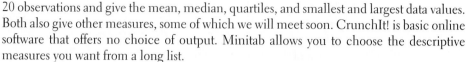

Figure 1.18 *Boxplots of the highway and city gas mileages for cars classified as two-seaters and as minicompacts by the Environmental Protection Agency.*

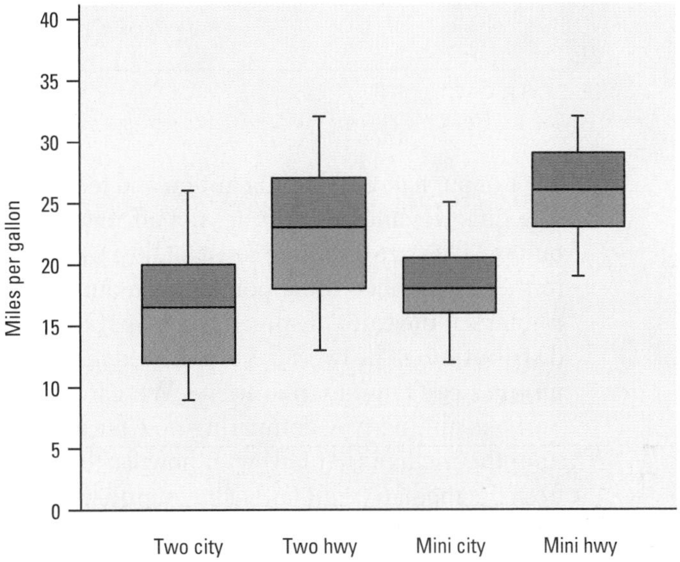

Because boxplots show less detail than histograms or stemplots, they are best used for side-by-side comparison of more than one distribution, as in Figure 1.18. When you look at a boxplot, first locate the median, which marks the center of the distribution. Then look at the spread. The quartiles show the spread of the middle half of the data, and the extremes (the smallest and largest observations) show the spread of the entire data set. We see at once that city mileages are lower than highway mileages. The minicompact cars have slightly higher median gas mileages than the two-seaters, and their mileages are markedly less variable. In particular, the low gas mileages of the Ferraris and Lamborghinis in the two-seater group pull the group minimum down.

The 1.5 × *IQR* Rule for Suspected Outliers

Look again at the stemplot of the distribution of tuition and fees for the 37 Virginia four-year colleges. Visualize the stemplot in Figure 1.4 (page 45) without the two-year college costs in the top row. You can check that the five-number summary is

9420 14,286 16,870 21,707.50 34,850

There is a clear outlier, the University of Richmond's $34,850. How shall we describe the spread of this distribution? The range is a bit misleading because of the high outlier. The distance between the quartiles (the range of the center half of the data) is a more resistant measure of spread. This distance is called the ***interquartile range***.

The Interquartile Range (IQR)

The **interquartile range (IQR)** is the distance between the first and third quartiles,

$$IQR = Q_3 - Q_1$$

For our data on Virginia tuition and fees, $IQR = 21{,}707.5 - 14{,}286 = 7421.5$. The quartiles and the IQR are not affected by changes in either tail of the distribution. They are therefore resistant, because changes in a few data points have no further effect once these points move outside the quartiles. However, **no single numerical measure of spread, such as IQR, is very useful for describing skewed distributions.** The two sides of a skewed distribution have different spreads, so one number can't summarize them. We can often detect skewness from the five-number summary by comparing how far the first quartile and the minimum are from the median (left tail) with how far the third quartile and the maximum are from the median (right tail). The interquartile range is mainly used as the basis for a rule of thumb for identifying suspected outliers.

The 1.5 × IQR *Rule for Outliers*

Call an observation a suspected outlier if it falls more than $1.5 \times IQR$ above the third quartile or below the first quartile.

Example 1.15 *Virginia tuition and fees data*
$1.5 \times IQR$ rule

For the Virginia four-year college data in Table 1.2 (page 45),

$$1.5 \times IQR = 1.5 \times 7421.5 = 11{,}132.25$$

Any values below $Q_1 - 1.5\ IQR = 14{,}286 - 11{,}132.25 = 3{,}153.75$ *or* above $Q_3 + 1.5\ IQR = 21{,}707.5 + 11{,}132.25 = 32{,}839.75$ are flagged as outliers. There are no low outliers, but the University of Richmond observation, 34,850, is a high outlier.

modified boxplot

Statistical software often uses the $1.5 \times IQR$ rule. Boxplots drawn by software are often ***modified boxplots*** that plot suspected outliers individually. Figure 1.19 is a modified boxplot of the Virginia four-year college tuition and fees data. The lines extend out from the central box only to the smallest and largest observations that are not flagged by the $1.5 \times IQR$ rule. The largest observation is plotted as an individual point.

Figure 1.19 *Boxplot of the Virginia four-year college tuition and fees data.*

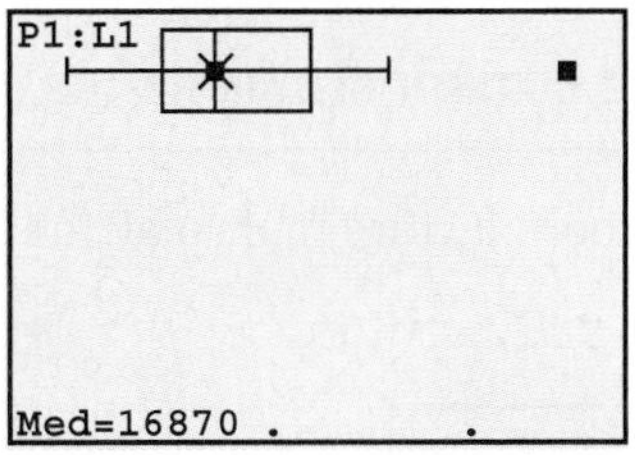

The stemplot in Figure 1.4 (page 45) and the modified boxplot in Figure 1.19 tell us much more about the distribution of Virginia college tuition and fees than the five-number summary or other numerical measures. The routine methods of statistics compute numerical measures and draw conclusions based on their values. These methods are very useful, and we will study them carefully in later chapters. But they should not be applied blindly, by feeding data to a computer program, because **statistical measures and methods based on them are generally meaningful only for distributions of sufficiently regular shape.** This principle will become clearer as we progress, but it is good to be aware at the beginning that quickly resorting to fancy calculations is the mark of a statistical amateur. Look, think, and choose your calculations selectively.

CAUTION

Technology Toolbox

Calculator boxplots and numerical summaries

The TI-83/84 and TI-89 can plot up to three boxplots in the same viewing window. Both calculators can also calculate the mean, median, quartiles, and other one-variable statistics for data stored in lists. In this example, we compare Barry Bonds to Babe Ruth, the "Sultan of Swat." Here are the numbers of home runs hit by Ruth in each of his seasons as a New York Yankee (1920 to 1934):

54	59	35	41	46	25	47	60	54	46	49	46	41	34	22

Bonds's home runs are shown in Exercise 1.32 (page 75).

1. Enter Bonds's home run data in L_1/list1 and Ruth's in L_2/list2.
2. Set up two statistics plots: Plot 1 to show a modified boxplot of Bonds's data and Plot 2 to show a modified boxplot of Ruth's data.

TI-83/84

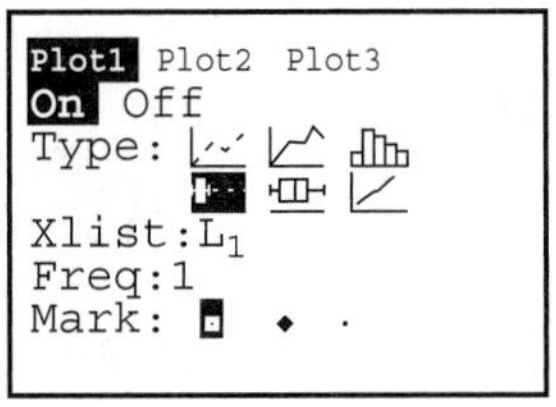

TI-89

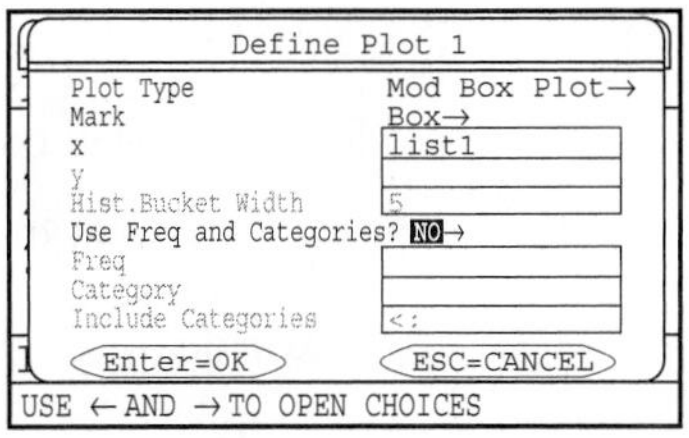

Technology Toolbox

Calculator boxplots and numerical summaries *(continued)*

3. Use the calculator's zoom feature to display the side-by-side boxplots.

- Press **ZOOM** and select `9:ZoomStat`.
- Press **F5** (ZoomData).

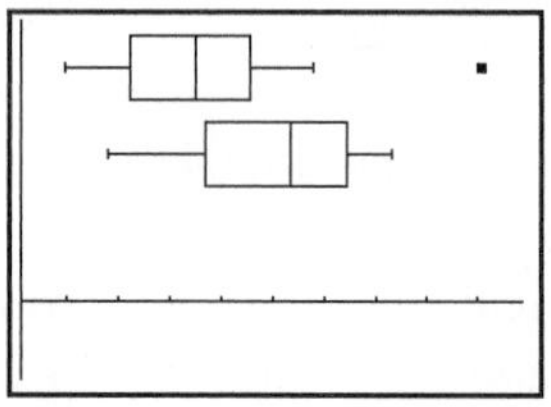

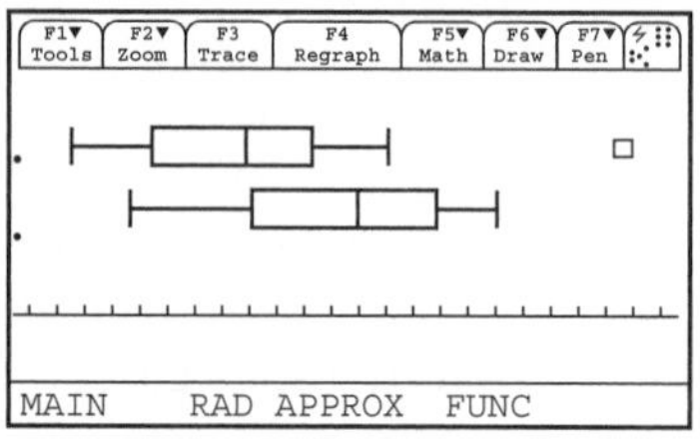

4. Calculate numerical summaries for each set of data.

- Press **STAT** **▶** (CALC) and select `1:1-Var Stats`
- Press **ENTER**. Now press **2nd** **1** (L1) and **ENTER**.
- Press **F4** (Calc) and choose `1:1-Var Stats`.
- Type list1 in the list box. Press **ENTER**.

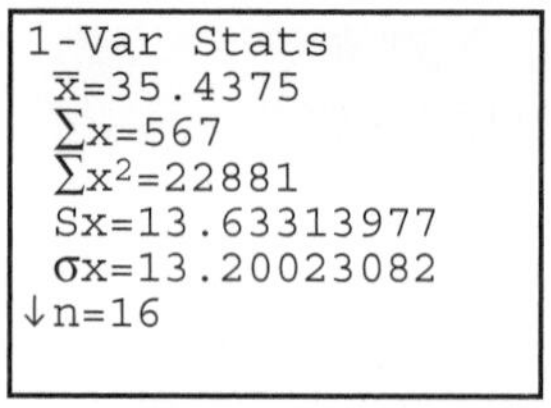

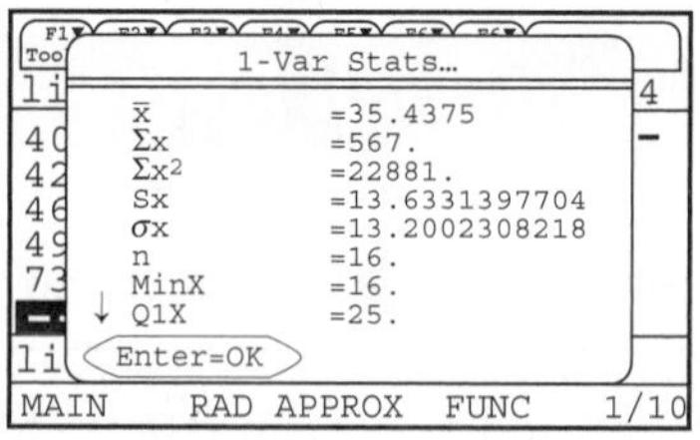

5. Notice the down-arrow on the left side of the display. Press **▼** to see Bonds's other statistics. Repeat the process to find the Babe's numerical summaries.

Exercises

1.33 SSHA scores, II Here are the scores on the Survey of Study Habits and Attitudes (SSHA) for 18 first-year college women:

154 109 137 115 152 140 154 178 101 103 126 126 137 165 165 129 200 148

and for 20 first-year college men:

108 140 114 91 180 115 126 92 169 146 109 132 75 88 113 151 70 115 187 104

(a) Make side-by-side boxplots to compare the distributions.

(b) Compute numerical summaries for these two distributions.

(c) Write a paragraph comparing the SSHA scores for men and women.

1.34 How old are presidents? Return to the data on presidential ages in Table 1.4 (page 57). In Exercise 1.11, you were asked to construct a histogram of the age data.

(a) From the shape of the histogram, do you expect the mean to be much less than the median, about the same as the median, or much greater than the median? Explain.

(b) Find the five-number summary and verify your expectation from (a).

(c) What is the range of the middle half of the ages of new presidents?

(d) Construct by hand a (modified) boxplot of the ages of new presidents.

(e) On your calculator, define Plot 1 to be a histogram using the list named PREZ that you created in the Technology Toolbox on page 59. Define Plot 2 to be a (modified) boxplot also using the list PREZ. Use the calculator's zoom command to generate a graph. To remove the overlap, adjust your viewing window so that Ymin = −6 and Ymax = 22. Then graph. Use TRACE to inspect values. Press the up and down cursor keys to toggle between plots. Is there an outlier? If so, who was it?

1.35 Is the interquartile range a resistant measure of spread? Give an example of a small data set that supports your answer.

1.36 Shopping spree, IV Figure 1.20 displays computer output for the data on amount spent by grocery shoppers in Exercise 1.6 (page 48).

Figure 1.20 *Numerical descriptions of the unrounded shopping data from the DataDesk and Minitab software, for Exercise 1.36.*

DataDesk

```
Session
Summaryof    spending
No Selector

Percentile  25

         Count   50
          Mean   34.7022
        Median   27.8550
        StdDev   21.6974
           Min    3.11000
           Max   93.3400
Lower ith %tile  19.2700
Upper ith %tile  45.4000
```

Minitab

```
Session
Descriptive Statistics
Variable      N     Mean   Median   TrMean    StDev   SEMean
spending     50    34.70    27.85    32.92    21.70     3.07

Variable    Min      Max       Q1       Q3
spending   3.11    93.34    19.06    45.72
```

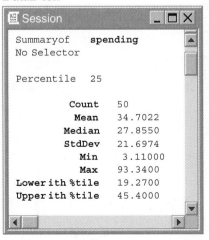

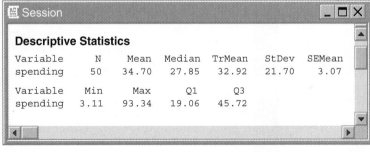

(a) Find the total amount spent by the shoppers.

(b) Make a boxplot from the computer output. Did you check for outliers?

1.37 Bonds's home runs

(a) Find the quartiles Q_1 and Q_3 for Barry Bonds's home run data. Refer to Exercise 1.32 (page 75).

(b) Use the $1.5 \times IQR$ rule for identifying outliers to see if Bonds's 73 home runs in 2001 is an outlier.

1.38 Senior citizens, III The stemplot you made for Exercise 1.23 (page 68) displays the distribution of the percents of residents aged 65 and over in the 50 states. Stemplots help you find the five-number summary because they arrange the observations in increasing order.

(a) Give the five-number summary of this distribution.

(b) Does the $1.5 \times IQR$ rule identify Alaska and Florida as suspected outliers? Does it also flag any other states?

Measuring Spread: The Standard Deviation

The five-number summary is not the most common numerical description of a distribution. That distinction belongs to the combination of the mean to measure center and the ***standard deviation*** to measure spread. The standard deviation measures spread by looking at how far the observations are from their mean.

The Variance s^2 and Standard Deviation s

The **variance** s^2 of a set of observations is the average of the squares of the deviations of the observations from their mean. In symbols, the variance of n observations $x_1, x_2, \ldots, x_n$ is

$$s^2 = \frac{(x_1 - \overline{x})^2 + (x_2 - \overline{x})^2 + \cdots + (x_n - \overline{x})^2}{n-1}$$

or, more compactly,

$$s^2 = \frac{1}{n-1} \Sigma(x_i - \overline{x})^2$$

The **standard deviation** s is the square root of the variance s^2:

$$s = \sqrt{\frac{1}{n-1} \Sigma(x_i - \overline{x})^2}$$

The idea behind the variance and the standard deviation as measures of spread is as follows. The deviations $x_i - \bar{x}$ display the spread of the values x_i about their mean $\bar{x}$. Some of these deviations will be positive and some negative because some of the observations fall on each side of the mean. In fact, *the sum of the deviations of the observations from their mean will always be zero.* Squaring the deviations makes them all positive, so that observations far from the mean in either direction have large positive squared deviations. The variance is the average squared deviation. Therefore, s^2 and s will be large if the observations are widely spread about their mean, and small if the observations are all close to the mean.

Example 1.16 ***Metabolic rate***
Standard deviation

A person's metabolic rate is the rate at which the body consumes energy. Metabolic rate is important in studies of weight gain, dieting, and exercise. Here are the metabolic rates of 7 men who took part in a study of dieting. (The units are calories per 24 hours. These are the same calories used to describe the energy content of foods.)

1792 1666 1362 1614 1460 1867 1439

Enter these data into your calculator or software and verify that

$$\bar{x} = 1600 \text{ calories} \qquad s = 189.24 \text{ calories}$$

Figure 1.21 plots these data as dots on the calorie scale, with their mean marked by the blue dot. The arrows mark two of the deviations from the mean. If you were calculating s by hand, you would find the first deviation as

$$x_1 - \bar{x} = 1792 - 1600 = 192$$

Figure 1.21 ***Metabolic rates for seven men, with the mean (blue dot) and the deviations of two observations from the mean indicated.***

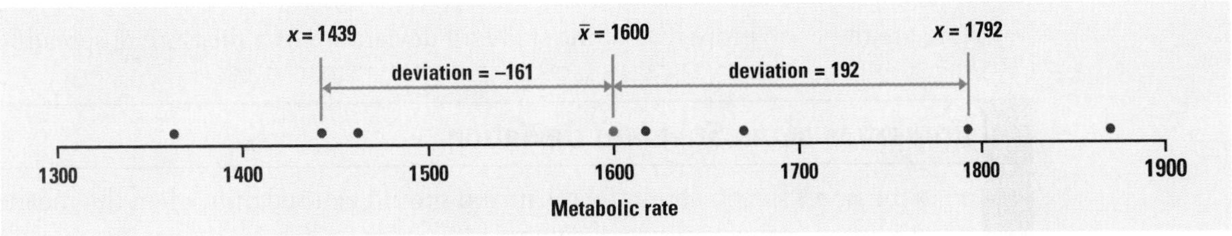

Exercise 1.41 (page 89) asks you to calculate the seven deviations, square them, and find s^2 and s directly from the deviations. Working one or two short examples by hand helps you understand how the standard deviation is obtained. In practice, you will use either software or a calculator that will find s from keyed-in data. The two software outputs in Figure 1.17 (page 77) both give the variance and standard deviation for the highway mileage data.

The idea of the variance is straightforward: it is the average of the squares of the deviations of the observations from their mean. The details we have just presented, however, raise some questions.

Why do we square the deviations? Why not just average the distances of the observations from their mean? There are two reasons, neither of them obvious. First, the sum of the squared deviations of any set of observations from their mean is the smallest such sum possible. The sum of the unsquared distances is always zero. So squared deviations point to the mean as center in a way that distances do not. Second, the standard deviation turns out to be the natural measure of spread for a particularly important class of symmetric unimodal distributions, the *Normal distributions*. We will meet the Normal distributions in the next chapter. We commented earlier that the usefulness of many statistical procedures is tied to distributions of particular shapes. This is distinctly true of the standard deviation.

Why do we emphasize the standard deviation rather than the variance? One reason is that s, not s^2, is the natural measure of spread for Normal distributions. There is also a more general reason to prefer s to s^2. Because the variance involves squaring the deviations, it does not have the same unit of measurement as the original observations. The variance of the metabolic rates, for example, is measured in squared calories. Taking the square root remedies this. The standard deviation s measures spread about the mean in the original scale.

Why do we "average" by dividing by $n - 1$ rather than n in calculating the variance? Because the sum of the deviations is always zero, the last deviation can be found once we know the other $n - 1$. So we are not averaging n unrelated numbers. Only $n - 1$ of the squared deviations can vary freely, and we average by dividing the total by $n - 1$. The number $n - 1$ is called the ***degrees of freedom*** of the variance or standard deviation. Many calculators offer a choice between dividing by n and dividing by $n - 1$, so be sure to use $n - 1$.

degrees of freedom

Properties of the Standard Deviation

Here are the basic properties of the standard deviation s as a measure of spread.

Properties of the Standard Deviation

- s measures spread about the mean and should be used only when the mean is chosen as the measure of center.
- $s = 0$ only when there is *no spread/variability*. This happens only when all observations have the same value. Otherwise, $s > 0$. As the observations become more spread out about their mean, s gets larger.
- s, like the mean $\bar{x}$, is not resistant. A few outliers can make s very large.

The use of squared deviations renders *s* even more sensitive than $\bar{x}$ to a few extreme observations. For example, dropping the Honda Insight from our list of two-seater cars reduces the mean highway mileage from 24.7 to 22.6 mpg. It cuts the standard deviation by more than half, from 10.8 mpg with the Insight to 5.3 mpg without it. Distributions with outliers and strongly skewed distributions have large standard deviations. The number *s* does not give much helpful information about such distributions.

Choosing Measures of Center and Spread

How do we choose between the five-number summary and $\bar{x}$ and *s* to describe the center and spread of a distribution? Because the two sides of a strongly skewed distribution have different spreads, no single number such as *s* describes the spread well. The five-number summary, with its two quartiles and two extremes, does a better job.

Choosing a Summary

The five-number summary is usually better than the mean and standard deviation for describing a skewed distribution or a distribution with strong outliers. Use $\bar{x}$ and *s* only for reasonably symmetric distributions that are free of outliers.

Example 1.17 *Investments*
Choosing summaries

A central principle in the study of investments is that taking bigger risks is rewarded by higher returns, at least on the average over long periods of time. It is usual in finance to measure risk by the standard deviation of returns, on the grounds that investments whose returns vary a lot from year to year are less predictable and therefore more risky than those whose returns don't vary much. Compare, for example, the approximate mean and standard deviation of the annual percent returns on American common stocks and U.S. Treasury bills over the period from 1950 to 2003:

Investment	Mean return	Standard deviation
Common stocks	13.2%	17.6%
Treasury bills	5.0%	2.9%

Stocks are risky. They went up more than 13% per year on the average during this period, but they dropped almost 28% in the worst year. The large standard deviation reflects the fact that stocks have produced both large gains and large losses. When you buy a Treasury bill, on the other hand, you are lending money to the government for one year. You know that the government will pay you back with interest. That is much less risky than buying stocks, so (on the average) you get a smaller return.

Are $\bar{x}$ and s good summaries for distributions of investment returns? Figure 1.22 displays stemplots of the annual returns for both investments. (Because stock returns are so much more spread out, a back-to-back stemplot does not work well. The stems in the stock stemplot are tens of percents; the stems for Treasury bills are percents. The lowest returns are −28% for stocks and 0.9% for bills.) You see that returns on Treasury bills have a right-skewed distribution. Convention in the financial world calls for $\bar{x}$ and s because some parts of investment theory use them. For describing this right-skewed distribution, however, the five-number summary would be more informative.

Figure 1.22 *Stemplots of annual returns for stocks and Treasury bills, 1950 to 2003. (a) Stock returns, in whole percents. (b) Treasury bill returns, in percents and tenths of a percent.*

Common Stocks

−2	8 1
−1	9 1 1 1 1 0
−0	9 6 4 3
0	0 0 0 1 2 3 8 9 9
1	1 3 3 4 4 6 6 6 7 8
2	0 1 1 2 3 4 4 4 5 7 7 9 9
3	0 1 1 2 3 4 6 7
4	5
5	0

(a)

Treasury bills

0	9
1	0 2 5 5 6 6 6 8
2	1 5 7 7 9
3	0 1 1 3 5 5 8 9 9
4	2 4 7 7 8
5	1 1 2 2 2 5 6 6 7 8 7 9
6	2 4 5 6 9
7	2 7 8
8	0 4 8
9	8
10	4 5
11	3
12	
13	
14	7

(b)

CAUTION

Remember that a graph gives the best overall picture of a distribution. Numerical measures of center and spread report specific facts about a distribution, but they do not describe its entire shape. Numerical summaries do not disclose the presence of multiple modes or gaps, for example. **Always plot your data.**

Exercises

1.39 Phosphate levels The level of various substances in the blood influences our health. Here are measurements of the level of phosphate in the blood of a patient, in milligrams of phosphate per deciliter of blood, made on 6 consecutive visits to a clinic:

5.6   5.2   4.6   4.9   5.7   6.4

A graph of only 6 observations gives little information, so we proceed to compute the mean and standard deviation.

(a) Find the mean from its definition. That is, find the sum of the 6 observations and divide by 6.

(b) Find the standard deviation from its definition. That is, find the deviations of each observation from the mean, square the deviations, then obtain the variance and the standard deviation. Example 1.16 shows the method.

(c) Now enter the data into your calculator to obtain $\bar{x}$ and s. Do the results agree with your hand calculations? Can you find a way to compute the standard deviation without using one-variable statistics?

1.40 Choosing measures of center and spread Which measure of center and spread should be used for the following data sets? In each case, write a sentence or two to explain your reasoning.

(a) The Treasury bill returns in Figure 1.22(b) (page 88).

(b) The 60 IQ scores of fifth-grade students in Example 1.6 (page 49).

(c) The 44 DRP test scores in Exercise 1.5 (page 48).

1.41 Metabolic rates Calculate the mean and standard deviation of the metabolic rates in Example 1.16 (page 85), showing each step in detail. First find the mean $\bar{x}$ by summing the 7 observations and dividing by 7. Then find each of the deviations $x_i - \bar{x}$ and their squares. Check that the deviations have sum 0. Calculate the variance as an average of the squared deviations (remember to divide by $n - 1$). Finally, obtain s as the square root of the variance.

1.42 Median and mean Create a set of 5 positive numbers (repeats allowed) that have median 10 and mean 7. What thought process did you use to create your numbers?

1.43 Contest This is a standard deviation contest. You must choose four numbers from the whole numbers 0 to 10, with repeats allowed.

(a) Choose four numbers that have the smallest possible standard deviation.

(b) Choose four numbers that have the largest possible standard deviation.

(c) Is more than one choice possible in either (a) or (b)? Explain.

1.44 Sum of deviations is zero Use the definition of the mean $\bar{x}$ to show that the sum of the deviations $x_i - \bar{x}$ of the observations from their mean is always zero. This is one reason why the variance and standard deviation use squared deviations.

Statistics in the courtroom In 1994 Digital Equipment Corporation (DEC) was sued by three individuals who claimed that DEC's computer keyboard caused repetitive stress injuries. Awards for economic loss were fairly easy to set, but deciding awards for pain and suffering was much more difficult. On appeal, Circuit Court Judge Jack Weinstein described ways to find a comparison group of similar cases. Then for the jury award to be considered reasonable, he ruled that it should not be more than two standard deviations away from the mean award of the comparison group. Any award outside this interval would be adjusted to be two standard deviations away from the mean.

"It's the new keyboard for the statistics lab. Once you learn how to use it, it will make computation of the standard deviation easier."

Changing the Unit of Measurement

The same variable can be recorded in different units of measurement. Americans commonly record distances in miles and temperatures in degrees Fahrenheit, while the rest of the world measures distances in kilometers and temperatures in degrees Celsius. Fortunately, it is easy to convert from one unit of measurement to another. This is true because a change in the measurement unit is a ***linear transformation*** of the measurements.

Linear Transformation

A **linear transformation** changes the original variable x into the new variable x_{new} given by an equation of the form

$$x_{\text{new}} = a + bx$$

Adding the constant a shifts all values of x upward or downward by the same amount. Multiplying by the positive constant b changes the size of the unit of measurement.

Example 1.18 *Miami Heat salaries*

Changing units

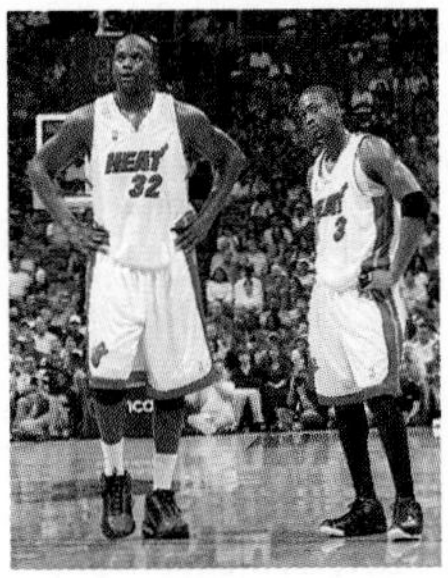

Table 1.7 gives the approximate base salaries of the 15 members of the Miami Heat basketball team for the year 2005. You can calculate that the mean is $\bar{x} = \$3.859$ million and that the median is $M = \$1.13$ million. No wonder professional basketball players have big houses!

Table 1.7 *Year 2005 salaries for Miami Heat players (in millions of dollars)*

Player	Salary	Player	Salary
Shaquille O'Neal	27.70	Christian Laettner	1.10
Eddie Jones	13.46	Steve Smith	1.10
Dwyane Wade	2.83	Shandon Anderson	0.87
Damon Jones	2.50	Keyon Dooling	0.75
Michael Doleac	2.40	Zhizhi Wang	0.75
Rasual Butler	1.20	Udonis Haslem	0.62
Dorell Wright	1.15	Alonzo Mourning	0.33
Qyntel Woods	1.13		

Source: www.hoopshype.com/salaries/miami.htm.

Figure 1.23(a) is a stemplot of the salaries of Miami Heat players, with millions as stems. The distribution is skewed to the right and there are two high outliers. The very high

Figure 1.23 *Minitab stemplots. (a) The salaries of Miami Heat players, with millions as stems. (b) When the same amount is added to every observation, the shape of the distribution remains unchanged. (c) Multiplying every observation by the same amount will either increase or decrease the spread by that amount.*

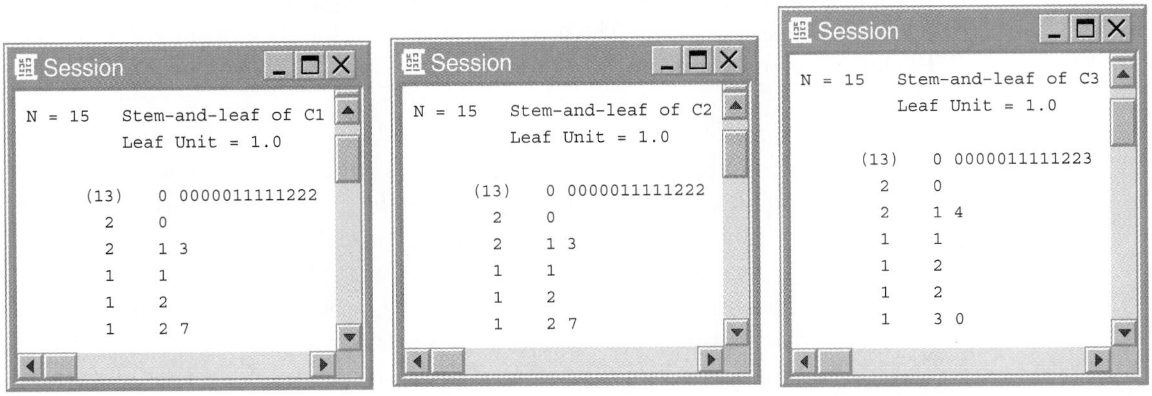

(a) (b) (c)

salaries of Eddie Jones and Shaquille O'Neal pull up the mean. Use your calculator to check that s = \$7.34 million, and that the five-number summary is

\$0.33 million \$0.75 million \$1.13 million \$2.50 million \$27.70 million

1. Suppose that each member of the team receives a \$100,000 bonus for winning the NBA Championship. How will this affect the shape, center, and spread of the distribution?

Since \$100,000 = \$0.1 million, each player's salary will increase by \$0.1 million. This linear transformation can be represented by $x_{new} = 0.1 + 1x$, where x_{new} is the salary after the bonus and x is the player's base salary. Increasing each value in Table 1.7 by 0.1 will also increase the mean by 0.1. That is, $\overline{x}_{new}$ = \$3.96 million. Likewise, the median salary will increase by 0.1 and become M = \$1.23 million.

What will happen to the spread of the distribution? The standard deviation of the Heat's salaries after the bonus is still s = \$7.34 million. With the bonus, the five-number summary becomes

\$0.43 million \$0.85 million \$1.23 million \$2.60 million \$27.80 million

Both before and after the salary bonus, the IQR for this distribution is \$1.75 million. *Adding a constant amount to each observation does not change the spread.* The shape of the distribution remains unchanged, as shown in Figure 1.23(b).

2. Suppose that, instead of receiving a \$100,000 bonus, each player is offered a 10% increase in his base salary. Alonzo Mourning, who is making a base salary of \$0.33 million, would receive an additional (0.10)(\$0.33 million) = \$0.033 million. To obtain his new salary, we could have used the linear transformation $x_{new} = 0 + 1.10x$, since multiplying the current salary (x) by 1.10 increases it by 10%. Increasing all 15 players' salaries in the same way results in the following list of values (in millions):

\$0.363	\$0.682	\$0.825	\$0.825	\$0.957	\$ 1.210	\$ 1.210	\$1.243
\$1.265	\$1.320	\$2.640	\$2.750	\$3.113	\$14.806	\$30.470	

Use your calculator to check that $\overline{x}_{new}$ = \$4.245 million, s_{new} = \$8.072 million, M_{new} = \$1.243 million, and the five-number summary for x_{new} is

\$0.363 million \$0.825 million \$1.243 million \$2.750 million \$30.470 million

Since \$3.859(1.10) = \$4.245 and \$1.13(1.10) = \$1.243, you can see that both measures of center (the mean and median) have increased by 10%. This time, the spread of the distribution has increased, too. Check for yourself that the standard deviation and the IQR have also increased by 10%. The stemplot in Figure 1.23(c) shows that the distribution of salaries is still right-skewed.

Linear transformations do not change the shape of a distribution. If measurements on a variable x have a right-skewed distribution, any new variable x_{new} obtained by a linear transformation $x_{new} = a + bx$ (for $b > 0$) will also have a right-skewed distribution. If the distribution of x is symmetric and unimodal, the distribution of x_{new} remains symmetric and unimodal.

Although a linear transformation preserves the basic shape of a distribution, the center and spread may change. Because linear transformations of measurement scales are common, we must be aware of their effect on numerical descriptive measures of center and spread. Fortunately, the changes follow a simple pattern.

Effect of a Linear Transformation

To see the **effect of a linear transformation** on measures of center and spread, apply these rules:

- Multiplying each observation by a positive number b multiplies both measures of center (mean and median) and measures of spread (interquartile range and standard deviation) by b.
- Adding the same number a (either positive, zero, or negative) to each observation adds a to measures of center and to quartiles but does not change measures of spread.

The measures of spread IQR and s do not change when we add the same number a to all of the observations because adding a constant changes the location of the distribution but leaves the spread unaltered. You can find the effect of a linear transformation $x_{\text{new}} = a + bx$ by combining these rules. For example, if x has mean $\overline{x}$, the transformed variable x_{new} has mean $a + b\overline{x}$.

Comparing Distributions

An experiment is carried out to compare the effectiveness of a new cholesterol-reducing drug with the one that is currently prescribed by most doctors. A survey is conducted to determine whether the proportion of males who are likely to vote for a political candidate is higher than the proportion of females who are likely to vote for the candidate. Students taking AP Calculus AB and AP Statistics are curious about which exam is harder. They have information on the distribution of scores earned on each exam from the year 2005. In each of these situations, we are interested in comparing distributions. This section presents some of the more common methods for making statistical comparisons. We also introduce the Data Analysis Toolbox. When you are investigating a statistical problem involving one or more sets of data, use the Data Analysis Toolbox to organize your thinking.

Data Analysis Toolbox

To answer a statistical question of interest involving one or more data sets, proceed as follows.

- **Data** Organize and examine the data. Answer the ***key questions*** from the Preliminary Chapter:

  1. **Who** are the individuals described by the data?
  2. **What** are the variables? In what units is each variable recorded?

Data Analysis Toolbox

(continued)

3. **Why** were the data gathered?
4. **When, where, how, and by whom** were the data produced?

- **Graphs** Construct appropriate graphical displays.
- **Numerical summaries** Calculate relevant summary statistics.
- **Interpretation** Discuss what the data, graphs, and numerical summaries tell you in the context of the problem. Answer the question!

Example 1.19 *The nation's report card*

Comparing categorical variables

The National Assessment of Educational Progress (NAEP), also known as "the Nation's Report Card," is a nationally representative and continuing assessment of what Americans know and can do in reading and math. NAEP results are based on samples of students. Table 1.8 shows the NAEP comparisons between Virginia students and the nation as a whole.

Table 1.8 *How Virginia students performed on national reading and math tests*

	Virginia	Nation
Grade 4 reading proficiency (%)	35	30
Grade 4 math proficiency (%)	36	31
Grade 8 reading proficiency (%)	36	30
Grade 8 math proficiency (%)	31	27

Source: Data from the Web site www.nces.ed.gov/nationsreportcard/.

- **Data**

1. **Who?** The individuals are students in grades 4 and 8.
2. **What?** The two variables for Virginia and the nation are the percent of individuals classified as proficient on the math test and the percent classified as proficient on the reading test.
3. **Why?** The National Center for Education Statistics is charged with periodically testing students in mathematics and reading. They are also charged with preparing reports for the public that enable comparisons of an individual's performance with that of students across the nation and in the state.
4. **When, where, how,** and **by whom?** The data were collected in 2003 by the National Center for Education Statistics, based in Washington, D.C.

- **Graphs** Figure 1.24 shows how Virginia's grade 4 and grade 8 students compare with the nation as a whole in reading and math proficiency.

Figure 1.24 *Bar chart comparing Virginia and the nation in reading and math proficiency.*

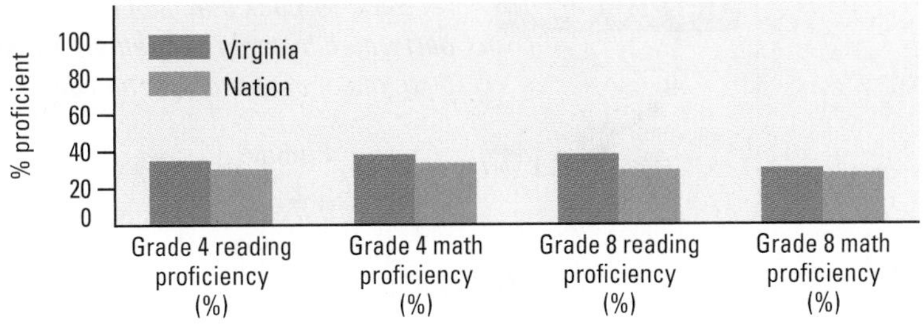

- **Numerical summaries** are the percents of students who test proficient. See Table 1.8 on the previous page.
- **Interpretation** Table 1.8 and Figure 1.24 show that higher percents of students were proficient in math and reading in Virginia than in the nation as a whole.

Example 1.20 *Swiss doctors*

Comparing quantitative variables: Data Analysis Toolbox

Do male doctors perform more cesarean sections (C-sections) than female doctors? A study in Switzerland examined the number of cesarean sections (surgical deliveries of babies) performed in a year by samples of male and female doctors.

- **Data** Here are the data for 15 male doctors, arranged from lowest to highest:

20	25	25	27	28	31	33	34
36	37	44	50	59	85	86	

The study also looked at 10 female doctors. The numbers of cesarean sections performed by these doctors (arranged in order) were

5	7	10	14	18	19	25	29	31	33

1. **Who?** The individuals are Swiss doctors.
2. **What?** The variable is the number of cesarean sections performed in a year.
3. **Why?** Researchers wanted to compare the number of cesarean sections performed by male and female doctors in Switzerland.
4. **When, where, how, and by whom?** The unpublished results of this study were provided to one of the co-authors in a private communication in the 1980s. The identities of the researchers are unknown.

- **Graphs** We can compare the number of cesarean sections performed by male and female doctors using a back-to-back stemplot and a side-by-side boxplot. Figure 1.25 shows the completed graphs. In the stemplot, the stems are listed in the middle and leaves are placed on the left for male doctors and on the right for female doctors. It is usual to have the leaves increase in value as they move away from the stem.

Figure 1.25 *(a) Back-to-back stemplot of the number of cesarean sections performed by male and female Swiss doctors. (b) Side-by-side boxplots of cesarean section data.*

Male		Female
	0	5 7
	1	0 4 8 9
8 7 5 5 0	2	5 9
7 6 4 3 1	3	1 3
4	4	
9 0	5	
	6	
	7	
6 5	8	

(a)

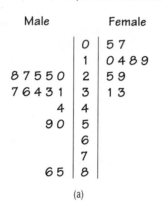

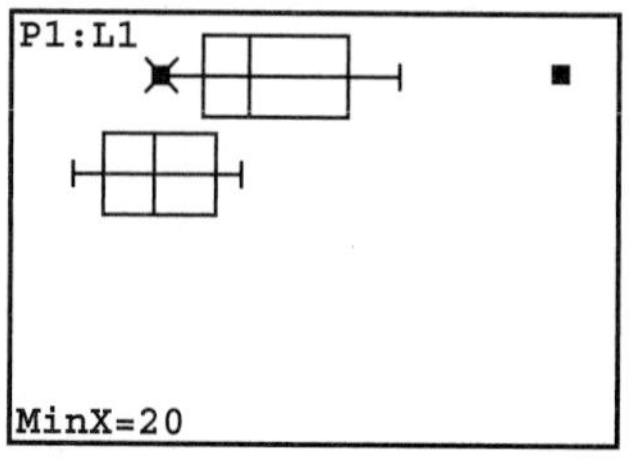

(b)

- **Numerical summaries** Here are summary statistics for the two distributions:

	$\bar{x}$	s	Min.	Q_1	M	Q_3	Max.	IQR
Male doctors	41.333	20.607	20	27	34	50	86	23
Female doctors	19.1	10.126	5	10	18.5	29	33	19

- **Interpretation** The distribution of the number of cesarean sections performed by female doctors is roughly symmetric. For the male doctors, the distribution is skewed to the right. *(Shape)*

  Two male physicians performed an unusually high number of cesarean sections, 85 and 86. You can use the $1.5 \times IQR$ rule to confirm that these two values are outliers, and that there are no other outliers in either data set. *(Outliers)*

  More than half of the female doctors in the study performed fewer than 20 cesarean sections in a year; 20 was the minimum number of cesarean sections performed by male doctors. The mean and median numbers of cesarean sections performed are higher for the male doctors. *(Center)*

  Both the standard deviation and the *IQR* for the male doctors are much larger than the corresponding statistics for the female doctors. So there is much greater variability in the number of cesarean sections performed by male doctors. *(Spread)*

  Due to the outliers in the male doctor data and the lack of symmetry of their distribution of cesareans, we should use the resistant medians and *IQR*s in our numerical comparisons. In Switzerland, it does seem that male doctors generally perform more cesarean sections each year (median = 34) than do female doctors (median = 18.5). In addition, male Swiss doctors are more variable in the number of cesarean sections performed each year ($IQR = 23$) than female Swiss doctors ($IQR = 19$). We may want to do more research on why this apparent discrepancy exists.

Exercises

1.45 Raising teachers' pay, I A school system employs teachers at salaries between $30,000 and $60,000. The teachers' union and the school board are negotiating the form of next year's increase in the salary schedule. Suppose that every teacher is given a flat $1000 raise.

(a) How much will the mean salary increase? The median salary?

(b) Will a flat $1000 raise increase the spread as measured by the distance between the quartiles?

(c) Will a flat $1000 raise increase the spread as measured by the standard deviation of the salaries?

1.46 Raising teachers' pay, II Suppose that the teachers in the previous exercise each receive a 5% raise. The amount of the raise will vary from $1500 to $3000, depending on present salary. Will a 5% across-the-board raise increase the spread of the distribution as measured by the distance between the quartiles? Do you think it will increase the standard deviation?

1.47 Which AP exam is easier: Calculus AB or Statistics? The table below gives the distribution of grades earned by students taking the AP Calculus AB and AP Statistics exams in 2005:

	5	4	3	2	1
Calculus AB	20.7%	19.5%	17.7%	16.7%	25.2%
Statistics	12.6%	22.8%	25.3%	19.2%	20.1%

(a) Make a graphical display to compare the exam grades for Calculus AB and Statistics.

(b) Write a few sentences comparing the two distributions of exam grades. Do you now know which exam is easier? Why or why not?

1.48 Get your hot dogs here! Face it. "A hot dog isn't a carrot stick." So said *Consumer Reports*, commenting on the low nutritional quality of the all-American frank. Table 1.9 on the next page shows the magazine's laboratory test results for calories and milligrams of sodium (mostly due to salt) in a number of major brands of hot dogs. There are three types: beef, "meat" (mainly pork and beef, but government regulations allow up to 15% poultry meat), and poultry. Because people concerned about their health may prefer low-calorie, low-sodium hot dogs, we ask: "Are there any systematic differences among the three types of hot dogs in these two variables?" Use side-by-side boxplots and numerical summaries to help you answer this question. Write a paragraph explaining your findings. Use the Data Analysis Toolbox (page 93) as a guide.

1.49 Who makes more? A manufacturing company is reviewing the salaries of its full-time employees below the executive level at a large plant. The clerical staff is almost entirely female, while a majority of the production workers and technical staff are male. As a result, the distributions of salaries for male and female employees may be quite different. Table 1.10 on the next page gives the frequencies and relative frequencies for women and men.

(a) Make histograms for these data, choosing a vertical scale that is most appropriate for comparing the two distributions.

Table 1.9 *Calories and sodium (milligrams) in three types of hot dogs*

Beef Hot Dogs		Meat Hot Dogs		Poultry Hot Dogs	
Calories	**Sodium**	**Calories**	**Sodium**	**Calories**	**Sodium**
186	495	173	458	129	430
181	477	191	506	132	375
176	425	182	473	102	396
149	322	190	545	106	383
184	482	172	496	94	387
190	587	147	360	102	542
158	370	146	387	87	359
139	322	139	386	99	357
175	479	175	507	170	528
148	375	136	393	113	513
152	330	179	405	135	426
111	300	153	372	142	513
141	386	107	144	86	358
153	401	195	511	143	581
190	645	135	405	152	588
157	440	140	428	146	522
131	317	138	339	144	545
149	319				
135	298				
132	253				

Source: *Consumer Reports*, June 1986, pp. 366–367.

Table 1.10 *Salary distributions of female and male workers in a large factory*

	Women		Men	
Salary ($1000)	**Number**	**%**	**Number**	**%**
10–15	89	11.8	26	1.1
15–20	192	25.4	221	9.0
20–25	236	31.2	677	27.6
25–30	111	14.7	823	33.6
30–35	86	11.4	365	14.9
35–40	25	3.3	182	7.4
40–45	11	1.5	91	3.7
45–50	3	0.4	33	1.4
50–55	2	0.3	19	0.8
55–60	0	0.0	11	0.4
60–65	0	0.0	0	0.0
65–70	1	0.1	3	0.1
Total	756	100.1	2451	100.0

(b) Describe the shapes of the salary distributions and the chief differences between them.

(c) Explain why the total for women is greater than 100%.

1.50 Linear transformations In each of the following settings, give the values of a and b for the linear transformation $x_{new} = a + bx$ that expresses the change in measurement units. Then explain how the transformation will affect the mean, the *IQR*, the median, and the standard deviation of the original distribution.

(a) You collect data on the power of car engines, measured in horsepower. Your teacher requires you to convert the power to watts. One horsepower is 746 watts.

(b) You measure the temperature (in degrees Fahrenheit) of your school's swimming pool at 20 different locations within the pool. Your swim team coach wants the summary statistics in degrees Celsius (°F = (9/5)°C + 32).

(c) Mrs. Swaynos has given a very difficult statistics test and is thinking about "curving" the grades. She decides to add 10 points to each student's score.

Section 1.2 Summary

A numerical summary of a distribution should report its **center** and its **spread,** or **variability**.

The mean $\bar{x}$ and the median M describe the center of a distribution in different ways. The mean is the average of the observations, and the median is the midpoint of the values.

When you use the median to indicate the center of a distribution, describe its spread by giving the **quartiles.** The **first quartile** Q_1 has about one-fourth of the observations below it, and the **third quartile** Q_3 has about three-fourths of the observations below it. An extreme observation is an **outlier** if it is smaller than $Q_1 - (1.5 \times IQR)$ or larger than $Q_3 + (1.5 \times IQR)$.

The **five-number summary** consists of the median, the quartiles, and the high and low extremes and provides a quick overall description of a distribution. The median describes the center, and the quartiles and extremes show the spread.

Boxplots based on the five-number summary are useful for comparing two or more distributions. The box spans the quartiles and shows the spread of the central half of the distribution. The median is marked within the box. Lines extend from the box to the smallest and the largest observations that are not outliers. Outliers are plotted as isolated points.

The **variance** s^2 and especially its square root, the **standard deviation** s, are common measures of spread about the mean as center. The standard deviation s is zero when there is no spread and gets larger as the spread increases.

The median is a **resistant** measure of center because it is relatively unaffected by extreme observations. The mean is nonresistant. Among measures of spread, the quartiles are resistant, but the standard deviation is not.

The mean and standard deviation are strongly influenced by outliers or skewness in a distribution. They are good descriptions for symmetric distributions and are most useful for the Normal distributions, which will be introduced in the next chapter.

The median and quartiles are not affected by outliers, and the two quartiles and two extremes describe the two sides of a distribution separately. The five-number summary is the preferred numerical summary for skewed distributions.

Linear transformations are quite useful for changing units of measurement. Linear transformations have the form $x_{new} = a + bx$. When you add a constant a to all the values in a data set, the mean and median increase by a. Measures of spread do not change. When you multiply all the values in a data set by a constant b, the mean, median, *IQR*, and standard deviation are multiplied by b.

Back-to-back stemplots and **side-by-side boxplots** are useful for comparing quantitative distributions.

Section 1.2 Exercises

1.51 Do girls study more than boys? We asked the students in three AP Statistics classes how many minutes they studied on a typical weeknight. Here are the responses of random samples of 30 girls and 30 boys from the classes:

Girls					Boys				
180	120	180	360	240	90	120	30	90	200
120	180	120	240	170	90	45	30	120	75
150	120	180	180	150	150	120	60	240	300
200	150	180	150	180	240	60	120	60	30
120	60	120	180	180	30	230	120	95	150
90	240	180	115	120	0	200	120	120	180

(a) Examine the data. Why are you not surprised that most responses are multiples of 10 minutes? We eliminated one student who claimed to study 30,000 minutes per night. Are there any other responses you consider suspicious?

(b) Make a back-to-back stemplot of these data. Report the approximate midpoints of both groups. Does it appear that girls study more than boys (or at least claim that they do)?

1.52 Educational attainment Table 1.11 on the next page shows the educational level achieved by U.S. adults aged 25 to 34 and by those aged 65 to 74. Compare the distributions of educational attainment graphically. Write a few sentences explaining what your display shows.

1.53 Logging in rain forests "Conservationists have despaired over destruction of tropical rainforest by logging, clearing, and burning." These words begin a report on a statistical study of the effects of logging in Borneo. Researchers compared forest plots that had never been logged (Group 1) with similar plots nearby that had been logged 1 year earlier (Group 2) and 8 years earlier (Group 3). All plots were 0.1 hectare in area. Here are the counts of trees for plots in each group:[18]

Group 1:	27	22	29	21	19	33	16	20	24	27	28	19
Group 2:	12	12	15	9	20	18	17	14	14	2	17	19
Group 3:	18	4	22	15	18	19	22	12	12			

(a) Give a complete comparison of the three distributions, using both graphs and numerical summaries.

Table 1.11 *Educational attainment by U.S. adults aged 25 to 34 and 65 to 74*

	Number of People (thousands)	
	Ages 25–34	**Ages 65–74**
Less than high school	5,063	4,546
High school graduate	11,380	6,737
Some college	7,613	2 481
Bachelor's degree	8,830	2,047
Advanced degree	2,943	1,413
Total	35,829	17,224

Source: Census Bureau, *Educational Attainment in the United States*, March 2004.

(b) To what extent has logging affected the count of trees?

(c) The researchers used an analysis based on $\bar{x}$ and s. Explain why this is reasonably well justified.

1.54 $\bar{x}$ and s are not enough The mean $\bar{x}$ and standard deviation s measure center and spread but are not a complete description of a distribution. Data sets with different shapes can have the same mean and standard deviation. To demonstrate this fact, use your calculator to find $\bar{x}$ and s for the following two small data sets. Then make a stemplot of each and comment on the shape of each distribution.

Data A:	9.14	8.14	8.74	8.77	9.26	8.10	6.13	3.10	9.13	7.26	4.74
Data B:	6.58	5.76	7.71	8.84	8.47	7.04	5.25	5.56	7.91	6.89	12.50

1.55 Sales of CDs In recent years, the Recording Industry Association of America (RIAA) has initiated legal action against individuals for illegally downloading copyrighted music from the Web. RIAA has targeted primarily college students, roughly aged 17 to 23, who access peer-to-peer (P2P) sites via fast Internet connections in their dorm rooms. The table below shows the sales from record labels, primarily CDs (87.8% of sales), from 1994 to 2003 for two different age groups.[19]

	1994	1995	1996	1997	1998	1999	2000	2001	2002	2003
15–34 years	44.8	44.7	44.7	42.3	39.4	35.7	36.0	36.1	34.2	32.3
Over 35	46.6	46.5	47.3	47.9	50.4	54.5	53.8	54.5	56.0	57.9

Make a graphical display to compare sales from record labels to these two consumer groups. Write a few sentences describing what the graph tells you

1.56 Linear transformation consequences A change of units that multiplies each unit by b, such as the change $x_{new} = 0 + 2.54x$ from inches x to centimeters x_{new}, multiplies our usual measures of spread by b. This is true of the IQR and standard deviation. What happens to the variance when we change units in this way?

1.57 Better corn Corn is an important animal food. Normal corn lacks certain amino acids, which are building blocks for protein. Plant scientists have developed new corn

varieties that have more of these amino acids. To test a new corn as an animal food, a group of 20 one-day-old male chicks was fed a ration containing the new corn. A control group of another 20 chicks was fed a ration that was identical except that it contained normal corn. Here are the weight gains (in grams) after 21 days:[20]

Normal corn

380	321	366	356
283	349	402	462
356	410	329	399
350	384	316	272
345	455	360	431

New corn

361	447	401	375
434	403	393	426
406	318	467	407
427	420	477	392
430	339	410	326

(a) Compute five-number summaries for the weight gains of the two groups of chicks. Then make boxplots to compare the two distributions. What do the data show about the effect of the new corn?

(b) The researchers actually reported means and standard deviations for the two groups of chicks. What are they? How much larger is the mean weight gain of chicks fed the new corn?

(c) The weights are given in grams. There are 28.35 grams in an ounce. Use the results of part (b) to compute the means and standard deviations of the weight gains measured in ounces.

1.58 Mean or median? Which measure of center, the mean or the median, should you use in each of the following situations?

(a) Middletown is considering imposing an income tax on citizens. The city government wants to know the average income of citizens so that it can estimate the total tax base.

(b) In a study of the standard of living of typical families in Middletown, a sociologist estimates the average family income in that city.

CASE CLOSED!

Nielsen ratings

Begin by reviewing the ratings data in the Nielsen ratings Case Study (page 37). Then answer each of the following questions in complete sentences. Be sure to communicate clearly enough for any of your classmates to understand what you are saying.

1. Construct by hand an appropriate graphical display for comparing the Nielsen ratings of the three networks. Write a few sentences describing what you see.

2. Calculate numerical summaries for the Nielsen ratings of the three networks. Which measures of center and spread would you choose to compare the distributions and why?
3. Determine whether there are any outliers in each of the three distributions. If there are outliers, identify them.
4. What does it mean to say that the mean percent of TV viewers watching a particular network is a nonresistant measure of center?
5. How would you rank the three networks based on your analysis?

Chapter Review

Summary

Data analysis is the art of describing data using graphs and numerical summaries. The purpose of data analysis is to describe the most important features of a set of data. This chapter introduces data analysis by presenting statistical ideas and tools for describing the distribution of a single variable. The figure below will help you organize the big ideas.

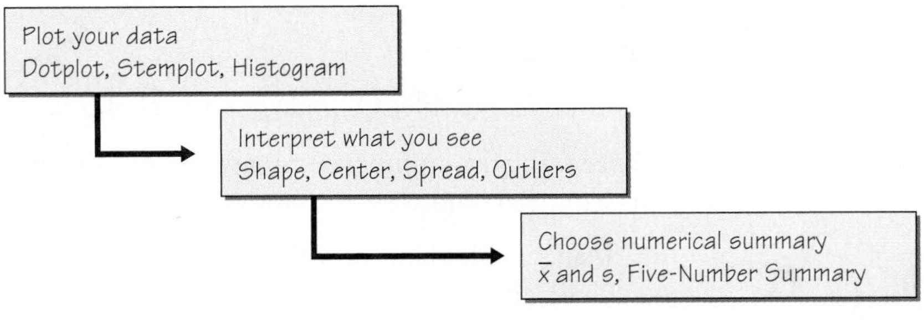

What You Should Have Learned

Here is a review list of the most important skills you should have acquired from your study of this chapter.

A. Displaying Distributions

1. Make a stemplot of the distribution of a quantitative variable. Trim the numbers or split stems as needed to make an effective stemplot.
2. Make a histogram of the distribution of a quantitative variable.
3. Construct and interpret an ogive of a set of quantitative data.

B. Inspecting Distributions (Quantitative Variables)

1. Look for the overall pattern and for major deviations from the pattern.
2. Assess from a dotplot, stemplot, or histogram whether the shape of a distribution is roughly symmetric, distinctly skewed, or neither. Assess whether the distribution has one or more major modes.
3. Describe the overall pattern by giving numerical measures of center and spread in addition to a verbal description of shape.
4. Decide which measures of center and spread are more appropriate: the mean and standard deviation (especially for symmetric distributions) or the five-number summary (especially for skewed distributions).
5. Recognize outliers.

C. Time Plots

1. Make a time plot of data, with the time of each observation on the horizontal axis and the value of the observed variable on the vertical axis.
2. Recognize strong trends or other patterns in a time plot.

D. Measuring Center

1. Find the mean $\bar{x}$ of a set of observations.
2. Find the median M of a set of observations.
3. Understand that the median is more resistant (less affected by extreme observations) than the mean. Recognize that skewness in a distribution moves the mean away from the median toward the long tail.

E. Measuring Spread

1. Find the quartiles Q_1 and Q_3 for a set of observations.
2. Give the five-number summary and draw a boxplot; assess center, spread, symmetry, and skewness from a boxplot. Determine outliers.
3. Using a calculator or software, find the standard deviation s for a set of observations.

4. Know the basic properties of s: $s \geq 0$ always; $s = 0$ only when all observations are identical; s increases as the spread increases; s has the same units as the original measurements; s is increased by outliers or skewness.

F. Changing Units of Measurement (Linear Transformations)

1. Determine the effect of a linear transformation on measures of center and spread.
2. Describe a change in units of measurement in terms of a linear transformation of the form $x_{new} = a + bx$.

G. Comparing Distributions

1. Use side-by-side bar graphs to compare distributions of categorical data.
2. Make back-to-back stemplots and side-by-side boxplots to compare distributions of quantitative variables.
3. Write narrative comparisons of the shape, center, spread, and outliers for two or more quantitative distributions.

Web Link

This site is an excellent source for available data:

www.census.gov/prod/ www/statistical-abstract.html

Chapter Review Exercises

1.59 Top companies Each year *Fortune* magazine lists the top 500 companies in the United States, ranked according to their total annual sales in dollars. Describe three other variables that could reasonably be used to measure the "size" of a company.

1.60 Density of the earth In 1798 the English scientist Henry Cavendish measured the density of the earth by careful work with a torsion balance. The variable recorded was the density of the earth as a multiple of the density of water. Here are Cavendish's 29 measurements:[21]

5.50	5.61	4.88	5.07	5.26	5.55	5.36	5.29	5.58	5.65
5.57	5.53	5.62	5.29	5.44	5.34	5.79	5.10	5.27	5.39
5.42	5.47	5.63	5.34	5.46	5.30	5.75	5.68	5.85	

Present these measurements graphically in a stemplot. Discuss the shape, center, and spread of the distribution. Are there any outliers? What is your estimate of the density of the earth based on these measurements?

1.61 Hummingbirds and tropical flowers Different varieties of the tropical flower *Heliconia* are fertilized by different species of hummingbirds. Over time, the lengths of the flowers and the forms of the hummingbirds' beaks have evolved to match each other. Here are data on the lengths in millimeters of three varieties of these flowers on the island of Dominica:[22]

H. bihai							
47.12	46.75	46.80	47.12	46.67	47.43	46.44	46.64
48.07	48.34	48.15	50.26	50.12	46.34	46.94	48.36
***H. caribaea* red**							
41.90	42.01	41.93	43.09	41.47	41.69	39.78	40.57
39.63	42.18	40.66	37.87	39.16	37.40	38.20	38.07
38.10	37.97	38.79	38.23	38.87	37.78	38.01	
***H. caribaea* yellow**							
36.78	37.02	36.52	36.11	36.03	35.45	38.13	37.10
35.17	36.82	36.66	35.68	36.03	34.57	34.63	

(a) Make boxplots to compare the three distributions. Report the five-number summaries along with your graphs. What are the most important differences among the three varieties of flower?

(b) Find $\bar{x}$ and s for each variety.

(c) Make a stemplot of each set of flower lengths. Do the distributions appear suitable for use of $\bar{x}$ and s as summaries?

(d) Starting from the $\bar{x}$ and s-values in millimeters, find the means and standard deviations in inches. (A millimeter is 1/1000 of a meter. A meter is 39.37 inches.)

1.62 Never on Sunday Figure 1.26 shows the distributions of number of births in Toronto, Canada, on each of the 365 days in a year, grouped by day of the week.[23] Based on these plots, give a more detailed description of how births depend on the day of the week.

Figure 1.26 *Side-by-side boxplots of the distributions of numbers of births in Toronto, Canada, for each day of the week during a year, for Exercise 1.62.*

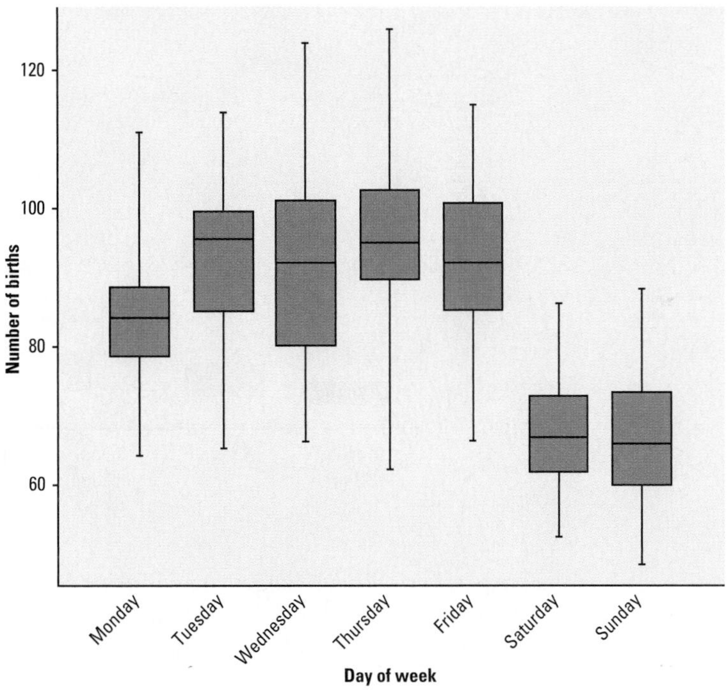

1.63 Presidential elections Here are the percents of the popular vote won by the successful candidate in each of the presidential elections from 1948 to 2004:

Year:	1948	1952	1956	1960	1964	1968	1972	1976	1980	1984	1988	1992	1996	2000	2004
Percent:	49.6	55.1	57.4	49.7	61.1	43.4	60.7	50.1	50.7	58.8	53.9	43.2	49.2	47.9	50.7

(a) Make a stemplot of the winners' percents. (Round to whole numbers and use split stems.)

(b) What is the median percent of the vote won by the successful candidate in presidential elections? (Work with the unrounded data.)

(c) Call an election a landslide if the winner's percent falls at or above the third quartile. Find the third quartile. Which elections were landslides?

1.64 Income and education level Each March, the Bureau of Labor Statistics compiles an Annual Demographic Supplement to its monthly Current Population Survey.[24] Data on about 71,067 individuals between the ages of 25 and 64 who were employed full-time in 2001 were collected in one of these surveys. The boxplots in Figure 1.27 compare the distributions of income for people with five levels of education. This figure is a variation

Figure 1.27 *Boxplots comparing the distributions of income for employed people aged 25 to 64 years with five different levels of education, for Exercise 1.64. The lines extend from the quartiles to the 5th and 95th percentiles.*

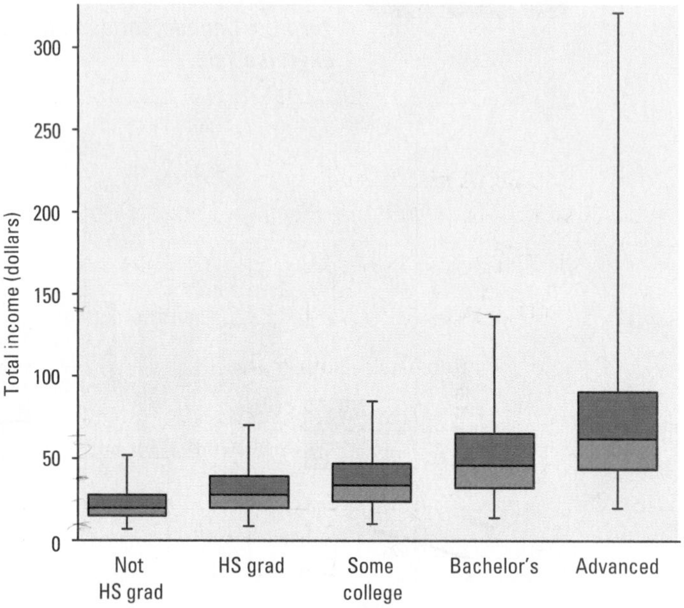

of the boxplot idea: because large data sets often contain very extreme observations, the lines extend from the central box only to the 5th and 95th percentiles. The data include 14,959 people whose highest level of education is a bachelor's degree.

(a) What is the position of the median in the ordered list of incomes (1 to 14,959) of people with a bachelor's degree? From the boxplot, about what is the median income?

(b) What is the position of the first and third quartiles in the ordered list of incomes for these people? About what are the numerical values of Q_1 and Q_3?

(c) You answered (a) and (b) from a boxplot that omits the lowest 5% and the highest 5% of incomes. Explain why leaving out these values has only a very small effect on the median and quartiles.

(d) About what are the positions of the 5th and 95th percentiles in the ordered list of incomes of the 14,959 people with a bachelor's degree? Incomes outside this range do not appear in the boxplot.

(e) About what are the numerical values of the 5th and 95th percentiles of income? (For comparison, the largest income among all 14,959 people was $481,720. That one person made this much tells us less about the group than does the 95th percentile.)

(f) Write a brief description of how the distribution of income changes with the highest level of education reached. Be sure to discuss center, spread, and skewness. Give some specifics read from the graphs to back up your statements.

1.65 Drive time Professor Moore, who lives a few miles outside a college town, records the time he takes to drive to the college each morning. Here are the times (in minutes) for 42 consecutive weekdays, with the dates in order along the rows:

8.25	7.83	8.30	8.42	8.50	8.67	8.17	9.00	9.00	8.17	7.92
9.00	8.50	9.00	7.75	7.92	8.00	8.08	8.42	8.75	8.08	9.75
8.33	7.83	7.92	8.58	7.83	8.42	7.75	7.42	6.75	7.42	8.50
8.67	10.17	8.75	8.58	8.67	9.17	9.08	8.83	8.67		

(a) Make a histogram of these drive times. Is the distribution roughly symmetric, clearly skewed, or neither? Are there any clear outliers?

(b) Construct an ogive for Professor Moore's drive times.

(c) Use your ogive from (b) to estimate the center and 90th percentile of the distribution.

(d) Use your ogive to estimate the percentile corresponding to a drive time of 8.00 minutes.

1.66 Computer use Mrs. Causey asked her students how much time they had spent using a computer during the previous week. Figure 1.28 is an ogive of her students' responses.

Figure 1.28 *Ogive of weekly computer use by Mrs. Causey's students, for Exercise 1.66.*

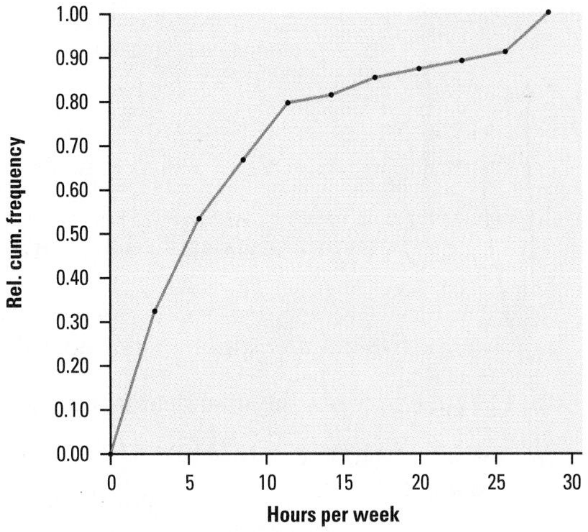

(a) Construct a relative frequency table based on the ogive. Then make a histogram.

(b) Estimate the median, Q_1, and Q_3 from the ogive. Then make a boxplot. Are there any outliers?

(c) At what percentile does a student who used her computer for 10 hours last week fall?

1.67 Wal-Mart stock The rate of return on a stock is its change in price plus any dividends paid. Rate of return is usually measured in percent of the starting value. We have data on the monthly rates of return for the stock of Wal-Mart stores for the years 1973 to 1991, the first 19 years Wal-Mart was listed on the New York Stock Exchange. There are 228 observations. Figure 1.29 displays output from statistical software that describes the distribution of these data. The stems in the stemplot are the tens digits of the percent returns. The leaves are the ones digits. The stemplot uses split stems to give a better display. The software gives high and low outliers separately from the stemplot rather than spreading out the stemplot to include them.

Figure 1.29 *Output from Minitab software describing the distribution of monthly returns from Wal-Mart stock, for Exercise 1.67.*

```
Session
Mean  =  3.064
Standard deviation  =  11.49

N  = 228   Median  =  3.4691
Quartiles  =  -2.950258,   8.4511

Decimal point is 1 place to the right of the colon

Low:  -34.04255  -31.25000  -27.06271  -26.61290


 -1 : 985
 -1 : 444443322222110000
 -0 : 99998877766666665555
 -0 : 44444444333333322222222222222111111100
  0 : 000001111111111112222222333333344444444
  0 : 55555555555555555555566666666666677777778888888888899999
  1 : 00000000011111111122233334444
  1 : 55566667889
  2 : 011334


High:  32.01923  41.80531  42.05607  57.89474  58.67769
```

(a) Give the five-number summary for monthly returns on Wal-Mart stock.

(b) Describe in words the main features of the distribution.

(c) If you had $1000 worth of Wal-Mart stock at the beginning of the best month during these 19 years, how much would your stock be worth at the end of the month? If you had $1000 worth of stock at the beginning of the worst month, how much would your stock be worth at the end of the month?

(d) Find the interquartile range (*IQR*) for the Wal-Mart data. Are there any outliers according to the $1.5 \times IQR$ criterion? Does it appear to you that the software uses this criterion in choosing which observations to report separately as outliers?

1.68 Jury awards A study of the size of jury awards in civil cases (such as injury, product liability, and medical malpractice) in Chicago showed that the median award was about $8000. But the mean award was about $69,000. Explain how a difference this big between the two measures of center can occur.

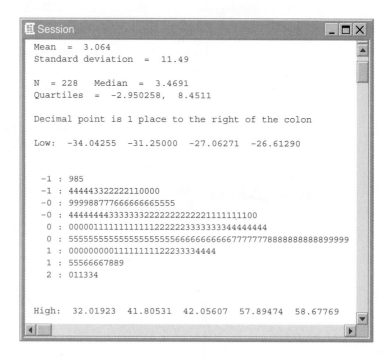

1.69 Women runners Women were allowed to enter the Boston Marathon in 1972. Here are the times (in minutes, rounded to the nearest minute) for the winning woman from 1972 to 2003:

Year	Time	Year	Time	Year	Time	Year	Time
1972	190	1980	154	1988	145	1996	147
1973	186	1981	147	1989	144	1997	146
1974	167	1982	150	1990	145	1998	143
1975	162	1983	143	1991	144	1999	143
1976	167	1984	149	1992	144	2000	146
1977	168	1985	154	1993	145	2001	144
1978	165	1986	145	1994	142	2002	141
1979	155	1987	146	1995	145	2003	145

Make a graph that shows change over time. What overall pattern do you see? Have times stopped improving in recent years? If so, when did improvement end?

1.70 Household incomes Rich and poor households differ in ways that go beyond income. Figure 1.30 displays histograms that compare the distributions of household size (number of people) for low-income and high-income households in 2002.[25] Low-income households had incomes less than $15,000, and high-income households had incomes of at least $100,000.

Figure 1.30 *The distributions of household size for households with incomes less than $15,000 and households with incomes of at least $100,000, for Exercise 1.70.*

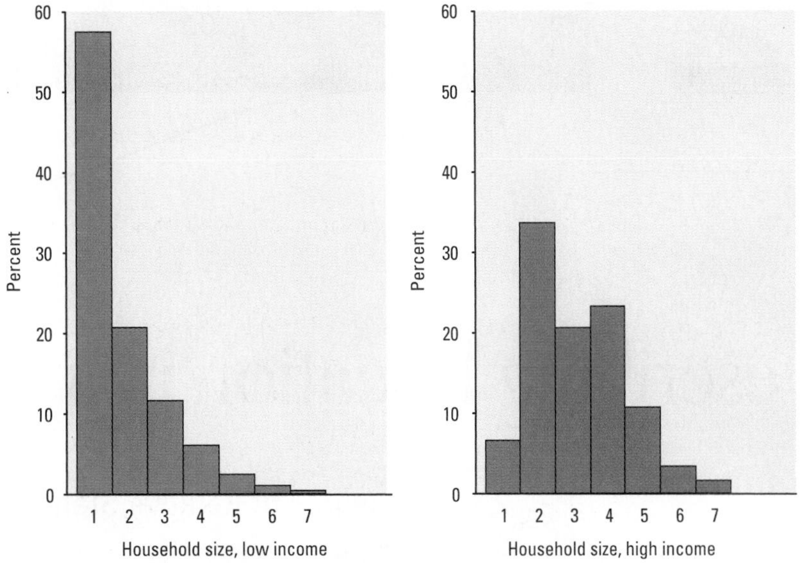

(a) About what percent of each group of households consisted of two people?

(b) What are the important differences between these two distributions? What do you think explains these differences?

CHAPTER

2

Describing Location in a Distribution

CASE STUDY

The new SAT

For many years, high school students across the United States have taken the SAT as part of the college admissions process. The SAT has undergone regular changes over time, but until recently, test takers have always received two scores: SAT Verbal and SAT Math. Motivated by the findings of a Blue Ribbon Panel in 1990 and threats by the mammoth University of California system to discontinue use of the SAT in its admissions process, the College Board decided to add a Writing section to the test. This section would include multiple-choice questions on grammar and usage from the old SAT II Writing Subject Test, as well as a timed, argumentative essay. At the same time, the existing Verbal and Math sections of the test underwent major content revisions. Analogies were removed from the Verbal section, which was renamed Critical Reading. In the Math section, quantitative comparison questions were eliminated, and new questions testing advanced algebra content were added.

In March 2005, the College Board administered the new SAT for the first time. Students, parents, teachers, high school counselors, and college admissions officers waited anxiously to hear about the results from this new exam. Would the scores on the new SAT be comparable to those from previous years? How would students perform on the new Writing section (and particularly on the timed essay)? In the past, boys had earned higher average scores than girls on both the Verbal and Math sections of the SAT. Would similar gender differences emerge on the new SAT?

By the end of this chapter, you will have developed the statistical tools you need to answer important questions about the new SAT.

Activity 2A

A fine-grained distribution

Materials: Sheet of grid paper; salt; can of spray paint; paint easel; newspapers

1. Place the grid paper on the easel with a horizontal fold as shown, at about a 45° angle to the horizontal. Provide a "lip" at the bottom to catch the salt. Place newspaper behind the grid paper and extending out on all sides so you will not get paint on the easel.

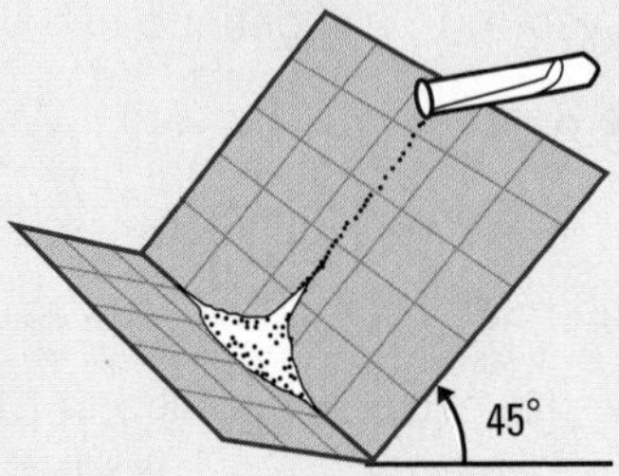

2. Pour a stream of salt slowly from a point near the middle of the top edge of the grid paper. The grains of salt will hop and skip their way down the grid as they collide with one another and bounce left and right. They will accumulate at the bottom, piled against the grid, with the smooth profile of a bell-shaped curve, known as a Normal distribution. We will learn about the Normal distribution in this chapter.
3. Now carefully spray the grid—salt and all—with paint. Then discard the salt. You should be able to easily measure the height of the curve at different places by simply counting lines on the grid, or you could approximate areas by counting small squares or portions of squares on the grid.

How could you get a tall, narrow curve? How could you get a short, broad curve? What factors might affect the height and breadth of the curve? From the members of the class, collect a set of Normal curves that differ from one another.

Activity 2B

Roll a Normal distribution

Materials: Several marbles, all the same size; two metersticks for a "ramp"; a ruled sheet of paper; a flat table about 4 feet long; carbon paper; Scotch Tape or masking tape

1. At one end of the table prop up the two metersticks in a V shape to provide a ramp for the marbles to roll down. Each marble will roll down

the chute, continue across the table, and fall off the table to the floor below. Make sure that the ramp is secure and that the tabletop does not have any grooves or obstructions.

2. Roll a marble down the ramp several times to get a good idea of the area of the floor where it will fall.

3. Center the ruled sheet of paper (see Figure 2.1) over this area, face up, with the bottom edge toward the table and parallel to the edge of the table. The ruled lines should go in the same direction as the marble's path. Tape the sheet securely to the floor. Place the sheet of carbon paper, carbon side down, over the ruled sheet.

4. Roll the marble for a class total of 200 times. The spots where it hits the floor will be recorded on the ruled paper as black dots. When the marble hits the floor, it will probably bounce, so try to catch it in midair after the impact so that you don't get any extra marks. After the first 100 rolls, replace the sheet of paper. This will make it easier for you to count the spots. Make sure that the second sheet is in exactly the same position as the first one.

5. When the marble has been rolled 200 times, make a histogram of the distribution of the points as follows. First, count the number of dots in each column. Then graph these numbers by drawing horizontal lines in the columns at the appropriate levels. Use the scale on the left-hand side of the sheet.

Figure 2.1 *Example of ruled sheet for Activity 2B.*

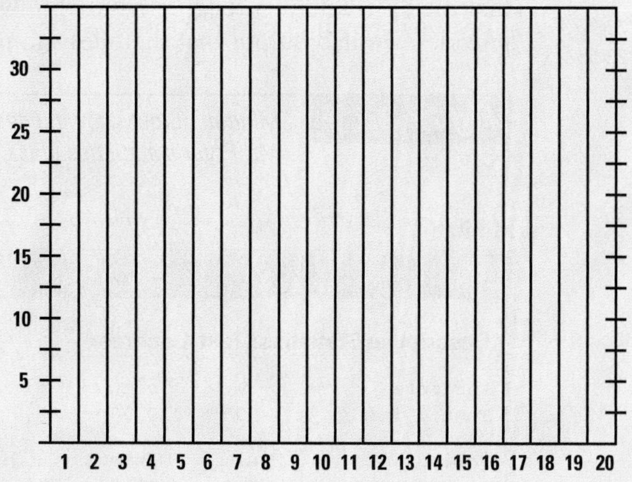

Introduction

Suppose that Jenny earns an 86 (out of 100) on her next statistics test. Should she be satisfied or disappointed with her performance? That depends on how her score compares to those of the other students who took the test. If 86 is the highest score, Jenny might be very pleased. Perhaps her teacher will "curve" the grades so that Jenny's 86 becomes an "A." But if Jenny's 86 falls below the "middle" of this set of test scores, she may not be so happy.

Section 2.1 focuses on describing the location of an individual within a distribution. We begin by using raw data to calculate two useful measures of location. One is based on the mean and standard deviation. The other is connected to the median and quartiles.

Sometimes it is helpful to use graphical models called *density curves* to describe the location of individuals within a distribution, rather than relying on actual data values. This is especially true when data fall in a bell-shaped pattern called a *Normal distribution*. Section 2.2 examines the properties of Normal distributions and shows you how to perform useful calculations with them.

2.1 Measures of Relative Standing and Density Curves

Here are the scores of all 25 students in Mr. Pryor's statistics class on their first test:

79	81	80	77	73	83	74	93	78	80	75	67	73
77	83	**86**	90	79	85	83	89	84	82	77	72	

The bold score is Jenny's 86. How did she perform on this test relative to her classmates?

6	7
7	2334
7	5777899
8	00123334
8	569
9	03

The stemplot in the margin displays this distribution of test scores. Notice that the distribution is roughly symmetric with no apparent outliers. Figure 2.2 provides Minitab output that includes summary statistics for the test score data.

Figure 2.2 *Minitab output showing descriptive statistics for the test scores of Mr. Pryor's statistics class.*

Minitab

```
Session

Descriptive Statistics: Test 1 scores

Variable          N     Mean    Median   TrMean   StDev  SE Mean
Test 1 scores    25    80.00    80.00    80.00     6.07     1.21

Variable       Minimum  Maximum     Q1       Q3
Test 1 scores   67.00    93.00    76.00    83.50
```

Where does Jenny's 86 fall relative to the center of this distribution? Since the mean and median are both 80, we can say that Jenny's result is "above average." But how much above average is it?

Measuring Relative Standing: *z*-Scores

One way to describe Jenny's position within the distribution of test scores is to tell how many standard deviations above or below the mean her score is. Since the mean is 80 and the standard deviation is about 6, Jenny's score of 86 is about one standard deviation above the mean. Converting scores like this from original values to standard deviation units is known as ***standardizing.*** To standardize a value, subtract the mean of the distribution and then divide by the standard deviation.

standardizing

Standardized Values and z-Scores

If x is an observation from a distribution that has known mean and standard deviation, the **standardized value** of x is

$$z = \frac{x - \text{mean}}{\text{standard deviation}}$$

A standardized value is often called a ***z*-score.**

A z-score tells us how many standard deviations away from the mean the original observation falls, and in which direction. Observations larger than the mean are positive when standardized, and observations smaller than the mean are negative.

Example 2.1 *Mr. Pryor's first test*
Standardizing scores

Jenny's score on the test was $x = 86$. Her *standardized* test score is

$$z = \frac{x - 80}{6.07} = \frac{86 - 80}{6.07} = 0.99$$

Katie earned the highest score in the class, 93. Her corresponding z-score is

$$z = \frac{93 - 80}{6.07} = 2.14$$

In other words, Katie's result is 2.14 standard deviations above the mean score for this test.

Norman got a 72 on this test. His standardized score is

$$z = \frac{72 - 80}{6.07} = -1.32$$

Norman's score is 1.32 standard deviations below the class mean of 80.

We can also use z-scores to compare the relative standing of individuals in different distributions, as the following example illustrates.

Example 2.2 ***Statistics or chemistry: which is easier?***
Using z-scores for comparison

The day after receiving her statistics test result of 86 from Mr. Pryor, Jenny earned an 82 on Mr. Goldstone's chemistry test. At first, she was disappointed. Then Mr. Goldstone told the class that the distribution of scores was fairly symmetric with a mean of 76 and a standard deviation of 4. Jenny quickly calculated her z-score:

$$z = \frac{82 - 76}{4} = 1.5$$

Her 82 in chemistry was 1.5 standard deviations above the mean score for the class. Since she only scored 1 standard deviation above the mean on the statistics test, Jenny actually did better in chemistry relative to the class.

We often standardize observations from symmetric distributions to express them on a common scale. We might, for example, compare the heights of two children of different ages by calculating their z-scores. The standardized heights tell us where each child stands in the distribution for his or her age group.

Exercises

2.1 SAT versus ACT Eleanor scores 680 on the mathematics part of the SAT. The distribution of SAT scores in a reference population is symmetric and single-peaked with mean 500 and standard deviation 100. Gerald takes the American College Testing (ACT) mathematics test and scores 27. ACT scores also follow a symmetric, single-peaked distribution but with mean 18 and standard deviation 6. Find the standardized scores for both students. Assuming that both tests measure the same kind of ability, who has the higher score?

2.2 Comparing batting averages Three landmarks of baseball achievement are Ty Cobb's batting average of .420 in 1911, Ted Williams's .406 in 1941, and George Brett's .390 in 1980. These batting averages cannot be compared directly because the distribution of major league batting averages has changed over the years. The distributions are quite symmetric, except for outliers such as Cobb, Williams, and Brett. While the mean batting average has been held roughly constant by rule changes and the balance between hitting and pitching, the standard deviation has dropped over time. Here are the facts:

Decade	Mean	Std. dev.
1910s	.266	.0371
1940s	.267	.0326
1970s	.261	.0317

Compute the standardized batting averages for Cobb, Williams, and Brett to compare how far each stood above his peers.[1]

2.3 Measuring bone density Individuals with low bone density (osteoporosis) have a high risk of broken bones (fractures). Physicians who are concerned about low bone density in patients can refer them for specialized testing. Currently, the most common method for testing bone density is dual-energy X-ray absorptiometry (DXA). A patient who undergoes a DXA test usually gets bone density results in grams per square centimeter (g/cm^2) and in standardized units.

Judy, who is 25 years old, has her bone density measured using DXA. Her results indicate a bone density in the hip of 948 g/cm^2 and a standardized score of $z = -1.45$. In the reference population[2] of 25-year-old women like Judy, the mean bone density in the hip is 956 g/cm^2.

(a) Judy has not taken a statistics class in a few years. Explain to her in simple language what the standardized score tells her about her bone density.

(b) Use the information provided to calculate the standard deviation of bone density in the reference population.

2.4 Comparing bone density Refer to the previous exercise. One of Judy's friends, Mary, has the bone density in her hip measured using DXA. Mary is 35 years old. Her bone density is also reported as 948 g/cm^2, but her standardized score is $z = 0.50$. The mean bone density in the hip for the reference population of 35-year-old women is 944 g/cm^2.

(a) Whose bones are healthier: Judy's or Mary's? Justify your answer.

(b) Calculate the standard deviation of the bone density in Mary's reference population. How does this compare with your answer to Exercise 2.3(b)? Are you surprised?

Measuring Relative Standing: Percentiles

We can also describe Jenny's performance on her first statistics test using percentiles. In Chapter 1, we defined the *p*th percentile of a distribution as the value with *p* percent of the observations less than or equal to it. Using the stemplot on page 116, we can see that Jenny's 86 is the 22nd highest score on this test, counting up from the lowest score. Since 22 of the 25 observations (88%) are at or below her score, Jenny scored at the 88th percentile.

Example 2.3 *Mr. Pryor's first test, continued*
Finding percentiles

For Norman, who got a 72, only 2 of the 25 test scores in the class are at or below his. Norman's percentile is computed as follows: 2/25=0.08, or 8%. So Norman scored at the 8th percentile on this test.

CAUTION

Katie's 93 puts her by definition at the 100th percentile, since 25 out of 25 test scores fall at or below her result. **Note that some people define the *p*th percentile of a distribution as the value with *p* percent of observations *below* it.** Using this alternative definition of percentile, it is never possible for an individual to fall at the 100th percentile. That is why you never see a standardized test score reported above the 99th percentile.

Two students scored an 80 on Mr. Pryor's first test. By our definition of percentile, 14 of the 25 scores in the class were at or below 80, so these two students are at the 56th percentile. If we used the alternative definition of percentile, these two students would fall at the 48th percentile (12 of 25 scores were *below* 80). Of course, since 80 is the median score, we prefer to think of it as being the 50th percentile. Calculating percentiles is not an exact science!

"I can't speak for everyone in the top one percent, but I'm fine."

You may have wondered by now whether there is a simple way to convert a z-score to a percentile. The percent of observations falling at or below a particular z-score depends on the shape of the distribution. An observation that is equal to the mean has a z-score of 0. In a heavily left-skewed distribution, where the mean will be less than the median, this observation will be somewhere below the 50th percentile (the median).

There is an interesting result that describes the percent of observations in *any* distribution that fall within a specified number of standard deviations of the mean. It is known as ***Chebyshev's inequality.***

Chebyshev's Inequality

In *any* distribution, the percent of observations falling within k standard deviations of the mean is *at least* $(100)\left(1-\frac{1}{k^2}\right)$.

If $k = 2$, for example, Chebyshev's inequality tells us that *at least* $(100)\left(1-\frac{1}{k^2}\right) = 75\%$ of the observations in any distribution must fall within 2 standard deviations of the mean. The following table displays the percent of observations within k standard deviations of the mean for several values of k.

k	*at least* $(100)\left(1-\frac{1}{k^2}\right)$ of observations within k standard deviations of mean
1	$(100)\left(1-\frac{1}{1^2}\right) = 0\%$
2	$(100)\left(1-\frac{1}{2^2}\right) = 75\%$

Going with 85% of the flow According to the *Los Angeles Times,* speed limits on California highways are set at the 85th percentile of vehicle speeds on those stretches of road. The idea, apparently, is that 85% of people behave in a safe and sane manner. The other 15% are reckless individuals who go too fast. To which group do you belong?

$$3 \quad (100)\left(1 - \frac{1}{3^2}\right) = 89\%$$

$$4 \quad (100)\left(1 - \frac{1}{4^2}\right) = 93.75\%$$

$$5 \quad (100)\left(1 - \frac{1}{5^2}\right) = 96\%$$

So it is fairly unusual to find an observation that is more than 5 standard deviations away from the mean in any distribution.

Chebyshev's inequality gives us some insight into how observations are distributed within distributions. It does not help us determine the percentile corresponding to a given z-score, however. For that, we need more advanced statistical models known as *density curves.*

Exercises

2.5 Unemployment in the states, I Each month the Bureau of Labor Statistics announces the unemployment rate for the previous month. Unemployment rates are economically important and politically sensitive. Unemployment may vary greatly among the states because types of work are unevenly distributed across the country. Table 2.1 presents the unemployment rates for each of the 50 states in May 2005.

Table 2.1 *Unemployment rates by state, May 2005*

State	Percent	State	Percent	State	Percent
Alabama	4.4	Louisiana	5.4	Ohio	6.1
Alaska	6.4	Maine	5.0	Oklahoma	4.5
Arizona	4.8	Maryland	4.2	Oregon	6.5
Arkansas	5.0	Massachusetts	4.8	Pennsylvania	4.8
California	5.3	Michigan	7.1	Rhode Island	4.5
Colorado	5.3	Minnesota	4.3	South Carolina	6.3
Connecticut	5.3	Mississippi	7.1	South Dakota	4.0
Delaware	4.1	Missouri	5.6	Tennessee	6.2
Florida	4.0	Montana	4.5	Texas	5.5
Georgia	5.2	Nebraska	4.0	Utah	4.9
Hawaii	2.7	Nevada	4.0	Vermont	3.1
Idaho	3.9	New Hampshire	3.6	Virginia	3.6
Illinois	5.8	New Jersey	3.9	Washington	5.7
Indiana	4.8	New Mexico	6.0	West Virginia	4.5
Iowa	4.8	New York	5.0	Wisconsin	4.7
Kansas	5.3	North Carolina	5.1	Wyoming	4.0
Kentucky	5.7	North Dakota	3.5		

Source: Bureau of Labor Statistics Web site, www.bls.gov.

(a) Make a histogram of these data. Be sure to label and scale your axes.

(b) Calculate numerical summaries for this data set. Describe the shape, center, and spread of the distribution of unemployment rates.

(c) Determine the percentile for Illinois. Explain in simple terms what this says about the unemployment rate in Illinois relative to the other states.

(d) Which state is at the 30th percentile? Calculate the z-score for this state.

(e) Compare the percent of state unemployment rates that fall within 1, 2, 3, 4, and 5 standard deviations of the mean with the percents guaranteed by Chebyshev's inequality (refer to the table on page 121).

2.6 Unemployment in the states, II Refer to the previous exercise. The December 2000 unemployment rates for the 50 states had a symmetric, single-peaked distribution with a mean of 3.47% and a standard deviation of about 1%. The unemployment rate for Illinois that month was 4.5%. There were 42 states with lower unemployment rates than Illinois.

(a) Write a sentence comparing the actual rates of unemployment in Illinois in December 2000 and May 2005.

(b) Compare the z-scores for the Illinois unemployment rate in these same two months in a sentence or two.

(c) Compare the percentiles for the Illinois unemployment rate in these same two months in a sentence or two.

2.7 PSAT scores In October 2004, about 1.4 million college-bound high school juniors took the Preliminary SAT (PSAT). The mean score on the Critical Reading test was 46.9, and the standard deviation was 10.9. Nationally, 5.2% of test takers earned a score of 65 or higher on the Critical Reading test's 20 to 80 scale.[3]

Scott was one of 50 junior boys to take the PSAT at his school. He scored 64 on the Critical Reading test. This placed Scott at the 68th percentile within the group of boys. Looking at all 50 boys' Critical Reading scores, the mean was 58.2, and the standard deviation was 9.4.

(a) Write a sentence or two comparing Scott's percentile among the national group of test takers and among the 50 boys at his school.

(b) Calculate and compare Scott's z-score among these same two groups of test takers.

(c) How well did the boys at Scott's school perform on the PSAT? Give appropriate evidence to support your answer.

(d) What does Chebyshev's inequality tell you about the performance of students nationally? At Scott's school?

2.8 Blood pressure Larry came home very excited after a visit to his doctor. He announced proudly to his wife, "My doctor says my blood pressure is at the 90th percentile among men like me. That means I'm better off than about 90% of similar men." How should his wife, who is a statistician, respond to Larry's statement?

Density Curves

In Chapter 1, we developed a kit of graphical and numerical tools that we call a "Data Analysis Toolbox" for describing distributions. In this chapter, we have explored some methods for measuring the relative standing of individuals within distributions. We now have a clear strategy for exploring data from a single quantitative variable.

1. Always plot your data: make a graph, usually a histogram or a stemplot.
2. Look for the overall pattern (shape, center, spread) and for striking deviations such as outliers.
3. Calculate a numerical summary to briefly describe center and spread.

Here is one more step to add to this strategy:

4. Sometimes the overall pattern of a large number of observations is so regular that we can describe it by a smooth curve. Doing so can help us describe the location of individual observations within a distribution.

Figure 2.3 is a histogram of the scores of all 947 seventh-grade students in Gary, Indiana, on the vocabulary part of the Iowa Test of Basic Skills.[4] Scores on this national test have a very regular distribution. The histogram is symmetric, and both tails fall off smoothly from a single center peak. There are no large gaps or obvious outliers. The smooth curve drawn through the tops of the histogram bars in Figure 2.3 is a good description of the overall pattern of the data.

Figure 2.3 *Histogram of the vocabulary scores of all seventh-grade students in Gary, Indiana.*

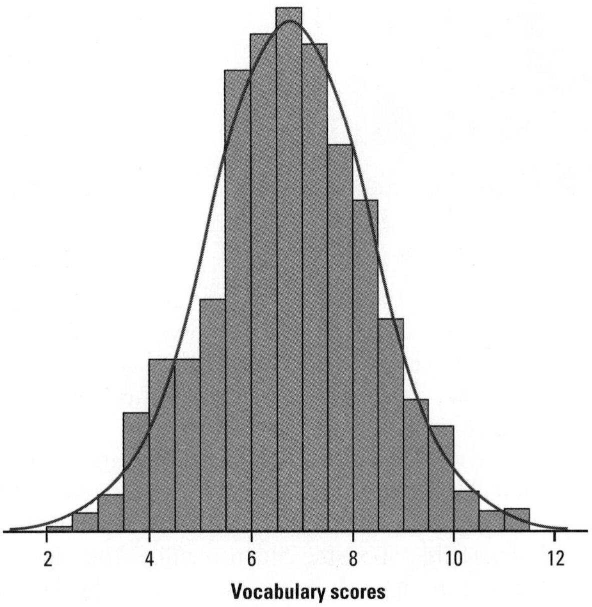

mathematical model

The curve is a ***mathematical model*** for the distribution. A mathematical model is an idealized description. It gives a compact picture of the overall pattern of the data but ignores minor irregularities as well as any outliers.

We will see that it is easier to work with the smooth curve in Figure 2.3 than with the histogram. The reason is that the histogram depends on our choice of classes, while with a little care we can use a curve that does not depend on any choices we make. Here's how we do it.

Example 2.4 *Seventh-grade vocabulary scores*

From histogram to density curve

Our eyes respond to the areas of the bars in a histogram. The bar areas represent proportions of the observations. Figure 2.4(a) is a copy of Figure 2.3 with the leftmost bars shaded blue. The area of the blue bars in Figure 2.4(a) represents the students with vocabulary scores of 6.0 or lower. There are 287 such students, who make up the proportion $287/947 = 0.303$ of all Gary seventh-graders. In other words, a score of 6.0 corresponds to about the 30th percentile.

Figure 2.4 *(a) The proportion of scores less than or equal to 6.0 from the histogram is 0.303. (b) The proportion of scores less than or equal to 6.0 from the density curve is 0.293.*

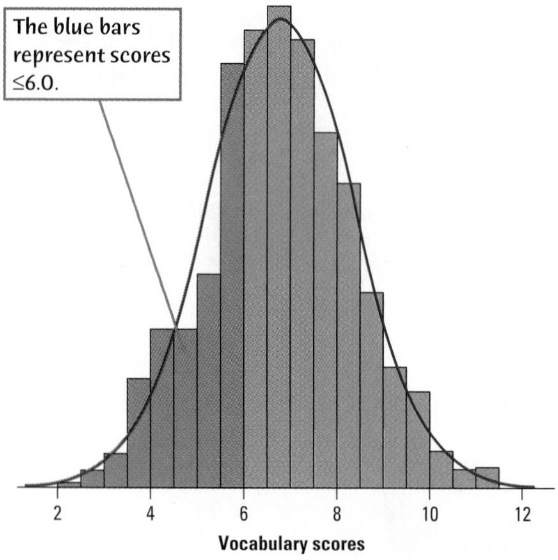

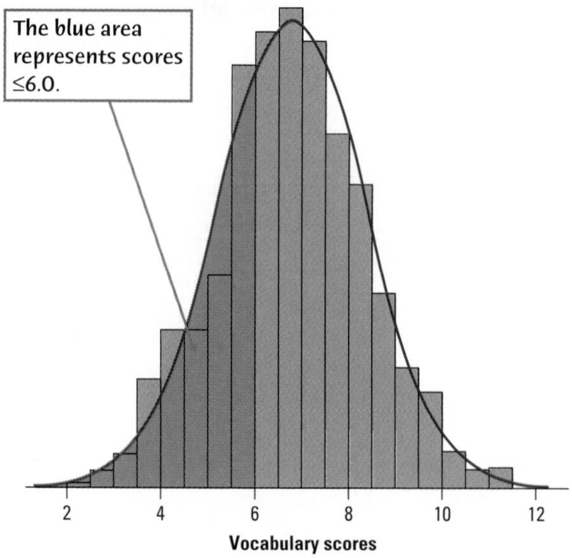

Now concentrate on the curve drawn through the bars. In Figure 2.4(b), the area under the curve to the left of 6.0 is shaded. Adjust the scale of the graph so that the total area under the curve is exactly 1. This area represents the proportion 1, that is, all the observations. Areas under the curve then represent proportions of the observations. The curve is now a ***density curve.*** The blue area under the density curve in Figure 2.4(b) represents the proportion of students with score 6.0 or lower. This area is 0.293, only 0.010 away from the histogram result. So our estimate based on the density curve is that a score of 6.0 falls at about the 29th percentile. You can see that areas under the density curve give quite good approximations of areas given by the histogram.

Density Curve

A **density curve** is a curve that

- is always on or above the horizontal axis, and
- has area exactly 1 underneath it.

A density curve describes the overall pattern of a distribution. The area under the curve and above any interval of values on the horizontal axis is the proportion of all observations that fall in that interval.

Normal curve

The density curve in Figures 2.3 and 2.4 is a ***Normal curve.*** Density curves, like distributions, come in many shapes. In later chapters, we will encounter important density curves that are skewed to the left or right and curves that may look like Normal curves but are not.

Example 2.5 *A left-skewed distribution*

Density curves and areas

Figure 2.5 shows the density curve for a distribution that is skewed to the left. The smooth curve makes the overall shape of the distribution clearly visible. The shaded area under the curve covers the range of values from 7 to 8. This area is 0.12. This means that the proportion 0.12 (12%) of all observations from this distribution have values between 7 and 8.

Figure 2.5 *The shaded area under this density curve is the proportion of observations taking values between 7 and 8.*

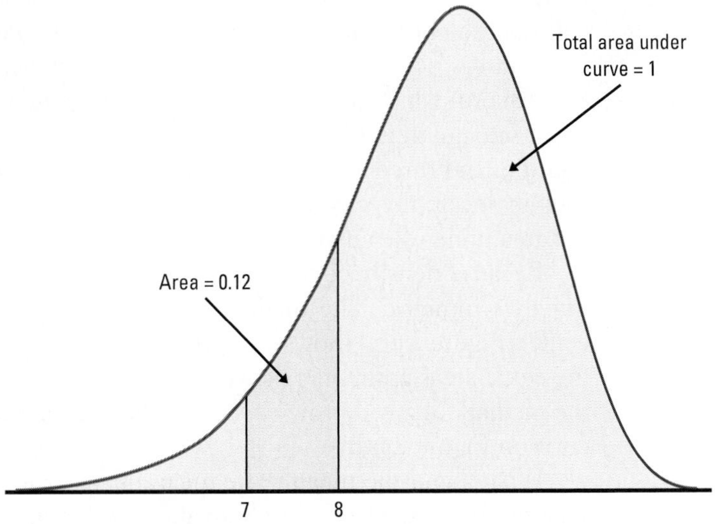

Figure 2.6 shows two density curves: a symmetric, Normal density curve and a right-skewed curve. The mean and median of each distribution, which will be discussed in more detail shortly, have been marked on the horizontal axis.

Figure 2.6 *(a) The median and mean of a symmetric density curve. (b) The median and mean of a right-skewed density curve.*

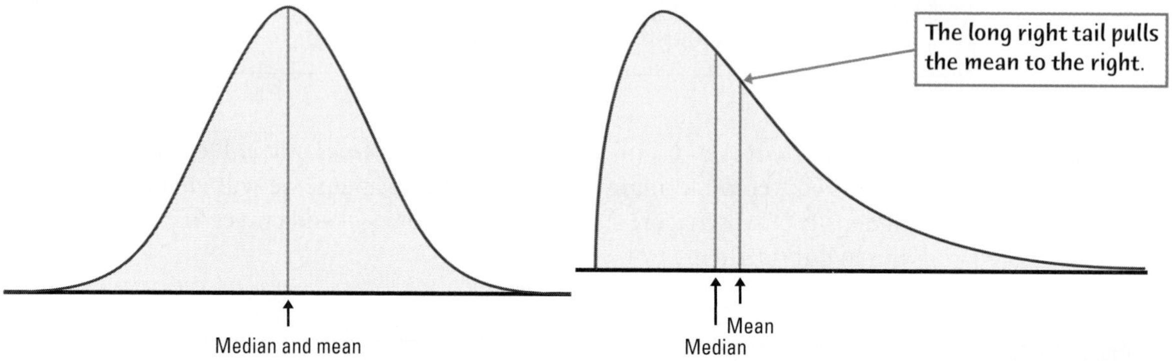

A density curve of the appropriate shape is often an adequate description of the overall pattern of a distribution. Outliers, which are deviations from the overall pattern, are not described by the curve. **Of course, no set of real data is exactly described by a density curve. The curve is an approximation that is easy to use and accurate enough for practical use.**

The Median and Mean of a Density Curve

Our measures of center and spread apply to density curves as well as to actual sets of observations. The median and quartiles are easy. Areas under a density curve represent proportions of the total number of observations. The median is the point with half the observations on either side. So the ***median of a density curve*** is the "equal-areas point," the point with half the area under the curve to its left and the remaining half of the area to its right. The quartiles divide the area under the curve into quarters. One-fourth of the area under the curve is to the left of the first quartile, and three-fourths of the area is to the left of the third quartile. You can roughly locate the median and quartiles of any density curve by eye by dividing the area under the curve into four equal parts.

Because density curves are idealized patterns, a symmetric density curve is exactly symmetric. The median of a symmetric density curve is therefore at its center. Figure 2.6(a) shows the median of a symmetric curve. It isn't so easy to spot the equal-areas point on a skewed curve. There are mathematical ways of finding the median for any density curve. We did that to mark the median on the skewed curve in Figure 2.6(b).

What about the mean? The mean of a set of observations is their arithmetic average. If we think of the observations as weights strung out along a thin rod, the mean is the point at which the rod would balance. This fact is also true of density curves. The ***mean of a density curve*** is the "balance point," the point at which the curve would balance if made of solid material. Figure 2.7 illustrates this fact about the mean. A symmetric curve balances at its center because the two sides are identical. *The mean and median of a symmetric density curve are equal*, as in Figure 2.6(a). We know that the mean of a skewed distribution is pulled toward the long tail. Figure 2.6(b) shows how the mean of a skewed density curve is pulled toward the long tail more than is the median. It's hard to locate the balance point by eye on a skewed curve. There are mathematical ways of calculating the mean for any density curve, so we are able to mark the mean as well as the median in Figure 2.6(b).

Figure 2.7 *The mean is the balance point of a density curve.*

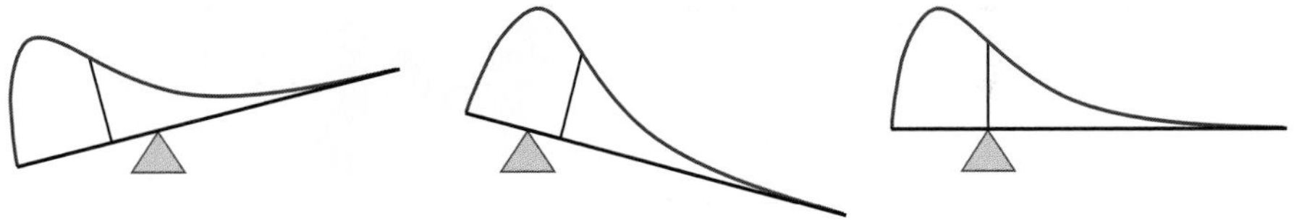

Median and Mean of a Density Curve

The **median of a density curve** is the "equal-areas point," the point that divides the area under the curve in half.

The **mean of a density curve** is the "balance point," at which the curve would balance if made of solid material.

The median and mean are the same for a symmetric density curve. They both lie at the center of the curve. The mean of a skewed curve is pulled away from the median in the direction of the long tail.

We can roughly locate the mean, median, and quartiles of any density curve by eye. This is not true of the standard deviation. When necessary, we can once again call on more advanced mathematics to learn the value of the standard deviation. The study of mathematical methods for doing calculations with density curves is part of theoretical statistics. Though we are concentrating on statistical practice, we often make use of the results of mathematical study.

Because a density curve is an idealized description of the distribution of data, we need to distinguish between the mean and standard deviation of the density curve and the mean $\bar{x}$ and standard deviation s computed from the actual observations. The usual notation for the ***mean of a density curve*** is μ (the Greek letter mu). We write the ***standard deviation of a density curve*** as σ (the Greek letter sigma).

mean μ

standard deviation σ

Exercises

2.9 Density curves Sketch density curves that might describe distributions with the following shapes:

(a) Symmetric, but with two peaks
(b) Single peak and skewed to the left

uniform distribution

2.10 A uniform distribution, I Figure 2.8 displays the density curve of a ***uniform distribution.*** The curve takes the constant value 1 over the interval from 0 to 1 and is 0 outside the range of values. This means that data described by this distribution take values that are uniformly spread between 0 and 1.

Figure 2.8 *The density curve of a uniform distribution, for Exercise 2.10.*

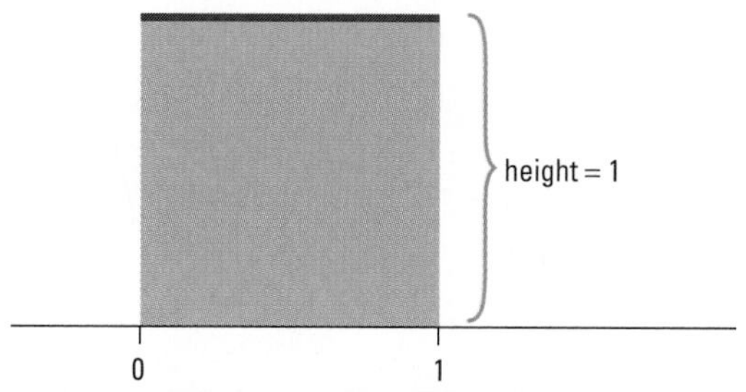

Use areas under this density curve to answer the following questions.

(a) Why is the total area under this curve equal to 1?
(b) What percent of the observations lie above 0.8?
(c) What percent of the observations lie below 0.6?
(d) What percent of the observations lie between 0.25 and 0.75?
(e) What is the mean μ for this distribution?

2.11 A uniform distribution, II Refer to the previous exercise. Can you construct a modified boxplot for the uniform distribution? If so, do it. If not, explain why not.

2.12 Finding means and medians Figure 2.9 displays three density curves, each with three points indicated. At which of these points on each curve do the mean and the median fall?

Figure 2.9 *Three density curves, for Exercise 2.12.*

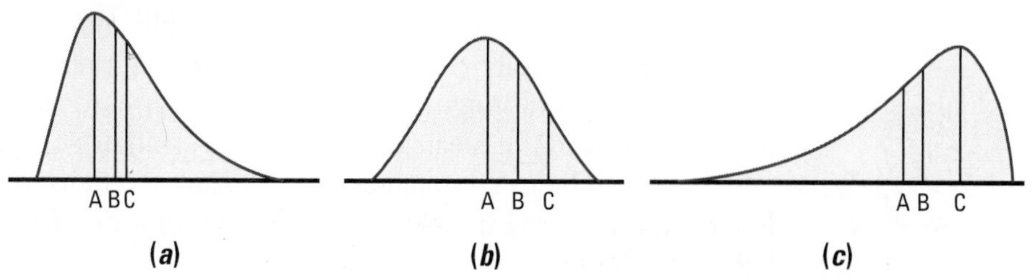

2.13 A weird density curve A line segment can be considered a density "curve," as shown in Exercise 2.10. A "broken-line" graph can also be considered a density curve. Figure 2.10 shows such a density curve.

Figure 2.10 *An unusual "broken-line" density curve, for Exercise 2.13.*

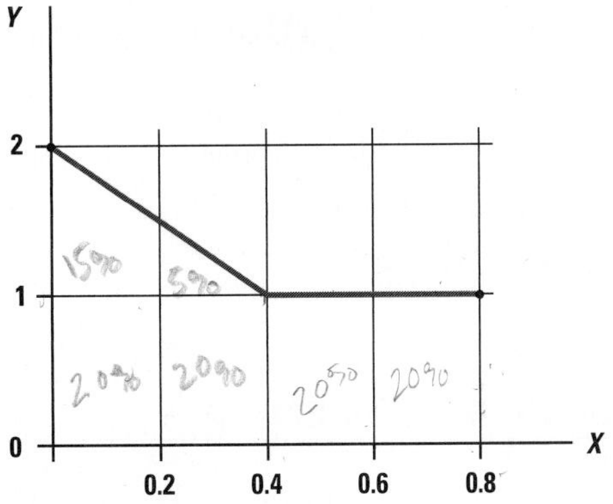

(a) Verify that the graph in Figure 2.10 is a valid density curve.

For each of the following, use areas under this density curve to find the proportion of observations within the given interval:

(b) $0.6 \le X \le 0.8$

(c) $0 \le X \le 0.4$

(d) $0 \le X \le 0.2$

(e) The median of this density curve is a point between $X = 0.2$ and $X = 0.4$. Explain why.

2.14 Roll a distribution In this exercise you will pretend to roll a regular, six-sided die 120 times. Each time you roll the die, you will record the result. The numbers 1, 2, 3, 4, 5, and 6 are called the ***outcomes*** of this chance phenomenon.

outcomes

In 120 rolls, how many of each number would you expect to roll? The TI-83/84 and TI-89 are useful devices for imitating chance behavior, especially in situations like this that involve performing many repetitions. Because you are only pretending to roll the die repeatedly, we call this process a ***simulation.*** There will be a more formal treatment of simulations in Chapter 6.

simulation

SIMULATIONS

- Begin by clearing L_1/list1 on your calculator.
- Use your calculator's random integer generator to generate 120 random whole numbers between 1 and 6 (inclusive), and then store these numbers in L_1/list1.
- On the TI-83/84, press MATH, choose PRB, then `5:RandInt.` On the TI-89, press CATALOG F3 and choose `randInt....`

- On the TI-83/84, complete the command `RandInt(1,6,120)` STO▶ L_1. On the TI-89, complete the command `tistat.randInt (1,6,120)` STO▶ `list1`.
- Set the viewing window as follows: $X[1, 7]_1$ by $Y[-5, 25]_5$.
- Specify a histogram using the data in L_1/list1.
- Then graph. Are you surprised?
- Repeat the simulation several times. You can recall and reuse the previous command by pressing 2nd ENTER. It's a good habit to clear L_1/list1 before you roll the die again.

(a) In theory, each number should come up 20 times. But in practice, there is chance variation, so the bars in the histogram will probably have different heights. Theoretically, what should the distribution look like?

(b) How is this distribution similar to the uniform distribution in Exercise 2.10? How is it different?

Section 2.1 Summary

Two ways of describing an individual's location within a distribution are **z-scores** and **percentiles.** To standardize any observation x, subtract the mean of the distribution and then divide by the standard deviation. The resulting z-score

$$z = \frac{x - \text{mean}}{\text{standard deviation}}$$

says how many standard deviations x lies from the distribution mean. An observation's percentile is the percent of the distribution that is at or below the value of that observation. We can also use z-scores and percentiles to compare the relative standing of individuals in different distributions.

Chebyshev's inequality gives us a useful rule of thumb for the percent of observations in any distribution that are within a specified number of standard deviations of the mean.

We can sometimes describe the overall pattern of a distribution by a **density curve.** A density curve always remains on or above the horizontal axis and has total area 1 underneath it. An area under a density curve gives the proportion of observations that fall in a range of values. A density curve is an idealized description of the overall pattern of a distribution that smooths out the irregularities in the actual data. Write the **mean of a density curve** as μ and the **standard deviation of a density curve** as σ to distinguish them from the mean $\bar{x}$ and the standard deviation s of the actual data.

The mean, the median, and the quartiles of a density curve can be located by eye. The mean μ is the balance point of the curve. The median divides the area under the curve in half. The quartiles, with the median, divide the area under the curve into quarters. The standard deviation σ cannot be located by eye on most density curves.

The mean and median are equal for symmetric density curves. The mean of a skewed curve is located farther toward the long tail than is the median.

Activity 2C
The Normal Curve Applet

The *Normal Curve Applet* allows you to investigate the properties of Normal distributions very easily. It is somewhat limited by the number of pixels available for use, so that it can't hit every value exactly. For the questions below, use the closest available values.

1. How accurate is 68–95–99.7? The 68–95–99.7 rule for Normal distributions is a useful approximation. But how accurate is this rule? On the applet, drag one flag past the other so that the applet shows the area under the curve between the two flags.
   (a) Place the flags one standard deviation on either side of the mean. What is the area between these two values? What does the 68–95–99.7 rule say this area is?
   (b) Repeat for locations two and three standard deviations on either side of the mean. Again compare the 68–95–99.7 rule with the area given by the applet.
2. How many standard deviations above or below the mean does the 40th percentile of the Normal distribution with mean 0 and standard deviation 1 fall? Describe how you used the applet to answer this question.

Exercises

2.23 Estimating standard deviations Figure 2.14 on the next page shows two Normal curves, both with mean 0. Approximately what is the standard deviation of each of these curves?

2.24 Men's heights The distribution of heights of adult American men is approximately Normal with mean 69 inches and standard deviation 2.5 inches. Draw a Normal curve on which this mean and standard deviation are correctly located. (*Hint*: Draw the curve first, locate the points where the curvature changes, then mark the horizontal axis.)

2.25 More on men's heights The distribution of heights of adult American men is approximately Normal with mean 69 inches and standard deviation 2.5 inches. Use the 68–95–99.7 rule to answer the following questions.

(a) What percent of men are taller than 74 inches?

(b) Between what heights do the middle 95% of men fall?

(c) What percent of men are shorter than 66.5 inches?

(d) A height of 71.5 inches corresponds to what percentile of adult male American heights?

Figure 2.14 *Two Normal curves with the same mean but different standard deviations, for Exercise 2.23.*

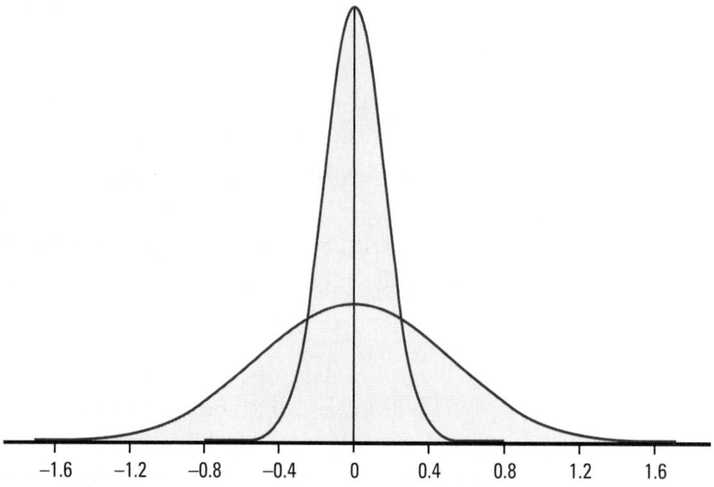

2.26 Potato chips The distribution of weights of 9-ounce bags of a particular brand of potato chips is approximately Normal with mean $\mu = 9.12$ ounces and standard deviation $\sigma = 0.15$ ounce.

(a) Draw an accurate sketch of the distribution of potato chip bag weights. Be sure to label the mean, as well as the points one, two, and three standard deviations away from the mean on the horizontal axis.

(b) A bag that weighs 8.97 ounces is at what percentile in this distribution?

(c) What percent of 9-ounce bags of this brand of potato chips weigh between 8.67 ounces and 9.27 ounces?

2.27 FLIP50 Refer to Exercise 2.22 (page 133). Run the program FLIP50 again.

(a) Use the Data Analysis Toolbox from Chapter 1 to describe your results.

(b) What percent of your observations fall within one standard deviation of the mean? Within two standard deviations? Within three standard deviations?

(c) How well do your data follow the 68–95–99.7 rule for Normal distributions? Explain.

2.28 A fine-grained distribution You can do this exercise if you spray-painted a Normal distribution in Activity 2A. On your "fine-grained distribution," first count the number of whole squares and parts of squares under the curve. Approximate as best you can. This represents the total area under the curve.

(a) Mark vertical lines at $\mu - 1\sigma$ and $\mu + 1\sigma$. Count the number of squares or parts of squares between these two vertical lines. Now divide the number of squares within one standard deviation of μ by the total number of squares under the curve and express your answer as a percent. How does this compare with 68%? Why would you expect your answer to differ somewhat from 68%?

(b) Count squares to determine the percent of area within 2σ of μ. How does your answer compare with 95%?

(c) Count squares to determine the percent of area within 3σ of μ. How does your answer compare with 99.7%?

The Standard Normal Distribution

As the 68–95–99.7 rule suggests, all Normal distributions share many common properties. In fact, all Normal distributions are the same if we measure in units of size σ about the mean μ as center. Changing to these units requires us to standardize, just as we did in Section 2.1:

$$z = \frac{x - \mu}{\sigma}$$

If the variable we standardize has a Normal distribution, then so does the new variable z. (This is because standardizing is a linear transformation, which, as we learned in Chapter 1, does not change the shape of a distribution.) This new distribution is called the ***standard Normal distribution.***

He said, she said The height and weight distributions in this chapter come from *actual* measurements in a government survey. Why not just ask people? When *asked* their weight, almost all women say they weigh less than they really do. Heavier men also underreport their weight, but lighter men claim to weigh more than the scale shows. We leave you to ponder the psychology of the two sexes. Just remember that "say so" is no replacement for measuring.

Standard Normal Distribution

The **standard Normal distribution** is the Normal distribution $N(0, 1)$ with mean 0 and standard deviation 1 (Figure 2.15).

Figure 2.15 *Standard Normal distribution.*

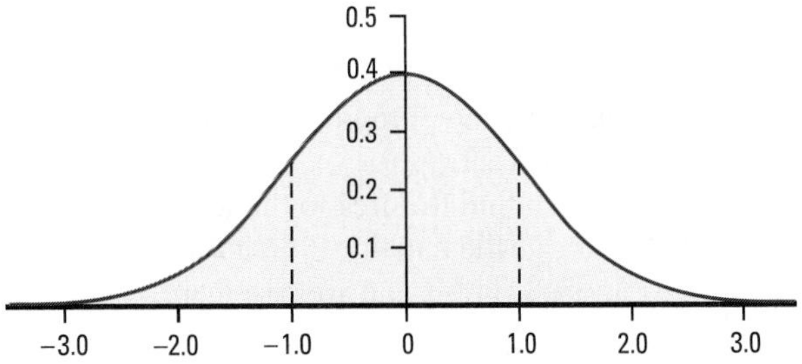

If a variable x has any Normal distribution $N(\mu, \sigma)$ with mean μ and standard deviation σ, then the standardized variable

$$z = \frac{x - \mu}{\sigma}$$

has the standard Normal distribution.

Standard Normal Calculations

An area under a density curve is a proportion of the observations in a distribution. Any question about what proportion of observations lie in some range of values can be answered by finding an area under the curve. Because all Normal distributions are the same when we standardize, we can find areas under any Normal curve from a single table, a table that gives areas under the curve for the standard Normal distribution. Table A, the ***standard Normal table***, gives areas under the standard Normal curve. You can find Table A inside the front cover.

The Standard Normal Table

Table A is a table of areas under the standard Normal curve. The table entry for each value z is the area under the curve to the left of z.

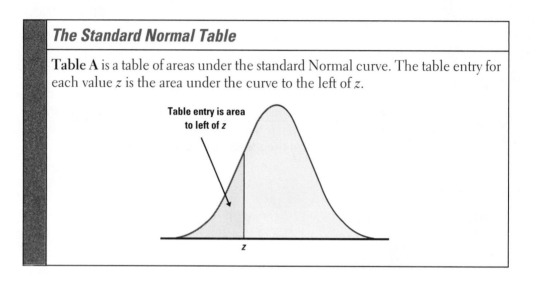

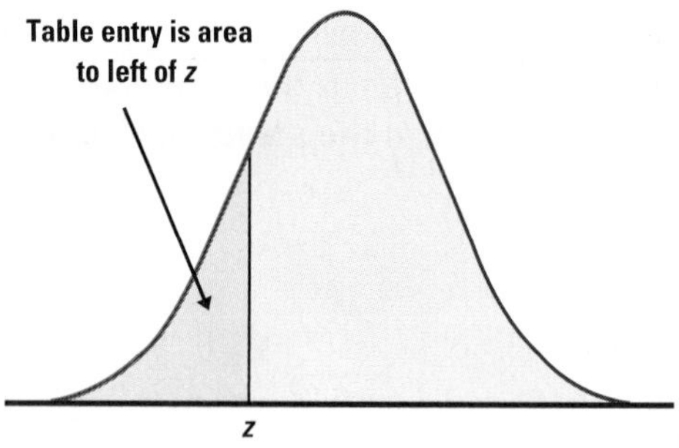

The next two examples show how to use the table.

Example 2.7 *Area to the left*
Using the standard Normal table

Problem: Find the proportion of observations from the standard Normal distribution that are less than 2.22.
Solution: To find the area to the left of 2.22, locate 2.2 in the left-hand column of Table A, then locate the remaining digit 2 as .02 in the top row. The entry opposite 2.2 and under .02 is 0.9868. This is the area we seek.

z	.00	.01	.02	.03
2.1	.9821	.9826	.9830	.9834
2.2	.9861	.9864	.9868	.9871
2.3	.9893	.9896	.9898	.9901

Figure 2.16 illustrates the relationship between the value $z = 2.22$ and the area 0.9868.

Figure 2.16 *The area under a standard Normal curve to the left of the point $z = 2.22$ is 0.9868.*

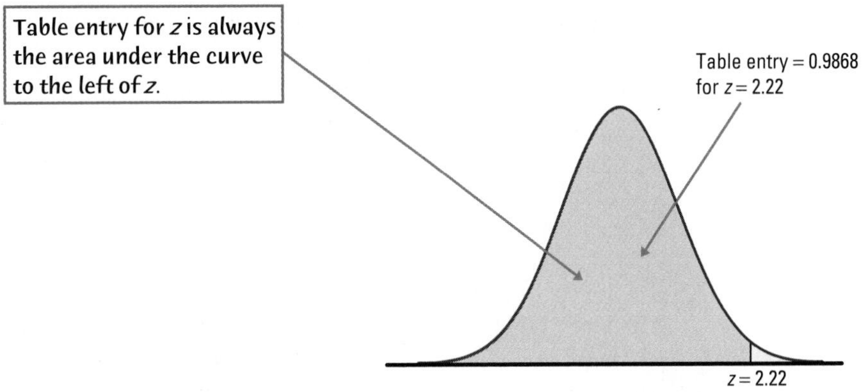

Example 2.8 *Area to the right*

More on using the standard Normal table

z	.04	.05	.06
−2.2	.0125	.0122	.0119
−2.1	.0162	.0158	.0154
−2.0	.0207	.0202	.0197

Problem: Find the proportion of observations from the standard Normal distribution that are greater than −2.15.

Solution: Enter Table A under $z = -2.15$. That is, find −2.1 in the left-hand column and .05 in the top row. The table entry is 0.0158. This is the area to the *left* of −2.15. Because the total area under the curve is 1, the area lying to the right of −2.15 is $1 - 0.0158 = 0.9842$. Figure 2.17 illustrates these areas.

Figure 2.17 *Areas under the standard Normal curve to the right and left of $z = -2.15$. Table A gives only areas to the left.*

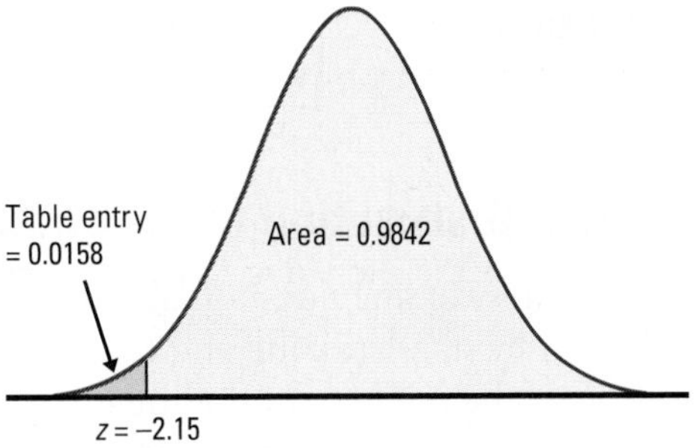

Caution! **A common student mistake is to look up a z-value in Table A and report the entry corresponding to that z-value, regardless of whether the problem asks for the area to the left or to the right of that z-value.** Always sketch the standard Normal curve, mark the z-value, and shade the area of interest. And before you finish, make sure your answer is reasonable in the context of the problem.

Exercises

2.29 Table A practice Use Table A to find the proportion of observations from a standard Normal distribution that satisfies each of the following statements. In each case, sketch a standard Normal curve and shade the area under the curve that is the answer to the question. Use the *Normal Curve Applet* to check your answers.

(a) $z < 2.85$

(b) $z > 2.85$

(c) $z > -1.66$

(d) $-1.66 < z < 2.85$

2.30 More Table A practice Use Table A to find the proportion of observations from a standard Normal distribution that satisfies each of the following statements. In each case, sketch a standard Normal curve and shade the area under the curve that is the answer to the question. Use the *Normal Curve Applet* to check your answers.

(a) $z < -2.46$

(b) $z > 2.46$

(c) $0.89 < z < 2.46$

(d) $-2.95 < z < -1.27$

Normal Distribution Calculations

We can answer any question about proportions of observations in a Normal distribution by standardizing and then using the standard Normal table. Here is an outline of the method for finding the proportion of the distribution in any region.

Solving Problems Involving Normal Distributions

Step 1: *State the problem* in terms of the observed variable x. Draw a picture of the distribution and shade the area of interest under the curve.

Step 2: *Standardize and draw a picture.* Standardize x to restate the problem in terms of a standard Normal variable z. Draw a picture to show the area of interest under the standard Normal curve.

Step 3: *Use the table.* Find the required area under the standard Normal curve, using Table A and the fact that the total area under the curve is 1.

Step 4: *Conclusion.* Write your conclusion in the context of the problem.

Example 2.9

Is cholesterol a problem for young boys?

Normal calculations

The level of cholesterol in the blood is important because high cholesterol levels may increase the risk of heart disease. The distribution of blood cholesterol levels in a large population of people of the same age and sex is roughly Normal. For 14-year-old boys,[6] the mean is $\mu = 170$ milligrams of cholesterol per deciliter of blood (mg/dl) and the standard deviation is $\sigma = 30$ mg/dl. Levels above 240 mg/dl may require medical attention. What percent of 14-year-old boys have more than 240 mg/dl of cholesterol?

Step 1: *State the problem.* Call the level of cholesterol in the blood x. The variable x has the $N(170, 30)$ distribution. We want the proportion of boys with cholesterol level $x > 240$. Sketch the distribution, mark the important points on the horizontal axis, and shade the area of interest. See Figure 2.18(a).

Figure 2.18a *(a) Cholesterol levels for 14-year-old boys who may require medical attention.*

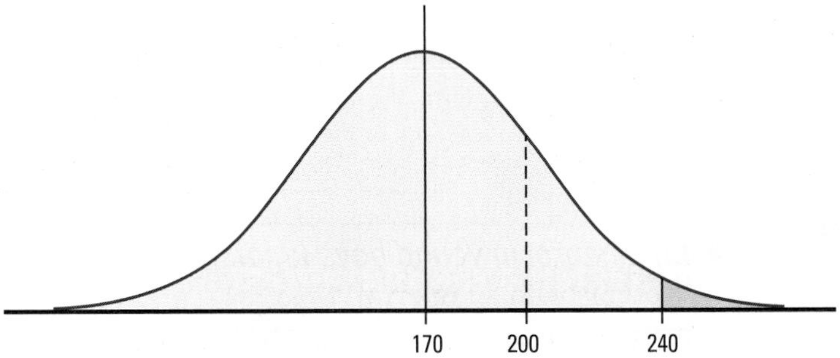

Step 2: *Standardize and draw a picture.* On both sides of the inequality, subtract the mean, then divide by the standard deviation, to turn x into a standard Normal z:

$$x > 240$$

$$\frac{x - 170}{30} > \frac{240 - 170}{30}$$

$$z > 2.33$$

Sketch a standard Normal curve, and shade the area of interest. See Figure 2.18(b).

Figure 2.18b *(b) Areas under the standard Normal curve.*

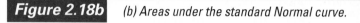

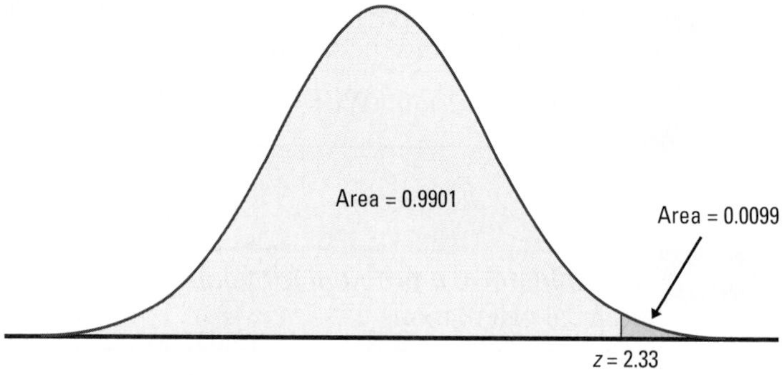

Step 3: *Use the table.* From Table A, we see that the proportion of observations less than 2.33 is 0.9901. About 99% of boys have cholesterol levels less than 240. The area to the right of 2.33 is therefore $1 - 0.9901 = 0.0099$. This is about 0.01, or 1%.

Step 4: *Conclusion.* (Write your conclusion in the context of the problem.) Only about 1% of 14-year-old boys have dangerously high cholesterol.

In a Normal distribution, the proportion of observations with $x > 240$ is the same as the proportion with $x \geq 240$. There is no *area* under the curve exactly above the point 240 on the horizontal axis, so the areas under the curve with $x > 240$ and $x \geq 240$ are the same. This isn't true of the actual data. There may be a boy with exactly 240 mg/dl of blood cholesterol. The Normal distribution is just an easy-to-use approximation, not a description of every detail in the actual data.

The key to doing a Normal calculation is to sketch the area you want, then match that area with the area that the table gives you. Here is another example.

Example 2.10 ***Cholesterol in young boys (again)***
Working with an interval

What percent of 14-year-old boys have blood cholesterol between 170 and 240 mg/dl?

Step 1: *State the problem.* We want the proportion of boys with $170 \leq x \leq 240$.

Step 2: *Standardize and draw a picture.*

$$170 \leq x \leq 240$$

$$\frac{170 - 170}{30} \leq \frac{x - 170}{30} \leq \frac{240 - 170}{30}$$

$$0 \leq z \leq 2.33$$

Sketch a standard Normal curve, and shade the area of interest. See Figure 2.19.

Figure 2.19 *Areas under the standard Normal curve.*

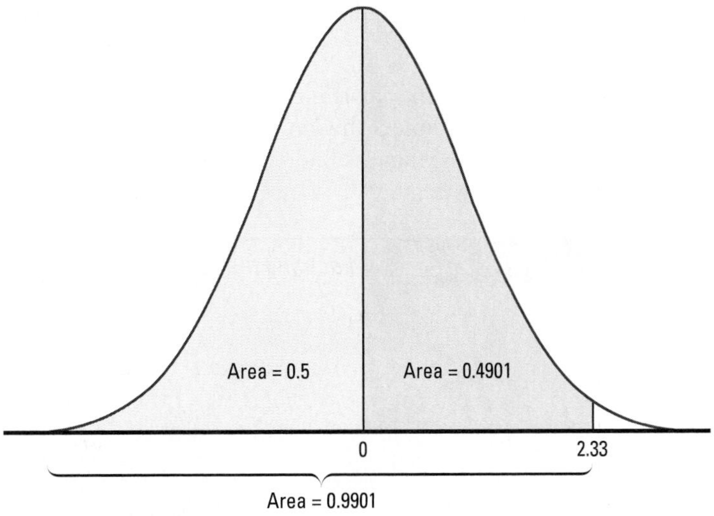

Step 3: *Use the table.* The area between 0 and 2.33 is the area to the left of 2.33 minus the area to the left of 0. Look at Figure 2.19 to check this. From Table A,

$$\begin{aligned}\text{area between 0 and 2.33} &= \text{area to the left of 2.33} - \text{area to the left of 0.00}\\ &= 0.9901 - 0.5000 = 0.4901\end{aligned}$$

Step 4: *Conclusion.* (State your conclusion in context.) About 49% of 14-year-old boys have cholesterol levels between 170 and 240 mg/dl.

What if we meet a z that falls outside the range covered by Table A? For example, the area to the left of $z = -4$ does not appear in the table. But since -4 is less than -3.4, this area is smaller than the entry for $z = -3.40$, which is 0.0003. There is very little area under the standard Normal curve outside the range covered by Table A. You can take this area to be zero with little loss of accuracy.

Finding a Value, Given a Proportion

Examples 2.9 and 2.10 illustrate the use of Table A to find what proportion of the observations satisfies some condition, such as "blood cholesterol between 170 and 240 mg/dl." We may instead want to find the observed value with a given proportion of the observations above or below it. To do this, use Table A backwards. Find the given proportion in the body of the table, read the corresponding z from the left column and top row, then "unstandardize" to get the observed value. Here is an example.

Example 2.11 *SAT Verbal scores*

Inverse Normal calculations

Scores on the SAT Verbal test in recent years follow approximately the N(505, 110) distribution. How high must a student score in order to place in the top 10% of all students taking the SAT?

Step 1: *State the problem and draw a picture.* We want to find the SAT score x with area 0.1 to its right under the Normal curve with mean $\mu = 505$ and standard deviation $\sigma = 110$. That's the same as finding the SAT score x with area 0.9 to its left. Figure 2.20 poses the question in graphical form.

Figure 2.20 *Locating the point on a Normal curve with area 0.10 to its right.*

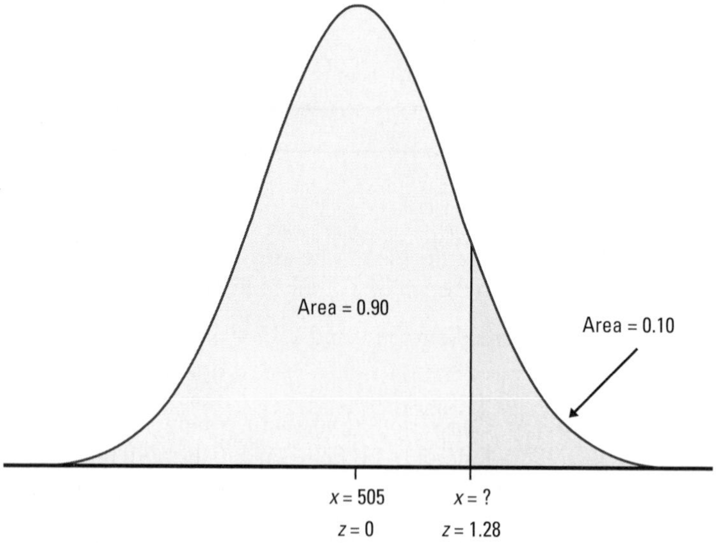

Because Table A gives the areas to the left of z-values, always state the problem in terms of the area to the left of x.

Step 2: *Use the table.* Look in the body of Table A for the entry closest to 0.9. It is 0.8997. This is the entry corresponding to $z = 1.28$. So $z = 1.28$ is the standardized value with area 0.9 to its left.

Step 3: *Unstandardize* to transform the solution from the z scale back to the original x scale. We know that the standardized value of the unknown x is $z = 1.28$. So x itself satisfies

$$\frac{x - 505}{110} = 1.28$$

Solving this equation for x gives

$$x = 505 + (1.28)(110) = 645.8$$

This equation should make sense: it finds the x that lies 1.28 standard deviations above the mean on this particular Normal curve. That is the "unstandardized" meaning of $z = 1.28$.

Step 4: *Conclusion.* We see that a student must score at least 646 to place in the highest 10%.

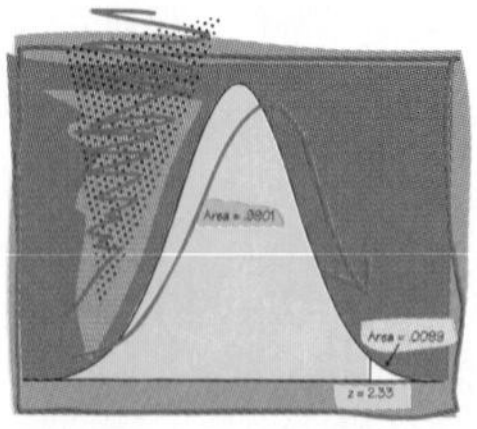

The bell curve? Does the distribution of human intelligence follow the "bell curve" of a Normal distribution? Scores on IQ tests do roughly follow a Normal distribution. That is because a test score is calculated from a person's answers in a way that is designed to produce a Normal distribution. To conclude that intelligence follows a bell curve, we must agree that the test scores directly measure intelligence. Many psychologists don't think there is one human characteristic that we can call "intelligence" and can measure by a single test score.

Exercises

2.31 How hard do locomotives pull? An important measure of the performance of a locomotive is its "adhesion," which is the locomotive's pulling force as a multiple of its weight. The adhesion of one 4400-horsepower diesel locomotive varies in actual use according to a Normal distribution with mean $\mu = 0.37$ and standard deviation $\sigma = 0.04$. For each part that follows, sketch and shade an appropriate Normal distribution. Then show your work.

(a) What proportion of adhesions measured in use are higher than 0.40?

(b) What proportion of adhesions are between 0.40 and 0.50?

(c) Improvements in the locomotive's computer controls change the distribution of adhesion to a Normal distribution with mean $\mu = 0.41$ and standard deviation $\sigma = 0.02$. Find the proportions in (a) and (b) after this improvement.

2.32 Table A in reverse Use Table A to find the value z from a standard Normal distribution that satisfies each of the following conditions. (Use the value of z from Table A that comes closest to satisfying the condition.) In each case, sketch a standard Normal curve with your value of z marked on the axis. Use the *Normal Curve Applet* to check your answers.

(a) The point z with 25% of the observations falling to the left of it.

(b) The point z with 40% of the observations falling to the right of it.

2.33 Length of pregnancies The length of human pregnancies from conception to birth varies according to a distribution that is approximately Normal with mean 266 days and standard deviation 16 days. For each part that follows, sketch and shade an appropriate Normal distribution. Then show your work.

(a) What percent of pregnancies last less than 240 days (that's about 8 months)?

(b) What percent of pregnancies last between 240 and 270 days (roughly between 8 months and 9 months)?

(c) How long do the longest 20% of pregnancies last?

2.34 IQ test scores Scores on the Wechsler Adult Intelligence Scale (a standard IQ test) for the 20 to 34 age group are approximately Normally distributed with $\mu = 110$ and $\sigma = 25$. For each part that follows, sketch and shade an appropriate Normal distribution. Then show your work.

(a) What percent of people aged 20 to 34 have IQ scores above 100?

(b) What percent have scores above 150?

(c) MENSA is an elite organization that admits as members people who score in the top 2% on IQ tests. What score on the Wechsler Adult Intelligence Scale would an individual have to earn to qualify for MENSA membership?

2.35 Locating quartiles The quartiles of any density curve are the points with area 0.25 and 0.75 to their left under the curve. Use Table A or the *Normal Curve Applet* to answer the following questions.

(a) What are the quartiles of a standard Normal distribution?

(b) How many standard deviations away from the mean do the quartiles lie in any Normal distribution? What are the quartiles for the lengths of human pregnancies in Exercise 2.33?

2.36 Brush your teeth The amount of time Ricardo spends brushing his teeth follows a Normal distribution with unknown mean and standard deviation. Ricardo spends less than one minute brushing his teeth about 40% of the time. He spends more than two minutes brushing his teeth 2% of the time. Use this information to determine the mean and standard deviation of this distribution.

Assessing Normality

The Normal distributions provide good models for some distributions of real data. Examples include the highway gas mileage of 2006 Corvette convertibles, statewide unemployment rates, and weights of 9-ounce bags of potato chips. The distributions of some other common variables are usually skewed and therefore distinctly non-Normal. Examples include economic variables such as personal income and gross sales of business firms, the survival times of cancer patients after treatment, and the service lifetime of mechanical or electronic components. While experience can suggest whether or not a Normal distribution is plausible in a particular case, it is risky to assume that a distribution is Normal without actually inspecting the data.

In the latter part of this course we will want to use various statistical inference procedures to try to answer questions that are important to us. These tests involve sampling people or objects and recording data to gain insights into the populations from which they come. Many of these procedures are based on the assumption that the host population is approximately Normally distributed. Consequently, we need to develop methods for assessing Normality.

Method 1 Construct a histogram or a stemplot. See if the graph is approximately bell-shaped and symmetric about the mean.

A histogram or stemplot can reveal distinctly non-Normal features of a distribution, such as outliers, pronounced skewness, or gaps and clusters. You can improve the effectiveness of these plots for assessing whether a distribution is Normal by marking the points $\bar{x}$, $\bar{x} \pm s$, and $\bar{x} \pm 2s$ on the horizontal axis. This gives the scale natural to Normal distributions. Then compare the count of observations in each interval with the 68–95–99.7 rule.

Example 2.12 ***Seventh-grade vocabulary scores (again)***
Assessing Normality

The histogram in Figure 2.3 (page 123) suggests that the distribution of the 947 Gary vocabulary scores is close to Normal. It is hard to assess by eye how close to Normal a histogram is. Let's use the 68–95–99.7 rule to check more closely. We enter the scores into a statistical software package and ask for the mean and standard deviation. The computer replies,

```
MEAN = 6.8585
STDEV = 1.5952
```

Now that we know that $\overline{x} = 6.8585$ and $s = 1.5952$, we check the 68–95–99.7 rule by finding the actual counts of Gary vocabulary scores in intervals of length s about the mean $\overline{x}$. The computer will also do this for us. Here are the counts:

1		21		129		331		318		125		21		1
	2.07		3.67		5.26		6.86		8.45		10.05		11.64	
	$\overline{x} - 3s$		$\overline{x} - 2s$		$\overline{x} - s$		$\overline{x}$		$\overline{x} + s$		$\overline{x} + 2s$		$\overline{x} + 3s$	

The distribution is very close to symmetric. It also follows the 68–95–99.7 rule closely: 68.5% of the scores (649 out of 947) are within one standard deviation of the mean, 95.4% (903 out of 947) are within two standard deviations, and 99.8% (945 out of 947) are within three standard deviations. These counts confirm that the Normal distribution with $\mu = 6.86$ and $\sigma = 1.595$ fits these data well.

Smaller data sets rarely fit the 68–95–99.7 rule as well as the Gary vocabulary scores. This is even true of observations taken from a larger population that really has a Normal distribution. There is more chance variation in small data sets.

Method 2 Construct a ***Normal probability plot***. A Normal probability plot provides a good assessment of the adequacy of the Normal model for a set of data. Most software packages, including Minitab, Fathom, and CrunchIt!, can construct Normal probability plots from entered data. The TI-83/84 and TI-89 will also make Normal probability plots. You will need to be able to produce a Normal probability plot (either with a calculator or with computer software) and interpret it.

Here is the basic idea of a Normal probability plot. The graphs produced by technology use more sophisticated versions of this idea.

1. Arrange the observed data values from smallest to largest. Record what percentile of the data each value occupies. For example, the smallest observation in a set of 20 is at the 5% point, the second smallest is at the 10% point, and so on.

2. Use the standard Normal distribution (Table A) to find the z-scores at these same percentiles. For example, $z = -1.645$ is the 5% point of the standard Normal distribution, and $z = -1.28$ is the 10% point.

3. Plot each data point x against the corresponding z. If the data distribution is close to Normal, the plotted points will lie close to some straight line. *Note: some software plots the x-values on the horizontal axis and the z-scores on the vertical axis, while other software does just the reverse.* You should be able to interpret the plot in either case.

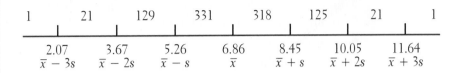

Use of Normal Probability Plots

If the points on a **Normal probability plot** lie close to a straight line, the plot indicates that the data are Normal. Systematic deviations from a straight line indicate a non-Normal distribution. Outliers appear as points that are far away from the overall pattern of the plot.

The next three examples show you how to interpret Normal probability plots.

Example 2.13 Connecting wires and wafers

Interpreting Normal probability plots

Let's return to Example 1.8 (page 54), which described the breaking strengths of connections between very fine wires and a semiconductor wafer. Figure 1.7 showed that the distribution of breaking strengths is symmetric and single-peaked. Now we ask: do the breaking strengths follow a Normal distribution?

Figure 2.21 is a Normal probability plot of the breaking strengths. Lay a transparent straightedge over the center of the plot to see that most points fall close to a straight line. A Normal distribution describes these points quite well. The only substantial deviations are short horizontal runs of points. Each run represents repeated observations having the same value—there are five measurements at 1150, for example. This phenomenon (called *granularity*) is a result of the limited precision of the measurements and does not represent an important deviation from Normality.

Figure 2.21 *Normal probability plot of the breaking strengths of wires bonded to a semiconductor wafer. This distribution has a Normal shape except for outliers in both tails.*

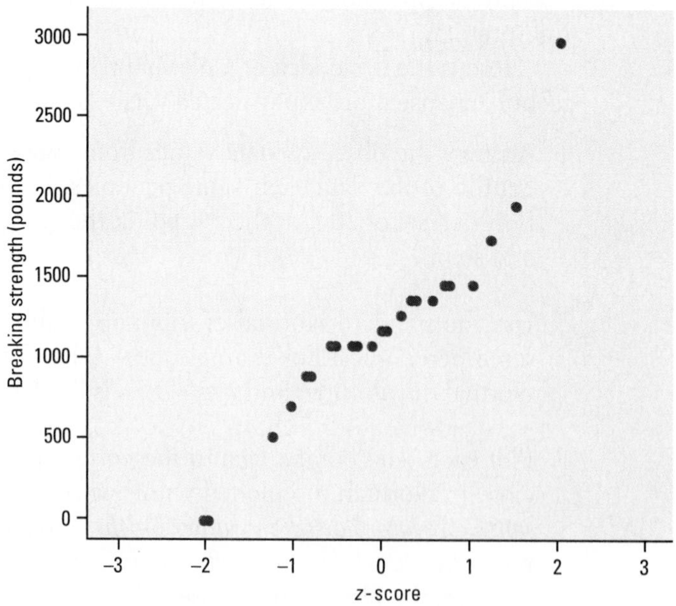

The high outlier at 3150 pounds lies above the line formed by the center of the data—it is farther out in the high direction than we expect Normal data to be. The two low outliers at 0 lie below the line—they are suspiciously far out in the low direction. These outliers were explained in Example 1.8.

Example 2.14 *Guinea pig survival*

Some non-Normal data

Table 2.2 gives the survival times in days of 72 guinea pigs after they were injected with tubercle bacilli in a medical experiment.

Table 2.2 *Survival times (days) of guinea pigs in a medical experiment*

43	45	53	56	56	57	58	66	67	73	74	79
80	80	81	81	81	82	83	83	84	88	89	91
91	92	92	97	99	99	100	100	101	102	102	102
103	104	107	108	109	113	114	118	121	123	126	128
137	138	139	144	145	147	156	162	174	178	179	184
191	198	211	214	243	249	329	380	403	511	522	598

Source: T. Bjerkedal, "Acquisition of resistance in guinea pigs infected with different doses of virulent tubercle bacilli," *American Journal of Hygiene,* 72 (1960), pp. 130–148.

Figure 2.22 is a Normal probability plot of the guinea pig survival times. Draw a line through the leftmost points, which correspond to the smaller observations. The larger observations fall systematically above this line. That is, the right-of-center observations have larger values than the Normal distribution. This Normal probability plot indicates that the guinea pig survival data are strongly right-skewed. *In a right-skewed distribution, the largest observations fall distinctly above a line drawn through the main body of points.* Similarly, left-skewness is evident when the smallest observations fall below the line. There appear to be no outliers in this set of data.

Figure 2.22 *Normal probability plot of the survival times of guinea pigs in a medical experiment. This distribution is skewed to the right.*

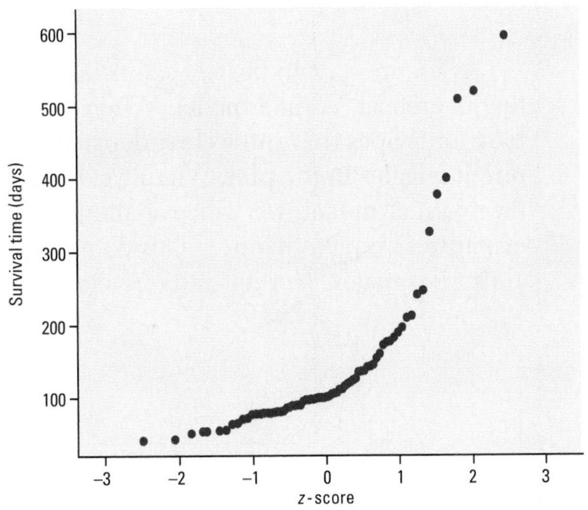

Example 2.15 *Acidity of rainwater*
Some Normal data

Researchers collected data on the acidity levels (measured by pH) in 105 samples of rainwater. Distilled water has pH 7.00. As water becomes more acidic, the pH goes down. The pH of rainwater is important to environmentalists because of the problem of acid rain.[7]

Figure 2.23 is a Normal probability plot of the 105 acidity (pH) measurements. The Normal probability plot makes it clear that a Normal distribution is a good description—there are only minor wiggles in a generally straight-line pattern.

Figure 2.23 *Normal probability plot of the acidity (pH) values of 105 samples of rainwater. This distribution is roughly Normal.*

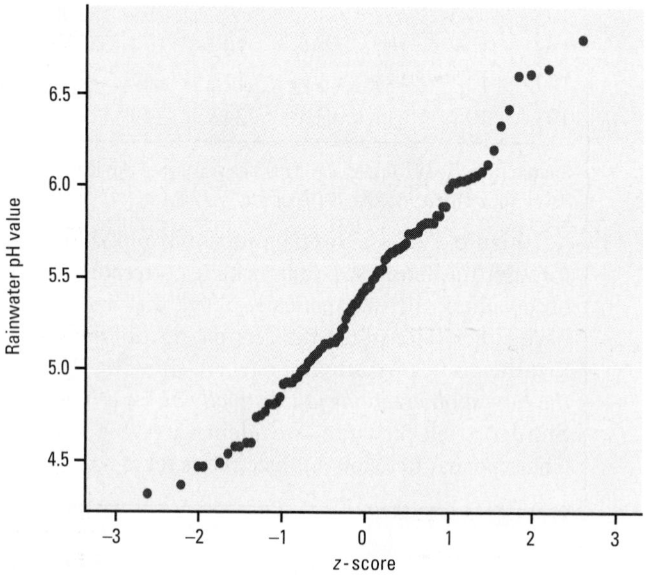

As Figure 2.23 indicates, real data almost always show some departure from the theoretical Normal model. **When you examine a Normal probability plot, look for shapes that show clear departures from Normality. Don't overreact to minor wiggles in the plot.** When we discuss statistical methods that are based on the Normal model, we will pay attention to the sensitivity of each method to departures from Normality. Many common methods work well as long as the data are approximately Normal and outliers are not present.

Technology Toolbox

Normal probability plots on the TI-83/84/89

If you ran the program FLIP50 in Exercise 2.22 (page 133), and you still have the data (100 numbers mostly in the 20s) in L1/list1, then use these data. If you have not entered the program and run it, take a few minutes to do that now. Duplicate this example with your data. Here is the histogram that was generated at the end of one run of this simulation on each calculator.

TI-83/84

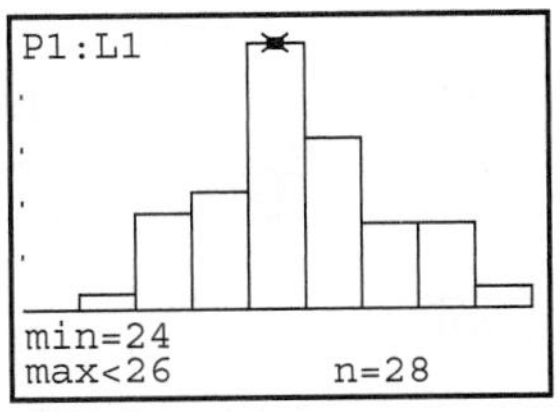

TI-89

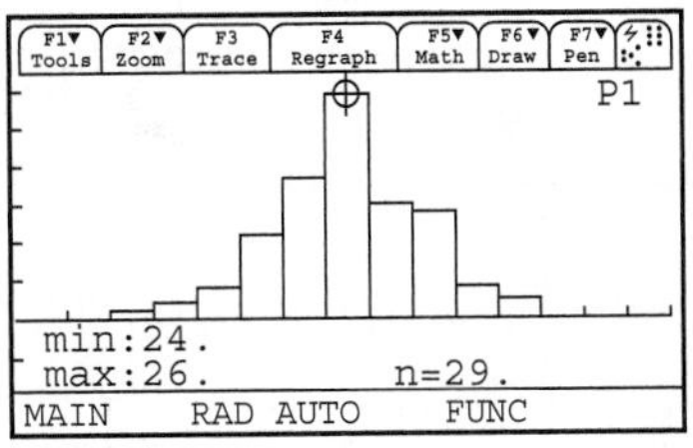

Ask for one-variable statistics:

- Press STAT, choose CALC, then `1:1-Var Stats` and 2nd 1 (L1)

This will give us the following:

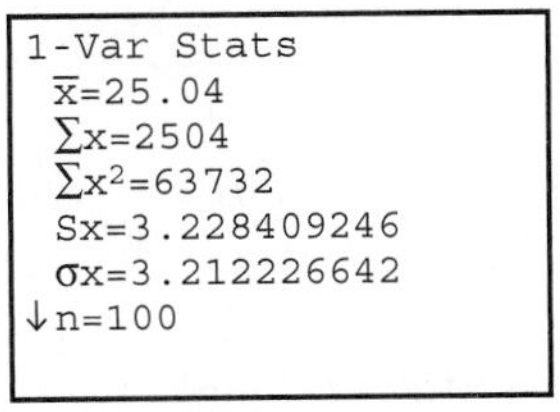

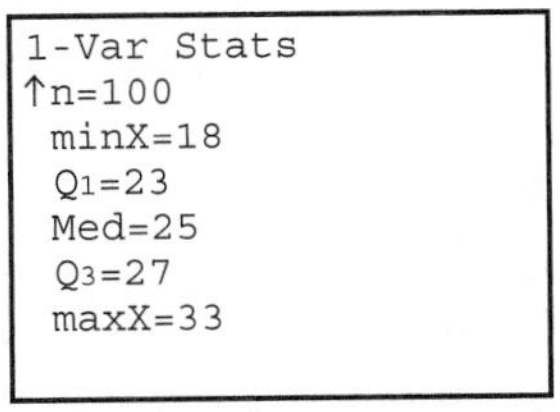

- In the Stats/List Editor, press; F4 (Calc) and choose `1:1-Var Stats` for list1.

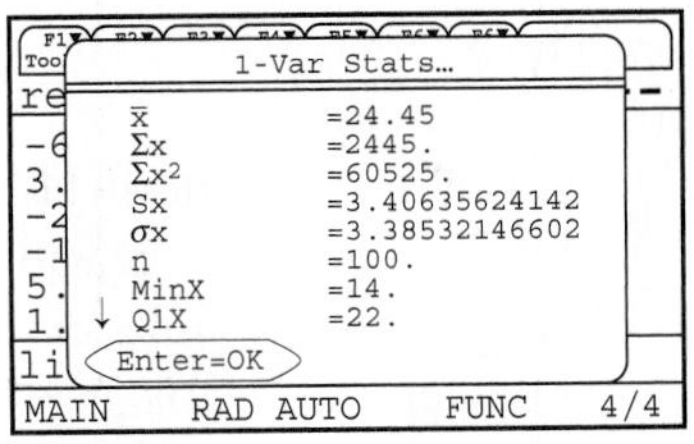

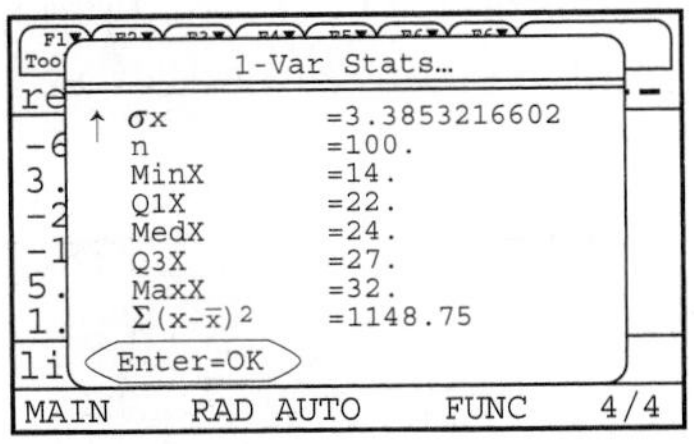

- Comparing the means and medians ($\bar{x} = 25.04$ versus $M = 25$ on the TI-83/84 and $\bar{x} = 24.45$ versus $M = 24$ on the TI-89) suggests that the distributions are fairly symmetric. Boxplots confirm the roughly symmetric shape.

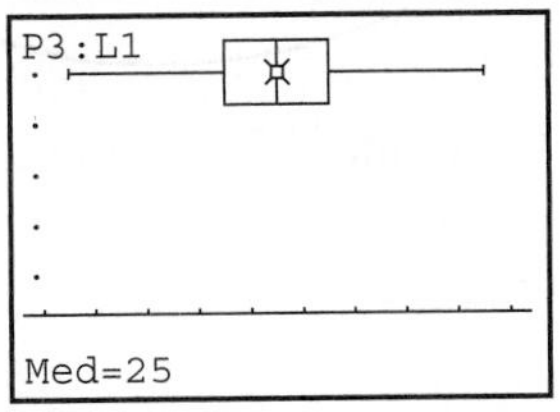

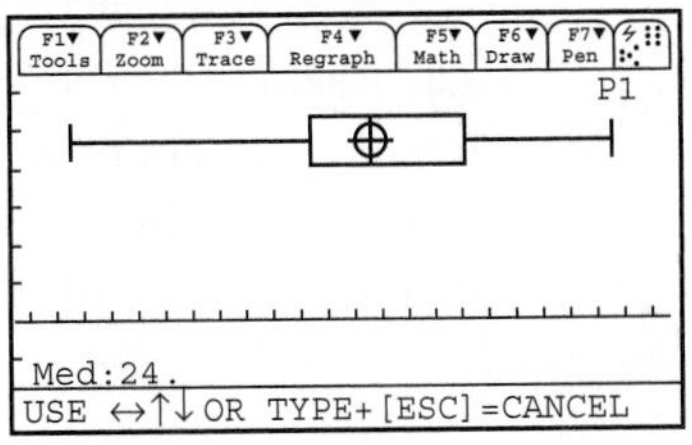

Technology Toolbox

Normal probability plots on the TI-83/84/89 *(continued)*

- To construct a Normal probability plot of the data *with the x-values on the horizontal axis*, define Plot 1 like this:

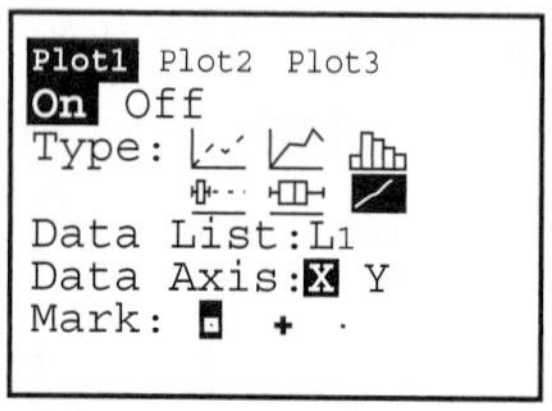

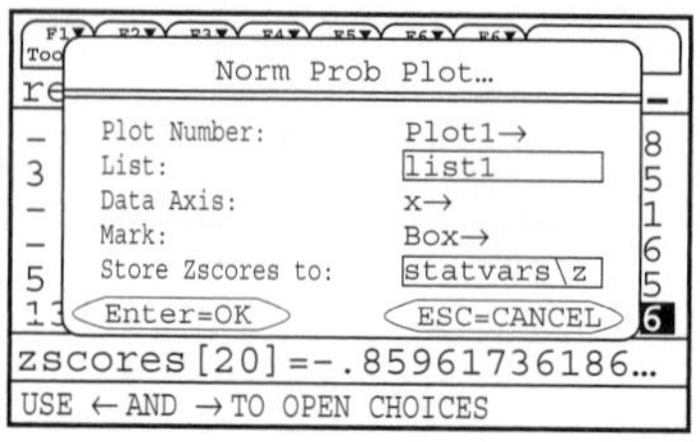

- Use ZoomStat (ZoomData on the TI-89) to see the finished graph.

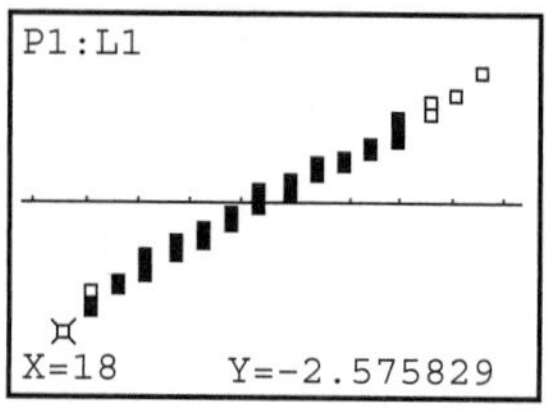

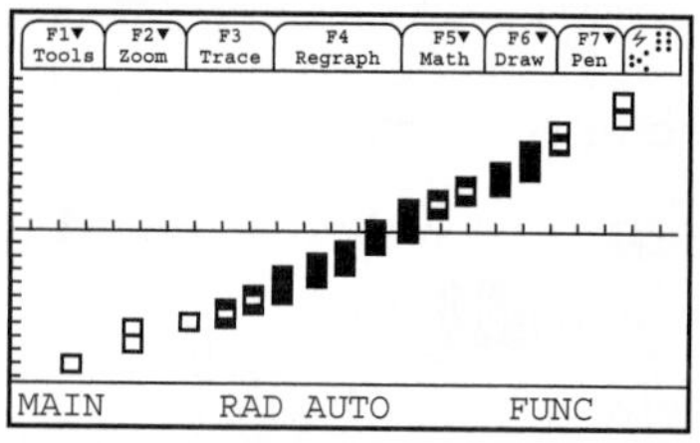

Interpretation: The Normal probability plot is quite linear, so it is reasonable to believe that the data follow a Normal distribution.

Comment: The cursor on the bottom TI-83/84 screen shot is on the point with an x-value of 18 and a z-score of -2.58. This is the point with the smallest x-value of the 100 data values generated by the program FLIP50. By our definition, that puts the value of $x = 18$ at the 1% point. The alternative definition of percentile (see page 119) would place $x = 18$ at the 0% point, since no data values are below this one. The Normal probability plot on the TI-83/84 and TI-89 uses a compromise: it puts this point at the 0.5% mark. If we look up an area to the left of 0.0050 (0.5%) in Table A, we find that $z = -2.58$.

Exercises

2.37 Taking stock Figure 2.24 (on the next page) is a Normal probability plot of the monthly percent returns for U.S. common stocks from June 1950 to June 2000. Because there are 601 observations, the individual points in the middle of the plot run together. In what way do monthly returns on stocks deviate from a Normal distribution?

Figure 2.24 *Normal probability plot of the percent returns on U.S. common stocks in 601 consecutive months, for Exercise 2.37.*

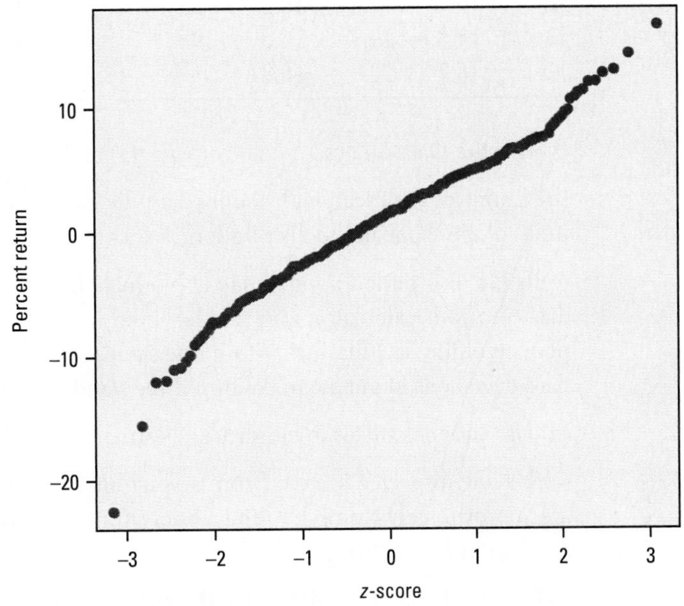

2.38 Carbon dioxide emissions Figure 2.25 is a Normal probability plot of the emissions of carbon dioxide per person in 48 countries.[8] In what ways is this distribution non-Normal?

Figure 2.25 *Normal probability plot of CO_2 emissions in 48 countries, for Exercise 2.38.*

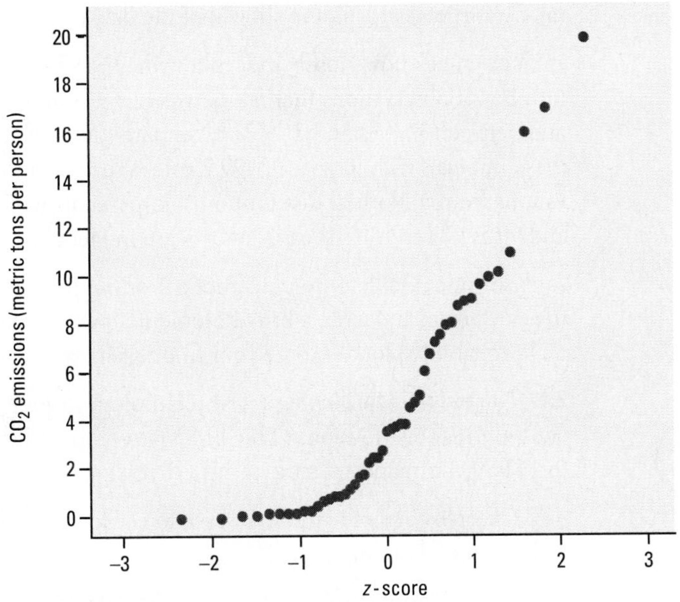

2.39 Great white sharks Here are the lengths in feet of 44 great white sharks:

18.7	12.3	18.6	16.4	15.7	18.3	14.6	15.8	14.9	17.6	12.1
16.4	16.7	17.8	16.2	12.6	17.8	13.8	12.2	15.2	14.7	12.4
13.2	15.8	14.3	16.6	9.4	18.2	13.2	13.6	15.3	16.1	13.5
19.1	16.2	22.8	16.8	13.6	13.2	15.7	19.7	18.7	13.2	16.8

(a) Use the Data Analysis Toolbox (page 93) to describe the distribution of these lengths.

(b) Compare the mean with the median. Does this comparison support your assessment in (a) of the shape of the distribution? Explain.

(c) Is the distribution approximately Normal? If you haven't done this already, enter the data into your calculator, and reorder them from smallest to largest. Then calculate the percent of the data that lies within one standard deviation of the mean. Within two standard deviations of the mean. Within three standard deviations of the mean.

(d) Use your calculator to construct a Normal probability plot. Interpret this plot.

(e) Having inspected the data from several different perspectives, do you think these data are approximately Normal? Write a brief summary of your assessment that combines your findings from (a) through (d).

2.40 Cavendish and the density of the earth Repeated careful measurements of the same physical quantity often have a distribution that is close to Normal. Look back at Cavendish's 29 measurements of the density of the earth from Exercise 1.60 (page 106):

5.50	5.61	4.88	5.07	5.26	5.55	5.36	5.29	5.58	5.65
5.57	5.53	5.62	5.29	5.44	5.34	5.79	5.10	5.27	5.39
5.42	5.47	5.63	5.34	5.46	5.30	5.75	5.68	5.85	

(a) Construct a stemplot to show that the data are reasonably symmetric.

(b) Now check how closely they follow the 68–95–99.7 rule. Find $\bar{x}$ and s, then count the number of observations that fall between $\bar{x} - s$ and $\bar{x} + s$, between $\bar{x} - 2s$ and $\bar{x} + 2s$, and between $\bar{x} - 3s$ and $\bar{x} + 3s$. Compare the percents of the 29 observations in each of these intervals with the 68–95–99.7 rule. We expect that, when we have only a few observations from a Normal distribution, the percents will show some deviation from 68, 95, and 99.7. Cavendish's measurements are in fact close to Normal.

(c) Use your calculator to construct a Normal probability plot for Cavendish's density of the earth data, and write a brief statement about the Normality of the data. Does the Normal probability plot reinforce your findings in (a) and (b)?

2.41 Normal random numbers Use technology to generate 100 observations from the standard Normal distribution. (The TI-83/84 command `randNorm(0,1,100)`→L_1 or the TI-89 command `tistat.randNorm(0,1,100)→list1` will do this.)

(a) Make a histogram of these observations. How does the shape of the histogram compare with a Normal density curve?

(b) Make a Normal probability plot of the data. Does the plot suggest any important deviations from Normality?

(c) Repeat the process of generating 100 observations and completing parts (a) and (b) several times. This should give you a better idea of how close to Normal the data will appear when we take a sample from a Normal population.

2.42 Uniform random numbers Use technology to generate 100 observations from the uniform distribution described in Exercise 2.10 (page 128). (The TI-83/84 command `rand(100)→L₁` or the TI-89 command `rand(100)→list1` will do this.)

(a) Make a histogram of these observations. How does the histogram compare with the density curve in Figure 2.8 (page 128)?

(b) Make a Normal probability plot of your data. According to this plot, how does the uniform distribution deviate from Normality?

Section 2.2 Summary

The **Normal distributions** are described by a special family of bell-shaped, symmetric density curves, called **Normal curves.** The mean μ and standard deviation σ completely specify a Normal distribution $N(\mu, \sigma)$. The mean is the center of the curve, and σ is the distance from μ to the inflection points on either side.

In particular, all Normal distributions satisfy the **68–95–99.7 rule,** which describes what percent of observations lie within one, two, and three standard deviations of the mean.

All Normal distributions are the same when measurements are standardized. If x has the $N(\mu,\sigma)$ distribution, then the **standardized variable** $z = (x - \mu)/\sigma$ has the **standard Normal distribution** $N(0, 1)$ with mean 0 and standard deviation 1.

Table A gives the proportions of standard Normal observations that are less than z for many values of z. By standardizing, we can use Table A for any Normal distribution. The *Normal Curve Applet* allows you to do Normal calculations quickly.

In order to perform certain inference procedures in later chapters, we will need to know that the data come from populations that are approximately Normally distributed. To assess Normality, one can observe the shape of histograms, stemplots, and boxplots and see how well the data fit the 68–95–99.7 rule for Normal distributions. Another good method for assessing Normality is to construct a **Normal probability plot.**

Section 2.2 Exercises

For Exercises 2.43 to 2.47, use Table A or the Normal Curve Applet.

2.43 Finding areas from *z*-scores Find the proportion of observations from a standard Normal distribution that fall in each of the following regions. In each case, sketch a standard Normal curve and shade the area representing the region.

(a) $z < 1.28$

(b) $z > -0.42$

(c) $-0.42 < z < 1.28$

(d) $-1.28 < z < -0.42$

2.44 Working backward: finding *z*-values

(a) Find the number z such that the proportion of observations that are less than z in a standard Normal distribution is 0.98.

(b) Find the number z such that 22% of all observations from a standard Normal distribution are greater than z.

2.45 A surprising calculation Changing the mean of a Normal distribution by a moderate amount (while keeping the standard deviation about the same) can greatly change the percent of observations in the tails. Suppose that a college is looking for applicants with SAT Math scores of 750 and above.

(a) In 2004, the scores of males on the SAT Math test followed the $N(537, 116)$ distribution. What percent of males scored 750 or better? Show your work.

(b) Females' SAT Math scores that year had the $N(501, 110)$ distribution. What percent of females scored 750 or better? Show your work. You see that the percent of males above 750 is almost three times the percent of females with such scores.

2.46 The stock market The annual rate of return on stock indexes (which combine many individual stocks) is approximately Normal. Since 1945, the Standard & Poor's 500 index has had a mean yearly return of 12%, with a standard deviation of 16.5%. Take this Normal distribution to be the distribution of yearly returns over a long period.

(a) In what range do the middle 95% of all yearly returns lie? Justify your answer.

(b) The market is down for the year if the return on the index is less than zero. In what proportion of years is the market down? Show your work.

(c) In what proportion of years does the index gain 25% or more? Show your work.

2.47 Deciles The deciles of any distribution are the points that mark off the lowest 10% and the highest 10%. The deciles of a density curve are therefore the points with area 0.1 and 0.9 to their left under the curve.

(a) What are the deciles of the standard Normal distribution?

(b) The heights of young women are approximately Normal with mean 64.5 inches and standard deviation 2.5 inches. What are the deciles of this distribution? Show your work.

2.48 Outliers in a Normal distribution The percent of the observations that are classified as outliers by the $1.5 \times IQR$ rule is the same in any Normal distribution. What is this percent? Show your method clearly.

2.49 Runners' heart rates Figure 2.26 is a Normal probability plot of the heart rates of 200 male runners after six minutes of exercise on a treadmill.[9] The distribution is close to Normal. How can you see this? Describe the nature of the small deviations from Normality that are visible in the plot.

Figure 2.26 *Normal probability plot of the heart rates of 200 male runners, for Exercise 2.49.*

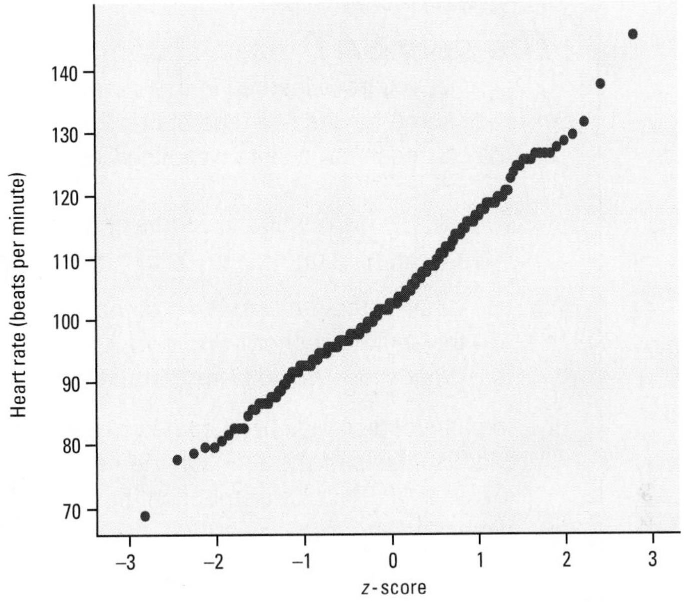

2.50 Weights aren't Normal The heights of people of the same sex and similar ages follow Normal distributions reasonably closely. Weights, on the other hand, are not Normally distributed. The weights of women aged 20 to 29 have mean 141.7 pounds and median 133.2 pounds. The first and third quartiles are 118.3 pounds and 157.3 pounds. What can you say about the shape of the weight distribution? Why?

Many pluggers fit the 'normal curve' when it comes to distribution of weight.

CASE CLOSED!

The new SAT

Using what you have learned in this chapter, you should now be ready to examine some results from the Spring 2005 administrations of the new SAT discussed in the chapter-opening Case Study (page 113).

1. Overall, the distribution of scores on the new Writing section followed a Normal distribution with mean 516 and standard deviation 115.
   (a) What percent of students earned scores between 600 and 700 on the Writing section? Show your work.
   (b) What score would place a student at the 65th percentile?
2. In past years, males have earned higher mean scores on the Verbal and Math sections of the SAT. On the new Writing section, male scores followed a $N(491, 110)$ distribution. Female scores followed a $N(502, 108)$ distribution.
   (a) What percent of male test takers earned scores below the mean score for female test takers?
   (b) What percent of female test takers earned scores above the mean score for male test takers?
   (c) What percent of male test takers earned scores above the 85th percentile of the female scores?
3. Here are the Writing scores on the new SAT for all test takers at a medium-sized private high school:

Males:	660	530	580	610	540	550	470	600	550	550	640	640
	540	520	430	580	490	710	670	570	700	520	530	630
	500	440	460	520	760	560	530	570	690	570	610	570
	620	700	720	630	720	530	690	690	530	580	540	520
Females:		500	690	420	480	630	640	640	560	490	630	620
		430	600	700	670	690	620	580	570	560	780	450
		660	580	500	560	580	580	530	630	560	630	610
		520	550	460	580	590	550					

   (a) Did males or females at this school perform better? Give appropriate graphical and numerical evidence to support your answer.
   (b) How do the scores of the males and females at this school compare with the national results? Write a few sentences discussing this issue.
   (c) Is the distribution of scores for the males at this school approximately Normal? How about the distribution of scores for the females at this school? Justify your answer with appropriate statistical evidence.

Chapter Review

Summary

This chapter focused on two big issues: describing an individual value's location within a distribution of data and modeling distributions with density curves. Both z-scores and percentiles provide easily calculated measures of relative standing for individuals. Density curves come in assorted shapes, but all share the property that the area beneath the curve is 1. We can use areas under density curves to estimate the proportion of individuals in a distribution whose values fall in a specified range. In the special case of Normally distributed data, we can use the standard Normal curve and Table A to calculate such areas. There are many real-world examples of Normal distributions. When you meet a new set of data, you can use the graphical and numerical tools discussed in this chapter to assess the Normality of the data.

What You Should Have Learned

Here is a review list of the most important skills you should have acquired from your study of this chapter.

A. Measures of Relative Standing

1. Find the standardized value (z-score) of an observation. Interpret z-scores in context.
2. Use percentiles to locate individual values within distributions of data.
3. Apply Chebyshev's inequality to a given distribution of data.

B. Density Curves

1. Know that areas under a density curve represent proportions of all observations and that the total area under a density curve is 1.
2. Approximately locate the median (equal-areas point) and the mean (balance point) on a density curve.
3. Know that the mean and median both lie at the center of a symmetric density curve and that the mean moves farther toward the long tail of a skewed curve.

C. Normal Distributions

1. Recognize the shape of Normal curves and be able to estimate both the mean and standard deviation from such a curve.
2. Use the 68–95–99.7 rule and symmetry to state what percent of the observations from a Normal distribution fall between two points when the points lie at the mean or one, two, or three standard deviations on either side of the mean.

3. Use the standard Normal distribution to calculate the proportion of values in a specified range and to determine a z-score from a percentile.
4. Given that a variable has the Normal distribution with mean μ and standard deviation σ, use Table A and your calculator to determine the proportion of values in a specified range.
5. Given that a variable has the Normal distribution with mean μ and standard deviation σ, calculate the point having a stated proportion of all values to the left or to the right of it.

D. Assessing Normality

1. Plot a histogram, stemplot, and/or boxplot to determine if a distribution is bell-shaped.
2. Determine the proportion of observations within one, two, and three standard deviations of the mean, and compare with the 68–95–99.7 rule for Normal distributions.
3. Construct and interpret Normal probability plots.

Web Links

You can get more information about standardized tests (like the PSAT, SAT, and ACT) and intelligence tests (like the Wechsler Adult Intelligence Scale) online.

College Board Web site	www.collegeboard.com
ACT Web site	www.act.org
MENSA Web site	www.us.mensa.org

Chapter Review Exercises

2.51 IQ scores for children The scores of a reference population on the Wechsler Intelligence Scale for Children (WISC) are Normally distributed with $\mu = 100$ and $\sigma = 15$. A school district classified children as "gifted" if their WISC score exceeded 135. There are 1300 sixth-graders in the school district. About how many of them are gifted? Show your work.

2.52 Understanding density curves Remember that it is areas under a density curve, not the height of the curve, that give proportions in a distribution. To illustrate this, sketch a density curve that has its peak at 0 on the horizontal axis but has greater area within 0.25 on either side of 1 than within 0.25 on either side of 0.

2.53 Are the data Normal? Scores on the ACT test for the 2004 high school graduating class had mean 20.9 and standard deviation 4.8. In all, 1,171,460 students in this class took the test, and 1,052,490 of them had scores of 27 or lower.[10] If the distribution of scores were Normal, what percent of scores would be 27 or lower? What percent of the actual scores were 27 or lower? Does the Normal distribution describe the actual data well?

2.54 Standardized test scores as percentiles Joey received a report that he scored in the 97th percentile on a national standardized reading test but in the 72nd percentile on the math portion of the test.

(a) Explain to Joey's grandmother, who knows no statistics, what these numbers mean.

(b) Can we determine Joey's z-scores for his reading and math performance? Why or why not?

2.55 Helmet sizes The army reports that the distribution of head circumference among soldiers is approximately Normal with mean 22.8 inches and standard deviation 1.1 inches. Helmets are mass-produced for all except the smallest 5% and the largest 5% of head sizes. Soldiers in the smallest or largest 5% get custom-made helmets. What head sizes get custom-made helmets? Show your work.

2.56 More weird density curves A certain density curve consists of a straight-line segment that begins at the origin, (0, 0), and has slope 1.

(a) Sketch the density curve. What are the coordinates of the right endpoint of the segment?

(b) Determine the median, the first quartile (Q_1), and the third quartile (Q_3).

(c) Relative to the median, where would you expect the mean of the distribution to be located?

(d) What percent of the observations lie below 0.5? Above 1.5?

2.57 Are the data Normal? A government report looked at the amount borrowed for college by students who graduated in 2000 and had taken out student loans.[11] The mean amount was $\overline{x} = \$17,776$ and the standard deviation was $s = \$12,034$. The quartiles were $Q_1 = \$9900$, $M = \$15,532$, and $Q_3 = \$22,500$.

(a) Compare the mean $\overline{x}$ and the median M. Also compare the distances of Q_1 and Q_3 from the median. Explain why both comparisons suggest that the distribution is right-skewed. Interpret this in the context of student loans.

(b) The right-skew increases the standard deviation. So a Normal distribution with the same mean and standard deviation would have a third quartile larger than the actual Q_3. Find the third quartile of the Normal distribution with $\mu = \$17,776$ and $\sigma = \$12,034$ and compare it with $Q_3 = \$22,500$.

2.58 Osteoporosis Osteoporosis is a condition in which the bones become brittle due to loss of minerals. To diagnose osteoporosis, an elaborate apparatus measures bone mineral density (BMD). BMD is usually reported in standardized form. The standardization is based on a population of healthy young adults. The World Health Organization (WHO) criterion for osteoporosis is a BMD score that is 2.5 standard deviations below the mean for young adults. BMD measurements in a population of people similar in age and sex roughly follow a Normal distribution.

(a) What percent of healthy young adults have osteoporosis by the WHO criterion?

(b) Women aged 70 to 79 are of course not young adults. The mean BMD in this age group is about -2 on the standard scale for young adults. Suppose that the standard deviation is the same as for young adults. What percent of this older population has osteoporosis?

2.59 Normal probability plots Figure 2.27 displays Normal probability plots for four different sets of data. Describe what each plot tells you about the Normality of the given data set.

Figure 2.27 *Four Normal probability plots, for Exercise 2.59.*

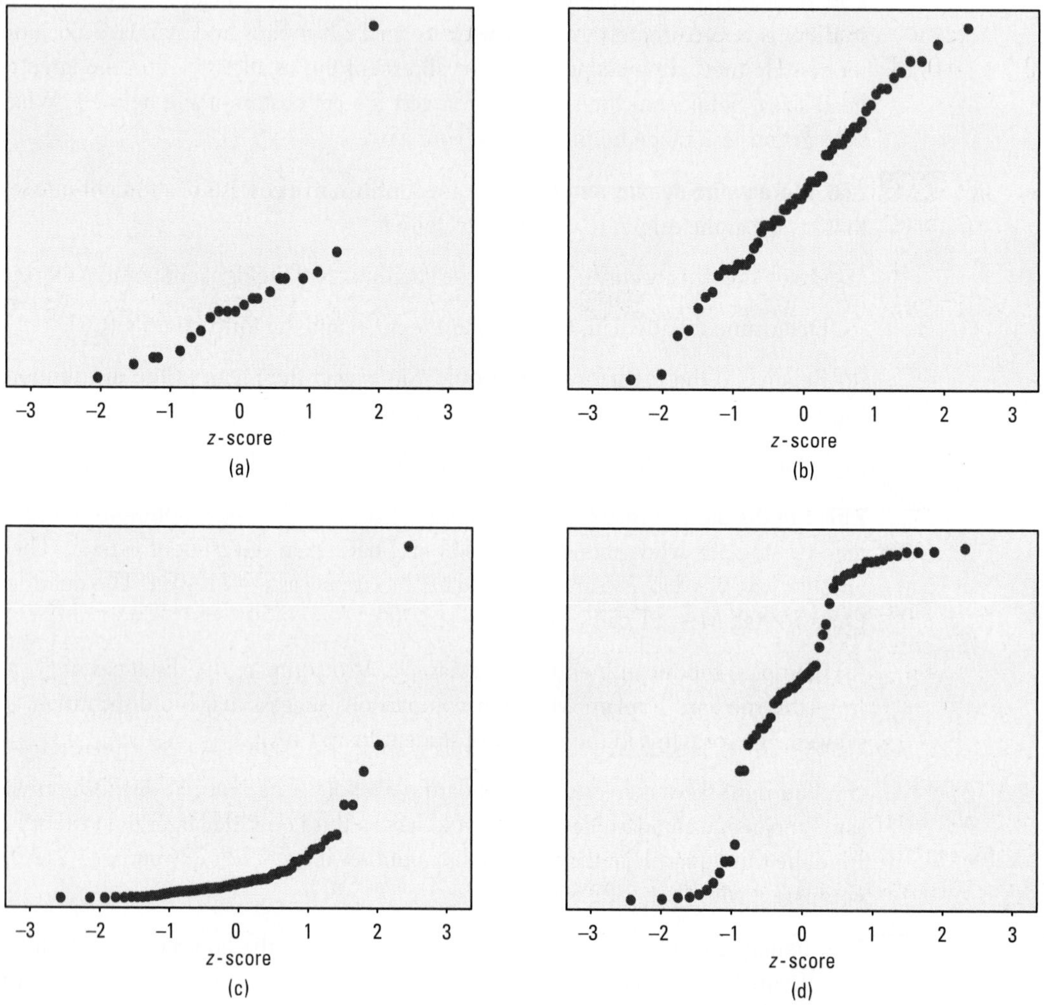

2.60 Grading managers Many companies "grade on a bell curve" to compare the performance of their managers and professional workers. This forces the use of some low performance ratings, so that not all workers are listed as "above average." Ford Motor Company's "performance management process" for a time assigned 10% A grades, 80% B grades, and 10% C grades to the company's 18,000 managers. Suppose that Ford's performance scores really are Normally distributed. This year, managers with scores less than 25 received C's and those with scores above 475 received A's. What are the mean and standard deviation of the scores? Show your work.

Technology Toolbox

Finding areas with ShadeNorm

The TI-83/84/89 can be used to find the area to the left or right of a point or above an interval in a Normal distribution without referring to a standard Normal table. Consider the WISC scores for children of Exercise 2.51. This distribution is N(100, 15). Suppose we want to find the percent of children whose WISC scores are above 125. Begin by specifying a viewing window as follows: $X[55, 145]_{12}$ and $Y[-0.008, 0.028]_{.01}$. You will generally need to experiment with the y settings to get a good graph.

TI-83/84

- Press 2nd VARS (DISTR), then choose DRAW and `1:ShadeNorm(`.
- Complete the command `ShadeNorm(125,1E99,100,15)` and press ENTER.

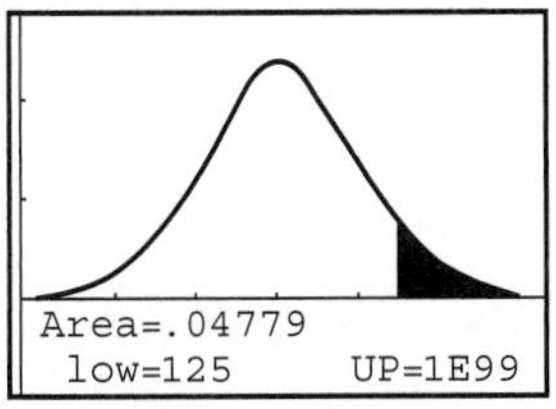

TI-89

- Press CATALOG F3 (Flash Apps) and choose `shadNorm(`.
- Complete the command `tistat.shadNorm(125,1E99,100,15)` and press ENTER.

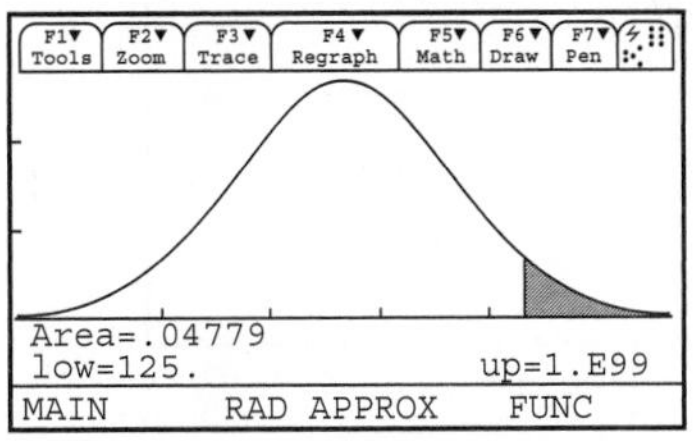

You must always specify an interval. An area in the right tail of the distribution would theoretically be the interval $(125, \infty)$. The calculator limitation dictates that we use a number that is at least 5 or 10 standard deviations to the right of the mean. To find the area to the left of 85, you would specify `ShadeNorm(−1E99,85,100,15)`. Or we could specify `ShadeNorm(0,85,100,15)` since WISC scores can't be negative. Both yield at least four-decimal-place accuracy. If you're using standard Normal values, then you need to specify only the endpoints of the interval; the mean 0 and standard deviation 1 will be understood. For example, use `ShadeNorm(1,2)` to find the area above the interval $z = 1$ to $z = 2$ to four decimal places. Compare this result to the 68–95–99.7 rule.

2.61 Made in the shade Use the calculator's ShadeNorm feature to find the following areas for the WISC scores of Exercise 2.51 (page 162), to four-decimal-place accuracy. Then write your findings in a sentence.

(a) The relative frequency of scores greater than 135.

(b) The relative frequency of scores lower than 75.

(c) Show two ways to find the percent of scores within two standard deviations of the mean.

Technology Toolbox

Finding areas with normalcdf

The normalcdf command on the TI-83/84/89 can be used to find areas under a Normal curve. This method has the advantage over ShadeNorm of being quicker, and the disadvantage of not providing a picture of the area it is finding. Here are the keystrokes for the WISC scores of Exercise 2.51:

TI-83/84

- Press 2nd VARS (DISTR) and choose `2:normalcdf(`.
- Complete the command `normalcdf(125,1E99,100,15)` and press ENTER.

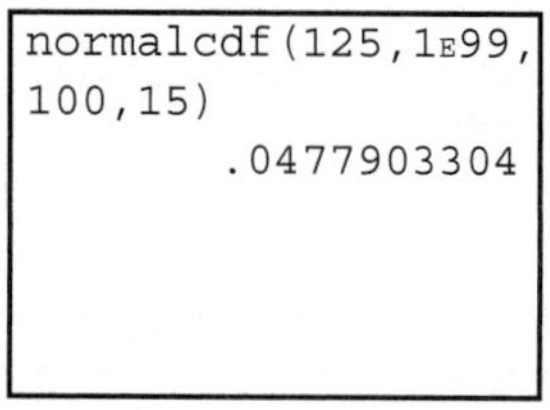

TI-89

- Press CATALOG F3 (Flash Apps) and choose `normCdf(`.
- Complete the command `tistat.normCdf(125,1E99,100,15)` and press ENTER.

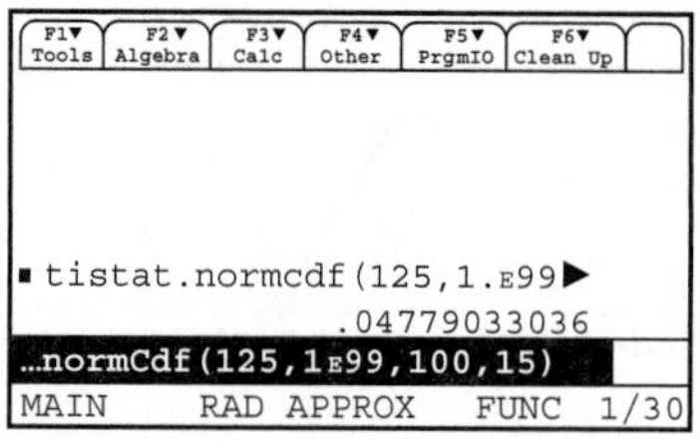

We can say that about 5% of the WISC scores are above 125. If the Normal values have already been standardized, then you need to specify only the left and right endpoints of the interval. For example, `normalcdf(-1,1)` returns 0.6827, meaning that the area from $z = -1$ to $z = 1$ is approximately 0.6827, correct to four decimal places.

2.62 Areas by calculator Use the calculator's normalcdf function to verify your answer to Exercise 2.53 (page 162).

Technology Toolbox

Finding values with invNorm

The TI-83/84/89 invNorm function calculates the raw or standardized Normal value corresponding to a known area under a Normal curve or a relative frequency. The following example uses the WISC scores, which have a N(100, 15) distribution. Here are the keystrokes:

TI-83/84

- Press **2nd** **VARS** (DISTR), then choose `3:invNorm(`.
- Complete the command `invNorm(.9,100,15)` and press **ENTER**. Compare this with the command `invNorm(.9)`.

TI-89

- Press Press **CATALOG** **F3** (Flash Apps) and choose `invNorm(`.
- Complete the command `tistat.invNorm(.9,100,15)` and press **ENTER**. Compare this with the command `tistat.invNorm(.9)`.

```
invNorm(.9,100,15)
        119.2232735
invNorm(.9)
        1.281551567
```

```
F1▼    F2▼      F3▼   F4▼    F5▼     F6▼
Tools  Algebra  Calc  Other  PrgmIO  Clean Up

■ tistat.invnorm(.9,100,15)
                        119.223
■ tistat.invnorm(.9)
                        1.28155
TIStat.invNorm(.9)
MAIN    RAD APPROX    FUNC  2/30
```

The first command finds that the WISC score that has 90% of the scores below it from the N(100, 15) distribution is $x = 119$. The second command finds that the standardized WISC score that has 90% of the scores below it is $z = 1.28$.

2.63 Inverse normal on the calculator Use the calculator's `invNorm` function to verify your answer to Exercise 2.55 (page 163).

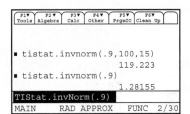

CHAPTER

3

Examining Relationships

CASE STUDY

Are baseballs "juiced"?

Home run mania hit the United States in 1998 when both Mark McGwire and Sammy Sosa broke the major league home run record of 61 home runs in one season. This mania was renewed in 2001 when Barry Bonds hit 73 home runs in one season. Are baseball players getting better at hitting home runs, or is something else going on? One claim is that the way baseballs are made has changed over the years and that the newer balls travel farther than the old ones. Some people say that modern baseballs are "juiced."

Table 3.1 shows the mean number of home runs hit per Major League Baseball game from 1960 to 2000. Do these data provide convincing evidence that baseballs have begun to fly farther over time?

Table 3.1 *Average number of home runs per game by year*

Year	Average	Year	Average	Year	Average	Year	Average
1960	1.722	1970	1.764	1980	1.467	1990	1.576
1961	1.909	1971	1.477	1981	1.278	1991	1.608
1962	1.851	1972	1.363	1982	1.604	1992	1.443
1963	1.670	1973	1.597	1983	1.565	1993	1.776
1964	1.699	1974	1.362	1984	1.548	1994	2.066
1965	1.656	1975	1.395	1985	1.713	1995	2.023
1966	1.698	1976	1.153	1986	1.813	1996	2.186
1967	1.419	1977	1.733	1987	2.118	1997	2.048
1968	1.228	1978	1.406	1988	1.514	1998	2.080
1969	1.603	1979	1.636	1989	1.464	1999	2.277
						2000	2.344

Source: C. Rist, "Foul ball? Unraveling the mystery of why it's so easy to hit a home run," *Discover,* May 2001, pp. 26–27.

Rawlings has had an exclusive contract to supply baseballs to Major League Baseball since 1977. Spaulding made baseballs for Major League Baseball for 100 years prior to that.

By the end of this chapter, you will have acquired the tools you need to make sense of the changing number of home runs hit over time.

Activity 3A

CSI stats: the case of the missing cookies

Materials: Metric ruler and meterstick; handprint and photo lineup (from Teacher Resource Binder) for each group of three to four students; one sheet of graph paper per student

On her desk in the math department office, Mrs. McGill keeps a large jar full of cookies for use in her statistics classes. Over the past few days, a few cookies have gone missing. Eager to catch the culprit, Mrs. McGill notes that the only people with access to her desk are the other math teachers at her school. So she asks her colleagues whether they have been making unauthorized withdrawals from the cookie jar. No one confesses to the crime.

But the next day, Mrs. McGill catches a break—she finds a clear handprint on the cookie jar. The careless culprit has left behind crucial evidence! At this point, Mrs. McGill calls in the CSI: Stats team (your class) to help her identify the prime suspect in "The Case of the Missing Cookies."

1. Measure the height and hand span of each member of your group to the nearest centimeter (cm). (Hand span is the maximum distance from the tip of the thumb to the tip of the pinkie finger on a person's fully stretched hand.)
2. Your teacher will make a data table on the board with two columns, labeled as follows:

Hand span (cm)	Height (cm)

Send a representative to record the data for each member of your group in the table.

3. Copy the data table onto your graph paper very near the left margin of the page. Next, you will make a graph of these data. Begin by constructing a set of coordinate axes. Allow plenty of space on the page for your graph. Label the horizontal axis "Hand span (cm)" and the vertical axis "Height (cm)."
4. Since neither hand span nor height can be close to 0 cm, we want to start our horizontal and vertical scales at larger numbers. Insert a "break in scale" symbol // on each axis very close to the origin (intersection point of the axes). Scale the horizontal axis in 0.5 cm

increments starting with 15 cm. Scale the vertical axis in 5 cm increments starting with 135 cm. Refer to the sketch below for comparison.

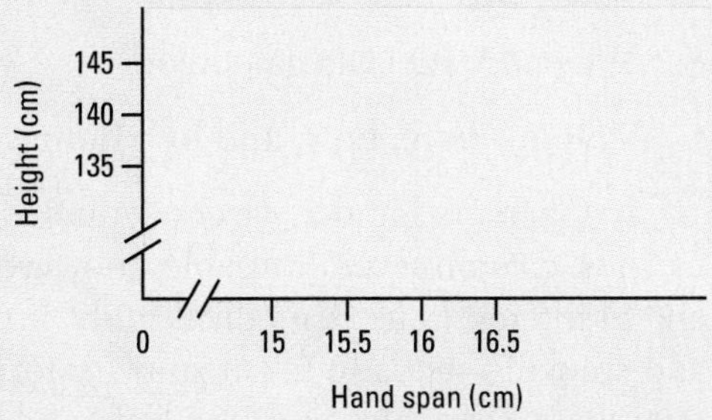

5. Plot each point from your class data table as accurately as you can on the graph. Compare your graph with those of your group members.
6. As a group, discuss what the graph tells you about the relationship between hand span and height. Summarize your observations in a sentence or two.
7. Ask your teacher for a copy of the handprint found at the scene and the photo lineup of the math department. Which math teacher does your group believe is the "prime suspect"? Justify your answer with appropriate statistical evidence.

Introduction

A medical study finds that short women are more likely to have heart attacks than women of average height, while tall women have the fewest heart attacks. An insurance group reports that heavier cars have fewer deaths per 100,000 vehicles than do lighter cars. These and many other statistical studies look at the *relationship between two variables.* Statistical relationships are overall tendencies, not ironclad rules. They allow individual exceptions. Although smokers on average die younger than nonsmokers, some people live to 90 while smoking three packs a day.

To understand a statistical relationship between two variables, we measure both variables on the same individuals. Often, we must examine other variables as well. To conclude that shorter women have higher risk from heart attacks, for example, the researchers had to eliminate the effect of other variables such as weight and exercise habits. In this chapter we begin our study of relationships between variables. One of our main themes is that **the relationship between two variables can be strongly influenced by other variables that are lurking in the background.**

When you examine the relationship between two or more variables, first ask the familiar key questions about the data:

- **Who** are the *individuals* described by the data?
- **What** are the *variables?*
- **Why** were the data gathered?
- **When, where, how, and by whom** were the data produced?

In Chapters 1 and 2, we concentrated mainly on quantitative variables. When we have data on several variables, however, categorical variables are often present and of interest to us. A medical study, for example, may record each subject's sex and smoking status (current smoker; former smoker; never smoked) along with quantitative variables such as weight and blood pressure. We may be interested in possible relationships between two quantitative variables (such as weight and blood pressure), between a categorical and a quantitative variable (such as sex and blood pressure), or between two categorical variables (such as sex and smoking status). This chapter will focus on relationships between quantitative variables. Categorical variables will play a larger role in the next chapter.

When you examine relationships among variables, a new question becomes important:

- Do you want simply to explore the nature of the relationship, or do you think that some of the variables help explain or even cause changes in the others? That is, are some of the variables ***response variables*** and others ***explanatory variables?***

Response Variable, Explanatory Variable

A **response variable** measures an outcome of a study. An **explanatory variable** helps explain or influences changes in a response variable.

You will often find explanatory variables called *independent variables*, and response variables called *dependent variables*. The idea behind this language is that the response variable depends on the explanatory variable. Because the words "independent" and "dependent" have other, unrelated meanings in statistics, we won't use them here.

It is easiest to identify explanatory and response variables when we specify values of one variable in order to see how it affects another variable.

Example 3.1 *Alcohol and body temperature*
Explanatory and response variables

Alcohol has many effects on the body. One effect is a drop in body temperature. To study this effect, researchers give several different amounts of alcohol to mice, then measure the change in each mouse's body temperature in the 15 minutes after taking the alcohol. Amount of alcohol is the explanatory variable, and change in body temperature is the response variable.

When we don't specify the values of either variable but just observe both variables, there may or may not be explanatory and response variables. Whether there are depends on how you plan to use the data.

Example 3.2 ***Are SAT Math and Verbal scores linked?***
Exploring a relationship

Jim wants to know how the mean 2005 SAT Math and Verbal scores in the 50 states are related to each other. He doesn't think that either score explains or causes the other. Jim has two related variables, and neither is an explanatory variable.

Julie looks at some data. She asks, "Can I predict a state's mean 2005 SAT Math score if I know its mean 2005 SAT Verbal score?" Julie is treating the Verbal score as the explanatory variable and the Math score as the response variable.

In Example 3.1 alcohol actually *causes* a change in body temperature. There is no cause-and-effect relationship between SAT Math and Verbal scores in Example 3.2. Because the scores are closely related, we can nonetheless use a state's 2005 SAT Verbal score to predict its Math score. We will learn how to do the prediction in Section 3.2. Prediction requires that we identify an explanatory variable and a response variable. Some other statistical techniques ignore this distinction. **Remember that calling one variable explanatory and the other response doesn't necessarily mean that changes in one *cause* changes in the other.**

CAUTION

Most statistical studies examine data on more than one variable. Fortunately, analysis of several-variable data builds on the tools used for examining individual variables. The principles that guide our work also remain the same:

- First plot the data, then compute numerical summaries.
- Look for overall patterns and deviations from those patterns.
- When the overall pattern is quite regular, use a compact mathematical model to describe it.

Exercises

3.1 Explanatory and response variables In each of the following situations, is it more reasonable to simply explore the relationship between the two variables or to view one of the variables as an explanatory variable and the other as a response variable? In the latter case, which is the explanatory variable and which is the response variable?

(a) The amount of time a student spends studying for a statistics exam and the grade on the exam

(b) The weight in kilograms and height in centimeters of a person

(c) Inches of rain in the growing season and the yield of corn in bushels per acre

(d) A student's grades in statistics and in French

(e) A family's income and the years of education their eldest child completes

3.2 The risks of obesity A study observes a large group of people over a 10-year period. The goal is to see if overweight and obese people are more likely to die during the decade than people who weigh less. Such studies can be misleading, because obese people are more likely to be inactive and to be poor. What are the explanatory and response variables in the study? What other variables are mentioned that may influence the relationship between the explanatory variable and the response variable?

3.3 Coral reefs How sensitive to changes in water temperature are coral reefs? To find out, measure the growth of corals in aquariums where the water temperature is controlled at different levels. Growth is measured by weighing the coral before and after the experiment. What are the explanatory and response variables? Are they categorical or quantitative?

3.4 Treating breast cancer Once the most common treatment for breast cancer was removal of the breast. It is now usual to remove only the tumor and nearby lymph nodes, followed by radiation. The change in policy was due to a large medical experiment that compared the two treatments. Some breast cancer patients, chosen at random, were given one or the other treatment. The patients were closely followed to see how long they lived following surgery. What are the explanatory and response variables? Are they categorical or quantitative?

3.1 Scatterplots and Correlation

The most effective way to display the relationship between two quantitative variables is a ***scatterplot.***

Example 3.3 *State SAT Math scores*
Constructing scatterplots

More than a million high school seniors take the SAT each year. We sometimes see state school systems "rated" by the average SAT scores of their seniors. This is not proper, because the percent of high school seniors who take the SAT varies from state to state. Let's examine the relationship between the percent of a state's high school seniors who took the exam in 2005 and the mean SAT Math score in the state that year.

We think that "percent taking" will help explain "mean score." Therefore, "percent taking" is the explanatory variable and "mean score" is the response variable. We want to see how mean score changes when percent taking changes, so we put percent taking (the explanatory variable) on the horizontal axis. Figure 3.1 is the scatterplot.[1] Each point represents a single state. In Colorado, for example, 26% took the SAT, and their mean SAT Math score was 560. Find 26 on the x (horizontal) axis and 560 on the y (vertical) axis.

Colorado appears as the point (26, 560) above 26 and to the right of 560. Figure 3.1 shows how to locate Colorado's point on the plot.

Figure 3.1 *Scatterplot of the mean SAT Math score in each state against the percent of that state's high school seniors who took the SAT in 2005. The dotted lines intersect at the point (26, 560), the data for Colorado.*

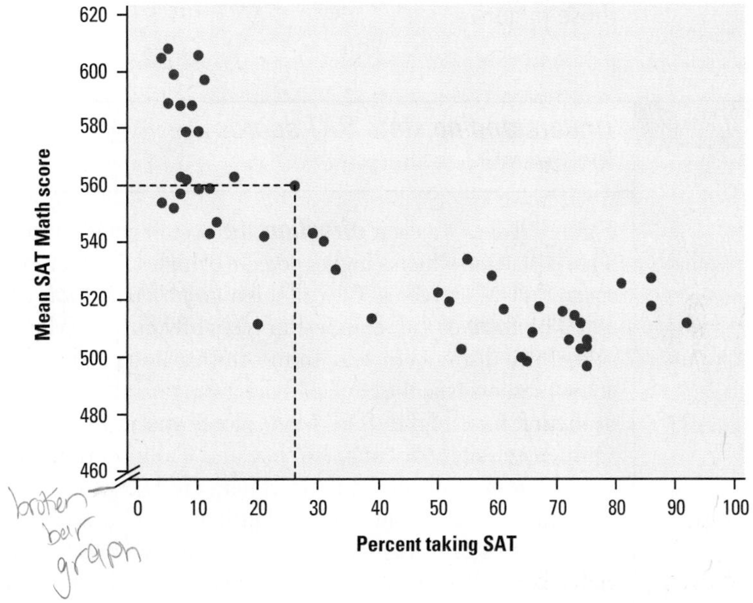

Scatterplot

A **scatterplot** shows the relationship between two quantitative variables measured on the same individuals. The values of one variable appear on the horizontal axis, and the values of the other variable appear on the vertical axis. Each individual in the data appears as the point in the plot fixed by the values of both variables for that individual.

Here are some tips for drawing scatterplots by hand:

1. Plot the explanatory variable, if there is one, on the horizontal axis (the x axis) of a scatterplot. As a reminder, we usually call the explanatory variable x and the response variable y. If there is no explanatory-response distinction, either variable can go on the horizontal axis.
2. Label both axes.
3. Scale the horizontal and vertical axes. The intervals must be uniform; that is, the distance between tick marks must be the same.

4. If you are given a grid, try to adopt a scale so that your plot uses the whole grid. Make your plot large enough so that the details can be easily seen. Don't compress the plot into one corner of the grid.

Interpreting Scatterplots

To interpret a scatterplot, look for patterns and any important deviations from those patterns.

Example 3.4 *Understanding state SAT scores*
Interpreting a scatterplot

Figure 3.1 shows a clear ***direction:*** the overall pattern moves from upper left to lower right. That is, states in which a higher percent of high school seniors take the SAT tend to have lower mean SAT Math scores. We call this a ***negative association*** between the two variables.

clusters

The ***form*** of the relationship is slightly curved. More important, most states fall into one of two distinct ***clusters.*** In the cluster at the right of the plot, more than half of high school seniors take the SAT and the average math scores are low. The states in the cluster at the left have higher SAT Math scores and no more than 30% of seniors take the test. Only three states (Arizona, Nevada, and California) lie in the gap between the clusters.

What explains the clusters? There are two widely used college entrance exams, the SAT and the American College Testing (ACT) exam. Each state favors one or the other. The left cluster in Figure 3.1 contains the ACT states, and the SAT states make up the right cluster. In ACT states, most students who take the SAT are applying to a selective college that requires SAT scores. This select group of students has a higher mean score than the much larger group of students who take the SAT in SAT states.

The ***strength*** of a relationship in a scatterplot is determined by how closely the points follow a clear form. The overall relationship in Figure 3.1 is only moderately strong—states with similar percents taking the SAT show quite a bit of scatter in their mean scores.

Do we see any deviations from the pattern? West Virginia, where 20% of high school seniors take the SAT, but the mean SAT Math score is only 511, stands out. This point can be described as an ***outlier.***

Interpreting a Scatterplot

In any graph of data, look for the **overall pattern** and for **striking deviations** from that pattern.

You can describe the overall pattern of a scatterplot by the **direction, form,** and **strength** of the relationship.

An important kind of deviation is an **outlier**, an individual value that falls outside the overall pattern of the relationship.

Here is an example of a stronger relationship with a clearer form.

Example 3.5

Beer and blood alcohol

A linear relationship

How well does the number of beers a student drinks predict his or her blood alcohol content (BAC)? Sixteen student volunteers at The Ohio State University drank a randomly assigned number of cans of beer. Thirty minutes later, a police officer measured their BAC. Here are the data:[2]

Student:	1	2	3	4	5	6	7	8
Beers:	5	2	9	8	3	7	3	5
BAC:	0.10	0.03	0.19	0.12	0.04	0.095	0.07	0.06
Student:	9	10	11	12	13	14	15	16
Beers:	3	5	4	6	5	7	1	4
BAC:	0.02	0.05	0.07	0.10	0.085	0.09	0.01	0.05

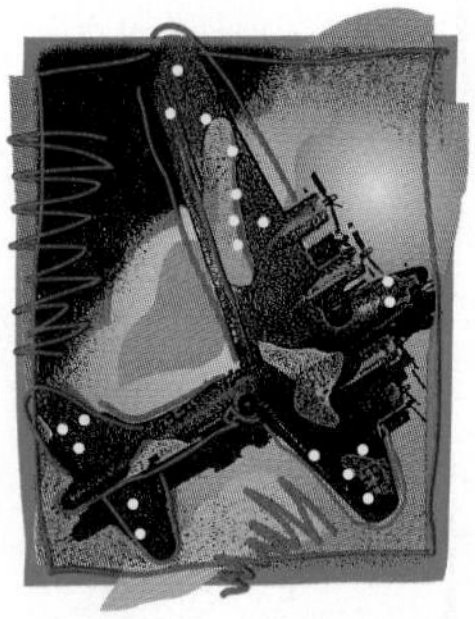

After you plot your data, think! The statistician Abraham Wald (1902–1950) worked on war problems during World War II. Wald invented some statistical methods that were military secrets until the war ended. Here is one of his simpler ideas. Asked where extra armor should be added to airplanes, Wald studied the location of enemy bullet holes in planes returning from combat. He plotted the locations on an outline of the plane. As data accumulated, most of the outline filled up. Put the armor in the few spots with no bullet holes, said Wald. That's where bullets hit the planes that didn't make it back.

The students were equally divided between men and women and differed in weight and usual drinking habits. Because of this variation, many students don't believe that number of drinks predicts BAC well. What do the data say?

The scatterplot in Figure 3.2 shows a fairly strong ***positive association.*** Generally, more beers consumed resulted in higher BAC. The form of the relationship is linear. That is, the points lie in a straight-line pattern. It is a fairly strong relationship because the points fall pretty close to a line, with relatively little scatter. If we know how many beers a student has consumed, we can predict BAC quite accurately from the scatterplot.

Figure 3.2 *Scatterplot of students' blood alcohol content against the number of cans of beer consumed.*

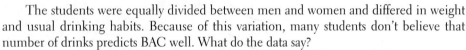

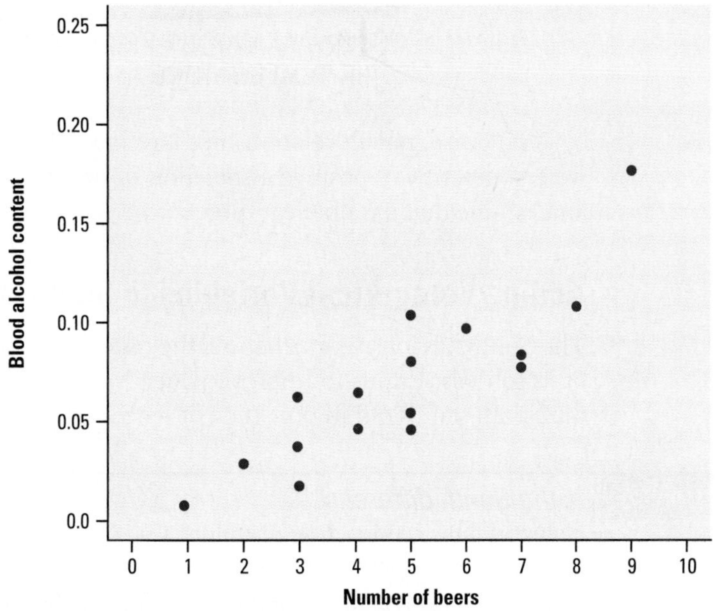

Positive Association, Negative Association

Two variables are **positively associated** when above-average values of one tend to accompany above-average values of the other, and below-average values also tend to occur together.

Two variables are **negatively associated** when above-average values of one tend to accompany below-average values of the other, and vice versa.

"SORRY, MA'AM, OUR PRICES GO UP WITH THE TEMPERATURE."

Of course, not all relationships have a simple form and a clear direction that we can describe as positive association or negative association. Exercise 3.9 gives an example that has no clear direction.

Adding Categorical Variables to Scatterplots

The South has long lagged behind the rest of the United States in the performance of its schools. Efforts to improve education have reduced the gap. We wonder if the South stands out in our study of state average SAT math scores.

Example 3.6 *Is the South different?*
Categorical variables in scatterplots

Figure 3.3 enhances the scatterplot in Figure 3.1 by plotting 12 southern states in blue. Most of the southern states blend in with the rest of the country. Several southern states do lie at the lower edges of their clusters. Florida, Georgia, South Carolina, and West Virginia have lower SAT Math scores than we would expect from the percent of their high school seniors who take the examination.

Figure 3.3 *Mean SAT Math scores and percent of high school seniors who take the test, by state, with the southern states highlighted.*

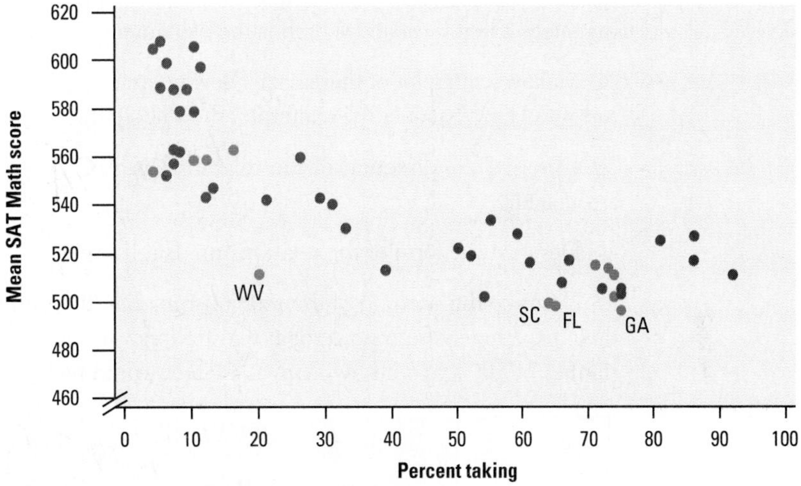

In Example 3.6, dividing the states into "southern" and "nonsouthern" introduces a third variable into the scatterplot. This is a categorical variable that has only two values. The two values are displayed by the two different plotting colors.

Categorical Variables in Scatterplots

To add a categorical variable to a scatterplot, use a different plotting color or symbol for each category.

Exercises

3.5 The endangered manatee, I Manatees are large, gentle sea creatures that live along the Florida coast. Many manatees are killed or injured by powerboats. Here are data on powerboat registrations (in thousands) and the number of manatees killed by boats in Florida in the years 1977 to 1990:

Year	Powerboat registrations (1000s)	Manatees killed	Year	Powerboat registrations (1000s)	Manatees killed
1977	447	13	1984	559	34
1978	460	21	1985	585	33
1979	481	24	1986	614	33

Year	Powerboat registrations (1000s)	Manatees killed	Year	Powerboat registrations (1000s)	Manatees killed
1980	498	16	1987	645	39
1981	513	24	1988	675	43
1982	512	20	1989	711	50
1983	526	15	1990	719	47

(a) We want to examine the relationship between number of powerboats and number of manatees killed by boats. Which is the explanatory variable?

(b) Make a scatterplot of these data. (Be sure to label the axes with the variable names, not just *x* and *y*.) What does the scatterplot show about the relationship between these variables?

(c) Describe the direction of the relationship. Are the variables positively or negatively associated?

(d) Describe the form of the relationship. Is it linear?

(e) Describe the strength of the relationship. Can the number of manatees killed be predicted accurately from powerboat registrations? If powerboat registrations remained constant at 719,000, about how many manatees would be killed by boats each year?

3.6 Bird colonies One of nature's patterns connects the percent of adult birds in a colony that return from the previous year and the number of new adults that join the colony. Here are data for 13 colonies of sparrowhawks:[3]

Percent returning	New adults	Percent returning	New adults	Percent returning	New adults
74	5	62	15	46	18
66	6	52	16	60	19
81	8	45	17	46	20
52	11	62	18	38	20
73	12				

(a) Plot the count of new birds (response) against the percent of returning birds (explanatory).

(b) Describe the direction, form, and strength of the relationship between number of new sparrowhawks in a colony and percent of returning adults.

(c) For short-lived birds, the association between these variables is positive: changes in weather and food supply drive the populations of new and returning birds up or down together. For long-lived territorial birds, on the other hand, the association is negative because returning birds claim their territories in the colony and don't leave room for new recruits. Which type of species is the sparrowhawk?

3.7 IQ and school grades Do students with higher IQ test scores tend to do better in school? Figure 3.4 is a scatterplot of IQ and school grade point average (GPA) for all 78 seventh-grade students in a rural midwestern school.[4]

Figure 3.4 *Scatterplot of school grade point average versus IQ test score for seventh-grade students.*

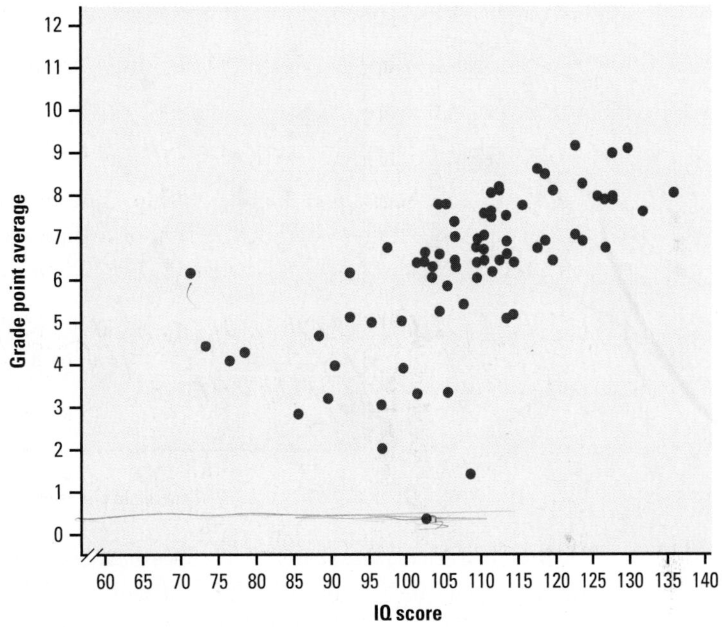

(a) Say in words what a positive association between IQ and GPA would mean. Does the plot show a positive association?

(b) What is the form of the relationship? Is it roughly linear? Is it very strong? Explain your answers.

(c) At the bottom of the plot are several points that we might call outliers. One student in particular has a very low GPA despite an average IQ score. What are the approximate IQ and GPA for this student?

3.8 Stocks versus T-bills What is the relationship between returns from buying Treasury bills and returns from` buying common stocks? To buy a Treasury bill is to make a short-term loan to the U.S. government. This is much less risky than buying stock in a company, so (on average) the returns on Treasury bills are lower than the returns on stocks. Figure 3.5 (on the next page) plots the annual returns on stocks for the years 1950 to 2003 against the returns on Treasury bills for the same years.

(a) The best year for stocks during this period was 1954. The worst year was 1974. About what were the returns on stocks in those two years?

(b) Treasury bills are a measure of the general level of interest rates. The years around 1980 saw very high interest rates. Treasury bill returns peaked in 1981. About what was the percent return that year?

(c) Some people say that high Treasury bill returns tend to go with low returns on stocks. Does such a pattern appear clearly in Figure 3.5? Does the plot have any clear pattern?

Figure 3.5 *Percent returns on common stocks and Treasury bills for the years 1950 to 2003.*

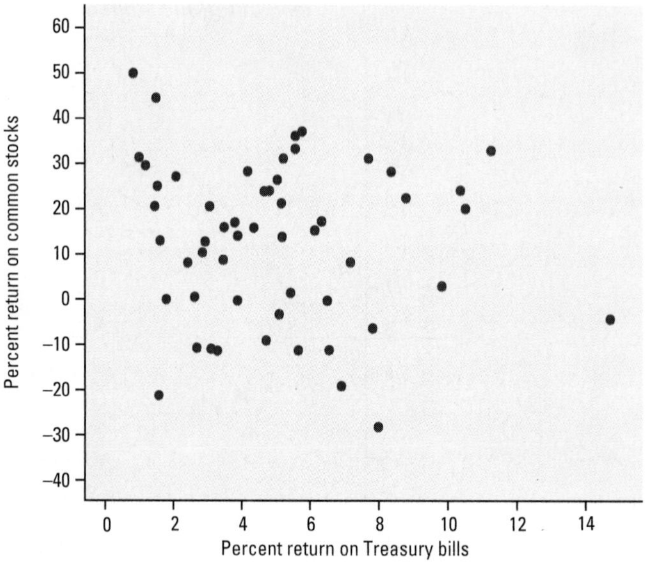

3.9 Does fast driving waste fuel? How does the fuel consumption of a car change as its speed increases? Here are data for a British Ford Escort. Speed is measured in kilometers per hour, and fuel consumption is measured in liters of gasoline used per 100 kilometers traveled.[5]

Speed (km/h)	Fuel used (liters/100 km)	Speed (km/h)	Fuel used (liters/100 km)
10	21.00	90	7.57
20	13.00	100	8.27
30	10.00	110	9.03
40	8.00	120	9.87
50	7.00	130	10.79
60	5.90	140	11.77
70	6.30	150	12.83
80	6.95		

(a) Make a scatterplot. (Which is the explanatory variable?)

(b) Describe the form of the relationship. Why is it not linear? Explain why the form of the relationship makes sense.

(c) It does not make sense to describe the variables as either positively associated or negatively associated. Why?

(d) Is the relationship reasonably strong or quite weak? Explain your answer.

3.10 Do heavier people burn more energy? Metabolic rate, the rate at which the body consumes energy, is important in studies of weight gain, dieting, and exercise. The table below gives data on the lean body mass and resting metabolic rate for 12 women and 7 men who are subjects in a study of dieting. Lean body mass, given in kilograms, is a person's weight leaving out all fat. Metabolic rate is measured in calories burned per 24 hours, the same calories used to describe the energy content of foods. The researchers believe that lean body mass is an important influence on metabolic rate.

Subject	Sex	Mass (kg)	Rate (cal)	Subject	Sex	Mass (kg)	Rate (cal)
1	M	62.0	1792	11	F	40.3	1189
2	M	62.9	1666	12	F	33.1	913
3	F	36.1	995	13	M	51.9	1460
4	F	54.6	1425	14	F	42.4	1124
5	F	48.5	1396	15	F	34.5	1052
6	F	42.0	1418	16	F	51.1	1347
7	M	47.4	1362	17	F	41.2	1204
8	F	50.6	1502	18	M	51.9	1867
9	F	42.0	1256	19	M	46.9	1439
10	M	48.7	1614				

(a) Make a scatterplot of the data for the female subjects. Which is the explanatory variable?

(b) Is the association between these variables positive or negative? What is the form of the relationship? How strong is the relationship?

(c) Now add the data for the male subjects to your graph, using a different color or a different plotting symbol. Does the pattern of relationship that you observed in (b) hold for men also? How do the male subjects as a group differ from the female subjects as a group?

Technology Toolbox

Making a calculator scatterplot

We will use the beer consumption and blood alcohol content data from Example 3.5 (page 177) to show how to construct a scatterplot on a TI-83/84/89.

- Enter the beer consumption values in L_1/list1 and the BAC values in L_2/list2.

TI-83/84

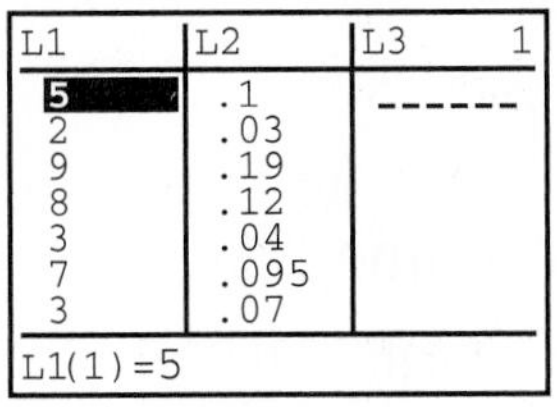

TI-89

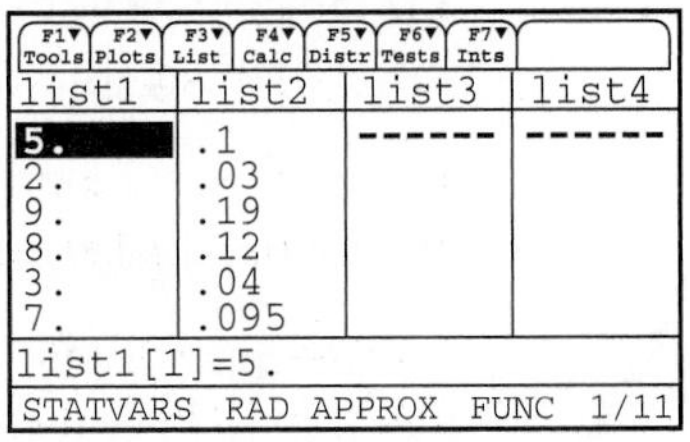

(continued)

Technology Toolbox

Making a calculator scatterplot *(continued)*

- Next, define a scatterplot in the statistics plot menu (press on the TI-89). Specify the settings shown.

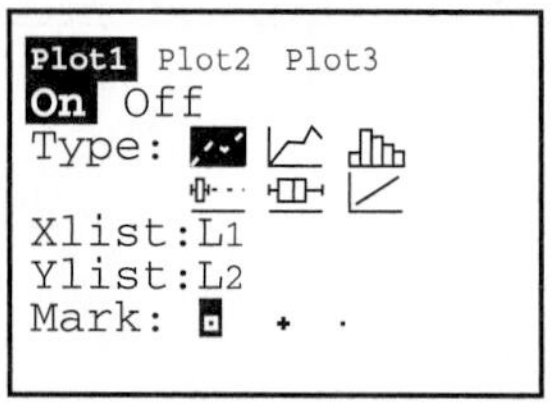

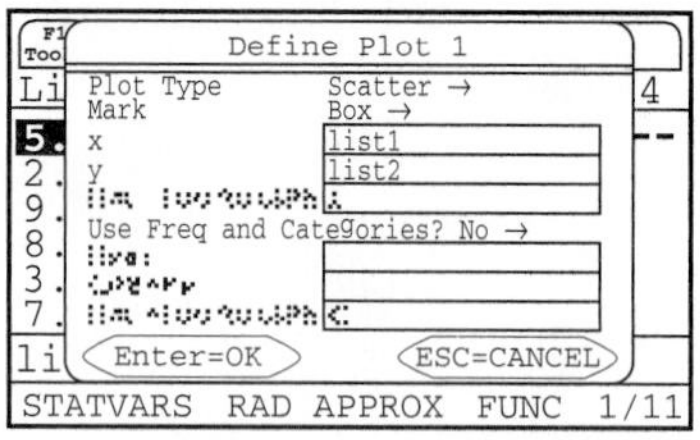

- Use ZoomStat (ZoomData on the TI-89) to obtain a graph. The calculator will set the window dimensions automatically by looking at the values in L_1/list1 and L_2/list2.

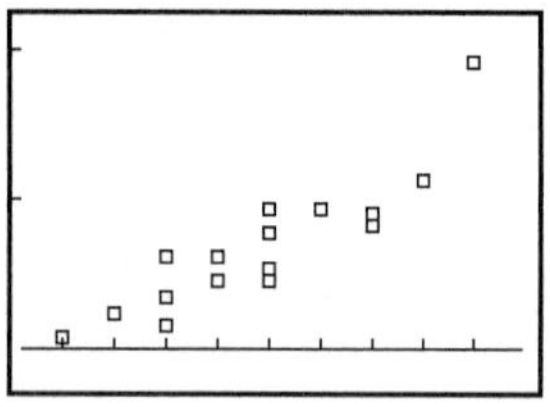

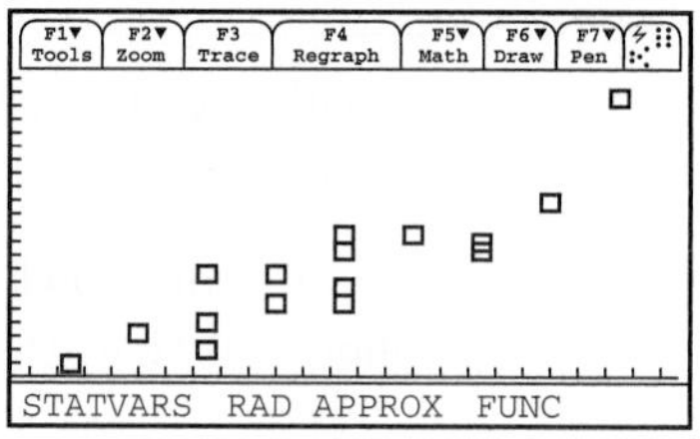

- Notice that there are no scales on the axes, and that the axes are not labeled. If you copy a scatterplot from your calculator onto your paper, make sure that you scale and label the axes. You can use TRACE to help you get started.

3.11 Scatterplot by calculator, I Rework Exercise 3.9 using your calculator. The command `seq(10X,X,1,15)`→ L_1 (→list1 on the TI-89) will assign the numbers 10, 20, . . . , 150 to L_1/list1. (Note that seq is found under 2nd / LIST / OPS on the TI-83/84 and under CATALOG on the TI-89.) Then enter the fuel data in L_2/list2. Define Plot 1 to be a scatterplot, and then use ZoomStat (ZoomData on the TI-89) to graph it. Verify your answers to Exercise 3.9.

3.12 Scatterplot by calculator, II Rework Exercise 3.10 using your calculator. Verify your answers to Exercise 3.10.

Measuring Linear Association: Correlation

CAUTION

A scatterplot displays the direction, form, and strength of the relationship between two quantitative variables. Linear relations are particularly important because a straight line is a simple pattern that is quite common. We say a linear relation is strong if the points lie close to a straight line, and weak if they are widely scattered about a line. **Our eyes are not good judges of how strong a linear relationship is.** The two scatterplots in Figure 3.6 depict exactly the same data, but the lower plot is drawn smaller in a large field. The lower plot seems to show a stronger linear relationship.

Figure 3.6 *Two scatterplots of the same data; the straight-line pattern in the lower plot appears stronger because of the surrounding empty space.*

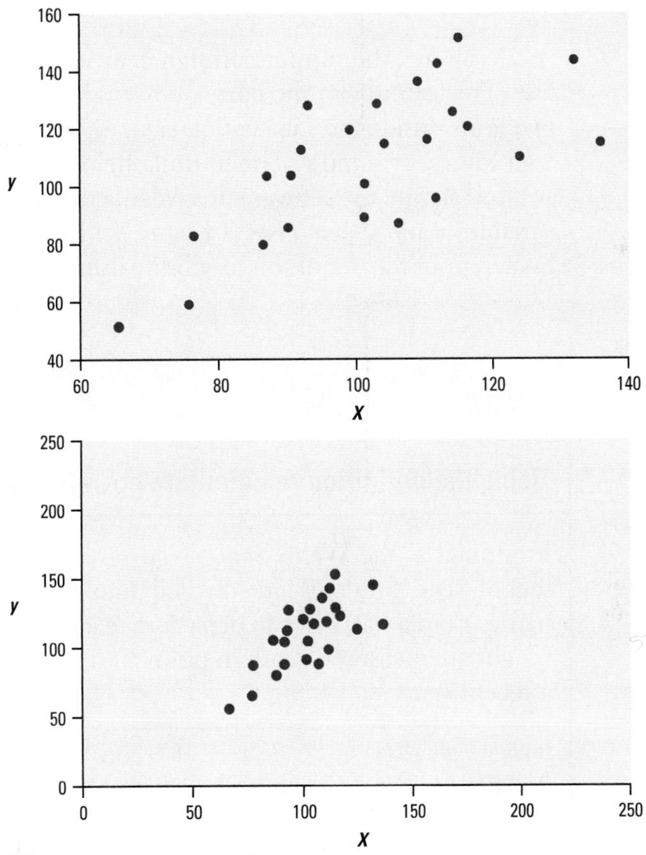

Our eyes can be fooled by changing the plotting scales or the amount of empty space around the cloud of points in a scatterplot.[6] We need to follow our strategy for data analysis by using a numerical measure to supplement the graph. ***Correlation*** is the measure we use.

Correlation r

The **correlation** measures the direction and strength of the linear relationship between two quantitative variables. Correlation is usually written as r.

Suppose that we have data on variables x and y for n individuals. The values for the first individual are x_1 and y_1, the values for the second individual are x_2 and y_2, and so on. The means and standard deviations of the two variables are $\bar{x}$ and s_x for the x-values, and $\bar{y}$ and s_y for the y-values. The correlation r between x and y is

$$r = \frac{1}{n-1} \sum \left(\frac{x_i - \bar{x}}{s_x}\right)\left(\frac{y_i - \bar{y}}{s_y}\right)$$

As always, the summation sign Σ means "add these terms for all the individuals." The formula for the correlation r is a bit complex. Notice, however, that the two terms inside the summation sign, $(x - \bar{x})/s_x$ and $(y - \bar{y})/s_y$, are just standardized values for x and y. The formula helps us see what correlation is, but in practice you should use software or a calculator that finds r from keyed-in values of two variables x and y. Exercises 3.13 and 3.14 ask you to calculate a correlation step-by-step from the definition to solidify its meaning.

Technology Toolbox

Using the definition to calculate correlation

Freshmen at the Webb Schools go on a backpacking trip at the start of each school year. Students are divided into hiking groups of size 8 by selecting names from a hat. Prior to departure, each student's body weight and backpack weight are measured (both in pounds). Here are data from one hiking group in a recent year:

Body weight (lb):	120	187	109	103	131	165	158	116
Backpack weight (lb):	26	30	26	24	29	35	31	28

We will use these data to show how to calculate the correlation using the definition and the list features of the TI-83/84/89.

- Begin by entering the body weights (x-values) in L_1/list1 and the backpack weights (y-values) in L_2/list2. Then calculate two-variable statistics for the x- and y-values. The calculator will remember all of the computed statistics until the next time you calculate one- or two-variable statistics.

TI-83/84

- Press STAT, choose `CALC`, then `2:2-Var Stats`.

TI-89

- In the Statistics/List Editor, press F4 and choose `2:2-Var Stats`.

- Complete the command `2-Var Stats L1, L2` and press **ENTER**.

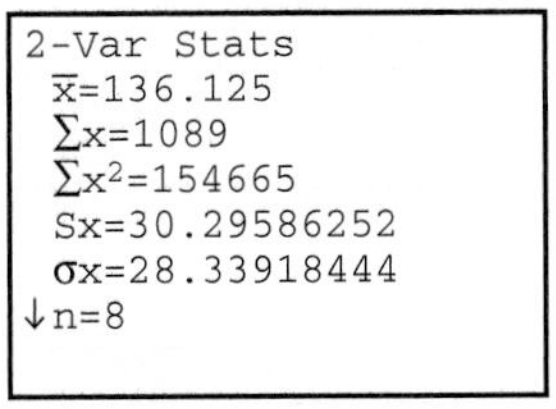

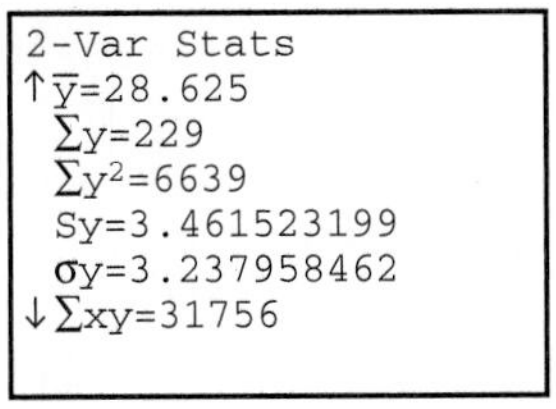

- In the new window, enter list1 as the Xlist and list2 as the Ylist, then press **ENTER**.

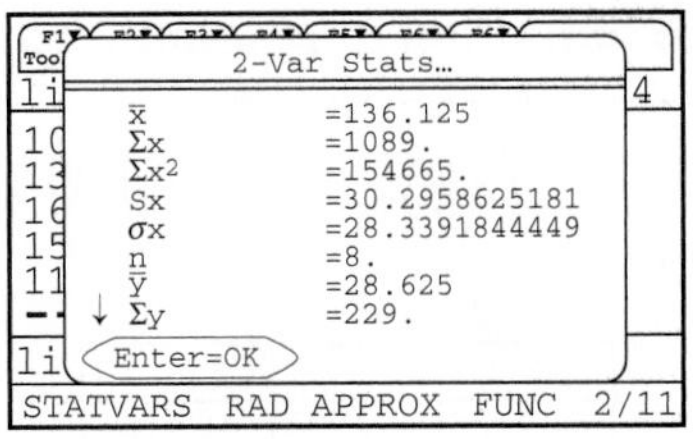

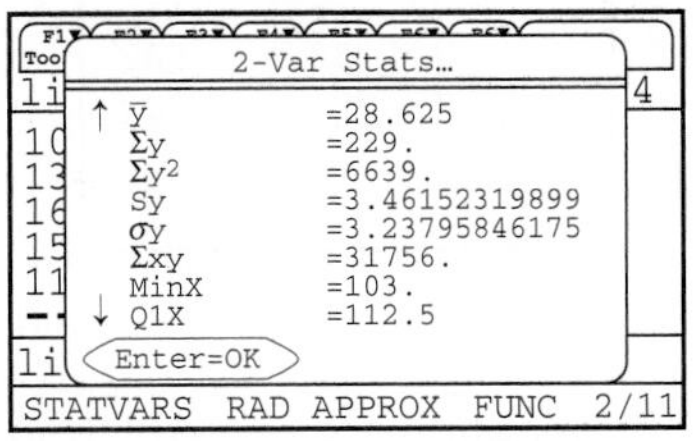

- Define $L_3 = ((L_1 - \bar{x})/s_x)$ and $L_4 = ((L_2 - \bar{y})/s_y)$

- Define list3 $= ((\text{list1} - \bar{x})/s_x)$ and list4 $= ((\text{list2} - \bar{y})/s_y)$

from the home screen as shown. Note that $\bar{x}$, $\bar{y}$, s_x, and s_y can be found under VARS/5:Statistics (in the VAR-LINK menu on the TI-89).

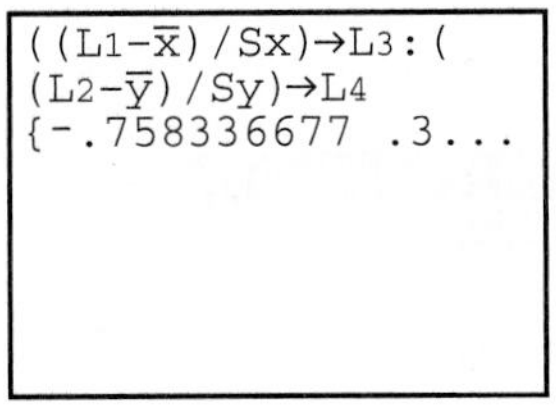

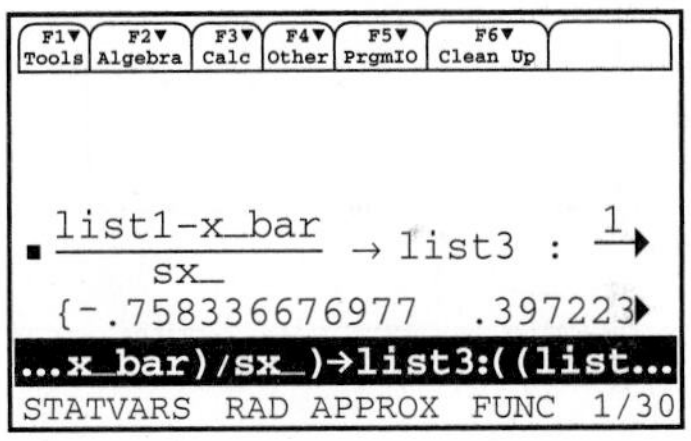

- Go into the Statistics/List Editor to look at the results. The first student listed has a body weight of 120 lb. and a backpack weight of 26 lb. In L_3, we see that his standardized body weight is $z = \dfrac{120 - 136.125}{30.296} = -0.53$. In other words, his weight is 0.53 standard deviations below the mean body weight for this group of 8 hikers. In L_4, we see that the z-score for his pack weight is $z = \dfrac{26 - 28.625}{3.462} = -0.76$. So his pack weight is 0.76 standard deviations below the mean backpack weight for the group.

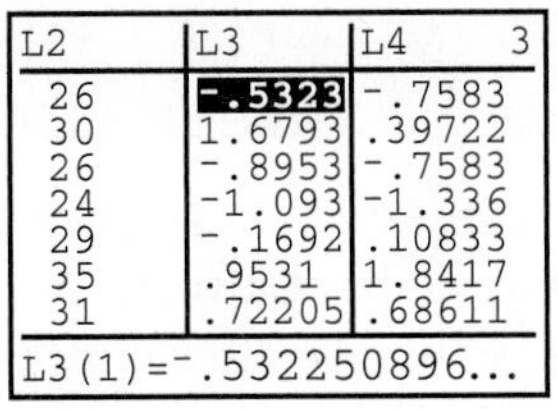

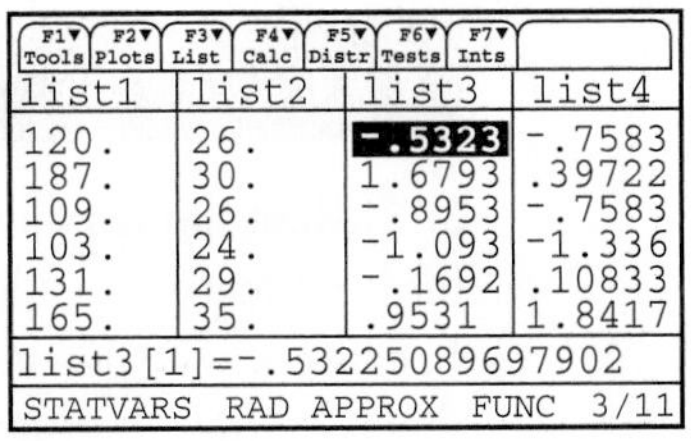

(continued)

Technology Toolbox

Using the definition to calculate correlation *(continued)*

- Define $L_5 = L_3{*}L_4$ (list5 = list3*list 4 on the TI-89). Notice that most of the values in L_5/list5 are positive.

L3	L4	L5
-.5323	-.7583	.40363
1.6793	.39722	.66705
-.8953	-.7583	.67897
-1.093	-1.336	1.4609
-.1692	.10833	-.0183
.9531	1.8417	1.7553
.72205	.68611	.49541

L5=L3*L4

list2	list3	list4	list5
26.	-.5323	-.7583	.40363
30.	1.6793	.39722	.66705
26.	-.8953	-.7583	.67897
24.	-1.093	-1.336	1.4609
29.	-.1692	.10833	-.0183
35.	.9531	1.8417	1.7553

list5=list3*list4

STATVARS RAD APPROX FUNC 5/11

- To finish calculating the correlation $r = \frac{1}{n-1}\sum\left(\frac{x-\bar{x}}{s_x}\right)\left(\frac{y-\bar{y}}{s_y}\right)$, we just need to add up the values in L_5/list5 and then to divide by 7. To do this, enter the command shown in the appropriate calculator screen. Press **ENTER** to see the correlation.

```
(1/(8-1))*sum(L5
)
            .7946926677
```

```
■ 1/(8-1) · sum(list5)
                  .794692667734
(1/(8-1))*sum(list5)
STATVARS RAD APPROX FUNC 1/30
```

The formula for *r* begins by standardizing the observations. In the Technology Toolbox, we began with body weight (x) and backpack weight (y) in pounds for each of the 8 hikers. Then we standardized the body weights and backpack weights and multiplied the resulting z-scores for each hiker. Standardized values have no units—in this example, they are no longer measured in pounds. The correlation r is an average of the products of the standardized body weight and backpack weight for the 8 hikers.

Exercises

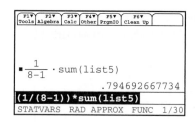

3.13 Meet the *Archaeopteryx* *Archaeopteryx* is an extinct beast having feathers like a bird but teeth and a long bony tail like a reptile. Only six fossil specimens are known. Because these specimens differ greatly in size, some scientists think they are different species rather than individuals from the same species. We will examine some data. If the specimens belong to the same species and differ in size because some are younger than others, there should be a positive linear relationship between the lengths of a pair of bones from all individuals. An outlier from this relationship would suggest a different species. Here are data on the lengths in centimeters of the femur (a leg bone) and the humerus (a bone in the upper arm) for the five specimens that preserve both bones:[7]

Femur:	38	56	59	64	74
Humerus:	41	63	70	72	84

(a) Make a scatterplot. Do you think that all five specimens come from the same species? Explain.

(b) Find the correlation r step-by-step. That is, find the mean and standard deviation of the femur lengths and of the humerus lengths. Then find the five standardized values for each variable and use the formula for r.

(c) Duplicate the steps in the Technology Toolbox to obtain the correlation for the *Archaeopteryx* data, and compare your result with that calculated by hand in (b).

3.14 Coffee and deforestation Coffee is a leading export from several developing countries. When coffee prices are high, farmers often clear forest to plant more coffee trees. Here are data on prices paid to coffee growers in Indonesia and the rate of deforestation in a national park that lies in a coffee-producing region for five years:[8]

Price (cents per pound)	Deforestation (percent)
29	0.49
40	1.59
54	1.69
55	1.82
72	3.10

(a) Make a scatterplot. Which is the explanatory variable? What kind of pattern does your plot show?

(b) Find the correlation r step-by-step. First find the mean and standard deviation of each variable. Then find the five standardized values for each variable. Finally, use the formula for r.

(c) Duplicate the steps in the Technology Toolbox to obtain the correlation for these data, and compare your result with that calculated by hand in (b).

Facts about Correlation

The formula for correlation helps us see that r is positive when there is a positive association between the variables. Height and weight, for example, have a positive association. People who are above average in height tend to also be above average in weight. Both the standardized height and the standardized weight are positive. People who are below average in height tend to also have below-average weight. Then both standardized height and standardized weight are negative. In both cases, the products in the formula for r are mostly positive and so r is positive. In the same way, we can see that r is negative when the association between x and y is negative. The following Activity lets you explore some important properties of the correlation.

Activity 3B

Correlation and Regression applet

Go to the book's Web site, www.whfreeman.com/tps3e and launch the *Correlation and Regression* applet.

1. You are going to use the *Correlation and Regression* applet to make scatterplots with 10 points that have correlation close to 0.7.

   (a) Stop after adding the first 2 points. What is the value of the correlation? Why does it have this value?

   (b) Continue making a lower-left to upper-right pattern of 10 points with correlation about $r = 0.7$. (You can drag points up or down to adjust r after you have 10 points.) Make a rough sketch of your scatterplot.

   (c) Make another scatterplot with 9 points in a vertical stack at the left of the plot. Add 1 point far to the right and move it until the correlation is close to 0.7. Make a rough sketch of your scatterplot.

   (d) Make yet another scatterplot with 10 points in a curved pattern that starts at the lower left, rises to the right, then falls again at the far right. Adjust the points up or down until you have a very smooth curve with correlation close to 0.7. Make a rough sketch of this scatterplot also.

Lesson: Many patterns can have the same correlation. Always plot your data before you trust a correlation.

2. Correlation is not resistant. Click on the scatterplot to create a group of 10 points in the lower-left corner of the scatterplot with a strong straight-line pattern (correlation about 0.9).

   (a) Add 1 point at the upper right that is in line with the first 10. How does the correlation change?

(b) Drag this last point down. How small can you make the correlation? Can you make the correlation negative?

(c) What lesson did you learn from part (b) about the effect of a single point on the correlation?

More detailed study of the correlation formula gives more detailed properties of r. Here is what you need to know in order to interpret correlation.

1. *Correlation makes no distinction between explanatory and response variables.* It makes no difference which variable you call x and which you call y in calculating the correlation.
2. Because r uses the standardized values of the observations, *r does not change when we change the units of measurement of x, y, or both.* Measuring height in centimeters rather than inches and weight in kilograms rather than pounds does not change the correlation between height and weight. The correlation r itself has no unit of measurement; it is just a number.
3. *Positive r indicates positive association between the variables, and negative r indicates negative association.*
4. *The correlation r is always a number between -1 and 1.* Values of r near 0 indicate a very weak linear relationship. The strength of the linear relationship increases as r moves away from 0 toward either -1 or 1. Values of r close to -1 or 1 indicate that the points in a scatterplot lie close to a straight line. The extreme values $r = -1$ and $r = 1$ occur *only* in the case of a perfect linear relationship, when the points lie exactly along a straight line.

Example 3.7 *From scatterplot to correlation*

Comparing correlation values

The scatterplots in Figure 3.7 (on the next page) illustrate how values of r closer to 1 or -1 correspond to stronger linear relationships. To make the meaning of r clearer, the standard deviations of both variables in these plots are equal, and the horizontal and vertical scales are the same. In general, it is not so easy to guess the value of r from the appearance of a scatterplot. Remember that changing the plotting scales in a scatterplot may mislead our eyes, but it does not change the correlation.

The real data we have examined also illustrate how correlation measures the strength and direction of linear relationships. Figure 3.2 (page 177) shows a rather strong positive linear relationship between number of beers consumed and blood alcohol content for a group of college students. The correlation is $r = 0.894$. Figure 3.1 (page 175) shows a slightly weaker but still quite strong negative association between percent of students taking the SAT and the mean SAT math score in a state. The correlation is $r = -0.845$.

Figure 3.7 *How correlation measures the strength of a linear relationship. Patterns closer to a straight line have correlations closer to 1 or −1.*

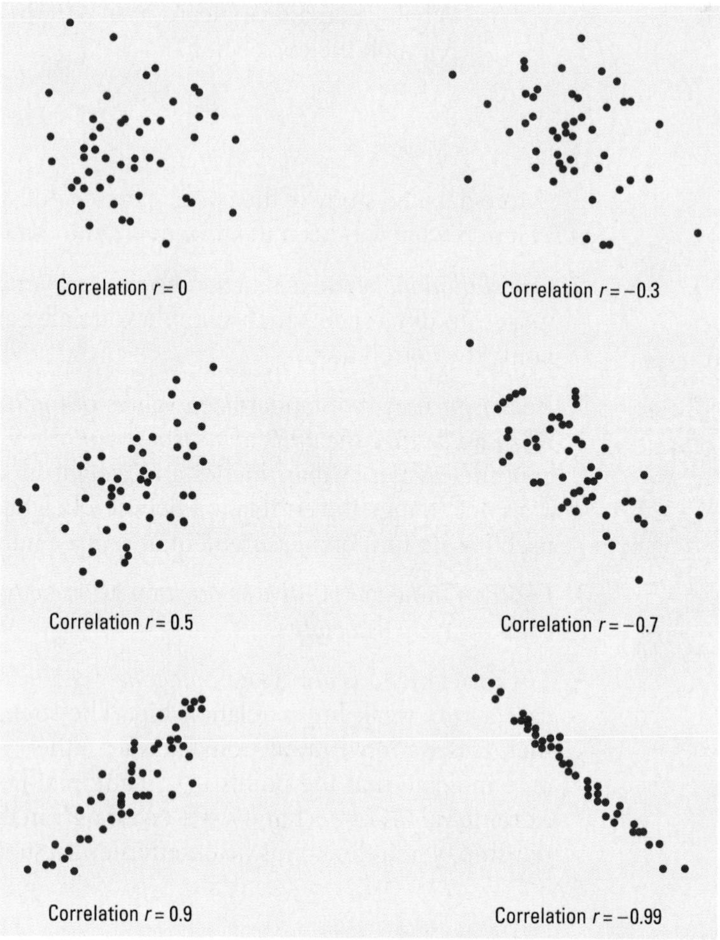

Describing the relationship between two variables is a more complex task than describing the distribution of one variable. Here are some more facts about correlation—cautions to keep in mind when you use r.

1. **Correlation requires that both variables be quantitative, so that it makes sense to do the arithmetic indicated by the formula for r.** We cannot calculate a correlation between the incomes of a group of people and what city they live in, because city is a categorical variable.

2. Correlation measures the strength of only the linear relationship between two variables. **Correlation does not describe curved relationships between variables, no matter how strong they are.** Exercise 3.20 illustrates this important fact.

3. **Like the mean and standard deviation, the correlation is not resistant: r is strongly affected by a few outlying observations.** Use r with caution when outliers appear in the scatterplot.

4. **Correlation is not a complete summary of two-variable data,** even when the relationship between the variables is linear. You should give the means and standard deviations of both x and y along with the correlation.

Because the formula for correlation uses the means and standard deviations, these measures are the proper choice to accompany a correlation. Here is an example in which understanding requires both means and correlation.

Example 3.8 *Scoring figure skaters*

Correlation doesn't tell the whole story

Until a scandal at the 2002 Olympics brought change, figure skating was scored by judges on a scale from 0.0 to 6.0. The scores were often controversial. We have the scores awarded by two judges, Pierre and Elena, for many skaters. How well do they agree? We calculate that the correlation between their scores is $r = 0.9$. But the mean of Pierre's scores is 0.8 point lower than Elena's mean.

These facts do not contradict each other. They are simply different kinds of information. The mean scores show that Pierre awards lower scores than Elena. But because Pierre gives *every* skater a score about 0.8 point lower than Elena, the correlation remains high. Adding the same number to all values of either x or y does not change the correlation. If both judges score the same skaters, the competition is scored consistently because Pierre and Elena agree on which performances are better than others. The high r shows their agreement. But if Pierre scores some skaters and Elena others, we must add 0.8 point to Pierre's scores to arrive at a fair comparison.

Of course, even giving means, standard deviations, and the correlation for "state SAT math scores" and "percent taking" will not point out the clusters in Figure 3.1 (page 175). Numerical summaries complement plots of data, but they do not replace them.

Exercises

3.15 Calories and salt in hot dogs Are hot dogs that are high in calories also high in salt? Figure 3.8 (on the next page) is a scatterplot of the calories and salt content (measured as milligrams of sodium) in 17 brands of meat hot dogs.[9]

(a) Roughly what are the lowest and highest calorie counts among these brands? Roughly what is the sodium level in the brands with the fewest and with the most calories?

(b) Does the scatterplot show a clear positive or negative association? Say in words what this association means about calories and salt in hot dogs.

(c) Are there any outliers? Is the relationship (ignoring any outliers) roughly linear in form? Still ignoring outliers, how strong would you say the relationship between calories and sodium is?

Figure 3.8 *Scatterplot of milligrams of sodium and calories in each of 17 brands of meat hot dogs, for Exercise 3.15.*

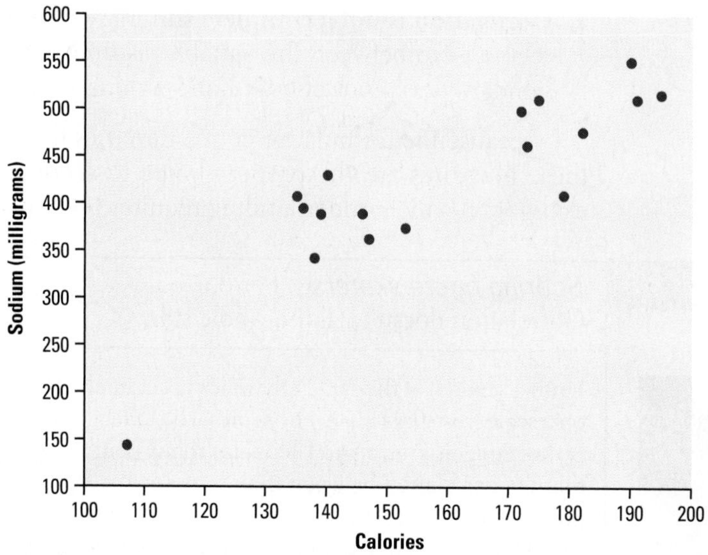

3.16 Thinking about correlation Figure 3.4 (page 181) is a scatterplot of school grade point average versus IQ score for 78 seventh-grade students.

(a) Is the correlation r for these data near -1, clearly negative but not near -1, near 0, clearly positive but not near 1, or near 1? Explain your answer.

(b) Refer to Figure 3.8 from the previous exercise. Is the correlation here closer to 1 than that for Figure 3.4, or closer to 0? Explain your answer.

(c) Both Figures 3.4 and 3.8 contain outliers. Removing the outliers will increase the correlation r in one figure and decrease r in the other figure. What happens in each figure, and why?

3.17 The effect of changing units Changing the units of measurement can dramatically alter the appearance of a scatterplot. Consider the following data:

x	-4	-4	-3	3	4	4
y	0.5	-0.6	-0.5	0.5	0.5	-0.6

(a) Enter the data into L_1/list1 and L_2/list2. Then use Plot 1 to define and plot the scatterplot. Use the box (□) as your plotting symbol.

(b) Use L_3/list3 and the technique described in the Technology Toolbox on page 186 to calculate the correlation.

(c) Define new variables $x^* = x/10$ and $y^* = 10y$, and enter these into L_4/list4 and L_5/list5 as follows: list4 = list1/10 and list5 = 10 × list2. Define Plot 2 to be a scatterplot with Xlist: list4 and Ylist: list5, and Mark: +. Plot both scatterplots at the same time, and on the same axes, using ZoomStat/ZoomData. The two plots are very different in appearance.

(d) Use L_6/list6 and the technique described in the Technology Toolbox to calculate the correlation between x^* and y^*. How are the two correlations (between x and y and between x^* and y^*) related? Explain why this isn't surprising.

3.18 Changing the correlation

(a) Use your calculator to find the correlation between the percent of returning birds and the number of new birds from the data in Exercise 3.6 (page 180).

(b) Make a scatterplot of the data with two new points added. Point A: 10% return, 25 new birds. Point B: 40% return, 5 new birds. Find two new correlations: for the original data plus Point A and for the original data plus Point B.

(c) Explain in terms of what correlation measures why adding Point A makes the correlation stronger (closer to -1) and adding Point B makes the correlation weaker (closer to 0).

3.19 Wives and husbands If women always married men who were 2 years older than themselves, what would be the correlation between the ages of husband and wife? (*Hint*: Draw a scatterplot for several ages.)

3.20 Strong association but no correlation The gas mileage of an automobile first increases and then decreases as the speed increases. Suppose that this relationship is very regular, as shown by the following data on speed (miles per hour) and mileage (miles per gallon):

Speed:	20	30	40	50	60
Mileage:	24	28	30	28	24

(a) Make a scatterplot of mileage versus speed.

(b) Show that the correlation between speed and mileage is $r = 0$. Explain why the correlation is 0 even though there is a strong relationship between speed and mileage.

Section 3.1 Summary

To study relationships between variables, we must measure the variables on the same group of individuals.

If we think that a variable x may explain or even cause changes in another variable y, we call x an **explanatory variable** and y a **response variable.**

A **scatterplot** displays the relationship between two quantitative variables measured on the same individuals. Mark values of one variable on the horizontal axis (x axis) and values of the other variable on the vertical axis (y axis). Plot each individual's data as a point on the graph.

- Always plot the explanatory variable, if there is one, on the x axis of a scatterplot. Plot the response variable on the y axis.
- Plot points with different colors or symbols to see the effect of a categorical variable in a scatterplot.

In examining a scatterplot, look for an overall pattern showing the **direction, form,** and **strength** of the relationship, and then for **outliers** or other deviations from this pattern.

- *Direction:* If the relationship has a clear direction, we speak of either **positive association** (high values of the two variables tend to occur together) or **negative association** (high values of one variable tend to occur with low values of the other variable).
- *Form:* Linear relationships, where the points show a straight-line pattern, are an important form of relationship between two variables. Be sure to look for curved relationships and clusters.
- *Strength:* The strength of a relationship is determined by how close the points in the scatterplot lie to a simple form such as a line.

The **correlation** r measures the strength and direction of the linear association between two quantitative variables x and y. Although you can calculate a correlation for any scatterplot, r measures only straight-line relationships.

Correlation indicates the direction of a linear relationship by its sign: $r > 0$ for a positive association and $r < 0$ for a negative association.

Correlation always satisfies $-1 \leq r \leq 1$ and indicates the strength of a relationship by how close it is to -1 or 1. Perfect correlation, $r = \pm 1$, occurs only when the points on a scatterplot lie exactly on a straight line.

Correlation ignores the distinction between explanatory and response variables. The value of r is not affected by changes in the unit of measurement of either variable. Correlation is not resistant, so outliers can greatly change the value of r.

Section 3.1 Exercises

3.21 Rich states, poor states One measure of a state's prosperity is the median income of its households. Another measure is the mean personal income per person in the state. Figure 3.9 (on the next page) is a scatterplot of these two variables, both measured in thousands of dollars. Because both variables have the same units, the plot uses equally spaced scales on both axes.[10]

(a) We have labeled the point for New York on the scatterplot. What are the approximate values of New York's median household income and mean income per person?

(b) Explain why you expect a positive association between these variables. Also explain why you expect household income to be generally higher than income per person.

(c) Nonetheless, the mean income per person in a state can be higher than the median household income. In fact, the District of Columbia has a median income of \$30,748 per household and mean income of \$33,435 per person. Explain why this can happen.

(d) Alaska is the state with the highest median household income. What is the approximate median household income in Alaska? We might call Alaska and the District of Columbia outliers in the scatterplot.

(e) Describe the direction, form, and strength of the relationship, ignoring the outliers.

Figure 3.9 *Scatterplot of mean income per person versus median household income for the states, for Exercise 3.21.*

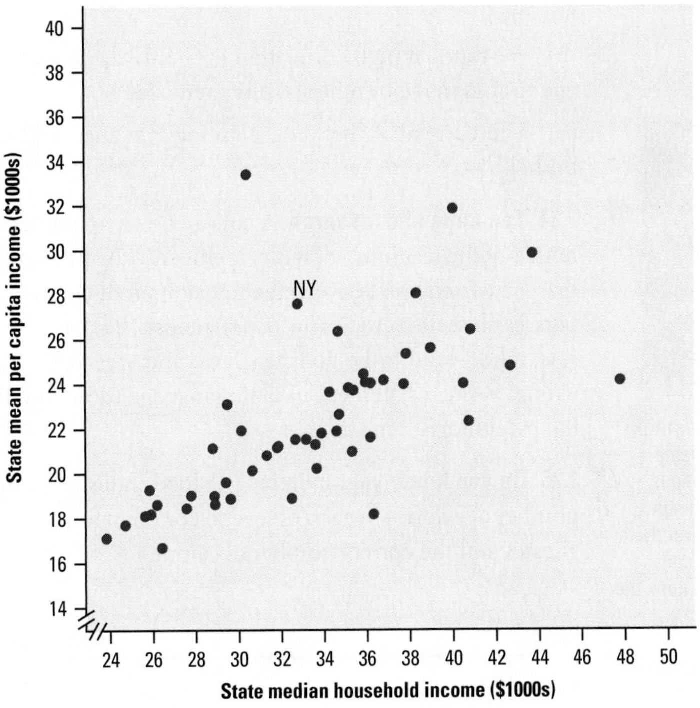

3.22 The professor swims, I Professor Moore swims 2000 yards regularly in a vain attempt to undo middle age. Here are his times (in minutes) and his pulse rate after swimming (in beats per minute) for 23 sessions in the pool:

Time:	34.12	35.72	34.72	34.05	34.13	35.72	36.17	35.57	35. 37
Pulse:	152	124	140	152	146	128	136	144	148
Time:	35.57	35.43	36.05	34.85	34.70	34.75	33.93	34.60	34. 00
Pulse:	144	136	124	148	144	140	156	136	148
Time:	34.35	35.62	35.68	35.28	35.97				
Pulse:	148	132	124	132	139				

(a) Make a scatterplot. (Which is the explanatory variable?)

(b) Is the association between these variables positive or negative? Explain why you expect the relationship to have this direction.

(c) Describe the form and strength of the relationship.

(d) Find the correlation r. Explain from looking at the scatterplot why this value of r is reasonable.

(e) Suppose that the times had been recorded in seconds. For example, the time 34.12 minutes would be 2047 seconds. How would the value of r change?

3.23 Sloppy writing about correlation Each of the following statements contains a blunder. Explain in each case what is wrong.

(a) "There is a high correlation between the gender of American workers and their income."

(b) "We found a high correlation ($r = 1.09$) between students' ratings of faculty teaching and ratings made by other faculty members."

(c) "The correlation between planting rate and yield of corn was found to be $r = 0.23$ bushel."

3.24 Teaching and research A college newspaper interviews a psychologist about student ratings of the teaching of faculty members. The psychologist says, "The evidence indicates that the correlation between the research productivity and teaching rating of faculty members is close to zero." The paper reports this as "Professor McDaniel said that good researchers tend to be poor teachers, and vice versa." Explain why the paper's report is wrong. Write a statement in plain language (don't use the word "correlation") to explain the psychologist's meaning.

Death from superstition? Is there a relationship between superstitious beliefs and bad things happening? Apparently there is. Chinese and Japanese people think the number 4 is unlucky because when pronounced it sounds like the word for "death." Sociologists looked at 15 years' worth of death certificates for Chinese Americans and Japanese Americans and for white Americans. Deaths from heart disease were notably higher on the fourth day of the month among Chinese and Japanese Americans, but not among whites. The sociologists think the explanation is increased stress on "unlucky days."

3.25 Do you know your calories? A food industry group asked 3368 people to guess the number of calories in each of several common foods. Here is a table of the average of their guesses and the correct number of calories:[11]

Food	Guessed calories	Correct calories
8 oz whole milk	196	159
5 oz spaghetti with tomato sauce	394	163
5 oz macaroni with cheese	350	269
One slice wheat bread	117	61
One slice white bread	136	76
2-oz candy bar	364	260
Saltine cracker	74	12
Medium-size apple	107	80
Medium-size potato	160	88
Cream-filled snack cake	419	160

(a) We think that how many calories a food actually has helps explain people's guesses of how many calories it has. With this in mind, make a scatterplot of these data. (Because both variables are measured in calories, you should use the same scale on both axes. Your plot will be square.)

(b) Describe the relationship.

(c) Calculate the correlation r (use your calculator). Explain why your r is reasonable based on the scatterplot.

(d) The guesses are all higher than the true calorie counts. Does this fact influence the correlation in any way? How would r change if every guess were 100 calories higher?

(e) The guesses are much too high for spaghetti and snack cake. Circle these points on your scatterplot. Calculate r for the other eight foods, leaving out these two points. Explain why r changed in the direction that it did.

3.26 Investment diversification A mutual funds company's newsletter says, "A well-diversified portfolio includes assets with low correlations." The newsletter includes a table of correlations between the returns on various classes of investments. For example, the correlation between municipal bonds and large-cap stocks is 0.50, and the correlation between municipal bonds and small-cap stocks is 0.21.[12]

(a) Rachel invests heavily in municipal bonds. She wants to diversify by adding an investment whose returns do not closely follow the returns on her bonds. Should she choose large-cap stocks or small-cap stocks for this purpose? Explain your answer.

(b) If Rachel wants an investment that tends to increase when the return on her bonds drops, what kind of correlation should she look for?

3.27 What affects correlation? Make a scatterplot of the following data:

x:	1	2	3	4	10	10
y:	1	3	3	5	1	11

Use your calculator to show that the correlation is 0.5. What feature of the data is responsible for reducing the correlation to this value despite a strong straight-line relationship between x and y in most of the observations?

3.28 Maximizing corn yields How much corn per acre should a farmer plant to obtain the highest yield? Too few plants will give a low yield. On the other hand, if there are too many plants, they will compete with each other for moisture and nutrients, and yields will fall. To find the best planting rate, plant at different rates on several plots of ground and measure the harvest. (Be sure to treat all the plots the same except for the planting rate.) Here are the data from such an experiment:[13]

Plants per acre	Yield (bushels per acre)			
12,000	150.1	113.0	118.4	142.6
16,000	166.9	120.7	135.2	149.8
20,000	165.3	130.1	139.6	149.9
24,000	134.7	138.4	156.1	
28,000	119.0	150.5		

(a) Make a scatterplot of these data.

(b) Describe the overall pattern of the relationship.

(c) Find the mean yield for each of the five planting rates. Plot each mean yield against its planting rate on your scatterplot and connect these five points with lines. This combination of numerical description and graphing makes the relationship clearer. What planting rate would you recommend to a farmer whose conditions were similar to those in the experiment?

3.2 Least-Squares Regression

Linear (straight-line) relationships between two quantitative variables are easy to understand and are quite common. In the previous section, we found linear relationships in settings as varied as sparrowhawk colonies, sodium and calories in hot dogs, and blood alcohol levels. Correlation measures the direction and strength of these relationships. When a scatterplot shows a linear relationship, we would like to summarize the overall pattern by drawing a line on the scatterplot. A ***regression line*** summarizes the relationship between two variables, but only in a specific setting: when one of the variables helps explain or predict the other. **Regression, unlike correlation, requires that we have an explanatory variable and a response variable.**

Regression Line

A **regression line** is a line that describes how a response variable y changes as an explanatory variable x changes. We often use a regression line to predict the value of y for a given value of x.

Example 3.9 *Does fidgeting keep you slim?*

Scatterplot, correlation, and regression line

Obesity is a growing problem around the world. Here is an account of a study that sheds some light on gaining weight.[14] We will use the Data Analysis Toolbox (page 93) with an additional category—Model—to organize our analysis.

Some people don't gain weight even when they overeat. Perhaps fidgeting and other "nonexercise activity" (NEA) explains why—some people may spontaneously increase nonexercise activity when fed more. Researchers deliberately overfed 16 healthy young adults for 8 weeks. They measured fat gain (in kilograms) and, as an explanatory variable, change in energy use (in calories) from activity other than deliberate exercise—fidgeting, daily living, and the like.

Data:

NEA change (cal):	−94	−57	−29	135	143	151	245	355
Fat gain (kg):	4.2	3.0	3.7	2.7	3.2	3.6	2.4	1.3
NEA change (cal):	392	473	486	535	571	580	620	690
Fat gain (kg):	3.8	1.7	1.6	2.2	1.0	0.4	2.3	1.1

1. *Who?* The individuals are 16 healthy young adults who participated in a study on overeating.
2. *What?* The explanatory variable is change in NEA (in calories), and the response variable is fat gain (kilograms).
3. *Why?* Researchers wondered whether changes in fidgeting and other NEA would help explain weight gain in individuals who overeat.
4. *When, where, how, and by whom?* The data come from a controlled experiment in which subjects were forced to overeat for an 8-week period. The results of the study were published in *Science* magazine in 1999.

Graphs: Figure 3.10 (on the next page) is a scatterplot of these data.
Numerical summaries: The correlation between NEA change and fat gain is $r = -0.7786$.
Model: Draw a line on the scatterplot to predict fat gain from change in NEA. We will do this in the next example.
Interpretation: The scatterplot shows a moderately strong negative linear association between NEA change and fat gain, with no outliers. People with larger increases in NEA do indeed gain less fat. A line drawn through the points will describe the overall pattern well.

CAUTION

1000 times larger, $b = -3.44$. **You can't say how important a relationship is by looking at how big the regression slope is.**

Prediction

We can use a regression line to predict the response y for a specific value of the explanatory variable x.

Example 3.11 *Predicting fat gain from NEA*

Using a regression line to make predictions

Based on the linear pattern, we want to predict the fat gain for an individual whose NEA increases by 400 calories when she overeats. To use the regression line to predict fat gain, go "up and over" on the graph in Figure 3.11. From 400 calories on the x axis, go up to the line and over to the y axis. The graph shows that the predicted gain in fat is a bit more than 2 kilograms.

If we have the equation of the line, it is faster and more accurate to substitute $x = 400$ in the equation. The predicted fat gain is

$$\text{fat gain} = 3.505 - 0.00344(400) = 2.13 \text{ kilograms}$$

The accuracy of predictions from a regression line depends on how much scatter about the line the data show. In Figure 3.11, fat gains for similar changes in NEA show a spread of 1 or 2 kilograms. The regression line summarizes the pattern but gives only roughly accurate predictions.

Example 3.12 *More prediction: fat gain and NEA*

When prediction is a bad idea

Can we predict the fat gain for someone whose NEA increases by 1500 calories when she overeats? We can certainly substitute 1500 calories into the equation of the line. The prediction is

$$\text{fat gain} = 3.505 - 0.00344(1500) = -1.66 \text{ kilograms}$$

That is, we predict that this individual *loses* fat when she overeats. This prediction makes no sense!

Look again at Figure 3.11. An NEA increase of 1500 calories is far outside the range of our data. We can't say whether increases this large ever occur, or whether the relationship remains linear at such extreme values. Predicting fat gain when NEA increases by 1500 calories is an ***extrapolation*** of the relationship beyond what the data show.

Extrapolation

Extrapolation is the use of a regression line for prediction outside the range of values of the explanatory variable x used to obtain the line. Such predictions are often not accurate.

Exercises

3.29 That's one big rat! Some data were collected on the weight of a male white laboratory rat following its birth. A scatterplot of the weight (in grams) and time since birth (in weeks) shows a fairly strong positive linear relationship. The linear regression equation

$$\text{weight} = 100 + 40(\text{time})$$

models the data fairly well.

(a) Interpret the slope of the regression line in this setting.

(b) Interpret the y intercept of the regression line in this setting.

(c) Draw a graph of this line between birth and 10 weeks of age.

(d) Would you be willing to use this line to predict the rat's weight at age 2 years? Do the prediction and think about the reasonableness of the result. (There are 454 grams in a pound. To help you assess the result, note that a medium-sized cat weighs about 10 pounds.)

3.30 IQ and reading scores Data on the IQ test scores and reading test scores for a group of fifth-grade children give the regression line

$$\text{reading score} = -33.4 + 0.882(\text{IQ score})$$

for predicting reading score from IQ score.

(a) Explain what the slope of this line tells you.

(b) Find the predicted reading scores for two children with IQ scores of 90 and 130, respectively.

(c) Draw a graph of the regression line for IQs between 90 and 130. (Be sure to show the scales for the x and y axes.)

(d) Interpret the y intercept of this line. Why doesn't this make any sense?

3.31 Penguins diving A study of king penguins looked for a relationship between how deep the penguins dive to seek food and how long they stay under water.[15] For all but the shallowest dives, there is a linear relationship that is different for different penguins. The study gives a scatterplot for one penguin titled "The Relation of Dive Duration (y) to Depth (x)." Duration y is measured in minutes and depth x is in meters. The report then says, "The regression equation for this bird is: $y = 2.69 + 0.0138x$."

(a) What is the slope of the regression line? Explain in specific language what this value says about this penguin's dives.

(b) According to the regression line, how long does a typical dive to a depth of 200 meters last?

(c) The dives varied from 40 meters to 300 meters in depth. Plot the regression line from $x = 40$ to $x = 300$.

(d) Does the y intercept of the regression line make any sense? If so, interpret it. If not, explain why not.

3.32 Acid rain Researchers studying acid rain measured the acidity of precipitation in a Colorado wilderness area for 150 consecutive weeks. Acidity is measured by pH. Lower pH values show higher acidity. The acid rain researchers observed a linear pattern over time. They reported that the regression line

$$\text{pH} = 5.43 - 0.0053(\text{weeks})$$

fit the data well.[16]

(a) What is the slope of the regression line? Explain clearly what this slope says about the change in the pH of the precipitation in this wilderness area.

(b) Draw a graph of this line over the duration of the study. Be sure to label and scale your axes.

(c) Does the y intercept of the regression line make sense in this setting? If so, interpret it. If not, explain why not.

(d) According to the regression line, what was the pH at the end of this study?

The Least-Squares Regression Line

In most cases, no line will pass exactly through all the points in a scatterplot. Different people will draw different lines by eye. We need a way to draw a regression line that doesn't depend on our guess as to where the line should go. Because we use the line to predict y from x, the prediction errors we make are errors in y, the vertical direction in the scatterplot. A *good regression line makes the vertical distances of the points from the line as small as possible.*

Figure 3.12 on the next page illustrates the idea. This plot shows three of the points from Figure 3.11, along with the line, on an expanded scale. The line passes above one of the points and below two of them. The three prediction errors appear as vertical line segments. For example, one subject had $x = -57$, a decrease of 57 calories in NEA. The line predicts a fat gain of 3.7 kilograms, but the actual fat gain for this subject was 3.0 kilograms. The prediction error is

$$\begin{aligned} \text{error} &= \text{observed response} - \text{predicted response} \\ &= 3.0 - 3.7 = -0.7 \text{ kilogram} \end{aligned}$$

There are many ways to make the collection of vertical distances "as small as possible." The most common is the "least-squares" method. The line in Figures 3.11 and 3.12 is in fact the ***least-squares regression line.***

> ***Least-Squares Regression Line***
>
> The **least-squares regression line** of y on x is the line that makes the sum of the squared vertical distances of the data points from the line as small as possible.

Figure 3.12 *The least-squares idea: make the errors in predicting y as small as possible by minimizing the sum of the squares of the vertical distances of the data points from the line.*

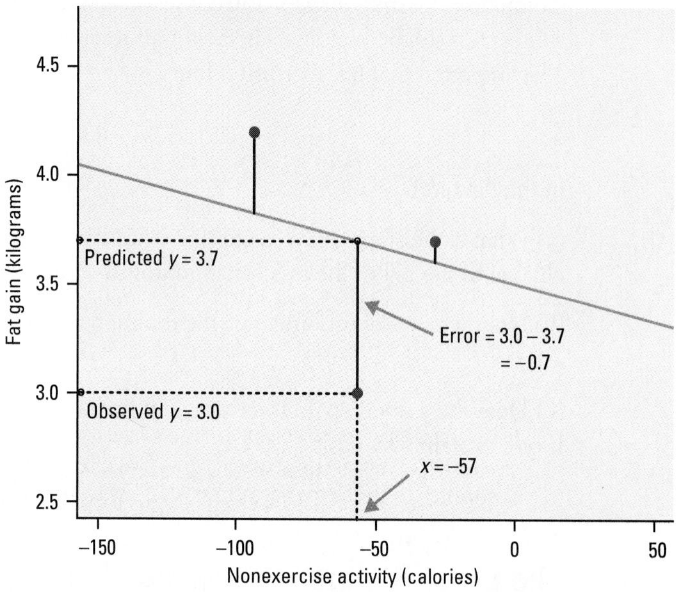

Figure 3.13 *The least-squares regression line results in the smallest possible sum of squared distances of the points from the line, 30.90. Fathom software allows you to explore this fact dynamically.*

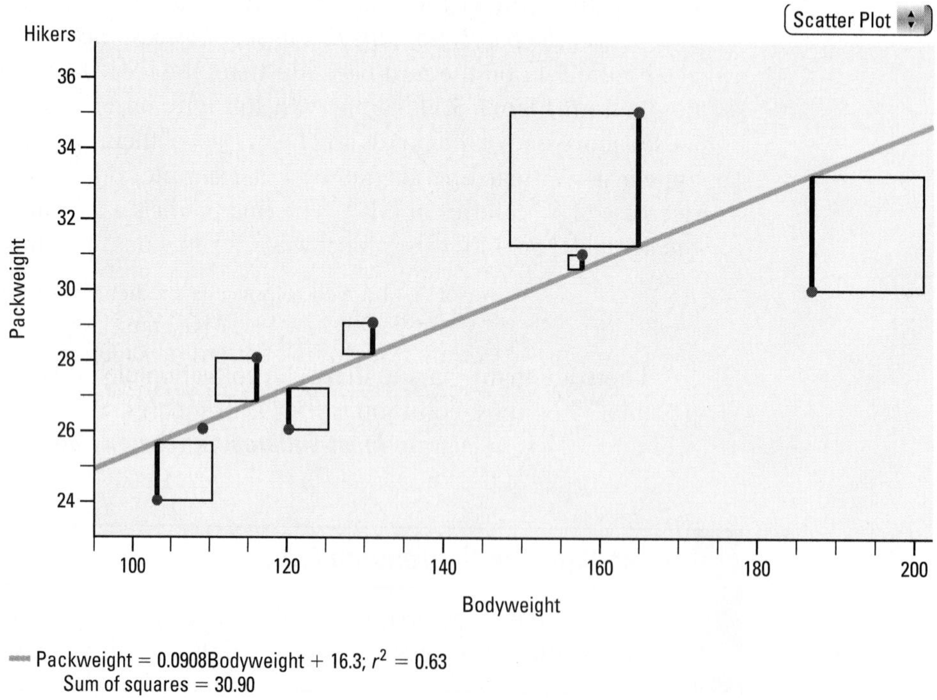

Figure 3.13 gives a geometric interpretation of the least-squares principle for the backpacker data from the Technology Toolbox in Section 3.1 (page 186). The least-squares regression line shown minimizes the sum of the squared vertical distances of the points from the line, 30.90. No other regression line would give a smaller sum of squared errors.

APPLET

Activity 3C

Investigating properties of the least-squares regression line

In this Activity, you will use the *Correlation and Regression* applet at the book's Web site, www.whfreeman.com/tps3e to explore some properties of the least-squares regression line.

1. Click on the scatterplot to create a group of 15 to 20 points from lower left to upper right with a clear positive straight-line pattern (correlation around 0.7). Click the "Draw line" button and use the mouse (right-click and drag) to draw a line through the middle of the cloud of points from lower left to upper right. Note the "thermometer" above the plot. The red portion is the sum of the squared vertical distances from the points in the plot to the least-squares line.

   The green portion shows by how much your line misses the smallest possible "sum of squares."

2. Use your mouse to adjust the slope and y intercept of your line to make the green portion of the thermometer as small as possible.
3. Now click the "Show least-squares line" box. How close did you get?
4. Click the "Draw line" button again to remove your line. Then click the "Show mean x and mean y lines" button. What do you notice?
5. Move points, one at a time, in your scatterplot to see if the result in Step 4 continues to hold true.
6. Create a new group of 10 points in the lower-left corner of the scatterplot with a strong straight-line pattern (correlation about 0.9). Click the "Show least-squares line" box to display the regression line. Record the equation of the line and the correlation.
7. Add one point at the upper right that is far from the other 10 points but exactly on the regression line. How does the addition of this point affect the equation of the line? The correlation?
8. Now use the mouse to drag this last point straight down. What happens to the least-squares line? The correlation?

One reason for the popularity of the least-squares regression line is that the problem of finding the equation of the line has a simple answer. We can give the ***equation of the least-squares regression line*** in terms of the means and standard deviations of the two variables and their correlation.

Equation of the Least-Squares Regression Line

We have data on an explanatory variable x and a response variable y for n individuals. From the data, calculate the means $\bar{x}$ and $\bar{y}$ and the standard deviations s_x and s_y of the two variables and their correlation r. The least-squares regression line is

$$\hat{y} = a + bx$$

with slope

$$b = r\frac{s_y}{s_x}$$

that passes through the point $(\bar{x}, \bar{y})$.

We write $\hat{y}$ (read "y hat") in the equation of the regression line to emphasize that the line gives a *predicted response* $\hat{y}$ for any x. Because of the scatter of points about the line, the predicted response will usually not be exactly the same as the actually *observed* response y.

Example 3.13 *Fat gain and NEA, continued*

Calculating the least-squares regression line

Use your calculator to verify that the mean and standard deviation of the 16 changes in NEA are $\bar{x} = 324.8$ calories and $s_x = 257.66$ calories and that the mean and standard deviation of the 16 fat gains are $\bar{y} = 2.388$ kg and $s_y = 1.1389$ kg.

The correlation between fat gain and NEA change is $r = -0.7786$. The least-squares regression line of fat gain y on NEA change x therefore has slope

$$b = r\frac{s_y}{s_x} = (-0.7786)\frac{1.1389}{257.66} = -0.00344 \text{ kg per calorie}$$

To find the y intercept, we use the fact that the least-squares line passes through $(\bar{x}, \bar{y})$:

$$y = a + bx$$

$$2.388 = a + (-0.00344)(324.8)$$

$$a = 3.505 \text{ kg}$$

The equation of the least-squares line is

$$\hat{y} = 3.505 - 0.00344x$$

CAUTION

When doing calculations like this by hand, you may need to carry extra decimal places in the preliminary calculations to get accurate values of the slope and y intercept. Using software or a calculator eliminates this worry.

Using Technology

In practice, you don't need to calculate the means, standard deviations, and correlation first. Software or your calculator will give the slope b and intercept a of the least-squares line from keyed-in values of the variables x and y. You can then concentrate on understanding and using the regression line.

Figure 3.14 displays the basic regression output for the NEA data from two statistical software packages: CrunchIt! and Minitab. Other software produces very similar output. Each output records the slope and y intercept of the least-squares line. The software also provides information that we don't yet need (or understand!), although we will use much of it later. (In fact, we left out part of the Minitab output.) Be sure that you can locate the slope and y intercept on both computer outputs. *Once you understand the statistical ideas, you can read and work with almost any software output.*

Figure 3.14 *Regression results for the nonexercise activity data from two statistical software packages. Other software produces similar output.*

CrunchIt!

Simple Linear Regression

Simple linear regression results:

Dependent Variable: Fat gain
Independent Variable: NEA
Fat gain = 3.505123 - 0.003441487 NEA
Sample size: 16
R (correlation coefficient) = -0.7786
R-sq = 0.6061492
Estimate of error standard deviation: 0.73985285

Parameter estimates:

Parameter	Estimate	Std. Err.	DF	T-Stat	P-Value
Intercept	3.505123	0.3036164	14	11.544577	<0.0001
Slope	-0.003441487	7.414096E-4	14	-4.641816	0.0004

Minitab

Session

Regression Analysis: Fat gain versus NEA

```
The regression equation is
Fat gain = 3.51 - 0.00344 NEA


Predictor        Coef     SE Coef      T      P
Constant       3.5051      0.3036  11.54  0.000
NEA        -0.0034415   0.0007414  -4.64  0.000


S = 0.739853   R-Sq = 60.6%   R-Sq(adj) = 57.8%
```

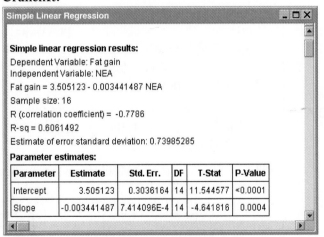

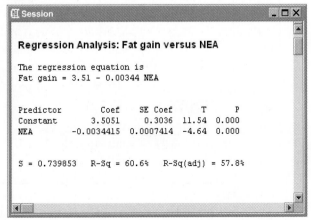

Technology Toolbox

Least-squares regression lines on the TI-83/84/89

We will use the fat gain and NEA data from Example 3.9 (page 200) to show how to use the TI-83/84/89 to determine the equation of the least-squares regression line.

- Enter the NEA change data into L_1/list1 and the fat gain data into L_2/list2.
- Define a scatterplot using L_1/list1 and L_2/list2, and then use ZoomStat (ZoomData) to plot the scatterplot.

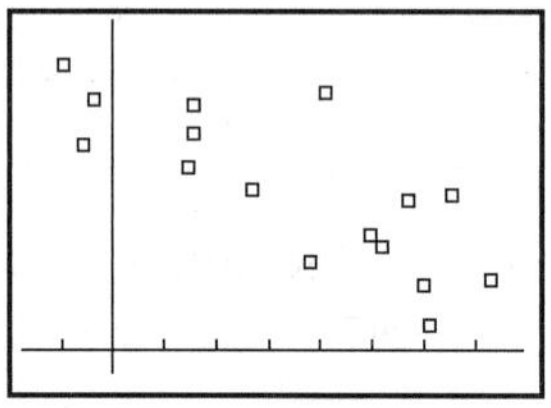

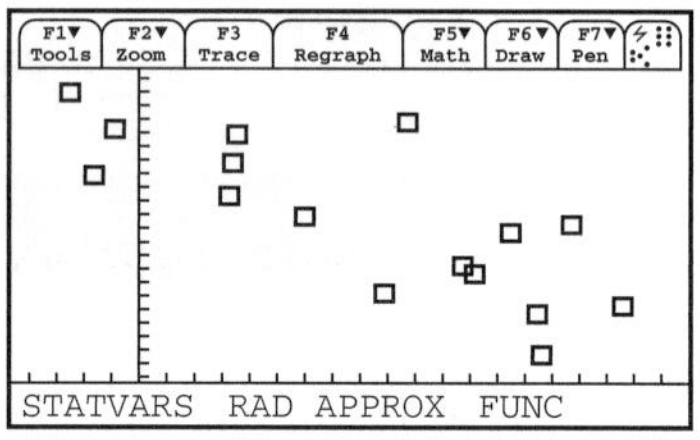

To determine the least-squares regression line:

TI-83/84

- Press **STAT**, choose CALC, then `8:LinReg (a+bx)`. Finish the command to read `LinReg(a+bx)L1,L2,Y1.` (Y1 is found under VARS/Y-VARS/1:Function.)

TI-89

- In the Statistics/List Editor, press **F4** (CALC), choose `3:Regressions`, then `1:LinReg(a+bx)`.
- Enter list1 for the Xlist, list2 for the Ylist, choose to store the RegEqn to y1(x), and press **ENTER**.

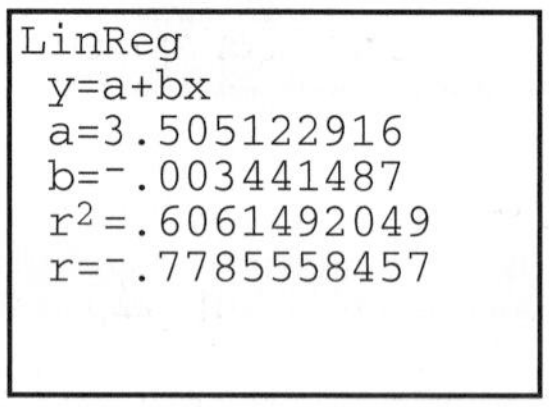

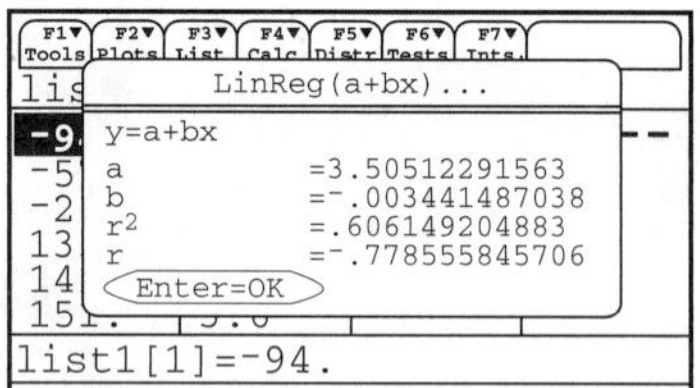

Note: If r^2 and r do not appear on your TI-83/84 screen, then do this one-time series of keystrokes: Press **2nd** **0** (CATALOG), scroll down to `DiagnosticOn`, and press **ENTER**. Press **ENTER** again to execute the command. The screen should say "Done." Then press **2nd** **ENTER** (ENTRY) to recall the regression command and **ENTER** again to calculate the least-squares line. The r^2- and r-values should now appear.

Technology Toolbox

Least-squares regression lines on the TI-83/84/89 *(continued)*

- Deselect all other equations in the Y=screen and press GRAPH (◆ F3 on the TI-89) to overlay the least-squares line on the scatterplot.

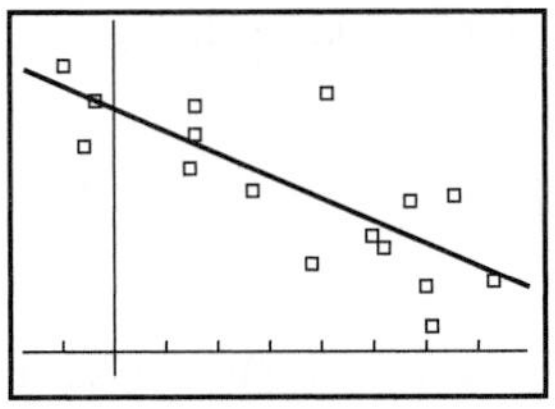

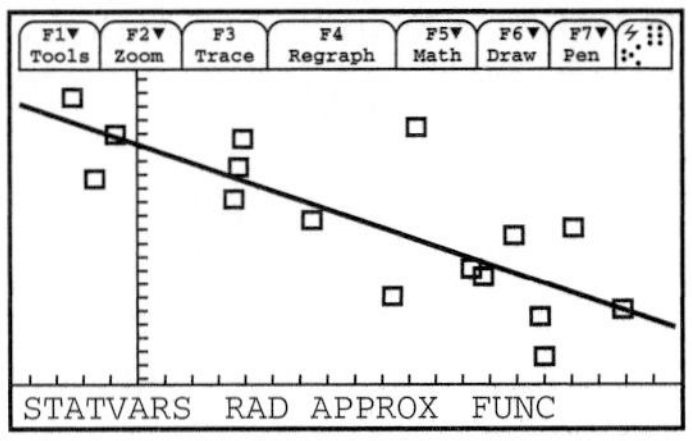

- Save these lists for later use. On the home screen, execute the command `L1→NEA:L2→FAT` (`list1→NEA:list2→FAT` on the TI-89).

Although the calculator will report the values for a and b to nine decimal places, we usually round off to fewer decimal places. You would write the equation as

$$\hat{y} = 3.505 - 0.00344x$$

When you write the equation, don't forget the hat symbol over the y; this means *predicted value.*

Exercises

3.33 The endangered manatee, II Exercise 3.5 (page 179) gives data on the number of powerboats registered in Florida and the number of manatees killed by boats in the years from 1977 to 1990.

(a) Use your calculator to make a scatterplot of these data.

(b) Find the equation of the least-squares line and overlay that line on your scatterplot.

(c) Predict the number of manatees that will be killed by boats in a year when 716,000 powerboats are registered.

(d) Here are four more years of manatee data, in the same form as in Exercise 3.5:

1991	716	53
1992	716	38
1993	716	35
1994	735	49

Add these points to your scatterplot. Florida took stronger measures to protect manatees during these years. Do you see any evidence that these measures succeeded?

(e) In part (c) you predicted manatee deaths in a year with 716,000 powerboat registrations. In fact, powerboat registrations were 716,000 for three years. Compare the mean manatee deaths in these three years with your prediction from part (c). How accurate was your prediction?

3.34 This is for the birds! Exercise 3.6 (page 180) gives data on the number of new birds y and percent of returning birds x for 13 sparrowhawk colonies. Enter the data into your calculator.

(a) Use your calculator's regression function to find the equation of the least-squares regression line.

(b) Use your calculator to find the mean and standard deviation of both x and y and their correlation r.

(c) Calculate the slope b and y intercept a of the regression line following the method of Example 3.13. Verify that your equation is the same as the one you obtained in part (a) except for slight rounding errors.

(d) Explain in words what the slope and y intercept of the regression line tell us.

(e) An ecologist uses the line to predict how many birds will join another colony of sparrowhawks, to which 60% of the adults from the previous year return. What is the prediction?

3.35 Predicting blood alcohol content, I Example 3.5 (page 177) describes a study that recorded data on number of beers consumed and blood alcohol content (BAC) for 16 students. Here is some partial computer output from Minitab relating to these data:

```
Predictor        Coef        StDev        T        P
Constant     -0.01270      0.01264    -1.00    0.332
Beers        0.017964     0.002402     7.48    0.000

S = 0.02044   R-Sq = 80.0%
```

(a) Use the computer output to write the equation of the least-squares line.

(b) Interpret the slope and y intercept of the equation in this setting.

(c) What blood alcohol level would your equation predict for a student who consumed 6 beers?

(d) The one student in the study who consumed 6 beers had a BAC of 0.10. What is your prediction error in part (c)?

3.36 Heating with natural gas, I The Sanchez household is about to install solar panels to reduce the cost of heating their house. In order to know how much the solar panels help, they record their consumption of natural gas before the panels are installed. Gas consumption is higher in cold weather, so the relationship between outside temperature and gas consumption is important. The Sanchez family collected data for 16 months. The response variable y is the average of the daily amounts of natural gas consumed during each month, in hundreds of cubic feet. The explanatory variable x is the average of the daily number of heating degree-days during each month. (Heating degree-days are the usual measure of demand for heating. One degree-day is accumulated for each degree a day's average temperature falls below 65°F. An average temperature of 20°F, for example, corresponds to 45 degree-days.)

Figure 3.15 is a scatterplot of the data with the least-squares regression line added.

Figure 3.15 *The Sanchez household gas consumption data, with a regression line for predicting gas consumption from average degree-days in a month.*

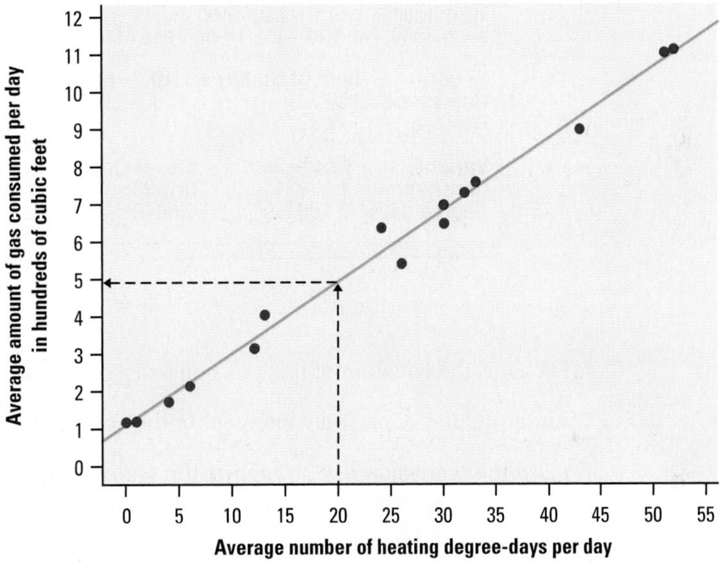

(a) Describe the direction, form, and strength of the relationship displayed in Figure 3.15.

(b) About how much gas does the regression line predict that the Sanchez family will use in a month that averages 20 degree-days per day?

(c) Explain why the regression line shown in Figure 3.15 is called the "least-squares line."

(d) How well does the least-squares line fit the data? Justify your answer.

3.37 Predicting blood alcohol content, II Here are two TI-84 screen shots produced using the data for Exercise 3.35.

```
2-Var Stats
 x̄=4.8125
 Σx=77
 Σx²=443
 Sx=2.1975365
 σx=2.127755566
↓n=16
```

```
2-Var Stats
↑n=16
 ȳ=.07375
 Σy=1.18
 Σy²=.11625
 Sy=.044139929
↓σy=.0427383025
```

Use the method of Example 3.13 to calculate the slope and y intercept of the least-squares regression line. Compare with your results from part (a) of Exercise 3.35.

3.38 Heating with natural gas, II Figure 3.16 provides computer output from DataDesk statistical software for the Sanchez family heating data of Exercise 3.36.

Figure 3.16 *Least-squares regression output from DataDesk for the gas consumption data, for Exercise 3.38.*

Dependent variable is: **Gas used**
No Selector
R squared = 99.1% R squared (adjusted) = 99.0%
s = 0.3389 with 16 − 2 = 14 degrees of freedom

Source	Sum of Squares	df	Mean Square	F-ratio
Regression	168.581	1	168.581	1468
Residual	1.60821	14	0.114872	

Variable	Coefficient	s.e. of Coeff	t-ratio	prob
Constant	1.08921	0.1389	7.84	≤0.0001
Degree-days	0.188999	0.0049	38.3	≤0.0001

(a) What is the equation of the least-squares regression line?

(b) Interpret the slope and y intercept in the context of this setting.

(c) Use the regression line to predict the amount of natural gas that the Sanchez family will consume in a month that averages 20 degree-days per day. Compare with your answer to Exercise 3.36(b).

(d) In one month of the study, the average degree-days was 30 and the Sanchez family used 640 cubic feet of gas. How much prediction error would there be in using the least-squares line to estimate the amount of natural gas consumed in this month?

How Well the Line Fits the Data: Residuals

One of the first principles of data analysis is to look for an overall pattern and also for striking deviations from the pattern. A regression line describes the overall pattern of a linear relationship between an explanatory variable and a response variable. We see deviations from this pattern by looking at the scatter of the data points about the regression line. The vertical distances from the points to the least-squares regression line are as small as possible, in the sense that they have the smallest possible sum of squares. Because they represent "left-over" variation in the response after fitting the regression line, these distances are called ***residuals.***

Residuals

A **residual** is the difference between an observed value of the response variable and the value predicted by the regression line. That is,

$$\begin{aligned} \text{residual} &= \text{observed } y - \text{predicted } y \\ &= y - \hat{y} \end{aligned}$$

Example 3.14 *Fat gain and NEA, again*

Examining residuals

Example 3.9 (page 200) describes measurements on 16 young people who volunteered to overeat for 8 weeks. Those whose nonexercise activity (NEA) rose substantially gained less fat than others. Figure 3.17 is a scatterplot of these data with the least-squares regression line superimposed. The equation of the least-squares line is

$$\text{fat gain} = 3.505 - 0.00344(\text{NEA change})$$

Figure 3.17 *Scatterplot of fat gain versus nonexercise activity, with the least-squares line.*

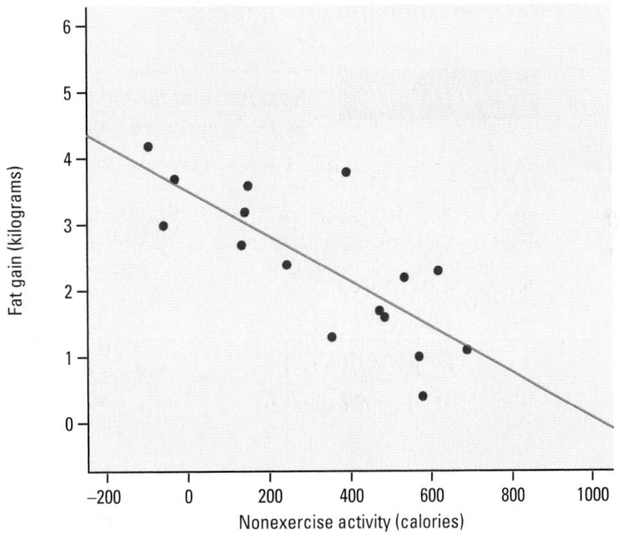

One subject's NEA rose by 135 calories. That subject gained 2.7 kilograms of fat. The predicted gain for 135 calories is

$$\hat{y} = 3.505 - 0.00344(135) = 3.04 \text{ kg}$$

The residual for this subject is therefore

$$\begin{aligned} \text{residual} &= \text{observed } y - \text{predicted } y \\ &= y - \hat{y} \\ &= 2.7 - 3.04 = -0.34 \text{ kg} \end{aligned}$$

The residual is negative because the data point lies below the line.

The 16 data points used in calculating the least-squares line produce 16 residuals. Rounded to two decimal places, they are

0.37	−0.70	0.10	−0.34	0.19	0.61	−0.26	−0.98
1.64	−0.18	−0.23	0.54	−0.54	−1.11	0.93	−0.03

Most graphing calculators and statistical software will calculate and store these residuals for you.

Because the residuals show how far the data fall from our regression line, examining the residuals helps assess how well the line describes the data. Although residuals can be calculated from any model that is fitted to the data, the residuals from the least-squares line have a special property: *the sum of the least-squares residuals is always zero.* You can check that the sum of the residuals in Example 3.14 is 0.01. The sum is not exactly 0 because we rounded to two decimal places.

You can see the residuals in the scatterplot of Figure 3.17 by looking at the vertical deviations of the points from the line. The ***residual plot*** in Figure 3.18 makes it easier to study the residuals by plotting them against the explanatory variable, change in NEA. Because the mean of the residuals is always zero, the horizontal line at zero in Figure 3.18 helps orient us. This "residual = 0" line corresponds to the regression line in Figure 3.17.

Figure 3.18 *Residual plot for the regression displayed in Figure 3.17. The line at y = 0 marks the mean of the residuals.*

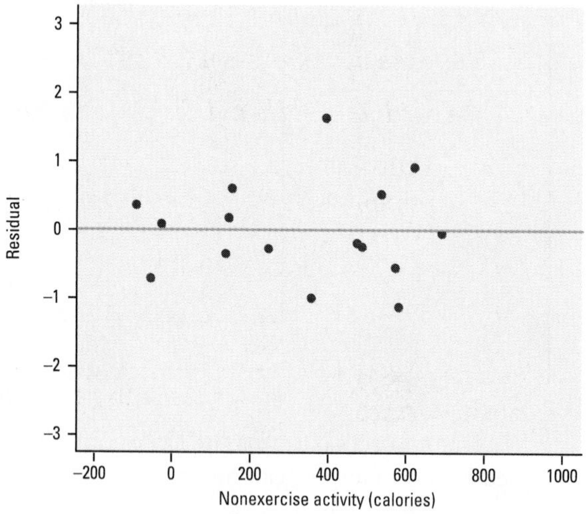

Residual Plots

A **residual plot** is a scatterplot of the regression residuals against the explanatory variable (or equivalently, against the predicted *y*-values). Residual plots help us assess how well a regression line fits the data.

CAUTION

You should be aware that **some computer software packages prefer to plot the residuals against the predicted values $\hat{y}$ instead of against the values of the explanatory variable.** The basic shape of the two plots is the same because $\hat{y}$ is linearly related to x.

The residual plot magnifies the deviations from the line to make patterns easier to see. If the regression line captures the overall pattern of the data, there should be *no pattern* in the residuals. The residual plot will look something like the simplified graph in Figure 3.19(a). That plot shows an unstructured scatter of points in a horizontal band centered at zero. Notice that the residuals in Figure 3.18 have such an irregular scatter about the "residual = 0" line.

Figure 3.19a *(a) The unstructured scatter of points indicates that the regression line fits the data well, so the line is a good model.*

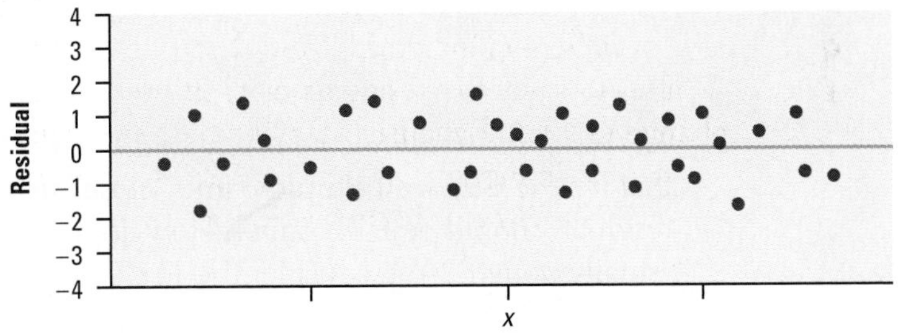

Here are two important things to look for when you examine a residual plot.

1. *The residual plot should show no obvious pattern.*
   - A curved pattern shows that the relationship is not linear. Figure 3.19(b) is a simplified example. A straight line may not be the best model for such data.

Figure 3.19b *(b) The residuals have a curved pattern, so a straight line may not be the best model.*

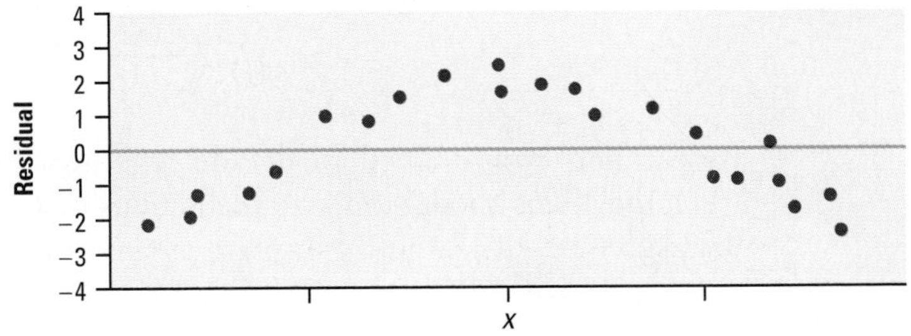

- Increasing (or decreasing) spread about the line as x increases indicates that prediction of y will be less accurate for larger x (for smaller x). Figure 3.19(c) is a simplified example of such a "fan-shaped" pattern.

Figure 3.19c *(c) The response variable y has more spread for larger values of the explanatory variable x, so prediction will be less accurate when x is large.*

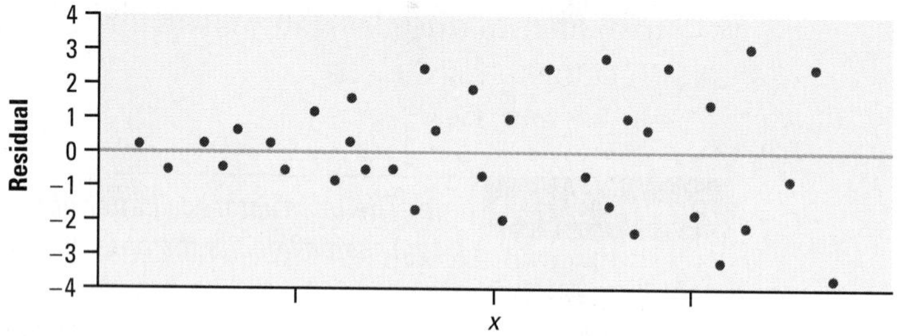

2. *The residuals should be relatively small in size.* A regression line in a model that fits the data well should come "close" to most of the points. That is, the residuals should be fairly small. How do we decide whether the residuals are "small enough"? We consider the size of a "typical" prediction error.

In Figure 3.18, for example, almost all of the residuals are between –0.7 and 0.7. For these individuals, the predicted fat gain from the least-squares line is within 0.7 kg of their actual fat gain during the study. That sounds pretty good. But the subjects gained only between 0.4 kg and 4.2 kg, so a prediction error of 0.7 kg is relatively large compared with the actual fat gain for an individual. The largest residual, 1.64, corresponds to a prediction error of 1.64 kg. This subject's actual fat gain was 3.8 kg, but the regression line predicted a fat gain of only 2.16 kg. That's a pretty large error, especially from the subject's perspective!

standard deviation of the residuals

A commonly used measure of typical prediction error is the ***standard deviation of the residuals,*** which is given by

$$s = \sqrt{\frac{\Sigma \text{residuals}^2}{n-2}}$$

For the NEA and fat gain data,

$$s = \sqrt{\frac{7.663}{14}} = 0.740$$

Researchers would need to decide whether they would feel comfortable using this linear model to make predictions that might be consistently "off" by 0.74 kg.

Technology Toolbox

Residual plots on the TI-83/84/89

Here is a procedure for calculating residuals on your TI-83/84/89 and then displaying a residual plot.

- Repeat all the steps in the Technology Toolbox on page 210 to reproduce the scatterplot and least-squares regression line for the fat gain and NEA data.

To graph the residual plot:

TI-83/84	TI-89
Define L_3/list3 as the predicted values from the regression equation.	
• With L_3 highlighted, enter the command `Y1(L1)` and press ENTER.	• With list3 highlighted, enter the command `y1(list1)` and press ENTER.
Define L_4/list4 as the observed y-value minus the predicted y-value.	
• With L_4 highlighted, enter the command `L2-L3` and press ENTER to show the residuals.	• With list4 highlighted, enter the command `list2-list3` and press ENTER to show the residuals.

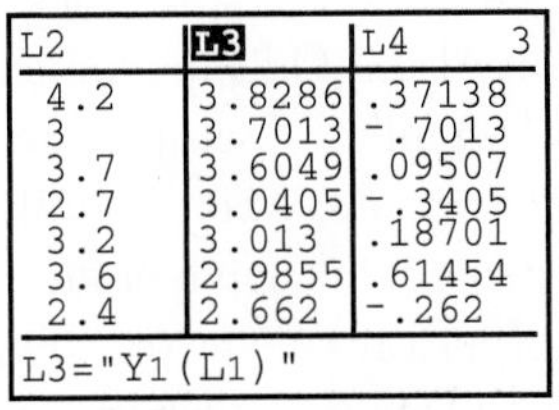

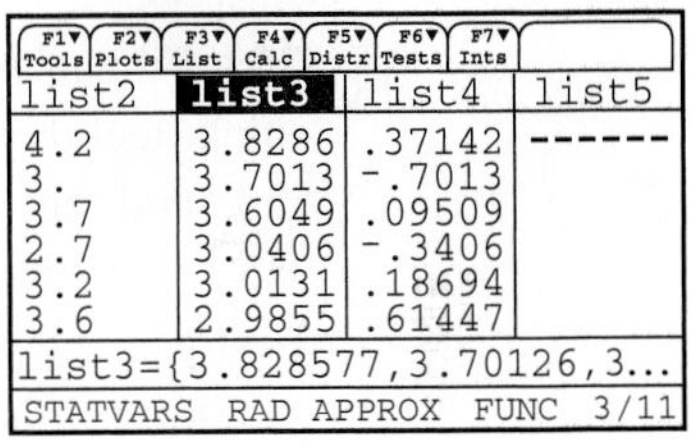

- Turn off Plot1 and deselect the regression equation. Specify Plot2 with L_1/list1 as the x variable and L_4/list4 as the y variable. Use ZoomStat (ZoomData) to see the residual plot.

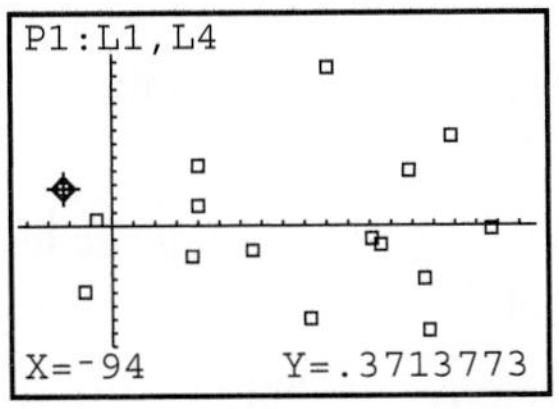

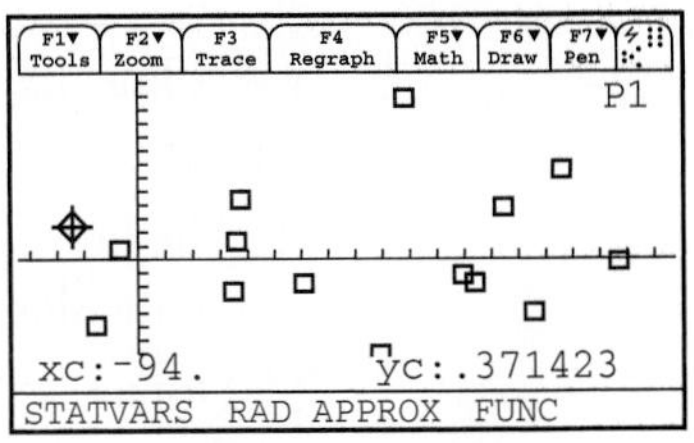

The x axis in the residual plot serves as a reference line, with points above this line corresponding to positive residuals and points below the line corresponding to negative residuals. We used TRACE to see the residual for the individual with an NEA change of -94 calories.

(continued)

Technology Toolbox

Residual plots on the TI-83/84/89 *(continued)*

- Finally, we have previously noted that an important property of residuals is that their mean is zero. Calculate one-variable statistics on the residuals list to verify that $\sum$residuals = 0 and that, consequently, the mean of the residuals is also 0.

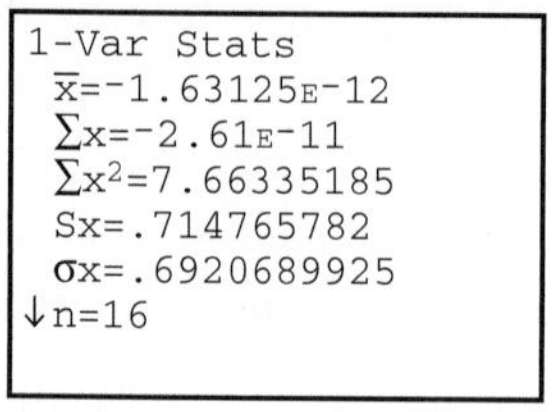

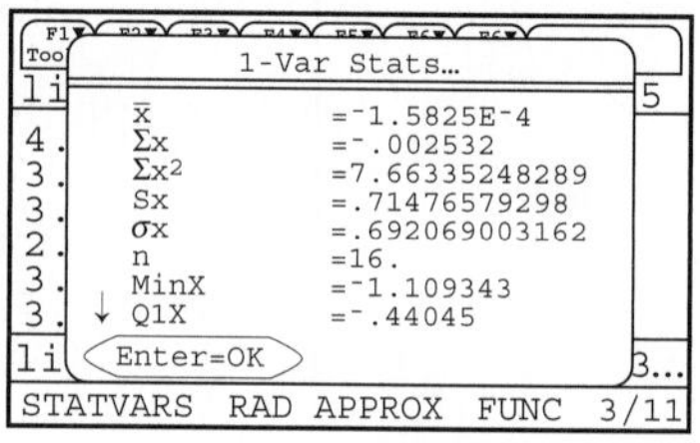

Note that the calculator is showing some roundoff error. You should recognize these peculiar-looking numbers as being equivalent to 0.

Exercises

3.39 Oil and residuals The Trans-Alaska Oil Pipeline is a tube formed from 1/2-inch-thick steel that carries oil across 800 miles of sensitive arctic and sub-arctic terrain. The pipe and the welds that join pipe segments were carefully examined before installation. How accurate are field measurements of the depth of small defects? Figure 3.20(a) on the next page compares the results of measurements on 100 defects made in the field with measurements of the same defects made in the laboratory.[17] The line $y = x$ is drawn on the scatterplot. Figure 3.20(b) is a residual plot for these data.

(a) Describe the overall pattern you see in the scatterplot, as well as any deviations from that pattern.

(b) If field and laboratory measurements all agree, then the points should fall on the $y = x$ line drawn on the plot, except for small variations in the measurements. But this is not the case. Explain.

(c) The line drawn on the scatterplot ($y = x$) is *not* the least-squares regression line. How would the slope and y intercept of the least-squares line compare? Justify your answer.

(d) Discuss what the residual plot tells you about how well the least-squares regression line fits the data.

3.40 Driving speed and fuel consumption Exercise 3.9 (page 182) gives data on the fuel consumption y of a car at various speeds x. Fuel consumption is measured in liters of gasoline per 100 kilometers driven and speed is measured in kilometers per hour. A statistical software package gives the least-squares regression line and also the residuals. The regression line is

$$\hat{y} = 11.058 - 0.01466x$$

Figure 3.20 *(a) Depths of small defects in pipe for the Trans-Alaska Oil Pipeline, measured in the field and in the laboratory, for Exercise 3.39.*

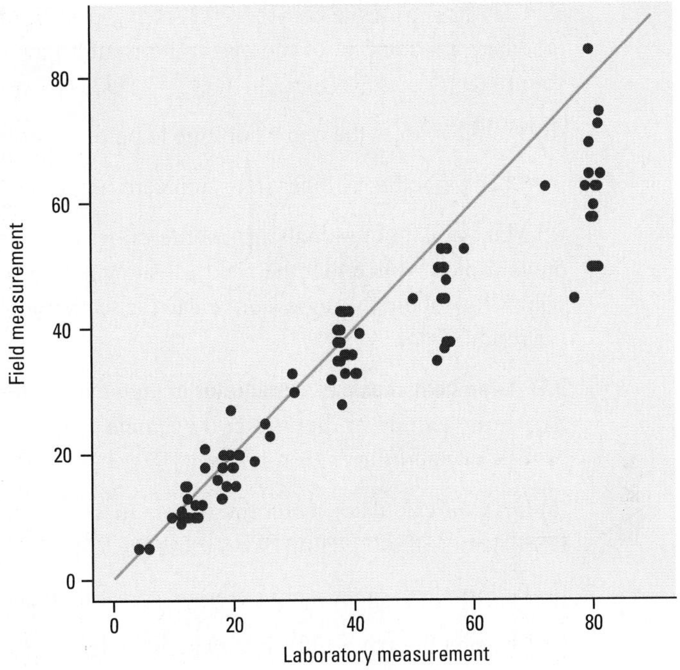

(b) Residual plot for the regression of field measurements of Alaska pipeline defects on laboratory measurements of the same defects, for Exercise 3.39.

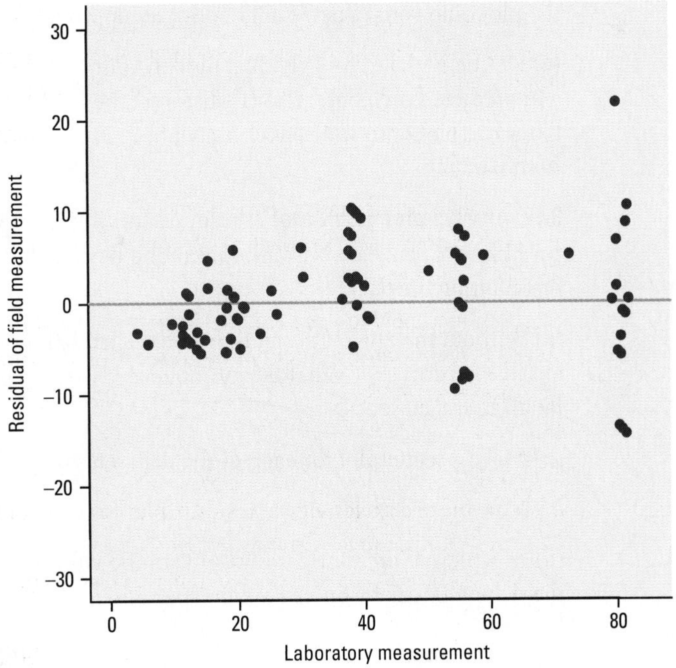

The residuals, in the same order as the observations, are

10.09	2.24	−0.62	−2.47	−3.33	−4.28	−3.73	−2.94
−2.17	−1.32	−0.42	0.57	1.64	2.76	3.97	

(a) Make a scatterplot of the observations and draw the least-squares regression line on your plot.

(b) Would you use the regression line to predict y from x? Justify your answer.

(c) Check that the residuals have sum zero (up to roundoff error).

(d) Make a plot of residuals against the values of x. Draw a horizontal line at height zero on your plot. Notice that the residuals show the same pattern about this line as the data points show about the regression line in the scatterplot in (a). What do you conclude about the residual plot?

3.41 Lean body mass as a predictor of metabolic rate Exercise 3.10 (page 183) provides data from a study of dieting for 12 female and 7 male subjects. We will explore the women's data further.

(a) On your calculator, enter the female mass data in L_1/list1 and the female metabolic rate data in L_2/list2. Define Plot 1 using the □ plotting symbol, and graph the scatterplot.

(b) Perform least-squares regression on your calculator and record the equation.

(c) Interpret the slope and y intercept of the least-squares line in context.

(d) Does the least-squares line provide an adequate model for the data? Calculate predicted y values in L_3/list3 and residuals in L_4/list4, as in the Technology Toolbox on page 219. Define Plot 2 to be a residual plot on your calculator with residuals on the vertical axis and lean body mass (x-values) on the horizontal axis. Use the □ plotting symbol. Copy the plot onto your paper. Label both axes appropriately.

(e) Define Plot 3 to be a residual plot on your calculator with residuals on the vertical axis and predicted metabolic rate (L_3/list3) on the horizontal axis. Use the □ plotting symbol. Copy the plot onto your paper. Label both axes. Compare this residual plot with the one from part (d).

3.42 Always plot your data! Table 3.2 (on the next page) presents four sets of data prepared by the statistician Frank Anscombe to illustrate the dangers of calculating without first plotting the data.

(a) Without making scatterplots, find the correlation and the least-squares regression line for all four data sets. What do you notice? Use the regression line to predict y for $x = 10$ for all four data sets.

(b) Make a scatterplot for each of the data sets and add the regression line to each plot.

(c) Now make a sketch of the residual plot for each of the four data sets.

(d) In which of the four cases would you be willing to use the regression line to describe the dependence of y on x? Explain your answer in each case.

Table 3.2 *Four data sets for exploring correlation and regression*

Data Set A

x	10	8	13	9	11	14	6	4	12	7	5
y	8.04	6.95	7.58	8.81	8.33	9.96	7.24	4.26	10.84	4.82	5.68

Data Set B

x	10	8	13	9	11	14	6	4	12	7	5
y	9.14	8.14	8.74	8.77	9.26	8.10	6.13	3.10	9.13	7.26	4.74

Data Set C

x	10	8	13	9	11	14	6	4	12	7	5
y	7.46	6.77	12.74	7.11	7.81	8.84	6.08	5.39	8.15	6.42	5.73

Data Set D

x	8	8	8	8	8	8	8	8	8	8	19
y	6.58	5.76	7.71	8.84	8.47	7.04	5.25	5.56	7.91	6.89	12.50

Source: Frank J. Anscombe, "Graphs in statistical analysis," *American Statistician,* 27 (1973), pp. 17–21.

How Well the Line Fits the Data: The Role of r^2 in Regression

A residual plot is a graphical tool for evaluating how well a linear model fits the data. There is also a numerical quantity that tells us how well the least-squares line does at predicting values of the response variable y. It is r^2, the ***coefficient of determination.*** Some computer packages call it "R-sq." You may have noticed this value in calculator and computer output for linear regression. For examples, look at the calculator screen shots in the Technology Toolbox on page 210 and the computer output in Figure 3.14 on page 209. Although it is true that this quantity is equal to the square of r, there is much more to this story.

When there is a linear relationship, some of the variation in y is accounted for by the fact that as x changes, it pulls y along the regression line with it. What if the least-squares regression line does *not* help predict the values of the response variable y as x changes? Then our "best guess" for the value of y at any x-value is simply the mean y-value, $\bar{y}$. The idea of r^2 is this: how much better is the least-squares line at predicting responses y than if we just used $\bar{y}$ as our prediction for every point?

Look at Figure 3.21 (on the next page), which shows the scatterplot of the fat gain and NEA data that we have studied throughout this section. The least-squares line is drawn on the plot in blue. Another line has been added in red: a horizontal line at the mean y-value, $\bar{y} = 2.3875$ kg.

For each point, we could ask: which comes closer to the actual y-value, the least-squares line or the horizontal line $y = \bar{y}$? Then we could count how many times each line is closer, and declare a "winner." But this approach doesn't take into account *how much better* one line does than the other.

We know that the least-squares line minimizes the sum of the squared residuals. For these data, $\sum \text{residual}^2 = \sum (y - \hat{y})^2 = 7.663$. Let's call this SSE, for sum

Figure 3.21 *Which does better at predicting the fat gain: the least-squares line (in blue) or the horizontal line at the mean y-value, y = 2.3875 kg (in red)? The coefficient of determination r^2 tells you how much better the least-squares line does.*

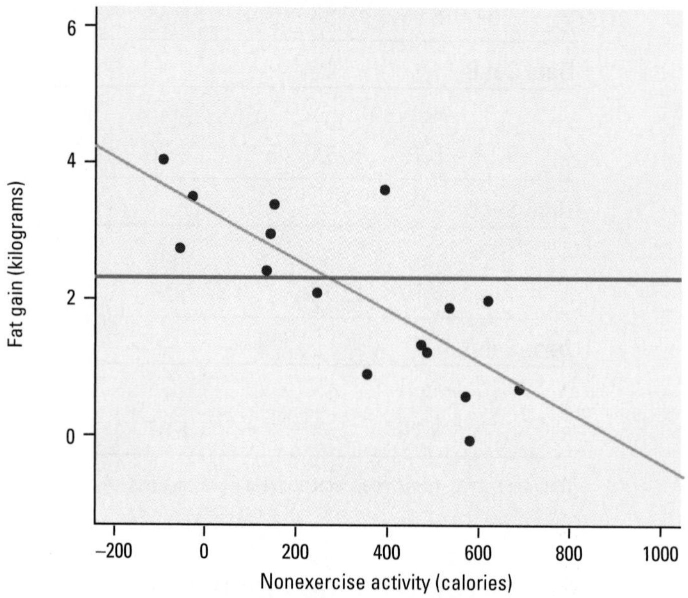

of squared errors. If we use $y = \bar{y}$ to make predictions, then our prediction errors are the vertical distances of the points away from the horizontal line. The sum of the squares of these numbers is $\text{SST} = \Sigma(y - \bar{y})^2 = 19.4575$. SST measures the total variation in the y-values (it's the variance of y times $n - 1$). The difference SST − SSE shows how much the least-squares line reduces the total variation in the responses y. In this case, SST − SSE = 19.4575 − 7.663 = 11.7945. We define

Recall from Chapter 1 that

$$s_y^2 = \sum \frac{(y - \bar{y})^2}{n - 1}$$

$$r^2 = \frac{\text{SST} - \text{SSE}}{\text{SST}} = 1 - \frac{\text{SSE}}{\text{SST}}$$

For the fat gain and NEA data,

$$r^2 = \frac{\text{SST} - \text{SSE}}{\text{SST}} = \frac{19.4575 - 7.663}{19.4575} = 0.606$$

We say "60.6% of the variation in fat gain is explained by the least-squares regression line relating fat gain and nonexercise activity."

The Coefficient of Determination: r^2 in Regression

The **coefficient of determination** r^2 is the fraction of the variation in the values of y that is explained by the least-squares regression line of y on x. We can calculate r^2 using the following formula:

$$r^2 = \frac{\text{SST} - \text{SSE}}{\text{SST}} = 1 - \frac{\text{SSE}}{\text{SST}}$$

where $\text{SSE} = \sum \text{residual}^2 = \sum (y - \hat{y})^2$ and $\text{SST} = \sum (y - \bar{y})^2$

If all of the points fall directly on the least-squares line, SSE = 0 and $r^2 = 1$. Then all of the variation in y is explained by the linear relationship with x. Since the least-squares line yields the smallest possible sum of squared prediction errors, SSE can never be more than SST, which is based on the line $y = \bar{y}$. In the worst-case scenario, the least-squares line does no better at predicting y than $y = \bar{y}$ does. Then SSE = SST and $r^2 = 0$. If $r^2 = 0.606$ (as in the fat gain and NEA example), then about 61% of the variation in y among the individual subjects is due to the straight-line relationship between y and x. The other 39% is individual variation among subjects that is not explained by the linear relationship.

It seems pretty remarkable that the coefficient of determination is actually the correlation squared. This fact provides an important connection between correlation and regression. When you report a regression, give r^2 as a measure of how successful the regression was in explaining the response. When you see a correlation, square it to get a better feel for the strength of the linear relationship.

Facts about Least-Squares Regression

One reason for the popularity of least-squares regression lines is that they have many convenient special properties. Here is a summary of several important facts about least-squares regression lines.

Fact 1. The distinction between explanatory and response variables is essential in regression. Least-squares regression makes the distances of the data points from the line small only in the y direction. If we reverse the roles of the two variables, we get a different least-squares regression line.

Example 3.15 ***Hubble's expanding universe***
Two different least-squares lines

Figure 3.22 (on the next page) is a scatterplot of data that played a central role in the discovery that the universe is expanding. They are the distances from earth of 24 spiral galaxies and the speed at which these galaxies are moving away from us, reported by the astronomer Edwin Hubble in 1929.[18] There is a positive linear relationship, $r = 0.7842$. More distant galaxies are moving away more rapidly. Astronomers believe that there is in fact a perfect linear relationship, and that the scatter is caused by imperfect measurements.

Figure 3.22 *Scatterplot of Hubble's data on the distance from earth of 24 galaxies and the velocity at which they are moving away from us. The two lines are the two least-squares regression lines: for predicting velocity from distance (green) and for predicting distance from velocity (dark blue).*

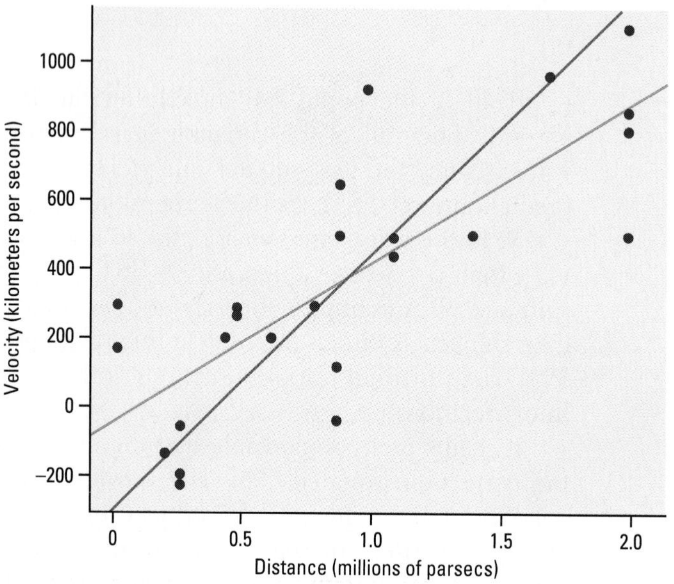

The two lines on the plot are two *different* least-squares regression lines. The green line is the least-squares line for predicting the velocity of a galaxy using its distance from earth. This line minimizes the sum of the squared vertical distances of points from the line. We might also use Hubble's data to predict another galaxy's distance from the earth using its velocity. Now the roles of the variables are reversed: velocity is the explanatory variable and distance is the response variable. The dark blue line in Figure 3.22 is the least-squares line for predicting distance from Earth using velocity. The two regression lines are not the same. **In the regression setting you must know clearly which variable is explanatory.**

CAUTION

Fact 2. There is a close connection between correlation and the slope of the least-squares line. The slope is

$$b = r\frac{s_y}{s_x}$$

This equation says that along the regression line, *a change of one standard deviation in x corresponds to a change of r standard deviations in y*. When the variables are perfectly correlated ($r = 1$ or $r = -1$), the change in the predicted response $\hat{y}$ is the same (in standard deviation units) as the change in x. Otherwise, because $-1 \le r \le 1$, the change in $\hat{y}$ is less than the change in x. As the correlation grows less strong, the prediction $\hat{y}$ moves less in response to changes in x.

Fact 3. The least-squares regression line of y on x always passes through the point $(\bar{x}, \bar{y})$. So the least-squares regression line of y on x is the line with slope rs_y/s_x that passes through the point $(\bar{x}, \bar{y})$.

Fact 4. The correlation r describes the strength of a straight-line relationship. In the regression setting, this description takes a specific form: *the square of the correlation, r^2, is the fraction of the variation in the values of y that is explained by the least-squares regression of y on x.*

Facts 2, 3, and 4 are special properties of least-squares regression. They are not true for other methods of fitting a line to data.

Exercises

3.43 Regression in the lab Gas chromatography is a technique used to detect very small amounts of a substance, for example, a contaminant in drinking water. Laboratories use regression to calibrate such techniques. The data below show the results of five measurements for each of four amounts of the substance being investigated.[19] The explanatory variable x is the amount of substance in the specimen, measured in nanograms (ng), units of 10^{-9} gram. The response variable y is the reading from the gas chromatograph.

Amount (ng)	Response				
0.25	6.25	7.98	6.54	6.37	7.96
1.00	29.7	30.0	30.1	29.5	29.1
5.00	211	204	212	213	205
20.00	929	905	922	928	919

(a) Make a scatterplot of the data. The relationship appears to be approximately linear, but the wide variation in the response values makes it hard to see the detail in this graph.

(b) Find the least-squares regression line for predicting reading from amount and add this line to your plot.

(c) Now compute the residuals and make a plot of the residuals against x. Describe carefully the pattern displayed by the residuals.

(d) What is r^2? What does this value say about the success of the linear regression in predicting gas chromatograph readings?

3.44 Class attendance and grades A study of class attendance and grades among first-year students at a state university showed that in general students who attended a higher percent of their classes earned higher grades. Class attendance explained 16% of the variation in grade index among the students. What is the numerical value of the correlation between percent of classes attended and grade index?

3.45 The professor swims, II Exercise 3.22 (page 197) gave Professor Moore's times (in minutes) to swim 2000 yards and his pulse rate after swimming (in beats per minute) for 23 sessions in the pool.

(a) A scatterplot shows a moderately strong negative linear relationship. Use your calculator or software to find the equation of the least-squares regression line.

(b) The next day's time is 34.30 minutes. Predict the professor's pulse rate. In fact, his pulse rate was 152. How accurate is your prediction?

(c) Suppose you were told only that the pulse rate was 152. You now want to predict swimming time. Find the equation of the least-squares regression line that is appropriate for this purpose. What is your prediction, and how accurate is it?

(d) Explain clearly, to someone who knows no statistics, why there are two different regression lines.

Regression toward the mean To "regress" means to go backward. Why are statistical methods for predicting a response from an explanatory variable called "regression"? Sir Francis Galton (1822–1911), who was the first to apply regression to biological and psychological data, looked at examples such as the heights of children versus the heights of their parents. He found that the taller-than-average parents tended to have children who were also taller than average but not as tall as their parents. Galton called this fact "regression toward the mean," and the name came to be applied to the statistical method.

3.46 Predicting fat, predicting NEA In Example 3.13 (page 208), we found that the least-squares regression line for predicting fat gain from nonexercise activity (NEA) was $\hat{y} = 3.505 - 0.00344x$. We might also use the data from the 16 subjects in this study to predict the change in NEA for another subject from that subject's fat gain when overfed for 8 weeks.

(a) Explain why you should *not* use the least-squares line calculated earlier to make such a prediction.

(b) Use the data from Example 3.9 (page 200) and your calculator to obtain the equation of the least-squares line that would be appropriate for predicting NEA from fat gain.

(c) Suppose the new subject's fat gain is 3.0 kg. One of the original 16 subjects had a fat gain of 3.0 kg, and that subject's NEA change was −57 calories. Explain why you should *not* just predict an NEA change of −57 calories for this new subject. What NEA change would you predict for this individual?

(d) Interpret the value of r^2 you obtained in part (b). How does this compare to the r^2-value we obtained earlier for the line $\hat{y} = 3.505 - 0.00344x$? Explain why this makes sense.

3.47 Predicting the stock market Some people think that the behavior of the stock market in January predicts its behavior for the rest of the year. Take the explanatory variable x to be the percent change in a stock market index in January and the response variable y to be the change in the index for the entire year. We expect a positive correlation between x and y because the change during January contributes to the full year's change. Calculation from data for the years 1960 to 1997 gives

$$\bar{x} = 1.75\% \qquad s_x = 5.36\% \qquad r = 0.596$$

$$\bar{y} = 9.07\% \qquad s_y = 15.35\%$$

(a) What percent of the observed variation in yearly changes in the index is explained by a straight-line relationship with the change during January?

(b) What is the equation of the least-squares line for predicting full-year change from January change?

(c) The mean change in January is $\bar{x} = 1.75\%$. Use your regression line to predict the change in the index in a year in which the index rises 1.75% in January. Why could you have given this result (up to roundoff error) without doing the calculation?

3.48 Beavers and beetles Ecologists sometimes find rather strange relationships in our environment. One study seems to show that beavers benefit beetles. The researchers laid out 23 circular plots, each four meters in diameter, in an area where beavers were cutting

down cottonwood trees. In each plot, they counted the number of stumps from trees cut by beavers and the number of clusters of beetle larvae. Here are the data:[20]

Stumps:	2	2	1	3	3	4	3	1	2	5	1	3
Beetle larvae:	10	30	12	24	36	40	43	11	27	56	18	40
Stumps:	2	1	2	2	1	1	4	1	2	1	4	
Beetle larvae:	25	8	21	14	16	6	54	9	13	14	50	

(a) Make a scatterplot that shows how the number of beaver-caused stumps influences the number of beetle larvae clusters. What does your plot show? (Ecologists think that the new sprouts from stumps are more tender than other cottonwood growth, so that beetles prefer them.)

(b) Find the least-squares regression line and draw it on your plot.

(c) Construct a residual plot. How well does the linear model fit the data?

(d) Interpret the r^2-value in the context of this problem.

Section 3.2 Summary

A **regression line** is a straight line that describes how a response variable y changes as an explanatory variable x changes. You can use a regression line to predict the value of y for any value of x by substituting this x into the equation of the line.

The **slope** b of a regression line $\hat{y} = a + bx$ is the rate at which the predicted response $\hat{y}$ changes along the line as the explanatory variable x changes. Specifically, b is the change in $\hat{y}$ when x increases by 1.

The **y intercept** a of a regression line $\hat{y} = a + bx$ is the predicted response $\hat{y}$ when the explanatory variable $x = 0$. This prediction is of no statistical use unless x can actually take values near 0.

Avoid **extrapolation,** the use of a regression line for prediction of values of the explanatory variable outside the range of the data from which the line was calculated.

The most common method of fitting a line to a scatterplot is least squares. The **least-squares regression line** is the straight line $\hat{y} = a + bx$ that minimizes the sum of the squares of the vertical distances of the observed points from the line.

The least-squares regression line of y on x is the line with slope $b = r(s_y/s_x)$ that passes through the point $(\bar{x}, \bar{y})$.

You can examine the fit of a regression line by studying the **residuals,** which are the differences between the observed and predicted values of y. Be on the lookout for points with unusually large residuals and also for nonlinear patterns and uneven variation about the residual $= 0$ line in the **residual plot.**

Correlation and regression are closely connected. The correlation r is the slope of the least-squares regression line when we measure both x and y in standardized units. The **coefficient of determination** r^2 is the fraction of the variance of one variable that is explained by least-squares regression on the other variable.

Section 3.2 Exercises

3.49 Measuring water quality Biochemical oxygen demand (BOD) measures organic pollutants in water by measuring the amount of oxygen consumed by microorganisms that break down these compounds. BOD is hard to measure accurately. Total organic carbon (TOC) is easy to measure, so it is common to measure TOC and use regression to predict BOD. A typical regression equation for water entering a city's water treatment plant is[21]

$$\text{BOD} = -55.43 + 1.507\ \text{TOC}$$

Both BOD and TOC are measured in milligrams per liter of water.

(a) What does the slope of this line say about the relationship between BOD and TOC?

(b) What is the predicted BOD when TOC = 0? Values of BOD less than 0 are impossible. Why do you think the prediction gives an impossible value?

3.50 IQ and school GPA Figure 3.4 (page 181) plots school grade point average (GPA) against IQ test score for 78 seventh-grade students. Calculation shows that the mean and standard deviation of the IQ scores are

$$\bar{x} = 108.9 \qquad s_x = 13.17$$

For the grade point averages, these values are

$$\bar{y} = 7.447 \qquad s_y = 2.10$$

The correlation between IQ and GPA is $r = 0.6337$.

(a) Find the equation of the least-squares line for predicting GPA from IQ.

(b) What percent of the observed variation in these students' GPAs can be explained by the linear relationship between GPA and IQ?

(c) One student has an IQ of 103 but a very low GPA of 0.53. What is the predicted GPA for a student with IQ = 103? What is the residual for this particular student?

3.51 Nahya infant weights A study of nutrition in developing countries collected data from the Egyptian village of Nahya. Here are the mean weights (in kilograms) for 170 infants in Nahya who were weighed each month during their first year of life:

Age (months):	1	2	3	4	5	6	7	8	9	10	11	12
Weight (kg):	4.3	5.1	5.7	6.3	6.8	7.1	7.2	7.2	7.2	7.2	7.5	7.8

(a) Plot the weight against time.

(b) A hasty user of statistics enters the data into software and computes the least-squares line without plotting the data. The result is

```
The regression equation is
Weight = 4.88 + 0.267 Age
```

Plot this line on your graph. Is it an acceptable summary of the overall pattern of growth? Remember that you can calculate the least-squares line for any set of two-variable data. It's up to you to decide if it makes sense to fit a line.

(c) Fortunately, the software also prints out the residuals from the least-squares line. In order of age along the rows, they are

−0.85	−0.31	0.02	0.35	0.58	0.62
0.45	0.18	−0.08	−0.35	−0.32	−0.29

Verify that the residuals have sum 0 (except for roundoff error). Plot the residuals against age and add a horizontal line at 0. Describe carefully the pattern that you see.

3.52 Keeping water clean Keeping water supplies clean requires regular measurement of levels of pollutants. The measurements are indirect—a typical analysis involves forming a dye by a chemical reaction with the dissolved pollutant, then passing light through the solution and measuring its "absorbance." To calibrate such measurements, the laboratory measures known standard solutions and uses regression to relate absorbance to pollutant concentration. This is usually done every day. Here is one series of data on the absorbance for different levels of nitrates. Nitrates are measured in milligrams per liter of water.[22]

Nitrates:	50	50	100	200	400	800	1200	1600	2000	2000
Absorbance:	7.0	7.5	12.8	24.0	47.0	93.0	138.0	183.0	230.0	226.0

(a) Chemical theory says that these data should lie on a straight line. If the correlation is not at least 0.997, something went wrong and the calibration procedure is repeated. Plot the data and find the correlation. Must the calibration be done again?

(b) What is the equation of the least-squares line for predicting absorbance from concentration? Interpret the slope and y intercept in the context of this problem.

(c) If the lab analyzed a specimen with 500 milligrams of nitrates per liter, what do you expect the absorbance to be?

(d) Do you expect your predicted absorbance to be very accurate? Give graphical and numerical evidence to support your answer.

3.53 A growing child Sarah's parents are concerned that she seems short for her age. Their doctor has the following record of Sarah's height:

Age (months):	36	48	51	54	57	60
Height (cm):	86	90	91	93	94	95

(a) Make a scatterplot of these data. Note the strong linear pattern.

(b) Using your calculator, find the equation of the least-squares regression line of height on age.

(c) Predict Sarah's height at 40 months and at 60 months. Use your results to draw the regression line on your scatterplot.

(d) What is Sarah's rate of growth, in centimeters per month? Normally growing girls gain about 6 cm in height between ages 4 (48 months) and 5 (60 months). What rate of growth is this in centimeters per month? Is Sarah growing more slowly than normal?

3.54 A nonsense prediction

(a) Use the least-squares regression line for the data in the previous exercise to predict Sarah's height at age 40 years (480 months). Your prediction is in centimeters. Convert it to inches using the fact that a centimeter is 0.3937 inch.

(b) The prediction is impossibly large. Explain why this happened.

3.55 What's my grade? In Professor Friedman's economics course the correlation between the students' total scores prior to the final examination and their final examination scores is $r = 0.6$. The pre-exam totals for all students in the course have mean 280 and standard deviation 30. The final-exam scores have mean 75 and standard deviation 8. Professor Friedman has lost Julie's final exam but knows that her total before the exam was 300. He decides to predict her final-exam score from her pre-exam total.

(a) Calculate and interpret the slope of the appropriate least-squares regression line for Professor Friedman's prediction. Do the same for the y intercept.

(b) Use the regression line to predict Julie's final-exam score.

(c) Julie doesn't think this method accurately predicts how well she did on the final exam. Determine r^2 and use the value you get to argue that her actual score could have been much higher (or much lower) than the predicted value.

3.56 Investing at home and overseas, I Investors ask about the relationship between returns on investments in the United States and on investments overseas. Table 3.3 gives the total returns on U.S. and overseas common stocks over a 27-year period. (The total return is change in price plus any dividends paid, converted into U.S. dollars. Both returns are averages over many individual stocks.)

Table 3.3 *Annual total return on overseas and U.S. stocks*

Year	Overseas % return	U.S. % return	Year	Overseas % return	U.S. % return	Year	Overseas % return	U.S. % return
1971	29.6	14.6	1980	22.6	32.3	1989	10.6	31.5
1972	36.3	18.9	1981	−2.3	−5.0	1990	−23.0	−3.1
1973	−14.9	−14.8	1982	−1.9	21.5	1991	12.8	30.4
1974	−23.2	−26.4	1983	23.7	22.4	1992	−12.1	7.6
1975	35.4	37.2	1984	7.4	6.1	1993	32.9	10.1
1976	2.5	23.6	1985	56.2	31.6	1994	6.2	1.3
1977	18.1	−7.4	1986	69.4	18.6	1995	11.2	37.6
1978	32.6	6.4	1987	24.6	5.1	1996	6.4	23.0
1979	4.8	18.2	1988	28.5	16.8	1997	2.1	33.4

Note: The U.S. returns are for the Standard & Poor's 500 index. The overseas returns are for the Morgan Stanley Europe, Australasia, Far East (EAFE) index.

(a) Make a scatterplot suitable for predicting overseas returns from U.S. returns.

(b) Find the correlation and r^2. Describe the relationship between U.S. and overseas returns in words, using r and r^2 to make your description more precise.

(c) Find the least-squares regression line for predicting overseas returns from U.S. returns. Draw the line on the scatterplot.

(d) In 1997, the return on U.S. stocks was 33.4%. Use the regression line to predict the return on overseas stocks. The actual overseas return was 2.1%. Are you confident that predictions using the regression line will be quite accurate? Why or why not?

(e) Circle the point that has the largest residual (either positive or negative). What year is this? Are there any other unusual points?

3.57 Investing at home and overseas, II Exercise 3.56 examined the relationship between returns on U.S. and overseas stocks. Investors also want to know what typical returns are and how much year-to-year variability (called "volatility" in finance) there is. Regression and correlation do not answer these questions.

(a) Find the five-number summaries for both U.S. and overseas returns, and make side-by-side boxplots to compare the two distributions.

(b) Were returns generally higher in the United States or overseas during this period? Explain your answer.

(c) Were returns more volatile (more variable) in the United States or overseas during this period? Explain your answer.

3.58 Will I bomb the final? We expect that students who do well on the midterm exam in a course will usually also do well on the final exam. Gary Smith of Pomona College looked at the exam scores of all 346 students who took his statistics class over a 10-year period.[23] The least-squares line for predicting final-exam score from midterm exam score was $\hat{y} = 46.6 + 0.41x$.

Octavio scores 10 points above the class mean on the midterm. How many points above the class mean do you predict that he will score on the final? (*Hint:* Use the fact that the least-squares line passes through the point $(\bar{x}, \bar{y})$ and the fact that Octavio's midterm score is $\bar{x} + 10$. This is an example of the phenomenon that gave "regression" its name: students who do well on the midterm will on the average do less well, but still above average, on the final.)

3.3 Correlation and Regression Wisdom

Correlation and regression are powerful tools for describing the relationship between two variables. When you use these tools, you must be aware of their limitations. You already know that:

CAUTION

- **Correlation and regression describe only linear relationships.** You can do the calculations for any relationship between two quantitative variables, but the results are useful only if the scatterplot shows a linear pattern.

CAUTION

- **Extrapolation** (using a model outside the range of the data) **often produces unreliable predictions.**

- **Correlation is not resistant.** Always plot your data and look for unusual observations before you interpret correlation.

Here are some other cautions to keep in mind when you apply correlation and regression or read accounts of their use.

Look for Outliers and Influential Observations

You already know that the correlation r is not resistant. One unusual point in a scatterplot can greatly change the value of r. Is the least-squares line resistant? The following example sheds some light on this question.

Example 3.16 *Gesell scores*

Identifying unusual points in regression

Does the age at which a child begins to talk predict later score on a test of mental ability? A study of the development of young children recorded the age in months at which each of 21 children spoke their first word and their Gesell Adaptive Score, the result of an aptitude test taken much later. The data appear in Table 3.4.

Table 3.4 *Age (months) at first word and Gesell score*

Child	Age	Score	Child	Age	Score	Child	Age	Score
1	15	95	8	11	100	15	11	102
2	26	71	9	8	104	16	10	100
3	10	83	10	20	94	17	12	105
4	9	91	11	7	113	18	42	57
5	15	102	12	9	96	19	17	121
6	20	87	13	10	83	20	11	86
7	18	93	14	11	84	21	10	100

Source: N. R. Draper and J. A. John, "Influential observations and outliers in regression," *Technometrics,* 23 (1981), pp. 21–26.

Data:

1. *Who?* The individuals are 21 young children.
2. *What?* The variables measured are age at first spoken word and later score on the Gesell Adaptive test.
3. *Why?* Researchers wondered whether the age at which a child first speaks can help predict the child's mental ability.
4. *When, where, how, and by whom?* These data were originally collected by L. M. Linde of UCLA but were first published by M. R. Mickey, O. J. Dunn, and V. Clark, "Note on the use of stepwise regression in detecting outliers," *Computers and Biomedical Research,* 1 (1967), pp. 105–111. The data have been used by several authors.

Graph: Figure 3.23 (on the next page) is a scatterplot, with age at first word as the explanatory variable x and Gesell score as the response variable y. Children 3 and 13, and also Children 16 and 21, have identical values of both variables. We use a different plotting symbol to show that one point stands for two individuals.

Figure 3.23 *Scatterplot of Gesell Adaptive Scores versus the age at first word for 21 children, from Table 3.4. The line is the least-squares regression line for predicting Gesell score from age at first word.*

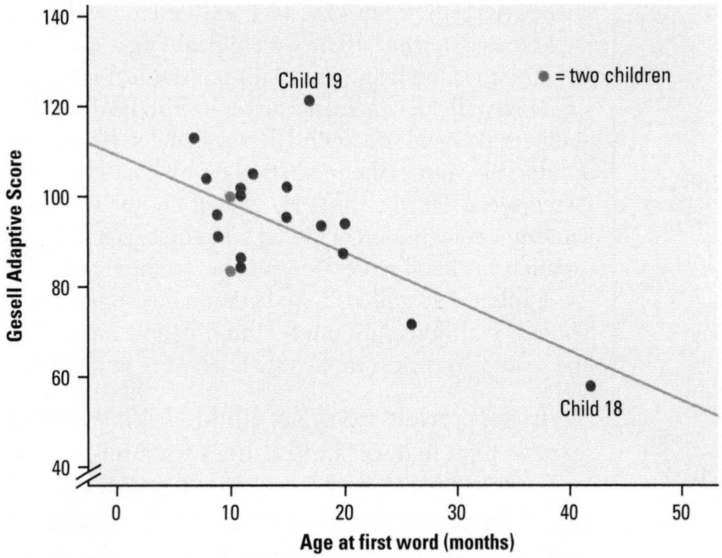

Numerical summaries: For these data, $\bar{x} = 14.381$, $s_x = 7.947$, $\bar{y} = 93.667$, $s_y = 13.987$, and $r = -0.640$.

Model: The line on the plot is the least-squares regression line of Gesell score on age at first word. Its equation is

$$\hat{y} = 109.8738 - 1.1270x$$

Figure 3.24 is a residual plot for these data.

Figure 3.24 *Residual plot for the regression of Gesell score on age at first word. Child 19 is an outlier, and Child 18 is an influential observation that does not have a large residual.*

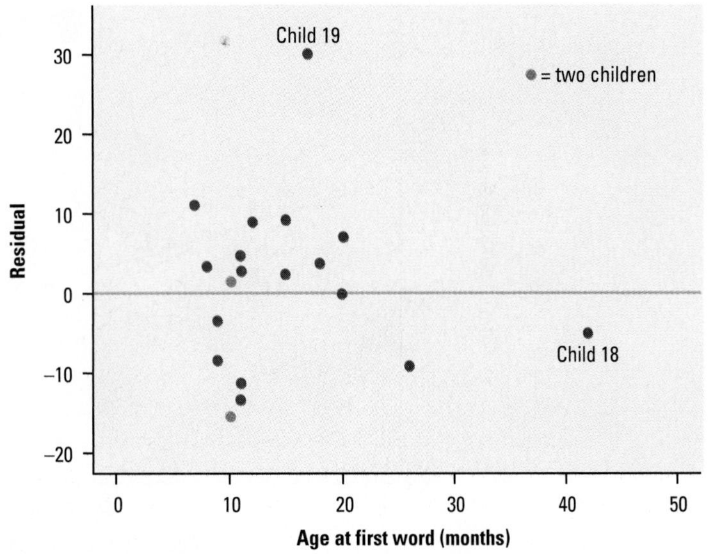

Interpretation: The scatterplot shows a negative association. That is, children who begin to speak later tend to have lower test scores than early talkers. The overall pattern is moderately linear ($r = -0.640$).

The slope of the regression line suggests that for every month older a child is when she begins to speak, her score on the Gesell test will decrease by about 1.13 points. According to the y intercept, 109.87, a child who first speaks at age 0 months would score about 109.9 on the Gesell test later. This nonsensical interpretation is due to extrapolation.

How well does the least-squares line fit the data? The residual plot shows a fairly "random" scatter of points around the "residual = 0" line with one very large positive residual (Child 19). Most of the prediction errors (that is, residuals) are 10 points or fewer on the Gesell score. Since $r^2 = 0.41$, 41% of the variation in Gesell scores is explained by the least-squares regression of Gesell score on age at first spoken word. That leaves 59% of the variation in Gesell scores unexplained by the linear relationship.

Children 18 and 19 have been marked on the scatterplot and on the residual plot. These two children are unusual in different ways. Child 19 lies far from the regression line. Child 18 is close to the line but far out in the x direction.

In the previous example, Child 19 is an *outlier in the y direction*, with a Gesell score so high that we should check for a mistake in recording it. In fact, the score is correct. Child 18 is an *outlier in the x direction*. This child began to speak much later than any of the other children. Because of its extreme position on the age scale, this point has a strong *influence* on the position of the regression line. Figure 3.25 adds a second regression line, calculated after leaving out Child 18. You can see that this one point moves the line quite a bit. (In fact, the equation of the new least-squares line is $\hat{y} = 105.63 - 0.779x$.)

Figure 3.25 *Two least-squares regression lines of Gesell score on age at first word. The green line is calculated from all the data. The dark blue line is calculated leaving out Child 18. Child 18 is an influential observation because leaving out this point moves the regression line quite a bit.*

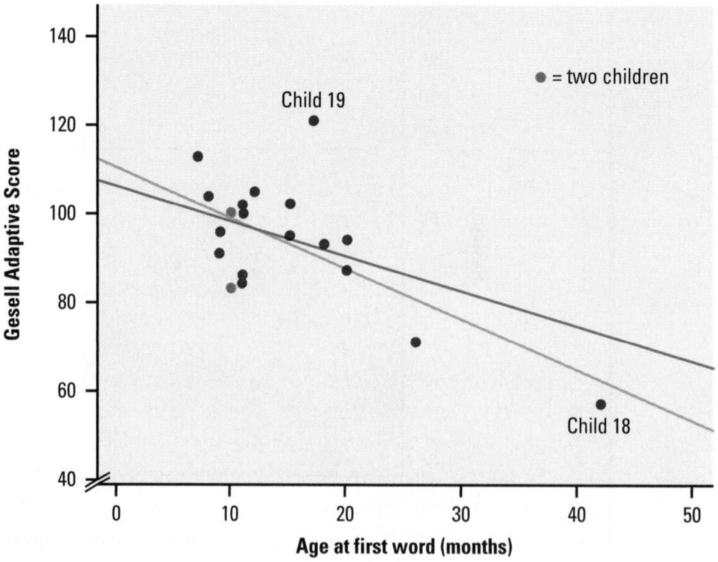

Least-squares lines make the sum of the squares of the vertical distances to the points as small as possible. A point that is extreme in the x direction with no other points near it pulls the line toward itself. We call such points ***influential.***

Outliers and Influential Observations in Regression

An **outlier** is an observation that lies outside the overall pattern of the other observations. Points that are outliers in the y direction of a scatterplot have large regression residuals, but other outliers need not have large residuals.

An observation is **influential** for a statistical calculation if removing it would markedly change the result of the calculation. Points that are outliers in the x direction of a scatterplot are often influential for the least-squares regression line.

We did not need the distinction between outliers and influential observations in Chapter 1. A single large salary that pulls up the mean salary $\bar{x}$ for a group of workers is an outlier because it lies far above the other salaries. It is also influential, because the mean changes when it is removed. In the regression setting, however, not all outliers are influential. The least-squares line is most likely to be heavily influenced by observations that are outliers in the x direction. The scatterplot will alert you to such observations. Influential points often have small residuals, because they pull the regression line toward themselves. If you just look at a residual plot, you may miss influential points.

The surest way to verify that a point is influential is to find the regression line both with and without the suspect point, as in Figure 3.25. If the line moves more than a small amount when the point is deleted, the point is influential.

Example 3.17 *Gesell scores, continued*
An influential observation

The strong influence of Child 18 makes the original regression of Gesell score on age at first word misleading. The original data have $r^2 = 0.41$. That is, the least-squares line relating age at which a child begins to talk with Gesell score explains 41% of the variation on this later test of mental ability. This relationship is strong enough to be interesting to parents. If we leave out Child 18, r^2 drops to only 11%. The apparent strength of the association was largely due to a single influential observation.

What should the child development researcher do? She must decide whether Child 18 is so slow to speak that this individual should not be allowed to influence the analysis. If she excludes Child 18, much of the evidence for a connection between the age at which a child begins to talk and later ability score vanishes. If she keeps Child 18, she needs data on other children who were also slow to begin talking, so that the analysis no longer depends so heavily on just one child.

Exercises

3.59 How many calories? Exercise 3.25 (page 198) gives data on the true calories in 10 foods and the average guesses made by a large group of people. In that exercise, you explored the influence of two outlying observations on the correlation.

(a) Make a scatterplot suitable for predicting guessed calories from true calories. Circle the points for spaghetti and snack cake on your plot. How do these points lie outside the linear pattern of the other 8 points?

(b) Use your calculator to find the least-squares regression line of guessed calories on true calories. Do this twice, first for all 10 data points and then leaving out spaghetti and snack cake.

(c) Plot both lines on your graph. (Make one dashed so that you can tell them apart.) Are spaghetti and snack cake, taken together, influential observations? Explain your answer.

3.60 Influential or not? The discussion of Example 3.17 shows that Child 18 in the Gesell data in Table 3.4 is an influential observation. Now we will examine the effect of Child 19, who is also an outlier in Figure 3.23.

(a) Find the least-squares regression line of Gesell score on age at first word, leaving out Child 19. Example 3.16 gives the regression line from all the children. Plot both lines on the same graph. (You do not have to make a scatterplot of all the points—just plot the two lines.) Would you call Child 19 very influential? Why?

(b) How does removing Child 19 change r^2 for this regression? Explain why r^2 changes in this direction when you drop Child 19.

3.61 More sparrowhawks Exercise 3.6 (page 180) asked you to examine the relationship between the percent of adult sparrowhawks in a colony that return from the previous year and the number of new adults that join the colony. Now we will use them to illustrate influence.

(a) Make a scatterplot of the data suitable for predicting new adults from percent of returning adults. Then add two new points. Point A: 10% return, 15 new birds. Point B: 60% return, 28 new birds. In which direction is each new point an outlier?

(b) Add three least-squares regression lines to your plot: for the original 13 colonies, for the original colonies plus Point A, and for the original colonies plus Point B. Which new point is more influential for the regression line? Explain in simple language why each new point moves the line in the way your graph shows.

3.62 I feel your pain "Empathy" means being able to understand what others feel. To see how the brain expresses empathy, researchers recruited 16 couples in their mid-twenties who were married or had been dating for two years. They zapped the man's hand with an electrode while the woman watched, and measured the activity in several parts of the woman's brain that would respond to her own pain. Brain activity was recorded as a fraction of the activity observed when the woman herself was zapped with the electrode. The women also completed a psychological test that measures empathy. Will women who are higher in empathy respond more strongly when their partner has a painful experience? Here are data for one brain region:[24]

Subject:	1	2	3	4	5	6	7	8
Empathy score:	38	53	41	55	56	61	62	48
Brain activity:	−0.120	0.392	0.005	0.369	0.016	0.415	0.107	0.506
Subject:	9	10	11	12	13	14	15	16
Empathy score:	43	47	56	65	19	61	32	105
Brain activity:	0.153	0.745	0.255	0.574	0.210	0.722	0.358	0.779

(a) Analyze these data in a way that will help answer the researchers' question. Follow the Data Analysis Toolbox.

(b) Is Subject 16 influential for the correlation? Justify your answer.

(c) Is Subject 16 influential for the least-squares line? Justify your answer.

Beware the Lurking Variable

Another caution is even more important: the relationship between two variables can often be understood only by taking other variables into account. ***Lurking variables*** can make a correlation or regression misleading.

Lurking Variable

A **lurking variable** is a variable that is not among the explanatory or response variables in a study and yet may influence the interpretation of relationships among those variables.

You should always think about possible lurking variables before you draw conclusions based on correlation or regression.

Example 3.18 *Is math the key to college success?*
Beware the lurking variable!

A College Board study of 15,941 high school graduates found a strong correlation between how much math minority students took in high school and their later success in college. News articles quoted the College Board as saying that "math is the gatekeeper for success in college."[25] Maybe so, but we should also think about lurking variables. Minority students from middle-class homes with educated parents no doubt take more high school math courses. They also are more likely to have a stable family, parents who emphasize education and can pay for college, and so on. These students would succeed in college even if they took fewer math courses. The family background of students is a lurking variable that probably explains much of the relationship between math courses and college success.

Example 3.19 *Imported goods and private health spending*

A nonsense correlation

Figure 3.26 displays a strong positive linear association. The correlation between these variables is $r = 0.9749$. Because $r^2 = 0.9504$, least-squares regression of y on x will explain 95% of the variation in the values of y.

Figure 3.26 *The relationship between private spending on health and the value of goods imported the same year.*

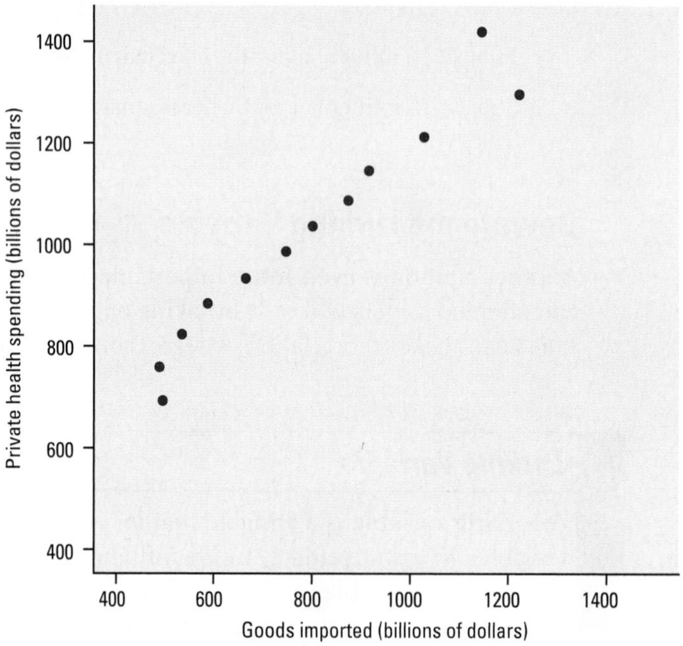

The explanatory variable in Figure 3.26 is the dollar value of goods imported into the United States in the years 1990 to 2001. The response variable is private spending on health in these years. There is no economic relationship between these variables. The strong association is due entirely to the fact that both imports and health spending grew rapidly in these years. The common year for each point is a lurking variable. Any two variables that both increase over time will show a strong association. This does not mean that one variable explains or influences the other.

Correlations such as that in Example 3.19 are sometimes called "nonsense correlations." The correlation is real. What is nonsense is the suggestion that the variables are directly related so that changing one of the variables *causes* changes in the other. The question of causation is important enough to merit separate treatment in the next chapter. For now, just remember that an association between two variables x and y can reflect many types of relationship among x, y, and one or more lurking variables. In other words, **association does not imply causation.**

Lurking variables sometimes create a correlation between x and y, as in Examples 3.18 and 3.19. They can also hide a true relationship between x and y, as the following example illustrates.

Example 3.20

Housing and health in Hull, England

Lurking variable hides relationship

A study of housing conditions in the city of Hull, England, measured a large number of variables for each of the wards in the city. Two of the variables were a measure x of overcrowding and a measure y of the lack of indoor toilets. Because x and y are both measures of inadequate housing, we expect a high correlation. In fact, the correlation was only $r = 0.08$. How can this be?

Investigation found that some poor wards had a lot of public housing. These wards had high values of x but low values of y because public housing always includes indoor toilets. Other poor wards lacked public housing, and these wards had high values of both x and y. Within wards of each type, there was a strong positive association between x and y. Analyzing all wards together ignored the lurking variable—amount of public housing—and hid the nature of the relationship between x and y.[26]

Figure 3.27 shows in simplified form how groups formed by a categorical lurking variable, as in the housing example, can make correlation and regression misleading. The groups appear as clusters of points in the scatterplot. There is a strong relationship between x and y within each of the clusters. In fact, $r = 0.85$ and $r = 0.91$ in the two clusters. However, because similar values of x correspond to quite different values of y in the two clusters, x alone is of little value for predicting y. The correlation for all the points together is only $r = 0.14$. This example is another reminder to plot the data instead of simply calculating numerical measures like the correlation.

Figure 3.27 *The variables in this scatterplot have a small correlation even though there is a strong correlation within each of the clusters.*

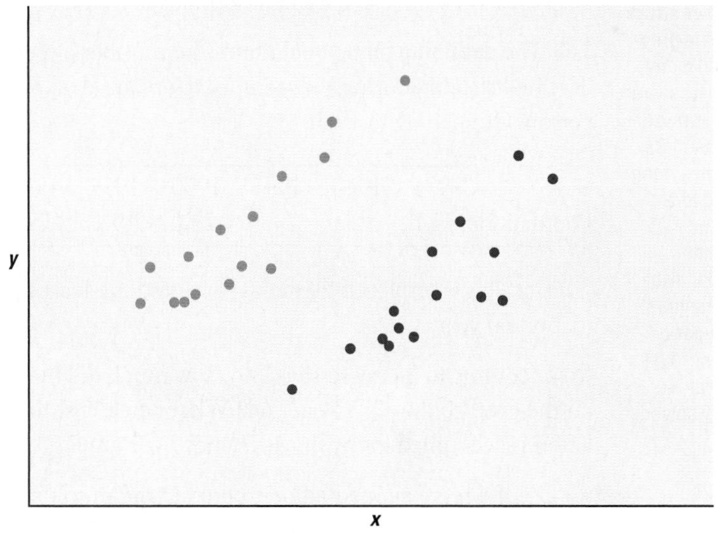

Beware Correlations Based on Averaged Data

Many regression or correlation studies work with averages or other measures that combine information from many individuals. For example, if we plot the average height of young children against their age in months, we will see a very strong positive association with correlation near 1. But individual children of the same age vary a great deal in height. A plot of height against age for individual children will show much more scatter and lower correlation than the plot of average height against age.

Correlations based on averages are usually too high when applied to individuals. This is another reminder that it is important to note exactly what variables were measured in a statistical study.

Exercises

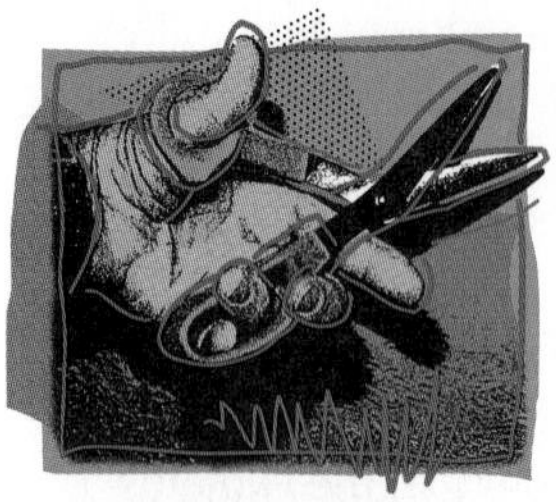

Do left-handers die early? Yes, said a study of 1000 deaths in California. Left-handed people died at an average age of 66 years; right-handers, at 75 years of age. Should left-handed people fear an early death? No—the lurking variable has struck again. Older people grew up in an era when many natural left-handers were forced to use their right hands. So right-handers are more common among older people, and left-handers are more common among the young. When we look at deaths, the left-handers who die are younger on the average because left-handers in general are younger. Mystery solved.

3.63 The link between health and income An article entitled "The Health and Wealth of Nations" says: "The positive correlation between health and income per capita is one of the best-known relations in international development. This correlation is commonly thought to reflect a causal link running from income to health. . . . Recently, however, another intriguing possibility has emerged: that the health-income correlation is partly explained by a causal link running the other way—from health to income."[27]

Explain how higher income in a nation can cause better health. Then explain how better health can cause higher income. There is no simple way to determine the direction of the link.

3.64 How to shorten a hospital stay A study shows that there is a positive correlation between the size of a hospital (measured by its number of beds x) and the median number of days y that patients remain in the hospital. Does this mean that you can shorten a hospital stay by choosing a small hospital? Explain.

3.65 The declining farm population The number of people living on American farms has declined steadily during this century. Here are data on the farm population (millions of persons) from 1935 to 1980:

Year:	1935	1940	1945	1950	1955	1960	1965	1970	1975	1980
Population:	32.1	30.5	24.4	23.0	19.1	15.6	12.4	9.7	8.9	7.2

(a) Make a scatterplot of these data and find the least-squares regression line of farm population on year.

(b) According to the regression line, how much did the farm population decline each year on the average during this period? What percent of the observed variation in farm population is accounted for by linear change over time?

(c) Use the regression equation to predict the number of people living on farms in 2010. Is this result reasonable? Why?

3.66 Predicting enrollment The mathematics department of a large state university must plan the number of sections and instructors required for its elementary courses. The department hopes that the number of students in these courses can be predicted from the number of first-year students, which is known before the new students actually choose courses. The table below contains data for several years.[28] The explanatory variable x is the number of first-year students. The response variable y is the number of students who enroll in elementary mathematics courses.

Year:	1993	1994	1995	1996	1997	1998	1999	2000
x:	4595	4827	4427	4258	3995	4330	4265	4351
y:	7364	7547	7099	6894	6572	7156	7232	7450

(a) Carry out a regression analysis to help answer the department's question. Follow the Data Analysis Toolbox.

(b) If you did not do so in (a), construct a plot of the residuals against the explanatory variable x. Now construct a plot of the residuals against time. Each plot raises a concern about the quality of the regression model you calculated in (a). Describe each concern clearly.

(c) Beginning in 1998, one of the schools in the university changed its program to require that entering students take another mathematics course. Explain how this change introduced a lurking variable that wasn't apparent from your work in (a), but that was evident in (b).

CAUTION

Note: **Many lurking variables change systematically over time.** One useful method for detecting lurking variables is therefore to **plot the residuals against the time order of the observations.**

3.67 Stock market indexes The Standard & Poor's 500 index is an average of the price of 500 stocks. There is a moderately strong correlation (roughly, $r = 0.6$) between how much this index changes in January and how much it changes during the entire year. If we looked instead at data on all 500 individual stocks, we would find a very different correlation. Would the correlation be higher or lower? Why?

3.68 Investing at home and overseas, III Exercise 3.56 (page 232) examined the relationship between returns on U.S. and overseas stocks. Return to the scatterplot and regression line for predicting overseas returns from U.S. returns.

(a) Circle the point that has the largest residual (either positive or negative). What year is this? Redo the regression without this point and add the new regression line to your plot. Was this observation very influential?

(b) Whenever we regress two variables that both change over time, we should plot the residuals against time as a check for time-related lurking variables. Make this plot for the stock returns data. Are there any suspicious patterns in the residuals?

Section 3.3 Summary

Correlation and regression must be interpreted with caution. Plot the data to be sure that the relationship is roughly linear and to detect **outliers.** Also look for **influential observations,** individual points that substantially change the correlation or the regression line. Outliers in x are often influential for the regression line.

Lurking variables may explain the relationship between the explanatory and response variables. Correlation and regression can be misleading if you ignore important lurking variables.

Remember that correlations based on averages are usually higher than correlations based on individual scores.

Section 3.3 Exercises

3.69 Heart attacks and hospitals If you need medical care, should you go to a hospital that handles many cases like yours? Figure 3.28 presents some data for heart attacks. The figure plots mortality (the proportion of patients who died) against the number of heart attack patients treated for a large group of hospitals in a recent year. The line on the plot is the least-squares regression line for predicting mortality from number of patients.

Figure 3.28 *Mortality of heart attack patients and number of heart attack cases treated for a large group of hospitals, for Exercise 3.69.*

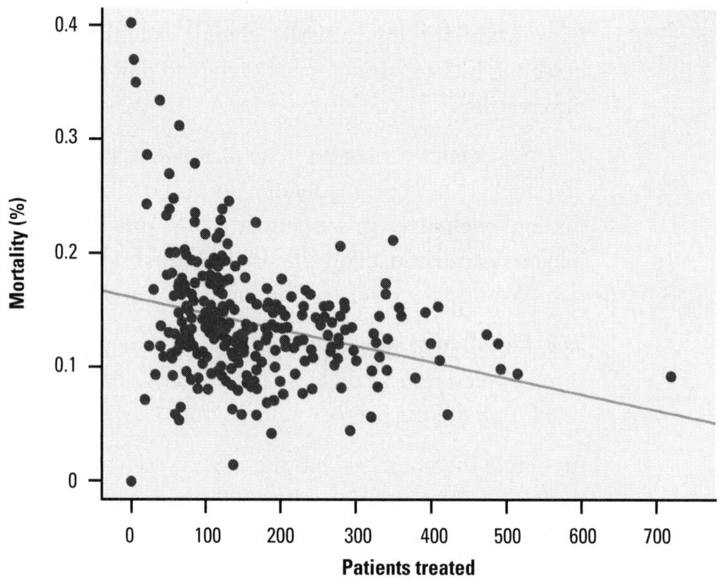

(a) Do the plot and regression generally support the thesis that mortality is lower at hospitals that treat more heart attacks? Is the relationship very strong?

(b) In what way is the pattern of the plot nonlinear? Does the nonlinearity strengthen or weaken the conclusion that heart attack patients should avoid hospitals that treat few heart attacks? Why?

3.70 Golf scores Here are the golf scores of 11 members of a women's golf team in two rounds of tournament play:

Player:	1	2	3	4	5	6	7	8	9	10	11
Round 1:	89	90	87	95	86	81	105	83	88	91	79
Round 2:	94	85	89	89	81	76	89	87	91	88	80

(a) Plot the data with the Round 1 scores on the x axis and the Round 2 scores on the y axis. There is a generally linear pattern except for one potentially influential observation. Circle this observation on your graph.

(b) Here are the equations of two least-squares lines. One of them is calculated from all 11 data points and the other omits the influential observation.

$$\hat{y} = 20.49 + 0.754x$$

$$\hat{y} = 50.01 + 0.410x$$

Draw both lines on your scatterplot. Which line omits the influential observation? How do you know this?

3.71 Better readers A study of elementary school children, ages 6 to 11, finds a high positive correlation between shoe size x and score y on a test of reading comprehension. What explains this correlation?

3.72 Managing diabetes People with diabetes must manage their blood sugar levels carefully. They measure their fasting plasma glucose (FPG) several times a day with a glucose meter. Another measurement, made at regular medical checkups, is called HbA. This is roughly the percent of red blood cells that have a glucose molecule attached. It measures average exposure to glucose over a period of several months. Table 3.5 gives data on both HbA and FPG for 18 diabetics five months after they had completed a diabetes education class.

Table 3.5 *Two measures of glucose level in diabetics*

Subject	HbA (%)	FPG (mg/ml)	Subject	HbA (%)	FPG (mg/ml)	Subject	HbA (%)	FPG (mg/ml)
1	6.1	141	7	7.5	96	13	10.6	103
2	6.3	158	8	7.7	78	14	10.7	172
3	6.4	112	9	7.9	148	15	10.7	359
4	6.8	153	10	8.7	172	16	11.2	145
5	7.0	134	11	9.4	200	17	13.7	147
6	7.1	95	12	10.4	271	18	19.3	255

Source: Debora L. Arsenau, "Comparison of diet management instruction for patients with non-insulin dependent diabetes mellitus: learning activity package vs. group instruction," MS thesis, Purdue University, 1993.

(a) Make a scatterplot with HbA as the explanatory variable. There is a positive linear relationship, but it is surprisingly weak.

(b) Subject 15 is an outlier in the y direction. Subject 18 is an outlier in the x direction. Find the correlation for all 18 subjects, for all except Subject 15, and for all except Subject 18. Are either or both of these subjects influential for the correlation? Explain in simple language why r changes in opposite directions when we remove each of these points.

(c) Add three regression lines for predicting FPG from HbA to your scatterplot: for all 18 subjects, for all except Subject 15, and for all except Subject 18. Is either Subject 15 or Subject 18 strongly influential for the least-squares line? Explain in simple language what features of the scatterplot explain the degree of influence.

3.73 Climate change Global warming has many indirect effects on climate. For example, the summer monsoon winds in the Arabian Sea bring rain to India and are critical for agriculture. As the climate warms and winter snow cover in the vast landmass of Europe and Asia decreases, the land heats more rapidly in the summer. This may increase the strength of the monsoon. Here are data on snow cover (in millions of square kilometers) and summer wind stress (in newtons per square meter):[29]

Snow cover	Wind stress	Snow cover	Wind stress	Snow cover	Wind stress
6.6	0.125	16.6	0.111	26.6	0.062
5.9	0.160	18.2	0.106	27.1	0.051
6.8	0.158	15.2	0.143	27.5	0.068
7.7	0.155	16.2	0.153	28.4	0.055
7.9	0.169	17.1	0.155	28.6	0.033
7.8	0.173	17.3	0.133	29.6	0.029
8.1	0.196	18.1	0.130	29.4	0.024

(a) Analyze these data to uncover the nature and strength of the effect of decreasing snow cover on wind stress. Follow the Data Analysis Toolbox.

(b) The report from which the data were taken is not clear about the time period that the data describe. Your work in (a) should include a scatterplot. That plot shows an odd pattern that correlation and regression don't describe. What is this pattern? On the basis of the scatterplot and rereading the report, we suspect that the data are for the months of May, June, and July over a period of 7 years. Why is the pattern in the graph consistent with this interpretation?

3.74 Economists' education and income There is a strong positive correlation between years of education and income for economists employed by business firms. (In particular, economists with doctorates earn more than economists with only a bachelor's degree.) There is also a strong positive correlation between years of education and income for economists employed by colleges and universities. But when all economists are considered, there is a negative correlation between education and income. The explanation for this is that business pays high salaries and employs mostly economists with bachelor's degrees, while colleges pay lower salaries and employ mostly economists with doctorates. Sketch a scatterplot with two groups of cases (business and academic) that illustrates how a strong positive correlation within each group and a negative overall correlation can occur together.

3.75 A computer game A multimedia statistics learning system includes a test of skill in using the computer's mouse. The software displays a circle at a random location on the computer screen. The subject tries to click in the circle with the mouse as quickly as possible. A new circle appears as soon as the subject clicks the old one. Table 3.6 gives data for one subject's trials, 20 with each hand. Distance is the distance from the cursor location to the center of the new circle, in units whose actual size depends on the size of the screen. Time is the time required to click in the new circle, in milliseconds.

(a) We suspect that time depends on distance. Make a scatterplot of time against distance, using separate symbols for each hand.

(b) Describe the pattern. How can you tell that the subject is right-handed?

(c) Find the regression line of time on distance separately for each hand. Draw these lines on your plot. Which regression does a better job of predicting time from distance? Give numerical measures that describe the success of the two regressions.

Table 3.6 *Reaction times in a computer game*

Time	Distance	Hand	Time	Distance	Hand
115	190.70	right	240	190.70	left
96	138.52	right	190	138.52	left
110	165.08	right	170	165.08	left
100	126.19	right	125	126.19	left
111	163.19	right	315	163.19	left
101	305.66	right	240	305.66	left
111	176.15	right	141	176.15	left
106	162.78	right	210	162.78	left
96	147.87	right	200	147.87	left
96	271.46	right	401	271.46	left
95	40.25	right	320	40.25	left
96	24.76	right	113	24.76	left
96	104.80	right	176	104.80	left
106	136.80	right	211	136.80	left
100	308.60	right	238	308.60	left
113	279.80	right	316	279.80	left
123	125.51	right	176	125.51	left
111	329.80	right	173	329.80	left
95	51.66	right	210	51.66	left
108	201.95	right	170	201.95	left

Source: P. Velleman, *ActivStats 2.0,* Addison-Wesley Interactive, 1997.

3.76 Using residuals It is possible that the subject in the previous exercise got better in later trials due to learning. It is also possible that he got worse due to fatigue. Plot the residuals from each regression against the time order of the trials (down the columns in Table 3.6). Is either of these systematic effects of time visible in the data?

CASE CLOSED!

Are baseballs "juiced"?

Return to the chapter-opening Case Study (page 169). With what you have learned about scatterplots, correlation, and linear regression, you are now ready to answer some questions about how the mean number of home runs hit per Major League Baseball game has changed over time.

1. Make a scatterplot of year versus the average number of home runs hit per game. Describe the pattern you see in the plot. Is there a linear relationship? Use your calculator to compute the correlation.

2. Look at the home run data since Rawlings started manufacturing baseballs for Major League Baseball in 1977. Make a scatterplot of these data. Describe the pattern you see. Find the correlation.

3. Fit a linear model to the home run data from 1977 to 2000 using year as the explanatory variable. Interpret the slope and *y* intercept of your linear model in context.

4. Construct a residual plot. Discuss what this plot tells you about how well your model fits the data.

5. Interpret the value of r^2 in the context of this problem.

6. Use your model to predict the mean number of home runs per game for 2001. Based on your answers to Questions 4 and 5, how accurate do you think this prediction will be?

7. The actual mean number of home runs hit per game in 2001 was 2.092. How good is the estimate that you found in your answer to the previous question?

8. Can we use these data to predict the mean number of home runs per game for 2020? Why or why not?

Chapter Review

Summary

Chapters 1 and 2 dealt with data analysis for a single variable. In this chapter, we have studied relations between two quantitative variables. Some of our analysis depends on whether one is an explanatory variable and the other a response variable.

Data analysis begins with graphs, then adds numerical summaries of specific aspects of the data. When the data show a regular pattern, we can use mathematical models to summarize this regularity. We should also examine any deviations from the pattern.

Scatterplots show the relationship, whether or not there is an explanatory-response distinction. Correlation describes the strength of a linear relationship, and least-squares regression fits a line to data that have an explanatory-response relation. Residual plots and r^2 help us assess how well the linear model fits the data.

Correlation and regression are powerful tools, but they have their limitations. Lurking variables can hide or alter the relationship between two variables. Outliers and influential points can drastically affect our interpretations of correlation and regression results.

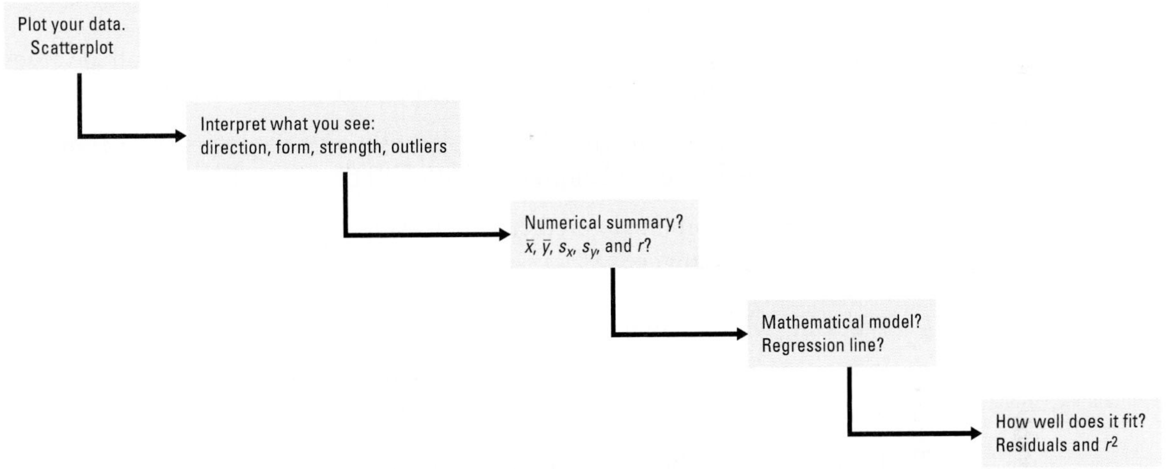

What You Should Have Learned

Here is a review list of the most important skills you should have gained from studying this chapter.

A. Data

1. Recognize whether each variable is quantitative or categorical.
2. Identify the explanatory and response variables in situations where one variable explains or influences another.

B. Scatterplots

1. Make a scatterplot to display the relationship between two quantitative variables. Place the explanatory variable (if any) on the horizontal scale of the plot.
2. Add a categorical variable to a scatterplot by using a different plotting symbol or color.

3. Describe the direction, form, and strength of the overall pattern of a scatterplot. In particular, recognize positive or negative association and linear (straight-line) patterns. Recognize outliers in a scatterplot.

C. Correlation

1. Using a calculator, find the correlation r between two quantitative variables.
2. Know the basic properties of correlation: r measures the strength and direction of linear relationships only; $-1 \leq r \leq 1$ always; $r = \pm 1$ only for perfect straight-line relations; r moves away from 0 toward ± 1 as the linear relation gets stronger.

D. Regression Lines

1. Explain what the slope b and the y intercept a mean in the equation $\hat{y} = a + bx$ of a regression line.
2. Using a calculator, find the least-squares regression line for predicting values of a response variable y from an explanatory variable x from data.
3. Find the slope and intercept of the least-squares regression line from the means and standard deviations of x and y and their correlation.
4. Use the regression line to predict y for a given x. Recognize extrapolation and be aware of its dangers.

E. Assessing Model Quality

1. Calculate the residuals and plot them against the explanatory variable x or against other variables. Recognize unusual patterns.
2. Use r^2 to describe how much of the variation in one variable can be accounted for by a straight-line relationship with another variable.
3. Recognize outliers and potentially influential observations from a scatterplot with the regression line drawn on it.

F. Interpreting Correlation and Regression

1. Understand that both r and the least-squares regression line can be strongly influenced by a few extreme observations.
2. Recognize possible lurking variables that may explain the correlation between two variables x and y.

Web Links

Try your hand at "guessing" correlations from scatterplots at www.stat.uiuc.edu/courses/stat100/cuwu/Games.html

Try the *Regression by Eye* Java applet for correlation and regression at www.ruf.rice.edu/~lane/stat_sim/reg_by_eye/

Chapter Review Exercises

3.77 Fighting fires Someone says, "There is a strong positive correlation between the number of firefighters at a fire and the amount of damage the fire does. So sending lots of firefighters just causes more damage." Why is this reasoning wrong?

3.78 Parents' heights Figure 3.29 is a scatterplot that displays the heights of 53 pairs of parents. The mother's height is plotted on the vertical axis and the father's height on the horizontal axis.[30]

Figure 3.29 *Scatterplot of the heights of the mother and father in 53 pairs of parents, for Exercise 3.78.*

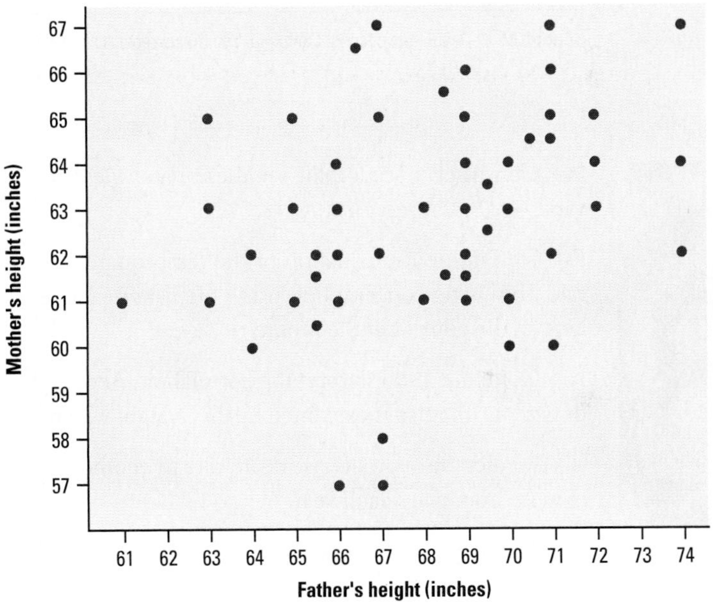

(a) What is the smallest height of any mother in the group? How many mothers have that height? What are the heights of the fathers in these pairs?

(b) What is the greatest height of any father in the group? How many fathers have that height? How tall are the mothers in these pairs?

(c) Are there clear explanatory and response variables, or could we freely choose which variable to plot horizontally?

(d) Say in words what a positive association between these variables means. The scatterplot shows a weak positive association. Why do we say the association is weak?

3.79 Is wine good for your heart? Table 3.7 (on the next page) gives data on average per capita wine consumption and heart disease death rates in 19 countries.

Table 3.7 *Average per capita wine consumption and heart disease*

Country	Alcohol from wine (liters/year)	Heart disease death rate (per 100,000)	Country	Alcohol from wine (liters/year)	Heart disease death rate (per 100,000)
Australia	2.5	211	Ireland	0.7	300
Austria	3.9	167	Italy	7.9	107
Belgium/Luxembourg	2.9	131	Netherlands	1.8	167
			New Zealand	1.9	266
Canada	2.4	191	Norway	0.8	227
Denmark	2.9	220	Spain	6.5	86
Finland	0.8	297	Sweden	1.6	207
France	9.1	71	Switzerland	5.8	115
Germany	2.7	172	United Kingdom	1.3	285
Iceland	0.8	211	United States	1.2	199

Source: M. H. Criqui, University of California, San Diego, reported in the *New York Times,* December 28, 1994.

(a) Construct a scatterplot for these data. Describe the relationship between the two variables.

(b) Determine the equation of the least-squares line for predicting heart disease death rate from wine consumption using the data in Table 3.7. Interpret the slope and y intercept in the context of the problem.

(c) Compute and interpret the correlation. About what percent of the variation in heart disease death rates is explained by the straight-line relationship with wine consumption?

(d) Predict the heart disease death rate in another country where adults average 4 liters of alcohol from wine each year.

(e) The slope of the least-squares line and the correlation in (b) and (c) are both negative. Is it possible for these two quantities to have opposite signs? Explain your answer.

3.80 Age and education in the states Because older people as a group have less education than younger people, we might suspect a relationship between the percent of state residents aged 65 and over and the percent who are not high school graduates. Figure 3.30 (on the next page) is a scatterplot of the data.

(a) There are at least two and perhaps three outliers in the plot. These states are Alaska, Utah, and Colorado. Give plausible reasons for why they might be outliers.

(b) If we ignore the outliers, does the relationship have a clear form and direction? Explain your answer.

(c) If we calculate the correlation with and without the three outliers, we get $r = 0.067$ and $r = 0.267$. Which of these is the correlation without the outliers? Explain your answer.

Figure 3.30 *Scatterplot of the percent of residents who are not high school graduates against the percent of residents aged 65 and over in the 50 states, for Exercise 3.80.*

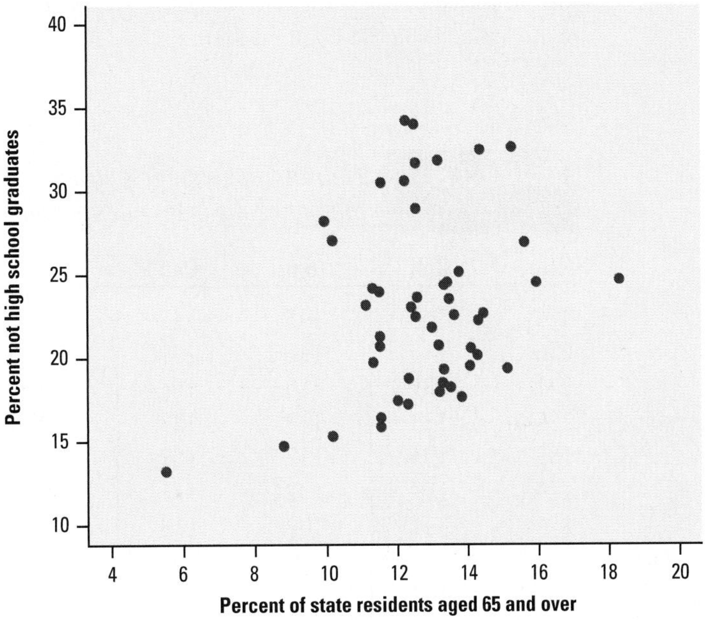

3.81 Competitive runners Good runners take more steps per second as they speed up. Here are the average numbers of steps per second for a group of top female runners at different speeds. The speeds are in feet per second.[31]

Speed (ft/s):	15.86	16.88	17.50	18.62	19.97	21.06	22.11
Steps per second:	3.05	3.12	3.17	3.25	3.36	3.46	3.55

(a) You want to predict steps per second from running speed. Make a scatterplot of the data with this goal in mind.

(b) Describe the pattern of the data and find the correlation.

(c) Find the least-squares regression line of steps per second on running speed. Draw this line on your scatterplot.

(d) Does running speed explain most of the variation in the number of steps a runner takes per second? Calculate r^2 and use it to answer this question.

(e) If you wanted to predict running speed from a runner's steps per second, would you use the same line? Explain your answer. Would r^2 stay the same?

3.82 Stride rate Refer to the previous exercise. There is a high positive correlation between steps per second and speed. Suppose that you had the full data set, which records steps per second for each runner separately at each speed. If you plotted each individual observation and computed the correlation, would you expect the correlation to be lower than, about the same as, or higher than the correlation for the published data? Why?

3.83 Listening to monkey calls The usual way to study the brain's response to sounds is to have individuals listen to "pure tones." The response to recognizable sounds may differ. To compare responses, researchers sedated macaque monkeys. They fed pure tones and monkey calls directly into their brains by inserting electrodes. Response to the stimulus was measured by the firing rate (electrical spikes per second) of neurons in various areas of the brain. Table 3.8 contains the responses for 37 neurons.

Table 3.8 *Neuron response (electrical spikes per second) to tones and monkey calls in macaque monkeys*

Tone	Call	Tone	Call	Tone	Call	Tone	Call
474	500	145	42	71	134	35	103
256	138	141	241	68	65	31	70
241	485	129	194	59	182	28	192
226	338	113	123	59	97	26	203
185	194	112	182	57	318	26	135
174	159	102	141	56	201	21	129
176	341	100	118	47	279	20	193
168	85	74	62	46	62	20	54
161	303	72	112	41	84	19	66
150	208						

(a) One notable finding is that responses to monkey calls are generally stronger than responses to pure tones. Give a numerical measure that supports this finding.

(b) Make a scatterplot of monkey call response against pure tone response (explanatory variable). Find the least-squares line and add it to your plot. Mark on your plot the point with the largest residual (positive or negative) and also the point that is an outlier in the x direction.

(c) How influential are each of these points on the correlation r? Justify your answer.

(d) How influential are each of these points on the regression line? Justify your answer.

3.84 Investing overseas One reason to invest abroad is that stock markets in different countries don't move in the same direction at the same time. When American stocks go down, foreign stocks may go up. So an investor who holds both bears less risk. That's the theory. Now we read, "The correlation between changes in American and European share prices has risen from 0.4 in the mid-1990s to 0.8 in 2000."[32]

(a) Explain to an investor who knows no statistics why this fact reduces the protection provided by buying European stocks.

(b) The same article goes on to say, "Crudely, that means the movements on Wall Street can explain 80% of price movements in Europe." Is this true? Why or why not?

3.85 Husbands and wives The mean height of American women in their early twenties is about 64.5 inches and the standard deviation is about 2.5 inches. The mean height of men the same age is about 68.5 inches, with standard deviation about 2.7 inches. If the correlation between the heights of husbands and wives is about $r = 0.5$, what is the slope of the regression line of the husband's height on the wife's height in young couples? Draw a graph of this regression line. Predict the height of the husband of a woman who is 67 inches tall.

3.86 The SAT essay: is longer better? Following the debut of the new SAT writing test in March 2005, Dr. Les Perelman, from the Massachusetts Institute of Technology, stirred controversy by reporting, "It appeared to me that regardless of what a student wrote, the longer the essay, the higher the score." He went on to say, "I have never found a quantifiable predictor in 25 years of grading that was anywhere as strong as this one. If you just graded them based on length without ever reading them, you'd be right over 90 percent of the time."[33] Table 3.9 shows the data set that Dr. Perelman used to draw his conclusions.

Carry out your own analysis of the data following the Data Analysis Toolbox. Then write a few sentences in response to each of Dr. Perelman's conclusions.

Table 3.9 *Length of essay and score for a sample of SAT essays*

Words	460	422	402	365	357	278	236	201	168	156	133
Score	6	6	5	5	6	5	4	4	4	3	2
Words	114	108	100	403	401	388	320	258	236	189	128
Score	2	1	1	5	6	6	5	4	4	3	2
Words	67	697	387	355	337	325	272	150	135		
Score	1	6	6	5	5	4	4	2	3		

Source: The College Board, www.collegeboard.com.

CHAPTER

4

More about Relationships between Two Variables

CASE STUDY

It's a matter of life and death

One sad fact about life is that we'll all die someday. Many adults plan ahead for their eventual passing by purchasing life insurance. Many different types of life insurance policies are available. Some provide coverage throughout an individual's life (whole life), while others last only for a specified number of years (term life). The policyholder makes regular payments (premiums) to the insurance company in return for the coverage. When the insured person dies, a payment is made to designated family members or other beneficiaries.

How do insurance companies decide how much to charge for life insurance? They rely on a staff of highly trained actuaries—people with expertise in probability, statistics, and advanced mathematics—to establish premiums. For an individual who wants to buy life insurance, the premium will depend on the type and amount of the policy as well as on personal characteristics like age, sex, and health status.

The following table shows monthly premiums for a 10-year term life insurance policy worth $1,000,000:[1]

Age	Monthly premium
40	$29
45	$46
50	$68
55	$106
60	$157
65	$257

How much would a 58-year-old expect to pay for such a policy? A 68-year-old? By the end of this chapter, you will have developed the tools you need to answer these questions.

Activity 4
Modeling the spread of cancer in the body

Materials: a regular six-sided die for each student; transparency grid; copy of grid for each student

Cancer begins with one cell, which divides into two cells.[2] Then these two cells divide and produce four cells. This process continues until there is some intervention such as radiation or chemotherapy to interrupt the spread of the disease or until the patient dies. In this activity you will simulate the spread of cancer cells in the body.

1. Select one student to represent the original cancer cell. That person rolls the die repeatedly, each roll representing a year. The number 5 will signal a cell division. When a 5 is rolled, a new student from the class will receive a die and join the original student (bad cell), so that there are now two cancer cells. These two students should be physically separated from the rest of the class, perhaps in a corner of the room.
2. As the die is rolled, another student will plot points on a transparency grid on the overhead projector. "Time," from 0 to 25 years, is marked on the horizontal axis, and the "Number of cancer cells," from 0 to 50, is on the vertical axis. The points on the grid will form a scatterplot.

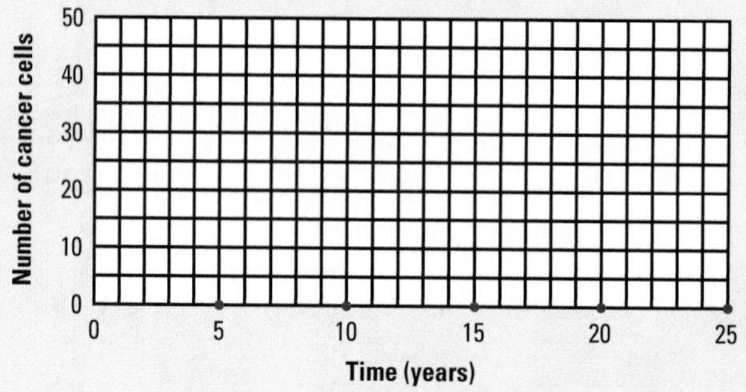

3. At a signal from the teacher, each "cancer cell" will roll his or her die. If anyone rolls the number 5, a new student from the class receives a die and joins the circle of cancer cells. The total number of cancer cells is counted, and the next point on the grid is plotted. The simulation continues until all students in the class have become cancer cells.
4. Copy the points shown on the transparency grid onto your copy of the grid.
5. Do the points show a pattern? If so, is the pattern linear? Is it a curved pattern? What mathematical function would best describe the pattern of points?

Keep your graph with the plotted points handy. We will analyze the results later in the chapter, after establishing some principles.

Introduction

In Chapter 3, we learned how to analyze relationships between two quantitative variables that showed a linear pattern. When two-variable data shows a nonlinear relationship, we must develop new techniques for finding an appropriate model. Section 4.1 describes several simple transformations of data that can straighten a nonlinear pattern. Once the data have been transformed to achieve linearity, we can use least-squares regression to generate a useful model for making predictions.

What about relationships between categorical variables? In Section 4.2, we examine graphical and numerical methods for describing such relationships.

As we noted in the previous chapter, association does not imply causation. Sometimes an observed association really does reflect cause and effect. In other cases, an association is explained by lurking variables, and the conclusion that x causes y is either wrong or not justified. We will tackle the issue of establishing causation in Section 4.3.

4.1 Transforming to Achieve Linearity

How is the weight of an animal's brain related to the weight of its body? Figure 4.1 (on the next page) is a scatterplot of brain weight against body weight for 96 species of mammals.[3] The line in the plot is the least-squares regression line for predicting brain weight from body weight. The outliers are interesting. We might say that dolphins and humans are smart, hippos are dumb, and elephants are just big. That's because dolphins and humans have larger brains than their body weights suggest, hippos have smaller brains, and the elephant is much heavier than any other mammal in both body and brain.

Example 4.1 ***Modeling mammal brain weight versus body weight***
Transforming data

The plot in Figure 4.1 is not very satisfactory. Most mammals are so small relative to elephants and hippos that their points overlap to form a blob in the lower-left corner of the plot. The correlation between brain weight and body weight is $r = 0.86$, but this is misleading. If we remove the elephant, the correlation for the other 95 species is $r = 0.50$. Figure 4.2 (on the next page) is a scatterplot of the data with the four outliers removed to allow a closer look at the other 92 observations. We can now see that the relationship is not linear. It bends to the right as body weight increases.

Biologists know that data on sizes often behave better if we take logarithms before doing more analysis. Figure 4.3 (page 261) plots the logarithm of brain weight against the logarithm of body weight for all 96 species. The effect is almost magical. There are no longer any extreme outliers or very influential observations. The pattern is very linear, with correlation $r = 0.96$. The vertical spread about the least-squares line is similar everywhere, so that predictions of brain weight from body weight will be about equally accurate for any body weight (in the log scale).

Figure 4.1 *Scatterplot and least-squares regression line of brain weight against body weight for 96 species of mammals.*

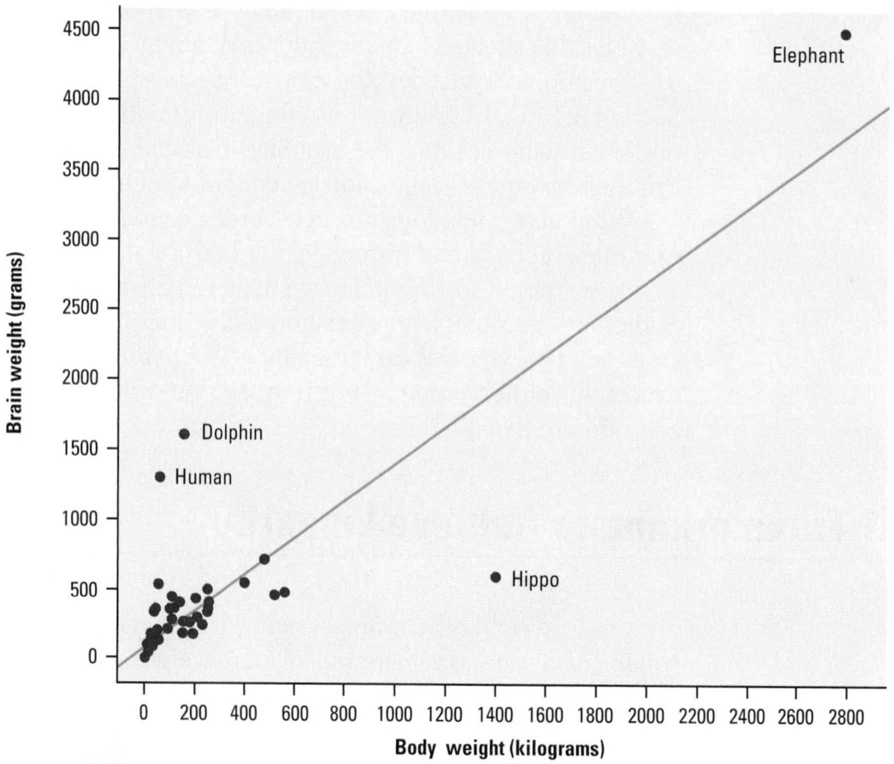

Figure 4.2 *Scatterplot of brain weight against body weight for mammals, with outliers removed.*

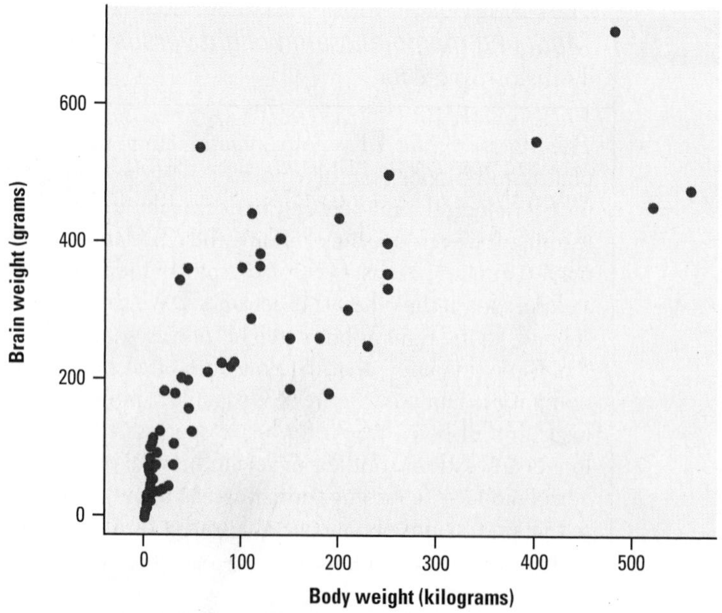

Figure 4.3 *Scatterplot and least-squares regression line of the logarithm of brain weight against the logarithm of body weight for 96 species of mammals.*

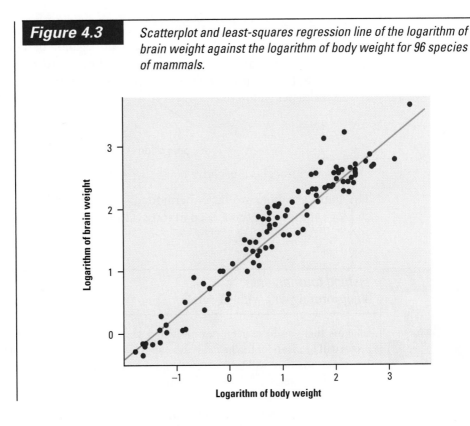

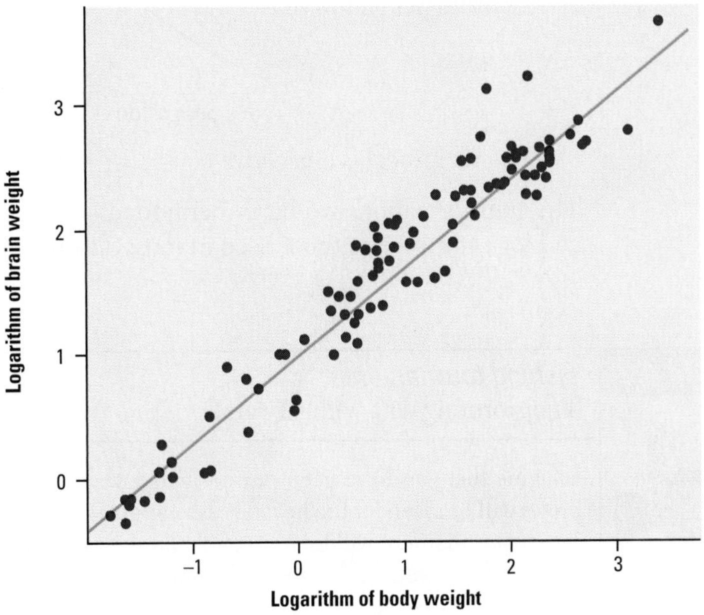

transforming
reexpressing

Example 4.1 shows that working with a function of our original measurements can greatly simplify statistical analysis. Applying a function such as the logarithm or square root to a quantitative variable is called ***transforming*** or ***reexpressing*** the data. We will see in this section that understanding how simple functions work helps us choose and use transformations to straighten nonlinear patterns.

First Steps in Transforming

Transforming data amounts to changing the scale of measurement that was used when the data were collected. We can choose to measure temperature in degrees Fahrenheit or in degrees Celsius, distance in miles or in kilometers. These changes of units are *linear transformations*, discussed in Chapter 1. Linear transformations cannot straighten a curved relationship between two variables. To do that, we resort to functions that are not linear. The logarithm function, applied in Example 4.1, is a nonlinear function. Here are some others.

- How should we measure the size of a sphere or of such roughly spherical objects as grains of sand or bubbles in a liquid? The size of a sphere can be expressed in terms of the diameter d, in terms of surface area (proportional to d^2), or in terms of volume (proportional to d^3). Any one of these powers of the diameter may be natural in a particular application.

- We commonly measure the fuel consumption t of a car in miles per gallon, which is how many miles the car travels on 1 gallon of fuel. Engineers prefer to measure in gallons per mile, which is how many gallons of fuel the car needs to travel 1 mile. This is a reciprocal transformation, $1/t$. A car that gets 25 miles per gallon uses

$$\frac{1}{\text{miles per gallon}} = \frac{1}{25} = 0.04 \text{ gallons per mile}$$

The reciprocal is a *negative power*: $1/t = t^{-1}$.

The transformations we have mentioned—linear, positive and negative powers, and logarithms—are those used in most statistical problems.

Example 4.2 *Fishing tournament*

Transforming data with powers

Imagine that you have been put in charge of organizing a fishing tournament in which prizes will be given for the heaviest fish caught. You know that many of the fish caught during the tournament will be measured and released. You are also aware that using delicate scales to try to weigh a fish that is flopping around in a moving boat will probably not yield very accurate results.

It would be much easier to measure the length of the fish while on the boat. What you need is a way to convert the length of the fish to its weight. You reason that since length is one-dimensional and weight (like volume) is three-dimensional, and since a fish 0 units long would weigh 0 pounds, the weight of a fish should be proportional to the cube of its length. Thus, a model of the form weight $= a \times \text{length}^3$ should work. You contact the nearby marine research laboratory, and they provide the average length (in centimeters) and weight (in grams) catch data for the Atlantic Ocean rockfish *Sebastes mentella* (Table 4.1). The lab

Table 4.1 *Average length and weight at different ages for Atlantic Ocean rockfish,* Sebastes mentella

Age (yr)	Length (cm)	Weight (g)	Age (yr)	Length (cm)	Weight (g)
1	5.2	2	11	28.2	318
2	8.5	8	12	29.6	371
3	11.5	21	13	30.8	455
4	14.3	38	14	32.0	504
5	16.8	69	15	33.0	518
6	19.2	117	16	34.0	537
7	21.3	148	17	34.9	651
8	23.3	190	18	36.4	719
9	25.0	264	19	37.1	726
10	26.7	293	20	37.7	810

Source: Gordon L. Swartzman and Stephen P. Kaluzny, *Ecological Simulation Primer*, Macmillan, 1987, p. 98.

also advises you that the model relationship between body length and weight has been found to be accurate for most fish species growing under normal feeding conditions. Figure 4.4 is a scatterplot of these weights versus lengths. The scatterplot shows a curved form.

Figure 4.4 *Scatterplot of Atlantic Ocean rockfish weight versus length.*

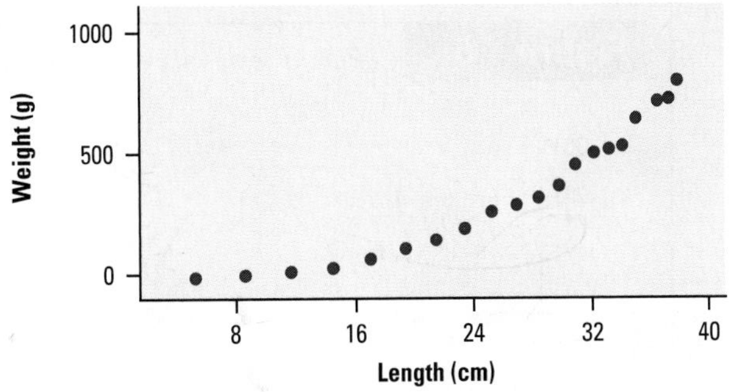

We have already decided on a model that makes sense in this context: weight = $a \times \text{length}^3$. What will happen if we cube the lengths in Table 4.1 and then graph weight versus length^3? Figure 4.5 gives us the answer. This transformation of the explanatory variable helps us produce a graph that is quite linear.

Figure 4.5 *The scatterplot of weight versus length^3 looks linear.*

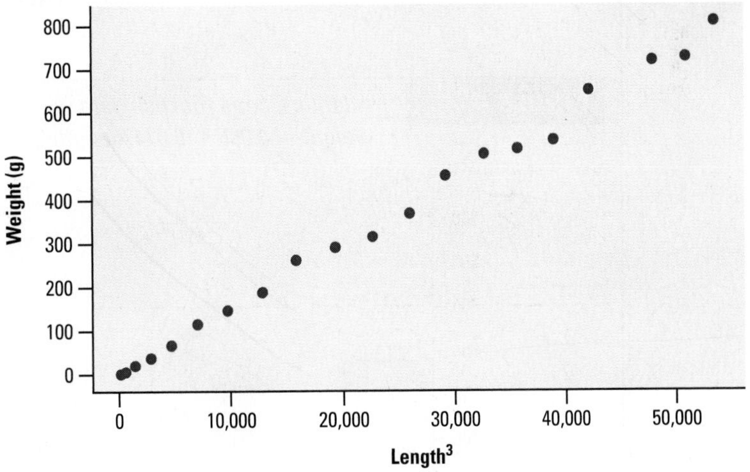

We perform a least-squares regression on the transformed points (length^3, weight). The resulting equation is

$$\text{weight} = 4.066 + 0.0147 \times \text{length}^3$$

with $r^2 = 0.995$. So 99.5% of the variation in the weight of Atlantic Ocean rockfish is accounted for by the linear regression of weight on length^3. Despite the very high r^2-value, it is still important to look at a residual plot. Figure 4.6 shows a slight pattern in the residuals for smaller values of length^3: for shorter fish, the residuals are all small negative numbers. So our model predicts a weight that is slightly higher than the actual weight of these fish. For heavier fish, the residuals are fairly randomly scattered. We should be safe using this model to predict the weight of fish caught in the tournament from the cube of their length.

Figure 4.6 *Plot of residuals versus length^3.*

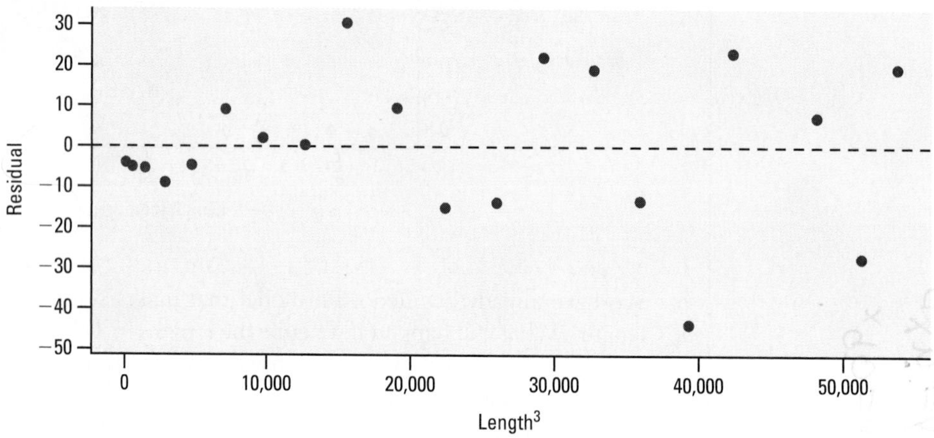

Figure 4.7 shows a scatterplot of the original data with the model weight = 4.066 + 0.0147 length^3 superimposed. The model appears to fit the original data well.

Figure 4.7 *Atlantic Ocean rockfish data with the model weight = 4.066 + 0.0147 length^3.*

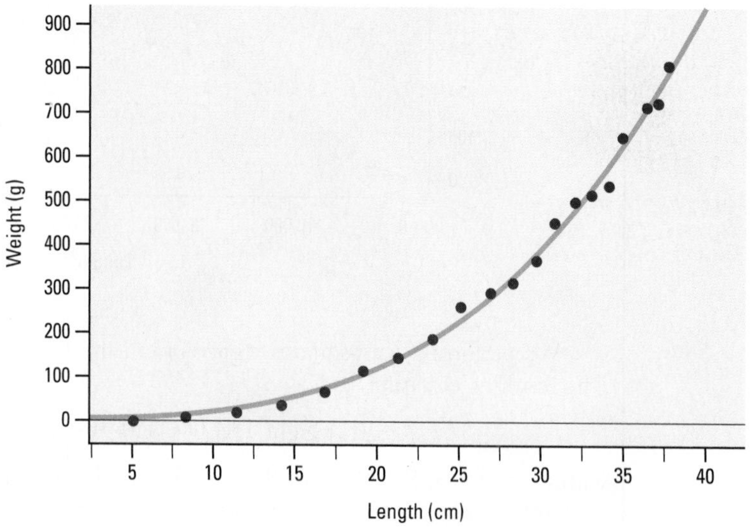

The original purpose for developing this model was to approximate the weight of a fish, given its length. Suppose your catch measured 36 centimeters. Our model predicts

$$\text{weight} = 4.066 + 0.0147(36)^3 = 689.9 \text{ grams}$$

If you entered a fishing contest, would you be comfortable with this procedure for determining the weights of the fish caught, and hence for determining the winner of the contest?

Exercises

4.1 Gone fishin' Return to the fishing tournament data of Example 4.2. Suppose that the underlying relationship between the length and weight of Atlantic Ocean rockfish is again weight $= a \times \text{length}^3$. There is another transformation that will achieve linearity. We can take the cube root of both sides of the equation:

$$\sqrt[3]{\text{weight}} = \sqrt[3]{a} \times \text{length}$$

If we plot the cube root of weight against length for the fish data, we should see a linear relationship.

(a) Make a scatterplot of the points (length, $\sqrt[3]{\text{weight}}$). Did this transformation achieve linearity?

(b) Use technology to calculate the least-squares regression equation for the transformed data. Interpret the slope and y intercept in the context of this problem.

(c) What does your model in (b) predict for the weight of a 36-centimeter fish? Show your work. Compare this with the prediction at the end of Example 4.2.

(d) Construct a residual plot. What does it tell you about how well your model fits the data?

(e) Interpret the value of r^2 for your model in (b).

4.2 The swinging pendulum, I Mrs. Hanrahan's precalculus class collected data on the length (in centimeters) of a pendulum and the time (in seconds) the pendulum took to complete one back-and-forth swing (called its period). Here are their data:

Length (cm)	Period (s)
16.5	0.777
17.5	0.839
19.5	0.912
22.5	0.878
28.5	1.004
31.5	1.087
34.5	1.129
37.5	1.111
43.5	1.290
46.5	1.371
106.5	2.115

(a) Make a scatterplot of the data with length as the explanatory variable. Describe what you see.

(b) At first, Mrs. Hanrahan's students thought that a linear model might describe the relationship between a pendulum's length and its period pretty well. After fitting a least-squares regression model to the data, they decided that a better model might be available. Perform this analysis yourself, and explain why the students didn't favor the linear model.

(c) A physics student in the class recalled that the period of a pendulum should be proportional to the square root of its length. Use this information to suggest a transformation that should achieve linearity. Then carry out your transformation.

(d) Find the least-squares regression equation for the transformed data in (c). How well does this model fit the transformed data? Justify your answer using a residual plot and r^2.

(e) In fact, the theoretical relationship between a pendulum's length and its period is

$$\text{period} = 2\pi\sqrt{\frac{\text{length}}{g}}$$

where g is acceleration due to gravity (in this case, $g = 980$ cm/s^2). How closely does the model you found in (d) agree with the theory?

(f) Use your model from (d) to predict the period of an 80-centimeter pendulum.

4.3 Boyle's law If you have taken a chemistry class, then you are probably familiar with Boyle's law: for gas in a confined space kept at a constant temperature, pressure times volume is a constant (in symbols, $PV = k$). Students collected the following data on pressure and volume using a syringe and a pressure probe.

Volume (cubic centimeters)	Pressure (atmospheres)
6	2.9589
8	2.4073
10	1.9905
12	1.7249
14	1.5288
16	1.3490
18	1.2223
20	1.1201

(a) Make a scatterplot suitable for predicting pressure from volume. Describe the direction, form, and strength of the relationship. Are there any outliers?

(b) If the true relationship between the pressure and volume of the gas is $PV = k$, we can divide both sides of this equation by V to obtain $P = k/V$, or $P = k(1/V)$. So if we graph pressure against the reciprocal of volume, we should see a linear relationship. Perform this transformation of the volume data, and graph P versus $1/V$. Did this transformation achieve linearity?

(c) Find the least-squares regression equation for the transformed data. How well does this model fit the transformed data? Justify your answer using a residual plot and r^2.

(d) Let's try a different transformation of the data. Take the reciprocal of the pressure values ($1/P$), and plot these against the original volume measurements. Find the least-squares regression equation. Why did this transformation achieve linearity?

(e) Use your models in parts (c) and (d) to predict the pressure in the syringe when the volume of gas is 15 cubic centimeters. How similar are the predictions?

4.4 The swinging pendulum, II Refer to the pendulum data of Exercise 4.2. This time, let's transform the data by squaring the period values.

(a) Make a scatterplot of period^2 versus length. Did this transformation achieve linearity?

(b) Find the least-squares regression equation for the transformed data. How well does this model fit the transformed data? Justify your answer using a residual plot and r^2.

(c) How similar is your model to the theoretical equation in Exercise 4.2(e)?

(d) Use your model to predict the period of an 80-centimeter pendulum. Compare with your answer from Exercise 4.2(f).

Transforming with Powers

Though simple in algebraic form and easy to compute with a calculator, the power and logarithm functions are varied in their behavior. It is natural to think of powers such as

$$\ldots, x^{-1}, x^{-1/2}, x^{1/2}, x, x^2, \ldots$$

as a hierarchy. Some facts about the family of power functions will help us choose transformations.

Figure 4.8 (on the next page) displays functions of the form $y = x^p$ for several values of p. What can we learn from this display?

- The graph of a linear function (power $p = 1$) is a straight line.
- Powers greater than 1 (like $p = 2$ and $p = 4$) give graphs that bend upward. The sharpness of the bend increases as p increases.
- Powers less than 1 but greater than 0 (like $p = 0.5$) give graphs that bend downward.
- Powers less than 0 (like $p = -0.5$ and $p = -1$) give graphs that decrease as x increases. Greater negative values of p result in graphs that decrease more quickly.
- Look at the $p = 0$ graph. You may be surprised that this is *not* the graph of $y = x^0$. Why not? The 0th power x^0 is just the constant 1, which is not very useful. The $p = 0$ entry in the figure is not constant; it is the logarithm, $\log x$. That is, *the logarithm fits into the hierarchy of power transformations at* $p = 0$.[4]

Figure 4.8 *The hierarchy of power functions. The logarithm function corresponds to $p = 0$.*

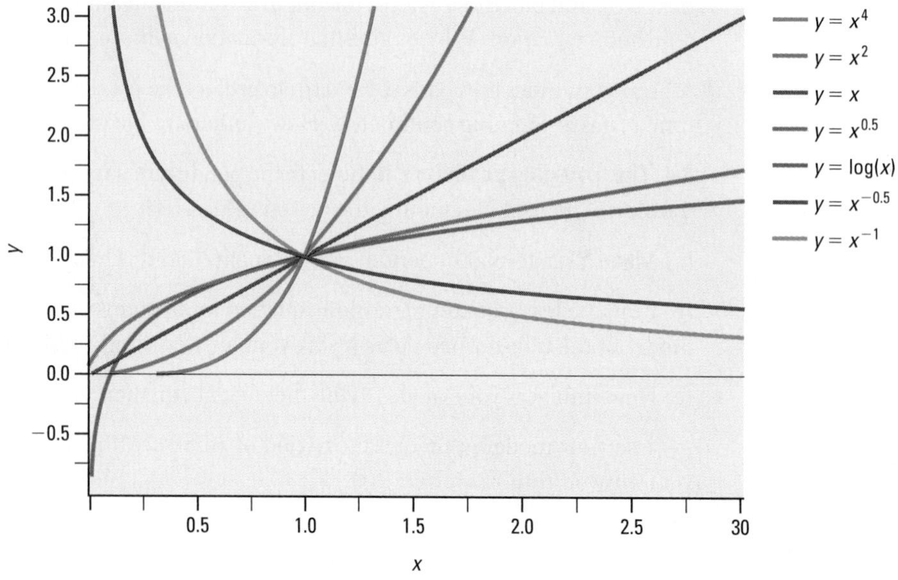

Example 4.3 *A country's GDP and life expectancy*

The hierarchy of power transformations

Figure 4.9(a) is a scatterplot of data from the World Bank.[5] The individuals are all the world's nations for which data are available. The explanatory variable x is a measure of how rich a country is: the gross domestic product (GDP) per person. GDP is the total value of the goods and services produced in a country, converted into dollars. The response variable y is life expectancy at birth.

Life expectancy increases in richer nations, but only up to a point. The pattern in Figure 4.9(a) at first rises rapidly as GDP increases but then levels out. Three African nations (Botswana, Gabon, and Namibia) are outliers with much lower life expectancy than the overall pattern suggests. Can we straighten the overall pattern by transforming?

Figures 4.9(b), (c), and (d) show the results of three transformations of GDP. The r-value in each figure is the correlation when the three outliers are omitted.

The square root, $\sqrt{x}$ when $p = 1/2$, reduces the curvature in the new scatterplot, but not enough. The logarithm $\log x$ ($p = 0$) straightens the pattern more, but it still bends to the right. The reciprocal square root $1/\sqrt{x}$, when $p = -1/2$, gives a pattern that is quite straight except for the outliers.

Example 4.3 shows the hierarchy of power transformations at work. As we moved through decreasing values of p, the scatterplots became straighter. But check that when we get to $p = -1$, using the reciprocal $1/x = x^{-1}$ bends the plot in the other direction. This "try it and see" approach isn't very satisfactory. That life expectancy depends linearly on $1/\sqrt{\text{GDP}}$ does not increase our understanding of

Figure 4.9 *The hierarchy of power transformations at work. The data are life expectancy and gross domestic product (GDP) for 115 nations. Panel (a) displays the original data. Panels (b), (c), and (d) transform GDP, using $\sqrt{x}$ (p = 1/2), log x (p = 0), and $1/\sqrt{x}$ (p = −1/2), respectively.*

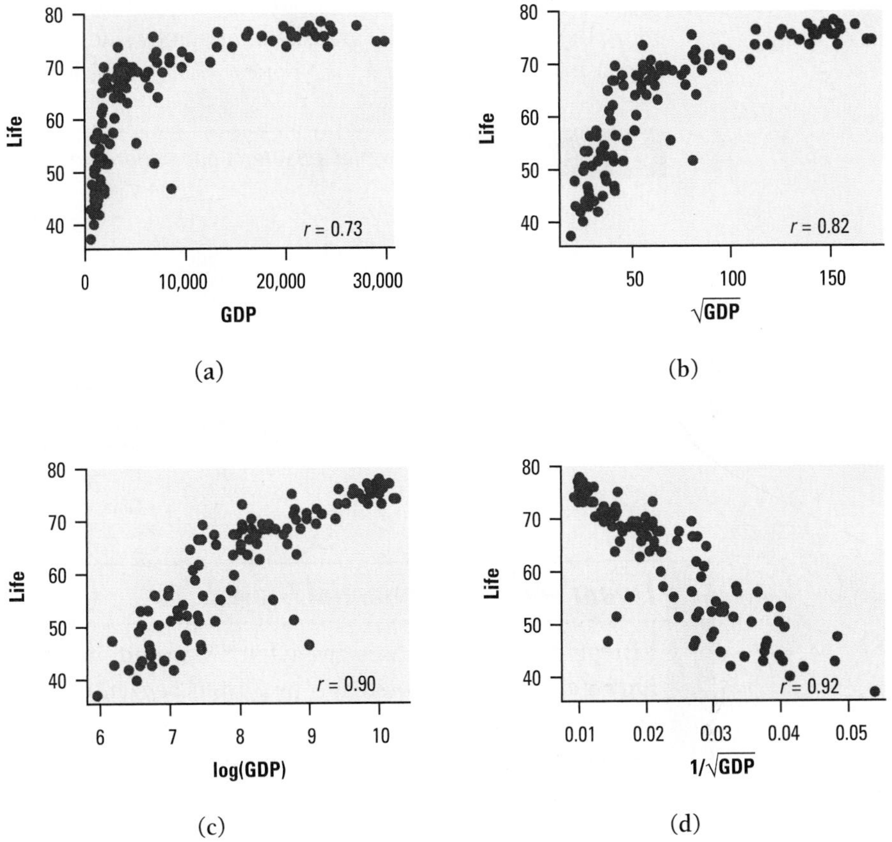

the relationship between the health and wealth of nations. *We don't recommend just pushing buttons on your calculator to try to straighten a scatterplot.*

Few people use the hierarchy of powers in practice. It is much more satisfactory to begin with a theory or mathematical model that we expect to describe a relationship. The transformation needed to make the relationship linear is then a consequence of the model. One of the most common models is ***exponential growth.***

Exponential Growth

In ***linear growth,*** a fixed increment is *added* to the variable in each equal time period. Exponential growth occurs when a variable is *multiplied* by a fixed number in each equal time period. To grasp the effect of multiplicative growth, consider a population of bacteria in which each bacterium splits into two each hour.

Beginning with a single bacterium, we have 2 after one hour, 4 at the end of two hours, 8 after three hours, then 16, 32, 64, 128, and so on. These first few numbers are deceiving. After 1 day of doubling each hour, there are 2^{24} (16,777,216) bacteria in the population. That number then doubles the next hour! Try successive multiplications by 2 on your calculator to see for yourself the very rapid increase after a slow start. Figure 4.10 shows the growth of the bacteria population over 24 hours. For the first 16 hours, the population is too small to rise visibly above the zero level on the graph. It is characteristic of exponential growth that the increase appears slow for a long period, then seems to explode.

Figure 4.10 *Growth of a bacteria population over a 24-hour period.*

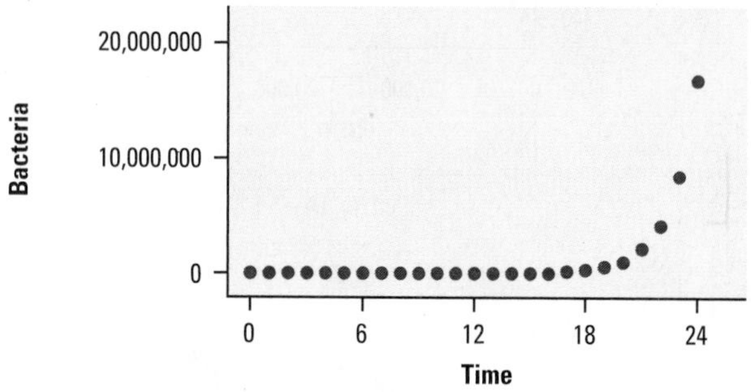

Linear versus Exponential Growth

Linear growth increases by a fixed *amount* in each equal time period. **Exponential growth** increases by a fixed *percent* of the previous total in each equal time period.

Populations of living things—like bacteria and the malignant cancer cells in Activity 4—tend to grow exponentially if not restrained by outside limits such as lack of food or space. More pleasantly, money also displays exponential growth when returns on an investment are compounded. Compounding means that last period's income earns income in the next period.

Example 4.4 *The growth of money*
Understanding exponential growth

A dollar invested at an annual rate of 6% turns into \$1.06 in a year. The original dollar remains and has earned \$0.06 in interest. That is, 6% annual interest means that any amount on deposit for the entire year is multiplied by 1.06. If the \$1.06 remains invested for a second year, the new amount is therefore 1.06×1.06, or 1.06^2. That is only \$1.12, but this in turn is multiplied by 1.06 during the third year, and so on. After x years, the dollar has become 1.06^x dollars.

If the Native Americans who sold Manhattan Island for \$24 in 1626 had deposited the \$24 in a savings account at 6% annual interest, they would now have almost \$80 billion.

Our savings accounts don't make us billionaires, because we don't stay around long enough. A century of growth at 6% per year turns \$24 into \$8143. That's 1.06^{100} times \$24. By 1826, two centuries after the sale, the account would hold a bit over \$2.7 million. Only after a patient 302 years do we finally reach \$1 billion. That's *real* money, but 302 years is a long time.

The count of bacteria after x hours is 2^x. The value of \$24 invested for x years at 6% interest is 24×1.06^x. Both are examples of the exponential growth model $y = ab^x$ for different constants a and b. In this model, the response y is multiplied by b in each time period.

Example 4.5 ***Moore's law and computer chips***
Exponential growth

Gordon Moore, one of the founders of Intel Corporation, predicted in 1965 that the number of transistors on an integrated circuit chip would double every 18 months. This is "Moore's law," one way to measure the revolution in computing. Here are data on the dates and number of transistors for Intel microprocessors:[6]

Processor	Date	Transistors	Processor	Date	Transistors
4004	1971	2,250	486 DX	1989	1,180,000
8008	1972	2,500	Pentium	1993	3,100,000
8080	1974	5,000	Pentium II	1997	7,500,000
8086	1978	29,000	Pentium III	1999	24,000,000
286	1982	120,000	Pentium 4	2000	42,000,000
386	1985	275,000			

Figure 4.11 shows the growth in the number of transistors on a computer chip from 1971 to 2000. Notice that we used "years since 1970" as the explanatory variable. We'll explain this a bit later.

Figure 4.11 *Scatterplot showing growth in the number of transistors on a computer chip from 1971 to 2000.*

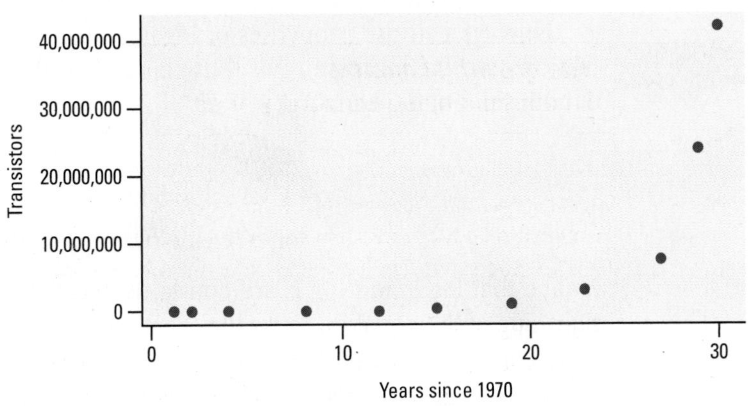

There is an increasing trend, but the overall pattern is not linear. The number of transistors on a computer chip has increased much faster than linear growth. The pattern of growth follows a smooth curve that looks a lot like an exponential curve. Is this exponential growth?

The Logarithm Transformation

The growth curve for the number of transistors on a computer chip does look somewhat like the exponential curve in Figure 4.10, but our eyes are not very good at comparing curves of roughly similar shape. We need a better way to check whether growth is exponential.

If an exponential model of the form $y = a \times b^x$ describes the relationship between x and y, we can use logarithms to transform the data to produce a linear relationship. But before we do the transformation, we need to review the properties of logarithms. The basic idea of a logarithm is this: $\log_2 8 = 3$ because 3 is the exponent to which 2 must be raised to yield 8. Here is a quick summary of the properties of logarithms.

Algebraic Properties of Logarithms

$$\log_b x = y \qquad \text{if and only if} \qquad b^y = x$$

The rules for logarithms are

1. $\log_b (MN) = \log_b M + \log_b N$
2. $\log_b (M/N) = \log_b M - \log_b N$
3. $\log_b X^p = p \log_b X$

These properties hold for all positive values of the base b except for $b = 1$. The two most commonly used bases are 10 and e (2.71828 . . .). When $b = 10$, we are dealing with "logarithms base 10," also called common logarithms. The notation for a common logarithm is $\log_{10} x$, or simply $\log x$. When the base is e, the resulting logarithm is called a natural logarithm. The corresponding notation is $\log_e x$, or more simply $\ln x$.

Now we can use properties of logarithms to carry out the required transformation: take the logarithm (we'll use base 10, but base e would work just as well) of both sides of the equation $y = ab^x$.

$$\begin{aligned} \log y &= \log(ab^x) \\ &= \log a + \log b^x && \text{using Rule 1} \\ &= \log a + (\log b)x && \text{using Rule 3} \end{aligned}$$

Notice that $\log a$ and $\log b$ are constants because a and b are constants. So the right side of the equation looks like a straight line. That is, *if our data are growing exponentially and we plot the logarithm (base 10 or base e) of y against x, we should observe a straight line for the transformed data.*

Example 4.6

Moore's law and computer chips (continued)
Transforming with logarithms

Figure 4.12 plots the natural logarithm (log base e or ln) of the number of transistors on a computer chip versus years since 1970 for the data of Example 4.5. This plot has a fairly linear form. It looks like our logarithm transformation achieved linearity.

Figure 4.12 *Scatterplot of* ln(number of transistors) *versus years since 1970.*

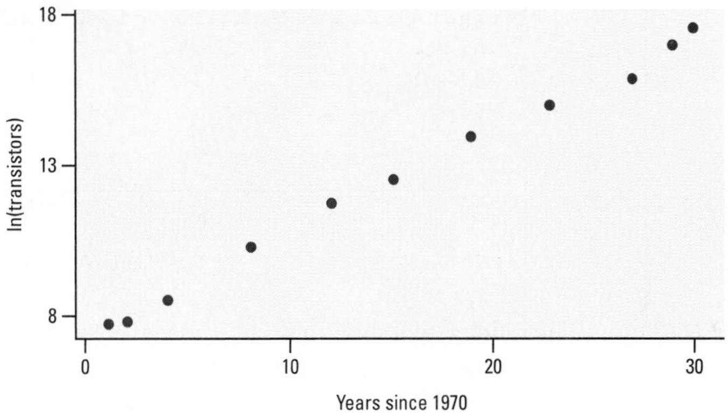

Applying least-squares regression to the transformed data, Minitab reports:

```
The regression equation is
ln(transistors) = 7.41 + 0.332 Years since 1970
Predictor            Coef       SE Coef        T       P
Constant             7.4078     0.1479      50.10   0.000
Years since 1970     0.331613   0.007941    41.76   0.000
S = 0.2735        R-Sq = 99.5%        R-Sq(adj) = 99.4%
```

As is usually the case, computer software tells us more than we want to know, but observe that the value of r^2 is 0.995. That means that 99.5% of the variation in ln(number of transistors) is explained by the least-squares regression model. That's pretty impressive. Let's continue. Figure 4.13 is a plot of the transformed data along with the fitted line.

Figure 4.13 *Plot of transformed transistor data with least-squares line.*

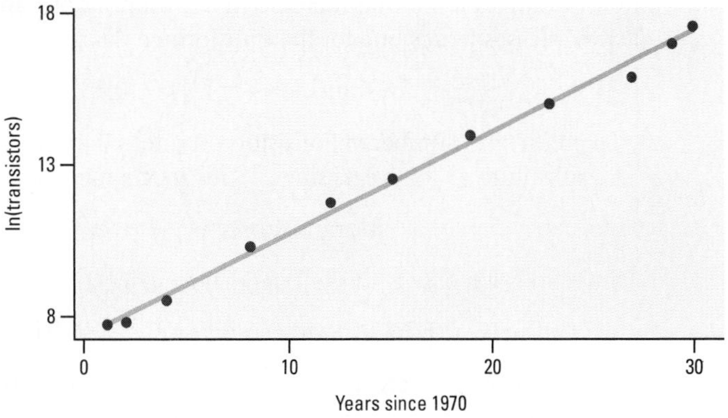

This appears to be a useful model for prediction purposes. Although the r^2-value is high, we should inspect the residual plot to further assess the quality of the model. Figure 4.14 is a residual plot.

Figure 4.14 *Residual plot for the transformed transistor data.*

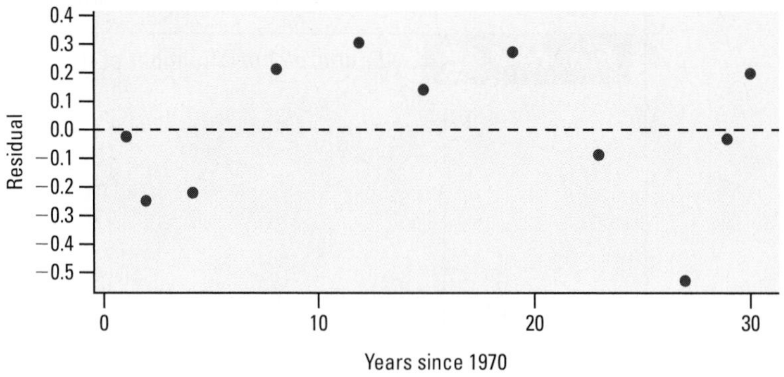

The residual plot shows a slight curved pattern, with the residuals going from negative to positive to negative as we move from left to right. But the residuals are small in size relative to the transformed y-values. We feel comfortable using this model to make predictions about the number of transistors on a computer chip.

Prediction in the Exponential Growth Model

Regression is often used for prediction. When we fit a least-squares regression line, we find the predicted response y for any value of the explanatory variable x by substituting our x-value into the equation of the line. In the case of exponential growth, the logarithms of the responses rather than the actual responses follow a linear pattern. To do prediction, we need to "undo" the logarithm transformation to return to the original units of measurement.

Example 4.7 *Predicting transistors on a computer chip in 2003*
An inverse transformation

Our examination of the increase in the number of transistors on a computer chip left us with a least-squares line for the transformed data of

$$\ln(\text{transistors}) = 7.41 + 0.332 \times \text{years since 1970}$$

To predict the number of transistors on Intel's Itanium 2 chip, which was released in 2003, we substitute 33 for "years since 1970" in the regression equation:

$$\ln(\text{transistors}) = 7.41 + 0.332 \times 33 = 18.366$$

Since ln is log base e, this tells us that

$$\text{transistors} = e^{18.366} = 94{,}678{,}737$$

So our model predicts about 95 million transistors on the Itanium 2 chip. In fact, this chip had about 410 million transistors!

An alternative method for predicting the number of transistors to the one shown in Example 4.7 uses algebraic properties of exponents. We could have used the definition of logarithm to change from

$$\ln(\text{transistors}) = 7.41 + 0.332(\text{years since 1970})$$

to

$$\text{transistors} = e^{7.41 + 0.332(\text{years since 1970})}$$

Using properties of exponents, we can simplify this as follows:

$$\text{transistors} = e^{7.41} \times e^{0.332(\text{years since 1970})}$$
$$\text{transistors} = 1652.426 \times (e^{0.332})^{(\text{years since 1970})}$$
$$\text{transistors} = 1652.426 \times (1.394)^{(\text{years since 1970})}$$

This equation is now in the familiar exponential format $y = ab^x$. To predict the number of transistors on the Itanium 2, we substitute 33 for "years since 1970" in the exponential equation:

$$\text{transistors} = 1652.426 \times (1.394)^{33} = 94{,}678{,}737$$

(*Note*: We used the unrounded values for $e^{7.41}$ and $e^{0.332}$ for the last calculation to avoid roundoff error.) Performing this messy algebra has the advantage of giving you the exponential model explicitly, rather than in terms of logarithms.

Make sure that you understand the big idea here. The necessary transformation is carried out by taking the logarithm of the response variable. Your calculator and most statistical software will calculate the logarithms of all the values of a variable with a single command. The essential property of the logarithm for our purposes is that *if a variable grows exponentially, its logarithm grows linearly.*

Example 4.8 ***Transforming bacteria counts***
Exact exponential growth

Figure 4.15 plots the logarithms of the bacteria counts in Figure 4.10 (page 270). Sure enough, exact exponential growth turns into an exact straight line when we plot the logarithms. After 15 hours, for example, the population contains $2^{15} = 32{,}768$ bacteria. The logarithm of 32,768 is 4.515, and this point appears above the 15-hour mark in Figure 4.15.

Figure 4.15 *Logarithms of the bacteria counts.*

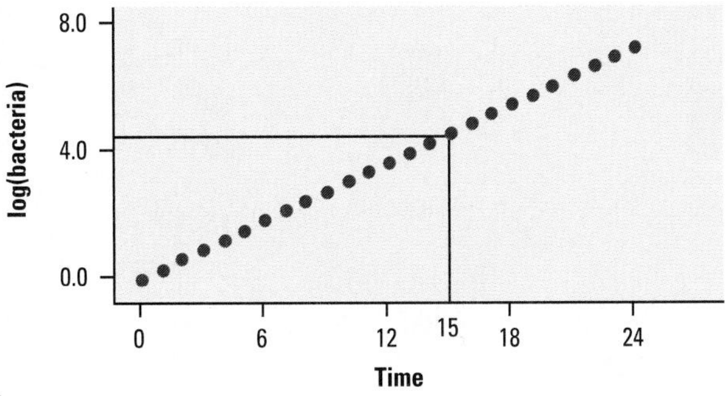

Exercises

4.5 Light through the water Some college students collected data on the intensity of light at various depths in a lake. Here are their data:

Depth (meters)	Light intensity (lumens)
5	168.00
6	120.42
7	86.31
8	61.87
9	44.34
10	31.78
11	22.78

(a) Make a scatterplot suitable for predicting light intensity from depth. Describe the form of the relationship.

(b) To verify that the decrease in light intensity follows an exponential model, calculate the ratios of light intensity at consecutive depths. Start with $120.42/168.00 = 0.717$. What do you conclude?

(c) Take the natural logarithm (ln) of the light intensity measurements and plot these values against the corresponding depths. Does this transformation achieve linearity?

(d) Calculate the least-squares regression equation for the transformed data. Interpret the slope and y intercept of this equation in this setting.

(e) Construct and interpret a residual plot.

(f) Perform the inverse transformation to express light intensity as an exponential function of depth in the lake. Display a scatterplot of the original data with the exponential model superimposed. Is your exponential function a satisfactory model for the data?

(g) Use your model to predict the light intensity at a depth of 22 meters. The actual light intensity reading at that depth was 0.58 lumens. Does this surprise you?

4.6 Gypsy moths Biological populations can grow exponentially if not restrained by predators or lack of food. The gypsy moth outbreaks that occasionally devastate the forests of the Northeast illustrate approximate exponential growth. It is easier to count the number of acres defoliated by the moths than to count the moths themselves. Here are data on an outbreak in Massachusetts:[7]

Year	Acres
1978	63,042
1979	226,260
1980	907,075
1981	2,826,095

(a) Plot the number of acres defoliated y against the year x. The pattern of growth appears exponential.

(b) Verify that y is being multiplied by about 4 each year by calculating the ratio of acres defoliated each year to the previous year. (Start with 1979 to 1978, when the ratio is 226,260/63,042 = 3.6.)

(c) Take the (base 10) logarithm of each number y and plot the logarithms against the year x. The linear pattern confirms that the growth is exponential.

(d) Verify that the least-squares line fitted to the transformed data is

$$\log y = -1094.51 + 0.5558 \times \text{year}$$

(e) Try to perform the inverse transformation (using the definition of logarithm as an exponent) to express acres as an exponential function of the year. What happens?

When the explanatory variable is years, transform the data to "years since" so that the values are smaller and don't create overflow problems when you perform the inverse transformation.

(f) Calculate the least-squares regression line of log(acres) on years since 1977.

(g) Construct and interpret a residual plot for $\log y$ on years since 1977.

(h) Perform the inverse transformation to express y as an exponential function. Display a scatterplot of the original data with the exponential curve model superimposed. Is your exponential function a satisfactory model for the data?

(i) Use your model to predict the number of acres defoliated in 1982.

(*Postscript:* A viral disease reduced the gypsy moth population between the readings in 1981 and 1982. The actual count of defoliated acres in 1982 was 1,383,265.)

4.7 More on Moore's law Refer to Examples 4.5 to 4.7 on Moore's law. Suppose that Moore's law is exactly correct. That is, the number of transistors is 2250 in year 1 (1971) and doubles every 18 months (1.5 years) thereafter.

(a) Write the model for predicting transistors in year x after 1970.

(b) What is the equation of the line that, according to your model in (a), connects the logarithm of transistors with x?

(c) Explain why a comparison of this line with the regression line from Example 4.6 shows that although transistor counts have grown exponentially, they have grown a bit more slowly than Moore's law predicts.

4.8 Gun violence (exact exponential growth) A paper in a scholarly journal once claimed (we are not making this up), "Every year since 1950, the number of American children gunned down has doubled."[8] To see that this is silly, suppose that in 1950 just 1 child was "gunned down" and suppose that the paper's claim is exactly right.

(a) Make a table of the number of children killed in each of the next 10 years, 1951 to 1960.

(b) Plot the number of deaths against the year and connect the points with a smooth curve. This is an exponential curve.

(c) The paper appeared in 1995, 45 years after 1950. How many children were killed in 1995, according to the paper?

(d) Take the logarithm of each of your counts from (a). Plot these logarithms against the year. You should get a straight line.

(e) From your graph in (d), find the approximate values of the slope b and the intercept a for the line. Use the equation $y = a + bx$ to predict the logarithm of the count for the 45th year. Check your result by taking the logarithm of the count you found in (c).

4.9 U.S. population The following table gives the resident population of the United States from 1790 to 2000, in millions of persons:

Date	Pop.	Date	Pop.	Date	Pop.	Date	Pop.
1790	3.9	1850	23.2	1910	92.0	1970	203.3
1800	5.3	1860	31.4	1920	105.7	1980	226.5
1810	7.2	1870	39.8	1930	122.8	1990	248.7
1820	9.6	1880	50.2	1940	131.7	2000	281.4
1830	12.9	1890	62.9	1950	151.3		
1840	17.1	1900	76.0	1960	179.3		

(a) Plot population against time since 1790. The growth of the American population appears roughly exponential.

(b) Plot the logarithms of population against time since 1790. The pattern of growth is now clear. An expert says that "the population of the United States increased exponentially from 1790 to about 1880. After 1880 growth was still approximately exponential, but at a slower rate." Explain how this description is obtained from the graph.

(c) Use the data from 1920 to 2000 to construct an exponential model for the purpose of predicting the population in 2010. Show your method clearly.

(d) Construct a residual plot for the transformed data. What is the value of r^2 for the transformed data?

(e) Use your model to predict the population in the year 2010. Do you think your prediction will be too low or too high? Explain.

4.10 Galileo's experiment Galileo studied motion by rolling balls down ramps. Newton later showed how Galileo's data fit his general laws of motion. Imagine that you are Galileo, without Newton's laws to guide you. He rolled a ball down a ramp with a horizontal shelf at the end of it so that the ball was moving horizontally when it started to fall. The ramp was placed at different heights above the floor, and Galileo measured the horizontal distance the ball traveled before it hit the floor. Here are Galileo's data. (We won't try to describe the obscure seventeenth-century units Galileo used to measure distance and height.)[9]

Distance	Height
1500	1000
1340	828
1328	800
1172	600
800	300

(a) Plot distance y against height x. The pattern is very regular, as befits data described by a physical law. We want to find distance as a function of height. That is, we want to transform x to straighten the graph.

(b) Think before you calculate: will powers x^p for $p < 1$ or $p > 1$ tend to straighten the graph? Why?

(c) Use the hierarchy of power transformations to try values of p until you find one that achieves linearity. What transformation do you suggest?

Technology Toolbox

Modeling exponential growth with the TI-83/84/89

Let's use the Moore's law data from Example 4.5 (page 271) to illustrate how to perform a logarithm transformation on the TI-83/84/89.

- Enter the coded "years since 1970" in L_1/list1 and the number of transistors in L_2/list2. Use ZoomStat/ZoomData to draw the scatterplot.

TI-83/84 **TI-89**

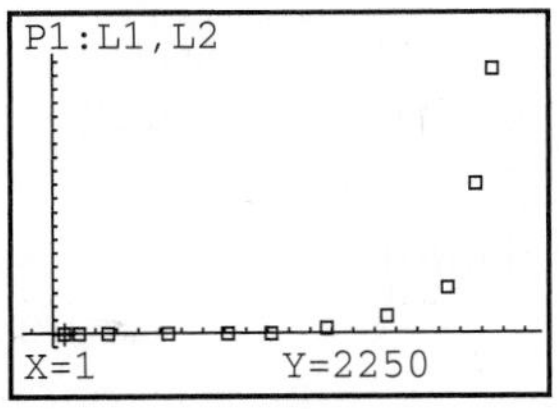

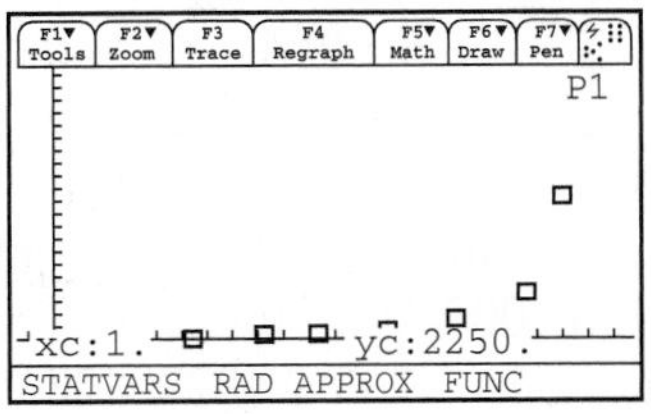

- Define L_3/list3 as the natural logarithm of L_2/list2. Then make a scatterplot of ln(transistors) versus year.

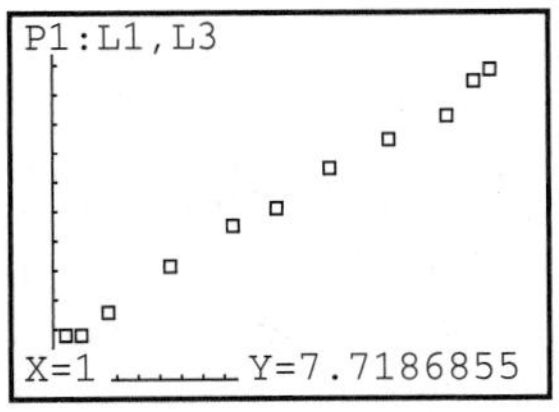

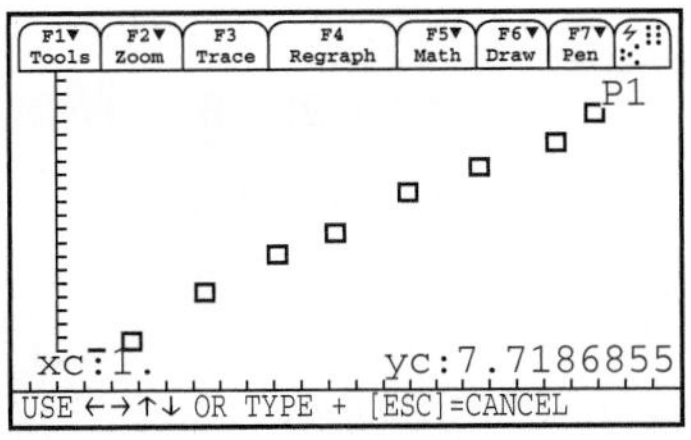

- Next, we perform the least-squares regression on the transformed data.

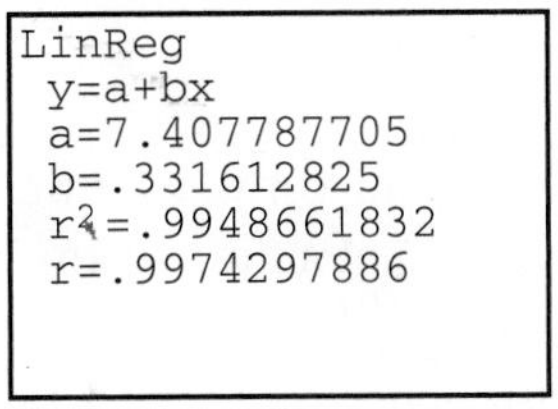

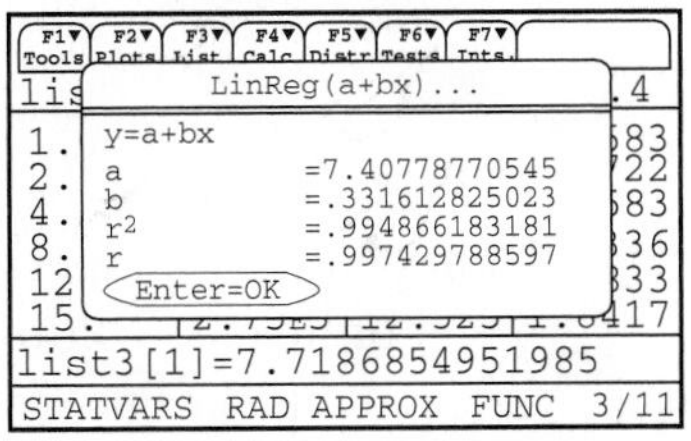

(continued)

Technology Toolbox

Modeling exponential growth with the TI-83/84/89 *(continued)*

- Here are the scatterplots with the least-squares line:

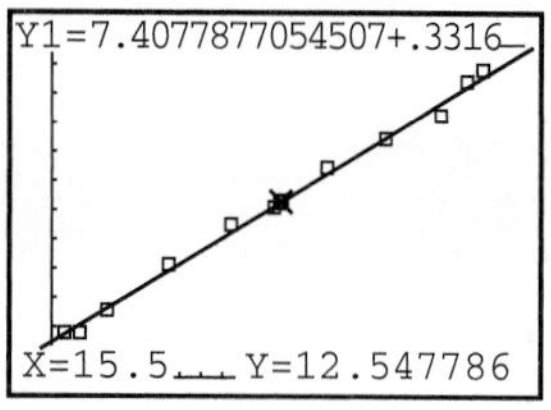

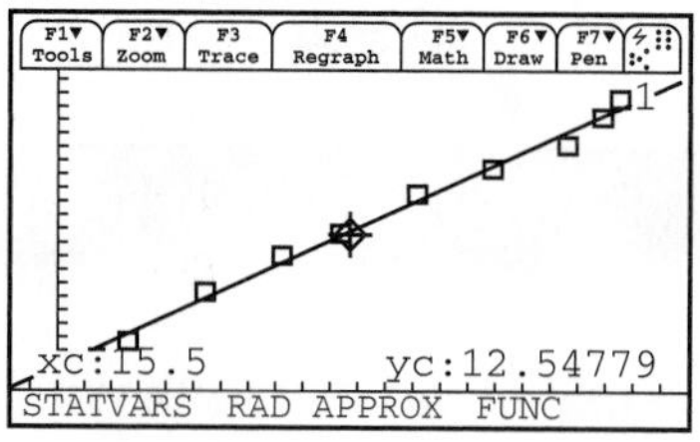

- Despite the high r^2-value, you should always inspect the residual plot. Be sure to plot the residuals (stored in the RESID list) versus L_1/list1.

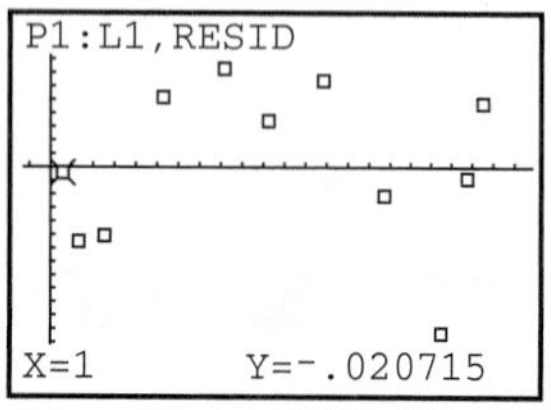

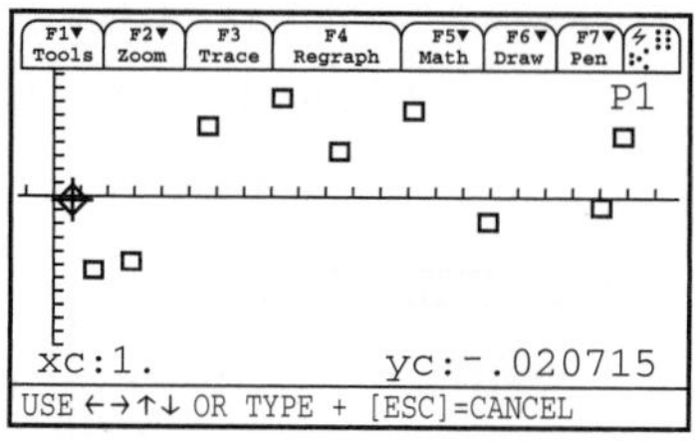

- Now we're ready to predict the number of transistors on the Itanium 2 chip, which was released by Intel in 2003. With the regression equation in Y1, define Y2 = e^(Y1). The predicted population is then Y2(33) = 93270173.5. Notice the large error due to roundoff in our prediction in Example 4.7 (page 274).

Power Law Models

When you visit a pizza parlor, you order a pizza by its diameter, say 10 inches, 12 inches, or 14 inches. But the amount you get to eat depends on the area of the pizza. The area of a circle is π times the square of its radius. So the area of a round pizza with diameter x is

$$\text{area} = \pi r^2 = \pi (x/2)^2 = \pi (x^2/4) = (\pi/4)x^2$$

power law model

This is a ***power law model*** of the form

$$y = ax^p$$

When we are dealing with things of the same general form, whether circles or fish or people, we expect area to go up with the square of a dimension such as diameter or height. Volume should go up with the cube of a linear dimension. That is, geometry tells us to expect power laws in some settings.

Biologists have found that many characteristics of living things are described quite closely by power laws. There are more mice than elephants, and more flies than mice—the abundance of species follows a power law with body weight as the explanatory variable. So do pulse rate, length of life, the number of eggs a bird lays, and so on. Sometimes the powers can be predicted from geometry, but sometimes they are mysterious. Why, for example, does the rate at which animals use energy go up as the 3/4 power of their body weight? Biologists call this relationship Kleiber's law. It has been found to work all the way from bacteria to whales. The search goes on for some physical or geometrical explanation for why life follows power laws. There is as yet no general explanation, but power laws are a good place to start in simplifying relationships for living things.

When we apply the logarithm transformation to the response variable y in an exponential growth model, we produce a linear relationship. To produce a linear relationship from a power law model, we apply the logarithm transformation to *both* variables. Here are the details:

1. The power law model is

$$y = ax^p$$

2. Take the logarithm of both sides of this equation. You see that

$$\log y = \log a + p \log x$$

   That is, taking the logarithm of both variables results in a linear relationship between $\log x$ and $\log y$.

3. Look carefully: the *power* p in the power law becomes the *slope* of the straight line that links $\log y$ to $\log x$.

Prediction in Power Law Models

If taking the logarithms of both variables produces a linear scatterplot, a power law is a reasonable model for the original data. We can even roughly estimate what power p the law involves by finding the least-squares regression line of $\log y$ on $\log x$ and using the slope of the regression line as an estimate of the power. Remember that the slope is only an estimate of the p in an underlying power model. The greater the scatter of the points in the scatterplot about the fitted line, the smaller our confidence that this estimate is accurate.

Example 4.9 ***Predicting brain weight from body weight***
Using a power model

The magical success of the logarithm transformation in Example 4.1 (page 259) would not surprise a biologist. We suspect that a power law governs this relationship. Least-squares regression for the scatterplot in Figure 4.3 (page 261) gives the line

$$\log y = 1.01 + 0.72 \log x$$

for predicting the logarithm of brain weight from the logarithm of body weight. To undo the logarithm transformation, remember that for logarithms with base 10, $y = 10^{\log y}$. We see that

$$\begin{aligned} y &= 10^{1.01 + 0.72 \log x} \\ &= 10^{1.01} \times 10^{0.72 \log x} \\ &= 10.2\,(10^{\log x})^{0.72} \end{aligned}$$

Because $10^{\log x} = x$, the estimated power model connecting predicted brain weight $\hat{y}$ with body weight x for mammals is

$$\hat{y} = 10.2x^{0.72}$$

Based on footprints and some other sketchy evidence, some people think that a large apelike animal, called Sasquatch or Bigfoot, lives in the Pacific Northwest. His weight is estimated to be about 280 pounds, or 127 kilograms. How big is Bigfoot's brain? Based on the power law estimated from data on other mammals, we predict

$$\begin{aligned} \hat{y} &= 10.2(127)^{0.72} \\ &= 10.2\,(32.7) \\ &= 333.7 \text{ grams} \end{aligned}$$

For comparison, gorillas have an average body weight of about 140 kilograms and an average brain weight of about 406 grams. Of course, Bigfoot may have a larger brain than his weight predicts—after all, he has avoided being captured, shot, or videotaped for many years!

Example 4.10 *What's a planet, anyway?*

Power law models from scratch

On July 31, 2005, a team of astronomers announced that they had discovered what appeared to be a new planet in our solar system. They had first observed this object almost two years earlier using a telescope at Caltech's Palomar Observatory in California. Originally named UB313, the potential planet is bigger than Pluto and has an average distance of about 9.5 billion miles from the sun. (For reference, Earth is about 93 million miles from the sun.) Could this new astronomical body, nicknamed Xena, be a new planet?

Data: At the time of Xena's discovery, there were nine known planets in our solar system. Here are data on the distance from the sun and period of revolution of those planets.[10]

Planet	Distance from sun (astronomical units)	Period of revolution (Earth years)
Mercury	0.387	0.241
Venus	0.723	0.615
Earth	1.000	1.000
Mars	1.524	1.881
Jupiter	5.203	11.862
Saturn	9.539	29.456
Uranus	19.191	84.070
Neptune	30.061	164.810
Pluto	39.529	248.530

- *Who?* The individuals are the nine planets in our current solar system.
- *What?* The explanatory variable is distance from the sun (in astronomical units), and the response variable is period of revolution (in Earth years).
- *Why?* To determine whether the newly discovered Sedna could be the tenth planet.
- *When, where, how, and by whom?* The existing planetary data have been around for quite some time. Data on Xena was first obtained in October 2003.

Graphs: Figure 4.16 is a scatterplot of the planetary data. There appears to be a strong curved relationship between distance from the sun and period of revolution.

Figure 4.16 *Scatterplot of planetary distance from the sun and period of revolution.*

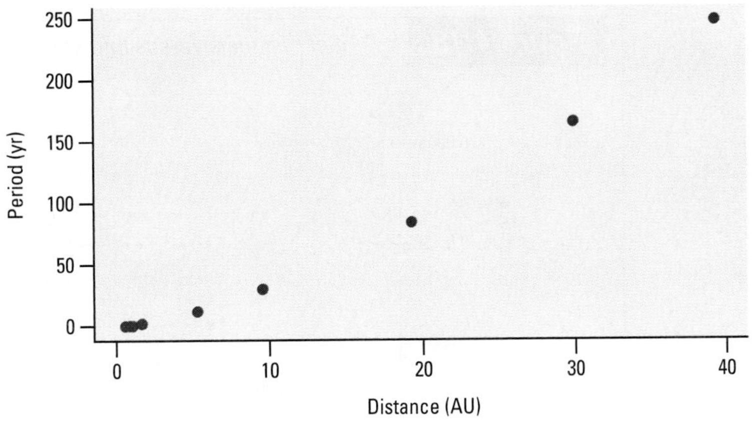

What type of transformation should we use to linearize the data? If the relationship between distance from the sun and period of revolution is *exponential*, then a plot of log(period) versus distance should be roughly linear. If the relationship between these variables follows a *power* model, then a plot of log(period) versus log(distance) should be fairly linear.

Figure 4.17 *(a) The scatterplot of ln(period) versus distance is not linear. (b) The scatterplot of ln(period) versus ln(distance) appears very linear.*

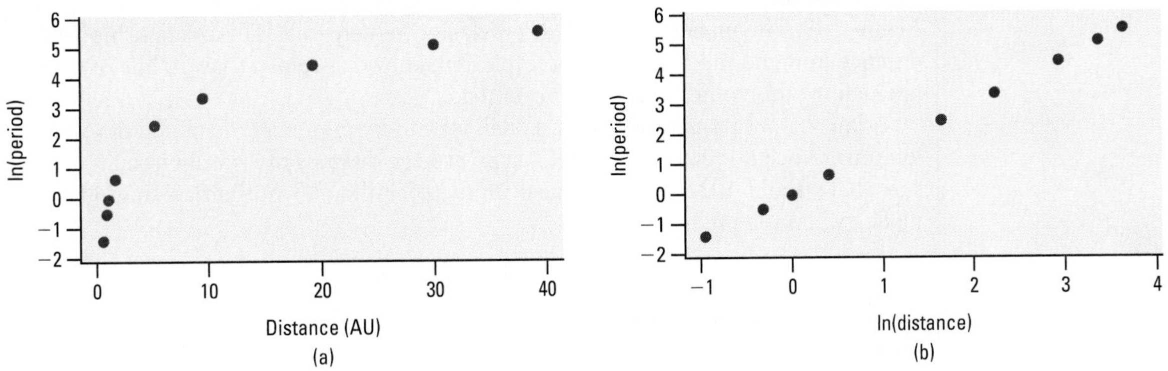

Figure 4.17(a) shows that an exponential model is not appropriate for these data. The logarithm transformation of the response variable made the new plot look even less linear than the original scatterplot. Figure 4.17(b) suggests that a power function would be appropriate for modeling these data since the transformation to ln(distance) and ln(period) produced a scatterplot with a linear form.

Numerical summaries: The correlation between ln(distance) and ln(period) is nearly 1.
Model: The least-squares regression equation is

$$\ln(\text{period}) = 0.000254 + 1.50 \ln(\text{distance})$$

with $r^2 \approx 1$. A plot of the residuals versus ln(distance) is shown in Figure 4.18. A curved pattern is apparent, but the size of the residuals relative to the values of ln(period) is very small.

Figure 4.18 *Plot of residuals versus ln(distance).*

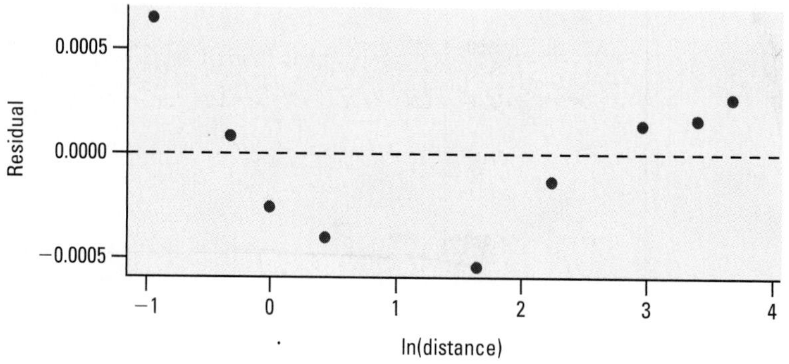

The last step is to perform an inverse transformation on the linear regression equation:

$$\ln(\text{period}) = 0.000254 + 1.50 \ln(\text{distance})$$
$$e^{\ln(\text{period})} = e^{0.000254 + 1.50\ln(\text{distance})}$$
$$\text{period} = e^{0.000254} \times e^{1.50\ln(\text{distance})}$$
$$\text{period} = 1.000e^{\ln(\text{distance})^{1.50}}$$
$$\text{period} = 1.000 \text{ distance}^{1.50}$$

This is the power model for the original data.

Interpretation: The scatterplot of the original data along with the power model appears in Figure 4.19. The model fits the planetary data extremely well. This relationship between distance from the sun and period of revolution, known as Kepler's law, is one of the most fascinating astronomical discoveries ever made.

Now that we have developed a model, we can try to predict the period of revolution for Xena. Since this potential planet is at an average distance of 9.5 billion miles from the sun, that's about 102.15 astronomical units (9.5 billion/93 million). Using our power model, we would predict a period of

$$\text{period} = 1.000(102.15)^{150} = 1032.4 \text{ years}$$

We wouldn't want to wait for Xena to make a full revolution to see if our prediction is accurate!

Figure 4.19 *Planetary data with power model.*

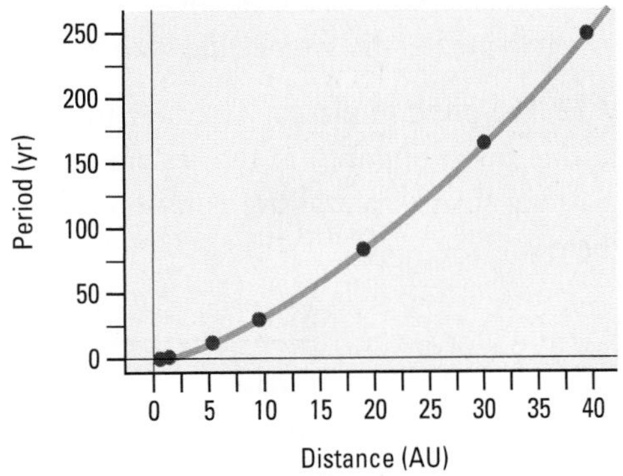

Postscript: In August 2006, the International Astronomical Union agreed on a new definition of planet. Both Pluto and UB313 (Xena) were classified as "dwarf planets."

Exercises

4.11 Body weight and lifetime Table 4.2 gives the average weight and average life span in captivity for several species of mammals.

Table 4.2 *Body weight and life span for several species of mammals*

Species	Weight (kg)	Life span (yr)	Species	Weight (kg)	Life span (yr)
Baboon	32	20	Guinea pig	1	4
Beaver	25	5	Hippopotamus	1400	41
Cat, domestic	2.5	12	Horse	480	20
Chimpanzee	45	20	Lion	180	15
Dog	8.5	12	Mouse, house	0.024	3
Elephant	2800	35	Pig, domestic	190	10
Goat, domestic	30	8	Red fox	6	7
Gorilla	140	20	Sheep, domestic	30	12
Grizzly bear	250	25			

Source: G. A. Sacher and E. F. Staffelt, "Relation of gestation time to body weight for placental mammals: implications for the theory of vertebrate growth," *American Naturalist,* 108 (1974), pp. 593–613. We found these data in F. L. Ramsey and D. W. Schafer, *The Statistical Sleuth: A Course in Methods of Data Analysis,* Duxbury, 1997.

(a) Use an appropriate transformation and make a plot to determine whether a power model would be appropriate. What do you conclude?

(b) Perform least-squares regression on the transformed data. Construct a residual plot and interpret the r^2-value.

(c) Carry out the inverse transformation to obtain a power model in the form $y = ax^p$.

(d) Use your fitted model to predict the average life span for humans (average weight 65 kilograms). Humans are an exception to the rule.

(e) Some writers on power laws in biology claim that life span depends on body weight according to a power law with power $p = 0.2$. Perform a transformation without using logarithms that should achieve linearity if the claim were true. Find the resulting model and use it to predict the average life span for humans as in (c).

4.12 The price of pizzas The new manager of a pizza restaurant wants to add variety to the pizza offerings at the restaurant. She also wants to determine if the prices for existing sizes of pizzas are consistent. Prices for plain (cheese only) pizzas are shown below:

Size	Diameter (inches)	Cost
Small	10	\$4.00
Medium	12	\$6.00
Large	14	\$8.00
Giant	18	\$14.00

(a) Would a power model or an exponential model be more appropriate for these data? Perform transformations and sketch graphs to justify your answer.

(b) Construct the model for these data that you chose in (a). Record the least-squares regression equation on the transformed data and show the inverse transformation to obtain your final model.

(c) Based on your analysis, would you advise the manager to adjust the price on any of the pizza sizes? If so, explain briefly.

(d) Use your model to suggest a price for a new "soccer team"–size pizza, with a 24-inch diameter.

(e) Suggest an alternative model to the one you calculated in (b) by performing a transformation that does not involve logarithms. What price does this new model suggest for the soccer team–size pizza?

4.13 How much weight is too much? The U.S. Department of Health and Human Services characterizes adults as "seriously overweight" if they meet the following criteria for height and weight (only a portion of the chart is reproduced here):

Height (ft in)	Height (in)	Severely overweight (lb)	Height (ft in)	Height (in)	Severely overweight (lb)
4′10″	58	138	5′8″	68	190
5′0″	60	148	6′0″	72	213
5′2″	62	158	6′2″	74	225
5′4″	64	169	6′4″	76	238
5′6″	66	179	6′6″	78	250

Weights are given in pounds, without clothes. Height is measured without shoes. There is no distinction between men and women; a note accompanying the table states, "The higher weights apply to people with more muscle and bone, such as many men." Despite any reservations you may have about the department's common standards for both genders, do the following:

(a) Without looking at the data, hypothesize a relationship between height and weight of U.S. adults. That is, write a general form of an equation that you believe will model the relationship.

(b) Which variable would you select as explanatory and which would be the response? Plot the data from the table.

(c) Perform a transformation to achieve linearity. Do a least-squares regression on the transformed data and interpret the value of r^2.

(d) Construct a residual plot of the transformed data. Interpret the residual plot.

(e) Perform the inverse transformation and write the equation for your model. Use your model to predict how many pounds a 5′10″ adult would have to weigh in order to be classified by the department as "seriously overweight." Do the same for a 7-foot-tall individual.

4.14 Heart weights of mammals Use the methods discussed in this section to analyze the following data on the hearts of various mammals.[11] Follow the Data Analysis Toolbox (page 93), as in Example 4.10.

Mammal	Heart weight (grams)	Length of cavity of left ventricle (centimeters)
Mouse	0.13	0.55
Rat	0.64	1.0
Rabbit	5.8	2.2
Dog	102	4.0
Sheep	210	6.5
Ox	2030	12.0
Horse	3900	16.0

Technology Toolbox

Power law modeling

- Enter the x data (explanatory) into L_1/list1 and the y data (response) into L_2/list2.
- Produce a scatterplot of y versus x. Confirm a nonlinear trend that could be modeled by a power function in the form $y = ax^p$.
- Define L_3/list3 to be the logarithm (base 10 or natural log) of L_1/list1 and define L_4/list4 to be the logarithm (base 10 or natural log) of L_2/list2.

(continued)

Technology Toolbox

Power law modeling *(continued)*

- Plot L_4/list4 versus L_3/list3. Verify that the pattern is approximately linear.
- Calculate the regression equation for the transformed data and store it in Y1. Remember that Y1 is really the logarithm of y. Check the r^2-value.
- Construct a residual plot, in the form of either RESID versus x or RESID versus predicted values (fits). Ideally, the points in a residual plot should be randomly scattered above and below the $y = 0$ reference line.
- Perform the inverse transformation to find the power function $y = ax^p$ that models the original data. Define Y2 to be (10^a)(x^b) or (e^a)(x^b), depending on the type of logarithm you used for the transformation. The calculator has stored the values of a and b for the most recent regression performed. Deselect Y1 and plot Y2 and the scatterplot for the original data together.
- To make a prediction for the value $x = k$, evaluate Y2(k) on the home screen.

Section 4.1 Summary

Nonlinear relationships between two quantitative variables can sometimes be changed into linear relationships by **transforming** one or both of the variables.

The most common transformations belong to the family of **power transformations** x^p. The logarithm $\log x$ fits into the hierarchy of power functions at position $p = 0$.

Transformation is particularly effective when there is reason to think that the data are governed by some mathematical model. The **exponential growth model** $y = ab^x$ becomes linear when we plot $\log y$ against x. The **power law model** $y = ax^p$ becomes linear when we plot $\log y$ against $\log x$.

We can fit exponential growth and power models to data by finding the least-squares regression line for the transformed data, then doing the inverse transformation.

Section 4.1 Exercises

4.15 Free-fallin' Some high school physics students dropped a ball and measured its height at various points along its descent. Table 4.3 shows the time since release (in seconds) and the distance the ball had fallen (in centimeters).

(a) Make a scatterplot suitable for predicting distance fallen from time since release. Describe the direction, form, and strength of the relationship.

Table 4.3 *Free-fall distance data*

Time (s)	Distance (cm)	Time (s)	Distance (cm)
0.16	12.1	0.57	150.2
0.24	29.8	0.61	182.2
0.25	32.7	0.61	189.4
0.30	42.8	0.68	220.4
0.30	44.2	0.72	254.0
0.32	55.8	0.72	261.0
0.36	63.5	0.83	334.6
0.36	65.1	0.88	375.5
0.50	124.6	0.89	399.1
0.50	129.7		

Source: *Data Analysis,* National Council of Teachers of Mathematics, 1990.

(b) If you have studied physics, then you probably know that the distance an object falls when released from a resting position is a function of time^2. Perform an appropriate transformation to achieve linearity under this assumption. Then find a least-squares regression model for the transformed data.

(c) Comment on the quality of your model in (b) by referring to a residual plot and r^2.

(d) Another possible transformation that could linearize the data is to use the square roots of the distance-fallen values. Make a scatterplot of the points (time, $\sqrt{\text{distance}}$) to see if this transformation works. Then find a least-squares regression model for the transformed data.

(e) Comment on the quality of your model in (d) by referring to a residual plot and r^2.

(f) Use the two models you obtained in (b) and (d) to predict the distance that the object had fallen after 0.47 seconds. Which prediction do you think is closer to the actual value? Why?

4.16 Determining tree biomass It is easy to measure the "diameter at breast height" of a tree. It's hard to measure the total "aboveground biomass" of a tree, because to do this you must cut and weigh the tree. The biomass is important for studies of ecology, so ecologists commonly estimate it using a power law. Combining data on 378 trees in tropical rain forests gives this relationship between biomass y measured in kilograms and diameter x measured in centimeters:[12]

$$\ln y = -2.00 + 2.42 \ln x$$

Note that the investigators chose to use natural logarithms, with base $e = 2.71828\ldots$, rather than common logarithms with base 10.

(a) Translate the line given into a power model. Use the fact that for natural logarithms,

$$y = e^{\ln y}$$

(b) Estimate the biomass of a tropical tree 30 centimeters in diameter.

4.17 Counting carnivores Ecologists look at data to learn about nature's patterns. One pattern they have found relates the size of a carnivore (body mass in kilograms) to how many of those carnivores there are in an area. The right measure of "how many" is to count carnivores per 10,000 kilograms of their prey in the area. Table 4.4 gives data for 25 carnivore species. Determine an appropriate model for predicting abundance from body mass. Follow the Data Analysis Toolbox (page 93).

Table 4.4 *Size and abundance of carnivores*

Carnivore species	Body mass (kg)	Abundance (per 10,000 kg of prey)	Carnivore species	Body mass (kg)	Abundance (per 10,000 kg of prey)
Least weasel	0.14	1656.49	Eurasian lynx	20.0	0.46
Ermine	0.16	406.66	Wild dog	25.0	1.61
Small Indian mongoose	0.55	514.84	Dhole	25.0	0.81
Pine marten	1.3	31.84	Snow leopard	40.0	1.89
Kit fox	2.02	15.96	Wolf	46.0	0.62
Channel Island fox	2.16	145.94	Leopard	46.5	6.17
Arctic fox	3.19	21.63	Cheetah	50.0	2.29
Red fox	4.6	32.21	Puma	51.9	0.94
Bobcat	10.0	9.75	Spotted hyena	58.6	0.68
Canadian lynx	11.2	4.79	Lion	142.0	3.40
European badger	13.0	7.35	Tiger	181.0	0.33
Coyote	13.0	11.65	Polar bear	310.0	0.60
Ethiopian wolf	14.5	2.70			

Source: Chris Carbone and John L. Gittleman, "A common rule for the scaling of carnivore density," *Science*, 295 (2002), pp. 2273–2276.

4.18 Crunch the data What is the relationship between the mature size of a species of fish and the size of the fish when they begin to reproduce? To answer this question, scientists collected data on 85 species of fish from many published studies. Go to the book's Web site, www.whfreeman.com/tps3e, and click on CrunchIt! Load the fish reproduction data set from Chapter 4 Resources. The first variable is the "asymptotic body length," the length (in centimeters) to which female fish will eventually grow. The second variable is the length (also in centimeters) at which 50% of females first reproduce. Explore the relationship between the variables carefully. Does any simple model describe the relationship well?

4.19 How mold grows, I Do mold colonies grow exponentially? In an investigation of the growth of molds, biologists inoculated flasks containing a growth medium with equal amounts of spores of the mold *Aspergillus nidulans*. They measured the size of a colony by analyzing how much remained of a radioactive tracer substance that is consumed by the mold as it grows. Each size measurement required destroying that colony, so that the data below refer to 30 separate colonies. To smooth the pattern, we take the mean size of the three colonies measured at each time.[13]

Hours	Colony sizes			Mean
0	1.25	1.60	0.85	1.23
3	1.18	1.05	1.32	1.18
6	0.80	1.01	1.02	0.94
9	1.28	1.46	2.37	1.70
12	2.12	2.09	2.17	2.13
15	4.18	3.94	3.85	3.99
18	9.95	7.42	9.68	9.02
21	16.36	13.66	12.78	14.27
24	25.01	36.82	39.83	33.89
36	138.34	116.84	111.60	122.26

(a) Graph the mean colony size against time. Then graph the logarithm of the mean colony size against time.

(b) On the basis of data such as these, microbiologists divide the growth of mold colonies into three phases that follow each other in time. Exponential growth occurs during only one of these phases. Briefly describe the three phases, making specific reference to the graphs to support your description.

(c) The exponential growth phase for these data lasts from about 6 hours to about 24 hours. Find the least-squares regression line of the logarithms of mean size on hours for only the data between 6 and 24 hours. Use this line to predict the size of a colony 10 hours after inoculation. (The line predicts the logarithm. You must obtain the size from its logarithm.)

4.20 How mold grows, II Find the correlation between the logarithm of mean size and hours for the data between 6 and 24 hours in the previous exercise. Make a scatterplot of the logarithms of the individual size measurements against hours for this same period and find the correlation. Why do we expect the second r to be smaller? Is it in fact smaller?

4.21 Muscle strength and body weight Bigger people are generally stronger than smaller people, though there's a lot of individual variation. Let's find a theoretical model.

(a) Body weight increases as the cube of height. The strength of a muscle increases with its cross-sectional area, which we expect to go up as the square of height. Put these together: what power law should describe how muscle strength increases with weight?

(b) Let's apply your result from (a). Graph the power law relation between strength and body weight for weights from (say) 1 to 1000. (Constants in the power law just reflect the units of measurement used, so we can ignore them.) Use the graph to explain why a person 1 million times as heavy as an ant can't lift a million times as much as an ant can lift.

4.22 Activity 4 revisited

(a) Using the data you and your class collected in the chapter-opening Activity (page 258), use transformation methods to construct an appropriate model. Follow the Data Analysis Toolbox.

(b) A theoretical analysis might begin as follows. The probability that an individual malignant cell reproduces is 1/6 each year. Let P = population of cancer cells at time t and let P_0 = population of cancer cells at time $t = 0$. At the end of Year 1, the population is $P = P_0 + (1/6)P_0 = P_0\,(7/6)$. At the end of Year 2, the population is $P = P_0\,(7/6) + P_0\,(1/6)(7/6) = P_0\,(7/6)^2$. Continue this line of reasoning to show that the growth equation after n years is $P = P_0\,(7/6)^n$.

(c) Enter the growth equation into your calculator as Y3, and plot it along with your exponential model calculated in (a). Specify a thick plotting line for one of the curves. How do the two exponential curves compare?

4.2 Relationships between Categorical Variables

We have concentrated on relationships in which at least the response variable is quantitative. Now we will describe relationships between two or more categorical variables. Some variables—such as sex, race, and occupation—are categorical by nature. Other categorical variables are created by grouping values of a quantitative variable into classes. Published data often appear in grouped form to save space. To analyze categorical data, we use the *counts* or *percents* of individuals that fall into various categories.

Example 4.11 *College students*

Organizing categorical variables

two-way table

row variable

column variable

Table 4.5 presents Census Bureau data describing the age and sex of college students. This is a ***two-way table*** because it describes two categorical variables. (Age is categorical here because the students are grouped into age categories.) Age group is the ***row variable*** because each row in the table describes students in one age group. Because age group has a natural order from youngest to oldest, the order of the rows reflects this order. Sex is the ***column variable*** because each column describes one sex. The entries in the table are the counts of students in each age-by-sex class.

Table 4.5 *College students by sex and age group, 2003 (thousands of persons)*

	Sex		
Age group	**Female**	**Male**	**Total**
15 to 17 years	89	61	150
18 to 24 years	5,668	4,697	10,365
25 to 34 years	1,904	1,589	3,494
35 years or older	1,660	970	2,630
Total	9,321	7,317	16,639

Source: From the October 2003 Current Population Survey, www.census.gov.

Marginal distributions

How can we best grasp the information contained in Table 4.5? First, *look at the distribution of each variable separately*. The distribution of a categorical variable says how often each outcome occurred. The "Total" column at the right of the table contains the totals for each of the rows. These row totals give the distribution of age (the row variable) among college students: 150,000 were 15 to 17 years old, 10,365,000 were 18 to 24 years old, and so on. In the same way, the "Total" row at the bottom of the table gives the distribution of sex. From the bottom row, we can calculate a striking and important fact: 56% of college students are women.

If the row and column totals are missing in a two-way table, you should calculate them. The distributions of sex alone and age alone are called ***marginal distributions*** because they appear at the bottom and right margins of the two-way table.

marginal distribution

If you check the row and column totals in Table 4.5, you will notice a few discrepancies. For example, the sum of the entries in the "25 to 34" row is 3493. The entry in the "Total" column for that row is 3494. The explanation is *roundoff error*. The table entries are in thousands of students and each is rounded to the nearest thousand. The Census Bureau obtained the "Total" entry by rounding the exact number of students aged 25 to 34 to the nearest thousand. The result was 3,494,000. Adding the row entries, each of which is already rounded, gives a slightly different result.

Percents are often more informative than counts, especially when we are making comparisons. We can display the marginal distribution of students' age groups in terms of percents by dividing each row total by the table total and converting to a percent.

Example 4.12 ***College students (continued)***

Calculating a marginal distribution

The percent of college students who are 18 to 24 years old is

$$\frac{\text{total age 18 to 24}}{\text{table total}} = \frac{10{,}365}{16{,}639} = 0.623 = 62.3\%$$

Are you surprised that only about 62% of students are in the traditional college age group? Do three more such calculations to obtain the marginal distribution of age group in percents:

	Age Group			
	15 to 17	18 to 24	25 to 34	35 or older
Percent	0.9	62.3	21.0	15.8

The total is 100% because everyone is in one of the four age categories.

Each marginal distribution from a two-way table is a distribution for a single categorical variable. As we saw in Chapter 1, we can use a bar graph or a pie chart to display such a distribution. Figure 4.20 (on the next page) is a bar graph of the distribution of age for college students.

In working with two-way tables, you must calculate lots of percents. Here's a tip to help decide what fraction gives the percent you want. Ask, "What group

Figure 4.20 *A bar graph of the distribution of age for college students. This is one of the marginal distributions for Table 4.5.*

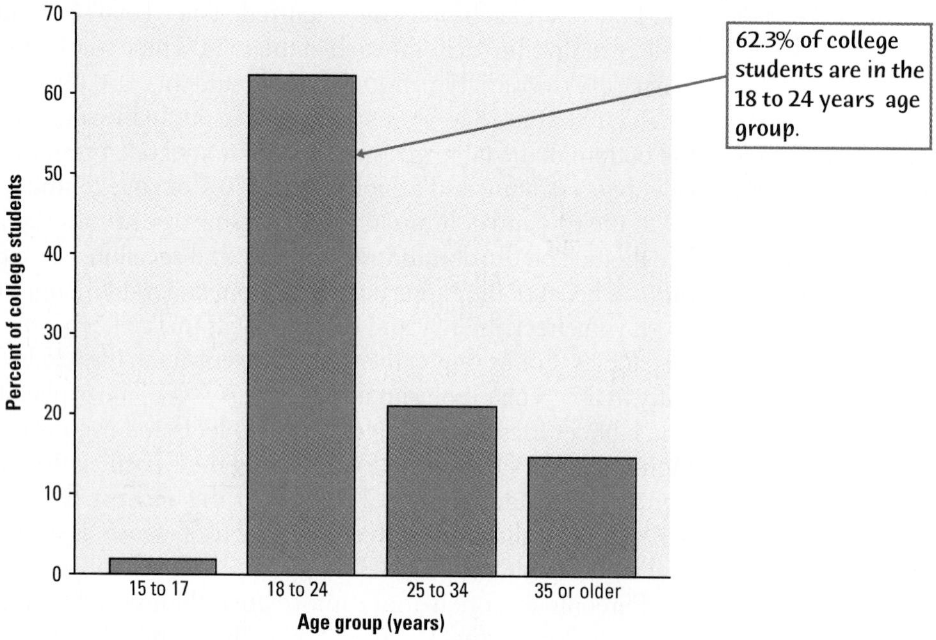

represents the total that I want a percent of?" The count for that group is the denominator of the fraction that leads to the percent. In Example 4.12, we wanted a percent of "college students," so the count of college students (the table total) is the denominator.

Describing Relationships

The marginal distributions of sex and of age from Table 4.5 do not tell us how the two variables are related. That information is in the body of the table. How can we describe the relationship between age and sex of college students? *To describe relationships among categorical variables, calculate appropriate percents from the counts given.* We use percents because counts are often hard to compare. For example, there are 5,668,000 female college students in the 18 to 24 years group, and only 1,660,000 in the 35 years and over group. Because there are many more students overall in the 18 to 24 group, these counts don't allow us to compare how prominent women are in the two age groups. In fact, women make up

$$\frac{5{,}668}{10{,}365} = 54.7\%$$

of the traditional college age group, 18 to 24 years. But they make up

$$\frac{1660}{2630} = 63.1\%$$

of students 35 years and older. Women are more likely than men to return to college after working for a number of years. That's an important part of the relationship between the sex and age of college students.

conditional distribution

When we compare the percent of women in two age groups, we are comparing two ***conditional distributions.***

Example 4.13 ***More college students***

Conditional distribution of sex, given age

If we know that a college student is 18 to 24 years old, we need only look at the "18 to 24 years" row in the two-way table, as in Table 4.5. To find the distribution of sex among only students in this age group, divide each count in the row by the row total, which is 10,365. The conditional distribution of sex, *given* that a student is 18 to 24 years old is

	Sex	
	Female	**Male**
Percent of 18 to 24 age group	54.7	45.3

The two percents add to 100% because all 18- to 24-year-old students are either female or male. We use the term "conditional" because these percents describe only students who satisfy the condition that they are between 18 and 24 years old. Find the conditional distribution of sex among students at least 35 years old in the same way, now looking only at the "35 years or older" row in Table 4.5:

	Sex	
	Female	**Male**
Percent of 35 or older age group	63.1	36.9

There is a separate conditional distribution of sex for each of the four age groups in Table 4.5. These distributions can differ from each other and also from the marginal distribution of sex among all college students.

Comparing conditional distributions reveals the nature of the association between the sex and age of college students. Because the variable "sex" has just two values, comparing conditional distributions just amounts to comparing the percents of women in the four age groups. The bar graph in Figure 4.21 (on the next page) makes the comparison visible. The heights of the bars do not add to 100% because they are not parts of a whole. Each bar describes a different age group. If age and sex were not related, all four bars would be the same height because the percent of women students would be the same in all age groups. As it is, the heights don't differ greatly but women are more common among the youngest and especially the oldest students.

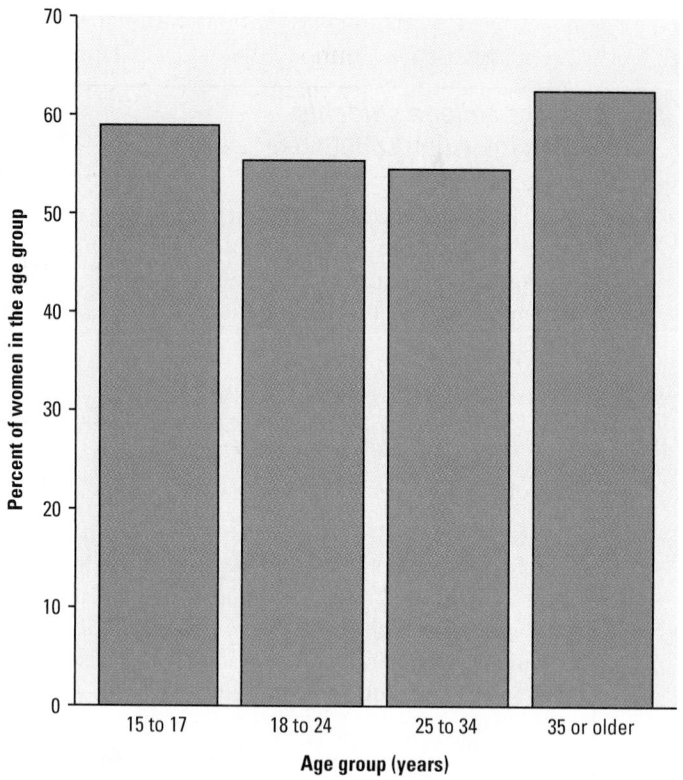

Figure 4.21 *Bar graph comparing the percent of female college students in four age groups. There are more women than men in all age groups, but the percent of women is highest among older students.*

Example 4.14 *College students again*

Conditional distribution of age, given sex

What is the distribution of age among female college students? Information about women students appears in the "Female" column of Table 4.5. So look only at this column. To find the conditional distribution of age, divide the count of women in each age group by the column total, which is 9321. Here is the distribution:

	Age Group			
	15 to 17	18 to 24	25 to 34	35 or older
Percent of women	1.0	60.8	20.4	17.8

Looking only at the "Male" column in the two-way table gives the conditional distribution of age for men:

	Age Group			
	15 to 17	18 to 24	25 to 34	35 or older
Percent of men	0.8	64.2	21.7	13.3

Comparing these two conditional distributions shows the relationship between sex and age in another form. Male students are more likely to be 18 to 24 years old and quite a bit less likely to be 35 or older.

Software will do these calculations for you. Most programs allow you to choose which conditional distributions you want to compare. The output in Figure 4.22 compares the two conditional distributions of sex given age and also the marginal distribution of age for all students. The percents in the first two columns agree (up to roundoff) with the displays in Example 4.14.

Figure 4.22 *CrunchIt! output of the two-way table of college students by age and sex, along with each entry as a percent of its row total. The percents in each row give the conditional distribution of sex for one age group, and the percents in the "Total" row give the marginal distribution of sex for all college students.*

CrunchIt!

Contingency Table with summary

Contingency table results:

Rows: Age group

Columns: Sex

Cell format
Count (Row percent)

	Female	Male	Total
15 to 17	89 (59.33%)	61 (40.67%)	150 (100.00%)
18 to 24	5668 (54.68%)	4697 (45.32%)	10365 (100.00%)
25 to 34	1904 (54.51%)	1589 (45.49%)	3493 (100.00%)
35 or older	1660 (63.12%)	970 (36.88%)	2630 (100.00%)
Total	9321 (56.02%)	7317 (43.98%)	16638 (100.00%)

CAUTION

No single graph (such as a scatterplot) portrays the form of the relationship between categorical variables. No single numerical measure (such as the correlation) summarizes the strength of the association. Bar graphs are flexible enough to be helpful, but you must think about what comparisons you want to display. For numerical measures, we rely on well-chosen percents. You must decide which percents you need. Here is a hint: *if there is a clear explanatory/response relationship, compare the conditional distributions of the response variable for the separate values of the explanatory variable.* If you think that sex influences the age at which students attend college, compare women and men as in Example 4.14.

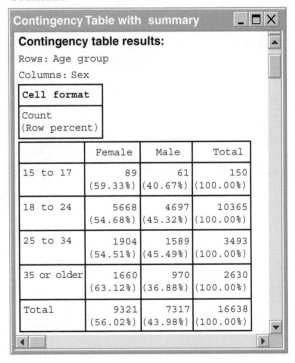

Exercises

4.23 Risks of playing soccer A study in Sweden looked at former elite soccer players, people who had played soccer but not at the elite level, and people of the same age who did not play soccer. Here is a two-way table that classifies these individuals by whether or not they had arthritis of the hip or knee by their midfifties:[14]

	Elite	Non-elite	Did not play
Arthritis	10	9	24
No arthritis	61	206	548

(a) How many people are described in the two-way table?

(b) How many of these people have arthritis of the hip or knee?

(c) Give the marginal distribution of participation in soccer, both as counts and as percents.

(d) Find the percent of each group who have arthritis. What do these percents say about the association between playing soccer and later arthritis?

4.24 Checking arithmetic Should the entries in the marginal distribution of Exercise 4.23(c) add to 100%? Do they? If not, why not?

4.25 Smoking by students and their parents Here are data from eight high schools on smoking among students and among their parents:[15]

	Neither parent smokes	One parent smokes	Both parents smoke
Student does not smoke	1168	1823	1380
Student smokes	188	416	400

(a) How many students are described in the two-way table?

(b) What percent of these students smoke?

(c) Give the marginal distribution of parents' smoking behavior, both in counts and in percents.

(d) Calculate three conditional distributions of students' smoking behavior: one for each of the three parental smoking categories. Describe the relationship between the smoking behaviors of students and their parents in a few sentences.

4.26 Python eggs How is the hatching of water python eggs influenced by the temperature of the snake's nest? Researchers assigned newly laid eggs to one of three temperatures: hot, neutral, or cold. Hot duplicates the extra warmth provided by the mother python, and cold duplicates the absence of the mother. Here are the data on the number of eggs and the number that hatched:[16]

	Cold	Neutral	Hot
Number of eggs	27	56	104
Number hatched	16	38	75

(a) Make a two-way table of temperature by outcome (hatched or not).

(b) Calculate the percent of eggs in each group that hatched. The researchers anticipated that eggs would not hatch in cold water. Do the data support that anticipation?

4.27 Majors for men and women in business A study of the career plans of young women and men sent questionnaires to all 722 members of the senior class in the College of Business Administration at the University of Illinois. One question asked which major within the business program the student had chosen. Here are the data from the students who responded:[17]

	Female	Male
Accounting	68	56
Administration	91	40
Economics	5	6
Finance	61	59

Smiling faces Women smile more than men. The same data that produce this fact allow us to link smiling to other variables in two-way tables. For example, add as the second variable whether or not the person thinks he or she is being observed. If yes, that's when women smile more. If no, there's no difference between women and men. Or take the second variable to be the person's occupation or social role. Within each occupational or social category, there is very little difference in smiling between women and men.

(a) Find the two conditional distributions of major, one for women and one for men. Based on your calculations, describe the differences between women and men with a graph and in words.

(b) What percent of the students did not respond to the questionnaire? The nonresponse weakens conclusions drawn from these data.

4.28 Marginal distributions aren't the whole story Here are the row and column totals for a two-way table with two rows and two columns:

a	b	50
c	d	50
60	40	100

Find *two different* sets of counts a, b, c, and d for the body of the table that give these same totals. This shows that the relationship between two variables cannot be obtained from the two individual distributions of the variables.

Simpson's Paradox

As is the case with quantitative variables, the effects of lurking variables can change or even reverse relationships between two categorical variables. Here is an example that demonstrates the surprises that can await the unsuspecting user of data.

Example 4.15 ***Do medical helicopters save lives?***
Simpson's paradox

Accident victims are sometimes taken by helicopter from the accident scene to a hospital. Helicopters save time. Do they also save lives? Let's compare the percents of accident

victims who die with helicopter evacuation and with the usual transport to a hospital by road. Here are data that illustrate a practical difficulty:[18]

	Helicopter	Road
Victim died	64	260
Victim survived	136	840
Total	**200**	**1100**

We see that 32% (64 out of 200) helicopter patients died, compared with only 24% (260 out of 1100) of the others. That seems discouraging.

The explanation is that the helicopter is sent mostly to serious accidents, so that the victims transported by helicopter are more often seriously injured. They are more likely to die with or without helicopter evacuation. Here are the same data broken down by the seriousness of the accident:

	Serious Accidents			Less Serious Accidents	
	Helicopter	**Road**		**Helicopter**	**Road**
Died	48	60	Died	16	200
Survived	52	40	Survived	84	800
Total	100	100	**Total**	100	1000

Inspect these tables to convince yourself that they describe the same 1300 accident victims as the original two-way table. For example, 200 were moved by helicopter, and 64 (48 + 16) of these died. Among victims of serious accidents, the helicopter saves 52% (52 out of 100) compared with 40% for road transport. If we look only at less serious accidents, 84% of those transported by helicopter survive, versus 80% of those transported by road. Both groups of victims have a higher survival rate when evacuated by helicopter.

At first, it seems paradoxical that the helicopter does better for both groups of victims but worse when all victims are lumped together. Examining the data makes the explanation clear. Half the helicopter transport patients are from serious accidents, compared with only 100 of the 1100 road transport patients. So the helicopter carries patients who are more likely to die. The seriousness of the accident was a lurking variable that, until we uncovered it, made the relationship between survival and mode of transport to a hospital hard to interpret. Example 4.15 illustrates ***Simpson's paradox.***

Simpson's Paradox

An association or comparison that holds for all of several groups can reverse direction when the data are combined to form a single group. This reversal is called **Simpson's paradox.**

The lurking variable in Simpson's paradox is categorical. That is, it breaks the individuals into groups, as when accident victims are classified as injured in a "serious accident" or a "less serious accident." Simpson's paradox is just an extreme form of the fact that observed associations can be misleading when there are lurking variables.

Exercises

4.29 Race and the death penalty Whether a convicted murderer gets the death penalty seems to be influenced by the race of the victim. Here are data on 326 cases in which the defendant was convicted of murder:[19]

	White Defendant			Black Defendant	
	White victim	Black victim		White victim	Black victim
Death	19	0	Death	11	6
Not	132	9	Not	52	97

(a) Use these data to make a two-way table of defendant's race (white or black) versus death penalty (yes or no).

(b) Show that Simpson's paradox holds: a higher percent of white defendants are sentenced to death overall, but for both black and white victims a higher percent of black defendants are sentenced to death.

(c) Use the data to explain why the paradox holds in language that a judge could understand.

4.30 College admissions paradox Upper Wabash Tech has two professional schools, business and law. Here are two-way tables of applicants to both schools, categorized by gender and admission decision. (Although these data are made up, similar situations occur in reality.)[20]

	Business			Law	
	Admit	Deny		Admit	Deny
Male	480	120	Male	10	90
Female	180	20	Female	100	200

(a) Make a two-way table of gender by admission decision for the two professional schools together by summing entries in the two-way tables above.

(b) From the two-way table, calculate the percent of male applicants who are admitted and the percent of female applicants who are admitted. Wabash admits a higher percent of male applicants.

(c) Now compute separately the percents of male and female applicants admitted by the business school and by the law school. Each school admits a higher percent of female applicants.

(d) This is Simpson's paradox: both schools admit a higher percent of the women who apply, but overall Wabash admits a lower percent of female applicants than of male applicants. Explain carefully, as if speaking to a skeptical reporter, how it can happen that Wabash appears to favor males when each school individually favors females.

"Yes, on the surface it would appear to be sex-bias but let us ask the following questions . . ."

Section 4.2 Summary

A **two-way table** of counts organizes data about two categorical variables. Values of the **row variable** label the rows that run across the table, and values of the **column variable** label the columns that run down the table. Two-way tables are often used to summarize large amounts of data by grouping outcomes into categories.

The row totals and column totals in a two-way table give the **marginal distributions** of the two individual variables. It is clearer to present these distributions as percents of the table total. Marginal distributions tell us nothing about the relationship between the variables.

To find the **conditional distribution** of the row variable for one specific value of the column variable, look only at that one column in the table. Find each entry in the column as a percent of the column total. There is a conditional distribution of the row variable for each column in the table. Comparing these conditional distributions is one way to describe the association between the row and the column variables. This approach is useful when the column variable is the explanatory variable. If the row variable is the explanatory variable, compare conditional distributions of the column variable for each value of the row variable.

Bar graphs are a flexible means of presenting categorical data. There is no single best way to describe an association between two categorical variables.

A comparison between two variables that holds for each individual value of a third variable can be reversed when the data for all values of the third variable are combined. This is **Simpson's paradox.** Simpson's paradox is an example of the effect of lurking variables on an observed association.

Attack of the killer TVs! Are kids in greater danger from TV sets or alligators? Alligator attacks make the news, but they aren't high on any count of causes of death and injury. In fact, the 28 children killed by falling TV sets in the United States between 1990 and 1997 is about twice the total number of people killed by alligators in Florida since 1948.

Section 4.2 Exercises

Marital status and job level *We sometimes hear that getting married is good for one's career. Table 4.6 presents data from one of the studies behind this generalization. To avoid gender effects, the investigators looked only at men. The data describe the marital status and the job level of all 8235 male managers and professionals employed by a large manufacturing firm. The firm assigns each position a grade that reflects the value of that particular job to the company. The authors of the study grouped the many job grades into quarters. Grade 1 contains jobs in the lowest quarter of the job grades, and Grade 4 contains those in the highest quarter. Exercises 4.31 to 4.35 are based on these data.*

Table 4.6 *Marital status and job grade*

	Marital Status				
Job grade	**Single**	**Married**	**Divorced**	**Widowed**	**Total**
1	58	874	15	8	955
2	222	3927	70	20	4239
3	50	2396	34	10	2490
4	7	533	7	4	551
Total	337	7730	126	42	8235

Source: Sanders Korenman and David Neumark, "Does marriage really make men more productive?" *Journal of Human Resources*, 26 (1991), pp. 282–307.

4.31 Marginal distributions Give (in percents) the two marginal distributions, for marital status and for job grade. Do each of your two sets of percents add to exactly 100%? If not, why not?

4.32 Percents What percent of single men hold Grade 1 jobs? What percent of Grade 1 jobs are held by single men?

4.33 Conditional distribution Give (in percents) the conditional distribution of job grade among single men. Should your percents add to 100% (up to roundoff error)?

4.34 Marital status and job grade One way to see the relationship is to look at who holds Grade 1 jobs.

(a) There are 874 married men with Grade 1 jobs, and only 58 single men with such jobs. Explain why these counts by themselves don't describe the relationship between marital status and job grade.

(b) Find the percent of men in each marital status group who have Grade 1 jobs. Then find the percent in each marital group who have Grade 4 jobs. What do these percents say about the relationship?

4.35 Association is not causation The data in Table 4.6 show that single men are more likely to hold lower-grade jobs than are married men. We should not conclude that single men can help their careers by getting married. What lurking variables might help explain the association between marital status and job grade?

4.36 Helping cocaine addicts Cocaine addiction is hard to break. Addicts need cocaine to feel any pleasure, so perhaps giving them an antidepressant drug will help. A 3-year study with

72 chronic cocaine users compared an antidepressant drug called desipramine with lithium and a placebo. (Lithium is a standard drug to treat cocaine addiction. A placebo is a dummy drug, used so that the effect of being in the study but not taking any drug can be seen.) One-third of the subjects, chosen at random, received each drug. Here are the results:[21]

	Desipramine	Lithium	Placebo
Relapse	10	18	20
No relapse	14	6	4
Total	24	24	24

(a) Compare the effectiveness of the three treatments in preventing relapse. Use percents and draw a bar graph.

(b) Do you think that this study gives good evidence that desipramine actually *causes* a reduction in relapses? Justify your answer.

4.37 Attitudes toward recycled products Recycling is supposed to save resources. Some people think recycled products are lower in quality than other products, a fact that makes recycling less practical. People who actually use a recycled product may have different opinions from those who don't use it. Here are data on attitudes toward coffee filters made of recycled paper among people who do and don't buy these filters:[22]

	Think the quality of the recycled product is:		
	Higher	The same	Lower
Buyers	20	7	9
Nonbuyers	29	25	43

(a) Find the marginal distribution of opinion about quality. Assuming that these people represent all users of coffee filters, what does this distribution tell us?

(b) How do the opinions of buyers and nonbuyers differ? Use conditional distributions as a basis for your answer. Can you conclude that using recycled filters *causes* more favorable opinions? (If so, giving away samples might increase sales.)

4.38 Baseball paradox Most baseball hitters perform differently against right-handed and left-handed pitching. Consider two players, Joe and Moe, both of whom bat right-handed. The table below records their performance against right-handed and left-handed pitchers:

Player	Pitcher	Hits	At-bats
Joe	Right	40	100
	Left	80	400
Moe	Right	120	400
	Left	10	100

(a) Make a two-way table of player (Joe or Moe) versus outcome (hit or no hit) by summing over both kinds of pitcher.

(b) Find the overall batting average (hits divided by total times at bat) for each player. Who has the higher batting average?

(c) Make a separate two-way table of player versus outcome for each kind of pitcher. From these tables, find the batting averages of Joe and Moe against right-handed pitching. Who does better? Do the same for left-handed pitching. Who does better?

(d) The manager doesn't believe that one player can hit better against both left-handers and right-handers yet have a lower overall batting average. Explain in simple language why this happens to Joe and Moe.

4.39 Obesity and health Recent studies have shown that earlier reports underestimated the health risks associated with being overweight. The error was due to overlooking lurking variables. In particular, smoking tends both to reduce weight and to lead to earlier death. Illustrate Simpson's paradox by a simplified version of this situation. That is, make up tables of overweight (yes or no) by early death (yes or no) by smoker (yes or no) such that

- overweight smokers and overweight nonsmokers both tend to die earlier than those not overweight,
- but when smokers and nonsmokers are combined into a two-way table of overweight by early death, persons who are not overweight tend to die earlier.

4.40 College degrees Here are data on the numbers of degrees earned in 2005–2006, as projected by the National Center for Education Statistics. The table entries are counts of degrees in thousands.[23]

	Female	Male
Associate's	431	244
Bachelor's	813	584
Master's	298	215
Professional	42	47
Doctor's	21	24

Describe briefly how the participation of women changes with level of degree. Follow the Data Analysis Toolbox (page 93).

4.3 Establishing Causation

When we study the relationship between two variables, we often hope to show that changes in the explanatory variable *cause* changes in the response variable. A strong association between two variables is not enough to draw conclusions about cause and effect. What ties between two variables (and others lurking in the background) can explain an observed association? What constitutes good evidence for causation? We begin our consideration of these questions with a set of examples. In each case, there is a clear association between an explanatory variable x and a response variable y.

Example 4.16 *Six interesting relationships*
Examining associations

The following are some examples of observed associations between x and y:

1. x = mother's body mass index
   y = daughter's body mass index
2. x = amount of the artificial sweetener saccharin in a rat's diet
   y = count of tumors in the rat's bladder
3. x = a high school senior's SAT score
   y = the student's first-year college grade point average
4. x = monthly flow of money into stock mutual funds
   y = monthly rate of return for the stock market
5. x = whether a person regularly attends religious services
   y = how long the person lives
6. x = the number of years of education a worker has
   y = the worker's income

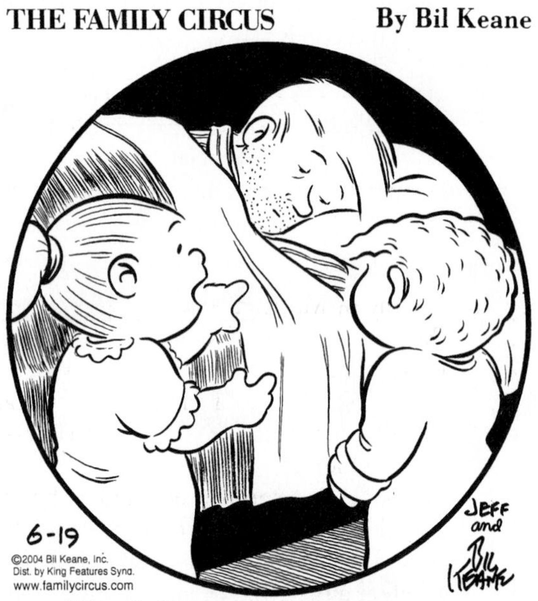

"See? Sleepin' makes Daddy's whiskers grow."

Explaining Association: Causation

Figure 4.23 shows in outline form how a variety of underlying links between variables can explain association. The dashed line represents an observed association between the variables x and y. Some associations are explained by ***causation***—a direct cause-and-effect link between the variables. The first diagram in Figure 4.23 shows "x causes y" by a solid arrow running from x to y.

causation

Figure 4.23 *Variables x and y show a strong association (dashed line). This association may be the result of any of several causal relationships (solid arrow). (a) Causation: Changes in x cause changes in y. (b) Common response: Changes in both x and y are caused by changes in a lurking variable z. (c) Confounding: The effect (if any) of x on y is confounded with the effect of a lurking variable z.*

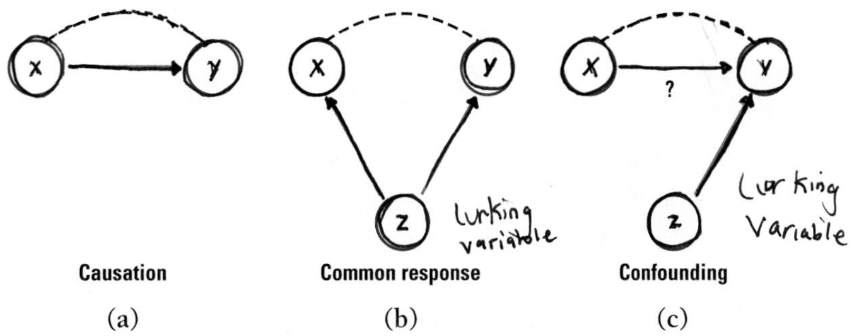

Example 4.17 *BMI in mothers and daughters; saccharin in rats*

Causation?

Items 1 and 2 in Example 4.16 are examples of direct causation. Thinking about these examples, however, shows that "causation" is not a simple idea.

1. A study of Mexican American girls aged 9 to 12 years recorded body mass index (BMI), a measure of weight relative to height, for both the girls and their mothers. People with high BMI are overweight or obese. The study also measured hours of television, minutes of physical activity, and intake of several kinds of food. The strongest correlation ($r = 0.506$) was between the BMI of daughters and the BMI of their mothers.[24]

   Body type is in part determined by heredity. Daughters inherit half their genes from their mothers. As a result, there is a direct causal link between the BMI of mothers and daughters. Yet the mothers' BMIs explain only 25.6% (that's r^2 again) of the variation among the daughters' BMIs. Other factors, such as diet and exercise, also influence BMI. **Even when direct causation is present, it is rarely a complete explanation of an association between two variables.**

2. The best evidence for causation comes from experiments that actually change x while holding all other factors fixed. If y changes, we have good reason to think that x caused the change in y. Experiments show conclusively that large amounts of saccharin in the diet cause bladder tumors in rats. Should we avoid saccharin as a replacement for sugar in food? Rats are not people. Although we can't experiment with people, studies of people who consume different amounts of saccharin show little association between saccharin and bladder tumors.[25] **Even well-established causal relations may not generalize to other settings.**

Explaining Association: Common Response

common response

"Beware the lurking variable" is good advice when thinking about an association between two variables. The second diagram in Figure 4.23 illustrates ***common response.*** The observed association between the variables x and y is explained by a lurking variable z. Both x and y change in response to changes in z. This common response creates an association even though there may be no direct causal link between x and y.

Example 4.18 ***SAT and GPA; mutual funds and the stock market***
Common response

The third and fourth items in Example 4.16 illustrate how common response can create an association.

3. Students who are smart and who have learned a lot tend to have both high SAT scores and high college grades. The positive correlation is explained by this common response to students' ability and knowledge.
4. There is a strong positive correlation between how much money individuals add to mutual funds each month and how well the stock market does the same month. Is the new money driving the market up? The correlation may be explained in part by common response to underlying investor sentiment: when optimism reigns, individuals send money to funds and large institutions also invest more. The institutions would drive up prices even if individuals did nothing. In addition, what causation there is may operate in the other direction: when the market is doing well, individuals rush to add money to their mutual funds.[26]

Explaining Association: Confounding

We noted in Example 4.17 that inheritance no doubt explains part of the association between the body mass indexes (BMIs) of daughters and their mothers. Can we use r or r^2 to say how much inheritance contributes to the daughters' BMIs? No. It may well be that mothers who are overweight also set an example of little exercise, poor eating habits, and lots of television. Their daughters pick up these habits to some extent, so the influence of heredity is mixed up with influences from the girls' environment. We call this mixing of influences ***confounding.***

Confounding

Two variables are **confounded** when their effects on a response variable cannot be distinguished from each other. The confounded variables may be either explanatory variables or lurking variables.

Confounding often prevents us from drawing conclusions about causation. The third diagram in Figure 4.23 illustrates confounding. Both the explanatory variable x and the lurking variable z may influence the response variable y. Because x is confounded with z, we cannot distinguish the influence of x from the influence of z. We cannot say how strong the direct effect of x on y is. In fact, it can be hard to say if x influences y at all.

Example 4.19 ***Religion and life span; education and income***
Confounding

The last two associations in Example 4.16 (Items 5 and 6) are explained in part by confounding.

5. Many studies have found that people who are active in their religion live longer than nonreligious people. But people who attend church or mosque or synagogue also take better care of themselves than nonattenders. They are less likely to smoke, more likely to exercise, and less likely to be overweight. The effects of these good habits are confounded with the direct effects of attending religious services.
6. It is likely that more education is a cause of higher income—many highly paid professions require advanced education. However, confounding is also present. People who have high ability and come from prosperous homes are more likely to get many years of education than people who are less able or poorer. Of course, people who start out able and rich are more likely to have high earnings even without much education. We can't say how much of the higher income of well-educated people is actually caused by their education.

Many observed associations are at least partly explained by lurking variables. Both common response and confounding involve the influence of a lurking variable (or variables) z on the response variable y. The distinction between these two types of relationships is less important than the common element—the influence of lurking variables. The most important lesson of these examples is one we have already emphasized: even a very strong association between two variables is not by itself good evidence that there is a cause-and-effect link between the variables.

Establishing Causation

How can a direct causal link between x and y be established? The best method of establishing causation is to conduct a carefully designed experiment in which the effects of possible lurking variables are controlled. Much of Chapter 5 is devoted to the art of designing convincing experiments.

Many of the sharpest disputes in which statistics plays a role involve questions of causation that cannot be settled by experiment. Does gun control reduce violent crime? Does living near power lines cause cancer? Has increased free trade helped to increase the gap between the incomes of more educated and less educated American workers? All of these questions have become public issues. All concern associations among variables. And all have this in common: they try to pinpoint cause and effect in a setting involving complex relations among many variables. Common response and confounding, along with the number of potential lurking variables, make observed associations misleading. Experiments are not possible for ethical or practical reasons. We can't assign some people to live near power lines or compare the same nation with and without free-trade agreements.

Example 4.20 *Do power lines increase the risk of leukemia?*

Relying on observational studies

Electric currents generate magnetic fields. So living with electricity exposes people to magnetic fields. Living near power lines increases exposure to these fields. Really strong fields can disturb living cells in laboratory studies. What about the weaker fields we experience if we live near power lines?

It isn't ethical to do experiments that expose children to magnetic fields. It's hard to compare cancer rates among children who happen to live in more and less exposed locations, because leukemia is rare and locations vary in many ways other than magnetic fields. We must rely on studies that compare children who have leukemia with children who don't.

A careful study of the effect of magnetic fields on children took five years and cost $5 million. The researchers compared 638 children who had leukemia and 620 who did not. They went into the homes and actually measured the magnetic fields in the children's bedrooms, in other rooms, and at the front door. They recorded facts about nearby power lines for the family home and also for the mother's residence when she was pregnant. *Result:* no evidence of more than a chance connection between magnetic fields and childhood leukemia.[27]

"No evidence" that magnetic fields are connected with childhood leukemia doesn't prove that there is no risk. It says only that a careful study could not find any risk that stands out from the play of chance that distributes leukemia cases across the landscape. Critics continue to argue that the study failed to measure some lurking variables, or that the children studied don't fairly represent all children. Nonetheless, a carefully designed study comparing children with and without leukemia is a great advance over haphazard and sometimes emotional counting of cancer cases.

Example 4.21 *Does smoking cause lung cancer?*

Establishing causation without experiments

Despite the difficulties, it is sometimes possible to build a strong case for causation in the absence of experiments. The evidence that smoking causes lung cancer is about as strong as nonexperimental evidence can be.

Doctors had long observed that most lung cancer patients were smokers. Comparison of smokers and similar nonsmokers showed a very strong association between smoking and death from lung cancer. Could the association be due to common response? Might there be, for example, a genetic factor that predisposes people both to nicotine addiction and to lung cancer? Smoking and lung cancer would then be positively associated even if smoking had no direct effect on the lungs. Or perhaps confounding is to blame. It might be that smokers live unhealthy lives in other ways (diet, alcohol, lack of exercise) and that some other habit confounded with smoking is a cause of lung cancer. How were these objections overcome?

Let's answer this question in general terms: what are the criteria for establishing causation when we cannot do an experiment?

- *The association is strong.* The association between smoking and lung cancer is very strong.
- *The association is consistent.* Many studies of different kinds of people in many countries link smoking to lung cancer. That reduces the chance that a lurking variable specific to one group or one study explains the association.
- *Larger values of the response variable are associated with stronger responses.* People who smoke more cigarettes per day or who smoke over a longer period get lung cancer more often. People who stop smoking reduce their risk.
- *The alleged cause precedes the effect in time.* Lung cancer develops after years of smoking. The number of men dying of lung cancer rose as smoking became more common, with a lag of about 30 years. Lung cancer kills more men than any other form of cancer. Lung cancer was rare among women until women began to smoke. Lung cancer in women rose along with smoking, again with a lag of about 30 years, and has now passed breast cancer as the leading cause of cancer death among women.
- *The alleged cause is plausible.* Experiments with animals show that tars from cigarette smoke do cause cancer.

Medical authorities do not hesitate to say that smoking causes lung cancer. The U.S. Surgeon General states that cigarette smoking is "the largest avoidable cause of death and disability in the United States."[28] The evidence for causation is overwhelming—but it is not as strong as the evidence provided by well-designed experiments. Conducting an experiment in which some subjects were forced to smoke and others were not allowed to would be unethical, however. In cases like this, observational studies are our best source of reliable information.

Section 4.3 Summary

The effect of lurking variables can operate through **common response** if changes in both the explanatory and response variables are caused by changes in lurking variables. **Confounding** of two variables (either explanatory or lurking variables) means that we cannot distinguish their effects on the response variable.

Be careful not to conclude that there is a cause-and-effect relationship between two variables just because they are strongly associated. The relationship could involve common response or confounding. The best evidence that an association is due to **causation** comes from an experiment in which the explanatory variable is directly changed and other influences on the response are controlled.

In the absence of experimental evidence, be cautious in accepting claims of causation. Good evidence of causation requires a strong association that appears consistently in many studies, a clear explanation for the alleged causal link, and careful examination of possible lurking variables.

Section 4.3 Exercises

For Exercises 4.41 through 4.48, state whether the relationship between the two variables involves causation, common response, or confounding. Identify possible lurking variable(s). Draw a diagram of the relationship in which each circle represents a variable. By each circle, write a brief description of the variable.

4.41 Does TV make you live longer? Measure the number of television sets per person x and the average life expectancy y for the world's nations. There is a high positive correlation: nations with many TV sets have higher life expectancies. Could we lengthen the lives of people in Rwanda by shipping them TV sets? Justify your answer.

4.42 Do artificial sweeteners cause weight gain? People who use artificial sweeteners in place of sugar tend to be heavier than people who use sugar. Does this mean that artificial sweeteners cause weight gain? Give a more plausible explanation for this association.

4.43 Does exposure to industrial chemicals cause miscarriages? A study showed that women who work in the production of computer chips have abnormally high numbers of miscarriages. The union claimed that exposure to chemicals used in production causes the miscarriages. Another possible explanation is that these workers spend most of their time standing up. Can we conclude that exposure to chemicals causes more miscarriages? Why or why not?

4.44 Does having more cars make you live longer? A serious study once found that people with two cars live longer than people who own only one car.[29] Owning three cars is even better, and so on. There is a substantial positive correlation between number of cars x and length of life y. What lurking variables might explain the association between number of cars owned and life span?

4.45 Are grades and TV watching linked? Children who watch many hours of television get lower grades in school on the average than those who watch less TV. Explain clearly why this fact does not show that watching TV causes poor grades. In particular, suggest some other variables that may be confounded with heavy TV viewing and may contribute to poor grades.

4.46 To earn more, get married? Data show that men who are married, and also divorced or widowed men, earn quite a bit more than men the same age who have never been married. This does not mean that a man can raise his income by getting married, because men who have never been married are different from married men in many ways other than marital status. Suggest several lurking variables that might help explain the association between marital status and income.

4.47 Raising SAT scores A study finds that high school students who take the SAT, enroll in an SAT coaching course, and then take the SAT a second time raise their SAT mathematics scores from a mean of 521 to a mean of 561.[30] What factors other than taking the course might explain this improvement?

4.48 How's your self-esteem? People who do well tend to feel good about themselves. Perhaps helping people feel good about themselves will help them do better in school and life. Raising self-esteem became for a time a goal in many schools. California even created a state commission to advance the cause. Can you think of explanations for the association between high self-esteem and good school performance other than "Self-esteem causes better work in school"?

CASE CLOSED!

It's a matter of life and death

Return to the chapter-opening Case Study (page 257) about life insurance. You should now be ready to answer the following questions.

1. Determining insurance premiums
   (a) Would a power model provide an appropriate description of the relationship between age and monthly premium? Transform the data and sketch a graph that will help answer this question.
   (b) Would an exponential model provide an appropriate description of the relationship between age and monthly premium? Transform the data and sketch a graph that will help answer this question.
   (c) Based on your answers to (a) and (b), use least-squares regression to fit the most appropriate type of model for these data. Perform the inverse transformation to write monthly premium as a function of age.
   (d) Use your model to predict the monthly premium for a 58-year-old who wants a $1 million, 10-year term life insurance policy. For a 68-year-old.
   (e) How comfortable do you feel about the predictions you made in (d)? Justify your answer using a residual plot and r^2.

(continued)

2. Death statistics

Here is a two-way table of number of deaths in the United States in three age groups from selected causes in 2003.[31] The entries are counts of deaths.

	15 to 24 years	25 to 44 years	45 to 64 years
Accidents	14,966	27,844	23,669
AIDS	171	6,879	5,917
Cancer	1,628	19,041	144,936
Heart disease	1,083	16,283	101,713
Homicide	5,148	7,367	2,756
Suicide	3,921	11,251	10,057
Total deaths	**33,022**	**128,924**	**437,058**

(a) Why don't the entries in each column add to the "Total deaths" count?
(b) Should you use counts or percents to compare the age groups? Explain.
(c) Construct the conditional distribution of cause of death for each age group. Then make a bar graph to display the results.
(d) Use your results in (c) to explain briefly how the leading causes of death change as people get older.

3. Stay fitter, live longer

A sign in a fitness center says, "Mortality is halved for men over 65 who walk at least 2 miles a day."

(a) Mortality is eventually 100% for everyone. What do you think "mortality is halved" means?
(b) Assuming that the claim is true, explain why this fact does not show that exercise *causes* lower mortality.

Chapter Review

Summary

In this chapter we learned how to use transformations to construct mathematical models for data that follow a curved pattern. By using powers and logarithms, we can often transform one or both variables to achieve a linear relationship. Many relationships between two quantitative variables can be accurately described by an exponential function ($y = ab^x$) or a power function ($y = ax^p$).

When both variables are categorical, there is no perfect graph for displaying the data, although bar graphs can be helpful. We describe the relationship by comparing percents.

A strong observed association between two variables may exist without a cause-and-effect link between them. You should always look for lurking variables that might affect the relationship.

What You Should Have Learned

Here is a review list of the most important skills you should have gained from studying this chapter.

A. Modeling Nonlinear Data

1. Use powers to transform nonlinear data to achieve linearity. Then fit a linear model to the transformed data.
2. Recognize that, when a variable is multiplied by a fixed number in each equal time period, exponential growth results and that, when one variable is proportional to a power of a second variable, a power law model results.
3. In the case of both exponential growth and power functions, perform a logarithmic transformation and obtain points that lie in a linear pattern. Then use least-squares regression on the transformed data. Carry out an inverse transformation to produce a curve that models the original data.
4. Know that deviations from the overall pattern are most easily examined by fitting a line to the transformed points and plotting the residuals from this line against the explanatory variable (or fitted values).

B. Relations in Categorical Data

1. From a two-way table of counts, find the marginal distributions of both variables by obtaining the row sums and column sums.
2. Describe the relationship between two categorical variables by computing and comparing percents. Often this involves comparing the conditional distributions of one variable for the different categories of the other variable. Construct bar graphs when appropriate.
3. Recognize Simpson's paradox and be able to explain it.

C. Establishing Causation

1. Recognize possible lurking variables that may help explain the observed association between two variables x and y.
2. Determine whether the relationship between two variables is most likely due to causation, common response, or confounding.
3. Understand that even a strong association does not mean that there is a cause-and-effect relationship between x and y.

Web Links

Moore's law on the Web www.intel.com/technology/silicon/index.htm

Chapter Review Exercises

4.49 TV and obesity Over the last 20 years there has developed a positive association between sales of television sets and the number of obese adolescents in the United States. Do more TVs cause more children to put on weight, or are there other factors involved? List some of the possible lurking variables.

4.50 Light intensity, I In physics class, the intensity of a 100-watt light bulb was measured by a sensing device at various distances from the light source, and the following data were collected. Note that a candela (cd) is a unit of luminous intensity in the International System of Units.

Distance (meters)	Intensity (candelas)
1.0	0.2965
1.1	0.2522
1.2	0.2055
1.3	0.1746
1.4	0.1534
1.5	0.1352
1.6	0.1145
1.7	0.1024
1.8	0.0923
1.9	0.0832
2.0	0.0734

(a) Plot the data. Describe what you see.

(b) Transform the data to achieve linearity. Then use linear regression to construct a model for the transformed data.

(c) Assess the quality of your linear model from (b) using r^2 and a residual plot.

(d) Use an inverse transformation to write an equation for predicting intensity directly from distance. Then plot the original data with this model on the same axes.

(e) What would you predict for the intensity of the 100-watt bulb at a distance of 2.1 meters? Show your work.

4.51 Light intensity, II Physics textbooks suggest that the relationship between light intensity and distance should follow an "inverse square law," that is, a power law model with $p = -2$. Let's examine that claim using the data from the previous exercise.

(a) Transform the distance measurements by squaring them and then taking their reciprocals. Now plot the points(1/distance^2, intensity). Did this transformation achieve linearity?

(b) Perform least-squares regression on the transformed data. Write the resulting equation that describes the relationship between intensity of light and distance from the light source.

(c) What would you predict for the intensity of a 100-watt bulb at a distance of 2.1 meters? Show your work.

(d) Compare your model from (b) and your prediction from (c) with the results of the previous exercise. Does an inverse square law describe the relationship well?

4.52 The power of herbal tea A group of college students believes that herbal tea has remarkable powers. To test this belief, they make weekly visits to a local nursing home, where they visit with the residents and serve them herbal tea. The nursing home staff reports that after several months many of the residents are more cheerful and healthy. A skeptical sociologist commends the students for their good deeds but scoffs at the idea that herbal tea helped the residents. Identify the explanatory and response variables in this informal study. Then explain what lurking variables account for the observed association.

4.53 Women and marital status The following two-way table describes the age and marital status of American women in 2003. The table entries are in thousands of women.

	Marital Status				
Age	**Single**	**Married**	**Widowed**	**Divorced**	**Total**
15–24	10,949	2,472	16	155	13,592
25–39	7,653	19,640	228	2,904	30,425
40–64	4,009	32,183	2,312	7,898	46,402
≥65	720	8,539	8,732	1,703	19,694
Total					110,115

(a) Calculate the sum of the entries in each column. Check that the total of these columns equals the row total.

(b) Give the marginal distribution of marital status for American women (use percents). Draw a bar graph to display this distribution.

(c) Compare the conditional distributions of marital status for women aged 15 to 24 and women aged 40 to 64. Briefly describe the most important differences between the two groups of women, and back up your description with percents.

(d) You are planning a magazine aimed at single women who have never been married. (That's what "single" means in government data.) Find the conditional distribution of ages among single women.

4.54 Follow the bouncing ball Students in Mr. Handford's class dropped a kickball beneath a motion detector. The detector recorded the height of the ball as it bounced up and down several times. Here are the heights of the ball at the highest point of each of the first few bounces:

Bounce number	Height (feet)
0	2.935
1	2.240
2	1.620
3	1.235
4	0.958
5	0.756

(a) Would a power model or an exponential model be more appropriate for predicting height from bounce number? Justify your answer by transforming the data with logarithms and plotting graphs.

(b) Find the least-squares regression line for the transformed data you chose in (a). Assess the quality of this model using a residual plot and r^2.

(c) Perform an inverse transformation to obtain an equation that gives height as a function of bounce number. Then use your model to predict the height of the ball's seventh bounce.

4.55 Ice cream and flu There is a negative correlation between the number of flu cases reported each week throughout the year and the amount of ice cream sold in that particular week. It's unlikely that ice cream prevents flu. What is a more plausible explanation for this observed correlation? Draw a diagram to illustrate your explanation.

4.56 Do angry people have more heart disease? People who get angry easily tend to have more heart disease. That's the conclusion of a study that followed a random sample of 12,986 people from three locations for about four years. All subjects were free of heart disease at the beginning of the study. The subjects took the Spielberger Trait Anger Scale test, which measures how prone a person is to sudden anger. Here are data for the 8474 people in the sample who had normal blood pressure.[32] CHD stands for "coronary heart disease." This includes people who had heart attacks and those who needed medical treatment for heart disease.

	Low anger	Moderate anger	High anger	Total
CHD	53	110	27	190
No CHD	3057	4621	606	8284
Total	3110	4731	633	8474

Do these data support the study's conclusion about the relationship between anger and heart disease? Follow the Data Analysis Toolbox (page 93) to help you answer this question.

4.57 Killing bacteria Expose marine bacteria to X-rays for time periods from 1 to 15 minutes. Here are the number of surviving bacteria (in hundreds) on a culture plate after each exposure time:[33]

Time *t*	Count *y*	Time *t*	Count *y*
1	355	9	56
2	211	10	38
3	197	11	36
4	166	12	32
5	142	13	21
6	106	14	19
7	104	15	15
8	60		

Theory suggests an exponential growth or decay model. Follow the Data Analysis Toolbox (page 93) to determine whether these data agree with the theory.

4.58 Smoking and staying alive In the mid-1970s, a medical study contacted randomly chosen people in a district in England. Here are data on the 1314 women contacted who were either current smokers or who had never smoked. The table classifies these women by their smoking status and age at the time of the survey and whether they were still alive 20 years later.[34]

	Age 18 to 44			Age 45 to 64			Age 65+	
	Smoker	Not		Smoker	Not		Smoker	Not
Dead	19	13	Dead	78	52	Dead	42	165
Alive	269	327	Alive	167	147	Alive	7	28

(a) Make a two-way table of smoking (yes or no) by dead or alive. What percent of the smokers stayed alive for 20 years? What percent of the nonsmokers survived? It seems surprising that a higher percent of smokers stayed alive.

(b) The age of the women at the time of the study is a lurking variable. Show that within each of the three age groups in the data, a higher percent of nonsmokers remained alive 20 years later. This is another example of Simpson's paradox.

(c) The study authors give this explanation: "Few of the older women (over 65 at the original survey) were smokers, but many of them had died by the time of follow-up." Compare the percent of smokers in the three age groups to verify the explanation.

PART I

Review Exercises

Communicate your thinking clearly in Exercises I.1 to I.10.

I.1 Long-term records from the Serengeti National Park in Tanzania show interesting ecological relationships. When wildebeest are more abundant, they graze the grass more heavily, so there are fewer fires and more trees grow. Lions feed more successfully when there are more trees, so the lion population increases. Here are data on one part of this cycle, wildebeest abundance (in thousands of animals) and the percent of the grass area burned in the same year:[1]

Wildebeest (1000s)	Percent burned	Wildebeest (1000s)	Percent burned
396	56	622	60
476	50	600	56
698	25	902	45
1049	16	1440	21
1178	7	1147	32
1200	5	1173	31
1302	7	1178	24
360	88	1253	24
444	88	1249	53
524	75		

To what extent do these data support the claim that more wildebeest reduce the percent of grasslands that are burned? How rapidly does burned area decrease as the number of wildebeest increases? Include a graph and suitable calculations. Follow the Data Analysis Toolbox.

I.2 What is the most important reason that students buy from catalogs? The answer may differ for different groups of students. Here are results for samples of American and East Asian students at a large midwestern university:[2]

	American	Asian
Save time	29	10
Easy	28	11
Low price	17	34
Live far from stores	11	4
No pressure to buy	10	3
Other reason	20	7
Total	**115**	**69**

(a) Give the marginal distribution of reasons for all students, in percents.

(b) Give the two conditional distributions of reasons, for American and for Asian students.

(c) What are the most important differences between the two groups of students? Justify your answer.

I.3 Biological measurements on the same species often follow a Normal distribution quite closely. The weights of seeds of a variety of winged bean are approximately Normal with mean 525 milligrams (mg) and standard deviation 110 mg.

(a) What percent of seeds weigh more than 500 mg? Show your method.

(b) If we discard the lightest 10% of these seeds, what is the smallest weight among the remaining seeds? Show your method.

I.4 Every 17 years, swarms of cicadas emerge from the ground in the eastern United States, live for about six weeks, then die. (There are several "broods," so we experience cicada eruptions more often than every 17 years.) There are so many cicadas that their dead bodies can serve as fertilizer and increase plant growth. In an experiment, a researcher added 10 cicadas under some plants in a natural plot of American bellflowers on the forest floor, leaving other plants undisturbed. One of the response variables was the size of seeds produced by the plants. Here are data (seed mass in milligrams) for 39 cicada plants and 33 undisturbed (control) plants:[3]

Cicada plants		Control plants	
0.237	0.277	0.212	0.188
0.109	0.209	0.261	0.265
0.261	0.227	0.203	0.241
0.276	0.234	0.215	0.285
0.239	0.266	0.178	0.244
0.238	0.210	0.290	0.253
0.218	0.263	0.268	0.190
0.351	0.245	0.246	0.145
0.317	0.310	0.241	0.253
0.192	0.201	0.263	0.170
0.241	0.142	0.135	0.155
0.238	0.277	0.257	0.266
0.171	0.235	0.198	0.212
0.255	0.296	0.190	0.253
0.296	0.217	0.249	0.220
0.295	0.193	0.196	0.140
0.305	0.257	0.247	
0.226	0.276		
0.223	0.229		
0.211			

Do the data support the idea that dead cicadas can serve as fertilizer? Give appropriate graphical and numerical evidence to support your conclusion. Follow the Data Analysis Toolbox.

The Nenana Ice Classic is an annual contest to guess the exact time in the spring thaw when a tripod erected on the frozen Tanana River near Nenana, Alaska, will fall through the ice. The 2005 jackpot prize was $285,000. The contest has been run since 1917. Table I.1 gives simplified data that record only the date on which the tripod fell each year. The earliest date so far is April 20. To make the data easier to use, the table gives the date each year in days starting with April 20. That is, April 20 is 1, April 21 is 2, and so on. You will need software or a graphing calculator to analyze these data in Exercises I.5 to I.7.

Table I.1 Days from April 20 for the Tanana River tripod to fall

Year	Day	Year	Day	Year	Day
1917	11	1947	14	1977	17
1918	22	1948	24	1978	11
1919	14	1949	25	1979	11
1920	22	1950	17	1980	10
1921	22	1951	11	1981	11
1922	23	1952	23	1982	21
1923	20	1953	10	1983	10
1924	22	1954	17	1984	20
1925	16	1955	20	1985	23
1926	7	1956	12	1986	19
1927	23	1957	16	1987	16
1928	17	1958	10	1988	8
1929	16	1959	19	1989	12
1930	19	1960	13	1990	5
1931	21	1961	16	1991	12
1932	12	1962	23	1992	25
1933	19	1963	16	1993	4
1934	11	1964	31	1994	10
1935	26	1965	18	1995	7
1936	11	1966	19	1996	16
1937	23	1967	15	1997	11
1938	17	1968	19	1998	1
1939	10	1969	9	1999	10
1940	1	1970	15	2000	12
1941	14	1971	19	2001	19
1942	11	1972	21	2002	18
1943	9	1973	15	2003	10
1944	15	1974	17	2004	5
1945	27	1975	21	2005	9
1946	16	1976	13		

Source: From the Nenana Ice Classic Web page, www.nenanaakiceclassic.com. See Raphael Sagarin and Fiorenza Micheli, "Climate change in nontraditional data sets," *Science*, 294 (2001), p. 811, for a careful discussion.

I.5 We have 89 years of data on the date of ice breakup on the Tanana River. Describe the distribution of the breakup date with both a graph or graphs and appropriate numerical summaries. What is the median date (month and day) for ice breakup?

I.6 Because of the high stakes, the falling of the tripod has been carefully observed for many years. If the date the tripod falls has been getting earlier, that may be evidence for the effects of global warming.

(a) Make a time plot of the date the tripod falls against year.

(b) There is a great deal of year-to-year variation. Fitting a regression line to the data may help us see the trend. Fit the least-squares line and add it to your time plot. What do you conclude?

(c) There is much variation about the line. Give a numerical description of how much of the year-to-year variation in ice breakup time is accounted for by the time trend represented by the regression line.

I.7 Side-by-side boxplots offer a different look at the data. Group the data into periods of roughly equal length: 1917 to 1939, 1940 to 1959, 1960 to 1979, and 1980 to 2005. Make boxplots to compare ice breakup dates in these four time periods. Write a brief description of what the plots show.

I.8 Many colleges offer online versions of courses that are also taught in the classroom. It often happens that the students who enroll in the online version do better than the classroom students on the course exams. Explain why this does not show that online instruction is more effective than classroom teaching. Suggest some differences between online and classroom students that might explain why online students do better.

I.9 Table I.2 gives data on the mean number of seeds produced in a year by several common tree species and the mean weight (in milligrams) of the seeds produced. (Some species appear twice because their seeds were counted in two locations.) We might expect that trees with heavy seeds produce fewer of them, but what mathematical model best describes the relationship? Use the Data Analysis Toolbox to help you answer this question.

Table I.2 *Count and weight of seeds produced by common tree species*

Tree species	Seed count	Seed weight (mg)
Paper birch	27,239	0.6
Yellow birch	12,158	1.6
White spruce	7,202	2.0
Engelmann spruce	3,671	3.3
Red spruce	5,051	3.4
Tulip tree	13,509	9.1
Ponderosa pine	2,667	37.7
White fir	5,196	40.0
Sugar maple	1,751	48.0
Sugar pine	1,159	216.0
American beech	463	247
American beech	1,892	247
Black oak	93	1,851
Scarlet oak	525	1,930
Red oak	411	2,475
Red oak	253	2,475
Pignut hickory	40	3,423
White oak	184	3,669
Chestnut oak	107	4,535

Source: Data from many studies compiled in D. F. Greene and E. A. Johnson, "Estimating the mean annual seed production of trees," *Ecology*, 75 (1994), pp. 642–647.

I.10 U.S. car sales change month by month because of car prices, the national economy, and other factors that affect the sales data. How about gas price? Does it also affect car sales in some manner? To answer this question, we analyze U.S. car sale data from January 2000 to June 2001, a period when gas prices were generally increasing, to see if we can find some relation between gas price and car sales.

(a) As gas price goes up, people may think more about gas mileage when they are deciding what car to buy. Figure I.1 shows side-by-side boxplots of gas mileage for each size class of car—subcompact, compact, midsize, and large. Write a paragraph comparing the distributions of gas mileage for the four car classes.

Figure I.1 *Boxplots of gas mileage for different classes of cars.*

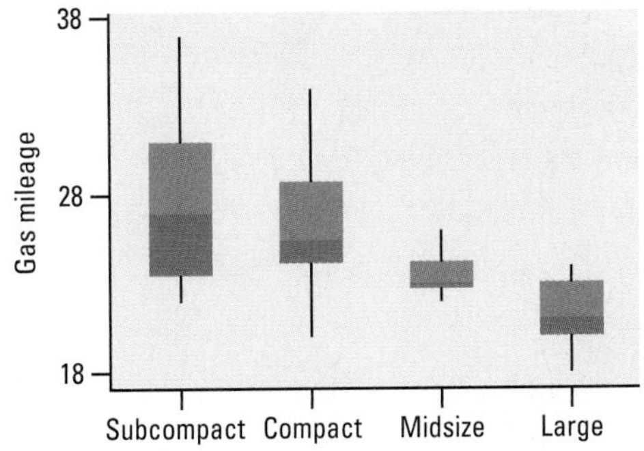

A reasonable theory is that as gas prices increase, people will buy smaller cars. Figure I.2 displays a scatterplot of the average price of regular unleaded gasoline and the percent of cars sold that month with gas mileage greater than 26.75 miles per gallon (mpg). The least-squares regression line has been added to the plot, and some computer regression output appears at the top of the graph.

Figure I.2 *Scatterplot of monthly gas prices and percent of cars purchased that month with gas mileage greater than 26.75 mpg.*

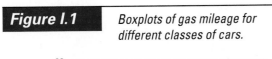

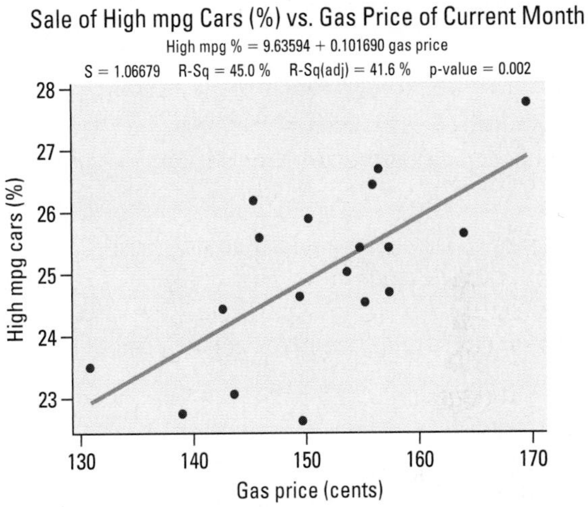

(b) Interpret the slope and y intercept of the least-squares regression line in the context of the problem.

(c) During one of the months in this study, the average gas price was \$1.50. Calculate the residual for that month. Show your method.

(d) Interpret the value of r^2 in the context of the problem.

PART II

Producing Data: Surveys, Observational Studies, and Experiments

CHAPTER

5

Producing Data

CASE STUDY

Can eating chocolate be good for you?

Chocolate and other products from cacao plants have been popular with people for centuries. The Mayan and Aztec cultures in Central America enjoyed cocoa, which they made from beans from the cacao tree's football-sized pods. In seventeenth-century France, chocolate was used to treat poor digestion, lung and heart conditions, and infectious diseases. We now know that chocolate contains vitamins, minerals, and flavonoids, which are compounds that help protect against heart disease. Contemporary studies have shown that chocolate acts like an antioxidant that protects cells from damage. Here are brief summaries of three recent studies.

Study I A University of Helsinki (Finland) study wanted to determine if chocolate consumption during pregnancy had an effect on infant temperament at age 6 months. Researchers began by asking 305 healthy pregnant women to rate their stress levels and chocolate consumption. Six months after birth, the researchers asked mothers to rate their infants' temperament, including smiling, laughter, and fear.[1]

Study II A study conducted by Norman Hollenberg, professor of medicine at Brigham and Women's Hospital and Harvard Medical School, involved 27 healthy people aged 18 to 72. Each subject consumed a cocoa beverage containing 900 milligrams of flavonols (a class of flavonoids) daily for five days. Using a finger cuff, blood flow was measured on the first and fifth days of the study.[2]

Study III Cardiologists at Athens Medical School in Greece wanted to test whether chocolate affected blood flow in the blood vessels. The researchers recruited 17 healthy young volunteers, who were given a 3.5-ounce bar of dark chocolate, either bittersweet or fake chocolate. On another day, the volunteers were switched. The subjects had no chocolate outside the study, and investigators did not know what each subject ate during the study. An ultrasound was taken of each volunteer's upper arm to see the functioning of the cells in the walls of the main artery.[3]

By the end of this chapter, you will be able to determine the strengths and weaknesses of each of these studies.

Activity 5A

Class survey

A class survey is a quick way to collect interesting data. Certainly there are things about the class as a group that you would like to know. Your task here is to construct a *draft* of a class survey, a questionnaire that would be used to gather data about the members of your class. Here are the steps to take:

1. As a class, discuss the questions you would like to include on the survey. In addition to what you want to ask, you should also consider how many questions you want to ask.
2. Once you have identified the topics, then work on the wording of the questions. Try to achieve as much consensus as possible.
3. Distribute one copy of the final draft of the survey to each student, but do not fill them out at this time. The surveys are to be put aside for the time being. As you complete this chapter, you will return to take another look at the survey you have constructed, make final adjustments, and then administer the survey to all of the members of your class. This survey should provide some interesting data that can be analyzed during the remainder of the course. As a starting point, here is a sample of a short survey:

Class Survey

Your answers to the questions below will help describe your class. DO NOT PUT YOUR NAME ON THIS PAPER. Your answers are completely private. They just help us describe the entire class.

1. Are you MALE or FEMALE? (Circle one.)
2. How many brothers and sisters do you have? ________
3. How tall are you in inches, to the nearest inch? ________
4. Estimate the number of pairs of shoes you own. ________
5. How much money in coins are you carrying right now? (Don't count any paper money, just coins.) ________
6. On a typical school night, how much time do you spend doing homework? (Answer in minutes. For example, 2 hours is 120 minutes.) ________
7. On a typical school night, how much time do you spend watching television? (Answer in minutes.) ________

Introduction

Exploratory data analysis seeks to discover and describe what data say by using graphs and numerical summaries. The conclusions we draw from data analysis apply to the specific data that we examine. Often, however, we want to answer questions about some large group of individuals. To get sound answers, we must produce data in a way that is designed to answer our questions.

Suppose our question is "What percent of American adults agree that the United Nations should continue to have its headquarters in the United States?" To answer the question, we interview American adults. We can't afford to ask all adults, so we put the question to a ***sample*** chosen to represent the entire adult population. How shall we choose a sample that truly represents the opinions of the entire population? Statistical designs for choosing samples are the topic of Section 5.1.

The distinction between experiments and observational studies that we discussed in the Preliminary Chapter is one of the most important ideas in statistics, and is worth repeating.

Observational Study versus Experiment

In an **observational study**, we observe individuals and measure variables of interest but do not attempt to influence the responses.

In an **experiment**, we deliberately impose some treatment on (that is, do something to) individuals in order to observe their responses.

Example 5.1 *Child care*
Observational study

A study of child care enrolled 1364 infants in 1991 and planned to follow them through their sixth year in school. In 2003, the researchers published an article finding that "the more time children spent in child care from birth to age four-and-a-half, the more adults tended to rate them, both at age four-and-a-half and at kindergarten, as less likely to get along with others, as more assertive, as disobedient, and as aggressive."[4]

This is an observational study. Parents made all child care decisions and the study did not attempt to influence them. "The study authors noted that their study was not designed to prove a cause and effect relationship. That is, the study cannot prove whether spending more time in child care *causes* children to have more problem behaviors." Perhaps employed parents who use child care are under stress and the children react to their parents' stress. Perhaps single parents are more likely to use child care. Perhaps parents are more likely to place in child care children who already have behavior problems. We can imagine an experiment that would remove these difficulties: from a large group of young children, choose some to be placed in child care and others to remain at home. This is an experiment because the treatment (child care or not) is imposed on the children. Of course, this particular experiment is neither practical nor ethical.

In Example 5.1, the variable "effect of child care" on behavior is *confounded* with other variables characteristic of families who use child care. Observational studies of the effect of one variable on another often fail because the explanatory variable is confounded with lurking variables. Because experiments allow us to pin down the effects of specific variables of interest to us, they are the preferred method for gaining knowledge in science, medicine, and industry.

5.1 Designing Samples

A political scientist wants to know what percent of the voting-age population consider themselves conservatives. An automaker hires a market research firm to learn what percent of adults aged 18 to 35 recall seeing television advertisements for a new sport utility vehicle. Government economists inquire about average household income. In all these cases, we want to gather information about a large group of individuals. We will not, as in an experiment, impose a treatment in order to observe the response. Also, time, cost, and inconvenience forbid contacting every individual. In such cases, we gather information about only part of the group in order to draw conclusions about the whole.

Population and Sample

The entire group of individuals that we want information about is called the **population.**

A **sample** is a part of the population that we actually examine in order to gather information.

You just don't understand A sample survey of journalists and scientists found quite a communications gap. Journalists think that scientists are arrogant, while scientists think that journalists are ignorant. We won't take sides, but here is one interesting result from the survey: 82% of the scientists agree that the "media do not understand statistics well enough to explain new findings" in medicine and other fields.

Notice that "population" is defined in terms of our desire for knowledge. If we wish to draw conclusions about all U.S. college students, that group is our population even if only local students are available for questioning. The sample is the part from which we draw conclusions about the whole. ***Sampling*** and conducting a ***census*** are two distinct ways of collecting data.

Sampling versus a Census

Sampling involves studying a part in order to gain information about the whole.

A **census** attempts to contact every individual in the entire population.

We want information on current unemployment and public opinion next week, not next year. Moreover, a carefully conducted sample is often more accurate than a census. Accountants, for example, sample a firm's inventory to verify the accuracy of the records. Attempting to count every last item in the warehouse would be not only expensive but inaccurate. Bored people do not count carefully.

sampling method

If conclusions based on a sample are to be valid for the entire population, a sound design for selecting the sample is required. ***Sampling method*** refers to the process used to choose the sample from the population. Poor sampling methods can produce misleading conclusions, as the following examples illustrate.

Example 5.2

Call-in opinion polls
Voluntary response

Television news programs like to conduct call-in polls of public opinion. The program announces a question and asks viewers to call one telephone number to respond "Yes" and another for "No." Telephone companies charge for these calls. The ABC network program *Nightline* once asked whether the United Nations should continue to have its headquarters in the United States. More than 186,000 callers responded, and 67% said "No."

People who spend the time and money to respond to call-in polls are not representative of the entire adult population. In fact, they tend to be the same people who call radio talk shows. People who feel strongly, especially those with strong negative opinions, are more likely to call. It is not surprising that a properly designed sample showed that 72% of adults want the UN to stay.[5]

Call-in opinion polls are an example of ***voluntary response sampling.*** A voluntary response sample can easily produce 67% "No" when the truth about the population is close to 72% "Yes." Example 5.3 also illustrates voluntary response sampling.

Voluntary Response Sample

A **voluntary response sample** consists of people who choose themselves by responding to a general appeal. Voluntary response samples are biased because people with strong opinions, especially negative opinions, are most likely to respond.

Example 5.3

Online polls
Voluntary response sampling

The American Family Association (AFA) is a conservative group that claims to stand for "traditional family values." It regularly posts online poll questions on its Web site—just click on a response to take part. Because the respondents are people who visit this site, the poll results always support AFA's positions. Well, almost always. In 2004, AFA's online poll asked about the heated issue of allowing same-sex marriage. Soon, email lists and social network sites favored mostly by young liberals pointed to the AFA poll. Almost 850,000 people responded, and 60% of them favored legalization of same-sex marriage. AFA claimed that homosexual rights groups had skewed its poll.

Online polls are now everywhere—some sites will even provide help in conducting your own online poll. As the AFA poll illustrates, you can't trust the results. People who take the trouble to respond to an open invitation are not representative of the entire adult population. That's true of regular visitors to AFA's site, of the activists who made a special effort to vote in the marriage poll, and of the people who bother to respond to write-in, call-in, or online polls in general.

Voluntary response is one common type of bad sampling method. Another is ***convenience sampling.***

Convenience Sampling

Choosing individuals who are easiest to reach is called **convenience sampling.**

Here is an example of convenience sampling.

Example 5.4

Interviewing at the mall
Convenience sampling

Manufacturers and advertising agencies often use interviews at shopping malls to gather information about the habits of consumers and the effectiveness of ads. A sample of mall shoppers is fast and cheap. "Mall interviewing is being propelled primarily as a budget issue," one expert told the *New York Times*. But people contacted at shopping malls are not representative of the entire U.S. population. They are richer, for example, and more likely to be teenagers or retired. Moreover, mall interviewers tend to select neat, safe-looking individuals from the stream of customers. Decisions based on mall interviews may not reflect the preferences of all consumers.[6]

Voluntary response sampling (Examples 5.2 and 5.3) and convenience sampling (Example 5.4) choose samples that are almost guaranteed not to represent the entire population. These sampling methods display ***bias,*** or systematic error.

Bias

The sampling method is **biased** if it systematically favors certain outcomes.

The remedy for bias in choosing a sample is to allow impersonal chance to do the choosing. In this section you have seen some sampling methods to avoid. In the next section you will see some examples of valid sampling methods.

Exercises

5.1 Students as customers A committee on community relations in a college town plans to survey local businesses about the importance of students as customers. From telephone book listings, the committee chooses 150 businesses at random. Of these, 73 return the questionnaire mailed by the committee. What is the population for this sample survey? What is the sample? What is the proportion of people who did not respond?

5.2 What is the population? For each of the following sampling situations, identify the population as exactly as possible. That is, say what kind of individuals the population consists of and say exactly which individuals fall in the population. If the information given is not complete, complete the description of the population in a reasonable way.

(a) Each week, the Gallup Poll questions a sample of about 1500 adult U.S. residents to determine national opinion on a wide variety of issues.

(b) The 2000 Census tried to gather basic information from every household in the United States. A "long form" requesting much additional information was sent to a sample of about 17% of households.

(c) A machinery manufacturer purchases voltage regulators from a supplier. There are reports that variation in the output voltage of the regulators is affecting the performance of the finished products. To assess the quality of the supplier's production, the manufacturer sends a sample of 5 regulators from the last shipment to a laboratory for study.

5.3 Teaching reading An educator wants to compare the effectiveness of computer software that teaches reading with that of a standard reading curriculum. He tests the reading ability of each student in a class of fourth-graders, then divides them into two groups. One group uses the computer regularly, while the other studies a standard curriculum. At the end of the year, he retests all the students and compares the increase in reading ability in the two groups. Is this an experiment? Why or why not? What are the explanatory and response variables?

5.4 The effects of propaganda In 1940, a psychologist conducted an experiment to study the effect of propaganda on attitude toward a foreign government. He administered a test of attitude toward the German government to a group of American students. After the students read German propaganda for several months, he tested them again to see if their attitudes had changed. Unfortunately, Germany attacked and conquered France while the experiment was in progress. Explain clearly why confounding makes it impossible to determine the effect of reading the propaganda.

5.5 Alcohol and heart attacks Many studies have found that people who drink alcohol in moderation have lower risk of heart attacks than either nondrinkers or heavy drinkers. Does alcohol consumption also improve survival after a heart attack? One study followed 1913 people who were hospitalized after severe heart attacks. In the year before their heart attack, 47% of these people did not drink, 36% drank moderately, and

17% drank heavily. After four years, fewer of the moderate drinkers had died.[7] Is this an observational study or an experiment? Why? What are the explanatory and response variables?

5.6 Are anesthetics safe? The National Halothane Study was a major investigation of the safety of anesthetics used in surgery. Records of over 850,000 operations performed in 34 major hospitals showed the following death rates for four common anesthetics:[8]

Anesthetic:	A	B	C	D
Death rate:	1.7%	1.7%	3.4%	1.9%

There seems to be a clear association between the anesthetic used and the death rate of patients. Anesthetic C appears to be dangerous.

(a) Explain why we call the National Halothane Study an observational study rather than an experiment, even though it compared the results of using different anesthetics in actual surgery.

(b) When the study looked at other variables that are confounded with a doctor's choice of anesthetic, it found that Anesthetic C was not causing extra deaths. Suggest several variables that are mixed up with what anesthetic a patient receives.

5.7 Call the shots A newspaper advertisement for *USA Today: The Television Show* once said: "Should handgun control be tougher? You call the shots in a special call-in poll tonight. If yes, call 1-900-720-6181. If no, call 1-900-720-6182. Charge is 50 cents for the first minute." Explain why this opinion poll is almost certainly biased.

5.8 Explain it to the congresswoman You are on the staff of a member of Congress who is considering a bill that would provide government-sponsored insurance for nursing-home care. You report that 1128 letters have been received on the issue, of which 871 oppose the legislation. "I'm surprised that most of my constituents oppose the bill. I thought it would be quite popular," says the congresswoman. Are you convinced that a majority of the voters oppose the bill? How would you explain the statistical issue to the congresswoman?

Simple Random Samples

In a voluntary response sample, people choose whether to respond. In a convenience sample, the interviewer makes the choice. In both cases, personal choice produces bias. The statistician's remedy is to allow chance to select the sample. A sample chosen by chance allows neither favoritism by the sampler nor self-selection by respondents. Choosing a sample by chance attacks bias by giving all individuals an equal chance to be chosen. Rich and poor, young and old, black and white, all have the same chance to be in the sample.

The simplest way to use chance to select a sample is to place names in a hat (the population) and draw out a handful (the sample). This is the idea of a ***simple random sample.***

Simple Random Sample

A **simple random sample (SRS)** of size n consists of n individuals from the population chosen in such a way that every set of n individuals has an equal chance to be the sample actually selected.

An SRS not only gives each individual an equal chance to be chosen (thus avoiding bias in the choice) but also gives every possible sample an equal chance to be chosen. There are other random sampling designs that give each individual, but not each sample, an equal chance. Exercise 5.32 describes one such design, called ***systematic random sampling.***

systematic random sampling

The idea of an SRS is to choose our sample by drawing names from a hat. In practice, computer software can choose an SRS almost instantly from a list of the individuals in the population. If you don't use a computer or calculator, you can randomize by using a table of ***random digits.***

Random Digits

A table of **random digits** is a long string of the digits 0, 1, 2, 3, 4, 5, 6, 7, 8, 9 with these two properties:

1. Each entry in the table is equally likely to be any of the 10 digits 0 through 9.
2. The entries are independent of each other. That is, knowledge of one part of the table gives no information about any other part.

Table B at the back of the book is a table of random digits. You can think of Table B as the result of asking an assistant (or a computer) to mix the digits 0 to 9 in a hat, draw one, then replace the digit drawn, mix again, draw a second digit, and so on. The assistant's mixing and drawing save us the work of mixing and drawing when we need to randomize. Table B begins with the digits 19223950340575628713. To make the table easier to read, the digits appear in groups of five and in numbered rows. The groups and rows have no meaning—the table is just a long list of randomly chosen digits. Because the digits in Table B are random:

- Each entry is equally likely to be any of the 10 possibilities 0, 1, . . . , 9.
- Each pair of entries is equally likely to be any of the 100 possible pairs 00, 01, . . . , 99.
- Each triple of entries is equally likely to be any of the 1000 possibilities 000, 001, . . . , 999, and so on.

These "equally likely" facts make it easy to use Table B to choose an SRS. Here is an example that shows how.

Example 5.5

Joan's accounting firm
How to choose an SRS

Joan's small accounting firm serves 30 business clients. Joan wants to interview a sample of 5 clients in detail to find ways to improve client satisfaction. To avoid bias, she chooses an SRS of size 5.

Step 1: Label. Give each client a numerical label, using as few digits as possible. Two digits are needed to label 30 clients, so we use labels

01, 02, 03, . . . , 29, 30

It is also correct to use labels 00 to 29 or even another choice of 30 two-digit labels. Here is the list of clients, with labels attached:

01	A-1 Plumbing	16	JL Records
02	Accent Printing	17	Johnson Commodities
03	Action Sport Shop	18	Keiser Construction
04	Anderson Construction	19	Liu's Chinese Restaurant
05	Bailey Trucking	20	MagicTan
06	Balloons Inc.	21	Peerless Machine
07	Bennett Hardware	22	Photo Arts
08	Best's Camera Shop	23	River City Books
09	Blue Print Specialties	24	Riverside Tavern
10	Central Tree Service	25	Rustic Boutique
11	Classic Flowers	26	Satellite Services
12	Computer Answers	27	Scotch Wash
13	Darlene's Dolls	28	Sewer's Center
14	Fleisch Realty	29	Tire Specialties
15	Hernandez Electronics	30	Von's Video Store

Step 2: Table. Enter Table B anywhere and read two-digit groups. Suppose we enter at line 130, which is

69051 64817 87174 09517 84534 06489 87201 97245

The first 10 two-digit groups in this line are

69 05 16 48 17 87 17 40 95 17

Each successive two-digit group is a label. The labels 00 and 31 to 99 are not used in this example, so we ignore them. The first 5 labels between 01 and 30 that we encounter in the table choose our sample. Of the first 10 labels in line 130, we strike through 5 because they are too high (over 30):

~~69~~ 05 16 ~~48~~ 17 ~~87~~ 17 ~~40~~ ~~95~~ 17

The others are 05, 16, 17, 17, and 17. The clients labeled 05, 16, and 17 go into the sample. Ignore the second and third 17s because that client is already in the sample.

Step 3: Stopping rule. Now run your finger across line 130 (and continue to line 131 if needed) until 5 clients are chosen.

~~84~~ ~~53~~ ~~40~~ ~~64~~ ~~89~~ ~~87~~ 20 19

Step 4: Identify sample. The sample is the clients labeled 05, 16, 17, 20, 19. These are Bailey Trucking, JL Records, Johnson Commodities, MagicTan, and Liu's Chinese Restaurant.

Choosing an SRS

Choose an SRS in four steps:

Step 1: **Label.** Assign a numerical label to every individual in the population.
Step 2: **Table.** Use Table B to select labels at random.
Step 3: **Stopping rule.** Indicate when you should stop sampling.
Step 4: **Identify sample.** Use the labels to identify subjects selected to be in the sample.

You can assign labels in any convenient manner, such as alphabetical order for names of people. Be certain that all labels have the same number of digits. Only then will all individuals have the same chance to be chosen. Use the shortest possible labels: one digit for a population of up to 10 members, two digits for 11 to 100 members, three digits for 101 to 1000 members, and so on. As standard practice, we recommend that you begin with label 1 (or 01 or 001, as needed). You can read digits from Table B in any order—across a row, down a column, and so on—because the table has no order. As standard practice, we recommend reading across rows. And don't always start at the same row when using Table B.

APPLET

Activity 5B
The Simple Random Sample applet

Randomly selecting a sample by hand is relatively easy. You can use a table of random digits or your calculator. If you have to select many individuals, this could be tedious. The *Simple Random Sample* applet at the book's Web site animates the selection of a simple random sample, and can handle larger samples.

1. Go to the textbook's Web site, www.whfreeman.com/tps3e. Click on the applet's link, and then choose the *Simple Random Sample* applet. You should see a screen like the one pictured on the next page.
2. In Example 5.5, you learned how to use the random digits table to randomly select 5 businesses from a population of size 30. Now we will use the applet to select our sample. Specify a population of 30 in the "Population = 1 to" box, and then click "Reset." The numbers from 1 to 30 appear in the population hopper. Then specify a sample size of 5 in the "Select a sample of size" box, and click "Sample."

Here are the results of randomly selecting 5 business clients from the population of 30.

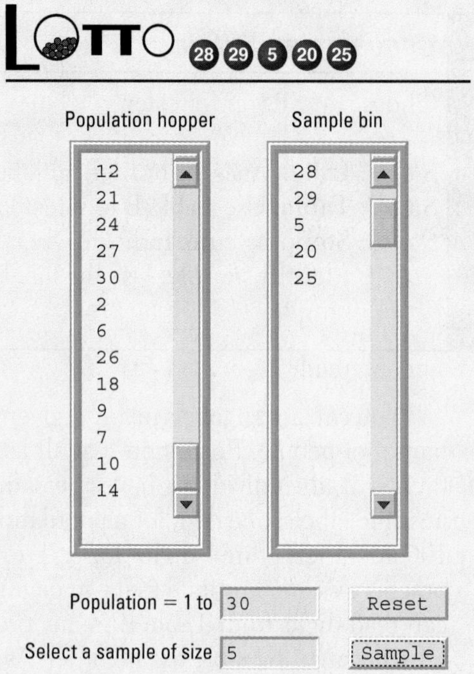

The sample bin shows the labels for an SRS of size 5. The 5 businesses selected are 28 Sewer's Center, 29 Tire Specialties, 5 Bailey Trucking, 20 MagicTan, and 25 Rustic Boutique.

3. Click "Reset," and then click "Sample" to select another SRS of size 5. How many numbers are left in the population hopper?

4. Which sample do you think is more likely: the sample that we got (28, 29, 5, 20, and 25) or the numbers 1, 2, 3, 4, and 5 (in that order)?

5. In the game Lotto, you choose 6 numbers in the range from 1 to 44. You can use this applet to let chance choose the 6 numbers for you. Enter 44 in the population box. Your population is now the numbers from 1 to 44. Click on "Reset." In the sample size box, enter 6. Then click on "Sample." Your 6 randomly selected numbers now appear in the sample bin. This is what happens when you buy a Lotto ticket and you tell the sales clerk that you want the computer to pick the numbers for you.

Other Sampling Methods

The general framework for a method that uses chance to choose a sample is a ***probability sample.***

Probability Sample

A **probability sample** is a sample chosen by chance. We must know what samples are possible and what chance, or probability, each possible sample has.

Some probability sampling methods (such as an SRS) give each member of the population an equal chance to be selected. This may not be true in more elaborate sampling methods. In every case, however, **the use of chance to select the sample is the essential principle of statistical sampling.**

Methods for sampling from large populations spread out over a wide area are usually more complex than an SRS. For example, it is common to sample important groups within the population separately and then combine these samples. This is the idea of a ***stratified random sample.***

Stratified Random Sample

To select a **stratified random sample,** first divide the population into groups of individuals, called **strata,** that are similar in some way that is important to the response. Then choose a separate SRS in each stratum and combine these SRSs to form the full sample.

"Stratum" is singular; "strata" is plural.

Define the strata, using information available before you choose your sample, so that each stratum groups individuals that are likely to be similar. For example, you might divide a population of high schools into public schools, Catholic schools, and other private schools. A stratified sample can give good information about each stratum separately as well as about the overall population. If the individuals in each stratum are less varied than the population as a whole, a stratified sample can produce better information about the population than an SRS of the same size. Think of the extreme case in which all the individuals in each stratum are identical—then just one individual from each stratum completely describes the population.

Example 5.6

Who wrote that song?
Stratified sampling

A radio station that broadcasts a piece of music owes a royalty to the composer. The organization of composers (called ASCAP) collects these royalties for all its members by charging stations a license fee for the right to play members' songs. ASCAP has four million songs in its catalog and collects $435 million in fees each year. How should ASCAP distribute this income among its members? By sampling: ASCAP tapes about 60,000 hours from the 53 million hours of local radio programs across the country each year.

Radio stations are stratified by type of community (metropolitan, rural), geographic location (New England, Pacific, etc.), and the size of the license fee paid to ASCAP, which reflects the size of the audience. In all, there are 432 strata. Tapes are made at random hours for randomly selected members of each stratum. The tapes are reviewed by experts who can recognize almost every piece of music ever written, and the composers are then paid according to their popularity.[9]

Another common type of probability sample is the ***cluster sample.***

Cluster Sampling

Cluster sampling divides the population into groups, or clusters. Some of these clusters are randomly selected. Then *all* individuals in the chosen clusters are selected to be in the sample.

Example 5.7 *What do AP students think?*
Cluster sampling

Suppose you want to survey opinions of AP Statistics students to see if they feel they have enough time on the free-response section of the AP exam. One option would be to begin with a list of all schools that had students who took the AP Statistics exam. A number of schools are randomly selected from the list of all schools represented, and every student at selected schools who took the AP exam is asked if he or she thought the time allowed to complete the free-response portion of the exam was adequate. The students interviewed would constitute a cluster sample.

As in Example 5.7, cluster sampling is often used for practical reasons. It is much easier to question all AP students in a sample of schools than to track down an SRS of individual AP Statistics students scattered across the country.

Be sure that you understand the distinction between stratified and cluster sampling. In stratified sampling, we study a random sample of individuals in every stratum. In cluster sampling, we study all the individuals in the chosen clusters and none of the individuals in other clusters.

Another common means of restricting random selection is to choose the sample in stages. This is usual practice for national samples of households or people. For example, data on employment and unemployment are gathered by the government's Current Population Survey, which conducts interviews in about 55,000 households each month. It is not practical to maintain a list of all U.S. households from which to select an SRS. Moreover, the cost of sending interviewers to the widely scattered households in an SRS would be too high. The Current Population Survey therefore uses a ***multistage sampling design.***

multistage sample

The final sample consists of clusters of nearby households that an interviewer can easily visit. Most opinion polls and other national samples are also multistage, though interviewing in most national samples today is done by telephone rather

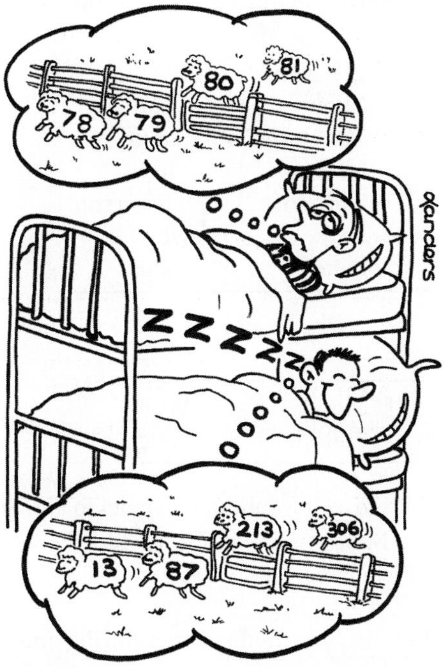

Statisticians fall asleep faster by taking a random sample of sheep.

than in person, eliminating the economic need for clustering. The Current Population Survey sampling design is roughly as follows:[10]

Stage 1: Divide the United States into 2007 geographical areas called Primary Sampling Units, or PSUs. Select a sample of 756 PSUs. This sample includes the 428 PSUs with the largest population and a stratified sample of 328 of the others.

Stage 2: Divide each PSU selected into smaller areas called "neighborhoods." Stratify the neighborhoods using ethnic and other information, and take a stratified sample of the neighborhoods in each PSU.

Stage 3: Sort the housing units in each neighborhood into clusters of four nearby units. Interview the households in a random sample of these clusters.

Analysis of data from sampling designs more complex than an SRS takes us beyond basic statistics. But the SRS is the building block of more elaborate designs, and analysis of other designs differs more in complexity of detail than in fundamental concepts.

Exercises

In the following exercises, use the four steps described on page 337 to take samples using Table B, the random digit table. In particular, write down the numbers you obtain from the table, and indicate for each number whether it is used or discarded.

5.9 Choose your sample You must choose an SRS of 10 of the 440 retail outlets in New York that sell your company's products. How would you label this population? Use Table B, starting at line 105, to choose your sample.

5.10 Spring break destinations A campus newspaper plans a major article on spring break destinations. The authors intend to call a few randomly chosen resorts at each destination to ask about their attitudes toward groups of students as guests. Here are the resorts listed in one city. The first step is to label the members of this population as shown.

01	Aloha Kai	08	Captiva	15	Palm Tree	22	Sea Shell
02	Anchor Down	09	Casa del Mar	16	Radisson	23	Silver Beach
03	Banana Bay	10	Coconuts	17	Ramada	24	Sunset Beach
04	Banyan Tree	11	Diplomat	18	Sandpiper	25	Tradewinds
05	Beach Castle	12	Holiday Inn	19	Sea Castle	26	Tropical Breeze
06	Best Western	13	Lime Tree	20	Sea Club	27	Tropical Shores
07	Cabana	14	Outrigger	21	Sea Grape	28	Veranda

Enter Table B at line 131, and choose three resorts.

5.11 Who goes to the convention? A club has 30 student members and 10 faculty members. The students are

Abel	Fisher	Huber	Miranda	Reinmann
Carson	Ghosh	Jimenez	Moskowitz	Santos
Chen	Griswold	Jones	Neyman	Shaw
David	Hein	Kim	O'Brien	Thompson
Deming	Hernandez	Klotz	Pearl	Utts
Elashoff	Holland	Liu	Potter	Varga

The faculty members are

Andrews	Fernandez	Kim	Moore	West
Besicovitch	Gupta	Lightman	Phillips	Yang

The club can send 4 students and 2 faculty members to a convention. It decides to choose those who will go by random selection. Use Table B, beginning at line 123, to choose a stratified random sample of 4 students and 2 faculty members.

5.12 Sampling by accountants Accountants often use stratified samples during audits to verify a company's records of such things as accounts receivable. The stratification is based on the dollar amount of the item and often includes 100% sampling of the largest items. One company reports 5000 accounts receivable. Of these, 100 are in amounts over \$50,000; 500 are in amounts between \$1000 and \$50,000; and the remaining 4400 are in amounts under \$1000. Using these groups as strata, you decide to verify all of the largest accounts and to sample 5% of the midsize accounts and 1% of the small accounts. How would you label the two strata from which you will sample? Use Table B, starting at line 115, to select only the first 5 accounts from each of these strata.

5.13 California polling A poll of opinion in California uses random digit dialing to choose telephone numbers at random. Numbers are selected separately within each California area code. The size of the sample in each area code is proportional to the population living there.

(a) What is the name for this kind of sampling method?

(b) California area codes, in rough order from north to south, are

530	707	916	209	415	925	510	650	408	831	805	559	760
661	818	213	626	323	562	709	310	949	909	858	619	

Another California survey does not call numbers in all area codes but starts with an SRS of 10 area codes. Choose such an SRS using Table B and starting at line 111.

5.14 High-speed Internet Laying fiber-optic cable is expensive, and so cable companies want to make sure that, if they extend their lines out to less dense suburban or rural areas, there will be sufficient demand and the work will be cost-effective. They decide to conduct a survey to determine the proportion of households in a rural subdivision that would buy the service. They select a sample of 5 blocks in the subdivision and survey each family that lives on one of those blocks.

(a) What is the name for this kind of sampling method?

(b) Suppose there are 65 blocks in the subdivision. Use the random digit table, beginning at line 142, or the *Simple Random Sample* applet to select 5 blocks to be sampled.

Cautions about Sample Surveys

Random selection eliminates bias in the choice of a sample from a list of the population. When the population consists of human beings, however, accurate information from a sample requires much more than a good sampling method. To begin, we need an accurate and complete list of the population. Because such a list is rarely available, most samples suffer from some degree of ***undercoverage.*** A sample survey of households, for example, will miss not only homeless people but prison inmates and students in dormitories. An opinion poll conducted by telephone will miss the 7% to 8% of American households without residential phones. The results of national sample surveys therefore have some bias if the people not covered—who most often are poor people—differ from the rest of the population.

A more serious source of bias in most sample surveys is ***nonresponse,*** which occurs when a selected individual cannot be contacted or refuses to cooperate. Nonresponse to sample surveys often reaches 30% or more, even with careful planning and several callbacks. Because nonresponse is higher in urban areas, most sample surveys substitute other people in the same area to avoid favoring rural areas in the final sample. If the people contacted differ from those who are rarely at home or who refuse to answer questions, some bias remains.

Undercoverage and Nonresponse

Undercoverage occurs when some groups in the population are left out of the process of choosing the sample.

Nonresponse occurs when an individual chosen for the sample can't be contacted or does not cooperate.

How bad is nonresponse? The Current Population Survey (CPS) has the lowest nonresponse rate of any poll we know: only about 4% of the households in the CPS sample refuse to take part and another 3% or 4% can't be contacted. People are more likely to respond to a government survey such as the CPS, and the CPS contacts its sample in person before doing later interviews by phone.

Example 5.8

Ring, ring, nobody home
Nonresponse

The General Social Survey (Figure 5.1) is the nation's most important social science research survey. The GSS also contacts its sample in person, and it is run by a university. Despite these advantages, its most recent survey had a 30% rate of nonresponse.

Figure 5.1 *Part of the subject index for the General Social Survey (GSS). The GSS has assessed attitudes on a wide variety of topics since 1972. Its continuity over time makes the GSS a valuable source for studies of changing attitudes.*

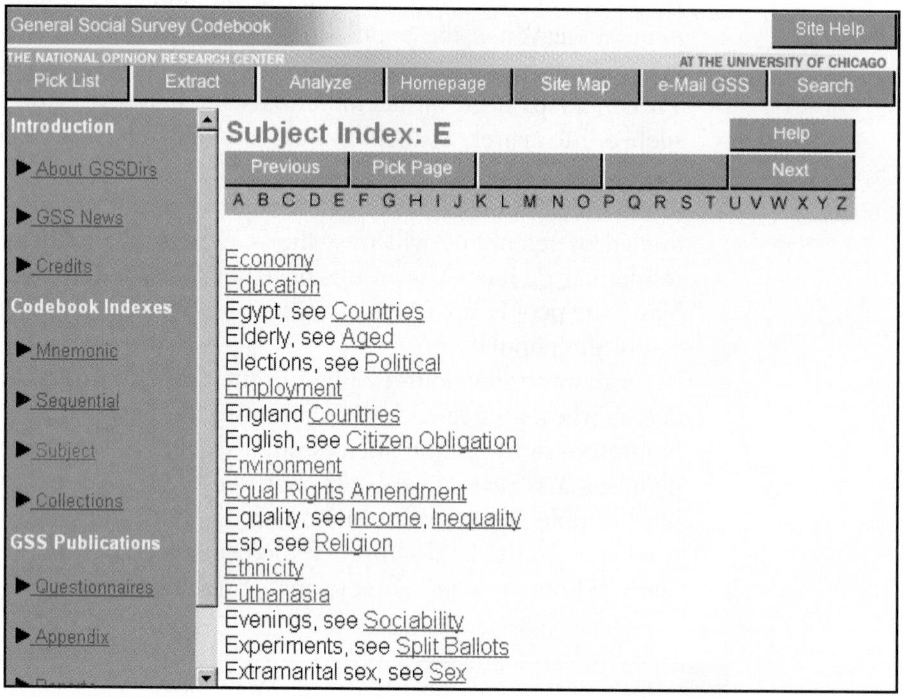

What about polls done by the media and by market research and opinion-polling firms? We don't know their rates of nonresponse, because they won't say. That itself is a bad sign. The Pew Research Center for the People and the Press imitated a careful telephone survey and published the results: out of 2879 households called, 1658 were never at home, refused, or would not finish the interview. That's a nonresponse rate of 58%.[11]

Most sample surveys, and almost all opinion polls, are now carried out by telephone. This and other details of the interview method can affect the results.

Example 5.9

Effect of interview method on poll results
Survey methods

A Pew Research Center Poll has asked about belief in God for many years. In response to the statement "I never doubt the existence of God," subjects are asked to choose from the responses

completely agree   mostly agree   mostly disagree   completely disagree

In 1990, subjects were interviewed in person and were handed a card with the four responses on it. In 1991, the poll switched to telephone interviews. In 1990, 60% said, "completely agree," in line with earlier years. In 1991, 71% completely agreed. The increase is probably explained by the effect of hearing "completely agree" read first by the interviewer.[12]

response bias

The behavior of the respondent or of the interviewer can cause ***response bias*** in sample results. Respondents may lie, especially if asked about illegal or unpopular behavior. The race or sex of the interviewer can influence responses to questions about race relations or attitudes toward feminism. Careful training of interviewers and careful supervision to avoid variation among the interviewers can greatly reduce response bias. Answers to questions that ask respondents to recall past events are often inaccurate because of faulty memory. For example, many people "telescope" events in the past, bringing them forward in memory to more recent time periods. "Have you visited a dentist in the last 6 months?" will often elicit a "Yes" from someone who last visited a dentist 8 months ago.[13]

Example 5.10 *Did you vote?*

Response bias

"One of the most frequently observed survey measurement errors is the overreporting of voting behavior."[14] People know they should vote, so those who didn't vote tend to save face by saying that they did. Here are the data from a typical sample of 663 people after an election:

		What They Said	
		I voted	I didn't
What They Did	Voted	358	13
	Didn't vote	120	172

You can see that 478 people (72%) said that they voted, but only 371 people (56%) actually did vote.

wording of questions

The ***wording of questions*** is the most important influence on the answers given to a sample survey. Confusing or leading questions can introduce strong bias, and even minor changes in wording can change a survey's outcome. Here are two examples.

Example 5.11 ***Should we ban disposable diapers?***
Wording of questions

A survey paid for by makers of disposable diapers found that 84% of the sample opposed banning disposable diapers. Here is the actual question:

> It is estimated that disposable diapers account for less than 2% of the trash in today's landfills. In contrast, beverage containers, third-class mail and yard wastes are estimated to account for about 21% of the trash in landfills. Given this, in your opinion, would it be fair to ban disposable diapers?[15]

This question gives information on only one side of an issue, then asks an opinion. That's a sure way to bias the responses. A different question that described how long disposable diapers take to decay and how many tons they contribute to landfills each year would draw a quite different response.

Example 5.12 ***Doubting the Holocaust***
Wording of questions

An opinion poll conducted in 1992 for the American Jewish Committee asked: "Does it seem possible or does it seem impossible to you that the Nazi extermination of the Jews never happened?" When 22% of the sample said "possible," the news media wondered how so many Americans could be uncertain that the Holocaust happened. Then a second poll asked the question in different words: "Does it seem possible to you that the Nazi extermination of the Jews never happened, or do you feel certain that it happened?" Now only 1% of the sample said "possible." The complicated wording of the first question confused many respondents.[16]

The statistical design of sample surveys is a science, but this science is only part of the art of sampling. Because of nonresponse, response bias, and the difficulty of posing clear and neutral questions, you should hesitate to fully trust reports about complicated issues based on surveys of large human populations. **Insist on knowing the exact questions asked, the rate of nonresponse, and the date and method of the survey before you trust a poll result.**

Inference about the Population

Despite the many practical difficulties in carrying out a sample survey, using chance to choose a sample does eliminate bias in the actual selection of the sample from the list of available individuals. But it is unlikely that results from a sample are exactly the same as for the entire population. Sample results, like the official unemployment rate obtained from the monthly Current Population Survey, are only estimates of the truth about the population. If we select two samples at random from the same population, we will draw different individuals. So the sample results will almost certainly differ somewhat. Two runs of the Current Population Survey would produce somewhat different unemployment rates. Properly designed samples avoid systematic bias, but their results are rarely exactly correct and they vary from sample to sample.

CAUTION

We can improve our results by knowing that **larger random samples give more accurate results than smaller samples.** By taking a very large sample, you can be confident that the sample result is very close to the truth about the population. The Current Population Survey's sample of 50,000 households estimates the national unemployment rate very accurately. Of course, only probability samples carry this guarantee. *Nightline*'s voluntary response sample is worthless even though 186,000 people called in. Using a probability sampling method and taking care to deal with practical difficulties reduce bias in a sample. The size of the sample then determines how close to the population truth the sample result is likely to fall.

Exercises

sampling frame

5.15 Sampling frame The list of individuals from which a sample is actually selected is called the ***sampling frame.*** Ideally, the frame should list every individual in the population, but in practice this is often difficult. A frame that leaves out part of the population is a common source of undercoverage.

(a) Suppose that a sample of households in a community is selected at random from the telephone directory. What households are omitted from this frame? What types of people do you think are likely to live in these households? These people will probably be underrepresented in the sample.

(b) It is more common in telephone surveys to use random digit dialing equipment that selects the last four digits of a telephone number at random after being given the exchange (the first three digits). Which of the households you mentioned in your answer to (a) will be included in the sampling frame by random digit dialing?

5.16 Ring-no-answer A common form of nonresponse in telephone surveys is "ring-no-answer." That is, a call is made to an active number but no one answers. The Italian National Statistical Institute looked at nonresponse to a government survey of households in Italy during the periods January 1 to Easter and July 1 to August 31. All calls were made between 7 and 10 P.M., but 21.4% gave "ring-no-answer" in one period versus 41.5% "ring-no-answer" in the other period.[17] Which period do you think had the higher rate of no answers? Why? Explain why a high rate of nonresponse makes sample results less reliable.

5.17 Question wording During the 2000 presidential campaign, the candidates debated what to do with the large government surplus. The Pew Research Center asked two questions of random samples of adults. Both questions stated that Social Security would be "fixed." Here are the uses suggested for the remaining surplus:

> Should the money be used for a tax cut, or should it be used to fund new government programs?
>
> Should the money be used for a tax cut, or should it be spent on programs for education, the environment, healthcare, crime-fighting and military defense?

One of these questions drew 60% favoring a tax cut; the other, only 22%. Which wording pulls respondents toward a tax cut? Why?

5.18 Grading the president A newspaper article about an opinion poll says that "43% of Americans approve of the president's overall job performance." Toward the end of the article, you read: "The poll is based on telephone interviews with 1210 adults from around the United States, excluding Alaska and Hawaii." What variable did this poll measure? What population do you think the newspaper wants information about? What was the sample? Are there any sources of bias in the sampling method used?

5.19 Advertisements targeted toward minors? The Excite Poll can be found online at poll.excite.com. The question appears on the screen, and you simply click buttons to vote "Yes," "No," "Not sure," or "Don't care." On February 17, 2004, the question was "Do you think that beer advertisements are targeted towards minors?" In all, 4316 (30%) said "Yes," another 8986 (62%) said "No," and the remaining 1182 were not sure or didn't care.

(a) What is the sample size for this poll?

(b) That's a much larger sample than standard sample surveys. In spite of this, we can't trust the result to give good information about any clearly defined population. Why?

5.20 Wording bias Comment on each of the following as a potential sample survey question. Is the question clear? Is it slanted toward a desired response?

(a) "Some cell phone users have developed brain cancer. Should all cell phones come with a warning label explaining the danger of using cell phones?"

(b) "Do you agree that a national system of health insurance should be favored because it would provide health insurance for everyone and would reduce administrative costs?"

(c) "In view of escalating environmental degradation and incipient resource depletion, would you favor economic incentives for recycling of resource-intensive consumer goods?"

Section 5.1 Summary

We can produce data intended to answer specific questions by sampling or experimentation. **Sampling** selects a part of a population of interest to represent the whole. **Experiments** are distinguished from **observational studies** such as sample surveys by the active imposition of some treatment on the subjects of the experiment.

A sample survey selects a **sample** from the **population** of all individuals about which we desire information. We base conclusions about the population on data about the sample.

The **sampling method** refers to the method used to select the sample from the population. **Probability sampling methods** use impersonal chance to select a sample.

The basic probability sample is a **simple random sample (SRS).** An SRS gives every possible sample of a given size the same chance to be chosen.

Choose an SRS by labeling the members of the population and using a **table of random digits** to select the sample. Software can automate this process.

To choose a **stratified random sample,** divide the population into **strata,** groups of similar individuals. Then choose a separate SRS from each stratum and combine them to form the full sample.

To choose a **cluster sample,** divide the population into groups, or clusters. Randomly select some of these clusters. All of the individuals in the chosen clusters are then selected to be in the sample.

Multistage samples select successively smaller groups within the population in stages, resulting in a sample consisting of clusters of individuals. Each stage may employ an SRS, a stratified sample, or another type of sample.

Failure to use probability sampling often results in **bias,** or systematic errors in the way the sample represents the population. **Voluntary response** samples, in which the respondents choose themselves, and **convenience samples,** in which individuals easiest to reach are chosen, are particularly prone to large bias.

In human populations, even probability samples can suffer from bias due to **undercoverage** or **nonresponse,** from **response bias** due to the behavior of the interviewer or the respondent, or from misleading results due to **poorly worded questions.**

Larger samples give more accurate results than smaller samples.

Section 5.1 Exercises

5.21 Describe the population For each of the following sampling situations, identify the population as exactly as possible. That is, say what kind of individuals the population consists of and say exactly which individuals fall in the population. If the information given is not complete, complete the description of the population in a reasonable way.

(a) An opinion poll contacts 1161 adults and then asks them, "Which political party do you think has better ideas for leading the country in the twenty-first century?"

(b) A sociologist wants to know the opinions of employed adult women about government funding for day care. She obtains a list of the 520 members of a local business and professional women's club and mails a questionnaire to 100 of these women selected at random.

(c) The American Community Survey (ACS) will replace the census "long form" starting with the 2010 Census. The main part of the ACS contacts 250,000 addresses by mail each month, with follow-up by phone and in person if there is no response. Each household answers questions about their housing, economic, and social status.

5.22 How many children? A teacher asks her class, "How many children are there in your family, including yourself?" The mean response is about 3 children. According to the 2000 census, families that have children average 1.86 children. Why is a sample like the teacher's class biased toward higher outcomes?

5.23 Testing chemicals A manufacturer of chemicals chooses 3 from each lot of 25 containers of a reagent to test for purity and potency. Below are the control numbers stamped on the bottles in the current lot. Use Table B at line 111 to choose an SRS of 3 of these bottles.

A1096	A1097	A1098	A1101	A1108
A1112	A1113	A1117	A2109	A2211
A2220	B0986	B1011	B1096	B1101
B1102	B1103	B1110	B1119	B1137
B1189	B1223	B1277	B1286	B1299

5.24 Increasing sample size Just before a presidential election, a national opinion-polling firm increases the size of its weekly sample from the usual 1500 people to 4000 people. Why do you think the firm does this?

5.25 Census tract The Census Bureau divides the entire country into "census tracts" that contain about 4000 people. Each tract is in turn divided into small "blocks," which in urban areas are bounded by local streets. An SRS of blocks from a census tract is often the next-to-last stage in a multistage sample. Figure 5.2 shows part of census tract 8051.12, in Cook County, Illinois, west of Chicago. The 44 blocks in this tract are divided into three "block groups." Group 1 contains 6 blocks numbered 1000 to 1005; Group 2 (outlined in Figure 5.2) contains 12 blocks numbered 2000 to 2011; Group 3 contains 26 blocks numbered 3000 to 3025. Use Table B, beginning at line 125, to choose an SRS of 5 of the 44 blocks in this census tract. Explain carefully how you labeled the blocks.

Figure 5.2 *Census blocks in Cook County, Illinois, for Exercise 5.25. The outlined area is a block group.*

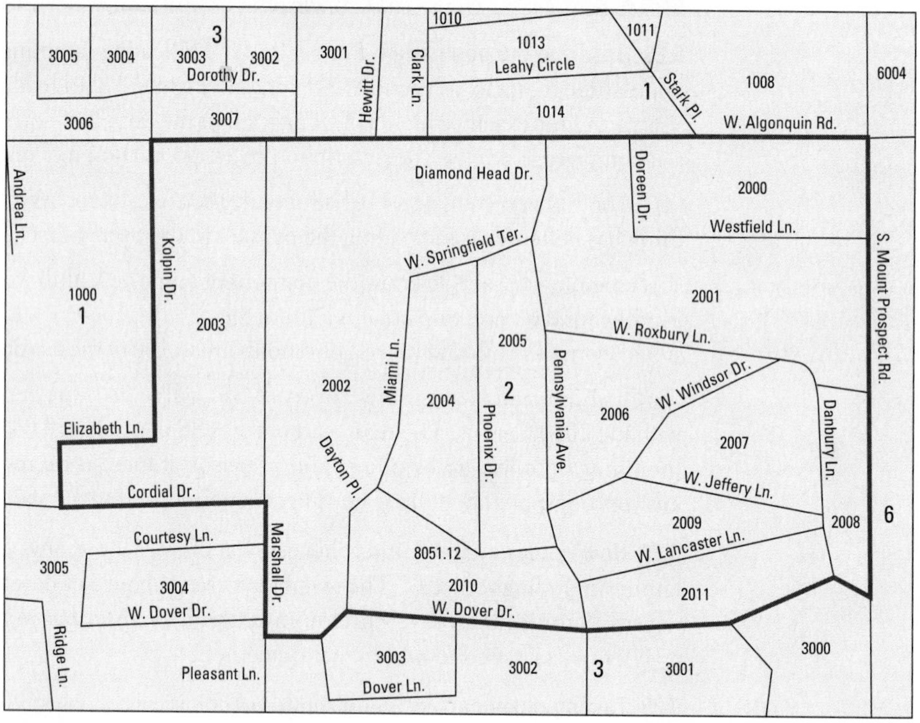

5.26 Random digits Which of the following statements are true of a table of random digits, and which are false? Briefly explain your answers.

(a) There are exactly four 0s in each row of 40 digits.

(b) Each pair of digits has 1 chance in 100 of being 00.

(c) The digits 0000 can never appear as a group, because this pattern is not random.

5.27 Is it an SRS? A corporation employs 2000 male and 500 female engineers. A stratified random sample of 200 male and 50 female engineers gives each engineer 1 chance in 10 to be chosen. This sample design gives every individual in the population the same chance to be chosen for the sample. Is it an SRS? Explain your answer.

5.28 Checking for bias Comment on each of the following as a potential sample survey question. Is the question clear? Is it slanted toward a desired response?

(a) Which of the following best represents your opinion on gun control?

1. The government should confiscate our guns.

2. We have the right to keep and bear arms.

(b) A freeze in nuclear weapons should be favored because it would begin a much-needed process to stop everyone in the world from building nuclear weapons now and reduce the possibility of nuclear war in the future. Do you agree or disagree?

5.29 Sampling error A *New York Times* opinion poll on women's issues contacted a sample of 1025 women and 472 men by randomly selecting telephone numbers. The *Times* publishes complete descriptions of its polling methods. Here is part of the description for this poll:

> In theory, in 19 cases out of 20 the results based on the entire sample will differ by no more than three percentage points in either direction from what would have been obtained by seeking out all adult Americans.
>
> The potential sampling error for smaller subgroups is larger. For example, for men it is plus or minus five percentage points.[18]

Explain why the margin of error is larger for conclusions about men alone than for conclusions about all adults.

5.30 Attitudes toward alcohol At a party there are 30 students over age 21 and 20 students under age 21. You choose at random 3 of those over 21 and separately choose at random 2 of those under 21 to interview about attitudes toward alcohol. You have given every student at the party the same chance to be interviewed: what is the chance? Why is your sample not an SRS?

5.31 What do schoolkids want? What are the most important goals of schoolchildren? Do girls and boys have different goals? Are goals different in urban, suburban, and rural areas? To find out, researchers wanted to ask children in the fourth, fifth, and sixth grades this question:

What would you most like to do at school?

A. *Make good grades.*

B. *Be good at sports.*

C. *Be popular.*

Because most children live in heavily populated urban and suburban areas, an SRS might contain few rural children. Moreover, it is too expensive to choose children at random from a large region—we must start by choosing schools rather than children. Describe a suitable sampling method for this study and explain the reasoning behind your choice of method.

5.32 Systematic random sample Sample surveys often use a systematic random sample to choose a sample of apartments in a large building or dwelling units in a block at the last stage of a multistage sample. Here is a description of how to choose a systematic sample.

Suppose that we must choose 4 addresses out of 100. Because 100/4 = 25, we can think of the list as four lists of 25 addresses. Choose 1 of the first 25 addresses at random using Table B. The sample contains this address and the addresses 25, 50, and 75 places down the list from it. If the table gives 13, for example, then the systematic random sample consists of the addresses numbered 13, 38, 63, and 88.

(a) Use Table B to choose a systematic random sample of 5 addresses from a list of 200. Enter the table at line 120.

(b) Like an SRS, a systematic random sample gives all individuals the same chance to be chosen. Explain why this is true. Then explain carefully why a systematic sample is nonetheless not an SRS.

Activity 5C

The class survey revisited

Each student should have a copy of the survey that the class constructed in Activity 5A at the beginning of the chapter. Now that you are experts on good and bad characteristics of survey questions, do the following:

1. Consider the questions in order. As you look at each item, see if the question contains bias. Does it advocate a position? Does the question contain any complicated words or phrasing that might be misinterpreted? Will any questions evoke response bias?
2. Make any changes that the group feels are needed. Remember that the survey should be *anonymous* (no names on the papers) so that students are assured that the class as a whole rather than themselves as individuals will be described.
3. Print the final version of the survey. Make one copy for each member of the class and an extra copy on which to tally the results.
4. Each student should complete the survey.
5. Place the completed surveys, upside down, in a pile. The last student finished should shuffle the pile of surveys to ensure anonymity.
6. Designate someone (the teacher?) to tally the responses as homework and prepare a cumulative summary. Give a copy of the results to each student in the class for later analysis.

5.2 Designing Experiments

Activity 5D
Good news for chocoholics!

Materials: One traditional Toll House cookie and one Hershey's dark chocolate cookie for each student; blindfolds; paper cups; water; class roll; calculator or table of random digits

Recent studies have found that chocolate has many health benefits. It can lower blood pressure, make blood vessels more flexible, and help prevent blood clots. Dark chocolate appears to have more health benefit potential than milk chocolate or white chocolate. Companies that make chocolate, like Mars and Hersheys, have moved quickly to take advantage of the good news about dark chocolate to boost sales and revenue. But consumers have to like the taste of the new, healthier dark chocolate.

In this Activity, you will compare the taste of Nestle's traditional Toll House chocolate chip cookies and cookies made with Hershey's new Special Dark Chips. First, you will need a supply of both kinds of cookies.

Prior to conducting the experiment, a list of students should be used to *randomly* select approximately half of the students. These randomly selected students will eat a traditional Toll House cookie first and the dark chocolate cookie second. The remaining students will receive a dark chocolate cookie first and a traditional cookie second. Important: *Subjects should NOT know which cookie they are eating at any given time.*

1. Decide on a method of providing the cookies so that they are as similar as possible (similar size, similar degree of baking, etc.). Perhaps you can ask the home economics teacher or class at your school to bake the two batches of cookies. Or perhaps a couple of parents or students in the class would volunteer to provide the cookies. The important thing is to make both batches of cookies as uniform (similar) as possible. The teacher, serving as the Chief Investigator, should label each container of cookies. Don't let anyone see the cookies or the labels.
2. Each student will, in turn, put on a blindfold (bandana or scarf). Then the teacher or designated assistant will give the subject a single cookie to sample. The kind of cookie (traditional Toll House or dark chocolate) will have been determined by the random selection process conducted earlier.
3. The subject will rinse his or her mouth with water for several seconds and swallow the water.
4. Give the subject one of the other type of cookie.

5. The subject will remove the blindfold and record his or her cookie preference (1 or 2) on a piece of paper.
6. The teacher will use the annotated list of students (showing which group they were in) to determine the cookie preference for each student and will share the results for the whole class. Which type of cookie do the majority of the class prefer?

A study is an experiment when we actually do something to individuals in order to observe the response. Here is the basic vocabulary of experiments.

Experimental Units, Subjects, Treatment

The individuals on which the experiment is done are the **experimental units.** When the units are human beings, they are called **subjects.** A specific experimental condition applied to the units is called a **treatment.**

factor

level

Because the purpose of an experiment is to reveal the response of one variable to changes in other variables, the distinction between explanatory and response variables is important. The explanatory variables in an experiment are often called ***factors.*** Many experiments study the joint effects of several factors. In such an experiment, each treatment is formed by combining a specific value (often called a ***level***) of each of the factors.

Example 5.13 *Effects of class size*

Experimental units, subjects, treatment

Do smaller classes in elementary school really benefit students in areas such as scores on standardized tests, staying in school, and going on to college? We might do an observational study that compares students who happened to be in smaller and larger classes in their early school years. Small classes are expensive, so they are more common in schools that serve richer communities. Students in small classes tend to also have other advantages: their schools have more resources; their parents are better educated, and so on. Confounding makes it impossible to isolate the effects of small classes.

The Tennessee STAR program was an experiment on the effects of class size. It has been called "one of the most important educational investigations ever carried out." The subjects were 6385 students who were beginning kindergarten. Each student was assigned to one of three treatments: regular class (22 to 25 students) with one teacher, regular class with a teacher and a full-time teacher's aide, and small class (13 to 17 students). These treatments are levels of a single factor, the type of class. The students stayed in the same type of class for four years, then all returned to regular classes. In later years, students from the small classes had higher scores on standardized tests, were less likely to fail a grade, had better high school grades, and so on. The benefits of small classes were greatest for minority students.[19]

Example 5.13 illustrates the big advantage of experiments over observational studies. *In principle, experiments can give good evidence for causation.* In an experiment, we study the specific factors we are interested in, while controlling the effects of lurking variables. All the students in the Tennessee STAR program followed the usual curriculum at their schools. Because students were assigned to different class types within their schools, school resources and family backgrounds were not confounded with class type. The only systematic difference was the type of class. When students from the small classes did better than those in the other two types, we can be confident that class size made the difference.

Example 5.14 *Advertising*
Response variables

What are the effects of repeated exposure to an advertising message? The answer may depend both on the length of the ad and on how often it is repeated. An experiment investigated this question using undergraduate students as *subjects*. All subjects viewed a 40-minute television program that included ads for a digital camera. Some subjects saw a 30-second commercial; others, a 90-second version. The same commercial was shown either 1, 3, or 5 times during the program.

This experiment has 2 *factors*: length of the commercial, with 2 levels, and repetitions, with 3 levels. The 6 combinations of one level of each factor form 6 *treatments*. Figure 5.3 shows the layout of the treatments. After viewing, all of the subjects answered questions about their recall of the ad, their attitude toward the camera, and their intention to purchase it. These are the *response variables*.[20]

Figure 5.3 *The treatments in the study of advertising for Example 5.14. Combining the levels of the two factors forms six treatments.*

		Factor B: Repetitions		
		1 time	3 times	5 times
Factor A: Length	30 seconds	1	2	3
	90 seconds	4	5	6

Example 5.14 shows how experiments allow us to study the combined effects of several factors. The interaction of several factors can produce effects that could not be predicted from looking at the effects of each factor alone. Perhaps longer commercials increase interest in a product, and more commercials also increase interest, but if we both make a commercial longer and show it more often, viewers get annoyed and their interest in the product drops. The two-factor experiment in Example 5.14 will help us find out.

Control

Laboratory experiments in science and engineering often have a simple design with only a single treatment, which is applied to all of the experimental units. The design of such an experiment can be outlined as

Treatment → Observe response

For example, we may place a heavy object on a support (treatment) and measure how much it bends (observation). We rely on the controlled environment of the laboratory to protect us from lurking variables. When experiments are conducted in the field or with living subjects, such simple designs can yield invalid data. That is, we cannot tell whether the response was due to the treatment or to lurking variables. A medical example will show what can go wrong.

Example 5.15 *Treating ulcers*
Placebo effect

"Gastric freezing" is a clever treatment for ulcers in the upper intestine. The patient swallows a deflated balloon with tubes attached, then a refrigerated liquid is pumped through the balloon for an hour. The idea is that cooling the stomach will reduce its production of acid and so relieve ulcers. An experiment reported in the *Journal of the American Medical Association* showed that gastric freezing did reduce acid production and relieve ulcer pain. The treatment was safe and easy and was widely used for several years. The design of the experiment was

Gastric freezing → Observe pain relief

placebo effect

The gastric-freezing experiment was poorly designed. The patients' response may have been due to the ***placebo effect.*** A placebo is a dummy treatment. Many patients respond favorably to any treatment, even a placebo. This may be due to trust in the doctor and expectations of a cure or simply to the fact that medical conditions often improve without treatment. The response to a dummy treatment is the placebo effect.

A later experiment divided ulcer patients into two groups. One group was treated by gastric freezing as before. The other group received a placebo treatment in which the liquid in the balloon was at body temperature rather than freezing. The results: 34% of the 82 patients in the treatment group improved, but so did 38% of the 78 patients in the placebo group. This and other properly designed experiments showed that gastric freezing was no better than a placebo, and its use was abandoned.[21]

The first gastric-freezing experiment gave misleading results because the effects of the explanatory variable were confounded with the placebo effect. We can defeat confounding by comparing two groups of patients, as in the second gastric-freezing experiment. The placebo effect and other lurking variables now operate on both groups. The only difference between the groups is the actual effect of gastric freezing. The group of patients who received a sham treatment is called a ***control group,*** because it enables us to control the effects of outside variables on the outcome. *Control is the first basic principle of statistical design of experiments. Comparison of several treatments in the same environment is the simplest form of control.*

control group

Without control, experimental results in medicine and the behavioral sciences can be dominated by such influences as the details of the experimental arrangement, the selection of subjects, and the placebo effect. The result is often bias, systematic favoritism toward one outcome. An uncontrolled study of a new medical therapy, for example, is biased in favor of finding the treatment effective because of the placebo effect. It should not surprise you to learn that uncontrolled studies in medicine give new therapies a much higher success rate than proper comparative experiments. Well-designed experiments, like the Tennessee STAR study and the second gastric-freezing study, usually compare several treatments.

Don't confuse control and control group. Control refers to the overall effort to minimize variability in the way the experimental units are obtained and treated.

Exercises

For each of the experimental situations described in Exercises 5.33 to 5.36, identify the experimental units or subjects, the factors, the treatments, and the response variables.

5.33 Growing in the shade Ability to grow in shade may help pines found in the dry forests of Arizona to resist drought. How well do these pines grow in shade? Investigators planted pine seedlings in a greenhouse in either full light, light reduced to 25% of normal by shade cloth, or light reduced to 5% of normal. At the end of the study, they dried the young trees and weighed them.

5.34 Internet telephone calls You can use your computer to make long-distance telephone calls over the Internet. How will the cost affect the use of this service? A university plans an experiment to find out. It will offer the service to all 350 students in one of its dormitories. Some students will pay a low flat rate. Others will pay higher rates at peak periods and very low rates off-peak. The university is interested in the amount and time of use and in the effect on the congestion of the network.

5.35 Improving response rate How can we reduce the rate of refusals in telephone surveys? Most people who answer at all listen to the interviewer's introductory remarks and then decide whether to continue. One study made telephone calls to randomly selected households to ask opinions about the next election. In some calls, the interviewer gave her name, in others she identified the university she was representing, and in still others she identified both herself and the university. For each type of call, the interviewer either did or did not offer to send a copy of the final survey results to the person interviewed. Do these differences in the introduction affect whether the interview is completed?

5.36 Sickle-cell disease Sickle-cell disease is an inherited disorder of the red blood cells that in the United States affects mostly blacks. It can cause severe pain and many complications. Can the drug hydroxyurea reduce the severe pain caused by sickle-cell disease? A study by the National Institutes of Health gave the drug to 150 sickle-cell sufferers and a placebo (a dummy medication) to another 150. The researchers then counted the episodes of pain reported by each subject.

5.37 Sham operation In the mid 1900s, a common treatment for angina (a disease marked by brief attacks of chest pain caused by insufficient oxygen to the heart) was called internal mammary ligation. In this procedure doctors made small incisions in the chest and tied knots in two arteries to try to increase blood flow to the heart. It was a popular procedure—90 percent of patients reported that it helped reduce pain. In 1960, Seattle cardiologist Dr. Leonard Cobb carried out an experiment where he compared ligation with a procedure in which he made incisions but did not tie off the arteries. This sham operation proved just as successful, and the ligation procedure was abandoned as a treatment for angina.

(a) What is the response variable in Dr. Cobb's experiment?

(b) Dr. Cobb showed that the sham operation was just as successful as ligation. What term do we use to describe the phenomenon that many subjects report good results from a pretend treatment?

(c) The ligation procedure is an example of the lack of an important property of a well-designed experiment. What is that property?

5.38 Eat well and exercise Most American adolescents don't eat well and don't exercise enough. Can middle schools increase physical activity among their students? Can they persuade students to eat better? Investigators designed a "physical activity intervention" to increase activity in physical education classes and during leisure periods throughout the school day. They also designed a "nutrition intervention" that improved school lunches and offered ideas for healthy home-packed lunches. Each participating school was randomly assigned to one of the interventions, both interventions, or no intervention. The investigators observed physical activity and lunchtime consumption of fat.

(a) What are the experimental units and the response variables in this experiment?

(b) How many factors are there? How many treatments? Use a diagram like that in Figure 5.3 (page 355) to lay out the treatments.

(c) How many experimental units are required for the experiment?

Replication

Even with control, there will still be natural variability among experimental units. If each treatment is assigned to only one unit, you won't know whether any *systematic* differences in responses were due to the treatments or to the natural variability in the units. We would like to see units within a treatment group responding similarly to one another, but differently from units in other treatment groups. Then we can be sure that the treatment groups really are responding differently from each other.

Example 5.16 ***Gastric freezing for ulcers***
Systematic differences

In the gastric-freezing experiment of Example 5.15 (page 356), there were no systematic differences between the gastric-freezing (treatment) group and the placebo (control) group. In the two groups, 34% and 38%, respectively, improved.

In the Tennessee STAR program experiment of Example 5.13 (page 354) 6385 students were assigned to one of three treatment groups. This is such a large number of subjects that systematic differences among the groups are easy to spot.

We cannot say that *any* specific difference in the performance of the students in the three groups is caused by the size of the class or number of teaching professionals. There would be some difference even if all three groups were treated the same, because the natural variability among children means that some are smarter than others. Chance assigns the smartest students to one group or another, so that there is a chance difference among groups. We would not trust an experiment with just one subject in each group, for example. The results would depend too much on which group got lucky and received the smartest child. If we assign many students to each group, however, *the effects of chance will average out.* There will be little difference in the performance of the three groups unless the size of the class or the number of teachers causes a difference.

Our second principle of statistical design of experiments is *replication: use enough subjects to reduce chance variation.* If one experiment is conducted and then the same or a similar experiment is independently conducted in a different location by different investigators, then this is also called replication. But that's not the kind of replication we mean here.

Randomization

The design of an experiment first describes the response variable or variables, the factors (explanatory variables), and the layout of the treatments, with comparison (control) as the leading principle. Figure 5.3 (page 355) illustrates this aspect of the design of a study of response to advertising. The third basic principle of design is *randomization: the rule used to assign the experimental units to the treatments.* Comparison of the effects of several treatments is valid only when all treatments are applied to similar groups of experimental units. If one corn variety is planted on more fertile ground, or if one cancer drug is given to more seriously ill patients, comparisons among treatments are meaningless. Systematic differences among the groups of experimental units in a comparative experiment cause bias. How can we assign experimental units to treatments in a way that is fair to all of the treatments?

Experimenters often attempt to match groups by elaborate balancing acts. Medical researchers, for example, try to match the patients in a "new drug" experimental group and a "standard drug" control group by age, sex, physical condition, smoker or not, and so on. Matching is helpful but not adequate in this situation—there are too many lurking variables that might affect the outcome. The experimenter is unable to measure some of these variables and will not think of others until after the experiment. Some important variables, such as how advanced a cancer patient's disease is, are so subjective that they can't be measured. Also, an experimenter might bias a study by, for example, assigning more advanced cancer cases to a promising new treatment in the unconscious hope that it will help them.

The statistician's remedy is to rely on chance to make an assignment that does not depend on any characteristic of the experimental units and that does not rely on the judgment of the experimenter in any way. The use of chance can be combined with matching, but the simplest design creates groups by chance alone. Here is an example.

Example 5.17 *Cell phones and driving, I*
Random assignment

Does talking on a hands-free cell phone distract drivers? Undergraduate students "drove" in a high-fidelity driving simulator equipped with a hands-free cell phone. The car ahead brakes: how quickly does the subject respond? Twenty students (the control group) simply drove. Another 20 (the experimental group) talked on the cell phone while driving.

This experiment has a single factor (cell phone use) with two levels. The researchers must divide the 40 student subjects into two groups of 20. To do this in a completely unbiased fashion, put the names of the 40 students in a hat, mix them up, and draw 20. These students form the experimental group, and the remaining 20 make up the control group. Figure 5.4 outlines the design of this experiment.[22]

Figure 5.4 *Outline of a randomized comparative experiment, for Example 5.17.*

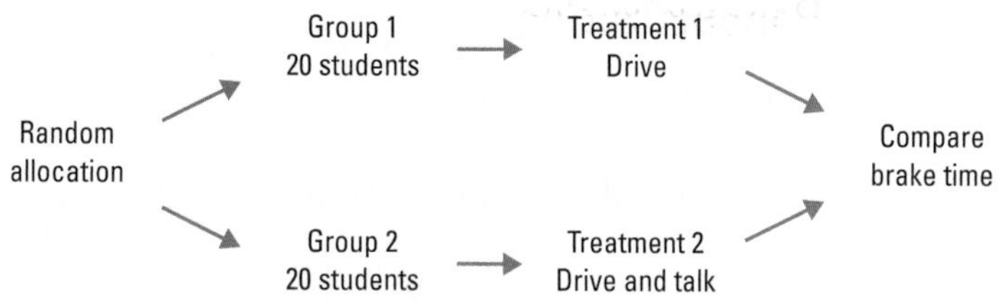

randomization

The use of chance to divide experimental units into groups is called ***randomization.*** The design in Figure 5.4 combines comparison and randomization to arrive at the simplest randomized comparative design. This "flowchart" outline presents all the essentials: randomization, the sizes of the groups and which treatment they receive, and the response variable. There are, as we will see later, statistical reasons for generally using treatment groups that are about equal in size.

Example 5.18 *The Physicians' Health Study*
Random assignment to treatments

Does regularly taking aspirin help protect people against heart attacks? The Physicians' Health Study was a medical experiment that helped answer this question. In fact, the Physicians' Health Study looked at the effects of two drugs: aspirin and beta-carotene. The body converts beta-carotene into vitamin A, which may help prevent some forms of cancer. The *subjects* were 21,996 male physicians. There were two *factors*, each having two levels: aspirin (yes or no) and beta-carotene (yes or no). Combinations of the levels of these factors form the four *treatments* shown in Figure 5.5. One-fourth of the subjects were assigned to each of these treatments.

Figure 5.5 *The treatments in the Physicians' Health Study.*

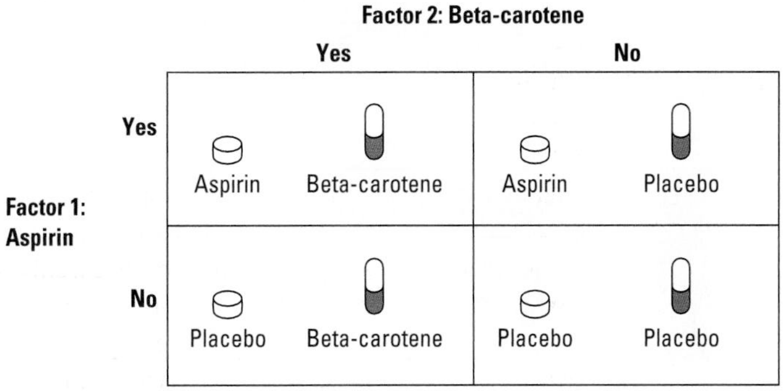

On odd-numbered days, the subjects took either a white tablet that contained aspirin or a dummy pill that looked and tasted like the aspirin but had no active ingredient (a placebo). On even-numbered days, they took either a blue capsule containing beta-carotene or a placebo. There were several response variables—the study looked for heart attacks, several kinds of cancer, and other medical outcomes. After several years, 239 of the placebo group but only 139 of the aspirin group had suffered heart attacks. This difference is large enough to give good evidence that taking aspirin does reduce heart attacks.[23] It did not appear, however, that beta-carotene had any effect.

Randomized Comparative Experiments

The logic behind the randomized comparative design in Figure 5.4 is as follows:

- Randomization produces two groups of subjects that we expect to be similar in all respects before the treatments are applied.
- Comparative design helps ensure that influences other than the cell phone operate equally on both groups.
- Therefore, differences in average brake reaction time must be due either to talking on the cell phone or to the play of chance in the random assignment of subjects to the two groups.

Principles of Experimental Design

The basic principles of statistical design of experiments are

1. **Control** the effects of lurking variables on the response, most simply by comparing two or more treatments.
2. **Replicate** each treatment on many units to reduce chance variation in the results.
3. **Randomize**—use impersonal chance to assign experimental units to treatments.

We hope to see a difference in the responses so large that it is unlikely to happen just because of chance variation. We can use the laws of probability, which give a mathematical description of chance behavior, to learn if the differences in treatment effects are larger than we would expect to see if only chance were operating. If they are, we call them ***statistically significant.***

Statistical Significance

An observed effect so large that it would rarely occur by chance is called **statistically significant.**

You will often see the phrase "statistically significant" in reports of investigations in many fields of study. It tells you that the investigators found good evidence for the effect they were seeking. The Physicians' Health Study, for example, reported statistically significant evidence that aspirin reduces the number of heart attacks compared with a placebo.

completely randomized design

When all experimental units are allocated at random among all treatments, the experiment is said to have a ***completely randomized design.*** The designs in Figures 5.4 and 5.5 are both completely randomized. The Physicians' Health Study had two factors, which combine to form the four treatments shown in Figure 5.5 (page 361). The study used a completely randomized design that assigned 5499 of the 21,996 subjects to each of the four treatments. Completely randomized designs can compare any number of treatments. The treatments can be formed by levels of a single factor or by more than one factor.

Example 5.19 *TV commercial*
Completely randomized design

Figure 5.3 (page 355) displays six treatments formed by the two factors in an experiment on response to a TV commercial. Suppose that we have 150 students who are willing to serve as subjects. We must assign 25 students at random to each group. Figure 5.6 outlines the completely randomized design.

To carry out the random assignment, label the 150 students 001 to 150. (Three digits are needed to label 150 subjects.) Enter Table B and read three-digit groups until you have selected 25 students to receive Treatment 1 (a 30-second ad shown once). If you start at line 140, the first few labels for Treatment 1 subjects are 129, 048, and 003.

Continue in Table B to select 25 more students to receive Treatment 2 (a 30-second ad shown 3 times). Then select another 25 for Treatment 3 and so on until you have assigned 125 of the 150 students to Treatments 1 through 5. The 25 students who remain get Treatment 6. The randomization is straightforward but very tedious to do by hand. The *Simple Random Sample* applet comes to our rescue. Alternatively, you can use your graphing calculator.

Figure 5.6 *Outline of a completely randomized design comparing six treatments, for Example 5.19.*

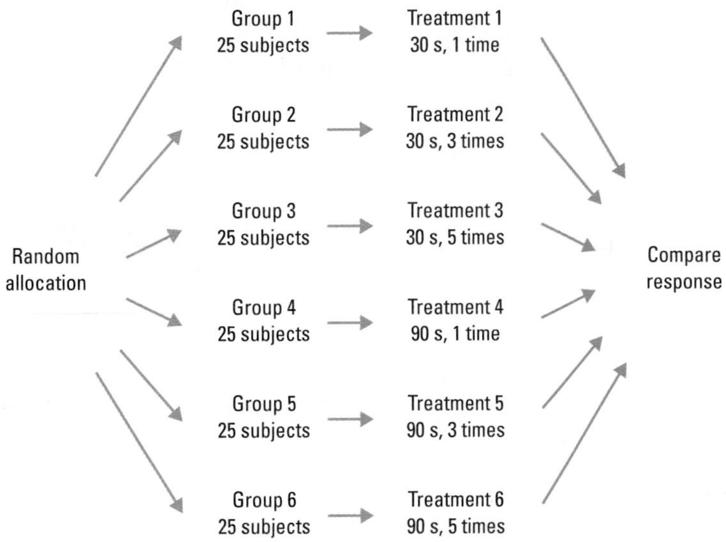

Activity 5E
Selecting random samples by calculator

In the experiment on the effect of cell phones on driving described in Example 5.17 (page 360), we need 20 students to be randomly selected from a group of 40 for the treatment group. On the TI-83/84, do the following steps:

1. `STAT/4:clrList/2ND/L`$_1$`/`**ENTER**. This clears any data in list L_1.
2. `MATH/PRB/5:randInt(`. Then complete the command to read `randInt(1,40,20)`$\rightarrow L_1$. You should see the first 5 numbers in a list of 20 numbers. To see the rest of the list, press the right-arrow key repeatedly. These 20 numbers have also been stored in list L_1 for convenience. Then match these 20 numbers with the corresponding student names to obtain the treatment group. If there are any duplicate numbers, you can use the command `randInt(1,40)` and press **ENTER** repeatedly to obtain random replacements, as needed. Which 20 students did you choose? The remaining 20 students make up the control group.

You can use the calculator to randomly assign experimental units to more than two groups without difficulty. Example 5.19 describes a randomized comparative experiment in which 150 students are randomly assigned to six groups of 25.

3. Use your calculator to randomly choose 25 out of 150 students to form the first group. Caution: Watch out for duplicate numbers. You may want to sketch an array of numbers from 1 to 150 and then cross out (or enter) numbers as they are selected. Which students are in your Group 1?

4. There are now 125 students who were not chosen. Randomly choose the 25 students for Group 2. If your previous calculator command was `randInt(1,150,25)→L1`, then you can quickly recall this command with `2nd/ENTRY/`ENTER. Again, make sure there are no duplicate numbers chosen. Which students were chosen for Group 2?

You can continue in this way until you have randomly assigned 25 students to each of the first five groups. The remaining 25 students not chosen would be assigned to Group 6.

Exercises

5.39 Treating prostate disease A large study used records from Canada's national health care system to compare the effectiveness of two ways to treat prostate disease. The two treatments are traditional surgery and a new method that does not require surgery. The records described many patients whose doctors had chosen each method. The study found that patients treated by the new method were significantly more likely to die within 8 years.[24]

(a) Further study of the data showed that this conclusion was wrong. The extra deaths among patients who got the new method could be explained by lurking variables. What lurking variables might be confounded with a doctor's choice of surgical or nonsurgical treatment?

(b) You have 300 prostate patients who are willing to serve as subjects in an experiment to compare the two methods. Use a diagram to outline the design of a randomized comparative experiment. (When using a diagram to outline the design of an experiment, be sure to indicate the size of the treatment groups and the response variable. The diagram in Example 5.17, page 360, is a model.)

5.40 Headache relief Doctors identify "chronic tension-type headaches" as headaches that occur almost daily for at least six months. Can antidepressant medications or stress management training reduce the number and severity of these headaches? Are both together more effective than either alone? Investigators compared four treatments: antidepressant alone, placebo alone, antidepressant plus stress management, and placebo plus stress management. Outline the design of the experiment. The headache sufferers named in the following table have agreed to participate in the study. Use a graphing calculator, software, or Table B at line 130 to randomly assign the subjects to the treatments.

01 Acosta	07 Duncan	13 Han	19 Liang	25 Padilla	31 Valasco
02 Asihiro	08 Durr	14 Howard	20 Maldonado	26 Plochman	32 Vaughn
03 Bennett	09 Edwards	15 Hruska	21 Marsden	27 Rosen	33 Wei
04 Bikalis	10 Farouk	16 Imrani	22 Montoya	28 Solomon	34 Wilder
05 Chen	11 Fleming	17 James	23 O'Brian	29 Trujillo	35 Willis
06 Clemente	12 George	18 Kaplan	24 Ogle	30 Tullock	36 Zhang

5.41 Layoffs and "survivor guilt" Workers who survive a layoff of other employees at their location may suffer from "survivor guilt." A study of survivor guilt and its effects used as subjects 120 students who were offered an opportunity to earn extra course credit by doing proofreading. Each subject worked in the same cubicle as another student, who was an accomplice of the experimenters. At a break midway through the work, one of three things happened:

Treatment 1: The accomplice was told to leave; it was explained that this was because she performed poorly.

Treatment 2: It was explained that unforeseen circumstances meant there was only enough work for one person. By "chance," the accomplice was chosen to be laid off.

Treatment 3: Both students continued to work after the break.

The subjects' work performance after the break was compared with performance before the break.[25]

(a) Outline the design of this completely randomized experiment.

(b) If you are using a graphing calculator or software, choose the subjects for Treatment 1. If not, use Table B at line 123 to choose the first four subjects for Treatment 1.

5.42 Smoking pot How does smoking marijuana affect willingness to work? Canadian researchers persuaded young adult men who used marijuana to live for 98 days in a "planned environment." The men earned money by weaving belts. They used their earnings to pay for meals and other consumption and could keep any money left over. One group smoked two potent marijuana cigarettes every evening. The other group smoked two weak marijuana cigarettes. All subjects could buy more cigarettes but were given strong or weak cigarettes, depending on their group. Did the weak and strong groups differ in work output and earnings?[26]

(a) Outline the design of this experiment.

(b) The following table lists the names of the 20 subjects. Use a graphing calculator, software, or Table B at line 131 to carry out the randomization your design requires.

01 Causey	05 Colton	09 Cuevas	13 Edwards	17 Frankum
02 French	06 Hankinson	10 Mathis	14 McGrath	18 Myers
03 Olsen	07 Pryor	11 Shenk	15 Shrader	19 Swaynos
04 Valenzuela	08 Vignolini	12 Waespe	16 Wenk	20 Zillgitt

5.43 Exercise and heart attacks Does regular exercise reduce the risk of a heart attack? Here are two ways to study this question.

1. A researcher finds 2000 men over 40 who exercise regularly and have not had heart attacks. She matches each with a similar man who does not exercise regularly, and she follows both groups for 5 years.

2. Another researcher finds 4000 men over 40 who have not had heart attacks and are willing to participate in a study. She assigns 2000 of the men to a regular program of supervised exercise. The other 2000 continue their usual habits. The researcher follows both groups for 5 years.

Explain clearly why the second design will produce more trustworthy data.

5.44 Growth of cataracts Eye cataracts are responsible for over 40% of blindness around the world. Can regularly drinking tea slow the growth of cataracts? We can't experiment on people, so we use rats as subjects. Researchers injected 18 young rats with a substance that causes cataracts. One group of the rats also received black tea extract; a second group received green tea extract; and a third got a placebo, a substance with no effect on the body. The response variable was the growth of cataracts over the next six weeks. Yes, both tea extracts did slow cataract growth.[27]

(a) Outline the design of this experiment.

(b) Use a graphing calculator, software, or Table B, starting at line 142, to assign rats to treatments.

"When Dr. Henderson comes in, everybody play dead."

Blocking

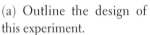

Suppose a fitness instructor believes that a certain exercise regimen will increase upper-body strength. He recruits students to test his theory by having them do as many push-ups as they can after they complete the training. You would expect a certain amount of inherent variability because some subjects will be able to do a lot of push-ups because they are stronger. Others will do fewer because they are not as strong. We try to control for these inherent differences by placing subjects into groups of similar individuals. One way to accomplish this is to separate the men and the women. We know that women tend to have less upper-body strength, so we would expect them, as a group, to be able to do fewer push-ups. By separating subjects by gender, we can reduce the effect of variation in strength on the number of push-ups. This is the idea behind a ***block design.***

Block Design

A **block** is a group of experimental units or subjects that are known before the experiment to be similar in some way that is expected to systematically affect the response to the treatments. In a **block design,** the random assignment of units to treatments is carried out separately within each block.

Block designs can have blocks of any size. A block design combines the idea of creating treatment groups as similar as possible with the principle of forming treatment groups at random. Blocks are another form of *control*. They control the effects of some outside variables by bringing those variables into the experiment to form the blocks. Here are some typical examples of block designs.

Example 5.20 *Comparing cancer therapies*
Block design

The progress of a type of cancer differs in women and men. A clinical experiment to compare three therapies for this cancer therefore treats gender as a blocking variable. Two separate randomizations are done, one assigning the female subjects to the treatments and the other assigning the male subjects. Figure 5.7 outlines the design of this experiment. Note that there is no randomization involved in making up the blocks. They are groups of subjects who differ in some way (gender in this case) that is apparent before the experiment begins.

Figure 5.7 *Outline of a block design for Example 5.20. The blocks consist of male and female subjects. The treatments are three therapies for cancer.*

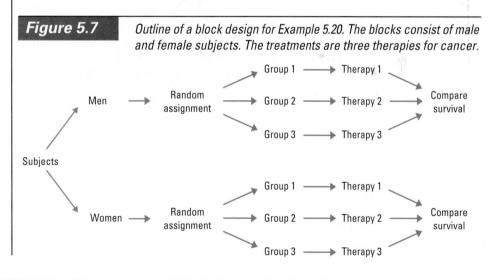

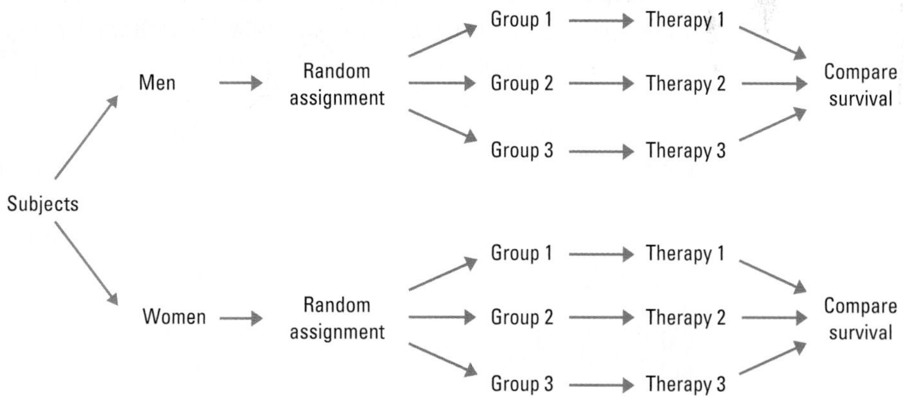

Example 5.21 *Soybeans*
Block design

The soil type and fertility of farmland differ by location. Because of this, a test of the effect of tillage type (two types) and pesticide application (three application schedules) on soybean yields uses small fields as blocks. Each block is divided into six plots, and the six treatments are randomly assigned to plots separately within each block.

Example 5.22 *Tennessee STAR*
Block design

The Tennessee STAR class size experiment (Example 5.13 on page 354) used a block design. It was important to compare different class types in the same school because the children in a school come from the same neighborhood, follow the same curriculum, and have the same school environment outside class. In all, 79 schools across Tennessee participated in the program. That is, there were 79 blocks. New kindergarten students were randomly placed in the three types of class separately within each school.

Blocks allow us to draw separate conclusions about each block, for example, about men and women in the cancer study in Example 5.20. Blocking also allows more precise overall conclusions because the systematic differences between men and women can be removed when we study the overall effects of the three therapies. The idea of blocking is an important additional principle of statistical design of experiments. **A wise experimenter will form blocks based on the most important unavoidable sources of variability among the experimental units.** Randomization will then average out the effects of the remaining variation and allow an unbiased comparison of the treatments. Hence the mantra: *control what you can, block on what you can't control, and randomize the rest.*

Matched Pairs Designs

Completely randomized designs are the simplest statistical designs for experiments. They illustrate clearly the principles of control, replication, and randomization. However, completely randomized designs are often inferior to more elaborate statistical designs. In particular, *matching the subjects in various ways can produce more precise results than simple randomization.*

matched pairs design

The simplest use of matching is a ***matched pairs design,*** which compares just two treatments. The subjects are matched in pairs. For example, an experiment to compare two advertisements for the same product might use pairs of subjects with the same age, sex, and income. The idea is that matched subjects are more similar than unmatched subjects, so that comparing responses within a number of pairs is more efficient than comparing the responses of groups of randomly assigned subjects. Randomization remains important: which one of a matched pair sees the first ad is decided at random. One common variation of the matched pairs design imposes both treatments on the same subjects, so that each subject serves as his or her own control. Here is an example.

Example 5.23 *Cell phones and driving, II*
Matched pairs

Example 5.17 (page 360) describes an experiment on the effects of talking on a cell phone while driving. The experiment compared two treatments: driving in a simulator and driving in a simulator while talking on a hands-free cell phone. The response variable is the time the driver takes to apply the brake when the car in front brakes suddenly. In Example 5.17, 40 student subjects were assigned at random, 20 students to each treatment. This is a completely randomized design, outlined in Figure 5.4 (page 360). Subjects differ in driving skill and reaction times. The completely randomized design relies on chance to create two similar groups of subjects.

In fact, the experimenters used a matched pairs design in which all subjects drove both with and without using the cell phone. They compared each individual's reaction times with and without the phone. If all subjects drove first with the phone and then without it, the effect of talking on the cell phone would be confounded with the fact that this is the first run in the simulator. The proper procedure requires that all subjects first be trained in using the simulator, that the *order* in which a subject drives with and without the phone be random, and that the two drives be on separate days to reduce carryover effects.

The completely randomized design uses chance to decide which 20 subjects will drive with the cell phone. The other 20 drive without it. The matched pairs design uses chance to decide which 20 subjects will drive first with and then without the cell phone. The other 20 drive first without and then with the phone.

Example 5.24 ***Cereal leaf beetles***
Matched pairs design

Are cereal leaf beetles more strongly attracted by the color yellow or by the color green? Agriculture researchers want to know because they detect the presence of the pests in farm fields by mounting sticky boards to trap insects that land on them. The board color should attract beetles as strongly as possible. We must design an experiment to compare yellow and green by mounting boards on poles in a large field of oats.

The experimental units are locations within the field far enough apart to represent independent observations. We erect a pole at each location to hold the boards. We might employ a completely randomized design in which we randomly select half the poles to receive a yellow board while the remaining poles receive green. The locations vary widely in the number of beetles present, however. For example, the alfalfa that borders the oats on one side is a natural host of the beetles, so locations near the alfalfa will have extra beetles. This variation among experimental units can hide the systematic effect of the board color.

It is more efficient to use a matched pairs design in which we mount boards of both colors on each pole. The observations (numbers of beetles trapped) are matched in pairs from the same poles. We compare the number of trapped beetles on a yellow board with the number trapped by the green board on the same pole. Because the boards are mounted one above the other, we select the color of the top board at random. Just toss a coin for each board—if the coin falls heads, the yellow board is mounted above the green board.

Matched pairs designs compare just two treatments. We choose *blocks* of two units that are as closely matched as possible. In Example 5.24, two boards on the same pole form a block. We assign one of the treatments to each unit by tossing a coin or reading odd and even digits from Table B. Alternatively, each block in a matched pairs design may consist of just one subject, who gets both treatments one after the other. Each subject serves as his or her own control. The *order* of

Is the placebo effect pseudoscience? Many believe that the placebo effect is purely psychological, with no physical basis (see Example 5.15, page 356). But a University of Michigan experiment published in the *Journal of Neuroscience* provides evidence that it is real. Fourteen healthy men, aged 20 to 30, were given a saltwater injection that caused pain to their jaws. They were then injected with a placebo and told that it was a painkiller. Researchers monitored their brain activity during the process. Each man's brain released more natural painkilling endorphins after the placebos were administered.

the treatments can influence the subject's response, so we randomize the order for each subject, again by a coin toss.

The matched pairs design of Example 5.24 uses the principles of control, replication, and randomization on several experimental units. However, the randomization is not complete (all locations randomly assigned to treatment groups) but is restricted to assigning the order of the boards at each location. The matched pairs design *reduces the effect of variation* among locations in the field by comparing the pair of boards at each location. *Matched pairs are an example of block designs.*

Cautions about Experimentation

The logic of a randomized comparative experiment depends on our ability to treat all the experimental units identically in every way except for the actual treatments being compared. Good experiments therefore require careful attention to details. For example, the subjects in the second gastric-freezing experiment (Example 5.15, page 356) all got the same medical attention during the study. Moreover, the study was ***double-blind***—neither the subjects themselves nor the medical personnel who measured their ulcer relief knew which treatment any subject had received. The double-blind method avoids unconscious bias by, for example, a doctor who doesn't think that "just a placebo" can benefit a patient.

Double-Blind Experiment

In a **double-blind experiment,** neither the subjects nor those who measure the response variable know which treatment a subject received.

Many—perhaps most—experiments have some weaknesses in detail. The environment of an experiment can influence the outcomes in unexpected ways. Although experiments are the gold standard for evidence of cause and effect, really convincing evidence usually requires that a number of studies in different places with different details produce similar results.

lack of realism

The most serious potential weakness of experiments is ***lack of realism.*** The subjects or treatments or setting of an experiment may not realistically duplicate the conditions we really want to study. Here are some examples of what can go wrong.

Example 5.25 *Placebo cigarettes?*
Lack of realism

A study of the effects of marijuana recruited young men who used marijuana. Some were randomly assigned to smoke marijuana cigarettes, while others were given placebo cigarettes. This failed: the control group recognized that their cigarettes were phony and complained loudly. It may be quite common for blindness to fail because the subjects can tell which treatment they are receiving.[28]

Example 5.26 *Response to advertising*
Lack of realism

A study compares two television advertisements by showing TV programs to student subjects. The students know it's "just an experiment." We can't be sure that the results apply to everyday television viewers. Many behavioral science experiments use as subjects students who know they are subjects in an experiment. That's not a realistic setting.

Example 5.27 *Genes and behavior*
Lack of realism

To study genetic influence on behavior, experimenters "knock out" a gene in one group of mice and compare their behavior with that of a control group of normal mice. The results of these experiments often don't agree as well as hoped, so investigators did exactly the same experiment with the same genetic strain of mice in Oregon, Alberta (Canada), and New York. Many results were very different. It appears that small differences in the lab environments have big effects on the behavior of the mice. Remember this the next time you read that our genes control our behavior.[29]

CAUTION

Lack of realism can limit our ability to apply the conclusions of an experiment to the settings of greater interest. Most experimenters want to generalize their conclusions to some setting wider than that of the actual experiment. **Statistical analysis of an experiment cannot tell us how far the results will generalize to other settings.** Nonetheless, the randomized comparative experiment, because of its ability to give convincing evidence for causation, is one of the most important ideas in statistics.

Exercises

5.45 Meditation for anxiety An experiment that claimed to show that meditation lowers anxiety proceeded as follows. The experimenter interviewed the subjects and rated their level of anxiety. Then the subjects were randomly assigned to two groups. The experimenter taught one group how to meditate and they meditated daily for a month. The other group was simply told to relax more. At the end of the month, the experimenter interviewed all the subjects again and rated their anxiety level. The meditation group now had less anxiety. Psychologists said that the results were suspect because the ratings were not blind. Explain what this means and how lack of blindness could bias the reported results.

5.46 Pain relief study Fizz Laboratories, a pharmaceutical company, has developed a new pain relief medication. Sixty patients suffering from arthritis and needing pain relief are available. Each patient will be treated and asked an hour later, "About what percent of pain relief did you experience?"

(a) Why should Fizz not simply administer the new drug and record the patients' responses?

(b) Outline the design of an experiment to compare the drug's effectiveness with that of aspirin and of a placebo.

(c) Should patients be told which drug they are receiving? How would this knowledge probably affect their reactions?

(d) If patients are not told which treatment they are receiving, the experiment is single-blind. Should this experiment be double-blind? Explain.

5.47 Comparing weight-loss treatments Twenty overweight females have agreed to participate in a study of the effectiveness of four weight-loss treatments: A, B, C, and D. The researcher first calculates how overweight each subject is by comparing the subject's actual weight with her "ideal" weight. The subjects and their excess weights in pounds are

Birnbaum	35	Hernandez	25	Moses	25	Smith	29
Brown	34	Jackson	33	Nevesky	39	Stall	33
Brunk	30	Kendall	28	Obrach	30	Tran	35
Cruz	34	Loren	32	Rodriguez	30	Wilansky	42
Deng	24	Mann	28	Santiago	27	Williams	22

The response variable is the weight lost after 8 weeks of treatment. Because a subject's excess weight will influence the response, a block design is appropriate.

(a) Arrange the subjects in order of increasing excess weight. Form five blocks of 4 subjects each by grouping the 4 least overweight, then the next 4, and so on.

(b) Use Table B to randomly assign the 4 subjects in each block to the 4 weight-loss treatments. Be sure to explain exactly how you used the table.

5.48 Carbon dioxide and tree growth The concentration of carbon dioxide (CO_2) in the atmosphere is increasing rapidly due to our use of fossil fuels. Because plants use CO_2 to fuel photosynthesis, more CO_2 may cause trees and other plants to grow faster. An elaborate apparatus allows researchers to pipe extra CO_2 to a 30-meter circle of forest. We want to compare the growth in base area of trees in treated and untreated areas to see if extra CO_2 does in fact increase growth. We can afford to treat three circular areas.[30]

(a) Describe the design of a completely randomized experiment using 6 well-separated 30-meter circular areas in a pine forest. Sketch the circles and carry out the randomization your design calls for.

(b) Areas within the forest may differ in soil fertility. Describe a matched pairs design using three pairs of circles that will reduce the extra variation due to different fertility. Sketch the circles and carry out the randomization your design calls for.

5.49 Does room temperature affect manual dexterity? An expert on worker performance is interested in the effect of room temperature on the performance of tasks requiring manual dexterity. She chooses temperatures of 70°F and 90°F as treatments. The response variable is the number of correct insertions, during a 30-minute period, in a peg-and-hole apparatus that requires the use of both hands simultaneously. Each subject is trained on the apparatus and then asked to make as many insertions as possible in 30 minutes of continuous effort.

(a) Outline a completely randomized design to compare dexterity at 70° and 90°. Twenty subjects are available.

(b) Because individuals differ greatly in dexterity, the wide variation in individual scores may hide the systematic effect of temperature unless there are many subjects in each group. Describe in detail the design of a matched pairs experiment in which each subject serves as his or her own control.

5.50 Calcium intake Calcium is important to the development of young girls. To study how the bodies of young girls process calcium, investigators used the setting of a summer camp. Calcium was given in Hawaiian Punch at either a high or a low level. The camp diet was otherwise the same for all girls. Suppose that there are 60 campers.

(a) Outline a completely randomized design for this experiment.

(b) Describe a matched pairs design in which each girl receives both levels of calcium (with a "washout period" between). What is the advantage of the matched pairs design over the completely randomized design?

(c) The same randomization can be used in different ways for both designs. Label the subjects 01 to 60. You must choose 30 of the 60. Use Table B at line 160 to choose just the first 5 of the 30. How are the 30 subjects chosen treated in the completely randomized design? How are they treated in the matched pairs design?

Section 5.2 Summary

In an experiment, one or more **treatments** are imposed on the **experimental units** or **subjects.** Each treatment is a combination of **levels** of the explanatory variables, which we call **factors.**

The **design** of an experiment refers to the choice of treatments and the manner in which the experimental units or subjects are assigned to the treatments.

The basic principles of statistical design of experiments are **control, replication,** and **randomization.**

The simplest form of control is **comparison.** Experiments should compare two or more treatments in order to prevent **confounding** the effect of a treatment with other influences, such as lurking variables.

Replication of the treatments on many units reduces the role of chance variation and makes the experiment more sensitive to differences among the treatments.

Randomization uses chance to assign subjects to the treatments. Randomization creates treatment groups that are similar (except for chance variation) before the treatments are applied. Randomization and comparison together prevent bias, or systematic favoritism, in experiments.

You can carry out randomization by giving numerical labels to the experimental units and using a **table of random digits** to choose treatment groups.

In addition to comparison, another form of control is to restrict randomization by forming **blocks** of experimental units. The units in each block are similar in some way that is important to the response. Randomization is then carried out separately within each block.

Matched pairs are a common form of blocking for comparing just two treatments. In some matched pairs designs, each subject receives both treatments in a random order. In others, the subjects are matched in pairs as closely as possible, and one subject in each pair receives one treatment, and the other subject receives the other treatment.

Good experiments require attention to detail as well as good statistical design. Many behavioral and medical experiments are **double-blind. Lack of realism** in an experiment can prevent us from generalizing its results.

The "**placebo effect**" is a term that doctors use to describe the phenomenon where patients get better because they expect the treatment to work even though they have taken a fake pill.

Section 5.2 Exercises

5.51 Does Saint-John's-wort relieve major depression? Here are some excerpts from the report of a study of this issue.[31] The study concluded that the herb is no more effective than a placebo.

(a) "Design: Randomized, double-blind, placebo-controlled clinical trial. . . ." Explain the meaning of each of the terms in this description.

(b) "Participants . . . were randomly assigned to receive either Saint-John's-wort extract ($n = 98$) or placebo ($n = 102$). . . . The primary outcome measure was the rate of change in the Hamilton Rating Scale for Depression over the treatment period." Based on this information, use a diagram to outline the design of this clinical trial.

5.52 Marketing to children If children are given more choices within a class of products, will they tend to prefer that product to a competing product that offers fewer choices? Marketers want to know. An experiment prepared three "choice sets" of beverages. The first contained two milk drinks and two fruit drinks. The second had the same two fruit drinks but four milk drinks. The third contained four fruit drinks but only the original two milk drinks. The researchers divided 210 children aged 4 to 12 years into 3 groups at random. They offered each group one of the choice sets. As each child chose a beverage to drink from the choice set presented, the researchers noted whether the choice was a milk drink or a fruit drink.

(a) What are the experimental units or subjects?

(b) What is the factor, and what are its levels?

(c) What is the response variable?

(d) The children's choice experiment has 210 subjects. Explain how you would assign labels to the 210 children in the actual experiment. Then use Table B at line 125 to choose only the first 5 children assigned to the first treatment.

5.53 Body temperature and surgery Surgery patients are often cold because the operating room is kept cool and the body's temperature regulation is disturbed by anesthetics. Will warming patients to maintain normal body temperature reduce infections after surgery? In one experiment, patients undergoing colon surgery received intravenous fluids from a warming machine and were covered with a blanket through which air circulated. For some patients, the fluid and the air were warmed; for others, they were not. The patients received identical treatment in all other respects.[32]

(a) Identify the experimental subjects, the factor and its levels, and the response variable.

(b) Draw a diagram to outline the design of a randomized comparative experiment for this study.

(c) The following subjects have given consent to participate in this study. Do the random assignment required by your design. (If you use Table B, begin at line 121.)

Abbott	Decker	Gutierrez	Lucero	Rosen
Adamson	Devlin	Howard	Masters	Sugiwara
Afifi	Engel	Hwang	McNeill	Thompson
Brown	Fluharty	Iselin	Morse	Travers
Cansico	Garcia	Janle	Ng	Turing
Chen	Gerson	Kaplan	Quinones	Ullmann
Cordoba	Green	Kim	Rivera	Williams
Curzakis	Gupta	Lattimore	Roberts	Wong

(d) To simplify the setup of the study, we might warm the fluids and air blanket for one operating team and not for another doing the same kind of surgery. Why might this design result in bias?

(e) The operating team did not know whether fluids and air blanket were heated, nor did the doctors who followed the patients after surgery. What is this practice called? Why was it used here?

5.54 Fabric science A maker of fabric for clothing is setting up a new line to "finish" the raw fabric. The line will use either metal rollers or natural-bristle rollers to raise the surface of the fabric; a dyeing-cycle time of either 30 or 40 minutes; and a temperature of either 150° or 175° Celsius. An experiment will compare all combinations of these choices. Three specimens of fabric will be subjected to each treatment and scored for quality.

(a) What are the factors and the treatments? How many individuals (fabric specimens) does the experiment require?

(b) Outline a completely randomized design for this experiment. (You need not actually do the randomization.)

5.55 Does calcium reduce blood pressure? You are participating in the design of a medical experiment to investigate whether a calcium supplement in the diet will reduce the blood pressure of middle-aged men. You have available 40 men with high blood pressure who are willing to serve as subjects.

(a) Outline an appropriate design for the experiment.

(b) The names of the subjects appear below. Use Table B, beginning at line 119, to do the randomization required by your design, and list the subjects to whom you will give the drug.

Alomar	Denman	Han	Liang	Rosen
Asihiro	Durr	Howard	Maldonado	Solomon
Bennett	Edwards	Hruska	Marsden	Tompkins
Bikalis	Farouk	Imrani	Moore	Townsend
Chen	Fratianna	James	O'Brian	Tullock
Clemente	George	Kaplan	Ogle	Underwood
Cranston	Green	Krushchev	Plochman	Willis
Curtis	Guillen	Lawless	Rodriguez	Zhang

(c) Preliminary work suggests that calcium may be effective and that the effect may be greater for black men than for white men. How could your design in (a) be modified to account for differences between black men and white men?

5.56 Treating low bone density Fractures of the spine are common and serious among women with advanced osteoporosis (low mineral density in the bones). Can taking strontium renelate help? A large medical trial assigned 1649 women to take either strontium renelate or a placebo each day. All of the subjects had osteoporosis and had had at least one fracture. All were taking calcium supplements and receiving standard medical care. The response variables were measurements of bone density and counts of new fractures over three years. The subjects were treated at 10 medical centers in 10 different countries.[33] Outline an appropriate design for this experiment. Explain why this is the proper design.

5.57 Placebo effect A survey of physicians found that some doctors give a placebo to a patient who complains of pain for which the physician can find no cause. If the patient's pain improves, these doctors conclude that it had no physical basis. The medical school researchers who conducted the survey claimed that these doctors do not understand the placebo effect. Why?

5.58 Will taking antioxidants help prevent colon cancer? People who eat lots of fruits and vegetables have lower rates of colon cancer than those who eat little of these foods. Fruits and vegetables are rich in "antioxidants" such as vitamins A, C, and E. Will taking antioxidants help prevent colon cancer? A clinical trial studied 864 people who were at risk of colon cancer. The subjects were divided into four groups: daily beta-carotene (which is made into vitamin A in the body), daily vitamins C and E, all three vitamins every day, and daily placebo. After four years, the researchers were surprised to find no significant difference in colon cancer among the groups.[34]

(a) What are the explanatory and response variables in this experiment?

(b) Outline the design of the experiment. Use your judgment in choosing the group sizes.

(c) Assign labels to the 864 subjects and use Table B, starting at line 118, to choose the first 5 subjects for the beta-carotene group.

(d) The study was double-blind. What does this mean?

(e) What does "no significant difference" mean in describing the outcome of the study?

(f) Suggest some lurking variables that could explain why people who eat lots of fruits and vegetables have lower rates of colon cancer. The experiment suggests that these variables, rather than the antioxidants, may be responsible for the observed benefits of fruits and vegetables.

5.59 Treating drunk drivers Once a person has been convicted of drunk driving, one purpose of court-mandated treatment or punishment is to prevent future offenses of the same kind. Suggest three different treatments that a court might require. Then outline the design of an experiment to compare their effectiveness. Be sure to specify the response variables you will measure.

5.60 Acculturation rating There are several psychological tests that measure the extent to which Mexican Americans are oriented toward Mexican/Spanish or Anglo/English culture. Two such tests are the Bicultural Inventory (BI) and the Acculturation Rating Scale for Mexican Americans (ARSMA). To study the correlation between the scores on these two tests, researchers will give both tests to a group of 22 Mexican Americans.

(a) Briefly describe a matched pairs design for this study. In particular, how will you use randomization in your design?

(b) You have an alphabetized list of the subjects (numbered 1 to 22). Carry out the randomization required by your design and report the result.

CASE CLOSED!

Can eating chocolate be good for you?

Use what you have learned in this chapter to resolve the Case Study that opened the chapter. Just as in a court proceeding, you can exclaim "Case Closed!" when you have finished.

Start by reviewing the information in the Case Study on page 327. Then answer each of the following questions in complete sentences. Be sure to communicate clearly enough for any of your classmates to understand what you are saying. First, here are the results of the three studies:

Study I The babies born to women who had been eating chocolate daily during pregnancy were found to be more active and "positively reactive"—a measure that the investigators said encompasses traits such as smiling and laughter. And the babies of stressed women who had regularly consumed chocolate showed less fear of new situations than babies of stressed women who abstained. Among those who reported they never/seldom ate chocolate, the mother's prenatal stress predicted more negative ratings of the infant's temperament. The researchers concluded that consuming chocolate produced subjective feelings of psychological well-being.

Study II Dr. Hollenberg's research indicates, preliminarily, that consuming high-flavonol cocoa helps regulate the synthesis of nitric oxide, a compound in the body that helps it maintain blood pressure and blood flow in vessel walls. The flavonols may also help vessels dilate and help keep platelets from clustering on the blood vessel walls. After five days,

(continued)

researchers measured what they called "significant improvement" in blood flow and the function of the cells that line the blood vessels.

Study III The investigators found that blood vessel function was improved when the subjects ate bittersweet dark chocolate, and that there were no such changes when they ate the placebo (fake chocolate). The favorable effects of dark chocolate lasted for a minimum of three hours.

1. **Study I**
   (a) Is Study 1 (effects of consumption of chocolate by pregnant women on infant temperament) an observational study or an experiment? Explain.
   (b) Is the conclusion of the researchers that consuming chocolate produced subjective feelings of psychological well-being justified? Explain.
2. **Study II**
   (a) Compare and contrast Study II with the gastric-freezing study in Example 5.15 (page 356). How are they alike, and how are they different?
   (b) Does Dr. Hollenberg's study describe a matched pairs design? Explain. (Be careful here.)
   (c) Describe a completely randomized design that includes a control group to investigate whether consuming flavonols for a week improves blood pressure and blood flow in vessel walls.
3. **Study III**
   (a) Why did the researchers not use a completely randomized design in Study III?
   (b) Identify the design that the investigators used in their study.
   (c) What advantage was gained by using the design they used rather than a completely randomized design?

Chapter Review

Summary

Designs for producing data are essential parts of statistics in practice. Random sampling and randomized comparative experiments are perhaps the most important statistical inventions of the last century. Both were slow to gain acceptance, and you will still see many voluntary response samples and uncontrolled experiments. This chapter has explained good techniques for producing data and has also explained why bad techniques often produce worthless data. The deliberate use of chance in producing data is a central idea in statistics. It allows use of the laws of probability to analyze data, as we will see in the following chapters.

What You Should Have Learned

Here are the major skills you should have gained from studying this chapter.

A. SAMPLING

1. Identify the population in a sampling situation.
2. Recognize bias due to voluntary response sampling and other inferior sampling methods.
3. Select a simple random sample (SRS) from a population.
4. Recognize cluster sampling and how it differs from other sampling methods.
5. Recognize the presence of undercoverage and nonresponse as sources of error in a sample survey. Recognize the effect of the wording of questions on the response.
6. Use random digits to select a stratified random sample from a population when the strata are identified.

B. EXPERIMENTS

1. Recognize whether a study is an observational study or an experiment.
2. Recognize bias due to confounding of explanatory variables with lurking variables in either an observational study or an experiment.
3. Identify the factors (explanatory variables), treatments, response variables, and experimental units or subjects in an experiment.
4. Outline the design of a completely randomized experiment using a diagram like those in Examples 5.17 (page 360) and 5.19 (page 362). The diagram in a specific case should show the sizes of the groups, the specific treatments, and the response variable(s).
5. Carry out the random assignment of subjects to groups in a completely randomized experiment.
6. Recognize the placebo effect. Recognize when the double-blind technique should be used.
7. Recognize a block design and when it would be appropriate. Know when a matched pairs design would be appropriate and how to design a matched pairs experiment.
8. Explain why a randomized comparative experiment can give good evidence for cause-and-effect relationships.

Chapter Review Exercises

5.61 Healthy cookies In Activity 5D, you compared the taste of cookies made with Hershey's new, allegedly healthier, dark chocolate chips and traditional Toll House chocolate chip cookies. The experimental cookies don't have to taste better than traditional cookies, but in order for consumers to buy them, the taste should be comparable to traditional cookies.

(a) What are the explanatory variable and the response variable? What values can the explanatory variable take?

(b) Describe the population for your study. Was the sample an SRS?

(c) Describe the design of the experiment. For example, was it completely randomized?

(d) What was done to control variability in this experiment?

(e) Discuss replication in this experiment.

(f) Was the experiment blind? Was it double-blind?

(g) Critique the experiment. For example, should a different design have been used? If you wanted to reach a conclusion about the cookie preference of all students in your school, what aspects of the experiment would you do differently, if any?

5.62 Ontario Health Survey The Ministry of Health in the province of Ontario, Canada, wants to know whether the national health care system is achieving its goals in the province. Much information about health care comes from patient records, but that source doesn't allow us to compare people who use health services with those who don't. So the Ministry of Health conducted the Ontario Health Survey, which interviewed a random sample of 61,239 people who live in the province of Ontario.[35]

(a) What is the population for this sample survey? What is the sample?

(b) The survey found that 76% of males and 86% of females in the sample had visited a general practitioner at least once in the past year. Do you think these estimates are close to the truth about the entire population? Why?

5.63 Study designs What is the name for each of these study designs?

(a) A study to compare two methods of preserving wood started with boards of southern white pine. Each board was ripped from end to end to form two edge-matched specimens. One was assigned to Method A; the other, to Method B.

(b) A survey on youth and smoking contacted by telephone 300 smokers and 300 nonsmokers, all 14 to 22 years of age.

(c) Does air pollution induce DNA mutations in mice? Starting with 40 male and 40 female mice, 20 of each sex were housed in a polluted industrial area downwind from a steel mill. The other 20 of each sex were housed at an unpolluted rural location 30 kilometers away.

5.64 Money talks Will cash bonuses speed unemployed people's return to work? The Illinois Department of Employment Security designed an experiment to find out. The subjects were 10,065 people aged 20 to 54 who were filing claims for unemployment insurance. Some were offered \$500 if they found a job within 11 weeks and held it for at least 4 months. Others could tell potential employers that the state would pay the employer \$500 for hiring them. A control group got neither kind of bonus.[36]

(a) Suggest a few response variables of interest to the state and outline the design of the experiment.

(b) How will you label the subjects for random assignment? Use Table B at line 167 to choose the first 3 subjects for the first treatment.

5.65 Coach, I need oxygen! We often see players on the sidelines of a football game inhaling oxygen. Their coaches think it will speed their recovery. We might measure recovery from intense exercise as follows. Have a football player run 100 yards three times in quick succession. Then allow three minutes to rest before running 100 yards again. Time the final run. Because players vary greatly in speed, you plan a matched pairs experiment using 25 football players as subjects. Discuss the design of such an experiment to investigate the effect of inhaling oxygen during the rest period.

5.66 Polling the faculty A labor organization wants to study the attitudes of college faculty members toward collective bargaining. These attitudes appear to differ depending on the type of college. The American Association of University Professors classifies colleges as follows:

Class I: Offer doctorate degrees and award at least 15 per year.
Class IIA: Award degrees above the bachelor's but are not in Class I.
Class IIB: Award no degrees beyond the bachelor's.
Class III: Two-year colleges.

Discuss the design of a sample of faculty from colleges in your state, with total sample size about 200.

5.67 New-drug study A drug manufacturer is studying how a new drug behaves in patients. Investigators compare two doses: 5 milligrams (mg) and 10 mg. The drug can be administered by injection, by a skin patch, or by intravenous drip. Concentration in the blood after 30 minutes (the response variable) may depend both on the dose and on the method of administration.

(a) Make a sketch that describes the treatments formed by combining dose and method. Then use a diagram to outline a completely randomized design for this two-factor experiment.

(b) "How many subjects?" is a tough issue. We will explain the basic ideas in Chapter 10. What can you say now about the advantage of using larger groups of subjects?

(c) The drug may behave differently in men and women. How would you modify your experimental design to take this into account?

5.68 Vitamin C for marathon runners An ultramarathon, as you might guess, is a footrace longer than the 26.2 miles of a marathon. Runners commonly develop respiratory infections after an ultramarathon. Will taking 600 milligrams of vitamin C daily reduce those infections? Researchers randomly assigned ultramarathon runners to receive either vitamin C or a placebo. Separately, they also randomly assigned these treatments to a group of nonrunners the same age as the runners. All subjects were watched for 14 days after the big race to see if infections developed.

(a) What is the name for this experimental design?

(b) Use a diagram to outline the design.

(c) The report of the study said:

> Sixty-eight percent of the runners in the placebo group reported the development of symptoms of upper respiratory tract infection after the race; this was significantly more than that reported by the vitamin C–supplemented group (33%).[37]

Explain to someone who knows no statistics why "significantly more" means there is good reason to think that vitamin C works.

5.69 Delivering the mail Does adding the five-digit postal zip code to an address really speed up delivery of letters? Does adding the four more digits that make up "zip + 4" speed delivery yet more? What about mailing a letter on Monday, Thursday, or Saturday? Describe the design of an experiment on the speed of first-class mail delivery. For simplicity, suppose that all letters go from you to a friend, so that the sending and receiving locations are fixed.

5.70 McDonald's versus Wendy's Do consumers prefer the taste of a cheeseburger from McDonald's or from Wendy's in a blind test in which neither burger is identified? Describe briefly the design of a matched pairs experiment to investigate this question.

5.71 Repairing knees in comfort Knee injuries are routinely repaired by arthroscopic surgery that does not require opening up the knee. Can we reduce patient discomfort by giving them a nonsteroidal anti-inflammatory drug (NSAID)? Eighty-three patients were placed in three groups. Group A received the NSAID both before and after the surgery. Group B was given a placebo before and the NSAID after. Group C received a placebo both before and after surgery. The patients recorded a pain score by answering questions one day after the surgery.[38]

(a) Outline the design of this experiment. You do not need to do the randomization that your design requires.

(b) You read that "the patients, physicians and physical therapists were blinded" during the study. What does this mean?

(c) You also read that "the pain scores for Group A were significantly lower than Group C but not significantly lower than Group B." What does this mean? What does this finding lead you to conclude about the use of NSAIDs?

5.72 Fresh air For years, scientists have suspected that a chemical in many household deodorizing products may cause short term lung problems. A 2006 study by scientists at the National Institutes of Health say they've found that people with relatively high blood concentrations of the substance—1,4-dichlorobenzene, an organic chemical—show signs of slightly reduced lung function. The study followed 953 Americans, average age 37, for six years.

(a) From the above description of this study, would you say it was an experiment or an observational study? Justify your answer.

(b) The number of subjects in this study is relatively large—almost 1,000 people. Does the large sample size lend credence to the notion that having household deodorizing products around is dangerous to your lungs? Justify your answer.

(c) A newspaper article announcing these results stated, "Other studies have had similar findings, which suggests that chronic exposure to a chemical in air fresheners can cause lung problems." Based on this NIH study, can we conclude that exposure to products that contain 1,4-dichlorobenzene causes lung problems? Why or why not?

PART

II Review Exercises

Communicate your thinking clearly in Exercises II.1 to II.10.

II.1 Is "tartar control" Crest toothpaste more effective at preventing tartar buildup on teeth than regular Crest toothpaste? Researchers recruit 210 volunteers (120 men and 90 women) who first get a free tooth cleaning. Then the researchers randomly assign 105 to the tartar control group and 105 to the regular toothpaste group. All 210 people are given unmarked tubes of the appropriate toothpaste and instructed on how often to brush their teeth each day. They are not told which group they are in, and the two kinds of toothpaste look the same. After 6 months, they are rated on tartar buildup by dentists who don't know who was in each group.

(a) Did the researchers use any form of blinding in this experiment? Justify your answer.

(b) Why did the researchers give all of the volunteers a free tooth cleaning?

(c) The researchers used a completely randomized design in this experiment. Describe a design that uses blocking to improve on the original design of the experiment. Give one reason why the block design might be preferable.

II.2 An opinion poll calls 2000 randomly chosen residential telephone numbers, then asks to speak with an adult member of the household. The interviewer asks, "How many movies have you watched in a movie theater in the past 12 months?" In all, 1131 people respond.

(a) Identify a potential source of bias related to the question being asked. Suggest a change that would help fix this problem.

(b) Identify a potential source of bias in this survey that is not related to question wording. Suggest a change that would help fix this problem.

II.3 It appears that people who drink alcohol in moderation have lower death rates than either people who drink heavily or people who do not drink at all. The protection offered by moderate drinking is concentrated among people over 50 and on deaths from heart disease. The Nurses' Health Study played an essential role in establishing these facts for women. This part of the study followed 85,709 female nurses for 12 years, during which time 2658 of the subjects died. The nurses completed a questionnaire that described their diet, including their use of alcohol. They were reexamined every 2 years. Conclusion: "As compared with nondrinkers and heavy drinkers, light-to-moderate drinkers had a significantly lower risk of death."[1]

(a) Was this study an experiment or an observational study? Justify your answer.

(b) What does "significantly lower risk of death" mean in simple language?

(c) Identify a lurking variable that might be confounded with how much a person drinks. Explain how the variable you chose could lead to confounding.

II.4 Few people want to eat discolored french fries. Potatoes are kept refrigerated before being cut for french fries to prevent spoiling and preserve flavor. But immediate processing of cold potatoes causes discoloring due to complex chemical reactions. The potatoes must therefore be brought to room temperature before processing. Design an experiment in which tasters will rate the color and flavor of french fries prepared from several groups of potatoes. The potatoes will be freshly picked or stored for a month at room temperature or stored for a month refrigerated. They will then be sliced and cooked either immediately or after an hour at room temperature.

(a) What are the factors and their levels, the treatments, and the response variables?

(b) Describe the design of your experiment. Write a few sentences explaining how you would implement your design.

(c) It is efficient to have each taster rate fries from all treatments. How will you use randomization in presenting fries to the tasters?

II.5 Here's the opening of a press release from June 2004: "Starbucks Corp. on Monday said it would roll out a line of blended coffee drinks intended to tap into the growing popularity of reduced-calorie and reduced-fat menu choices for Americans." You wonder if Starbucks customers like the new "Mocha Frappuccino® Light" as well as the regular version of this coffee.

(a) Describe a matched pairs design using 40 regular Starbucks customers that would help answer this question. Be sure to discuss blinding.

(b) Explain how you would use the partial table of random digits below to do the randomization that your design requires. Then use your method to assign treatments to the first 3 subjects. Show your work clearly on your paper.

07511	88915	41267	16853
84569	79367	32337	03316
81486	69487	60513	09297
00412	71238	27649	39950

(c) Would it have been better to use a completely randomized design instead of a matched pairs design? Justify your answer.

II.6 A large university wants to gather student opinion about parking for students on campus. It isn't practical to contact all enrolled students.

(a) Give an example of a way to choose a sample of students that is poor practice because it depends on voluntary response.

(b) Give an example of a bad way to choose a sample that doesn't use voluntary response.

(c) Now suggest a method of obtaining a sample that should lead to more reliable results.

II.7 You read in a magazine that "nonphysical treatments such as meditation and prayer have been shown to be effective in controlled scientific studies for such ailments as high blood pressure, insomnia, ulcers, and asthma."

a) Explain in simple language what the article means by "controlled scientific studies."

(b) Why can such studies provide good evidence that meditation and prayer are effective treatments for some medical problems?

II.8 Elephants sometimes damage crops in Africa. It turns out that elephants dislike bees. They recognize beehives in areas where they are common and avoid them. Can this behavior be used to keep elephants away from trees? A group in Kenya placed active beehives in some trees and empty beehives in others. Will elephant damage be less in trees with hives? Will even empty hives keep elephants away?[2]

(a) Describe the design of an experiment to answer these questions using 72 acacia trees. Write a few sentences about how you would implement your design.

(b) Explain the purpose of randomization in the context of this experiment.

(c) Should any form of blinding be used in this experiment? Justify your answer.

II.9 Observational studies had suggested that vitamin E reduces the risk of heart disease. Careful experiments showed that vitamin E has no effect, at least for women. A commentary in the *Journal of the American Medical Association* said:

> Thus, vitamin E enters the category of therapies that were promising in epidemiologic and observational studies but failed to deliver in adequately powered randomized controlled trials. As in other studies, the "healthy user" bias must be considered, ie, the healthy lifestyle behaviors that characterize individuals who care enough about their health to take various supplements are actually responsible for the better health, but this is minimized with the rigorous trial design.[3]

A friend who knows no statistics asks you to explain this.

(a) What is the difference between observational studies and experiments?

(b) Explain what is meant by a "randomized controlled trial."

(c) How does "healthy user bias" explain how people who take vitamin E supplements have better health in observational studies but not in controlled experiments?

II.10 Stores advertise price reductions to attract customers. What type of price cut is most attractive?

Experiments with more than one factor allow insight into interactions between the factors. A study of the attractiveness of advertised price discounts had two factors: percent of all foods on sale (25%, 50%, 75%, or 100%) and whether the discount was stated precisely as "60% off" or as a range, "50% to 70% off." Subjects rated the attractiveness of the sale on a scale of 1 to 7.

(a) Describe a completely randomized design using 200 student subjects.

(b) Explain how you would use the partial table of random digits below to assign subjects to treatment groups. Then use your method to select the first 3 subjects for one of the treatment groups. Show your work clearly on your paper.

45740	41807	65561	33302
07051	93623	18132	09547
12975	13258	13048	45144
72321	81940	00360	02428

(c) Figure II.1 shows the mean ratings for the eight treatments formed from the two factors.[4] Based on these results, write a careful description of how percent on sale and precise discount versus range of discounts influence the attractiveness of a sale.

Figure II.1 *Mean responses to eight treatments in an experiment with two factors, showing interaction between the factors, for Exercise II.10.*

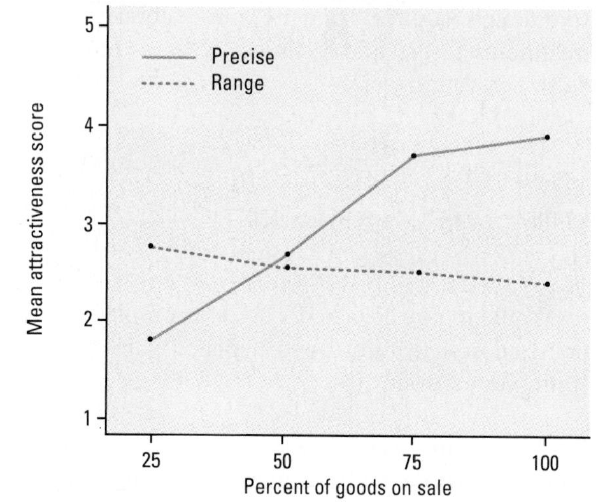

PART III

Probability and Random Variables: Foundations for Inference

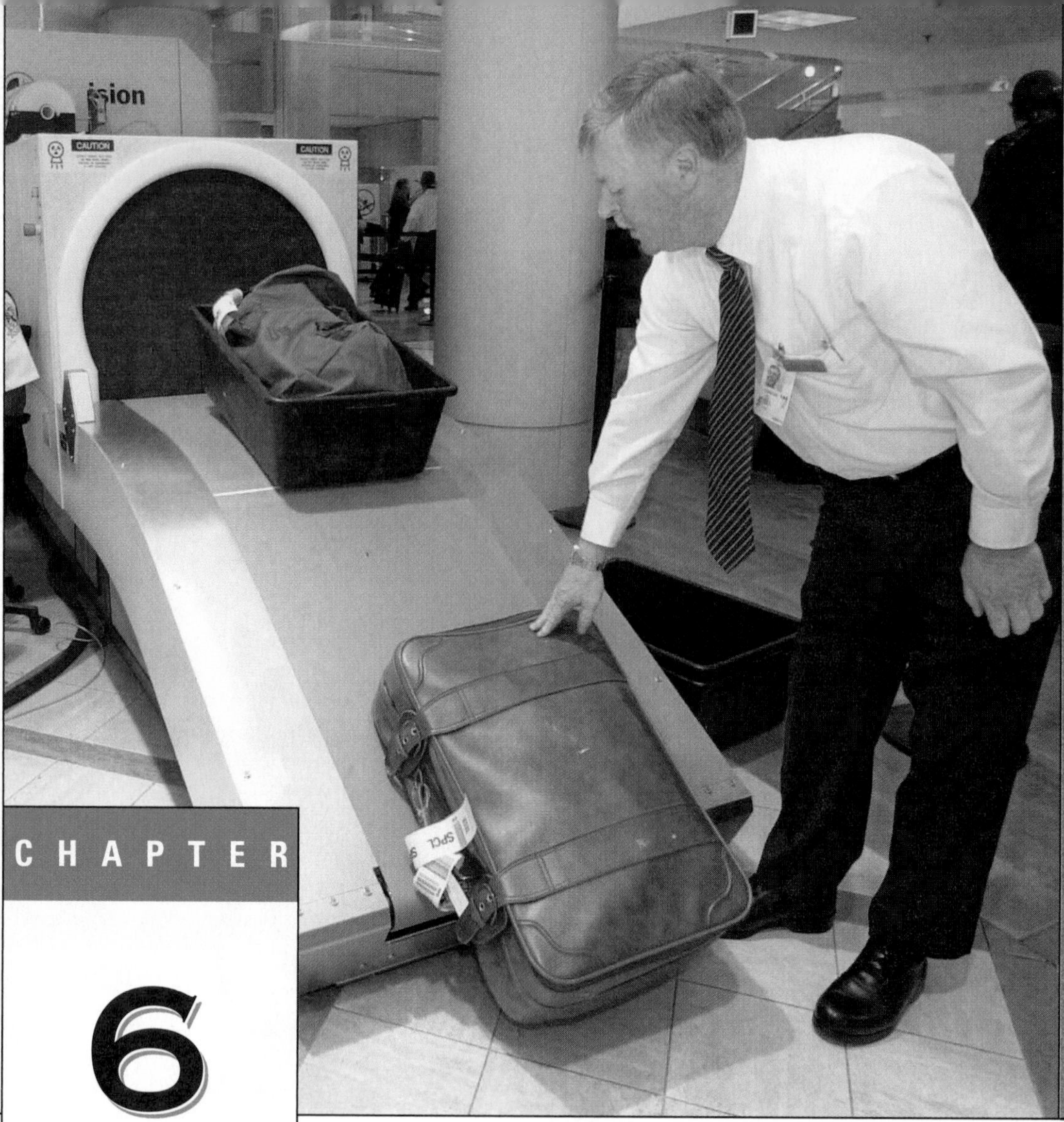

CHAPTER

6

Probability and Simulation: The Study of Randomness

CASE STUDY

False alarms at airports are an explosive issue

Our lives are full of false-positives. When the smoke alarm goes off because you've burned something on the stove, that's a false-positive. It's positive because an alarm goes off alerting you to a danger. It's false because your house is not actually burning down. When the metal in your sneakers sets off the metal detector at the airport, that's a false-positive, too: the metal the machine sensed in your shoes might have been a knife or a gun, but it wasn't.

After the World Trade Center and Pentagon tragedies of September 11, 2001, the Transportation Security Administration (TSA) was established and charged with installing a screening system for airports that would detect weapons and bombs on individuals or in baggage. Since January 1, 2003, TSA has been screening all checked luggage.

The machines for checking baggage tend to be large, about the size of an SUV, and costly, over $1 million per machine. Unfortunately, the technology is not perfect. Shampoo, for example, which has the same density as certain explosives, can be mistaken for explosives and generate a false-positive. Other items that produce false-positives are certain food items (like cheese or chocolate), books, deodorant sticks, toothpaste, golf balls, and ticking objects like electric toothbrushes! The machines also flag luggage that has items the scanner can't see through, such as laptop computers, camera equipment, and cell phones. TSA screeners will hand-search bags that register a positive reading.

Screening luggage at airports raises a host of questions. By the end of this chapter you will have developed the necessary statistical methods to answer important questions about false-positives and false-negatives and their implications.

Activity 6A

Austin and Sara's game

Materials: Graphing calculator; table of random digits

1. Suppose Austin picks a random whole number from 1 to 5 twice and adds them together. And suppose Sara picks a random whole number from 1 to 10. High score wins. What would you guess is the proportion of times Austin would win, if this game were played many times? What would you guess is the proportion of times Sara would win?

2. Get a partner and play a few rounds of the game, using the digits from line 101 in Table B (back inside cover of your text) in the order that they appear in line 101. Read the digit 0 as the number 10. For example, the first block of digits is 19223. Austin picks two numbers first. His numbers are 1 and 2. Note that we ignore the 9 and go on to the next digit because the numbers he chooses have to be in the 1 to 5 range. Then the next digit (the second 2) is Sara's number. Since Austin's $1 + 2 = 3$ is greater than Sara's 2, Austin wins the first game. Record the results. Continue in this fashion, reading left to right in row 101 and discarding digits as necessary, until you come to the end of the row. At this point, is either player pulling ahead, or are they about even?

3. Carefully enter the following commands in the home screen on your calculator. `randInt(1,5,200)→L1:randInt(1,5,200)→L2:randInt(1,10,200)→L3:((L1+L2)>L3)→L4:sum(L4)/200.` The command `randInt` is found under `MATH/PRB/5:randInt` on the TI-83/84, and under **CATALOG** **F3** (Flash Apps) on the TI-89. This sequence of commands tells the calculator to pretend to play 200 times. (This is called a *simulation*, which we will define more formally soon.) Press **ENTER** to carry out the simulation, and record the results. Are you surprised? Press **ENTER** again to simulate 200 more games, and record the results. Continue to press **ENTER** until you have 25 sets of repetitions for a total of 5,000 games.

4. Enter your 25 results in list L_4. Define a viewing window to be $X[0.4,0.6]_{0.025}$ and $Y[-4,14]_{.1}$. Then plot a histogram with a boxplot superimposed. TRACE to find the center of both plots. Based on what you see, what is your prediction for the proportion of times Austin would win if this game were played many times?

5. Do `1-Var Stats` and look at the mean and median for your data. Do they tend to agree with your prediction in Step 4?

6. Now let's consider the game from Sara's perspective. Do you think that Sara's chances of winning over the long term are the same as Austin's? What minor change to the calculator instructions in Step 3 would simulate 200 plays of the game and report the proportion of times Sara wins? Make this change, and then press ENTER repeatedly until you get 25 results. Calculate the mean of the 25 decimals. Are you surprised?
7. Would you like to modify your estimate of the proportion of times Sara wins? If so, what is your new estimate?

Introduction

Toss a coin 10 times. What is the likelihood of a run of 3 or more consecutive heads or tails? A couple plans to have children until they have a girl or until they have four children, whichever comes first. What are the chances that they will have a girl among their children? An airline knows from past experience that a certain percent of customers who have purchased tickets will not show up to board the airplane. If the airline "overbooks" a particular flight (that is, sells more tickets than they have seats), what are the chances that the airline will encounter more ticketed passengers than they have seats for? There are three methods we can use to answer questions like these involving chance:

1. Try to estimate the likelihood of a result of interest by actually observing the random phenomenon many times and calculating the relative frequency of the results.
2. *probability model* Develop a ***probability model*** and use it to calculate a theoretical answer.
3. Start with a model that, in some fashion, reflects the truth about the random phenomenon, and then develop a plan for imitating—or simulating—a number of repetitions of the procedure.

Option 1 is slow, sometimes costly, and often impractical or logistically difficult. Option 2 requires that we know something about the rules of probability and therefore may not be feasible. Option 3, constructing a simulation, is usually quicker than repeating the real procedure, especially if we can use the TI-83/84/89 or a computer. Perhaps most importantly, simulation methods allow us to get reasonably accurate results and answer questions that are more difficult when done with formal mathematical analysis.

In this chapter, we will begin our study of chance phenomena by first learning some fundamental simulation techniques. Once these techniques are mastered, you will see that many, if not most, problems involving chance can be solved by simulation methods.

Probability is the branch of mathematics that describes the pattern of chance outcomes. When we produce data by random sampling or randomized comparative experiments, the laws of probability answer the question "What would happen if we did this many times?" Sections 6.2 and 6.3 present the laws of probability in the form of a probability model.

Probability calculations are the basis for inference. The tools you acquire in this chapter will help you describe the behavior of statistics from random samples and randomized comparative experiments in later chapters. Even our brief acquaintance with probability will enable us to answer questions like these:

- If we know the blood types of a man and a woman, what can we say about the blood types of their future children?
- Give a test for the AIDS virus to the employees of a small company. What is the chance of at least one positive test if all the people tested are free of the virus?
- An opinion poll asks a sample of 1500 adults what they consider the most serious problem facing our schools. How often will the percent of those who answer "drugs" come within two percentage points of the truth about the entire population?

6.1 Simulation

You already have some experience with several simulations. In Exercise 2.14 (page 129) you used the calculator to simulate rolling a six-sided die many times. Then you compared the resulting distribution to a uniform distribution. The calculator program FLIP50 in Exercise 2.22 (page 133) simulates flipping a fair coin 50 times. In this section, you will learn how to build simulations to answer various questions involving chance.

Here are some situations where simulation is helpful.

Example 6.1 *Simulations on a grand scale*
Applications of simulation

SIMULATIONS

(a) *Cleaning up the environment.* In the 1970s Allied Chemical was cited for dumping a waste chemical called Kepone into the James River. Kepone killed fish and other aquatic life and leeched into the river bottom, where it became a long-term environmental problem. If the pollution were confined to the river, that would be bad enough. But the James feeds into the Chesapeake Bay, which in turn leads to the Atlantic Ocean. Scientists who were studying the environmental impact of a Kepone-type incident wanted to better understand the dynamics, so they created a very complicated simulation model. The simulation allowed them to vary factors such as the rate of introduction of the chemical into the river and then observe what happened downstream.

(b) *Training soldiers for warfare.* Live military field exercises are expensive, hazardous, difficult to control, and frequently constrained by location and weather. So, in 2004, the army spent \$1.8 billion to develop and train soldiers in a variety of tasks, from operating a huge ship to engaging in hand-to-hand combat, all in a virtual-reality world. The goals are low casualties and victory in war. The Hampton Roads area in Virginia is becoming the nation's center for military applications of modeling and simulation. Old Dominion University, in Norfolk, Virginia, is developing a "Battle Lab" as part of its Virginia Modeling, Analysis, and Simulation Center in Suffolk, Virginia. The lab is designed to test new methods in military warfare.

(c) *Pilot and driver training.* For years, pilots have used computer simulators to learn how to operate the latest technology built into new airliners. They practice landings and takeoffs and what to do if they encounter hazardous conditions like severe storms and wind sheer. The insurance industry, in an effort to reduce the number of accidents and injuries due to drunk driving, takes driver simulators from school to school. The simulator can show scenes on the screens as they would be seen by someone under the influence of alcohol. It shows reduced reaction times as well as the distance needed to stop a vehicle once braking begins. The experience provides a powerful message against driving while intoxicated.

Example 6.2 *A girl in the family*

Setting up a simulation

Suppose we are interested in estimating the likelihood of a couple's having a girl among their first four children. Let a flip of a fair coin represent a birth, with heads corresponding to a girl and tails a boy. Assuming that girls and boys are equally likely to occur on any birth, the coin flip is an accurate imitation of the situation. Flip the coin until a head appears or until the coin has been flipped 4 times, whichever comes first. The appearance of a head within the first 4 flips corresponds to the couple's having a girl among their first four children.

If this coin-flipping procedure is repeated many times, to represent the births in a large number of families, then the proportion of times that a head appears within the first 4 flips should be a good estimate of the true likelihood of the couple's having a girl.

A single die (one of a pair of dice) could also be used to simulate the birth of a son or daughter. Let an even number of spots (called pips) represent a girl, and let an odd number of spots represent a boy.

Simulation

The imitation of chance behavior, based on a model that accurately reflects the phenomenon under consideration, is called a **simulation.**

Simulation is an effective tool for finding likelihoods of complex results once we have a trustworthy model. As you saw in Activity 6A, we can use random digits from a table, graphing calculator, or computer software to simulate many repetitions quickly. The proportion of repetitions on which a result occurs will eventually be close to its true likelihood, so simulation can give good estimates of probabilities. The art of random digit simulation can be illustrated by a series of examples.

Example 6.3 *Simulation steps*

Simulation mechanics

Step 1: **State the problem or describe the random phenomenon.** Toss a coin 10 times. What is the likelihood of a run of at least 3 consecutive heads or 3 consecutive tails?

Step 2: **State the assumptions.** There are two:

- A head or a tail is equally likely to occur on each toss.
- Tosses are independent of each other (that is, what happens on one toss will not influence the next toss).

Step 3: **Assign digits to represent outcomes.** In a random number table, such as Table B, even digits and odd digits occur with the same long-term relative frequency, 50%. Here is one assignment of digits for coin tossing:

- One digit simulates one toss of the coin.
- Odd digits represent heads; even digits represent tails.

Successive digits in the table simulate independent tosses.

Step 4: **Simulate many repetitions.** Looking at 10 consecutive digits in Table B simulates one repetition. Read many groups of 10 digits from the table to simulate many repetitions. Be sure to keep track of whether or not the event we want (a run of at least 3 heads or at least 3 tails) occurs on each repetition.

Here are the first three repetitions, starting at line 101 in Table B. Runs of 3 or more heads or tails have been underlined.

Digits:	1 9 2 2 3	9 5 0 3 4	0 5 7 5 6	2 8 7 1 3	9 6 4 0 9	1 2 5 3 1
Heads/tails:	H H T T H	H H T H T	T H H H T	T T H H H	H T T T H	H T H H H
Run of 3:		YES		YES		YES

Twenty-two additional repetitions were done for a total of 25 repetitions; 23 of them did have a run of 3 or more heads or 3 or more tails.

Step 5: **State your conclusions.** We estimate the probability of a run of size 3 by the proportion

$$\text{estimated probability} = \frac{23}{25} = 0.92$$

Of course, 25 repetitions are not enough to be confident that our estimate is accurate. Now that we understand how to do the simulation, we can tell a computer to do many thousands of repetitions (or ***trials***). A long simulation (or mathematical analysis) finds that the true probability is about 0.826.

trials

Once you have gained some experience in simulation, establishing a correspondence between random numbers and possible outcomes is usually the hardest part and must be done carefully. Although coin tossing may not fascinate you, the model in Example 6.3 is typical of many probability problems. The results of one toss have no effect or influence over the next coin toss. For this reason, we say that the coin tosses are ***independent.*** We will define independent more formally in Section 6.2. Coin tosses, as well as dice rolls, are said to be *independent trials*

independent

because the tosses all have the same possible outcomes and probabilities. Shooting 10 free throws and observing the sexes of 10 children have similar models and are simulated in much the same way.

The idea is to state the basic structure of the random phenomenon and then use simulation to move from this model to the probabilities of more complicated events. The model is based on opinion and past experience. If the model does not correctly describe the random phenomenon, the probabilities derived from it by simulation will also be incorrect.

Step 3 (assigning digits) can usually be done in several different ways, but some assignments are more efficient than others. Here are some examples of this step.

Example 6.4 ***Assigning digits***
Simulation mechanics

(a) Choose a person at random from a group of which 70% are employed. One digit simulates one person:

$$0, 1, 2, 3, 4, 5, 6 = \text{employed}$$
$$7, 8, 9 = \text{not employed}$$

It doesn't matter which 3 digits are assigned to "not employed" as long as they are distinct. The following correspondence is also satisfactory:

$$00, 01, \ldots, 69 = \text{employed}$$
$$70, 71, \ldots, 99 = \text{not employed}$$

This assignment is less efficient, however, because it requires twice as many digits and ten times as many numbers.

(b) Choose one person at random from a group of which 73% are employed. Now two digits simulate one person:

$$00, 01, 02, \ldots, 72 = \text{employed}$$
$$73, 74, 75, \ldots, 99 = \text{not employed}$$

We assigned 73 of the 100 two-digit pairs to "employed" to get probability 0.73. Representing "employed" by 01, 02, . . . , 73 would also be correct.

(c) Choose one person at random from a group of which 50% are employed, 20% are unemployed, and 30% are not in the labor force. There are now three possible outcomes, but the principle is the same. One digit simulates one person:

$$0, 1, 2, 3, 4 = \text{employed}$$
$$5, 6 = \text{unemployed}$$
$$7, 8, 9 = \text{not in the labor force}$$

Another valid assignment of digits might be

$$0, 1 = \text{unemployed}$$
$$2, 3, 4 = \text{not in the labor force}$$
$$5, 6, 7, 8, 9 = \text{employed}$$

What is important is the number of digits assigned to each outcome, not the order of the digits.

As Example 6.4 shows, simulation methods work just as easily when outcomes are not equally likely. Consider the following slightly more complicated example.

Example 6.5

Frozen yogurt sales
Using the random digit table

Orders of frozen yogurt flavors (based on sales) have the following relative frequencies: 38% chocolate, 42% vanilla, and 20% strawberry. We want to simulate customers entering the store and ordering yogurt.

Step 1: **State the problem or describe the random phenomenon.** How would you simulate 10 frozen yogurt sales based on this recent history?

Step 2: **State the assumptions.** Orders of frozen yogurt flavors, based on sales, have the following relative frequencies: 38% chocolate, 42% vanilla, and 20% strawberry. We also assume that customers order one flavor only, and that customers' choices of flavors do not influence one another.

Step 3: **Assign digits to represent outcomes.** Instead of considering the random number table to be made up of single digits, we now consider it to be made up of pairs of digits. Thus we may assign the numbers in the random number table as follows:

- 00 to 37 to correspond to the outcome chocolate (C)
- 38 to 79 to correspond to the outcome vanilla (V)
- 80 to 99 to correspond to the outcome strawberry (S)

Step 4: **Simulate many repetitions.** The sequence of random numbers (starting at the 21st column of row 112 in Table B) is as follows:

19352 73089 84898 45785

This yields the following two-digit numbers:

19 35 27 30 89 84 89 84 57 85

which correspond to the outcomes

C C C C S S S S V S

Step 5: **State your conclusions.** The problem only asked for the process, but let's take a look at our results. We estimate the probability of an order for chocolate to be 4/10 = 0.4, an order for vanilla to be 1/10 = 0.1, and an order for strawberry to be 5/10 = 0.5. *Once again, we need to point out that 10 repetitions are not enough to be confident that our estimates are accurate.*

Example 6.6

A girl or four
Random digit table

A couple plans to have children until they have a girl or until they have four children, whichever comes first. We will show how to use random digits to estimate the likelihood that they will have a girl.

The model is the same as for coin tossing. We will assume that each child has probability 0.5 of being a girl and 0.5 of being a boy, and the sexes of successive children are independent.

Assigning digits is also easy. One digit simulates the sex of one child:

$$0, 1, 2, 3, 4 = \text{girl}$$
$$5, 6, 7, 8, 9 = \text{boy}$$

To simulate one repetition of this childbearing strategy, read digits from Table B until the couple has either a girl or four children. Notice that the number of digits needed to simulate one repetition can vary from 1 to 4. Here is the simulation, using line 130 of Table B. To interpret the digits, G for girl and B for boy are written under them, space separates repetitions, and under each repetition "+" indicates that a girl was born and "−" indicates that one was not.

690	51	64	81	7871	74	0
BBG	BG	BG	BG	BBBG	BG	G
+	+	+	+	+	+	+
951	784	53	4	0	64	8987
BBG	BBG	BG	G	G	BG	BBBB
+	+	+	+	+	+	−

In these 14 repetitions, a girl was born 13 times. Our estimate of the probability that this strategy will produce a girl is therefore

$$\text{estimated probability} = \frac{13}{14} = 0.93$$

Some fairly complicated mathematics shows that if our probability model is correct, the actual probability of having a girl is 0.938. Our simulated answer came quite close. Unless the couple is unlucky, they will succeed in having a girl.

Exercises

6.1 Establishing a correspondence State how you would use the following aids to establish a correspondence in a simulation that involves a 75% chance:

(a) a coin

(b) a six-sided die

(c) a random digit table (Table B)

(d) a standard deck of playing cards

6.2 The clever coins Suppose you left your statistics textbook and calculator in your locker, and you need to simulate a random phenomenon that has a 25% chance of a desired outcome. You discover two nickels in your pocket that are left over from your lunch money. Describe how you could use the two coins to set up your simulation.

6.3 Abolish evening exams? Suppose that 84% of a university's students favor abolishing evening exams. You ask 10 students chosen at random. What is the likelihood that all 10 favor abolishing evening exams?

(a) Describe how you would pose this question to 10 students independently of each other. How would you model the procedure?

(b) Assign digits to represent the answers "Yes" and "No."

(c) Simulate 5 repetitions, starting at line 129 of Table B. Then combine your results with those of the rest of your class. What is your estimate of the likelihood of the desired result?

6.4 Shooting free throws A basketball player makes 70% of her free throws in a long season. In a tournament game she shoots 5 free throws late in the game and misses 3 of them. The fans think she was nervous, but the misses may simply be chance. You will shed some light by estimating a probability.

(a) Describe how to simulate a single shot if the probability of making each shot is 0.7. Then describe how to simulate 5 independent shots.

(b) Simulate 50 repetitions of the 5 shots and record the number missed on each repetition. Use Table B, starting at line 125. What is the approximate likelihood that the player will miss 3 or more of the 5 shots?

6.5 A political poll, I An opinion poll selects adult Americans at random and asks them, "Which political party, Democratic or Republican, do you think is better able to manage the economy?" Explain carefully how you would assign digits from Table B to simulate the response of one person in each of the following situations.

(a) Of all adult Americans, 50% would choose the Democrats and 50% the Republicans.

(b) Of all adult Americans, 60% would choose the Democrats and 40% the Republicans.

(c) Of all adult Americans, 40% would choose the Democrats, 40% would choose the Republicans, and 20% would be undecided.

(d) Of all adult Americans, 53% would choose the Democrats and 47% the Republicans.

6.6 A political poll, II Use Table B to simulate the responses of 10 independently chosen adults in each of the four situations of Exercise 6.5.

(a) For situation (a), use line 110.

(b) For situation (b), use line 111.

(c) For situation (c), use line 112.

(d) For situation (d), use line 113.

Simulations with the Calculator or Computer

The calculator and computer can be extremely useful in conducting simulations because they can be easily programmed to quickly perform a large number of repetitions. Study the reasoning and the steps involved in the following example so that you may become adept at using the capabilities of the TI-83/84/89 to design and carry out simulations.

Example 6.7

Randomizing with the calculator

Simulation mechanics

The command `randInt` can be used to generate random digits between any two specified values. Here are three applications.

The command `randInt(0,9,5)` generates 5 random integers between 0 and 9. This could serve as a block of 5 random digits in the random number table. The command `randInt(1,6,7)` could be used to simulate rolling a die 7 times. Generating 10 two-digit numbers between 00 and 99 from Example 6.5 could be done with the command `randInt(0,99,10)`.

```
randInt(0,9,5)
      {5 6 5 7 1}
randInt(1,6,7)
    {5 6 5 5 3 4 1}
randInt(0,99,10)

{81 23 86 2 40...
```

```
F1▼   F2▼     F3▼  F4▼   F5     F6▼
Tools Algebra Calc Other PrgmIO Clean Up
■ tistat.randint(0,9,5)
          {6. 2. 2. 4. 7.}
■ tistat.randint(1,6,7)
          {3. 3. 4. 1. 5. 4. ▸
■ tistat.randint(0,99,10)
          {96. 87. 52. 34. 63▸
TIStat.randInt(0,99,10)
MAIN      RAD AUTO      FUNC
```

Using the statistical software package Minitab, the following commands will generate a set of 10 random numbers in the range 00 to 99 and store these numbers in column C1. From the Menu bar, pull down "Calc," select "Random," followed by "Integer." Specify 10 rows of data to be stored in column C1. Enter a minimum value of 0 and a maximum value of 99. Numbers like the following will appear in column C1 in the worksheet:

C1									
38	93	14	30	50	92	16	18	84	20

When you combine the power and simplicity of simulations with the power of technology, you have formidable tools for answering questions involving chance behavior.

Activity 6B

Is this discrimination?

A gentleman sent the following letter to the ombudsman at his local newspaper. "The company I worked for recently laid off 10 of its sales staff, including me, due to budget cuts. But after talking with one of my former fellow workers, we both realized that 6 of the 10 people they fired were older than 55, while a large proportion of the younger sales staff—who are paid less—kept their jobs. How can I find out if I have an age-discrimination case, and where can I turn for help?" What the gentleman is asking is whether this can reasonably be attributed to chance. We learn from the Bureau of Labor Statistics that 24% of all sales people in the last census were 55 or older.

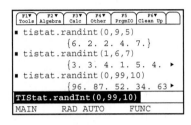

Here's the plan. In this investigation, you will use your calculator's random number generator to conduct 20 repetitions of a simulation. Place the results in a frequency table. Then estimate the relative frequency that 6 or more sales people in a randomly selected group of 10 are 55 years old or older.

1. Let digits 1 to 100 represent the salespeople. Let digits 1 to 24 represent salespeople 55 or older, and let 25 to 100 represent salespeople younger than 55. Now randomly select 10 salespeople:

   randInt(1,100,10)→L_1:SortA(L_1)

2. Look at your sample of 10 salespeople (in list L_1) and count the number of salespeople 55 and older (numbers 1 to 24). Record a tally mark in the appropriate column of your frequency table.

3. Repeat Steps 1 and 2 for a total of 20 repetitions (20 tally marks). It will go faster if you edit the above command to read

   randInt(1,100,10)→L_1:SortA(L_1):($L_1 \leq 24$)→L_2:sum(L_2)

   Then all you have to do is keep pressing **ENTER** and recording the appropriate tally mark.

Number of salespeople 55 or older	Frequency
0	
1	
2	
3	
4	
5	
6	
7	
8	
9	
10	

4. Where should the center of the distribution be? Where is the center of your sample?

5. Calculate the relative frequency that 6 or more of the 10 salespeople laid off are 55 or older.
6. Combine your results with those of your classmates to obtain a more accurate relative frequency. Theory tells us that, in the long run, only about 1.6 times in 100 would you see 6 or more people 55 or older out of 10 if only chance were involved. This is unlikely to happen by chance alone. The gentleman appears to have a case. Compare your relative frequency with the theoretical relative frequency of 0.016.

Example 6.8 *Carly Patterson, Olympic gymnastics gold medalist*
Simulating with technology

SIMULATIONS

Carly Patterson, then only 16 years old, won the gold medal in all-around gymnastics at the 2004 Summer Olympics in Athens, Greece. A competitor's total score is determined by adding the scores for four events: vault, parallel bars, balance beam, and floor exercise. Suppose we know that the eventual silver medalist, Russia's Svetlana Khorkina, earned a total score of 38.211 (her actual total in Athens). Suppose also that Carly's scores in 100 previous meets leading up to the Olympics have been approximately Normally distributed with means and standard deviations as shown in the table below. We will further assume that Carly's performance at the Olympics will follow the same pattern as her pre-Olympic meets. This is a reasonable assumption for world-class athletes like Carly.

Event	Mean	Standard deviation
Vault	9.314	0.216
Parallel bars	9.553	0.122
Balance beam	9.461	0.203
Floor exercise	9.543	0.099

We want to know Carly's chances of beating the total score of the current leader, Khorkina. One way to simulate this situation is to randomly sample Carly's scores from 100 competitions for each of the four events and to store those scores in lists: vault in L_1, parallel bars in L_2, balance beam in L_3, and floor exercise in L_4. Then add the corresponding entries in lists L_1 through L_4 and store the total scores in L_5. Next, calculate the mean of the 100 total scores in L_5. That will be an estimate of Carly's all-around performance in the Olympics. Here are the details:

`randNorm(9.314,0.216,100)` $\rightarrow L_1$	simulates 100 of Carly's vault scores
`randNorm(9.553,0.122,100)` $\rightarrow L_2$	simulates 100 parallel bars scores
`randNorm(9.461,0.203,100)` $\rightarrow L_3$	simulates 100 balance beam scores
`randNorm(9.543,0.099,100)` $\rightarrow L_4$	simulates 100 floor exercise scores
$(L_1 + L_2 + L_3 + L_4) \rightarrow L_5$	simulated total score

Calculating one-variable statistics on L_5 tells us that Carly's simulated mean total score is 37.854 with a standard deviation of 0.207. Using the 68–95–99.7 rule, we would expect about 95% of Carly's total scores to be between 37.44 and 38.268. A score at the top end of that range would beat Khorkina's total score of 38.211. In fact, Carly exceeded expectations by earning 9.375 on the vault, 9.575 on the parallel bars, 9.725 on the balance beam, and 9.712 on the floor exercise, for a total of 38.387. With that total score, she won the gold medal, the first Olympic gold medal in all-around gymnastics for the U.S.A. since Mary Lou Retton won all-around gold at the 1984 Olympics in Los Angeles.

Exercises

6.7 A girl or four Use your calculator to simulate a couple's having children until they have a girl or until they have four children, whichever comes first. (See Example 6.6.) Use the simulation to estimate the probability that they will have a girl among their no-more-than-four children. How do your calculator results compare with those of Example 6.6?

6.8 World Series Suppose that in a particular year the American League baseball team is considered to have a 60% chance of beating the National League team in any given World Series game. (This assumption ignores any possible home field advantage, which is probably not very realistic.) To win the World Series, a team must win 4 out of 7 games in the series. Further assume that the outcome of each game is not influenced by the outcome of any other game (that is, who wins one game is independent of who wins any other game).

(a) Use simulation methods to approximate the number of games that would have to be played in order to determine the world champion.

(b) The so-called home field advantage is one factor that might be an explanatory variable in determining the winner of a game. What are some other possible factors?

6.9 Tennis racquets Professional tennis players bring multiple racquets to each match. They know that high string tension, the force with which they hit the ball, and occasional "racquet abuse" are all reasons why racquets break during a match. Brian Lob's coach tells him that he has a 15% chance of breaking a racquet in any given match. How many matches, on average, can Brian expect to play until he breaks a racquet and needs to use a backup? Use simulation methods to answer this question.

6.10 Simulating families Imagine a large group of families, each having 4 children. You want to use simulation to estimate the percent of families that have exactly 0 girls, exactly 1 girl, exactly 2 girls, exactly 3 girls, and exactly 4 girls. Assume that the outcomes (girl) and (boy) occur with the same frequency in the long run.

(a) Describe a correspondence between random digits and outcomes that one could use to perform this simulation.

(b) Conduct 40 repetitions of the simulation, and then calculate the percents specified above.

Don't shoot Bambi In recent years the deer population in many areas, and particularly in urban areas, has increased to the extent that the deer have become a nuisance. In search of food, they have been emboldened to graze in vegetable gardens and flowerbeds and eat prized plants down to the ground. One sophisticated mathematical modeling program was created to find the optimal strategy for thinning the herd. After specifying assumptions from population biology, the program surprised researchers by declaring that, over time, the most effective method of reducing the deer population would be to "harvest" (kill) the young.

6.11 The airport van, I Your company operates a van service from the airport to downtown hotels. Each van carries 7 passengers. Many passengers who reserve seats don't show up—in fact, 25 times out of 100, in the long run, a passenger will fail to appear. Assume that whether a passenger shows up has no relation to any other passenger showing up. If you allow 9 reservations for each van, what is the probability that more than 7 passengers will appear? Carry out a simulation to answer this question.

6.12 Should I guess? A multiple-choice test is scored as follows: For each question you answer correctly, you get 4 points. For each question you answer incorrectly, you lose 1 point. For simplicity, suppose that there are 10 multiple-choice questions with four choices for each question.

(a) Suppose Jack doesn't know the answers to any of the questions, and he guesses on each one. Use simulation methods to determine Jack's score.

(b) Sam is in the same boat. He's clueless, so he leaves all 10 questions blank. Who do you expect to score higher, Jack or Sam? Be careful.

Section 6.1 Summary

There are times when actually carrying out an experiment, sample survey, or observational study is too costly, too slow, or simply impractical. In situations like these, a carefully designed **simulation** can provide approximate answers to our questions. A *probability model* can be used to calculate a theoretical probability.

A simulation is an imitation of chance behavior, most often carried out with random numbers representing independent trials.

Here are the **steps of a simulation**:

1. State the problem or describe the random phenomenon.
2. State the assumptions.
3. Assign digits to represent outcomes.
4. Simulate many repetitions (trials).
5. State your conclusions.

Calculators, like the TI-83/84/89, and computers are particularly useful for conducting simulations because they can perform many repetitions quickly.

Section 6.1 Exercises

6.13 Game of chance, I Amarillo Slim is a card sharp who likes to play the following game. Draw 2 cards from a standard deck of 52 cards. If at least one of the cards is a heart, you win $1. If neither card is a heart, you lose $1.

(a) Describe a correspondence between random numbers and possible outcomes in this game.

(b) Simulate playing the game for 25 rounds. Shuffle the cards after each round. See if you can beat Amarillo Slim at his own game. Remember to write down the results of each game. When you finish, combine your results with those of 3 other students to obtain a total of 100 trials. Report your cumulative proportion of wins. Do you think this is a "fair" game? That is, do both you and Slim have an equal chance of winning?

6.14 Game of chance, II A certain game of chance is based on randomly selecting three numbers from 00 to 99 (inclusive), allowing repetitions, and adding the numbers. A person wins the game if the resulting sum is a multiple of 5.

(a) Describe your scheme for assigning random numbers to outcomes in this game.

(b) Use simulation to estimate the proportion of times a person wins the game.

6.15 The birthday problem Use your calculator and the simulation method to determine the chances of at least 2 students with the same birthday in a class of 23 unrelated students. Determine the chances of at least 2 people having the same birthday in a room of 41 people. What assumptions are you using in your simulations?

6.16 Batter up! Suppose a major league baseball player has a current batting average of .320. Note that batting average = (number of hits)/(number of at-bats).

(a) Describe an assignment of random numbers to possible results in order to simulate the player's next 20 at-bats.

(b) Carry out the simulation for 20 repetitions, and report your results. What is the relative frequency of at-bats in which the player gets a hit?

(c) Compare your simulated experimental results with the player's actual batting average of .320.

6.17 Nuclear safety A nuclear reactor is equipped with two independent automatic shutdown systems to shut down the reactor when the core temperature reaches the danger level. Neither system is perfect. System A shuts down the reactor 90% of the time when the danger level is reached. System B does so 80% of the time. The reactor is shut down if either system works.

(a) Explain how to simulate the response of System A to a dangerous temperature level.

(b) Explain how to simulate the response of System B to a dangerous temperature level.

(c) Both systems are in operation simultaneously. Combine your answers to (a) and (b) to simulate the response of both systems to a dangerous temperature level. Explain why you cannot use the same entry in Table B to simulate both responses.

(d) Now simulate 100 trials of the reactor's response to an emergency of this kind. Estimate the probability that it will shut down. This probability is higher than the probability that either system working alone will shut down the reactor.

6.18 Spreading a rumor On a small island there are 25 inhabitants. One of these inhabitants, named Jack, starts a rumor that spreads around the isle. Any person who hears the rumor continues spreading it until he or she meets someone who has heard the story before. At that point, the person stops spreading it, since nobody likes to spread stale news.

(a) Do you think that all 25 inhabitants will eventually hear the rumor or will the rumor die out before that happens? Estimate the proportion of inhabitants who will hear the rumor.

(b) In the first time increment, Jack randomly selects one of the other inhabitants, named Jill, to tell the rumor to. In the second time increment, both Jack and Jill each randomly select one of the remaining 24 inhabitants to tell the rumor to. (*Note*: They could conceivably pick each other again.) In the next time increment, there are 4 rumor spreaders, and so on. If a randomly selected person has already heard the rumor, that rumor teller stops spreading the rumor. Design a record-keeping chart, and simulate this procedure. Use your TI-83/84/89 to help with the random selection. Continue until all 25 inhabitants hear the rumor or the rumor dies out. How many inhabitants out of 25 eventually heard the rumor?

(c) Combine your results with those of other students in the class. What is the mean number of inhabitants who hear the rumor?

6.19 Shaq The basketball player Shaquille O'Neal makes about half of his free throws over an entire season. We will use the calculator to simulate 100 free throws shot independently by a player who makes 50% of his attempts. We let the number 1 represent the outcome "Hit" and 0 represent a "Miss."

(a) Enter the command `randInt(0,1,100)→SHAQ.` (randInt is found in the CATALOG under Flash Apps on the TI-89.) This tells the calculator to randomly select a hit (1) or a miss (0). Do this 100 times in succession, and store the results in the list named SHAQ.

(b) What percent of the 100 shots are hits?

(c) Examine the sequence of hits and misses. How long was the longest run of shots made? Of shots missed? (Sequences of random outcomes often show runs longer than we might expect.)

6.20 Simulating an opinion poll A recent opinion poll showed that about 73% of married women agree that their husbands do at least their fair share of household chores. Suppose that this is exactly true. Choosing a married woman at random then has a 73% chance of getting one who agrees that her husband does his share. Use software or your calculator to simulate choosing many women independently. (In most software, the key phrase to look for is *Bernoulli trials*. This is the technical term for independent trials with Yes/No outcomes. Our outcomes here are "Agree" or not.)

(a) Simulate drawing 20 women, then 80 women, then 320 women. What proportion agree in each case? We expect (but because of chance variation we can't be sure) that the proportion will be closer to 0.73 when more trials are examined.

(b) Simulate drawing 20 women 10 times and record the percents in each trial who agree. Then simulate drawing 320 women 10 times and again record the 10 percents. Which set of 10 results is less variable? Write a statement about the relationship between the number of trials and the variability in the results.

6.2 Probability Models

Activity 6C

The spinning wheel

Materials: Margarine tub spinner or graphing calculator or table of random numbers

Imagine a spinner with three sectors, all the same size, marked 1, 2, and 3 as shown.

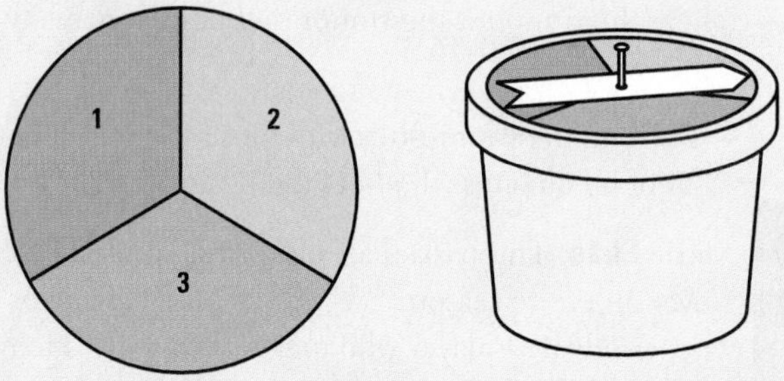

The investigation consists of spinning the spinner three times and recording the numbers as they occur (for example, 123). We want to determine the proportion of times that *at least one digit occurs in its correct position.* For example, in the number 123, all of the digits are in their proper positions, but in the number 331, none are. For this Activity, use a spinner like the one in the illustration, a table of random digits, or your calculator.

1. Guess the proportion of times at least one digit will occur in its proper place.
2. To use your calculator to randomly generate the three-digit number, enter the command `randInt(1,3,3)`. Continue to press ENTER to generate more three-digit numbers. Use a tally mark to record the results in a table like the one below. Do 20 trials and then calculate the relative frequency for the event "at least one digit in the correct position."

At least one digit in the correct position	
Not	

To use a random number table, select a row and (discarding digits 4 to 9 and 0) record digits in the 1 to 3 range in groups of three.

3. Combine your results with those of your classmates to obtain as many trials as possible (at least 100 randomly generated three-digit numbers; 200 would be better).
4. Count the number of times at least one digit occurred in its correct position, and calculate the proportion.
5. Your teacher has a program, SPIN123, that implements the procedure for the TI-83/84/89. The key step uses the calculator's Boolean logic to count the number of "hits." Enter the program or link it from a classmate or your teacher. Execute the program for 25, 50, and 100 repetitions. Compare the calculator results with the results you obtained in Steps 2 to 4.

Later in the chapter we will calculate the theoretical probability of this event happening, so keep your data handy so that you can compare the theoretical probability with your experimental results.

The Idea of Probability

The mathematics of probability begins with the observed fact that some phenomena are random—that is, the relative frequencies of their outcomes seem to settle down to fixed values in the long run. Recall the coin tossing in Example P.9 (page 21). The relative frequency of heads is quite erratic in 2 or 5 or 10 tosses. But after several thousand tosses it remains stable, changing very little over further thousands of tosses. The big idea is this: *chance behavior is unpredictable in the short run but has a regular and predictable pattern in the long run.*

In Example P.9 (page 21) we saw that the proportion of heads in many tosses of a balanced coin eventually gets close to 0.5. But does the actual count of heads get close to one-half the number of tosses? Let's find out.

Activity 6D

Proportion of heads versus count of heads

Go to the book's Web site, www.whfreeman.com/tps3e and launch the *Probability* applet.

1. Set the "Probability of heads" to 0.5 and the number of tosses to 40.
2. After 40 tosses,

   (a) What is the proportion of heads?

   (b) What is the count of heads?

(c) What is the difference between the count of heads and 20 (one-half the number of tosses)?

You can extend the number of tosses by clicking "Toss" again to get 40 more. *Important:* Don't click "Reset" during this Activity.

3. Keep going to 120 tosses. Again record the proportion and count of heads and the difference between the count and 60 (half the number of tosses).
4. Keep going. Stop at 240 tosses and again at 480 tosses to record the same facts.
5. Although it may take a long time, the laws of probability say that the proportion of heads will always get close to 0.5. They also say that the difference between the count of heads and half the number of tosses will always grow without limit. Did you find this to be the case?

The Language of Probability

"Random" in statistics is not a synonym for "haphazard" but a description of a kind of order that emerges only in the long run. We often encounter the unpredictable side of randomness in our everyday experience, but we rarely see enough repetitions of the same random phenomenon to observe the long-term regularity that probability describes. You can see that regularity emerging in Activity 6D. In the very long run, the proportion of tosses that give a head is 0.5. This is the intuitive idea of probability. Probability 0.5 means "occurs half the time in a very large number of trials."

We might suspect that a coin has probability 0.5 of coming up heads just because the coin has two sides. As Exercise 6.21 (page 410) illustrates, such suspicions are not always correct. The idea of probability is ***empirical.*** That is, it is based on observation rather than theorizing. Probability describes what happens in very many trials, and we must actually observe many trials to pin down a probability. In the case of tossing a coin, some diligent people have in fact made thousands of tosses.

empirical

Example 6.9 *Some coin tossers*
Long-term relative frequency

The French naturalist Count Buffon (1707–1788) tossed a coin 4040 times. Result: 2048 heads, or proportion 2048/4040 = 0.5069 for heads.

Around 1900, the English statistician Karl Pearson heroically tossed a coin 24,000 times. Result: 12,012 heads, a proportion of 0.5005.

While imprisoned by the Germans during World War II, the South African mathematician John Kerrich tossed a coin 10,000 times. Result: 5067 heads, a proportion of 0.5067.

Randomness and Probability

We call a phenomenon **random** if individual outcomes are uncertain but there is nonetheless a regular distribution of outcomes in a large number of repetitions.

The **probability** of any outcome of a random phenomenon is the proportion of times the outcome would occur in a very long series of repetitions. That is, probability is long-term relative frequency.

Does God play dice? Few things in the world are truly random in the sense that no amount of information will allow us to predict the outcome. We could in principle apply the laws of physics to a specific toss of a coin, for example, and calculate whether it will land heads or tails. But randomness does rule events inside individual atoms. Albert Einstein didn't like this feature of the new quantum theory. "I shall never believe that God plays dice with the world," said the great scientist. Eighty years later, it appears that Einstein was wrong.

Thinking about Randomness

That some things are random is an observed fact about the world. The outcome of a coin toss, the time between emissions of particles by a radioactive source, and the sexes of the next litter of lab rats are all random. So is the outcome of a random sample or a randomized experiment. Probability theory is the branch of mathematics that describes random behavior. Of course, we can never observe a probability exactly. We could always continue tossing the coin, for example. Mathematical probability is an idealization based on imagining what would happen in an indefinitely long series of trials.

The best way to understand randomness is to observe random behavior—not only the long-run regularity but the unpredictable results of short runs. You can do this with physical devices, as in Exercises 6.21, 6.22, and 6.25 (pages 410–411), but computer simulations of random behavior allow faster exploration. As you explore randomness, remember:

- You must have a long series of *independent* trials. That is, the outcome of one trial must not influence the outcome of any other. Imagine a crooked gambling house where the operator of a roulette wheel can stop it where she chooses—she can prevent the proportion of "red" from settling down to a fixed number. These trials are not independent.
- The idea of probability is empirical. Computer simulations start with given probabilities and imitate random behavior, but we can estimate a real-world probability only by actually observing many trials.
- Nonetheless, computer simulations are very useful because we need to see the results of many trials. In situations such as coin tossing, the proportion of an outcome often requires several hundred trials to settle down to the probability of that outcome. The kinds of physical random devices suggested in the exercises are too slow for this. Conducting only a few trials will give only a rough estimate of a probability.

The Uses of Probability

Probability theory originated in the study of games of chance. Tossing dice, dealing shuffled cards, and spinning a roulette wheel are examples of deliberate randomization that are similar to random sampling. Although games of chance are ancient, they were not studied by mathematicians until the sixteenth and seventeenth centuries. It is only a mild simplification to say that probability as a branch of mathematics arose when seventeenth-century French gamblers asked the mathematicians Blaise Pascal and Pierre de Fermat for help. Gambling is still with us, in casinos and state lotteries. We will make use of games of chance as simple examples that illustrate the principles of probability.

Really random digits For purists, the RAND Corporation long ago published a book titled *One Million Random Digits*. The book lists 1,000,000 digits that were produced by a very elaborate physical randomization and really are random. An employee of RAND once commented that this is not the most boring book that RAND has ever published.

"MR. WILSON SAYS MOST ACCIDENTS HAPPEN WITHIN FIVE MILES OF YOUR HOME. MAYBE WE SHOULD MOVE."

Careful measurements in astronomy and surveying led to further advances in probability in the eighteenth and nineteenth centuries because the results of repeated measurements are random and can be described by distributions much like those arising from random sampling. Similar distributions appear in data on human life span (mortality tables) and in data on lengths or weights in a population of skulls, leaves, or cockroaches.[1] In the twenty-first century, we employ the mathematics of probability to describe the flow of traffic through a highway system, a telephone interchange, or a computer processor; the genetic makeup of individuals or populations; the energy states of subatomic particles; the spread of epidemics or rumors; and the rate of return on risky investments. Although we are interested in probability because of its usefulness in statistics, the mathematics of chance is important in many fields of study.

Exercises

6.21 Pennies spinning Hold a penny upright on its edge under your forefinger on a hard surface, then snap it with your other forefinger so that it spins for some time before falling. Based on 50 spins, estimate the probability of heads.

6.22 A game of chance Ginny and Fred play a game of "Heads or Tails." They toss a coin four times. Ginny wins a dollar from Fred for each head and pays Fred a dollar for each tail—that is, she wins or loses the difference between the number of heads and the number of tails. For example, if a game results in one head and three tails, Ginny loses \$2. You can check that Ginny's possible outcomes are

$$\{-4, -2, 0, 2, 4\}$$

Assign probabilities to these outcomes by playing the game 20 times and using the proportions of the outcomes as estimates of the probabilities. If possible, combine your trials with those of other students to obtain long-run proportions that are closer to the probabilities.

6.23 Random digits The table of random digits (Table B) was produced by a random mechanism that gives each digit probability 0.1 of being a 0. What proportion of the first 200 digits in the table are 0s? This proportion is an estimate, based on 200 repetitions, of the true probability, which in this case is known to be 0.1.

6.24 Matching probabilities Probability is a measure of how likely an event is to occur. Match each statement about an event with one of the probabilities that follow. (The probability is usually a much more exact measure of likelihood than is the verbal statement.)

$$0, 0.01, 0.3, 0.6, 0.99, 1$$

(a) This event is impossible. It can never occur.

(b) This event is certain. It will occur on every trial of the random phenomenon.

(c) This event is very unlikely, but it will occur once in a while in a long sequence of trials.

(d) This event will occur more often than not.

6.25 Colors of M&M's It is reasonable to think that packages of M&M's Milk Chocolate Candies are filled at the factory with candies chosen at random from the very large number produced. So a package of M&M's contains a number of repetitions of a random phenomenon: choosing a candy at random and noting its color. What is the probability that an M&M's Milk Chocolate Candy is blue? To find out, buy one or more packs. How many candies did you examine? How many were blue? What is your estimate of the probability that a randomly chosen candy is blue?

6.26 How many tosses to get a head? When we toss a penny, experience shows that the probability (long-term proportion) of a head is close to 1/2. Suppose now that we toss the penny repeatedly until we get a head. What is the probability that the first head comes up in an odd number of tosses (1, 3, 5, and so on)? To find out, repeat this procedure 50 times, and keep a record of the number of tosses needed to get a head on each of your 50 trials.

(a) From your experiment, estimate the probability of a head on the first toss. What value should we expect this probability to have?

(b) Use your results to estimate the probability that the first head appears on an odd-numbered toss.

6.27 Winning a baseball game A study of the home field advantage in baseball found that over the period from 1969 to 1989 the league champions won 63% of their home games.[2] The two league champions meet in the World Series. Would you use the study results to assign probability 0.63 to the event that the home team wins in a World Series game? Explain your answer.

6.28 Probability and poker hands Many Internet sites give the probabilities of being dealt various five-card poker hands. For example, the probability of being dealt two pairs is approximately 1/21. Explain in simple language what "probability 1/21" means. Also explain why it does *not* mean that in 21 deals you will get exactly one two-pair hand.

Probability Models

Earlier chapters gave mathematical models for linear relationships (in the form of the equation of a line) and for some distributions of data (in the form of Normal density curves). Now we must give a mathematical description or model for randomness. To see how to proceed, think first about a very simple random phenomenon, tossing a coin once. When we toss a coin, we cannot know the outcome in advance. What do we know? We are willing to say that the outcome will be either heads or tails. We believe that each of these outcomes has probability 1/2. This description of coin tossing has two parts:

- A list of possible outcomes.
- A probability for each outcome.

Such a description is the basis for all probability models. Here is the basic vocabulary we use.

Probability Models

The **sample space S** of a random phenomenon is the set of all possible outcomes.

An **event** is any outcome or a set of outcomes of a random phenomenon. That is, an event is a subset of the sample space.

A **probability model** is a mathematical description of a random phenomenon consisting of two parts: a sample space S and a way of assigning probabilities to events.

To specify S, we must state what constitutes an individual outcome and then state which outcomes can occur. The sample space S can be very simple or very complex. When we toss a coin once, there are only two outcomes, heads and tails. The sample space is S = {H, T}. If we draw a random sample of 50,000 U.S. households, as the Current Population Survey does, the sample space contains all possible choices of 50,000 of the 113 million households in the country. This S is extremely large. Each member of S is a possible sample, which explains the term *sample space.*

Example 6.10 *Random phenomena*
Sample space

(a) Rolling two dice is a common way to lose money in casinos. There are 36 possible outcomes when we roll two dice and record the up-faces in order (first die, second die). Figure 6.1 displays these outcomes. They make up the sample space S.

Figure 6.1 *The 36 possible outcomes in rolling two dice, for Example 6.10.*

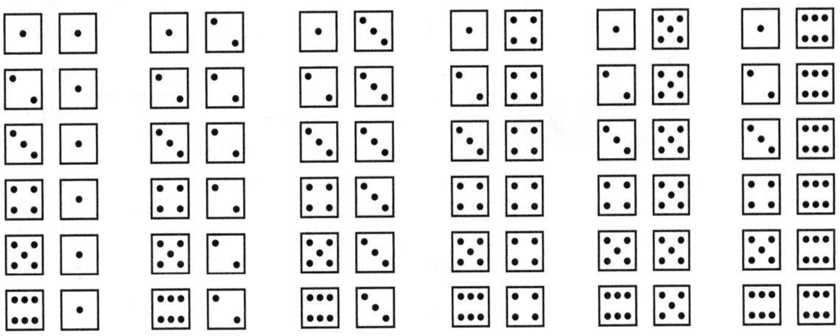

"Roll a 5" is an event, call it A, that contains four of these 36 outcomes:

A = {  }

(b) Gamblers care only about the number of pips on the up-faces of the dice. The sample space for rolling two dice and counting the pips is

$$S = \{2, 3, 4, 5, 6, 7, 8, 9, 10, 11, 12\}$$

Comparing this S with Figure 6.1 reminds us that we can change S by changing the detailed description of the random phenomenon we are describing.

(c) Let your pencil point fall blindly into Table B of random digits; record the value of the digit it lands on. The possible outcomes are

$$S = \{0, 1, 2, 3, 4, 5, 6, 7, 8, 9\}$$

(d) Toss a coin four times and record the results. That's a bit vague. To be exact, record the results of each of the four tosses in order. A typical outcome is then HTTH. Counting shows that there are 16 possible outcomes. The sample space S is the set of all 16 strings of four H's and T's.

Suppose that our only interest is the number of heads in four tosses. Now we can be exact in a simpler fashion. The random phenomenon is to toss a coin four times and count the number of heads. The sample space contains only five outcomes:

$$S = \{0, 1, 2, 3, 4\}$$

This example also illustrates the importance of carefully specifying what constitutes an individual outcome.

(e) Flip a coin and roll a die. Possible outcomes are a head (H) followed by any of the digits 1 to 6, or a tail (T) followed by any of the digits 1 to 6. The sample space contains 12 outcomes:

$$S = \{H1, H2, H3, H4, H5, H6, T1, T2, T3, T4, T5, T6\}$$

tree diagram

Being able to properly enumerate the outcomes in a sample space will be critical to determining probabilities. Three techniques are very helpful in making sure you don't accidentally overlook any outcomes. The first is called a ***tree diagram*** because it resembles the branches of a tree. The first action in Example 6.10(e) is to toss a coin. To construct the tree diagram, begin with a point and draw a line from the point to H and a second line from the point to T. The second action is to roll a die; there are six possible faces that can come up on the die. So draw a line from each of H and T to these six outcomes. Notice that an outcome is one of the 12 paths through the tree. See Figure 6.2.

Figure 6.2 *Tree diagram.*

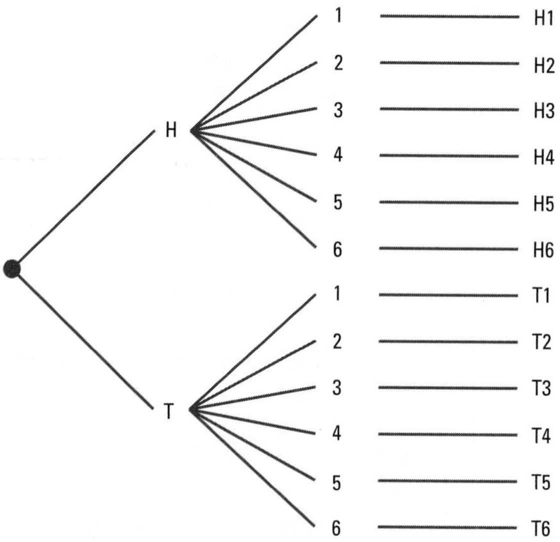

The second technique is to make use of the following rule.

Multiplication Principle

If you can do one task in n_1 number of ways and a second task in n_2 number of ways, then both tasks can be done in $n_1 \times n_2$ number of ways.

In Example 6.10(d), there are 2 possible results for each coin toss. Applying the multiplication principle, for four coin tosses there are $2 \times 2 \times 2 \times 2 = 16$ possible outcomes in the sample space. To see why this is true, just sketch a tree diagram.

The third technique is to make an organized list of all the possible outcomes. The sample space in Example 6.10(d) is organized easily.

Example 6.11 *Flip four coins*

Sample space as an organized list

Listing all 16 outcomes when you flip a coin four times in succession requires a scheme or systematic method so that you don't leave out any possibilities. One technique is to list all the ways you can obtain 0 heads, then list all the ways you can get 1 head, 2 heads, 3 heads, and finally all 4 heads. Here is the list:

0 heads	1 head	2 heads	3 heads	4 heads
TTTT	HTTT	HHTT	HHHT	HHHH
	THTT	HTHT	HHTH	
	TTHT	HTTH	HTHH	
	TTTH	THHT	THHH	
		THTH		
		TTHH		

Although these examples may seem remote from the practice of statistics, the connection is surprisingly close. Suppose that in the course of conducting an opinion poll you select four people at random from a large population and ask each if he or she favors reducing federal spending on low-interest student loans. The possible outcomes—the sample space—are the answers "Yes" or "No." Similarly, the possible outcomes of an SRS of 1500 people are the same in principle as the possible outcomes of tossing a coin 1500 times. One of the great advantages of mathematics is that the essential features of quite different phenomena can be described by the same mathematical model.

Of course, some sample spaces are simply too large to allow all of the possible outcomes to be listed, as the next example shows.

Example 6.12 ***Generate a random decimal number***

Nondiscrete sample space

Many computing systems have a function that will generate a random number between 0 and 1. The sample space is

$$S = \{\text{all numbers between 0 and 1}\}$$

This S is a mathematical idealization. Any specific random number generator produces numbers with some limited number of decimal places so that, strictly speaking, not all numbers between 0 and 1 are possible outcomes.

If you are selecting objects from a collection of distinct choices, such as drawing playing cards from a standard deck of 52 cards, then much depends on whether each choice is exactly like the previous choice. If you are selecting random digits by drawing numbered slips of paper from a hat, and you want all 10 digits to be equally likely to be selected each draw, then after you draw a digit and record it, you must put it back into the hat. Then the second draw will be exactly like the first. This is referred to as ***sampling with replacement.*** If you do not replace the slips you draw, however, there are only nine choices for the second slip picked, and eight for the third. This is called ***sampling without replacement.*** So if the question is "How many three-digit numbers can you make?" the answer is, by the multiplication principle, $10 \times 10 \times 10 = 1000$, if you are sampling with replacement. On the other hand, there are $10 \times 9 \times 8 = 720$ different ways

sampling with replacement

sampling without replacement

to construct a three-digit number if you are sampling without replacement. You should be able to determine from the context of the problem whether the selection is with or without replacement, and this will help you properly identify the sample space.

Exercises

6.29 Describe the sample space In each of the following situations, describe a sample space S for the random phenomenon. In some cases, you have some freedom in your choice of S.

(a) A seed is planted in the ground. It either germinates or fails to grow.

(b) A patient with a usually fatal form of cancer is given a new treatment. The response variable is the length of time that the patient lives after treatment.

(c) A student enrolls in a statistics course and at the end of the semester receives a letter grade.

(d) A basketball player shoots four free throws. You record the sequence of hits and misses.

(e) A basketball player shoots four free throws. You record the number of baskets she makes.

6.30 Describe the sample space In each of the following situations, describe a sample space S for the random phenomenon. In some cases you have some freedom in specifying S, especially in setting the largest and the smallest value in S.

(a) Choose a student in your class at random. Ask how much time that student spent studying during the past 24 hours.

(b) The Physicians' Health Study asked 11,000 physicians to take an aspirin every other day and observed how many of them had a heart attack in a five-year period.

(c) In a test of a new package design, you drop a carton of a dozen eggs from a height of 1 foot and count the number of broken eggs.

(d) Choose a student in your class at random. Ask how much cash that student is carrying.

(e) A nutrition researcher feeds a new diet to a young male white rat. The response variable is the weight (in grams) that the rat gains in 8 weeks.

6.31 Calories in hot dogs Give a reasonable sample space for the number of calories in a hot dog. (Table 1.9 on page 98 contains some typical values to guide you.)

6.32 Listing outcomes, I For each of the following, use a tree diagram or the multiplication principle to determine the number of outcomes in the sample space. Then write the sample space using set notation.

(a) Toss 2 coins.

(b) Toss 3 coins.

(c) Toss 5 coins.

6.33 Listing outcomes, II For each of the following, use a tree diagram or the multiplication principle to determine the number of outcomes in the sample space.

(a) Suppose a county license tag has a four-digit number for identification. If any digit can occupy any of the four positions, how many county license tags can you have?

(b) If the county license tags described in (a) do not allow duplicate digits, how many county license tags can you have?

(c) Suppose the county license tags described in (a) can have up to four digits. How many county license tags will this scheme allow?

6.34 Spin-123 Refer to Activity 6C (page 406).

(a) Determine the number of outcomes in the sample space.

(b) List the outcomes in the sample space.

6.35 Rolling two dice Figure 6.1 (page 413) showed the 36 outcomes when we roll two dice. Another way to summarize these results is to make a table like this:

Number of ways	Sum	Outcomes
1	2	1,1
2	3	1,2 2,1
...	...	...

(a) Complete the table.

(b) In how many ways can you get an even sum?

(c) In how many ways can you get a sum of 5? A sum of 8?

(d) Describe any patterns that you see in the table.

6.36 Pick a card Suppose you select a card from a standard deck of 52 playing cards. In how many ways can the selected card be

(a) a red card?

(b) a heart?

(c) a queen and a heart?

(d) a queen or a heart?

(e) a queen that is not a heart?

Probability Rules

The true probability of any outcome—say, "roll a 5 when we toss two dice"—can be found only by actually tossing two dice many times, and then only approximately. How then can we describe probability mathematically? Rather than try to give "correct" probabilities, we start by laying down facts that must be true for any assignment of probabilities. These facts follow from the idea of probability as "the long-run proportion of repetitions on which an event occurs."

1. **Any probability is a number between 0 and 1.** Any proportion is a number between 0 and 1, so any probability is also a number between 0 and 1. An event with probability 0 never occurs, and an event with probability 1 occurs on every trial. An event with probability 0.5 occurs in half the trials in the long run.
2. **The sum of the probabilities of all possible outcomes must equal 1.** Note that some outcome must occur on every trial.
3. **If two events have no outcomes in common, the probability that one or the other occurs is the sum of their individual probabilities.** If one event occurs in 40% of all trials, a different event occurs in 25% of all trials, and the two can never occur together, then one or the other occurs on 65% of all trials because 40% + 25% = 65%.
4. **The probability that an event does not occur is 1 minus the probability that the event does occur.** If an event occurs in 70% of all trials, it fails to occur in the other 30%. The probability that an event occurs and the probability that it does not occur always add to 100%, or 1.

We can use mathematical notation to state Facts 1 to 4 more concisely. Capital letters near the beginning of the alphabet denote events. If A is any event, we write its probability as $P(A)$. Here are our probability facts in formal language. As you apply these rules, remember that they are just another form of intuitively true facts about long-run proportions of trials.

Probability Rules

Rule 1. The probability $P(A)$ of any event A satisfies $0 \le P(A) \le 1$.

Rule 2. If S is the sample space in a probability model, then $P(S) = 1$.

Rule 3. Two events A and B are **disjoint** (also called "mutually exclusive") if they have no outcomes in common and so can never occur simultaneously. If A and B are disjoint,

$$P(A \text{ or } B) = P(A) + P(B)$$

This is the **addition rule** for disjoint events.

Rule 4. The **complement** of any event A is the event that A does not occur, written as A^c. The **complement rule** states that

$$P(A^c) = 1 - P(A)$$

union Sometimes we use set notation to describe events. The event $A \cup B$, read "A ***union*** B," is the set of all outcomes that are either in A or in B. So $A \cup B$ is just another way to indicate the event "A or B." We will use "and" and "$\cap$," and "or" and "$\cup$" interchangeably. The symbol $\varnothing$ is used for the ***empty event,*** that is, the event that has no outcomes in it. If two events A and B are disjoint (mutually exclusive), we can write $A \cap B = \varnothing$, read "A ***intersect*** B is empty."

empty event

intersect

You may find it helpful to draw a picture to remind yourself of the meaning of complements and disjoint events. A picture like Figure 6.3 that shows the sample space S as a rectangular area and events as areas within S is called a ***Venn diagram.*** The events A and B in Figure 6.3 are disjoint because they do not overlap; that is, they have no outcomes in common. Their intersection is the empty event, ∅. Their union consists of the two shaded regions.

Venn diagram

Figure 6.3 *Venn diagram showing disjoint (mutually exclusive) events* A *and* B.

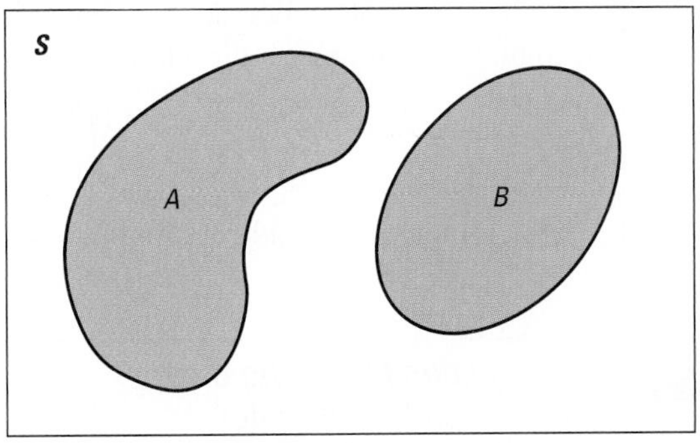

The complement A^c in Figure 6.4 contains exactly the outcomes that are not in A. Note that we could write $A \cup A^c = S$ and $A \cap A^c = \varnothing$.

Figure 6.4 *Venn diagram showing the complement* A^c *of an event* A.

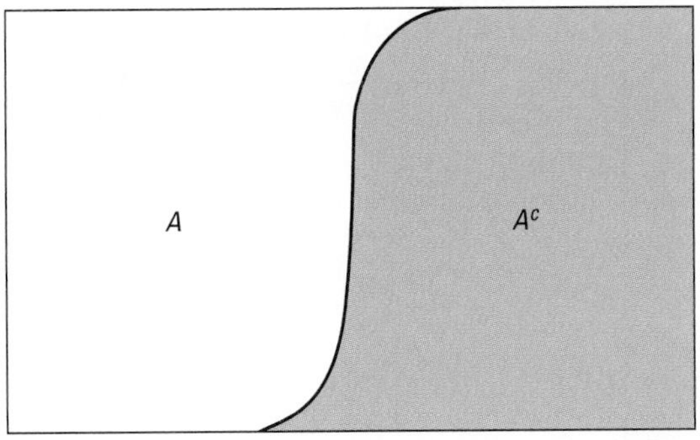

Example 6.13 ***Distance learning***

Complement rule

Distance learning courses are rapidly gaining popularity among college students. Choose at random an undergraduate taking a distance learning course for credit, and record the student's age. "At random" means that we give every such student the same chance to be the one we choose. This is an SRS of size 1. The probability of any age group is just the proportion of all distance learners in that age group—if we drew many students, this is the proportion we would get. Here is the probability model:[3]

Age group (yr):	18 to 23	24 to 29	30 to 39	40 or over
Probability:	0.57	0.17	0.14	0.12

Each probability is a number between 0 and 1. The probabilities add to 1 because these outcomes together make up the sample space S.

The probability that the student we draw is not in the traditional undergraduate age range of 18 to 23 years is, by the complement rule,

$$\begin{aligned} P(\text{not 18 to 23 years}) &= 1 - P(\text{18 to 23 years}) \\ &= 1 - 0.57 = 0.43 \end{aligned}$$

That is, if 57% of distance learners are 18 to 23 years old, then the remaining 43% are not in this age group.

The events "30 to 39 years" and "40 years or over" are disjoint because no student can be in both age groups. So the addition rule says that

$$\begin{aligned} P(\text{30 years or over}) &= P(\text{30 to 39 years}) + P(\text{40 years or over}) \\ &= 0.14 + 0.12 = 0.26 \end{aligned}$$

That is, 26% of undergraduates in distance learning courses are at least 30 years old.

Example 6.14 ***Probabilities for rolling dice***

Applying probability rules

Figure 6.1 (page 413) displays the 36 possible outcomes of rolling two dice. What probabilities should we assign to these outcomes?

Casino dice are carefully made. Their spots are not hollowed out, which would give the faces different weights, but are filled with white plastic of the same density as the colored plastic of the body. For casino dice it is reasonable to assign the same probability to each of the 36 outcomes in Figure 6.1. Because all 36 outcomes together must have probability 1 (Rule 2), each outcome must have probability 1/36.

Gamblers are often interested in the sum of the pips on the up-faces. What is the probability of rolling a 5? Because the event "roll a 5" contains the four outcomes displayed in Example 6.10(a), the addition rule (Rule 3) says that its probability is

P(roll a 5) = P(⚀ ⚃) + P(⚂ ⚁) + P(⚁ ⚂) + P(⚃ ⚀)

$$= \frac{1}{36} + \frac{1}{36} + \frac{1}{36} + \frac{1}{36}$$

$$= \frac{4}{36} = 0.111$$

What about the probability of rolling a 7? In Figure 6.1 you will find six outcomes for which the sum of the pips is 7. The probability is 6/36, or about 0.167.

Assigning Probabilities: Finite Number of Outcomes

Examples 6.13 and 6.14 illustrate one way to assign probabilities to events: assign a probability to every individual outcome, and then add these probabilities to find the probability of any event. If such an assignment is to satisfy the rules of probability, the probabilities of all the individual outcomes must sum to exactly 1.

Probabilities in a Finite Sample Space

Assign a probability to each individual outcome. These probabilities must be numbers between 0 and 1 and must have sum 1.

The probability of any event is the sum of the probabilities of the outcomes making up the event.

Example 6.15 *Benford's law*

Applying probability rules

Faked numbers in tax returns, payment records, invoices, expense account claims, and many other settings often display patterns that aren't present in legitimate records. Some patterns, like too many round numbers, are obvious and easily avoided by a clever crook. Others are more subtle. It is a striking fact that the first digits of numbers in legitimate records often follow a distribution known as *Benford's law*. Here it is (note that a first digit can't be 0):[4]

First digit:	1	2	3	4	5	6	7	8	9
Probability:	0.301	0.176	0.125	0.097	0.079	0.067	0.058	0.051	0.046

Benford's law usually applies to the first digits of the sizes of similar quantities, such as invoices, expense account claims, and county populations. Investigators can detect fraud by comparing these probabilities with the first digits in records such as invoices paid by a business.

(a) Consider the events

$$A = \{\text{first digit is } 1\}$$
$$B = \{\text{first digit is 6 or greater}\}$$

From the table of probabilities,

$$P(A) = P(1) = 0.301$$
$$P(B) = P(6) + P(7) + P(8) + P(9)$$
$$= 0.067 + 0.058 + 0.051 + 0.046 = 0.222$$

(b) The probability that a first digit is anything other than a 1 is, by the complement rule,

$$P(A^c) = 1 - P(A)$$
$$= 1 - 0.301 = 0.699$$

(c) The events A and B are disjoint, so the probability that a first digit either is 1 or is 6 or greater is, by the addition rule,

$$P(A \text{ or } B) = P(A) + P(B)$$
$$= 0.301 + 0.222 = 0.523$$

CAUTION

Be careful to apply the addition rule only to disjoint events.

(d) Check that the probability of the event C that a first digit is odd is

$$P(C) = P(1) + P(3) + P(5) + P(7) + P(9) = 0.609$$

(e) The probability

$$P(B \text{ or } C) = P(1) + P(3) + P(5) + P(6) + P(7) + P(8) + P(9) = 0.727$$

is *not* the sum of $P(B)$ and $P(C)$, because events B and C are not disjoint. Outcomes 7 and 9 are common to both events.

Assigning Probabilities: Equally Likely Outcomes

Assigning correct probabilities to individual outcomes often requires long observation of the random phenomenon. In some special circumstances, however, we are willing to assume that individual outcomes are equally likely because of some balance in the phenomenon. Ordinary coins have a physical balance that should make heads and tails equally likely, for example, and the table of random digits comes from a deliberate randomization.

Example 6.16 *Random digits*

Applying probability rules

You might think that first digits are distributed "at random" among the digits 1 to 9. The 9 possible outcomes would then be equally likely. The sample space for a single digit is

$$S = \{1, 2, 3, 4, 5, 6, 7, 8, 9\}$$

Because the total probability must be 1, the probability of each of the 9 outcomes must be 1/9. That is, the assignment of probabilities to outcomes is

First digit:	1	2	3	4	5	6	7	8	9
Probability:	1/9	1/9	1/9	1/9	1/9	1/9	1/9	1/9	1/9

The probability of the event B that a randomly chosen first digit is 6 or greater is

$$P(B) = P(6) + P(7) + P(8) + P(9)$$
$$= \frac{1}{9} + \frac{1}{9} + \frac{1}{9} + \frac{1}{9} = \frac{4}{9} = 0.444$$

Compare this with the Benford's law probability in Example 6.15. A crook who fakes data by using "random" digits will end up with too many first digits 6 or greater and too few 1s and 2s.

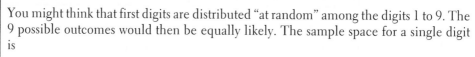

Equally likely? A game of bridge begins by dealing all 52 cards in the deck to the four players, 13 to each. If the deck is well shuffled, all of the immense number of possible hands will be equally likely. But don't expect the hands that appear in newspaper bridge columns to reflect the equally likely probability model. Writers on bridge choose "interesting" hands—especially those that lead to high bids that are rare in actual play.

In Example 6.16 all outcomes have the same probability. Because there are 9 equally likely outcomes, each must have probability 1/9. Because exactly 4 of the 9 equally likely outcomes are 6 or greater, the probability of this event is 4/9. In the special situation where all outcomes are equally likely, we have a simple rule for assigning probabilities to events.

Equally Likely Outcomes

If a random phenomenon has k possible outcomes that are all equally likely, then each individual outcome has probability $1/k$. The probability of any event A is

$$P(A) = \frac{\text{count of outcomes in A}}{\text{count of outcomes in S}} = \frac{\text{count of outcomes in A}}{k}$$

Most random phenomena do not have equally likely outcomes, so the general rule for finite sample spaces is more important than the special rule for equally likely outcomes.

Exercises

6.37 Blood types All human blood can be typed as one of O, A, B, or AB, but the distribution of the types varies a bit with race. Here is the distribution of the blood type of a randomly chosen black American:

Blood type:	O	A	B	AB
Probability:	0.49	0.27	0.20	?

(a) What is the probability of type AB blood? Why?

(b) Maria has type B blood. She can safely receive blood transfusions from people with blood types O and B. What is the probability that a randomly chosen black American can donate blood to Maria?

6.38 Distribution of M&M colors If you draw an M&M candy at random from a bag of the candies, the candy you draw will have one of six colors. The probability of drawing each color depends on the proportion of each color among all candies made.

(a) The table below gives the probability of each color for a randomly chosen milk chocolate M&M:

Color:	Brown	Red	Yellow	Green	Orange	Blue
Probability:	0.13	0.13	0.14	0.16	0.20	?

What must be the probability of drawing a blue candy?

(b) The probabilities for peanut M&Ms are a bit different. Here they are:

Color:	Brown	Red	Yellow	Green	Orange	Blue
Probability:	0.12	0.12	0.15	0.15	0.23	?

What is the probability that a peanut M&M chosen at random is blue?

(c) What is the probability that a milk chocolate M&M is any of red, yellow, or orange? What is the probability that a peanut M&M has one of these colors?

6.39 Heart disease and cancer Government data assign a single cause for each death that occurs in the United States. The data show that the probability is 0.45 that a randomly chosen death was due to cardiovascular (mainly heart) disease, and 0.22 that it was due to cancer. What is the probability that a death was due either to cardiovascular disease or to cancer? What is the probability that the death was due to some other cause?

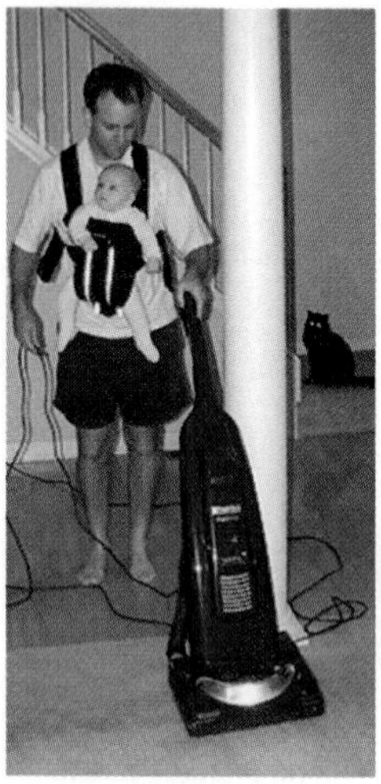

6.40 Do husbands do their share? The *New York Times* (August 21, 1989) reported a poll that interviewed a random sample of 1025 women. The married women in the sample were asked whether their husbands did their fair share of household chores. Here are the results:

Outcome	Probability
Does more than his fair share	0.12
Does his fair share	0.61
Does less than his fair share	?

These proportions are probabilities for the random phenomenon of choosing a married woman at random and asking her opinion.

(a) What must be the probability that the woman chosen says that her husband does less than his fair share? Why?

(b) The event "I think my husband does at least his fair share" contains the first two outcomes. What is its probability?

6.41 Academic rank Select a first-year college student at random and ask what his or her academic rank was in high school. Here are the probabilities, based on proportions from a large sample survey of first-year students:

Rank:	Top 20%	Second 20%	Third 20%	Fourth 20%	Lowest 20%
Probability:	0.41	0.23	0.29	0.06	0.01

(a) What is the sum of these probabilities? Why do you expect the sum to have this value?

(b) What is the probability that a randomly chosen first-year college student was not in the top 20% of his or her high school class?

(c) What is the probability that a first-year student was in the top 40% in high school?

6.42 Spin 123 again Refer to the experiment described in Activity 6C (page 406) and Exercise 6.34 (page 417).

(a) Determine the theoretical probability that at least one digit will occur in its correct place.

(b) Compare the theoretical probability with your experimental (empirical) results.

6.43 Tetrahedral dice Psychologists sometimes use tetrahedral dice to study our intuition about chance behavior. A regular tetrahedron is a pyramid (think of Egypt) with four identical faces, each a triangle with all sides equal in length. Label the four faces of a tetrahedral die with 1, 2, 3, and 4 spots.

(a) Give a probability model for rolling such a die and recording the number of spots on the down-face. Explain why you think your model is at least close to correct.

(b) Give a probability model for rolling two such dice. That is, write down all possible outcomes and give a probability to each. What is the probability that the sum of the down-faces is 5?

6.44 Benford's law Example 6.15 (page 421) states that the first digits of numbers in legitimate records often follow a distribution known as Benford's law. Here is the distribution:

First digit:	1	2	3	4	5	6	7	8	9
Probability:	0.301	0.176	0.125	0.097	0.079	0.067	0.058	0.051	0.046

It was shown in Example 6.15 that

$$P(A) = P(\text{first digit is } 1) = 0.301$$
$$P(B) = P(\text{first digit is 6 or greater}) = 0.222$$
$$P(C) = P(\text{first digit is odd}) = 0.609$$

We will define event D to be {first digit is less than 4}. Using the union and intersection notation, find the following probabilities.

(a) $P(D)$

(b) $P(B \cup D)$

(c) $P(D^c)$

(d) $P(C \cap D)$

(e) $P(B \cap C)$

Independence and the Multiplication Rule

Rule 3, the addition rule for disjoint events, describes the probability that *one or the other* of two events A and B will occur in the special situation when A and B cannot occur together because they are disjoint. Now we will describe the probability that *both* events A and B occur, again only in a special situation. More general rules appear in Section 6.3.

Suppose that you toss a balanced coin twice. You are counting heads, so two events of interest are

$$A = \text{first toss is a head}$$
$$B = \text{second toss is a head}$$

The events A and B are not disjoint. They occur together whenever both tosses give heads. We want to compute the probability of the event "A and B" that both tosses are heads. The Venn diagram in Figure 6.5 illustrates the event "A and B" as the overlapping area that is common to both A and B.

Figure 6.5 *Venn diagram showing the event* {A *and* B}.

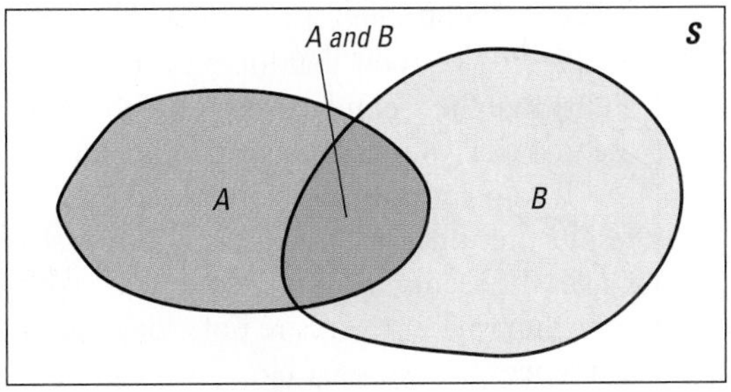

The coin tossing of Buffon, Pearson, and Kerrich described in Example 6.9 makes us willing to assign probability 1/2 to a head when we toss a coin. So

$$P(A) = 0.5$$
$$P(B) = 0.5$$

What is $P(A \text{ and } B)$? Our common sense says that it is 1/4. The first toss will give a head half the time and then the second will give a head on half of those trials, so both tosses will give heads on $1/2 \times 1/2 = 1/4$ of all trials in the long run. This reasoning assumes that the second toss still has probability 1/2 of a head after the first has given a head. This is true—we can verify it by performing many trials of two tosses and observing the proportion of heads on the second toss after the first toss has produced a head. We say that the events "head on the first toss" and "head on the second toss" are ***independent.*** Here is our next probability rule.

The Multiplication Rule for Independent Events

Rule 5. Two events A and B are **independent** if knowing that one occurs does not change the probability that the other occurs. If A and B are independent,

$$P(A \text{ and } B) = P(A)P(B)$$

This is the **multiplication rule** for independent events.

A more precise definition of independence appears in Section 6.3. In practice, independence is usually *assumed* as part of a probability model when we want to describe random phenomena that seem to be physically unrelated to each other.

Example 6.17 *Independent or not independent?*
Applying the definition

Because a coin has no memory and most coin tossers cannot influence the fall of the coin, it is safe to assume that successive coin tosses are independent. For a balanced coin this means that after we see the outcome of the first toss, we still assign probability 1/2 to heads on the second toss.

On the other hand, the colors of successive cards dealt from the same deck (sampling without replacement) are not independent. A standard 52-card deck contains 26 red and 26 black cards. For the first card dealt from a shuffled deck, the probability of a red card is $26/52 = 0.50$ because the 52 possible cards are equally likely. Once we see that the first card is red, we know that there are only 25 reds among the remaining 51 cards. The probability that the second card is red is therefore only $25/51 = 0.49$. Knowing the outcome of the first deal changes the probability for the second.

If a doctor measures your blood pressure twice, it is reasonable to assume that the two results are independent because the first result does not influence the instrument that makes the second reading. But if you take an IQ test or other mental test twice in succession, the two test scores are not independent. The learning that occurs on the first attempt influences how you score on your second attempt.

When independence is part of a probability model, the multiplication rule applies. Here is an example.

Example 6.18 ***Mendel's peas***
Applying the definition of independence

Gregor Mendel used garden peas in some of the experiments that revealed that inheritance operates randomly. The seed color of Mendel's peas can be either green or yellow. Two parent plants are "crossed" (one pollinates the other) to produce seeds. Each parent plant carries two genes for seed color, and each of these genes has probability 1/2 of being passed to a seed. The two genes that the seed receives, one from each parent, determine its color. The parents contribute their genes independently of each other.

Suppose that both parents carry the G and the Y genes. The seed will be green if both parents contribute a G gene; otherwise it will be yellow. If M is the event that the male contributes a G gene and F is the event that the female contributes a G gene, then the probability of a green seed is

$$\begin{aligned} P(M \text{ and } F) &= P(M)P(F) \\ &= (0.5)(0.5) = 0.25 \end{aligned}$$

In the long run, 1/4 of all seeds produced by crossing these plants will be green.

CAUTION

The multiplication rule applies only to independent events; you cannot use it if events are not independent. Here is a distressing example of misuse of the multiplication rule.

Example 6.19 ***Sudden infant death***
Multiplication rule

Sudden infant death syndrome (SIDS) causes babies to die suddenly (often in their cribs), but as yet no cause is known. Deaths from SIDS have been greatly reduced by placing babies on their backs.

When more than one SIDS death occurs in a family, the parents are sometimes accused. One "expert witness" popular with prosecutors in England told juries that there is only a 1 in 72 million chance that two children in the same family could have died naturally. Here is his calculation: The rate of SIDS in a nonsmoking middle-class family is 1 in 8500. So the probability of two deaths is

$$\frac{1}{8500} \times \frac{1}{8500} = \frac{1}{72{,}250{,}000}$$

Several women were convicted of murder on this basis, without any direct evidence that they harmed their children.

As the Royal Statistical Society said, this reasoning is nonsense. It assumes that SIDS deaths in the same family are independent events. The cause of SIDS is unknown: "There may well be unknown genetic or environmental factors that predispose families to SIDS, so that a second case within the family becomes more likely."[5] The British government decided to review the cases of 258 parents convicted of murdering their babies.

The multiplication rule $P(A \text{ and } B) = P(A) \times P(B)$ holds if A and B are independent but not otherwise. The addition rule $P(A \text{ or } B) = P(A) + P(B)$ holds if A and B are disjoint but not otherwise. Resist the temptation to use these simple formulas when the circumstances that justify them are not present. You must also be certain not to confuse disjointness and independence. **Disjoint events cannot be independent.** If A and B are disjoint, then the fact that A occurs tells us that B cannot occur—look again at Figure 6.4 (page 419). Unlike disjointness of complements, independence cannot be pictured by a Venn diagram, because it involves the probabilities of the events rather than just the outcomes that make up the events.

More Applications of the Probability Rules

If two events A and B are independent, then their complements A^c and B^c are also independent, and A^c is independent of B. Suppose, for example, that 75% of all registered voters in a suburban district are Republicans. If an opinion poll interviews two voters chosen independently, the probability that the first is a Republican and the second is not a Republican is $(0.75)(0.25) = 0.1875$. The multiplication rule also extends to collections of more than two events, provided that all are independent. Independence of events A, B, and C means that no information about any one or any two can change the probability of the remaining events. The formal definition is a bit messy. Fortunately, independence is usually assumed in setting up a probability model. We can then use the multiplication rule freely, as in this example.

Example 6.20 *Six degrees of separation*
Using independence

The "six degrees of separation" idea says that any two people on earth are linked by a chain of friends of friends of friends no more than six steps long. Does this work online? That is, if we give you the name and location of someone you don't know, can you get a message to that person by sending an email to an acquaintance who you think is closer to your target and asking him or her to pass on the message?

As you might guess, success depends on how willing the people in the chain are to take the next step. Here's a model based on a large study: after the first step (you send the message to someone you know), the probability that the message is passed on is 0.37 and steps are independent.[6] Take A_i to be the event that the *i*th person after you in the chain passes on the message.

The probability that the next two links pass on the message is

$$\begin{aligned} P(A_1 \text{ and } A_2) &= P(A_1)P(A_2) \\ &= 0.37 \times 0.37 = 0.1369 \end{aligned}$$

If there really are six degrees of separation between you and a specific person you don't know, you need at most five people after you to pass on the message. The probability that five people will all pass on the message is

$$P(A_1 \text{ and } A_2 \text{ and } A_3 \text{ and } A_4 \text{ and } A_5) = P(A_1)P(A_2)P(A_3)P(A_4)P(A_5)$$
$$= 0.37 \times 0.37 \times 0.37 \times 0.37 \times 0.37$$
$$= 0.37^5 = 0.0069$$

Even if five people in a row keep your message moving, the fifth person may not succeed in reaching your target. In the study, only 384 of 24,163 six-person chains were successful. It's very unlikely that you can pass a message by way of friends of friends of friends successfully.

By combining the rules we have learned, we can compute probabilities for rather complex events. Here is an example.

Example 6.21 *AIDS testing*

Independence and the complement rule

Many people who come to clinics to be tested for HIV, the virus that causes AIDS, don't come back to learn the test results. Clinics now use "rapid HIV tests" that give a result in a few minutes. Applied to people who have no HIV antibodies, one rapid test has probability about 0.004 of producing a false-positive (that is, of falsely indicating that antibodies are present).[7] If a clinic tests 200 people who are free of HIV antibodies, what is the probability that at least one false-positive will occur?

It is reasonable to assume as part of the probability model that the test results for different individuals are independent. The probability that the test is positive for a single person is 0.004, so the probability of a negative result is $1 - 0.004 = 0.996$ by the complement rule. The probability of at least one false-positive among the 200 people tested is therefore

$$P(\text{at least one positive}) = 1 - P(\text{no positives})$$
$$= 1 - P(200 \text{ negatives})$$
$$= 1 - 0.996^{200}$$
$$= 1 - 0.4486 = 0.5514$$

The probability is greater than 1/2 that at least 1 of the 200 people will test positive for HIV even though no one has the virus.

Exercises

6.45 A battle plan A general can plan a campaign to fight one major battle or three small battles. He believes that he has probability 0.6 of winning the large battle and probability 0.8 of winning each of the small battles. Victories or defeats in the small battles are independent. The general must win either the large battle or all three small battles to win the campaign. Which strategy should he choose?

6.46 Defective chips An automobile manufacturer buys computer chips from a supplier. The supplier sends a shipment containing 5% defective chips. Each chip chosen from this shipment has probability 0.05 of being defective, and each automobile uses 12 chips selected independently. What is the probability that all 12 chips in a car will work properly?

6.47 College-educated laborers? Government data show that 26% of the civilian labor force have at least 4 years of college and that 16% of the labor force work as laborers or operators of machines or vehicles. Can you conclude that because $(0.26)(0.16) = 0.0416$, about 4% of the labor force are college-educated laborers or operators? Explain your answer.

6.48 College student demographics Choose at random a college student at least 15 years of age. We are interested in the events

$$A = \{\text{The person chosen is male}\}$$
$$B = \{\text{The person chosen is 25 years old or older}\}$$

Government data recorded in Table 4.5 (page 292) allow us to assign probabilities to these events.

(a) Explain why $P(A) = 0.44$.

(b) Find $P(B)$.

(c) Find the probability that the person chosen is a male at least 25 years old, $P(A \text{ and } B)$. Are the events A and B independent?

6.49 Bright lights? A string of Christmas lights contains 20 lights. The lights are wired in series, so that if any light fails the whole string will go dark. Each light has probability 0.02 of failing during a 3-year period. The lights fail independently of each other. What is the probability that the string of lights will remain bright for 3 years?

6.50 Detecting steroids An athlete suspected of having used steroids is given two tests that operate independently of each other. Test A has probability 0.9 of being positive if steroids have been used. Test B has probability 0.8 of being positive if steroids have been used. What is the probability that neither test is positive if steroids have been used?

6.51 Telephone success Most sample surveys use random digit dialing equipment to call residential telephone numbers at random. The telephone polling firm Zogby International reports that the probability that a call reaches a live person is 0.2.[8] Calls are independent.

(a) A polling firm places 5 calls. What is the probability that none of them reaches a person?

(b) When calls are made to New York City, the probability of reaching a person is only 0.08. What is the probability that none of 5 calls made to New York City reaches a person?

6.52 Pick 3 A state lottery's Pick 3 game asks players to choose a three-digit number, 000 to 999. The state chooses the winning three-digit number at random, so that each number has probability 1/1000. You win if the winning number contains the digits in your number, in any order:

(a) Your number is 123. What is your probability of winning?

(b) Your number is 112. What is your probability of winning?

Section 6.2 Summary

A **random phenomenon** has outcomes that we cannot predict but that nonetheless have a regular distribution in very many repetitions.

The **probability** of an event is the proportion of times the event occurs in many repeated trials of a random phenomenon.

A **probability model** for a random phenomenon consists of a sample space S and an assignment of probabilities.

The **sample space S** is the set of all possible outcomes of the random phenomenon. Sets of outcomes are called **events.** A number $P(A)$ is assigned to an event A as its probability.

A **tree diagram** begins with a point and draws a line to each possible outcome. Additional lines are drawn to each new outcome, and so on. The result looks like a tree with branches. An outcome in a sample space is one of the paths through the tree.

The **multiplication principle** says if you can do one task in n_1 number of ways and a second task in n_2 number of ways, then both tasks can be done in $n_1 \times n_2$ number of ways.

Sampling with replacement requires that objects selected from distinct choices be replaced before the next selection. Probabilities are the same for each draw. In **sampling without replacement,** probabilities change for each new selection.

The **complement A^c** of an event A consists of exactly the outcomes that are not in A. Events A and B are **disjoint** (mutually exclusive) if they have no outcomes in common. Events A and B are **independent** if knowing that one event occurs does not change the probability we would assign to the other event.

$\varnothing$ is the empty event. There are no outcomes in the empty event.

A **Venn diagram** show events as disjoint or intersecting regions.

Any assignment of probability must obey the rules that state the basic properties of probability:

1. $0 \leq P(A) \leq 1$ for any event A.
2. $P(S) = 1$ for the sample space S.
3. **Addition rule:** If events A and B are **disjoint**, then $P(A \text{ or } B) = P(A \cup B) = P(A) + P(B)$.
4. **Complement rule:** For any event A, $P(A^c) = 1 - P(A)$.
5. **Multiplication rule:** If events A and B are **independent,** then $P(A \text{ and } B) = P(A \cap B) = P(A)P(B)$.

Section 6.2 Exercises

6.53 Student survey Choose a student at random from a large statistics class. Give a reasonable sample space S for answers to each of the following questions. (In some cases you may have some freedom in specifying S.)

(a) Are you right-handed or left-handed?

(b) What is your height in centimeters? (One inch is 2.54 centimeters.)

(c) How much money in coins (not bills) are you carrying?

(d) How many minutes did you study last night?

6.54 A study subject's fitness A randomly chosen subject arrives for a study of exercise and fitness. Describe a sample space for each of the following. (In some cases, you may have some freedom in your choice of S.)

(a) The subject is either female or male.

(b) After 10 minutes on an exercise bicycle, you ask the subject to rate his or her effort on the Rate of Perceived Exertion (RPE) scale. The RPE scale ranges in whole-number steps from 6 (no exertion at all) to 20 (maximal exertion).

(c) You measure VO2, the maximum volume of oxygen consumed per minute during exercise. VO2 is generally between 2.5 and 6 liters per minute.

(d) You measure the maximum heart rate (beats per minute).

6.55 Does it satisfy the rules of probability? In each of the following situations, state whether or not the given assignment of probabilities to individual outcomes is legitimate (that is, satisfies the rules of probability). If not, give specific reasons for your answer.

(a) Roll a die and record the count of spots on the up-face: $P(1) = 0$, $P(2) = 1/6$, $P(3) = 1/3$, $P(4) = 1/3$, $P(5) = 1/6$, $P(6) = 0$.

(b) Choose a college student at random and record sex and enrollment status: $P(\text{female full-time}) = 0.56$, $P(\text{female part-time}) = 0.24$, $P(\text{male full-time}) = 0.44$, $P(\text{male part-time}) = 0.17$.

(c) Deal a card from a shuffled deck: $P(\text{clubs}) = 12/52$, $P(\text{diamonds}) = 12/52$, $P(\text{hearts}) = 12/52$, $P(\text{spades}) = 16/52$.

6.56 Disjoint versus independent events This exercise explores the relationship between mutually exclusive and independent events.

(a) Assume that events A and B are non-empty, independent events. Show that A and B must intersect (that is, that $A \cap B \neq \emptyset$).

(b) Use the results of (a) to argue that if A and B are disjoint, then they cannot be independent.

(c) Find an example of two events that are neither disjoint nor independent.

6.57 New Census categories The 2000 Census allowed each person to choose one or more from a long list of races. That is, in the eyes of the Census Bureau, you belong to whatever race or races you say you belong to. "Hispanic/Latino" is a separate category;

Hispanics may be of any race. If we choose a resident of the United States at random, the 2000 Census gives these probabilities:

Race	Hispanic	Not Hispanic
Asian	0.000	0.036
Black	0.003	0.121
White	0.060	0.691
Other	0.062	0.027

Let A be the event that a randomly chosen American is Hispanic, and let *B* be the event that the person chosen is white.

(a) Verify that the table gives a legitimate assignment of probabilities.

(b) What is *P*(A)?

(c) Describe B^c in words and find $P(B^c)$ by the complement rule.

(d) Express "the person chosen is a non-Hispanic white" in terms of events A and *B*. What is the probability of this event?

6.58 Being Hispanic Exercise 6.57 assigns probabilities for the ethnic background of a randomly chosen resident of the United States. Let A be the event that the person chosen is Hispanic, and let *B* be the event that he or she is white. Are events A and *B* independent? How do you know?

6.59 Preparing for the GMAT A company that offers courses to prepare would-be MBA students for the Graduate Management Admission Test (GMAT) finds that 40% of its customers are currently undergraduate students and 60% are college graduates. After completing the course, 50% of the undergraduates and 70% of the graduates achieve scores of at least 600 on the GMAT.

(a) What proportion of customers are undergraduates and score at least 600? What proportion of customers are graduates and score at least 600?

(b) What proportion of all customers score at least 600 on the GMAT?

6.60 Austin and Sara's game Refer to Activity 6A on page 390.

(a) What are Austin's possible choices for his two numbers? Make a vertical list of the pairs of numbers Austin might pick. Here's a start:

1, 1
1, 2

Write Sara's possible choices in a second column. Begin drawing a tree diagram to show the possible outcomes.

(b) How many outcomes are there in the sample space?

(c) In a third column, write the number of ways Austin's numbers add to greater than Sara's number.

(d) What is the probability that Austin wins? Add the numbers in your third column and divide by your answer in (b). How close did your results from your simulation in Activity 6A come to the theoretical probability?

(e) In a fourth column, write the number of ways Sara's number is greater than the sum of Austin's two numbers.

(f) What is the probability that Sara wins? Add the numbers in your fourth column and divide by your answer in (b). How close did your results from your simulation come to the theoretical probability?

(g) What is the probability of a tie? Is it true that $P(\text{Austin wins}) + P(\text{Sara wins}) + P(\text{tie}) = 1$?

6.61 Using a table to find probabilities The type of medical care a patient receives may vary with the age of the patient. A large study of women who had a breast lump investigated whether or not each woman received a mammogram and a biopsy when the lump was discovered. Here are some probabilities estimated by the study. The entries in the table are the probabilities that both of two events occur; for example, 0.321 is the probability that a patient is under 65 years of age and the tests were done. The four probabilities in the table have sum 1 because the table lists all possible outcomes.

	Tests Done?	
Age	**Yes**	**No**
Under 65	0.321	0.124
65 or over	0.365	0.190

(a) What is the probability that a patient in this study is under 65? That a patient is 65 or over?

(b) What is the probability that the tests were done for a patient? That they were not done?

(c) Are the events A = {patient was 65 or older} and B = {the tests were done} independent? Were the tests omitted on older patients more or less frequently than would be the case if testing were independent of age?

6.62 Roulette A roulette wheel has 38 slots, numbered 0, 00, and 1 to 36. The slots 0 and 00 are colored green, 18 of the others are red, and 18 are black. The dealer spins the wheel and at the same time rolls a small ball along the wheel in the opposite direction. The wheel is carefully balanced so that the ball is equally likely to land in any slot when the wheel slows. Gamblers can bet on various combinations of numbers and colors.

"NO! YOU MAY NOT OIL THE WHEEL."

(a) What is the probability that the ball will land in any one slot?

(b) If you bet on "red," you win if the ball lands in a red slot. What is the probability of winning?

(c) The slot numbers are laid out on a board on which gamblers place their bets. One column of numbers on the board contains all multiples of 3, that is, 3, 6, 9, . . . , 36. You place a "column bet" that wins if any of these numbers comes up. What is your probability of winning?

6.63 Which is most likely? A six-sided die has four green and two red faces and is balanced so that each face is equally likely to come up. The die will be rolled several times. You must choose one of the following three sequences of colors; you will win \$25 if the first rolls of the die give the sequence that you have chosen.

RGRRR

RGRRRG

GRRRRR

Which sequence do you choose? Explain your choice.[9]

6.64 Albinism in genetics The gene for albinism in humans is recessive. That is, carriers of this gene have probability 1/2 of passing it to a child, and the child is albino only if both parents pass the albinism gene. Parents pass their genes independently of each other. If both parents carry the albinism gene, what is the probability that their first child is albino? If they have two children (who inherit independently of each other), what is the probability that both are albino? That neither is albino?

6.3 General Probability Rules

In this section we will consider some additional laws that govern any assignment of probabilities. The purpose of learning more laws of probability is to be able to give probability models for more complex random phenomena. We have already met and used five rules.

Rules of Probability

Rule 1. $0 \le P(A) \le 1$ for any event A.

Rule 2. $P(S) = 1$.

Rule 3. **Addition rule:** If A and B are disjoint events, then

$$P(A \text{ or } B) = P(A) + P(B)$$

Rule 4. **Complement rule:** For any event A,

$$P(A^c) = 1 - P(A)$$

Rule 5. **Multiplication rule:** If A and B are independent events, then

$$P(A \text{ and } B) = P(A)P(B)$$

General Addition Rules

If A and B are disjoint events, then $P(A \text{ or } B) = P(A) + P(B)$. What if there are more than two events, or if the events are not disjoint? These circumstances are covered by more general addition rules for probability.

Union

The **union** of any collection of events is the event that at least one of the collection occurs.

For two events A and B, $A \cup B$ is the event that A or B or both occur. From the addition rule for two disjoint events, we can obtain rules for more general unions. Suppose first that we have several events—say A, B, and C—that are disjoint in pairs. That is, no two can occur simultaneously. The Venn diagram in Figure 6.6 illustrates three disjoint events.

Figure 6.6 *The addition rule for disjoint events:* P(A *or* B *or* C) = P(A) + P(B) + P(C) *when events* A, B, *and* C *are disjoint.*

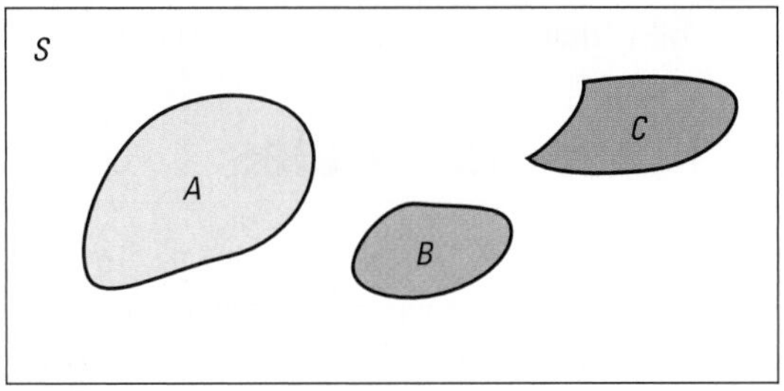

The addition rule for two disjoint events extends to the following law.

Addition Rule for Disjoint Events

If events A, B, and C are disjoint in the sense that no two have any outcomes in common, then

$$P(A \text{ or } B \text{ or } C) = P(A) + P(B) + P(C)$$

This rule extends to any number of disjoint events.

Example 6.22 *Uniform distribution*
Addition rule for disjoint events

Generate a random number X between 0 and 1. What is the probability that the first digit will be odd? We will learn in Chapter 7 that the variable X has the density curve of a uniform distribution (see Exercise 2.10, page 128). This density curve has constant height 1 between 0 and 1 and is 0 elsewhere. The event that the first digit of X is odd is the union of five disjoint events. These events are

$$0.10 \leq X < 0.20$$
$$0.30 \leq X < 0.40$$
$$0.50 \leq X < 0.60$$
$$0.70 \leq X < 0.80$$
$$0.90 \leq X < 1.00$$

Figure 6.7 illustrates the probabilities of these events as areas under the density curve. Each has probability 0.1 equal to its length. The union of the five therefore has probability equal to the sum, or 0.5. As we should expect, a random number is equally likely to begin with an odd or an even digit.

Figure 6.7 *The probability that the first digit of a random number is odd is the sum of the probabilities of the 5 disjoint events shown, for Example 6.22.*

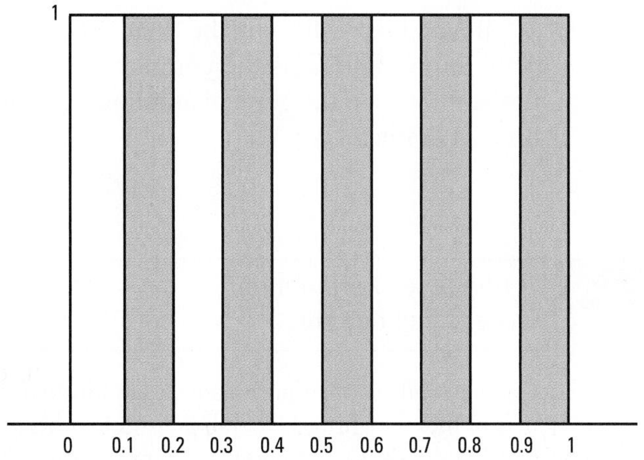

If events A and B are not disjoint, they can occur simultaneously. The probability of their union is then *less* than the sum of their probabilities. As Figure 6.8 suggests, the outcomes common to both are counted twice when we add probabilities, so we must subtract this probability once. Here is the addition rule for the union of any two events, disjoint or not.

Figure 6.8 *The general addition rule for the union of two events:* P(A *or* B) = P(A) + P(B) − P(A *and* B) *for any events* A *and* B.

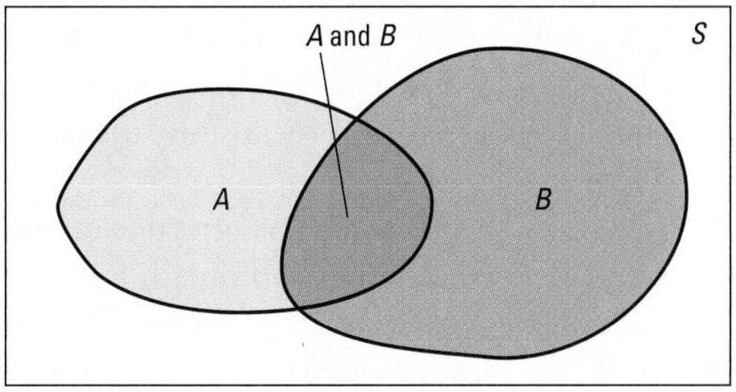

General Addition Rule for Unions of Two Events

For any two events A and B,

$$P(A \text{ or } B) = P(A) + P(B) - P(A \text{ and } B)$$

Equivalently,

$$P(A \cup B) = P(A) + P(B) - P(A \cap B)$$

If A and B are disjoint, the event $A \cap B$ that both occur has no outcomes in it. This *empty event* $\varnothing$ is the complement of the sample space S and must have probability 0. So the general addition rule includes Rule 3, the addition rule for disjoint events.

Example 6.23 *Probability of promotion*
General addition rule

personal probability

Deborah and Matthew are anxiously awaiting word on whether they have been made partners of their law firm. Deborah guesses that her probability of making partner is 0.7 and that Matthew's is 0.5. (These are ***personal probabilities*** reflecting Deborah's assessment of chance.) This assignment of probabilities does not give us enough information to compute the probability that at least one of the two is promoted. In particular, adding the individual probabilities of promotion gives the impossible result 1.2. If Deborah also guesses that the probability that *both* she and Matthew are made partners is 0.3, then by the addition rule for unions

$$P(\text{at least one is promoted}) = 0.7 + 0.5 - 0.3 = 0.9$$

The probability that *neither* is promoted is then 0.1 by the complement rule.

Venn diagrams are a great help in finding probabilities for unions, because you can just think of adding and subtracting areas. Figure 6.9 shows some events and their probabilities for Example 6.23. What is the probability that Deborah is promoted and Matthew is not? The Venn diagram shows that this is the probability that Deborah is promoted minus the probability that both are promoted, $0.7 - 0.3 = 0.4$. Similarly, the probability that Matthew is promoted and Deborah is not is $0.5 - 0.3 = 0.2$. The four probabilities that appear in the figure add to 1 because they refer to four disjoint events whose union is the entire sample space.

joint event

joint probability

The simultaneous occurrence of two events, such as A = Deborah is promoted and B = Matthew is promoted, is called a ***joint event.*** The probability of a joint event, such as $P(\text{Deborah is promoted and Matthew is promoted}) = P(A \text{ and } B)$, is called a ***joint probability.*** Determining joint probabilities when you have equally likely outcomes can be as easy as counting outcomes. For most situations, however, we will need more powerful methods, which will be developed later.

Figure 6.9 *Venn diagram and probabilities, for Example 6.23.*

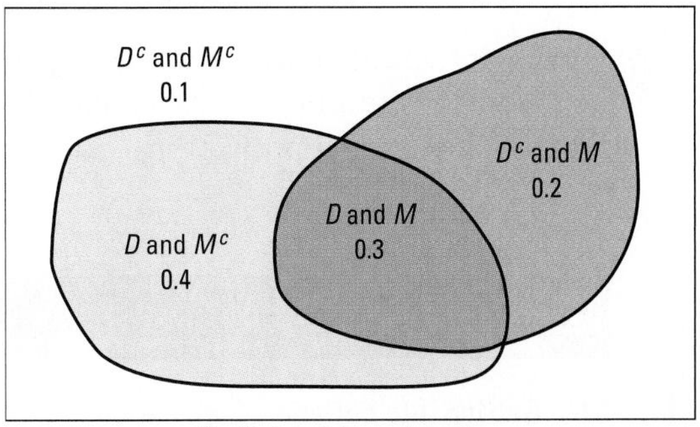

D = Deborah is made partner
M = Matthew is made partner

Here's another way to work with joint events. We have two variables. One variable is employee, which has two values: Deborah and Matthew. The other variable is promotion, which also has two values: promoted and not promoted.

$$D = \{\text{Deborah promoted}\}$$
$$D^c = \{\text{Deborah not promoted}\}$$
$$M = \{\text{Matthew promoted}\}$$
$$M^c = \{\text{Matthew not promoted}\}$$

We can construct a table and write in the probabilities that Deborah guesses:

	Matthew		
Deborah	**Promoted**	**Not promoted**	**Total**
Promoted	0.3		0.7
Not promoted			
Total	0.5		1

The rows and columns have to add to the totals shown, so we can fill in the rest of the table to produce the completed table:

	Matthew		
Deborah	**Promoted**	**Not promoted**	**Total**
Promoted	0.3	0.4	0.7
Not promoted	0.2	0.1	0.3
Total	0.5	0.5	1

The four entries in the body of the table are the probabilities of the joint events of interest:

$P(D \text{ and } M) = P(\text{Deborah and Matthew are both promoted}) = 0.3$
$P(D \text{ and } M^c) = P(\text{Deborah is promoted and Matthew is not promoted}) = 0.4$
$P(D^c \text{ and } M) = P(\text{Deborah is not promoted and Matthew is promoted}) = 0.2$
$P(D^c \text{ and } M^c) = P(\text{Deborah is not promoted and Matthew is not promoted}) = 0.1$

Note that these joint probabilities add to 1.

We will continue our discussion of tables and joint events in the next subsection, on conditional probability.

Poker mania raises young gambling addicts A national survey showed a huge increase in card-playing among males aged 14 to 22. The number of youths reporting that they gambled in card games at least once a week jumped from 6.2% in 2003 to 11.4% in 2004. Four percent reported that they stole money from relatives to gamble, which is a sign of addiction. The vast majority of poker players are male. At some schools, it's estimated that half or more of the males are regular poker players. State and local councils on compulsive gambling report that poker playing ranks along with hip-hop and video games as a symbol of American youth culture.

Exercises

6.65 Getting into college Zack has applied to both Princeton and Stanford. He thinks the probability that Princeton will admit him is 0.4, the probability that Stanford will admit him is 0.5, and the probability that both will admit him is 0.2.

(a) Make a Venn diagram marked with the given probabilities.

(b) What is the probability that neither university admits Zack?

(c) What is the probability that he gets into Stanford but not Princeton?

6.66 Prosperity and education Call a household prosperous if its income exceeds \$100,000. Call the household educated if the householder completed college. Select an American household at random, and let A be the event that the selected household is prosperous and B the event that it is educated. According to the Current Population Survey, $P(A) = 0.138$, $P(B) = 0.261$, and the probability that a household is both prosperous and educated is $P(A \cap B) = 0.082$. What is the probability $P(A \cup B)$ that the household selected is either prosperous or educated?

6.67 Show the relation Draw a Venn diagram that shows the relation between events A and B in Exercise 6.66. Indicate each of the following events on your diagram and use the information in Exercise 6.66 to calculate the probability of each event. Finally, describe in words what each event is.

(a) A and B

(b) A and B^c

(c) A^c and B

(d) A^c and B^c

6.68 Finding a job Stephanie is graduating from college. She has studied biology, chemistry, and computing and hopes to work as a forensic scientist and apply her science background to crime investigation. Late one night she thinks about some jobs she has applied for. Let A, B, and C be the events that Stephanie is offered a job by

A = the Connecticut Office of the Chief Medical Examiner

B = the New Jersey Division of Criminal Justice

C = the federal Disaster Mortuary Operations Response Team

Stephanie writes down her personal probabilities for being offered these jobs:

$P(A) = 0.6$ $\quad$ $P(B) = 0.4$ $\quad$ $P(C) = 0.2$

$P(A \cap B) = 0.1$ $\quad$ $P(A \cap C) = 0.05$ $\quad$ $P(B \cap C) = 0.05$

$P(A \cap B \cap C) = 0$

(a) Make a Venn diagram of the events A, B, and C. As in Figure 6.9 (page 439) mark the probabilities of every intersection involving these events and their complements.

(b) What is the probability that Stephanie is offered at least one of the three jobs?

(c) What is the probability that Stephanie is offered both the Connecticut and New Jersey jobs but not the federal job?

6.69 Caffeine in the diet Common sources of caffeine are coffee, tea, and cola drinks. Suppose that

55% of adults drink coffee

25% of adults drink tea

45% of adults drink cola

and also that

15% drink both coffee and tea

5% drink all three beverages

25% drink both coffee and cola

5% drink only tea

Draw a Venn diagram marked with this information. Use it along with the addition rules to answer the following questions.

(a) What percent of adults drink only cola?

(b) What percent drink none of these beverages?

6.70 Tastes in music, I Musical styles other than rock and pop are becoming more popular. A survey of college students finds that 40% like country music, 30% like gospel music, and 10% like both.

(a) Make a Venn diagram with these results.

(b) What percent of college students like country but not gospel?

(c) What percent like neither country nor gospel?

Conditional Probability

The probability we assign to an event can change if we know that some other event has occurred. This idea is the key to many applications of probability.

Example 6.24 *Amarillo Slim wants an ace*
Conditional probability

Slim is a professional poker player. He stares at the dealer, who prepares to deal. What is the probability that the card dealt to Slim is an ace? There are 52 cards in the deck. Because the deck was carefully shuffled, the next card dealt is equally likely to be any of the cards. Four of the 52 cards are aces. So

$$P(\text{ace}) = \frac{4}{52} = \frac{1}{13}$$

This calculation assumes that Slim knows nothing about any cards already dealt. Suppose now that he is looking at 4 cards already in his hand, and that 1 of them is an ace. He knows nothing about the other 48 cards except that exactly 3 aces are among them. Slim's probability of being dealt an ace, *given what he knows*, is now

$$P(\text{ace} \mid 1 \text{ ace in 4 visible cards}) = \frac{3}{48} = \frac{1}{16}$$

Knowing that there is 1 ace among the 4 cards Slim can see changes the probability that the next card dealt is an ace.

conditional probability

The new notation $P(A \mid B)$ is a ***conditional probability.*** That is, it gives the probability of one event (the next card dealt is an ace) under the condition that we know another event (exactly 1 of the 4 visible cards is an ace). You can read the bar | as "given the information that."

In Example 6.24 we could find probabilities because we were willing to use an equally likely probability model for a shuffled deck of cards. Here is an example based on data.

Example 6.25 *University grades*
Conditional probability

Students at the University of New Harmony received 10,000 course grades last semester. Table 6.1 breaks down these grades by which school of the university taught the course. The schools are Liberal Arts, Engineering and Physical Sciences, and Health and Human Services. (Table 6.1 is based closely on grade distributions at an actual university, simplified a bit for clarity.)[10]

Table 6.1 *Grades awarded at a university, by school*

	Grade Level			
	A	B	Below B	Total
Liberal Arts	2,142	1,890	2,268	**6,300**
Engineering and Physical Sciences	368	432	800	**1,600**
Health and Human Services	882	630	588	**2,100**
Total	**3,392**	**2,952**	**3,656**	**10,000**

It is common knowledge that college grades are lower in engineering and the physical sciences (EPS) than in liberal arts and social sciences. Consider the two events

$$A = \text{the grade comes from an EPS course}$$
$$B = \text{the grade is below a B}$$

There are 10,000 grades, of which 3656 are below B. Choosing at random gives each grade an equal chance, so the probability of choosing a grade below a B is

$$P(B) = \frac{3656}{10{,}000} = 0.3656$$

To find the *conditional* probability that a grade is below a B, *given the information* that it comes from the EPS school, look only at the "Engineering and Physical Sciences" row. The EPS grades are all in this row, so the information given says that only this row is relevant. The conditional probability is

$$P(B \mid A) = \frac{800}{1600} = 0.5$$

The conditional probability that a grade is below a B, given that we know it comes from an EPS course, is much higher than the probability for a randomly chosen grade.

It is easy to confuse probabilities and conditional probabilities involving the same events. For example, Table 6.1 says that

$$P(A) = \frac{1600}{10{,}000} = 0.16$$

$$P(A \text{ and } B) = \frac{800}{10{,}000} = 0.08$$

$$P(B \mid A) = \frac{800}{1600} = 0.5$$

Be sure you understand how we found these three probabilities. There is a relationship among these probabilities. The probability that a grade is both from EPS *and* below a B is the product of the probabilities that it is from EPS and that it is below a B, *given* that it is from EPS. That is,

$$\begin{aligned} P(A \text{ and } B) &= P(A) \times P(B \mid A) \\ &= \frac{1600}{10{,}000} \times \frac{800}{1600} \\ &= \frac{800}{10{,}000} = 0.08 \qquad \text{(as before)} \end{aligned}$$

Try to think your way through this in words: first, the grade is from EPS; then, given that it is from EPS, it is below B. We have just discovered the fundamental multiplication rule of probability.

General Multiplication Rule for Any Two Events

The joint probability that events A and B both happen can be found by

$$P(A \cap B) = P(A)P(B \mid A)$$

Here $P(B \mid A)$ is the conditional probability that B occurs, given the information that A occurs.

In words, this rule says that for both of two events to occur, first one must occur and then, given that the first event has occurred, the second must occur. Notice also that there is a kind of symmetry here. Since $A \cap B = B \cap A$, we could write the general multiplication rule as $P(A \cap B) = P(B)P(A \mid B)$.

Example 6.26 *Downloading music*
Multiplication rule

The multiplication rule is just common sense made formal. For example, 29% of Internet users download music files, and 67% of downloaders say they don't care if the music is copyrighted.[11] So the percent of Internet users who download music (event A) *and* don't care about copyright (event B) is 67% of the 29% who download, or

$$(0.67)(0.29) = 0.1943 = 19.43\%$$

The multiplication rule expresses this as

$$\begin{aligned} P(A \text{ and } B) &= P(A) \times P(B \mid A) \\ &= (0.29)(0.67) = 0.1943 \end{aligned}$$

Example 6.27 *Slim wants diamonds*
Conditional probability

Slim is still at the poker table. At the moment, he wants very much to draw 2 diamonds in a row. As he sits at the table looking at his hand and at the upturned cards on the table, Slim sees 11 cards. Of these, 4 are diamonds. The full deck contains 13 diamonds among its 52 cards, so 9 of the 41 unseen cards are diamonds. To find Slim's probability of drawing 2 diamonds, first calculate

$$P(\text{first card diamond}) = \frac{9}{41}$$

$$P(\text{second card diamond} \mid \text{first card diamond}) = \frac{8}{40}$$

Slim finds both probabilities by counting cards. The probability that the first card drawn is a diamond is 9/41 because 9 of the 41 unseen cards are diamonds. If the first card is a diamond, that leaves 8 diamonds among the 40 remaining cards. So the conditional probability of another diamond is 8/40. The multiplication rule now says that

$$P(\text{both cards diamonds}) = \frac{9}{41} \times \frac{8}{40} = 0.044$$

Slim will need luck to draw his diamonds.

If we know $P(A)$ and $P(A \text{ and } B)$, we can rearrange the general multiplication rule to produce a *definition* of the ***conditional probability*** $P(B \mid A)$ in terms of unconditional probabilities.

Conditional Probability

When $P(A) > 0$, the **conditional probability** of B, given A, is

$$P(B \mid A) = \frac{P(A \text{ and } B)}{P(A)}$$

CAUTION

Be sure to keep in mind the distinct roles in $P(B \mid A)$ of the event B whose probability we are computing and the event A that represents the information we are given. The conditional probability $P(B|A)$ makes no sense if the event A can never occur, so we require that $P(A) > 0$ whenever we talk about $P(B \mid A)$.

Example 6.28 *The University of New Harmony*

Finding conditional probabilities

What is the conditional probability that a grade at the University of New Harmony is an A, given that it comes from a liberal arts course? We see from Table 6.1 that

$$P(\text{liberal arts course}) = \frac{6300}{10{,}000} = 0.63$$

$$P(\text{A grade and liberal arts course}) = \frac{2142}{10{,}000} = 0.2142$$

The definition of conditional probability therefore says that

$$P(\text{A grade} \mid \text{liberal arts course}) = \frac{P(\text{A grade and liberal arts course})}{P(\text{liberal arts course})} = \frac{0.2142}{0.63} = 0.34$$

Exercises

6.71 Cars and light trucks Motor vehicles sold to individuals are classified as either cars or light trucks (including SUVs) and as either domestic or imported. In early 2004, 69% of vehicles sold were light trucks, 78% were domestic, and 55% were domestic light trucks. Let A be the event that a vehicle is a light truck and B the event that it is imported. Write each of the following events in terms of events A and B and give its probability.

(a) The vehicle is a car.

(b) The vehicle is an imported car.

(c) Given that a vehicle is imported, what is the conditional probability that it is a car?

(d) Are the events "vehicle is a car" and "vehicle is imported" independent? Justify your answer.

6.72 Pay at the pump At a self-service gas station, 40% of the customers pump regular gas, 35% pump midgrade, and 25% pump premium gas. Of those who pump regular, 30% pay at least \$20. Of those who pump midgrade, 50% pay at least \$20. And of those who pump premium, 60% pay at least \$20. What is the probability that the next customer pays at least \$20?

6.73 Paying for premium In the setting of Exercise 6.72, what percent of customers who pay at least \$20 pump premium? (Write this as a conditional probability.)

6.74 Women managers Choose an employed person at random. Let A be the event that the person chosen is a woman and B the event that the person holds a managerial or professional job. Government data tell us that $P(A) = 0.46$ and the probability of managerial and professional jobs among women is $P(B \mid A) = 0.32$. Find the probability that a randomly chosen employed person is a woman holding a managerial or professional position.

6.75 Buying from Japan Functional Robotics Corporation buys electrical controllers from a Japanese supplier. The company's treasurer thinks that there is probability 0.4 that the dollar will fall in value against the Japanese yen in the next month. The treasurer also believes that if the dollar falls there is probability 0.8 that the supplier will demand renegotiation of the contract. What probability has the treasurer assigned to the event that the dollar falls and the supplier demands renegotiation?

6.76 The probability of a flush A poker player holds a flush when all 5 cards in the hand belong to the same suit. We will find the probability of a flush when 5 cards are dealt. Remember that a deck contains 52 cards, 13 of each suit, and that when the deck is well shuffled, each card dealt is equally likely to be any of those that remain in the deck.

(a) We will concentrate on spades. What is the probability that the first card dealt is a spade? What is the conditional probability that the second card is a spade, given that the first is a spade?

(b) Continue to count the remaining cards to find the conditional probabilities of a spade on the third, the fourth, and the fifth card, given in each case that all previous cards are spades.

(c) The probability of being dealt 5 spades is the product of the five probabilities you have found. Why? What is this probability?

(d) The probability of being dealt 5 hearts or 5 diamonds or 5 clubs is the same as the probability of being dealt 5 spades. What is the probability of being dealt a flush?

6.77 The probability of a royal flush A royal flush is the highest hand possible in poker. It consists of the ace, king, queen, jack, and ten of the same suit. Modify the outline given in Exercise 6.76 to find the probability of being dealt a royal flush in a five-card deal.

6.78 Tastes in music, II Musical styles other than rock and pop are becoming more popular. A survey of college students finds that 40% like country music, 30% like gospel music, and 10% like both.

(a) What is the conditional probability that a student likes gospel music if we know that he or she likes country music?

(b) What is the conditional probability that a student who does not like country music likes gospel music? (A Venn diagram may help you.)

Extended Multiplication Rules

The definition of conditional probability reminds us that in principle all probabilities, including conditional probabilities, can be found from the assignment of probabilities to events that describe a random phenomenon. More often, however, conditional probabilities are part of the information given to us in a probability model, and the multiplication rule is used to compute $P(A \text{ and } B)$. This rule extends to more than two events.

The union of a collection of events is the event that *any* of them occur. Here is the corresponding term for the event that *all* of them occur.

Intersection

The **intersection** of any collection of events is the event that *all* of the events occur.

To extend the multiplication rule to the probability that all of several events occur, the key is to condition each event on the occurrence of *all* of the preceding events. For example, the intersection of three events A, B, and C has probability

$$P(A \text{ and } B \text{ and } C) = P(A)P(B \mid A)P(C \mid A \text{ and } B)$$

Example 6.29 ***The future of high school athletes***
Extended multiplication rule

Only 5% of male high school basketball, baseball, and football players go on to play at the college level. Of these, only 1.7% enter major league professional sports. About 40% of the athletes who compete in college and then reach the pros have a career of more than 3 years.[12] Define these events:

$$A = \{\text{competes in college}\}$$
$$B = \{\text{competes professionally}\}$$
$$C = \{\text{pro career longer than 3 years}\}$$

What is the probability that a high school athlete competes in college and then goes on to have a pro career of more than 3 years? We know that

$$P(A) = 0.05$$
$$P(B \mid A) = 0.017$$
$$P(C \mid A \text{ and } B) = 0.400$$

The probability we want is therefore

$$\begin{aligned} P(A \text{ and } B \text{ and } C) &= P(A)P(B \mid A)P(C \mid A \text{ and } B) \\ &= 0.05 \times 0.017 \times 0.40 = 0.00034 \end{aligned}$$

Only about 3 of every 10,000 high school athletes can expect to compete in college and have a professional career of more than 3 years. High school students would be wise to concentrate on studies rather than on unrealistic hopes of earning a fortune from pro sports.

Tree Diagrams Revisited

Probability problems often require us to combine several of the basic rules into a more elaborate calculation. Here is an example that illustrates how to solve problems that have several stages.

Example 6.30 ***Online chat rooms***
Tree diagram

Online chat rooms are dominated by the young. Teens are the biggest users. If we look only at adult Internet users (aged 18 and over), 47% of the 18 to 29 age group chat, as do 21% of those aged 30 to 49 and just 7% of those 50 and over. To learn what percent of all Internet users participate in chat rooms, we also need the age breakdown of users. Here it is: 29% of adult Internet users are aged 18 to 29 (event A_1), another 47% are 30 to 49 (event A_2), and the remaining 24% are 50 and over (event A_3).[13]

What is the probability that a randomly chosen user of the Internet participates in chat rooms (event C)? To find out, use the tree diagram in Figure 6.10 to organize your thinking. Each segment in the tree is one stage of the problem. Each complete branch shows a path through the two stages. The probability written on each segment is the conditional probability of an Internet user following that segment, given that he or she has reached the node from which it branches.

Figure 6.10 *Tree diagram for Example 6.30. The probability P(C) is the sum of the probabilities of the three branches marked with asterisks (*).*

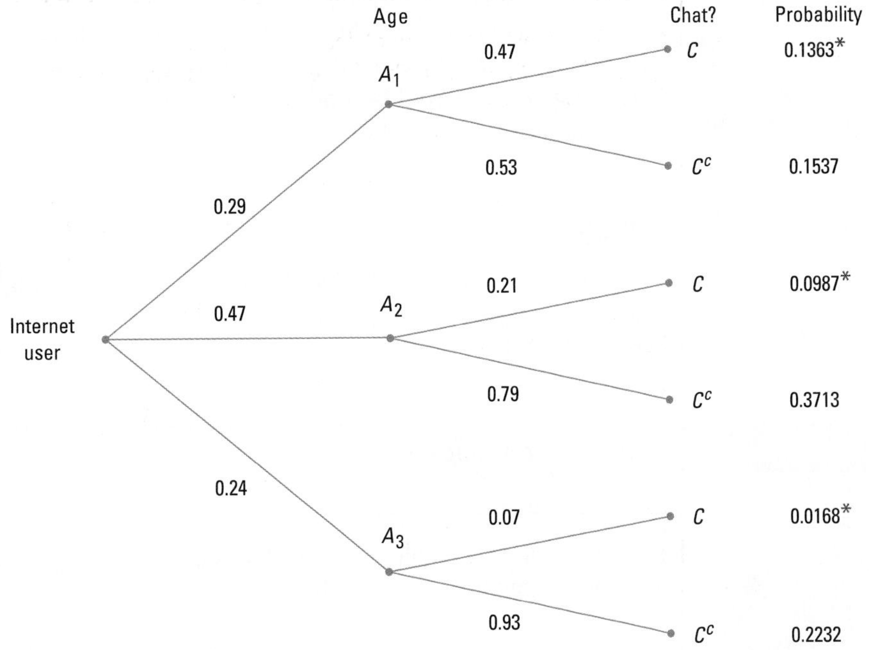

Starting at the left, an Internet user falls into one of the three age groups. The probabilities of these groups

$$P(A_1) = 0.29 \qquad P(A_2) = 0.47 \qquad P(A_3) = 0.24$$

mark the leftmost branches in the tree. Conditional on being 18 to 29 years old, the probability of participating in chat is $P(C \mid A_1) = 0.47$. So the conditional probability of *not* participating is

$$P(C^c \mid A_1) = 1 - 0.47 = 0.53$$

These conditional probabilities mark the paths branching out from the A_1 node in Figure 6.10. The other two age group nodes similarly lead to two branches marked with the conditional probabilities of chatting or not. The probabilities on the branches from any node add to 1 because they cover all possibilities, given that this node was reached.

There are three disjoint paths to C, one for each age group. By the addition rule, $P(C)$ is the sum of their probabilities. The probability of reaching C through the 18 to 29 age group is

$$\begin{aligned} P(C \text{ and } A_1) &= P(A_1)\,P(C \mid A_1) \\ &= 0.29 \times 0.47 = 0.1363 \end{aligned}$$

Follow the paths to C through the other two age groups. The probabilities of these paths are

$$P(C \text{ and } A_2) = P(A_2)P(C \mid A_2) = (0.47)(0.21) = 0.0987$$
$$P(C \text{ and } A_3) = P(A_3)P(C \mid A_3) = (0.24)(0.07) = 0.0168$$

The final result is

$$P(C) = 0.1363 + 0.0987 + 0.0168 = 0.2518$$

About 25% of all adult Internet users take part in chat rooms.

It takes longer to explain a tree diagram than it does to use it. Once you have understood a problem well enough to draw the tree, the rest is easy. Tree diagrams combine the addition and multiplication rules. The multiplication rule says that *the probability of reaching the end of any complete branch is the product of the probabilities written on its segments.* The probability of any outcome, such as the event C that an adult Internet user takes part in chat rooms, is then found by adding the probabilities of all branches that are part of that event.

Bayes's Rule

There is another kind of probability question that we might ask in the context of studies concerning online chat. What percent of adult chat room participants are aged 18 to 29?

Example 6.31 *Chat room participants*
Bayes's rule

In the notation of Example 6.30 this is the conditional probability $P(A_1 \mid C)$. Start from the definition of conditional probability and then apply the results of Example 6.30:

$$P(A_1 \mid C) = \frac{P(A_1 \text{ and } C)}{P(C)}$$

$$= \frac{0.1363}{0.2518} = 0.5413$$

Over half of adults who participate in chat rooms are between 18 and 29 years old. Compare this conditional probability with the original information (unconditional) that 29% of adult Internet users are between 18 and 29 years old. Knowing that a person chats increases the probability that he or she is young.

We know the probabilities $P(A_1)$, $P(A_2)$, and $P(A_3)$ that give the age distribution of adult Internet users. We also know the conditional probabilities $P(C \mid A_1)$, $P(C \mid A_2)$, and $P(C \mid A_3)$ that a person from each age group chats. Example 6.30 shows how to use this information to calculate $P(C)$. The method can be summarized in a single expression that adds the probabilities of the three paths to C in the tree diagram:

$$P(C) = P(A_1)P(C \mid A_1) + P(A_2)P(C \mid A_2) + P(A_3)P(C \mid A_3)$$

Independence Again

The conditional probability $P(B \mid A)$ is generally not equal to the unconditional probability $P(B)$. That is because the occurrence of event A generally gives us some additional information about whether or not event B occurs. If knowing that A occurs gives no additional information about B, then A and B are independent events. The formal definition of independence is expressed in terms of conditional probability.

Independent Events

Two events A and B that both have positive probability are **independent** if

$$P(B \mid A) = P(B)$$

This definition makes precise the informal description of independence given in Section 6.2. We now see that the multiplication rule for independent events, $P(A \text{ and } B) = P(A)P(B)$, is a special case of the general multiplication rule, $P(A \text{ and } B) = P(A)P(B \mid A)$, just as the addition rule for disjoint events is a special case of the general addition rule.

What are the odds? Gamblers often express chance in terms of odds rather than probability. Odds of *A* to *B* against an outcome means that the probability of that outcome is $B/(A+B)$. So "odds of 5 to 1" is another way of saying "probability of 1/6." A probability is always between 0 and 1, but odds range from 0 to infinity. Although odds are mainly used in gambling, they give us a way to make very small probabilities clearer. "Odds of 999 to 1" may be easier to understand than "probability 0.001."

Exercises

6.79 Educated and prosperous This is a continuation of Exercise 6.66 (page 440). What is the conditional probability that a household is prosperous, given that it is educated? Explain why your result shows that events A and B are not independent.

6.80 IRS returns In 2000, the Internal Revenue Service received 129,075,000 individual tax returns. Of these, 10,855,000 reported an adjusted gross income of at least \$100,000, and 240,000 reported at least \$1 million.[14] If you know that a randomly chosen return shows an income of \$100,000 or more, what is the conditional probability that the income is at least \$1 million?

6.81 Surgery risks You have torn a tendon and are facing surgery to repair it. The orthopedic surgeon explains the risks to you. Infection occurs in 3% of such operations, the repair fails in 14%, and both infection and failure occur together in 1%. What percent of these operations succeed and are free from infection?

6.82 HIV testing, I Enzyme immunoassay (EIA) tests are used to screen blood specimens for the presence of antibodies to HIV, the virus that causes AIDS. Antibodies indicate the presence of the virus. The test is quite accurate but is not always correct. Here are approximate probabilities of positive and negative EIA outcomes when the blood tested does and does not actually contain antibodies to HIV.[15]

	Test Result	
	+	−
Antibodies present	0.9985	0.0015
Antibodies absent	0.006	0.994

Suppose that 1% of a large population carries antibodies to HIV in their blood.

(a) Draw a tree diagram for selecting a person from this population (outcomes: antibodies present or absent) and for testing his or her blood (outcomes: EIA positive or negative).

(b) What is the probability that the EIA test is positive for a randomly chosen person from this population?

(c) What is the probability that a person has the antibody, given that the EIA test is positive?

(*Comment:* This exercise illustrates a fact that is important when considering proposals for widespread testing for HIV, illegal drugs, or agents of biological warfare: if the condition being tested is uncommon in the population, many positives will be false-positives.)

6.83 HIV testing, II The previous exercise gives data on the results of EIA tests for the presence of antibodies to HIV. Repeat part (c) of that exercise for two different populations:

(a) Blood donors are prescreened for HIV risk factors, so perhaps only 0.1% (0.001) of this population carries HIV antibodies.

(b) Clients of a drug rehab clinic are a high-risk group, so perhaps 10% of this population carries HIV antibodies.

(c) What general lesson do your calculations illustrate?

6.84 Who studies education? The probability that a randomly chosen student at the University of New Harmony is a woman is 0.6. The probability that the student is studying education is 0.15. The conditional probability that the student is a woman, given that the student is studying education, is 0.8. What is the conditional probability that the student is studying education, given that she is a woman?

Section 6.3 Summary

The **complement** A^c of an event A contains all outcomes that are not in A. The **union** $A \cup B$ of events A and B contains all outcomes in A, in B, or in both A and B. The **intersection** $A \cap B$ contains all outcomes that are in both A and B, but not outcomes in A alone or B alone.

The essential general rules of elementary probability are

Legitimate values: $0 \le P(A) \le 1$ for any event A
Total probability 1: $P(S) = 1$
Complement rule: $P(A^c) = 1 - P(A)$
General addition rule: $P(A \cup B) = P(A) + P(B) - P(A \cap B)$
Multiplication rule: $P(A \cap B) = P(A)P(B \mid A)$

The **conditional probability** $P(B \mid A)$ of an event B, given an event A, is defined by

$$P(B \mid A) = \frac{P(A \cap B)}{P(A)}$$

when $P(A) > 0$. In practice, conditional probabilities are most often found from directly available information.

If A and B are **disjoint** (mutually exclusive), then $P(A \cap B) = 0$. The general addition rule for unions then becomes the special addition rule, $P(A \cup B) = P(A) + P(B)$.

A and B are **independent** when $P(B \mid A) = P(B)$. The multiplication rule for intersections then becomes $P(A \cap B) = P(A)P(B)$.

A **Venn diagram,** together with the general addition rule, can be helpful in finding probabilities of the union of two events $P(A \cup B)$ or the joint probability $P(A \cap B)$. The joint probability $P(A \cap B)$ can also be found using the general multiplication rule: $P(A \cap B) = P(A)P(B \mid A) = P(B)P(A \mid B)$.

Constructing a table is a good approach for determining a conditional probability.

In problems with several stages, draw a **tree diagram** to organize use of the multiplication and addition rules.

Section 6.3 Exercises

6.85 Inspecting switches A shipment contains 10,000 switches. Of these, 1000 are bad. An inspector draws switches at random, so that each switch has the same chance to be drawn.

(a) Draw one switch. What is the probability that the switch you draw is bad? What is the probability that it is not bad?

(b) Suppose the first switch drawn is bad. How many switches remain? How many of them are bad? Draw a second switch at random. What is the conditional probability that this switch is bad?

(c) Answer the questions in (b) again, but now suppose that the first switch drawn is not bad.

(*Comment:* Knowing the result of the first trial changes the conditional probability for the second trial, so the trials are not independent. But because the shipment is large, the probabilities change very little. The trials are almost independent.)

6.86 Nobel Prize winners The numbers of Nobel Prize laureates in selected sciences, 1901 to 2003, are shown in the following table by location of award-winning research:[16]

Country	Physics	Chemistry	Physiology/medicine
United States	74	51	90
United Kingdom	21	27	28
Germany	19	29	15
France	11	7	7
Soviet Union	9	1	2
Japan	4	4	0

If a laureate is selected at random, what is the probability that

(a) his or her award was in chemistry?

(b) the award was won by someone working in the United States?

(c) the awardee was working in the United States, given that the award was for physiology/medicine?

(d) the award was for physiology/medicine, given that the awardee was working in the United States?

(e) Interpret each of your results in parts (a) through (d) in terms of percents.

6.87 Academic degrees, I Here are the counts (in thousands) of earned degrees in the United States in the 2005–2006 academic year, classified by level and by the sex of the degree recipient:[17]

	Bachelor's	Master's	Professional	Doctorate	Total
Female	784	276	39	20	1119
Male	559	197	44	25	825
Total	1343	473	83	45	1944

(a) If you choose a degree recipient at random, what is the probability that the person you choose is a woman?

(b) What is the conditional probability that you choose a woman, given that the person chosen received a professional degree?

(c) Are the events "choose a woman" and "choose a professional degree recipient" independent? How do you know?

6.88 Pick two cards The suit of 13 hearts (A, 2 to 10, J, Q, K) from a standard deck of cards is placed in a hat. The cards are thoroughly mixed and a student reaches into the hat and selects two cards without replacement.

(a) What is the probability that the first card selected is the jack?

(b) Given that the first card selected is the jack, what is the probability that the second card is the 5?

(c) What is the probability of selecting the jack on the first draw and then the 5?

(d) What is the probability that both cards selected are greater than 5 (when the ace is considered "low")?

6.89 Academic degrees, II Exercise 6.87 gives the counts (in thousands) of earned degrees in the United States in the 2005–2006 academic year. Use these data to answer the following questions.

(a) What is the probability that a randomly chosen degree recipient is a man?

(b) What is the conditional probability that the person chosen received a bachelor's degree, given that he is a man?

(c) Use the multiplication rule to find the joint probability of choosing a male bachelor's degree recipient. Check your result by finding this probability directly from the table of counts.

6.90 Teenage drivers An insurance company has the following information about drivers aged 16 to 18 years: 20% are involved in accidents each year; 10% in this age group are A students; among those involved in an accident, 5% are A students.

(a) Let A be the event that a young driver is an A student and C the event that a young driver is involved in an accident this year. State the information given in terms of probabilities and conditional probabilities for the events A and C.

(b) What is the probability that a randomly chosen young driver is an A student and is involved in an accident?

6.91 More on teenage drivers Use your work from Exercise 6.90 to find the percent of A students who are involved in accidents. (Start by expressing this as a conditional probability.)

6.92 If the dollar does not fall? Suppose that in Exercise 6.75 (page 446) the treasurer also feels that, if the dollar does not fall, there is probability 0.2 that the Japanese supplier will demand that the contract be renegotiated. What is the probability that the supplier will demand renegotiation?

6.93 Multiple-choice exam strategies An examination consists of multiple-choice questions, each having five possible answers. Linda estimates that she has probability 0.75 of knowing the answer to any question that may be asked. If she does not know the answer, she will guess, with conditional probability 1/5 of being correct. What is the probability that Linda gives the correct answer to a question? (Draw a tree diagram to guide the calculation.)

6.94 Election math The voters in a large city are 40% white, 40% black, and 20% Hispanic. (Hispanics may be of any race in official statistics, but in this case we are speaking of political blocks.) A black mayoral candidate anticipates attracting 30% of the white vote, 90% of the black vote, and 50% of the Hispanic vote. Draw a tree diagram with probabilities for the race (white, black, or Hispanic) and vote (for or against the candidate) of a randomly chosen voter. What percent of the overall vote does the candidate expect to get?

6.95 More exam strategies In the setting of Exercise 6.93, find the conditional probability that Linda knows the answer, given that she supplies the correct answer. (*Hint:* Use the result of Exercise 6.93 and the definition of conditional probability.)

6.96 Geometric probability Choose a point at random in the graph of the square with sides $0 \le x \le 1$ and $0 \le y \le 1$. Since the area of the square is 1 unit, the probability that the point falls in any region within the square is the area of that region. Let X be the x coordinate and Y the y coordinate of the point chosen. Find the conditional probability $P(Y < 1/2 \mid Y > X)$. (*Hint:* Draw a diagram of the square and the events $Y < 1/2$ and $Y > X$.)

CASE CLOSED!

False alarms at airports

False-positives are disruptive. Having one's luggage flagged as possibly carrying a bomb stigmatizes innocent passengers as potential terrorist suspects. False-positives at the airport lead to embarrassment and humiliation as well as delays. *False positives are costly.* Hand-searches of airport luggage mean more wage-earning workers, raising the total cost of airport security to $5 to $7 per passenger. More important, *false-positives undermine the effectiveness of hand-searches.* If there are only a few alarms a day, screeners can investigate them thoroughly. If there are dozens, screeners will feel pressured to hurry and may become desensitized to the routine and miss something important.

A recent Government Accounting Office report found that the detection of weapons or bombs has not improved much since 2003. In a separate Federal Aviation Administration test of the effectiveness of airport screening systems, 40 percent of explosives, 30 percent of guns, and 70 percent of knives planted by government agents made it through such security checkpoints.

Using what you have learned about simulations and probability, you should now be able to answer some questions about false-positives and false-negatives in connection with the screening of luggage at airports. Refer to the Case Study at the beginning of this chapter (page 389).

1. If a false alarm is a false-positive, what is a false-negative?
2. Which is the more serious error in airport baggage screening, a false-positive or a false-negative? Justify your answer.
3. It is said that the occurrence of false-positives in airport screenings has been about 30%. What does that mean? Write this as a probability statement.
4. Assume that on average 1 suitcase in 10,000 has a bomb in it. Find the probability that a suitcase with a bomb would be detected. What's the probability that a piece of luggage that has a bomb in it would escape detection?
5. Find the probability that no alarm is sounded for a suitcase that has no bomb.

Chapter Review

Summary

Probability describes the pattern of chance outcomes. Probability calculations provide the basis for inference. When data are produced by random sampling or randomized comparative experiments, the laws of probability answer the question "What would happen if we did this very many times?" Probability is used to describe the long-term regularity that results from many repetitions of the same random phenomenon. The reasoning of statistical inference rests on asking, "How often would this method give a correct answer if I used it very many times?" This chapter developed a probability model, including rules and tools that will help you describe the behavior of statistics from random samples in later chapters.

What You Should Have Learned

Here are the major skills you should have gained from studying this chapter.

A. Simulations

1. Recognize that many random phenomena can be investigated by means of a carefully designed simulation.
2. Use the following steps to construct and run a simulation:
   a. State the problem or describe the random phenomenon.
   b. State the assumptions.
   c. Assign digits to represent outcomes.
   d. Simulate many repetitions.
   e. Calculate relative frequencies and state your conclusions.
3. Use a random number table, the TI-83/84/89, or a computer utility such as Minitab, DataDesk, or a spreadsheet to conduct simulations.

B. Probability Rules

1. Describe the sample space of a random phenomenon. For a finite number of outcomes, use the multiplication principle to determine the number of outcomes, and use counting techniques, Venn diagrams, and tree diagrams to determine simple probabilities. For the continuous case, use geometric areas to find probabilities (areas under simple density curves) of events (intervals on the horizontal axis).
2. Know the probability rules and be able to apply them to determine probabilities of defined events. In particular, determine if a given assignment of probabilities is valid.

3. Determine if two events are disjoint, complementary, or independent. Find unions and intersections of two or more events.
4. Use Venn diagrams to picture relationships among several events.
5. Use the general addition rule to find probabilities that involve intersecting events.
6. Understand the idea of independence. Judge when it is reasonable to assume independence as part of a probability model.
7. Use the multiplication rule for independent events to find the probability that all of several independent events occur.
8. Use the multiplication rule for independent events in combination with other probability rules to find the probabilities of complex events.
9. Understand the idea of conditional probability. Find conditional probabilities for individuals chosen at random from a table of counts of possible outcomes.
10. Use the general multiplication rule to find the joint probability $P(A \cap B)$ from $P(A)$ and the conditional probability $P(B \mid A)$.
11. Construct tree diagrams to organize the use of the multiplication and addition rules to solve problems with several stages.

Web Links

The following site from the National Center for Education Statistics simulates rolling two dice as many times as you specify and then plots the results as a histogram with very nice graphics: nces.ed.gov/nceskids/probability/

The following site provides a discussion of the famous birthday problem with an interactive applet: www.mste.uiuc.edu/java/java/birthday/birthday.html

Chapter Review Exercises

6.97 A spinner game of chance A game of chance is based on spinning a 1–10 spinner like the one shown in the illustration two times in succession. The player wins if the larger of the two numbers is greater than 5.

(a) What constitutes a single run of this game? What are the possible outcomes resulting in "win" or "lose"?

(b) Describe a correspondence between random digits from a random number table and outcomes in the game.

(c) Describe a technique using the `randInt` command on the TI-83/84/89 to simulate the result of a single run of the game.

(d) Use either the random number table or your calculator to simulate 20 trials. Report the proportion of times you win the game. Then combine your results with those of other students to obtain results for a large number of trials.

6.98 Gauging the demand for cheesecake The owner of a bakery knows that the daily demand for a highly perishable cheesecake is as follows:

Number/day:	0	1	2	3	4	5
Relative frequency:	0.05	0.15	0.25	0.25	0.20	0.10

(a) Use simulation to find the demand for the cheesecake on 30 consecutive business days.

(b) Suppose that it costs the baker \$5 to produce a cheesecake, and that the unused cheesecakes must be discarded at the end of the business day. Suppose also that the selling price of a cheesecake is \$13. Use simulation to estimate the number of cheesecakes that he should produce each day in order to maximize his profit.

6.99 Hot streaks in foul shooting Joey is interested in investigating so-called hot streaks in foul shooting among basketball players. He's a fan of Carla, who has been making approximately 80% of her free throws. Specifically, Joey wants to use simulation methods to determine Carla's longest run of baskets on average, for 20 consecutive free throws.

(a) Describe a correspondence between random numbers and outcomes.

(b) What will constitute 1 repetition in this simulation? Carry out 20 repetitions and record the longest run for each repetition. Combine your results with those of 4 other students to obtain at least 100 repetitions.

(c) What is the mean run length? Are you surprised? Determine the five-number summary for the data.

(d) Construct a histogram of the results.

6.100 The wine tasters, I Two wine tasters rate each wine they taste on a scale of 1 to 5. From data on their ratings of a large number of wines, we obtain the following probabilities for both tasters' ratings of a randomly chosen wine:

	Taster 2				
Taster 1	**1**	**2**	**3**	**4**	**5**
1	0.03	0.02	0.01	0.00	0.00
2	0.02	0.08	0.05	0.02	0.01
3	0.01	0.05	0.25	0.05	0.01
4	0.00	0.02	0.05	0.20	0.02
5	0.00	0.01	0.01	0.02	0.06

(a) Why is this a legitimate assignment of probabilities to outcomes?

(b) What is the probability that the tasters agree when rating a wine?

(c) What is the probability that Taster 1 rates a wine higher than 3?

(d) What is the probability that Taster 2 rates a wine higher than 3?

6.101 Are you my (blood) type? All human blood can be "ABO-typed" as one of O, A, B, or AB, but the distribution of the types varies a bit among groups of people. Here is the distribution of blood types for a randomly chosen person in the United States:

Blood type:	O	A	B	AB
U.S. probability:	0.45	0.40	0.11	?

(a) What is the probability of type AB blood in the United States?

(b) An individual with type B blood can safely receive transfusions only from persons with type B or type O blood. What is the probability that the husband of a woman with type B blood is an acceptable blood donor for her?

(c) What is the probability that in a randomly chosen couple the wife has type B blood and the husband has type A?

(d) What is the probability that one of a randomly chosen couple has type A blood and the other has type B?

(e) What is the probability that at least one of a randomly chosen couple has type O blood?

6.102 Blood types in China The distribution of blood types in China differs from the U.S. distribution given in the previous exercise:

Blood type:	O	A	B	AB
China probability:	0.35	0.27	0.26	0.12

Choose two people, one American and one Chinese at random, and independently of each other.

(a) What is the probability that both have type O blood?

(b) What is the probability that both have the same blood type?

6.103 Income and savings A sample survey chooses a sample of households and measures their annual income and their savings. Some events of interest are

A = the household chosen has income at least \$100,000

C = the household chosen has at least \$50,000 in savings

Based on this sample survey, we estimate that $P(A) = 0.07$ and $P(C) = 0.2$.

(a) We want to find the probability that a household either has income at least \$100,000 or has savings at least \$50,000. Explain why we do not have enough information to find this probability. What additional information is needed?

(b) We want to find the probability that a household has income at least \$100,000 and savings at least \$50,000. Explain why we do not have enough information to find this probability. What additional information is needed?

6.104 Screening job applicants A company retains a psychologist to assess whether job applicants are suited for assembly-line work. The psychologist classifies applicants as A (well suited), *B* (marginal), or *C* (not suited). The company is concerned about event *D*: an employee leaves the company within a year of being hired. Data on all people hired in the past 5 years give these probabilities:

$P(A) = 0.4$	$P(B) = 0.3$	$P(C) = 0.3$
$P(A \text{ and } D) = 0.1$	$P(B \text{ and } D) = 0.1$	$P(C \text{ and } D) = 0.2$

Sketch a Venn diagram of the events A, *B*, *C*, and *D* and mark on your diagram the probabilities of all combinations of psychological assessment and leaving (or not) within a year. What is $P(D)$, the probability that an employee leaves within a year?

6.105 At the gym Many conditional probability calculations are just common sense made automatic. For example, 10% of adults belong to health clubs, and 40% of these health club members go to the club at least twice a week. What percent of all adults go to a health club at least twice a week? Write the information in terms of probabilities and use the general multiplication rule.

6.106 Toss two coins Independence of events is not always obvious. Toss two balanced coins independently. The four possible combinations of heads and tails in order each have probability 0.25. The events

A = head on the first toss

B = both tosses have the same outcome

may seem intuitively related. Show that $P(B \mid A) = P(B)$, so that A and *B* are in fact independent.

6.107 The wine tasters, II In the setting of Exercise 6.100, Taster 1's rating for a wine is 3. What is the conditional probability that Taster 2's rating is higher than 3?

6.108 Poll on sensitive issues It is difficult to conduct sample surveys on sensitive issues because many people will not answer questions if the answers might embarrass them. *Randomized response* is an effective way to guarantee anonymity while collecting information on topics such as student cheating or sexual behavior. Here is the idea. To ask a sample of students whether they have plagiarized a term paper while in college, have each student toss a coin in private. If the coin lands "heads" and they have not plagiarized, they are to answer "No." Otherwise they are to give "Yes" as their answer. Only the student knows whether the answer reflects the truth or just the coin toss, but the researchers can use a proper random sample with follow-up for nonresponse and other good sampling practices.

Suppose that in fact the probability is 0.3 that a randomly chosen student has plagiarized a paper. Draw a tree diagram in which the first stage is tossing the coin and the second is the truth about plagiarism. The outcome at the end of each branch is the answer given to the randomized-response question. What is the probability of a "No" answer in the randomized-response poll? If the probability of plagiarism were 0.2, what would be the probability of a "No" response on the poll? Now suppose that you get 39% "No" answers in a randomized-response poll of a large sample of students at your college. What do you estimate to be the percent of the population who have plagiarized a paper?

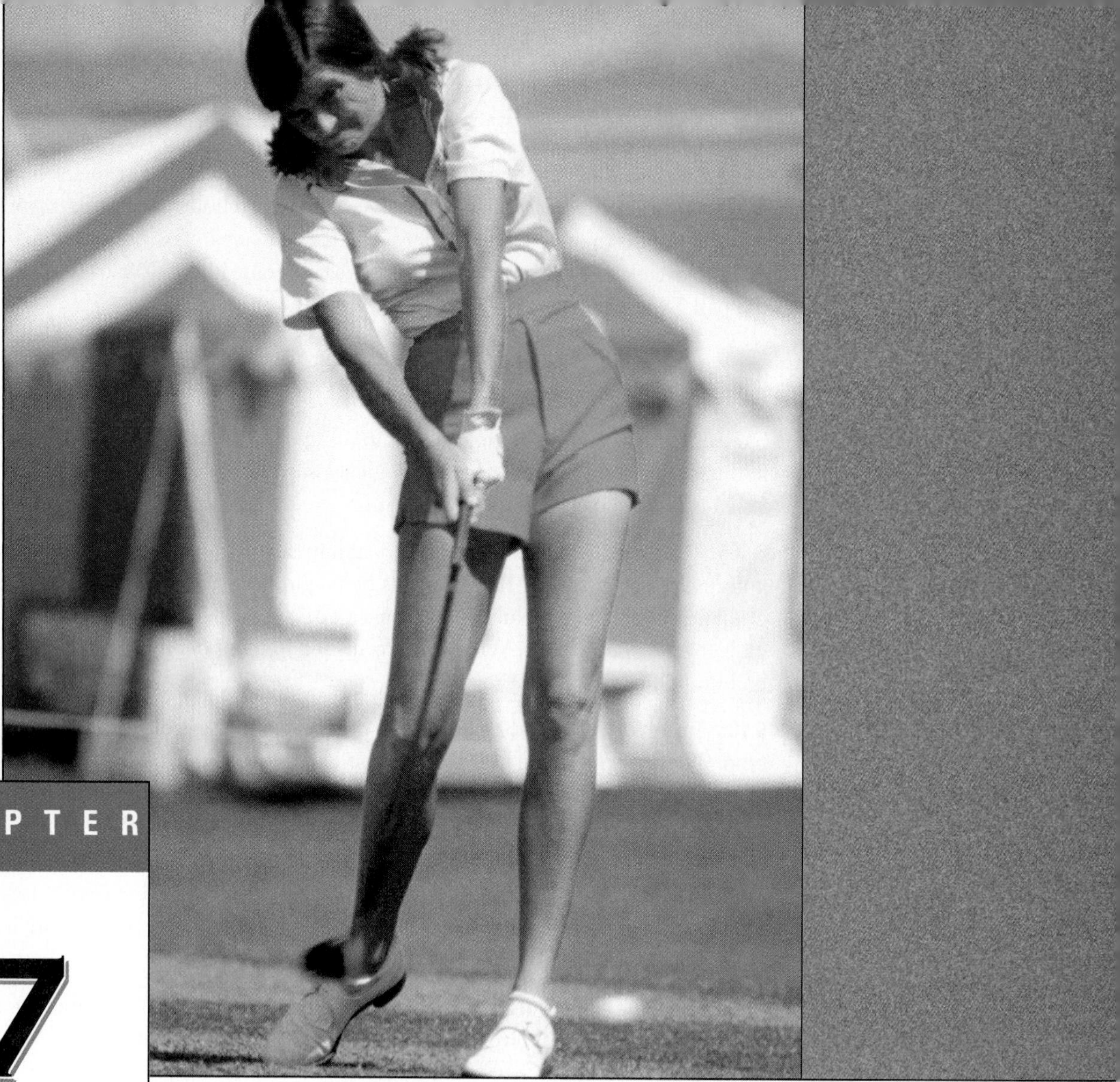

CHAPTER

7

Random Variables

CASE STUDY

Lost income and the courts

Jane Blaylock joined the Ladies Professional Golf Association (LPGA) in 1969 and by 1972 had become the leading money winner on the tour. During the Bluegrass Invitational Tournament, she was disqualified for an alleged rules infraction. The LPGA appointed a committee of her competitors, who suspended her from the next tournament, the Carling Open. Blaylock sued for damages and expenses under the Sherman Antitrust Act, which says that individuals cannot be prevented by their peers from working in their profession because it would lessen competition. She won but then had to come up with a method of determining how much money she might reasonably have made if she had been allowed to play in the next tournament. This is a difficult issue in most antitrust cases but was particularly problematic for a professional golfer, who might play well one day and poorly another day. Her task was challenging. She would have to use a measure that the judge and the jury would understand and that would be sufficiently convincing for a ruling in her favor. She and her legal team used a statistical procedure called the *expected value,* which we will study in this chapter. Using data from the nine most recent tournaments that Blaylock played in prior to the disqualification, they estimated the probability that she would achieve various scores based on her past performance. The scores for players who won money ranged from 209 (the tournament winner) to 232. To simplify things, the 24 possible scores were reduced to 8, where 210, for example, represents 209, 210, and 211. Here is a table that summarizes her possible outcomes and the probabilities calculated for each of the outcomes:

Possible outcomes X:	210	213	216	219	222	225	228	231
Probabilities, $P(X)$:	0.07	0.16	0.23	0.24	0.17	0.09	0.03	0.01

The probability is 0.07 that she would score 209, 210, or 211; and so forth. Using these numbers, her expected score was calculated to be approximately 218. Had she played in the tournament, her 218 would have earned her $1427.50. Not only was the jury persuaded, but they also believed that Blaylock might well have won the tournament, so they awarded her first-place money, $4500. This amount was then tripled to $13,500 to cover legal expenses, according to the provisions of the Sherman act. Statistics to the rescue.

Activity 7A
The game of craps

Materials: Pair of dice for each pair of students

The game of craps is one of the most famous (or notorious) of all gambling games played with dice. In this game, the player rolls a pair of six-sided dice, and the sum of the numbers that turn up on the two faces is noted. If the sum is 7 or 11, the player wins immediately. If the sum is 2, 3, or 12, the player loses immediately. If any other sum is obtained, the player continues to throw the dice until he either wins by repeating the first sum he obtained or loses by rolling a 7. Your mission in this activity is to estimate the probability of a player winning at craps. But first, let's get a feel for the game. For this activity, your class will be divided into groups of two. Your instructor will provide a pair of dice for each group of two students.

1. In your group of two students, play a total of 20 games of craps. One person will roll the dice; the other will keep track of the sums and record the end result (win or lose). If you like, you can switch jobs after 10 games have been completed. How many times out of 20 does the player win? What is the relative frequency (that is, percent, written as a decimal) of wins?
2. Combine your results with those of all the other two-student groups in the class. What is the relative frequency of wins for the entire class?
3. Use simulation techniques to represent 20 games of craps, using either the table of random digits or the random number generating feature of your TI-83/84/89. What is the relative frequency of wins based on the 20 simulations? How does this number compare to the relative frequency you found in Step 2?
4. One of the ways you can win at craps is to roll a sum of 7 or 11 on your first roll. Using your results and those of your fellow students, determine the number of times a player won by rolling a sum of 7 on the first roll. What is the relative frequency of rolling a sum of 7? Repeat these calculations for a sum of 11. Which of these sums appears more likely to occur than the other, based on the class results?
5. One of the ways you can lose at craps is to roll a sum of 2, 3, or 12 on your first roll. Using your results and those of your fellow students, determine the number of times a player lost by rolling a sum of 2 on the first roll. What is the relative frequency of rolling a sum of 2? Repeat these calculations for a sum of 3 and a sum of 12. Which of these sums appears more likely to occur than the others, based on the class results?

6. Clearly, the key quantity of interest in craps is the sum of the numbers on the two dice. Let's try to get a better idea of how this sum behaves in general by conducting a simulation. First, determine how you would simulate the roll of a single fair die. (*Hint:* Just use digits 1 to 6 and ignore the others.) Then determine how you would simulate a roll of two fair dice. Using this model, simulate 36 rolls of a pair of dice and determine the relative frequency of each of the possible sums. Alternatively, use the applet at the Web site nces.ed.gov/nceskids/probability/.
7. Construct a relative frequency histogram of the relative frequency results in Step 6. What is the approximate shape of the distribution? What sum appears most likely to occur? Which appears least likely to occur?
8. From the relative frequency data in Step 6, compute the relative frequency of winning and the relative frequency of losing on your first roll in craps. How do these simulated results compare with what the class obtained?

The fortune cookie works
In the March 30, 2005, Powerball lottery, 110 players got 5 of the 6 numbers correct and won almost $19 million. The lottery keeps a $25 million reserve for such odd situations. Normally, only 4 or 5 second-prize winners would be expected. The Multi-State Lottery Association, which runs Powerball, suspected fraud. But Powerball tickets are sold in 29 states, and lottery officials were befuddled. After some sleuthing, officials discovered that the 110 players got their numbers from the same batch of fortune cookies, which came from Wonton Food's Long Island City factory.

Introduction

Sample spaces need not consist of numbers. When we toss four coins, we can record the outcome as a string of heads and tails, such as HTTH. In statistics, however, we are most often interested in numerical outcomes such as the count of heads in the four tosses. It is convenient to use a shorthand notation: Let X be the number of heads. If our outcome is HTTH, then $X = 2$. If the next outcome is TTTH, the value of X changes to $X = 1$. The possible values of X are 0, 1, 2, 3, and 4. Tossing a coin four times will give X one of these possible values. Tossing four more times will give X another and probably different value. We call X a ***random variable*** because its values vary when the coin tossing is repeated.

Random Variable

A **random variable** is a variable whose value is a numerical outcome of a random phenomenon.

We usually denote random variables by capital letters near the end of the alphabet, such as X or Y. Of course, the random variables of greatest interest to us are outcomes such as the mean $\bar{x}$ of a random sample, for which we will keep the familiar notation.[1] As we progress from general rules of probability toward statistical inference, we will concentrate on random variables. When a random variable X describes a random phenomenon, the sample space S just lists the possible values of the random variable. We usually do not mention S separately. There remains the second part of any probability model, the assignment of probabilities to events. In this section, we will learn two ways of assigning probabilities to the values of a random variable. The two types of probability models that result will dominate our application of probability to statistical inference.

7.1 Discrete and Continuous Random Variables

Discrete Random Variables

We have learned several rules of probability but only one method of assigning probabilities: state the probabilities of the individual outcomes and assign probabilities to events by summing over the outcomes. The outcome probabilities must be between 0 and 1 and have sum 1. When the outcomes are numerical, they are values of a random variable. We will call such random variables having probability assigned in this way ***discrete random variables.***[2]

Discrete Random Variable

A **discrete random variable** X has a countable number of possible values. The **probability distribution** of a discrete random variable X lists the values and their probabilities:

Value of X:	x_1	x_2	x_3	...	x_k
Probability:	p_1	p_2	p_3	...	p_k

The probabilities p_i must satisfy two requirements:

1. Every probability p_i is a number between 0 and 1.
2. The sum of the probabilities is 1: $p_1 + p_2 + \ldots + p_k = 1$.

Find the probability of any event by adding the probabilities p_i of the particular values x_i that make up the event.

Example 7.1 *Getting good grades*
Finding discrete probabilities

North Carolina State University posts the grade distributions for its courses online. Students in Statistics 101 in the fall 2003 semester received 21% A's, 43% B's, 30% C's, 5% D's, and 1% F's. Choose a Statistics 101 student at random. To "choose at random" means to give every student the same chance to be chosen. The student's grade on a four-point scale (with A = 4) is a random variable X.

The value of X changes when we repeatedly choose students at random, but it is always one of 0, 1, 2, 3, or 4. Here is the distribution of X:

Value of X:	0	1	2	3	4
Probability:	0.01	0.05	0.30	0.43	0.21

The probability that the student got a B or better is the sum of the probabilities of an A and a B. In the language of random variables,

$$P(X \geq 3) = P(X = 3) + P(X = 4)$$
$$= 0.43 + 0.21 = 0.64$$

probability histogram

We can use histograms to show probability distributions as well as distributions of data. Figure 7.1 displays ***probability histograms*** that compare the probability model for random digits (Example 6.16, page 422) with the model given by Benford's law (Example 6.15, page 421). The height of each bar shows the probability of the outcome at its base. Because the heights are probabilities, they add to 1. As usual, all the bars in a histogram have the same width. So the areas of the bars also display the assignment of probability to outcomes. Think of these histograms as idealized pictures of the results of very many trials. The histograms make it easy to quickly compare the two distributions.

Figure 7.1 *Probability histograms for (a) random digits 0 to 9 and (b) Benford's law. The height of each bar shows the probability assigned to a single outcome.*

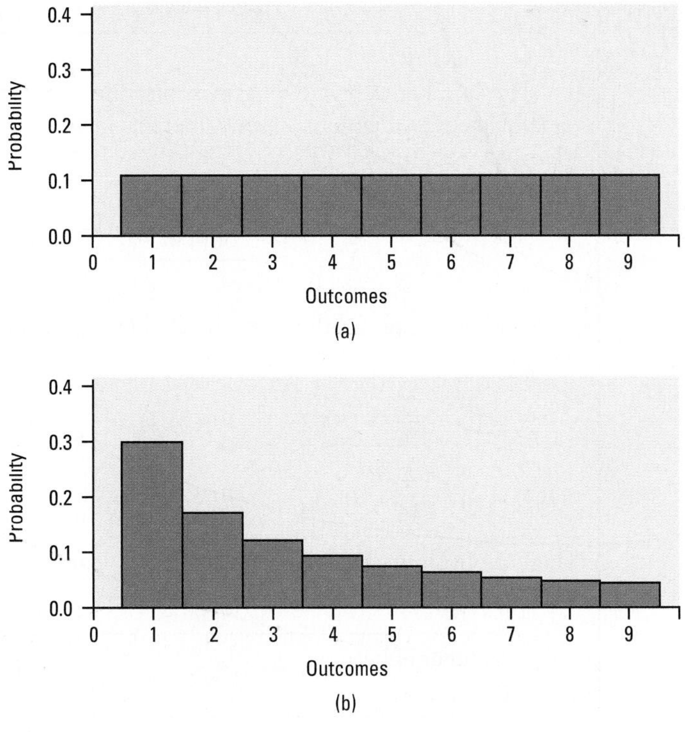

Example 7.2

Tossing coins
Values of a random variable

What is the probability distribution of the discrete random variable X that counts the number of heads in four tosses of a coin? We can derive this distribution if we make two reasonable assumptions:

1. The coin is balanced, so each toss is equally likely to give H or T.
2. The coin has no memory, so tosses are independent.

The outcome of four tosses is a sequence of heads and tails such as HTTH. There are 16 possible outcomes in all. Figure 7.2 lists these outcomes along with the value of X for each outcome. The multiplication rule for independent events tells us that, for example,

$$P(\text{HTTH}) = \frac{1}{2} \times \frac{1}{2} \times \frac{1}{2} \times \frac{1}{2} = \frac{1}{16}$$

Each of the 16 possible outcomes similarly has probability 1/16. That is, these outcomes are equally likely.

Figure 7.2 *Possible outcomes in four tosses of a coin, for Example 7.2. The random variable X is the number of heads.*

		HHTT		
		HTHT		
	HTTT	HTTH	HHHT	
	THTT	THHT	HHTH	
	TTHT	THTH	HTHH	
TTTT	TTTH	TTHH	THHH	HHHH
$X = 0$	$X = 1$	$X = 2$	$X = 3$	$X = 4$

The number of heads X has possible values 0, 1, 2, 3, and 4. These values are not equally likely. As Figure 7.2 shows, there is only one way that $X = 0$ can occur: namely, when the outcome is TTTT. So $P(X = 0) = 1/16$. But the event $\{X = 2\}$ can occur in six different ways, so that

$$P(X = 2) = \frac{\text{count of ways } X = 2 \text{ can occur}}{16} = \frac{6}{16}$$

We can find the probability of each value of X from Figure 7.2 in the same way. Here is the result:

$$P(X = 0) = \frac{1}{16} = 0.0625 \qquad P(X = 1) = \frac{4}{16} = 0.25 \qquad P(X = 2) = \frac{6}{16} = 0.375$$

$$P(X = 3) = \frac{4}{16} = 0.25 \qquad P(X = 4) = \frac{1}{16} = 0.0625$$

These probabilities have sum 1, so this is a legitimate probability distribution. In table form the distribution is

Number of heads:	0	1	2	3	4
Probability:	0.0625	0.25	0.375	0.25	0.0625

Figure 7.3 is a probability histogram for this distribution. The probability distribution is exactly symmetric. It is an idealization of the relative frequency distribution of the number of heads after many tosses of four coins, which would be nearly symmetric but is unlikely to be exactly symmetric.

Figure 7.3 *Probability histogram for the number of heads in four tosses of a coin, for Example 7.2.*

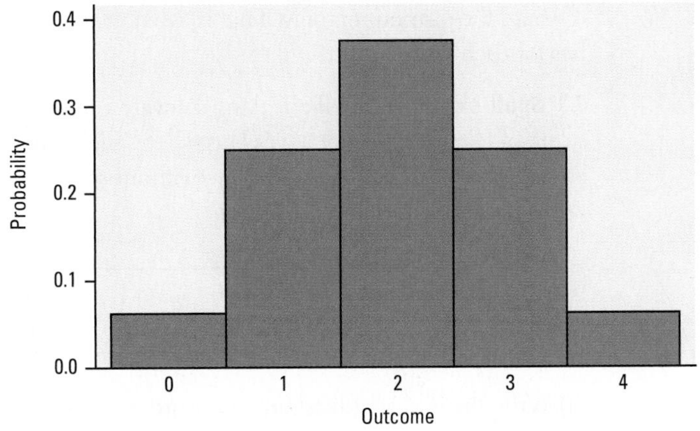

Any event involving the number of heads observed can be expressed in terms of X, and its probability can be found from the distribution of X. For example, the probability of tossing at most two heads is

$$\begin{aligned} P(X \leq 2) &= P(X = 0) + P(X = 1) + P(X = 2) \\ &= 0.0625 + 0.25 + 0.375 \\ &= 0.6875 \end{aligned}$$

The probability of at least one head is most simply found by use of the complement rule:

$$\begin{aligned} P(X \geq 1) &= 1 - P(X = 0) \\ &= 1 - 0.0625 = 0.9375 \end{aligned}$$

Tossing a coin n *times is similar to choosing an SRS of size* n *from a large population and asking a yes-or-no question.* We will extend the results of Example 7.2 when we return to sampling distributions in the next two chapters.

Exercises

7.1 Roll of the die If a carefully made die is rolled once, it is reasonable to assign probability 1/6 to each of the six faces.

(a) What is the probability of rolling a number less than 3?

(b) Use your TI-83/84/89 to simulate rolling a die 100 times, and assign the values to L_1/list1. Sort the list in ascending order, and then count the outcomes that are either 1s or 2s. Record the relative frequency.

(c) Repeat part (b) four more times, and then average the five relative frequencies. Is this number close to your result in (a)?

7.2 Three children, I A couple plans to have three children. There are 8 possible arrangements of girls and boys. For example, GGB means the first two children are girls and the third child is a boy. All 8 arrangements are (approximately) equally likely.

(a) Write down all 8 arrangements of the sexes of three children. What is the probability of any one of these arrangements?

(b) Let X be the number of girls the couple has. What is the probability that $X = 2$?

(c) Find the distribution of X. That is, what values can X take, and what are the probabilities for each value?

7.3 Spell-checking Spell-checking software catches "nonword errors," which result in a string of letters that is not a word, as when "the" is typed as "teh." When undergraduates are asked to write a 250-word essay (without spell-checking), the number X of nonword errors has the following distribution:

Value of X:	0	1	2	3	4
Probability:	0.1	0.2	0.3	0.3	0.1

(a) Write the event "at least one nonword error" in terms of X. What is the probability of this event?

(b) Describe the event $X \leq 2$ in words. What is its probability? What is the probability that $X < 2$?

7.4 Housing in San Jose, I How do rented housing units differ from units occupied by their owners? Here are the distributions of the number of rooms for owner-occupied units and renter-occupied units in San Jose, California:[3]

	Number of Rooms									
	1	2	3	4	5	6	7	8	9	10
Owned	0.003	0.002	0.023	0.104	0.210	0.224	0.197	0.149	0.053	0.035
Rented	0.008	0.027	0.287	0.363	0.164	0.093	0.039	0.013	0.003	0.003

Make probability histograms of these two distributions, using the same scales. What are the most important differences between the distributions of owner-occupied and rented housing units?

7.5 Housing in San Jose, II Let the random variable X be the number of rooms in a randomly chosen owner-occupied housing unit in San Jose, California. Exercise 7.4 gives the distribution of X.

(a) Express "the unit has five or more rooms" in terms of X. What is the probability of this event?

(b) Express the event $\{X > 5\}$ in words. What is its probability?

(c) What important fact about discrete random variables does comparing your answers to (a) and (b) illustrate?

7.6 Pass on the message Example 6.20 (page 428) gives a model for an attempt to send an email message to someone you don't know by sending an email to an acquaintance who you think is closer to your target and asking him or her to pass on the message. At each step after you send the initial email, the probability is 0.37 that the recipient will pass on your message. Steps are independent. Let T be the total length of the chain. That is, if your friend does not relay your message, $T = 2$ because the chain stops at the second person. If your friend passes it on but the next person does not, $T = 3$, and so on.

(a) What is $P(T = 2)$? What is $P(T = 3)$?

(b) Express in words the meaning of $P(T \leq 4)$. What is the value of this probability?

Continuous Random Variables

When we use the table of random digits to select a digit between 0 and 9, the result is a discrete random variable. The probability model assigns probability 1/10 to each of the 10 possible outcomes, as Figure 7.1(a) on page 467 shows. Suppose that we want to choose a number at random between 0 and 1, allowing any number between 0 and 1 as the outcome. Software random number generators will do this. You can visualize such a random number by thinking of a spinner (Figure 7.4) that turns freely on its axis and slowly comes to a stop. The pointer can come to rest anywhere on a circle that is marked from 0 to 1. The sample space is now an entire interval of numbers:

$$S = \{\text{all numbers } x \text{ such that } 0 \leq x \leq 1\}$$

Figure 7.4 *A spinner that generates a random number between 0 and 1.*

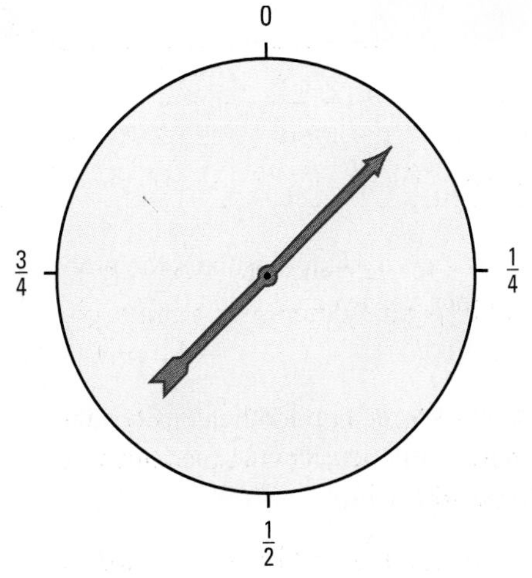

How can we assign probabilities to such events as $0.3 \leq x \leq 0.7$? As in the case of selecting a random digit, we would like all possible outcomes to be equally likely. But we cannot assign probabilities to each individual value of x and then sum, because there are infinitely many possible values. Instead, we use a new way of assigning probabilities directly to events as *areas under a density curve*. Any density curve has area exactly 1 underneath it, corresponding to total probability 1.

Example 7.3

Random numbers and the uniform distribution

Areas under a density curve

The random number generator will spread its output uniformly across the entire interval from 0 to 1 as we allow it to generate a long sequence of numbers. The results of many trials are represented by the density curve of a *uniform distribution* (Figure 7.5). This density curve has height 1 over the interval from 0 to 1. The area under the density curve is 1, and the probability of any event is the area under the density curve and above the event in question.

Figure 7.5 *Assigning probability for generating a random number between 0 and 1, for Example 7.3. The probability of any interval of numbers is the area above the interval and under the curve.*

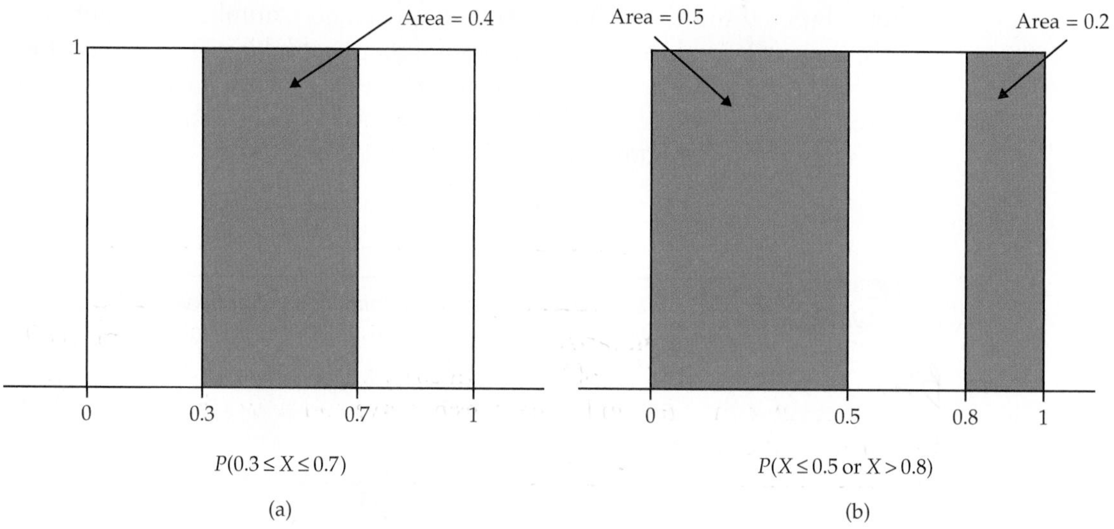

As Figure 7.5(a) illustrates, the probability that the random number generator produces a number X between 0.3 and 0.7 is

$$P(0.3 \leq X \leq 0.7) = 0.4$$

because the area under the density curve and above the interval from 0.3 to 0.7 is 0.4. The height of the density curve is 1 and the area of a rectangle is the product of height and length, so the probability of any interval of outcomes is just the length of the interval. So,

$$P(X \leq 0.5) = 0.5$$
$$P(X > 0.8) = 0.2$$
$$P(X \leq 0.5 \text{ or } X > 0.8) = 0.7$$

Notice that the last event consists of two nonoverlapping intervals, so the total area above the event is found by adding two areas, as illustrated by Figure 7.5(b). This assignment of probabilities obeys all of our rules for probability.

Probability as area under a density curve is a second important way of assigning probabilities to events. Figure 7.6 illustrates this idea in general form. We call X in Example 7.3 a ***continuous random variable*** because its values are not isolated numbers but an entire interval of numbers.

Figure 7.6 *The probability distribution of a continuous random variable assigns probabilities as area under a density curve.*

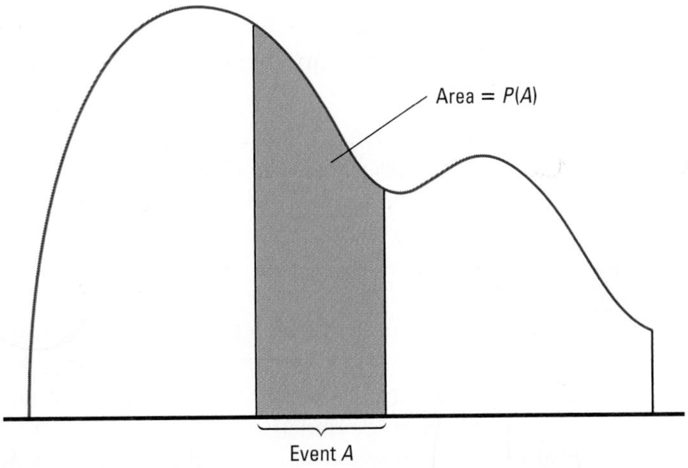

Continuous Random Variable

A **continuous random variable** X takes all values in an interval of numbers. The **probability distribution** of X is described by a density curve. The probability of any event is the area under the density curve and above the values of X that make up the event.

CAUTION

The probability model for a continuous random variable assigns probabilities to intervals of outcomes rather than to individual outcomes. In fact, **all continuous probability distributions assign probability 0 to every individual outcome!** Only intervals of values have positive probability. To see that this is true, consider a specific outcome such as $P(X = 0.8)$ in Example 7.3. The probability of any interval is the same as its length. The point 0.8 has no length, so its probability is 0. Although this fact may seem odd at first glance, it does make intuitive as well as mathematical sense. The random number generator produces a number between 0.79 and 0.81 with probability 0.02. An outcome between 0.799 and 0.801 has probability 0.002, and a result between 0.7999 and 0.8001 has probability 0.0002. Continuing to hone in on 0.8, you can see why an outcome exactly equal to 0.8 should have probability 0. Because there is no probability exactly at $X = 0.8$, the two events $\{X > 0.8\}$ and $\{X \geq 0.8\}$ have the same probability. We can ignore the distinction between $>$ and $\geq$ when finding probabilities for continuous (but not discrete) random variables.

Normal Distributions as Probability Distributions

The density curves that are most familiar to us are the Normal curves. (We discussed Normal curves in Section 2.2.) Because any density curve describes an assignment of probabilities, *Normal distributions are probability distributions.* Recall that $N(\mu, \sigma)$ is our shorthand notation for the Normal distribution having mean μ and standard deviation σ. In the language of random variables, if X has the $N(\mu, \sigma)$ distribution, then the standardized variable

$$Z = \frac{X - \mu}{\sigma}$$

is a standard Normal random variable having the distribution $N(0, 1)$.

Example 7.4 *Cheating in school*
Continuous random variables

Students are reluctant to report cheating by other students. A sample survey puts this question to an SRS of 400 undergraduates: "You witness two students cheating on a quiz. Do you go to the professor?" Suppose that if we could ask all undergraduates, 12% would answer "Yes."[4]

The proportion $p = 0.12$ is a *parameter* that describes the population of all undergraduates. The proportion $\hat{p}$ of the sample who answer "Yes" is a *statistic* used to estimate p. The statistic $\hat{p}$ is a random variable, because repeating the SRS would give a different sample of 400 undergraduates and a different value of $\hat{p}$. We will see in the next chapter that $\hat{p}$ has approximately the $N(0.12, 0.016)$ distribution. The mean 0.12 of this distribution is the same as the population parameter. The standard deviation is controlled mainly by the size of the sample.

What is the probability that the survey result differs from the truth about the population by more than 2 percentage points? Because $p = 0.12$, the survey misses by more than 2 percentage points if $\hat{p} < 0.10$ or $\hat{p} > 0.14$. Figure 7.7 shows this probability as an area under a Normal density curve. You can find it using software or by standardizing and using Table A. Let's start with the complement rule,

$$P(\hat{p} < 0.10 \text{ or } \hat{p} > 0.14) = 1 - P(0.10 \leq \hat{p} \leq 0.14)$$

From Table A,

$$\begin{aligned} P(0.10 \leq \hat{p} \leq 0.14) &= P\left(\frac{0.10 - 0.12}{0.016} \leq \frac{\hat{p} - 0.12}{0.016} \leq \frac{0.14 - 0.12}{0.016}\right) \\ &= P(-1.25 \leq Z \leq 1.25) \\ &= 0.8944 - 0.1056 = 0.7888 \end{aligned}$$

The probability we seek is therefore $1 - 0.7888 = 0.2112$. About 21% of sample results will be off by more than two percentage points. The arrangement of this calculation is familiar from our earlier work with Normal distributions. Only the language of probability is new.

Figure 7.7 *Probability as area under a Normal density curve, for Example 7.4.*

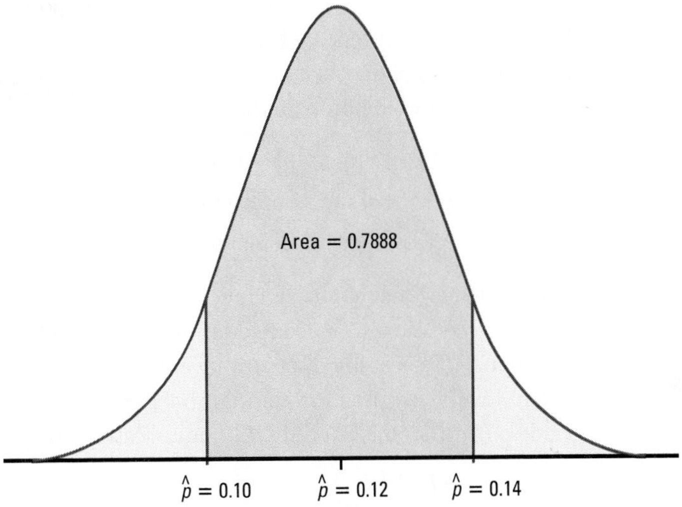

Exercises

7.7 Continuous random variable, I Let the random variable X be a random number with the uniform density curve in Figure 7.5 (page 472). Find the following probabilities:

(a) $P(X < 0.49)$

(b) $P(X \leq 0.49)$

(c) $P(X \geq 0.27)$

(d) $P(0.27 < X < 1.27)$

(e) $P(0.1 \leq X \leq 0.2 \text{ or } 0.8 \leq X \leq 0.9)$

(f) The probability that X is not in the interval from 0.3 to 0.8.

(g) $P(X = 0.5)$

7.8 Continuous random variable, II Let X be a random number with a uniform distribution between 0 and 1 as described in Example 7.3 and Figure 7.5 (page 472). Find the following probabilities:

(a) $P(0 \leq X \leq 0.4)$

(b) $P(0.4 \leq X \leq 1)$

(c) $P(0.3 \leq X \leq 0.5)$

(d) $P(0.3 < X < 0.5)$

(e) $P(0.226 \leq X \leq 0.713)$

(f) What important fact about continuous random variables does comparing your answers to (c) and (d) illustrate?

7.9 Violence in schools, I An SRS of 400 American adults is asked, "What do you think is the most serious problem facing our schools?" Suppose that in fact 40% of all adults would answer "violence" if asked this question. The proportion $\hat{p}$ of the sample who answer "violence" will vary in repeated sampling. In fact, we can assign probabilities to values of $\hat{p}$ using the Normal density curve with mean 0.4 and standard deviation 0.024. Use this density curve to find the probabilities of the following events:

(a) At least 45% of the sample believes that violence is the schools' most serious problem.

(b) Less than 35% of the sample believes that violence is the most serious problem.

(c) The sample proportion is between 0.35 and 0.45.

7.10 Violence in schools, II How could you design a simulation to answer part (b) of Exercise 7.9? What we need to do is simulate 400 observations from the N(0.4, 0.024) distribution. This is easily done on the calculator. Here's one way. Clear L_1/list1 and enter the following commands (randNorm is found under the MATH/PRB menu on the TI-83/84, and in the CATALOG under FlashApps on the TI-89):

TI-83/84	TI-89
• `randNorm(0.4,.024,400)→L₁`	• `tistat.randNorm(0.4,.024,400)→list1`

This will select 400 random observations from the N(0.4, 0.024) distribution.

TI-83/84	TI-89
• `SortA(L₁)`	• `SortA list1`

This will sort the 400 observations in L_1/list1 in ascending order.

Then scroll through L_1/list1. How many entries (observations) are less than 0.35? What is the relative frequency of this event? Compare the results of your simulation with your answer to Exercise 7.9(b).

Section 7.1 Summary

The previous chapter included a general discussion of the idea of probability and the properties of probability models. Two very useful specific types of probability models are distributions of discrete and continuous random variables. In our study of statistics we will employ only these two types of probability models.

A **random variable** is a variable taking numerical values determined by the outcome of a random phenomenon. The **probability distribution** of a random variable X tells us what the possible values of X are and what probabilities are assigned to those values.

A random variable X and its distribution can be discrete or continuous.

A **discrete random variable** has a countable number of possible values. The probability distribution assigns each of these values a probability between 0 and 1 such that the sum of all the probabilities is exactly 1. The probability of any event is the sum of the probabilities of all the values that make up the event.

A **continuous random variable** takes all values in some interval of numbers. A density curve describes the probability distribution of a continuous random variable. The probability of any event is the area under the curve above the values that make up the event.

Normal distributions are one type of continuous probability distribution.

You can picture a probability distribution by drawing a probability histogram in the discrete case or by graphing the density curve in the continuous case.

When you work problems, get in the habit of first identifying the random variable of interest. X = number of ______ for discrete random variables, and X = amount of ______ for continuous random variables.

Section 7.1 Exercises

7.11 Rolling two dice Some games of chance rely on tossing two dice. Each die has six faces, marked with 1, 2, . . . , 6 spots called pips. The dice used in casinos are carefully balanced so that each face is equally likely to come up. When two dice are tossed, each of the 36 possible pairs of faces is equally likely to come up. The outcome of interest to a gambler is the sum of the pips on the two up-faces. Call this random variable X.

(a) Write down all 36 possible pairs of faces.

(b) If all pairs have the same probability, what must be the probability of each pair?

(c) Define the random variable X. Then write the value of X next to each pair of faces and use this information with the result of (b) to give the probability distribution of X. Draw a probability histogram to display the distribution.

(d) One bet available in craps wins if a 7 or 11 comes up on the next roll of two dice. What is the probability of rolling a 7 or 11 on the next roll? Compare your answer with your results (relative frequency) in Activity 7A, Step 4.

(e) After the dice are rolled the first time, several bets lose if a 7 is then rolled. If any outcome other than a 7 occurs, these bets either win or continue to the next roll. What is the probability that anything other than a 7 is rolled?

7.12 Size of American households, I In government data, a household consists of all occupants of a dwelling unit, while a family consists of two or more persons who live together and are related by blood or marriage. So all families form households, but some households are not families. Here are the distributions of household size and family size in the United States:

	Number of Persons						
	1	2	3	4	5	6	7
Household probability	0.25	0.32	0.17	0.15	0.07	0.03	0.01
Family probability	0	0.42	0.23	0.21	0.09	0.03	0.02

(a) Verify that each is a legitimate discrete probability distribution.

(b) Make probability histograms for these two discrete distributions, using the same scales. What are the most important differences between the sizes of households and families?

7.13 Size of American households, II Choose an American household at random and let the random variable Y be the number of persons living in the household. Exercise 7.12 gives the distribution of Y.

(a) Express "more than one person lives in this household" in terms of Y. What is the probability of this event?

(b) What is $P(2 < Y \leq 4)$?

(c) What is $P(Y \neq 2)$?

7.14 Car ownership Choose an American household at random and let the random variable X be the number of cars (including SUVs and light trucks) they own. Here is the probability model if we ignore the few households that own more than 5 cars:

Number of cars X:	0	1	2	3	4	5
Probability:	0.09	0.36	0.35	0.13	0.05	0.02

(a) Verify that this is a legitimate discrete distribution. Display the distribution in a probability histogram.

(b) Say in words what the event $\{X \geq 1\}$ is. Find $P(X \geq 1)$.

(c) A housing company builds houses with two-car garages. What percent of households have more cars than the garage can hold?

7.15 Education levels A study of education followed a large group of fifth-grade children to see how many years of school they eventually completed. Let X be the highest year of school that a randomly chosen fifth-grader completes. (Students who go on to college are included in the outcome $X = 12$.) The study found this probability distribution for X:

Years:	4	5	6	7	8	9	10	11	12
Probability:	0.010	0.007	0.007	0.013	0.032	0.068	0.070	0.041	0.752

(a) Check that this is a legitimate discrete probability distribution.

(b) What percent of fifth-graders eventually finished twelfth grade?

(c) Find $P(X \geq 6)$.

(d) Find $P(X > 6)$.

(e) What values of X make up the event "the student completed at least one year of high school"? (High school begins with the ninth grade.) What is the probability of this event?

7.16 How student fees are used Weary of the low turnout in student elections, a college administration decides to choose an SRS of three students to form an advisory board that represents student opinion. Suppose that 40% of all students oppose the use of student fees to fund student interest groups and that the opinions of the three students on the board are independent. Then the probability is 0.4 that each opposes the funding of interest groups.

(a) Call the three students A, B, and C. What is the probability that A and B support funding and C opposes it?

(b) List all possible combinations of opinions that can be held by Students A, B, and C. (*Hint:* There are eight possibilities.) Then give the probability of each of these outcomes. Note that they are not equally likely.

(c) Let the random variable X be the number of student representatives who oppose the funding of interest groups. Give the probability distribution of X.

(d) Express the event "a majority of the advisory board opposes funding" in terms of X and find its probability.

7.17 A uniform distribution Many random number generators allow users to specify the range of the random numbers to be produced. Suppose that you specify that the range is to be $0 \le Y \le 2$. Then the density curve of the outcomes has constant height between 0 and 2, and height 0 elsewhere.

(a) What is the height of the density curve between 0 and 2? Draw a graph of the density curve.

(b) Use your graph from (a) and the fact that probability is area under the curve to find $P(Y \le 1)$.

(c) Find $P(0.5 < Y < 1.3)$.

(d) Find $P(Y \ge 0.8)$.

7.18 The sum of two random decimals Generate two random numbers between 0 and 1 and take Y to be their sum. Then Y is a continuous random variable that can take any value between 0 and 2. The density curve of Y is the triangle shown in Figure 7.8.

Figure 7.8 *The density curve for the sum of two random numbers, for Exercise 7.18. This continuous random variable takes values between 0 and 2.*

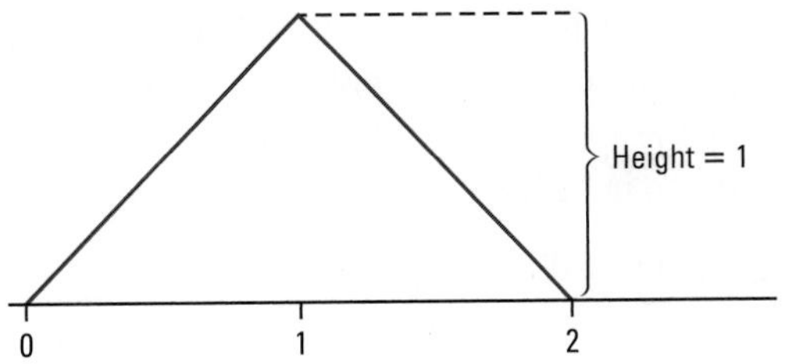

(a) Verify that the area under this curve is 1.

(b) What is the probability that Y is less than 1? (Sketch the density curve, shade the area that represents the probability, and then find that area. Do this for (c) also.)

(c) What is the probability that Y is less than 0.5?

(d) Use simulation methods to answer the questions in (b) and (c). Here's one way using the TI-83/84/89. Clear L_1/list1, L_2/list2, and L_3/list3 and enter these commands:

TI-83/84	TI-89	
`rand(200)→L1`	`tistat.rand83 (200)→list1`	Generates 200 random numbers and stores them in L_1/list1
`rand(200)→L2`	`tistat.rand83 (200)→list2`	Generates 200 random numbers and stores them in L_2/list2
`L1+L2→L3`	`list1+list2→ list3`	Adds the first number in L_1/list1 and the first number in L_2/list2 and stores the sum in L_3/list3, and so forth, from $i = 1$ to $i = 200$
`SortA(L3)`	`SortA list3`	Sorts the sums in L_3/list3 in ascending order

Now simply scroll through L_3/list3 and count the number of sums that satisfy the conditions stated in (b) and (c), and determine the relative frequencies.

7.19 Picturing a distribution This is a continuation of the previous exercise. If you carried out the simulation in 7.18(d), you can picture the distribution as follows: Deselect any active functions in the $Y =$ screen, and turn off all STAT PLOTs. Define Plot1 to be a histogram using L_3/list3. On the TI-89, set the Hist. Bucket Width at 0.1. Set WINDOW dimensions as follows: $X[0,2]_{0.1}$ and $Y[-6,25]_5$. Then press GRAPH. Does the resulting histogram resemble the triangle in Figure 7.8? Can you imagine the triangle superimposed on top of the histogram? Of course, some bars will be too short and others will be too long, but this is due to chance variation. To overlay the triangle, define Y_1 to be

- TI-83/84: $Y_1 = (25X)(X \geq 0 \text{ and } X \leq 1) + (-25X + 50)(X \geq 1 \text{ and } X \leq 2)$
- TI-89: when $(x \geq 0 \text{ and } x \leq 2, \text{when } (x \leq 1, 25x, -25x + 50), 0)$

Then press GRAPH again. How well does this "curve" fit your histogram?

7.20 Joggers, I An opinion poll asks an SRS of 1500 adults, "Do you happen to jog?" Suppose that the population proportion who jog is $p = 0.15$. To estimate p, we use the proportion $\hat{p}$ in the sample who answer "Yes." The statistic $\hat{p}$ is a random variable that is approximately Normally distributed with mean $\mu = 0.15$ and standard deviation $\sigma = 0.0092$. Find the following probabilities:

(a) $P(\hat{p} \geq 0.16)$

(b) $P(0.14 \leq \hat{p} \leq 0.16)$

7.21 Joggers, II Describe the details of a simulation you could carry out to approximate an answer to Exercise 7.20(a). Then carry out the simulation. About how many repetitions do you need to get a result close to your answer to Exercise 7.20(a)?

7.22 Weird dice Nonstandard dice can produce interesting distributions of outcomes. You have two balanced, six-sided dice. One is a standard die, with faces having 1, 2, 3, 4, 5, and 6 spots. The other die has three faces with 0 spots and three faces with 6 spots. Find the probability distribution for the total number of spots Y on the up-faces when you roll these two dice.

7.2 Means and Variances of Random Variables

Activity 7B

Means of random variables

To see how means of random variables work, consider a random variable that takes values {1, 1, 2, 3, 5, 8}. Do the following.

1. Calculate the mean μ of the population.
2. Make a list of all of the samples of size 2 from this population. (*Caution:* Notice that in our population, the first two values are the same. To distinguish them from one another, we will use subscripts: 1_a and 1_b.) As a check, you should have 15 subsets of size 2. Here's the beginning of our list:

Sample number	Sample	$\bar{x}$
1	1_a, 1_b	1
2	1_a, 2	1.5
etc.		

3. Find the mean of the 15 $\bar{x}$-values in the third column. Compare this with the population mean that you calculated in Step 1.
4. Repeat Steps 1 to 3 for a different (but still small) population of your choice. Now compare your results with those of other students in your class.
5. Write a brief statement that describes what you discovered.

Probability is the mathematical language that describes the long-run regular behavior of random phenomena. The probability distribution of a random variable is an idealized relative frequency distribution. The probability histograms and density curves that picture probability distributions resemble our earlier pictures of distributions of data. In describing data, we moved from graphs to numerical measures such as means and standard deviations. Now we will make the same move to expand our descriptions of the distributions of random variables. We can speak of the mean winnings in a game of chance or the standard deviation of the randomly varying number of calls a travel agency receives in an hour. In this section we will learn more about how to compute these descriptive measures and about the laws they obey.

The Mean of a Random Variable

The mean $\bar{x}$ of a set of observations is their ordinary average. The mean of a discrete random variable X is also an average of the possible values of X, but with an essential change to take into account the fact that not all outcomes need be equally likely. An example will show what we must do.

Example 7.5 *The Tri-State Pick 3*
Expected value

Most states and Canadian provinces have government-sponsored lotteries. Here is a simple lottery wager, from the Tri-State Pick 3 game that New Hampshire shares with Maine and Vermont. You choose a three-digit number; the state chooses a three-digit winning number at random and pays you \$500 if your number is chosen. Because there are 1000 three-digit numbers, you have probability 1/1000 of winning. Taking X to be the amount your ticket pays you, the probability distribution of X is

Payoff X:	\$0	\$500
Probability:	0.999	0.001

What is your average payoff from many tickets? The ordinary average of the two possible outcomes \$0 and \$500 is \$250, but that makes no sense as the average because \$500 is much less likely than \$0. In the long run you receive \$500 once in every 1000 tickets and \$0 on the remaining 999 of 1000 tickets. The long-run average payoff is

$$\$500\,\frac{1}{1000} + \$0\,\frac{999}{1000} = \$0.50$$

or fifty cents. That number is the mean of the random variable X. (Tickets cost \$1, so in the long run the state keeps half the money you wager.)

If you play Tri-State Pick 3 several times, we would as usual call the mean of the actual amounts you win $\bar{x}$. The mean in Example 7.5 is a different quantity—it is the long-run average winnings you expect if you play a very large number of times. Just as probabilities are an idealized description of long-run proportions, the mean of a probability distribution describes the long-run average outcome. We can't call this mean $\bar{x}$, so we need a different symbol. The common symbol for the ***mean of a probability distribution*** is μ, the Greek letter mu. We used μ in Chapter 2 for the mean of a Normal distribution, so this is not a new notation. We will often be interested in several random variables, each having a different probability distribution with a different mean. To remind ourselves that we are talking about the mean of X we often write μ_X rather than simply μ. In Example 7.5, $\mu_X = \$0.50$. Notice that, as often happens, the mean is not a possible value of X. You will often find the mean of a random variable X called the ***expected value*** of X. This term can be misleading, for we don't necessarily expect one observation on X to be close to the expected value of X.

mean μ

expected value

The **mean of any discrete random variable** is found just as in Example 7.5. It is an average of the possible outcomes, but a weighted average in which each outcome is weighted by its probability. Because the probabilities add to 1, we have total weight 1 to distribute among the outcomes. An outcome that occurs half the time has probability one-half and so gets one-half the weight in calculating the mean. Here is the general definition.

Mean of a Discrete Random Variable

Suppose that X is a discrete random variable whose distribution is

Value of X:	x_1	x_2	x_3	...	x_k
Probability:	p_1	p_2	p_3	...	p_k

To find the **mean** of X, multiply each possible value by its probability, then add all the products:

$$\begin{aligned}\mu_X &= x_1p_1 + x_2p_2 + \ldots + x_kp_k \\ &= \sum x_ip_i\end{aligned}$$

Example 7.6

Benford's law

Calculating the expected first digit

If first digits in a set of data appear "at random," the nine possible digits 1 to 9 all have the same probability. The probability distribution of the first digit X is then

First digit X:	1	2	3	4	5	6	7	8	9
Probability:	1/9	1/9	1/9	1/9	1/9	1/9	1/9	1/9	1/9

The mean of this distribution is

$$\begin{aligned}\mu_X &= 1 \times \frac{1}{9} + 2 \times \frac{1}{9} + 3 \times \frac{1}{9} + 4 \times \frac{1}{9} + 5 \times \frac{1}{9} + 6 \times \frac{1}{9} + 7 \times \frac{1}{9} + 8 \times \frac{1}{9} + 9 \times \frac{1}{9} \\ &= 45 \times \frac{1}{9} = 5\end{aligned}$$

If, on the other hand, the data obey Benford's law, the distribution of the first digit V is

First digit V:	1	2	3	4	5	6	7	8	9
Probability:	0.301	0.176	0.125	0.097	0.079	0.067	0.058	0.051	0.046

The mean of V is

$$\begin{aligned}\mu_V &= (1)(0.301) + (2)(0.176) + (3)(0.125) + (4)(0.097) + (5)(0.079) + (6)(0.067) \\ &\quad + (7)(0.058) + (8)(0.051) + (9)(0.046) \\ &= 3.441\end{aligned}$$

The different means reflect the greater probability of smaller first digits under Benford's law.

Figure 7.9 on the next page locates the means of X and V on the two probability histograms. Because the discrete uniform distribution in Figure 7.9(a) is symmetric, the mean lies at the center of symmetry. We can't locate the mean of the right-skewed distribution in Figure 7.9(b) by eye—calculation is needed.

Figure 7.9 *Locating the mean of a discrete random variable on the probability histogram for (a) digits between 1 and 9 chosen at random and (b) digits between 1 and 9 chosen from records that obey Benford's law.*

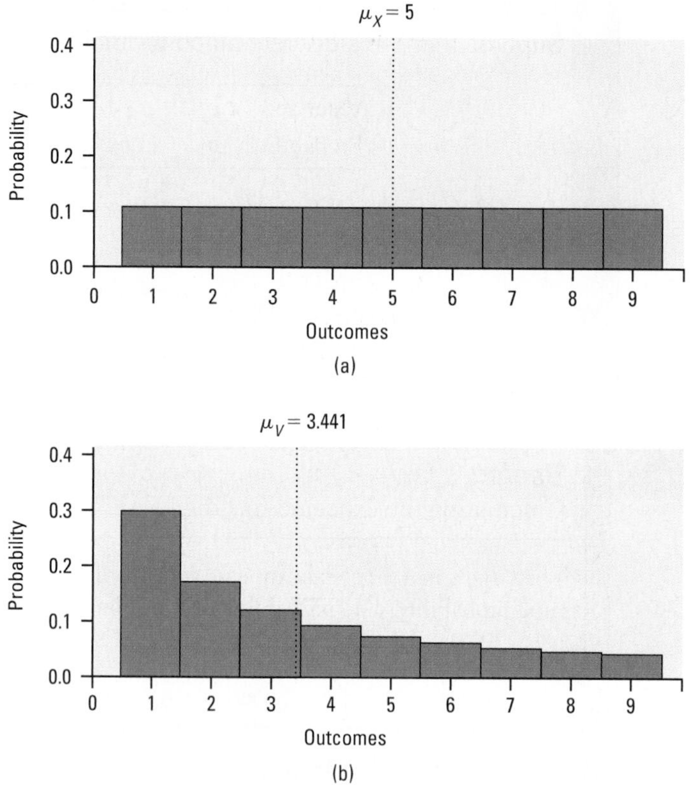

What about continuous random variables? The probability distribution of a continuous random variable X is described by a density curve. Chapter 2 showed how to find the mean of the distribution: it is the point at which the area under the density curve would balance if it were made out of solid material. The mean lies at the center of symmetric density curves such as the Normal curves. Exact calculation of the mean of a distribution with a skewed density curve requires advanced mathematics.[5]

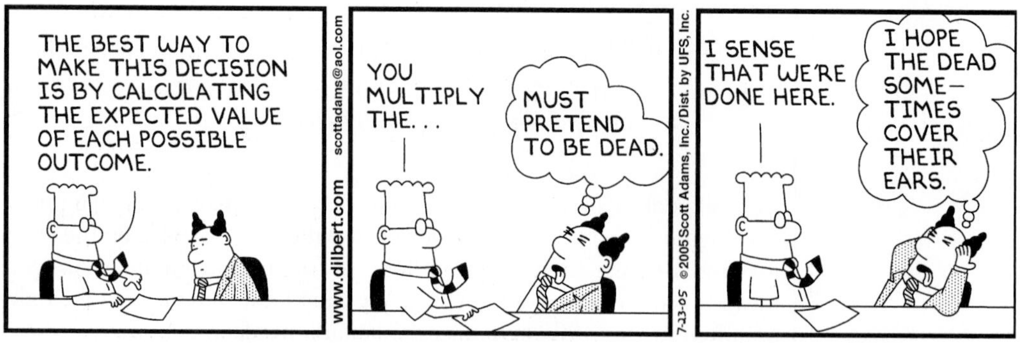

The Variance of a Random Variable

The mean is a measure of the center of a distribution. Even the most basic numerical description requires in addition a measure of the spread or variability of the distribution. The variance and the standard deviation are the measures of spread that accompany the choice of the mean to measure center. Just as for the mean, we need a distinct symbol to distinguish the variance of a random variable from the variance s^2 of a data set. We write the variance of a random variable X as σ_X^2. Once again the subscript reminds us which variable we have in mind. The definition of the variance σ_X^2 of a random variable is similar to the definition of the sample variance s^2 given in Chapter 1. That is, the variance is an average of the squared deviation $(X - \mu_X)^2$ of the variable X from its mean μ_X. As for the mean, the average we use is a weighted average in which each outcome is weighted by its probability in order to take account of outcomes that are not equally likely. Calculating this weighted average is straightforward for discrete random variables but requires advanced mathematics in the continuous case. Here is the definition.

Variance of a Discrete Random Variable

Suppose that X is a discrete random variable whose distribution is

Value of X:	x_1	x_2	x_3	...	x_k
Probability:	p_1	p_2	p_3	...	p_k

and that μ is the mean of X. The **variance** of X is

$$\sigma_X^2 = (x_1 - \mu_X)^2 p_1 + (x_2 - \mu_X)^2 p_2 + \ldots + (x_k - \mu_X)^2 p_k$$
$$= \sum (x_i - \mu_X)^2 p_i$$

The **standard deviation** σ_X of X is the square root of the variance.

Example 7.7 *Linda sells cars*
Variance of a random variable

Linda is a sales associate at a large auto dealership. She motivates herself by using probability estimates of her sales. For a sunny Saturday in April, she estimates her car sales as follows:

Cars sold:	0	1	2	3
Probability:	0.3	0.4	0.2	0.1

We can find the mean and variance of X by arranging the calculation in the form of a table. Both μ_X and σ_X^2 are sums of columns in this table.

x_i	p_i	$x_i p_i$	$(x_i - \mu_X)^2 p_i$
0	0.3	0.0	$(0 - 1.1)^2(0.3) = 0.363$
1	0.4	0.4	$(1 - 1.1)^2(0.4) = 0.004$
2	0.2	0.4	$(2 - 1.1)^2(0.2) = 0.162$
3	0.1	0.3	$(3 - 1.1)^2(0.1) = 0.361$
		$\mu_X = 1.1$	$\sigma_X^2 = 0.890$

We see that $\sigma_X^2 = 0.89$. The standard deviation of X is $\sigma_X = \sqrt{0.89} = 0.943$. The standard deviation is a measure of the variability of the number of cars Linda sells. As in the case of distributions for data, the standard deviation of a probability distribution is easiest to understand for Normal distributions.

Exercises

7.23 Three children, II Refer to Exercise 7.2 (page 470). Use the definition to find the expected number of girls and the standard deviation of the random variable.

7.24 Grade distribution, I Example 7.1 (page 466) gives the distribution of grades (A = 4, B = 3, and so on) in a large class as

Grade:	0	1	2	3	4
Probability:	0.01	0.05	0.30	0.43	0.21

Find the average (that is, the mean) grade in this course.

7.25 Owned and rented housing, I How do rented housing units differ from units occupied by their owners? Exercise 7.4 (page 470) gives the distributions of the number of rooms for owner-occupied units and renter-occupied units in San Jose, California. Find the mean number of rooms for both types of housing unit. How do the means reflect the differences between the distributions that you found in Exercise 7.4?

7.26 Pick 3 The Tri-State Pick 3 lottery game offers a choice of several bets. You choose a three-digit number. The lottery commission announces the winning three-digit number, chosen at random, at the end of each day. The "box" pays \$83.33 if the number you choose has the same digits as the winning number, in any order. Find the expected payoff for a \$1 bet on the box. (Assume that you chose a number having three different digits.)

7.27 Keno Keno is a favorite game in casinos, and similar games are popular with the states that operate lotteries. Balls numbered 1 to 80 are tumbled in a machine as the bets are placed, then 20 of the balls are chosen at random. Players select numbers by marking a card. The simplest of the many wagers available is "Mark 1 Number." Your payoff is \$3 on a \$1 bet if the number you select is one of those chosen. Because 20 of 80 numbers are chosen, your probability of winning is 20/80, or 0.25.

(a) What is the probability distribution (the outcomes and their probabilities) of the payoff X on a single play?

(b) What is the mean payoff μ_X?

(c) In the long run, how much does the casino keep from each dollar bet?

7.28 Grade distribution, II Find the standard deviation σ_X of the distribution of grades in Exercise 7.24.

7.29 Households and families Exercise 7.12 (page 477) gives the distributions of the number of people in households and in families in the United States. An important difference is that many households consist of one person living alone, whereas a family must have at least two members. Some households may contain families along with other people, and so will be larger than the family. These differences make it hard to compare the distributions without calculations. Find the mean and standard deviation of both household size and family size. Combine these with your descriptions from Exercise 7.12 to give a comparison of the two distributions.

7.30 Owned and rented housing, II Which of the two distributions for room counts in Exercises 7.4 (page 470) and 7.25 appears more spread out in the probability histograms? Why? Find the standard deviation for both distributions. The standard deviation provides a numerical measure of spread.

Statistical Estimation and the Law of Large Numbers

We would like to estimate the mean height μ of the population of all American women between the ages of 18 and 24 years. This μ is the mean μ_X of the random variable X obtained by choosing a young woman at random and measuring her height. To estimate μ, we choose an SRS of young women and use the sample mean $\bar{x}$ to estimate the unknown population mean μ. Statistics obtained from probability samples are random variables because their values would vary in repeated sampling. The ***sampling distributions*** of statistics are just the probability distributions of these random variables. We will study sampling distributions in Chapter 9.

sampling distributions

It seems reasonable to use $\bar{x}$ to estimate μ. An SRS should fairly represent the population, so the mean $\bar{x}$ of the sample should be somewhere near the mean μ of the population. Of course, we don't expect $\bar{x}$ to be exactly equal to μ, and we realize that if we choose another SRS, the luck of the draw will probably produce a different $\bar{x}$.

If $\bar{x}$ is rarely exactly right and varies from sample to sample, why is it nonetheless a reasonable estimate of the population mean μ? If we keep on adding observations to our random sample, the statistic $\bar{x}$ is *guaranteed* to get as close as we wish to the parameter μ and then stay that close. We have the comfort of knowing that if we can afford to keep on measuring more young women, eventually we will estimate the mean height of all young women very accurately. This remarkable fact is called the ***law of large numbers.*** It is remarkable because it holds for *any* population, not just for some special class such as Normal distributions.

Law of Large Numbers

Draw independent observations at random from any population with finite mean μ. Decide how accurately you would like to estimate μ. As the number of observations drawn increases, the mean $\bar{x}$ of the observed values eventually approaches the mean μ of the population as closely as you specified and then stays that close.

The behavior of $\bar{x}$ is similar to the idea of probability. In the long run, the *proportion* of outcomes taking any value gets close to the *probability* of that value, and the *average outcome* gets close to the distribution mean. The figure in Example P.9 (page 21) shows how proportions approach probability in one example. Here is an example of how sample means approach the distribution mean.

Example 7.8

Heights of young women
Law of large numbers

The distribution of the heights of all young women is close to the Normal distribution with mean 64.5 inches and standard deviation 2.5 inches. Suppose that $\mu = 64.5$ were exactly true. Figure 7.10 shows the behavior of the mean height $\bar{x}$ of n women chosen at random from a population whose heights follow the N(64.5, 2.5) distribution. The graph plots the values of $\bar{x}$ as we add women to our sample. The first woman drawn had height 64.21 inches, so the line starts there. The second had height 64.35 inches, so for $n = 2$ the mean is

$$\bar{x} = \frac{64.21 + 64.35}{2} = 64.28$$

This is the second point on the line in the graph.

Figure 7.10 *The law of large numbers in action, for Example 7.8. As we increase the size of our sample, the sample mean $\bar{x}$ always approaches the mean μ of the population.*

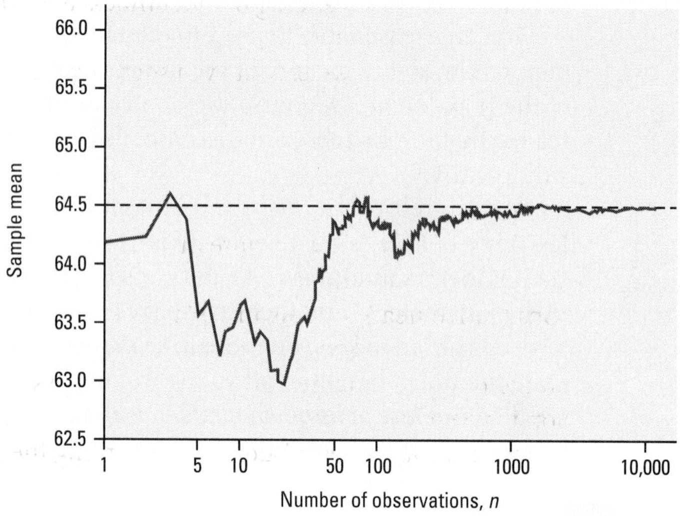

At first, the graph shows that the mean of the sample changes as we take more observations. Eventually, however, the mean of the observations gets close to the population mean $\mu = 64.5$ and settles down at that value. The law of large numbers says that this *always* happens.

The mean μ of a random variable is the average value of the variable in two senses. By its definition, μ is the average of the possible values, weighted by their probability of occurring. The law of large numbers says that μ is also the long-run average of many independent observations on the variable. The law

High-tech gambling
There are more than 450,000 slot machines in the United States. Once upon a time, you put in a coin and pulled the lever to spin three wheels, each with 20 symbols. No longer. Now the machines are video games with flashy graphics and outcomes produced by random number generators. Machines can accept many coins at once, can pay off on a bewildering variety of outcomes, and can be networked to allow common jackpots. Gamblers still search for systems, but in the long run the law of large numbers guarantees the house its 5% profit.

of large numbers can be proved mathematically starting from the basic laws of probability.

Thinking about the Law of Large Numbers

The law of large numbers says broadly that the average results of many independent observations are stable and predictable. The gamblers in a casino may win or lose, but the casino will win in the long run because the law of large numbers says what the average outcome of many thousands of bets will be. An insurance company deciding how much to charge for life insurance and a fast-food restaurant deciding how many beef patties to prepare also rely on the fact that averaging over many individuals produces a stable result. It is worth the effort to think a bit more closely about so important a fact.

The "law of small numbers"

Both the rules of probability and the law of large numbers describe the regular behavior of chance phenomena *in the long run*. Psychologists have discovered that our intuitive understanding of randomness is quite different from the true laws of chance.[6] For example, most people believe in an incorrect "law of small numbers." That is, we expect even short sequences of random events to show the kind of average behavior that in fact appears only in the long run.

Try this. Write down a sequence of heads and tails that you think imitates 10 tosses of a balanced coin. How long was the longest string (called a run) of consecutive heads or consecutive tails in your tosses? Most people will write a sequence with no runs of more than two consecutive heads or tails. Longer runs don't seem "random" to us. In fact, the probability of a run of three or more consecutive heads or tails in 10 tosses is greater than 0.5078.[7] The runs of consecutive heads or consecutive tails that appear in real coin tossing (and that are predicted by the mathematics of probability) seem surprising to us. Because we don't expect to see long runs, we may conclude that the coin tosses are not independent or that some influence is disturbing the random behavior of the coin.

Example 7.9 *The "hot hand" in basketball*

No "law of small numbers"

Belief in the law of small numbers influences behavior. If a basketball player makes several consecutive shots, both the fans and her teammates believe that she has a "hot hand" and is more likely to make the next shot. This is doubtful. Careful study suggests that runs of baskets made or missed are no more frequent in basketball than would be expected if each shot were independent of the player's previous shots. Baskets made or missed are just like heads and tails in tossing a coin. (Of course, some players make 30% of their shots in the long run and others make 50%, so a coin toss may not be the best model for basketball shots.) Our perception of hot or cold streaks simply shows that we don't perceive random behavior very well.[8]

Gamblers often follow the hot-hand theory, betting that a run will continue. At other times, however, they draw the opposite conclusion when confronted with a run of outcomes. If a coin gives 10 straight heads, some gamblers feel that it must now produce some extra tails to get back to the average of half heads and half tails. Not so. If the next 10,000 tosses give about 50% tails, those 10 straight heads will be swamped by the later thousands of heads and tails. No compensation is needed to get back to the average in the long run. Remember that it is only in the long run that the regularity described by probability and the law of large numbers takes over.

CAUTION

Our intuition doesn't do a good job of distinguishing random behavior from systematic influences. This is also true when we look at data. We need statistical inference to supplement exploratory analysis of data because probability calculations can help verify that what we see in the data is more than a random pattern.

How large is a large number?

The law of large numbers says that the actual mean outcome of many trials gets close to the distribution mean μ as more trials are made. It doesn't say how many trials are needed to guarantee a mean outcome close to μ. That depends on the *variability* of the random outcomes. The more variable the outcomes, the more trials are needed to ensure that the mean outcome $\bar{x}$ is close to the distribution mean μ. Casinos understand this: the outcomes of games of chance are variable enough to hold the interest of gamblers. Only the casino plays often enough to rely on the law of large numbers. Gamblers get entertainment; the casino has a business.

Exercises

SIMULATIONS

7.31 Law of large numbers simulation This exercise is based on Example 7.8 and uses the TI-83/84/89 to simulate the law of large numbers and the sampling process. Begin by clearing L_1/list1, L_2/list2, L_3/list3, and L_4/list4. Then enter the commands from the following table.

Specify Plot1 as follows: xyLine (2nd Type icon on the TI-83/84); Xlist: L_1/list1; Ylist: L_4/list4; Mark: . Set the viewing WINDOW as follows: $X[1,10]_{10}$. To set the Y dimensions, scan the values in L_4/list4. Or start with $Y[60,69]_1$ and adjust as necessary. Press GRAPH.

In the WINDOW screen, change Xmax to 100, and press GRAPH again. In your own words, write a short description of the principle that this exercise demonstrates.

TI-83/84	TI-89	
`seq(X,X,1,200)→L1`	`seq(X,X,1,200)→list1`	Enters the positive integers 1 to 200 into L_1/list1 (for seq, look under `2nd/LIST/OPS` on the TI-83/84 and under `CATALOG` on the TI-89).
`randNorm(64.5,2.5,200)→L2`	`tistat.randNorm(64.5,2.5,200)→list2`	Generates 200 random heights (in inches) from the $N(64.5, 2.5)$ distribution and stores these values in L_2/list2 (for `randNorm`, look under `MATH/PRB` on the TI-83/84 and under `CATALOG` on the TI-89).
`cumSum(L2)→L3`	`cumSum(list2)→list3`	Provides a cumulative sum of the observations and stores these values in L_3/list3 (for cumSum, look under `2nd/LIST/OPS` on the TI-83/84 and under `CATALOG` on the TI-89).
`L3/L1→L4`	`list3/list1→list4`	Calculates the average heights of the women and stores these values in L_4/list4

7.32 Is red hot?

(a) A gambler knows that red and black are equally likely to occur on each spin of a roulette wheel. He observes five consecutive reds and bets heavily on red at the next spin. Asked why, he says that "red is hot" and that the run of reds is likely to continue. Explain to the gambler what is wrong with this reasoning.

(b) After hearing you explain why red and black remain equally probable after five reds on the roulette wheel, the gambler moves to a poker game. He is dealt five straight red cards. He remembers what you said and assumes that the next card dealt in the same hand is equally likely to be red or black. Is the gambler right or wrong? Why?

7.33 A game of chance One consequence of the law of large numbers is that once we have a probability distribution for a random variable, we can find its mean by simulating many outcomes and averaging them. The law of large numbers says that if we take enough outcomes, their average value is sure to approach the mean of the distribution.

I have a little bet to offer you. Toss a coin 10 times. If there is no run of three or more straight heads or tails in the 10 outcomes, I'll pay you \$2. If there is a run of three or more, you pay me just \$1. Surely you will want to take advantage of me and play this game?

Simulate enough plays of this game (the outcomes are +\$2 if you win and −\$1 if you lose) to estimate the mean outcome. Is it to your advantage to play?

7.34 Overdue for a hit Retired baseball player Tony Gwynn got a hit about 35% of the time over an entire season. After he failed to hit safely in six straight at-bats, a TV commentator said, "Tony is due for a hit by the law of averages." Is that right? Why?

7.35 Roll the dice This exercise uses the *Expected Value* applet.

(a) What is the expected result when you roll a single die? Specify 40 rolls, and then click on "roll the dice." After the last roll, click on "show mean." Does your expected sum agree with the green mean? What happens to the plotted points as the number of rolls increases? If the picture is not clear, click on "roll the dice" for a new total of 80 rolls. Describe what you see in a sentence. What is the name for this principle?

(b) Click "reset," specify 40 rolls, and then click "more dice" once so that 2 dice show in the left box. Carry out the steps in (a) above for two dice.

(c) Reset the applet, and carry out the same routine for 3, then 4, and then 5 dice.

(d) Construct a table with number of dice, 1 to 5, in the first column, and expected value in the second column. What is the greatest number of dice possible for this applet? What is the expected value when rolling that many dice? Describe the pattern you see in the table.

7.36 Applying Benford's law To use Benford's law (Example 7.6, page 483) to detect faked invoice amounts, compare the distribution of first digits among the invoices with the Benford's law distribution. The process is more effective when you have very many invoices from the vendor in question rather than just a few. Explain why this is true.

Activity 7C
Combining random variables

The purpose of this activity is to gain some insight into what happens when we combine random variables. Consider the following very simple random variables and their probability distributions.

X:	1	2	5
P(X):	0.2	0.5	0.3

Y:	2	4
P(Y):	0.7	0.3

1. Use the definitions of mean and variance of a random variable on pages 483 and 485 to find the mean and variance for both X and Y.

2. Build a new random variable, which will be the sum of the random variables X and Y. Determine all of the possible values for X + Y. (*Hint:* The multiplication rule tells us that there are 6 possible values of X + Y, some of which may be the same.) To simplify things, begin with a table like the one below.

X + *Y*		*Y*	*Y*
		2	4
X	1		
X	2		
X	5		

3. Calculate the probabilities for each of the values of $X + Y$. Then make a probability distribution table for $X + Y$. For example, the smallest value of $X + Y$ is 3, and $P(X + Y = 3) = P(X = 1 \text{ and } Y = 2) = P(X = 1) \times P(Y = 2 \mid X = 1) = (0.2)(0.7) = 0.14$. Verify that the sum of the probabilities is 1.

4. Use the distribution from Step 3 and the definition of the mean of a random variable to calculate the mean of $X + Y$. Verify that *the mean of the sum is the sum of the means:* $\mu_{X+Y} = \mu_X + \mu_Y$.

5. Using the probability distribution table for $X + Y$ constructed in Step 3 and the definition, calculate the variance of the sum, $\text{var}(X + Y)$. Verify that this is the same as $\text{var}(X) + \text{var}(Y)$. Thus, for these random variables X and Y, *the variance of the sum of X and Y is the sum of the variances.*

6. Repeat Steps 1 to 5 for the difference $X - Y$. For example, the smallest difference is $1 - 2 = -1$ and $P(X - Y = -3) = P(X = 1 \text{ and } Y = 4) = P(X = 1) \times P(Y = 4 \mid X = 1) = (0.2)(0.3) = 0.06$. Begin with a table similar to the table in Step 2. Verify that *the mean of the difference is the difference of the means:* $\mu_{X-Y} = \mu_X - \mu_Y$.

7. Using the probability distribution table for $X - Y$ and the definition, calculate the variance of the difference, $\text{var}(X - Y)$. Verify that this is the same as $\text{var}(X) + \text{var}(Y)$. Conclude that for these two random variables, $\text{var}(X \pm Y) = \text{var}(X) + \text{var}(Y)$.

Note that this activity has demonstrated that certain results for means and variances hold true for this particular pair of random variables X and Y. To prove that the rules always holds when the hypotheses are satisfied requires showing that the results hold in the general case.

Rules for Means

You are studying flaws in the painted finish of refrigerators made by your firm. Dimples and paint sags are two kinds of surface flaw. Not all refrigerators have the same number of dimples: many have none, some have one, some two, and so on. You ask for the average number of imperfections on a refrigerator. How many total imperfections of both kinds (on the average) are there on a refrigerator? That's easy: if the average number of dimples is 0.7 and the average number of sags is 1.4, then counting both gives an average of $0.7 + 1.4 = 2.1$ flaws.

In more formal language, the number of dimples on a refrigerator is a random variable X that takes values 0, 1, 2, and so on. X varies as we inspect one refrigerator after another. Only the mean number of dimples $\mu_X = 0.7$ was reported to you. The number of paint sags is a second random variable Y having mean $\mu_Y = 1.4$. (You see how the subscripts keep straight which variable we are talking about.) The total number of both dimples and sags is the sum $X + Y$. That sum is another random variable that varies from refrigerator to refrigerator. Its mean μ_{X+Y} is the average number of dimples and sags together and is just the sum of the individual means μ_X and μ_Y. That is an important rule for how means of random variables behave.

Here's another rule. The crickets living in a field have mean length 1.2 inches. What is the mean in centimeters? There are 2.54 centimeters in an inch, so the length of a cricket in centimeters is 2.54 times its length in inches. If we multiply every observation by 2.54, we also multiply their average by 2.54. The mean in centimeters must be 2.54×1.2, or about 3.05 centimeters. More formally, the length in inches of a cricket chosen at random from the field is a random variable X with mean μ_X. The length in centimeters is 2.54X, and this new random variable has mean $2.54\mu_X$.

The point of these examples and Activity 7C is that means of random variables behave like averages of one variable data. Here are the rules we need.

Rules for Means

Rule 1. If X is a random variable and a and b are fixed numbers, then

$$\mu_{a+bX} = a + b\mu_X$$

Rule 2. If X and Y are random variables, then

$$\mu_{X+Y} = \mu_X + \mu_Y$$

Here is an example that applies these rules.

Example 7.10 ***Linda sells cars and trucks***

Applying rules for means

In Example 7.7 we saw that the number X of cars that Linda hopes to sell has distribution

Cars sold:	0	1	2	3
Probability:	0.3	0.4	0.2	0.1

Linda's estimate of her truck and SUV sales is

Vehicles sold:	0	1	2
Probability:	0.4	0.5	0.1

Take X to be the number of cars Linda sells and Y the number of trucks and SUVs. The means of these random variables are

$$\begin{aligned}\mu_X &= (0)(0.3) + (1)(0.4) + (2)(0.2) + (3)(0.1)\\ &= 1.1 \text{ cars}\end{aligned}$$

$$\begin{aligned}\mu_Y &= (0)(0.4) + (1)(0.5) + (2)(0.1)\\ &= 0.7 \text{ trucks and SUVs}\end{aligned}$$

At her commission rate of 25% of gross profit on each vehicle she sells, Linda expects to earn \$350 for each car sold and \$400 for each truck or SUV sold. So her earnings are

$$Z = 350X + 400Y$$

Combining Rules 1 and 2, her mean earnings are

$$\begin{aligned}\mu_Z &= 350\mu_X + 400\mu_Y \\ &= (350)(1.1) + (400)(0.7) = \$665\end{aligned}$$

This is Linda's best estimate of her earnings for the day.

Rules for Variances

What are the facts for variances that parallel Rules 1 and 2 for means? The mean of a sum of random variables is always the sum of their means, but this addition rule is not always true for variances. To understand why, take X to be the percent of a family's after-tax income that is spent and Y the percent that is saved. When X increases, Y decreases by the same amount. Though X and Y may vary widely from year to year, their sum X + Y is always 100% and does not vary at all. It is the association between the variables X and Y that prevents their variances from adding. If random variables are independent, this kind of association between their values is ruled out and their variances do add. Two random variables X and Y are *independent* if knowing that any event involving X alone did or did not occur tells us nothing about the occurrence of any event involving Y alone. Probability models often assume independence when the random variables describe outcomes that appear unrelated to each other. You should ask in each instance whether the assumption of independence seems reasonable. Here are the rules for variances.

Rules for Variances

Rule 1. If X is a random variable and a and b are fixed numbers, then

$$\sigma^2_{a+bX} = b^2\sigma^2_X$$

Rule 2. If X and Y are *independent* random variables, then

$$\begin{aligned}\sigma^2_{X+Y} &= \sigma^2_X + \sigma^2_Y \\ \sigma^2_{X-Y} &= \sigma^2_X + \sigma^2_Y\end{aligned}$$

This is the **addition rule for variances of independent random variables.**

Notice that because a variance is the average of *squared* deviations from the mean, multiplying X by a constant b multiplies σ^2_X by the *square* of the constant. Adding a constant a to a random variable changes its mean but does not change its variability. The variance of $X + a$ is therefore the same as the variance of X. Because the square of -1 is 1, the addition rule says that the variance of a difference is the *sum* of the variances. For independent random variables, the difference $X - Y$ is more variable than either X or Y alone because variations in both X and Y contribute to variation in their difference.

As with data, we prefer the standard deviation of a random variable to the variance as a measure of variability. The addition rule for variances implies that **standard deviations do *not* generally add. Standard deviations are most easily combined by using the rules for variances rather than by giving separate rules for standard deviations.** For example, the standard deviations of 2X and $-2X$ are both equal to $2\sigma_X$ because this is the square root of the variance $4\sigma_X^2$.

Example 7.11 *Winning the lottery*

Applying rules for variances

The payoff X of a $1 ticket in the Tri-State Pick 3 game is $500 with probability 1/1000 and $0 the rest of the time. Here is the combined calculation of mean and variance:

x_i	p_i	$x_i p_i$	$(x_i - \mu_X)^2 p_i$
0	0.999	0	$(0 - 0.5)^2 (0.999) = 0.24975$
500	0.001	0.5	$(500 - 0.5)^2 (0.001) = 249.50025$
		$\mu_X = 0.5$	$\sigma_X^2 = 249.75$

The standard deviation is $\sigma_X = \sqrt{249.75} = \$15.80$. It is usual for games of chance to have large standard deviations, because large variability makes gambling exciting.

If you buy a Pick 3 ticket, your winnings are $W = X - 1$ because the dollar you paid for the ticket must be subtracted from the payoff. By the rules for means, the mean amount you win is

$$\mu_W = \mu_X - 1 = -\$0.50$$

That is, you lose an average of 50 cents on a ticket. The rules for variances remind us that the variance and standard deviation of the winnings $W = X - 1$ are the same as those of X. Subtracting a fixed number changes the mean but not the variance.

Suppose now that you buy a $1 ticket on each of two different days. The payoffs X and Y on the two tickets are independent because separate drawings are held each day. Your total payoff $X + Y$ has mean

$$\mu_{X+Y} = \mu_X + \mu_Y = \$0.50 + \$0.50 = \$1.00$$

Because X and Y are independent, the variance of $X + Y$ is

$$\sigma_{X+Y}^2 = \sigma_X^2 + \sigma_Y^2 = 249.75 + 249.75 = 499.50$$

The standard deviation of the total payoff is

$$\sigma_{X+Y} = \sqrt{499.5} = \$22.35$$

This is not the same as the sum of the individual standard deviations, which is $15.80 + $15.80 = $31.60. *Variances of independent random variables add; standard deviations do not.*

If you buy a ticket every day (365 tickets in a year), your mean payoff is the sum of 365 daily payoffs. That's 365 times 50 cents, or $182.50. Of course, it costs $365 to play, so the state's mean take from a daily Pick 3 player is $182.50. Results for individual players

Example 7.12 ***SAT scores***
The role of independence

A college uses SAT scores as one criterion for admission. Experience has shown that the distribution of SAT scores among its entire population of applicants is such that

SAT Math score X	$\mu_X = 519$	$\sigma_X = 115$
SAT Verbal score Y	$\mu_Y = 507$	$\sigma_Y = 111$

What are the mean and standard deviation of the total score X + Y among students applying to this college?

The mean overall SAT score is

$$\mu_{X+Y} = \mu_X + \mu_Y = 519 + 507 = 1026$$

The variance and standard deviation of the total *cannot be computed* from the information given. SAT Verbal and Math scores are not independent, because students who score high on one exam tend to score high on the other also. Therefore, Rule 2 does not apply.

Example 7.13 ***Investing in stocks and T-bills***
Independence

Zadie has invested 20% of her funds in Treasury bills and 80% in an "index fund" that represents all U.S. common stocks. The rate of return of an investment over a time period is the percent change in the price during the time period, plus any income received. If X is the annual return on T-bills and Y the annual return on stocks, the portfolio rate of return R is

$$R = 0.2X + 0.8Y$$

The returns X and Y are random variables because they vary from year to year. Based on annual returns between 1950 and 2000, we have[9]

X = annual return on T-bills	$\mu_X = 5.0\%$	$\sigma_X = 2.9\%$
Y = annual return on stocks	$\mu_Y = 13.2\%$	$\sigma_Y = 17.6\%$

Stocks had higher returns than T-bills on the average, but the standard deviations show that returns on stocks varied much more from year to year. That is, the risk of investing in stocks is greater than the risk for T-bills because their returns are less predictable.

For the return R on Zadie's portfolio of 20% T-bills and 80% stocks,

$$\begin{aligned} R &= 0.2X + 0.8Y \\ \mu_R &= 0.2\mu_X + 0.8\mu_Y \\ &= (0.2 \times 5.0) + (0.8 \times 13.2) = 11.56\% \end{aligned}$$

Finding the variance of the portfolio return is beyond the scope of this text, because returns on T-bills and returns on stocks are not independent. The portfolio has a smaller mean return than an all-stock portfolio, but it is also less risky.

Combining Normal Random Variables

So far, we have concentrated on finding rules for means and variances of random variables. If a random variable is Normally distributed, we can use its mean and variance to compute probabilities. Example 7.4 (page 474) shows the method. What if we combine two Normal random variables?

Any linear combination of independent Normal random variables is also Normally distributed. That is, if X and Y are independent Normal random variables and a and b are any fixed numbers, $aX + bY$ is also Normally distributed. In particular, the sum or difference of independent Normal random variables has a Normal distribution. The mean and standard deviation of $aX + bY$ are found as usual from the addition rules for means and variances. These facts are often used in statistical calculations.

Example 7.14 *A round of golf*

Combining random variables

Tom and George are playing in the club golf tournament. Their scores vary as they play the course repeatedly. Tom's score X has the N(110, 10) distribution, and George's score Y varies from round to round according to the N(100, 8) distribution. If they play independently, what is the probability that Tom will score lower than George and thus do better in the tournament? The difference X − Y between their scores is Normally distributed, with mean and variance

$$\mu_{X-Y} = \mu_X - \mu_Y = 110 - 100 = 10$$
$$\sigma^2_{X-Y} = \sigma^2_X + \sigma^2_Y = 10^2 + 8^2 = 164$$

Because $\sqrt{164} = 12.8$, X − Y has the N(10, 12.8) distribution. Figure 7.11 illustrates the probability computation:

$$\begin{aligned} P(X < Y) &= P(X - Y < 0) \\ &= P\left(\frac{(X-Y) - 10}{12.8} < \frac{0 - 10}{12.8}\right) \\ &= P(Z < -0.78) = 0.2177 \end{aligned}$$

Figure 7.11 *The Normal probability calculation, for Example 7.14.*

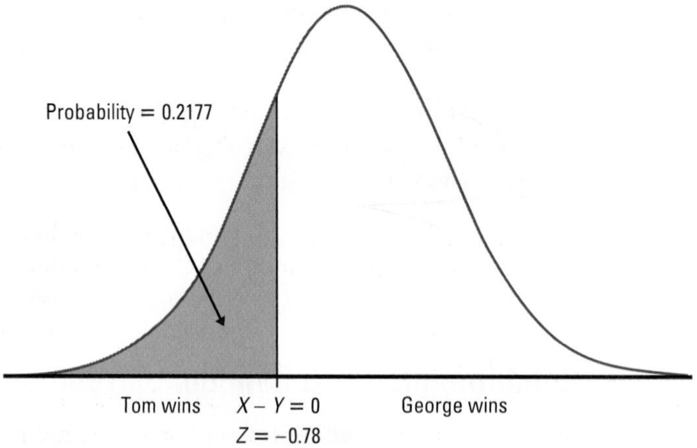

Although George's score is 10 strokes lower on average, Tom will have the lower score in about one of every five matches.

Exercises

7.37 Checking rules This exercise extends Activity 7C (page 492). Recall that the probability distribution for a random variable X was given as

X:	1	2	5
$P(X)$:	0.2	0.5	0.3

(a) Let $a = 2$ and $b = 3$. Determine the probability distribution function for the new random variable $a + bX$.

(b) Use the definitions (pages 483 and 485) to find the mean and variance of $a + bX$.

(c) Now use Rule 1 for means (page 494) to find the mean and variance of $a + bX$. Verify that your results in (b) and (c) are the same.

(d) Verify that Rule 1 for variances holds for this example. That is, show that var$(2 + 3X) = 9$ var(X).

(e) Which method do you prefer: using the definition or using the rules? Why?

7.38 Checking independence, I For each of the following situations, would you expect the random variables X and Y to be independent? Explain your answers.

(a) X is the rainfall (in inches) on November 6 of this year, and Y is the rainfall at the same location on November 6 of next year.

(b) X is the amount of rainfall today, and Y is the rainfall at the same location tomorrow.

(c) X is today's rainfall at the airport in Orlando, Florida, and Y is today's rainfall at Disney World just outside Orlando.

Some people have all the luck
In June 2005, Donna Goeppert won a million dollar jackpot playing the Pennsylvania Lottery. The odds of winning a lottery like the one in Pennsylvania are 1.44 million to 1. But what is extraordinary is that Ms. Goeppert had previously won $1 million playing the lottery earlier in the year. The odds of winning twice vary, depending on how many tickets are scratched. A university professor in Pennsylvania estimated that if you played 100 tickets, the odds of winning the lottery twice are about 419 million to 1.

7.39 Checking independence, II In which of the following games of chance would you be willing to assume independence of X and Y in making a probability model? Explain your answer in each case.

(a) In blackjack, you are dealt two cards and examine the total points X on the cards (face cards count 10 points). You can choose to be dealt another card and compete based on the total points Y on all three cards.

(b) In craps, the betting is based on successive rolls of two dice. X is the sum of the faces on the first roll, and Y is the sum of the faces on the next roll.

7.40 Chemical reactions Laboratory data show that the time required to complete two chemical reactions in a production process varies. The first reaction has a mean time of 40 minutes and a standard deviation of 2 minutes; the second has a mean time of 25 minutes and a standard deviation of 1 minute. The two reactions are run in sequence during production. There is a fixed period of 5 minutes between them as the product of the first reaction is pumped into the vessel where the second reaction will take place. What is the mean time required for the entire process?

7.41 Time and motion A time and motion study measures the time required for an assembly-line worker to perform a repetitive task. The data show that the time required to bring a part from a bin to its position on an automobile chassis varies from car to car with mean 11 seconds and standard deviation 2 seconds. The time required to attach the part to the chassis varies with mean 20 seconds and standard deviation 4 seconds.

(a) What is the mean time required for the entire operation of positioning and attaching the part?

(b) If the variation in the worker's performance is reduced by better training, the standard deviations will decrease. Will this decrease change the mean you found in (a) if the mean times for the two steps remain as before?

(c) The study finds that the times required for the two steps are independent. A part that takes a long time to position, for example, does not take more or less time to attach than other parts. Find the standard deviation of the time required for the two-step assembly operation.

7.42 Electronic circuit The design of an electronic circuit calls for a 100-ohm resistor and a 250-ohm resistor connected in series so that their resistances add. The components used are not perfectly uniform, so that the actual resistances vary independently according to Normal distributions. The resistance of 100-ohm resistors has mean 100 ohms and standard deviation 2.5 ohms, while that of 250-ohm resistors has mean 250 ohms and standard deviation 2.8 ohms.

(a) What is the distribution of the total resistance of the two components in series?

(b) What is the probability that the total resistance lies between 345 and 355 ohms? Show your work.

Section 7.2 Summary

The probability distribution of a random variable X, like a distribution of data, has a **mean** μ_X and a **standard deviation** σ_X.

The **mean** μ is the balance point of the probability histogram or density curve. If X is discrete with possible values x_i having probabilities p_i, the mean is the average of the values of X, each weighted by its probability:

$$\mu_X = x_1p_1 + x_2p_2 + \ldots + x_kp_k$$

The **variance** σ_X^2 is the average squared deviation of the values of the variable from their mean. For a discrete random variable,

$$\sigma_X^2 = (x_1 - \mu)^2\, p_1 + (x_2 - \mu)^2\, p_2 + \ldots + (x_k - \mu)^2\, p_k$$

The **standard deviation** σ_X is the square root of the variance. The standard deviation measures the variability of the distribution about the mean. It is easiest to interpret for Normal distributions.

The mean and variance of a continuous random variable can be computed from the density curve, but to do so requires more advanced mathematics.

The **law of large numbers** says that the average of the values of X observed in many trials must approach μ.

The means and variances of random variables obey the following rules. If a and b are fixed numbers, then

$$\mu_{a+bX} = a + b\mu_X$$
$$\sigma^2_{a+bX} = b^2\sigma^2_X$$

If X and Y are any two random variables, then

$$\mu_{X+Y} = \mu_X + \mu_Y$$

and if X and Y are *independent*, then

$$\sigma^2_{X+Y} = \sigma^2_X + \sigma^2_Y$$
$$\sigma^2_{X-Y} = \sigma^2_X + \sigma^2_Y$$

Any linear combination of independent Normal random variables is also Normally distributed.

Section 7.2 Exercises

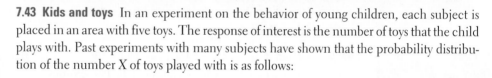

7.43 Kids and toys In an experiment on the behavior of young children, each subject is placed in an area with five toys. The response of interest is the number of toys that the child plays with. Past experiments with many subjects have shown that the probability distribution of the number X of toys played with is as follows:

Number of toys x_i:	0	1	2	3	4	5
Probability p_i:	0.03	0.16	0.30	0.23	0.17	0.11

(a) Calculate the mean μ_X and the standard deviation σ_X.

(b) Describe the details of a simulation you could carry out to approximate the mean number of toys μ_X and the standard deviation σ_X. Then carry out your simulation. Are the mean and standard deviation produced from your simulation close to the values you calculated in (a)?

7.44 Buying stock You purchase a hot stock for \$1000. The stock either gains 30% or loses 25% each day, each with probability 0.5. Its returns on consecutive days are independent of each other. You plan to sell the stock after two days.

(a) What are the possible values of the stock after two days, and what is the probability for each value? What is the probability that the stock is worth more after two days than the \$1000 you paid for it?

(b) What is the mean value of the stock after two days? You see that these two criteria give different answers to the question "Should I invest?"

7.45 SSHA The academic motivation and study habits of female students as a group are better than those of males. The Survey of Study Habits and Attitudes (SSHA) is a psychological test that measures these factors. The distribution of SSHA scores among the women at a college has mean 120 and standard deviation 28, and the distribution of scores among male students has mean 105 and standard deviation 35. You select a single male student and a single female student at random and give them the SSHA test.

(a) Explain why it is reasonable to assume that the scores of the two students are independent.

(b) What are the mean and standard deviation of the difference (female minus male) between their scores?

(c) From the information given, can you find the probability that the woman chose scores higher than the man? If so, find this probability. If not, explain why you cannot.

7.46 Weird dice again You have two balanced, six-sided dice. The first has 1, 3, 4, 5, 6, and 8 spots on its six faces. The second die has 1, 2, 2, 3, 3, and 4 spots on its faces.

(a) What is the mean number of spots on the up-face when you roll each of these dice?

(b) Write the probability model for the outcomes when you roll both dice independently. From this, find the probability distribution of the sum of the spots on the up-faces of the two dice.

(c) Find the mean number of spots on the two up-faces in two ways: from the distribution you found in (b) and by applying the addition rule to your results in (a). You should, of course, get the same answer.

7.47 A glass act, I In a process for manufacturing glassware, glass stems are sealed by heating them in a flame. The temperature of the flame varies a bit. Here is the distribution of the temperature X measured in degrees Celsius:

Temperature:	540°	545°	550°	555°	560°
Probability:	0.1	0.25	0.3	0.25	0.1

(a) Find the mean temperature μ_X and the standard deviation σ_X.

(b) The target temperature is 550°C. What are the mean and standard deviation of the number of degrees off target, $X - 550$?

(c) A manager asks for results in degrees Fahrenheit. The conversion of X into degrees Fahrenheit is given by

$$Y = \frac{9}{5}X + 32$$

What are the mean μ_Y and the standard deviation σ_Y of the temperature of the flame in the Fahrenheit scale?

7.48 A glass act, II In continuation of the previous exercise, describe the details of a simulation you could carry out to approximate the mean temperature and the standard deviation in degrees Celsius. Then carry out your simulation. Are the mean and standard deviation produced from your simulation close to the values you calculated in 7.47(a)?

7.49 His and her earnings A study of working couples measures the income X of the husband and the income Y of the wife in a large number of couples in which both partners are employed. Suppose that you knew the means μ_X and μ_Y and the variances σ_X^2 and σ_Y^2 of both variables in the population.

(a) Is it reasonable to take the mean of the total income X + Y to be $\mu_X + \mu_Y$? Explain your answer.

(b) Is it reasonable to take the variance of the total income to be $\sigma_X^2 + \sigma_Y^2$? Explain your answer.

7.50 Applying torque A machine fastens plastic screw-on caps onto containers of motor oil. If the machine applies more torque than the cap can withstand, the cap will break. Both the torque applied and the strength of the caps vary. The capping-machine torque has the Normal distribution with mean 7 inch-pounds and standard deviation 0.9 inch-pounds. The cap strength (the torque that would break the cap) has the Normal distribution with mean 10 inch-pounds and standard deviation 1.2 inch-pounds.

(a) Explain why it is reasonable to assume that the cap strength and the torque applied by the machine are independent.

(b) What is the probability that a cap will break while being fastened by the capping machine? Show your work.

7.51 Estimated sales Examples 7.7 (page 485) and 7.10 (page 494) concern a probability projection of Linda's sales of cars and trucks.

(a) Find the variance and standard deviation of Linda's estimated sales Y of trucks using the distribution and mean from Example 7.10.

(b) Because car and truck sales are not closely linked, Linda is willing to assume that car and truck sales vary independently. Combine your result from (a) with the results for car sales from Example 7.10 to obtain the standard deviation of the total sales X + Y.

(c) Find the standard deviation of Linda's estimated earnings, $Z = 350X + 400Y$.

7.52 Loser buys the pizza Leona and Fred are friendly competitors in high school. Both are about to take the ACT college entrance examination. They agree that if one of them scores 5 or more points better than the other, the loser will buy the winner a pizza. Suppose that in fact Fred and Leona have equal ability, so that each score varies Normally with mean 24 and standard deviation 2. (The variation is due to luck in guessing and the accident of the specific questions being familiar to the student.) The two scores are independent. What is the probability that the scores differ by 5 or more points in either direction?

CASE CLOSED!

Lost income and the courts

In the Case Study (page 463), the table of possible outcomes and probabilities of those outcomes was given as

Possible outcomes X:	210	213	216	219	222	225	228	231
$P(X)$:	0.07	0.16	0.23	0.24	0.17	0.09	0.03	0.01

1. Identify the random variable of interest. $X = ?$
2. Is this a valid probability model? Support your answer.
3. Verify that the probability distribution table yields an expected score of about 218.
4. Determine the variance and standard deviation of X.
5. What is the probability that Blaylock's score would be 218 or less? No more than 220? Between 209 and 218, inclusive?

Chapter Review

Summary

A random variable defines what is counted or measured in a statistics application. If the random variable X is a count, such as the number of heads in four tosses of a coin, then X is discrete, and its distribution can be pictured as a histogram. If X is measured, as in the number of inches of rainfall in Richmond in April, then X is continuous, and its distribution is pictured as a density curve. Among the continuous random variables, the Normal random variable is the most important. The Normal distribution, first introduced in Chapter 2, is revisited, with emphasis this time on it as a probability distribution. The mean and variance of a random variable are calculated, and rules for the sum or difference of two random variables are developed.

What You Should Have Learned

Here is a checklist of the major skills you should have acquired by studying this chapter.

A. Random Variables

1. Recognize and define a discrete random variable, and construct a probability distribution table and a probability histogram for the random variable.

2. Recognize and define a continuous random variable, and determine probabilities of events as areas under density curves.

3. Given a Normal random variable, use the standard Normal table or a graphing calculator to find probabilities of events as areas under the standard Normal distribution curve.

B. Means and Variances of Random Variables

1. Calculate the mean and variance of a discrete random variable. Find the expected payout in a raffle or similar game of chance.

2. Use simulation methods and the law of large numbers to approximate the mean of a distribution.

3. Use rules for means and rules for variances to solve problems involving sums, differences, and linear combinations of random variables.

Chapter Review Exercises

7.53 Gold medal gymnast Rework Example 6.8 (page 401) about Carly Patterson using what you now know about combining random variables.

7.54 Two-finger morra Ann and Bob are playing the game two-finger morra. Each player shows either one or two fingers and at the same time calls out a guess for the number of fingers the other player will show. If a player guesses correctly and the other player does not, the player wins a number of dollars equal to the total number of fingers shown by both players. If both or neither guesses correctly, no money changes hands. On each play both Ann and Bob choose one of the following options:

Choice	Show	Guess
A	1	1
B	1	2
C	2	1
D	2	2

(a) Give the sample space S by writing all possible choices for both players on a single play of this game.

(b) Let X be Ann's winnings on a play. (If Ann loses \$2, then $X = -2$; when no money changes hands, $X = 0$.) Write the value of the random variable X next to each of the outcomes you listed in (a). This is another choice of sample space.

(c) Now assume that Ann and Bob choose independently of each other. Moreover, they both play so that all four choices listed above are equally likely. Find the probability distribution of X.

(d) If the game is fair, X should have mean zero. Does it? What is the standard deviation of X?

Insurance. *The business of selling insurance is based on probability and the law of large numbers. Consumers (including businesses) buy insurance because we all face risks that are unlikely but carry high cost. Think of a fire destroying your home. So we form a group to share the risk: we all pay a small amount, and the insurance policy pays a large amount to those few of us whose homes burn down. The insurance company sells many policies, so it can rely on the law of large numbers. Exercises 7.55 to 7.58 explore aspects of insurance.*

7.55 Life insurance, I A life insurance company sells a term insurance policy to a 21-year-old male that pays \$100,000 if the insured dies within the next 5 years. The probability that a randomly chosen male will die each year can be found in mortality tables. The company collects a premium of \$250 each year as payment for the insurance. The amount X that the company earns on this policy is \$250 per year, less the \$100,000 that it must pay if the insured dies. Here is the distribution of X. Fill in the missing probability in the table and calculate the mean profit μ_X.

Age at death:	21	22	23	24	25	≥26
Profit:	−\$99,750	−\$99,500	−\$99,250	−\$99,000	−\$98,750	\$1250
Probability:	0.00183	0.00186	0.00189	0.00191	0.00193	

7.56 Life insurance, II It would be quite risky for you to insure the life of a 21-year-old friend under the terms of the previous exercise. There is a high probability that your friend would live and you would gain \$1250 in premiums. But if he were to die, you would lose almost \$100,000. Explain carefully why selling insurance is not risky for an insurance company that insures many thousands of 21-year-old men.

7.57 Life insurance, III The risk of an investment is often measured by the standard deviation of the return on the investment. The more variable the return is (the larger σ is), the riskier the investment. We can measure the great risk of insuring a single person's life in Exercise 7.55 by computing the standard deviation of the income X that the insurer will receive. Find σ_X, using the distribution and mean found in Exercise 7.55.

7.58 Life insurance, IV The risk of insuring one person's life is reduced if we insure many people. Use the result of the previous exercise and rules for means and variances to answer the following questions.

(a) Suppose that we insure two 21-year-old males, and that their ages at death are independent. If X and Y are the insurer's income from the two insurance policies, the insurer's average income on the two policies is

$$Z = \frac{X + Y}{2} = 0.5X + 0.5Y$$

Find the mean and standard deviation of Z. You see that the mean income is the same as for a single policy but the standard deviation is less.

(b) If four 21-year-old men are insured, the insurer's average income is

$$Z = \frac{1}{4}(X_1 + X_2 + X_3 + X_4)$$

where X_i is the income from insuring one man. The X_i are independent and each has the same distribution as before. Find the mean and standard deviation of Z. Compare your results with the results of (a). We see that averaging over many insured individuals reduces risk.

7.59 Auto emissions The amount of nitrogen oxides (NOX) present in the exhaust of a particular type of car varies from car to car according to the Normal distribution with mean 1.4 grams per mile (g/mi) and standard deviation 0.3 g/mi. Two cars of this type are tested. One has 1.1 g/mi of NOX; the other has 1.9. The test station attendant finds this much variation between two similar cars surprising. If X and Y are independent NOX levels for cars of this type, find the probability

$$P(X - Y \geq 0.8 \text{ or } X - Y \leq -0.8)$$

that the difference is at least as large as the value the attendant observed.

7.60 Making a profit Rotter Partners is planning a major investment. The amount of profit X is uncertain, but a probabilistic estimate gives the following distribution (in millions of dollars):

Profit:	1	1.5	2	4	10
Probability:	0.1	0.2	0.4	0.2	0.1

(a) Find the mean profit μ_X and the standard deviation of the profit.

(b) Rotter Partners owes its source of capital a fee of $200,000 plus 10% of the profits X. So the firm actually retains

$$Y = 0.9X - 0.2$$

from the investment. Find the mean and standard deviation of Y.

7.61 A balanced scale You have two scales for measuring weights in a chemistry lab. Both scales give answers that vary a bit in repeated weighings of the same item. If the true weight of a compound is 2.00 grams (g), the first scale produces readings X that have mean 2.000 g and standard deviation 0.002 g. The second scale's readings Y have mean 2.001 g and standard deviation 0.001 g.

(a) What are the mean and standard deviation of the difference $Y - X$ between the readings? (The readings X and Y are independent.)

(b) You measure once with each scale and average the readings. Your result is $Z = (X + Y)/2$. What are μ_Z and σ_Z? Is the average Z more or less variable than the reading Y of the less-variable scale?

7.62 It's a girl! A couple plans to have children until they have a girl or until they have four children, whichever comes first. Example 6.6 (page 396) estimated the probability that they will have a girl among their children. Now we ask a different question: How many children, on the average, will couples who follow this plan have?

(a) To answer this question, construct a simulation similar to that in Example 6.6 but this time keep track of the number of children in each repetition. Carry out 25 repetitions and then average the results to estimate the expected value.

(b) Construct the probability distribution table for the random variable X = number of children.

(c) Use the table from (b) to calculate the expected value of X. Compare this number with the result from your simulation in (a).

7.63 Slim again Amarillo Slim is back and he's got another deal for you. We have a fair coin (heads and tails each have probability 1/2). Toss it twice. If two heads come up, you win. If you get any other result, you get another chance: toss the coin twice more, and if you get two heads, you win. If you fail to get two heads on the second try, you lose. You pay a dollar to play. If you win, you get your dollar back plus another dollar.

(a) Explain how to simulate one play of this game using Table B. How could you simulate one play using your calculator?

(b) Simulate 50 plays, using Table B or your calculator. Use your simulation to estimate the expected value of the game.

(c) There are two outcomes in this game: win or lose. Let the random variable X be the (monetary) outcome. What are the two values X can take? Calculate the actual probabilities of each value of X. Then calculate μ_X. How does this compare with your estimate from the simulation in (b)?

7.64 A mechanical assembly A mechanical assembly (Figure 7.12) consists of a rod with a bearing at each end. The total length of the assembly is the sum X + Y + Z of the rod length X and the lengths Y and Z of the bearings. These lengths vary from part to part in production, independently of each other and with Normal distributions. The rod length X has mean 11.2 inches and standard deviation 0.002 inch, while each bearing length Y and Z has mean 0.4 inch and standard deviation 0.001 inch.

Figure 7.12 *Sketch of a mechanical assembly, for Exercise 7.64.*

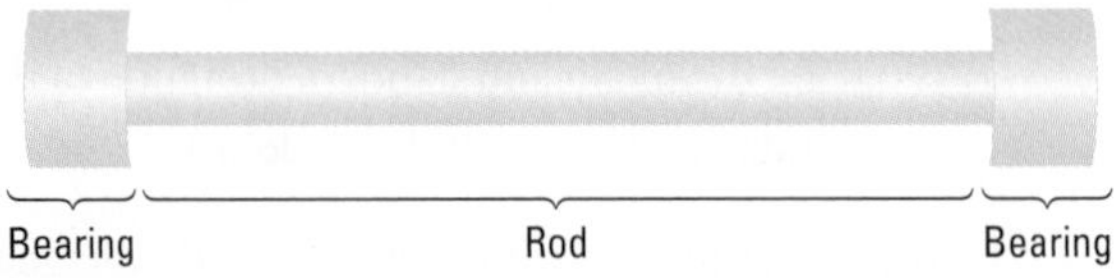

(a) According to the 68–95–99.7 rule, about 95% of all rods have lengths in the range 11.2 $\pm\, d_1$ inches. What is the value of d_1? Similarly, about 95% of the bearing lengths fall in the range of $0.4 \pm d_2$. What is the value of d_2?

(b) It is common practice in industry to state the "natural tolerance" of parts in the form used in (a). An engineer who knows no statistics thinks that tolerances add, so that the natural tolerance for the total length of the assembly (rod and two bearings) is $12 \pm d$ inches, where $d = d_1 + 2d_2$. Find the standard deviation of the total length $X + Y + Z$. Then find the value d such that about 95% of all assemblies have lengths in the range $12 \pm d$. Was the engineer correct?

CHAPTER

8

The Binomial and Geometric Distributions

CASE STUDY

Psychic probability

Researchers in parapsychology must be both wary of tricksters and able to discount instances of luck when investigating evidence of psychic ability. This is especially true in the testing of individuals who claim to possess these extraordinary abilities. In one such test of psychic ability, researchers claim to have eliminated the possibility of trickery by a self-proclaimed psychic and report that his ability exceeded that dictated by chance.[1]

The subject of this study was the controversial psychic Olof Jonsson. The test was conducted using a computer, a video-game paddle, an internal random number generator, and a modified version of a computer program known as the *ESPerciser*.

For each trial, the *ESPerciser* program began with a screen with four blank rectangles and the words "IMPRESSION PERIOD—PRESS BUTTON WHEN READY." After the subject clicked the paddle button, the program would display four black-and-white symbols (star, wave, cross, and circle) in a random order on the screen. This was followed by a screen with all four symbols and the words "USE PADDLE TO POINT, PRESS BUTTON TO SELECT." The subject was supposed to select the symbol he thought the computer had picked. After the subject had selected one of the symbols, the program asked the subject to indicate his level of confidence in his selection as low, medium, or high. If the subject's guess matched the symbol chosen by the computer, the word "HIT" was repeatedly flashed in large multicolored letters, along with sound effects.

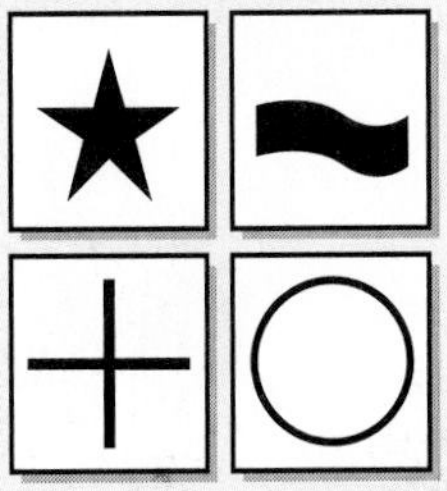

The experiment was conducted in three sessions on three consecutive days. Each session consisted of four runs of 24 trials for an experiment total of 288 trials. For each run, there was a 50% chance that the computer was in a precognition mode, in which the computer selected the symbol after the subject made a guess, and there was a 50% chance that it was in a clairvoyance mode, in which the computer made its selection at the start of each trial. The subject was not informed of the mode until after the run was completed.

We want to determine if the subject, Olof Jonsson, obtained results significantly better than chance in this experiment. After you have studied this chapter, you will have the skills needed to answer this question.

Activity 8A

Everyone's worst nightmare

One particularly stressful night, you dream that you are now in college and that the professor has announced a pop quiz. The quiz topic is ancient Ugaritic language and literature during the thirteenth century B.C. To make matters worse, the quiz is written in Farsi. Of course, you are totally unprepared. In your dream, you are the only student in the class because everyone else was exempt from taking the quiz. The professor notices that you are sweating profusely, and he correctly surmises that you are not prepared. He says, "Since you must guess at all the answers anyway, I won't even bother providing you with the questions. It saves paper!"

Now pretend that you are living this nightmare (except that the results of the quiz won't count).

1. Number your paper from 1 to 10 as shown. There are five choices for each question: A, B, C, D, or E. Select an answer to each question.

Question	Answer
1	
2	
3	
4	
5	
6	
7	
8	
9	
10	

2. Your teacher will display the correct answers on the overhead when everyone has finished. Score your own paper.
3. What kind of results should you expect to see for the entire population of unprepared students in your class? Display a graph of this expected distribution of results. Label it clearly. Would the distribution look different if there were only four choices? Explain.
4. Design a *simulation* to determine the distribution of the number of correct guesses if you had five choices. Use the random digits table or your calculator. What will the digits represent? Describe how you will run the simulation. Perform 30 trials of your simulation. Record the number of correct guesses for each trial by creating a frequency distribution table.

Introduction

In practice, we frequently encounter random phenomena where there are two outcomes of interest. Here are some examples:

- We use a coin toss to see which of the two football teams gets the choice of kicking off or receiving to begin the game.
- A basketball player shoots a free throw; the outcomes of interest are {she makes the shot; she misses}.
- A young couple prepares for their first child; the possible outcomes are {boy; girl}.
- A quality control inspector selects an automotive part coming off the assembly line; he is interested in whether or not the part meets production requirements.

In this chapter we will explore two important classes of distributions—the binomial distributions and the geometric distributions—and learn some of their properties. We will use what we have learned about probability and random variables from previous chapters, with the view toward completing the necessary foundation to study inference.

8.1 The Binomial Distributions

In Activity 8A, we simulated guessing on a five-choice, ten-item multiple-choice quiz. Spinning a five-sector spinner with equally likely choices A to E would produce similar results. The characterizing features of this setting are as follows: A *trial* consists of spinning the spinner once. We are interested in two outcomes in each trial: the result of the spin is the correct answer to the question (success) or it is not (failure). Typically we are interested in the number of successes. We will spin the spinner 10 times. The spins are *independent* in the sense that the outcome of one spin has no influence on the outcome of the next spin. And last, the probability of success (0.2) is the same for each spin (trial). A situation where these four conditions are satisfied is said to be a ***binomial setting.***

The Binomial Setting

1. Each observation falls into one of just two categories, which for convenience we call "success" or "failure."
2. There is a fixed number n of observations.
3. The n observations are all **independent.** That is, knowing the result of one observation tells you nothing about the other observations.
4. The probability of success, call it p, is the same for each observation.

If you are presented with a random phenomenon, it is important to be able to recognize it as a binomial setting or a geometric setting (covered in the next section) or neither. If you can verify that each of these four conditions is satisfied, you will be able to make use of known properties of binomial situations to gain more insights.

binomial random variable

If data are produced in a binomial setting, then the random variable X = number of successes is called a ***binomial random variable,*** and the probability distribution of X is called a ***binomial distribution.***

Binomial Distribution

The distribution of the count X of successes in the binomial setting is the **binomial distribution** with parameters n and p. The parameter n is the number of observations, and p is the probability of a success on any one observation. The possible values of X are the whole numbers from 0 to n. As an abbreviation, we say that X is $B(n, p)$.

CAUTION

The binomial distributions are an important class of discrete probability distributions. Later in this section we will learn how to assign probabilities to outcomes and how to find the mean and standard deviation of binomial distributions. **The most important skill for using binomial distributions is the ability to recognize situations to which they do and don't apply.**

Example 8.1 *Blood types*
Binomial setting?

Blood type is inherited. If both parents carry genes for the O and A blood types, each child has probability 0.25 of getting two O genes and so of having blood type O. Different children inherit independently of each other. The number of O blood types among 5 children of these parents is the count X of successes in 5 independent observations with probability 0.25 of a success on each observation. So X has the binomial distribution with $n = 5$ and $p = 0.25$. We say that X is $B(5, 0.25)$.

Example 8.2 *Dealing cards*
Binomial setting?

Deal 10 cards from a shuffled deck and count the number X of red cards. There are 10 observations, and each gives either a red or a black card. A "success" is a red card. But the observations are not independent. If the first card is black, the second is more likely to be red because there are more red cards than black cards left in the deck. The count X does *not* have a binomial distribution.

CLOSE TO HOME JOHN McPHERSON

A diabolical new testing technique: math essay questions.

Binomial Distributions in Statistical Sampling

The binomial distributions are important in statistics when we wish to make inferences about the proportion p of "successes" in a population. Here is a typical example.

Example 8.3 *Inspecting switches*

Binomial setting?

An engineer chooses an SRS of 10 switches from a shipment of 10,000 switches. Suppose that (unknown to the engineer) 10% of the switches in the shipment are bad. The engineer counts the number X of bad switches in the sample.

This is not quite a binomial setting. Just as removing one card in Example 8.2 changed the makeup of the deck, removing one switch changes the proportion of bad switches remaining in the shipment. So the state of the second switch chosen is not independent of the first. But removing one switch from a shipment of 10,000 changes the makeup of the remaining 9999 switches very little. In practice, the distribution of X is very close to the binomial distribution with $n = 10$ and $p = 0.1$.

Example 8.3 shows how we can use the binomial distributions in the statistical setting of selecting an SRS. When the population is much larger than the sample, a count of successes in an SRS of size n has approximately the binomial distribution with n equal to the sample size and p equal to the proportion of successes in the population.

Sampling Distribution of a Count

Choose an SRS of size n from a population with proportion p of successes. When the population is much larger than the sample, the count X of successes in the sample has approximately the binomial distribution with parameters n and p.

Example 8.4 *Aircraft engine reliability*
Binomial setting?

Engineers define reliability as the probability that an item will perform its function under specific conditions for a specific period of time. If an aircraft engine turbine has probability 0.999 of performing properly for an hour of flight, the number of turbines in a fleet of 350 engines that fly for an hour without failure has the $B(350, 0.999)$ distribution. This binomial distribution is obtained by assuming, as seems reasonable, that the turbines fail independently of each other. A common cause of failure, such as sabotage, would destroy the independence and make the binomial model inappropriate.

Was he good or was he lucky?
When a baseball player hits .300, everyone applauds. A .300 hitter gets a hit in 30% of times at bat. Could a .300 year just be luck? Typical major leaguers bat about 500 times a season and hit about .260. A hitter's successive tries seem to be independent, so we have a binomial setting. From this model, we can calculate or simulate the probability of hitting .300. It is about 0.025. Out of 100 run-of-the-mill major league hitters, two or three each year will bat .300 because they were lucky.

Exercises

In each of Exercises 8.1 to 8.5, X is a count. Does X have a binomial distribution? Give your reasons in each case.

8.1 Quality control An auto manufacturer chooses one car from each hour's production for a detailed quality inspection. One variable recorded is the count X of finish defects (dimples, ripples, etc.) in the car's paint.

8.2 Polling a jury pool The pool of potential jurors for a murder case contains 100 persons chosen at random from the adult residents of a large city. Each person in the pool is asked whether he or she opposes the death penalty; X is the number who say "Yes."

8.3 Random digit dialing When an opinion poll calls residential telephone numbers at random, only 20% of the calls reach a live person. You watch the random dialing machine make 15 calls. X is the number that reach a live person.

8.4 Logging in At peak periods, 15% of attempted log-ins to an email service fail. Log-in attempts are independent and each has the same probability of failing. Darci logs in repeatedly until she succeeds. X is the number of the log-in attempt that finally succeeds.

8.5 Computer instruction A student studies binomial distributions using computer-assisted instruction. After the lesson, the computer presents 10 problems. The student solves each problem and enters her answer. The computer gives additional instruction between problems if the answer is wrong. The count X is the number of problems that the student gets right.

8.6 I can't relax Opinion polls find that 14% of Americans "never have time to relax."[2] If you take an SRS of 500 adults, what is the approximate distribution of the number in your sample who say they never have time to relax? Justify your answer.

Binomial Formulas

We can find a formula for the probability that a binomial random variable takes any value by adding probabilities for the different ways of getting exactly that many successes in n observations. Here is the example we will use to show the idea.

Example 8.5 *Inheriting blood type*
Developing a formula

Each child born to a particular set of parents has probability 0.25 of having blood type O. If these parents have 5 children, what is the probability that exactly 2 of them have type O blood? The count of children with type O blood is a binomial random variable X with $n = 5$ tries and probability $p = 0.25$ of a success on each try. We want $P(X = 2)$.

Because the method doesn't depend on the specific example, let's use "S" for success and "F" for failure for short. Do the work in two steps.

Step 1: Find the probability that a specific 2 of the 5 tries give successes, say the first and the third. This is the outcome SFSFF. Here's how to find the probability of this outcome:

- The probability that the first try is a success is 0.25. That is, in many repetitions, we succeed on the first try 25% of the time.
- Out of all the repetitions with a success on the first try, 75% have a failure on the second try. So the proportion of repetitions on which the first two tries are SF is (0.25)(0.75). We can multiply here because the tries are independent. That is, the first try has no influence on the second.
- Keep going: Of these repetitions, the proportion 0.25 have S on the third try. So the probability of SFS is (0.25)(0.75)(0.25). After two more tries, the probability of SFSFF is the product of the try-by-try probabilities:

$$(0.25)(0.75)(0.25)(0.75)(0.75) = (0.25)^2(0.75)^3$$

Step 2: Observe that the probability of *any one* arrangement of 2 S's and 3 F's has this same probability. That's true because we multiply together 0.25 twice and 0.75 three times whenever we have 2 S's and 3 F's. The probability that $X = 2$ is the probability of getting 2 S's and 3 F's in any arrangement whatsoever. Here are all the possible arrangements:

SSFFF	SFSFF	SFFSF	SFFFS	FSSFF
FSFSF	FSFFS	FFSSF	FFSFS	FFFSS

There are 10 of them, all with the same probability. The overall probability of 2 successes is therefore

$$P(X = 2) = 10(0.25)^2(0.75)^3 = 0.2637$$

The pattern of this calculation works for any binomial probability. To use it, we need to be able to count the number of arrangements of k successes in n observations without actually listing them. We use the following fact to do the counting:

Binomial Coefficient

The number of ways of arranging k successes among n observations is given by the **binomial coefficient**

$$\binom{n}{k} = \frac{n!}{k!(n-k)!}$$

for $k = 0, 1, 2, \ldots, n$.

The formula for binomial coefficients uses the factorial notation. For any positive whole number n, its factorial $n!$ is

$$n! = n \times (n-1) \times (n-2) \times \ldots \times 3 \times 2 \times 1$$

Also, $0! = 1$.

Notice that the larger of the two factorials in the denominator of a binomial coefficient will cancel much of the $n!$ in the numerator. For example, the binomial coefficient we need for Example 8.5 is

$$\binom{5}{2} = \frac{5!}{2!3!}$$
$$= \frac{(5)(4)(3)(2)(1)}{(2)(1) \times (3)(2)(1)}$$
$$= \frac{(5)(4)}{(2)(1)} = \frac{20}{2} = 10$$

The notation $\binom{n}{k}$ is not related to the fraction $\frac{n}{k}$. A helpful way to remember its meaning is to read it as "binomial coefficient n choose k." Binomial coefficients have many uses in mathematics, but we are interested in them only as an aid to finding binomial probabilities. The binomial coefficient $\binom{n}{k}$ counts the number of ways in which k successes can be distributed among n observations. The ***binomial probability*** $P(X = k)$ is this count multiplied by the probability of any specific arrangement of the k successes. Here is the formula we seek:

Binomial Probability

If X has the binomial distribution with n observations and probability p of success on each observation, the possible values of X are $0, 1, 2, \ldots, n$. If k is any one of these values,

$$P(X = k) = \binom{n}{k} p^k (1-p)^{n-k}$$

Example 8.6

Defective switches

Applying the formula

The number X of switches that fail inspection in Example 8.3 has approximately the binomial distribution with $n = 10$ and $p = 0.1$. The probability that no more than 1 switch fails is

$$\begin{aligned} P(X \leq 1) &= P(X = 1) + P(X = 0) \\ &= \binom{10}{1}(0.1)^1(0.9)^9 + \binom{10}{0}(0.1)^0(0.9)^{10} \\ &= \frac{10!}{1!9!}(0.1)(0.3874) + \frac{10!}{0!10!}(1)(0.3487) \\ &= (10)(0.1)(0.3874) + (1)(1)(0.3487) \\ &= 0.3874 + 0.3487 \\ &= 0.7361 \end{aligned}$$

Notice that the calculation uses the facts that $0! = 1$ and that $a^0 = 1$ for any number a other than 0.

Exercises

In each of the following exercises, you are to use the binomial probability formula to answer the question. Begin with the formula, and show substitution into the formula.

8.7 Blood types The count X of children with type O blood among 5 children whose parents carry genes for both the O and the A blood types is $B(5, 0.25)$. See Example 8.1 on page 514. Use the binomial probability formula to find $P(X = 3)$.

8.8 Broccoli plants Suppose you purchase a bundle of 10 bare-root broccoli plants. The sales clerk tells you that on average you can expect 5% of the plants to die before producing any broccoli. Assume that the bundle is a random sample of plants. Use the binomial formula to find the probability that you will lose at most 1 of the broccoli plants.

8.9 More on blood types Refer to Exercise 8.7. Let X = number of children who have type O blood. Then X has the $B(5, 0.25)$ distribution. Use the binomial probability formula to find the probability that at least one of the children in Exercise 8.7 has type O blood. (*Hint*: Do not calculate more than one binomial formula.)

8.10 Do our athletes graduate? A university claims that 80% of its basketball players get degrees. An investigation examines the fate of all 20 players who entered the program over a period of several years that ended six years ago. Of these players, 11 graduated and the remaining 9 are no longer in school. If the university's claim is true, the number of players among the 20 who graduate should have the binomial distribution with $n = 20$ and $p = 0.8$. Use the binomial probability formula to answer the following questions.

(a) What is the probability that exactly 11 out of 20 players graduate?

(b) What is the probability that all 20 graduate?

(c) What is the probability that not all of the 20 graduate?

8.11 Hispanic representation A factory employs several thousand workers, of whom 30% are Hispanic. If the 15 members of the union executive committee were chosen from the workers at random, the number of Hispanics on the committee would have the binomial distribution with $n = 15$ and $p = 0.3$.

(a) What is the probability that exactly 3 members of the committee are Hispanic?

(b) What is the probability that none of the committee members are Hispanic?

8.12 Scuba diving The mailing list of an agency that markets scuba-diving trips to the Florida Keys contains 70% males and 30% females. The agency calls 30 people chosen at random from its list.

(a) What is the probability that 20 of the 30 are men?

(b) What is the probability that the first woman is reached on the fourth call? (That is, the first 4 calls give MMMF.)

Finding Binomial Probabilities

In practice, you will rarely have to use the formula for calculating the probability that a binomial random variable takes any of its values. The TI-83/84/89 and most statistical software packages calculate binomial probabilities.

Example 8.7 *Inspecting switches*
Calculating a binomial probability

A quality engineer selects an SRS of 10 switches from a large shipment for detailed inspection. Unknown to the engineer, 10% of the switches in the shipment fail to meet the specifications. What is the probability that no more than 1 of the 10 switches in the sample fail inspection?

The count X of bad switches in the sample has approximately the $B(10, 0.1)$ distribution. Figure 8.1 is a probability histogram for this distribution.

The distribution is strongly skewed. Although X can take any whole-number value from 0 to 10, the probabilities of values larger than 5 are so small that they do not appear in the histogram. We want to calculate

$$P(X \leq 1) = P(X = 0) + P(X = 1)$$

when X is $B(10, 0.1)$. The TI-83/84 command `binompdf(n,p,X)` and the TI-89 command `tistat.binomPdf(n,p,X)` calculate the binomial probability of the value X. The suffix `pdf` stands for ***probability distribution function.*** We met the probability distribution in Chapter 7.

The `binompdf` command is found under **2nd** `(DISTR)/0:binompdf` on the TI-83/84. On the TI-89, it's in the CATALOG under Flash Apps. The TI-83/84 command `binompdf(10,.1,0)` and the TI-89 command `tistat.binomPdf(10,.1,0)` calculate the binomial probability that X = 0 to be 0.3486784401. The command `binompdf(10,.1,1)` returns the probability 0.387420489. Thus,

$$P(X \leq 1) = P(X = 0) + P(X = 1)$$
$$= 0.3487 + 0.3874 = 0.7361$$

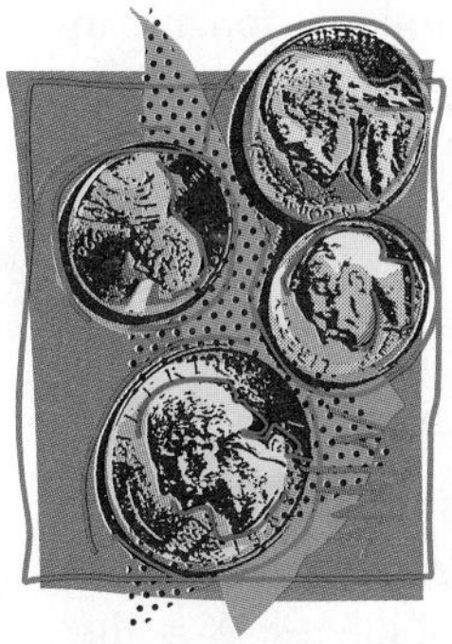

What looks random? Toss a coin six times and record heads (H) or tails (T) on each toss. Which of these outcomes is more probable: HTHTTH or TTTHHH? Almost everyone says that HTHTTH is more probable, because TTTHHH does not "look random." In fact, both are equally probable. That heads has probability 0.5 says that about half of a very long sequence of tosses will be heads. It doesn't say that heads and tails must come close to alternating in the short run. The coin doesn't know what past outcomes were, and it can't try to create a balanced sequence.

Figure 8.1 *Probability histogram for the binomial distribution with* n = *10 and* p = *0.1, for Example 8.7.*

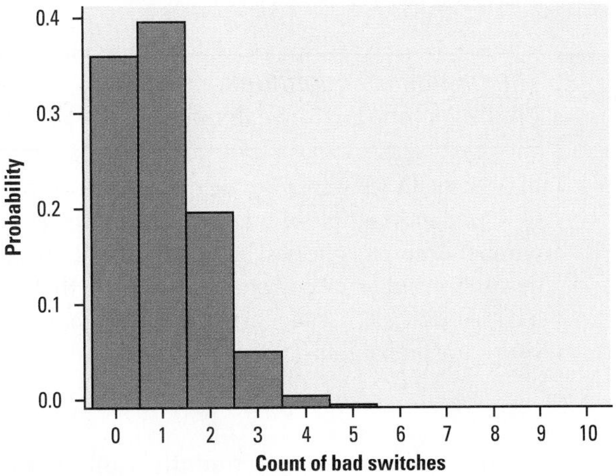

About 74% of all samples will contain no more than 1 bad switch. A sample of size 10 cannot be trusted to alert the engineer to the presence of unacceptable items in the shipment.

pdf

Given a discrete random variable X, the **probability distribution function (pdf)** assigns a probability to each value of X. The probabilities must satisfy the rules for probabilities given in Chapter 6.

Example 8.8 *Corinne's free throws*

Calculating binomial probabilities

Corinne is a basketball player who makes 75% of her free throws over the course of a season. In a key game, Corinne shoots 12 free throws and makes only 7 of them. The fans think that she failed because she was nervous. Is it unusual for Corinne to perform this poorly? To answer this question, assume that free throws are independent with probability 0.75 of a success on each shot. (Studies of long sequences of free throws have found no evidence that making a free throw affects a later shot.) The number X of baskets (successes) in 12 attempts has the $B(12, 0.75)$ distribution.

We want the probability of making a basket on at most 7 free throws. This is

$$\begin{aligned} P(X \leq 7) &= P(X = 0) + P(X = 1) + P(X = 2) + \ldots + P(X = 7) \\ &= 0.0000 + 0.0000 + 0.0000 + 0.0004 + 0.0024 + 0.0115 + \\ &\quad 0.0401 + 0.1032 \\ &= 0.1576 \end{aligned}$$

Corinne will make at most 7 of her 12 free throws about 16% of the time, or roughly in one of every six games. While below her average level, this performance is well within the range of the usual chance variation in her shooting.

Example 8.9 ***The nightmare continues***
Finding a binomial by calculator

In Activity 8A we wanted to determine the distribution of the number of correct guesses on a 10-item multiple-choice quiz. In this case, the random variable of interest, X = the number of correct guesses, has the $B(10, 0.2)$ distribution. We want to find the probability that the number of correct guesses is 4, that is, $P(X = 4)$. The TI-83/84 command `binompdf(10,.2,4)` and the TI-89 command `tistat.binomPdf(10,.2,4)` return the probability 0.088.

In applications we frequently want to find the probability that a random variable takes a range of values. The ***cumulative binomial probability*** is useful in these cases.

cdf

Given a random variable X, the **cumulative distribution function** (cdf) of X calculates the sum of the probabilities for 0, 1, 2, . . . , up to the value X. That is, it calculates the probability of obtaining at most X successes in n trials.

For the count X of defective switches in Example 8.6 on page 519, the command `binomcdf(10,.1,1)` and the TI-89 command `tistat.binomCdf(10,.1,1)` output 0.7360989303 for the cumulative probability $P(X \leq 1)$.

Example 8.10 ***Is Corinne in a slump?***
Binomial probability by calculator

In Example 8.8 Corinne shoots $n = 12$ free throws and makes only 7 of them. Since she is a 75% free-throw shooter ($p = 0.75$), we wanted to know if it was unusual for Corinne to perform this poorly. If X = number of baskets made on free throws, then X has the $B(12, 0.75)$ distribution, and we need to find the probability that she makes at most 7 of her free throws, that is, $P(X \leq 7)$. The TI-83/84 command `binomcdf(12,.75,7)` and the TI-89 command `tistat.binomCdf(12,.75,7)` calculate the cumulative probability $P(X \leq 7)$ to be 0.1576436761. We round the answer to four decimal places and report that the probability that Corinne makes *at most* 7 of her 12 free throws is 0.1576.

The pdf table for Corinne's shots looks like this:

X:	0	1	2	3	4	5	6
P(X):	0.000	0.000	0.000	0.000	0.002	0.011	0.040
X:	7	8	9	10	11	12	
P(X):	0.103	0.194	0.258	0.232	0.127	0.032	

If we denote the cumulative distribution function by $F(X)$, we can record the cumulative sum of the probabilities in a third row of the table:

X:	0	1	2	3	4	5	6
P(X):	0.000	0.000	0.000	0.000	0.002	0.011	0.040
F(X):	0.000	0.000	0.000	0.000	0.003	0.014	0.054
X:	7	8	9	10	11	12	
P(X):	0.103	0.194	0.258	0.232	0.127	0.032	
F(X):	0.158	0.351	0.609	0.842	0.968	1	

Notice that terms sometimes don't appear to add up as they should. The cumulative function $F(4)$, for example, should equal $P(0) + P(1) + P(2) + P(3) + P(4)$. Of course, the culprit is roundoff error. With your calculator, enter the integers 0 to 12 into L_1/list1, the corresponding binomial probabilities into L_2/list2, and use the command `binomcdf(12,.75,L₁)→L₃` (`tistat.binomCdf(12,.75, list1)→list3` on the TI-89) to enter the cumulative probabilities into L_3/list3.

In addition to being helpful in answering questions involving wording such as "find the probability that the number of successes is at most 6," the cdf is also particularly useful for calculating the probability that the number of successes is *more* than a certain number. This calculation uses the complement rule:

$$P(X > k) = 1 - P(X \leq k) \quad k = 0, 1, 2, 3, 4, \ldots$$

Exercises

Use your calculator's binomial pdf or cdf commands to find the following probabilities.

8.13 Inheriting blood type This exercise is a continuation of Exercise 8.9 (page 519). Each child born to a particular set of parents has probability 0.25 of having type O blood. Suppose these parents have 5 children. Let X = number of children who have type O blood. Then X is $B(5, 0.25)$.

(a) What is the probability that exactly 2 of the children have type O blood?

(b) Make a table for the pdf of the random variable X. Then use your calculator to find the probabilities of all possible values of X, and complete the table.

(c) Verify that the sum of the probabilities is 1.

(d) Construct a histogram of the pdf.

(e) Use the calculator to find the cumulative probabilities, and add these values to your pdf table. Then construct a cumulative distribution histogram. How is this histogram different from the histogram for Corinne's free throws?

8.14 Guessing on a true-false quiz Suppose that James guesses on each question of a 50-item true-false quiz. Find the probability that James passes if

(a) a score of 25 or more correct is needed to pass.

(b) a score of 30 or more correct is needed to pass.

(c) a score of 32 or more correct is needed to pass.

8.15 Guessing on a multiple-choice quiz Suppose that Erin guesses on each question of a multiple-choice quiz.

(a) If each question has four different choices, find the probability that Erin gets 1 or more correct answers on a 10-item quiz.

(b) If the quiz consists of three questions, Question 1 has 3 possible answers, Question 2 has 4 possible answers, and Question 3 has 5 possible answers, find the probability that Erin gets 1 or more correct answers.

8.16 Dad's in the pokey According to a 2000 study by the Bureau of Justice Statistics, approximately 2% of the nation's 72 million children had a parent behind bars—nearly 1.5 million minors. Suppose that 100 children are randomly selected. Let X be the number of children who have an incarcerated parent.

(a) Does X satisfy the requirements for a binomial setting? Explain. If $X = B(n, p)$, what are n and p?

(b) Describe $P(X = 0)$ in words. Then find $P(X = 0)$ and $P(X = 1)$.

(c) What is the probability that 2 or more of the 100 children have a parent behind bars?

8.17 Graduation rate for athletes See Exercise 8.10 (page 519). The number of athletes who graduate is $B(20, 0.8)$. Use your calculator to answer the three questions posed in Exercise 8.10.

8.18 Marital status Among employed women, 25% have never been married. Select 10 employed women at random.

(a) The number in your sample who have never been married has a binomial distribution. What are n and p?

(b) What is the probability that exactly 2 of the 10 women in your sample have never been married?

(c) What is the probability that 2 or fewer have never been married?

Binomial Mean and Standard Deviation

If a count X has the binomial distribution based on n observations with probability p of success, what is its mean μ? We can guess the answer. If a basketball player makes 75% of her free throws, the mean number made in 12 tries should be 75% of 12, or 9. In general, the mean of a binomial distribution should be $\mu = np$. To derive the expressions for the mean and standard deviation in the general case, let X represent the number of successes in a single trial. Then X takes two values, 1 (for success) and 0 (for failure). We'll let p be the probability of success on a single trial, and introduce $q = 1 - p$ as the probability of failure. This is common notation. Then the probability distribution is simply

X:	0	1
P(X):	q	p

The expected value for this one trial is

$$E(X) = \mu_X = 0(q) + 1(p) = p$$

The variance is

$$\sigma_X^2 = (0 - p)^2 q + (1 - p)^2 p = p^2 q + pq^2 = pq(p + q) = pq$$

Now define a new random variable Y to be the number of successes in n independent trials. Then $Y = X_1 + X_2 + \ldots + X_n$. Using the rules for means and variances of linear combinations of independent random variables, we can say that

$$\begin{aligned} \mu_Y &= \mu_{(X_1 + X_2 + \ldots + X_n)} = \mu_{X_1} + \mu_{X_2} + \ldots + \mu_{X_n} \\ &= p + p + \ldots + p \\ &= np \end{aligned}$$

and

$$\begin{aligned} \sigma_Y^2 &= \sigma^2_{X_1 + X_2 + \ldots + X_n} = \sigma^2_{X_1} + \sigma^2_{X_2} + \ldots + \sigma^2_{X_n} \\ &= pq + pq + \ldots + pq \\ &= npq = np(1 - p) \end{aligned}$$

and the standard deviation of Y is $\sqrt{np(1 - p)}$. Here is what we have shown:

Mean and Standard Deviation of a Binomial Random Variable

If a count X has the binomial distribution with number of observations n and probability of success p, the mean and standard deviation of X are

$$\mu = np$$

$$\sigma = \sqrt{np(1 - p)}$$

Be careful. These short formulas are good only for binomial distributions. They can't be used for other discrete random variables.

Example 8.11 *Bad switches*
Calculating the mean and standard deviation

Continuing Examples 8.6 and 8.7, the count X of bad switches is binomial with $n = 10$ and $p = 0.1$. This is the sampling distribution the engineer would see if she drew all possible SRSs of 10 switches from the shipment and recorded the value of X for each sample.

The mean and standard deviation of the binomial distribution are

$$\begin{aligned} \mu &= np \\ &= (10)(0.1) = 1 \\ \sigma &= \sqrt{np(1-p)} \\ &= \sqrt{(10)(0.1)(0.9)} = \sqrt{0.9} = 0.9487 \end{aligned}$$

The mean is marked on the probability histogram in Figure 8.2.

Figure 8.2 *Probability histogram for the binomial distribution with* n = *10 and* p = *0.1, for Example 8.11.*

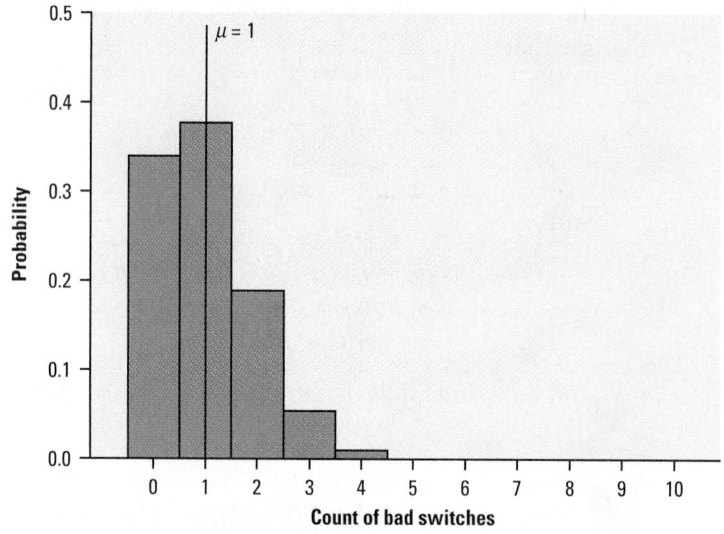

The Normal Approximation to Binomial Distributions

The formula for binomial probabilities becomes awkward as the number of trials n increases. You can use software or a statistical calculator to handle some problems for which the formula is not practical. Here is another alternative: as the number of trials n gets larger, the binomial distribution gets close to a Normal distribution. When n is large, we can use Normal probability calculations to approximate hard-to-calculate binomial probabilities.

Example 8.12 *Attitudes toward shopping*
A formidable computation

Are attitudes toward shopping changing? Sample surveys show that fewer people enjoy shopping than in the past. A survey asked a nationwide random sample of 2500 adults if they agreed or disagreed that "I like buying new clothes, but shopping is often frustrating and time-consuming."[3] The population that the poll wants to draw conclusions about is all U.S. residents aged 18 and over. Suppose that in fact 60% of all adult U.S. residents would say "Agree" if asked the same question. What is the probability that 1520 or more of the sample agree?

Because there are more than 218 million adults, we can take the responses of 2500 randomly chosen adults to be independent. So the number in our sample who agree that shopping is frustrating is a random variable X having the binomial distribution with $n = 2500$ and $p = 0.6$. To find the probability that at least 1520 of the people in the sample find shopping frustrating, we must add the binomial probabilities of all outcomes from $X = 1520$ to $X = 2500$. This isn't practical. Here are three ways to do this problem.

1. Statistical software can do the calculation. The result is

$$\begin{aligned} P(X \geq 1520) &= 1 - P(X \leq 1519) \\ &= 1 - 0.7869 \\ &= 0.2131 \end{aligned}$$

This answer is exactly correct to four decimal places.

2. We can simulate a large number of repetitions of the sample. Figure 8.3 on the next page displays a histogram of the counts X from 1000 samples of size 2500 when the truth about the population is $p = 0.6$. Because 221 of these 1000 samples have X at least 1520, the probability estimated from the simulation is

$$P(X \geq 1520) = \frac{221}{1000} = 0.221$$

This answer is only approximately correct. The law of large numbers says that the results of such simulations always get closer to the true probability as we simulate larger numbers of samples.

3. Both of the previous methods require software. Instead, look at the Normal curve in Figure 8.3. This is the density curve of the Normal distribution with the same mean and standard deviation as the binomial variable X:

$$\mu = np = (2500)(0.6) = 1500$$
$$\sigma = \sqrt{np(1-p)} = \sqrt{(2500)(0.6)(0.4)} = 24.49$$

As the figure shows, this Normal distribution approximates the binomial distribution quite well. So we can do a Normal calculation.

Figure 8.3 *Histogram of 1000 binomial counts (n = 2500, p = 0.6) and the Normal density curve that approximates this binomial distribution.*

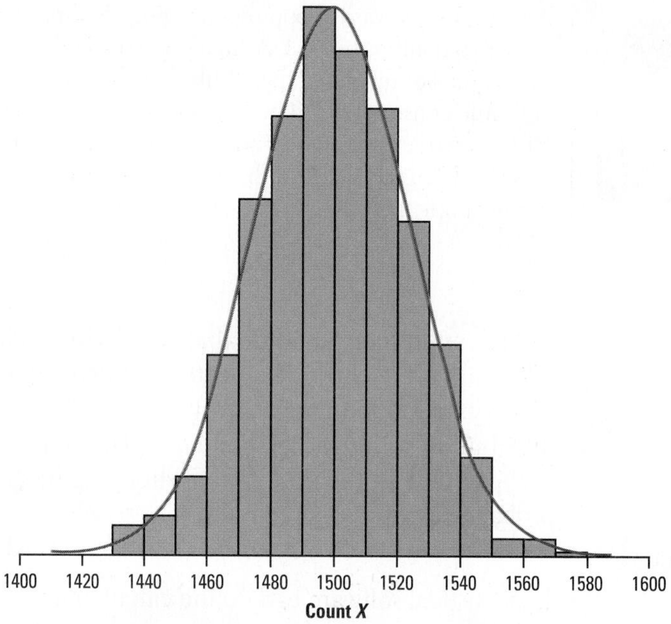

Example 8.13 *Normal calculation of a binomial probability*

The Normal approximation

If we act as though the count X has the N(1500, 24.49) distribution, here is the probability we want, using Table A:

$$P(X \geq 1520) = P\left(\frac{X - 1500}{24.49} \geq \frac{1520 - 1500}{24.49}\right)$$
$$= P(Z \geq 0.82)$$
$$= 1 - 0.7939 = 0.2061$$

The Normal approximation 0.2061 differs from the software result 0.2131 by only 0.007.

Normal Approximation for Binomial Distributions

Suppose that a count X has the binomial distribution with n trials and success probability p. When n is large, the distribution of X is approximately Normal, $N(np, \sqrt{np(1-p)})$. As a rule of thumb, we will use the Normal approximation when n and p satisfy $np \geq 10$ and $n(1-p) \geq 10$.

The Normal approximation is easy to remember because it says that X is Normal with its binomial mean and standard deviation. The accuracy of the Normal approximation improves as the sample size n increases. It is most accurate for any fixed n when p is close to 1/2 and least accurate when p is near 0 or 1 and the distributions are skewed. The *Normal Approximation to Binomial* applet shows in visual form how well the Normal approximation fits the binomial distribution for any n and p. Whether or not you use the Normal approximation should depend on how accurate your calculations need to be. For most statistical purposes great accuracy is not required. Our "rule of thumb" for use of the Normal approximation reflects this judgment.

Exercises

8.19 Attitudes on shopping Refer to Example 8.12 (page 527) on attitudes toward shopping.

(a) Verify that the rule of thumb conditions are satisfied for using the Normal approximation to the binomial distribution.

(b) Use your calculator and the cumulative binomial function to verify that the exact answer for the probability that at least 1520 people in the sample find shopping frustrating is 0.2131. What is the probability, correct to six decimal places?

(c) What is the probability that at most 1468 people in the sample would agree with the statement that shopping is frustrating?

8.20 Using Benford's law According to Benford's law (Example 7.6, page 483) the probability that the first digit of the amount of a randomly chosen invoice is a 1 or a 2 is 0.477. You examine 90 invoices from a vendor and find that 29 have first digits 1 or 2. If Benford's law holds, the count of 1s and 2s will have the binomial distribution with $n = 90$ and $p = 0.477$. Too few 1s and 2s suggests fraud. What is the approximate probability of 29 or fewer if the invoices follow Benford's law? Do you suspect that the invoice amounts are not genuine?

8.21 Do our athletes graduate?

(a) Find the mean number of graduates out of 20 players in the setting of Exercise 8.10 (page 519).

(b) Find the standard deviation σ of the count X.

(c) Suppose that the 20 players came from a population of which $p = 0.9$ graduated. What is the standard deviation σ of the count of graduates? If $p = 0.99$, what is σ? What does your work show about the behavior of the standard deviation of a binomial distribution as the probability p of success gets closer to 1?

8.22 Mark McGwire's home runs In 1998, Mark McGwire of the St. Louis Cardinals hit 70 home runs, a new major league record. Was this feat as surprising as most of us thought? In the three seasons before 1998, McGwire hit a home run in 11.6% of his times at bat. He went to bat 509 times in 1998. McGwire's home run count in 509 times at bat has approximately the binomial distribution with $n = 509$ and $p = 0.116$. What is the mean number of home runs he will hit in 509 times at bat? What is the probability of 70 or more home runs? (Use the Normal approximation.)

8.23 Polling Many local polls of public opinion use samples of size 400 to 800. Consider a poll of 400 adults in Richmond that asks the question "Do you approve of President George W. Bush's response to the World Trade Center terrorist attacks in September 2001?" Suppose we know that President Bush's approval rating on this issue nationally is 92% a week after the incident.

(a) What is the random variable X? Is X binomial? Explain.

(b) Calculate the binomial probability that at most 358 of the 400 adults in the Richmond poll answer "Yes" to this question.

(c) Find the expected number of people in the sample who indicate approval. Find the standard deviation of X.

(d) Perform a Normal approximation to answer the question in (b), and compare the results of the binomial calculation and the Normal approximation. Is the Normal approximation satisfactory?

8.24 Checking for survey errors One way of checking the effect of undercoverage, nonresponse, and other sources of error in a sample survey is to compare the sample with known facts about the population. About 12% of American adults are black. The number X of blacks in a random sample of 1500 adults should therefore vary with the binomial ($n = 1500$, $p = 0.12$) distribution.

(a) What are the mean and standard deviation of X?

(b) Use the Normal approximation to find the probability that the sample will contain between 165 and 195 blacks. Be sure to check that you can safely use the approximation.

Binomial Distribution with the Calculator

The following Technology Toolbox summarizes some important calculator techniques when working in a binomial setting:

Technology Toolbox

Exploring binomial distributions

For illustration purposes, we will use the sample of $n = 10$ switches with probability $p = 0.10$ of a defective switch from Example 8.6 (page 519). The random variable X is the number of defective switches (success) and $X = B(10, 0.1)$. To have the calculator make the probability distribution table and plot a histogram for the distribution of defective switches in a sample of 10 switches, proceed as follows:

Technology Toolbox

Exploring binomial distributions *(continued)*

TI-83/84

1. Enter the values of X into list L_1, either through the STAT/EDIT mode or by entering the command `seq (X, X, 0, 10, 1)`→L_1. (The `seq` command is under 2nd/LIST/OPS/5:seq. The syntax is the first X is the function, the second X is the counting variable, the next two numbers define the starting and ending values, and the last number is the increment.)
2. Enter the binomial probabilities into list L_2. Highlight L_2 and press 2nd VARS (DISTR)/ `C:binompdf(`. Then complete the command: `binompdf (10,0.1,L1)`. Note that the largest probability listed is about 0.3874. This will help us define our viewing window.

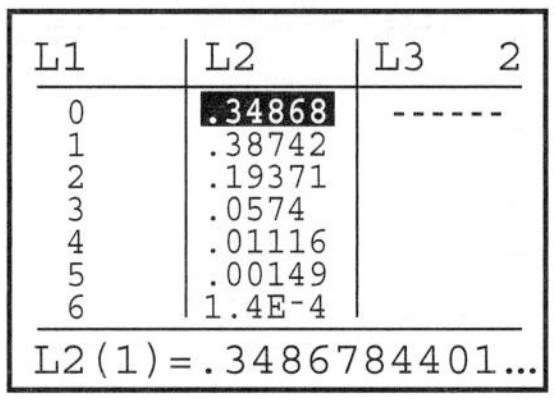

3. Deselect or delete any active defined functions in the Y = window.
4. Define Plot1 to be a histogram with Xlist: L_1 and Freq:L_2
5. Set the viewing window to be $X[0,11]_1$ and $Y[-.15,.5]_1$. Press TRACE and use the left and right cursor keys to inspect heights of various bars in the histogram.

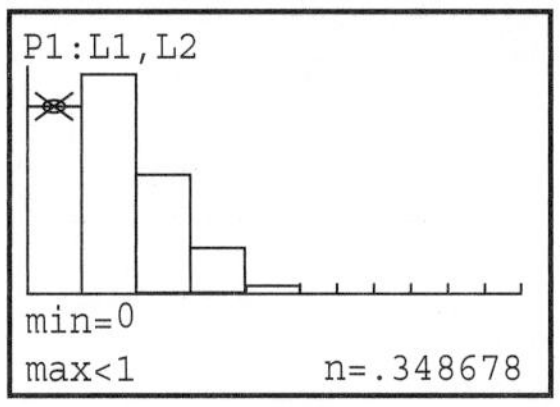

TI-89

1. Enter the values of X into list1, either through the Statistics/List Editor or by entering the command `seq (X, X, 0, 10, 1) →List1`. (The `seq` command is in the CATALOG. The syntax is: the first X is the function, the second X is the counting variable, the next two numbers define the starting and ending values, and the last number is the increment.)
2. Highlight list2 and press CATALOG, type F3 (Flash Apps), choose `binomPdf(`. Press ENTER and complete the command: `tistat.binomPdf(10, 0.1, list1)` and then ENTER.

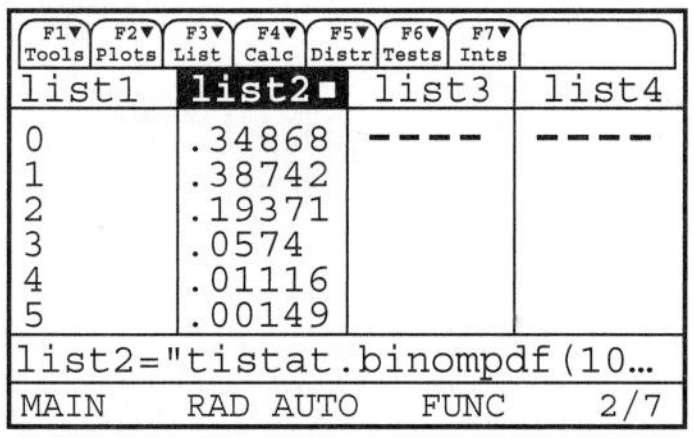

3. Deselect or delete any active defined functions in the Y = window.
4. Define Plot1 to be a histogram using list1 for *x* and list2 for frequency.
5. Define the viewing window: ◆ F2 (WINDOW). Specify $X[0,11]_1$ and $Y[-.15,.5]_1$. Press ◆ F3. Here is the histogram for the pdf.

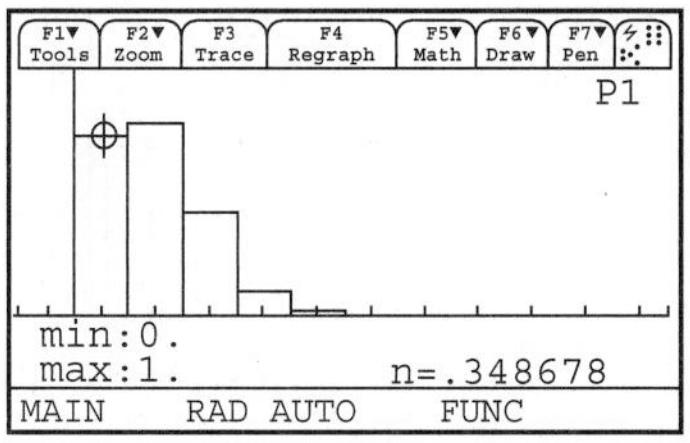

Outcomes larger than 6 do not have probability exactly 0, but their probabilities are so small that the rounded vaues are 0.0000. Verify that the sum of the probabilities is 1.

(continued)

Technology Toolbox

Exploring binomial distributions *(continued)*

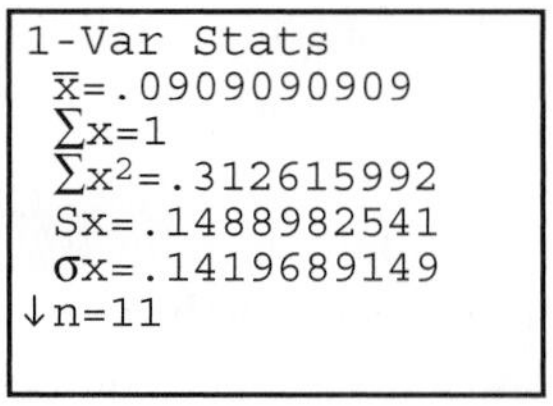

6. To calculate the cumulative probabilities, highlight list L_3. Press 2nd DISTR, and select `A:binomcdf(`. Complete the command: `binomcdf (10,.1,L1)`. Press ENTER. The cumulative probabilities are in L_3.

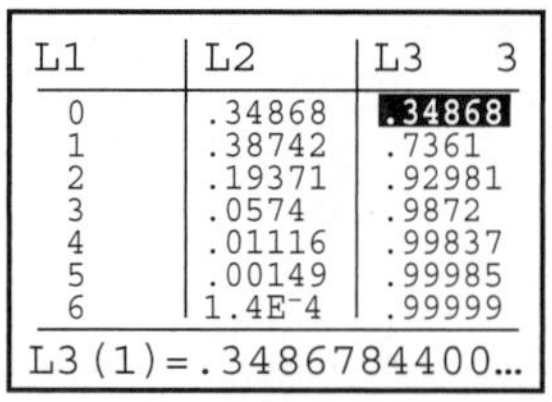

7. Turn off Plot1 and turn on Plot2. Define Plot2 to be a histogram with Xlist: L_1 and Ylist:L_3. In the viewing window, set Ymin $= -.3$ and Ymax $= 1.2$. Here is the histogram for the cdf:

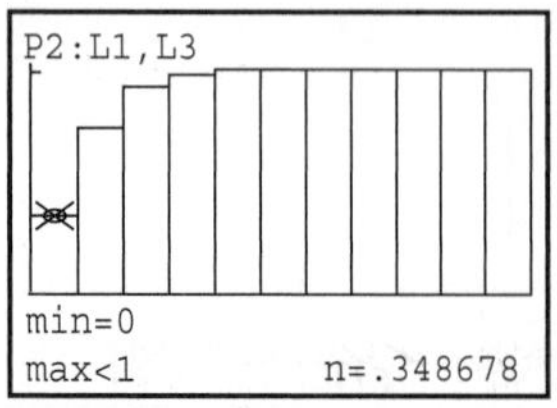

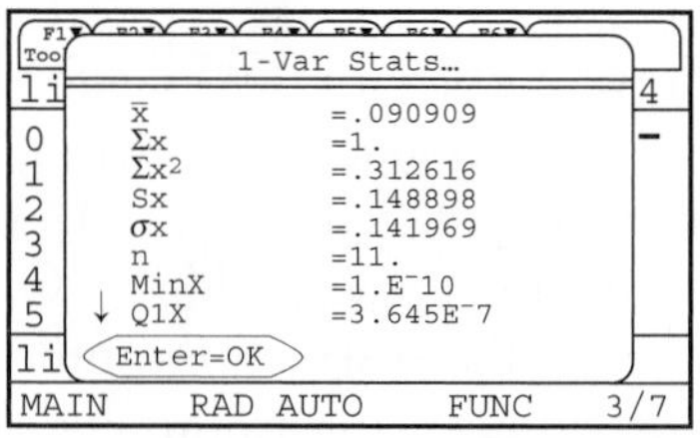

6. To calculate the cumulative distribution values, highlight list3 and press CATALOG F3 (FlashApps), and choose `binomCdf(`. Press ENTER and complete the command: `tistat.binomCdf (10,.1,list1)` and then ENTER.

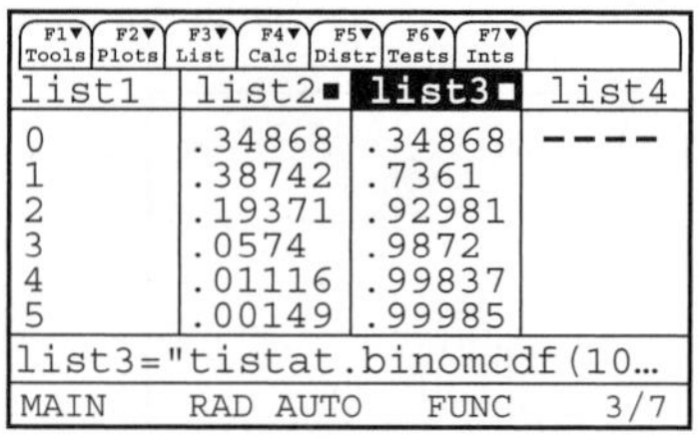

7. Turn off Plot1 (F2 (Plot Setup)), highlight Plot1, then F4 (✓) to deselect Plot1. Define Plot2 to be a histogram, except this time specify list3 for the frequency. In the viewing window, set Ymin $= -.3$ and Ymax $= 1.2$. Here is the histogram for the cdf:

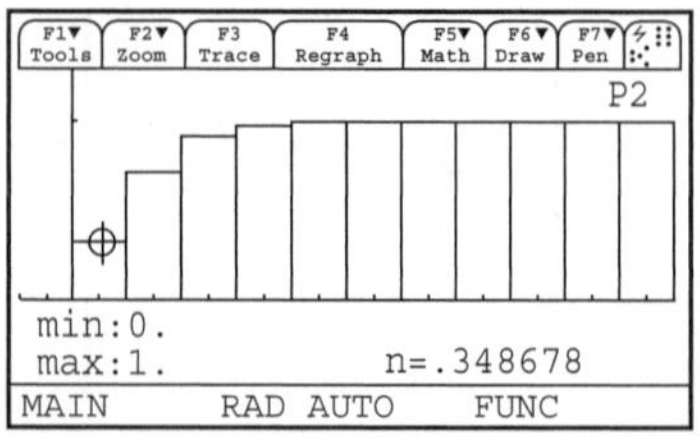

Simulating Binomial Events

In order to simulate a binomial event, you need to know how the random variable X and "success" are defined, the probability of success, and the number of trials. But if you know these things, you can apply the rules learned in this section to calculate the probabilities of events exactly. Nevertheless, being able to simulate a binomial event can give credence to results obtained by applying formulas and rules when the results may be less than convincing to someone who knows no statistics.

Example 8.14 *Corinne's free throws*

Simulating binomial events

SIMULATIONS

Recall that Corinne's free throw percent was 75% (see Example 8.8). In a particular game, she had 12 attempts and she made only 7. The question was "How unusual was it for Corinne to make at most 7 shots out of 12 attempts?" In Example 8.8, we calculated this binomial probability to be $P(X \leq 7) = 0.1576$. Now we will use the calculator to simulate 12 attempted shots and we will count the number of hits (baskets). Let X = number of hits in 12 free-throw attempts. Note that the probability of "success" is 0.75. To set up the simulation, we will assign the digit 0 to a miss and a 1 to a hit. The command `randBin(1,.75,12)` simulates 12 free-throw attempts. In the long run, this random function will select the number "1" 75% of the time and the number "0" 25% of the time.

Here are the results of one simulated game on the TI-83/84: the first three shots were hits, the fourth was a miss, the fifth was a hit, and the next two were misses, and so forth. If we repeated this many times and counted the proportion of times Corinne had 7 or fewer hits, that would give an estimate of the probability $P(X \leq 7)$ that Corinne made at most 7 of her 12 attempts. One way to automate this more is to assign these results to list L_1/list1 and then sum the entries in the list. Enter the TI-83/84 command `rand-Bin(1,.75,12)→L1:sum(L1)`. For the TI-89, press **CATALOG** **F3** (Flash Apps) and select `randbin(` and then complete the command: `tistat.randbin(1,.75,12)`.

TI-83

```
randBin(1,.75,12
)
{1 1 1 0 1 0 0…
```

TI-89

```
F1▼   F2▼     F3▼  F4▼   F5      F6▼
Tools Algebra Calc Other PrgmIO Clean Up

▪ tistat.randbin(1,.75,12)▶
  {0.  1.  1.  0.  1.  0.▶
…▪randBin(1,.75,12)→list1
MAIN    RAD AUTO    FUNC    1/30
```

```
randBin(1,.75,12
)→L1:sum(L1)
                10
```

```
F1▼   F2▼     F3▼  F4▼   F5      F6▼
Tools Algebra Calc Other PrgmIO Clean Up

▪ tistat.randbin(1,.75,12)▶
                       10.
…▪75,12)→list1:sum(list1)
MAIN    RAD AUTO    FUNC    1/30
```

Continue pressing the **ENTER** key until you have 10 numbers. Record these numbers.

This makes 10 repetitions (that is, simulates 10 games) for both calculators. (The TI-89 results were 10, 10, 10, 9, 10, 12, 9, 7, 11, 9.) So far Corinne has made 7 or fewer shots in 1 out of 10 games, for a relative frequency of 0.10. Compare this with the binomial probabiliy of 0.1576 for this event. Continue to press **ENTER** to simulate 10 more games.

```
randBin(1,.75,12)
→L1:sum(L1)
                10
                 9
                 8
                 7
                 8
```

```
                11
                11
                 9
                 9
                 8
```

```
F1▾   F2▾     F3▾  F4▾   F5     F6▾
Tools Algebra Calc Other PrgmIO Clean Up
■ tistat.randbin(1,.75,12)▸
                          7.
■ tistat.randbin(1,.75,12)▸
                         11.
■ tistat.randbin(1,.75,12)▸
                          9.
...■75,12)→list1:sum(list1)
MAIN    RAD AUTO    FUNC   10/30
```

Calculate the relative frequency for 20 games and so on. According to the law of large numbers, these relative frequencies should get closer to 0.1576 as the number of simulated games increases. Continue in this fashion until you have simulated 50 games. Are your cumulative results close to 0.1576?

Exercises

8.25 Corinne's free throws Use lists L_1/list1 and L_2/list2 on your calculator to construct a pdf for Corinne's free-throw probabilities. (Refer to Examples 8.8 and 8.10 on pages 521 and 522.) Use the random variable X = number of baskets made on free throws. Then on both calculators, execute the command `cumSum(L2)→L3` (`cumSum` is found under `2nd/LIST/OPS/ 6:cumSum` on the TI-83/84; `cumSum` is found in the `CATALOG` on the TI-89). What do you think this command does? Then use the `binomcdf` command to enter the cumulative probabilities into list L_4/list4. Compare L_3/list3 and L_4/list4. Are they the same?

SIMULATIONS

8.26 Simulating defective switches

(a) Use the TI-83/84/89 command `randBin(10,.1,10)→L1:1-Var Stats L1` (list1 on the T1-89) to simulate randomly selecting a sample of 10 switches from the B(10, 0.1) distribution. (Refer to Examples 8.6 and 8.7, and the Technology Toolbox on pages 519, 520 and 530–532.) Press **ENTER**. This command will store these 10 results to L_1/list1, and display descriptive statistics. Record the mean number of defective switches among the 10. Continue to press **ENTER** until you have simulated 10 samples of size $n = 10$ and recorded values of $\bar{x}$. Calculate the mean of the $\bar{x}$'s and the sample standard deviation, s_x. Compare the mean of the 10 values of $\bar{x}$ with the population mean, $\mu = 1$. Calculate the value of $\sigma/\sqrt{n} = 0.9487/\sqrt{n}$. Compare this result with s_x.

(b) Repeat the steps in (a) for samples of size 25, and then for samples of size 50.

(c) What effect, if any, does the number of switches sampled have on the mean number of switches? As the sample size increases from 10 to 25 and then to 50, what is happening to the spread of $\bar{x}$ values?

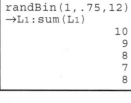

8.27 Eggs and salmonella Sara read that 1 out of 4 eggs contains salmonella bacteria. So she never uses more than 3 eggs in cooking. Assume that eggs do or don't contain salmonella independently of each other.

(a) What is the distribution of the number of contaminated eggs when Sara uses 3 eggs that are chosen at random?

(b) Construct a simulation to determine the probability that at least one of Sara's eggs contains salmonella.

(c) Calculate the exact probability that at least one of Sara's eggs contains salmonella. Compare this result with your simulation in (b).

8.28 Student indebtedness According to a 2006 study conducted by the Department of Education's National Center for Education Statistics, about 65% of college students who graduated in the 2003–2004 school year did so after getting student loans. For students who took out loans, the average debt was $19,202, most of which came through federal loan programs. Assume that a randomly selected student has probability 0.65 of having a student loan. Randomly select 30 college students.

(a) Use simulation methods to determine the probability that more than 24 of the 30 students have student loans.

(b) Use your calculator to find the probability that more than 24 of the 30 students have student loans, and compare this (theoretical) probability with your simulation results.

(c) Use a Normal approximation to find the probability that more than 24 of the 30 students have student loans. Then compare the probabilities you obtained in (b) and (c). Explain why a Normal approximation would not be advisable in this situation.

8.29 The random digits table Each entry in a table of random digits like Table B has probability 0.1 of being a zero.

(a) Find the probability of finding exactly 4 zeros in a line 40 digits long.

(b) What is the probability that a group of five digits from the table will contain at least 1 zero?

8.30 Drawing poker chips There are 50 poker chips in a container, 25 of which are red, 15 white, and 10 blue. You draw a chip without looking 25 times, each time returning the chip to the container.

(a) What is the expected number of white chips you will draw in 25 draws?

(b) What is the standard deviation of the number of blue chips that you will draw?

(c) Simulate 25 draws by hand or by calculator. Repeat the process as many times as you think necessary.

(d) Based on your answers to parts (a) to (c), is it likely or unlikely that you will draw 9 or fewer blue chips?

(e) Is it likely or unlikely that you will draw 15 or fewer blue chips?

Section 8.1 Summary

A count X of successes has a binomial distribution in the **binomial setting:** there are n observations; the observations are independent of each other; each observation results in a success or a failure; and each observation has the same probability p of a success.

If X has the binomial distribution with parameters n and p, the possible values of X are the whole numbers 0, 1, 2, . . . , n. The **binomial probability** of observing k successes in n trials is

$$P(X = k) = \binom{n}{k} p^k(1 - p)^{n-k}$$

Binomial probabilities in practice are best found using calculators or software.

The **binomial coefficient**

$$\binom{n}{k} = \frac{n!}{k!(n-k)!}$$

counts the number of ways k successes can be arranged among n observations. Here the **factorial** $n!$ is

$$n! = n \times (n-1) \times (n-2) \times \ldots \times 3 \times 2 \times 1$$

for positive whole numbers n, and $0! = 1$.

Given a random variable X, the **probability distribution function** (pdf) assigns a probability to each value of X. For each value of X, the **cumulative distribution function** (cdf) assigns the sum of the probabilities for values less than or equal to X.

The **mean** and **standard deviation** of a binomial count X are

$$\mu = np$$
$$\sigma = \sqrt{np(1-p)}$$

The **Normal approximation** to the binomial distribution says that if X is a count having the binomial distribution with parameters n and p, then when n is large, X is approximately $N(np, \sqrt{np(1-p)})$. We will use this approximation when $np \geq 10$ and $n(1 - p) \geq 10$.

Section 8.1 Exercises

8.31 Binomial setting? In which of these two sports settings is a binomial distribution most likely to be at least approximately correct? Explain your answer.

(a) A National Football League kicker has made 80% of his field goal attempts in the past. This season he attempts 20 field goals. The attempts differ widely in distance, angle, wind, and so on.

(b) A National Basketball Association player has made 80% of his free-throw attempts in the past. This season he takes 150 free throws. Basketball free throws are always attempted from 15 feet away with no interference from other players.

8.32 Binomial setting? In each situation below, is it reasonable to use a binomial distribution for the random variable X? Give reasons for your answer in each case.

(a) In December 2005, a CBS News/*New York Times* Poll of 1155 randomly selected adults asked if they "approved or disapproved" of the way George W. Bush is handling the situation in Iraq. X is the number who approved.

(b) On a bright October day, Canada geese arrive to foul the pond at an apartment complex at the average rate of 12 geese per hour; X is the number of geese that arrive in the next three hours.

(c) Joe buys a ticket in his state's Pick 3 lottery game every week; X is the number of times in a year that he wins a prize.

8.33 Random digits Ten lines in the random digits table (Table B) contain 400 digits. Dan says that the count of zeros in these lines is approximately Normal.

(a) Select 10 lines in the random digits table, and for each line, count the number of zeros. Fill in a table like the following:

Line number:										
Count of zeros:										

Construct a dotplot and a boxplot. Calculate the mean and standard deviation for this distribution. Then describe the shape of the distribution. In particular, does it appear to be approximately Normal or is it skewed?

(b) Let X be the number of zeros in one line in the random digits table. Explain why X is a binomial random variable. What is n? What is p?

(c) If you included more than 10 rows from the random digits table, how would that affect the shape of the distribution? Would you expect the center and spread to change? Take additional rows—you decide how many—and expand your table. Describe the effect of including more than 10 rows on the shape of the resulting distribution.

(d) Write a sentence or two to explain why you think Dan is right or wrong.

8.34 Testing ESP In a test for ESP (extrasensory perception), a subject is told that cards the experimenter can see but the subject cannot contain either a star, a circle, a wave, or a square. As the experimenter looks at each of 20 cards in turn, the subject names the shape on the card. A subject who is just guessing has probability 0.25 of guessing correctly on each card.

(a) The count of correct guesses in 20 cards has a binomial distribution. What are n and p?

(b) What is the mean number of correct guesses in many repetitions?

(c) What is the probability of exactly 5 correct guesses?

8.35 Activity Follow-up Refer to Activity 8A (page 512). Determine the theoretical probability of getting

(a) no questions correct.

(b) 1 question correct.

(c) 2 questions correct.

(d) 3 or fewer questions correct.

(e) more than 3 questions correct.

(f) all 10 questions correct.

(g) Use your results from (a) through (f) to construct a probability distribution table.

(h) Prior to taking the quiz, what was your *expected number* of correct answers? That is, how many questions out of 10 would you have expected to answer correctly by guessing alone?

8.36 Lie detectors A federal report finds that lie detector tests given to truthful persons have probability about 0.2 of suggesting that the person is deceptive.[4]

(a) A company asks 12 job applicants about thefts from previous employers, using a lie detector to assess their truthfulness. Suppose that all 12 answer truthfully. What is the probability that the lie detector says all 12 are truthful? What is the probability that the lie detector says at least 1 is deceptive?

(b) Among 12 truthful persons, what is the mean number who will be classified as deceptive? What is the standard deviation of this number?

(c) What is the probability that the number classified as deceptive is less than the mean?

8.37 Simulating graduation Refer to Exercise 8.10 (page 519). Construct a simulation to estimate the probability that at most 11 of the 20 basketball players graduated. Describe the design of your simulation, including the correspondence between digits and outcomes in the simulation and the number of repetitions you carried out. Report your results.

8.38 Planning a survey You are planning a sample survey of small businesses in your area. You will choose an SRS of businesses listed in the telephone book's Yellow Pages. Experience shows that only about half the businesses you contact will respond.

(a) If you contact 150 businesses, it is reasonable to use the binomial distribution with $n = 150$ and $p = 0.5$ for the number X who respond. Explain why.

(b) What is the expected number (the mean) who will respond?

(c) What is the probability that 70 or fewer will respond? (Use the Normal approximation.)

(d) How large a sample must you take to increase the mean number of respondents to 100?

8.39 On the Web What kinds of Web sites do males aged 18 to 34 visit most often? Pornographic sites take first place, but about 50% of male Internet users in this age group visit an auction site such as eBay at least once a month.[5] Interview a random sample of 12 male Internet users aged 18 to 34.

(a) What is the distribution of the number who have visited an online auction site in the past month?

(b) What is the probability that exactly 8 of the 12 have visited an auction site in the past month? Find the probability that at least 8 of the 12 have visited an auction site in the past month.

8.40 AIDS test A test for the presence of antibodies to the AIDS virus in blood has probability 0.99 of detecting the antibodies when they are present. Suppose that during a year 20 units of blood with AIDS antibodies pass through a blood bank.

(a) Take X to be the number of these 20 units that the test detects. What is the distribution of X?

(b) What is the probability that the test detects all 20 contaminated units? What is the probability that at least 1 unit is not detected?

(c) What is the mean number of units among the 20 that will be detected? What is the standard deviation of the number detected?

8.2 The Geometric Distributions

Activity 8B

Mrs. Hathaway's homework offer

Mrs. Hathaway, the AP Statistics teacher, always assigns 10 problems for homework. One day, she decides to make an unusual offer to her class. "My little cherubs," she says, "I have a proposition for you. Instead of giving you the typical ten terrific textbook teasers, I would gladly allow probability to play a pivotal part in the process. When class begins each day, I will select a student at random using my trusty calculator. Then, I will give the lucky student the opportunity to guess the day of the week on which one of my many friends was born. If the chosen student guesses correctly, then I will assign only one homework problem that night. If, on the other hand, your representative gives the wrong day of the week, he or she will try to guess the day on which another Hathaway friend was born. This time, a correct answer will net you two homework problems. We will continue this little game until the chosen one's guess matches the day on which one of my acquaintances was born. I will then assign you a number of homework questions equal to the number of guesses made by your chosen spokesperson. What say you?"

1. Before you make a decision about Mrs. Hathaway's offer, try the birthday game for yourself. Play several times before you draw any conclusions.

For each of Mrs. Hathaway's friends, the lucky student has a 1/7 chance of correctly guessing his or her day of birth. Note that the trials (birthday guesses) are independent. The game continues until the first correct guess is made. In statistical language, we count the number of trials (birthday guesses) up to and including the first "success" (birthday match). If we let X = the number of guesses the student makes until he or she matches a Hathaway friend's day of birth, then X is a *geometric random variable*. Mrs. Hathaway's birthday challenge is an example of a geometric probability problem. We will learn more about geometric probabilities and distributions in this section.

2. What is the theoretical probability that Mrs. Hathaway assigns exactly10 homework problems as a result of a randomly selected student playing the birthday game?
3. Find the theoretical probability that Mrs. Hathaway assigns fewer than the typical 10 homework problems as a result of a randomly selected student playing the birthday game.
4. Find the theoretical probability that Mrs. Hathaway assigns more than the typical 10 homework problems.

5. Here is a set of results from playing Mrs. Hathaway's birthday game for 100 days. What do you notice?

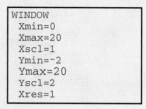

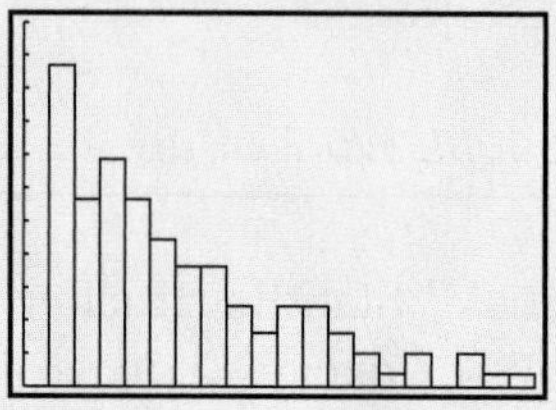

```
WINDOW
 Xmin=0
 Xmax=20
 Xscl=1
 Ymin=-2
 Ymax=20
 Yscl=2
 Xres=1
```

In the case of a binomial random variable, the number of trials is fixed beforehand, and the binomial variable X counts the number of successes in that fixed number of trials. If there are n trials, then the possible values of X are 0, 1, 2, . . . , n. By way of comparison, there are situations in which the goal is to obtain a fixed number of successes. In particular, if the goal is to obtain one success, a random variable X can be defined that counts the number of trials needed to obtain that first success. A random variable that satisfies the above description is called *geometric,* and the distribution produced by this random variable is called a ***geometric distribution.*** The possible values of a geometric random variable are 1, 2, 3, . . . , that is, an infinite set, because it is theoretically possible to proceed indefinitely without ever obtaining a success. Consider the following situations:

geometric distribution

- Flip a coin until you get a head.
- Roll a die until you get a 3.
- In basketball, attempt a three-point shot until you make a basket.

Notice that all of these situations involve counting the number of trials until an event of interest happens. We are now ready to characterize the ***geometric setting.***

A random variable X is geometric provided that the following conditions are met:

The Geometric Setting

1. Each observation falls into one of just two categories, which for convenience we call "success" or "failure."
2. The observations are all **independent.**
3. The probability of a success, call it p, is the same for each observation.
4. The variable of interest is the number of trials required to obtain the first success.

Example 8.15 *Roll a die*

Geometric setting?

A game consists of rolling a single die. The event of interest is rolling a 3; this event is called a success. The random variable is defined as X = the number of trials until a 3 occurs. To verify that this is a geometric setting, note that rolling a 3 will represent a success, and rolling any other number will represent a failure. The probability of rolling a 3 on each roll is the same: 1/6. The observations are independent. A trial consists of rolling the die once. We roll the die until a 3 appears. Since all of the requirements are satisfied, this game describes a geometric setting.

Example 8.16 *Draw an ace*
Geometric setting?

Suppose you repeatedly draw cards without replacement from a deck of 52 cards until you draw an ace. There are two categories of interest: ace = success; not ace = failure. But is the probability of success the same for each trial? No. The probability of an ace on the first card is 4/52. If you don't draw an ace on the first card, then the probability of an ace on the second card is 4/51. Since the result of the first draw affects probabilities on the second draw (and on all successive draws required), the trials are not independent. So this is not a geometric setting.

Using the setting of Example 8.15, let's calculate some probabilities.

$$X = 1{:}\ P(X = 1) = P(\text{success on first roll}) = 1/6$$

$$\begin{aligned} X = 2{:}\ P(X = 2) &= P(\text{success on second roll}) \\ &= P(\text{failure on first roll and success on second roll}) \\ &= P(\text{failure on first roll}) \times P(\text{success on second roll}) \\ &= (5/6) \times (1/6) \end{aligned}$$

(since trials are independent).

$$\begin{aligned} X = 3{:}\ P(X = 3) &= P(\text{failure on first roll}) \\ &\quad \times P(\text{failure on second roll}) \times P(\text{success on third roll}) \\ &= (5/6) \times (5/6) \times (1/6) \end{aligned}$$

Continue the process. The pattern suggests that a general formula for the variable X is

$$P(X = n) = (5/6)^{n-1}(1/6)$$

Now we can state the following principle:

Rule for Calculating Geometric Probabilities

If X has a geometric distribution with probability p of success and $(1 - p)$ of failure on each observation, the possible values of X are 1, 2, 3, ... If n is any one of these values, the probability that the first success occurs on the nth trial is

$$P(X = n) = (1 - p)^{n-1}p$$

Although the setting for the geometric distribution is very similar to the binomial setting, there are some striking differences. In rolling a die, for example, it is possible that you will have to roll the die many times before you roll a 3. In fact, it is theoretically possible to roll the die forever without rolling a 3 (although the probability gets closer and closer to 0 the longer you roll the die without getting a 3). The probability of observing the first 3 on the fiftieth roll of the die is $P(X = 50) = 0.0000$.

A probability distribution table for the geometric random variable is strange indeed because it never ends; that is, the number of table entries is infinite. The rule for calculating geometric probabilities shown above can be used to construct the table:

X:	1	2	3	4	5	6	7 . . .
P(X):	p	$(1-p)p$	$(1-p)^2p$	$(1-p)^3p$	$(1-p)^4p$	$(1-p)^5p$	$(1-p)^6p$ . . .

The probabilities (that is, the entries in the second row) are the terms of a geometric sequence (hence the name for this random variable). You may recall from your study of algebra that the general form for a geometric sequence is

$$a, ar, ar^2, ar^3, \ldots, ar^{n-1}, \ldots$$

where a is the first term, r is the ratio of one term in the sequence to the next, and the nth term is ar^{n-1}. You may also recall that even though the sequence continues forever, and even though you could never finish adding the terms, the sequence does have a sum (one of the implausible truths of the infinite!). This sum is

$$\frac{a}{1-r}$$

In order for the geometric random variable to have a valid pdf, the probabilities in the second row of the table must add to 1. Using the formula for the sum of a geometric sequence, we have

$$\sum_{i=1}^{\infty} P(x_i) = p + (1-p)p + (1-p)^2p + \ldots$$
$$= \frac{p}{1-(1-p)} = \frac{p}{p} = 1$$

Example 8.17 *Roll a die*

Probability distribution table

The rule for calculating geometric probabilities can be used to construct a probability distribution table for X = number of rolls of a die until a 3 occurs:

X:	1	2	3	4	5	6	7	. . .
P(X):	0.1667	0.1389	0.1157	0.0965	0.0804	0. 0670	0.0558	. . .

Here's one way to find these probabilities with your calculator:

1. Enter the probability of success, `1/6`. Press ENTER.
2. Enter `*(5/6)` and press ENTER.
3. Continue to press ENTER repeatedly.

```
1/6
        .1666666667
Ans*(5/6)
        .1388888889
        .1157407407
        .0964506173
        .0803755144
```

Verify that the entries in the second row are as shown:

X:	1	2	3	4
P(X):	1/6	5/36	25/216	125/1296

Figure 8.4 is a graph of the distribution of X. As you might expect, the probability distribution histogram is strongly skewed to the right with a peak at the leftmost value, 1. It is easy to see why this must be so, since the height of each bar after the first is the height of the previous bar times the probability of failure $1 - p$. Since you're multiplying the height of each bar by a number less than 1, each new bar will be shorter than the previous bar, and hence the histogram will be right-skewed. Always.

Figure 8.4 *Probability histogram for the geometric distribution, for Example 8.17.*

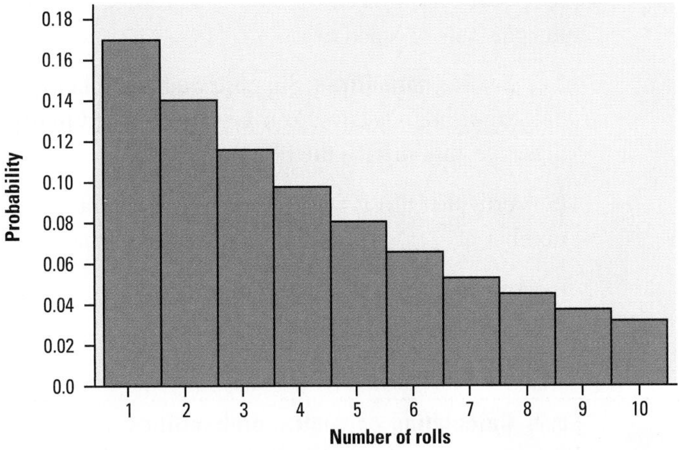

Exercises

8.41 Geometric setting? For each of the following, determine if the random phenomenon describes a geometric setting. If it does, describe the two events of interest (success and failure) and what constitutes a trial, and state the probability of success on each trial. If the random variable is not geometric, identify a condition of the geometric setting that is not satisfied.

(a) Flip a coin until you observe a tail.

(b) Record the number of times a player makes both shots in a one-and-one foul-shooting situation. (In this situation, you get to attempt a second shot only if you make your first shot.)

(c) Draw a card from a deck, observe the card, and replace the card within the deck. Count the number of times you draw a card in this manner until you observe a jack.

(d) Buy a Match 6 lottery ticket every day until you win the lottery. (In a Match 6 lottery, a player chooses 6 different numbers from the set $\{1, 2, 3, \ldots, 44\}$. A lottery representative draws 6 different numbers from this set. To win, the player must match all 6 numbers, in any order.)

(e) There are 10 red marbles and 5 blue marbles in a jar. You reach in and, without looking, select a marble. You want to know how many marbles you will have to draw (without replacement), on average, in order to be sure that you have 3 red marbles.

8.42 Roll a prime An experiment consists of rolling a die until a prime number (2, 3, or 5) is observed. Let X = number of rolls required to get the first prime number.

(a) Verify that X has a geometric distribution.

(b) Construct a probability distribution table that includes at least five entries for the probabilities of X. Record probabilities to four decimal places.

(c) Construct a graph of the pdf of X.

(d) Compute the cdf of X and plot its histogram.

(e) Use the formula for the sum of a geometric sequence to show that the probabilities in the pdf table of X add to 1.

8.43 Testing hard drives Suppose we have data that suggest that 3% of a company's hard disk drives are defective. You have been asked to determine the probability that the first defective hard drive is the fifth unit tested.

(a) Verify that this is a geometric setting. Identify the random variable; that is, write X = number of ________ and fill in the blank. What constitutes a success in this situation?

(b) Answer the original question: what is the probability that the first defective hard drive is the fifth unit tested?

(c) Find the first four entries in the table of the pdf for the random variable X.

8.44 Calculating geometric probabilities For each of the parts of Exercise 8.41 that describes a geometric setting, find the probability that $X = 4$.

The Expected Value and Other Properties of the Geometric Random Variable

If you were flipping a fair coin, how many times would you expect to have to flip the coin in order to observe the first head? If you were rolling a die, how many times would you expect to have to roll the die in order to observe the first 3? If you said 2 coin tosses and 6 rolls of the die, then your intuition is serving you well. To derive an expression for the mean (expected value) of a geometric random variable, we begin with the probability distribution table. The notation will be simplified if we let p = probability of success and let q = probability of failure. Then $q = 1 - p$ and the probability distribution table looks like this:

X:	1	2	3	4	...
P(X):	p	pq	pq^2	pq^3	...

The mean (expected value) of X is calculated as follows:

$$\begin{aligned}\mu_X &= 1(p) + 2(pq) + 3(pq^2) + 4(pq^3) + \ldots \\ &= p(1 + 2q + 3q^2 + 4q^3 + \ldots)\end{aligned}$$

Multiplying both sides by q, we have

$$q\mu_X = p(q + 2q^2 + 3q^3 + 4q^4 + \ldots)$$

Now subtract this equation from the previous equation, and group like terms on the right.

$$\mu_X - q\mu_X = p(1 + q + q^2 + q^3 + \ldots)$$

$$\mu_X(1 - q) = p\left(\frac{1}{1-q}\right)$$

$$\mu_X = \frac{p}{(1-q)^2} = \frac{p}{p^2} = \frac{1}{p}$$

Deriving the variance and standard deviation of the geometric random variable X is a bit more work and would take us too far afield.

Here are the facts:

The Mean and Standard Deviation of a Geometric Random Variable

If X is a geometric random variable with probability of success p on each trial, then the **mean,** or **expected value,** of the random variable, that is, the expected number of trials required to get the first success, is $\mu = 1/p$. The variance of X is $(1 - p)/p^2$.

Example 8.18 *Arcade game*

Mean and standard deviation

Glenn likes the game at the state fair where you toss a coin into a saucer. You win if the coin comes to rest in the saucer without sliding off. Glenn has played this game many times and has determined that on average he wins 1 out of every 12 times he plays. He believes that his chances of winning are the same for each toss. He has no reason to think that his tosses are not independent. Let X be the number of tosses until a win. Glenn believes that this describes a geometric setting.

Since $E(X) = 12 = 1/p$, the probability of success on any given trial is

$$p = 1/12 = 0.0833$$

The variance of X is

$$\sigma_X^2 = \frac{1-p}{p^2} = \frac{11/12}{1/144} = 132$$

And the standard deviation is $\sigma_X \cong 11.5$.

There is another interesting result that relates to the probability that it takes more than a certain number of trials to achieve success. Here are the steps:

$$\begin{aligned} P(X > n) &= 1 - P(X \leq n) \\ &= 1 - (p + qp + q^2p + \ldots + q^{n-1}p) \\ &= 1 - p(1 + q + q^2 + \ldots + q^{n-1}) \\ &= 1 - p\left(\frac{1-q^n}{1-q}\right) \\ &= 1 - p\left(\frac{1-q^n}{p}\right) \\ &= 1 - (1 - q^n) \\ &= q^n = (1-p)^n \end{aligned}$$

We summarize as follows:

$P(X > n)$

The probability that it takes *more* than n trials to see the first success is

$$P(X > n) = (1-p)^n$$

Example 8.19 *Rolling and tossing*

Applying the formula

Roll a die until a 3 is observed. The probability that it takes more than 6 rolls to observe a 3 is

$$P(X > 6) = (1-p)^n = (5/6)^6 \cong 0.335$$

Let Y be the number of Glenn's coin tosses until a coin stays in the saucer (see Example 8.18). The expected number is 12. The probability that it takes more than 12 tosses to win a stuffed animal is

$$P(X > 12) = (11/12)^{12} \cong 0.352$$

The probability that it takes more than 24 tosses to win a stuffed animal is

$$P(X > 24) = (11/12)^{24} \cong 0.124$$

The following Technology Toolbox summarizes some calculator techniques when working in a geometric setting:

Technology Toolbox

Exploring geometric distributions

For illustration purposes, we will use the roll of a die with $n = 6$ equally likely outcomes and probability $p = 1/6$ of rolling a 3, from Example 8.15 (page 540). The random variable X is the number of rolls until a 3 is observed.

To have the calculator calculate the probability distribution table and plot a histogram for the distribution, proceed as follows:

TI-83/84

1. Enter the numbers 1 to 10 in list L_1. Next, enter the probabilities into L_2 by first highlighting L_2. Then press **2nd** **VARS** (DISTR). Scroll down and select D: geometpdf(. Complete the command: geometpdf $(1/6, L_1)$, and press **ENTER**. Here are the results:

L1	L2	L3 2
1	.16667	------
2	.13889	
3	.11574	
4	.09645	
5	.08038	
6	.06698	
7	.05582	

L2(1)=.1666666666…

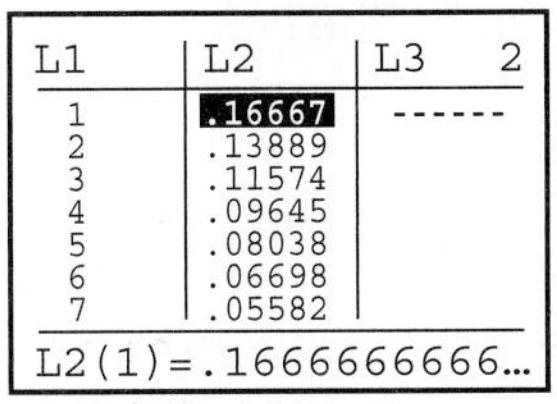

2. Specify the dimensions of an appropriate viewing window. Scanning the list of values gives you insight into reasonable dimensions for the window. Specify $X[0,11]_1$ and $Y[-.05,.2]_1$.

3. When you define a histogram for Plot1, specify Xlist:L_1 and Freq:L_2. The resulting plot shows that the distribution is strongly right-skewed.

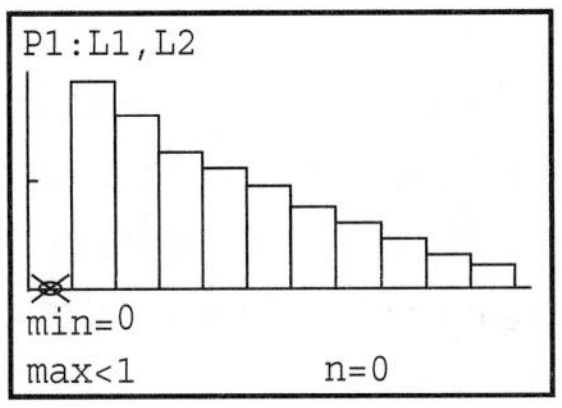

TI-89

1. Enter the numbers 1 to 10 in list1. Next, enter the probabilities into list2 by first highlighting list2. Then press **CATALOG** **F3** (Flash Apps) and scroll down to select geomPdf(. Complete the command: tistat.geomPdf $(1/6, list1)$. Here are the results:

list1	list2■	list3	list4
1	.16667	----	----
2	.13889		
3	.11574		
4	.09645		
5	.08038		
6	.06698		

list2[1]=.16666666666667

MAIN   RAD AUTO   FUNC   2/7

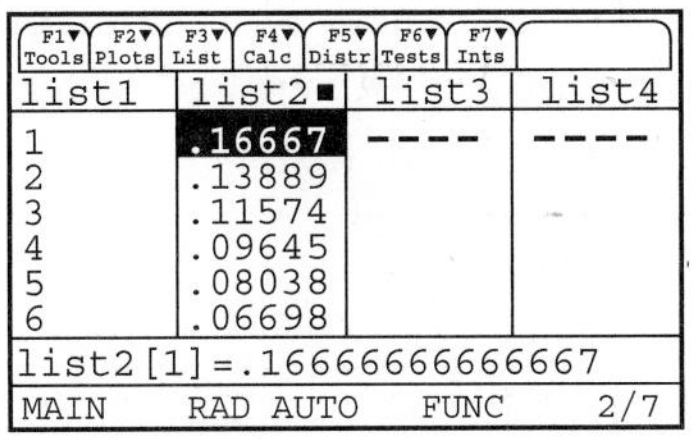

2. Specify these dimensions for the viewing window: $X[0,11]_1$ and $Y[-.05,.2]_1$.

3. From the Statistics/List Editor, press **F2** (Plots). Select 1: Plot Setup. Define Plot 1 to be a histogram using list1 for the X-values and list2 for the frequency. To plot the histogram, press ◆ **GRAPH**. Here is the pdf histogram:

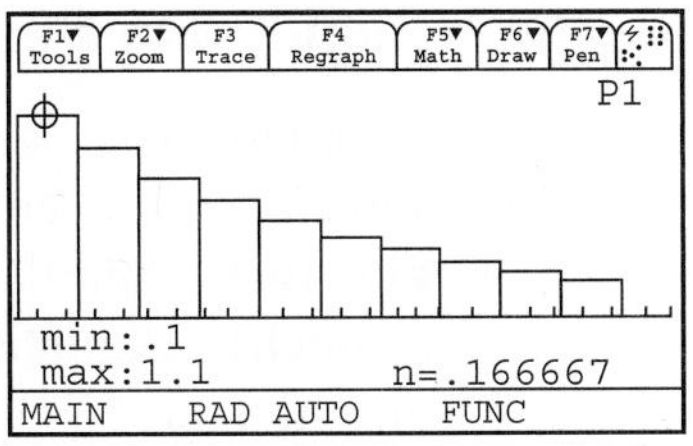

(continued)

Technology Toolbox

Exploring geometric distributions *(continued)*

4. Next install the cdf as list L_3. In the STAT/Edit window, place the cursor on list L_3. Enter the formula geometcdf (1/6, L_1) and press ENTER.

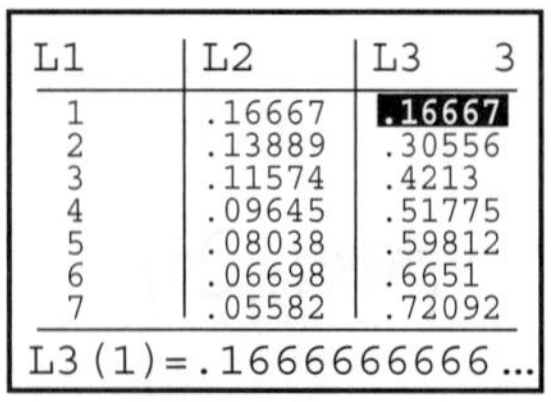

4. To calculate the cumulative distribution values, highlight list3 and press CATALOG F3 (Flash Apps), then scroll down and select `geomCdf(`. Complete the command `tistat.geomCdf (1/6, list1)`. Here is the geometric cdf:

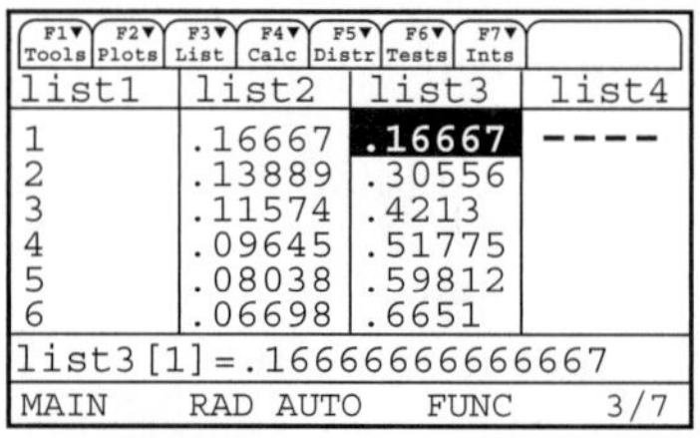

5. To plot the cumulative distribution histogram, first specify the viewing window: $X[0,11]_1$ and $Y[-.3,1]_1$. Then deselect Plot1 and define Plot2 to be a histogram with Xlist:L_1 and Freq:L_3. Press GRAPH. Here is the cdf histogram:

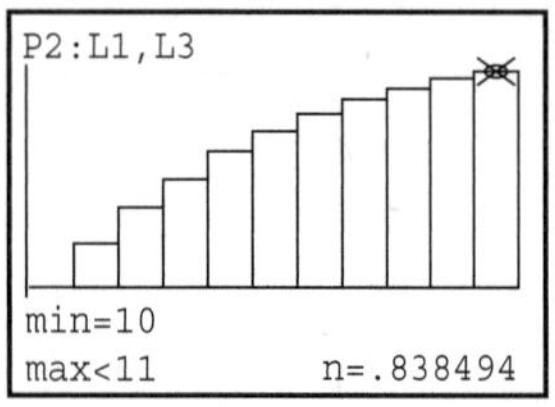

5. Deselect Plot 1. Then define Plot 2 to be a histogram using list3 for the frequency. Here is the histogram for the cdf:

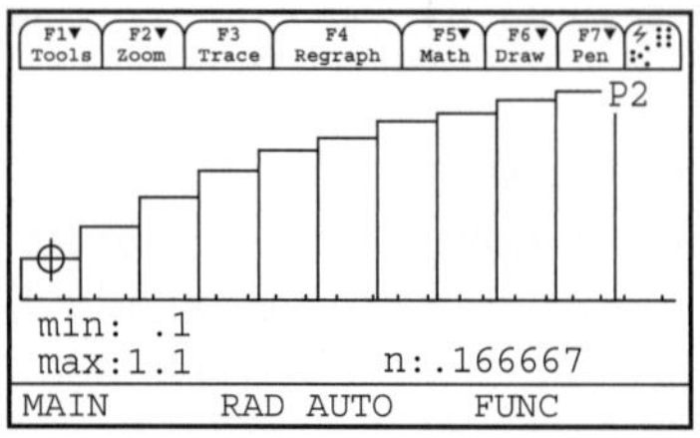

Simulating Geometric Situations

Geometric simulations are frequently called "waiting time" simulations because you continue to conduct trials and wait until a "success" is observed. Conducting a geometric simulation by hand is generally pretty easy but tedious. Conducting a geometric simulation by calculator or computer usually takes more effort initially, but the payoff is that you can quickly run many repetitions and often get results that are quite respectable. Here is an example.

Example 8.20 *Show me the money!*

Applying means and standard deviations

In 1986–1987, Cheerios™ cereal boxes displayed a dollar bill on the front of the box and a cartoon character who said, "Free \$1 bill in every 20th box."

Here is a simulation to determine the number of boxes of Cheerios you would expect to buy in order to get one of the "free" dollar bills.

Let a two-digit number, 00 to 99, represent a box of Cheerios, and let the digits 01 to 05 represent a box of Cheerios with a \$1 bill in it. The digits 00 and 06 to 99 represent boxes without the \$1 bill. Starting at the third block of digits in line 127 of Table B, we select digits in pairs:

23	33	06	43	59	40	08	61	69	25
85	11	73	60	71	15	68	91	42	27
06	56	51	43	74	13	35	24	93	67
81	98	28	72	09	36	75	95	89	84
68	28	82	29	13	18	63	84	43	**03**

So in our first run of this simulation, we had to buy 50 boxes of Cheerios until we found one with a \$1 bill in it! If you don't usually buy Cheerios, would this promotion induce you to buy a box in hopes of getting one with a dollar in it?

Why did it take so many boxes (50) to achieve success? Since the probability of success on a single trial is $p = 1/20 = 0.05$, we know that the mean (expected value) is $E(X) = 1/p = 20$, so a value of 50 in our simulation seems high. But the variance is

$$\sigma_X^2 = \frac{1-p}{p^2} = \frac{0.95}{0.0025} = 380$$

and

$$\sigma_X = \sqrt{380} \cong 19.49$$

Our simulated result of 50 is about 1.5 standard deviations to the right of the mean, 20. So perhaps we should not be too surprised with our 50. The geometric distribution has a considerable amount of variability.

Exercises

8.45 Flip a coin Flip a coin until a head appears.

(a) Identify the random variable X.

(b) Construct the pdf table for X. Then plot the probability histogram.

(c) Compute the cdf and plot its histogram.

8.46 Arcade game Refer to Example 8.19 (page 546).

(a) Use the formula for calculating $P(X > n)$ on page 546 to find the probability that it takes more than 10 tosses until Glenn wins a stuffed animal.

(b) Find the answer to (a) by calculating the probability of the complementary event: $1 - P(X \leq 10)$. Your results should agree, of course.

(*Note:* The formula for $P(X > n)$ is not practically important since there are other ways to answer the question. But it's a nice little result, and it's quite easy to derive.)

8.47 Roll a die

(a) Plot the cumulative distribution histogram for the die-rolling activity described in Example 8.15 (page 540) with the pdf table in Example 8.17 (page 542).

(b) Find the probability that it takes more than 10 rolls to observe a 3.

(c) Find the smallest positive integer k for which $P(X \leq k) > 0.99$.

8.48 Language skills The State Department is trying to identify an individual who speaks Farsi to fill a foreign embassy position. They have determined that 4% of the applicant pool are fluent in Farsi.

(a) If applicants are contacted randomly, how many individuals can they expect to interview in order to find one who is fluent in Farsi?

(b) What is the probability that they will have to interview more than 25 until they find one who speaks Farsi? More than 40?

8.49 Shooting free throws A basketball player makes 80% of her free throws. We put her on the free-throw line and ask her to shoot free throws until she misses one. Let X = the number of free throws the player takes until she misses.

(a) What assumption do you need to make in order for the geometric model to apply? With this assumption, verify that X has a geometric distribution. What action constitutes "success" in this context?

(b) What is the probability that the player will make 5 shots before she misses?

(c) What is the probability that she will make at most 5 shots before she misses?

8.50 Game of chance Three friends each toss a coin. The odd man wins; that is, if one coin comes up different from the other two, that person wins that round. If the coins all match, then no one wins and they toss again. We're interested in the number of times the players will have to toss the coins until someone wins.

(a) What is the probability that no one will win on a given coin toss?

(b) Define a success as "someone wins on a given coin toss." What is the probability of a success?

(c) Define the random variable of interest: X = number of __________. Is X binomial? Geometric? Justify your answer.

(d) Construct a probability distribution table for X. Then extend your table by the addition of cumulative probabilities in a third row.

(e) What is the probability that it takes no more than 2 rounds for someone to win?

(f) What is the probability that it takes more than 4 rounds for someone to win?

(g) What is the expected number of tosses needed for someone to win?

(h) Use the `randInt` function on your calculator to simulate 25 rounds of play. Then calculate the relative frequencies for X = 1, 2, 3, ... Compare the results of your simulation with the theoretical probabilities you calculated in (d).

Section 8.2 Summary

A count X of successes has a **geometric distribution** in the geometric setting if the following conditions are satisfied: each observation results in a success or a failure; observations are independent; each observation has the same probability p of success; and X counts the number of trials required to obtain the first success. A geometric random variable differs from a binomial variable because in the geometric setting the number of trials varies and the desired number of defined successes (1) is fixed in advance.

If X has the geometric distribution with probability of success p, the possible values of X are the positive integers 1, 2, 3, ... The **geometric probability** that X takes any value is

$$P(X = n) = (1 - p)^{n-1}p$$

The **mean** (expected value) of a geometric count X is $1/p$.
The standard deviation is

$$\sqrt{\frac{(1 - p)}{p^2}}$$

The probability that it takes more than n trials to see the first success is

$$P(X > n) = (1 - p)^n$$

Section 8.2 Exercises

8.51 Drawing marbles, I There are 20 red marbles, 10 blue marbles, and 5 white marbles in a jar. Select a marble without looking, note the color, and then replace the marble in the jar. We're interested in the number of marbles you would have to draw in order to be sure you have a red marble.

(a) Is this a binomial or a geometric setting? Explain your choice, and write a description of the random variable X.

(b) Calculate the probability of drawing a red marble on the second draw. Calculate the probability of drawing a red marble by the second draw. Calculate the probability that it would take more than 2 draws to get a red marble.

(c) What single calculator command will install the first 20 values of X into L_1/list1? What single command will install the corresponding probabilities into L_2/list2? What single command will install the cumulative probabilities into L_3/list3? Enter these commands in the home screen. Copy this information from your calculator onto your paper to make an expanded probability distribution table (with the cdf as the third row).

(d) Construct a probability distribution histogram as STAT PLOT1, and then construct a cumulative distribution histogram as STAT PLOT2.

8.52 Drawing marbles, II This is a continuation of Exercise 8.51. Given the jar containing red, white, and blue marbles, Joey thinks a more interesting problem would be to find the number of marbles you would have to draw, without replacing them in the jar, to be sure that you have 2 red marbles.

(a) Does this activity describe a geometric setting? Why or why not?

(b) Would your answer to (a) change if the marble was replaced after each draw? Explain.

(c) Design and carry out a simulation to determine the number of marbles you would have to draw, with replacement, until you get 2 red marbles. Compare the results from your simulation with the results from the previous exercise.

8.53 Multiple choice Carla makes random guesses on the multiple-choice test in Activity 8A (page 512), which has five choices for each question. We want to know how many questions Carla answers until she gets one correct.

(a) Define a success in this context, and define the random variable X of interest. What is the probability of success?

(b) What is the probability that Carla's first correct answer occurs on Question 5?

(c) What is the probability that it takes more than 4 questions before Carla answers one correctly?

(d) Construct a probability distribution table for X.

(e) If Carla took a test like this test many times and randomly guessed at each question, what would be the average number of questions she would have to answer before she answered one correctly?

8.54 It's a boy! In some cultures, it is considered very important to have a son to carry on the family name. Suppose that a couple in one of these cultures plans to have children until they have exactly one son.

(a) Find the average number of children per family in such a culture.

(b) What is the expected number of girls in this family?

(c) Describe a simulation that could be used to find approximate answers to the questions in (a) and (b).

8.55 Family planning, I Example 6.6 (page 396) used simulation techniques to explore the following situation: a couple plan to have children until they have a girl or until they have four children, whichever comes first.

(a) List the outcomes in the sample space for this investigation. What event represents a success?

(b) Let X = the number of boys in this family. What values can X take? Use an appropriate probability rule to calculate the probability for each value of X, and make a probability distribution table for X. Then show that the sum of the probabilities is 1.

(c) Let Y = the number of children produced in this family until a girl is produced. Show that Y starts out as a geometric distribution but then is stopped abruptly. Make a probability distribution table for Y.

(d) What is the expected number of children for this couple?

(e) What is the probability that this couple will have more than the expected number of children?

(f) At the end of Example 6.6 (page 396), it states that the probability of having a girl in this situation is 0.938. How can you prove this?

8.56 Family planning, II This is a continuation of Exercise 8.55. A couple plan to have children until they have a girl or until they have four children, whichever comes first. Use the random number table (Table B), beginning on line 130, to simulate 25 repetitions of this childbearing strategy. As in Example 6.6, since a girl and boy are equally likely, let the digits 0 to 4 represent a girl, and let digits 5 to 9 represent a boy. Write the digits in a string until you observe a girl, write B or G under each digit, and write the number of children noted at the bottom. The first two repetitions would be recorded as

6	9	0		5	1
B	B	G		B	G
	3			2	

Then find the mean of the 25 repetitions. How do your results compare with the theoretical expected value of 1.8 children?

8.57 Family planning, III This is a continuation of Exercises 8.55 and 8.56. Devise a simulation procedure for the calculator to approximate the expected number of children. List the steps and commands you use as well as the number of repetitions and the results. Alternatively, incorporate these steps into a calculator program similar to the programs SPIN123 (page 406) and FLIP50 (page 133).

8.58 Making the connection This exercise provides visual reinforcement of the relationship between the probability of success and the mean (expected value) of a geometric random variable.

(a) Begin by completing the table below, where X = probability of success and Y = expected value.

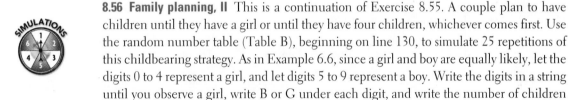

X:	0.10	0.20	0.30	0.40	0.50	0.60	0.70	0.80	0.90
Y:									

(b) Make a scatterplot of the points (X, Y).

(c) Enter the data into your calculator, and transform the data assuming a power function model.

(d) Remember that the purpose of transforming data is to make the data points linear so that the method of least squares can be employed. Sketch the plot of the transformed data.

(e) What is the correlation r for the transformed data?

(f) Write the equation of the power function. Draw the power function curve on your scatterplot.

(g) Briefly explain the connection between this curve and what you have learned about the expected value of a geometric random variable.

CASE CLOSED!

Psychic probability

The results of the experiment described in the Case Study (page 511) are presented in the following table:

Data breakdown	Trials	Hits
All trials	288	88
High confidence	165	55
Medium confidence	48	12
Low confidence	75	21
Precognition mode	72	27
Clairvoyance mode	216	61

(a) Compute the proportion of successes for all trials and separately for the high-confidence, medium-confidence, and low-confidence trials using the data from the table. If we assume that the trials are independent and the subject is simply guessing, what would the expected proportion of successes be in each case? Comment on the observed versus expected proportions.

(b) If the computer selected one of the four black-and-white symbols independently and at random on each trial, is the overall result of 88 hits in 288 trials evidence of psychic ability? Use an appropriate statistical procedure to justify your answer.

(c) The paper by the researchers does not indicate whether the trials were independent or whether the *ESPerciser* functions like a deck of 24 cards with trials like draws from the deck without replacement. In (a), we assumed that the trials were independent (draws from a deck of cards with replacement). We now investigate what the effect might be if trials were like draws of cards without replacement. To keep things simple, assume that we have a deck with only four cards, one of each symbol. What is the probability of making a correct guess on the first, second, third, and fourth draws if you are told what happens on each draw (that is, after each guess, you are shown the actual card)?

(d) Under the same assumptions as in (b), if you are just guessing, how many guesses would you expect to get correct? (*Hint:* This expectation is just the sum of the probabilities of guessing correctly on the first, second, third, and fourth draws.) How does this compare with the number you would expect if draws were made with replacement? What are the implications of these results on the experiment conducted with Olof Jonsson?

(e) According to the researchers, the computer randomly selected the precognition mode with equal probability for each run of 24 trials. There were 12 runs (4 runs on each of the three days) in the experiment. From the table, we can see that 3 of the runs (72 trials/24 trials per run) were in the precognition mode. How likely is it to get as few as 3 runs out of 12 in the precognition mode? Comment on the experimental results in view of your findings.

Chapter Review

Summary

The previous chapter introduced discrete and continuous random variables and described methods for finding means and variances, as well as rules for means and variances. This chapter focused on two important classes of discrete random variables, each of which involves two outcomes or events of interest. Both require independent trials and the same probability of success on each trial. The **binomial** random variable requires a fixed number of trials; the **geometric** random variable has the property that the number of trials varies. Both the binomial and the geometric settings occur sufficiently often in applications that they deserve special attention.

What You Should Have Learned

Here is a checklist of the major skills you should have acquired by studying this chapter:

A. BINOMIAL

1. Identify a random variable as binomial by verifying four conditions: two outcomes (success and failure); fixed number of trials; independent trials; and the same probability of success for each trial.
2. Use technology or the formula to determine binomial probabilities and to construct probability distribution tables and histograms.
3. Calculate cumulative distribution functions for binomial random variables, and construct cumulative distribution tables and histograms.
4. Calculate means (expected values) and standard deviations of binomial random variables.
5. Use a Normal approximation to the binomial distribution to compute probabilities.

B. GEOMETRIC

1. Identify a random variable as geometric by verifying four conditions: two outcomes (success and failure); independent trials; the same probability of success for each trial; and the count of interest is the number of trials required to get the first success.
2. Use formulas or technology to determine geometric probabilities and to construct probability distribution tables and histograms.
3. Calculate cumulative distribution functions for geometric random variables, and construct cumulative distribution tables and histograms.
4. Calculate expected values and standard deviations of geometric random variables.

Web Links

The applet at www.stat.berkeley.edu/~stark/Java/Html/BinHist.htm calculates binomial probabilities and shows the corresponding histogram. There is an option of having a Normal curve superimposed.

Chapter Review Exercises

8.59 Binomial setting? In each of the following cases, decide whether or not a binomial distribution is an appropriate model, and give your reasons.

(a) You want to know what percent of married people believe that mothers of young children should not be employed outside the home. You plan to interview 50 people, and for the sake of convenience you decide to interview both the husband and wife in 25 married couples. The random variable X is the number among the 50 persons interviewed who think mothers should not be employed.

(b) You are interested in attitudes toward drinking among the 75 members of a fraternity. You choose 25 members at random to interview. One question is "Have you had five or more drinks at one time during the last week?" Suppose that in fact 20% of the 75 members would say "Yes." Explain why you cannot safely use the $B(25, 0.2)$ distribution for the count X in your sample who say "Yes."

8.60 How many cars? Twenty percent of American households own three or more motor vehicles. You choose 12 households at random.

(a) What is the probability that none of the chosen households owns three or more vehicles? What is the probability that at least one household owns three or more vehicles?

(b) What are the mean and standard deviation of the number of households in your sample that own three or more vehicles?

(c) What is the probability that your sample count is greater than the mean?

8.61 Seven brothers! There's a movie classic entitled *Seven Brides for Seven Brothers*. Even if these brothers had a few sisters, this many brothers is unusual. We will assume that there are no sisters.

(a) Let X = number of boys in a family of 7 children. Assume that sons and daughters are equally likely outcomes. Do you think the distribution of X will be skewed left, symmetric, or skewed right? The answer to this question depends on what fact?

(b) Use the `binompdf` command to construct a pdf table for X. Then construct a probability distribution histogram and a cumulative distribution histogram for X. Keep a written record of your numerical results as they are produced by your calculator, as well as sketches of the histograms.

(c) What is the probability that all of the 7 children are boys?

8.62 Get a head Suppose we toss a penny repeatedly until we get a head. We want to determine the probability that the first head comes up in an *odd* number of tosses (1, 3, 5, and so on).

(a) Toss a penny until the first head occurs, and repeat the procedure 50 times. Keep a record of the results of the first toss and of the number of tosses needed to get a head on each of your 50 repetitions.

(b) Based on the result of your first toss in the 50 repetitions, estimate the probability of getting a head on the first toss.

(c) Use your 50 repetitions to estimate the probability that the first head appears on an odd-numbered toss.

8.63 Faith and healing A higher percent of southerners believe in God and prayer, according to a 1998 study by the University of North Carolina's Institute for Research in Social Science. The survey was conducted by means of telephone interviews with 844 adults in 12 southern states and 413 adults in other states. One of the findings was that 46% of southerners believe they have been healed by prayer, compared with 28% of others. Assume that the results of the UNC survey are true for the region. Suppose that 20 southerners are selected at random and asked if they believe they have been healed by prayer. Find the probability that the number who answer "Yes" to this question is

(a) exactly 10.

(b) between 10 and 15.

(c) over 75% of the 20.

(d) less than 8.

8.64 Tooth decay and gum disease Dentists are increasingly concerned about the growing trend of local school districts to grant soft drink companies exclusive rights to install soda pop machines in schools in return for money—usually millions—that goes directly into school coffers. According to a recent study by the National Soft Drink Association, 62% of schools nationally already have such contracts. This comes at a time when dentists are seeing an alarming increase in horribly decayed teeth and eroded enamel in the mouths of teenagers and young adults. With ready access to soft drinks, children tend to drink them all day. That, combined with no opportunity to brush, leads to disaster, dentists say. Suppose that 20 schools around the country are randomly selected and asked if they have a soft drink contract. Find the probability that the number of "Yes" answers is

(a) exactly 8.

(b) at most 8.

(c) at least 4.

(d) between 4 and 12, inclusive.

(e) Identify the random variable of interest, X. Then write the probability distribution table for X.

(f) Draw a probability histogram for X.

8.65 False-positives in testing for HIV A rapid test for the presence in the blood of antibodies to HIV, the virus that causes AIDS, gives a positive result with probability about 0.004 when a person who is free of HIV antibodies is tested. A clinic tests 1000 people who are all free of HIV antibodies.

(a) What is the distribution of the number of positive tests?

(b) What is the mean number of positive tests?

(c) You cannot safely use the Normal approximation for this distribution. Explain why.

8.66 A simulation surprise In Example 6.5 (page 396) the strawberry flavor of frozen yogurt accounted for 20% of sales. Yet when we simulated 10 customers, half of them ordered strawberry. That's quite a difference. At that time, you didn't have the tools to calculate the probability of observing results as surprising as these. Now you do. Find the probability of observing 5 or more orders for strawberry frozen yogurt among 10 customers. Then write a sentence or two that puts the probability you found in context.

8.67 First hit of the season Suppose that Roberto, a well-known major league baseball player, finished last season with a .325 batting average. He wants to calculate the probability that he will get his first hit of this new season in his first at-bat. You define a success as getting a hit and define the random variable X = number of at-bats until Roberto gets his first hit.

(a) What is the probability that Roberto will get a hit on his first at-bat (that is, that $X = 1$)?

(b) What is the probability that it will take him at most 3 at-bats to get his first hit?

(c) What is the probability that it will take him more than 4 at-bats to get his first hit?

(d) Roberto wants to know the expected number of at-bats until he gets a hit. What would you tell him?

(e) Enter the first 10 values of X into L_1/list1, the corresponding geometric probabilities into L_2/list2, and the cumulative probabilities into L_3/list3.

(f) Construct a probability distribution histogram as STAT PLOT1, and then construct a cumulative distribution histogram as STAT PLOT2.

You show this analysis to Roberto, and he is so impressed he gives you two free tickets to his first game.

8.68 Quality control Many manufacturing companies use statistical techniques to ensure that the products they make meet standards. One common way to do this is to take a random sample of products at regular intervals throughout the production shift. Assuming that the process is working properly, the mean measurements from these random samples will vary Normally around the target mean μ, with a standard deviation of σ.

(a) If the process is working properly, what is the probability that 4 out of 5 consecutive sample means fall within the interval $(\mu - \sigma, \mu + \sigma)$?

(b) If the process is working properly, what is the probability that the first sample mean that is greater than $\mu + 2\sigma$ is the one from the fourth sample taken?

CHAPTER

9

Sampling Distributions

CASE STUDY

Building better batteries

Everyone wants to have the latest technological gadget. That's why iPods, digital cameras, PDAs, Game Boys, and camera phones have sold millions of units. These devices require lots of power and can drain traditional alkaline batteries quickly. Battery manufacturers are constantly searching for ways to build longer-lasting batteries. In July 2005, Panasonic began marketing its new Oxyride battery in the United States. According to the results of preliminary testing, Oxyride batteries produced more power and lasted up to twice as long as alkaline batteries.[1]

Battery manufacturers must constantly measure battery lifetimes to ensure that their production process is working properly. Because testing a battery's lifetime requires the battery to be drained completely, the manufacturer wants to test as few batteries as possible. As part of the quality control process, the manufacturer selects a sample of batteries to test at regular intervals throughout production. By looking at the results from the sample, the manufacturer can determine whether the entire batch of batteries produced meets specifications.

At a particular battery production plant, when the process is working properly, AA batteries last an average of 17 hours with a standard deviation of 0.8 hour. Quality control inspectors select a random sample of 30 batteries during each hour of production and then drain them under conditions that mimic normal use. Here are the lifetimes (in hours) of the batteries from one such sample:

16.91	18.83	17.58	15.84	17.42	17.65	16.63	16.84	15.63	16.37
15.80	15.93	15.81	17.45	16.85	16.33	16.22	16.59	17.13	17.10
16.96	16.40	17.35	16.37	15.98	16.52	17.04	17.07	15.73	16.74

Do these data suggest that the process is working properly?

In this chapter, you will develop the tools you need to help answer questions like this.

Activity 9A

Young women's heights

Materials: Several 3″ × 3″ or 3″ × 5″ Post-it Notes.

In Example 7.8 (page 488), we saw that the height of young women varies approximately according to the N(64.5, 2.5) distribution. That is to say, the population of young women is Normally distributed with mean $\mu = 64.5$ inches and standard deviation $\sigma = 2.5$ inches. The random variable measured (call it X) is the height of a randomly selected young woman. In this Activity you will use the TI-83/84/89 to sample from this distribution and then use Post-it Notes to construct a distribution of averages.

1. If we choose one woman at random, the heights we get in repeated choices follow the $N(64.5, 2.5)$ distribution. On your calculator, go into the Statistics/List Editor and clear L_1/list1. Simulate the heights of 100 randomly selected young women and store these heights in L_1/list1 as follows:
   - Place your cursor at the top of L_1/list1 (on the list name, not below it).
   - TI-83/84: Press MATH, choose PRB, choose `6:randNorm(`.

     TI-89: Press F4, choose `4:Probability`, choose `6:randNorm(`.
   - Complete the command: `randNorm(64.5,2.5,100)` and press ENTER.
2. Plot a histogram of the 100 heights as follows. Deselect active functions in the Y = window, and turn off all STAT PLOTS. Set WINDOW dimensions to $X[57,72]_{2.5}$ and $Y[-10,45]_5$ to extend three standard deviations to either side of the mean, 64.5. Define Plot 1 to be a histogram using the heights in L_1/list1. (You must set the Hist. Bucket Width to 2.5 in the TI-89 Plot Setup.) Press GRAPH (on the TI-89, press ◆ F3) to plot the histogram. Describe the approximate shape of your histogram. Is it fairly symmetric or clearly skewed?
3. Approximately how many heights should there be within 3σ of the mean (that is, between 57 and 72)? Use TRACE to count the number of heights within 3σ. How many heights should there be within 1σ of the mean? Within 2σ of the mean? Again use TRACE to find these counts, and compare them with the numbers you would expect.
4. Use 1-Var Stats to find the mean, median, and standard deviation for your data. Compare $\bar{x}$ with the population mean $\mu = 64.5$. Compare the sample standard deviation s with $\sigma = 2.5$. How do the mean and median for your 100 heights compare? Recall that the closer the mean and the median are, the more symmetric the distribution.

5. Define Plot 2 to be a boxplot using L_1/list1, and then press GRAPH again. The boxplot will be plotted above the histogram. Does the boxplot appear symmetric? How close is the median in the boxplot to the mean of the histogram? Based on the appearance of the histogram and the boxplot, and a comparison of the mean and median, would you say that the distribution is nonsymmetric, moderately symmetric, or very symmetric?

6. Repeat Steps 1 to 5 two or three more times. Each time, record the mean $\bar{x}$, median, and standard deviation s.

7. In large print write the mean $\bar{x}$ for each sample on a different Post-it Note. Next, you will build a "Post-it Note histogram" of the distribution of the sample means $\bar{x}$. The teacher will draw an x axis on the blackboard, with tick marks indicating different mean heights. When instructed, go to the blackboard and stick each of your notes directly above the tick mark that is closest to the mean written on the note. When the Post-it Note histogram is complete, answer the following questions:

   (a) What is the approximate shape of the distribution of $\bar{x}$?

   (b) Where is the center of the distribution of $\bar{x}$? How does this center compare with the mean of heights of the population of *all* young women?

   (c) Roughly, how does the spread of the distribution of $\bar{x}$ compare with the spread of the original distribution ($\sigma = 2.5$)?

8. While someone calls out the values of $\bar{x}$ from the Post-it Notes, enter these values into L_2/list 2 in your calculator. Turn off Plot 1 and define Plot 3 to be a boxplot of the $\bar{x}$ data. How do these distributions of X and $\bar{x}$ compare visually? Use 1-Var Stats to calculate the standard deviation $s_{\bar{x}}$ for the distribution of $\bar{x}$. Compare this value with $2.5/\sqrt{100} = 0.25$.

9. Fill in the blanks in the following statement with a function of μ or σ: "The distribution of $\bar{x}$ is approximately Normal with mean $\mu(\bar{x})=$ ________ and standard deviation $\sigma(\bar{x})$ = ________."

Introduction

The reasoning of statistical inference rests on asking, "How often would this method give a correct answer if I used it very many times?" If it doesn't make sense to imagine repeatedly producing your data in the same circumstances, statistical inference is not possible.[2] Exploratory data analysis makes sense for any data, but formal inference does not. Even experts can disagree about how widely statistical inference should be used. But all agree that inference is most secure when we produce data by random sampling or randomized comparative experiments. The reason is that when we use chance to choose respondents or assign subjects, the laws of probability answer the question "What would happen if we did this many times?" The purpose of this chapter is to prepare for the study of statistical inference by looking at the probability distributions of some very common statistics: sample proportions and sample means.

9.1 Sampling Distributions

How much on the average do American households earn? The government's Current Population Survey contacted a sample of 113,146 households in March 2005. Their mean income in 2004 was $\bar{x}$ = \$60,528.[3] That \$60,528 describes the sample, but we use it to estimate the mean income of all households. We must now take care to keep straight whether a number describes a sample or a population. Here is the vocabulary we use.

Parameter, Statistic

A **parameter** is a number that describes the population. In statistical practice, the value of a parameter is not known, because we cannot examine the entire population.

A **statistic** is a number that can be computed from the sample data without making use of any unknown parameters. In practice, we often use a statistic to estimate an unknown parameter.

Example 9.1 *Making money*
Statistic versus parameter: means

The mean income of the sample of households contacted by the Current Population Survey was $\bar{x}$ = \$60,528. The number \$60,528 is a *statistic* because it describes this one Current Population Survey sample. The population that the poll wants to draw conclusions about is all 113 million U.S. households. The *parameter* of interest is the mean income of all of these households. We don't know the value of this parameter.

population mean μ
sample mean $\bar{x}$

Remember: statistics come from samples, and parameters come from populations. As long as we were just doing data analysis, the distinction between population and sample was not important. Now, however, it is essential. The notation we use must reflect this distinction. We write μ (the Greek letter mu) for the ***mean of a population.*** This is a fixed parameter that is unknown when we use a sample for inference. The ***mean of the sample*** is the familiar $\bar{x}$, the average of the observations in the sample. The sample mean $\bar{x}$ from a sample or an experiment is an estimate of the mean μ of the underlying population.

sampling variability

How can $\bar{x}$, based on a sample of only a few of the 113 million American households, be an accurate estimate of $\bar{x}$? After all, a second random sample taken at the same time would choose different households and no doubt produce a different value of $\bar{x}$. This basic fact is called ***sampling variability:*** the value of a statistic varies in repeated random sampling.

Example 9.2 ***Do you believe in ghosts?***
Statistic versus parameter: proportions

The Gallup Poll asked a random sample of 515 U.S. adults whether they believe in ghosts. Of the respondents, 160 said "Yes."[4] So the proportion of the sample who say they believe in ghosts is

$$\hat{p} = \frac{160}{515} = 0.31$$

The number 0.31 is a *statistic*. We can use it to estimate our *parameter* of interest: p, the proportion of all U.S. adults who believe in ghosts.

population proportion p

sample proportion $\hat{p}$

We use p to represent a ***population proportion.*** The ***sample proportion*** $\hat{p}$ is used to estimate the unknown parameter p. Based on the sample survey of Example 9.2, we might conclude that the proportion of all U.S. adults who believe in ghosts *is* 0.31. That would be a mistake. After all, a second random sample of 515 adults would probably yield a different value of $\hat{p}$. Sampling variability strikes again!

Sampling Variability

To understand why sampling variability is not fatal, we ask, "What would happen if we took many samples?" Here's how to answer that question:

- Take a large number of samples from the same population.
- Calculate the sample mean $\bar{x}$ or sample proportion $\hat{p}$ for each sample.
- Make a histogram of the values of $\bar{x}$ or $\hat{p}$.
- Examine the distribution displayed in the histogram for shape, center, and spread, as well as outliers or other deviations.

In practice it is too expensive to take many samples from a population like all adult U.S. residents. But we can imitate many samples by using simulation.

Example 9.3 ***Baggage check!***
Simulating sampling variability

Thousands of travelers pass through Guadalajara airport each day. Before leaving the airport, each passenger must pass through the Customs inspection area. Customs officials want to be sure that passengers do not bring illegal items into the country. But they do not have time to search every traveler's luggage. Instead, they require each person to press a button. Either a red or a green bulb lights up. If the red light shows, the passenger will be searched by Customs agents. A green light means "go ahead." Customs officers claim that the probability that the light turns green on any press of the button is 0.70.

We will simulate drawing simple random samples (SRSs) of size 100 from the population of travelers passing through Guadalajara airport. The parameter of interest is the proportion of travelers who get a green light at the Customs station. Assuming the Customs officials are telling the truth, we know that $p = 0.70$.

We can imitate the population by a huge table of random digits, such as Table B, with each entry standing for a traveler. Seven of the ten digits (say 0 to 6) stand for passengers who get a green light at Customs. The remaining three digits, 7 to 9, stand for those who get a red light and are searched. Because all digits in a random number table are equally likely, this assignment produces a population proportion of passengers who get the green light equal to $p = 0.7$. We then imitate an SRS of 100 travelers from the population by taking 100 consecutive digits from Table B. The statistic $\hat{p}$ is the proportion of the digits from 0 through 6 in the sample.

For example, if we begin at line 101 in Table B:

GRGGG	RGGGG	GGRGG	GRRGG	...
19223	95034	05756	28713	...

71 of the first 100 entries are from 0 through 6, so $\hat{p} = 71/100 = 0.71$. A second SRS based on the next 100 entries in Table B gives a different result, $\hat{p} = 0.62$. The two sample results are different, and neither is equal to the true population value $\hat{p} = 0.7$. That's sampling variability.

Simulation is a powerful tool for studying chance. It is much faster to use Table B than to actually draw repeated SRSs, and much faster yet to use technology to produce random digits. Figure 9.1 is the histogram of values of $\hat{p}$ from 1000

Figure 9.1 *The distribution of the sample proportion $\hat{p}$ from SRSs of size 100 drawn from a population with population proportion $p = 0.7$. The histogram shows the results of drawing 1000 SRSs.*

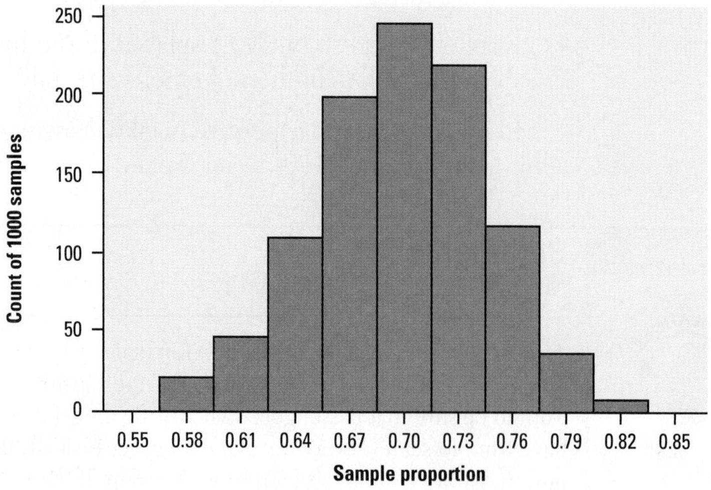

separate SRSs of size 100 drawn from a population with $p = 0.7$. This histogram shows what would happen if we drew many samples. It approximates the ***sampling distribution*** *of* $\hat{p}$.

Sampling Distribution

The **sampling distribution** of a statistic is the distribution of values taken by the statistic in all possible samples of the same size from the same population.

Strictly speaking, the sampling distribution is the ideal pattern that would emerge if we looked at all possible samples of size 100 from our population. A distribution obtained from a fixed number of trials, like the 1000 trials in Figure 9.1, is only an approximation to the sampling distribution. One of the uses of probability in statistics is to obtain exact sampling distributions without simulation. The interpretation of a sampling distribution is the same, however, whether we obtain it by simulation or by the mathematics of probability.

Example 9.4

Random digits

An exact sampling distribution

The population used to construct the random number table (Table B) can be described by the probability distribution shown in Figure 9.2.

Figure 9.2 *The probability distribution used to construct Table B, for Example 9.4.*

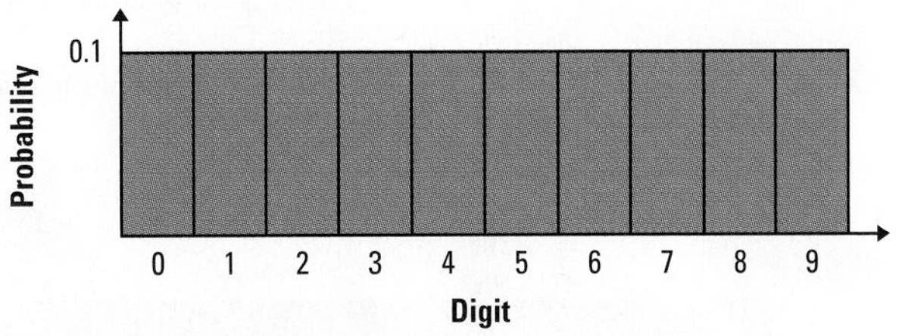

Consider the process of taking an SRS of size 2 from this population and computing $\bar{x}$ for the sample. We could perform a simulation to get a rough picture of the sampling distribution of $\bar{x}$. But in this case, we can construct the actual sampling distribution. Figure 9.3 (page 568) displays the values of $\bar{x}$ for all 100 possible samples of two random digits.

The distribution of $\bar{x}$ can be summarized by the histogram shown in Figure 9.4 (page 568). Since this graph displays *all* possible values of $\bar{x}$ from SRSs of size $n = 2$ from the population, it *is* the sampling distribution of $\bar{x}$.

Figure 9.3 *The values of $\bar{x}$ in all possible samples of two random digits, for Example 9.4.*

First digit \ Second digit	0	1	2	3	4	5	6	7	8	9
0	$\bar{x} = 0$	$\bar{x} = 0.5$	$\bar{x} = 1$	$\bar{x} = 1.5$	$\bar{x} = 2$	$\bar{x} = 2.5$	$\bar{x} = 3$	$\bar{x} = 3.5$	$\bar{x} = 4$	$\bar{x} = 4.5$
1	$\bar{x} = 0.5$	$\bar{x} = 1$	$\bar{x} = 1.5$	$\bar{x} = 2$	$\bar{x} = 2.5$	$\bar{x} = 3$	$\bar{x} = 3.5$	$\bar{x} = 4$	$\bar{x} = 4.5$	$\bar{x} = 5$
2	$\bar{x} = 1$	$\bar{x} = 1.5$	$\bar{x} = 2$	$\bar{x} = 2.5$	$\bar{x} = 3$	$\bar{x} = 3.5$	$\bar{x} = 4$	$\bar{x} = 4.5$	$\bar{x} = 5$	$\bar{x} = 5.5$
3	$\bar{x} = 1.5$	$\bar{x} = 2$	$\bar{x} = 2.5$	$\bar{x} = 3$	$\bar{x} = 3.5$	$\bar{x} = 4$	$\bar{x} = 4.5$	$\bar{x} = 5$	$\bar{x} = 5.5$	$\bar{x} = 6$
4	$\bar{x} = 2$	$\bar{x} = 2.5$	$\bar{x} = 3$	$\bar{x} = 3.5$	$\bar{x} = 4$	$\bar{x} = 4.5$	$\bar{x} = 5$	$\bar{x} = 5.5$	$\bar{x} = 6$	$\bar{x} = 6.5$
5	$\bar{x} = 2.5$	$\bar{x} = 3$	$\bar{x} = 3.5$	$\bar{x} = 4$	$\bar{x} = 4.5$	$\bar{x} = 5$	$\bar{x} = 5.5$	$\bar{x} = 6$	$\bar{x} = 6.5$	$\bar{x} = 7$
6	$\bar{x} = 3$	$\bar{x} = 3.5$	$\bar{x} = 4$	$\bar{x} = 4.5$	$\bar{x} = 5$	$\bar{x} = 5.5$	$\bar{x} = 6$	$\bar{x} = 6.5$	$\bar{x} = 7$	$\bar{x} = 7.5$
7	$\bar{x} = 3.5$	$\bar{x} = 4$	$\bar{x} = 4.5$	$\bar{x} = 5$	$\bar{x} = 5.5$	$\bar{x} = 6$	$\bar{x} = 6.5$	$\bar{x} = 7$	$\bar{x} = 7.5$	$\bar{x} = 8$
8	$\bar{x} = 4$	$\bar{x} = 4.5$	$\bar{x} = 5$	$\bar{x} = 5.5$	$\bar{x} = 6$	$\bar{x} = 6.5$	$\bar{x} = 7$	$\bar{x} = 7.5$	$\bar{x} = 8$	$\bar{x} = 8.5$
9	$\bar{x} = 4.5$	$\bar{x} = 5$	$\bar{x} = 5.5$	$\bar{x} = 6$	$\bar{x} = 6.5$	$\bar{x} = 7$	$\bar{x} = 7.5$	$\bar{x} = 8$	$\bar{x} = 8.5$	$\bar{x} = 9$

Figure 9.4 *The sampling distribution of $\bar{x}$ for samples of size $n = 2$, for Example 9.4.*

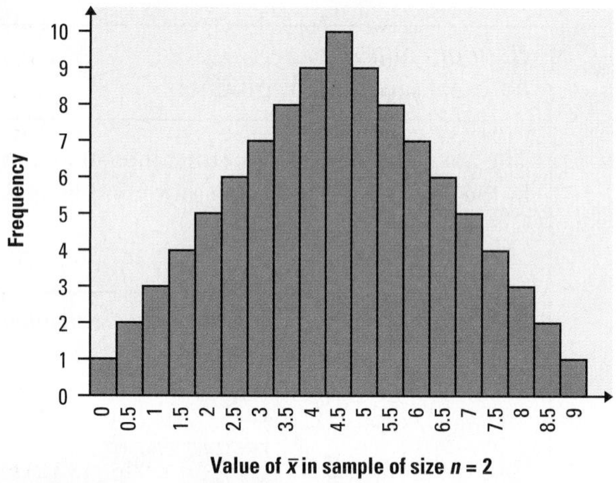

Exercises

For each boldface number in Exercises 9.1 and 9.2, (1) state whether it is a parameter or a statistic and (2) use appropriate notation to describe each number; for example, $p = 0.65$.

9.1 Ball bearings and unemployment

(a) The ball bearings in a large container have mean diameter **2.5003** centimeters (cm). This is within the specifications for acceptance of the container by the purchaser. By chance, an inspector chooses 100 bearings from the container that have mean diameter **2.5009** cm. Because this is outside the specified limits, the container is mistakenly rejected.

(b) The Bureau of Labor Statistics last month interviewed 60,000 members of the U.S. labor force, of whom **7.2%** were unemployed.

9.2 Telemarketing and well-fed rats

(a) A telemarketing firm in Los Angeles uses a device that dials residential telephone numbers in that city at random. Of the first 100 numbers dialed, **48%** are unlisted. This is not surprising, because **52%** of all Los Angeles residential phones are unlisted.

(b) A researcher carries out a randomized comparative experiment with young rats to investigate the effects of a toxic compound in food. She feeds the control group a normal diet. The experimental group receives a diet with 2500 parts per million of the toxic material. After 8 weeks, the mean weight gain is **335** grams for the control group and **289** grams for the experimental group.

Exercises 9.3 through 9.5 ask you to use simulations to study sampling distributions.

9.3 Murphy's Law and tumbling toast If a piece of toast falls off your breakfast plate, is it more likely to land with the buttered side down? According to Murphy's Law (the assumption that if anything can go wrong, it will), the answer is "Yes." Most scientists would argue that by the laws of probability, the toast is equally likely to land butter-side up or butter-side down. Robert Matthews, science correspondent of the *Sunday Telegraph*, disagrees. He claims that when toast falls off a plate that is being carried at a "typical height," the toast has just enough time to rotate once (landing butter-side down) before it lands. To test his claim, Mr. Matthews has arranged for 150,000 students in Great Britain to carry out an experiment with tumbling toast.[5]

Assuming scientists are correct, the proportion of times that the toast will land butter-side down is $p = 0.5$. We can use a coin toss to simulate the experiment. Let heads represent the toast landing butter-side down.

(a) Toss a coin 20 times and record the proportion of heads obtained, $\hat{p}$ = (number of heads)/20. Explain how your result relates to the tumbling-toast experiment.

(b) Repeat this sampling process 10 times. Make a histogram of the 10 values of $\hat{p}$. Is the center of this distribution close to 0.5?

(c) Ten repetitions give a very crude approximation to the sampling distribution. Pool your work with that of other students to obtain several hundred repetitions. Make a histogram of all the values of $\hat{p}$. Is the center close to 0.5? Is the shape approximately Normal?

(d) How much sampling variability is present? That is, how much do your values of $\hat{p}$ based on samples of size 20 differ from the actual population proportion, $p = 0.5$?

(e) Why do you think Mr. Matthews is asking so many students to participate in his experiment?

9.4 More tumbling toast Use your calculator to replicate Exercise 9.3 as follows. The command `randBin(20,.5)` simulates tossing a coin 20 times. The output is the number of heads in 20 tosses. The command `randBin(20,.5,10)/20` simulates 10 repetitions of tossing a coin 20 times and finding the proportions of heads. Go into your Statistics/List Editor and place your cursor on the top of L_1/list1. Execute the command `randBin(20,.5,10)/20` as follows:

- TI-83/84: Press MATH, choose PRB, choose `7:randBin(`. Complete command and press ENTER.
- TI-89: Press F4, choose `4:Probability`, choose `7:randBin(`. Complete command and press ENTER.

(a) Plot a histogram of the 10 values of $\hat{p}$. Set WINDOW parameters to $X[-0.05,1.05]_{0.1}$ and $Y[-2,6]_1$ and then TRACE. Is the center of the histogram close to 0.5? Do this several times to see if you get similar results each time.

(b) Increase the number of repetitions to 100. The command should read `randBin(20,.5,100)/20`. Execute the command (be patient!) and then plot a histogram using these 100 values. Don't change the XMIN and XMAX values, but do adjust the Y-values to $Y[-20, 50]_{10}$ to accommodate the taller bars. Is the center close to 0.5? Describe the shape of the distribution.

(c) Define Plot 2 to be a boxplot using L_1/list1, and TRACE again. How close is the median (in the boxplot) to the mean (balance point) of the histogram?

(d) Note that we didn't increase the sample size, only the number of repetitions. Did the spread of the distribution change? What would you change to decrease the spread of the distribution?

9.5 Sampling test scores, I Let us illustrate the idea of a sampling distribution of $\bar{x}$ in the case of a very small sample from a very small population. The population is the scores of 10 students on an exam:

Student:	0	1	2	3	4	5	6	7	8	9
Score:	82	62	80	58	72	73	65	66	74	62

The parameter of interest is the mean score in this population, which is 69.4. The sample is an SRS drawn from the population. Because the students are labeled 0 to 9, a single random digit from Table B chooses 1 student for the sample.

(a) Use Table B to draw an SRS of size $n = 4$ from this population. Write the four scores in your sample and calculate the mean $\bar{x}$ of the sample scores. This statistic is an estimate of the population parameter.

(b) Repeat this process 10 times. Make a histogram of the 10 values of $\bar{x}$. You are constructing the sampling distribution of $\bar{x}$. Is the center of your histogram close to 69.4?

(c) Ten repetitions give a very crude approximation to the sampling distribution. Pool your work with that of other students—using different parts of Table B—to obtain several hundred repetitions. Make a histogram of all the values of $\bar{x}$. Is the center close to 69.4? Describe the shape of the distribution. This histogram is a better approximation to the sampling distribution.

9.6 Sampling test scores, II Refer to the previous exercise.

(a) It is possible to construct the actual sampling distribution of $\overline{x}$ for samples of size n = 2 taken from this population. (Refer to Example 9.4.) Draw this sampling distribution.

(b) Compare the sampling distributions of $\overline{x}$ for samples of size 2 and size 4. Are the shapes, centers, and spreads similar or different?

Describing Sampling Distributions

We can use the tools of data analysis to describe any distribution. Let's apply these tools in the world of television.

Example 9.5 *Are you a* Survivor *fan?*
Describing the sampling distribution of $\hat{p}$

Television executives and companies who advertise on TV are interested in how many viewers watch particular television shows. According to 2005 Nielsen ratings, *Survivor: Guatemala* was one of the most-watched television shows in the United States during every week that it aired. Suppose that the true proportion of U.S. adults who watched *Survivor: Guatemala* is $p = 0.37$. Figure 9.5 shows the results of drawing 1000 SRSs of size $n = 100$ from a population with $p = 0.37$.

Figure 9.5 *Proportions of samples who watched* Survivor: Guatemala *in samples of size* n = *100, for Example 9.5.*

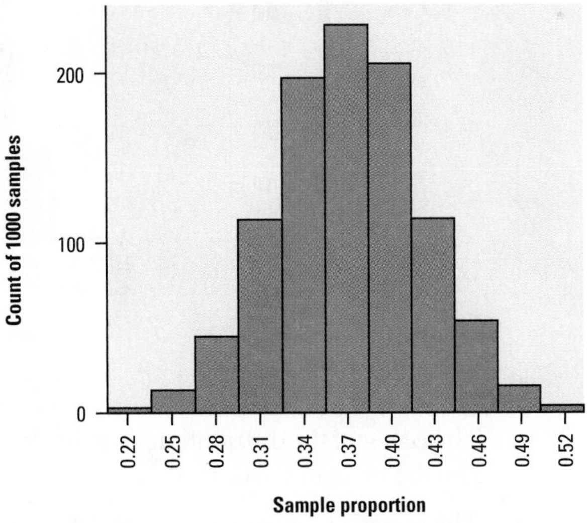

From the figure, we can see that:

- The overall *shape* of the distribution is symmetric and approximately Normal.
- The *center* of the distribution is very close to the true value $p = 0.37$ for the population from which the samples were drawn. In fact, the mean of the 1000 sample proportions is 0.372 and their median is exactly 0.370.

- The values of $\hat{p}$ have a large *spread*. They range from 0.22 to 0.535. Because the distribution is close to Normal, we can use the standard deviation to describe its spread. The standard deviation is about 0.05.
- There are no *outliers* or other important deviations from the overall pattern.

Figure 9.5 shows that a sample of 100 people often gave a $\hat{p}$ quite far from the population parameter $p = 0.37$. That is, a sample of 100 people does not produce a trustworthy estimate of the population proportion. That is why a Gallup Poll asked, not 100, but 1000 people whether they had watched *Survivor*. Let's repeat our simulation, this time taking 1000 SRSs of size 1000 from a population with proportion $p = 0.37$ who have watched *Survivor: Guatemala*.

Figure 9.6 displays the distribution of the 1000 values of $\hat{p}$ from these new samples. Figure 9.6 uses the same horizontal scale as Figure 9.5 to make comparison easy.

Figure 9.6 *The approximate sampling distribution of the sample proportion $\hat{p}$ from SRSs of size 1000 drawn from a population with population proportion $p = 0.37$. The histogram shows the results of 1000 SRSs. The scale is the same as in Figure 9.5.*

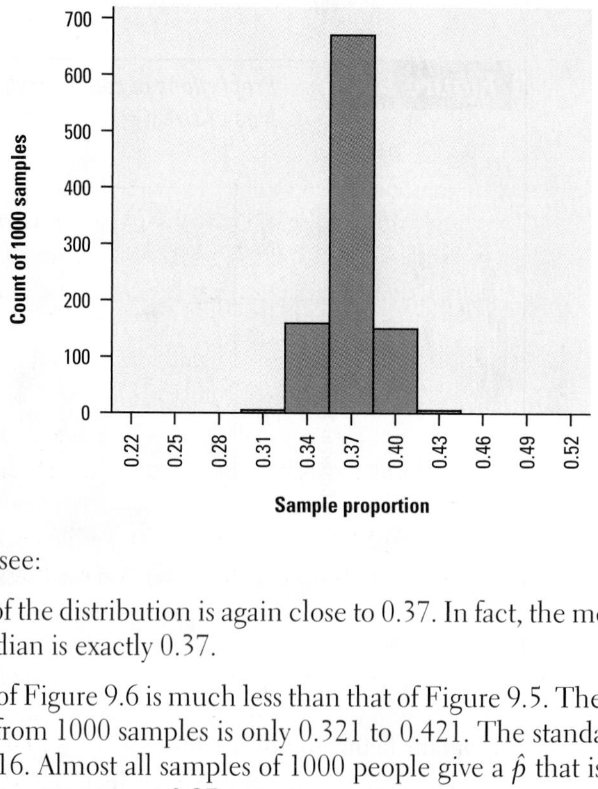

Here's what we see:

- The *center* of the distribution is again close to 0.37. In fact, the mean is 0.3697 and the median is exactly 0.37.
- The *spread* of Figure 9.6 is much less than that of Figure 9.5. The range of the values of $\hat{p}$ from 1000 samples is only 0.321 to 0.421. The standard deviation is about 0.016. Almost all samples of 1000 people give a $\hat{p}$ that is close to the population parameter $p = 0.37$.
- Because the values of $\hat{p}$ cluster so tightly about 0.37, it is hard to see the *shape* of the distribution in Figure 9.6. Figure 9.7 displays the same 1000 values of

$\hat{p}$ on an expanded scale that makes the shape clearer. The distribution is again approximately Normal in shape.

Figure 9.7 *The approximate sampling distribution from Figure 9.6, for samples of size 1000, redrawn on an expanded scale to better display the shape.*

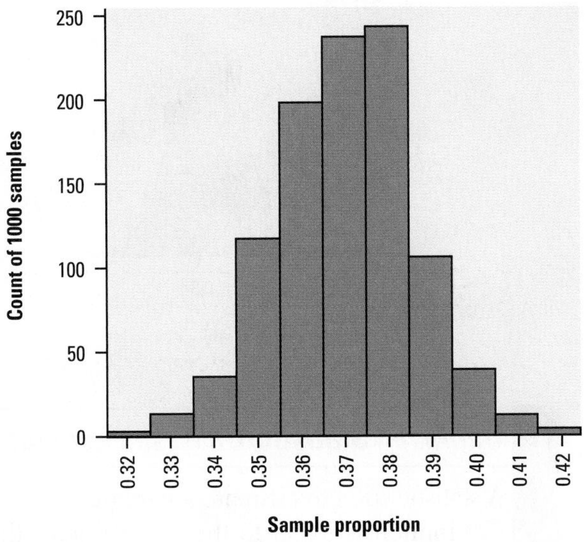

The appearance of the approximate sampling distributions in Figures 9.5 to 9.7 is a consequence of random sampling. Haphazard sampling does not give such regular and predictable results. When randomization is used in a design for producing data, statistics computed from the data have a definite pattern of behavior over many repetitions, even though the result of a single repetition is uncertain.

The Bias of a Statistic

The fact that statistics from random samples have definite sampling distributions allows a more careful answer to the question of how trustworthy a statistic is as an estimate of a parameter. Figure 9.8 (on the next page) shows the two sampling distributions of $\hat{p}$ for samples of 100 people and samples of 1000 people, side by side and drawn to the same scale. Both distributions are approximately Normal, so we have also drawn Normal curves for both. How trustworthy is the sample proportion $\hat{p}$ as an estimator of the population proportion p in each case?

bias Sampling distributions allow us to describe ***bias*** more precisely by speaking of the bias of a statistic rather than bias in a sampling method. Bias concerns the center of the sampling distribution. The centers of the approximate sampling distributions in Figure 9.8 are very close to the true value of the population parameter. Those distributions show the results of 1000 samples. In fact, the mean of the sampling distribution (think of taking *all* possible samples, not just 1000 samples) is *exactly* equal to 0.37, the parameter in the population.

Figure 9.8 *The approximate sampling distributions for sample proportions $\hat{p}$ for SRSs of two sizes drawn from a population with $p = 0.37$. (a) Sample size 100. (b) Sample size 1000. Both statistics are unbiased because the means of their distributions equal the true population value $p = 0.37$. The statistic from the larger sample is less variable.*

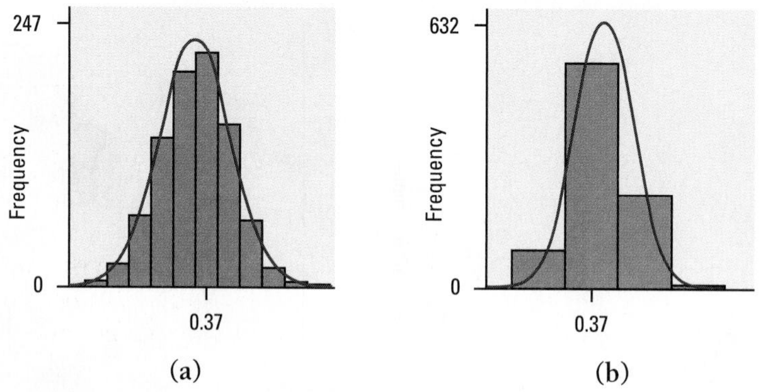

Unbiased Statistic/Unbiased Estimator

A statistic used to estimate a parameter is **unbiased** if the mean of its sampling distribution is equal to the true value of the parameter being estimated. The statistic is called an **unbiased estimator** of the parameter.

An *unbiased statistic* will sometimes fall above the true value of the parameter and sometimes below if we take many samples. Because its sampling distribution is centered at the true value, however, there is no systematic tendency to overestimate or underestimate the parameter. This makes the idea of lack of bias in the sense of "no favoritism" more precise. The sample proportion $\hat{p}$ from an SRS is an *unbiased estimator* of the population proportion p. If we draw an SRS from a population in which 37% have watched *Survivor: Guatemala,* the mean of the sampling distribution of $\hat{p}$ is 0.37. If we draw an SRS from a population in which 50% have seen *Survivor: Guatemala,* the mean of the sampling distribution of $\hat{p}$ is then 0.5.

The Variability of a Statistic

The statistics whose approximate sampling distributions appear in Figure 9.8 are both unbiased. That is, both distributions are centered at 0.37, the true population proportion. The sample proportion $\hat{p}$ from a random sample of any size is an unbiased estimator of the parameter p. Larger samples have a clear advantage, however. They are much more likely to produce an estimate close to the true value of the parameter because there is much less variability among large samples than among small samples.

Example 9.6 ***The statistics have spoken***
Describing sampling variability

The approximate sampling distribution of $\hat{p}$ for samples of size 100, shown in Figure 9.8(a), is close to the Normal distribution with mean 0.37 and standard deviation 0.05. Recall the 68–95–99.7 rule for Normal distributions. It says that 95% of values of $\hat{p}$ will fall within two standard deviations of the mean of the distribution, $p = 0.37$. So 95% of all samples will estimate $\hat{p}$ as

$$\text{mean} \pm (2 \times \text{standard deviation}) = 0.37 \pm (2 \times 0.05) = 0.37 \pm 0.1$$

If, in fact, 37% of U.S. adults have seen *Survivor: Guatemala,* the estimates from repeated SRSs of size 100 will usually fall between 27% and 47%. That's not very precise.

For samples of size 1000, Figure 9.8(b) shows that the standard deviation is only about 0.01. So 95% of these samples will give an estimate within about 0.02 of the true parameter, that is, between 0.35 and 0.39. An SRS of size 1000 can be trusted to give sample estimates that are very close to the truth about the entire population.

In Section 9.2 we will give the standard deviation of $\hat{p}$ for any size sample. We will then see Example 9.6 as part of a general rule that shows exactly how the variability of sample results decreases for larger samples. One important and surprising fact is that the spread of the sampling distribution does *not* depend very much on the size of the *population.*

Variability of a Statistic

The **variability of a statistic** is described by the spread of its sampling distribution. This spread is determined by the sampling design and the size of the sample. Larger samples give smaller spread. As long as the population is much larger than the sample (say, at least 10 times as large), the spread of the sampling distribution is approximately the same for any population size.

Why does the size of the population have little influence on the behavior of statistics from random samples? To see that this is plausible, imagine sampling harvested corn by thrusting a scoop into a lot of corn kernels. The scoop doesn't know whether it is surrounded by a bag of corn or by an entire truckload. As long as the corn is well mixed (so that the scoop selects a random sample), the variability of the result depends only on the size of the scoop.

The fact that the variability of sample results is controlled by the size of the sample has important consequences for sampling design. A statistic from an SRS of size 2500 from the more than 300 million residents of the United States is just as precise as an SRS of size 2500 from the 750,000 inhabitants of San Francisco. This is good news for designers of national samples but bad news for those who want accurate information about the citizens of San Francisco. If both use an SRS, both must use the same size sample to obtain equally trustworthy results.

Bias and Variability

We can think of the true value of the population parameter as the bull's-eye on a target and of the sample statistic as an arrow fired at the target. Both bias and variability describe what happens when we take many shots at the target. *Bias* means that our aim is off and we consistently miss the bull's-eye in the same direction. Our sample values do not center on the population value. *High variability* means that repeated shots are widely scattered on the target. Repeated samples do not give very similar results. Figure 9.9 shows this target illustration of the two types of error.

Figure 9.9 *Bias and variability. (a) High bias, low variability. (b) Low bias, high variability. (c) High bias, high variability. (d) The ideal: low bias, low variability.*

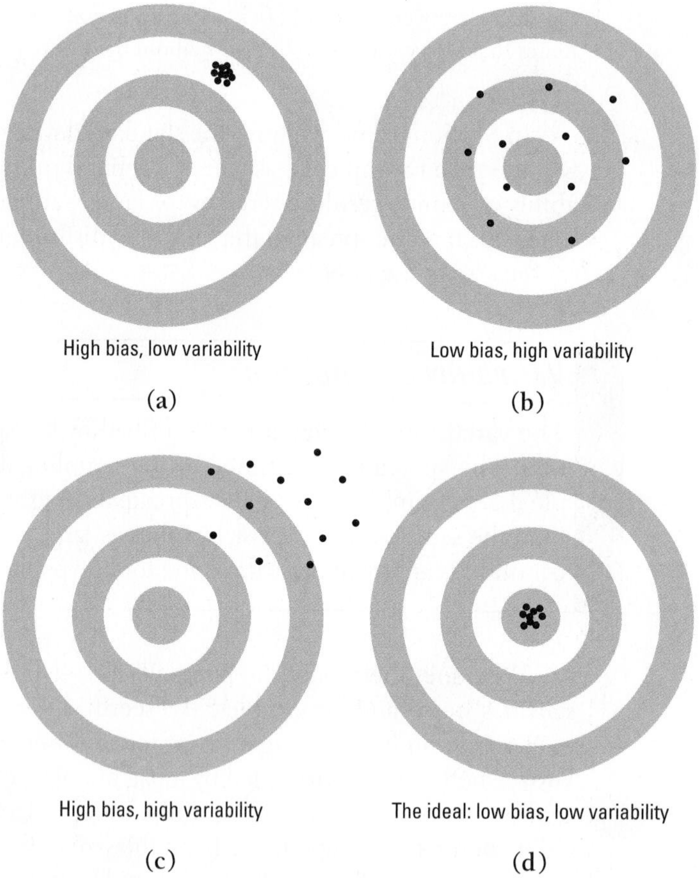

Notice that low variability (shots are close together) can accompany high bias (shots are consistently away from the bull's-eye in one direction). And low bias (shots center on the bull's-eye) can accompany high variability (shots are widely scattered). *Properly chosen statistics computed from random samples of sufficient size will have low bias and low variability.*

Exercises

9.7 Guinea pigs Here, again, are the survival times of 72 guinea pigs from the medical experiment described in Example 2.14 (page 151). Consider these 72 animals to be the population of interest.

43	45	53	56	56	57	58	66	67	73	74	79
80	80	81	81	81	82	83	83	84	88	89	91
91	92	92	97	99	99	100	100	101	102	102	102
103	104	107	108	109	113	114	118	121	123	126	128
137	138	139	144	145	147	156	162	174	178	179	184
191	198	211	214	243	249	329	380	403	511	522	598

(a) Make a histogram of the 72 survival times. This is the population distribution. It is strongly skewed to the right.

(b) Find the mean of the 72 survival times. This is the population mean μ. Mark μ on the x axis of your histogram.

SIMULATIONS

(c) Label the members of the population 01 to 72 and use Table B to choose an SRS of size $n = 12$. What is the mean survival time $\bar{x}$ for your sample? Mark the value of $\bar{x}$ with a point on the axis of your histogram from (a).

(d) Choose four more SRSs of size 12, using different parts of Table B. Find $\bar{x}$ for each sample and mark the values on the axis of your histogram from (a). Would you be surprised if all five $\bar{x}$'s fell on the same side of μ? Why?

(e) If you chose all possible SRSs of size 12 from this population and made a histogram of the $\bar{x}$-values, where would you expect the center of this sampling distribution to lie?

(f) Pool your results with those of your classmates to construct a histogram of the $\bar{x}$-values you obtained. Describe the shape, center, and spread of this distribution. Is the histogram approximately Normal?

9.8 Bearing down The table below contains the results of simulating on a computer 100 repetitions of drawing an SRS of size 200 from a large lot of ball bearings. Ten percent of the bearings in the lot do not conform to the specifications. That is, $p = 0.10$ for this population. The numbers in the table are the counts of nonconforming bearings in each sample of 200.

17	23	18	27	15	17	18	13	16	18	20	15	18	16	21	17	18	19	16	23
20	18	18	17	19	13	27	22	23	26	17	13	16	14	24	22	16	21	24	21
30	24	17	14	16	16	17	24	21	16	17	23	18	23	22	24	23	23	20	19
20	18	20	25	16	24	24	24	15	22	22	16	28	15	22	9	19	16	19	19
25	24	20	15	21	25	24	19	19	20	28	18	17	17	25	17	17	18	19	18

(a) Make a table that shows how often each count occurs. For each count in your table, give the corresponding value of the sample proportion $\hat{p} = \text{count}/200$. Then draw a histogram for the values of the statistic $\hat{p}$.

(b) Describe the shape of the distribution.

(c) Find the mean of the 100 observations of $\hat{p}$. Mark the mean on your histogram to show its center. Does the statistic $\hat{p}$ appear to have large or small bias as an estimate of the population proportion p?

(d) The sampling distribution of $\hat{p}$ is the distribution of the values of $\hat{p}$ from all possible samples of size 200 from this population. What is the mean of this distribution?

(e) If we repeatedly selected SRSs of size 1000 instead of 200 from this same population, what would be the mean of the sampling distribution of the sample proportion $\hat{p}$? Would the spread be larger, smaller, or about the same when compared with the spread of your histogram in (a)?

9.9 IRS audits The Internal Revenue Service plans to examine an SRS of individual federal income tax returns from each state. One variable of interest is the proportion of returns claiming itemized deductions. The total number of tax returns in each state varies from over 15 million in California to about 240,000 in Wyoming.

(a) Will the sampling variability of the sample proportion change from state to state if an SRS of 2000 tax returns is selected in each state? Explain your answer.

(b) Will the sampling variability of the sample proportion change from state to state if an SRS of 1% of all tax returns is selected in each state? Explain your answer.

9.10 Bias and variability Figure 9.10 shows histograms of four sampling distributions of statistics intended to estimate the same parameter. Label each distribution relative to the others as having large or small bias and as having large or small variability.

Figure 9.10 *Which of these sampling distributions displays large or small bias and large or small variability?*

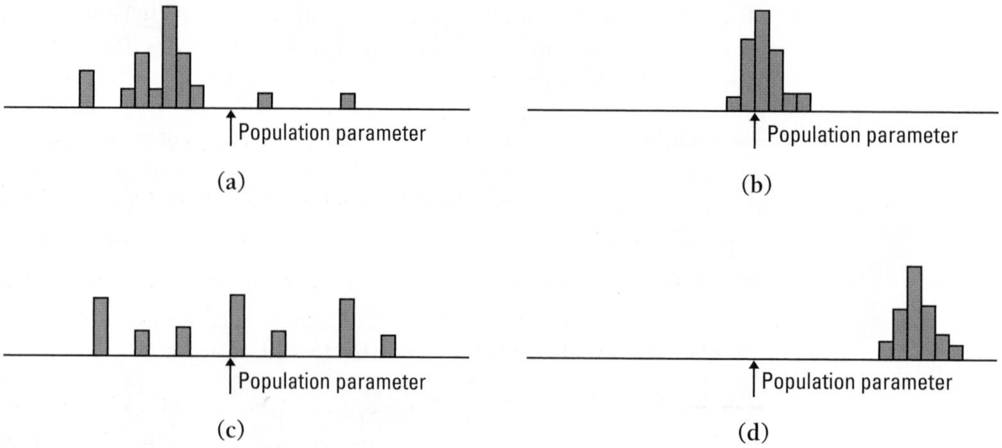

Section 9.1 Summary

A number that describes a population is called a **parameter.** A number that can be computed from the sample data is called a **statistic.** The purpose of sampling or experimentation is usually to use statistics to make statements about unknown parameters.

A statistic produced from a probability sample or randomized experiment has a **sampling distribution** that describes how the statistic varies in repeated data production. The sampling distribution answers the question "What would happen if we repeated the sample or experiment many times?" Formal statistical inference is based on the sampling distributions of statistics.

A statistic as an estimator of a parameter may suffer from **bias** or from high **variability.** Bias means that the center of the sampling distribution is not equal to the true value of the parameter. The variability of the statistic is described by the spread of its sampling distribution.

Properly chosen statistics from randomized data production designs have no bias resulting from the way the sample is selected or the way the experimental units are assigned to treatments. The variability of the statistic is determined by the size of the sample or by the size of the experimental groups. Statistics from larger samples have less variability.

Section 9.1 Exercises

In Exercises 9.11 and 9.12, (1) state whether each boldface number is a parameter or a statistic, and (2) use appropriate notation to describe each number.

9.11 Small classes in school The Tennessee STAR experiment randomly assigned children to regular or small classes during their first four years of school. When these children reached high school, **40.2%** of blacks from small classes took the ACT or SAT college entrance exams. Only **31.7%** of blacks from regular classes took one of these exams.

9.12 How tall? A random sample of female college students has a mean height of **64.5** inches, which is greater than the **63**-inch mean height of all adult American women.

9.13 A sample of teens A study of the health of teenagers plans to measure the blood cholesterol level of an SRS of youths aged 13 to 16. The researchers will report the mean $\bar{x}$ from their sample as an estimate of the mean cholesterol level μ in this population.

(a) Explain to someone who knows no statistics what it means to say that $\bar{x}$ is an unbiased estimator of μ.

(b) The sample result $\bar{x}$ is an unbiased estimator of the population mean μ no matter what size SRS the study chooses. Explain to someone who knows no statistics why a large sample gives more trustworthy results than a small sample.

9.14 Bad eggs An entomologist samples a field for egg masses of a harmful insect by placing a yard-square frame at random locations and examining the ground within the frame carefully. He wants to estimate the proportion of square yards in which egg masses are present. Suppose that in a large field egg masses are present in 20% of all possible yard-square areas. That is, $p = 0.2$ in this population.

SIMULATIONS

(a) Use Table B to simulate the presence or absence of egg masses in each square yard of an SRS of 10 square yards from the field. Be sure to explain clearly which digits you used to represent the presence and the absence of egg masses. What proportion of your 10 sample areas had egg masses? This is the statistic $\hat{p}$.

(b) Repeat (a) with different lines from Table B, until you have simulated the results of 20 SRSs of size 10. What proportion of the square yards in each of your 20 samples had egg masses? Make a stemplot from these 20 values to display the distribution of your 20 observations on $\hat{p}$. What is the mean of this distribution? What is its shape?

(c) If you looked at all possible SRSs of size 10, rather than just 20 SRSs, what would be the mean of the values of $\hat{p}$? This is the mean of the sampling distribution of $\hat{p}$.

(d) In another field, 40% of all square-yard areas contain egg masses. What is the mean of the sampling distribution of $\hat{p}$ in samples from this field?

9.15 Rolling the dice, I Consider the population of all rolls of a fair, six-sided die.

(a) Draw a histogram that shows the population distribution. Find the mean μ and standard deviation σ of this population.

(b) If you took an SRS of size $n = 2$ (with replacement) from this population, what would you actually be doing?

(c) List all possible SRSs of size 2 from this population, and compute $\bar{x}$ for each sample.

(d) Draw the sampling distribution of $\bar{x}$ for samples of size $n = 2$. Describe its shape, center, and spread. How do these characteristics compare with those of the population distribution?

9.16 Rolling the dice, II In Exercise 9.15, you constructed the sampling distribution of $\bar{x}$ in samples of size $n = 2$ from the population of rolls of a fair, six-sided die. What would happen if we increased the sample size to $n = 3$? For starters, it would take you a long time to list all possible SRSs for $n = 3$. Instead, you can use your calculator to simulate rolling the die three times.

(a) Generate L_1/list1 using the command `randInt(1,6,100)+randInt(1,6,100)+randInt(1,6,100).`

This will run 100 simulations of rolling the die three times and calculating the sum of the three rolls.

(b) Define L_2/list2 as $L_1/3$ (list1/3). Now L_2/list2 contains the values of $\bar{x}$ for the 100 simulations.

(c) Plot a histogram of the $\bar{x}$-values.

9.17 School vouchers A national opinion poll recently estimated that 44% ($\hat{p} = 0.44$) of all adults agree that parents of school-age children should be given vouchers good for education at any public or private school of their choice. The polling organization used a probability sampling method for which the sample proportion $\hat{p}$ has a Normal distribution with standard deviation about 0.015. If a sample were drawn by the same method from the state of New Jersey (population 8.7 million) instead of from the entire United States (population about 300 million), would this standard deviation be larger, about the same, or smaller? Explain your answer.

9.18 Simulating *Survivor* Suppose the true proportion of U.S. adults who have watched *Survivor: Guatemala* is 0.41. Your teacher will provide a calculator program that simulates sampling from this population.

(a) In the program, what digits are assigned to U.S. adults? What digits are assigned to U.S. adults who say they have watched *Survivor: Guatemala*? Does the program output a count of adults who answer "Yes," a percent, or a proportion?

(b) Execute the program and specify 5 trials (sample size = 5). Do this 10 times, and record the 10 numbers.

(c) Execute the program 10 more times, specifying a sample size of 25. Record the 10 results for sample size = 25.

(d) Execute the program 10 more times, specifying a sample size of 100. Record the 10 results for sample size = 100.

(e) Enter the 10 outputs for sample size = 5 in L_1/list1, the 10 results for sample size = 25 in L_2/list2, and the 10 results for sample size = 100 in L_3/list3. Then do 1-Var Stats for L_1/list 1, L_2/list2, and L_3/list3, and record the means and sample standard deviations s_x for each sample size. Complete the sentence "As the sample size increases, the variability _______________."

9.2 Sample Proportions

What proportion of U.S. teens know that 1492 was the year in which Columbus "discovered" America? A Gallup Poll found that 210 out of a random sample of 501 American teens aged 13 to 17 knew this historically important date.[6] The sample proportion

$$\hat{p} = \frac{210}{501} = 0.42$$

is the statistic that we use to gain information about the unknown population parameter p. We may say that "42% of U.S. teenagers know that Columbus discovered America in 1492." Statistical recipes work with proportions expressed as decimals, so 42% becomes 0.42.

The Sampling Distribution of a Sample Proportion $\hat{p}$

How good is the statistic $\hat{p}$ as an estimate of the parameter p? To find out, we ask, "What would happen if we took many samples?" The ***sampling distribution of*** $\hat{p}$ answers this question. How do we determine the center, shape, and spread of the sampling distribution of $\hat{p}$? By making an important connection between proportions and counts. We want to estimate the proportion of "successes" in the population. We take an SRS from the population of interest. Our estimator is the sample proportion of successes:

$$\hat{p} = \frac{\text{count of "successes" in sample}}{\text{size of sample}} = \frac{X}{n}$$

Since values of X and $\hat{p}$ will vary in repeated samples, both X and $\hat{p}$ are random variables. Provided that the population is much larger than the sample (say at least 10 times), the count X will follow a binomial distribution. The proportion $\hat{p}$ does not have a binomial distribution.

From Chapter 8, we know that

$$\mu_X = np \quad \text{and} \quad \sigma_X = \sqrt{np(1-p)}$$

give the mean and standard deviation of the random variable X. Since $\hat{p} = X/n = (1/n)X$, we can use the rules from Chapter 7 to find the mean and standard deviation of the random variable $\hat{p}$. Recall that if $Y = a + bX$, then $\mu_Y = a + b\mu_X$ and $\sigma_Y = b\sigma_X$. In this case, $\hat{p} = 0 + (1/n)X$, so

$$\mu_{\hat{p}} = 0 + \frac{1}{n}np = p$$

$$\sigma_{\hat{p}} = \frac{1}{n}\sqrt{np(1-p)} = \sqrt{\frac{np(1-p)}{n^2}} = \sqrt{\frac{p(1-p)}{n}}$$

Sampling Distribution of a Sample Proportion

Choose an SRS of size n from a large population with population proportion p having some characteristic of interest. Let $\hat{p}$ be the proportion of the sample having that characteristic. Then:

- The **mean** of the sampling distribution of $\hat{p}$ is exactly p.
- The **standard deviation** of the sampling distribution of $\hat{p}$ is

$$\sqrt{\frac{p(1-p)}{n}}$$

Because the mean of the sampling distribution of $\hat{p}$ is always equal to the parameter p, the sample proportion $\hat{p}$ is an unbiased estimator of p. The standard deviation of $\hat{p}$ gets smaller as the sample size n increases because n appears in the denominator of the formula for the standard deviation. That is, $\hat{p}$ is less variable in larger samples. What is more, the formula shows just how quickly the standard deviation decreases as n increases. The sample size n is under the square root sign, so to cut the standard deviation in half, we must take a sample four times as large, not just twice as large.

The formula for the standard deviation of $\hat{p}$ doesn't apply when the sample is a large part of the population. You can't use this recipe if you choose an SRS of 50 of the 100 people in a class, for example. In practice, we usually take a sample only when the population is large. Otherwise, we could examine the entire population. Here is a practical guide.[7]

Rule of Thumb 1

Use the recipe for the standard deviation of $\hat{p}$ only when the population is at least 10 times as large as the sample; that is, when $N \geq 10n$.

Using the Normal Approximation for $\hat{p}$

What about the shape of the sampling distribution of $\hat{p}$? In the simulation examples in Section 9.1, we found that the sampling distribution of $\hat{p}$ is *approximately Normal* and is closer to a Normal distribution when the sample size n is large. For example, if we sample 100 individuals, the only possible values of $\hat{p}$ are 0, 1/100, 2/100, and so on. The statistic has only 101 possible values, so its distribution cannot be exactly Normal. The accuracy of the Normal approximation improves as the sample size n increases. For a fixed sample size n, the Normal approximation is most accurate when p is close to 1/2, and least accurate when p is near 0 or 1. If $p = 1$, for example, then $\hat{p} = 1$ in every sample because every individual in the population has the characteristic we are counting. The Normal approximation is no good at all when $p = 1$ or $p = 0$. Here is a rule of thumb that ensures that Normal calculations are accurate enough for most statistical purposes. Unlike the first rule of thumb, this one rules out some settings of practical interest.

Rule of Thumb 2

We will use the Normal approximation to the sampling distribution of $\hat{p}$ for values of n and p that satisfy $np \geq 10$ and $n(1 - p) \geq 10$.

Using what we have learned about the sampling distribution of $\hat{p}$, we can determine the likelihood of obtaining an SRS in which $\hat{p}$ is close to p. This is especially useful to college admissions officers, as the following example shows.

Example 9.7 *Applying to college*

Normal calculations involving $\hat{p}$

A polling organization asks an SRS of 1500 first-year college students whether they applied for admission to any other college. In fact, 35% of all first-year students applied to colleges besides the one they are attending. What is the probability that the random sample of 1500 students will give a result within 2 percentage points of this true value?

We have an SRS of size $n = 1500$ drawn from a population in which the proportion $p = 0.35$ applied to other colleges. The sampling distribution of $\hat{p}$ has mean $\mu_{\hat{p}} = 0.35$. What about its standard deviation? By the first "rule of thumb," the population must contain at least $10 \times 1500 = 15{,}000$ people for us to use the standard deviation formula we derived. There are over 1.7 million first-year college students, so

$$\sigma_{\hat{p}} = \sqrt{\frac{p(1-p)}{n}} = \sqrt{\frac{(0.35)(0.65)}{1500}} = 0.0123$$

Can we use a Normal distribution to approximate the sampling distribution of $\hat{p}$? Checking the second "rule of thumb": $np = (1500)(0.35) = 525$ and $n(1-p) = (1500)(0.65) = 975$. Both are much larger than 10, so the Normal approximation will be quite accurate.

We want to find the probability that $\hat{p}$ falls between 0.33 and 0.37 (within 2 percentage points, or 0.02, of 0.35). This is a Normal distribution calculation. Figure 9.11 shows the Normal distribution that approximates the sampling distribution of $\hat{p}$. The area of the green region corresponds to the probability that $0.33 \le \hat{p} \le 0.37$.

Figure 9.11 *The Normal approximation to the sampling distribution of $\hat{p}$, for Example 9.7.*

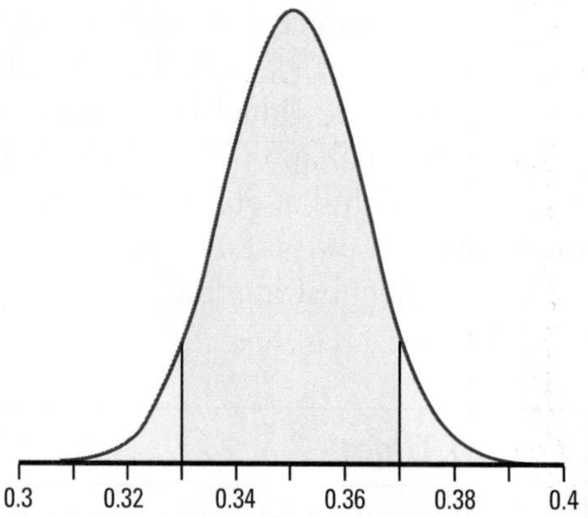

Step 1: Standardize $\hat{p}$ by subtracting its mean 0.35 and dividing by its standard deviation 0.0123. That produces a new statistic that has the standard Normal distribution. It is usual to call such a statistic z:

$$z = \frac{\hat{p} - 0.35}{0.0123}$$

Step 2: Find the standardized values (z-scores) of $\hat{p} = 0.33$ and $\hat{p} = 0.37$. For $\hat{p} = 0.33$:

$$z = \frac{0.33 - 0.35}{0.0123} = -1.63$$

For $\hat{p} = 0.37$:

$$z = \frac{0.37 - 0.35}{0.0123} = 1.63$$

Step 3: Draw a picture of the area under the standard Normal curve corresponding to these standardized values (Figure 9.12). Then use Table A to find the green area. Here is the calculation:

$$P(0.33 \le \hat{p} \le 0.37) = P(-1.63 \le Z \le 1.63) = 0.9484 - 0.0516 = 0.8968$$

Figure 9.12 *Probabilities as areas under the standard Normal curve, for Example 9.7.*

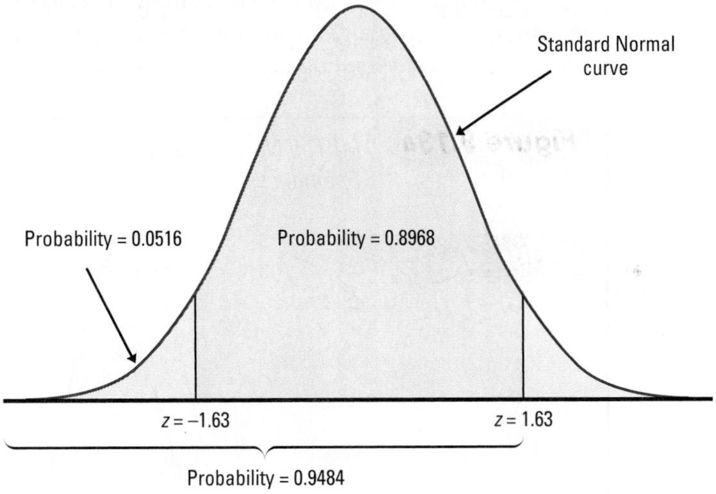

We see that almost 90% of all samples will give a result within 2 percentage points of the truth about the population.

The outline of the calculation in Example 9.7 is familiar from Chapter 2, but the language of probability is new. The sampling distribution of $\hat{p}$ gives probabilities for its values, so the entries in Table A are now probabilities. We used a brief notation that is common in statistics. The capital P in $P(0.33 \le \hat{p} \le 0.37)$ stands for "probability." The expression inside the parentheses tells us what event we are finding the probability of. This entire expression is a short way of writing "the probability that the value of $\hat{p}$ is between 0.33 and 0.37."

Example 9.8 *Survey undercoverage?*
More Normal calculations

One way of checking the effect of undercoverage, nonresponse, and other sources of error in a sample survey is to compare the sample with known facts about the population. About 11% of American adults are black. The proportion $\hat{p}$ of blacks in an SRS of 1500 adults should therefore be close to 0.11. It is unlikely to be exactly 0.11 because of sampling variability. If a national sample contains only 9.2% blacks, should we suspect that the sampling procedure is somehow underrepresenting blacks? We will find the probability that a sample contains no more than 9.2% blacks when the population is 11% black.

The mean of the sampling distribution of $\hat{p}$ is $p = 0.11$. Since the population of all black American adults is larger than $10 \times 1500 = 15{,}000$, the standard deviation of $\hat{p}$ is

$$\sqrt{\frac{p(1-p)}{n}} = \sqrt{\frac{(0.11)(0.89)}{1500}} = 0.00808$$

(by Rule of Thumb 1). Next, we check to see that $np = (1500)(0.11) = 165$ and $n(1-p) = (1500)(0.89) = 1335$. So Rule of Thumb 2 tells us that we can use the Normal approximation to the sampling distribution of $\hat{p}$. Figure 9.13(a) shows the Normal distribution with the area corresponding to $\hat{p} \le 0.092$ shaded.

Figure 9.13a *(a) The Normal approximation to the sampling distribution of $\hat{p}$, for Example 9.8.*

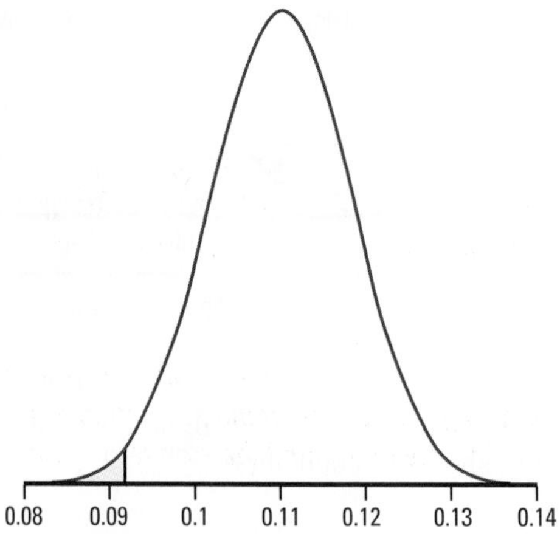

Step 1: Standardize $\hat{p}$.

$$z = \frac{\hat{p} - 0.11}{0.00808}$$

has the standard Normal distribution.

Step 2: Find the standardized value (z-score) of $\hat{p} = 0.092$.

$$z = \frac{0.092 - 0.11}{0.00808} = -2.23$$

Step 3: Draw a picture of the area under the standard Normal curve corresponding to the standardized value (Figure 9.13(b)). Then use Table A to find the shaded area.

$$P(\hat{p} \leq 0.092) = P(Z \leq -2.23) = 0.0129$$

Only 1.29% of all samples would have so few blacks. Because it is unlikely that a sample would include so few blacks, we have good reason to suspect that the sampling procedure underrepresents blacks.

Figure 9.13b *(b) The probability as an area under the standard Normal curve, for Example 9.8.*

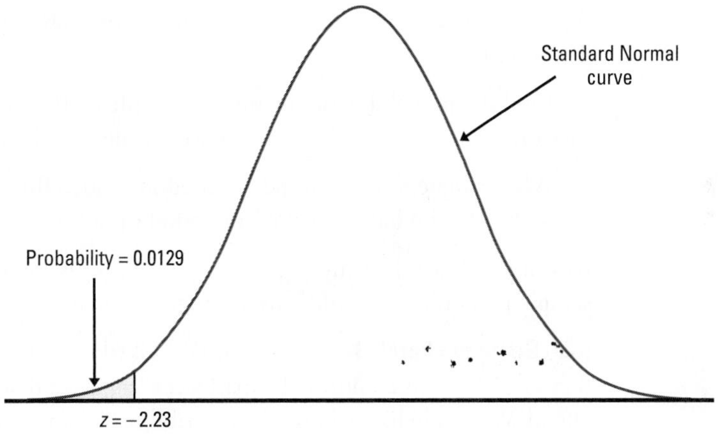

Figure 9.14 summarizes the facts that we have learned about the sampling distribution of $\hat{p}$ in a form that helps you remember the big idea of a sampling distribution.

Figure 9.14 *Select a large SRS from a population in which the proportion p are successes. The sampling distribution of the proportion $\hat{p}$ of successes in the sample is approximately Normal. The mean is p and the standard deviation is $\sqrt{p(1-p)/n}$.*

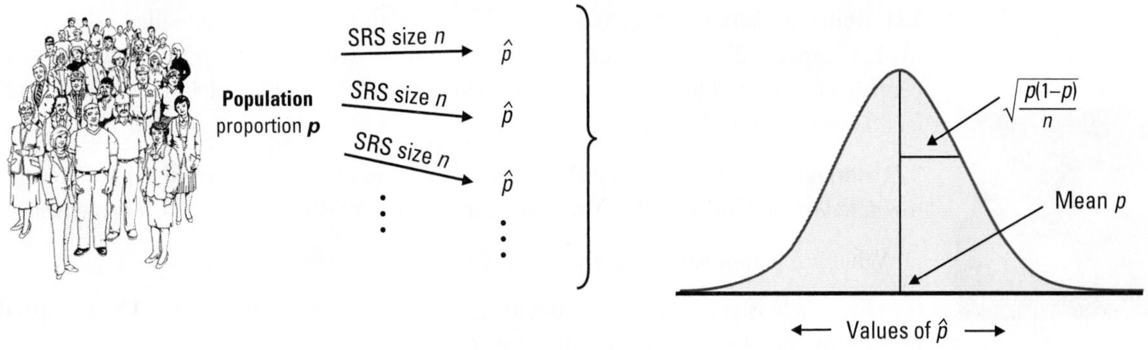

Exercises

9.19 Do you drink the cereal milk? A *USA Today* poll asked a random sample of 1012 U.S. adults what they do with the milk in the bowl after they have eaten the cereal. Of the respondents, 67% said that they drink it. Suppose that 70% of U.S. adults actually drink the cereal milk.

(a) Find the mean and standard deviation of the proportion $\hat{p}$ of the sample who say they drink the cereal milk.

(b) Explain why you can use the formula for the standard deviation of $\hat{p}$ in this setting (Rule of Thumb 1).

(c) Check that you can use the Normal approximation for the distribution of $\hat{p}$ (Rule of Thumb 2).

(d) Find the probability of obtaining a sample of 1012 adults in which 67% or fewer say they drink the cereal milk. Do you have any doubts about the result of this poll?

(e) What sample size would be required to reduce the standard deviation of the sample proportion to one-half the value you found in (a)?

(f) If the pollsters had surveyed 1012 teenagers instead of 1012 adults, do you think the sample proportion $\hat{p}$ would have been greater than, equal to, or less than 0.67? Explain.

9.20 Going to church, I The Gallup Poll asked a random sample of 1785 adults whether they attended church during the past week. Suppose that 40% of the adult population did attend. We would like to know the probability that an SRS of size 1785 would come within plus or minus 3 percentage points of this true value.

(a) If $\hat{p}$ is the proportion of the sample who did attend church, what is the mean of the sampling distribution of $\hat{p}$? What is its standard deviation?

(b) Explain why you can use the formula for the standard deviation of $\hat{p}$ in this setting (Rule of Thumb 1).

(c) Check that you can use the Normal approximation for the distribution of $\hat{p}$ (Rule of Thumb 2).

(d) Find the probability that $\hat{p}$ takes a value between 0.37 and 0.43. Will an SRS of size 1785 usually give a result $\hat{p}$ within plus or minus 3 percentage points of the true population proportion? Explain.

9.21 Going to church, II Suppose that 40% of the adult population attended church last week. Exercise 9.20 asks the probability that $\hat{p}$ from an SRS estimates $p = 0.4$ within 3 percentage points. Find this probability for SRSs of sizes 300, 1200, and 4800. What general fact do your results illustrate?

9.22 Harley motorcycles Harley-Davidson motorcycles make up 14% of all the motorcycles registered in the United States. You plan to interview an SRS of 500 motorcycle owners.

(a) What is the approximate distribution of your sample who own Harleys?

(b) How likely is your sample to contain 20% or more who own Harleys? Do a Normal probability calculation to answer this question.

(c) How likely is your sample to contain at least 15% who own Harleys? Do a Normal probability calculation to answer this question.

9.23 On-time shipping Your mail-order company advertises that it ships 90% of its orders within three working days. You select an SRS of 100 of the 5000 orders received in the past week for an audit. The audit reveals that 86 of these orders were shipped on time.

(a) What is the sample proportion of orders shipped on time?

(b) If the company really ships 90% of its orders on time, what is the probability that the proportion in an SRS of 100 orders is as small as the proportion in your sample or smaller?

(c) A critic says, "Aha! You claim 90%, but in your sample the on-time percent is lower than that. So the 90% claim is wrong." Explain in simple language why your probability calculation in (b) shows that the result of the sample does not refute the 90% claim.

9.24 Students on diets A sample survey interviews an SRS of 267 college women. Suppose (as is roughly true) that 70% of college women have been on a diet within the past 12 months. What is the probability that 75% or more of the women in the sample have been on a diet? Show your work.

Section 9.2 Summary

When we want information about the **population proportion** p of individuals with some special characteristic, we often take an SRS and use the **sample proportion** $\hat{p}$ to estimate the unknown parameter p.

The **sampling distribution** of $\hat{p}$ describes how the statistic varies in all possible samples from the population.

The **mean** of the sampling distribution is equal to the population proportion p. That is, $\hat{p}$ is an unbiased estimator of p.

The **standard deviation** of the sampling distribution is $\sqrt{p(1-p)/n}$ for an SRS of size n. This formula can be used if the population is at least 10 times as large as the sample.

The standard deviation of $\hat{p}$ gets smaller as the sample size n gets larger. Because of the square root, a sample four times larger is needed to cut the standard deviation in half.

When the sample size n is large, the sampling distribution of $\hat{p}$ is close to a Normal distribution with mean p and standard deviation $\sqrt{p(1-p)/n}$. In practice, use this **Normal approximation** when both $np \geq 10$ and $n(1-p) \geq 10$.

Section 9.2 Exercises

9.25 Do you jog? The Gallup Poll once asked a random sample of 1540 adults, "Do you happen to jog?" Suppose that in fact 15% of all adults jog.

(a) Find the mean and standard deviation of the proportion $\hat{p}$ of the sample who jog. (Assume the sample is an SRS.)

(b) Explain why you can use the formula for the standard deviation of $\hat{p}$ in this setting.

(c) Check that you can use the Normal approximation for the distribution of $\hat{p}$.

(d) Find the probability that between 13% and 17% of the sample jog.

(e) What sample size would be required to reduce the standard deviation of the sample proportion to one-third the value you found in (a)?

9.26 More jogging! Suppose that 15% of all adults jog. Exercise 9.25 asks the probability that the sample proportion $\hat{p}$ from an SRS estimates $p = 0.15$ within 2 percentage points. Find this probability for SRSs of sizes 200, 800, and 3200. What general conclusion can you draw from your calculations?

9.27 Underage drinking The Harvard College Alcohol Study finds that 67% of college students support efforts to "crack down on underage drinking." The study took a random sample of almost 15,000 students, so the population proportion who support a crackdown is close to $p = 0.67$.[8] The administration of a local college surveys an SRS of 100 students and finds that 62 support a crackdown on underage drinking.

(a) What is the sample proportion who support a crackdown on underage drinking?

(b) If in fact the proportion of all students attending this college who support a crackdown is the same as the national 67%, what is the probability that the proportion in an SRS of 100 students is as small or smaller than the result of the administration's sample?

(c) A writer in the college's student paper says that "support for a crackdown is lower at our school than nationally." Write a short letter to the editor explaining why the survey does not support this conclusion.

Rigging the lottery We have all seen televised lottery drawings in which numbered balls bubble about and are randomly popped out by air pressure. How might we rig such a drawing? In 1980, when the Pennsylvania lottery used just three balls, a drawing was rigged by the host and several stagehands. They injected paint into all balls bearing 8 of the 10 digits. This weighed them down and guaranteed that all three balls for the winning number would have the remaining 2 digits. The perps then bet on all combinations of these digits. When 6-6-6 popped out, they won $1.2 million. Yes, they were caught.

9.28 Unlisted numbers According to a market research firm, 52% of all residential telephone numbers in Los Angeles are unlisted. A telephone sales firm uses random digit dialing equipment that dials residential numbers at random, whether or not they are listed in the telephone directory. The firm calls 500 numbers in Los Angeles.

(a) What are the mean and standard deviation of the proportion of unlisted numbers in the sample?

(b) What is the probability that at least half the numbers dialed are unlisted? (Remember to check that you can use the Normal approximation.)

9.29 Multiple-choice tests Here is a simple probability model for multiple-choice tests. Suppose that a student has probability p of correctly answering a question chosen at random from a universe of possible questions. (A good student has a higher p than a poor student.) The correctness of an answer to any specific question doesn't depend on other questions. A test contains n questions. Then the proportion of correct answers that a student gives is a sample proportion $\hat{p}$ from an SRS of size n drawn from a population with population proportion p.

(a) Julie is a good student for whom $p = 0.75$. Find the probability that Julie scores 70% or lower on a 100-question test.

(b) If the test contains 250 questions, what is the probability that Julie will score 70% or lower?

(c) How many questions must the test contain in order to reduce the standard deviation of Julie's proportion of correct answers to one-fourth its value for a 100-item test?

(d) Laura is a weaker student for whom $p = 0.6$. Does the answer you gave in (c) for the standard deviation of Julie's score apply to Laura's standard deviation also? Explain.

9.30 Rules of thumb Explain why you cannot use the methods of this section to find the following probabilities.

(a) A factory employs 3000 unionized workers, of whom 30% are Hispanic. The 15-member union executive committee contains 3 Hispanics. What would be the probability of 3 or fewer Hispanics if the executive committee were chosen at random from all the workers?

(b) A university is concerned about the academic standing of its intercollegiate athletes. A study committee chooses an SRS of 50 of the 316 athletes to interview in detail. Suppose that in fact 40% of the athletes have been told by coaches to neglect their studies on at least one occasion. What is the probability that at least 15 in the sample are among this group?

(c) Use what you learned in Chapter 8 to find the probability described in part (a).

9.3 Sample Means

Sample proportions arise most often when we are interested in categorical variables. We then ask questions like "What proportion of U.S. adults have watched *Survivor: Guatemala*?" or "What percent of the adult population attended church last week?" When we record quantitative variables—the income of a household, the lifetime of a car brake pad, the blood pressure of a patient—we are interested in other statistics, such as the median or mean or standard deviation of the variable. Because sample means are just averages of observations, they are among the most common statistics. This section describes the sampling distribution of the mean of the responses in an SRS.

Example 9.9 *Bull market or bear market?*

Sampling distribution of $\overline{x}$

A basic principle of investment is that diversification reduces risk. That is, buying several securities rather than just one reduces the variability of the return on an investment. Figure 9.15 (on the next page) illustrates this principle in the case of common stocks listed on the New York Stock Exchange. Figure 9.15(a) shows the distribution of returns for all 1815 stocks listed on the Exchange for the entire year 1987.[9] This was a year of extreme swings in stock prices, including a record loss of over 20% in a single day. The mean return for all 1815 stocks was $\mu = -3.5\%$, and the distribution shows a very wide spread.

Figure 9.15(b) shows the distribution of returns for all possible portfolios that invested equal amounts in each of 5 stocks. A portfolio is just a sample of 5 stocks, and its return is the average return for the 5 stocks chosen. The mean return for all portfolios is still -3.5%, but the variation among portfolios is much less than the variation among individual stocks. For example, 11% of all individual stocks had a loss of more than 40%, but only 1% of the portfolios had a loss that large.

Figure 9.15 *(a) The distribution of returns for New York Stock Exchange common stocks in 1987. (b) The distributions of returns for portfolios of five stocks in 1987.*

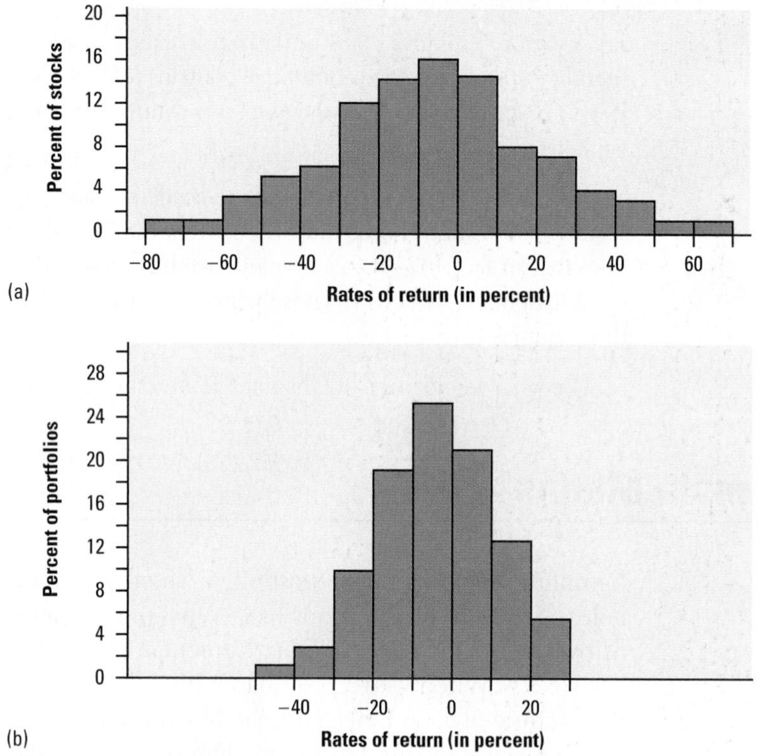

Source: Figure 9.15 is taken with permission from John K. Ford, "A method for grading 1987 stock recommendations," *American Association of Individual Investors Journal,* March 1988, pp. 16–17.

The histograms in Figure 9.15 emphasize a principle that we will make precise in this section:

- Means of random samples are *less variable* than individual observations.

More detailed examination of the distributions would point to a second principle:

- Means of random samples are *more Normal* than individual observations.

These two facts contribute to the popularity of sample means in statistical inference.

The Mean and the Standard Deviation of $\bar{x}$

The sampling distribution of $\bar{x}$ is the distribution of the values of $\bar{x}$ in all possible samples of the same size from the population. Figure 9.15(a) shows the distribution of a population, with mean $\mu = -3.5\%$. Figure 9.15(b) is the sampling distribution of the sample mean $\bar{x}$ from all samples of size $n = 5$ from this population. The mean of all the values of $\bar{x}$ is again -3.5%, but the values of $\bar{x}$ are less spread out than the individual values in the population. This is an example of a general fact.

Mean and Standard Deviation of a Sample Mean

Suppose that $\bar{x}$ is the mean of an SRS of size n drawn from a large population with mean μ and standard deviation σ. Then the **mean** of the sampling distribution of $\bar{x}$ is $\mu_{\bar{x}} = \mu$ and its **standard deviation** is $\sigma_{\bar{x}} = \sigma / \sqrt{n}$.

The behavior of $\bar{x}$ in repeated samples is much like that of the sample proportion $\hat{p}$:

- The sample mean $\bar{x}$ is an unbiased estimator of the population mean μ.
- The values of $\bar{x}$ are less spread out for larger samples. Their standard deviation decreases at the rate $\sqrt{n}$, so you must take a sample four times as large to cut the standard deviation of $\bar{x}$ in half.
- You should use the recipe $\sigma / \sqrt{n}$ for the standard deviation of $\bar{x}$ only when the population is at least 10 times as large as the sample. This is almost always the case in practice.

Notice that *these facts about the mean and standard deviation of $\bar{x}$ are true no matter what the population distribution looks like.*

Example 9.10 *Young women's heights*

Describing the sampling distribution of $\bar{x}$

The height of young women varies approximately according to the N(64.5, 2.5) distribution. This is a population distribution with $\mu = 64.5$ and $\sigma = 2.5$. If we choose one young woman at random, the heights we get in repeated choices follow this distribution. That is, the distribution of the population is also the distribution of one observation chosen at random. So we can think of the population distribution as a distribution of probabilities, just like a sampling distribution.

Now measure the height of an SRS of 10 young women. The sampling distribution of their sample mean height $\bar{x}$ will have mean $\mu_{\bar{x}} = \mu = 64.5$ inches and standard deviation

$$\sigma_{\bar{x}} = \frac{\sigma}{\sqrt{n}} = \frac{2.5}{\sqrt{10}} = 0.79 \text{ inch}$$

The heights of individual women vary widely about the population mean, but the average height of a sample of 10 women is less variable.

In Activity 9A, you plotted the distribution of $\bar{x}$ for samples of size $n = 100$, so the standard deviation of $\bar{x}$ is $\sigma / \sqrt{100} = 2.5/10 = 0.25$. How close did your class come to this number?

We have described the mean and standard deviation of the sampling distribution of a sample mean $\bar{x}$, but not its shape. The shape of the distribution of $\bar{x}$ depends on the shape of the population distribution. In particular, if the population distribution is Normal, then so is the distribution of the sample mean.

Sampling Distribution of a Sample Mean from a Normal Population

Draw an SRS of size n from a population that has the Normal distribution with mean μ and standard deviation σ. Then the sample mean $\bar{x}$ has the Normal distribution with mean μ and standard deviation $\sigma / \sqrt{n}$.

We already knew the mean and standard deviation of the sampling distribution. All that we have added now is the Normal shape. In Activity 9A, we began with a Normal distribution, $N(64.5, 2.5)$. The center (mean) of the approximate sampling distribution of $\bar{x}$ should have been very close to the mean of the population: 64.5 inches. Was it? The spread of the distribution of $\bar{x}$ should have been very close to $\sigma / \sqrt{n}$. Was it? The reason that you don't observe exact agreement is sampling variability.

Example 9.11 *More on young women's heights*
Calculating probabilities involving $\bar{x}$

What is the probability that a randomly selected young woman is taller than 66.5 inches? What is the probability that the mean height of an SRS of 10 young women is greater than 66.5 inches? We can answer both of these questions using Normal calculations.

If we let X be the height of a randomly selected young woman, then the random variable X follows a Normal distribution with $\mu = 64.5$ inches and $\sigma = 2.5$ inches. To find $P(X > 66.5)$, we first standardize the values of X by setting

$$Z = \frac{X - \mu}{\sigma}$$

The random variable Z follows the standard Normal distribution. When $X = 66.5$,

$$Z = \frac{66.5 - 64.5}{2.5} = 0.80$$

From Table A,

$$P(X > 66.5) = P(Z > 0.80) = 1 - 0.7881 = 0.2119$$

The probability of choosing a young woman at random whose height exceeds 66.5 inches is about 0.21.

Now let's take an SRS of 10 young women from this population and compute $\bar{x}$ for the sample. In Example 9.10, we saw that in repeated samples of size $n = 10$, the values of $\bar{x}$ will follow the $N(64.5, 0.79)$ distribution. To find the probability that $\bar{x} > 66.5$ inches, we start by standardizing:

$$z = \frac{\bar{x} - \mu_{\bar{x}}}{\sigma_{\bar{x}}}$$

A sample mean of 66.5 inches yields a z-score of

$$z = \frac{66.5 - 64.5}{0.79} = 2.53$$

Finally,

$$P(\bar{x} > 66.5) = P(Z > 2.53) = 1 - 0.9943 = 0.0057$$

It is very unlikely (less than a 1% chance) that we would draw an SRS of 10 young women whose average height exceeds 66.5 inches.

Figure 9.16 compares the population distribution and the sampling distribution of $\bar{x}$. It also shows the areas corresponding to the probabilities that we just computed.

Figure 9.16 *The sampling distribution of the mean height $\bar{x}$ for samples of 10 young women compared with the distribution of the height of a single woman chosen at random, for Example 9.11.*

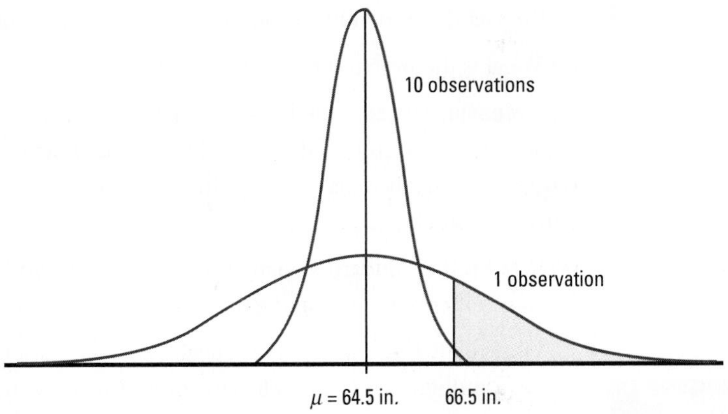

The fact that averages of several observations are less variable than individual observations is important in many settings. For example, it is common practice to repeat a careful measurement several times and report the average of the results. Think of the results of n repeated measurements as an SRS from the population of outcomes we would get if we repeated the measurement forever. The average of the n results (the sample mean $\bar{x}$) is less variable than a single measurement.

Exercises

9.31 Bull market or bear market? Investors remember 1987 as the year stocks lost 20% of their value in a single day. For 1987 as a whole, the mean return of all common stocks on the New York Stock Exchange was $\mu = -3.5\%$. (That is, these stocks lost an average of 3.5% of their value in 1987.) The standard deviation of the returns was about $\sigma = 26\%$. Figure 9.15(a) (page 592) shows the distribution of returns. Figure 9.15(b) is the sampling distribution of the mean returns $\bar{x}$ for all possible samples of 5 stocks.

(a) What are the mean and the standard deviation of the distribution in Figure 9.15(b)?

(b) Assuming that the population distribution of returns on individual common stocks is Normal, what is the probability that a randomly chosen stock showed a return of at least 5% in 1987? Show your work.

(c) Assuming that the population distribution of returns on individual common stocks is Normal, what is the probability that a randomly chosen portfolio of 5 stocks showed a return of at least 5% in 1987? Show your work.

(d) What percent of 5-stock portfolios lost money in 1987? Show your work.

9.32 ACT scores The scores of individual students on the American College Testing (ACT) composite college entrance examination have a Normal distribution with mean 18.6 and standard deviation 5.9.

(a) What is the probability that a single student randomly chosen from all those taking the test scores 21 or higher?

(b) Now take an SRS of 50 students who took the test. What are the mean and standard deviation of the average (sample mean) score for the 50 students? Do your results depend on the fact that individual scores have a Normal distribution?

(c) What is the probability that the mean score $\bar{x}$ of these students is 21 or higher?

9.33 Measurements in the lab Juan makes a measurement in a chemistry laboratory and records the result in his lab report. The standard deviation of students' lab measurements is $\sigma = 10$ milligrams. Juan repeats the measurement 3 times and records the mean $\bar{x}$ of his 3 measurements.

(a) What is the standard deviation of Juan's mean result? (That is, if Juan kept on making 3 measurements and averaging them, what would be the standard deviation of all his $\bar{x}$'s?)

(b) How many times must Juan repeat the measurement to reduce the standard deviation of $\bar{x}$ to 3 milligrams? Explain to someone who knows no statistics the advantage of reporting the average of several measurements rather than the result of a single measurement.

9.34 Lightning strikes The number of lightning strikes on a square kilometer of open ground in a year has mean 6 and standard deviation 2.4. (These values are typical of much of the United States.) The National Lightning Detection Network uses automatic sensors to watch for lightning in a sample of 10 square kilometers.

(a) What are the mean and standard deviation of $\bar{x}$, the mean number of strikes per square kilometer?

(b) Can you calculate the probability tha $\bar{x} < 5$? If so, do it. If not, explain why not.

Activity 9B
Sampling pennies

Materials: For a week or so prior to this experiment, you should collect 25 pennies from current circulation. You should bring to class your 25 pennies, as well as 2 nickels, 2 dimes, 1 quarter, and a small container such as a Styrofoam coffee cup or a margarine tub.

1. This Activity begins by plotting the distribution of ages (in years) of the pennies you have brought to class.[10] Sketch a density curve that you think will capture the shape of the distribution of ages of the pennies.
2. Make a table of years, beginning with the current year and counting backward. Make the second column the age of the penny. For the age, subtract the date on the penny from the current year. Make the third column the frequency, and use tally marks to record the number of pennies of each age. For example, if it is 2007:

Year	Age	Frequency
2007	0	| | |
2006	1	|
2005	2	| | | |
. . .		

3. Put your 25 pennies in a cup, and randomly select 5 pennies. Find the average age of the 5 pennies in your sample, and record the mean age as $\bar{x}$ (5). If you are in a small class (fewer than about 15), you should repeat this step. Replace the pennies in the cup, stir so they are randomly distributed, and then repeat the process.
4. Repeat Step 3, except this time randomly select 10 pennies. Calculate the average age of the sample of 10 pennies, and record this as $\bar{x}$ (10). If your class is small, do this twice to obtain two means.
5. Repeat Step 3 but take all 25 pennies. Record the mean age as $\bar{x}$ (25).
6. Select a flat surface (or clear a space on the floor), and use masking tape to make a number line (axis) with ages marked from 0 to about 30 on the axis. Each interval should be a little more than the width of a penny. You should stack your 25 pennies on the axis according to age. When everyone has done this, look at the shape of the histogram. Are you surprised? How would you explain the shape?
7. Make a second axis on the floor, and label it 0, 0.5, 1, 1.5, etc. This time, use nickels to plot the means for penny samples of size 5. What is the shape of this histogram for the distribution of $\bar{x}$ (5)?
8. Make a third histogram for the means of penny samples of size 10. Use dimes to make this histogram.
9. Finally, use the quarters to make a histogram of the distribution of means for penny samples of size 25. Describe the shape of this histogram. Are you surprised?
10. Write a short description of what you have discovered by doing this Activity.

The Central Limit Theorem

Although many populations have roughly Normal distributions, very few indeed are exactly Normal. What happens to $\bar{x}$ when the population distribution is not Normal? In Activity 9B, the distribution of ages of pennies should have been right-skewed, but as the sample size increased from 1 to 5 to 10 and then to 25, the distribution should have gotten closer and closer to a Normal distribution. This is true no matter what shape the population distribution has, as long as the population has a finite standard deviation σ. This famous fact of probability is called the ***central limit theorem.*** It is much more useful than the fact that the distribution of $\bar{x}$ is exactly Normal if the population is exactly Normal.

Central Limit Theorem

Draw an SRS of size n from any population whatsoever with mean μ and finite standard deviation σ. When n is large, the sampling distribution of the sample mean $\bar{x}$ is close to the Normal distribution $N(\mu, \sigma/\sqrt{n})$ with mean μ and standard deviation $\sigma/\sqrt{n}$.

How large a sample size n is needed for $\bar{x}$ to be close to Normal depends on the population distribution. More observations are required if the shape of the population distribution is far from Normal.

Example 9.12 *A skewed population distribution*

The central limit theorem in action

The *Central Limit Theorem* applet at the book's Web site allows you to watch the central limit theorem in action. Figure 9.17 on the next page presents snapshots from the applet. Figure 9.17(a) shows the density curve of a single observation, that is, of the population. The distribution is strongly right-skewed, and the most probable outcomes are near 0. The mean μ of this distribution is 1, and its standard deviation σ is also 1. This particular distribution is called an *exponential distribution.* Exponential distributions are used as models for the service lifetime of electronic components and for the time required to serve a customer or repair a machine.

Figures 9.17(b), (c), and (d) are the density curves of the sample means of 2, 10, and 25 observations from this population. As n increases, the shape becomes more Normal. The mean remains at $\mu = 1$, and the standard deviation decreases, taking the value $1/\sqrt{n}$. The density curve for 10 observations is still somewhat skewed to the right but already resembles a Normal curve having $\mu = 1$ and $\sigma = 1/\sqrt{10} = 0.32$. The density curve for $n = 25$ is even more Normal. The contrast between the shapes of the population distribution and the distribution of the mean of 10 or 25 observations is striking.

Figure 9.17 *The central limit theorem in action: the distribution of sample means $\bar{x}$ from a strongly non-Normal population becomes more Normal as the sample size increases. (a) The distribution of 1 observation. (b) The distribution of $\bar{x}$ for 2 observations. (c) The distribution of $\bar{x}$ for 10 observations. (d) The distribution of $\bar{x}$ for 25 observations.*

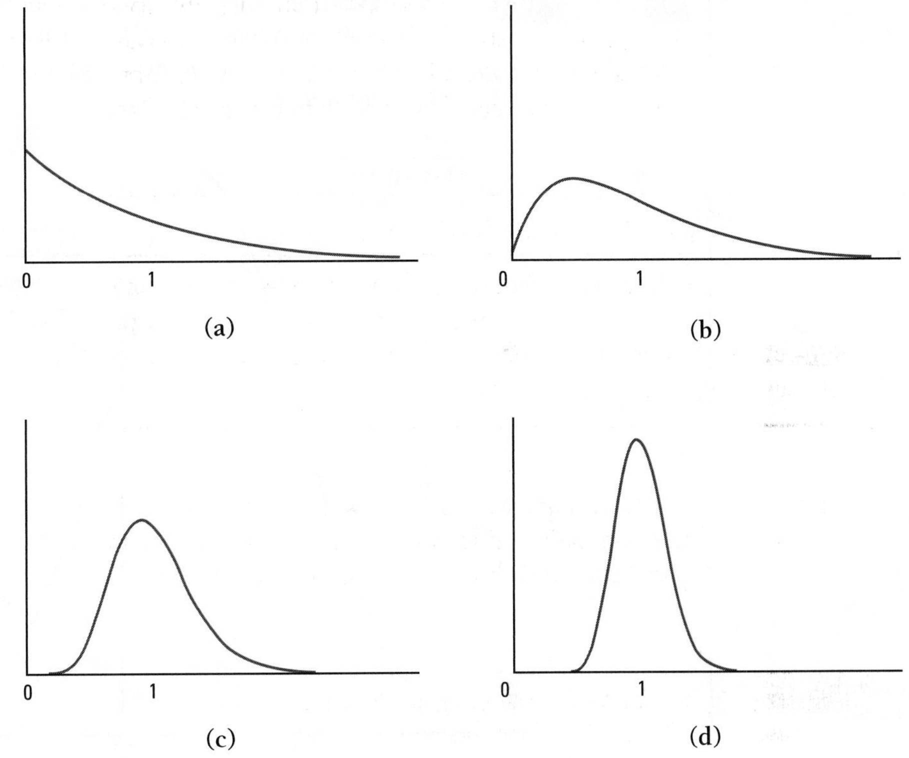

Let's use Normal calculations based on the central limit theorem to answer a question about the very non-Normal distribution in Figure 9.17(a).

Example 9.13 *Servicing air conditioners*

Calculations using the central limit theorem

The time that a technician requires to perform preventive maintenance on an air-conditioning unit is governed by the exponential distribution whose density curve appears in Figure 9.17(a). The mean time is $\mu = 1$ hour and the standard deviation is $\sigma = 1$ hour. Your company has a contract to maintain 70 of these units in an apartment building. You must schedule technicians' time for a visit to this building. Is it safe to budget an average of 1.1 hours for each unit? Or should you budget an average of 1.25 hours?

The central limit theorem says that the sample mean time $\bar{x}$ (in hours) spent working on 70 units has approximately the Normal distribution with mean equal to the population mean $\mu = 1$ hour and standard deviation

$$\frac{\sigma}{\sqrt{70}} = \frac{1}{\sqrt{70}} = 0.120 \text{ hour}$$

The distribution of $\bar{x}$ is therefore approximately $N(1, 0.120)$. This Normal curve is the solid curve in Figure 9.18.

Figure 9.18 *The exact distribution (dotted) and the Normal approximation from the central limit theorem (solid) for the average time needed to maintain an air conditioner, for Example 9.13. The probability we want is the area to the right of 1.1.*

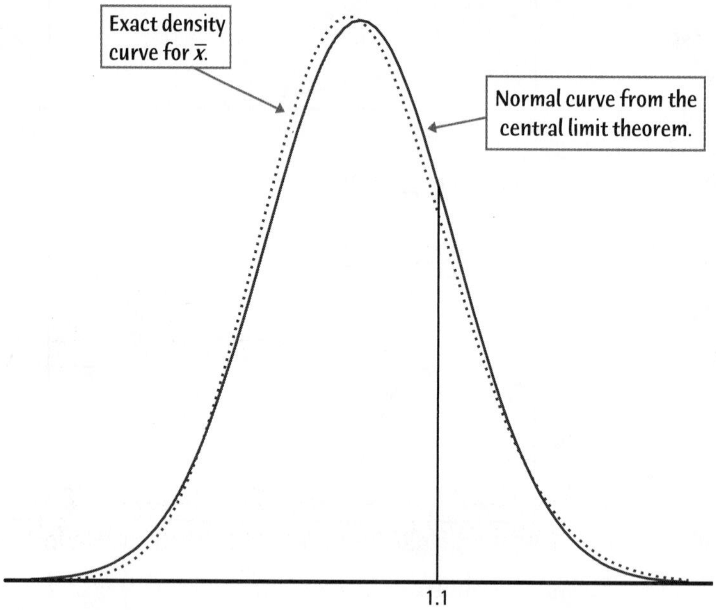

To determine whether it is safe to budget 1.1 hours, on average, the probability we want is $P(\bar{x} > 1.1)$. Since

$$Z = \frac{\bar{x} - \mu_{\bar{x}}}{\sigma_{\bar{x}}} = \frac{1.1 - 1}{0.120} = 0.83$$

the probability that the work will exceed the time allotted is

$$P(\bar{x} > 1.1) = P(Z > 0.83) = 1 - 0.7967 = 0.2033$$

This is the area to the right of 1.1 under the solid Normal curve in Figure 9.18. The exactly correct probability is the area under the dotted density curve in the figure. It is 0.1977. The central limit theorem Normal approximation is off by only about 0.006.

If you budget an average of 1.25 hours, a similar calculation yields

$$P(\bar{x} > 1.25 \text{ hours}) = 0.0182$$

What can we conclude? If you budget 1.1 hours per unit, there is a 20% chance that the technicians will not complete the work in the building within the budgeted time. This chance drops to 2% if you budget 1.25 hours. You therefore budget 1.25 hours per unit.

Figure 9.19 summarizes the facts about the sampling distribution of $\bar{x}$. It reminds us of the big idea of a sampling distribution. Keep taking random samples of size n from a population with mean μ. Find the sample mean $\bar{x}$ for each sample. Collect all the $\bar{x}$'s and display their distribution. That's the sampling distribution of $\bar{x}$. Sampling distributions are the key to understanding statistical inference. Keep this figure in mind for future reference.

Figure 9.19 *The sampling distribution of a sample mean $\bar{x}$ has mean μ and standard deviation $\sigma/\sqrt{n}$. The distribution is Normal if the population distribution is Normal; it is approximately Normal for large samples in any case.*

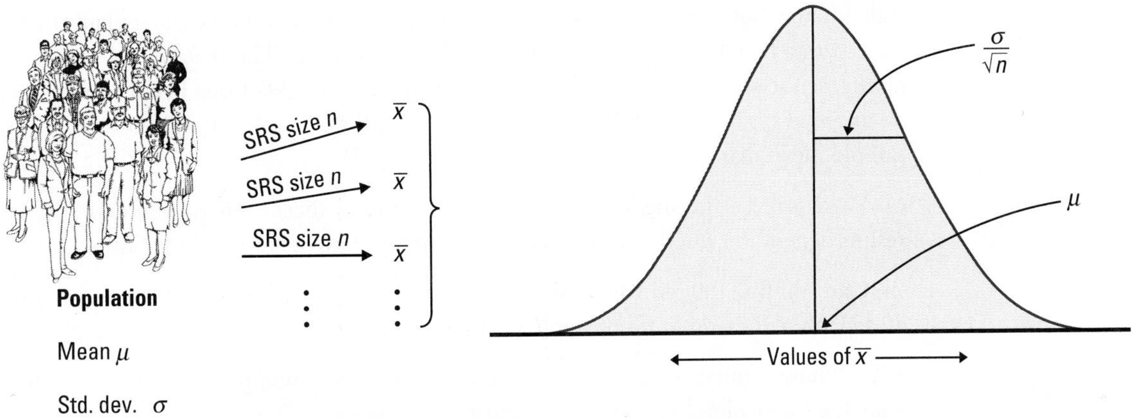

Exercises

9.35 Bad carpet The number of flaws per square yard in a type of carpet material varies with mean 1.6 flaws per square yard and standard deviation 1.2 flaws per square yard. The population distribution cannot be Normal, because a count takes only whole-number values. An inspector studies 200 square yards of the material, records the number of flaws found in each square yard, and calculates $\bar{x}$, the mean number of flaws per square yard inspected. Use the central limit theorem to find the approximate probability that the mean number of flaws exceeds 2 per square yard. Show your work.

9.36 Investing for retirement The distribution of annual returns on common stocks is roughly symmetric, but extreme observations are more frequent than in a Normal distribution. Because the distribution is not strongly non-Normal, the mean return over even a moderate number of years is close to Normal. From 1950 to 2004, annual returns (not adjusted for inflation) on common stocks have had mean 13.2% and standard deviation 17.5%. Andrew plans to retire in 40 years and is considering investing in stocks. What is the probability (assuming that the past pattern of variation continues) that the mean annual return on common stocks over the next 40 years will exceed 15%? What is the probability that the mean return will be less than 10%?

9.37 Airline passengers get heavier In response to the increasing weight of airline passengers, the Federal Aviation Administration (FAA) in 2003 told airlines to assume that

passengers average 190 pounds in the summer, including clothes and carry-on baggage. But passengers vary, and the FAA did not specify a standard deviation. A reasonable standard deviation is 35 pounds. Weights are not Normally distributed, especially when the population includes both men and women, but they are not very non-Normal. A commuter plane carries 20 passengers.

(a) Can you calculate the probability that a randomly selected passenger weighs more than 200 pounds? If so, do it. If not, explain why not.

(b) Can you calculate the probability that the total weight of the passengers on the flight exceeds 4000 pounds? (*Hint:* Restate the problem in terms of the mean weight.) If so, do it. If not, explain why not.

9.38 Making auto parts An automatic grinding machine in an auto parts plant prepares axles with a target diameter $\mu = 40.125$ millimeters (mm). The machine has some variability, so the standard deviation of the diameters is $\sigma = 0.002$ mm. The machine operator inspects a sample of 4 axles each hour for quality control purposes and records the sample mean diameter.

(a) What will be the mean and standard deviation of the numbers recorded? Do your results depend on whether or not the axle diameters have a Normal distribution?

(b) Can you find the probability that an SRS of 4 axles has a mean diameter greater than 40.127 mm? If so, do it. If not, explain why not.

9.39 Glucose testing, I Sheila's doctor is concerned that she may suffer from gestational diabetes (high blood glucose levels during pregnancy). There is variation both in the actual glucose level and in the test that measures the level. A patient is classified as having gestational diabetes if the glucose level is above 140 milligrams per deciliter (mg/dl) one hour after a sugary drink. Sheila's measured glucose level one hour after the sugary drink varies according to the Normal distribution with $\mu = 125$ mg/dl and $\sigma = 10$ mg/dl.

(a) If a single glucose measurement is made, what is the probability that Sheila is diagnosed as having gestational diabetes? Show your work.

(b) If measurements are made on 4 separate days and the mean result is compared with the criterion 140 mg/dl, what is the probability that Sheila is diagnosed as having gestational diabetes?

(c) Did you use the central limit theorem for either part (a) or part (b)? Explain.

9.40 Glucose testing, II Refer to the previous exercise. What is the level L such that there is probability only 0.05 that the mean glucose level of 4 test results on Sheila falls above L? (*Hint:* This requires a backward Normal calculation. See Chapter 2 if you need to review.)

Section 9.3 Summary

When we want information about the **population mean** μ for some variable, we often take an SRS and use the **sample mean** $\bar{x}$ to estimate the unknown parameter μ.

The **sampling distribution of** $\bar{x}$ describes how the statistic $\bar{x}$ varies in *all* possible samples from the population.

- The **mean** of the sampling distribution is μ, so that $\bar{x}$ is an unbiased estimator of μ.
- The **standard deviation** of the sampling distribution of $\bar{x}$ is $\sigma/\sqrt{n}$ for an SRS of size n if the population has standard deviation σ. This formula can be used if the population is at least 10 times as large as the sample.
- If the population has a Normal distribution, so does $\bar{x}$.
- The **central limit theorem** states that for large n the sampling distribution of $\bar{x}$ is approximately Normal for any population with finite standard deviation σ. The mean and standard deviation of the Normal distribution are the mean μ and standard deviation $\sigma/\sqrt{n}$ of $\bar{x}$ itself.

Section 9.3 Exercises

9.41 Bottling cola A bottling company uses a filling machine to fill plastic bottles with cola. The bottles are supposed to contain 300 milliliters (ml). In fact, the contents vary according to a Normal distribution with mean $\mu = 298$ ml and standard deviation $\sigma = 3$ ml.

(a) What is the probability that an individual bottle contains less than 295 ml? Show your work.

(b) What is the probability that the mean contents of the bottles in a six-pack is less than 295 ml? Show your work.

9.42 Stop the car! A company that owns and services a fleet of cars for its sales force has found that the service lifetime of disc brake pads varies from car to car according to a Normal distribution with mean $\mu = 55{,}000$ miles and standard deviation $\sigma = 4500$ miles. The company installs a new brand of brake pads on 8 cars.

(a) If the new brand has the same lifetime distribution as the previous type, what is the distribution of the sample mean lifetime for the 8 cars?

(b) The average life of the pads on these 8 cars turns out to be $\bar{x} = 51{,}800$ miles. What is the probability that the sample mean lifetime is 51,800 miles or less if the lifetime distribution is unchanged? (The company takes this probability as evidence that the average lifetime of the new brand of pads is less than 55,000 miles.)

9.43 What a wreck! The number of traffic accidents per week at an intersection varies with mean 2.2 and standard deviation 1.4. The number of accidents in a week must be a whole number, so the population distribution is not Normal.

(a) Let $\bar{x}$ be the mean number of accidents per week at the intersection during a year (52 weeks). What is the approximate distribution of $\bar{x}$ according to the central limit theorem?

(b) What is the approximate probability that $\bar{x}$ is less than 2? Show your work.

(c) What is the approximate probability that there are fewer than 100 accidents at the intersection in a year? (*Hint*: Restate this event in terms of $\bar{x}$.)

9.44 Testing kindergarten children Children in kindergarten are sometimes given the Ravin Progressive Matrices Test (RPMT) to assess their readiness for learning. Experience at Southwark Elementary School suggests that the RPMT scores for its kindergarten pupils have mean 13.6 and standard deviation 3.1. The distribution is close to Normal. Mr. Lewis has 22 children in his kindergarten class this year. He suspects that their RPMT scores will be unusually low because the test was interrupted by a fire drill. To check this suspicion, he wants to find the level L such that there is probability only 0.05 that the mean score of 22 children falls below L when the usual Southwark distribution remains true. What is the value of L? (*Hint:* This requires a backward Normal calculation. See Chapter 2 if you need to review.)

9.45 Insurance The idea of insurance is that we all face risks that are unlikely but carry high cost. Think of a fire destroying your home. So we form a group to share the risk: we all pay a small amount, and the insurance policy pays a large amount to those few of us whose homes burn down. An insurance company looks at the records for millions of homeowners and sees that the mean loss from fire in a year is $\mu = \$250$ per person. (Most of us have no loss, but a few lose their homes. The \$250 is the average loss.) The company plans to sell fire insurance for \$250 plus enough to cover its costs and profit. Explain clearly why it would be a poor practice to sell only 12 policies. Then explain why selling thousands of such policies is a safe business practice.

9.46 More on insurance The insurance company in the previous exercise sees that in the entire population of homeowners, the mean loss from fire is $\mu = \$250$ and the standard deviation of the loss is $\sigma = \$300$. The distribution of losses is strongly right-skewed: many policies have \$0 loss, but a few have large losses. If the company sells 10,000 policies, what is the approximate probability that the average loss will be greater than \$260? Show your method.

CASE CLOSED!

Building better batteries

Return to the Case Study on page 561. You should now be ready to answer some questions about the battery production process described there.

1. Assuming the process is working properly ($\mu = 17$ and $\sigma = 0.8$), what are the shape, center, and spread of the distribution of sample means for random samples of 30 batteries? Justify each of your answers.
2. For the random sample of 30 batteries, $\bar{x} = 16.70$ hours. Calculate the probability of obtaining a sample with a mean lifetime of 16.70 hours or less if the production process is working properly. Show your work.
3. Based on your answer to Question 2, do you believe that the process is working properly? Justify your answer.

The plant manager also wants to know what proportion of all the batteries produced that day lasted less than 16 hours, which he has declared "unsuitable." From past experience, about 10% of batteries made at the plant are unsuitable. If he can feel relatively certain that the proportion of unsuitable batteries p produced that day is less than 0.10, he might take the chance of shipping the whole batch of batteries to customers.

4. Assuming that the actual proportion of unsuitable batteries produced that day is 0.10, what are the shape, center, and spread of the distribution of sample proportions for random samples of 480 batteries? Justify each of your answers.
5. On this particular day, a total of 480 batteries were tested during 16 hours of production. Forty of these batteries were unsuitable. Calculate the probability of obtaining a sample in which such a small proportion of batteries is unsuitable if $p = 0.10$. Show your work.
6. Based on your answer to Question 5, what advice would you give the plant manager? Why?

Chapter Review

Summary

This chapter lays the foundations for the study of statistical inference. Statistical inference uses data to draw conclusions about the population from which the data come. What is special about inference is that the conclusions include a statement, in the language of probability, about how reliable they are. The statement gives a probability that answers the question "What would happen if I used this method very many times?"

This chapter introduced sampling distributions of statistics. A sampling distribution describes the values a statistic would take in very many repetitions of a sample or an experiment under the same conditions. Understanding that idea is the key to understanding statistical inference. The chapter gave details about the sampling distributions of two important statistics: a sample proportion $\hat{p}$ and a sample mean $\bar{x}$. These statistics behave in much the same way. In particular, their sampling distributions are approximately Normal if the sample is large. This is a main reason why Normal distributions are so important in statistics. We can use everything we know about Normal distributions to study the sampling distributions of proportions and means.

What You Should Have Learned

Here is a checklist of the major skills you should have acquired by studying this chapter.

A. Sampling Distributions

1. Identify parameters and statistics in a sample or experiment.
2. Recognize the fact of sampling variability: a statistic will take different values when you repeat a sample or experiment.
3. Interpret a sampling distribution as describing the values taken by a statistic in all possible repetitions of a sample or experiment under the same conditions.
4. Describe the bias and variability of a statistic in terms of the mean and spread of its sampling distribution.
5. Understand that the variability of a statistic is controlled by the size of the sample. Statistics from larger samples are less variable.

B. Sample Proportions

1. Recognize when a problem involves a sample proportion $\hat{p}$.
2. Find the mean and standard deviation of the sampling distribution of a sample proportion $\hat{p}$ for an SRS of size n from a population having population proportion p.
3. Know that the standard deviation (spread) of the sampling distribution of $\hat{p}$ gets smaller at the rate $\sqrt{n}$ as the sample size n gets larger.
4. Recognize when you can use the Normal approximation to the sampling distribution of $\hat{p}$. Use the Normal approximation to calculate probabilities that concern $\hat{p}$.

C. Sample Means

1. Recognize when a problem involves the mean $\bar{x}$ of a sample.
2. Find the mean and standard deviation of the sampling distribution of a sample mean $\bar{x}$ from an SRS of size n when the mean μ and standard deviation σ of the population are known.
3. Know that the standard deviation (spread) of the sampling distribution of $\bar{x}$ gets smaller at the rate $\sqrt{n}$ as the sample size n gets larger.
4. Understand that $\bar{x}$ has approximately a Normal distribution when the sample is large (central limit theorem). Use this Normal distribution to calculate probabilities involving $\bar{x}$.

Web Links

You can explore sampling distributions of sample means and sample proportions using Java applets at these two Web sites:

- Sampling distributions applet, www.ruf.rice.edu/~ane/stat_sim/sampling_dist/
- Reese's Pieces and sampling pennies applets, www.rossmanchance.com/applets/

Chapter Review Exercises

9.47 Republican voters Voter registration records show that **68%** of all voters in Indianapolis are registered as Republicans. To test whether the numbers dialed by a random digit dialing device really are random, you use the device to call 150 randomly chosen residential telephones in Indianapolis. Of the registered voters contacted, **73%** are registered Republicans.

(a) Is each of the boldface numbers a parameter or a statistic? Give the appropriate notation for each.

(b) What are the mean and the standard deviation of the sample proportion of registered Republicans in samples of size 150 from Indianapolis?

(c) Find the probability of obtaining an SRS of size 150 from the population of Indianapolis voters in which 73% or more are registered Republicans. How well is your random digit device working?

9.48 Baggage check! In Example 9.3 (page 565), we performed a simulation to determine what proportion of a sample of 100 travelers would get the "green light" in Customs at Guadalajara airport. Suppose the Customs agents say that the probability that the light shows green is 0.7 on each push of the button. You observe 100 passengers at the Customs "stoplight." Only 65 get a green light. Does this give you reason to doubt the Customs officials?

(a) Use your calculator to simulate 50 groups of 100 passengers activating the Customs light. Generate L_1/list1 with the command `randBin(100,0.7,50)/100`. L_1/list1 will contain 50 values of $\hat{p}$, the proportion of the 100 passengers who got a green light.

(b) Sort L_1/list1 in descending order. In how many of the 50 simulations did you obtain a value of $\hat{p}$ that is less than or equal to 0.65? Do you believe the Customs agents?

(c) Describe the shape, center, and spread of the sampling distribution of $\hat{p}$ for samples of $n = 100$ passengers.

(d) Use the sampling distribution from part (c) to find the probability of getting a sample proportion of 0.65 or less if $p = 0.7$ is actually true. How does this compare with the results of your simulation in part (b)?

(e) Repeat parts (c) and (d) for samples of size $n = 1000$ passengers.

9.49 IQ tests The Wechsler Adult Intelligence Scale (WAIS) is a common "IQ test" for adults. The distribution of WAIS scores for persons over 16 years of age is approximately Normal with mean 100 and standard deviation 15.

(a) What is the probability that a randomly chosen individual has a WAIS score of 105 or higher? Show your work.

(b) What are the mean and standard deviation of the sampling distribution of the average WAIS score $\bar{x}$ for an SRS of 60 people?

(c) What is the probability that the average WAIS score of an SRS of 60 people is 105 or higher? Show your work.

(d) Would your answers to any of (a), (b), or (c) be affected if the distribution of WAIS scores in the adult population were distinctly non-Normal?

9.50 Polling women Suppose that 47% of all adult women think they do not get enough time for themselves. An opinion poll interviews 1025 randomly chosen women and records the sample proportion who feel they don't get enough time for themselves.

(a) Describe the sampling distribution of $\hat{p}$.

(b) The truth about the population is $p = 0.47$. In what range will the middle 95% of all sample results fall?

(c) What is the probability that the poll gets a sample in which fewer than 45% say they do not get enough time for themselves? Show your method.

9.51 Detecting gypsy moths The gypsy moth is a serious threat to oak and aspen trees. A state agriculture department places traps throughout the state to detect the moths. When traps are checked periodically, the mean number of moths trapped is only 0.5, but some traps have several moths. The distribution of moth counts is discrete and strongly skewed, with standard deviation 0.7.

(a) What are the mean and standard deviation of the average number of moths $\bar{x}$ in 50 traps?

(b) Use the central limit theorem to find the probability that the average number of moths in 50 traps is greater than 0.6.

9.52 High school dropouts High school dropouts make up 20.2% of all Americans aged 18 to 24. A vocational school that wants to attract dropouts mails an advertising flyer to 25,000 persons between the ages of 18 and 24.

(a) If the mailing list can be considered a random sample of the population, what is the mean number of high school dropouts who will receive the flyer?

(b) What is the probability that at least 5000 dropouts will receive the flyer? Show your method.

9.53 Online photos Suppose (as is roughly true) that 20% of all Internet users have posted photos online. A sample survey interviews an SRS of 1555 Internet users.

(a) Describe the shape, center, and spread of the distribution of the proportion in the sample who have posted photos online.

(b) What is the probability that 300 or fewer people in the sample have posted photos online? Use the distribution in (a) to help answer this question.

9.54 Pollutants in auto exhausts, I The level of nitrogen oxides (NOX) in the exhaust of cars of a particular model varies Normally with mean 0.2 grams per mile (g/mi) and standard deviation 0.05 g/mi. Governments regulations call for NOX emissions to be no higher than 0.3 g/mi.

(a) What is the probability that a single car of this model fails to meet the NOX requirement? Show your work.

(b) A company has 25 cars of this model in its fleet. What is the probability that the average NOX level $\bar{x}$ of these cars is above the 0.3 g/mi limit?

9.55 Pollutants in auto exhausts, II Refer to the previous exercise. What is the level L such that the probability that the average NOX level $\bar{x}$ for the fleet is greater than L is only 0.01? Show your method clearly.

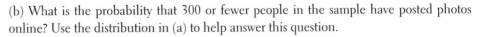

9.56 How many people in a car? A study of rush-hour traffic in San Francisco counts the number of people in each car entering a freeway at a suburban interchange. Suppose that this count has mean 1.5 and standard deviation 0.75 in the population of all cars that enter at this interchange during rush hours.

(a) Could the exact distribution of the count be Normal? Why or why not?

(b) Traffic engineers estimate that the capacity of the interchange is 700 cars per hour. According to the central limit theorem, what is the approximate distribution of the mean number of persons $\bar{x}$ in 700 randomly selected cars at this interchange?

(c) What is the probability that 700 cars will carry more than 1075 people? (*Hint:* Restate this event in terms of the mean number of people $\bar{x}$ per car.) Show your work.

9.57 Student drinking The Harvard College Alcohol Study interviewed an SRS of 14,941 college students about their drinking habits. Suppose that half of all college students "drink to get drunk" at least once in a while. That is, $p = 0.5$.

(a) What are the mean and the standard deviation of the proportion $\hat{p}$ of the sample who drink to get drunk?

(b) Use the Normal approximation to find the probability that $\hat{p}$ is between 0.49 and 0.51.

9.58 Low-birth-weight babies A newborn baby has extremely low birth weight (ELBW) if it weighs less than 1000 grams. A study of the health of such children in later years examined a random sample of 219 children. Their mean weight at birth was $\bar{x} = 810$ grams.

(a) This sample mean is an unbiased estimator of the mean weight μ in the population of all ELBW babies. Explain in simple terms what this means.

(b) Do you think the population distribution of birth weights among ELBW babies is roughly Normal? Justify your answer.

(c) Do you think the distribution of mean birth weights $\bar{x}$ in samples of 219 ELBW babies is roughly Normal? Justify your answer.

PART III

Review Exercises

Communicate your thinking clearly in Exercises III.1 to III.10.

III.1 Spelling errors in a text are either "nonword errors" or "word errors." A nonword error produces a string of letters that is not a word, as when "the" is typed as "teh." Word errors produce the wrong word, as when "loose" is typed as "lose." Nonword errors make up 25% of all errors. A human proofreader will catch 90% of nonword errors and 70% of word errors.

(a) What percent of all errors will the proofreader catch? Show your work.

(b) Of all errors that the proofreader catches, what percent are nonword errors? Show your work.

(c) You ask a fellow student to proofread an essay in which you have deliberately made 10 word errors. Missing 3 or more out of 10 errors seems a poor performance. What is the probability that your fellow student misses 3 or more of the 10 word errors that you made?

III.2 In the language of government statistics, you are "in the labor force" if you are available for work and either working or actively seeking work. The unemployment rate is the proportion of the labor force (not of the entire population) who are unemployed. Here are data from the Current Population Survey for the civilian population aged 25 years and over in 2004. The table entries are counts in thousands of people.

Highest education	Total pop.	In labor force	Employed
Didn't finish high school	27,669	12,470	11,408
High school but no college	59,860	37,834	35,857
Less than bachelor's degree	47,556	34,439	32,977
College graduate	51,582	40,390	39,293

(a) Find the unemployment rate for people with each level of education. How does the unemployment rate change with education?

(b) What is the probability that a randomly chosen person 25 years of age or older is in the labor force? Show your work.

(c) If you know that a randomly chosen person 25 years of age or older is a college graduate, what is the probability that he or she is in the labor force? Show your work.

(c) Are the events "in the labor force" and "college graduate" independent? Justify your answer.

III.3 Choose a young person (aged 19 to 25) at random and ask, "In the past seven days, how many days did you watch television?" Call the response X for short. Here is a probability model for the response:[1]

Days X	0	1	2	3	4	5	6	7
Probability	0.04	0.03	0.06	0.08	0.09	0.08	0.05	???

(a) What is the probability that $X = 7$? Justify your answer.

(b) Calculate the mean of the random variable X. Interpret this value in context.

(c) Suppose that you asked 100 randomly selected young people (aged 19 to 25) to respond to the question and found that the mean $\bar{x}$ of their responses was 4.96. Would this result surprise you? Justify your answer.

III.4 Does delaying oral practice hinder learning a foreign language? Researchers randomly assigned 23 beginning students of Russian to begin speaking practice immediately and another 23 to delay speaking for 4 weeks. At the end of the semester, both groups took a test of comprehension of spoken Russian. Suppose that in the population of all beginning students, the test scores for early speaking vary according to a Normal distribution with mean 32 and standard deviation 6. For the delayed speakers, the scores follow a Normal distribution with mean 29 and standard deviation 5.

(a) Determine the probability that the mean score $\bar{x}$ for the delayed speakers in this experiment would exceed 32. Show your work.

(b) Explain the purpose of the random assignment in the context of this experiment.

(c) Find the probability that the experiment will show (misleadingly) that the mean score for delayed speaking is at least as large as that for early speaking. Show your work.

III.5 According to the Census Bureau, 12.1% of American adults (aged 18 and over) are Hispanics. An opinion poll plans to contact an SRS of 1200 adults.

(a) Calculate the mean and standard deviation of the number of Hispanics in such samples. Show your work.

(b) According to the 68–95–99.7 rule, what range will include the counts of Hispanics in 95% of all such samples?

(c) How large a sample is required to make the mean number of Hispanics at least 200?

III.6 The Census Bureau says that the 10 most common names in the United States are (in order) Smith, Johnson, Williams, Jones, Brown, Davis, Miller, Wilson, Moore, and Taylor. These names account for 5.6% of all U.S. residents. Out of curiosity, you look at the authors of the textbooks for your current courses. There are 9 authors in all. Would you be surprised if none of the names of these authors were among the 10 most common? Give a probability to support your answer and explain the reasoning behind your calculation.

III.7 Below are histograms of the values taken by three sample statistics in several hundred samples from the same population. The true value of the population parameter is marked with an arrow on each histogram.

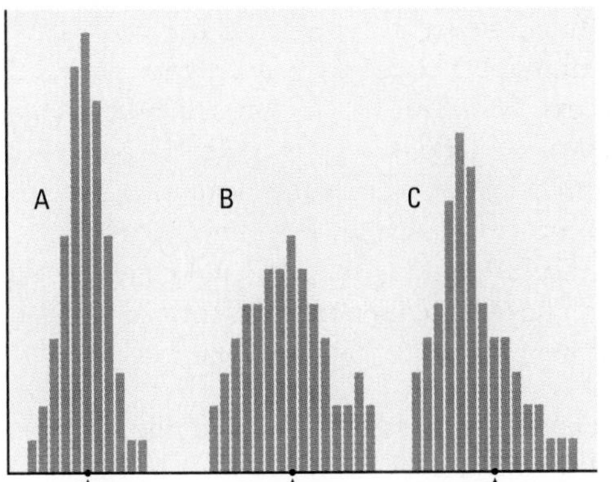

Which statistic would provide the best estimate of the parameter? Be sure to address both variability and bias in your explanation.

III.8 Suppose (as is roughly true) that 88% of college men and 82% of college women were employed last summer. A sample survey interviews independent SRSs of 500 college men and 500 college women.

(a) Find the probability that fewer than 85% of the men in the sample were employed last summer. Show your work.

(b) Explain the purpose of random sampling in the design of this survey.

(c) Determine the probability that a higher proportion of women than men in the sample worked last summer. Show your work.

III.9 Of the 16,701 degrees in mathematics given by U.S. colleges and universities in the most recent year, 73% were bachelor's degrees, 21% were master's degrees, and the rest were doctorates. Moreover, women earned 48% of the bachelor's degrees, 42% of the master's degrees, and 29% of the doctorates.[2]

(a) How many of the mathematics degrees given in this year were earned by women? Justify your answer.

(b) Are the events "degree earned by a woman" and "degree was a master's degree" independent? Justify your answer using appropriate probabilities.

(c) If you choose 2 of the 16,701 mathematics degrees at random, what is the probability that at least 1 of the 2 degrees was earned by a woman? Show your work.

III.10 According to the National Campaign for Hearing Health, deafness is the most common birth defect. Audiologists estimate that 3 out of every 1000 babies are born with some kind of hearing loss. Early identification of hearing loss through a screening test shortly after birth and subsequent intervention can have a significant impact on a child's cognitive development, as well as financial consequences. Thus, universal newborn hearing screening (UNHS) calls for high-performing, accurate, and inexpensive testing equipment.

Studies conducted at the University of Utah (2000) were undertaken to assess the adequacy of a new miniature screening device called the Handtronix-OtoScreener.[3] The accuracy of this new device was judged using as "truth" the results found using standard equipment. Results are shown in the following table:

		Baby's Hearing	
		Loss	**Normal**
Test	Loss	54	6
Result	Normal	4	36

(a) Sensitivity and specificity are measures of the accuracy of a diagnostic procedure. For the Handtronix-OtoScreener, sensitivity is defined as the probability that the test shows a hearing loss when there really is a hearing loss. Specificity is defined as the probability that the test shows that hearing is normal when hearing is really normal. Estimate the "sensitivity" and the "specificity" of this new small audiometer using the data in the table. Show your work.

(b) It is estimated that the total number of births in the United States each year is about 4,008,083, and the prevalence of hearing loss is 3 per 1000. Based on these figures, how many babies each year can be expected to have a hearing loss? Suppose all these babies are screened for a hearing loss. Based on the table above, estimate the number of babies with a hearing loss that will be missed by this new screening device.

(c) Still assume that the prevalence of hearing loss in infants is 3 per 1000, that is, 0.003. Suppose we choose a baby randomly among all newborns and have the hearing tested using the Handtronix-OtoScreener. If the test shows that the baby has a hearing loss, what is the estimated probability that the baby really has a hearing loss? Show your work.

PART

IV

Inference: Conclusions with Confidence

CHAPTER

10

Estimating with Confidence

CASE STUDY

Need help? Give us a call!

If your cable television goes out, you phone the cable company to get it fixed. Does a real person answer your call? These days, probably not. It is far more likely that you will get an automated response. You will probably be offered several options, such as: to order cable service, press 1; for questions about your bill, press 2; to add new channels, press 3; (and finally) to speak with a customer service agent, press 4. Callers will get frustrated if they have to wait too long before speaking to a live voice. So companies try hard to minimize the time required to connect to a customer service representative.

A large bank decided to study the call response times in its customer service department. The bank's goal was to have a representative answer an incoming call within 30 seconds. Here are the results from a random sample of 300 calls to the bank's customer service center. Table 10.1 shows the time from the first ring until a customer service agent answers. By the end of this chapter, you will learn how to use these sample data to provide estimates of how well the bank is doing in meeting its call response goal.

Table 10.1 *Call center response times (seconds)*

59	13	2	24	11	18	38	12	46	17	77	12	46
44	4	74	41	22	25	7	10	46	78	14	6	9
122	8	16	15	17	17	9	15	24	70	9	9	10
32	9	68	8	10	41	13	17	50	12	82	97	33
76	56	42	19	14	21	12	44	63	5	21	11	47
8	12	4	111	37	12	24	43	37	27	65	32	3
9	26	5	10	30	27	21	14	19	44	49	10	24
11	10	22	43	70	27	10	32	96	11	29	7	28
22	17	9	24	15	14	34	5	38	29	16	65	6
5	58	17	7	44	14	16	4	46	32	52	75	11
11	17	31	8	36	25	14	85	4	46	23	58	5
54	28	6	46	4	28	11	111	6	3	83	27	6
83	27	2	56	26	21	276	14	30	8	7	12	4
29	21	23	4	14	23	22	19	66	51	60	14	111
20	7	7	87	22	11	53	20	14	41	30	7	10
11	9	9	101	55	18	20	77	14	13	11	22	15
2	14	20	83	25	10	34	23	21	5	14	22	10
68	8	70	56	8	26	7	15	7	9	144	11	109
20	4	16	20	124	16	16	47	97	27	61	35	18
22	244	19	10	6	43	20	77	22	7	33	67	20
4	28	5	7	118	18	1	35	78	35	71	85	24
333	50	11	12	11	13	19	16	91	4	63	14	22
43	25	18	55	13	11	6	13	4	3	17	11	6
17												

Source: Haipeng Shen, "Nonparametric regression for problems involving lognormal distributions," PhD thesis, University of Pennsylvania, 2003. Thanks to Haipeng Shen and Larry Brown for sharing the data.

Activity 10A

Read any good books lately?

Materials: Book provided by your teacher, random digits table (Table B) or graphing calculator

What makes a book easier or more difficult to read? An interesting story line, intriguing characters, and an appealing writing style certainly help. But from an educator's standpoint, the readability of a book or article depends mainly on word and sentence length. In this Activity, you will learn some methods for evaluating readability. Work with one or two partners, as directed by your teacher.

1. The first step in most standard readability tests is the selection of a random sample of words from the book or article. Use a random digits table (Table B) or your graphing calculator to select a page at random from the book that your teacher has provided. Check to see that the chosen page contains at least 100 words. If not, randomly select another page. Your sample will consist of the first 100 words on the chosen page.

2. For each of the 100 words in your sample, count the number of letters. Record the data.

3. Make a graph that displays the frequency of the *number of letters per word* for your sample. Describe what you see in a few sentences.

4. Calculate the mean $\bar{x}$ and standard deviation s of the number of letters per word. Now construct an interval as follows:

$$\bar{x} \pm 2\frac{s}{\sqrt{n}}$$

   This interval, based on the sample, gives a range of plausible values for the population mean μ, the average number of letters per word in the book. In this chapter, you will learn why this is called a *confidence interval.*

5. Start with the first complete sentence on the page that you randomly selected. Count the number of words in the sentence. Repeat this process for the next 29 sentences, so that you have a total of 30 sentence lengths. (*Note:* You may have to continue to the next page before you have counted 30 sentences.)

6. Repeat Steps 3 and 4 for the *number of words per sentence* in this sample.
7. Compare your results with those of your classmates. Which book do you think would be most difficult to read? Easiest to read? Explain.

Introduction

How long can you expect a AA battery to last? What proportion of college undergraduates have engaged in binge drinking? Is caffeine dependence real? These are the types of questions we would like to be able to answer. It wouldn't be practical to determine the lifetime of *every* AA battery, to ask *all* college undergraduates about their drinking habits, or to conduct an experiment with *every* potential caffeine-dependent individual in the population. Instead, we select a sample of individuals (batteries, college students, caffeine-dependent people) to represent the population, and we collect data from those individuals. We want to infer from the sample data some conclusion about the population. That is the goal of ***statistical inference.***

Statistical Inference

Statistical inference provides methods for drawing conclusions about a population from sample data.

We cannot be certain that our conclusions are correct—a different sample might lead to different conclusions. Statistical inference uses the language of probability to express the strength of our conclusions. Probability allows us to take chance variation into account and to correct our judgment by calculation.

confidence intervals

significance tests

In this chapter and the next, we will meet the two most common types of formal statistical inference. This chapter concerns ***confidence intervals*** for estimating the value of a population parameter. Chapter 11 presents ***significance tests,*** which assess the evidence for a claim about a population. Both types of inference are based on the sampling distributions of statistics. That is, both report probabilities that state *what would happen if we used the inference method many times.*

CAUTION

The methods of formal inference require the long-run regular behavior that probability describes. Inference is most reliable when the data are produced by a properly randomized design. **When you use statistical inference, you are acting as if the data are a random sample or come from a randomized experiment.**

If this is not true, your conclusions may be open to challenge. Formal inference cannot remedy basic flaws in producing data, such as voluntary response samples and uncontrolled experiments. Use the common sense developed in your study of the first nine chapters of this book. Proceed to formal inference only when you are satisfied that the data deserve such analysis.

In this chapter and the next, we concentrate on the reasoning of inference. To see the reasoning more clearly, we start by pretending that the world is simpler than it is. In particular, we act as if we know that the population standard deviation σ has a particular value, even though we don't know the value of μ. Once we have introduced the basic ideas, we'll remove this unrealistic requirement.

We will begin by presenting an oversimplified technique for estimating a population mean. Then we will show how to modify this technique to make it practically useful. Finally, we will provide a method for estimating a population proportion. There are libraries—both of books and of computer software—full of more elaborate statistical techniques. Informed use of any of these methods requires an understanding of the underlying reasoning. *A computer will do the arithmetic, but you must still exercise judgment based on understanding.*

10.1 Confidence Intervals: The Basics

How do colleges and universities attempt to attract qualified students? Most schools publish statistical information about their incoming freshman classes—high school GPAs, class ranks, number of honors/AP courses taken, and standardized test scores—in their viewbooks and on their Web sites. High school students (and their parents) use this information to decide where to apply. Magazines like *U.S. News & World Report* rank colleges and universities, in part based on the academic strength of their enrolled students. Highly selective schools are constantly searching for creative ways to secure the best students.

Example 10.1 *IQ and admissions*
Estimating an unknown parameter

The admissions director at Big City University has a novel idea. He proposes using the IQ scores of current students as a marketing tool. The university agrees to provide him with enough money to administer IQ tests to 50 students. So the director gives the IQ test to an SRS of 50 of the university's 5000 freshmen. The mean IQ score for the sample is $\bar{x} = 112$. What can the director say about the mean score μ of the population of all 5000 freshmen?

Is the mean IQ score μ of all Big City University freshmen exactly 112? Probably not. But the law of large numbers tells us that the sample mean $\bar{x}$ from a large SRS will be close to the unknown population mean μ. Because $\bar{x} = 112$, we guess that μ is "somewhere around 112." How close to 112 is μ likely to be? To answer

Beautiful theory, ugly facts "Science is organized common sense where many a beautiful theory was killed by an ugly fact." So said Thomas Huxley. The job of statistical inference is to draw conclusions based on ugly facts. Educational theorists, for example, long pushed the "whole-language" approach to teaching reading and belittled the perceived need to break words into phonetic sounds, called "phonics." In 2000, a national panel reviewed ugly facts from 52 randomized studies. Conclusion: no matter what theory says, phonics is essential in teaching reading.

this question, we ask another: *How would the sample mean $\bar{x}$ vary if we took many samples of 50 freshmen from this same population?*

The sampling distribution of $\bar{x}$ describes how the values of $\bar{x}$ vary in repeated samples. Recall the essential facts about this sampling distribution from Chapter 9:

- The mean of the sampling distribution of $\bar{x}$ is the same as the unknown mean μ of the entire population.
- The standard deviation of $\bar{x}$ for an SRS of 50 freshmen is $\sigma/\sqrt{50}$, where σ is the standard deviation of individual IQ scores among all Big City University freshmen. Suppose we know that these IQ scores have standard deviation $\sigma = 15$. Then the standard deviation of $\bar{x}$ is

$$\frac{\sigma}{\sqrt{n}} = \frac{15}{\sqrt{50}} \approx 2.1$$

- The central limit theorem tells us that the mean $\bar{x}$ of 50 scores has a distribution that is close to Normal.

Putting these facts together gives us the reasoning of statistical estimation in a nutshell.

1. To estimate the unknown population mean μ, use the mean $\bar{x}$ of our random sample.
2. Although $\bar{x}$ is an unbiased estimate of μ, it will rarely be exactly equal to μ, so our estimate has some error.
3. In repeated samples, the values of $\bar{x}$ follow (approximately) a Normal distribution with mean μ and standard deviation 2.1.
4. The 95 part of the 68–95–99.7 rule for Normal distributions says that in about 95% of all samples, the mean IQ score $\bar{x}$ for the sample will be within 4.2 (that's two standard deviations) of the population mean μ.
5. Whenever $\bar{x}$ is within 4.2 points of μ, μ is within 4.2 points of $\bar{x}$. This happens in about 95% of all possible samples. So the unknown μ lies between $\bar{x} - 4.2$ and $\bar{x} + 4.2$ in about 95% of all samples.
6. If we estimate that μ lies somewhere in the interval from

$$\bar{x} - 4.2 = 112 - 4.2 = 107.8$$

to

$$\bar{x} + 4.2 = 112 + 4.2 = 116.2$$

we would be calculating this interval using a method that "captures" the true μ in about 95% of all possible samples.

The big idea is that the sampling distribution of $\bar{x}$ tells us how big the error is likely to be when we use $\bar{x}$ to estimate μ. A few pictures and graphs should help to clarify the idea.

Example 10.2 ***Statistical estimation in pictures***
The idea of a confidence interval

Figures 10.1 to 10.3 illustrate the reasoning of estimation in graphical form. Begin with Figure 10.1, and follow the flow from left to right. Starting with the population, imagine taking many SRSs of 50 Big City University freshmen. The first sample has mean IQ score $\bar{x} = 112$, the second has $\bar{x} = 109$, the third has $\bar{x} = 114$, and so on. If we collect all these sample means and display their distribution, we get a Normal-shaped distribution with mean equal to the unknown μ and standard deviation 2.1.

Next, look at Figure 10.2. This is the same as the Normal curve in Figure 10.1. The standard deviation of this curve is 2.1. The 68–95–99.7 rule says that in 95% of all samples, $\bar{x}$ lies within two standard deviations of the population mean μ. That's the same as saying that the interval $\bar{x} \pm 4.2$ captures μ in 95% of all samples.

Finally, Figure 10.3 puts it all together. Starting with the population, imagine taking many SRSs of 50 freshmen. The recipe $\bar{x} \pm 4.2$ gives an interval based on each sample; *95% of these intervals capture the unknown population mean* μ.

Figure 10.1 *The sampling distribution of the mean score $\bar{x}$ of an SRS of 50 Big City University freshmen on an IQ test.*

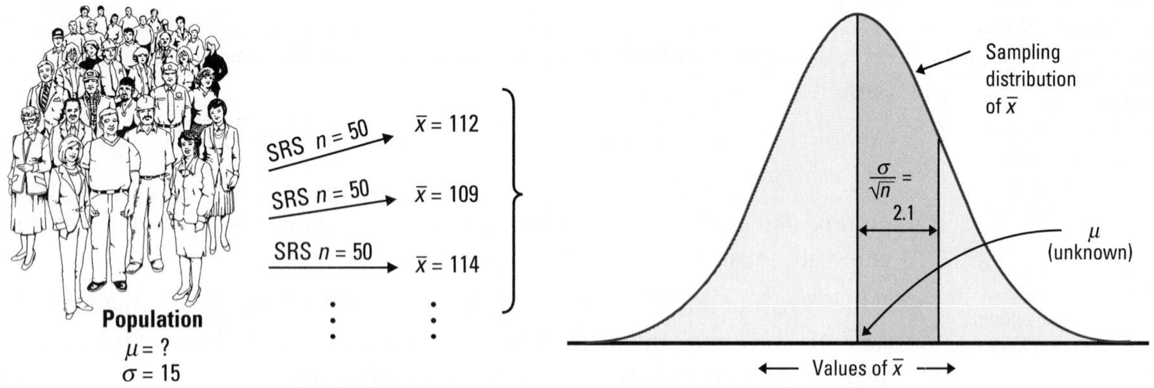

Figure 10.2 *In 95% of all samples, $\bar{x}$ lies within ± 4.2 of the unknown population mean μ. So μ also lies within ± 4.2 of $\bar{x}$ in those samples.*

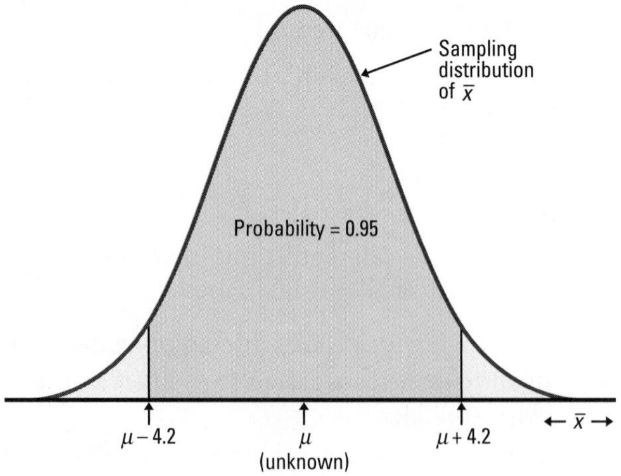

Figure 10.3 *To say that $\bar{x} \pm 4.2$ is a 95% confidence interval for the population mean μ is to say that, in repeated samples, 95% of these intervals capture μ.*

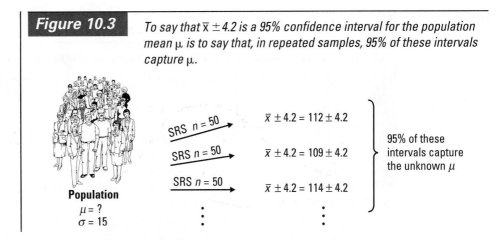

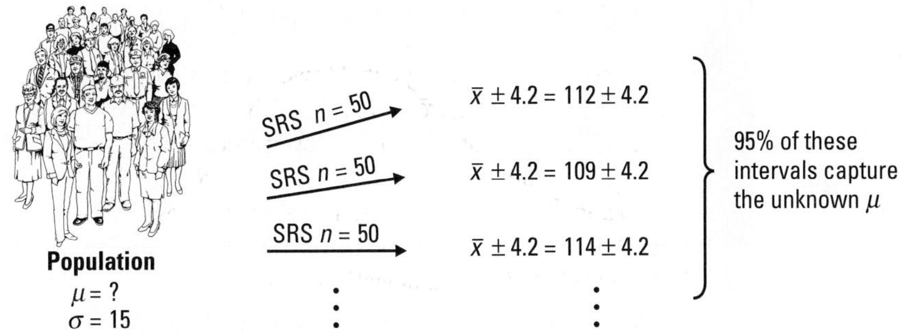

We have learned that in 95% of all samples of 50 Big City University freshmen, the interval $\bar{x} \pm 4.2$ will contain the true population mean μ. The language of statistical inference uses this fact about what would happen in *many* samples to express our confidence in the results of any *one* sample.

Example 10.3 *Conclusion: IQ at Big City University*

Interpreting a confidence interval

Our sample of 50 freshmen gave $\bar{x} = 112$. The resulting interval is 112 ± 4.2, which can be written as (107.8, 116.2). We say that we are 95% *confident* that the unknown mean IQ score for all Big City University freshmen is between 107.8 and 116.2.

Be sure you understand the basis for our confidence. There are only two possibilities:

1. The interval between 107.8 and 116.2 contains the true μ.
2. Our SRS was one of the few samples for which $\bar{x}$ is not within 4.2 points of the true μ. Only 5% of all samples give such inaccurate results.

We cannot know whether our sample is one of the 95% for which the interval $\bar{x} \pm 4.2$ catches μ, or whether it is one of the unlucky 5%. The statement that we are "95% confident" that the unknown μ lies between 107.8 and 116.2 is shorthand for saying, "We got these numbers by a method that gives correct results 95% of the time."

The interval of numbers $\bar{x} \pm 4.2$ is called a 95% ***confidence interval*** for μ. The ***confidence level*** is 95%. Like most confidence intervals we will meet, this one has the form

$$\text{estimate} \pm \text{margin of error}$$

margin of error

The estimate ($\bar{x}$ in this case) is our best guess for the value of the unknown parameter. The ***margin of error*** ± 4.2 shows how accurate we believe our guess is, based on the variability of the estimate. This is a 95% confidence interval because it catches the unknown μ in 95% of all possible samples.

Confidence Interval and Confidence Level

A **level C confidence interval** for a parameter has two parts:

- A **confidence interval** calculated from the data, usually of the form

  estimate ± margin of error

- A **confidence level C**, which gives the probability that the interval will capture the true parameter value in repeated samples. That is, the confidence level is the success rate for the method.

We can choose the confidence level, usually 90% or higher because we usually want to be quite sure of our conclusions. We will use C to stand for the confidence level in decimal form. For example, a 95% confidence level corresponds to C = 0.95.

Figure 10.3 is one way to picture the idea of a 95% confidence interval. Figure 10.4 illustrates the idea in a different form. Study these figures carefully. If you understand what they say, you have mastered one of the big ideas of statistics.

Figure 10.4 *Twenty-five samples from the same population gave these 95% confidence intervals. In the long run, 95% of all samples give an interval that contains the population mean μ.*

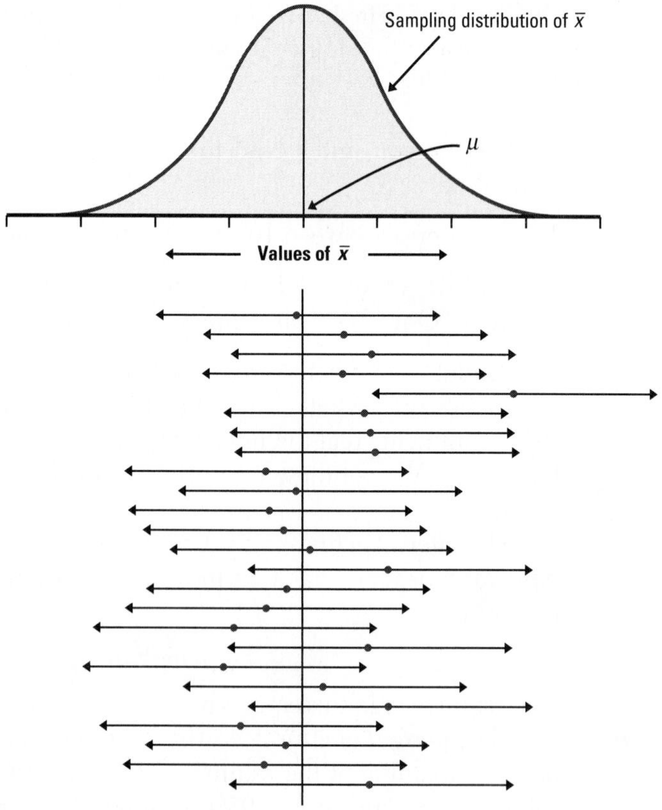

Figure 10.4 shows the result of drawing many SRSs from the same population and calculating a 95% confidence interval from each sample. The center of each interval is at $\bar{x}$ and therefore varies from sample to sample. The sampling distribution of $\bar{x}$ appears at the top of the figure to show the long-term pattern of this variation. The 95% confidence intervals from 25 SRSs appear on page 622. Here's what you should notice:

- The center $\bar{x}$ of each interval is marked by a dot.
- The arrows on either side of the dot span the confidence interval. The distance from the dot to the end of an arrow is the margin of error for that interval.
- 24 of these 25 intervals (that's 96%) cover the true value of μ. If we took all possible samples, 95% of the resulting confidence intervals would contain μ.

Activity 10B
Confidence Interval Applet

The *Confidence Interval* applet at the book's Web site, www.whfreeman.com/tps3e, animates Figure 10.4. You can use the applet to watch confidence intervals from one sample after another capture or fail to capture the true parameter.

1. Go to the Web site and launch the *Confidence Interval* applet.

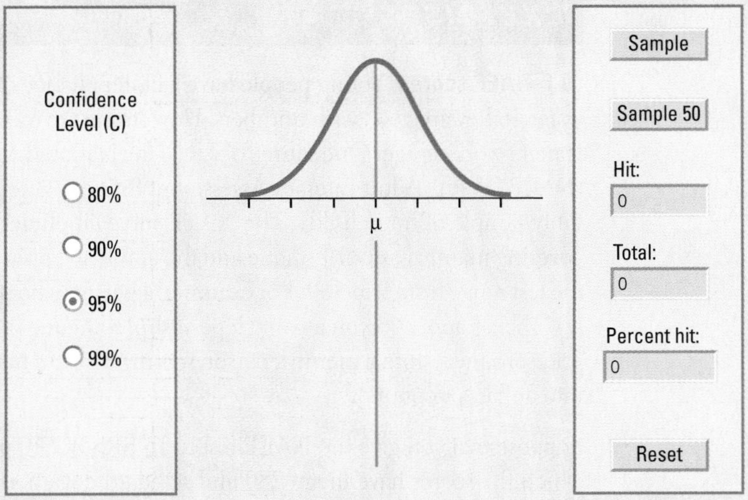

2. Exploring 80% confidence intervals
   (a) Set the confidence level to 80%. Click "Sample" to choose an SRS and calculate the confidence interval. Do this 10 times to simulate

10 SRSs with their 10 confidence intervals. How many of the intervals captured the true mean μ? How many missed? Can you predict whether the next interval you generate will hit or miss?

(b) Reset the applet. Click "Sample 50" to choose 50 SRSs and calculate the confidence intervals. How many hit? Keep clicking "Sample 50" and record the percent of hits among 100, 200, 300, 400, 500, 600, 700, 800, 900, and 1000 SRSs. What do you notice?

3. What confidence means

(a) Reset the applet. Set the confidence level to 90%. Click "Sample 50" repeatedly until you have 1000 samples. What percent of the 1000 confidence intervals captured the true μ?

(b) Do not reset the applet yet! Click on 95% confidence. How do the intervals change? What about the percent hits? Click on 99% confidence and answer the same two questions.

(c) Now reset the applet. Choose 95% confidence. Generate 1000 SRSs and record the percent of intervals that capture μ.

(d) Repeat part (c) for 99% confidence.

4. Write a paragraph describing what you learned from this exploration.

Exercises

10.1 NAEP scores Young people have a better chance of full-time employment and good wages if they are good with numbers. How strong are the quantitative skills of young Americans of working age? One source of data is the National Assessment of Educational Progress (NAEP) Young Adult Literacy Assessment Survey, which is based on a nationwide probability sample of households. The NAEP survey includes a short test of quantitative skills, covering mainly basic arithmetic and the ability to apply it to realistic problems. Scores on the test range from 0 to 500. For example, a person who scores 233 can add the amounts of two checks appearing on a bank deposit slip; someone scoring 325 can also determine the price of a meal from a menu; a person scoring 375 can transform a price in cents per ounce into dollars per pound.[1]

Suppose that you give the NAEP test to an SRS of 840 people from a large population in which the scores have mean 280 and standard deviation $\sigma = 60$. The mean $\bar{x}$ of the 840 scores will vary if you take repeated samples.

(a) Describe the shape, center, and spread of the sampling distribution of $\bar{x}$.

(b) Sketch the Normal curve that describes how $\bar{x}$ varies in many samples from this population. Mark its mean and the values one, two, and three standard deviations on either side of the mean.

(c) According to the 68–95–99.7 rule, about 95% of all the values of $\bar{x}$ fall within ________ of the mean of this curve. What is the missing number? Call it m for "margin of error." Shade the region from the mean minus m to the mean plus m on the axis of your sketch.

(d) Whenever $\bar{x}$ falls in the region you shaded, the true value of the population mean, $\mu = 280$, lies in the confidence interval between $\bar{x} - m$ and $\bar{x} + m$. Draw the confidence interval below your sketch for one value of $\bar{x}$ inside the shaded region and one value of $\bar{x}$ outside the shaded region. (Use Figure 10.4 as a model for the drawing.)

(e) In what percent of all samples will the true mean $\mu = 280$ be covered by the confidence interval $\bar{x} \pm m$?

10.2 Explaining confidence, I The admissions director from Big City University found that (107.8, 116.2) is a 95% confidence interval for the mean IQ score of all freshmen. Comment on whether each of the following explanations is correct.

(a) There is a 95% probability (chance) that the interval from 107.8 to 116.2 contains μ.

(b) There is a 95% chance that the interval (107.8, 116.2) contains $\bar{x}$.

(c) This interval was constructed using a method that captures the true mean in 95% of all possible samples.

(d) 95% of all possible samples will contain the interval (107.8, 116.2).

(e) The probability that the interval (107.8, 116.2) captures μ is either 0 or 1, but we don't know which.

10.3 Explaining confidence, II A student reads that a 95% confidence interval for the mean NAEP quantitative score for men aged 21 to 25 is 267.8 to 276.2. Asked to explain the meaning of this interval, the student says, "95% of all young men have scores between 267.8 and 276.2." Is the student right? Justify your answer.

10.4 Auto emissions Oxides of nitrogen (called NOX for short) emitted by cars and trucks are important contributors to air pollution. The amount of NOX emitted by a particular model varies from vehicle to vehicle. For one light truck model, NOX emissions vary with mean μ that is unknown and standard deviation $\sigma = 0.4$ grams per mile. You test an SRS of 50 of these trucks. The sample mean NOX level $\bar{x}$ estimates the unknown μ. You will get different values of $\bar{x}$ if you repeat your sampling.

(a) Describe the shape, center, and spread of the sampling distribution of $\bar{x}$.

(b) Sketch the Normal curve for the sampling distribution of $\bar{x}$. Mark its mean and the values one, two, and three standard deviations on either side of the mean.

(c) According to the 68–95–99.7 rule, about 95% of all values of $\bar{x}$ lie within a distance m of the mean of the sampling distribution. What is m? Shade the region on the axis of your sketch that is within m of the mean.

(d) Whenever $\bar{x}$ falls in the region you shaded, the unknown population mean μ lies in the confidence interval $\bar{x} \pm m$. For what percent of all possible samples does this happen?

(e) Following the style of Figure 10.4, draw the confidence intervals below your sketch for two values of $\bar{x}$, one that falls within the shaded region and one that falls outside it.

10.5 Losing weight A Gallup Poll in November 2002 found that 51% of the people in its sample said "Yes" when asked, "Would you like to lose weight?" Gallup announced: "For results based on the total sample of national adults, one can say with 95% confidence that the margin of (sampling) error is ±3 percentage points."[2]

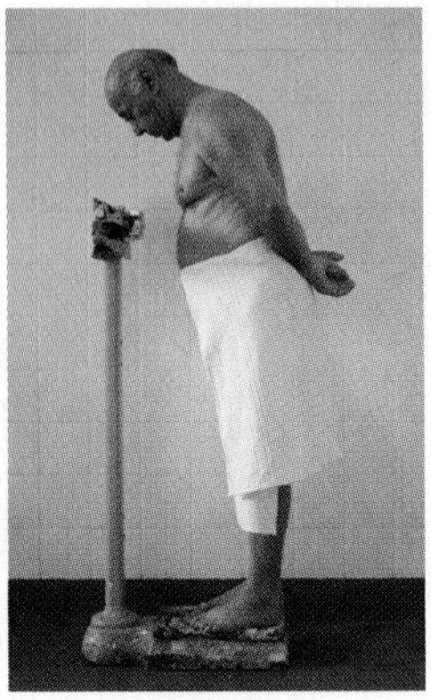

(a) What is the 95% confidence interval for the percent of all adults who want to lose weight?

(b) Explain clearly to someone who knows no statistics why we can't just say that 51% of all adults want to lose weight.

(c) Then explain clearly what "95% confidence" means.

(d) As Gallup indicates, the 3% margin of error for this poll includes only sampling variability (what they call "sampling error"). What other potential sources of error (Gallup calls these "nonsampling errors") could affect the accuracy of the 51% estimate?

10.6 Taller girls Based on information from the National Center for Health Statistics, the heights of 10-year-old girls closely follow a Normal distribution with mean 54.5 inches and standard deviation 2.7 inches. The heights of 10-year-old boys follow a Normal distribution with mean 54.1 inches and standard deviation 2.4 inches. A particular fourth-grade class has 10 girls and 7 boys. The children's heights are recorded on their tenth birthdays.

(a) Treat the students in this class as random samples of 10-year-old boys and girls. Use random variables to calculate the probability that the mean height of the girls is greater than the mean height of the boys. Show your work.

(b) Design a calculator simulation to estimate the probability described in (a). Describe your simulation method clearly enough so that a classmate could carry it out without further explanation from you. Then carry out your simulation. Interpret your results.

Confidence Interval for a Population Mean (When σ Is Known)

We can now show you how to calculate a ***level C confidence interval*** for the unknown mean μ of a population. The calculation of the interval depends on three important conditions:

1. **SRS** Our method of calculation assumes that the data come from an SRS of size n from the population of interest. Other types of random samples (stratified or cluster, say) might give a better representation of the population than an SRS, but they would require more complex calculations than the ones we'll use.
2. **Normality** The construction of the interval depends on the fact that the sampling distribution of the sample mean $\bar{x}$ is at least approximately Normal. This distribution is exactly Normal if the population distribution is Normal. When the population distribution is not Normal, the central limit theorem tells us that the sampling distribution of $\bar{x}$ will be approximately Normal if n is sufficiently large (say, at least 30).

3. **Independence** The procedure for calculating a confidence interval assumes that individual observations are independent. Independent observations are required for us to use the formula $\sigma_{\bar{x}} = \sigma/\sqrt{n}$ from Chapter 9. We will be safe if data are produced by sampling *with replacement* from the population. But that is rarely the case. To keep the calculations reasonably accurate when we sample *without replacement* from a finite population, we should sample no more than 10% of the population. In symbols, we need $N \geq 10n$.

CAUTION

Be sure to check that these *conditions for constructing a confidence interval for* μ are satisfied before you perform any calculations.

Conditions for Constructing a Confidence Interval for μ

The construction of a confidence interval for a population mean μ is appropriate when

- the data come from an SRS from the population of interest (**SRS**),
- the sampling distribution of $\bar{x}$ is approximately Normal (**Normality**), and
- individual observations are independent; when sampling without replacement, the population size N is at least 10 times the sample size n. (**Independence**).

Our construction of a 95% confidence interval for the mean IQ score of Big City University freshmen began by noting that, in any Normal distribution, about 95% of the observations fall within two standard deviations of its mean. To construct a level C confidence interval, we want to catch the central probability C under a Normal curve. To do that, we must go out z^* standard deviations on either side of the mean. Since any Normal distribution can be standardized, we can get the value z^* from the standard Normal table. Here is an example of how to find z^*.

Example 10.4 *Finding* z^*

Using Table A

z	.07	.08	.09
1.1	.8790	.8810	.8830
1.2	.8980	.8997	.9015
1.3	.9147	.9162	.9177

To construct an 80% confidence interval, we must catch the central 80% of the Normal sampling distribution of $\bar{x}$. In catching the central 80% we leave out 20%, or 10% in each tail. So z^* is the point with area 0.1 to its right (and 0.9 to its left) under the standard Normal curve. Search the body of Table A to find the point with area 0.9 to its left. The closest entry is $z^* = 1.28$. There is area 0.8 under the standard Normal curve between -1.28 and 1.28. Figure 10.5 (page 628) shows how z^* is related to areas under the curve.

Figure 10.5 *The central probability 0.8 under a standard Normal curve lies between −1.28 and 1.28. That is, there is area 0.1 to the right of 1.28 under the curve.*

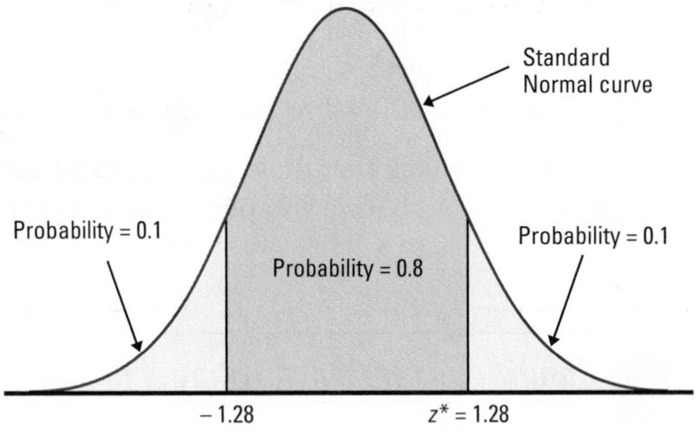

Figure 10.6 shows the general situation for any confidence level C. If we catch the central area C, the leftover tail area is 1 − C, or (1 − C)/2 for each of the *upper* and *lower tails*. You can find z^* for any C by searching Table A. Here are the results for the most common confidence levels:

Confidence level	Tail area	z^*
90%	0.05	1.645
95%	0.025	1.960
99%	0.005	2.576

Notice that for 95% confidence we use $z^* = 1.960$. This is more exact than the approximate value $z^* = 2$ given by the 68–95–99.7 rule. Values z^* that mark off a specified area under the standard Normal curve are often called ***critical values*** of the distribution.

Figure 10.6 *In general, the central probability* C *under a standard Normal curve lies between* $-z^*$ *and* z^**. Because* z^* *has area (1 −* C*)/2 to its right under the curve, we call it the upper (1 −* C*)/2 critical value.*

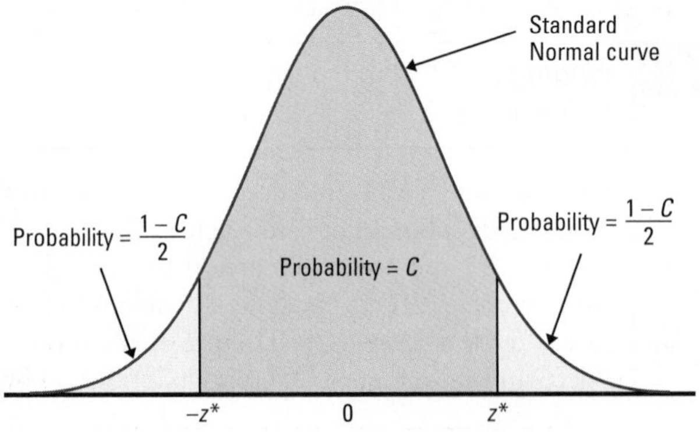

Critical Values

The number z^* with probability p lying to its right under the standard Normal curve is called the upper p **critical value** of the standard Normal distribution.

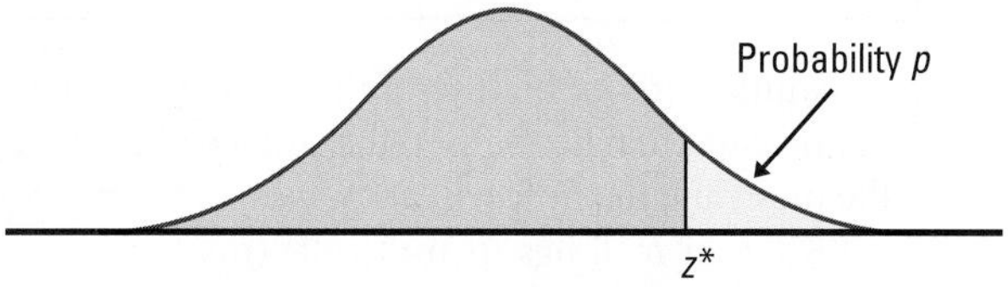

Here's the thinking that leads to the level C confidence interval:

- Under any Normal curve, the area between the point z^* standard deviations below its mean and the point z^* standard deviations above its mean is C.
- The standard deviation of the sampling distribution of $\bar{x}$ is $\sigma/\sqrt{n}$, and its mean is the population mean μ. So there is probability C that the observed sample mean $\bar{x}$ takes a value between

$$\mu - z^*\frac{\sigma}{\sqrt{n}} \quad \text{and} \quad \mu + z^*\frac{\sigma}{\sqrt{n}}$$

- Whenever this happens, the population mean μ is contained between

$$\bar{x} - z^*\frac{\sigma}{\sqrt{n}} \quad \text{and} \quad \bar{x} + z^*\frac{\sigma}{\sqrt{n}}$$

That is our confidence interval. The estimate of the unknown μ is $\bar{x}$, and the margin of error is $z^*\sigma/\sqrt{n}$.

Confidence Interval for a Population Mean (σ Known)

Choose an SRS of size n from a population having unknown mean μ and known standard deviation σ. A **level C confidence interval** for μ is

$$\bar{x} \pm z^*\frac{\sigma}{\sqrt{n}}$$

Here z^* is the value that determines area C between $-z^*$ and z^* under the standard Normal curve. This interval is exact when the population distribution is Normal and is approximately correct for large n in other cases.

Now we are ready to calculate a confidence interval from scratch. The following example shows you how to do it.

Example 10.5 *Video screen tension, I*

Constructing a confidence interval for μ

A manufacturer of high-resolution video terminals must control the tension on the mesh of fine wires that lies behind the surface of the viewing screen. Too much tension will tear the mesh and too little will allow wrinkles. The tension is measured by an electrical device with output readings in millivolts (mV). Some variation is inherent in the production process. Careful study has shown that when the process is operating properly, the standard deviation of the tension readings is $\sigma = 43$ mV. Here are the tension readings from an SRS of 20 screens from a single day's production:

269.5	297.0	269.6	283.3	304.8	280.4	233.5	257.4	317.5	327.4
264.7	307.7	310.0	343.3	328.1	342.6	338.8	340.1	374.6	336.1

Construct and interpret a 90% confidence interval for the mean tension μ of all the screens produced on this day.

Step 1: Parameter *Identify the population of interest and the parameter you want to draw conclusions about.* The population of interest is all of the video terminals produced on the day in question. We want to estimate μ, the mean tension for all of these screens.

Step 2: Conditions *Choose the appropriate inference procedure. Verify the conditions for using it.* Since we know σ, we should use the confidence interval for a population mean that was just introduced to estimate μ. Now we must check that the three required conditions are met.

- **SRS** We are told that the data come from an SRS of 20 screens from the population of all screens produced that day.
- **Normality** Is the sampling distribution of $\bar{x}$ approximately Normal? Past experience suggests that the tension readings of screens produced on a single day follow a Normal distribution quite closely. If the population distribution is Normal, the sampling distribution of $\bar{x}$ will be Normal. But what if the population distribution weren't known to be Normal?

The sample size is too small ($n = 20$) to use the central limit theorem. So we examine the sample data. A boxplot of the tension readings, Figure 10.7(a), shows no outliers or strong skewness. The Normal probability plot in Figure 10.7(b) tells us that the sample data are approximately Normally distributed. These data give us no reason to doubt the Normality of the population from which they came.

- **Independence** Since we are sampling without replacement, we must assume that at least $(10)(20) = 200$ video terminals were produced on this day.

Step 3: Calculations *If the conditions are met, carry out the inference procedure.* You can check that the mean tension reading for the 20 screens in our sample is $\bar{x} = 306.3$ mV. The confidence interval formula is $\bar{x} \pm z^*\sigma/\sqrt{n}$. For a 90% confidence level, the critical value is $z^* = 1.645$. So the 90% confidence interval for μ is

$$\bar{x} \pm z^*\frac{\sigma}{\sqrt{n}} = 306.3 \pm 1.645\frac{43}{\sqrt{20}} = 306.3 \pm 15.8 = (290.5, 322.1)$$

Figure 10.7 *(a) A boxplot and (b) a Normal probability plot of the video screen tension readings, for Example 10.5.*

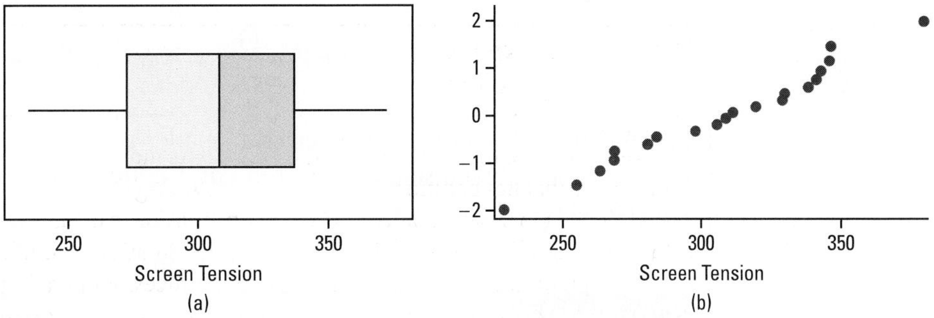

Step 4: **Interpretation** *Interpret your results in the context of the problem.* We are 90% confident that the true mean tension in the entire batch of video terminals produced that day is between 290.5 and 322.1 mV. (Notice that our *conclusion* about the plausible values for μ *connects* directly to our work in Step 3 and is given in the *context* of this problem. Be sure to include all three of these elements in Step 4.)

We will use the four-step process of Example 10.5 throughout our study of statistical inference. You can think of this general structure as your ***Inference Toolbox.*** Specific inference procedures (tools) will be added to your toolbox for use in a variety of settings. Examples that use the Inference Toolbox will be marked with a toolbox icon.

Inference Toolbox

Confidence intervals

To construct a confidence interval:

Step 1: **Parameter** Identify the population of interest and the parameter you want to draw conclusions about.
Step 2: **Conditions** Choose the appropriate inference procedure. Verify conditions for using it.
Step 3: **Calculations** If the conditions are met, carry out the inference procedure.

$$\text{confidence interval} = \text{estimate} \pm \text{margin of error}$$

Step 4: **Interpretation** Interpret your results in the context of the problem. Remember the "three C's": conclusion, connection, and context.

Suppose that a single computer screen had a tension reading of 306.3 mV, the same value as the sample mean in Example 10.5. Repeating the calculation with $n = 1$ shows that the 90% confidence interval based on a single measurement is

$$\bar{x} \pm z^* \frac{\sigma}{\sqrt{n}} = 306.3 \pm 1.645 \frac{43}{\sqrt{1}} = 306.3 \pm 70.7 = (235.6, 377.0)$$

The mean of 20 measurements gives a smaller margin of error and therefore a shorter interval than a single measurement. Figure 10.8 illustrates the gain from using 20 observations.

Figure 10.8 *Confidence intervals for* n = *20 and* n = *1, for Example 10.5. Larger samples give shorter intervals.*

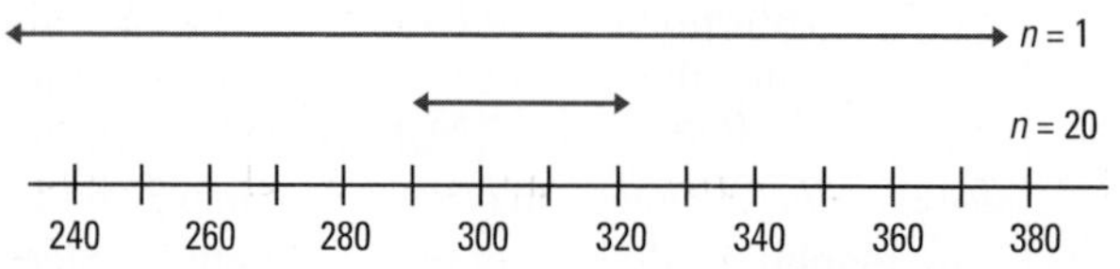

Exercises

10.7 Finding z^*, I Use Table A of standard Normal probabilities to find z^* for confidence level 97.5%. Start by making a copy of Figure 10.5 with C = 0.975 that shows how much area is left in each tail when the central area is 0.975.

10.8 Finding z^*, II Use Table A of standard Normal probabilities to find z^* for confidence level 94%. Start by making a copy of Figure 10.5 with C = 0.94 that shows how much area is left in each tail when the central area is 0.4.

10.9 IQ test scores Here are the IQ test scores of 31 seventh-grade girls in a midwestern school district:[3]

114	100	104	89	102	91	114	114	103	105	
108	130	120	132	111	128	118	119	86	72	
111	103	74	112	107	103	98	96	112	112	93

Treat these girls as an SRS of all seventh-grade girls in the school district. Suppose that the standard deviation of IQ scores in this population is known to be $\sigma = 15$.

(a) Construct and interpret a 99% confidence interval for the mean IQ score in the population. Follow the Inference Toolbox.

(b) If these girls had actually come from one teacher's class within the school district, how would your work in part (a) have changed? Be specific.

10.10 Analyzing pharmaceuticals A manufacturer of pharmaceutical products analyzes a specimen from each batch of a product to verify the concentration of the active ingredient.

The chemical analysis is not perfectly precise. Repeated measurements on the same specimen give slightly different results. The results of repeated measurements follow a Normal distribution quite closely. The analysis procedure has no bias, so the mean μ of the population of all measurements is the true concentration in the specimen. From previous experience, the standard deviation of this distribution is known to be $\sigma = 0.0068$ grams per liter. The laboratory analyzes each specimen three times and reports the mean result.

Three analyses of one specimen give concentrations

0.8403	0.8363	0.8447

(a) What is a pharmaceutical product, and why would such an analysis be important?

(b) Construct and interpret a 99% confidence interval for the true concentration μ. Use the Inference Toolbox as a guide.

(c) Explain what "99% confident" means. (That is, interpret the confidence level.)

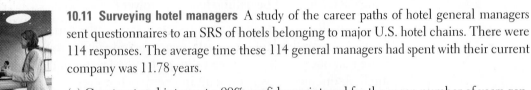

10.11 Surveying hotel managers A study of the career paths of hotel general managers sent questionnaires to an SRS of hotels belonging to major U.S. hotel chains. There were 114 responses. The average time these 114 general managers had spent with their current company was 11.78 years.

(a) Construct and interpret a 99% confidence interval for the mean number of years general managers of major-chain hotels have spent with their current company. (Take it as known that the standard deviation of time with the company for all general managers is 3.2 years.) Use the Inference Toolbox as a guide.

(b) In fact, the questionnaires were mailed to 160 hotels. How might this additional information affect your work in part (a)? Be specific.

10.12 Sisters and brothers How strongly do physical characteristics of sisters and brothers correlate? Here are data on the heights (in inches) of 11 adult pairs:[4]

Brother:	71	68	66	67	70	71	70	73	72	65	66
Sister:	69	64	65	63	65	62	65	64	66	59	62

(a) Construct a scatterplot using brother's height as the explanatory variable. Describe what you see.

(b) Use your calculator to compute the least-squares regression line for predicting sister's height from brother's height. Interpret the slope of the least-squares line in context.

(c) Damien is 70 inches tall. Predict the height of his sister Tonya.

(d) Do you expect your prediction in (c) to be very accurate? Give appropriate evidence to support your answer.

How Confidence Intervals Behave

The confidence interval $\bar{x} \pm z^*\sigma/\sqrt{n}$ for the mean of a population illustrates several important properties that are shared by all confidence intervals in common use. The user chooses the confidence level, and the margin of error follows from this choice. We would like high confidence and also a small margin of error. High confidence says that our method almost always gives correct answers. A small margin of error says that we have pinned down the parameter quite precisely. The margin of error is

$$\text{margin of error} = z^* \frac{\sigma}{\sqrt{n}}$$

This expression has z^* and σ in the numerator and $\sqrt{n}$ in the denominator. So the margin of error gets smaller when

- z^* gets smaller. Smaller z^* is the same as smaller confidence level C (look at Figure 10.6, on page 628, again). There is a trade-off between the confidence level and the margin of error. To obtain a smaller margin of error from the same data, you must be willing to accept lower confidence.
- σ gets smaller. The standard deviation σ measures the variation in the population. You can think of the variation among individuals in the population as noise that obscures the average value μ. It is easier to pin down μ when σ is small. Reducing σ is very difficult in practice, however.
- n gets larger. Increasing the sample size n reduces the margin of error for any fixed confidence level. Because we take the square root of n, we must take four times as many observations in order to cut the margin of error in half.

Example 10.6 *Video screen tension, II*
Changing the confidence level

Suppose that the manufacturer in Example 10.5 wants 99% confidence rather than 90%. The critical value for 99% confidence is $z^* = 2.57$ (from Table A). The 99% confidence interval for μ based on an SRS of 20 video terminals with mean $\bar{x} = 306.3$ is

$$\bar{x} \pm z^* \frac{\sigma}{\sqrt{n}} = 306.3 \pm 2.57 \frac{43}{\sqrt{20}} = 306.3 \pm 24.7 = (281.6, 331.0)$$

Demanding 99% confidence instead of 90% confidence has increased the margin of error from 15.8 to 24.7. Figure 10.9 compares these two measurements.

Figure 10.9 *90% and 99% confidence intervals, for Example 10.6. Higher confidence requires a longer interval.*

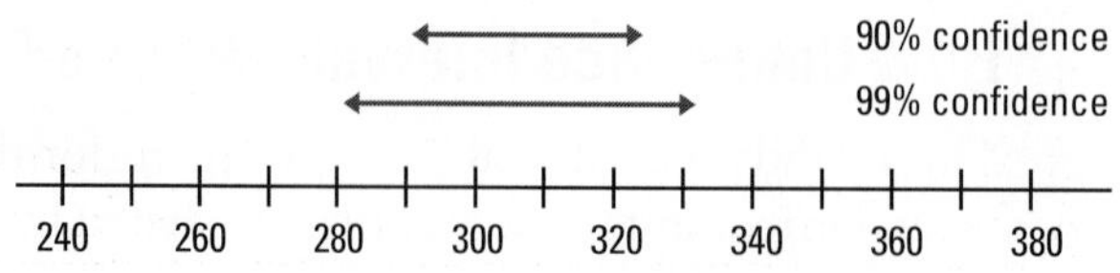

Determining Sample Size

A wise user of statistics never plans data collection without planning the inference at the same time. You can arrange to have both high confidence and a small margin of error by taking enough observations. The margin of error m of the confidence interval for the population mean μ is $m = z^*\sigma/\sqrt{n}$. To determine the ***sample size for a desired margin of error m***, substitute the value of z^* for your desired confidence level, set the expression for m less than or equal to the specified margin of error, and solve the inequality for n. The procedure is best illustrated with an example.

Example 10.7 *How many monkeys?*
Determining sample size from margin of error

Researchers would like to estimate the mean cholesterol level μ of a particular variety of monkey that is often used in laboratory experiments. They would like their estimate to be within 1 milligram per deciliter of blood (mg/dl) of the true value of μ at a 95% confidence level. A previous study involving this variety of monkey suggests that the standard deviation of cholesterol level is about $\sigma = 5$ mg/dl. Obtaining monkeys is time-consuming and expensive, so the researchers want to know the minimum number of monkeys they will need to generate a satisfactory estimate.

For 95% confidence, Table A gives $z^* = 1.96$. We know that $\sigma = 5$. Set the margin of error to be at most 1:

$$z^* \frac{\sigma}{\sqrt{n}} \leq 1$$

$$1.96 \frac{5}{\sqrt{n}} \leq 1$$

$$\sqrt{n} \geq \frac{(1.96)(5)}{1}$$

$$\sqrt{n} \geq 9.8$$

$$n \geq 96.04 \quad \text{so take } n = 97$$

Because 96 monkeys would give a slightly larger margin of error than desired, and 97 monkeys would give a slightly smaller margin of error, the researchers would need 97 monkeys to estimate the cholesterol levels to their satisfaction. **Always round up to the next whole number when finding n.** On learning the cost of getting this many monkeys, the researchers might want to consider studying rats instead!

Here is the principle:

Sample Size for a Desired Margin of Error

To determine the sample size n that will yield a confidence interval for a population mean with a specified margin of error m, set the expression for the margin of error to be less than or equal to m and solve for n:

$$z^* \frac{\sigma}{\sqrt{n}} \leq m$$

Taking observations costs time and money. The required sample size may be impossibly expensive. **Notice that it is the size of the sample that determines the margin of error. The size of the population does not influence the sample size we need.** (This is true as long as the population is much larger than the sample.)

Some Cautions

Any method of inference is correct only in specific circumstances. If statistical procedures carried warning labels like those on drugs, most inference methods would have long labels indeed. Our handy recipe $\bar{x} \pm z^*\sigma/\sqrt{n}$ for estimating an unknown population mean comes with the following list of warnings for the user:

- **The data must be an SRS from the population.** We are safest if we actually carried out the random selection of an SRS. We are not in great danger if the data can plausibly be thought of as observations taken at random from a population. That is the case in Exercise 10.10 (page 632), where we have in mind the population resulting from a very large number of repeated analyses of the same specimen. Remember, though, that in some cases an attempt to choose an SRS can be frustrated by nonresponse and other practical problems.
- **Different methods are needed for different designs.** The confidence interval formula isn't correct for probability samples more complex than an SRS. Correct methods for other designs are available. We will not discuss confidence intervals based on multistage, cluster, or stratified random samples. If you plan such samples, be sure that you (or your statistical consultant) know how to carry out the inference you desire.
- **There is no correct method for inference from data haphazardly collected with bias of unknown size.** Fancy formulas cannot rescue badly produced data.
- **Outliers can distort the results.** Because $\bar{x}$ is strongly influenced by a few extreme observations, outliers can have a large effect on the confidence interval. You should search for outliers and try to correct them or justify their removal before computing the interval. If the outliers cannot be removed, ask your statistical consultant about procedures that are not sensitive to outliers.
- **The shape of the population distribution matters.** If the sample size is small and the population is not Normal, the true confidence level will be different from the value C used in computing the interval. Examine your data carefully for skewness and other signs of non-Normality. The interval relies only on the distribution of $\bar{x}$, which even for quite small sample sizes is much closer to Normal than the individual observations. When $n \geq 15$, the confidence level is not greatly disturbed by non-Normal populations unless extreme outliers or quite strong skewness are present. We will discuss this issue in more detail in the next section.
- **You must know the standard deviation σ of the population.** This unrealistic requirement renders the interval $\bar{x} \pm z^*\sigma/\sqrt{n}$ of little use in statistical practice. We will learn in the next section what to do when σ is unknown.

The most important caution concerning confidence intervals is a consequence of the first of these warnings. **The margin of error in a confidence interval covers only random sampling errors.** The margin of error is obtained from the sampling distribution and indicates how much error can be expected because of chance variation in randomized data production. Practical difficulties, such as undercoverage and nonresponse in a sample survey, can cause additional errors that may be larger than the random sampling error. Remember this unpleasant fact when reading the results of an opinion poll or other sample survey. The practical conduct of the survey influences the trustworthiness of its results in ways that are not included in the announced margin of error.

Every inference procedure that we will meet has its own list of warnings. Because many of the warnings are similar to those above, we will not print the full warning label each time. It is easy to state (from the mathematics of probability) conditions under which a method of inference is exactly correct. These conditions are never fully met in practice. For example, no population is exactly Normal. Deciding when a statistical procedure should be used in practice often requires judgment assisted by exploratory analysis of the data.

Finally, you should understand what statistical confidence does not say. We are 95% confident that the mean IQ score for all Big City University freshmen lies between 107.8 and 116.2. That is, these numbers were calculated by a method that gives correct results in 95% of all possible samples. We cannot say that the probability is 95% that the true mean falls between 107.8 and 116.2. No randomness remains after we draw one particular sample and get from it one particular interval. The true mean either is or is not between 107.8 and 116.2. The probability calculations of standard statistical inference describe how often the *method* gives correct answers.

Exercises

10.13 Sample size and margin of error Find the margin of error for 90% confidence in Example 10.5 (page 630) if the manufacturer measures the tension of 80 video terminals. Check that your result is half as large as the margin of error based on 20 measurements from the Example.

10.14 A balanced scale? To assess the accuracy of a laboratory scale, a standard weight known to weigh 10 grams is weighed repeatedly. The scale readings are Normally distributed with unknown mean (this mean is 10 grams if the scale has no bias). The standard deviation of the scale readings is known to be 0.0002 gram.

(a) The weight is weighed five times. The mean result is 10.0023 grams. Construct and interpret a 98% confidence interval for the mean of repeated measurements of the weight.

(b) How many measurements must be averaged to get a margin of error of ± 0.0001 with 98% confidence? Show your work.

10.15 Confidence level and margin of error High school students who take the SAT Mathematics exam a second time generally score higher than on their first try. The change in score has a Normal distribution with standard deviation $\sigma = 50$. A random sample of 1000 students gain an average of $\bar{x} = 22$ points on their second try.[5]

(a) Construct and interpret a 95% confidence interval for the mean score gain μ in the population.

(b) Calculate the 90% and 99% confidence intervals for μ.

(c) Make a sketch like Figure 10.9 (page 634) to compare these three intervals. How does increasing the confidence level affect the length of the confidence interval?

10.16 The SAT, again Refer to the previous exercise. Complete part (a) if you have not done so already.

(a) Suppose that the same result, $\bar{x} = 22$ had come from a sample of 250 students. Calculate the 95% confidence interval for the population mean μ in this case.

(b) Then suppose that a sample of 4000 students had produced the sample mean $\bar{x} = 22$. Again compute the 95% confidence interval for μ.

(c) What are the margins of error for samples of size 250, 1000, and 4000? How does increasing the sample size affect the margin of error of a confidence interval?

(d) How large a sample of high school students would be needed to estimate the mean change in SAT score μ to within ± 2 points with 95% confidence? Show your work.

10.17 A talk show opinion poll A radio talk show invites listeners to enter a dispute about a proposed pay increase for city council members. "What yearly pay do you think council members should get? Call us with your number." In all, 958 people call. The mean pay they suggest is $\bar{x} = \$8740$ per year, and the standard deviation of the responses is $s = \$1125$. For a large sample such as this, s is very close to the unknown population σ. The station calculates the 95% confidence interval for the mean pay μ that all citizens would propose for council members to be \$8669 to \$8811.

(a) Is the station's calculation correct?

(b) Does their conclusion describe the population of all the city's citizens? Explain your answer.

10.18 The 2004 presidential election A closely contested presidential election pitted George W. Bush against John Kerry in 2004. A poll taken immediately before the 2004 election showed that 51% of the sample intended to vote for Kerry. The polling organization announced that they were 95% confident that the sample result was within ± 2 points of the true percent of all voters who favored Kerry.

(a) Explain in plain language to someone who knows no statistics what "95% confident" means in this announcement.

(b) The poll showed Kerry leading. Yet the polling organization said the election was too close to call. Explain why.

(c) On hearing of the poll, a nervous politician asked, "What is the probability that over half the voters prefer Kerry?" A statistician replied that this question can't be answered from the poll results, and that it doesn't even make sense to talk about such a probability. Explain why.

Section 10.1 Summary

A **confidence interval** uses sample data to estimate an unknown population parameter with an indication of how accurate the estimate is and of how confident we are that the result is correct.

Any confidence interval has two parts: an interval computed from the data and a confidence level. The interval often has the form

$$\text{estimate} \pm \text{margin of error}$$

The **confidence level** states the probability that the method will give a correct answer. That is, if you use many 95% confidence intervals, in the long run 95% of your intervals will contain the true parameter value. You do not know whether a 95% confidence interval calculated from a particular set of data contains the true parameter value.

A **level C confidence interval** for the mean μ of a Normal population with known standard deviation σ, based on an SRS of size n, is given by

$$\bar{x} \pm z^* \frac{\sigma}{\sqrt{n}}$$

The **critical value** z^* is chosen so that the area under the standard Normal curve between $-z^*$ and z^* is C. Because of the central limit theorem, this interval is approximately correct for large samples when the population is not Normal.

A specific confidence interval formula is correct only under specific conditions. The most important conditions for inference about a population mean are: **SRS** from the population of interest, **Normality** of the sampling distribution, and **independence** of observations. Remember that the margin of error for a confidence interval includes only sampling variability, not other sources of error like nonresponse and undercoverage.

Use the **Inference Toolbox** (page 631) as a guide when you construct a confidence interval.

Other things being equal, the **margin of error** of a confidence interval gets smaller as

- the confidence level C decreases (z^* gets smaller),
- the population standard deviation σ decreases, and
- the sample size n increases.

The sample size required to obtain a confidence interval with specified margin of error m for the mean $\bar{x}$ is found by setting

$$z^* \frac{\sigma}{\sqrt{n}} \leq m$$

and solving for n, where z^* is the critical value for the desired level of confidence. Always round n up when you use this formula.

Section 10.1 Exercises

10.19 Constructing a confidence interval A study based on a sample of size 25 reported a mean of 76 with a margin of error of 12 for 95% confidence.

(a) What conditions must be satisfied before you would feel comfortable reporting the results of this study in a scientific journal?

(b) Give the 95% confidence interval. Explain what "margin of error of 12" means.

(c) Explain the meaning of "95% confidence."

10.20 California SAT scores Suppose you want to estimate the mean SAT Math score for the more than 385,000 high school seniors in California. Only about 49% of California students take the SAT. These self-selected seniors are planning to attend college and so are not representative of all California seniors. You know better than to make inferences about the population based on such sample data. At considerable effort and expense, you give the test to a simple random sample (SRS) of 500 California high school seniors. The mean for your sample is $\bar{x} = 461$. Assume that $\sigma = 100$.

(a) What can you say about the mean score μ in the population of all 385,000 seniors? Construct and interpret a 99% confidence interval to answer this question.

(b) How large a sample would you need to estimate the mean score μ within 5 points at a 99% confidence level? Show your work.

10.21 Making money A newspaper advertisement seeks people who are willing to work for a telemarketing company. The ad states, "Earn between \$500 and \$1000 per week." Do you think that the ad is describing a confidence interval? Explain your answer.

10.22 Sampling children's heights Consider the following two scenarios: (a) Take a simple random sample of 100 students from a large elementary school with children in grades kindergarten through fifth grade. (b) Take a simple random sample of 100 third-graders from the same school. For each of these samples, you will measure the height of each child in the sample. Which sample should have the smaller margin of error for 95% confidence? Justify your answer.

10.23 Prayer in the schools A *New York Times*/CBS News Poll asked the question "Do you favor an amendment to the Constitution that would permit organized prayer in public schools?" Sixty-six percent of the sample answered "Yes." The article describing the poll says that it "is based on telephone interviews conducted from Sept. 13 to Sept. 18 with 1,664 adults around the United States, excluding Alaska and Hawaii. . . . the telephone numbers were formed by random digits, thus permitting access to both listed and unlisted residential numbers."

(a) The article gives the margin of error as 3 percentage points. Make a 99% confidence statement about the percent of all adults who favor a school prayer amendment.

(b) The news article goes on to say: "The theoretical errors do not take into account a margin of additional error resulting from the various practical difficulties in taking any survey of public opinion." List some of the "practical difficulties" that may cause errors in addition to the ±3% margin of error. Pay particular attention to the news article's description of the sampling method.

10.24 Healing of skin wounds Biologists studying the healing of skin wounds measured the rate at which new cells closed a razor cut made in the skin of an anesthetized newt. Here are data from 18 newts, measured in micrometers (millionths of a meter) per hour:[6]

29	27	34	40	22	28	14	35	26
35	12	30	23	18	11	22	23	33

(a) Make a boxplot of the healing rates. It is difficult to assess Normality from 18 observations, but look for outliers or extreme skewness. Now make a Normal probability plot. What do you find?

(b) Scientists usually assume that animal subjects are SRSs from their species or genetic type. Treat these newts as an SRS and suppose you know that the standard deviation of healing rates for this species of newt is 8 micrometers per hour. Construct and interpret a 90% confidence interval for the mean healing rate for the species.

(c) A friend who knows almost no statistics follows the formula $\bar{x} \pm z^* \sigma / \sqrt{n}$ in a biology lab manual to get a 95% confidence interval for the mean. Is her interval wider or narrower than yours? Explain to her why it makes sense that higher confidence changes the length of the interval.

10.25 More newts! How large a sample would enable you to estimate the mean healing rate of skin wounds in newts (see previous exercise) within a margin of error of 1 micrometer per hour with 90% confidence?

10.26 Confidence intervals on the TI-83/84/89

(a) You can use your TI-83/84/89 to find the confidence interval for the mean rate of healing for newts in Exercise 10.24. Follow the Technology Toolbox below.

(b) If you have summary statistics but not raw data, you would select "Stats" as your input method and then provide the sample mean $\bar{x}$, the population standard deviation σ, and the number n of observations. Use your calculator in this way to find the confidence interval for the mean number of years hotel managers have spent with their current company in Exercise 10.11 (page 633). **Calculating the confidence interval is only one part of the inference process. Follow the steps in the Inference Toolbox (page 631).**

Technology Toolbox

Calculator confidence intervals

You can use your TI-83/84/89 to construct confidence intervals, using either data stored in a list or summary statistics. In Exercise 10.10 (pages 632–633), for example, we would begin by entering the three specimen concentrations into L_1/list1.

(continued)

Technology Toolbox

Calculator confidence intervals *(continued)*

TI-83/84

- Press STAT, choose TESTS, and then choose 7:ZInterval . . .
- Adjust your settings as shown.
- Then choose "Calculate."

```
ZInterval
 Inpt:Data Stats
 σ:.0068
 List:L1
 Freq:1
 C-Level:.99
 Calculate
```

```
ZInterval
 (.83032,.85055)
 x̄=.8404333333
 Sx=.004201587
 n=3
```

TI-89

- Run the Stats/List Editor APP
- Press 2nd F2 ([F7]), choose 1:ZInterval . . .
- Choose Input Method = Data.
- Adjust your settings as shown. Press ENTER.

```
Z Interval
σ:        .0068
List:     list1
Freq:     1
C Level:  .99
Enter=OK   ESC=CANCEL
list1[4]=
TYPE+[ENTER]=OK AND[ESC]=CANCEL
```

```
Z Interval
C Int  =(.8303,.8505)
x̄      =.840433
ME     =.010113
Sx     =.004202
n      =3.
σ      =.0068
Enter=OK
list1[4]=
MAIN   RAD AUTO   FUNC   1/1
```

10.2 Estimating a Population Mean

How can we construct a confidence interval for an unknown population mean μ when we don't know the population standard deviation σ? In the previous section, we made the unrealistic assumption that we knew the value of σ. In practice, σ is unknown. We must estimate σ from the data even though we are primarily interested in μ. The need to estimate σ changes some computational details of confidence intervals for μ, but not their interpretation.

As before, we need to verify three important conditions before we estimate a population mean. When we do inference in practice, verifying the conditions is often a bit more complicated.

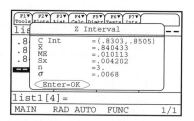

Conditions for Inference about a Population Mean

- **SRS** Our data are a simple random sample (SRS) of size n from the population of interest or come from a randomized experiment. This condition is very important.
- **Normality** Observations from the population have a Normal distribution with mean μ and standard deviation σ. In practice, it is enough that the distribution be symmetric and single-peaked unless the sample is very small. Both μ and σ are unknown parameters.
- **Independence** The method for calculating a confidence interval assumes that individual observations are independent. To keep the calculations reasonably accurate when we sample without replacement from a finite population, we should verify that the population size is at least 10 times the sample size ($N \geq 10n$).

In this setting, the sample mean $\bar{x}$ has the Normal distribution with mean μ and standard deviation $\sigma/\sqrt{n}$. Because we don't know σ, we estimate it by the sample standard deviation s. We then estimate the standard deviation of $\bar{x}$ by $s/\sqrt{n}$. This quantity is called the ***standard error*** of the sample mean $\bar{x}$.

Standard Error

When the standard deviation of a statistic is estimated from the data, the result is called the **standard error** of the statistic. The standard error of the sample mean $\bar{x}$ is $s/\sqrt{n}$.

The *t* Distributions

When we know the value of σ, we base a confidence interval for μ on the sampling distribution of $\bar{x}$, which has a Normal distribution if the Normality condition is satisfied. In that case, we find a critical value z^* using the standard Normal curve and our desired level of confidence. When we do not know σ, we substitute the standard error $s/\sqrt{n}$ of $\bar{x}$ for its standard deviation $\sigma/\sqrt{n}$. The distribution of the resulting statistic, t, is not Normal. It is a ***t distribution.***

t distribution

Unlike the standard Normal distribution, there is a different t distribution for each sample size n. We specify a particular t distribution by giving its ***degrees of freedom (df).*** When we perform inference about μ using a t distribution, the appropriate degrees of freedom is $df = n - 1$. This is because we are using the sample standard deviation s in our calculations; in Chapter 1 we briefly described why s has $n - 1$ degrees of freedom. There are other t statistics with different degrees of freedom that we will meet later. We will write the t distribution with k degrees of freedom as $t(k)$ for short. We also will refer to the standard Normal distribution as the ***z distribution.***

degrees of freedom (df)

z distribution

Activity 10C

Comparing the z and t Distributions

In this Activity, you will use the TI-83/84/89 to compare the standard Normal distribution and several *t* distribution curves with different degrees of freedom.

1. Begin by clearing any functions in Y_1, Y_2, and Y_3. Turn off all STAT PLOTS and clear the graphics screen (ClrDraw).
2. Define Y_1 = normalpdf(X). Change the graph style to a thick line. (On the TI-83/84, move to the left of Y_1 and press **ENTER**. On the TI-89, press **2nd** **F1** ([F6]) and choose 4: Thick.)
3. Next define Y_2 = tpdf(X, 2). Note that tpdf is found under the DISTR menu on the TI-83/84 and in the CATALOG on the TI-89 under Flash Apps. The second parameter, 2, specifies the degrees of freedom.
4. Set your WINDOW to $X[-3,3]_1$ and $Y[-0.1,0.4]_{0.1}$ and then GRAPH. Sketch the graphs of the two curves on your paper and write a brief description of their similarities and differences.

TI-83/84 **TI-89**

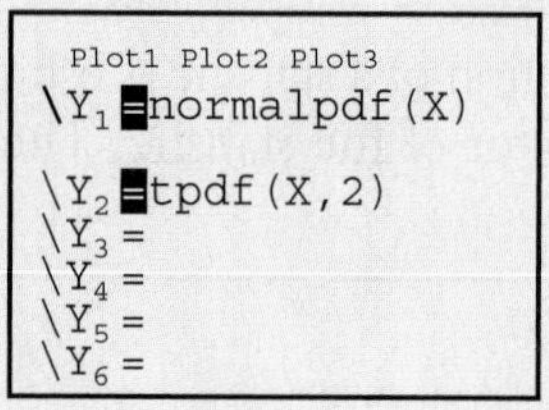

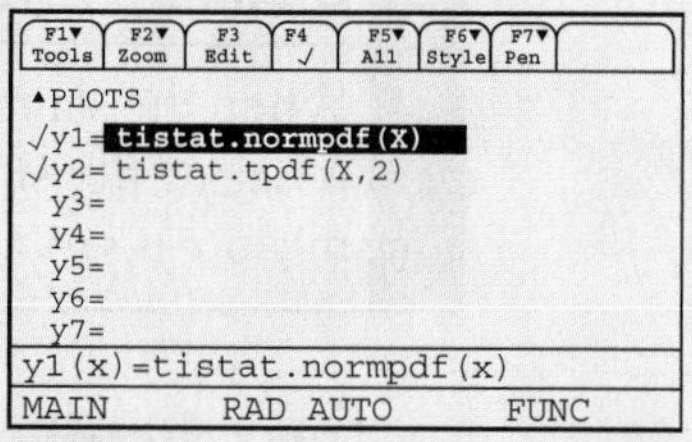

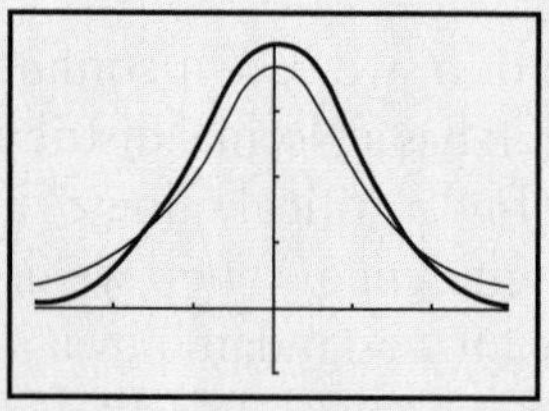

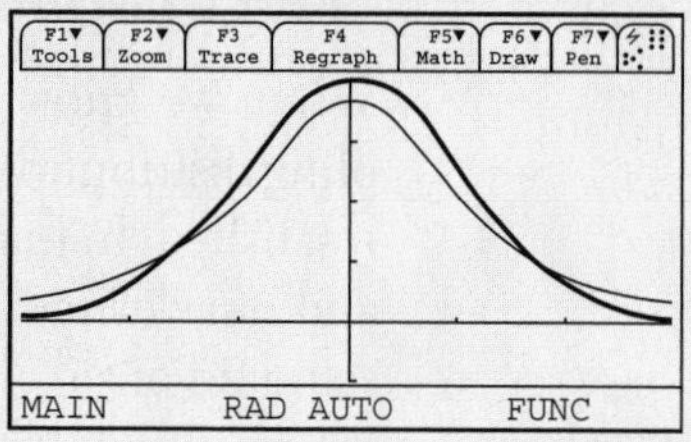

5. Change the graph style for Y_1 to a dotted line. Deselect Y_2 and define Y_3 = tpdf (X, 9). GRAPH these two functions. How do these two curves compare?
6. Deselect Y_3 and define Y_4 = tpdf(X, 30). GRAPH these two functions. What appears to be happening to the shape of the *t* distribution curve as the number of degrees of freedom increases?

Better statistics, better beer The *t* distribution and the *t* inference procedures were invented by William S. Gosset (1876–1937). Gosset worked for the Guinness brewery, and his goal in life was to make better beer. He used his new *t* procedures to find the best varieties of barley and hops. Gosset's statistical work helped him become head brewer, a more interesting title than professor of statistics.

Figure 10.10 compares the density curves of the standard Normal distribution and the *t* distributions with 2 and 9 degrees of freedom. Activity 10C and Figure 10.10 illustrate these facts about the *t* distributions:

- The density curves of the *t* distributions are similar in shape to the standard Normal curve. They are symmetric about zero, single-peaked, and bell-shaped.
- The spread of the *t* distributions is a bit greater than that of the standard Normal distribution. The *t* distributions in Figure 10.10 have more area (probability) in the tails and less in the center than does the standard Normal. This is true because substituting the estimate s for the fixed parameter σ introduces more variation.
- As the degrees of freedom k increase, the $t(k)$ density curve approaches the $N(0, 1)$ curve ever more closely. This happens because s estimates σ more accurately as the sample size increases. So using s in place of σ causes little extra variation when the sample is large.

Figure 10.10 *Density curves for the t distributions with 2 and 9 degrees of freedom and for the standard Normal distribution. All are symmetric with center 0. The t distributions have more probability in the tails than does the standard Normal.*

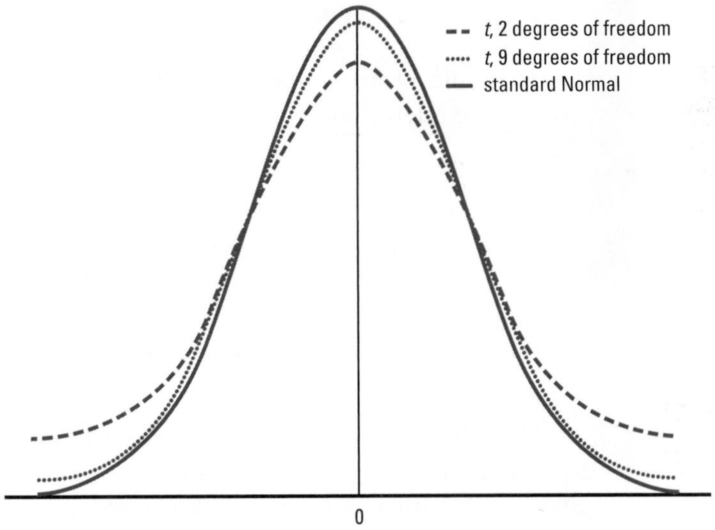

Table C in the back of the book gives critical values t^* for the *t* distributions. Each row in the table contains critical values for one of the *t* distributions; the degrees of freedom appear at the left of the row. For confidence intervals, several of the more common confidence levels C (in percent) are given at the bottom of the table. By looking down any column, you can check that the *t* critical values approach the Normal values as the degrees of freedom increase. As in the case of the table of standard Normal probabilities, statistical software often makes Table C unnecessary.

Example 10.8

Finding t^*
Using Table C

	Upper-tail probability p			
df	.05	.025	.02	.01
10	1.812	2.228	2.359	2.764
11	1.796	2.201	2.328	2.718
12	1.782	2.179	2.303	2.681
z^*	1.645	1.960	2.054	2.326
	90%	95%	96%	98%
	Confidence level C			

Suppose you want to construct a 95% confidence interval for the mean μ of a population based on an SRS of size $n = 12$. What critical value t^* should you use?

In Table C, we consult the row corresponding to df $= n - 1 = 11$. We move across that row to the entry that is directly above 95% confidence level on the bottom of the chart. The desired critical value is $t^* = 2.201$. Notice that the corresponding critical value from the z distribution is $z^* = 1.96$.

The One-Sample *t* Confidence Intervals

To construct a confidence interval for μ based on a sample from a Normal population with unknown σ, replace the standard deviation $\sigma / \sqrt{n}$ of $\bar{x}$ by its standard error $s / \sqrt{n}$ in the formula $\bar{x} \pm z^* \sigma / \sqrt{n}$ of Section 10.1. Use critical values from the t distribution with $n - 1$ degrees of freedom in place of the z critical values. The ***one-sample t interval*** is similar in both reasoning and computational detail to the z interval of Section 10.1. So we will now pay more attention to questions about using these methods in practice.

The One-Sample t Interval

Draw an SRS of size n from a population having unknown mean μ. A level C confidence interval for μ is

$$\bar{x} \pm t^* \frac{s}{\sqrt{n}}$$

where t^* is the critical value for the $t(n - 1)$ distribution. This interval is exactly correct when the population distribution is Normal and is approximately correct for large n in other cases.

The following example shows you how to construct a confidence interval for a population mean when σ is unknown. You should recognize the four-step process as the Inference Toolbox developed in Section 10.1.

Example 10.9

Auto pollution
Constructing a one-sample t interval for μ

Environmentalists, government officials, and vehicle manufacturers are all interested in studying the auto exhaust emissions produced by motor vehicles. The major pollutants in auto exhaust from gasoline engines are hydrocarbons, monoxide, and nitrogen oxides (NOX). Table 10.2 gives the NOX levels (in grams per mile) for a random sample of light-duty engines of the same type.

Table 10.2 *Amount of nitrogen oxides (NOX) emitted by light-duty engines (grams/mile)*

1.28	1.17	1.16	1.08	0.60	1.32	1.24	0.71	0.49	1.38	1.20	0.78
0.95	2.20	1.78	1.83	1.26	1.73	1.31	1.80	1.15	0.97	1.12	0.72
1.31	1.45	1.22	1.32	1.47	1.44	0.51	1.49	1.33	0.86	0.57	1.79
2.27	1.87	2.94	1.16	1.45	1.51	1.47	1.06	2.01	1.39		

Source: T. J. Lorenzen, "Determining statistical characteristics of a vehicle emissions audit procedure," *Technometrics,* 22 (1980), pp. 483–493.

Construct and interpret a 95% confidence interval for the mean amount of NOX emitted by light-duty engines of this type.

Step 1: **Parameter** The population of interest is all light-duty engines of this type. We want to estimate μ, the mean amount of the pollutant NOX emitted, for all of these engines.

Step 2: **Conditions** Since we do not know σ, we should construct a one-sample t interval for the mean NOX level μ if the conditions are satisfied.

- **SRS** The data come from a "random sample" of 46 engines from the population of all light-duty engines of this type. We are not told that the sample is an SRS, however. Our calculations may be slightly off if a different sampling method was used.
- **Normality** Is the population distribution of NOX emissions Normal? We do not know from the problem statement. Let's examine the sample data.

Figure 10.11 shows a Minitab stemplot and boxplot of the data. The distribution of NOX values in the sample is fairly symmetric if we ignore the one extremely high value in the stemplot. But the boxplot shows three high values that are outliers. Figure 10.12 (on the next page) shows a Normal probability plot from a TI-83/84 calculator. The plot is somewhat linear, although the one engine with the extremely high NOX reading is obvious once again. So we have mild concerns about the Normality of the population distribution of NOX readings. We'll proceed with caution and examine the impact of the outliers later.

Figure 10.11 *Minitab stemplot and boxplot of NOX emissions in a sample of 46 light-duty engines, for Example 10.9. Note the roughly symmetric shape and the high outliers.*

```
Stem-and-leaf of NOX        N = 46

Leaf Unit  =  0.10

   3   0  455
   7   0  6777
  10   0  899
  17   1  0011111
 (12)  1  222223333333
  17   1  4444445
  10   1  777
   7   1  888
   4   2  0
   3   2  22
   1   2
   1   2
   1   2  9
```

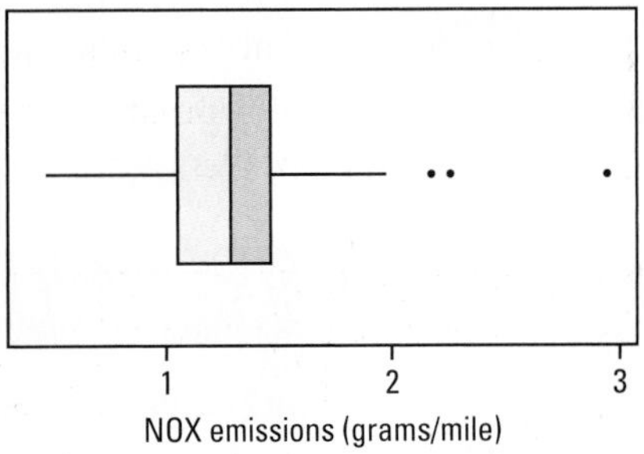

Figure 10.12 *Calculator Normal probability plot of the NOX sample data, for Example 10.9. If the data are Normally distributed, the plot will be approximately linear.*

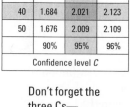

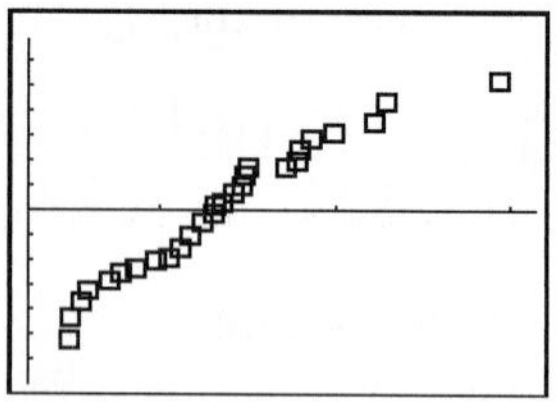

- **Independence** We must assume that there are at least (10)(46) = 460 light-duty engines of this type since we are sampling without replacement.

Upper-tail probability p			
df	.05	.025	.02
30	1.697	2.042	2.147
40	1.684	2.021	2.123
50	1.676	2.009	2.109
	90%	95%	96%
Confidence level C			

Step 3: **Calculations** Check that the mean NOX emission reading for the 46 light-duty engines in our sample is $\bar{x} = 1.329$ grams per mile. The confidence interval formula is $\bar{x} \pm t^* s/\sqrt{n}$. We use the t distribution with df = 46 − 1 = 45. Unfortunately, there is no row corresponding to 45 degrees of freedom in Table C. Instead, we use the df = 40 row, which will yield a higher critical value t^*, and thus a wider confidence interval. At a 95% confidence level, the critical value is $t^* = 2.021$. So the 95% confidence interval for μ is

$$\bar{x} \pm t^* \frac{s}{\sqrt{n}} = 1.329 \pm 2.021 \frac{0.484}{\sqrt{46}} = 1.329 \pm 0.144 = (1.185, 1.473)$$

Step 4: **Interpretation** We are 95% confident that the true mean level of nitrogen oxides emitted by this type of light-duty engine is between 1.185 and 1.473 grams/mile.

Don't forget the three Cs—conclusion, connection, context!

The one-sample t confidence interval has the form

$$\text{estimate} \pm t^* \text{SE}_{\text{estimate}}$$

where "SE" stands for "standard error." We use Table C to determine the correct value of t^* for a given confidence interval. All we need to know are the confidence level C and the degrees of freedom (df). **When the actual df does not appear in Table C, use the greatest df available that is less than your desired df.** This guarantees a wider confidence interval than we need to justify a given confidence level.

CAUTION

Exercises

10.27 Standard error Writers in some fields often summarize data by giving $\bar{x}$ and its standard error rather than $\bar{x}$ and s. The standard error of the sample mean $\bar{x}$ is often abbreviated as SEM.

(a) A medical study finds that $\bar{x} = 114.9$ and $s = 9.3$ for the seated systolic blood pressure of the 27 members of one treatment group. What is the standard error of the mean?

(b) Biologists studying the levels of several compounds in shrimp embryos reported their results in a table, with the note, "Values are means ± SEM for three independent samples." The table entry for the compound ATP was 0.84 ± 0.01. The researchers made three measurements of ATP, which had $\bar{x} = 0.84$. What was the sample standard deviation s for these measurements?

10.28 Finding critical values What critical value t^* from Table C should be used for a confidence interval for the mean of the population in each of the following situations?

(a) A 90% confidence interval based on $n = 12$ observations.

(b) A 95% confidence interval from an SRS of 30 observations.

(c) An 80% confidence interval from a sample of size 18.

10.29 Upper tail probabilities in the *z* and *t* distributions This exercise uses the TI-83/84/89 to compare upper tail probabilities in the standard Normal and several representative t distributions. Begin by clearing the graphics screen (`ClrDraw`). Set your WINDOW to $X[-3,3]_1$ and $Y[-0.1,0.4]_{0.1}$.

(a) Enter `ShadeNorm(2,100)` to shade the area under the standard Normal curve to the right of $z = 2$. Record this area (probability), rounded to four decimal places.

(b) Make a table like the one following.

df	$P(t > 2)$	Absolute difference
2		
10		
30		
50		
100		

(c) Clear the graphics screen (`ClrDraw`). Enter `Shade_t(2,100,2)`. The syntax is Shade_t(leftendpoint, rightendpoint, df). Round the probability value to four decimal places, and enter it in the second column of the table. Calculate the absolute value of the difference between the upper tail probabilities for the Normal curve and the $t(2)$ curve, and enter this value in column 3.

(d) Calculate the areas to the right of $t = 2$ for the $t(10)$, $t(30)$, $t(50)$, and $t(100)$ curves, and record your answers in the table.

(e) Describe what happens to the area to the right of $t = 2$ under the $t(k)$ distribution as the degrees of freedom increase.

10.30 Vitamin C content In 1996, the U.S. Agency for International Development provided 238,300 metric tons of corn-soy blend (CSB) for development programs and emergency relief in countries throughout the world. CSB is a highly nutritious, low-cost fortified food that is partially precooked and can be incorporated into different food preparations by the recipients. As part of a study to evaluate appropriate vitamin C levels in this commodity, measurements were taken on samples of CSB produced in a factory.[7] The following data are the amounts of vitamin C, measured in milligrams per 100 grams (mg/100 g) of blend (dry basis), for a random sample of size 8 from a production run:

26	31	23	22	11	22	14	31

(a) What conditions must be satisfied in order to make inferences about μ, the mean vitamin C content of the CSB produced during this run? Determine whether each of the conditions is met in this case.

(b) If the conditions are satisfied, construct and interpret a 95% confidence interval for μ using the Inference Toolbox. If not, explain why it would not be wise to calculate the interval.

10.31 Give it some gas! Computers in some vehicles calculate various quantities related to performance. One of these is fuel efficiency, or gas mileage, usually expressed as miles per gallon (mpg). For one vehicle equipped in this way, the mpg was recorded each time the gas tank was filled and the computer was then reset.[8] Here are the mpg values for a random sample of 20 of these records:

15.8	13.6	15.6	19.1	22.4	15.6	22.5	17.2	19.4	22.6
19.4	18.0	14.6	18.7	21.0	14.8	22.6	21.5	14.3	20.9

(a) Describe the distribution using graphical methods and summarize the results in a few sentences.

(b) Is it appropriate to use methods based on Normal distributions to analyze these data? Explain why or why not.

(c) Find and report the mean, standard deviation, standard error, and the margin of error for 95% confidence.

(d) Construct and interpret a 95% confidence interval for μ, the mean mpg for this vehicle based on these data.

(e) Do you think that this interval would apply to other similar vehicles? Explain why or why not.

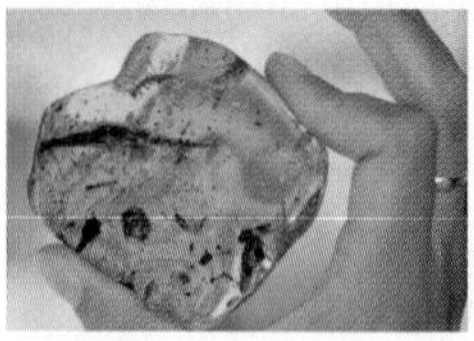

10.32 Ancient air The composition of the earth's atmosphere may have changed over time. To try to discover the nature of the atmosphere long ago, we can examine the gas in bubbles trapped inside ancient amber. Amber is a tree resin that has hardened and been trapped in rocks. The gas in bubbles within amber should be a sample of the atmosphere at the time the amber was formed. Measurements on specimens of amber from the late Cretaceous era (75 to 95 million years ago) give these percents of nitrogen:[9]

63.4	65.0	64.4	63.3	54.8	64.5	60.8	49.1	51.0

Assume (this is not yet agreed on by experts) that these observations are an SRS from the late Cretaceous atmosphere.

(a) Graph the data, and comment on skewness and outliers. The t procedures will be only approximate for these data.

(b) Construct and interpret a 95% confidence interval for the mean percent of nitrogen in ancient air.

Paired *t* Procedures

Comparative studies are more convincing than single-sample investigations. For that reason, one-sample inference is less common than comparative inference. A common design to compare two treatments makes use of one-sample procedures. Recall from Chapter 5 that in a *matched pairs design*, subjects are matched in pairs and each treatment is given to one subject in each pair. Alternatively, each subject receives both treatments in some order. A coin toss can be used to assign the treatments to the two subjects in each pair or to determine the order in which an individual subject receives the two treatments. Another situation calling for ***paired t procedures*** is before-and-after observations on the same subjects.

Paired t Procedures

To compare the responses to the two treatments in a matched pairs design or before-and-after measurements on the same subjects, apply one-sample t procedures to the observed differences.

The parameter μ in a paired t procedure is

- the mean difference in the responses to the two treatments within matched pairs of subjects in the entire population (when subjects are matched in pairs), or
- the mean difference in response to the two treatments for individuals in the population (when the same subject receives both treatments), or
- the mean difference between before-and-after measurements for all individuals in the population (for before-and-after observations on the same individuals).

Example 10.10 illustrates use of the paired t procedures.

Example 10.10 *Is caffeine dependence real?*
The paired t procedures

Our subjects are 11 people diagnosed as being dependent on caffeine. Each subject was barred from coffee, colas, and other substances containing caffeine. Instead, they took capsules containing their normal caffeine intake. During a different time period, they took placebo capsules. The order in which subjects took caffeine and the placebo was randomized. Table 10.3 (on the next page) contains data on two of several tests given to the subjects. "Depression" is the score on the Beck Depression Inventory. Higher scores show more symptoms of depression. "Beats" is the number of beats per minute the subject achieved when asked to press a button 200 times as quickly as possible. We are interested in whether being deprived of caffeine affects these outcomes. Let's construct and interpret a 90% confidence interval for the mean change in depression score. As always, we follow the Inference Toolbox format.

Step 1: Parameter The population of interest is all people who are dependent on caffeine. We want to estimate the mean difference $\mu_{\text{DIFF}} = \mu_{\text{PLACEBO - CAFFEINE}}$ in depression

Table 10.3 *Results of a caffeine-deprivation study*

Subject	Depression (caffeine)	Depression (placebo)	Number of beats (caffeine)	Number of beats (placebo)
1	5	16	281	201
2	5	23	284	262
3	4	5	300	283
4	3	7	421	290
5	8	14	240	259
6	5	24	294	291
7	0	6	377	354
8	0	3	345	346
9	2	15	303	283
10	11	12	340	391
11	1	0	408	411

Source: E. C. Strain et al., "Caffeine dependence syndrome: evidence from case histories and experimental evaluation," *Journal of the American Medical Association,* 272 (1994), pp. 1604–1607.

score that would be reported if all individuals in the population took both the caffeine capsule and the placebo.

Step 2: **Conditions** If the conditions are met, we will use one-sample t procedures to construct a confidence interval for μ_{DIFF} since the population standard deviation σ of the differences in depression scores is unknown.

- **SRS** These data probably do not come from an SRS from the population of interest. Subjects in such experiments tend to be volunteers. They may not be truly representative of the population of interest. As a result, we may have trouble generalizing the results of this study to the population of caffeine-dependent people. However, since the researchers randomly assigned the order in which each subject took the placebo and the caffeine tablet, any consistent differences we observe in subjects' responses should be due to the treatments.
- **Normality** Is the population distribution of differences Normal? We don't know this, so we examine the sample data. Figure 10.13 shows a Fathom boxplot and Normal probability plot (Fathom software calls this a Normal quantile plot) of the differences (placebo − caffeine) in depression scores for our 11 subjects. There are no obvious outliers or other departures from Normality in the sample data. We are given no reason to doubt the Normality of the population of differences.
- **Independence** We need to feel comfortable that the difference (placebo − caffeine) values for the 11 subjects are independent. This seems reasonable based on the design of the experiment. Note that the two depression scores for each subject are *not* independent, however. We would expect the same individual to show similar tendencies under both treatments.

Step 3: **Calculations** For our 11 subjects, $\bar{x}_{\text{PLACEBO} - \text{CAFFEINE}} = \bar{x}_{\text{DIFF}} = 7.364$ and $s_{\text{DIFF}} = 6.918$. The t critical value for a 90% confidence interval with $11 - 1 = 10$ degrees of freedom is $t^* = 1.812$. So the desired confidence interval is

$$\bar{x}_{\text{DIFF}} \pm t^* \frac{s_{\text{DIFF}}}{\sqrt{n}} = 7.364 \pm 1.812 \frac{6.918}{\sqrt{11}} = 7.364 \pm 3.780 = (3.584, 11.144)$$

Figure 10.13 *Fathom boxplot and Normal probability plot (Fathom software calls it a Normal quantile plot) of the differences in Beck Depression Inventory scores for the subjects in the matched pairs experiment, for Example 10.10.*

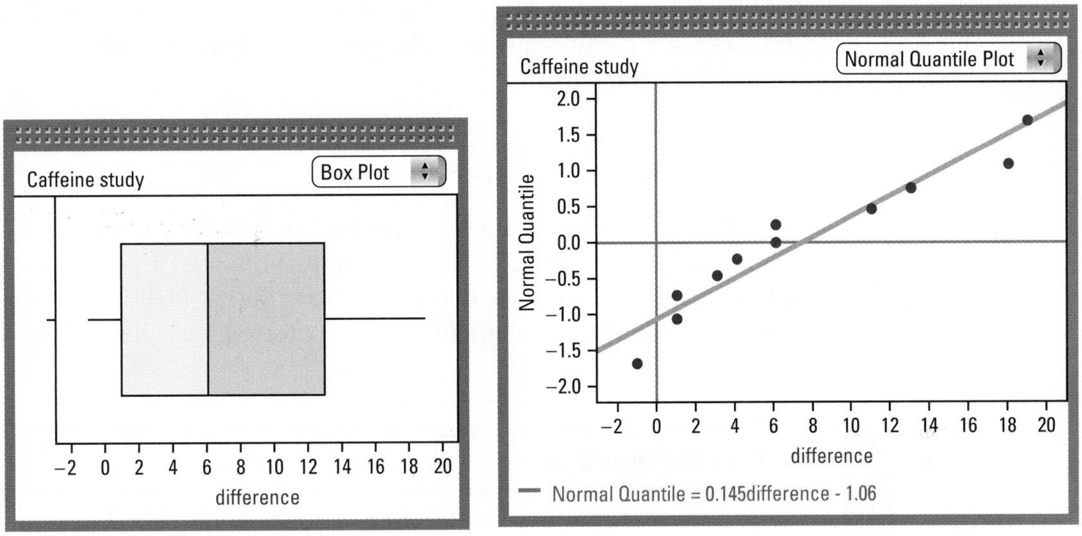

Step 4: **Interpretation** We are 90% confident that the actual mean difference in depression score for the population is between 3.584 and 11.144 points. That is, we estimate that caffeine-dependent individuals would score, on average, between 3.6 and 11.1 points higher on the Beck Depression Inventory when they are given a placebo instead of caffeine. This study provides evidence that withholding caffeine from caffeine-dependent individuals may lead to depression. However, the fact that the subjects in this study were not an SRS from the population of interest prevents us from generalizing any further.

Many studies that require the use of paired t procedures involve individuals who are *not* chosen at random from the population of interest, like the caffeine-dependent subjects in the previous example. In such cases, we may not be able to generalize our findings to the population of interest. By randomly assigning the treatments, however, the researchers helped ensure that the sizable mean difference in depression scores could be attributed to the caffeine treatment. *Random selection* of individuals for a statistical study allows us to generalize the results of that study to a larger population. *Random assignment* of treatments to subjects in an experiment lets us investigate whether there is evidence of a treatment effect, which might suggest that the treatment caused the observed difference. **Be sure you understand the different purposes of random selection and random assignment!**

CAUTION

Example 10.10 shows how to turn paired data into single-sample data by taking differences within each pair. We are making inferences about a single population, the population of differences for all caffeine-dependent individuals. It would be incorrect to ignore the pairing and analyze the data as if we had two samples, one from subjects who took a placebo pill and the other from subjects who took a caffeine pill. Inference procedures for two samples assume that the samples are selected independently of each other. This assumption does not hold when the same subjects are measured twice. **The proper analysis depends on the design used to produce the data.**

Robustness of *t* Procedures

The t confidence interval is exactly correct when the distribution of the population is exactly Normal. No real data are exactly Normal. The usefulness of the t procedures in practice therefore depends on how strongly they are affected by lack of Normality. Procedures that are not strongly affected are called ***robust.***

Robust Procedures

An inference procedure is called **robust** if the probability calculations involved in that procedure remain fairly accurate when a condition for use of the procedure is violated. For confidence intervals, this means that the stated confidence level is still fairly accurate.

If outliers are present in the sample data, then the population may not be Normal. The t procedures are *not* robust against outliers, because $\bar{x}$ and s are not resistant to outliers.

Example 10.11 *More NOX emissions*

t procedures are *not* robust against outliers

In Example 10.9, we constructed a confidence interval for the mean level of NOX emitted by a specific type of light-duty car engine. One of the 46 engines in our sample emitted an unusually high amount (2.94 grams/mile) of NOX. If we remove that single data point, $\bar{x} = 1.293$ and $s = 0.424$ for the remaining 45 sample values. The confidence interval based on this sample of 45 engines would be

$$\bar{x} \pm t^* \frac{s}{\sqrt{n}} = 1.293 \pm 2.021 \frac{0.424}{\sqrt{45}} = 1.293 \pm 0.128 = (1.165, 1.421)$$

Our new confidence interval is narrower and is centered at a lower value than our original interval of 1.185 to 1.473.

Can we be 95% confident in either of these intervals? Not really, because only intervals constructed in this way from a Normal population distribution would capture the true μ in 95% of all possible samples of engines. The outlier suggests that the population distribution of NOX emissions may not be Normal.

As Example 10.11 illustrates, a single outlier can have a drastic effect on the results of the t procedures. Moreover, the presence of an outlier prevents us from claiming 95% confidence, since the probability calculations involved depend on a Normal distribution.

Fortunately, the t procedures are quite robust against non-Normality of the population when there are no outliers, especially when the distribution is roughly symmetric. Larger samples improve the accuracy of critical values from the t distributions when the population is not Normal. The main reason for this is the central limit theorem. The t statistic is based on the sample mean $\bar{x}$, which becomes more nearly Normal as the sample size gets larger even when the population does not have a Normal distribution.

Always make a plot to check for skewness and outliers before you use the t procedures for small samples. For most purposes, you can safely use the one-sample t procedures when $n \geq 15$ unless an outlier or quite strong skewness is present. Here are practical guidelines for inference on a single mean.[10]

Using the t Procedures

- Except in the case of small samples, the assumption that the data are an SRS from the population of interest is more important than the assumption that the population distribution is Normal.
- *Sample size less than 15.* Use t procedures if the data are close to Normal. If the data are clearly non-Normal or if outliers are present, do not use t procedures.
- *Sample size at least 15.* The t procedures can be used except in the presence of outliers or strong skewness.
- *Large samples.* The t procedures can be used even for clearly skewed distributions when the sample is large, say $n \geq 30$.

If your sample data would give a biased estimate for some reason, then you shouldn't bother computing a t interval. Or if the data you have *is* the entire population of interest, then there's no need to perform inference.

Example 10.12 *The elderly, lightning strikes, and Shakespeare*
Can we use t?

Figure 10.14 (on the next page) shows histograms of three different data sets.

- Figure 10.14(a) is a histogram of the percent of each state's residents who are at least 65 years of age. We have data on the entire population of 50 states, so formal inference makes no sense. We can calculate the exact mean for the population. There is no uncertainty due to having only a sample from the population, and no need for a confidence interval.
- Figure 10.14(b) shows the time of the first lightning strike each day in a mountain region in Colorado. The data contain more than 70 observations that have a symmetric

Figure 10.14 *Can we use* t *procedures for these data? (a) Percent of residents aged 65 and over in the states. No: this is an entire population, not a sample. (b) Times of first lightning strike each day at a site in Colorado. Yes: there are over 70 observations with a symmetric distribution. (c) Word lengths in Shakespeare's plays.* Yes, if the sample size is large enough *to overcome the right-skewness.*

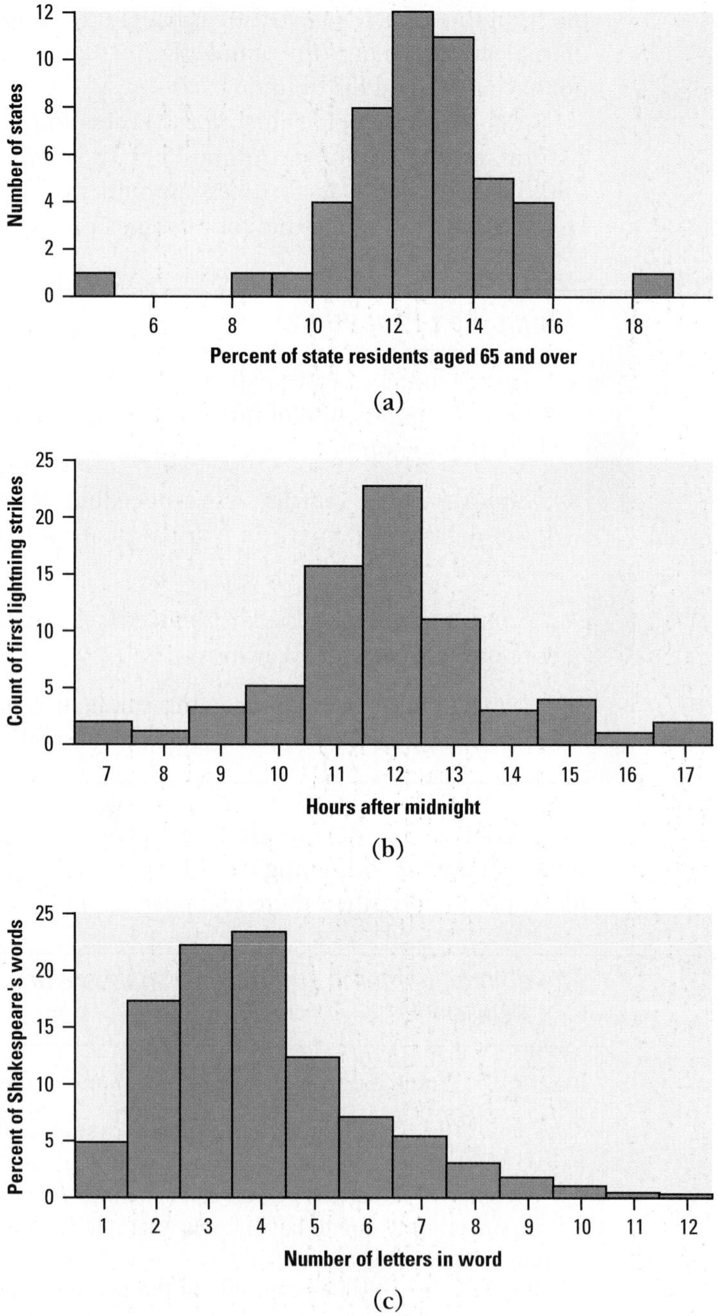

distribution. You can use the t procedures to draw conclusions about the mean time of a day's first lightning strike with complete confidence.

- Figure 10.14(c) shows that the distribution of word lengths in Shakespeare's plays is skewed to the right. We aren't told how large the sample is. You can use the t procedures for a distribution like this if the sample size is roughly 30 or larger.

Exercises

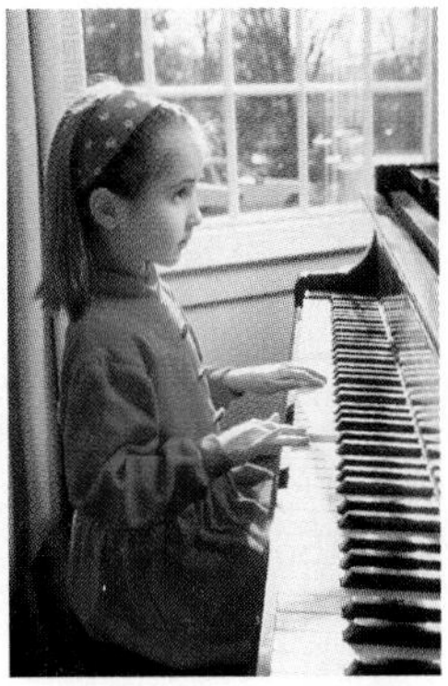

10.33 Does playing the piano make you smarter? Do piano lessons improve the spatial-temporal reasoning of preschool children? Neurobiological arguments suggest that this may be true. A study designed to test this hypothesis measured the spatial-temporal reasoning of 34 preschool children before and after six months of piano lessons.[11] (The study also included children who took computer lessons and a control group; but we are not concerned with those here.) The changes in the reasoning scores are

2	5	7	−2	2	7	4	1	0	7	3	4
3	4	9	4	5	2	9	6	0	3	6	−1
3	4	6	7	−2	7	−3	3	4	4		

(a) Display the data and summarize the distribution.

(b) Construct and interpret a 95% confidence interval for the mean improvement in reasoning scores.

(c) Can we conclude that playing the piano causes an improvement in spatial-temporal reasoning in children? Justify your answer.

10.34 Caffeine dependence

(a) The study in Example 10.10 (page 651) was double-blind. What does this mean?

(b) Examine the differences in beats per minute with and without caffeine with graphs and numerical summaries. Write a few sentences describing what you see.

(c) Is it appropriate to use t procedures to construct a 90% confidence interval for the mean difference in beats per minute? If so, do it. If not, explain why not.

10.35 The moon made me do it! Many people believe that the moon influences the actions of some individuals. A study of dementia patients in nursing homes recorded various types of disruptive behaviors every day for 12 weeks. Days were classified as moon days if they were in a three-day period centered at the day of the full moon. For each patient, the average number of disruptive behaviors was computed for moon days and for all other days. The data for the 15 subjects whose behaviors were classified as aggressive are presented in Table 10.4 (on the next page).

(a) Construct and interpret a 95% confidence interval for the mean difference in aggressive behaviors on moon days and other days among dementia patients.

(b) Can we conclude that the moon is causing dementia patients to become more aggressive? Why or why not?

Table 10.4 *Aggressive behaviors of dementia patients*

Patient	Moon days	Other days	Patient	Moon days	Other days
1	3.33	0.27	9	6.00	1.59
2	3.67	0.59	10	4.33	0.60
3	2.67	0.32	11	3.33	0.65
4	3.33	0.19	12	0.67	0.69
5	3.33	1.26	13	1.33	1.26
6	3.67	0.11	14	0.33	0.23
7	4.67	0.30	15	2.00	0.38
8	2.67	0.40			

Source: These data were collected as part of a larger study of dementia patients conducted by Nancy Edwards, School of Nursing, and Alan Beck, School of Veterinary Medicine, Purdue University.

23	0
24	0
25	
26	5
27	
28	7
29	
30	259
31	399
32	033677
33	0236

10.36 Should we use *t*? In each of the following situations, discuss whether it would be appropriate to construct a t interval to estimate the population mean.

(a) We want to estimate the average age at which U.S. presidents have died. So we obtain a list of all U.S. presidents who have died and their ages at death.

(b) How much time do students spend on the Internet? We collect data from the 32 members of our AP Statistics class and calculate the mean amount of time that each student spent on the Internet yesterday.

(c) The stemplot in the margin shows the force required to pull apart 20 pieces of Douglas fir.

Section 10.2 Summary

Confidence intervals for the mean μ of a Normal population are based on the sample mean $\bar{x}$ of an SRS. Because of the central limit theorem, the resulting procedures are approximately correct for other population distributions when the sample is large.

When we know σ, we use the z critical value and the standard Normal distribution to help calculate confidence intervals. In practice, we do not know σ. Replace the standard deviation $\sigma/\sqrt{n}$ of $\bar{x}$ by the **standard error** $s/\sqrt{n}$ and use the **t distribution** with $n - 1$ **degrees of freedom.**

There is a t distribution for every positive degrees of freedom k. All are symmetric distributions similar in shape to the standard Normal distribution. The $t(k)$ distribution approaches the N(0, 1) distribution as k increases.

An exact level C **confidence interval** for the mean μ of a Normal population is

$$\bar{x} \pm t^* \frac{s}{\sqrt{n}}$$

where t^* is the critical value of the $t(n - 1)$ distribution.

Use this one-sample procedure to analyze **paired data** by using as the data the observed difference for each pair.

The t procedures are relatively **robust** when the population is non-Normal, especially for larger sample sizes. The t procedures are useful for non-Normal data when $n \geq 15$ unless the data show outliers or strong skewness.

Section 10.2 Exercises

10.37 More critical values What critical value t^* from Table C would you use for a confidence interval for the mean of the population in each of the following situations?

(a) A 95% confidence interval based on $n = 10$ observations.

(b) A 99% confidence interval from an SRS of 20 observations.

(c) An 80% confidence interval from a sample of size 7.

10.38 Red wine is good for the heart Observational studies suggest that moderate use of alcohol reduces heart attacks, and that red wine may have special benefits. One reason may be that red wine contains polyphenols, substances that do good things to cholesterol in the blood and so may reduce the risk of heart attacks. In an experiment, healthy men were assigned at random to several groups. One group of 9 men drank half a bottle of red wine each day for two weeks. The level of polyphenols in their blood was measured before and after the two-week period. Here are the percent changes in level:[12]

3.5	8.1	7.4	4.0	0.7	4.9	8.4	7.0	5.5

(a) Make appropriate graphical displays to assess the Normality condition. What do you conclude?

(b) Construct and interpret a 90% confidence interval for the mean percent change μ in the population of interest. Be sure to define the population clearly.

(c) Explain why the data are paired in this situation based on the design of the study. Is this a matched pairs experiment?

10.39 A big-toe problem, I Hallux abducto valgus (call it HAV) is a deformation of the big toe that is uncommon in youth and often requires surgery. Doctors used X-rays to measure the angle (in degrees) of deformity in 38 consecutive patients under the age of 21 who came to a medical center for surgery to correct HAV. The angle is a measure of the seriousness of the deformity. Here are the data:[13]

28	32	25	34	38	26	25	18	30	26	28	13	20
21	17	16	21	23	14	32	25	21	22	20	18	26
16	30	30	20	50	25	26	28	31	38	32	21	

We are willing to consider these patients as a random sample of young patients who require HAV surgery. Give a 95% confidence interval for the mean HAV angle in the population of all such patients. Follow the Inference Toolbox.

10.40 A big-toe problem, II The data in the previous problem follow a Normal distribution quite closely except for one patient with HAV angle 50 degrees, a high outlier.

(a) Find the 95% confidence interval for the population mean based on the 37 patients who remain after you drop the outlier.

(b) Compare your interval in (a) with your interval from the previous problem. What is the most important effect of removing the outlier?

10.41 Spanish teachers' workshop The National Endowment for the Humanities sponsors summer institutes to improve the skills of high school language teachers. One institute hosted 20 Spanish teachers for four weeks. At the beginning of the period, the teachers took the Modern Language Association's listening test of understanding of spoken Spanish. After four weeks of immersion in Spanish in and out of class, they took the listening test again. (The actual spoken Spanish in the two tests was different, so that simply taking the first test should not improve the score on the second test.) Table 10.5 gives the pretest and posttest scores. The maximum possible score on the test is 36.

Table 10.5 *MLA listening scores for 20 Spanish teachers*

Subject	Pretest	Posttest	Subject	Pretest	Posttest
1	30	29	11	30	32
2	28	30	12	29	28
3	31	32	13	31	34
4	26	30	14	29	32
5	20	16	15	34	32
6	30	25	16	20	27
7	34	31	17	26	28
8	15	18	18	25	29
9	28	33	19	31	32
10	20	25	20	29	32

Source: Data provided by Joseph A. Wipf, Department of Foreign Languages and Literatures, Purdue University.

(a) Make a graphical check for outliers or strong skewness in the data. What do you conclude?

(b) Construct and interpret a 90% confidence interval for the mean increase in listening score due to attending the summer institute.

(c) Can you conclude that attending the institute causes Spanish teachers to improve their listening skills? Why or why not?

10.42 ADHD and organization, I In a study of children with attention deficit hyperactivity disorder (ADHD), parents were asked to rate their child on a variety of items related to how well their child performs different tasks.[14] One item was "has difficulty organizing work," rated on a five-point scale of 0 to 4, with 0 corresponding to "not at all" and 4, corresponding to "very much." The mean rating for 282 boys with ADHD was reported as 2.22 with a standard deviation of 1.03.

(a) Do you think that these data are Normally distributed? Explain why or why not.

(b) Is it appropriate to use the methods of this section to compute a 99% confidence interval? Explain why or why not.

(c) If appropriate, calculate the 99% confidence interval. Write a sentence explaining the interval and the meaning of the 99% confidence level.

(d) The boys in this study were all evaluated at the Western Psychiatric Institute and Clinic at the University of Pittsburgh. To what extent do you think the results could be generalized to boys with ADHD in other locations?

10.43 ADHD and organization, II Refer to the previous exercise. Compute the 90% and 95% confidence intervals. Display the three intervals graphically and write a short explanation of the effect of the confidence level on the width of the interval.

10.44 Does nature heal best? Differences of electric potential occur naturally from point to point on a body's skin. Is the natural electric field strength best for helping wounds to heal? If so, changing the field will slow healing. The research subjects are anesthetized newts.

Make a razor cut in both hind limbs. Let one heal naturally (the control). Use an electrode to change the electric field in the other to half its normal value. After two hours, measure the healing rate. Table 10.6 gives the healing rates (in micrometers per hour) for 14 newts.

Table 10.6 *Healing rates (micrometers per hour) for newts*

Newt	Experimental limb	Control limb	Difference in healing	Newt	Experimental limb	Control limb	Difference in healing
13	24	25	−1	20	33	36	−3
14	23	13	10	21	28	35	−7
15	47	44	3	22	28	38	−10
16	42	45	−3	23	21	43	−22
17	26	57	−31	24	27	31	−4
18	46	42	4	25	25	26	−1
19	38	50	−12	26	45	48	−3

Source: D. D. S. Iglesia, E. J. Cragoe, Jr., and J. W. Vanable, "Electric field strength and epithelization in the newt (Notophthalmus viridescens)," *Journal of Experimental Zoology*, 274 (1996), pp. 56–62.

(a) As is usual, the paper did not report these raw data. Readers are expected to be able to interpret the summaries that the paper did report. The paper summarized the differences in the table above as "-5.71 ± 2.82" and said, "All values are expressed as means ± standard error of the mean." Show carefully where the numbers −5.71 and 2.82 come from.

(b) Use the summary information from part (a) to give a 90% confidence interval for the amount by which changing the field changes the rate of healing. Then explain in a sentence what it means to say that you are "90% confident" of your result.

(c) Check the conditions that are required for calculating the interval in part (b)—after the fact! Do you still feel comfortable about your conclusions?

Technology Toolbox

t intervals

Confidence intervals using *t* procedures can be constructed on the TI-83/84/89, thus avoiding the use of tables. Here is a brief summary of the techniques using the healing-rates data from

(continued)

Technology Toolbox

t intervals *(continued)*

Exercise 10.44. For reference, the differences in healing rates for the 14 newts (in micrometers per hour) are

−1	10	3	−3	−31	4	−12	−3	−7	−10	−22	−4	−1	−3

Enter these data in L_1/list1.

On the TI-83/84, all inference routines are found under STAT/TESTS. On the TI-89, all inference routines can be accessed from inside the Stats/List Editor APP. Choose F7 (Ints) for confidence intervals.

To determine a confidence interval for these data:

TI-83/84

- Choose 8:TInterval.

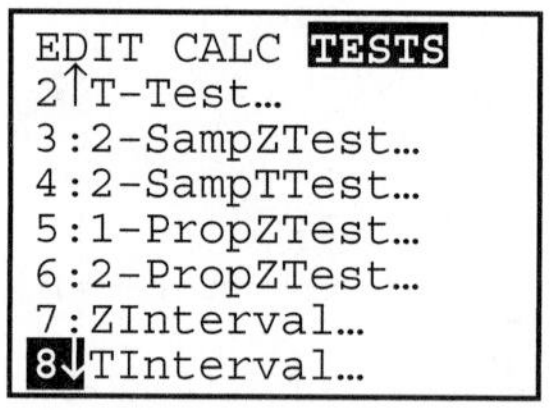

TI-89

- Choose 2:TInterval.

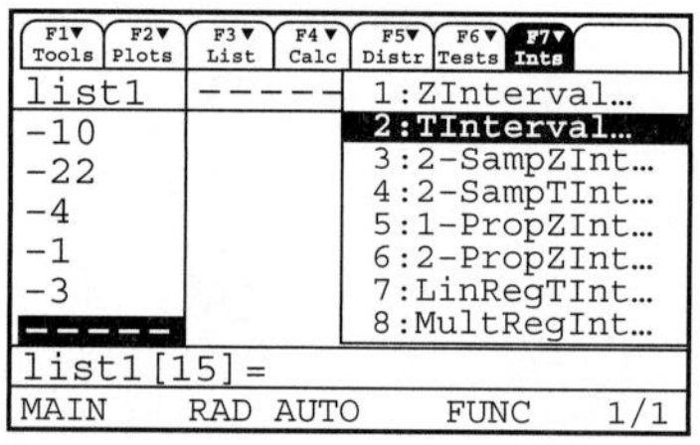

- Choose "Data" (not "Stats") and adjust the TInterval screen as shown.

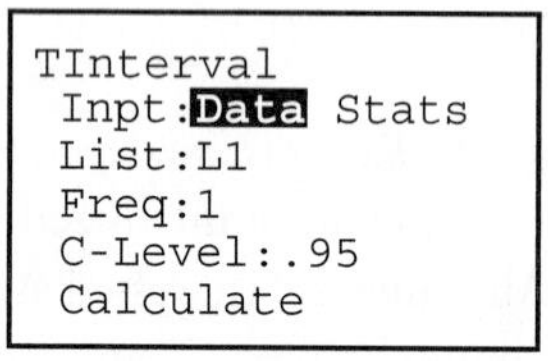

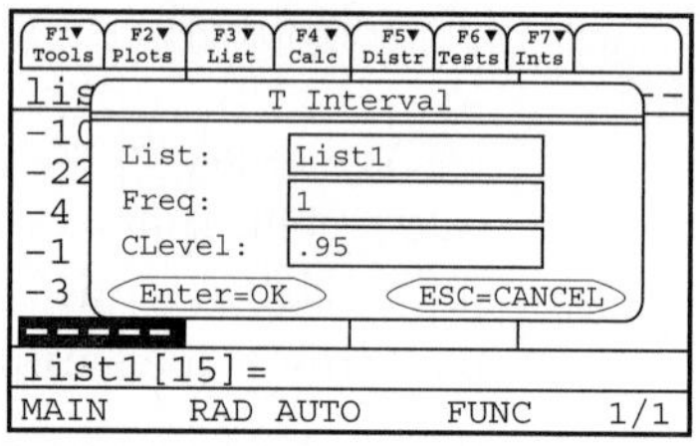

- Select "Calculate" and press ENTER.

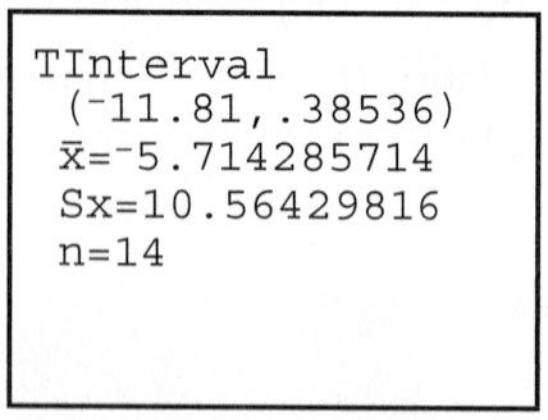

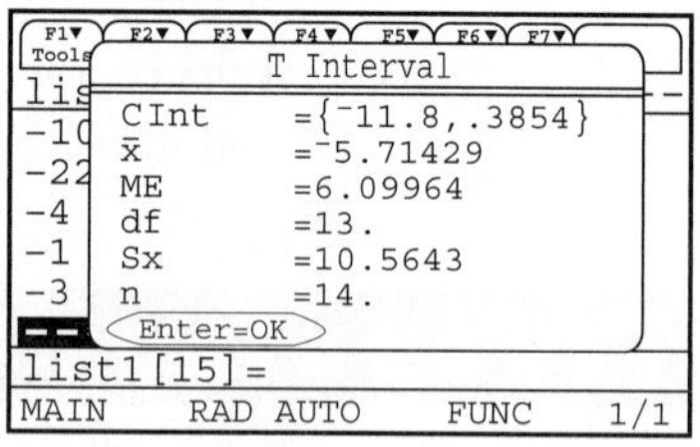

The results tell us that the 95% confidence interval for the true mean population difference in healing rate is between −11.81 and 0.385 micrometers per hour. If the researchers wanted to keep the 95% confidence level but wanted a shorter, more precise confidence interval, they would need to use more newts in the experiment (that is, increase the sample size n).

10.3 Estimating a Population Proportion

Activity 10D

Give me a kiss!

Materials: Bag of plain Hershey's Kisses (enough for one Kiss per student)

When you flip a fair coin, it is equally likely to land "heads" or "tails." Do plain Hershey's Kisses behave in the same way? In this activity, you will toss a Hershey's Kiss several times and observe whether it comes to rest on its side (S) or on its base (B). The question you are trying to answer is: what proportion of the time does a tossed Hershey's Kiss settle on its base?

1. Before you begin, make a guess about what will happen. If you could toss your Hershey's Kiss over and over and over, what proportion of all tosses do you think would settle on the base?
2. Toss your Hershey's Kiss 50 times. Record the result of each toss (S or B) in a table like the one shown. In the third column, calculate the proportion of base landings you have obtained so far.

Toss	Outcome	Cumulative proportion of B's
1	B	1/1 = 1.00
2	S	1/2 = 0.50
3	S	1/3 = 0.33
⋮	⋮	⋮

3. Make a scatterplot with the number of tosses on the horizontal axis and the cumulative proportion of B's on the vertical axis. Connect consecutive points with a line segment. Does the overall proportion of B's seem to be approaching a single value?
4. **SRS** Your set of 50 tosses can be thought of as a simple random sample from the population of all possible tosses of your Hershey's Kiss. The parameter p is the (unknown) population proportion of tosses that would land on the base. What is your best estimate for p? It's $\hat{p}$, the proportion of B's in your 50 tosses. Record your value of $\hat{p}$. How does it compare with the conjecture you made in Step 1?
5. If you tossed your Hershey's Kiss 50 more times (don't do it!), would you expect to get the same value of $\hat{p}$? In Chapter 9, we learned that the values of $\hat{p}$ in repeated samples could be described by a sampling distribution. The mean of the sampling distribution $\mu_{\hat{p}}$ is equal to the population proportion p. How far will your sample proportion $\hat{p}$ be from the true value p? If the sampling distribution is approximately Normal,

then the 68–95–99.7 rule tells us that about 95% of all $\hat{p}$-values will be within two standard deviations of p.

6. **Normality** The sampling distribution of $\hat{p}$ will be approximately Normal if $n\hat{p} \geq 10$ and $n(1 - \hat{p}) \geq 10$. Verify that the Normality condition is satisfied for your sample.
7. **Independence** What assumption would you have to make about your Kiss in order to claim that the outcomes of the 50 tosses are independent?
8. Estimate the standard deviation of the sampling distribution by computing $\sqrt{\frac{\hat{p}(1-\hat{p})}{n}}$ using your value of $\hat{p}$. This is the formula we developed in Chapter 9 but with p replaced by $\hat{p}$.
9. Construct the interval $\hat{p} \pm 2\sqrt{\frac{\hat{p}(1-\hat{p})}{n}}$ based on your sample of 50 tosses. This is called a *confidence interval* for p.
10. Your teacher will draw a number line with a scale marked off from 0 to 1 that has tick marks every 0.05 units. Draw your confidence interval above the number line. Your classmates will do the same. Do most of the intervals overlap? If so, what values are contained in all of the overlapping intervals?
11. About 95% of the time, the sample proportion $\hat{p}$ of base-landing tosses will be within two standard deviations of the actual population proportion of base-landing tosses of a Hershey's Kiss. But if $\hat{p}$ is within two standard deviations of p, then p is within two standard deviations of $\hat{p}$. So about 95% of the time, the interval $\hat{p} \pm 2\sqrt{\frac{\hat{p}(1-\hat{p})}{n}}$ will contain the true proportion p.
12. There is no way to know whether the confidence interval you constructed in Step 9 actually "catches" the true proportion p of times that your Hershey's Kiss will land on its base. What we can say is that the method you used in Step 9 will succeed in capturing the unknown population parameter about 95% of the time. Likewise, we expect that about 95% of all the confidence intervals drawn by the members of your class in Step 10 will capture p.

Our discussion of statistical inference to this point has concerned making inferences about population means. But we often want to answer questions about the proportion of some outcome in the population. Here are some examples.

- What proportion of U.S. adults are unemployed right now?
- What proportion of teenagers have a computer with Internet access in their bedroom?

- What proportion of college students pray on a daily basis?
- What proportion of preteens have a cell phone?
- What proportion of Californians approve of President Bush's handling of the situation in Iraq?

Example 10.13 ***Binge drinking in college***
Estimating a population proportion

Alcohol abuse has been described by college presidents as the number one problem on campus, and it is an important cause of death in young adults. How common is it? A 2001 survey of 10,904 U.S. college students collected information on drinking behavior and alcohol-related problems.[15] The researchers defined "frequent binge drinking" as having five or more drinks in a row three or more times in the past two weeks. According to this definition, 2486 students were classified as frequent binge drinkers. That's 22.8% of the sample. Based on these data, what can we say about the proportion of all college students who have engaged in frequent binge drinking?

Who is a smoker? When estimating a proportion p, be sure you know what counts as a "success." The news says that 20% of adolescents smoke. Shocking. It turns out that this is the percent who smoked at least once in the past month. If we say that a smoker is someone who smoked on at least 20 of the past 30 days and smoked at least half a pack on those days, fewer than 4% of adolescents qualify.

We are interested in the unknown proportion p of a population that has some outcome. For convenience, call the outcome we are looking for a "success." In Example 10.13, the population is U.S. college students, and the parameter p is the proportion who have engaged in frequent binge drinking during the two weeks prior to the survey. To estimate p, the Harvard School of Public Health sent surveys to a random sample of undergraduate students at 120 colleges and universities. Of the 10,904 students who responded, 2486 were frequent binge drinkers. The statistic that estimates the parameter p is the sample proportion

$$\hat{p} = \frac{\text{number of success in the sample}}{\text{number of individuals in the sample}}$$

$$= \frac{2486}{10{,}904} = 0.228$$

Conditions for Inference about a Proportion

As always, inference is based on the sampling distribution of a statistic. We described the sampling distribution of a sample proportion $\hat{p}$ in Section 9.2. Here is a brief review of its important properties:

Center The mean is p. That is, the sample proportion $\hat{p}$ is an unbiased estimator of the population proportion p.

Spread The standard deviation of $\hat{p}$ is $\sqrt{p(1-p)/n}$, provided that the population is at least 10 times as large as the sample.

Shape If the sample size is large enough that both np and $n(1-p)$ are at least 10, the distribution of $\hat{p}$ is approximately Normal.

Figure 10.15 (on the next page) displays this sampling distribution.

In practice, of course, we don't know the value of p. (If we did, we wouldn't need to construct a confidence interval for it!) So we cannot check whether np

Figure 10.15 *Select a large SRS from a population that contains proportion* p *of successes. The sampling distribution of the proportion* $\hat{p}$ *of successes in the sample is approximately Normal. The mean is* p *and the standard deviation is* $\sqrt{p(1-p)/n}$.

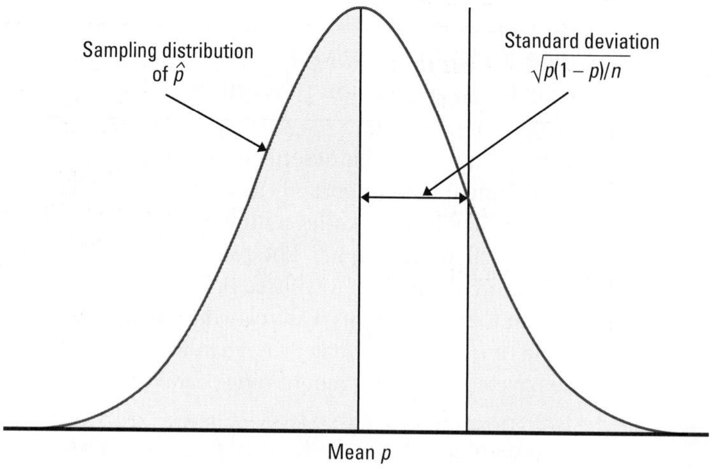

and $n(1 - p)$ are 10 or greater. In large samples, $\hat{p}$ will be close to p. Therefore, we replace p by $\hat{p}$ in determining the values of np and $n(1 - p)$. We also replace the standard deviation by the ***standard error of*** $\hat{p}$

standard error of $\hat{p}$

$$\text{SE} = \sqrt{\frac{\hat{p}\,(1-\hat{p})}{n}}$$

to get a confidence interval of the form

$$\text{estimate} \pm z^{*}\text{SE}_{\text{estimate}}$$

The requirements for using the z procedures for calculating a confidence interval about a population proportion are stated in terms of $\hat{p}$.

Conditions for Inference about a Proportion

- **SRS** The data are an SRS from the population of interest.
- **Normality** For a confidence interval, n is so large that both the count of successes $n\hat{p}$ and the count of failures $n(1 - \hat{p})$ are 10 or more.
- **Independence** Individual observations are independent. When sampling without replacement, the population is at least 10 times as large as the sample.

If you have a small sample or a sampling design more complex than an SRS, you can still do inference, but the details are more complicated. Get expert advice.

Example 10.14 *More on binge drinking in college*

Are the conditions met?

We want to use the Harvard School of Public Health survey data to give a confidence interval for the proportion of college students who have engaged in frequent binge drinking. Does the sample meet the requirements for inference?

- **SRS** The sampling design was in fact a complex stratified random sample, and the survey used inference procedures for that design. The overall effect is close to an SRS, however.
- **Normality** The counts of "Yes" and "No" responses are much greater than 10:

$$n\hat{p} = (10{,}904)(0.228) = 2486$$

$$n(1 - \hat{p}) = 10{,}904(1 - 0.228) = 8418$$

- **Independence** The number of college undergraduates (the population) is much larger than 10 times the sample size, $n = 10{,}904$.

The second and third requirements are easily met. The first requirement, that the sample be an SRS, is only approximately met.

A Confidence Interval for p: Using z Procedures

Here are the z procedures for constructing a confidence interval about p when our conditions are satisfied.

Confidence Interval for a Population Proportion

Draw an SRS of size n from a large population with unknown proportion p of successes. An approximate level C confidence interval for p is

$$\hat{p} \pm z^* \sqrt{\frac{\hat{p}(1 - \hat{p})}{n}}$$

where z^* is the upper $(1 - C)/2$ critical value of the standard Normal distribution.

Example 10.15 *Estimating risky behavior*

Calculating a confidence interval for p

The Harvard School of Public Health survey found that 2486 of a sample of 10,904 college undergraduates said they had engaged in frequent binge drinking. That is, $\hat{p} = 0.228$. We will act as if the sample were an SRS.

A 99% confidence interval for the proportion p of all college undergraduates who admit to frequent binge drinking uses the critical value $z^* = 2.576$. (Look in the bottom row of Table C for critical values from the standard Normal distribution.) The confidence interval is

$$\hat{p} \pm z^* \sqrt{\frac{\hat{p}(1-\hat{p})}{n}} = 0.228 \pm 2.576 \sqrt{\frac{(0.228)(0.772)}{10{,}904}}$$

$$= 0.228 \pm 0.010$$

$$= (0.218, 0.238)$$

We are 99% confident that the proportion of college undergraduates who engaged in frequent binge drinking lies between 0.218 and 0.238.

CAUTION

Remember that the margin of error in this confidence interval includes only random sampling error! There are other sources of error that are not taken into account. This survey uses a design that is more complicated than an SRS, but we treated the data as if they were an SRS. As is the case with many surveys, we are forced to assume that the respondents answered truthfully. If they didn't, then our estimate may be biased. We also have the typical problem of nonresponse. The response rate for this survey was over 50%, an excellent rate for surveys of this type. Do the students who didn't respond have different drinking habits from those who did? If so, this is another source of bias.

Exercises

For Exercises 10.45 to 10.47, (a) Describe the population of interest and explain in words what the parameter p *is. (b) Give the numerical value of the statistic* $\hat{p}$ *that estimates* p*. (c) Determine whether each of the conditions is met for calculating a confidence interval for the population proportion* p.

10.45 Rating dorm food Tonya wants to estimate what proportion of the students in her dormitory like the dorm food. She interviews an SRS of 50 of the 175 students living in the dormitory. She finds that 14 think the dorm food is good.

10.46 High tuition costs Glenn wonders what proportion of the students at his school think that tuition is too high. He interviews an SRS of 50 of the 2400 students at his college. Thirty-eight of those interviewed think tuition is too high.

10.47 AIDS and risk factors In the National AIDS Behavioral Surveys sample of 2673 adult heterosexuals, 0.2% had both received a blood transfusion and had a sexual partner from a group at high risk of AIDS. We want to estimate the proportion p in the population who share these two risk factors.

10.48 Plagiarism and the Internet The National Survey of Student Engagement found that 87% of students report that their peers at least "sometimes" copy information from the Internet in their reports without citing the source.[16] Assume that the sample size is 430,000.

(a) Find the margin of error for 95% confidence.

(b) Here are some items from the report that summarizes the survey. More than 430,000 students from 730 four-year colleges and universities participated. The average response rate was 43% and ranged from 15% to 89%. Institutions pay a participation fee of between $3000 and $7500 based on the size of their undergraduate enrollment. Are these issues part of the margin of error you found in (a)? What impact might these issues have on the survey results?

10.49 Abstain from drinking! In the Harvard School of Public Health survey, 2105 of the 10,904 respondents were classified as abstainers (nondrinkers).

(a) Define the population and parameter of interest.

(b) Confirm that the conditions for constructing a confidence interval for the true population proportion p are satisfied.

(c) Calculate a 99% confidence interval for p.

(d) Interpret the interval in the context of this problem.

10.50 The millennium begins with optimism In January 2000, a Gallup Poll asked a random sample of 1633 adults, "In general, are you satisfied or dissatisfied with the way things are going in the United States at this time?" It found that 1127 said that they were satisfied. Write a short report of this finding, as if you were writing for a newspaper. Be sure to include a margin of error.

Putting It All Together: The Inference Toolbox

Taken together, Examples 10.14 and 10.15 show you how to construct a confidence interval for an unknown population proportion p. You can organize the inference process using the four-step Inference Toolbox, as the next example illustrates.

Example 10.16 *Teens say sex can wait*
Confidence interval for p

The 2004 Gallup Youth Survey asked a random sample of 439 U.S. teens aged 13 to 17 whether they thought young people should wait to have sex until marriage.[17] 246 said "Yes." Let's construct a 95% confidence interval for the proportion of all teens who would say "Yes" if asked this question.

Step 1: Parameter The population is all 13- to 17-year-olds in the United States. The parameter of interest is p, the actual proportion of this age group who would say that they thought young people should wait to have sex until they get married.

Step 2: Conditions We should use z procedures to estimate p if the conditions are satisfied.

- **SRS** The sample was not actually a simple random sample of U.S. teens. However, the 439 respondents were chosen using a method that is designed to ensure a representative sample. If this is not actually the case, we will have trouble generalizing our results.
- **Normality** We check the counts of "successes" and "failures":

$$n\hat{p} = 246 \geq 10 \text{ and } n(1 - \hat{p}) = 193 \geq 10$$

- **Independence** There are at least $(10)(439) = 4390$ teens aged 13 to 17 in the United States.

Step 3: Calculations A 95% confidence interval for p is given by

$$\hat{p} \pm z^* \sqrt{\frac{\hat{p}(1 - \hat{p})}{n}} = 0.56 \pm 1.96 \sqrt{\frac{(0.56)(0.44)}{439}}$$

$$= 0.56 \pm 0.046 = (0.514, 0.606)$$

Step 4: Interpretation We can say with 95% confidence that the true proportion of 13- to 17-year-olds in the United States who would say that teens should wait until marriage to have sex is between 0.514 and 0.606. Again, this conclusion depends on whether our sample of 439 teens is truly representative of the population.

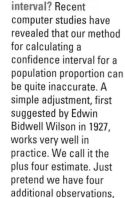

A better confidence interval? Recent computer studies have revealed that our method for calculating a confidence interval for a population proportion can be quite inaccurate. A simple adjustment, first suggested by Edwin Bidwell Wilson in 1927, works very well in practice. We call it the plus four estimate. Just pretend we have four additional observations, two of which are successes and two of which are failures. The resulting Wilson estimate is $\tilde{p} = \frac{(X + 2)}{(n + 4)}$.

But do teens' opinions about sex match their behaviors? According to the Gallup survey report:

> There's strong evidence that teens are putting thoughts about abstinence into action—or inaction. Sexual activity among high school students has been declining for the last 13 years, according to U.S. Centers for Disease Control and Prevention. In 1991, 54% of teens reported having had sexual intercourse; that figure had fallen to 47% in 2003.

The question remains: are the teens who were surveyed telling the truth? This is a question that statistical inference cannot answer.

Choosing the Sample Size

In planning a study, we may want to choose a sample size that will allow us to estimate the parameter within a given margin of error. We saw earlier how to do this for a population mean. The method is similar for estimating a population proportion.

The margin of error in the approximate confidence interval for p is

$$m = z^* \sqrt{\frac{\hat{p}(1-\hat{p})}{n}}$$

Here z^* is the critical value for the level of confidence we want. Because the margin of error involves the sample proportion of successes $\hat{p}$, we need to guess this value when choosing n. Call our guess p^*. Here are two ways to get p^*:

1. Use a guess p^* based on a pilot study or on past experience with similar studies. You should do several calculations that cover the range of $\hat{p}$-values you might get.
2. Use $p^* = 0.5$ as the guess. The margin of error m is largest when $\hat{p} = 0.5$, so this guess is conservative in the sense that if we get any other $\hat{p}$ when we do our study, we will get a margin of error smaller than planned.

Once you have a guess p^*, the formula for the margin of error can be solved to give the sample size n needed. Here is the result.

Sample Size for Desired Margin of Error

To determine the sample size n that will yield a level C confidence interval for a population proportion p with a specified margin of error m, set the following expression for the margin of error to be less than or equal to m, and solve for n:

$$z^* \sqrt{\frac{p^*(1-p^*)}{n}} \le m$$

where p^* is a guessed value for the sample proportion. The margin of error will be less than or equal to m if you take the guess p^* to be 0.5.

Which method for finding the guess p^* should you use? The n you get doesn't change much when you change p^* as long as p^* is not too far from 0.5. So use the conservative guess $p^* = 0.5$ if you expect the true p to be roughly between 0.3 and 0.7. If the true $\hat{p}$ is close to 0 or 1, using $p^* = 0.5$ as your guess will give a sample much larger than you need. So try to use a better guess from a pilot study when you suspect that $\hat{p}$ will be less than 0.3 or greater than 0.7.

Example 10.17 *Customer satisfaction*
Determining sample size

A company has received complaints about its customer service. They intend to hire a consultant to carry out a survey of customers. Before contacting the consultant, the company president wants some idea of the sample size that she will be required to pay for. One critical question is the degree of satisfaction with the company's customer service, measured on a five-point scale. The president wants to estimate the proportion p of customers who are satisfied (that is, who choose either "satisfied" or "very satisfied," the two highest levels on the five-point scale). She decides that she wants the estimate to be within 3% (0.03) at a 95% confidence level.

Since we have no idea of the true proportion p of satisfied customers, we decide to use $p^* = 0.5$. The sample size required is given by

$$z^* \sqrt{\frac{p^*(1-p^*)}{n}} \leq m$$

$$1.96 \sqrt{\frac{(0.5)(0.5)}{n}} \leq 0.03$$

$$\frac{(1.96)(0.5)}{0.03} \leq \sqrt{n}$$

$$32.667 \leq \sqrt{n}$$

$$n \geq 1067.11$$

So we round up to 1068 respondents to ensure that the margin of error is no more than 3%.

News reports frequently describe the results of surveys with sample sizes between 1000 and 1500 and a margin of error of about 3%. These surveys generally use sampling procedures more complicated than simple random sampling, so the calculation of confidence intervals is more involved than what we have studied in this section. The calculations of Example 10.17 nonetheless show in principle how such surveys are planned.

Exercises

10.51 Do job applicants lie? When trying to hire managers and executives, companies sometimes verify the academic credentials described by the applicants. One company that performs these checks summarized their findings for a six-month period. Of the 84 applicants whose credentials were checked, 15 lied about having a degree.[18]

(a) Find the proportion of applicants who lied about having a degree, and find the standard error.

(b) Consider these data to be a random sample of credentials from a large collection of similar applicants. Calculate and interpret a 90% confidence interval for the true proportion of applicants who lie about having a degree.

10.52 Equality for women? Have efforts to promote equality for women gone far enough in the United States? A poll on this issue by the cable network MSNBC contacted 1019 adults. A newspaper article about the poll said, "Results have a margin of sampling error of plus or minus 3 percentage points."[19]

(a) Overall, 54% of the sample (550 of 1019 people) answered "Yes." Construct and interpret a 95% confidence interval for the proportion in the adult population who would say "Yes" if asked. Is the report's claim about the margin of error roughly correct? (Assume that the sample is an SRS.)

(b) The news article said that 65% of men, but only 43% of women, think that efforts to promote equality have gone far enough. Explain why we do not have enough information to give confidence intervals for men and women separately.

(c) Would a 95% confidence interval for women alone have a margin of error less than 0.03, about equal to 0.03, or greater than 0.03? Why? You see that the news article's statement about the margin of error for poll results is a bit misleading.

10.53 Surveying students You are planning a survey of students at a large university to determine what proportion favor an increase in student fees to support an expansion of the student newspaper. Using records provided by the registrar, you can select a random sample of students. You will ask each student in the sample whether he or she is in favor of the proposed increase. Your budget will allow a sample of 100 students.

(a) For a sample of 100, construct a table of the margins of error for 95% confidence intervals when $\hat{p}$ takes the values 0.1, 0.2, 0.3, 0.4, 0.5, 0.6, 0.7, 0.8, and 0.9.

(b) A former editor of the student newspaper offers to provide funds for a sample of size 500. Repeat the margin of error calculations in (a) for the larger sample size. Then write a short thank-you note to the former editor describing how the larger sample size will improve the results of the survey.

10.54 School vouchers A national opinion poll found that 44% of all American adults agree that parents should be given vouchers good for education at any public or private school of their choice. The result was based on a small sample. How large an SRS is required to obtain a margin of error of 0.03 (that is, ±3%) in a 95% confidence interval?

(a) Answer this question using the previous poll's result as the guessed value p^*.

(b) Do the problem again using the conservative guess $p^* = 0.5$. By how much do the two sample sizes differ?

10.55 Teens and their TV sets According to a February 2005 Gallup Poll report, 64% of teens aged 13 to 17 have TVs in their rooms. Here is the footnote to this report:

> These results are based on telephone interviews with a randomly selected national sample of 1,028 teenagers in the Gallup Poll Panel of households, aged 13 to 17, conducted Jan. 17 to Feb. 6, 2005. For results based on this sample, one can say with 95% confidence that the maximum error attributable to sampling and other random effects is ±3 percentage points. In addition to sampling error, question wording and practical difficulties in conducting surveys can introduce error or bias into the findings of public opinion polls.

(a) State and interpret the 95% confidence interval described in the footnote.

(b) Explain the second sentence to someone who knows no statistics.

(c) Give an example of a "practical difficulty" that could result in biased results for this survey.

10.56 Starting a night club A college student organization wants to start a nightclub for students under the age of 21. To assess support for this proposal, they will select an SRS of students and ask each respondent if he or she would patronize this type of establishment. They expect that about 70% of the student body would respond favorably.

(a) What sample size is required to obtain a 90% confidence interval with an approximate margin of error of 0.04?

(b) Suppose that 50% of the sample responds favorably. Calculate the margin of error of the 90% confidence interval.

Technology Toolbox

Estimating a population proportion

The TI-83/84/89 can be used to construct a confidence interval for an unknown population proportion. Let's revisit Example 10.16.

Of $n = 439$ teens surveyed, $X = 246$ said they thought young people should wait to have sex until after marriage.

To construct a confidence interval:

TI-83/84

- Press **STAT**, then choose TESTS and A:1-PropZInt.

TI-89

- In the Statistics/List Editor, press **2nd** **F2** ([**F7**]) and choose 5:1-PropZInt.

- When the 1-PropZInt screen appears, enter $x = 246$, $n = 439$, and confidence level 0.95.

```
1-PropZInt
 x:246
 n:439
 C-Level:.95
 Calculate
```

```
F1 Tools  F2 Plots  F3 List  F4 Calc  F5 Distr  F6 Tests  F7 Ints
lis   1-Proportion Z Interval          4
--    Successes, x: 246                --
      n:            439
      C Level:      .95
      Enter=OK      ESC=CANCEL
list2[1]=
TYPE+[ENTER]=OK AND[ESC]=CANCEL
```

- Highlight "Calculate" and press **ENTER**. The 95% confidence interval for p is reported, along with the sample proportion $\hat{p}$ and the sample size, as shown here.

```
1-PropZInt
 (.51393,.60679)
 p̂=.5603644647
 n=439
```

```
F1 Tools  F2 Plots  F3 List  F4 Calc  F5 Distr  F6 Tests  F7 Ints
lis   1-Proportion Z Interval          4
--    C Int     ={.5139,.6068}         --
      p_hat     =.560364464692
      ME        =.04642987549
      n         =439.
      Enter=OK
list1[1]=
STATVARS RAD APPROX   FUNC   1/11
```

Section 10.3 Summary

Confidence intervals for a population proportion p when the data are an SRS of size n are based on the sample proportion $\hat{p}$.

When n is large, $\hat{p}$ has approximately the Normal distribution with mean p and standard deviation $\sqrt{p(1-p)/n}$.

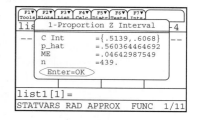

The level C **confidence interval for p** is

$$\hat{p} \pm z^* \sqrt{\frac{\hat{p}(1-\hat{p})}{n}}$$

where z^* is the upper $(1 - C)/2$ critical value from the standard Normal distribution.

This inference procedure is approximately correct when the data come from an SRS of size n, the sample is large enough to satisfy $n\hat{p} \geq 10$ and $n(1 - \hat{p}) \geq 10$, and the population is at least 10 times as large as the sample.

The sample size needed to obtain a confidence interval with approximate margin of error m for a population proportion involves solving

$$z^* \sqrt{\frac{p^*(1-p^*)}{n}} \leq m$$

for n, where p^* is a guessed value for the sample proportion $\hat{p}$, and z^* is the critical value for the level of confidence you want. If you use $p^* = 0.5$ in this formula, the margin of error of the interval will be less than or equal to m no matter what the value of $\hat{p}$ is.

Section 10.3 Exercises

10.57 Running red lights A random digit dialing telephone survey of 880 drivers asked, "Recalling the last ten traffic lights you drove through, how many of them were red when you entered the intersections?" Of the 880 respondents, 171 admitted that at least one light had been red.[20]

(a) Calculate and interpret a 95% confidence interval for the population proportion.

(b) Nonresponse is a practical problem for this survey—only 21.6% of calls that reached a live person were completed. Another practical problem is that people may not give truthful answers. What is the likely direction of the bias: do you think more or fewer than 171 of the 880 respondents really ran a red light? Why?

10.58 Customer satisfaction An automobile manufacturer would like to know what proportion of its customers are not satisfied with the service provided by the local dealer. The customer relations department will survey a random sample of customers and compute a 99% confidence interval for the proportion who are not satisfied.

(a) Past studies suggest that this proportion will be about 0.2. Find the sample size needed if the margin of error of the confidence interval is to be about 0.015.

(b) When the sample is actually contacted, 10% of the sample say they are not satisfied. What is the margin of error of the 99% confidence interval?

10.59 Drunken cyclists? In the United States approximately 900 people die in bicycle accidents each year. One study examined the records of 1711 bicyclists aged 15 or older who were fatally injured in bicycle accidents were tested for alcohol. Of these, 542 tested positive for alcohol (blood alcohol concentration of 0.01% or higher).[21]

(a) Find a 99% confidence interval for p. Follow the Inference Toolbox.

(b) Can you conclude from your statistical analysis of this study that alcohol causes fatal bicycle accidents? Explain.

10.60 Can you taste PTC? PTC is a substance that has a strong bitter taste for some people and is tasteless for others. The ability to taste PTC is inherited. About 75% of Italians can taste PTC, for example. You want to estimate the proportion of Americans with at least one Italian grandparent who can taste PTC. Starting with the 75% estimate for Italians, how large a sample must you test in order to estimate the proportion of PTC tasters within ±0.04 with 95% confidence?

10.61 Long sermons The National Congregations Study collected data in a one-hour interview with a key informant—that is, a minister, priest, rabbi, or other staff person or leader.[22] One question asked concerned the length of the typical sermon. For this question 390 out of 1191 congregations reported that the typical sermon lasted more than 30 minutes.

(a) Estimate the true population proportion for this question with a 95% confidence interval.

(b) There were 1236 congregations surveyed in this study. Calculate the nonresponse rate for this question. Does this influence how you interpret the results? Write a short discussion of this issue.

(c) The respondents to this question were not asked to use a stopwatch to record the lengths of a random sample of sermons at their congregations. They responded based on their impressions of the sermons. Do you think that ministers, priests, rabbis, or other staff persons or leaders might perceive sermon lengths differently from the people listening to the sermons? Discuss how your ideas would influence your interpretation of the results of this study.

10.62 Gambling and the NCAA Gambling is an issue of great concern to those involved in intercollegiate athletics. Because of this concern, the National Collegiate Athletic Association (NCAA) surveyed student-athletes concerning their gambling-related behaviors.[23] There were 5594 Division I male athletes in the survey. Of these, 3547 reported participation in some gambling behavior. This includes playing cards, betting on games of skill, buying lottery tickets, betting on sports, and similar activities.

(a) Find the sample proportion and the margin of error for 90% confidence. Explain in simple terms the meaning of the 90%.

(b) Because of the way that the study was designed to protect the anonymity of the student-athletes who responded, it was not possible to calculate the number of students who were asked to respond but did not. Does this fact affect the way that you interpret the results? Write a short paragraph explaining your answer.

10.63 A television poll A television news program conducts a call-in poll about a proposed city ban on handgun ownership. Of the 2372 calls, 1921 oppose the ban. The station, following recommended practice, makes a confidence statement: "81% of the Channel 13 Pulse Poll sample opposed the ban. We can be 95% confident that the true proportion of citizens opposing a handgun ban is within 1.6% of the sample result." Is this conclusion justified?

10.64 Do college students pray? Social scientists asked 127 undergraduate students "from courses in psychology and communications" about prayer and found that 107 prayed at least a few times a year.[24]

(a) Calculate and interpret a 99% confidence interval for the proportion p of all students who pray.

(b) To use any inference procedure, we must be willing to regard these 127 students, as far as their religious behavior goes, as an SRS from the population of all undergraduate students. Do you think it is reasonable to do this? Why or why not?

CASE CLOSED!

Need help? Give us a call!

Refer to the chapter-opening Case Study on page 615. The bank manager has hired you as a statistical consultant. She would like you to examine whether the bank's customer service agents generally met the goal of answering incoming calls within 30 seconds.

1. Construct an appropriate graphical display for the call pickup times. Then calculate numerical summary statistics. Write a few sentences describing the distribution of response times.
2. Calculate and interpret a 95% confidence interval for the mean call pickup time at this bank's customer service department.
3. Calculate and interpret a 95% confidence interval for the proportion of calls coming into this bank's customer service department that were answered within 30 seconds.
4. Compose a brief report of your findings for the bank manager. Be sure to explain the meaning of "95% confidence" in plain English when you discuss your confidence intervals.

Chapter Review

Summary

Statistical inference draws conclusions about a population on the basis of sample data and uses probability to indicate how reliable the conclusions are. A confidence interval estimates an unknown parameter.

The probabilities in confidence intervals tell us what would happen if we used the calculation method for the interval very many times. A confidence level is the probability that the recipe for a confidence interval actually produces an interval that contains the unknown parameter. A 95% confidence interval gives a correct result 95% of the time when we use it repeatedly.

Figure 10.16 presents the idea of a confidence interval in picture form. This idea will reappear throughout the rest of the book. We will have much to say about many statistical methods and their use in practice. In every case, the basic reasoning of confidence intervals remains the same.

Figure 10.16 *The idea of a confidence interval. See if you can write a paragraph describing the contents of this figure.*

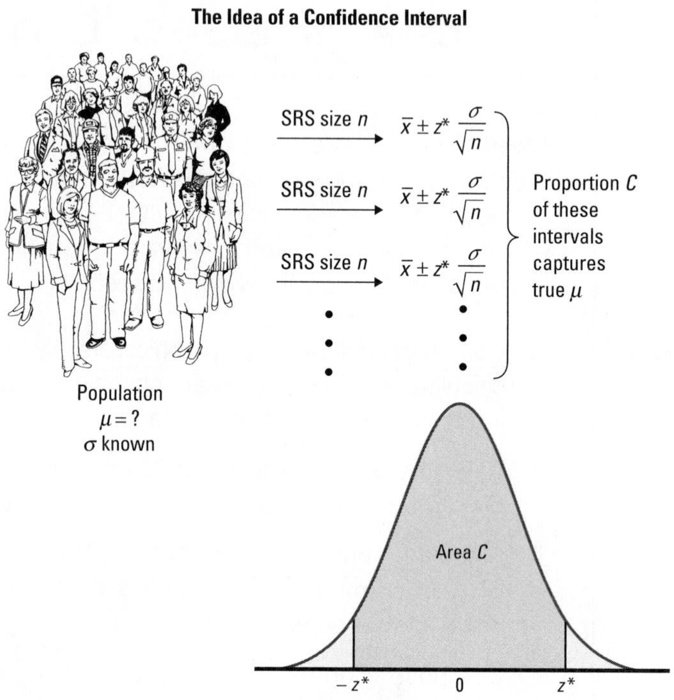

Section 10.1 concentrated on the reasoning behind forming confidence intervals for a population mean μ in the unrealistic setting where the population standard deviation σ is known. Section 10.2 presented t confidence intervals for estimating μ when σ is unknown. Matched pairs and other paired-data designs use one-sample t procedures because you first create a single sample by taking the differences of the responses within each pair.

The t procedures require that the data come from a simple random sample (SRS) and that the distribution of the population be Normal (or the sample size be at least 30). One reason for the wide use of t procedures is that they are robust. They are not very strongly affected by lack of Normality. If you can't regard your data as a random sample, however, the results of inference may be of little value.

Inference about population proportions is based on sample proportions. We rely on the fact that a sample proportion has a distribution that is close to Normal unless the sample is small. The z confidence interval in Section 10.3

works well when the samples are large enough (but not more than 10% of the population).

Before you use any of the inference methods in this chapter, be sure to verify whether each of the three conditions—SRS, Normality, and independence—is satisfied.

What You Should Have Learned

Here is a checklist of the major skills you should have acquired by studying this chapter.

A. Confidence Intervals for μ (σ known)

1. State in nontechnical language what is meant by "95% confidence" or other statements of confidence in statistical reports.
2. Calculate a confidence interval for the mean μ of a Normal population with known standard deviation σ, using the recipe $\bar{x} \pm z^*\sigma/\sqrt{n}$.
3. Recognize when you can safely use this confidence interval and when the data collection design or a small sample from a skewed population makes it inaccurate. Understand that the margin of error does not include the effects of undercoverage, nonresponse, or other practical difficulties.
4. Understand how the margin of error of a confidence interval changes with the sample size and the level of confidence C.
5. Find the sample size required to obtain a confidence interval of specified margin of error m when the confidence level and other information are given.

B. Confidence Intervals for μ (σ unknown)

1. Use the t procedures to obtain a confidence interval at a stated level of confidence for the mean μ of a population.
2. Recognize when the t interval is appropriate in practice, in particular that it is quite robust against lack of Normality but is influenced by outliers.
3. Also recognize when the design of the study, outliers, or a small sample from a skewed distribution make the t procedures risky.
4. Recognize paired-data designs and use the one-sample t procedures to obtain confidence intervals for such data.

C. Confidence Intervals for p

1. Use the z procedure to give a confidence interval for a population proportion p.
2. Check that you can safely use the z procedures in a particular setting.
3. Determine the sample size required to obtain a level C confidence interval with a specified margin of error.

Web Links

For more information about the Harvard School of Public Health's College Alcohol Study, visit www.hsph.harvard.edu/cas/

For other interesting poll results, try the Gallup Poll Web site, www.gallup.com

Chapter Review Exercises

10.65 Engine crankshafts Here are measurements (in millimeters) of a critical dimension on a sample of auto engine crankshafts:

224.120	224.001	224.017	223.982	223.989	223.961	223.960	224.089
223.987	223.976	223.902	223.980	224.098	224.057	223.913	223.999

The data come from a production process that is known to have standard deviation $\sigma = 0.060$ mm. The process mean is supposed to be $\mu = 224$ mm but can drift away from this target during production.

(a) We expect the distribution of the dimension to be close to Normal. Make a plot of these data and describe the shape of the distribution.

(b) Construct and interpret a 95% confidence interval for the process mean at the time these crankshafts were produced.

(c) Explain what "95% confident" means in this setting.

(d) How large a sample of crankshafts would be needed to estimate the mean μ within ± 0.020 mm with 95% confidence? Show your work.

10.66 We love football! A recent Gallup Poll conducted telephone interviews with a random sample of adults aged 18 and older. Data were obtained for 1000 people. Of these, 37% said that football is their favorite sport to watch on television.

(a) Explain to someone who knows no statistics why we can't just say that 37% of adults would say that football is their favorite sport to watch on television.

(b) Construct and interpret a 95% confidence interval for the population proportion p.

(c) The poll announced a margin of error of ± 3 percentage points. Do you agree?

(d) This poll was taken in December, an exciting part of the football season. Do you think a similar poll conducted in June might produce different results? Explain why or why not.

10.67 Calcium and blood pressure In a randomized comparative experiment on the effect of calcium in the diet on blood pressure, researchers divided 54 healthy white males at random into two groups. One group received calcium; the other, a placebo. At the beginning of the study, the researchers measured many variables on the subjects. The paper reporting the study gives $\bar{x} = 114.9$ and $s = 9.3$ for the seated systolic blood pressure of the 27 members of the placebo group.

(a) Calculate and interpret a 99% confidence interval for the mean blood pressure in the population from which the subjects were recruited.

(b) What conditions about the population and the study design are required by the procedure you used in (a)? Which of these conditions are important for the validity of the procedure in this case?

10.68 Eat more food, gain more weight If we increase our food intake, we generally gain weight. Nutrition scientists can calculate the amount of weight gain that would be associated with a given increase in calories. In one study, 16 nonobese adults, aged 25 to 36 years, were fed 1000 calories per day in excess of the calories needed to maintain a stable body weight. The subjects maintained this diet for 8 weeks, so they consumed a total of 56,000 extra calories.[25] According to theory, 3500 extra calories will translate into a weight gain of 1 pound. Therefore, we expect each of these subjects to gain 56,000/3500 = 16 pounds (lb). Here are the weights before and after the 8-week period expressed in kilograms (kg):

Subject:	1	2	3	4	5	6	7	8
Weight before:	55.7	54.9	59.6	62.3	74.2	75.6	70.7	53.3
Weight after:	61.7	58.8	66.0	66.2	79.0	82.3	74.3	59.3
Subject:	9	10	11	12	13	14	15	16
Weight before:	73.3	63.4	68.1	73.7	91.7	55.9	61.7	57.8
Weight after:	79.1	66.0	73.4	76.9	93.1	63.0	68.2	60.3

(a) Calculate and interpret a 95% confidence interval for the mean weight gain μ in the population of interest.

(b) Convert the confidence interval in part (a) from kilograms to pounds.

(c) Do you believe that the theory stating an expected average weight gain of 16 pounds is correct? Explain why or why not.

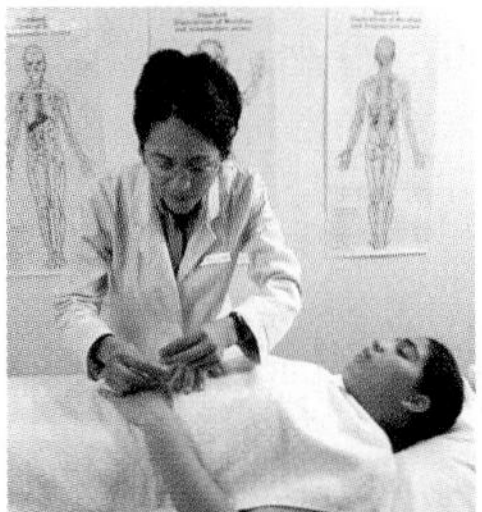

10.69 Alternative medicine A nationwide random survey of 1500 adults asked about attitudes toward "alternative medicine" such as acupuncture, massage therapy, and herbal therapy. Among the respondents, 660 said they would use alternative medicine if traditional medicine was not producing the results they wanted.[26]

(a) Construct and interpret a 95% confidence interval for the proportion of all adults who would use alternative medicine.

(b) Write a short paragraph for a news report based on the survey results.

10.70 Canadians and doctor-assisted suicide A Gallup Poll asked a sample of Canadian adults if they thought the law should allow doctors to end the life of the patient who is in great pain and near death if the patient makes a request in writing. The poll included 270 people in Quebec, 221 of whom agreed that doctor-assisted suicide should be allowed.[27]

(a) What is the margin of error for a 99% confidence interval for the proportion of all Quebec adults who would allow doctor-assisted suicide?

(b) How large a sample is needed to get the common ± 3 percentage points margin of error?

10.71 The cost of Internet access How much do users pay for Internet service? Here are the monthly fees (in dollars) paid by a random sample of 50 users of commercial Internet service providers in August 2000:[28]

20	40	22	22	21	21	20	10	20	20
20	13	18	50	20	18	15	8	22	25
22	10	20	22	22	21	15	23	30	12
9	20	40	22	29	19	15	20	20	20
20	15	19	21	14	22	21	35	20	22

(a) Display these data graphically. Describe the distribution of fees paid for Internet access.

(b) The t procedures are still approximately correct in this case. Explain why.

(c) Calculate and interpret a 90% confidence interval for the mean monthly cost of Internet access in August 2000.

10.72 Longer-lasting batteries A company that produces AA batteries makes the following statistical announcement: "Our AA batteries last 7.5 hours plus or minus 20 minutes, and our confidence in that interval is 95%."[29] They tested a random sample of 40 batteries using a special device designed to imitate real-world use and recorded data on battery lifetime.

(a) Determine the sample mean and standard deviation.

(b) A reporter translates the statistical announcement into "plain English" as follows: "If you buy one of this company's AA batteries, there is a 95% chance that it will last between 430 and 470 minutes." Comment on this interpretation.

(c) Your friend, who has just started studying statistics, claims that if you select 40 more AA batteries at random from those manufactured by this company, there is a 95% probability that the mean lifetime will fall between 430 and 470 minutes. Do you agree?

(d) Give a statistically correct interpretation of the confidence interval that could be published in a newspaper report.

10.73 Do you go to church? A Gallup Poll asked a sample of 1785 adults, "Did you, yourself, happen to attend church or synagogue in the last 7 days?" Of the respondents, 750 said "Yes." Suppose (it is not, in fact, true) that Gallup's sample was an SRS of all American adults.

(a) Calculate and interpret a 99% confidence interval for the proportion of all adults who attended church or synagogue during the week preceding the poll.

(b) How large a sample would be required to obtain a margin of error of 0.01 in a 99% confidence interval for the proportion who attend church or synagogue? (Use the conservative guess $p^* = 0.5$, and explain why this method is reasonable in this situation.)

10.74 Healthy bones Dual-energy X-ray absorptiometry (DXA) is a technique for measuring bone health. One of the most common measures is total body bone mineral content (TBBMC). A highly skilled operator is required to take the measurements. Recently a new DXA machine was purchased by a lab and two operators were trained to take the measurements. TBBMC for eight individuals was measured by both operators.[30] The units are grams (g). A comparison of the measurements for the two operators provides a check on the training they received and allows us to determine if one of the operators is producing measurements that are consistently higher than the other. Here are the data:

	Subject							
Operator	**1**	**2**	**3**	**4**	**5**	**6**	**7**	**8**
1	1.328	1.342	1.075	1.228	0.939	1.004	1.178	1.286
2	1.323	1.322	1.073	1.233	0.934	1.019	1.184	1.304

(a) Take the difference between the TBBMC recorded by Operator 1 and the TBBMC by Operator 2. Describe the distribution of these differences.

(b) Use a 95% confidence interval to estimate the mean difference in TBBMC readings in the population of interest based on the data.

(c) The eight subjects in this sample were not a random sample. In fact, they were friends of the researchers whose ages and heights were similar to these of the types of people who would be measured with the DXA technique. Comment on the appropriateness of this procedure for selecting a sample, and discuss any consequences regarding the interpretation of the confidence interval.

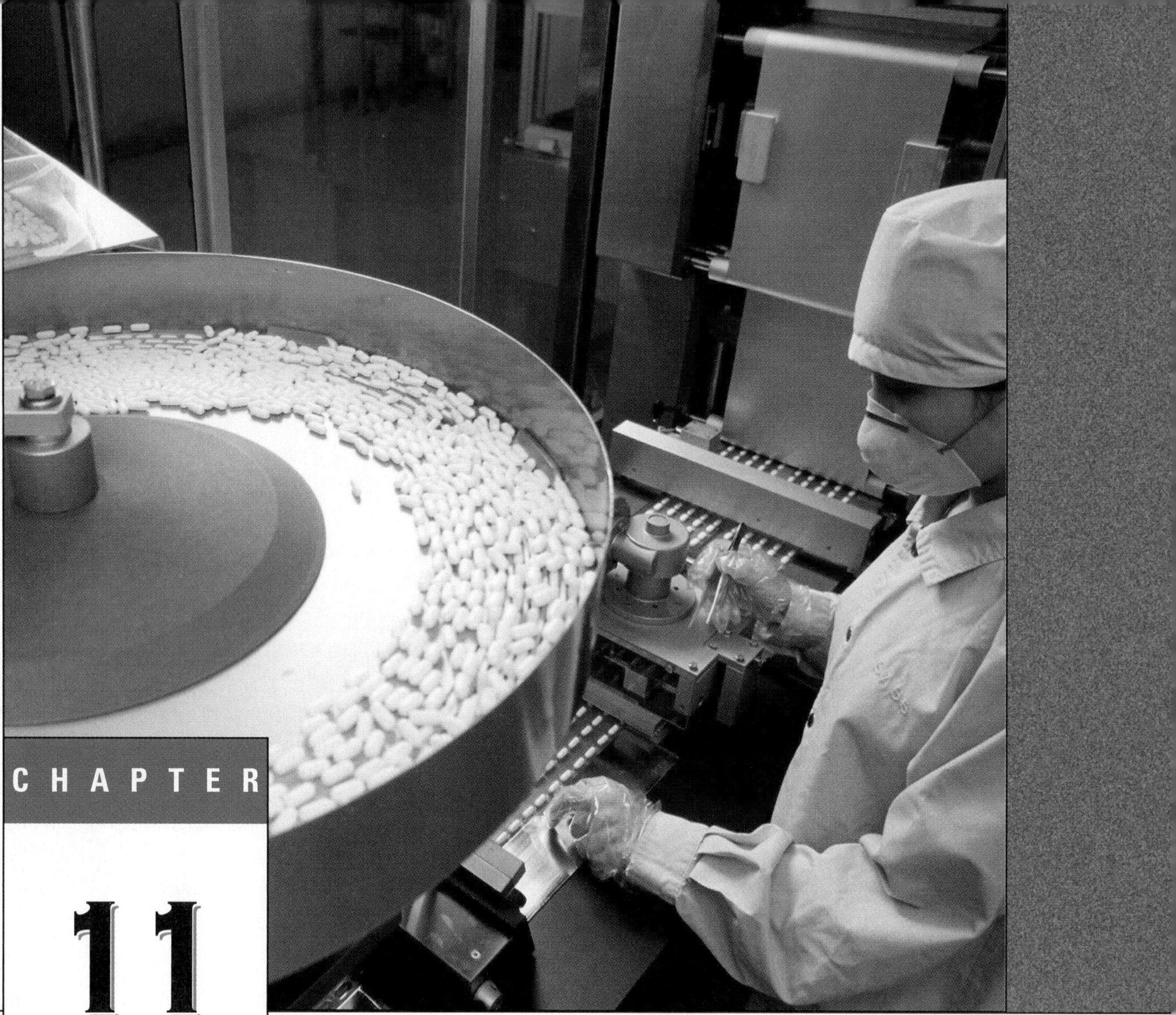

CHAPTER

11

Testing a Claim

CASE STUDY

I'm getting a headache!

The makers of Aspro brand aspirin want to be sure that their tablets contain the right amount of active ingredient (acetylsalicylic acid). So they inspect a sample of 36 tablets from a batch of production. When the production process is working properly, Aspro tablets have an average of $\mu = 320$ milligrams (mg) of acetylsalicylic acid with a standard deviation of $\sigma = 3$ mg. Here are the amounts (in mg) of acetylsalicylic acid in the 36 selected tablets:

319	328	321	324	322	320	324	321	320	324	322	317
321	320	322	318	326	316	316	326	325	320	316	319
319	321	322	319	326	320	324	320	318	321	322	318

What do these data tell us about the mean acetylsalicylic acid content of the tablets in this batch? Should the company distribute these tablets to drugstores or dispose of the entire batch? By the end of this chapter, you'll have all the statistical tools you'll need to answer these questions.

Activity 11A

Pick a card

Materials: Three decks of playing cards, provided by your teacher

Your teacher has what appear to be three standard decks of playing cards. In this Activity, your task is to try to determine the distribution of black and red cards in each of the decks.

1. As a class, make a guess about the proportion of red cards in the first deck. You may use your experience with playing cards, but you may not look at the cards in this deck. Write your class's guess (we will call it a "hypothesis") on the board in the form $p =$ ______.
2. The teacher will have one student draw a card, note the color, and replace the card in the deck. Write RED and BLACK on the board and use a tally mark to record the result of the first draw. After the deck has been shuffled, a second student should draw a card and tally the result. Continue until your class reaches a consensus on the proportion p of red cards in the deck. Then write this revised class estimate about the proportion p on the board as Deck 1: $p =$ ________.
3. The teacher will introduce a second deck and pose the same question about the proportion of red cards in this deck. Repeat the process described in Step 2.
4. The teacher will introduce a third deck and pose the same question about the proportion of red cards. But this time, since we know that larger sample sizes give better results, your teacher will have each student draw 2 cards at the same time. After looking at them, the student will record the proportion of red cards $\hat{p}$ in the sample ($\hat{p} = 0, \hat{p} = 0.5$, or $\hat{p} = 1$) on the board. Replace the cards in the deck, shuffle, and proceed to the next student. After five or six students have drawn their samples and recorded $\hat{p}$, ask whether the data suggest that the true proportion of red cards is different from 0.5. If not, then collect enough data until the class is ready to make an "alternative hypothesis" about the proportion of red cards in the deck or to conclude that $p = 0.5$. If you decide to make an alternative hypothesis, choose from the following:

$p < 0.5$ if you really think there are *fewer* red cards than black cards

$p > 0.5$ if you really think there are *more* red cards than black cards

$p \neq 0.5$ if you can't decide between the first two choices but you still think that there are more cards of one color than the other

5. If you agreed on an alternative hypothesis, your task now is to try to gather evidence against the original hypothesis that $p = 0.5$. Since larger samples reduce the variability of an estimate, each student should draw 5 cards this time. After at least 10 students have recorded their sample proportion $\hat{p}$ of red cards, calculate the mean of these

estimates of p. What do you conclude about whether the original hypothesis or your alternative hypothesis is more believable? Why?

In most statistical situations, we have no way of determining the true population parameters. So for this Activity to model real-world inference procedures, the true proportions of red cards in the three decks would not be known. The decision about whether to reveal this information is left to your teacher.

Introduction

Confidence intervals are one of the two most common types of statistical inference. Use a confidence interval when your goal is to estimate a population parameter. The second common type of inference, called *significance tests*, has a different goal: to assess the evidence provided by data about some claim concerning a population. Here is the reasoning of statistical tests in a nutshell.

Example 11.1 *I'm a great free-throw shooter!*
Significance tests: the big idea

I claim that I make 80% of my basketball free throws. To test my claim, you ask me to shoot 20 free throws. I make only 8 of the 20. "Aha!" you say. "Someone who makes 80% of his free throws would almost never make only 8 out of 20. So I don't believe your claim."

Your reasoning is based on asking what would happen if my claim were true and we repeated the sample of 20 free throws many times—I would almost never make as few as 8. This outcome is so unlikely that it gives strong evidence that my claim is not true.

You can say how strong the evidence against my claim is by giving the probability that I would make as few as 8 out of 20 free throws if I really make 80% in the long run. This probability is 0.0001. I would make as few as 8 of 20 only once in 10,000 tries in the long run if my claim to make 80% is true. The small probability convinces you that my claim is false.

Significance tests use an elaborate vocabulary, but the basic idea is simple: *an outcome that would rarely happen if a claim were true is good evidence that the claim is not true.*

APPLET

Activity 11B
I'm a great free-throw shooter!

The *Reasoning of a Statistical Test* applet on the book's Web site (www.whfreeman.com/tps3e) animates Example 11.1. That Example asks if a basketball player's actual performance gives evidence against the claim that he or she makes 80% of free throws. The parameter in question is the proportion p of free throws that the player will make if he or she shoots free throws forever. The population is all free throws the player will ever shoot. The "null hypothesis" is always the same, that the player makes 80% of shots taken: we write this as H_0: $p = 0.80$.

The applet does not do a formal statistical test. Instead, it allows you to ask the player to shoot until you are reasonably confident that the true percent of hits is or is not very close to 80%. I claim that I make 80% of my free throws. To test my claim, we go to the gym and I shoot 20 free throws.

1. Set the applet to take 20 shots. Check "Show null hypothesis" so that my claim is visible in the graph.
2. Click "Shoot." How many of the 20 shots did I make? Are you convinced that I really make less than 80%?
3. If you are not convinced, click "Shoot" again for 20 more shots. Keep going until *either* you are convinced that I don't make 80% of my shots *or* it appears that my true percent made is pretty close to 80%. How many shots did you watch me shoot? How many did I make? What did you conclude?
4. Click "Show true %" to reveal the truth. Was your conclusion correct?

Comment: You see why statistical tests say how strong the evidence is *against* some claim. If I make only 10 of 40 shots, you are pretty sure I can't make 80% in the long run. But even if I make exactly 80 of 100, my true long-term percent might be 78% or 81% instead of 80%. It's hard to be convinced that I make exactly 80%.

11.1 Significance Tests: The Basics

A significance test is a formal procedure for comparing observed data with a hypothesis whose truth we want to assess. The hypothesis is a statement about a population parameter, like the population mean μ or population proportion p. The results of a test are expressed in terms of a probability that measures how well the data and the hypothesis agree. The reasoning of statistical tests, like that of confidence intervals, is based on asking what would happen if we repeated the sampling or experiment many times. As in the previous chapter, we begin with the unrealistic assumption that we know σ, the population standard deviation.

Example 11.2 *Call the paramedics!*
Testing a claim: getting started

Vehicle accidents can result in serious injuries to drivers and passengers. When they do, someone usually calls 911. Police, firefighters, and paramedics respond to these

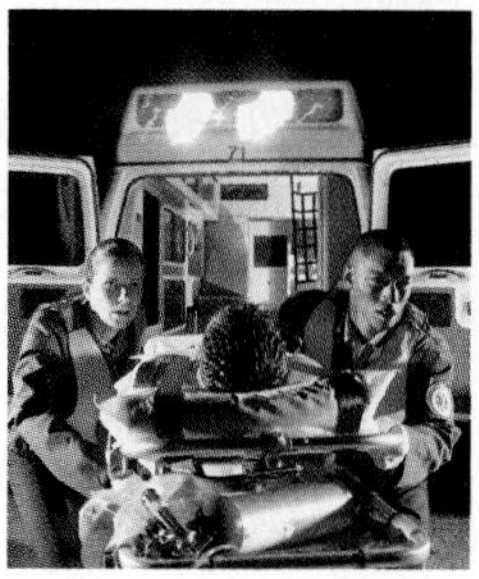

emergency calls as quickly as possible. Slow response times can have serious consequences for accident victims. In case of life-threatening injuries, victims generally need medical attention within 8 minutes of the crash.

Several cities have begun to monitor paramedic response times. In one such city, the mean response time to all accidents involving life-threatening injuries last year was $\mu = 6.7$ minutes with a standard deviation of $\sigma = 2$ minutes. The city manager shares this information with emergency personnel and encourages them to "do better" next year. At the end of the following year, the city manager selects a simple random sample (SRS) of 400 calls involving life-threatening injuries and examines the response times. For this sample, the mean response time was $\bar{x} = 6.48$ minutes. Do these data provide good evidence that response times have decreased since last year?

Should the city manager in Example 11.2 congratulate the city's paramedics for responding faster this year? Not yet. After all, sample results vary. Maybe the response times haven't improved at all, and $\bar{x} = 6.48$ minutes is simply a result of sampling variability. The reasoning we use here is the same as in Example 11.1. We make a claim and ask if the data give evidence *against* it. We would like to conclude that the mean response time has decreased, so the claim we test is that response times have *not* decreased. In that case, the mean response time for the population of all calls involving life-threatening injury this year would be $\mu = 6.7$ minutes. (We will assume that $\sigma = 2$ minutes for this year's calls, too.)

- If the claim that $\mu = 6.7$ minutes is true, the sampling distribution of $\bar{x}$ from 400 calls will be approximately Normal (by the central limit theorem) with mean $\mu = 6.7$ minutes and standard deviation

$$\frac{\sigma}{\sqrt{n}} = \frac{2}{\sqrt{400}} = 0.10 \text{ minutes}$$

Figure 11.1 (on the next page) shows this sampling distribution. We can judge whether any observed $\bar{x}$ is surprising by locating it on this distribution.

- Suppose that the city manager had sampled 400 response times for which $\bar{x} = 6.61$ minutes. It is clear from Figure 11.1 that an $\bar{x}$ this close to 6.7 minutes could occur just by chance when the population mean is $\mu = 6.7$. Such a result does not provide convincing evidence of a decrease in response times.

- In fact, the city manager's sample yielded $\bar{x} = 6.48$ minutes. That's much farther from $\mu = 6.7$ in Figure 11.1—so far that an observed value this small would rarely occur by chance if the true μ were 6.7 minutes. This observed value is good evidence that the true μ is less than 6.7 minutes, that is, that the average response time decreased this year. Maybe the city manager should offer congratulations to the city's paramedics for a job well done.

This outline of the reasoning of statistical tests omits many important details. However, if you don't understand the underlying reasoning, the fine details won't make much sense anyway!

Figure 11.1 *If response times have not decreased, the mean response time $\bar{x}$ for 400 calls will have this sampling distribution. If the result had been $\bar{x} = 6.61$ minutes, that could easily happen just by chance. But the actual result was $\bar{x} = 6.48$ minutes. That's so far out on the Normal curve that it's good evidence of a decrease in paramedics' response times.*

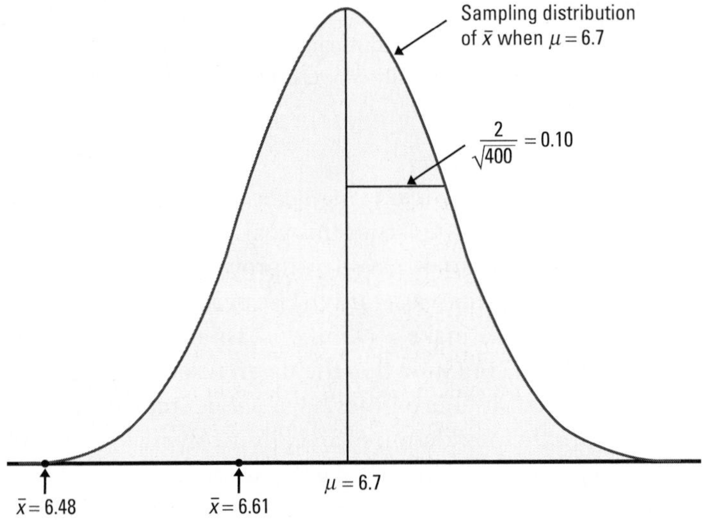

Exercises

11.1 Student attitudes, I The Survey of Study Habits and Attitudes (SSHA) is a psychological test that measures students' attitudes toward school and study habits. Scores range from 0 to 200. The mean score for U.S. college students is about 115, and the standard deviation is about 30. A teacher suspects that older students have better attitudes toward school. She gives the SSHA to a random sample of 25 students at her college who are at least 30 years of age. Assume that scores in the population of older students are Normally distributed with standard deviation $\sigma = 30$.

(a) Carefully define the parameter μ in this setting.

(b) We seek evidence *against* the claim that $\mu = 115$. What is the sampling distribution of the mean score $\bar{x}$ of a sample of 25 older students if the null hypothesis is true? Make a sketch of the Normal curve for this distribution. (Sketch a Normal curve, then mark the axis using what you know about locating the mean and standard deviation on a Normal curve.)

(c) Suppose that the sample data give $\bar{x} = 118.6$. Mark this point on the axis of your sketch. In fact, the result was $\bar{x} = 125.7$. Mark this point on your sketch. Using your sketch, explain in simple language why one result is good evidence that the mean score of all older students is greater than 115 and why the other outcome is not.

(d) Did we *need* to assume that the distribution of SSHA scores for older students is Normal in this problem? Justify your answer.

(e) Can we generalize our findings about SSHA scores to the population of all older students in U.S. colleges? Explain why or why not.

11.2 Anemia, I Hemoglobin is a protein in red blood cells that carries oxygen from the lungs to body tissues. People with less than 12 grams of hemoglobin per deciliter of blood (g/dl) are anemic. A public health official in Jordan suspects that the mean μ for all children in Jordan is less than 12. He measures a sample of 50 children. Suppose we know that the hemoglobin level for all children of this age follows a Normal distribution with standard deviation $\sigma = 1.6$ g/dl.

(a) Carefully define the parameter μ in this case.

(b) We seek evidence *against* the claim that $\mu = 12$. What is the sampling distribution of $\bar{x}$ of a sample of 50 children if $\mu = 12$? Make a sketch of the Normal curve for this distribution. (Sketch a Normal curve, then mark the axis using what you know about locating the mean and standard deviation on a Normal curve.)

(c) The sample mean was $\bar{x} = 11.3$. Mark this outcome on the sampling distribution. Also mark the outcome $\bar{x} = 11.8$ of a different study of 50 children. Explain carefully from your sketch why one of these outcomes is good evidence that μ is lower than 12, and also why the other outcome is not good evidence for this conclusion.

(d) Did we *need* to know that the distribution of hemoglobin level for children this age is Normal? Justify your answer.

(e) Can we generalize our findings about hemoglobin levels to the population of all children this age in Jordan? Explain why or why not.

Stating Hypotheses

A statistical test starts with a careful statement of the claims we want to compare. In Example 11.2, we asked whether the accident response time data are likely if, in fact, there is no decrease in paramedics' response times. Because the reasoning of tests looks for evidence *against* a claim, we start with the claim we seek evidence against, such as "no decrease in response time." This claim is our ***null hypothesis.***

Null and Alternative Hypotheses

The statement being tested in a significance test is called the **null hypothesis.** The significance test is designed to assess the strength of the evidence *against* the null hypothesis. Usually the null hypothesis is a statement of "no effect," "no difference," or no change from historical values.

The claim about the population that we are trying to find evidence *for* is the **alternative hypothesis.**

We abbreviate the null hypothesis as H_0 and the alternative hypothesis as H_a. In Example 11.2, we are seeking evidence of a decrease in response time this year. The null hypothesis says "no decrease" on the average in the large population of all calls involving life-threatening injury this year. The alternative hypothesis says "there is a decrease." So the hypotheses are

$$H_0\colon \mu = 6.7 \text{ minutes}$$

$$H_a\colon \mu < 6.7 \text{ minutes}$$

one-sided alternative

where μ is the mean response time to all calls involving life-threatening injury in the city this year. The alternative hypothesis is ***one-sided*** because we are interested only in deviations from the null hypothesis in one direction.

CAUTION

Hypotheses always refer to some population, not to a particular outcome. Be sure to state H_0 and H_a in terms of a population parameter. Because H_a expresses the effect that we hope to find evidence *for*, it is often easier to begin by stating H_a and then set up H_0 as the statement that the hoped-for effect is not present. Stating H_a is not always straightforward. It is not always clear, in particular, whether H_a should be one-sided or ***two-sided.*** Here is an example in which the alternative hypothesis is two-sided.

two-sided alternative

Example 11.3 *Studying job satisfaction*

Stating hypotheses

Does the job satisfaction of assembly workers differ when their work is machine-paced rather than self-paced? One study chose 18 subjects at random from a group of people who assembled electronic devices. Half of the subjects were assigned at random to each of two groups. Both groups did similar assembly work, but one work setup allowed workers to pace themselves, and the other featured an assembly line that moved at fixed time intervals so that the workers were paced by machine. After two weeks, all subjects took the Job Diagnosis Survey (JDS), a test of job satisfaction. Then they switched work setups and took the JDS again after two more weeks. This is a matched pairs design. The response variable is the difference in JDS scores, self-paced minus machine-paced.[1]

The parameter of interest is the mean μ of the differences in JDS scores in the population of all assembly workers. The null hypothesis says that there is no difference in the scores, that is,

$$H_0\colon \mu = 0$$

The authors of the study wanted to know if the two work conditions have different levels of job satisfaction. They did not specify the direction of the difference. The alternative hypothesis is therefore two-sided; that is, either $\mu < 0$ or $\mu > 0$. For simplicity, we write this as

$$H_a\colon \mu \neq 0$$

CAUTION

The alternative hypothesis should express the hopes or suspicions we have before we see the data. It is cheating to first look at the data and then frame H_a to fit what the data show. Thus, the fact that the workers in the study of Example

11.3 were more satisfied with self-paced work should not influence our choice of H_a. If you do not have a specific direction firmly in mind in advance, use a two-sided alternative.

Honest hypotheses? Chinese and Japanese, for whom the number 4 is unlucky, die more often on the fourth day of the month than on other days. The authors of a study did a statistical test of the claim that the fourth day has more deaths than other days and found good evidence in favor of this claim. Can we trust this? Not if the authors looked at all days, picked the one with the most deaths, then made "this day is different" the claim to be tested. A critic raised that issue, and the authors replied: No, we had Day 4 in mind in advance, so our test was legitimate.

Exercises

11.3 Student attitudes and anemia: stating hypotheses

(a) State appropriate null and alternative hypotheses for the study of older students' attitudes described in Exercise 11.1 (page 690).

(b) State appropriate null and alternative hypotheses for the anemia study described in Exercise 11.2 (page 691).

11.4 State your claims, I Each of the following situations calls for a significance test. State the appropriate null hypothesis H_0 and alternative hypothesis H_a in each case. Be sure to define your parameter each time.

(a) Larry's car averages 26 miles per gallon on the highway. He switches to a new brand of motor oil that is advertised to increase gas mileage. After driving 3000 highway miles with the new oil, he wants to determine if the average gas mileage has increased.

(b) A May 2005 Gallup Poll report on a national survey of 1028 teenagers revealed that 72% of teens said they rarely or never argue with their friends.[2] You wonder whether this national result would be true in your school. So you conduct your own survey of a random sample of students at your school.

11.5 State your claims, II Each of the following situations calls for a significance test. State the appropriate null hypothesis H_0 and alternative hypothesis H_a in each case. Be sure to define your parameter each time.

(a) In the setting of Example 11.2 (page 688), the city manager also noted that paramedics arrived within 8 minutes after 78% of all calls involving life-threatening injury last year. Based on this year's random sample of 400 calls, she wants to determine whether the paramedics have arrived within 8 minutes more frequently this year.

(b) A national study reports that households spend an average of 30% of their food expenditures in restaurants. A restaurant association in your area wonders if the national results apply locally. They interview a sample of households and ask about their total food budget and the amount spent in restaurants.

11.6 Wrong hypotheses Here are several situations where there is an incorrect application of the ideas presented in this section. Explain what is wrong in each situation and why it is wrong.

(a) A change is made that should improve student satisfaction with the parking situation at your school. The null hypothesis, that there is an improvement, is tested versus the alternative, that there is no change.

(b) A researcher tests the following null hypothesis: H_0: $\bar{x} = 10$.

(c) A climatologist wants to test the null hypothesis that it will rain tomorrow.

Conditions for Significance Tests

In Chapter 10, we introduced three conditions that should be satisfied before we construct a confidence interval about an unknown population mean or proportion: SRS from the population of interest, Normality, and independent observations. These same three conditions must be verified before performing a significance test about a population mean or proportion. As in the previous chapter, the details for checking the Normality condition are different for means and proportions:

For means	population distribution Normal or large sample size ($n \geq 30$)
For proportions	$np \geq 10$ and $n(1-p) \geq 10$

Example 11.4 ***Call the paramedics! (continued)***
Checking conditions

Before we attempt to carry out a significance test about the mean response time in Example 11.2 (page 688), we should confirm that the three important conditions are met.

- **SRS** The city manager took a simple random sample of 400 calls involving life-threatening accidents from the population of such calls this year.
- **Normality** The population distribution of paramedic response times may not follow a Normal distribution. But the sample size ($n = 400$) is large enough to ensure Normality of the sampling distribution of $\overline{x}$ (by the central limit theorem).
- **Independence** Since the city manager is sampling calls without replacement, we must assume that there were at least $(10)(400) = 4000$ calls that involved life-threatening injuries in the city this year.

All three conditions are met, so we can proceed to the calculations.

Test Statistics

test statistic A significance test uses data in the form of a ***test statistic.*** Here are some principles that apply to most tests:

- The test is based on a statistic that compares the value of the parameter as stated in the null hypothesis with an estimate of the parameter from the sample data.
- Values of the estimate far from the parameter value in the direction specified by the alternative hypothesis give evidence against H_0.
- To assess how far the estimate is from the parameter, standardize the estimate. In many common situations, the test statistic has the form

$$\text{test statistic} = \frac{\text{estimate} - \text{hypothesized value}}{\text{standard deviation of the estimate}}$$

Example 11.5 *Call the paramedics! (continued)*
Calculating the test statistic

In Example 11.2 (page 688), the null hypothesis is H_0: $\mu = 6.7$ minutes, and the estimate of μ is $\bar{x} = 6.48$ minutes. Since we are pretending that $\sigma = 2$ minutes for the distribution of response times this year, our test statistic is

$$z = \frac{\text{estimate} - \text{hypothesized value}}{\text{standard deviation of the estimate}}$$

$$= \frac{\bar{x} - \mu_0}{\sigma/\sqrt{n}}$$

where μ_0 is the value of μ specified by the null hypothesis. The test statistic z says how far $\bar{x}$ is from μ_0 in standard deviation units. For Example 11.2,

$$z = \frac{6.48 - 6.70}{2/\sqrt{400}} = -2.20$$

Because the sample result is over two standard deviations below the hypothesized mean 6.7, it gives good evidence that the mean response time this year is not 6.7 minutes but, rather, less than 6.7 minutes.

P-values

The null hypothesis H_0 states the claim we are seeking evidence against. The test statistic measures how much the sample data diverge from the null hypothesis. If the test statistic is large and is in the direction suggested by the alternative hypothesis H_a, we have data that would be unlikely if H_0 were true. We make "unlikely" precise by calculating a probability, called a ***P-value.***

P-value

The probability, computed assuming that H_0 is true, that the observed outcome would take a value as extreme as or more extreme than that actually observed is called the **P-value** of the test. The smaller the *P*-value is, the stronger the evidence against H_0 provided by the data.

Small *P*-values are evidence against H_0 because they say that the observed result is unlikely to occur when H_0 is true. Large *P*-values fail to give evidence against H_0.

Example 11.6 *Call the paramedics! (continued)*
Computing the *P*-value

A random sample of 400 call response times in Example 11.2 (page 688) yielded $\bar{x} = 6.48$ minutes. This is somewhat far from the hypothesized value H_0: $\mu = 6.7$ minutes. The test statistic says just how far, in the standard scale,

$$z = \frac{6.48 - 6.70}{2 / \sqrt{400}} = -2.20$$

The alternative H_a says that $\mu < 6.7$ minutes, so negative values of z favor H_a over H_0.

The *P*-value is the probability of getting a sample result at least as extreme as the one we did if H_0 were true. In other words, the *P*-value is $P(\bar{x} \le 6.48)$ calculated assuming $\mu = 6.7$. Figure 11.2(a) shows the sampling distribution of $\bar{x}$ when $\mu = 6.7$. The shaded area under the curve is the *P*-value of the sample result $\bar{x} = 6.48$.

Figure 11.2 *(a) The P-value for the one-sided test in Example 11.6. (b) The P-value for the value z = −2.20 of the test statistic. The P-value is the probability (when H_0 is true) that z takes a value as small or smaller than the actually observed value.*

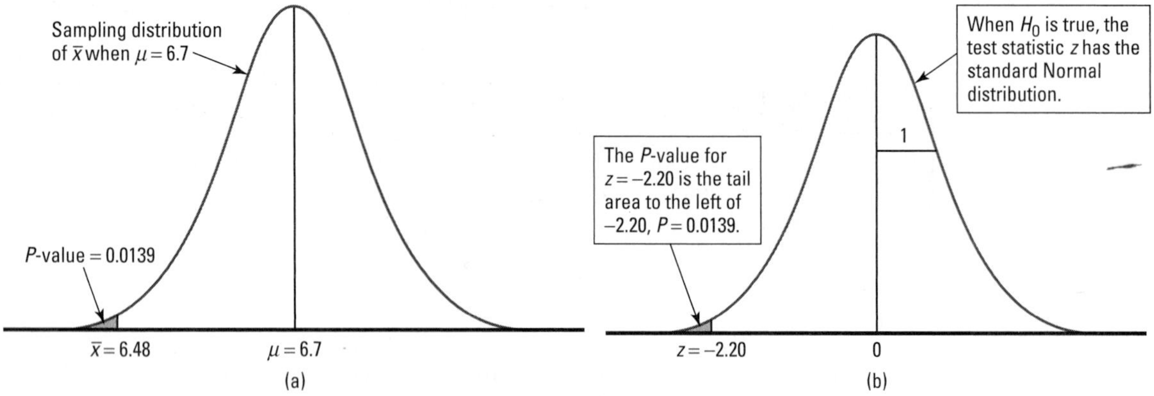

Figure 11.2(b) shows this same *P*-value on the standard Normal curve, which displays the distribution of the z statistic. The *P*-value is $P(Z \le -2.20)$. Using Table A, we find that this probability is 0.0139. So if H_0 is true, and the mean response time this year is still 6.7

minutes, there is about a 1.4% chance that the city manager would obtain a sample of 400 calls with a mean response time of 6.48 minutes or less. The small *P*-value provides strong evidence against H_0 and in favor of the alternative H_a: $\mu < 6.7$ minutes.

The *P*-value in Example 11.6 is the probability of getting a z less than or equal to the observed $z = -2.20$. The alternative hypothesis sets the direction that counts as evidence against H_0. If the alternative is two-sided, both directions count. Here is an example that shows the *P*-value for a two-sided test.

Example 11.7 ***Job satisfaction***
Calculating a two-sided *P*-value

Suppose we know that differences in job satisfaction scores in Example 11.3 (page 692) follow a Normal distribution with standard deviation $\sigma = 60$. If there is no difference in job satisfaction between the two work environments, the mean is $\mu = 0$. This is H_0. The alternative hypothesis says simply "there is a difference," H_a: $\mu \neq 0$.

Data from 18 workers gave $\bar{x} = 17$. That is, these workers preferred the self-paced environment on average. The test statistic is

$$z = \frac{\bar{x} - \sigma}{\sigma/\sqrt{n}}$$

$$= \frac{17 - 0}{60/\sqrt{18}} = 1.20$$

Because the alternative is two-sided, the *P*-value is the probability of getting a z *at least as far from 0 in either direction as the observed* $z = 1.20$. As always, calculate the *P*-value taking H_0 to be true. When H_0 is true, $\mu = 0$, and z has the standard Normal distribution. Figure 11.3 shows the *P*-value as an area under the standard Normal curve. It is

$$P\text{-value} = P(Z < -1.20 \text{ or } Z > 1.20)) = 2P(Z < -1.20)$$

$$= (2)(0.1151) = 0.2302$$

Values as far from 0 as $\bar{x} = 17$ would happen 23% of the time when the true population mean is $\mu = 0$. An outcome that would occur so often when H_0 is true is not good evidence against H_0.

Figure 11.3 *The P-value for the two-sided test in Example 11.7. The observed value of the test statistic is z = 1.20.*

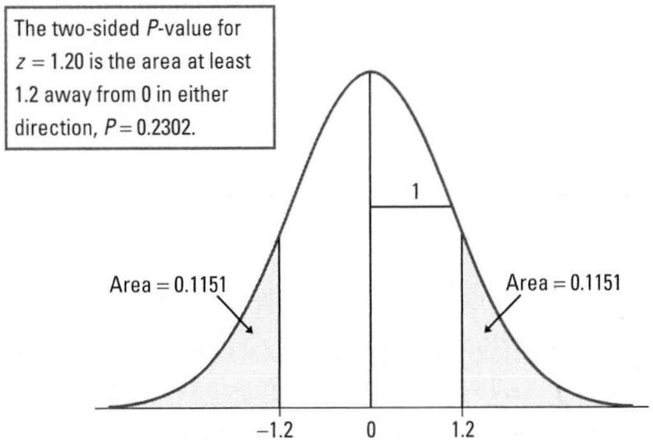
The two-sided P-value for z = 1.20 is the area at least 1.2 away from 0 in either direction, P = 0.2302.
1
Area = 0.1151
Area = 0.1151
−1.2
0
1.2

The *P-value of a Significance Test* applet at the book's Web site www.whfreeman.com/tps3e automates the work of finding *P*-values. The applet even displays *P*-values as areas under a Normal curve, just like Figures 11.2 and 11.3.

Exercises

11.7 Job satisfaction: the conditions Refer to Example 11.7.

(a) Verify that the three important conditions are satisfied.

(b) Was it necessary to know that differences in job satisfaction follow a Normal distribution? Why or why not?

11.8 Job satisfaction with a larger sample Suppose that the job satisfaction study had produced exactly the same outcome $\bar{x} = 17$ as in Example 11.7, but from a sample of 75 workers rather than just 18 workers.

(a) Now is it necessary to know that differences in job satisfaction follow a Normal distribution? Why or why not?

(b) Calculate the test statistic z and its two-sided *P*-value.

(c) Do the data give good evidence that the population mean is not zero? Justify your answer.

11.9 Student attitudes, II Return to Exercise 11.1 (page 690). Start with the picture you drew there, and then do the following.

(a) Shade the area under the curve that is the *P*-value for $\bar{x} = 118.6$. Then calculate the test statistic and the *P*-value.

(b) Shade differently the area under the curve that is the *P*-value for $\bar{x} = 125.7$. Then calculate the test statistic and the *P*-value.

(c) Explain what each of the *P*-values in parts (a) and (b) tells us about the evidence against the null hypothesis.

11.10 Anemia, II Return to Exercise 11.2 (page 691). Start with the picture you drew there, and then do the following:

(a) Shade the area under the curve that is the *P*-value for $\bar{x} = 11.3$. Then calculate the test statistic and the *P*-value.

(b) Shade differently the area under the curve that is the *P*-value for $\bar{x} = 11.8$. Then calculate the test statistic and the *P*-value.

(c) Explain what each of the *P*-values in parts (a) and (b) tells us about the evidence against the null hypothesis.

11.11 Coffee sales Weekly sales of regular ground coffee at a supermarket have in the recent past varied according to a Normal distribution with mean $\mu = 354$ units per week and standard deviation $\sigma = 33$ units. The store reduces the price by 5%. Sales in the next three weeks are 405, 378, and 411 units. Is this good evidence that average sales are now higher? The hypotheses are

$$H_0: \mu = 354$$
$$H_a: \mu > 354$$

Assume that the standard deviation of the population of weekly sales remains $\sigma = 33$.

(a) Find the value of $\bar{x}$.

(b) Sketch the Normal curve for the sampling distribution of $\bar{x}$ when H_0 is true. Why is the sampling distribution Normal? Are the other conditions satisfied?

(c) Shade the area that represents the *P*-value for the observed outcome. Calculate the test statistic and the *P*-value.

(d) Do you think there is convincing evidence that mean sales are higher? Explain.

11.12 Finding *P*-values A test of the null hypothesis H_0: $\mu = 0$ gives test statistic $z = 1.6$.

(a) What is the *P*-value if the alternative is H_a: $\mu > 0$?

(b) What is the *P*-value if the alternative is H_a: $\mu < 0$?

(c) What is the *P*-value if the alternative is H_a: $\mu \neq 0$?

Statistical Significance

We sometimes take one final step to assess the evidence against H_0. We can compare the *P*-value with a fixed value that we regard as decisive. This amounts to announcing in advance how much evidence against H_0 we will insist on. The decisive value of *P* is called the ***significance level.*** We write it as α, the Greek letter alpha. If we choose $\alpha = 0.05$, we are requiring that the data give evidence against H_0 so strong that it would happen no more than 5% of the time (1 time in 20 samples in the long run) when H_0 is true. If we choose $\alpha = 0.01$, we are insisting on stronger evidence against H_0, evidence so strong that it would appear only 1% of the time (1 time in 100 samples) if H_0 is in fact true.

significance level

Statistical Significance

If the *P*-value is as small as or smaller than alpha, we say that the data are **statistically significant at level α.**

"Significant" in the statistical sense does not mean "important." It means simply "not likely to happen just by chance." The significance level α makes "not likely" more exact. Significance at level 0.01 is often expressed by the statement "The results were significant ($P < 0.01$)." Here *P* stands for the *P*-value. The *P*-value is more informative than a statement of significance because it allows us to assess significance at any level we choose. For example, a result with $P = 0.03$ is significant at the $\alpha = 0.05$ level but is not significant at the $\alpha = 0.01$ level.

Example 11.8 ***Paramedics and job satisfaction***
Determining statistical significance

In Example 11.6 (page 696) concerning paramedic response times, we found the *P*-value to be $P = 0.0139$. This result is statistically significant at the $\alpha = 0.05$ level since $P < 0.05$. But it is not significant at the $\alpha = 0.01$ level since $P > 0.01$. See Figure 11.4.

Figure 11.4 *An outcome with P-value P is significant at all levels α at or above P and is not significant at smaller levels α.*

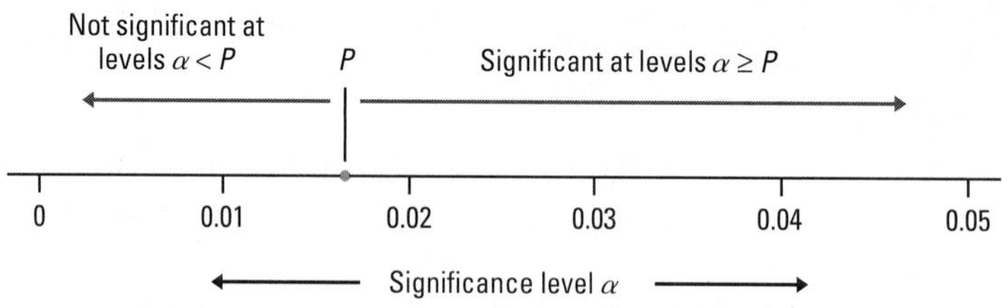

In Example 11.7 (page 697) about job satisfaction, the *P*-value was 0.2302. This outcome is not statistically significant at the $\alpha = 0.10$ or any smaller α level.

In practice, the most commonly used significance level is $\alpha = 0.05$. This is mainly due to Sir Ronald A. Fisher, a famous statistician who worked on agricultural experiments in England during the early twentieth century. Fisher was the first to suggest deliberately using random assignment in an experiment. In a paper published in 1926, Fisher wrote that

> it is convenient to draw the line at about the level at which we can say: "Either there is something in the treatment, or a coincidence has occurred such as does not occur more than once in twenty trials."[3]

Sometimes it may be preferable to choose $\alpha = 0.01$ or $\alpha = 0.10$, for reasons we will discuss in the next section.

Interpreting Results in Context

Remember the three C's: conclusion, connection, and context.

The final step in performing a significance test is to draw a conclusion about the competing claims you were testing. As with confidence intervals, your *conclusion* should have a clear *connection* to your calculations and should be stated in the *context* of the problem. In significance testing, there are two accepted methods for drawing conclusions: one based on *P*-values and another based on statistical significance. Both methods describe the strength of the evidence against the null hypothesis H_0.

reject H_0

fail to reject H_0

We have already illustrated the *P*-value approach in Examples 11.6 (page 696) and 11.7 (page 697). Our other option is to make one of two decisions about the null hypothesis H_0—***reject H_0*** or ***fail to reject*** H_0—based on whether our result is statistically significant. We will reject H_0 if our sample result is too unlikely to have occurred by chance assuming H_0 is true. In other words, we will reject H_0 if our result is statistically significant at the given α level. If our sample result could plausibly have happened by chance assuming H_0 is true, we will fail to reject H_0. That is, we will fail to reject H_0 if our result is not significant at the given α level.

Example 11.9 ***Paramedics and job satisfaction (continued)***
Deciding to reject or fail to reject H_0

In Example 11.6 (page 696), we calculated the P-value for the city manager's study of paramedic response times as $P = 0.0139$. If we were using an $\alpha = 0.05$ significance level, we would reject H_0: $\mu = 6.7$ minutes (conclusion) since our P-value, 0.0139, is less than $\alpha = 0.05$ (connection). It appears that the mean response time to all life-threatening calls this year is less than last year's average of 6.7 minutes (context).

For the job satisfaction study, the P-value was 0.2302. Using an $\alpha = 0.05$ significance level, we would fail to reject H_0: $\mu = 0$ since $0.2302 > \alpha = 0.05$. It is possible that the mean difference in job satisfaction scores for workers in a self-paced versus machine-paced environment is 0.

CAUTION

Warning: if you are going to draw a conclusion based on statistical significance, then the significance level α should be stated before the data are produced. Otherwise, a deceptive user of statistics might set an α level after the data have been analyzed in an obvious attempt to manipulate the conclusion. This is just as inappropriate as choosing an alternative hypothesis to be one-sided in a particular direction *after* looking at the data.

A P-value is more informative than a "reject" or "fail to reject" conclusion at a given significance level. For instance, a P-value of 0.0139 allows us to reject H_0 at the $\alpha = 0.05$ significance level. But the P-value, 0.0139, gives a better sense of how strong the evidence against H_0 is. *The P-value is the smallest α level at which the data are significant.* Knowing the P-value allows us to assess significance at any level. However, interpreting the P-value is more challenging than making a decision about H_0 based on statistical significance. A well-trained user of statistics should be able to handle either approach.

Exercises

11.13 Statistical significance Explain in plain language why a significance test that is significant at the 1% ($\alpha = 0.01$) level must always be significant at the 5% ($\alpha = 0.05$) level. If a test is significant at the 5% level, what can you say about its significance at the 1% level?

11.14 Nicotine in cigarettes To determine whether the mean nicotine content of a brand of cigarettes is greater than the advertised value of 1.4 milligrams, a health advocacy group tests H_0: $\mu = 1.4$ versus H_a: $\mu > 1.4$. The calculated value of the test statistic is $z = 2.42$.

(a) Is the result significant at the 5% level? Why or why not?

(b) Is the result significant at the 1% level? Why or why not?

(c) What decision would you make about H_0 in part (a)? Part (b)? Explain.

11.15 Significance tests You will perform a significance test of H_0: $\mu = 0$ versus H_a: $\mu > 0$.

(a) What values of z would lead you to reject H_0 at the 5% ($\alpha = 0.05$) significance level?

(b) If the alternative hypothesis was H_a: $\mu \neq 0$, what values of z would lead you to reject H_0 at the 5% significance level?

(c) Explain why your answers to parts (a) and (b) are different.

11.16 Testing a random number generator, I A certain random number generator is supposed to produce random numbers that are uniformly distributed on the interval from 0 to 1. If this is true, the numbers generated come from a population with $\mu = 0.5$ and $\sigma = 0.2887$. A command to generate 100 random numbers gives outcomes with mean $\bar{x} = 0.4365$. Assume that the population σ remains fixed. We want to test H_0: $\mu = 0.5$ versus H_a: $\mu \neq 0.5$.

(a) Calculate the value of the z test statistic and the P-value.

(b) Is the result significant at the 5% level ($\alpha = 0.05$)? Why or why not?

(c) Is the result significant at the 1% level ($\alpha = 0.01$)? Why or why not?

(d) What decision would you make about H_0 in part (b)? Part (c)? Explain.

11.17 Testing a random number generator, II The `rand` function on the TI-83/84 (MATH / PRB / 1:rand) and the `rand83` function on the TI-89 (CATALOG/F3) generate a pseudorandom real number in the interval [0, 1)—that is, in the interval $0 \leq X < 1$. The command `rand(100)` (`tistat.rand83(100)` on the TI-89) generates 100 random real numbers in the interval [0, 1). Describe how you would use your calculator to carry out a simulation like the one described in the previous exercise. (*Hint:* Store the 100 values in a list.) As in Exercise 11.16, take $\sigma = 0.2887$. Then carry out your plan and answer the questions in Exercise 11.16.

11.18 *P*-values and statistical significance A research report described the results of two studies. The *P*-value for the first study was 0.049; for the second it was 0.00002. Write a few sentences comparing what these *P*-values tell you about statistical significance in each of the two studies.

Section 11.1 Summary

A **significance test** assesses the evidence provided by data against a **null hypothesis H_0** in favor of an **alternative hypothesis H_a**.

The hypotheses are stated in terms of population parameters. Usually, H_0 is a statement that no effect is present, and H_a says that a parameter differs from its null value (the null hypothesis value) in a specific direction **(one-sided alternative)** or in either direction **(two-sided alternative)**.

As with confidence intervals, you should verify that the three conditions—SRS, Normality, and independence—are met before you proceed to calculations.

The essential reasoning of a significance test is as follows. Suppose that the null hypothesis is true. If we repeated our data production many times, would we often get data as inconsistent with H_0 as the data we actually have? If the data are unlikely when H_0 is true, they provide evidence against H_0.

A test is based on a **test statistic.** The **P-value** is the probability, computed supposing H_0 to be true, that the test statistic will take a value at least as extreme as that actually observed. Small *P*-values indicate strong evidence against H_0. Calculating *P*-values requires knowledge of the sampling distribution of the test statistic when H_0 is true.

If the P-value is as small as or smaller than a specified value α, the data are **statistically significant** at significance level α. In that case, we can **reject** H_0. If the P-value is larger than α, we **fail to reject** H_0. An alternative to making a decision about the null hypothesis in your conclusion is to interpret the P-value in context.

Down with driver ed! Who could object to driver-training courses in schools? The killjoy who looks at data, that's who. Careful studies show no significant effect of driver training on the behavior of teenage drivers. Because many states allow those who take driver ed to get a license at a younger age, the programs may actually increase accidents and road deaths by increasing the number of young and risky drivers.

Section 11.1 Exercises

11.19 Research and hypotheses For each of the following research questions, (1) describe an appropriate method for producing data to answer the question, and (2) state an appropriate pair of hypotheses (null and alternative) to test.

(a) The mean area of the several thousand similar apartments in a new development is advertised to be 1250 square feet. A tenant group thinks the apartments are smaller than advertised.

(b) Last year, a company's service technicians took an average of 1.8 hours to respond to trouble calls from customers who had purchased service contracts. The company president wants to know whether response times have changed this year.

(c) Simon reads a newspaper report claiming that 12% of all adults in the U.S. are left-handed. He wonders if 12% of the students at his large public high school are left-handed.

11.20 Spending on housing The Census Bureau reports that households spend an average of 31% of their total spending on housing. A homebuilders association in Cleveland believes that this average is lower in their area. They interview a sample of 40 households in the Cleveland metropolitan area to learn what percent of their spending goes toward housing. Take μ to be the mean percent of spending devoted to housing among all Cleveland households. We want to test the hypotheses

$$H_0: \mu = 31\%$$

$$H_a: \mu < 31\%$$

Assume that the population standard deviation is $\sigma = 9.6\%$.

(a) What is the sampling distribution of the mean percent $\bar{x}$ that the sample spends on housing if the null hypothesis is true? Sketch the density curve of the sampling distribution. Be sure to locate μ and σ correctly.

(b) Suppose that the study finds $\bar{x} = 30.2\%$ for the 40 households in the sample. Mark this point on the axis in your sketch. Then suppose that the study result is $\bar{x} = 27.6\%$. Mark this point on your sketch. Referring to your sketch, explain in simple language why one result is good evidence that average Cleveland spending on housing is less than 31%, whereas the other result is not.

(c) Shade the area under the curve that gives the P-value for the result $\bar{x} = 30.2\%$. (Note that we are looking for evidence that spending is less than the null hypothesis states.) Now shade differently the area under the curve that represents the P-value for $\bar{x} = 27.6\%$. Find these two P-values.

(d) What conclusion would you draw for each of the two results at the $\alpha = 0.05$ level? At the $\alpha = 0.01$ level? Justify your answers.

11.21 Statistical significance: one-sided and two-sided tests You are performing a significance test of H_0: $\mu = 0$ based on an SRS of 20 observations from a Normal population.

(a) If the alternative hypothesis is H_a: $\mu \neq 0$, what values of the z statistic are significant at the $\alpha = 0.005$ level? Show your work.

(b) If the alternative hypothesis is H_a: $\mu > 0$, what values of the z statistic are significant at the $\alpha = 0.005$ level? Show your work.

11.22 Interpreting statistical significance Asked to explain the meaning of "statistically significant at the $\alpha = 0.05$ level," a student says: "This means that the probability that the null hypothesis is true is less than 0.05." Is this explanation correct? Why or why not?

11.23 Interpreting *P*-values Suppose you perform a significance test of H_0: $\mu = 15$ versus the two-sided alternative and obtain a *P*-value of 0.082.

(a) Explain to someone who knows no statistics what this *P*-value means.

(b) What decision would you make about the null hypothesis? Why?

(c) Suppose you decide to reject H_0. What is the probability that you are wrong (H_0 is true)?

11.24 The Supreme Court speaks Court cases in such areas as employment discrimination often involve statistical evidence. The Supreme Court has said that z-scores beyond $z^* = 2$ or 3 are generally considered convincing statistical evidence. For a two-sided test, what significance level α corresponds to $z^* = 2$? To $z^* = 3$?

11.25 Thinking about conditions Explain in your own words why each of the three conditions—SRS, Normality, and independence—is important when performing inference about a population mean μ.

11.26 Diet and diabetes Does eating more fiber reduce the blood cholesterol level of patients with diabetes? A randomized clinical trial compared normal and high-fiber diets. Here is part of the researchers' conclusion:

> The high-fiber diet reduced plasma total cholesterol concentrations by 6.7 percent ($P = 0.02$), triglyceride concentrations by 10.2 percent ($P = 0.02$), and very-low-density lipoprotein concentrations by 12.5 percent ($P = 0.01$).[4]

A doctor who knows little statistics says that a drop of 6.7% in cholesterol isn't a lot—maybe it's just an accident due to the chance assignment of patients to the two diets. Explain to the doctor in simple language how "$P = 0.02$" answers this objection.

11.2 Carrying Out Significance Tests

Although the reasoning of significance testing isn't simple, carrying out a test is. The process is very similar to the one we followed when constructing a confidence interval. With a few minor changes, the four-step Inference Toolbox will once again guide us through the inference procedure.

Inference Toolbox

Significance tests

To test a claim about an unknown population parameter:

Step 1: Hypotheses Identify the population of interest and the parameter you want to draw conclusions about. State hypotheses.

Step 2: Conditions Choose the appropriate inference procedure. Verify the conditions for using it.

Step 3: Calculations If the conditions are met, carry out the inference procedure.

- Calculate the **test statistic.**
- Find the ***P*-value.**

Step 4: Interpretation Interpret your results in the context of the problem.

- Interpret the *P*-value *or* make a decision about H_0 using statistical significance.
- Don't forget the 3 C's: conclusion, connection, and context.

Once you have completed the first two steps, your calculator or computer can do Step 3 by following a built-in procedure. Here is the calculation step for carrying out a significance test about the population mean μ in the unrealistic setting when σ is known.

z Test for a Population Mean

To test the hypothesis H_0: $\mu = \mu_0$ based on an SRS of size n from a population with unknown mean μ and known standard deviation σ, compute the **one-sample *z* statistic**

$$z = \frac{\bar{x} - \mu_0}{\sigma / \sqrt{n}}$$

In terms of a variable Z having the standard Normal distribution, the *P*-value for a test of H_0 against

H_a: $\mu > \mu_0$ is $P(Z \geq z)$

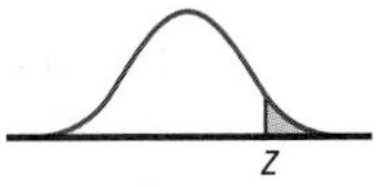

H_a: $\mu < \mu_0$ is $P(Z \leq z)$

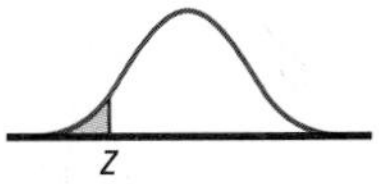

H_a: $\mu \neq \mu_0$ is $P(Z \geq |z|)$

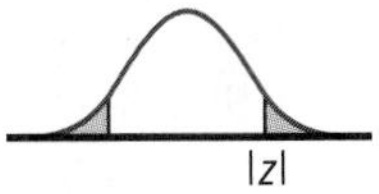

These *P*-values are exact if the population distribution is Normal and are approximately correct for large n in other cases.

Here is an example illustrating the use of the z test for a population mean.

Example 11.10 *Executives' blood pressures*

z test for a population mean

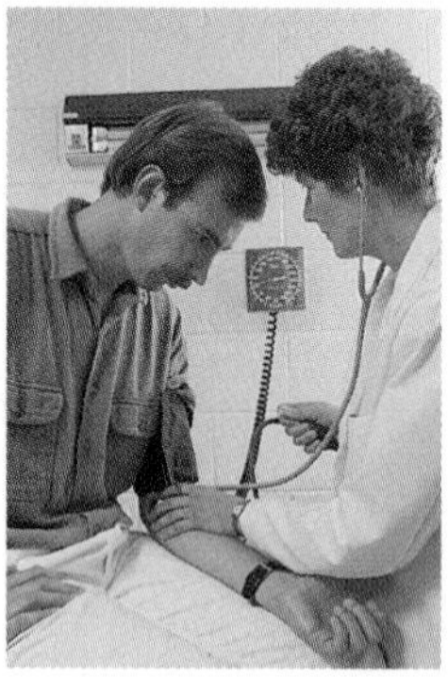

The medical director of a large company is concerned about the effects of stress on the company's younger executives. According to the National Center for Health Statistics, the mean systolic blood pressure for males 35 to 44 years of age is 128, and the standard deviation in this population is 15. The medical director examines the medical records of 72 male executives in this age group and finds that their mean systolic blood pressure is $\bar{x} = 129.93$. Is this evidence that the mean blood pressure for all the company's younger male executives is different from the national average? (As usual in this chapter, we make the unrealistic assumption that the population standard deviation is known, in this case that executives have the same σ as the general population.)

Step 1: Hypotheses The population of interest is all middle-aged male executives in this company. We want to test a claim about the mean blood pressure μ for these executives. The null hypothesis is "no difference" from the national mean $\mu = 128$. The alternative is two-sided because the medical director did not have a particular direction in mind before examining the data. So the hypotheses about the unknown mean μ of the executive population are

H_0: $\mu = 128$ Younger male company executives' mean blood pressure is 128.

H_a: $\mu \neq 128$ Younger male company executives' mean blood pressure differs from the national mean of 128.

Step 2: Conditions Since σ is known, we will use a one-sample z test for a population mean. Now we check conditions.

- **SRS** The z test assumes that the 72 executives in the sample are an SRS from the population of all younger male executives in the company. We should check this assumption by asking how the data were produced. If medical records are available only for executives with recent medical problems, for example, the data are of little value for our purpose. It turns out that all executives are given a free annual medical exam and that the medical director selected 72 exam results at random.
- **Normality** We do not know that the population distribution of blood pressures among the company executives is Normally distributed. But the large sample size ($n = 72$) guarantees that the sampling distribution of $\bar{x}$ will be approximately Normal (by the central limit theorem).
- **Independence** Since the medical director is selecting executives without replacement, we must assume that there are at least $(10)(72) = 720$ middle-aged male executives in this large company.

Step 3: Calculations

- **Test statistic** The one-sample z statistic is

$$z = \frac{\bar{x} - \mu_0}{\sigma/\sqrt{n}} = \frac{129.93 - 128}{15/\sqrt{72}} = 1.09$$

- ***P*-value** You should still draw a picture to help find the *P*-value, but now you can sketch the standard Normal curve with the observed value of z. Figure 11.5 shows that the *P*-value is the probability that a standard Normal variable Z takes an absolute value of 1.09 or greater. From Table A we find that this probability is

$$P\text{-value} = 2P(Z \geq 1.09) = 2(1 - 0.8621) = 0.2758$$

Figure 11.5 *The P-value for the two-sided test in Example 11.10.*

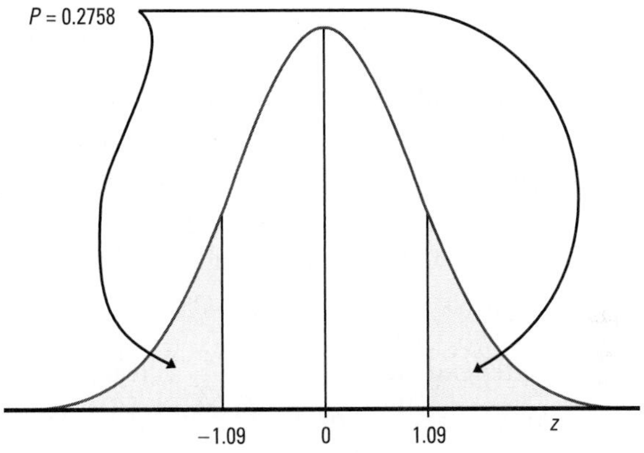

Step 4: **Interpretation** More than 27% of the time, an SRS of size 72 from the general male population would have a mean blood pressure at least as far from 128 as that of the executive sample. The observed $\overline{x} = 129.93$ is therefore not good evidence that middle-aged male executives' blood pressures differ from the national average.

The data in Example 11.10 do *not* establish that the mean blood pressure μ for this company's middle-aged male executives *is* 128. We sought evidence that μ differed from 128 and failed to find convincing evidence. That is all we can say. No doubt the mean blood pressure of the entire executive population is not exactly equal to 128. A large enough sample would give evidence of the difference, even if it is very small. Significance tests assess the evidence *against* H_0. If the evidence is strong, we can confidently reject H_0 in favor of the alternative. **Failing to find evidence against H_0 means only that the data are consistent with H_0, not that we have clear evidence that H_0 is true.**

CAUTION

In Example 11.10, we performed a *two-sided,* or ***two-tailed,*** test, because our alternative hypothesis was two-sided. We used the *P*-value approach to draw a conclusion in Step 4. The next example illustrates a *one-sided,* or ***one-tailed,*** test in which we draw a conclusion about H_0 using statistical significance.

two-tailed test

one-tailed test

Example 11.11 *A health promotion program*

One-tailed test and statistical significance

The company medical director of Example 11.10 institutes a health promotion campaign to encourage employees to exercise more and eat a healthier diet. One measure of the effectiveness of such a program is a drop in blood pressure. The director chooses a random sample of 50 employees and compares their blood pressures from physical exams given before the campaign and again a year later. The mean change in systolic blood pressure for these $n = 50$ employees is $\bar{x} = -6$. We take the population standard deviation to be $\sigma = 20$. The director decides to use an $\alpha = 0.05$ significance level.

Step 1: **Hypotheses** We want to know if the health campaign reduced blood pressure on average in the population of all employees at this large company. Taking μ to be the mean change in blood pressure for all employees, we test

$$H_0: \mu = 0$$
$$H_a: \mu < 0$$

Step 2: **Conditions** Since σ is known, we will use a one-sample z test for a population mean. Now we check conditions.

- **SRS** The medical director took a "random sample" of 50 company employees. Our calculation method assumes that an SRS was taken.
- **Normality** The large sample size ($n = 50$) guarantees approximate Normality of the sampling distribution of $\bar{x}$, even if the population distribution of change in blood pressure isn't Normal.
- **Independence** There must be at least $(10)(50) = 500$ employees in this large company since the medical director is sampling without replacement.

Step 3: **Calculations**

- **Test statistic** The one-sample z test is appropriate:

$$z = \frac{\bar{x} - \mu_0}{\sigma/\sqrt{n}} = \frac{-6-0}{20/\sqrt{50}} = -2.12$$

- **P-value** Because H_a is one-sided on the low side, large negative values of z count against H_0. See Figure 11.6. From Table A, we find that the P-value is

$$P = P(Z \leq -2.12) = 0.0170$$

Figure 11.6 *The P-value for the one-sided test in Example 11.11.*

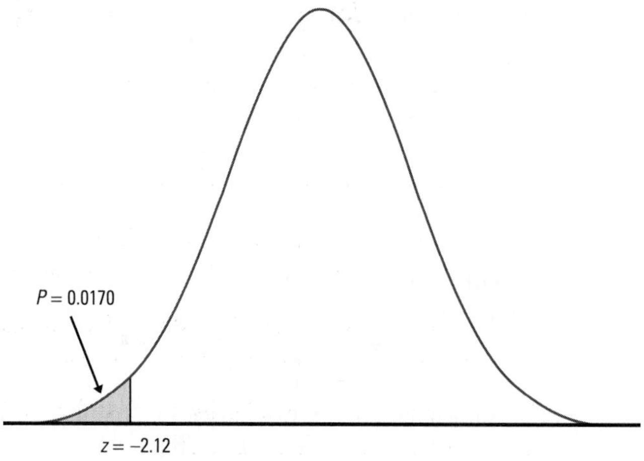

Step 4: Interpretation Since our *P*-value, 0.0170, is less than $\alpha = 0.05$, this result is statistically significant. We reject H_0 and conclude that the mean difference in blood pressure readings from before and after the campaign among this company's employees is negative. In other words, the data suggest that employees' blood pressure readings have decreased on average.

Our conclusion in Example 11.11 is cautious. We would like to conclude that the health campaign *caused* the drop in mean blood pressure. But there may be other possible explanations. Suppose the local television station runs a series on the risk of heart attacks and the value of better diet and exercise. Many employees may improve their health habits even without encouragement from the company. Only a randomized comparative experiment protects against such lurking variables. The medical director preferred to launch a company-wide campaign that appealed to all employees. This may be a good medical decision, but the lack of a control group weakens the statistical conclusion.

When you interpret your results in context (Step 4 of the Inference Toolbox), be sure to link your comments directly to your *P*-value or significance level. Do not simply say "reject H_0." Provide a basis for any decision that you make about the claim expressed in your hypotheses.

Exercises

11.27 Water quality An environmentalist group collects a liter of water from each of 45 randomly chosen locations along a stream and measures the amount of dissolved oxygen in each specimen. The mean is 4.62 milligrams per liter (mg/L). Is this strong evidence that the stream has a mean dissolved oxygen content of less than 5 mg per liter? (Suppose we know that dissolved oxygen varies among locations with $\sigma = 0.92$ mg/L.) Follow the Inference Toolbox.

11.28 Improving your SAT score We suspect that students will generally score higher the second time they take the SAT Mathematics exam than on their first attempt. Suppose we know that the changes in score (second try minus first try) have population standard deviation $\sigma = 50$. Here are the results for 46 randomly chosen high school students:

−30	24	47	70	−62	55	−41	−32	128	−11
−43	122	−10	56	32	−30	−28	−19	1	17
57	−14	−58	77	27	−33	51	17	−67	29
94	−11	2	12	−53	−49	49	8	−24	96
120	2	−33	−2	−39	99				

(a) Construct graphical displays and calculate numerical summaries for these data. Write a few sentences about the distribution of changes in SAT Math scores.

(b) Based on your work in part (a), do you believe that the population of differences in SAT Math score is Normally distributed? Why or why not?

(c) Do these data give good evidence that the mean change in the population is greater than 0? Follow the Inference Toolbox.

11.29 Pressing pills A drug manufacturer forms tablets by compressing a granular material that contains the active ingredient and various fillers. The hardness of a sample from each batch of tablets produced is measured in order to control the compression process. The target values for the hardness are $\mu = 11.5$ and $\sigma = 0.2$. The hardness data for a sample of 20 tablets are

11.627	11.613	11.493	11.602	11.360	11.374	11.592	11.458	11.552	11.463
11.383	11.715	11.485	11.509	11.429	11.477	11.570	11.623	11.472	11.531

(a) We are not told that the distribution of hardness measurements in the population is Normal, and the sample size is too small ($n = 20$) for the central limit theorem to help us. In cases like this, the best we can do is to examine the sample data. Make appropriate graphs and calculate numerical summaries for determining whether these data could have come from a Normal population. Write a few sentences explaining your conclusion.

(b) Is there significant evidence at the 5% level that the mean hardness of the tablets is different from the target value? Use the Inference Toolbox.

11.30 Filling cola bottles Bottles of a popular cola are supposed to contain 300 milliliters (ml) of cola. There is some variation from bottle to bottle because the filling machinery is not perfectly precise. The distribution of the contents is Normal with standard deviation $\sigma = 3$ ml. An inspector who suspects that the bottler is underfilling measures the contents of six randomly selected bottles from a single day's production. The results are

299.4	297.7	301.0	298.9	300.2	297.0

Is this convincing evidence that the mean amount of cola in all the bottles filled that day is less than the advertised 300 ml? Follow the Inference Toolbox.

Tests from Confidence Intervals

A 95% confidence interval captures the true value of μ in 95% of all samples. If we are 95% confident that the true μ lies in our interval, we are also confident that values of μ that fall outside our interval are incompatible with the data. That sounds like the conclusion of a significance test. In fact, there is an intimate connection between 95% confidence and significance at the 5% level. The same connection holds between 99% confidence intervals and significance at the 1% level, and so on.

Confidence Intervals and Two-Sided Tests

A level α two-sided significance test rejects a hypothesis H_0: $\mu = \mu_0$ exactly when the value μ_0 falls outside a level $1 - \alpha$ confidence interval for μ.

duality

The following example demonstrates the link between two-sided significance tests and confidence intervals (sometimes called ***duality***).

Example 11.12 *Analyzing drugs*

Duality: confidence intervals and significance tests

The Deely Laboratory analyzes specimens of a drug to determine the concentration of the active ingredient. Such chemical analyses are not perfectly precise. Repeated measurements on the same specimen will give slightly different results. The results of repeated measurements follow a Normal distribution quite closely. The analysis procedure has no bias, so the mean μ of the population of all measurements is the true concentration of the specimen. The standard deviation of this distribution is a property of the analysis method and is known to be $\sigma = 0.0068$ grams per liter. The laboratory analyzes each specimen three times and reports the mean result.

A client sends a specimen for which the concentration of active ingredient is supposed to be 0.86%. Deely's three analyses give concentrations

0.8403	0.8363	0.8447

Is there significant evidence at the 1% level that the true concentration is not 0.86%? This calls for a test of the hypotheses

$$H_0: \mu = 0.86$$
$$H_a: \mu \neq 0.86$$

We will carry out the test twice, first with the usual significance test and then from a 99% confidence interval. You should fill in the gaps in the four steps of the Inference Toolbox as you read on.

- *The significance test* The mean of the three analyses is $\bar{x} = 0.8404$. The one-sample z test statistic is therefore

$$z = \frac{\bar{x} - \mu_0}{\sigma / \sqrt{n}} = \frac{0.8404 - 0.86}{0.0068 / \sqrt{3}} = -4.99$$

Because the alternative is two-sided, the P-value is

$$P = 2P(Z \leq -4.99)$$

We cannot find this probability in Table A. The largest negative value of z in Table A is -3.49. All we can say from Table A is that P is less than $2P(Z \leq -3.49) = 2(1-0.9998) = 0.0004$. A calculator or computer could be used to give an accurate P-value. However, because the P- value is clearly less than the company's standard of 1%, we reject H_0.

- *The confidence interval* The 99% confidence interval for μ is

$$\bar{x} \pm z^* \frac{\sigma}{\sqrt{n}} = 0.8404 \pm 2.576 \frac{0.0068}{\sqrt{3}}$$
$$= 0.8404 \pm 0.0101$$
$$= (0.8303, 0.8505)$$

The hypothesized value $\mu_0 = 0.86$ falls outside this 99% confidence interval. According to this interval, $\mu = 0.86$ is not a plausible value for the mean concentration of active ingredient in this specimen. We can therefore reject H_0: $\mu = 0.86$ at the 1% significance level.

What if the null hypothesis in Example 11.12 had been H_0: $\mu = 0.85$? In that case, we would not be able to reject H_0: $\mu = 0.85$ in favor of the two-sided alternative H_a: $\mu \neq 0.85$, because 0.85 lies inside the 99% confidence interval for μ. Figure 11.7 shows both cases.

Figure 11.7 *Values of μ falling outside a 99% confidence interval can be rejected at the 1% significance level. Values falling inside the interval cannot be rejected.*

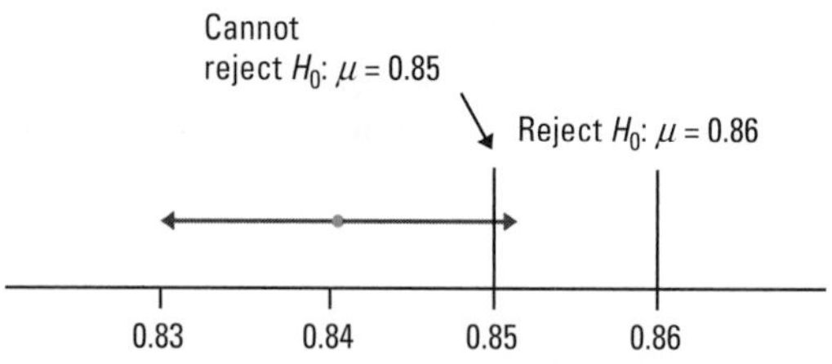

Exercises

11.31 Significance tests and confidence intervals The *P*-value for a two-sided test of the null hypothesis H_0: $\mu = 10$ is 0.06.

(a) Does the 95% confidence interval include 10? Why or why not?

(b) Does the 90% confidence interval include 10? Why or why not?

11.32 Confidence intervals and significance tests A 95% confidence interval for a population mean is 31.5 ± 3.5.

(a) Can you reject the null hypothesis that $\mu = 34$ at the 5% significance level? Why or why not?

(b) Can you reject the null hypothesis that $\mu = 36$ at the 5% significance level? Why or why not?

11.33 Radon detectors Radon is a colorless, odorless gas that is naturally released by rocks and soils and may concentrate in tightly closed houses. Because radon is slightly radioactive, there is some concern that it may be a health hazard. Radon detectors are sold to homeowners worried about this risk, but the detectors may be inaccurate. University researchers placed a random sample of 12 detectors in a chamber where they were exposed to 105 picocuries per liter of radon over 3 days. Here are the readings given by the detectors:[5]

91.9	97.8	111.4	122.3	105.4	95.0
103.8	99.6	96.6	119.3	104.8	101.7

Assume that repeated readings using detectors of this type follow a Normal distribution with $\sigma = 9$.

(a) Construct and interpret a 90% confidence interval for the mean reading μ for this type of detector.

(b) Is there significant evidence at the 10% level that the mean reading differs from the true value 105? State hypotheses and base a test on your confidence interval from (a).

11.34 One-sided tests and confidence intervals The *P*-value of a one-sided test of H_0: $\mu = 30$ is 0.04.

(a) Would the 95% confidence interval for μ include 30? Explain.

(b) Would the 90% confidence interval for μ include 30? Explain.

Section 11.2 Summary

Significance tests for the hypothesis H_0: $\mu = \mu_0$ concerning the unknown mean μ of a population are based on the **one-sample z statistic**

$$z = \frac{\bar{x} - \mu_0}{\sigma/\sqrt{n}}$$

The z test assumes an SRS of size n, known population standard deviation σ, independent observations, and either a Normal population or a large sample. *P*-values are computed from the standard Normal distribution (Table A). Use the Inference Toolbox as a guide when you perform a significance test.

Confidence intervals and two-sided significance tests are closely connected, provided that the significance level for the test and the confidence level for the interval add to 100%.

Section 11.2 Exercises

11.35 Calcium and pregnancy The level of calcium in the blood in healthy young adults varies with mean about 9.5 grams per deciliter and standard deviation about $\sigma = 0.4$. A clinic in rural Guatemala measures the blood calcium level of 160 healthy pregnant women at their first visit for prenatal care. The mean is $\bar{x} = 9.57$. Is this an indication that the mean calcium level in the population from which these women come differs from 9.5?

(a) Check the conditions for performing inference in this setting. Describe any concerns you may have about the data production.

(b) Carry out a significance test at the $\alpha = 0.05$ significance level. Report your conclusion.

(c) Construct and interpret a 95% confidence interval for the mean calcium level μ in the population. What additional information does this interval provide over the test in (b)?

Note: Based on our work in this problem, we are confident that μ lies quite close to 9.5. This illustrates the fact that **a test based on a large sample (n = 160 here) will often declare a small deviation from H_0 to be statistically significant.**

11.36 Is this milk watered down? Cobra Cheese Company buys milk from several suppliers. Cobra suspects that some producers are adding water to their milk to increase their profits. Excess water can be detected by measuring the freezing point of the milk. The freezing temperature of natural milk varies Normally, with mean $\mu = -0.545°$ Celsius (C) and standard deviation $\sigma = 0.008°C$. Added water raises the freezing temperature toward 0°C, the freezing point of water. Cobra's laboratory manager measures the freezing temperature of a random sample of five containers of milk from one producer. The mean measurement is $\bar{x} = -0.538°C$. Is this good evidence that the producer is adding water to the milk? Provide appropriate statistical evidence to justify your answer.

11.37 Connecting confidence intervals and significance tests, I The *P*-value for a two-sided test of the null hypothesis H_0: $\mu = 30$ is 0.09.

(a) Does the 95% confidence interval include the value 30? Why?

(b) Does the 90% confidence interval include the value 30? Why?

11.38 Connecting confidence intervals and significance tests, II A 90% confidence interval for a population mean is (12, 15).

(a) Can you reject the null hypothesis that $\mu = 13$ against the two-sided alternative at the 10% significance level? Why?

(b) Can you reject the null hypothesis that $\mu = 13$ against a one-sided alternative at the 10% significance level? Why?

(c) Can you reject the null hypothesis that $\mu = 10$ against the two-sided alternative at the 10% significance level? Why?

(d) Can you reject the null hypothesis that $\mu = 10$ against a one-sided alternative at the 10% significance level? Why?

11.39 California SAT scores In a discussion of SAT scores, someone comments: "Because only a minority of high school students take the test, the scores overestimate the ability of typical high school seniors. The mean SAT Mathematics score is about 475, but I think that if all seniors took the test, the mean score would be no more than 450." You arrange to give the test to an SRS of 500 seniors from California. These students had a mean score of $\bar{x} = 461$. Assume that the population standard deviation is $\sigma = 100$. Is this good evidence against the claim that the mean for all California seniors is no more than 450? Give appropriate statistical evidence to justify your answer.

11.40 Cockroaches An understanding of cockroach biology may lead to an effective control strategy for these annoying insects. Researchers studying the absorption of sugar by insects feed cockroaches a diet containing measured amounts of a particular sugar. After 10 hours, the cockroaches are killed and the concentration of the sugar in various body parts is determined by a chemical analysis. The paper that reports the research states that a 95% confidence interval for the mean amount (in milligrams) of the sugar in the hindguts of the cockroaches is 4.2 ± 2.3.[6]

(a) Does this paper give evidence that the mean amount of sugar in the hindguts under these conditions is not equal to 7 mg? State H_0 and H_a and base a test on the confidence interval.

(b) Would the hypothesis that $\mu = 5$ mg be rejected at the 5% level in favor of a two-sided alternative?

APPLET

11.41 *Statistical Significance* applet Go to the *Statistical Significance* applet on the book's Web site (www.whfreeman.com/tps3e). This applet illustrates statistical tests with a fixed significance level when sampling from a Normal population with known standard deviation.

(a) Open the applet and keep the default settings for the null ($\mu = 0$) and alternative ($\mu > 0$) hypotheses, the sample size ($n = 10$), the standard deviation ($\sigma = 1$), and the significance level ($\alpha = 0.05$). In the "I have data, and the observed $\bar{x}$ is $\bar{x}$ =" box, enter the value 1. Is the difference between $\bar{x}$ and μ_0 significant at the 5% level?

(b) Repeat (a) for $\bar{x}$ equal to 0.1, 0.2, 0.3, 0.4, 0.5, 0.6, 0.7, 0.8, and 0.9. Make a table giving $\bar{x}$ and the results of the significance tests. What do you conclude?

(c) Repeat parts (a) and (b) with significance level $\alpha = 0.01$. How does the choice of α change the statistical significance of the values of $\bar{x}$?

Technology Toolbox

Performing significance tests

The TI-83/84 and TI-89 can be used to conduct one-sample z tests, using either data stored in a list or summary statistics.

TI-83/84

- Press STAT and choose TESTS and 1:Z-Test to access the Z-Test screen, as shown.
- Choose "stats" as the input method.

```
EDIT CALC TESTS
1:Z-Test...
2:T-Test...
3:2-SampZTest...
4:2-SampTTest...
5:1-PropZTest...
6:2-PropZTest...
7↓ZInterval...
```

TI-89

- Press 2nd F1 ([F6]) and choose 1:Z-Test.
- Choose "Stats" as the input method.

```
F1▼   F2▼   F3▼  F4▼  F5▼   F6▼   F7▼
Tools Plots List Calc Distr Tests Ints
list1  lis 1:Z-Test...
------ --- 2:T-Test...
           3:2-SampZTest...
           4:2-SampTTest...
           5:1-PropZTest...
           6:2-PropZTest...
           7:Chi2 GOF...
           8↓Chi2 2-way...
list2[1]=
TYPE OR USE ←→↑↓ +[ENTER]OR[ESC]
```

For the executives' blood pressures of Example 11.10 (page 706), for example, you would enter 128 for the null hypothesized mean μ_0. Next enter 15 for σ, 129.93 for $\bar{x}$, and 72 for n. Select $\neq \mu_0$ for the alternative hypothesis, and choose "Calculate."

```
Z-Test
 Inpt:Data Stats
 μ0:128
 σ:15
 x̄:129.93
 n:72
 μ:≠μ0 <μ0 >μ0
 Calculate Draw
```

```
                 Z Test
 μ0:              128
 σ:               15
 x̄:               129.93
 n:               72
 Alternate Hyp:   μ≠μ0→
 Results:         Calculate →
  Enter=OK             ESC=CANCEL
USE ← AND → TO OPEN CHOICES
```

(continued)

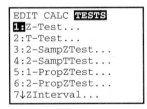

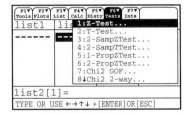

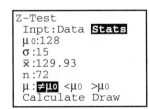

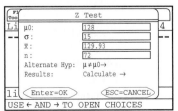

Technology Toolbox

Performing significance tests *(continued)*

Here are the results of the z test: $z = -1.09$ and P-value $= 0.2749$. Notice the slight difference from the P-value obtained using Table A in Example 11.10 ($P = 0.2758$).

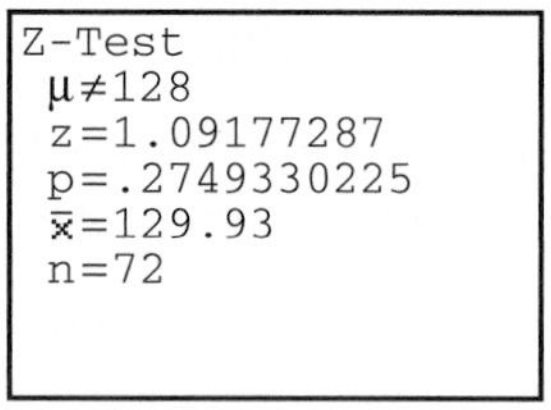

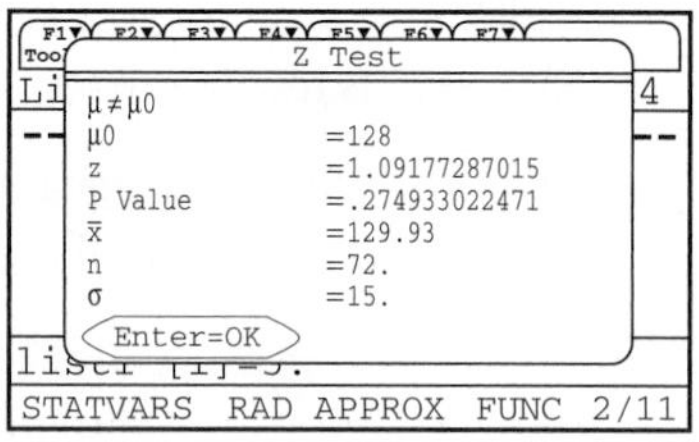

If you select "Draw" instead of "Calculate," the two critical areas under the Normal curve will be shaded and the z statistic and P-value will be displayed.

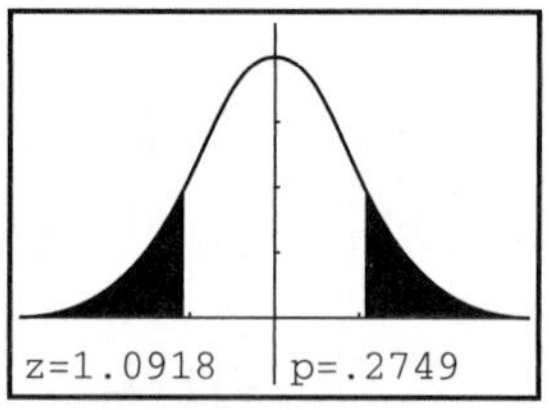

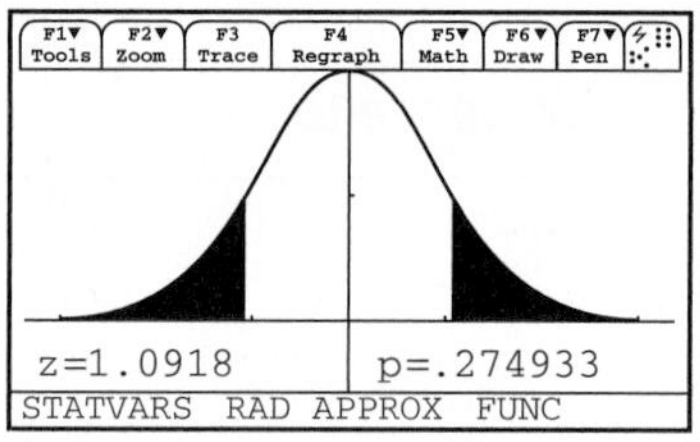

11.42 Calculator significance tests

(a) Follow the method of the Technology Toolbox to find the z statistic and P-value for the test in Exercise 11.36. Compare the results with your manual calculations.

(b) If you have data, enter the data into any list and then specify a z test. In Exercise 11.29 (page 710), for example, you would enter the 20 hardness values into L_1/list1. Then from the Z-Test screen, specify "Data" and enter the appropriate values.

11.3 Use and Abuse of Tests

Significance tests are widely used in reporting the results of research in many fields. New drugs require significant evidence of effectiveness and safety. Courts ask about statistical significance in hearing discrimination cases. Marketers want to know whether a new ad campaign significantly outperforms the old one, and medical researchers want to know whether a new therapy performs significantly better. In all these uses, statistical significance is valued because it points to an effect that is unlikely to occur simply by chance.

Carrying out a significance test is often quite simple, especially if you get a *P*-value effortlessly from a calculator or computer. Using tests wisely is not so simple. Here are some points to keep in mind when using or interpreting significance tests.

Choosing a Level of Significance

The purpose of a test of significance is to give a clear statement of the degree of evidence provided by the sample against the null hypothesis. The *P*-value does this. But how small a *P*-value is convincing evidence against the null hypothesis? This depends mainly on two circumstances:

- *How plausible is H_0?* If H_0 represents an assumption that the people you must convince have believed for years, strong evidence (small *P*) will be needed to persuade them.
- *What are the consequences of rejecting H_0?* If rejecting H_0 in favor of H_a means making an expensive changeover from one type of product packaging to another, you need strong evidence that the new packaging will boost sales.

These criteria are a bit subjective. Different people will insist on different levels of significance. Giving the *P*-value allows each of us to decide individually if the evidence is sufficiently strong.

Users of statistics have often emphasized standard levels of significance such as 10%, 5%, and 1%. For example, courts have tended to accept 5% as a standard in discrimination cases.[7] This emphasis reflects the time when tables of critical values rather than software dominated statistical practice. The 5% level $\alpha = 0.05$ is particularly common. **There is no sharp border between "statistically significant" and "statistically insignificant," only increasingly strong evidence as the *P*-value decreases.** There is no practical distinction between the *P*-values 0.049 and 0.051. It makes no sense to treat $P < 0.05$ as a universal rule for what is significant.

Statistical Significance and Practical Importance

When a null hypothesis ("no effect" or "no difference") can be rejected at the usual levels ($\alpha = 0.05$ or $\alpha = 0.01$), there is good evidence that an effect is present. But that effect may be very small. When large samples are available, even tiny deviations from the null hypothesis will be significant.

Example 11.13 *Wound healing time*

Statistical significance and practical importance

Suppose we're testing a new antibacterial cream, "Formulation NS," on a small cut made on the inner forearm. We know from previous research that with no medication, the mean healing time (defined as the time for the scab to fall off) is 7.6 days with a standard deviation of 1.4 days. The claim we want to test here is that Formulation NS speeds healing. We will use a 5% significance level.

Procedure: We cut 25 volunteer college students and apply Formulation NS to the wounds. The mean healing time for these subjects is $\bar{x} = 7.1$ days. We will assume that $\sigma = 1.4$ days.

Discussion: We want to test a claim about the mean healing time μ in the population of people who treat cuts with Formulation NS. Our hypotheses are

$$H_0: \mu = 7.6 \text{ days}$$

$$H_a: \mu < 7.6 \text{ days}$$

We carry out a z test and find that $z = -1.79$ and the P-value is 0.0367. Since $0.0367 < \alpha = 0.05$, we reject H_0 and conclude that Formulation NS's healing effect is statistically significant. However, this result is not practically important. Having your scab fall off half a day sooner is no big deal.

CAUTION

Remember the wise saying: **Statistical significance is not the same thing as practical importance.** Exercise 11.44 (page 722) demonstrates in detail the effect on P of increasing the sample size.

The remedy for attaching too much importance to statistical significance is to pay attention to the actual data as well as to the P-value. Plot your data and examine them carefully. Are there outliers or other deviations from a consistent pattern? A few outlying observations can produce highly significant results if you blindly apply common significance tests. Outliers can also destroy the significance of otherwise-convincing data. **The foolish user of statistics who feeds the data to a calculator or computer without exploratory analysis will often be embarrassed.** Is the effect you are seeking visible in your plots? If not, ask yourself if the effect is large enough to be practically important. Give a confidence interval for the parameter in which you are interested. A confidence interval actually estimates the size of an effect rather than simply asking if it is too large to reasonably occur by chance alone. Confidence intervals are not used as often as they should be, while tests of significance are perhaps overused.

CAUTION

Don't Ignore Lack of Significance

There is a tendency to infer that there is no effect whenever a P-value fails to attain the usual 5% standard. A provocative editorial in the *British Medical Journal* entitled "Absence of Evidence Is Not Evidence of Absence" deals with this issue.[8] Here is one of the examples they cite.

Example 11.14 *Reducing HIV transmission*

Interpreting lack of significance

In an experiment to compare methods for reducing transmission of HIV, subjects were randomly assigned to a treatment group and a control group. Result: the treatment group and the control group had the same rate of HIV infection. Researchers described this as an "incident rate ratio" of 1.00. A ratio above 1.00 would mean that there was a greater rate of HIV infection in the treatment group, while a ratio below 1.00 would indicate a greater rate of HIV infection in the control group. The 95% confidence interval for the incident rate ratio was reported as 0.63 to 1.58.[9] Saying that the treatment has no effect on HIV infection is misleading. The confidence interval for the incident rate ratio indicates that the treatment may be able to achieve a 37% decrease in infection; it might also produce a 58% increase in infection. Clearly, more data are needed to distinguish between these possibilities.

The situation can be worse. Research in some fields has rarely been published unless significance at the 5% level is attained.

Example 11.15 *Psychology journals*

Not significant = not published

A survey of four journals published by the American Psychological Association showed that of 294 articles using statistical tests, only 8 reported results that did not attain the 5% significance level.[10]

In some areas of research, small effects that are detectable only with large sample sizes can be of great practical significance. Data accumulated from a large number of patients taking a new drug may be needed before we can conclude that there are life-threatening consequences for a small number of people. **When planning a study, verify that the test you plan to use has a high probability of detecting an effect of the size you hope to find.**

Statistical Inference Is Not Valid for All Sets of Data

Badly designed surveys or experiments often produce invalid results. Formal statistical inference cannot correct basic flaws in the design. Each test is valid only in certain circumstances, with properly produced data being particularly important. The z test, for example, should bear the same warning label that we attached on page 636 to the z confidence interval. Similar warnings accompany the other tests that we will learn.

Example 11.16 *Does music increase worker productivity?*

When inference isn't valid

Hawthorne effect

You wonder whether background music would improve the productivity of the staff who process mail orders in your business. After discussing the idea with the workers, you add music and find a statistically significant increase. You should not be impressed. In fact, almost any change in the work environment together with knowledge that a study is under way will produce a short-term productivity increase. This is the ***Hawthorne effect***, named after the Western Electric manufacturing plant where it was first noted.

The significance test correctly informs you that an increase has occurred that is larger than would often arise by chance alone. It does not tell you what other than chance caused the increase. The most plausible explanation is that workers change their behavior when they know they are being studied. Your experiment was uncontrolled, so the significant result cannot be interpreted. A randomized comparative experiment would isolate the actual effect of background music and so make significance meaningful.

CAUTION

Significance tests and confidence intervals are based on the laws of probability. Randomization in sampling or experimentation ensures that these laws apply. **But we must often analyze data that do not arise from randomized samples or experiments. To apply statistical inference to such data, we must have confidence in the use of probability to describe the data.** The diameters of successive holes bored in auto engine blocks during production, for example, may behave like a random sample from a Normal distribution. We can check this probability model by examining the data. If the model appears correct, we can apply the recipes of this chapter to do inference about the process mean diameter μ. Always ask how the data were produced, and don't be too impressed by *P*-values on a printout until you are confident that the data deserve a formal analysis.

Beware of Multiple Analyses

Statistical significance ought to mean that you have found an effect that you were looking for. The reasoning behind statistical significance works well if you decide what effect you are seeking, design a study to search for it, and use a test of significance to weigh the evidence you get. In other settings, significance may have little meaning.

Example 11.17 *Cell phones and brain cancer*
Don't search for significance!

Might the radiation from cell phones be harmful to users? Many studies have found little or no connection between using cell phones and various illnesses. Here is part of a news account of one study:

> A hospital study that compared brain cancer patients and a similar group without brain cancer found no statistically significant association between cell phone use and a group of brain cancers known as gliomas. But when 20 types of glioma were considered separately an association was found between phone use and one rare form. Puzzlingly, however, this risk appeared to decrease rather than increase with greater mobile phone use.[11]

Think for a moment. Suppose that the 20 null hypotheses for these 20 significance tests are all true. Then each test has a 5% chance of being significant at the 5% level. That's what $\alpha = 0.05$ means: results this extreme occur only 5% of the time just by chance when the null hypothesis is true. Because 5% is 1/20, we expect about 1 of 20 tests to give a significant result just by chance. Running one test and reaching the $\alpha = 0.05$ level is reasonably good evidence that you have found something; running 20 tests and reaching that level only once is not.

Dropping out An experiment found that weight loss is significantly more effective than exercise for reducing high cholesterol and high blood pressure. The 170 subjects were randomly assigned to a weight-loss program, an exercise program, or a control group. Only 111 of the 170 subjects completed their assigned treatment, and the analysis used data from these 111. Did the dropouts create bias? Always ask about details of the data before trusting inference.

Searching data for suggestive patterns is certainly legitimate. Exploratory data analysis is an important aspect of statistics. But the reasoning of formal inference does not apply when your search for a striking effect in the data is successful. The remedy is clear. Once you have a hypothesis, design a study to search specifically for the effect you now think is there. If the result of this study is statistically significant, you have real evidence.

Section 11.3 Summary

P-values are more informative than the reject-or-not result of a fixed level α test. Beware of placing too much weight on traditional values of α, such as $\alpha = 0.05$.

Very small effects can be highly significant (small *P*), especially when a test is based on a large sample. A statistically significant effect need not be practically important. Plot the data to display the effect you are seeking, and use confidence intervals to estimate the actual value of parameters.

On the other hand, lack of significance does not imply that H_0 is true, especially when the test is based on just a few observations.

Significance tests are not always valid. Faulty data collection, outliers in the data, and testing a hypothesis on the same data that suggested the hypothesis can invalidate a test. Many tests run at once will probably produce some significant results by chance alone, even if all the null hypotheses are true.

Section 11.3 Exercises

11.43 Is it significant? In the absence of special preparation, SAT Math scores in recent years have varied Normally with mean $\mu = 518$ and $\sigma = 114$. One hundred students go through a rigorous training program designed to raise their SAT Math scores by improving their mathematics skills. Either by hand or using the *P-value of a Significance Test* applet, carry out a test of

$$H_0: \mu = 518$$
$$H_a: \mu > 518$$

(with $\sigma = 114$) in each of the following situations:

(a) The students' average score is $\bar{x} = 536.7$. Is this result significant at the 5% level?

(b) The average score is $\bar{x} = 536.8$. Is this result significant at the 5% level?

The difference between the two outcomes in (a) and (b) is of no importance. Beware attempts to treat $\alpha = 0.05$ as sacred.

11.44 Coaching and the SAT, I Suppose that SAT Math scores in the absence of coaching vary Normally with mean $\mu = 518$ and $\sigma = 100$. Suppose that coaching may change μ but does not change σ. A coaching service finds that a sample of students it has coached have mean score $\bar{x} = 522$. An increase in the SAT Math score from 518 to 522 is of no importance in seeking admission to college. But this service may still be able to advertise that its customers "score significantly higher." To see this, calculate the *P*-value for a test of

$$H_0: \mu = 518$$
$$H_a: \mu > 518$$

by hand or using the *P-value of a Significance Test* applet in each of the following situations:

(a) The service coaches 100 students. Their SAT Math scores average $\bar{x} = 522$.

(b) By the next year, the service has coached 1000 students. For these students, $\bar{x} = 522$.

(c) An advertising campaign brings the number of students coached to 10,000. Their average score is still $\bar{x} = 522$.

Should tests be banned? Significance tests don't tell us how large or how important an effect is. Research in psychology has emphasized tests, so much so that some think their weaknesses should ban them from use. The American Psychological Association asked a group of experts. They said to use anything that sheds light on a study, particularly more data analysis and confidence intervals. But: "The task force does not support any action that could be interpreted as banning the use of null hypothesis significance testing or *P*-values in psychological research and publication."

11.45 Coaching and the SAT, II Give a 99% confidence interval for the mean SAT Math score μ after coaching in each part of the previous exercise. For large samples, the confidence interval tells us, "Yes, the mean score is higher than 518 after coaching, but only by a small amount."

11.46 Ages of presidents Joe is writing a report on the backgrounds of American presidents. He looks up the ages of all the presidents when they entered office. Because Joe took a statistics course, he uses these numbers to perform a significance test about the mean age of all U.S. presidents. This makes no sense. Why not?

11.47 Do you have ESP? A researcher looking for evidence of extrasensory perception (ESP) tests 500 subjects. Four of these subjects do significantly better ($P < 0.01$) than random guessing.

(a) Is it proper to conclude that these four people have ESP? Explain your answer.

(b) What should the researcher now do to test whether any of these four subjects have ESP?

11.48 What is significance good for? Which of the following questions does a test of significance answer? Justify your answer.

(a) Is the sample or experiment properly designed?

(b) Is the observed effect due to chance?

(c) Is the observed effect important?

11.4 Using Inference to Make Decisions

In Section 11.1, we presented significance tests as methods for assessing the strength of evidence against the null hypothesis. Most users of statistics think of tests this way. But signs of another way of thinking were present in our discussion of significance tests with fixed significance level α. A level of significance α chosen in advance points to the outcome of the test as a *decision*. If the *P*-value is less than α, we reject H_0 in favor of H_a. Otherwise, we fail to reject H_0. There is a big distinction between measuring the strength of evidence and making a decision. Many statisticians feel that making decisions is too grand a goal for statistics alone. Decision makers should take account of the results of statistical studies, but their decisions should be based on many factors that are hard to reduce to numbers.

Sometimes, however, we really are concerned about making a decision or choosing an action based on our evaluation of the data. Here's one such example.

Example 11.18 ***Perfect potatoes***
Inference as decision

A potato chip producer and a supplier of potatoes agree that each shipment of potatoes must meet certain quality standards. If less than 8% of the potatoes in the shipment have "blemishes," the producer will accept the entire truckload. Otherwise, the truck will be sent away to get another load of potatoes from the supplier. It isn't practical to inspect every potato in a given shipment. Instead, the producer inspects a sample of potatoes. On the basis of the sample results, the potato chip producer uses a significance test to decide whether to accept or reject the shipment.

Type I and Type II Errors

When we make a decision based on a significance test, we hope that our decision will be correct. But sometimes it will be wrong. There are two types of incorrect decisions. We can reject the null hypothesis when it's actually true. Or we can fail to reject a false null hypothesis. To distinguish these two types of errors, we give them specific names.

Type I and Type II Errors

If we reject H_0 when H_0 is actually true, we have committed a **Type I error.**

If we fail to reject H_0 when H_0 is false, we have committed a **Type II error.**

The possibilities are summed up in Figure 11.8 (on the next page). If H_0 is true, our decision is either correct (if we fail to reject H_0) or is a Type I error. If H_0 is false (H_a is true), our decision is either correct or is a Type II error. Only one error is possible at one time.

It is important to be able to describe Type I and Type II errors in the context of a decision-making problem. Considering the consequences of each of these types of error is also important, as the following examples show.

Figure 11.8 *The two types of error in testing hypotheses.*

Decision based on sample	Truth about the population: H_0 true	Truth about the population: H_a true
Reject H_0	Type I error	Correct decision
Fail to reject H_0	Correct decision	Type II error

Example 11.19 *Perfect potatoes: possible errors*

Interpreting Type I and II errors

In Example 11.18, the potato chip producer wants to test the hypotheses

$$H_0: p = 0.08$$
$$H_a: p < 0.08$$

where p is the actual proportion of potatoes with blemishes in a given truckload. Let's examine the two types of error that the producer could make and the consequences of each.

- Type I error: reject H_0 when H_0 is true

Description: The producer concludes that the proportion of potatoes with blemishes in the shipment is less than 0.08 when actually 8% (or more) of the potatoes have blemishes. *Consequences:* The producer accepts the truckload of potatoes and uses them to make potato chips. More chips will be made from blemished potatoes, which may upset consumers.

- Type II error: fail to reject H_0 when H_0 is false

Description: The producer decides that the proportion of potatoes with blemishes could be 0.08 (or higher), but actually fewer than 8% of the potatoes in the shipment have blemishes. *Consequences:* The potato chip producer sends the truckload of acceptable potatoes away, which may result in lost revenue for the supplier. In addition, the producer will have to wait on another shipment of potatoes before producing the next batch of potato chips.

Example 11.20 *Awful accidents*

Consequences of Type I and II errors

In Section 11.1, we looked at paramedic response times following serious accidents in a major city. The city manager wanted to test the hypotheses

$$H_0: \mu = 6.7 \text{ minutes}$$
$$H_a: \mu < 6.7 \text{ minutes}$$

where μ was the mean response time to all calls involving life-threatening injuries this year.

- Type I error: reject H_0 when H_0 is true

Description: The city manager concludes that the mean response time this year is less than 6.7 minutes (last year's average) when in fact the mean response time is still 6.7 minutes (or higher).
Consequences: The city manager believes that paramedic response times have improved when they really haven't. This could result in additional loss of life for accident victims.

- Type II error: fail to reject H_0 when H_0 is false

Description: The city manager decides that the paramedics' mean response time this year is still 6.7 minutes (or higher) when it is actually less than 6.7 minutes.
Consequences: The city manager may take action to decrease paramedic response times when such action is unnecessary. This could result in considerable expense for the city, as well as some disgruntled paramedics.

Which is more serious: a Type I error or a Type II error? That depends on the circumstances, as the previous examples illustrate. For the potato chip producer of Example 11.19, a Type I error could result in decreased sales. A Type II error may not have much impact if additional shipments of potatoes can be obtained fairly easily. The potato supplier sees things quite differently. A Type II error would result in lost revenue from the truckload of potatoes. That's bad for business! The supplier would prefer a Type I error, as long as the producer didn't subsequently discover the mistake and seek a new supplier.

Error Probabilities

We can assess any rule for making decisions by looking at the probabilities of the two types of error. This is in keeping with the idea that statistical inference is based on asking, "What would happen if I did this many times?" We cannot (without inspecting the whole truckload) guarantee that good shipments of potatoes will never be rejected and bad shipments will never be accepted. But by random sampling and the laws of probability, we can say what the probabilities of both kinds of error are.

Example 11.21 *Awful accidents (continued)*
Type I and II error probabilities

For the paramedic response times of Example 11.20, we were testing

$$H_0\colon \mu = 6.7 \text{ minutes}$$

$$H_a\colon \mu < 6.7 \text{ minutes}$$

Suppose that the city manager decides to carry out this test using a 5% significance level ($\alpha = 0.05$). A Type I error is to reject H_0 when in fact $\mu = 6.7$ minutes.

What about Type II errors? Because there are many values of μ that satisfy $H_a: \mu < 6.7$ minutes, we will concentrate on one value. The city manager decides that she really wants the significance test to detect a decrease in the mean response time to 6.4 minutes or less. So a particular Type II error is to fail to reject H_0 when $\mu = 6.4$ minutes.

Figure 11.9 shows how the two probabilities of error are obtained from the two sampling distributions of $\bar{x}$, for $\mu = 6.7$ and $\mu = 6.4$. When $\mu = 6.7$, H_0 is true and to reject H_0 is a Type I error. If our sample results in a value of $\bar{x}$ that is much smaller than 6.7, we will reject H_0. How small would $\bar{x}$ need to be? The 5% significance level tells us to count results that could happen less than 5% of the time by chance if H_0 is true as evidence that H_0 is false. Look at the right-hand sampling distribution. The light blue area is 5%. Values of $\bar{x}$ to the left of the point marked "Critical value of $\bar{x}$" will cause us to reject H_0 even though H_0 is true. This will happen in 5% of all possible samples.

When $\mu = 6.4$ minutes, H_a is true and to fail to reject H_0 is a Type II error. In the previous paragraph, we decided to *reject* H_0 if our sample yielded a value of $\bar{x}$ less than the critical value marked on the horizontal axis of Figure 11.9. So we would *fail* to reject H_0 if the sample mean $\bar{x}$ falls to the right of this critical value. Look at the left-hand sampling distribution. The green area represents the probability of failing to reject H_0 when H_0 is false. This is the probability of a Type II error.

Figure 11.9 *The two error probabilities for Example 11.21. The probability of a Type I error (light blue shaded area) is the probability of rejecting $H_0: \mu = 6.7$ minutes when in fact $\mu = 6.7$ minutes. The probability of a Type II error (green shaded area) is the probability of failing to reject H_0 when in fact $\mu = 6.4$ minutes.*

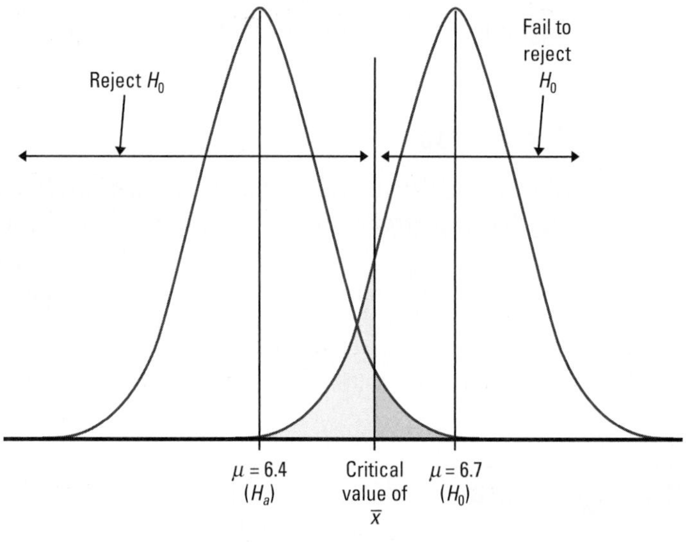

The probability of a Type I error is the probability of rejecting H_0 when it is really true. As Example 11.21 shows, this is exactly the significance level of the test.

Significance and Type I Error

The significance level α of any fixed level test is the probability of a Type I error. That is, α is the probability that the test will reject the null hypothesis H_0 when H_0 is in fact true.

The probability of a Type II error (sometimes called β) for the particular alternative $\mu = 6.4$ minutes in Example 11.21 is the probability that the test will fail to reject H_0 when μ has this alternative value. This is the probability that the sample mean $\bar{x}$ falls to the right of the critical value of $\bar{x}$ in Figure 11.9, calculated assuming $\mu = 6.4$. This probability is *not* $1 - 0.05$, because the probability 0.05 was found assuming that $\mu = 6.7$. Calculating Type II error by hand is possible but unpleasant. It's better to let technology do the work for you.

Example 11.22 *Awful accidents (one last time)*

Calculating Type II error probability

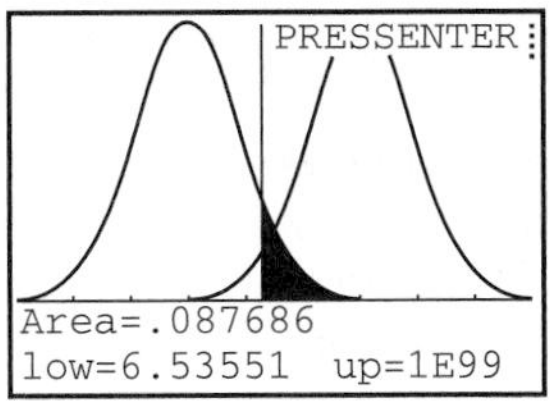

The TI-83/84 screen shot in the margin reproduces the two sampling distributions of Figure 11.9. This time, only the area corresponding to the Type II error probability is shaded. We can see that the probability β of making a Type II error (failing to reject H_0: $\mu = 6.7$ when actually $\mu = 6.4$) is about 8.8%. The city manager must decide whether this is an acceptable chance of failing to detect that the paramedics' mean response time has dropped to 6.4 minutes.

Where did the 6.53551 value for "low" in the screen shot come from? This is the critical value of $\bar{x}$ from Example 11.21. If $\bar{x}$ for our sample is less than 6.53551, we would reject H_0. If $\bar{x} \geq 6.53551$, we would fail to reject H_0. Since H_0 *is false* in this setting, rejecting H_0 is the correct decision. Failing to reject H_0 is a Type II error, which will occur in about 8.8% of all possible samples of 400 response times.

Exercises

11.49 Awful accidents Another way for the city manager of Example 11.20 to measure response times is to look at the *proportion* of calls for which paramedics arrived within 8 minutes. Last year, paramedics arrived on scene 75% of the time within 8 minutes. The city manager wants to determine whether they have done significantly better this year.

(a) State appropriate null and alternative hypotheses for the city manager to test.

(b) Describe a Type I and a Type II error in this setting.

(c) Explain the consequences of each type of error.

(d) Which is more serious: a Type I error or a Type II error? Justify your answer.

(e) If you sustain a life-threatening injury due to a vehicle accident, you want to receive medical treatment as quickly as possible. Which of the two significance tests—H_0: $\mu = 6.7$ versus H_a: $\mu < 6.7$ or the one from part (a) of this exercise—would you be more interested in? Justify your answer.

11.50 Blood pressure screening Your company markets a computerized device for detecting high blood pressure. The device measures an individual's blood pressure once per hour at a randomly selected time throughout a 12-hour period. Then it calculates the mean systolic (top number) pressure for the sample of measurements. Based on the sample results, the device determines whether there is significant evidence that the individual's actual mean systolic pressure is greater than 130. If so, it recommends that the person seek medical attention.

(a) State appropriate null and alternative hypotheses in this setting.

(b) Describe a Type I and a Type II error.

(c) The program can be adjusted to decrease one error probability at the cost of an increase in the other error probability. Which error probability would you choose to make smaller, and why?

11.51 Catalog sales You want to see if the redesign of the cover of a mail-order catalog will increase sales. A very large number of customers will receive the original catalog, and a random sample of customers will receive the one with the new cover. Based on past sales, you are willing to assume that the sales from the new catalog will follow a Normal distribution with $\sigma = 60$ dollars and that the mean for the original catalog will be $\mu = 40$ dollars. You decide to use a sample size of $n = 1000$. You wish to test

$$H_0: \mu = 40$$
$$H_a: \mu > 40$$

at the 1% significance level ($\alpha = 0.01$).

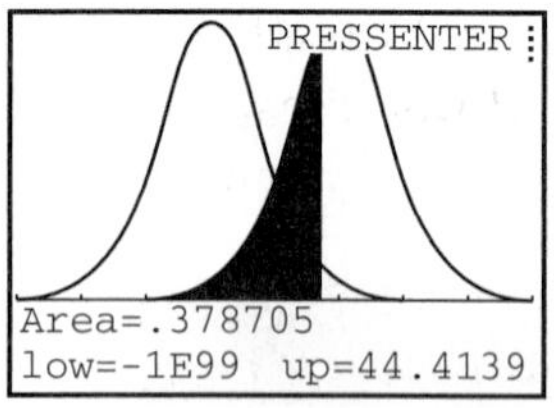

(a) Describe a Type I error in this setting and the consequences of making a Type I error.

(b) Describe a Type II error in this setting and the consequences of making a Type II error.

(c) Which type of error is more serious in this case? Explain.

(d) Using the information provided, including the TI-83/84 calculator screen shot in the margin, determine the probability of each of the two types of error. (Refer to Example 11.21.)

(e) Try to determine where the value 44.4139 came from.

11.52 Beetles in the wood An outbreak of the mountain pine beetle has affected several types of pine trees in British Columbia. The beetle leaves behind a fungus that produces blue-colored stains in the wood. Some consumers might worry that lumber obtained from the blue-stained trees is weaker as a result of the effects of the fungus. A Canadian company performed a test on the breaking strength of blue-stained wood.[12] They measured the mean breaking strength of a sample of 100 pieces of blue-stained pine. The target breaking strength of lumber made from healthy pine trees is 10,000 pounds per square inch (psi).

(a) State appropriate null and alternative hypotheses for the company to test.

(b) Describe a Type I and a Type II error, and give the consequences of each.

(c) Which type of error is more serious? Why?

Power

A significance test makes a Type II error when it fails to reject a null hypothesis that really is false. A high probability of a Type II error for a particular alternative means that the test is not sensitive enough to usually detect that alternative. Calculations of the probability of Type II errors are therefore useful even if you don't think that a statistical test should be viewed as making a decision. In the significance test setting, it is more common to report the probability that a test *does* reject H_0 when an alternative is true. This probability is called the ***power*** of the test against that specific alternative. The higher this probability is, the more sensitive the test is.

Power and Type II Error

The probability that a fixed level α significance test will reject H_0 when a particular alternative value of the parameter is true is called the **power** of the test against that alternative.

The power of a test against any alternative is 1 minus the probability of a Type II error for that alternative; that is, power $= 1 - \beta$.

Calculations of power are essentially the same as calculations of the probability of Type II error (which means we should use technology!). In Example 11.22 (page 727), the power is the probability of rejecting H_0: $\mu = 6.7$ when $\mu = 6.4$. It is $1 - 0.0877$, or 0.9123.

Calculations of *P*-values and calculations of power both say what would happen if we repeated the test many times. A *P*-value describes what would happen supposing that the null hypothesis is true. Power describes what would happen supposing that a particular alternative is true.

When planning a study that will include a significance test, a careful user of statistics decides what alternative values of the parameter the test should detect and checks that the power is adequate. The power depends on which particular parameter value in H_a we are interested in. To calculate power, we must fix an α so that there is a fixed rule for rejecting H_0. The following Activity will give you a chance to investigate what affects the power of a test.

Activity 11C

Exercise is good!

APPLET

Can a 6-month exercise program increase the total body bone mineral content (TBBMC) of young women? A team of researchers is planning a study to examine this question. Based on the results of a previous study, they are willing to assume that $\sigma = 2$ for the percent change in TBBMC over the 6-month period. A change in TBBMC of 1% would be considered important, and the researchers would like to have a reasonable chance of detecting a change this large or larger. How many subjects should they include in their study?

Fish, fishermen, and power Are the stocks of cod in the ocean off eastern Canada declining? Studies over many years failed to find significant evidence of a decline. These studies had low power—that is, they might fail to find a decline even if one was present. When it became clear that the cod were vanishing, quotas on fishing ravaged the economy in parts of Canada. If the earlier studies had had high power, they would likely have seen the decline. Quick action might have reduced the economic and environmental costs.

The null hypothesis is that the exercise program has no effect on TBBMC. In other words, the mean percent change is zero. The alternative is that exercise is beneficial; that is, the mean change is positive. Formally, we have

H_0: $\mu = 0$ (mean percent change in TBBMC is 0)
H_a: $\mu > 0$ (mean percent change in TBBMC is positive)

The alternative of interest is $\mu = 1\%$.

In this Activity, you will use the *Power* applet at the book's Web site, www.whfreeman.com/tps3e, to see what affects the power of a significance test.

1. Enter the values: $\mu_0 = 0$, $\sigma = 2$, one-sided test, alternative $\mu = 1$, $n = 25$, and $\alpha = 0.05$. Find the power and explain in simple terms what this number means.
2. Change the sample size to $n = 50$ and calculate the power again. What effect did increasing the sample size have on the power of the test to detect $\mu = 1$? Why?
3. Change the sample size back to $n = 25$. Decrease the significance level to $\alpha = 0.01$. Then calculate the power. What effect did decreasing the significance level have? Why?
4. Change the significance level back to 5%. Decrease the standard deviation to 1.5. What effect does this have on the power of the test to detect $\mu = 1$? Why?
5. Change the standard deviation back to 2 and increase the alternative value of μ to 1.3. What effect does this have on the power of the test to detect $\mu = 1$? Why?
6. Write a few sentences summarizing what you have learned about how you can increase the power of a significance test.

Increasing the Power

High power is desirable. Along with 95% confidence intervals and 5% significance tests, 80% power is becoming a standard. Many U.S. government agencies that provide research funds require that the sample size for the funded studies be sufficient to detect important results 80% of the time using a significance test with $\alpha = 0.05$.

Suppose you have performed a power calculation and found that the power is too small. What can you do to increase it? Here are four ways:

- Increase α. A test at the 5% significance level will have a greater chance of rejecting the alternative than a 1% test because the strength of evidence required for rejection is less.

- Consider a particular alternative that is farther away from μ_0. Values of μ that are in H_a but lie close to the hypothesized value μ_0 are harder to detect (lower power) than values of μ that are far from μ_0.
- Increase the sample size. More data will provide more information about $\overline{x}$, so we will have a better chance of distinguishing values of μ.
- Decrease σ. This has the same effect as increasing the sample size: more information about $\overline{x}$. Improving the measurement process and restricting attention to a subpopulation are two common ways to decrease σ.

Power calculations are important in planning studies. Using a significance test with low power makes it unlikely that you will find a significant effect even if the truth is far from the null hypothesis. A null hypothesis that is in fact false can become widely believed if repeated attempts to find evidence against it fail because of low power. Our best advice for maximizing the power of a test is to choose as high an α level (Type I error probability) as you are willing to risk *and* as large a sample size as you can afford.

Exercises

11.53 Opening a restaurant You are thinking about opening a restaurant and are searching for a good location. From research you have done, you know that the mean income of those living near the restaurant must be over \$85,000 to support the type of upscale restaurant you wish to open. You decide to take a simple random sample of 50 people living near one potential location. Based on the mean income of this sample, you will decide whether to open a restaurant there. A number of similar studies have shown that σ = \$5000.[13]

(a) State appropriate null and alternative hypotheses. Be sure to define your parameter.

(b) Describe the two types of errors that you might make. Identify which is a Type I error and which is a Type II error.

(c) Which of the two types of error is more serious? Explain.

(d) If you had to choose one of the "standard" significance levels for your significance test, would you choose $\alpha = 0.01, 0.05$, or 0.10? Justify your choice.

(e) Based on your choice in part (d), if the mean income in a certain area is \$87,000, how likely are you to open a restaurant in that area? What is β? (Use the TYPE2 program provided by your teacher.)

11.54 Salty potato chips The mean salt content of a certain type of potato chip is supposed to be 2.0 milligrams (mg). The salt content of these chips varies Normally with standard deviation $\sigma = 0.1$ mg. From each batch produced, an inspector takes a sample of 50 chips and measures the salt content of each chip. The inspector rejects the entire batch if the sample mean salt content is significantly different from 2 mg at the 5% significance level.

(a) What null and alternative hypotheses is the inspector testing? Be sure to define your parameter.

(b) Explain what a Type I error would mean in this setting. What's the probability of making a Type I error?

(c) Explain what a Type II error would mean in this setting. Use the TYPE2 program provided by your teacher to find β if $\mu = 2.05$.

(d) What is the power of the test to detect $\mu = 2.05$?

(e) What is the power of the test to detect $\mu = 1.95$? Why does this make sense?

(f) If the inspector used a 10% significance level instead of a 5% significance level, how would this affect the probability of a Type I error? A Type II error? The power of the test?

(g) Would you recommend a 1% significance level, a 5% significance level, or a 10% significance level to the company? Justify your answer.

11.55 More power to you! You are reviewing a research proposal that includes a section on sample size justification. A careful reading of this section indicates that the power is 20% for detecting an effect that most people would consider important. Write a short explanation of what this means and make a recommendation on whether the study should be run.

11.56 Power and the alternative A one-sided test of the null hypothesis $\mu = 50$ versus the alternative $\mu = 70$ has power equal to 0.5. Will the power for the alternative $\mu = 80$ be higher or lower than 0.5? Justify your answer.

11.57 Choose the right distribution You must decide which of two probability distributions a discrete random variable X has. We will call the probability distributions p_0 and p_1. Here are the probabilities they assign to the possible values of X:

X:	0	1	2	3	4	5	6
p_0:	0.1	0.1	0.1	0.1	0.2	0.1	0.3
p_1:	0.3	0.2	0.1	0.1	0.1	0.1	0.1

(a) Verify that both p_0 and p_1 are legitimate probability distributions.

You make a single observation on X and use it to test

$$H_0: p_0 \text{ is correct}$$
$$H_a: p_1 \text{ is correct}$$

One possible decision rule is to reject H_0 only if $X = 0$ or $X = 1$.

(b) Find the probability of making a Type I error, that is, the probability that you reject H_0 when p_0 is the correct distribution.

(c) Find the probability of making a Type II error. Show your work.

11.58 *Power* applet Go to the book's Web site and launch the *Power* applet. Redo Exercises 11.53 and 11.54 using the applet. In a few sentences, describe what the calculator program TYPE2 does that the applet does not do.

Section 11.4 Summary

When we use statistical tests to make decisions, we view H_0 and H_a as two competing hypotheses that we must decide between. Our decision is based on comparing the *P*-value with a fixed significance level α. If the *P*-value is less than α, we reject H_0. Otherwise, we fail to reject H_0.

A **Type I error** occurs if we reject H_0 when it is in fact true. A **Type II error** occurs if we fail to reject H_0 when it is actually false.

The **power** of a significance test measures its ability to detect an alternative hypothesis. The power against a specific alternative is the probability that the test will reject H_0 when the alternative is true.

In a fixed level α significance test, the significance level α is the probability of a Type I error, and the power against a specific alternative is 1 minus the probability of a Type II error for that alternative.

Increasing the size of the sample increases the power (reduces the probability of a Type II error) when the significance level remains fixed. We can also increase the power of a test by using a higher significance level (say, $\alpha = 0.10$ instead of $\alpha = 0.05$).

Section 11.4 Exercises

11.59 Power and sample size Two studies are identical in all respects except for the sample sizes. Will the study with the larger sample size have more or less power than the one with the smaller sample size? Explain your answer in terms that could be understood by someone with very little knowledge of statistics.

11.60 Strong pipes A large high school needs to replace all of its water pipes as part of a major construction project. Such water pipes frequently experience water pressures of up to 100 pounds per square inch (psi). To be safe, the construction company requires that any pipes it will use have a mean breaking strength of greater than 120 pounds per square inch (psi). A water pipe supplier wants to provide the 100 pipes that will be needed for the project. They claim that their pipes are strong enough to meet the 120 psi standard. The supplier has 1500 pipes currently available for use. Before deciding whether to use this supplier's water pipes, the construction manager wants to test the breaking strength of a random sample of pipes from among the 1500 that are currently available. From previous experience with this company's pipes, we can assume that $\sigma = 8$ psi.

(a) State appropriate null and alternative hypotheses to test the supplier's claim.

(b) The supplier offers to provide 25 pipes for testing. Will this give the test in part (a) enough power to detect pipes that have a mean breaking strength of 125 pounds at the 5% significance level? Use technology to calculate the power.

(c) After some discussion, the supplier agrees to provide 40 pipes for testing. Describe a method for choosing a random sample of 40 pipes from the 1500 that are available using Table B of random digits. Use line 140 to carry out your method.

(d) Describe a Type I and a Type II error in this situation, and describe the consequences of each.

(e) Would you recommend a significance level of 0.01, 0.05, or 0.10 for this test? Justify your choice.

(f) If you were the construction company's manager, would you prefer the test about the mean breaking strength of the supplier's pipes or a test about the proportion of the supplier's pipes that have a breaking strength less than or equal to 120 psi? Explain.

11.61 Heavy bags Captain Ben flies small passenger jets. These jets carry 50 passengers, plus their luggage. On a full flight, these jets will perform properly as long as the total weight of passengers' checked baggage does not exceed 5000 pounds. Ben is concerned that passengers on a particular flight have brought unusually heavy bags. He selects a random sample of 10 passengers and weighs their checked baggage. Based on the results from this sample, he must decide whether it is safe to take off.

(a) Ben wants to perform a test to determine whether the mean weight μ of passengers' luggage on this flight is too heavy. State appropriate null and alternative hypotheses.

(b) Describe a Type I and a Type II error in this setting, and give the consequences of each.

(c) If you had to choose one of the "standard" significance levels for your significance test, would you choose $\alpha = 0.01, 0.05$, or 0.10? Justify your choice.

(d) Discuss any concerns you have about how the data were produced.

11.62 Power on either side The power for a two-sided test of the null hypothesis $\mu = 0$ versus the specific alternative $\mu = 10$ is 0.82.

(a) What is the probability of a Type II error in this setting?

(b) What is the power of the test versus the specific alternative $\mu = -10$? Explain your answer.

(c) Would the power of the test versus $\mu = -6$ be larger or smaller than the power you calculated in part (b)? Explain.

11.63 Choosing alpha You are the statistical expert on a team that is planning a study. After you have made a careful presentation of the mechanics of significance testing, one of the team members suggests using $\alpha = 0.50$ for this study because you would be more likely to obtain statistically significant results with this choice. Explain in simple terms why this would not be a good use of statistical methods.

11.64 Filling cola bottles: power Exercise 11.30 (page 710) concerns a test about the mean contents of cola bottles. The hypotheses are

$$H_0: \mu = 300 \text{ milliliters}$$

$$H_a: \mu < 300 \text{ milliliters}$$

The sample size is $n = 6$, and the population is assumed to have a Normal distribution with $\sigma = 3$. We'll use a 5% significance level. Power calculations help us see how large a shortfall in the bottle contents the test can be expected to detect.

(a) Explain to someone who knows little statistics what a Type I error, a Type II error, and the power of the test represent in this situation.

(b) Use program TYPE2 or the *Power* applet to find the power of this test against the alternative $\mu = 299$.

(c) Find the power against the alternative $\mu = 295$.

(d) Explain two different ways to increase the power of the test to detect $\mu = 295$. Calculate the power for each suggestion you make.

CASE CLOSED!

I'm getting a headache!

Refer to the chapter-opening Case Study (page 685). Use what you have learned about significance tests and confidence intervals to answer the following questions.

1. Does it matter what method was used to select the sample of 36 tablets? Explain.
2. What null and alternative hypotheses would you recommend testing? Be sure to justify your choice of a one-sided or two-sided test.
3. Describe a Type I and a Type II error in this setting, and give a possible consequence of each.
4. Which significance level would you recommend: $\alpha = 0.10$, 0.05, or 0.01? Justify your answer.
5. Carry out a significance test. What conclusion would you draw about this batch of tablets?
6. Construct a confidence interval to estimate the true mean acetylsalicylic acid content in this batch of Aspro tablets. Use a confidence level that corresponds to the significance level you chose in Question 4.

In fact, these data were generated from a Normal distribution with $\mu = 321$ mg and $\sigma = 3$ mg.

7. Use technology to calculate the power of your test from Question 5. Then explain two ways that the company could have increased the ability of the test to detect a 1 mg difference from the target of $\mu = 320$ mg.
8. Write a brief report to Aspro executives summarizing your findings.

Chapter Review

Summary

A significance test shows how strong the evidence is for some claim about a parameter. In significance testing, the null hypothesis often says "no effect" or "no difference." The alternative hypothesis typically reflects what we hope to find evidence for.

A *P*-value is the probability that the test would produce a result at least as extreme as the observed result if the null hypothesis really were true. That is, a *P*-value tells us how surprising the observed outcome is. Very surprising outcomes (small *P*-values) are good evidence that the null hypothesis is not true.

Figure 11.10 presents the idea of a significance test in picture form.

Figure 11.10 *The idea of a significance test in picture form.*

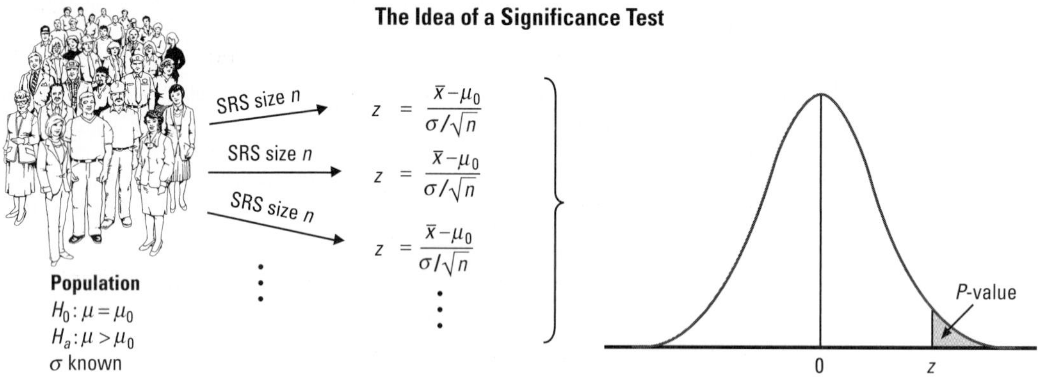

An alternative to the *P*-value way of thinking is to use a significance test to make a decision based on a fixed significance level α. If we decide to reject H_0 when H_0 is true, we have committed a Type I error. The probability of making a Type I error is equal to the significance level α of the test. If we fail to reject H_0 when H_0 is false, we have made a Type II error. The power of a significance test is the probability that the test will detect a specific value of the alternative hypothesis. Power $= 1 - P(\text{Type II error})$. We can increase the power of a test by increasing the sample size or by increasing our willingness to make a Type I error.

Section 11.1 presented the underlying logic and structure of significance tests. In Section 11.2, we looked at how to carry out significance tests about a population mean μ in the unrealistic setting when σ is known. Several important cautions for the wise use of significance tests were discussed in Section 11.3. Some of the implications of using significance tests to make decisions were examined in Section 11.4.

What You Should Have Learned

Here is a checklist of the major skills you should have acquired by studying this chapter.

Significance Tests for μ (σ known)

1. State the null and alternative hypotheses in a testing situation when the parameter in question is a population mean μ.
2. Explain in nontechnical language the meaning of the *P*-value when you are given the numerical value of *P* for a test.
3. Calculate the one-sample z statistic and the *P*-value for both one-sided and two-sided tests about the mean μ of a Normal population.
4. Assess statistical significance at standard levels α by comparing P to α.
5. Recognize that significance testing does not measure the size or importance of an effect.
6. Recognize when you can use the z test and when the data collection design or a small sample from a skewed population makes it inappropriate.
7. Explain Type I error, Type II error, and power in a significance-testing problem.

Web Links

Significance tests and confidence intervals appear frequently in reports of studies from many fields. Most scientific journals have Web sites that display at least summaries of articles appearing in the journals.

The British Medical Journal www.bmj.bmjjournals.com

Journal of the American Medical Association www.jama.ama-assn.org

American Psychological Association www.apa.org/journals/

Chapter Review Exercises

11.65 Stating hypotheses State the appropriate null and alternative hypotheses in each of the following cases.

(a) Census Bureau data show that the mean household income in the area served by a shopping mall is \$72,500 per year. A market research firm questions shoppers at the mall to find out whether the mean household income of mall shoppers is higher than that of the general population.

(b) Mr. Starnes believes that fewer than 75% of the students at his school completed their math homework last night. The math teachers inspect the homework assignments from a random sample of students at the school to help Mr. Starnes test his claim.

(c) Experiments on learning in animals sometimes measure how long it takes a mouse to find its way through a maze. The mean time is 20 seconds for one particular maze. A researcher thinks that playing rap music will cause the mice to complete the maze faster. She measures how long each of 12 mice takes with the rap music as a stimulus.

11.66 Corn yield The mean yield of corn in the United States is about 120 bushels per acre. A survey of 40 farmers this year gives a sample mean yield of $\bar{x} = 123.8$ bushels per acre. We want to know whether this is good evidence that the national mean this year is not 120 bushels per acre. Assume that the farmers surveyed are an SRS from the population of all commercial corn growers and that the standard deviation of the yield in this population is $\sigma = 10$ bushels per acre. Are you convinced that the population mean is not 120 bushels per acre? Is your conclusion correct if the distribution of corn yields is somewhat non-Normal? Why?

11.67 Significance tests and confidence intervals The *P*-value of a one-sided test of the null hypothesis H_0: $\mu = 15$ is 0.02.

(a) Does the 99% confidence interval for μ include 15? Why or why not?

(b) Does the 95% confidence interval for μ include 15? Why or why not?

11.68 Analyzing study results A study with 12 subjects reported a result that failed to achieve statistical significance at the 5% level. The *P*-value was 0.052. Write a short summary of how you would interpret these findings.

11.69 Student study times A student group claims that first-year students at a university must study 2.5 hours per night during the school week. A skeptic suspects that they study less than that on the average. A class survey finds that the average study time claimed by 269 students is $\bar{x} = 137$ minutes. Regard these students as a random sample of all first-year students and suppose we know that study times follow a Normal distribution with standard deviation 65 minutes. What conclusion would you draw? Give appropriate evidence to support your conclusion.

11.70 CEO pay A study of the pay of corporate chief executive officers (CEOs) examined the increase in cash compensation of the CEOs of 104 companies, adjusted for inflation, in a recent year. The mean increase in real compensation was $\bar{x} = 6.9\%$, and the standard deviation of the increases was $s = 55\%$. Is this good evidence that the mean real compensation μ of all CEOs increased that year? (Because the sample size is large, the sample s is close to the population σ, so take $\sigma = 55\%$.)

11.71 Why are large samples better? Statisticians prefer large samples. Describe briefly the effect of increasing the size of a sample (or the number of subjects in an experiment) on each of the following:

(a) The margin of error of a 95% confidence interval.

(b) The P-value of a test, when H_0 is false and all facts about the population remain unchanged as n increases.

(c) The power of a fixed level α test, when α, the alternative hypothesis, and all facts about the population remain unchanged.

11.72 Workers' earnings The Bureau of Labor Statistics generally uses 90% confidence in its reports. One report gives a 90% confidence interval for the mean hourly earnings of American workers in 2000 as \$15.49 to \$16.11. This result was calculated from the National Compensation Survey, a multistage probability sample of businesses.

(a) Would a 95% confidence interval be wider or narrower? Explain.

(b) Would the null hypothesis that the year 2000 mean hourly earnings of all workers was \$16 be rejected at the 10% significance level in favor of the two-sided alternative? What about the null hypothesis that the mean was \$15? Justify your answers.

11.73 Strong chairs? A company that manufactures classroom chairs for high school students claims that the mean breaking strength of the chairs they make is 300 pounds. From years of production, they have seen that $\sigma = 15$ pounds. One of the chairs collapsed beneath a 220-pound student last week. You wonder whether the manufacturer is exaggerating the breaking strength of their chairs.

(a) State null and alternative hypotheses in words and symbols.

(b) Describe a Type I error and a Type II error in this situation. Which is more serious?

(c) There are 30 chairs in your classroom. You decide to determine the breaking strength of each chair and then to find the mean of those values. What values of $\bar{x}$ would cause you to reject H_0 at the 5% significance level?

(d) If the truth is that $\mu = 290$ pounds, use technology to find the probability that you will make a Type II error.

(e) Explain two ways that you could improve the power of this test.

11.74 Significance and sample size A study with 5000 subjects reported a result that was statistically significant at the 5% level. Explain why this result might not be particularly large or important.

CHAPTER 12

Significance Tests in Practice

CASE STUDY

Do you have a fever?

Sometimes when you're sick, your forehead feels really warm. You might have a fever. How can you find out whether you do? By taking your temperature, of course. But what temperature should the thermometer show if you're healthy? Is this temperature the same for everyone?

In the early 1990s, researchers conducted a study to determine whether the "accepted" value for normal body temperature, 98.6°F, is accurate. They used an oral thermometer to measure the temperatures of 140 healthy men and women aged 18 to 40 several times over a three-day period. The resulting data consisted of 700 temperature readings.

Exploratory data analysis revealed several interesting facts about this data set:[1]

- The distribution of all 700 temperature readings was approximately Normal.
- The mean temperature was $\bar{x} = 98.25$°F.
- The standard deviation of the temperature readings was $s = 0.73$°F.
- 62.3% of the temperature readings were less than 98.6°F.

Based on the results of this study, can we conclude that "normal" body temperature in the population of healthy 18- to 40-year-olds is *not* 98.6°F? By the end of this chapter, you will have developed the necessary tools to answer this question.

Activity 12
Is one side of a coin heavier?

Materials: 20 pennies for each student

To use a coin to randomly determine an outcome, most people would flip the coin. Is that equivalent to holding the coin vertically on a tabletop and then spinning the coin with a quick flick of your finger? In this Activity, we will try a third variation. We will stand pennies on edge and then bang the table to make the pennies fall. We are interested in the proportion of times the pennies fall heads up. If the pennies are equally heavy on both sides of the coin, then it would be reasonable to expect the long-term proportion p of heads to be about 0.5. We state the following hypotheses:

$$H_0: p = 0.5$$
$$H_a: p \neq 0.5$$

1. Stand 20 pennies on edge on a horizontal tabletop. Take your time—this may take a steady hand and some patience.
2. Bang the table just hard enough to make all of the pennies fall.
3. Count the number of pennies that fall heads up.
4. Using your data, construct and interpret a 95% confidence interval for the true proportion of heads. Follow the Inference Toolbox.
5. This part should be done as a class activity. Draw a horizontal line at the top of the blackboard, and mark a scale wide enough to accommodate each student's confidence interval. Then below this scaled line, each student can draw his or her confidence interval. These intervals should vary somewhat. Looking at all of the confidence intervals, make a conjecture about the 95% confidence interval for the whole class.
6. Combine your results with those of other students in the class to obtain the overall proportion of pennies that landed heads up. Use the cumulative data collected by the whole class to calculate the 95% confidence interval, and compare this interval with the interval conjectured in Step 5.
7. Each student should compare his or her confidence interval with the confidence interval for the whole class. Which confidence interval do you prefer, and why? What accounts for the difference in the width?
8. Do you think it is likely, by chance alone, to obtain results like those you actually observed if H_0 is true? Why or why not?

Keep these results handy. As soon as we develop the necessary theory, you will test to see if your results are statistically significant.

Introduction

With the principles of testing claims in hand, we proceed to practice. We begin by dropping the unrealistic assumption that we know the population standard deviation σ when testing claims about a population mean. As with confidence intervals, this leads to the use of t distributions when carrying out significance tests about μ. Section 12.1 provides the details.

In Section 12.2, we examine how to perform a significance test about an unknown population proportion p. When the Normality condition is satisfied, we can use the familiar z test from Chapter 11 with a few minor modifications.

Don't forget the cautions about uses and abuses of significance tests from Section 11.3 as you learn the new methods of this chapter. Also remember that you could be committing a Type I or a Type II error whenever you use a significance test to make a decision. In short, think before you calculate!

12.1 Tests about a Population Mean

What would cause the head brewer of the famous Guinness brewery in Dublin, Ireland, not only to use statistics but also to invent new statistical methods? The search for better beer, of course.

William S. Gosset (1876–1937), fresh from Oxford University, joined Guinness as a brewer in 1899. He soon became involved in experiments and in statistics to understand the data from those experiments. What are the best varieties of barley and hops for brewing? How should they be grown, dried, and stored? The results of the field experiments, as you can guess, varied. Gosset faced the problem we noted in using the z test to introduce the reasoning of statistical tests: he didn't know the population standard deviation σ. He observed that just replacing σ by s in the z statistic

$$z = \frac{\overline{x} - \mu}{\sigma/\sqrt{n}}$$

and calling the result roughly Normal wasn't accurate enough.

After much work, Gosset developed what we now call the t distributions. His new t test identified the best barley variety, and Guinness promptly bought up all the available seed. Guinness allowed Gosset to publish his discoveries, but not under his own name. He used the name "Student," so Gosset's t test is sometimes called "Student's t" in his honor.

The One-Sample t *Statistic and the* t *Distributions*

Draw an SRS of size n from a population that has the Normal distribution with mean μ and standard deviation σ. The **one-sample t statistic**

$$t = \frac{\overline{x} - \mu}{s/\sqrt{n}}$$

has the t distribution with $n - 1$ degrees of freedom.

The t statistic has the same interpretation as any standardized statistic: it says how far $\overline{x}$ is from its mean μ in standard deviation units. In Section 10.2, we used Table C to find critical values for constructing confidence intervals about an unknown population mean μ. Now we show how to calculate P-values for a significance test about μ using Table C.

Example 12.1 *Determining* P*-values*

Using the "t table"

Upper tail probability p			
df	.05	.025	.02
18	1.734	2.101	2.214
19	1.729	2.093	2.205
20	1.725	2.086	2.197

Upper tail probability p			
df	.005	.0025	.001
29	2.756	3.038	3.396
30	2.750	3.030	3.385
40	2.704	2.971	3.307

Suppose you carry out a significance test of H_0: $\mu = 5$ versus H_a: $\mu > 5$ based on a sample of size $n = 20$ and obtain $t = 1.81$. In Table C, you would find the row corresponding to df $= 20 - 1 = 19$. The t statistic falls between the values 1.729 and 2.093. If you look at the top of the corresponding columns in Table C, you'll find that the "Upper tail probability p" is between 0.025 and 0.05. Since you are performing a one-tailed test with the alternative H_a: $\mu > 5$, only positive values of t count as evidence against H_0. So the "Upper tail probability p" is the P-value of the test. You can say that the P-value is between 0.025 and 0.05.

What if you were performing a test of H_0: $\mu = 5$ versus H_a: $\mu \neq 5$ based on a sample size of $n = 37$ and obtained $t = -3.17$? Table C shows only areas to the *right* of specified *positive* values of t. Since this is a two-tailed test, you are interested in the probability of getting a value of t less than -3.17 or greater than 3.17. Since df $= 37 - 1 = 36$ is not available on the table, use df $= 30$. This will result in a higher P-value than if you used df $= 36$. Move across the df $= 30$ row, and notice that $t = 3.17$ falls between 3.030 and 3.385. The corresponding "Upper tail probability p" is between 0.001 and 0.0025. For this two-tailed test, the P-value is between $2(0.001) = 0.002$ and $2(0.0025) = 0.005$.

Exercises

12.1 Using the "*t* table," I What critical value t^* from Table C satisfies each of the following conditions?

(a) The t distribution with 5 degrees of freedom has probability 0.05 to the right of t^*.

(b) The t distribution with 21 degrees of freedom has probability 0.99 to the left of t^*.

12.2 Using the "*t* table," II What critical value t^* from Table C satisfies each of the following conditions?

(a) The one-sample t statistic from a sample of 15 observations has probability 0.025 to the right of t^*.

(b) The one-sample t statistic from an SRS of 20 observations has probability 0.75 to the left of t^*.

12.3 One-sided *t* test The one-sample t statistic for testing

$$H_0: \mu = 0$$
$$H_a: \mu > 0$$

from a sample of $n = 15$ observations has the value $t = 1.82$.

(a) What are the degrees of freedom for this statistic?

(b) Give the two critical values t^* from Table C that bracket t. What are the upper tail probabilities p for these two entries?

(c) Between what two values does the P-value of the test fall?

(d) Is the value $t = 1.82$ significant at the 5% level? Is it significant at the 1% level?

12.4 Two-sided *t* test The one-sample t statistic from a sample of $n = 25$ observations for the two-sided test of

$$H_0: \mu = 64$$
$$H_a: \mu \neq 64$$

has the value $t = 1.12$.

(a) What are the degrees of freedom for t?

(b) Locate the two critical values t^* from Table C that bracket t. What are the upper tail probabilities p for these two values?

(c) Between what two values does the P-value of the test fall? (Note that H_a is two-sided.)

(d) Is the value $t = 1.12$ statistically significant at the 10% level? At the 5% level?

The One-Sample *t* Test

In tests as in confidence intervals, we allow for unknown σ by using the standard error and replacing z by t. Here are the details.

The One-Sample t Test

Draw an SRS of size n from a population having unknown mean μ. To test the hypothesis H_0: $\mu = \mu_0$ based on an SRS of size n, compute the **one-sample *t* statistic**

$$t = \frac{\bar{x} - \mu_0}{s/\sqrt{n}}$$

In terms of a random variable T having the $t(n-1)$ distribution, the P-value for a test of H_0 against

H_a: $\mu > \mu_0$ is $P(T \geq t)$

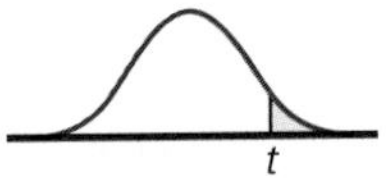

H_a: $\mu < \mu_0$ is $P(T \leq t)$

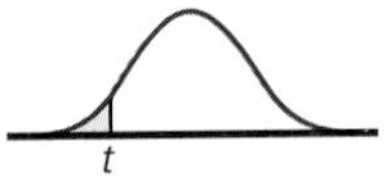

H_a: $\mu \neq \mu_0$ is $2P(T \geq |t|)$

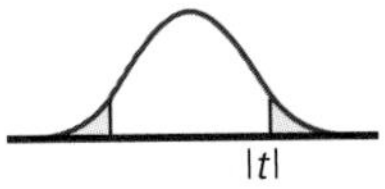

These P-values are exact if the population distribution is Normal and are approximately correct for large n in other cases.

Now we can do a realistic analysis of data produced to test a claim about an unknown population mean. Once again, we follow the steps in our Inference Toolbox.

Example 12.2 *Sweet cola*

Performing a one-sample t test

Diet colas use artificial sweeteners to avoid sugar. These sweeteners gradually lose their sweetness over time. Manufacturers therefore test new colas for loss of sweetness before marketing them. Trained tasters sip the cola along with drinks of standard sweetness and score the cola on a "sweetness scale" of 1 to 10. The cola is then stored for a month at high temperature to imitate the effect of four months' storage at room temperature. Each taster scores the cola again after storage. Our data are the differences (score before storage minus score after storage) in the tasters' scores. The bigger these differences, the bigger the loss of sweetness. Here are the sweetness losses for a new cola, as measured by 10 trained tasters:

2.0	0.4	0.7	2.0	−0.4	2.2	−1.3	1.2	1.1	2.3

Most are positive. That is, most tasters found a loss of sweetness. But the losses are small, and two tasters (the negative scores) thought the cola gained sweetness. *Are these data good evidence that the cola lost sweetness in storage?*

Step 1: Hypotheses Tasters vary in their perception of sweetness loss. So we ask the question in terms of the mean loss $\mu_{\text{BEFORE-AFTER}} = \mu_{\text{DIFF}}$ for a large population of tasters. The null hypothesis is "no loss" and the alternative hypothesis says "there is a loss."

H_0: $\mu_{\text{DIFF}} = 0$ The mean sweetness loss for the population of tasters is 0.

H_a: $\mu_{\text{DIFF}} > 0$ The mean sweetness loss for the population of tasters is positive. The cola seems to be losing sweetness in storage.

Step 2: Conditions Since we do not know the standard deviation of sweetness loss in the population of tasters, we must use a one-sample t test. Now we must check the three required conditions.

- **SRS** We must be willing to treat our 10 tasters as an SRS from the population of tasters if we want to draw conclusions about tasters in general. The tasters all have the same training. So even though we don't actually have an SRS from the population of interest, we are willing to act as if we did. This is a matter of judgment.
- **Normality** The assumption that the population distribution is Normal cannot be effectively checked with only 10 observations. In part, the researchers rely on experience with similar variables. They also look at the data. We can construct a stemplot and a Normal probability plot of the sweetness loss data (notice that the stem 0 must appear twice, to display the difference between −0.9 and 0 and between 0 and 0.9):

−1	3
−0	4
0	4 7
1	1 2
2	0 0 2 3

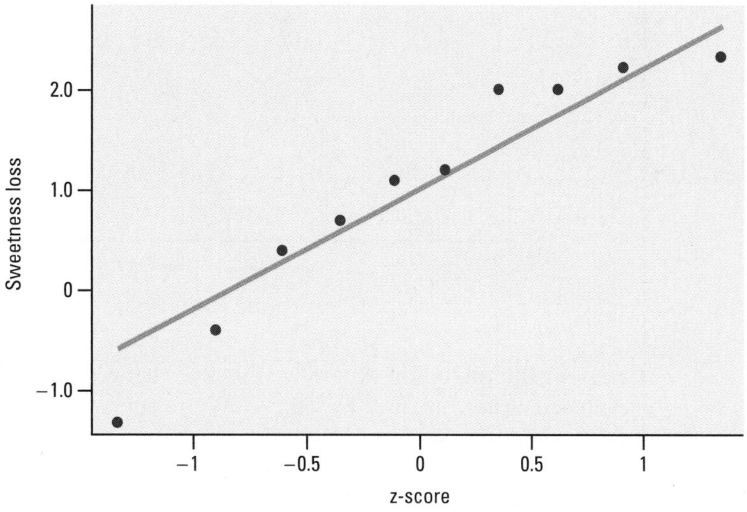

The distribution is left-skewed, but there are no gaps or outliers or other signs of non-Normal behavior. So we proceed with caution.

- **Independence** We need to know that the 10 sweetness loss scores can be viewed as independent observations. There is no reason to doubt this. Note, however, that the two sweetness scores (before and after storage) given by each taster are *not* independent.

Step 3: Calculations

- **Test statistic** The basic statistics are

$$\bar{x}_{\text{DIFF}} = 1.02 \text{ and } s_{\text{DIFF}} = 1.196$$

The one-sample t statistic is

$$t = \frac{\bar{x}_{\text{DIFF}} - \mu_0}{s_{\text{DIFF}}/\sqrt{n}} = \frac{1.02 - 0}{1.196/\sqrt{10}} = 2.70$$

- **P-value** The P-value for $t = 2.70$ is the area to the right of 2.70 under the t distribution curve with degrees of freedom $n - 1 = 9$. Figure 12.1 shows this area. We can't find the exact value of P without a calculator or computer. But we can pin P between two values by using Table C. Search the df = 9 row of Table C for entries that bracket $t = 2.70$. Because the observed t lies between 2.398 and 2.821, the P-value lies between 0.01 and 0.02. Technology gives the more exact result $P = 0.0123$.

Upper tail probability p			
df	.02	.01	.005
8	2.449	2.896	3.355
9	2.398	2.821	3.250
10	2.359	2.764	3.169

Figure 12.1 *The P-value for the one-sided t test of Example 12.2.*

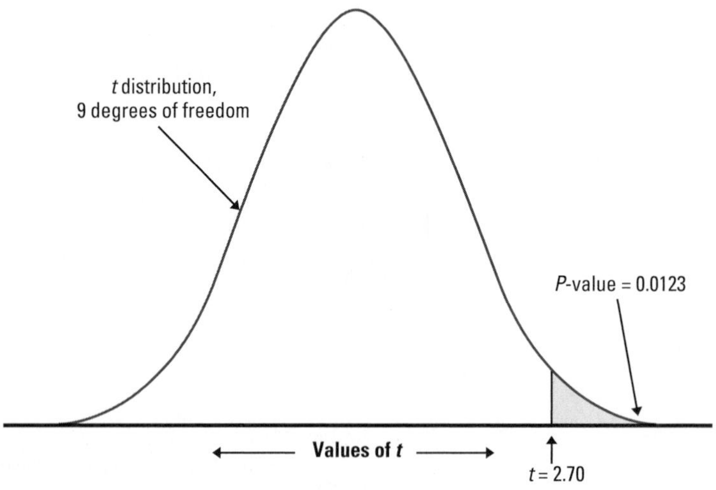

Step 4: **Interpretation** A P-value this low (between 0.01 and 0.02) gives quite strong evidence against the null hypothesis. We reject H_0 and conclude that the cola has lost sweetness during storage.

Notice the linkage between the P-value computed in Step 3 and the conclusion drawn in Step 4. If you keep in mind the three C's—conclusion, connection, context—then you will include all the important elements in your interpretation of results.

Because the t procedures are so common, all statistical software systems will do the calculations for you. Figure 12.2 shows the output from four statistical packages: Data Desk, Minitab, CrunchIt!, and Fathom. In each case, we entered the 10 sweetness losses as values of a variable called "cola" and asked for the one-sample t test of H_0: $\mu = 0$ against H_a: $\mu > 0$. The four outputs report slightly different information, but all include the basic facts: $\bar{x} = 1.02$, $t = 2.70$, $P = 0.012$. These are the results we found in Example 12.2.

Figure 12.2 *Output for the one-sample t test of Example 12.2 from four statistical software packages. You can easily locate the basic results in output from any statistical software.*

DataDesk

```
cola:
Test Ho: mu (cola) = 0 vs Ha: mu (cola) > 0
Sample Mean = 1.02000 t-Statistic = 2.697 w/9 df
Reject Ho at Alpha = 0.0500
p = 0.0123
```

Minitab

Session

```
TEST  OF  MU  =  0.000  VS  MU  G.  T.  0.000
          N     MEAN     STDEV   SE MEA    T    P  VALUE
cola      10    1.02     1.196     0.3    2.7     0.012
```

CrunchIt!

One sample T statistics

T-test results:
μ - mean of Variable
$H_0 : \mu = 0$
$H_A : \mu > 0$

Variable	Sample Mean	Std. Err	DF	T-Stat	P-value
cola	1.02	0.37824154	9	2.6966896	0.0123

Fathom

Test of sweetness — Test Mean

Attribute (numeric): SweetLoss

Ho: population mean of **SweetLoss** equals **0**
Ha: population mean of **SweetLoss is greater than 0**

Count:	**10**
Mean:	**1.02**
Std dev:	**1.1961**
Std error:	**0.378242**
Student's t:	**2.697**
DF:	**9**
P-value:	**0.012**

Here's an example involving a two-tailed test.

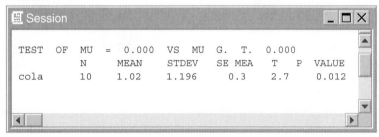

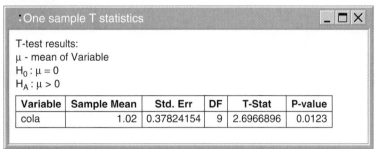

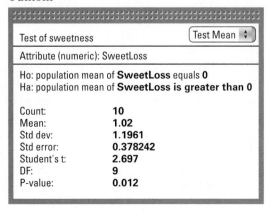

Example 12.3 *Diversify or be sued!*
Two-tailed t test

An investor with a stock portfolio worth several hundred thousand dollars sued his broker because lack of diversification in his portfolio led to poor performance (low returns). Table 12.1 gives the rates of return for the 39 months that the account was managed by the broker. An arbitration panel compared these returns with the average of the Standard & Poor's 500 stock index for the same period. Consider the 39 monthly returns as a random sample from the monthly returns the broker would generate if he managed the account forever. Are these returns compatible with a population mean of $\mu = 0.95\%$, the S&P 500 average?

Table 12.1 *Monthly rates of return on a portfolio (percent)*

−8.36	1.63	−2.27	−2.93	−2.70	−2.93	−9.14	−2.64	6.82	−2.35	−3.58	6.13	7.00
−15.25	−8.66	−1.03	−9.16	−1.25	−1.22	−10.27	−5.11	−0.80	−1.44	1.28	−0.65	4.34
12.22	−7.21	−0.09	7.34	5.04	−7.24	−2.14	−1.01	−1.41	12.03	−2.56	4.33	2.35

Source: C. Don Wiggins, "The legal perils of 'underdiversification'—a case study," *Personal Financial Planning,* 1, No. 6 (1999), pp. 16–18.

Step 1: **Hypotheses** Our hypotheses are

$$H_0: \mu = 0.95$$
$$H_a: \mu \neq 0.95$$

where μ is the mean return for all possible months that the broker could manage this account.

Step 2: **Conditions** Since we don't know σ, we should perform a one-sample t test. Checking conditions:

- **SRS** We are told to view this sample of 39 months as a random sample from all possible months during which the broker could have managed the account.
- **Normality** Figure 12.3(a) gives a histogram and Figure 12.3(b) gives a Normal probability plot for these data. There are no outliers and the distribution shows no strong skewness. We are comfortable assuming that the population distribution of returns is Normal. (Alternatively, since the sample size is large, $n = 39$, the central limit theorem confirms the approximate Normality of the sampling distribution of $\bar{x}$.)
- **Independence** We must be willing to treat the 39 monthly returns as independent observations from the (hypothetical) population of months in which this broker could have managed the account. This is a matter of judgment.

Step 3: **Calculations**

- **Test statistic** The one-sample t statistic is

$$t = \frac{\bar{x} - \mu_0}{s/\sqrt{n}} = \frac{-1.10 - 0.95}{5.99/\sqrt{39}} = -2.14$$

Figure 12.3 *(a) Histogram and (b) Normal probability plot of monthly percent returns on a portfolio, for Example 12.3.*

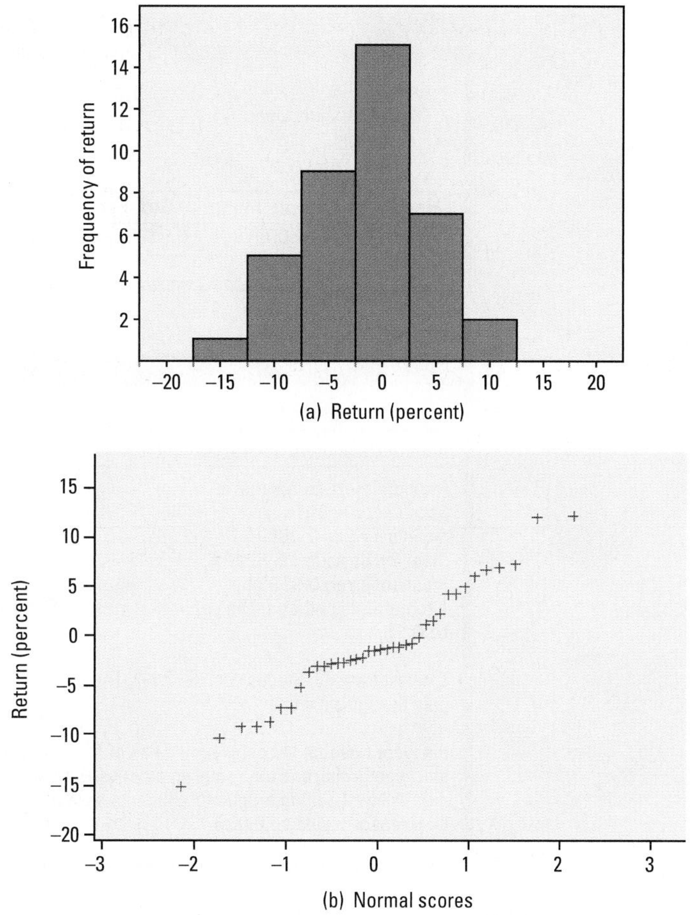

Upper tail probability p			
df	.05	.025	.02
29	1.699	2.045	2.150
30	1.697	2.042	2.147
40	1.684	2.021	2.123

- ***P*-value** Although df $= 39 - 1 = 38$, we use the more conservative df $= 30$ from Table C. For $t = 2.14$, the upper tail probability is between 0.02 and 0.025. Since we are performing a two-tailed test, our *P*-value is between $2(0.02) = 0.04$ and $2(0.025) = 0.05$.

Step 4: **Interpretation** Here is one way to report the conclusion: The mean monthly return on investment for this client's account was $\bar{x} = -1.1\%$. This differs significantly from the performance of the S&P 500 for the same period ($t = -2.14$, $P < 0.05$).

Figure 12.4 (on the next page) shows CrunchIt!, Fathom, and Minitab outputs for Example 12.3. Output from other software would look similar. Notice that the software reports the *P*-value as $P = 0.039$. This is slightly smaller than the *P*-value we calculated using Table C because we used df $= 30$ and the computer used df $= 38$. But our conclusion is the same: the mean return on the client's account differs significantly from that of the S&P 500. Now let's evaluate the return on the client's account with a confidence interval.

Figure 12.4 *CrunchIt!, Fathom, and Minitab output for Example 12.3.*

CrunchIt!

:One sample T statistics

T-test results:
μ - mean of Variable
$H_0 : \mu = 0.95$
$H_A : \mu$ not equal 0.95

Variable	Sample Mean	Std. Err.	DF	T-Stat	P-value
Return	−1.0997436	0.95930994	38	−2.1366856	0.0391

Fathom

Test of Stock portfolio — Test Mean

Attribute (numeric): Return_percent

Attribute: **Return_percent**
Sample count: **39**
Sample mean: **−1.09974**
Standard deviation: **5.99089**
Standard error: **0.95931**
Alternative hypothesis: The population mean of **Return_percent is not equal to 0.95**

The test statistic, Student's t, is – **2.137**. There are **38** degrees of freedom (one less than the sample size).

If it were true that the population mean of **Return_percent** were equal to 0.95 (the null hypothesis), and the sampling process were performed repeatedly, the probability of getting a value for Students's t **with an absolute value this great or greater** would be **0.039.**

Minitab

Session

```
Test of mu = 0.950 vs mu not = 0.950

Variable    N    Mean    StDev   SE Mean      T       P
RETURN     39  -1.100    5.991     0.959  -2.14   0.039
```

Example 12.4 *Estimating mean stock return*
Confidence intervals clarify!

The mean monthly return on the client's portfolio was $\bar{x} = -1.1\%$, and the standard deviation was $s = 5.99\%$. Our resulting 95% confidence interval is:

$$\bar{x} \pm t^* \frac{s}{\sqrt{n}} = -1.1 \pm 2.042 \frac{5.99}{\sqrt{39}} = -1.1 \pm 1.96 = (-3.06, 0.86)$$

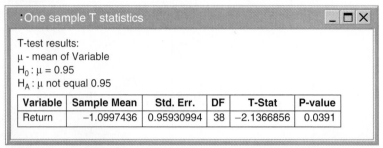

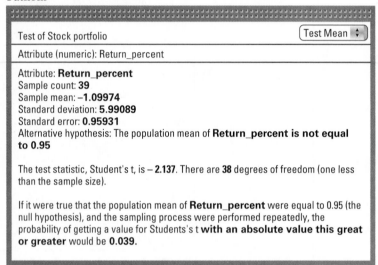

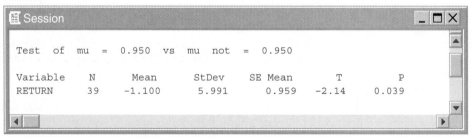

Notice that we used df = 30 in Table C to get $t^* = 2.042$. Figure 12.5 gives the Minitab, Excel, Fathom, and CrunchIt! outputs for a 95% confidence interval for the population mean μ. Note that Excel gives the margin of error next to the label "Confidence Level(95.0%)" rather than the actual confidence interval. We see that the 95% confidence interval using the more accurate df = 38 is (−3.04, 0.84), which is slightly narrower than the one we obtained with df = 30.

Figure 12.5 *Minitab, Excel, Fathom, and CrunchIt! output for Example 12.4.*

Minitab

Session

Variable	N	Mean	StDev	SE Mean	95.0% C.I.
RETURN	39	-1.100	5.991	0.959	(-3.042, 0.842)

Fathom

Estimate of Stock portfolio — Estimate Mean

Attribute (numeric): Return_percent

Interval estimate for population mean of **Return_percent**

Count:	**39**
Mean:	**−1.09974**
Std dev:	**5.99089**
Std error:	**0.95931**
Confidence level:	**95.0%**
Estimate:	**−1.09974 +/− 1.94202**
Range:	**−3.04176 to 0.842278**

Excel

Mean	−1.09974359
Standard Error	0.95930991
Standard Deviation	5.990888471
Count	39
Confidence Level(95.0%)	1.942021452

CrunchIt!

:One sample T statistics

95% Confidence interval results:

μ - mean of Variable

Variable	Sample Mean	Std. Err.	DF	L. Limit	U. Limit
Return	−1.0997436	0.95930994	38	−3.041765	0.84227777

Because the S&P 500 return, 0.95%, falls outside this interval, we know that μ differs significantly from 0.95% at the $\alpha = 0.05$ level. In fact, the confidence interval suggests that the broker's management of this account had a long-term mean somewhere between a loss of 3.04% and a gain of 0.84% per month. Since the S&P 500 showed a mean gain of 0.95% during this time period, we can say with 95% confidence that the underperformance of this portfolio is between 0.11% and 3.99% per month. This estimate helps determine the compensation owed to the investor.

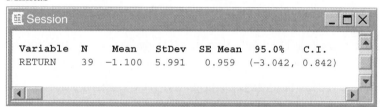

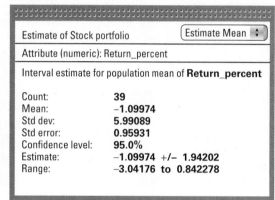

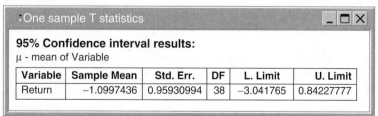

Catching cheaters A certification test for surgeons asks 277 multiple-choice questions. Smith and Jones have 193 common right answers and 53 identical wrong choices. The computer flags their 246 identical answers as evidence of possible cheating. They sue. The court wants to know how unlikely it is that exams this similar would occur just by chance. That is, the court wants a *P*-value. Statisticians offer several *P*-values based on different models for the exam-taking process. They all say that results this similar would almost never happen just by chance. Smith and Jones fail the exam.

Exercises

12.5 Healthy bones Here are estimates of the daily intakes of calcium (in milligrams) for 38 women between the ages of 18 and 24 years who participated in a study of women's bone health:

808	882	1062	970	909	802	374	416	784	997
651	716	438	1420	1425	948	1050	976	572	403
626	774	1253	549	1325	446	465	1269	671	696
1156	684	1933	748	1203	2433	1255	1100		

Suppose that the recommended daily allowance (RDA) of calcium for women in this age range is 1200 milligrams. Doctors involved in the study suspected that participating subjects had significantly lower calcium intakes than the RDA.

(a) Test the doctors' claim at the $\alpha = 0.05$ significance level. Follow the Inference Toolbox.

(b) Remove any outliers from the data and run the test in (a) again. Explain any differences in your results.

12.6 Heat through glass How well materials conduct heat matters when designing houses, for example. Conductivity is measured in terms of watts of heat power transmitted per square meter of surface per degree Celsius of temperature difference on the two sides of the material. In these units, glass has conductivity about 1. The National Institute of Standards and Technology provides exact data on properties of materials. Here are 11 measurements of the heat conductivity of a particular type of glass:[2]

1.11	1.07	1.11	1.07	1.12	1.08	1.08	1.18	1.18	1.18	1.12

Is there evidence that the conductivity of this type of glass is greater than 1? Carry out an appropriate test. Then give an interval estimate of the mean conductivity.

12.7 Eating and weight gain Return to Exercise 10.68 (page 681). Test the null hypothesis that the mean weight gain is 16 lb. Be sure to justify the alternative hypothesis that you choose.

12.8 NEAT Nonexercise activity thermogenesis (NEAT) provides a partial explanation for the results you found in the previous exercise. NEAT is energy burned by fidgeting, maintenance of posture, spontaneous muscle contraction, and other activities of daily living. In the study of the previous exercise, the 16 subjects increased their NEAT by 328 calories per day in response to the additional food intake. The standard deviation was 256.

(a) Test the null hypothesis that there was no change in NEAT versus the two-sided alternative. Summarize the results of your test in a few sentences.

(b) Calculate and interpret a 95% confidence interval for the change in NEAT. Discuss the additional information provided by the confidence interval that is not evident from the results of the significance test.

Paired *t* Tests

In the taste test of Example 12.2 (page 746), the same 10 tasters rated before and after sweetness. Since the data were paired by taster, we performed a one-sample t test on the differences. That is, we used a *paired t test.* Here's another example that illustrates the use of a paired t test.

Example 12.5 ***Floral scents and learning***
Performing a paired t test

TOOLBOX

We hear that listening to Mozart improves students' performance on tests. Perhaps pleasant odors have a similar effect. To test this idea, 21 subjects worked a paper-and-pencil maze while wearing a mask. The mask was either unscented or carried a floral scent. The response variable is their average time on three trials. Each subject worked the maze with both masks, in a random order. The randomization is important because subjects tend to improve their times as they work a maze repeatedly. Table 12.2 gives the subjects' average times with both masks.

Table 12.2 *Average time to complete a maze*

Subject	Unscented (seconds)	Scented (seconds)	Difference	Subject	Unscented (seconds)	Scented (seconds)	Difference
1	30.60	37.97	−7.37	12	58.93	83.50	−24.57
2	48.43	51.57	−3.14	13	54.47	38.30	16.17
3	60.77	56.67	4.10	14	43.53	51.37	−7.84
4	36.07	40.47	−4.40	15	37.93	29.33	8.60
5	68.47	49.00	19.47	16	43.50	54.27	−10.77
6	32.43	43.23	−10.80	17	87.70	62.73	24.97
7	43.70	44.57	−0.87	18	53.53	58.00	−4.47
8	37.10	28.40	8.70	19	64.30	52.40	11.90
9	31.17	28.23	2.94	20	47.37	53.63	−6.26
10	51.23	68.47	−17.24	21	53.67	47.00	6.67
11	65.40	51.10	14.30				

Source: A. R. Hirsch and L. H. Johnston, "Odors and learning," *Journal of Neurological and Orthopedic Medicine and Surgery,* 17 (1996), pp. 119–126.

To analyze these data, subtract the scented time from the unscented time for each subject. The 21 differences form a single sample. They appear in the "Difference" columns in Table 12.2. The first subject, for example, was 7.37 seconds slower wearing the scented mask, so the difference is negative. Positive differences show that the subject did better when wearing the scented mask.

Step 1: Hypotheses To assess whether the floral scent significantly improved performance, we test

$$H_0: \mu = 0$$

$$H_a: \mu > 0$$

Here μ is the mean difference in the population from which the subjects were drawn. The null hypothesis says that no improvement occurs, and H_a says that unscented times are longer than scented times on average.

Step 2: Conditions We do not know the standard deviation of the population of differences, so we should perform a paired t test. Next, we check the required conditions.

- **SRS** The data come from a randomized, matched pairs experiment. But we can generalize the results of this study to the population of interest only if we view our 21 subjects as an SRS from the population.
- **Normality** A stemplot and Normal probability plot of the differences show that the distribution is symmetric and appears reasonably Normal in shape. We have no reason to question the Normality of the population of differences.

Figure 12.6 *(a) Stemplot of the differences in time to complete a maze for 21 subjects, for Example 12.5. The data are rounded to the nearest whole second. (b) Normal probability plot of the differences in maze completion time for the 21 subjects.*

−2	5
−1	7 1 1
−0	8 7 6 4 4 3 1
0	3 4 7 9 9
1	2 4 6 9
2	5

(a)

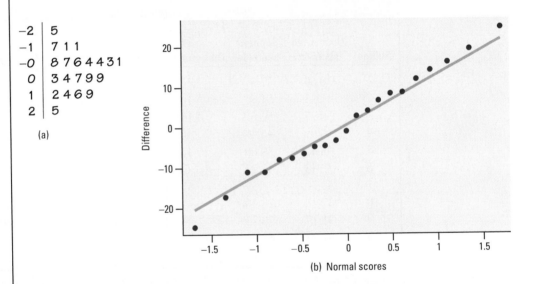

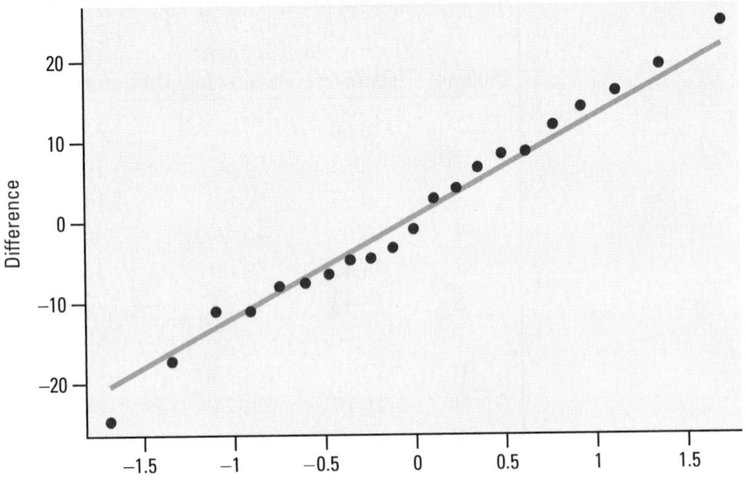

(b) Normal scores

- **Independence** We are willing to assume that the 21 subjects' differences in average maze completion times with the two types of masks are independent observations. Of course, an individual subject's average times with scented and unscented masks are *not* independent observations.

Step 3: Calculations

- **Test statistic** The 21 differences have $\bar{x} = 0.9567$ and $s = 12.5479$. The one-sample t statistic is therefore

$$t = \frac{\bar{x} - \mu_0}{s/\sqrt{n}} = \frac{0.9567 - 0}{12.5479/\sqrt{21}} = 0.349$$

- ***P*-value** Table C shows that 0.349 is less than the 0.25 critical value of the $t(20)$ distribution. The *P*-value is therefore greater than 0.25. Statistical software gives the value $P = 0.3652$.

Step 4: **Interpretation** The data do not support the claim that floral scents improve performance. The average improvement is small, just 0.96 second compared with the 50 seconds that the average subject took when wearing the unscented mask. This small improvement is not statistically significant at even the 25% level.

Figure 12.7 shows the output for Example 12.5 from Minitab, DataDesk, Excel, CrunchIt!, and Fathom. In each case, we entered the data and asked for the one-sided paired t test. The five outputs report slightly different information, but all include the basic facts: $t = 0.349$, $P = 0.365$.

Figure 12.7 *Computer output for the matched pairs t test of Example 12.5 from five different programs.*

Minitab

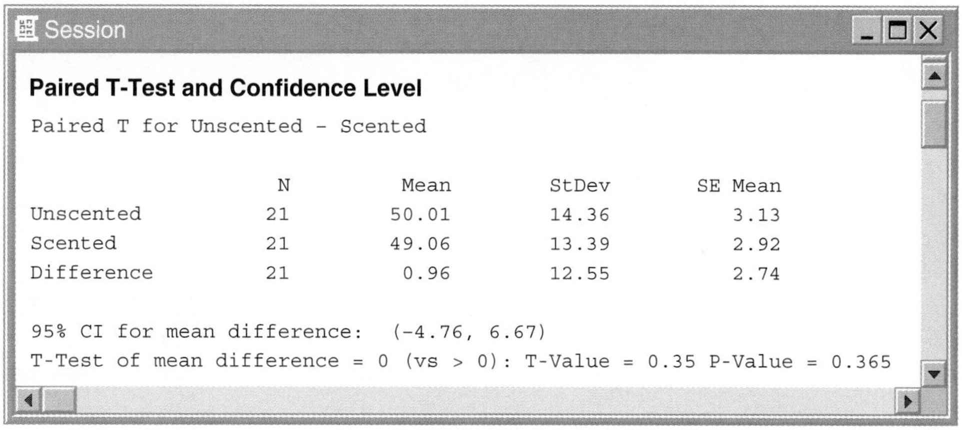

DataDesk

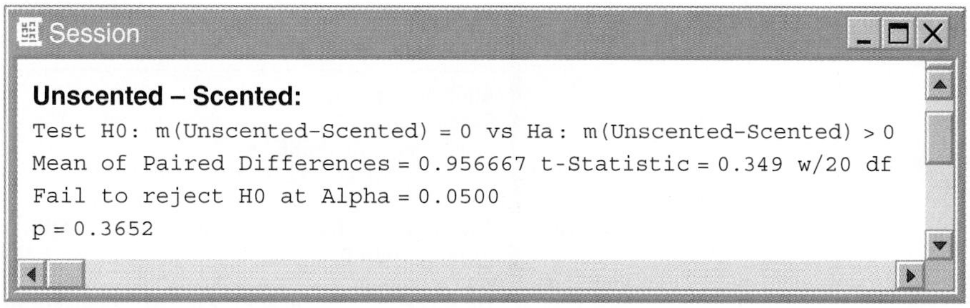

More about the One-Sample *t* Test: Robustness and Power

Recall from Section 10.2 that the t procedures are robust against non-Normality of the population except when outliers or strong skewness are present. (Skewness

Figure 12.7 *(continued)*

Excel

Session

t-Test: Paired Two Sample for Means

	Variable 1	Variable 2
Mean	50.01429	49.05762
Variance	206.3097	179.1748
Observations	21	21
Pearson Correlation	0.593026	
Hypothesized Mean Difference	0	
df	20	
t Stat	0.349381	
P(T<=t) one-tail	0.365227	
t Critical one-tail	1.724718	
P(T<=t) two-tail	0.730455	
t Critical two-tail	2.085962	

CrunchIt!

:One sample T statistics

Paired T statistics

Hypothesis test results:

$\mu_1 - \mu_2$: mean of the paired difference between Unscented and Scented

$H_0 : \mu_1 - \mu_2 = 0$

$H_A : \mu_1 - \mu_2 > 0$

Difference	Sample Diff.	Std. Err.	DF	T-Stat	P-value
Unscented – Scented	0.95666665	2.7381723	20	0.34938148	0.3652

Fathom

Test of maze times — Test Mean

Attribute (numeric): diff

Attribute: **diff**
Sample count: **21**
Sample mean: **0.956667 s**
Standard deviation: **12.5479 s**
Standard error: **2.73817 s**
Alternative hypothesis: The population mean of **diff is greater than 0**

The test statistic, Student's t, is **0.3494**. There are **20** degrees of freedom (one less than the sample size).

If it were true that the population mean of **diff** were equal to **0** (the null hypothesis), and the sampling process were performed repeatedly, the probability of getting a value for Student's t **this great or greater** would be **0.37.**

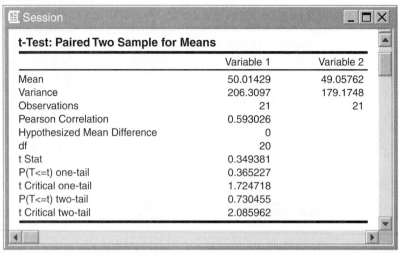

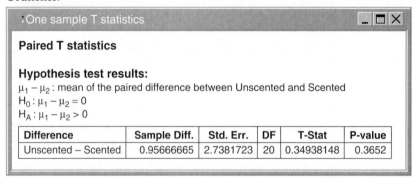

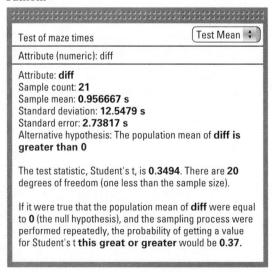

is more serious than other kinds of non-Normality.) As the sample size increases, the central limit theorem ensures that the distribution of the sample mean $\bar{x}$ becomes more nearly Normal and that the t distribution becomes more accurate for calculating P-values. Review the guidelines in the box "Using the t Procedures" on page 655.

The power of a statistical test measures its ability to detect deviations from the null hypothesis. In practice we carry out the test in the hope of showing that the null hypothesis is false, so higher power is important. The power of the one-sample t test against a specific alternative value of the population mean μ is the probability that the test will reject the null hypothesis when the mean has this alternative value.

Calculation of the exact power of the t test takes into account the estimation of σ by s and is a bit complex. But an approximate calculation that acts as if σ were known is usually adequate for planning a study. In that case, we can use the calculator program TYPE2 or the *Power* applet to estimate the power of the test. Whether we use a z test or a t test, we can commit a Type I error or a Type II error. The following Example illustrates these ideas for Example 12.3.

Example 12.6 ***Diversify or be sued***
Power, Type I, and Type II errors

In Example 12.3 (page 750), a client sued his broker because his investment portfolio lost money while the S&P 500 showed a mean return of 0.95%. We performed a significance test of H_0: $\mu = 0.95\%$ versus H_a: $\mu \neq 0.95\%$ using a sample of 39 monthly returns.

A Type I error in this setting would be rejecting H_0: $\mu = 0.95\%$ when this null hypothesis is actually true. In practical terms, we would be concluding that the mean monthly return for the client's portfolio is not the same as the overall market return. Given this information, the judge would probably rule in the client's favor and give him a monetary award to which he was not entitled. The probability of making a Type I error is the significance level α of our test.

Committing a Type II error would mean that we failed to reject H_0 when it's actually false. In that case, we would decide that the mean return on the client's portfolio in the long term could be 0.95%, the same for the overall market. The judge would rule against the client, giving him no monetary settlement even though he deserves one.

With $n = 39$ and $\alpha = 0.05$, we can calculate the power of the test to detect a specific value of the alternative hypothesis, like $\mu = 0\%$. Using the TYPE2 program on a TI-84, we obtained a power of 0.17. The probability of making a Type II error in this setting is $1 - 0.17 = 0.83$. This test has only a 25% chance of detecting a difference as large as 0.95% from the mean return of the S&P 500. To increase the power, we could increase our willingness to make a Type I error (α), increase the sample size n, or both.

Exercises

12.9 Right versus left, I The design of controls and instruments affects how easily people can use them. A student project investigated this effect by asking 25 right-handed students to turn a knob (with their right hands) that moved an indicator by screw action. There were two identical instruments, one with a right-hand thread (the knob turns clockwise) and the other with a left-hand thread (the knob turns counterclockwise). The following table gives the times in seconds each subject took to move the indicator a fixed distance:[3]

Subject	Right thread	Left thread	Subject	Right thread	Left thread
1	113	137	14	107	87
2	105	105	15	118	166
3	130	133	16	103	146
4	101	108	17	111	123
5	138	115	18	104	135
6	118	170	19	111	112
7	87	103	20	89	93
8	116	145	21	78	76
9	75	78	22	100	116
10	96	107	23	89	78
11	122	84	24	85	101
12	103	148	25	88	123
13	116	147			

(a) Each of the 25 students used both instruments. Discuss briefly how you would use randomization in arranging the experiment.

(b) The project designers hoped to show that right-handed people find right-hand threads easier to use. Carry out a significance test at the 5% significance level to investigate this claim.

(c) Describe a Type I error and a Type II error in this situation, and give a possible consequence of each.

(d) How likely is the significance test in part (b) to detect a mean difference of 5 seconds between the left-hand and right-hand times? Is this sufficient power?

12.10 Right versus left, II

(a) Construct and interpret a 90% confidence interval for the mean time advantage of right-hand over left-hand threads in the setting of the previous exercise.

(b) Do you think that the time saved would be of practical importance if the task were performed many times—for example, by an assembly-line worker? To help answer this question, find the mean time for right-hand threads as a percent of the mean time for left-hand threads.

12.11 Mutual funds performance Do "index funds" that simply buy and hold all the stocks in one of the stock market indexes, such as the Standard & Poor's 500 stock index, perform better than actively managed mutual funds? Compare the percent total return (price change plus dividends) of a large actively managed fund with that of the Vanguard 500 Index Fund for the 24 years from 1977 to 2000. Vanguard did better by an average of 2.83% per year, and the standard deviation of the 24 annual differences was 11.65%. Is there convincing evidence that the index fund does better? Give appropriate statistical evidence to support your conclusion.

12.12 Measuring acculturation The Acculturation Rating Scale for Mexican Americans (ARSMA) and the Bicultural Inventory (BI) both measure the extent to which Mexican Americans have adopted Anglo/English culture. These tests were compared by administering both tests to 22 Mexican Americans. Both tests have the same range of scores (1.00 to 5.00) and are scaled to have similar means for the groups used to develop them. There was a high correlation between the two scores, giving evidence that both are measuring the same characteristics. The researchers wanted to know whether the population mean difference in scores for the two tests is 0. The differences in scores (ARSMA − BI) for the 22 participants had $\bar{x} = 0.2519$ and $s = 0.2767$.[4]

(a) Describe briefly how to arrange the administration of the two tests to the subjects, including randomization.

(b) Carry out a significance test to help answer the researchers' question.

(c) Construct and interpret a 95% confidence interval for the mean difference in scores on the two tests.

12.13 No-fee credit card offer, I A bank wonders whether omitting the annual credit card fee for customers who charge at least \$2400 in a year would increase the amount charged on its credit cards. The bank makes this offer to an SRS of 200 of its credit card customers. It then compares how much these customers charge this year with the amount that they charged last year. The mean increase is \$332, and the standard deviation is \$108.

(a) Is there significant evidence at the 1% level that the mean amount charged increases under the no-fee offer? Give appropriate statistical evidence to support your conclusion.

(b) Construct and interpret a 99% confidence interval for the mean amount of the increase.

(c) The distribution of the amount charged is skewed to the right, but outliers are prevented by the credit limit that the bank enforces on each card. Use of the t procedures is justified in this case even though the population distribution is not Normal. Explain why.

(d) A critic points out that the customers would probably have charged more this year than last even without the new offer, because the economy is more prosperous and interest rates are lower. Briefly describe the design of an experiment to study the effect of the no-fee offer that would avoid this criticism.

12.14 No-fee credit card offer, II The bank in the previous exercise tested a new idea on a sample of 200 customers. The bank wants to be quite certain of detecting a mean increase of $\mu = \$100$ in the amount charged, at the $\alpha = 0.01$ significance level. Perhaps a sample of only $n = 50$ customers would accomplish this.

(a) Describe a Type I error and a Type II error in this setting. Which is more serious?

(b) Assume that $\mu = \$100$ (the given alternative) and that $\sigma = \$108$ (an estimate from the data in Exercise 12.13). Find the power using technology. Would you recommend that the bank do a test on 50 customers, or should more customers be included? Explain.

Section 12.1 Summary

Significance tests for the mean μ of a Normal population are based on the sample mean $\bar{x}$ of an SRS of size n. Thanks to the central limit theorem, the resulting procedures are approximately correct for other population distributions when the sample is large.

When we know σ, we use the z statistic and the standard Normal distribution. In practice, we do not know σ. Then we use the **one-sample *t* statistic**

$$t = \frac{\bar{x} - \mu_0}{s/\sqrt{n}}$$

The t statistic has the t distribution with $n - 1$ degrees of freedom.

Significance tests for H_0: $\mu = \mu_0$ are based on the t statistic. Use P-values or fixed significance levels from the $t(n-1)$ distribution.

Use a **paired *t* test** to analyze paired data by first taking the difference within each pair to produce a single sample. Then use one-sample t procedures.

Calculating the **power of a *t* test** is complicated, so we usually rely on technology.

Section 12.1 Exercises

12.15 Using Table C The one-sample t statistic for a test of

$$H_0: \mu = 10$$
$$H_a: \mu < 10$$

based on $n = 10$ observations has the value $t = -2.25$.

(a) What are the degrees of freedom for this statistic?

(b) Between what two probabilities from Table C does the P-value of the test fall?

12.16 Significance You are testing H_0: $\mu = 0$ against H_a: $\mu \neq 0$ based on an SRS of 20 observations from a Normal population. What values of the t statistic are statistically significant at the $\alpha = 0.005$ level?

12.17 Sample size and significance For a sample of size 5, a test of a null hypothesis versus a two-sided alternative gives $t = 2.45$.

(a) Is the test result significant at the 5% level? Draw a sketch of the appropriate t distribution and illustrate your calculation with this sketch.

(b) Now assume that the same test statistic was obtained for a sample size of $n = 10$. Assess the statistical significance of the result and illustrate the calculation with a sketch. How did the statistical significance change with the sample size? Explain your answer.

12.18 Growing tomatoes An agricultural field trial compares the yield of two varieties of tomatoes for commercial use. The researchers divide in half each of 10 small plots of land in different locations and plant each tomato variety on one-half of each plot. After harvest,

they compare the yields in pounds per plant at each location. The 10 differences (Variety A − Variety B) give $\bar{x} = 0.34$ and $s = 0.83$. Is there convincing evidence that Variety A has the higher mean yield?

(a) Describe in words what the parameter μ is in this setting.

(b) Perform a significance test to answer the question. Follow the Inference Toolbox.

12.19 The power of tomatoes The tomato experts who carried out the field trial described in the previous exercise suspect that the large *P*-value there is due to low power. They would like to be able to detect a mean difference in yields of 0.5 pound per plant at the 0.05 significance level. Based on the previous study, use 0.83 as an estimate of both the population σ and the value of s in future samples.

(a) Describe a Type I and a Type II error in this setting. Which is more serious?

(b) Use technology to find the power of the test with $n = 10$ against the alternative $\mu = 0.5$.

(c) If the sample size is increased to $n = 25$ plots of land, use technology to find the power against the same alternative.

(d) Give two other ways to increase the power of the test besides increasing the sample size.

12.20 Paying high prices? A retailer entered into an exclusive agreement with a supplier who guaranteed to provide all products at competitive prices. The retailer eventually began to purchase supplies from other vendors who offered better prices. The original supplier filed a lawsuit claiming violation of the agreement. In defense, the retailer had an audit performed on a random sample of invoices. For each audited invoice, all purchases made from other suppliers were examined and compared with those offered by the original supplier. For each invoice, the percent of purchases for which an alternative supplier offered a lower price than the original supplier was recorded.[5] Here is some computer output relating to these data:

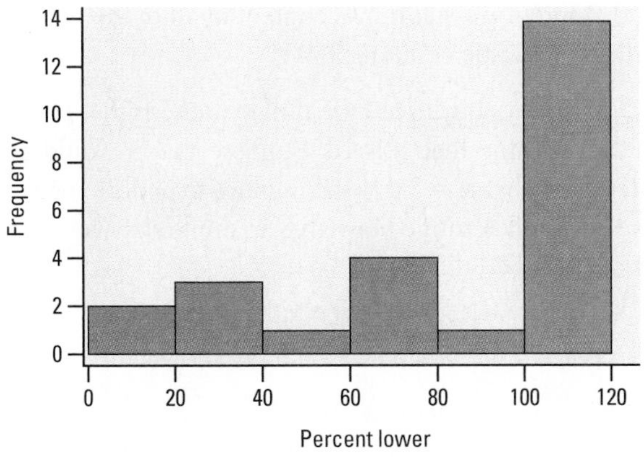

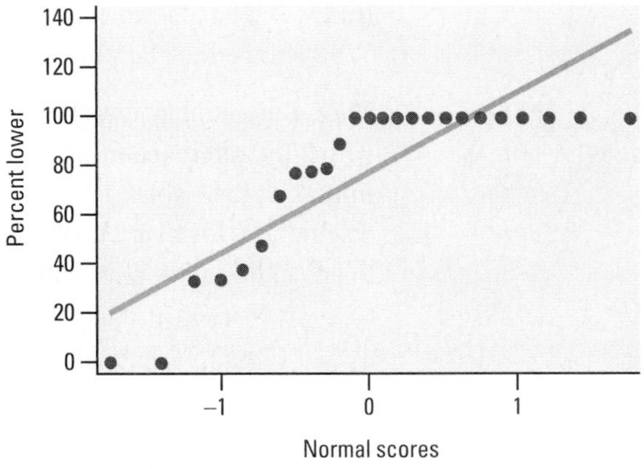

Summary statistics:

Column	n	Mean	Variance	Std. Dev.	Std. Err.	Median	Range	Min	Max	Q1	Q3
pctlower	25	77.76	1067.7733	32.6768	6.5353603	100	100	0	100	68	100

Do the sample invoices suggest that the original supplier's prices are not competitive on average? Give appropriate evidence to support your answer.

12.21 Vitamin C content The researchers studying vitamin C in CSB in Exercise 10.30 (page 649) were also interested in a similar commodity called wheat-soy blend (WSB). A major concern was the possibility that some of the vitamin C content would be destroyed as a result of storage and shipment of the commodity to its final destination. The researchers specially marked a collection of bags at the factory and took a sample from each of these to determine the vitamin C content. Five months later in Haiti they found the specially marked bags and took samples. The data consist of two vitamin C measures for each bag, one at the time of production in the factory and the other five months later in Haiti. The units are mg/100 g as in Exercise 10.30. Here are the data:

Factory	Haiti	Factory	Haiti	Factory	Haiti
44	40	45	38	39	43
50	37	32	40	52	38
48	39	47	35	45	38
44	35	40	38	37	38
42	35	38	34	38	41
47	41	41	35	44	40
49	37	43	37	43	35
50	37	40	34	39	38
39	34	37	40	44	36

(a) Examine the question of interest to these researchers. Provide appropriate statistical evidence to justify your conclusion.

(b) Estimate the loss in vitamin C content over the five-month period. Use a 95% confidence level.

(c) Do these data provide evidence that the mean vitamin C content of all of the bags of WSB shipped to Haiti differs from the target value of 40 mg/100 g?

APPLET

12.22 Determining power You are designing a study to test the null hypothesis that $\mu = 0$ versus the alternative that μ is positive. Assume that σ is 10. Suppose that it would be important to be able to detect the alternative that $\mu = 2$. Use technology to perform power calculations for a variety of sample sizes, and determine how large a sample size you need to detect this alternative with power of at least 0.80.

Technology Toolbox

t Tests

You can perform a t test using either raw data or summary statistics on the TI-83/84/89. Let's use the data on sweetness loss of cola from Example 12.2 (page 746) to carry out a test of H_0: $\mu = 0$ (no loss in sweetness) versus H_a: $\mu > 0$ (there is a sweetness loss during storage). Enter the data in L_1/list1. Then go to STAT/TESTS (Tests menu on the TI-89). Choose 2: T-Test. Adjust your settings as shown.

TI-83/84 **TI-89**

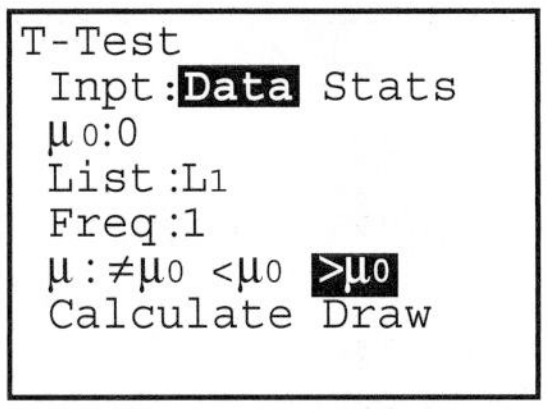

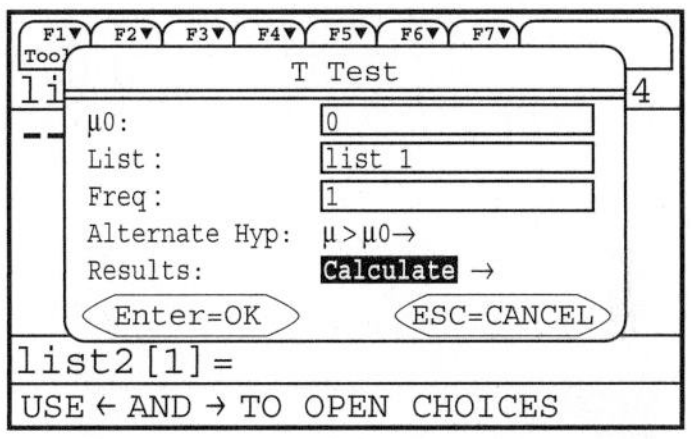

If you select "Calculate," the following screen appears:

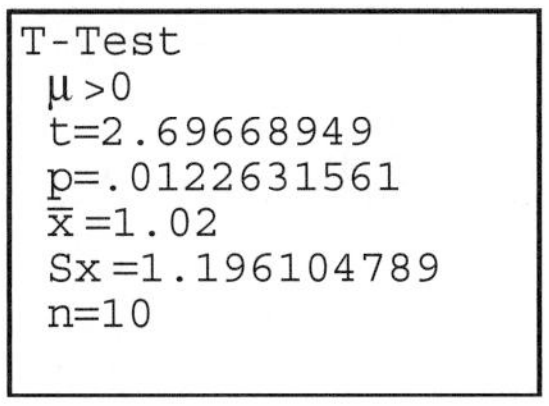

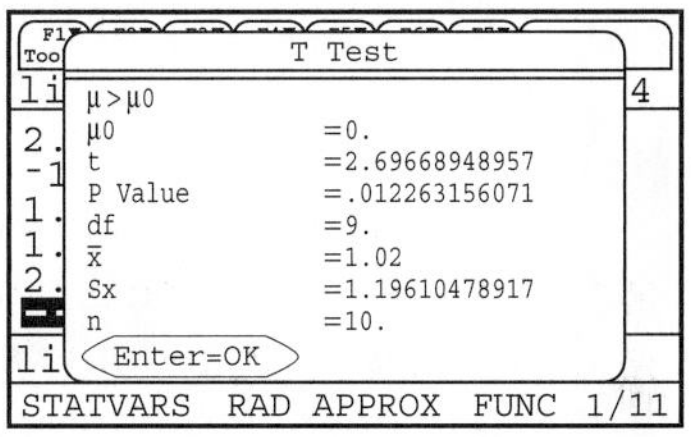

The test statistic is $t = 2.70$ and the P-value is 0.0123.

If you specify "Draw," you see a $t(9)$ distribution curve with the upper tail shaded. The P-value is 0.0123.

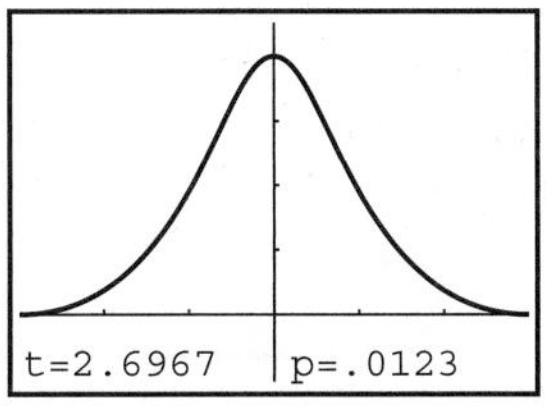

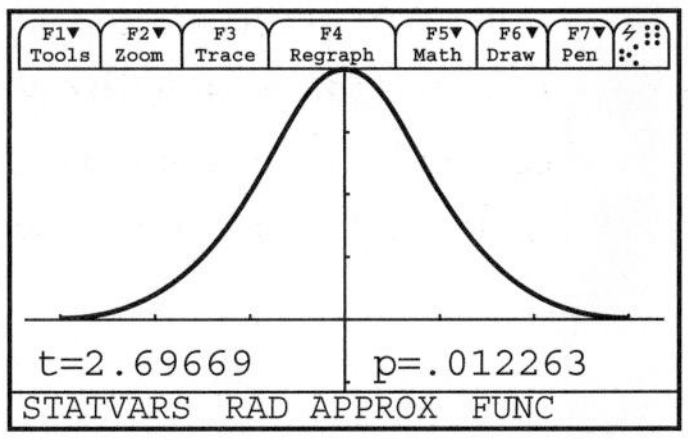

If you are given summary statistics instead of the original data, you would select the option "Stats" instead of "Data."

12.2 Tests about a Population Proportion

When the three important conditions are met—SRS, Normality, and independence—the sampling distribution of $\hat{p}$ is approximately Normal with mean $\mu_{\hat{p}} = p$ and standard deviation $\sigma_{\hat{p}} = \sqrt{p(1-p)/n}$. For confidence intervals, we substitute $\hat{p}$ for p in the last expression to obtain the standard error. When performing a significance test, however, the null hypothesis specifies a value for p, which we will call p_0. We assume that this value is correct when performing our calculations. If we standardize $\hat{p}$ by subtracting its mean p_0 and dividing by its standard deviation, we obtain the following z statistic:

$$z = \frac{\text{estimate} - \text{hypothesized parameter value}}{\text{standard deviation of estimate}} = \frac{\hat{p} - p_0}{\sqrt{\dfrac{p_0(1-p_0)}{n}}}$$

Here is a summary of the details for a ***one-proportion z test.***

The One-Proportion z Test

Choose an SRS of size n from a large population with unknown proportion p of successes. To test the hypothesis H_0: $p = p_0$, compute the **z statistic**

$$z = \frac{\hat{p} - p_0}{\sqrt{\dfrac{p_0(1-p_0)}{n}}}$$

In terms of a random variable Z having the standard Normal distribution, the approximate P-value for a test of H_0 against

H_a: $p > p_0$ is $P(Z \geq z)$

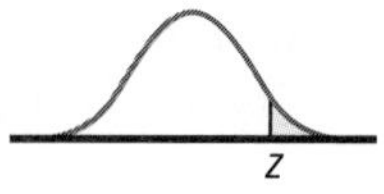

H_a: $p < p_0$ is $P(Z \leq z)$

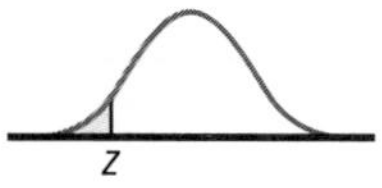

H_a: $p \neq p_0$ is $2P(Z \geq |z|)$

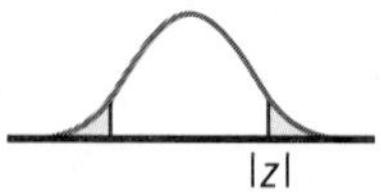

Normality condition: Use this test when the expected number of successes np_0 and failures $n(1 - p_0)$ are both at least 10.

large-sample test

Many people call this test a ***large-sample test*** because it is based on a Normal approximation to the sampling distribution of $\hat{p}$ that becomes more accurate as the sample size increases. If the Normality condition is not satisfied, or if the population is less than 10 times as large as the sample, other procedures should be used. Here is an example of the one-proportion z test.

Example 12.7 *Work stress*

Performing a one-proportion z test

According to the National Institute for Occupational Safety and Health, job stress poses a major threat to the health of workers. A national survey of restaurant employees found that 75% said that work stress had a negative impact on their personal lives.[6] A random sample of 100 employees from a large restaurant chain finds that 68 answer "Yes" when asked, "Does work stress have a negative impact on your personal life?" Is this good reason to think that the proportion of all employees in this chain who would say "Yes" differs from the national proportion $p_0 = 0.75$?

Step 1: **Hypotheses** We want to test a claim about p, the true proportion of this chain's employees who would say that work stress has a negative impact on their personal lives. Our hypotheses are

$$H_0: p = 0.75$$

$$H_a: p \neq 0.75$$

Step 2: **Conditions** We should use a one-proportion z test if the conditions are met.

- **SRS** We are told that a "random sample" of the restaurant chain's employees were surveyed. If the sampling design was not an SRS, our calculations may not be accurate.
- **Normality** The expected number of "Yes" and "No" responses are $(100)(0.75) = 75$ and $(100)(0.25) = 25$, respectively. Both are at least 10, so we can use the z test.
- **Independence** Since we are sampling without replacement, this "large chain" must have at least $(10)(100) = 1000$ employees.

Step 3: **Calculations**

- **Test statistic** The one-proportion z statistic is

$$z = \frac{\hat{p} - p_0}{\sqrt{\dfrac{p_0(1-p_0)}{n}}} = \frac{0.68 - 0.75}{\sqrt{\dfrac{(0.75)(0.25)}{100}}} = -1.62$$

- ***P*-value** Figure 12.8 (on the next page) displays the P-value as an area under the standard Normal curve for this two-tailed test. From Table A, we find $P(Z \leq -1.62) = 0.0526$. The P-value is the area in both tails, $P = 2 \times 0.0526 = 0.1052$.

Step 4: **Interpretation** There is over a 10% chance of obtaining a sample result as unusual as or even more unusual than we did ($\hat{p} = 0.68$) when the null hypothesis is true. We have insufficient evidence to suggest that the proportion of this chain restaurant's employees who suffer from work stress is different from the national survey result, 0.75.

Figure 12.8 *The P-value for the two-sided test of Example 12.7.*

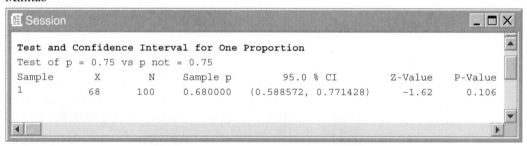

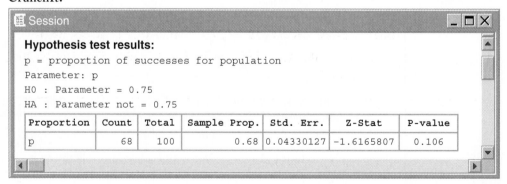

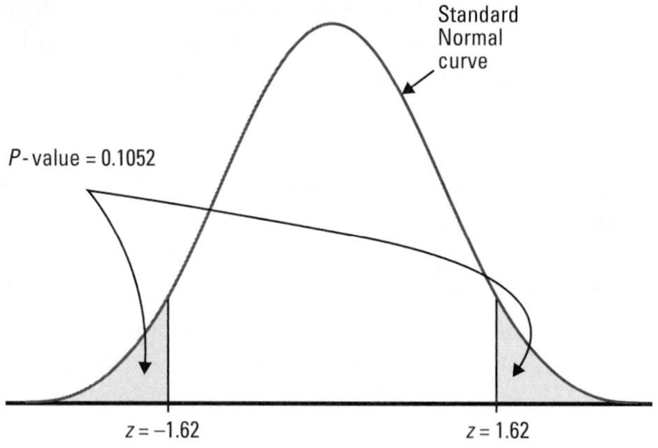

Figure 12.9 gives computer output from Minitab and CrunchIt! software for Example 12.7. *Note that for some entries software gives many more digits than we need.* You should decide how many digits are important for your analysis.

Figure 12.9 *Minitab and CrunchIt! output for Example 12.7.*

Minitab

Session

```
Test and Confidence Interval for One Proportion
Test of p = 0.75 vs p not = 0.75
Sample       X       N     Sample p        95.0 % CI              Z-Value    P-Value
1           68     100     0.680000    (0.588572, 0.771428)        -1.62      0.106
```

CrunchIt!

Session

Hypothesis test results:

```
p = proportion of successes for population
Parameter: p
H0 : Parameter = 0.75
HA : Parameter not = 0.75
```

Proportion	Count	Total	Sample Prop.	Std. Err.	Z-Stat	P-value
p	68	100	0.68	0.04330127	-1.6165807	0.106

In Example 12.7 we arbitrarily chose a "Yes" answer to the survey question as a "success" and a "No" answer as a "failure." Suppose we reversed the choice. If 68 respondents said "Yes," then the other $100 - 68 = 32$ people said "No." Let's repeat the significance test with "No" representing "success." The national comparison value for the significance test is now 0.25, the proportion in the national survey who responded "No."

Example 12.8 ***Work stress, again***
Definition of "success" doesn't matter

A random sample of 100 workers from a large chain restaurant were asked whether or not work stress had a negative impact on their personal lives. Thirty-two of them responded "No." A large national survey reported that 25% of restaurant workers did not feel that stress exerted a negative impact. We test the hypotheses

$$H_0: p = 0.25$$

$$H_a: p \neq 0.25$$

The test statistic is

$$z = \frac{\hat{p} - p_0}{\sqrt{\dfrac{p_0(1 - p_0)}{n}}} = \frac{0.32 - 0.25}{\sqrt{\dfrac{(0.25)(0.75)}{100}}} = 1.62$$

Using Table A, we find that the (two-tailed) P-value is 0.1052.

When we interchanged "Yes" and "No" (or success and failure), we simply changed the sign of the test statistic z. The P-value remained the same. These facts are true in general. Our conclusion does not depend on an arbitrary choice of success and failure.

The results of our significance test have limited use in this example, as in many cases of inference about a single parameter. Of course, we do not expect the experience of the restaurant workers to be *exactly* the same as that of the workers in the national survey. *If the sample of restaurant workers is sufficiently large, we will have sufficient power to detect a very small difference.* On the other hand, **if our sample size is very small, we may be unable to detect differences that could be very important.** For these reasons, we prefer to include a confidence interval as part of our analysis.

CAUTION

Confidence Intervals Provide Additional Information

To see what other values of p are compatible with the sample results, we will calculate a confidence interval.

Example 12.9 ***Estimating work stress***
Confidence intervals give more information

The restaurant worker survey in Example 12.7 found that 68 of a random sample of 100 employees agreed that work stress had a negative impact on their personal lives. We checked the three important conditions for performing the significance test earlier. Before we construct a confidence interval for the population proportion p, we should verify that both $n\hat{p}$ and $n(1 - \hat{p})$ are at least 10. Since the number of successes and failures in the sample are 68 and 32, respectively, we can proceed with the calculation.

Our 95% confidence interval is

$$\hat{p} \pm z^* \sqrt{\frac{\hat{p}(1-\hat{p})}{n}} = 0.68 \pm 1.96 \sqrt{\frac{(0.68)(0.32)}{100}}$$
$$= 0.68 \pm 0.09$$
$$= (0.59, 0.77)$$

We are 95% confident that between 59% and 77% of the restaurant chain's employees feel that work stress is damaging their personal lives.

The confidence interval of Example 12.9 is much more informative than the significance test of Example 12.7. We have determined the values of p that are consistent with the observed results. **Note that the standard error used for the confidence interval is estimated from the data, whereas the denominator for the test statistic z is based on the value assumed in the null hypothesis.** A consequence of this fact is that the correspondence between a two-tailed significance test and a confidence interval for a population proportion is no longer exact. However, the correspondence is still very close. The confidence interval (0.59, 0.77) gives an approximate range of p_0's that would not be rejected by a test at the $\alpha = 0.05$ significance level. We would not be surprised if the true proportion of restaurant workers who would say that work stress has a negative impact on their lives was as low as 60% or as high as 75%.

Section 12.2 Summary

Tests of H_0: $p = p_0$ are based on the **z statistic**

$$z = \frac{\hat{p} - p_0}{\sqrt{\frac{p_0(1-p_0)}{n}}}$$

with P-values calculated from the standard Normal distribution.

The **one-proportion z test** is approximately correct when the population is at least 10 times as large as the sample and the sample is large enough to satisfy $np_0 \geq 10$ and $n(1 - p_0) \geq 10$.

Confidence intervals provide additional information that significance tests do not—namely, a range of plausible values for the true population parameter p. For inference about a population proportion, confidence intervals and two-tailed tests do not have the perfect correspondence that we saw with inference about a population mean. But the relationship is similar.

Section 12.2 Exercises

12.23 Checking conditions In which of the following situations can you safely use the methods of this section for a significance test? Explain your answers.

(a) You toss a coin 10 times in order to test the hypothesis H_0: $p = 0.5$ that the coin is balanced.

(b) A college president says that "99% of the alumni support my firing of Coach Boggs." You contact an SRS of 200 of the college's 15,000 living alumni to test the hypothesis H_0: $p = 0.99$.

(c) Do a majority of the 250 students in a statistics course agree that knowing statistics will help them in their future careers? You interview an SRS of 20 students to test H_0: $p = 0.5$.

12.24 We want to be rich In a recent year, 73% of first-year college students responding to a national survey identified "being very well-off financially" as an important personal goal. A state university finds that 132 of an SRS of 200 of its first-year students say that this goal is important.

(a) Is there good evidence that the proportion of all first-year students at this university who think being very well-off is important differs from the national value, 73%? Carry out a significance test to help answer this question.

(b) Construct and interpret a 95% confidence interval for the proportion of all first-year students at the university who would identify being well-off as an important personal goal. What additional information does the confidence interval provide?

12.25 Hack-a-Shaq Any Miami Heat fan or archrival knows the team's very large "SHAQilles heel"—the free-throw shooting of the NBA's most dominant center, Shaquille O'Neal. Over his NBA career, Shaq has made 53.3% of his free throws.

Shaquille O'Neal worked in the off-season with an assistant coach on his free-throw technique. During the first two games of the next season, Shaq made 26 out of 39 free throws.

(a) Do these results provide evidence that Shaq has significantly improved his free-throw shooting? Justify your answer with appropriate evidence.

(b) Describe a Type I error and a Type II error in this situation.

(c) Suppose that Shaq has actually improved his free-throw shooting percent to 60%. What is the probability that you will correctly reject the claim that $p = 0.533$? Use a 5% significance level. (Use technology to help you.)

(d) Find the probability of a Type I error and a Type II error.

12.26 Side effects An experiment on the side effects of pain relievers assigned arthritis patients to one of several over-the-counter pain medications. Of the 440 patients who took one brand of pain reliever, 23 suffered some "adverse symptom."

(a) Does this experiment provide strong evidence that fewer than 10% of patients who take this medication have adverse symptoms? Perform an appropriate test.

(b) Describe a Type I error and a Type II error in this situation. Which is more serious?

12.27 Work stress and power Refer to Example 12.7 (page 767) concerning the work stress of a restaurant chain's employees.

(a) Describe a Type I error and a Type II error in this situation, and give potential consequences of each.

(b) Calculate the power of the test to detect the specific alternative $p = 0.80$ at the $\alpha = 0.005$ significance level. (Use technology!)

(c) Suppose you obtained the same results from a survey with twice as many respondents ($n = 200$). Repeat part (b). How does doubling the sample size affect the power of the test?

(d) How would your answer to part (b) change if you used a 1% significance level? A 10% significance level?

12.28 Activity 12 analysis

(a) Use the proportion of heads you obtained from your 20 coins in Activity 12 to test the null hypothesis H_0: $p = 0.5$ against the two-sided alternative. Report the *P*-value and state your conclusion.

(b) Use the proportion of heads for your entire class to test the same null and alternative hypotheses as in (a). Compare your results with those you obtained in (a).

12.29 Risky behavior The National AIDS Behavioral Surveys interviewed a sample of adults in the cities where AIDS is most common. This sample included 803 heterosexuals who reported having more than one sexual partner in the past year. We can consider this an SRS of size 803 from the population of all heterosexuals in high-risk cities who have multiple partners. These people risk infection with the AIDS virus. Yet 304 of the respondents said they never use condoms. Is this strong evidence that more than one-third of this population never use condoms? Give appropriate evidence to support your answer.

Technology Toolbox

One-proportion *z* test

The TI-83/84/89 can be used to test a claim about a population proportion. Let's revisit Example 12.7 (page 767) concerning work stress.

In a random sample of size $n = 100$, $X = 68$ people said "Yes" when asked whether work stress negatively impacted their personal lives. To perform a significance test:

TI-83/84

- Press STAT, then choose TESTS and 5:1-PropZTest

TI-89

- In the Statistics/List Editor, press 2nd F1 ([F6]) and choose 5:1-PropZTest.

- On the 1-PropZTest screen, enter the values $p_0 = 0.75$, $x = 68$, and $n = 100$. Specify the alternative hypothesis as "prop $\neq p_0$."

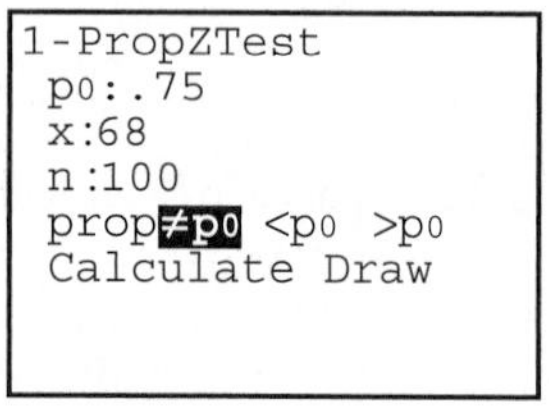

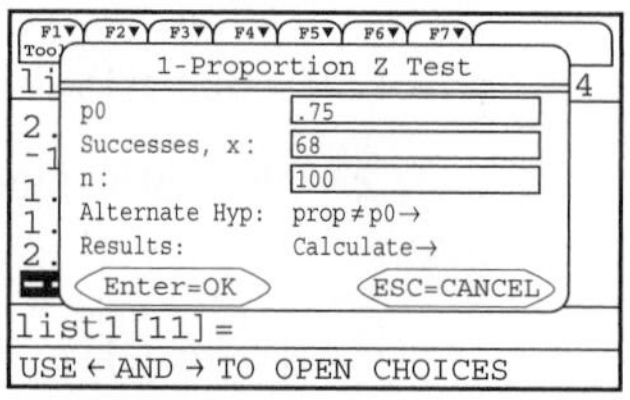

(continued)

Technology Toolbox

One-proportion *z* test *(continued)*

- If you select the "Calculate" choice and press ENTER, you will see that the z statistic is -1.62 and the P-value is 0.106.

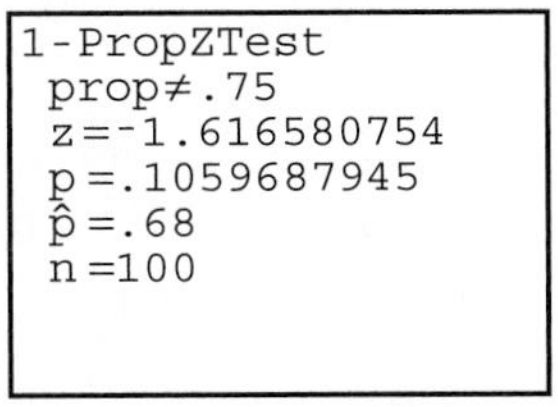

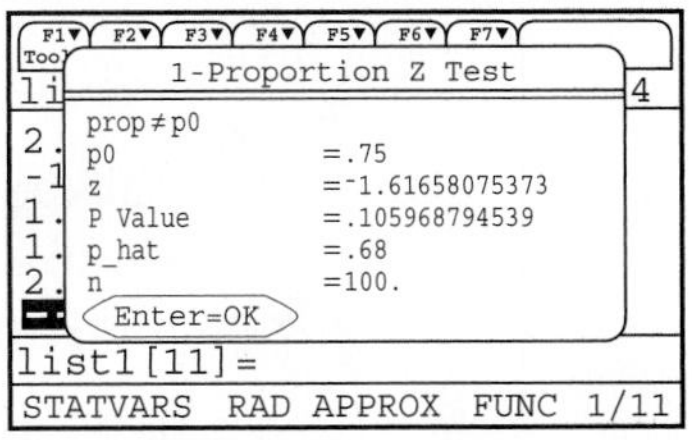

- If you select the "Draw" option, you will see the screen shown here. Compare these results with those in Example 12.7.

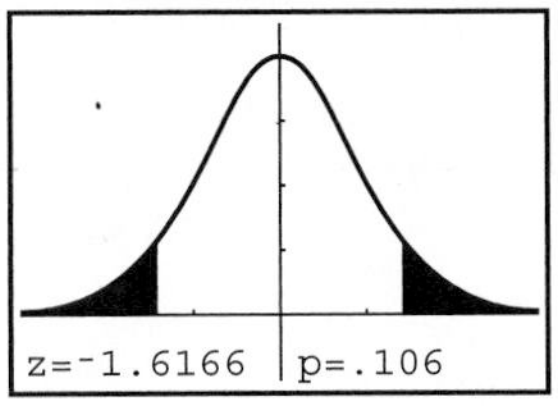

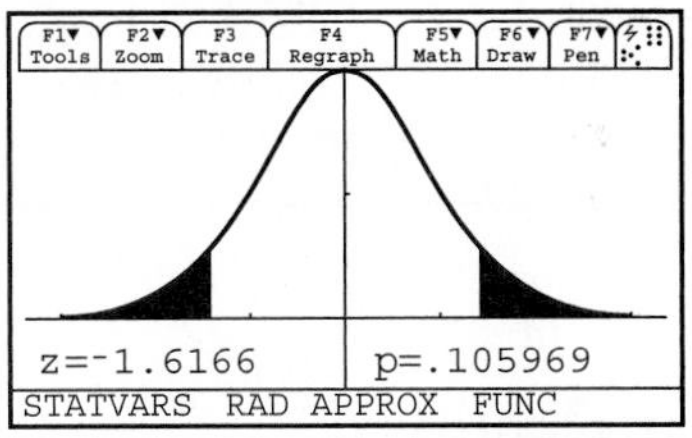

12.30 College food Tonya, Frank, and Sarah are investigating student attitudes toward college food for an assignment in their introductory statistics class. Based on comments overheard from other students, they believe that fewer than 1 in 3 students like college food. To test this hypothesis, each selects an SRS of students who regularly eat in the cafeteria and asks them if they like college food. Fourteen in Tonya's SRS of 50 replied "Yes," while 98 in Frank's sample of 350 and 140 in Sarah's sample of 500 said they like college food. Use your calculator to perform a test of significance on all three results and fill in a table like this:

X	n	$\hat{p}$	z	P-value
14	50			
98	350			
140	500			

Describe your findings in a short narrative.

CASE CLOSED!

Do you have a fever?

At the beginning of the chapter, we presented some details of a study concerning normal human body temperature. Return to the Case Study (page 741) to refresh yourself on the details.

(continued)

1. Use a significance test to determine whether we can conclude based on this study that "normal" body temperature in the population of healthy 18- to 40-year-olds differs from 98.6°F.
2. Construct and interpret a 95% confidence interval for the mean body temperature μ in the population of healthy 18- to 40-year-olds. What additional information does the confidence interval provide that the significance test in Question 1 did not?
3. If "normal" body temperature really is 98.6°F, we would expect that about half of all healthy 18- to 40-year-olds would have a body temperature less than 98.6°F. Carry out an appropriate test of this claim.
4. Use a 95% confidence interval to gain some additional information about the proportion described in Question 3.
5. Discuss any concerns you have about the design of this study that might affect your conclusions.

Chapter Review

Summary

Section 12.1 discussed the one-sample t test for a population mean when σ is unknown. We saw that paired data require a one-sample t test on the differences in each pair. In Section 12.2, we examined significance tests about a population proportion p. We noted that confidence intervals give more information than a mere reject or fail to reject significance test.

All of the methods of this chapter work well when the three important conditions—SRS, Normality, and independence—are satisfied. Be sure to verify these conditions before you proceed to calculations.

What You Should Have Learned

Here is a checklist of the major skills you should have acquired by studying this chapter.

A. One-Sample t Test for μ (σ unknown)

1. Carry out a t test for the hypothesis that a population mean μ has a specified value against either a one-sided or a two-sided alternative. Use Table C of t critical values to approximate the P-value or carry out a fixed α test.
2. Recognize when the t procedures are appropriate in practice, in particular that they are quite robust against lack of Normality but are influenced by skewness and outliers.

3. Also recognize when the design of the study, outliers, or a small sample from a skewed distribution make the t procedures risky.
4. Recognize paired data and use the t procedures to perform significance tests for such data.

B. Inference about One Proportion

1. Use the z statistic to carry out a test of significance for the hypothesis H_0: $p = p_0$ about a population proportion p against either a one-sided or a two-sided alternative.
2. Check that you can safely use the one-proportion z test in a particular setting.

Web Links

The *Journal of Statistics Education* has a wonderful data archive at www.amstat.org/publications/jse/jse_data_archive.html

Chapter Review Exercises

12.31 What's wrong? Explain what is wrong with each of the following:

(a) An approximate 95% confidence interval for an unknown proportion p is $\hat{p}$ plus or minus its standard error.

(b) You can use a significance test to evaluate the hypothesis H_0: $\hat{p} = 0.3$ versus the two-sided alternative.

(c) The large-sample significance test for a population proportion is based on a t statistic with $n - 1$ degrees of freedom.

12.32 Attitudes toward nuclear power A Gallup Poll on energy use asked 512 randomly selected adults if they favored "increasing the use of nuclear power as a major source of energy." Gallup reported that 225 said "Yes." Does this poll give good evidence that fewer than half of all adults favor increased use of nuclear power? Give appropriate statistical evidence to support your conclusion.

12.33 Store sales You will have complete sales information for last month in a week, but right now you have data from a random sample of 40 stores. The mean change in sales in the sample is +3.8% and the standard deviation of the changes is 12%.

(a) Are average sales for all stores different from last month? Carry out an appropriate test to help answer this question.

(b) If the test gives strong evidence against the null hypothesis, would you conclude that store sales are up in every one of your stores? Explain your answer.

12.34 The placebo effect The placebo effect is particularly strong in patients with Parkinson's disease. To understand the workings of the placebo effect, scientists made chemical measurements at a key point in the brain when patients received a placebo that they thought was an active drug and also when no treatment was given.[7] The same six patients were measured both with and without the placebo, at different times.

(a) Explain why the proper procedure to compare the mean response to placebo with control (no treatment) is a paired *t* test.

(b) The six differences (treatment minus control) had $\bar{x} = -0.326$ and $s = 0.181$. Is there significant evidence of a difference between treatment and control? Justify your answer.

12.35 Coffee preferences One-sample procedures for proportions, like those for means, are used to analyze data from matched pairs designs. Here is an example.

Each of 50 subjects tastes two unmarked cups of coffee and says which he or she prefers. One cup in each pair contains instant coffee; the other, fresh-brewed coffee. Thirty-one of the subjects prefer the fresh-brewed coffee. Take p to be the proportion of the population who would prefer fresh-brewed coffee in a blind tasting.

(a) Test the claim that a majority of people prefer the taste of fresh-brewed coffee. Is your result significant at the 5% level? What is your practical conclusion?

(b) Find a 90% confidence interval for p.

(c) When you do an experiment like this, in what order should you present the two cups of coffee to the subjects?

12.36 Hotel managers' personalities Successful hotel managers must have personality characteristics often thought of as feminine (such as "compassionate") as well as those often thought of as masculine (such as "forceful"). The Bem Sex-Role Inventory (BSRI) is a personality test that gives separate ratings for female and male stereotypes, both on a scale of 1 to 7.

Here are summary statistics for a sample of 148 general managers of three-star and four-star hotels.[8] The data come from a comprehensive mailing to these hotels. The response rate was 48%, which is good for mail surveys of this kind. Although nonresponse remains an issue, users of statistics usually act as if they have an SRS when the response rate is "good enough."

	Masculinity score	Femininity score
Mean	$\bar{x} = 5.91$	$\bar{x} = 5.29$
Standard deviation	$s = 0.57$	$s = 0.75$

The mean BSRI scores for the general male population are $\mu = 4.88$ for masculinity and $\mu = 5.19$ for femininity. (It does seem odd that the femininity score is higher, but such is the world of personality tests. The two scales are separate.) Is there evidence that hotel managers on the average score higher in masculinity than the general male population? Is there evidence that they also score higher in femininity? Show your work.

12.37 Sharks Great white sharks are big and hungry. Here are the lengths in feet of 44 great whites:[9]

18.7	12.3	18.6	16.4	15.7	18.3	14.6	15.8	14.9	17.6	12.1
16.4	16.7	17.8	16.2	12.6	17.8	13.8	12.2	15.2	14.7	12.4
13.2	15.8	14.3	16.6	9.4	18.2	13.2	13.6	15.3	16.1	13.5
19.1	16.2	22.8	16.8	13.6	13.2	15.7	19.7	18.7	13.2	16.8

(a) Examine these data for shape, center, spread, and outliers. The distribution is reasonably Normal except for one outlier in each direction. Because these are not extreme and preserve the symmetry of the distribution, use of the t procedures is safe with 44 observations.

(b) Construct and interpret a 95% confidence interval for the mean length of great white sharks. Based on this interval, is there significant evidence at the 5% level to reject the claim "Great white sharks average 20 feet in length"? Justify your answer.

(c) It isn't clear exactly what parameter μ you estimated in (b). What information do you need to say what μ is?

12.38 Is the Belgian euro coin "fair"? Two Polish math professors and their students spun a Belgian euro coin 250 times. It landed "heads" 140 times. One of the professors concluded that the coin was minted asymmetrically. A representative from the Belgian mint said the result was just chance.[10] Do your own analysis. Write a few sentences explaining your conclusion.

CHAPTER 13

Comparing Two Population Parameters

CASE STUDY

Fast-food frenzy!

Over $70 billion is spent each year in the drive-thru lanes of America's fast-food restaurants. Having fast, accurate, and friendly service at a drive-thru window translates directly into revenue for the restaurant. According to Jack Greenberg, the CEO of McDonald's, sales increase 1% for every six seconds saved at the drive-thru. So industry executives, stockholders, and analysts closely follow the ratings of fast-food drive-thru lanes that appear annually in *QSR,* a publication that reports on the quick-service restaurant industry.

The 2004 *QSR* magazine drive-thru study involved a total of 11,073 visits to restaurants in the 25 largest fast-food chains in all 50 states. Visits occurred during the lunch hours of 11:00 A.M. to 2:30 P.M. or during the dinner hours of 4:00 P.M. to 7:00 P.M. Four aspects of drive-thru operations were measured at each visit: speed, accuracy, menuboard appearance, and speaker clarity. The speed and accuracy measurements are particularly important since together they make up 80% of the final overall rating.

During each visit, the researcher ordered a modified main item (for example, a hamburger with no pickles), a side item, and a drink. If any item was not received as ordered, or if the restaurant failed to give the correct change or supply a straw and a napkin, then the order was considered "inaccurate." Service time, which is the time from when the car stopped at the speaker to when the entire order was received, was measured each visit.[1]

What is a typical service time at a fast-food drive-thru? How accurately are drive-thru orders filled? Which fast-food restaurant has the lowest average service time? The highest proportion of correctly filled orders? How can fast-food restaurant owners improve drive-thru service time and accuracy? By the end of the chapter, you should have acquired the tools to help answer questions like these.

Activity 13
Paper airplane experiment

Materials: Two paper airplane pattern sheets (in the Teacher Resource Binder), scissors, masking tape, 25-meter steel tape measure, graphing calculator

The Experiment The purpose of this Activity is to see which of two prototype paper airplane models flies farther. Specifically, the object is to determine the average distance flown for each prototype plane and to compare these average distances flown. The null hypothesis will be that there is no difference between the average distance flown by Prototype A and the average distance flown by Prototype B. So H_0: $\mu_A = \mu_B$. Equivalently, we could write H_0: $\mu_A - \mu_B = 0$. What form should the alternative hypothesis take? (Remember that the alternative hypothesis should be stated before you conduct the experiment.)

The Task Your task, as a class, is to design an experiment to determine which of the two prototype paper airplanes flies the farthest. Then you will carry out your plan and gather the necessary data. You should explore the data both numerically and graphically prior to conducting formal inference. As is typical in real-world settings, we don't know the population standard deviation σ for either prototype, so we can't use z procedures to conduct a significance test.

We will develop methods in this chapter that will enable you to calculate a test statistic to help answer the question about which prototype paper airplane flies the farthest. We will also calculate confidence intervals for the true difference in mean flight distances. Keep your data handy so that you can perform this analysis later.

Introduction

Which of two popular drugs—Lipitor or Pravachol—helps lower "bad cholesterol" more? Researchers designed an experiment, called the PROVE-IT Study, to find out. They used about 4000 people with heart disease as subjects. These individuals were randomly assigned to one of two treatment groups: Lipitor or Pravachol. At the end of the study, researchers compared the mean "bad cholesterol levels" for the two groups. For the Pravachol subjects, the mean was 95 milligrams per deciliter of blood; for the Lipitor subjects, it was 62 milligrams per deciliter.[2] Is this difference statistically significant? This is a question about *comparing two means.*

The researchers also compared the proportion of subjects in each group who died, had a heart attack, or suffered other serious consequences within two years. For those using Pravachol, the proportion was 0.263; for those using Lipitor, it was 0.224. Could such a difference have occurred purely by chance? This is a question about *comparing two proportions.*

How do small businesses that fail differ from those that succeed? Business school researchers compare the asset to liability ratios of two samples of firms started in 2000, one sample of failed businesses and one of firms that are still going after two years. This observational study compares two random samples, one from each of two different populations.

Comparing two populations or two treatments is one of the most common situations encountered in statistical practice. We call such situations ***two-sample problems.***

Two-Sample Problems

- The goal of inference is to compare the responses to two treatments or to compare the characteristics of two populations.
- We have a separate sample from each treatment or each population.
- The responses of each group are independent of those in the other group.

A two-sample problem can arise from a randomized comparative experiment that randomly divides subjects into two groups and exposes each group to a different treatment, like the PROVE-IT Study. Comparing random samples separately selected from two populations, like the successful and failed small businesses, is also a two-sample problem. Unlike the matched pairs designs studied earlier, there is no matching of the units in the two samples and the two samples can be of different sizes. Inference procedures for two-sample data differ from those for matched pairs. Here are some typical two-sample problems.

Example 13.1 ***Comparing means and proportions***
Two-sample problems

(a) Who is more likely to binge drink: male or female college students? The Harvard School of Public Health surveys random samples of male and female undergraduates at four-year colleges and universities about their drinking behaviors. This is an observational study designed to compare the proportion of all undergraduate males who binge drink and the proportion of all undergraduate females who binge drink.
(b) A bank wants to know which of two incentive plans will most increase the use of its credit cards. It offers each incentive to a random sample of credit card customers and compares the amount charged during the following six months. This is a randomized experiment designed to compare the mean amount spent under each of the two incentive "treatments."

13.1 Comparing Two Means

We can examine two-sample data graphically by comparing dotplots or stemplots (for small samples) or histograms or boxplots (for larger samples). Now we will apply the ideas of formal inference in this setting. When both population distributions are symmetric, and especially when they are at least approximately Normal, a comparison of the mean responses in the two populations is the most common goal of inference. Here are the conditions that must be satisfied.

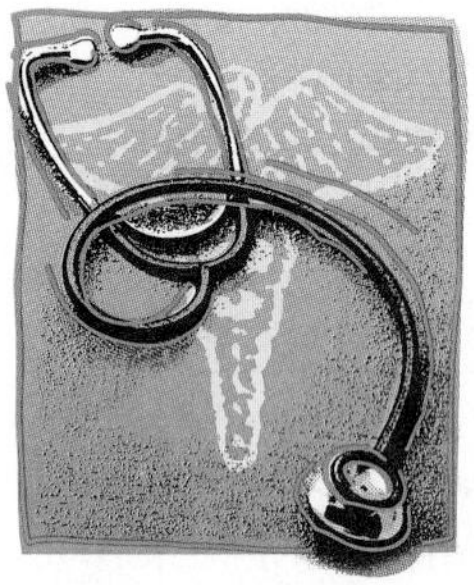

Sounds good—but no comparison Most women have mammograms to check for breast cancer once they reach middle age. Could a fancier test do a better job of finding cancers early? PET scans are a fancier (and more expensive) test. Doctors used PET scans on 14 women with tumors and got the detailed diagnosis right in 12 cases. That's promising. But there were no controls, and 14 cases are not statistically significant. Medical standards require randomized comparative experiments and statistically significant results. Only then can we be confident that the fancy test really is better.

Conditions for Comparing Two Means

- **SRS** We have two SRSs, from two distinct populations. This allows us to generalize our findings. We measure the same variable for both samples.
- **Normality** Both populations are Normally distributed. In practice, it is enough that the distributions have similar shapes and that the data have no strong outliers. The means and standard deviations of the populations are unknown.
- **Independence** The samples are independent. That is, one sample has no influence on the other. Paired observations violate independence, for example. When sampling without replacement from two distinct populations, each population must be at least 10 times as large as the corresponding sample size (in symbols, $N \geq 10n$).

Here is the notation we will use to describe the two populations and samples, since the variable may have different distributions in the two populations:

		Parameters			Statistics	
Population	**Variable**	**Mean**	**Standard deviation**	**Sample size**	**Mean**	**Standard deviation**
1	x_1	μ_1	σ_1	n_1	$\bar{x}_1$	s_1
2	x_2	μ_2	σ_2	n_2	$\bar{x}_2$	s_2

There are four unknown parameters, the two means and the two standard deviations. We want to compare the two population means, either by giving a confidence interval for their difference $\mu_1 - \mu_2$ or by testing the hypothesis of no difference, H_0: $\mu_1 = \mu_2$.

We use the sample means and standard deviations to estimate the unknown parameters. To do inference about the difference $\mu_1 - \mu_2$ between the means of the two populations, we start from the difference $\bar{x}_1 - \bar{x}_2$ between the means of the two samples.

Example 13.2 *Calcium and blood pressure*

Comparing two means

Does increasing the amount of calcium in our diet reduce blood pressure? Examination of a large sample of people revealed a relationship between calcium intake and blood pressure. The relationship was strongest for black men. Such observational studies do not establish causation. Researchers therefore designed a randomized comparative experiment.

The subjects in part of the experiment were 21 healthy black men. A randomly chosen group of 10 of the men received a calcium supplement for 12 weeks. The control group of 11 men received a placebo pill that looked identical. The experiment was double-blind. The response variable is the decrease in systolic (top number) blood pressure for a subject after 12 weeks, in millimeters of mercury. An increase appears as a negative response.[3]

Take Group 1 to be the calcium group and Group 2 the placebo group. Here are the data for the 10 men in Group 1 (calcium),

7	−4	18	17	−3	−5	1	10	11	−2

and for the 11 men in Group 2 (placebo),

−1	12	−1	−3	3	−5	5	2	−11	−1	−3

From the data, calculate the summary statistics:

Group	Treatment	n	$\bar{x}$	s
1	Calcium	10	5.000	8.743
2	Placebo	11	−0.273	5.901

The calcium group shows a drop in blood pressure, $\bar{x}_1 = 5.000$, while the placebo group shows a small increase, $\bar{x}_2 = -0.273$. Is this outcome good evidence that calcium decreases blood pressure in the entire population of healthy black men more than a placebo does?

This example fits the two-sample setting. Since we want to test a claim, we will perform a significance test. Our Inference Toolbox provides the procedure.

Step 1: **Hypotheses** We write hypotheses in terms of the mean decreases we would see in the entire population μ_1 for men taking calcium for 12 weeks and μ_2 for men taking a placebo. The hypotheses are

$$H_0: \mu_1 = \mu_2 \qquad \text{or, equivalently,} \qquad H_0: \mu_1 - \mu_2 = 0$$
$$H_a: \mu_1 > \mu_2 \qquad\qquad\qquad\qquad H_a: \mu_1 - \mu_2 > 0$$

Step 2: **Conditions** We do not yet know what procedure to use, but let's check the required conditions.

- **SRS** The 21 subjects in this experiment were not obtained by random selection from a larger population. As a result, it may be difficult to generalize our findings to all healthy black men. However, the random assignment of subjects to treatments should help ensure that any significant difference in mean blood pressure between the two groups is due to the treatments.

- **Normality** Although the samples are small, we check for serious non-Normality by examining the data. Inspection of the data reveals that there are no outliers (you should confirm this). The Normal probability plots in Figure 13.1 give a more detailed picture. The placebo responses appear roughly Normal. The calcium group has an irregular distribution, which is not unusual when we have only a few observations. There are no outliers and no departures from Normality that prevent use of t procedures.

Figure 13.1 *Normal probability plots of the decrease in blood pressure for subjects in the calcium and placebo groups, for Example 13.2.*

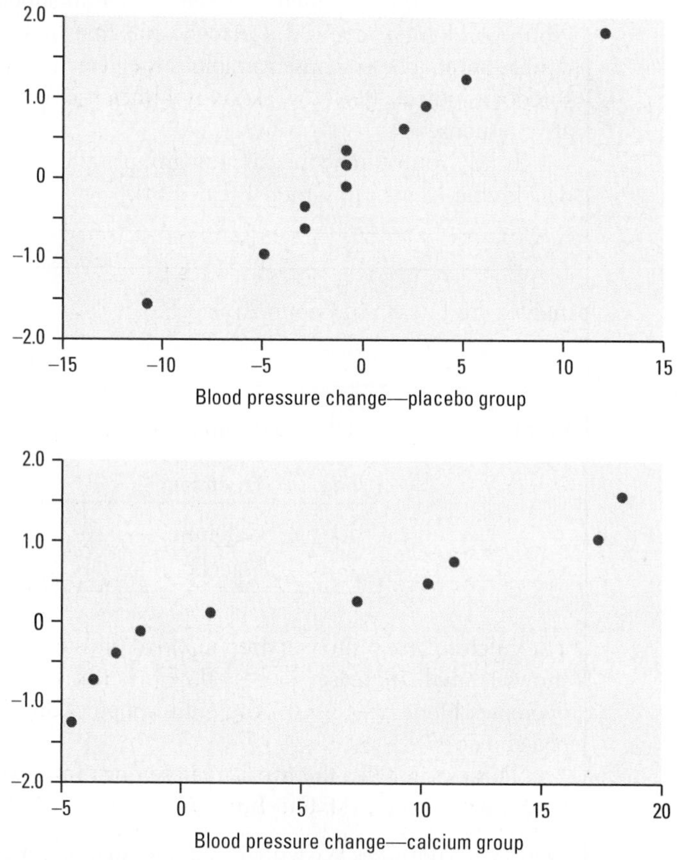

- **Independence** Because of the randomization, we are willing to regard the calcium and placebo groups as two independent samples. We are not sampling without replacement from a population of interest in this case.

We will continue with Steps 3 and 4 in Example 13.3.

The natural estimator of the difference $\mu_1 - \mu_2$ is the difference between the sample means:

$$\bar{x}_1 - \bar{x}_2 = 5.000 - (-0.273) = 5.273$$

This statistic measures the average advantage of calcium over a placebo. In order to use it for inference, we must know its sampling distribution. We will discuss the sampling distribution of $\bar{x}_1 - \bar{x}_2$ shortly.

Exercises

13.1 Toyota or Nissan? Are Toyota or Nissan owners more satisfied with their vehicles? Let's design a study to find out. We'll select a random sample of 400 Toyota owners and a separate random sample of 400 Nissan owners. Then we'll ask each individual in the sample: "Would you say that you are generally satisfied with your (Toyota/Nissan) vehicle?"

(a) Is this a problem about comparing means or comparing proportions? Explain.

(b) What type of study design is being used to produce data?

13.2 Computer gaming Do experienced computer game players earn higher scores when they play with someone present to cheer them on or when they play alone? Fifty teenagers who are experienced at playing a particular computer game have volunteered for a study. We randomly assign 25 of them to play the game alone and the other 25 to play the game with a supporter present. Each player's score is recorded.

(a) Is this a problem about comparing means or comparing proportions? Explain.

(b) What type of study design is being used to produce data?

13.3 Effective textbooks The following situations require inference about a mean or means. Identify each as involving a single sample, paired data, or two independent samples. Justify your choice.

(a) An education researcher wants to learn whether it is more effective to put questions before or after introducing a new concept in an elementary school mathematics text. He prepares two text segments that teach the concept, one with motivating questions before and the other with review questions after. He uses each text segment to teach a separate group of children. The researcher compares the scores of the groups on a test on the material.

(b) Another researcher approaches the same issue differently. She prepares text segments on two unrelated topics. Each segment comes in two versions, one with questions before and the other with questions after. The subjects are a single group of children. Each child studies both topics, one (chosen at random) with questions before and the other with questions after. The researcher compares test scores for each child on the two topics to see which topic he or she learned better.

13.4 Testing chemicals The following situations require inference about a mean or means. Identify each as involving a single sample, paired data, or two independent samples. Justify your choice.

(a) To check a new method of measuring the concentration of a chemical solution, a chemist obtains a solution of known concentration from the National Institute of Standards and Technology. She pours the solution in equal amounts into 20 beakers and then measures the concentration of the solution in each beaker with the new method. She checks for bias by comparing the mean result of her 20 measurements with the known concentration of the solution.

(b) Another chemist is checking the same new method of measuring concentration. He has no reference solution, but a familiar technique for measuring concentration is available. He wants to know if the new and old methods agree. He takes a solution of unknown concentration and pours it in equal amounts into 20 beakers. He measures the concentration in 10 of the beakers with the new method and in the other 10 with the old method.

13.5 Teaching reading, I An educator believes that new reading activities in the classroom will help elementary school pupils improve their reading ability. She arranges for a third-grade class of 21 students to follow these activities for an 8-week period. A control classroom of 23 third-graders follows the same curriculum without the activities. At the end of the 8 weeks, all students are given the Degree of Reading Power (DRP) test, which measures the aspects of reading ability that the treatment is designed to improve. Here are the data:[4]

Treatment					Control				
24	43	58	71	43	42	43	55	26	62
49	61	44	67	49	37	33	41	19	54
53	56	59	52	62	20	85	46	10	17
54	57	33	46	43	60	53	42	37	42
57					55	28	48		

(a) State appropriate hypotheses for testing the educator's belief. Be sure to define your parameter(s).

(b) Examine the data graphically and numerically. Write a few sentences comparing the DRP scores of the two groups of students.

(c) Although this study is an experiment, its design is not ideal because it had to be done in a school without disrupting classes. What aspect of good experimental design is missing?

13.6 Does breast-feeding weaken bones? Breast-feeding mothers secrete calcium into their milk. Some of the calcium may come from their bones, so mothers may lose bone mineral. Researchers compared 47 breast-feeding women with 22 women of similar age who were neither pregnant nor lactating. They measured the percent change in the mineral content of the women's spines over three months. Here are the data:[5]

Breast-feeding women						Other women					
−4.7	−2.5	−4.9	−2.7	−0.8	−5.3	2.4	0.0	0.9	−0.2	1.0	1.7
−8.3	−2.1	−6.8	−4.3	2.2	−7.8	2.9	−0.6	1.1	−0.1	−0.4	0.3
−3.1	−1.0	−6.5	−1.8	−5.2	−5.7	1.2	−1.6	−0.1	−1.5	0.7	−0.4
−7.0	−2.2	−6.5	−1.0	−3.0	−3.6	2.2	−0.4	−2.2	−0.1		
−5.2	−2.0	−2.1	−5.6	−4.4	−3.3						
−4.0	−4.9	−4.7	−3.8	−5.9	−2.5						
−0.3	−6.2	−6.8	1.7	0.3	−2.3						
0.4	−5.3	0.2	−2.2	−5.1							

(a) What two populations and parameters did the investigators want to compare? State appropriate hypotheses.

(b) Examine the data graphically and numerically. Write a few sentences comparing the change in bone mineral content for the two groups.

(c) Can we conclude that breast-feeding causes a mother's bones to weaken? Why or why not?

The Two-Sample *z* Statistic

Here are the facts about the sampling distribution of the difference $\bar{x}_1 - \bar{x}_2$ between the two sample means of independent SRSs. These facts can be derived using the mathematics of probability or demonstrated by simulation.

- The mean of $\bar{x}_1 - \bar{x}_2$ is $\mu_1 - \mu_2$. That is, the difference of sample means is an unbiased estimator of the difference of population means.
- The variance of the difference is the sum of the variances of $\bar{x}_1 - \bar{x}_2$, which is

$$\frac{\sigma_1^2}{n_1} + \frac{\sigma_2^2}{n_2}$$

  Note that the variances add because the samples are *independent*. The standard deviations do not.
- If the two population distributions are both Normal, then the distribution of $\bar{x}_1 - \bar{x}_2$ is also Normal.

When the statistic $\bar{x}_1 - \bar{x}_2$ has a Normal distribution, we can standardize it to obtain a standard Normal z statistic. Here is that new ***two-sample z statistic.***

Two-Sample z Statistic

Suppose that $\bar{x}_1$ is the mean of an SRS of size n_1 drawn from a Normally distributed population with mean μ_1 and standard deviation σ_1, and that $\bar{x}_2$ is the mean of an SRS of size n_2 drawn from a Normally distributed population with mean μ_2 and standard deviation σ_2. Then the **two-sample z statistic**

$$z = \frac{\bar{x}_1 - \bar{x}_2 - (\mu_1 - \mu_2)}{\sqrt{\dfrac{\sigma_1^2}{n_1} + \dfrac{\sigma_2^2}{n_2}}}$$

has the standard Normal distribution.

In the very unlikely event that both population standard deviations are known, the two-sample z statistic is the basis for inference about $\mu_1 - \mu_2$. Since this is rarely the case, we move immediately to the more useful t procedures.

The Two-Sample *t* Procedures

Because we don't know the population standard deviations, we estimate them by the standard deviations from our two samples. The result is the standard error, or estimated standard deviation, of the difference in sample means:

$$\text{SE} = \sqrt{\frac{s_1^2}{n_1} + \frac{s_2^2}{n_2}}$$

When we standardize our estimate $(\bar{x}_1 - \bar{x}_2)$, the result is the ***two-sample t statistic:***

$$t = \frac{(\bar{x}_1 - \bar{x}_2) - (\mu_1 - \mu_2)}{\sqrt{\frac{s_1^2}{n_1} + \frac{s_2^2}{n_2}}}$$

The statistic t has the same interpretation as any z or t statistic: it says how far $\bar{x}_1 - \bar{x}_2$ is from its mean in standard deviation units.

The two-sample t statistic has approximately a t distribution. It does not have exactly a t distribution even if the populations are both exactly Normal. In practice, however, the approximation is very accurate. There is a catch: we must calculate the degrees of freedom of the distribution we want to use from the data by a somewhat messy formula. Moreover, the degrees of freedom need not be a whole number. There are two practical options for using the two-sample t procedures:

Option 1: With technology, use the statistic t with accurate critical values from the approximating t distribution.

Option 2: Without technology, use the statistic t with critical values from the t distribution with degrees of freedom equal to the smaller of $n_1 - 1$ and $n_2 - 1$. These procedures are always conservative for any two Normal populations.

Most statistical software systems and graphing calculators use the two-sample t statistic with Option 1 for two-sample problems unless the user requests another method. Using this option without technology is a bit complicated. We will therefore present the second, simpler, option first. Here is a description of the Option 2 procedures that includes a statement of just how they are "conservative."

The Two-Sample t Procedures

Draw an SRS of size n_1 from a Normal population with unknown mean μ_1, and draw an independent SRS of size n_2 from another Normal population with unknown mean μ_2. The **level C confidence interval** for $\mu_1 - \mu_2$ given by

$$(\bar{x}_1 - \bar{x}_2) \pm t^* \sqrt{\frac{s_1^2}{n_1} + \frac{s_2^2}{n_2}}$$

has confidence level *at least* C no matter what the population standard deviations may be. Here t^* is the critical value for the $t(k)$ distribution with k equal to the smaller of $n_1 - 1$ and $n_2 - 1$.

To test the hypothesis H_0: $\mu_1 = \mu_2$, compute the **two-sample *t* statistic**

$$t = \frac{\bar{x}_1 - \bar{x}_2}{\sqrt{\frac{s_1^2}{n_1} + \frac{s_2^2}{n_2}}}$$

and use a *P*-value or critical values for the $t(k)$ distribution. The true *P*-value or fixed significance level will always be *equal to or less than* the value calculated from $t(k)$ no matter what values the unknown population standard deviations have.

These two-sample t procedures always err on the safe side, reporting *higher* P-values and *lower* confidence than may actually be true. The gap between what is reported and the truth is quite small unless the sample sizes are both small and unequal. As the sample sizes increase, probability values based on t with degrees of freedom equal to the smaller of $n_1 - 1$ and $n_2 - 1$ become more accurate.[6] The following examples illustrate the two-sample t procedures.

Example 13.3 *Calcium and blood pressure, continued*

Two-sample t test

TOOLBOX

The medical researchers in Example 13.2 can use the two-sample t procedures to compare calcium with a placebo.

Step 3: **Calculations**

- **Test statistic** The two-sample t statistic is

$$t = \frac{\bar{x}_1 - \bar{x}_2}{\sqrt{\frac{s_1^2}{n_1} + \frac{s_2^2}{n_2}}} = \frac{5.000 - (-0.0273)}{\sqrt{\frac{8.743^2}{10} + \frac{5.901^2}{11}}} = \frac{5.273}{3.2878} = 1.604$$

- **P-value** There are 9 degrees of freedom, the smaller of $n_1 - 1 = 9$ and $n_2 - 1 = 10$. Because H_a counts only positive values of t as evidence against H_0, the P-value is the area to the right of $t = 1.604$ under the $t(9)$ curve. Figure 13.2 illustrates this P-value. Table C shows that it lies between 0.05 and 0.10.

Upper tail probability p			
df	.10	.05	.025
8	1.397	1.860	2.306
9	1.383	1.833	2.262
10	1.372	1.812	2.228

Figure 13.2 *The P-value for Example 13.3 using the conservative method, which leads to the t distribution with 9 degrees of freedom.*

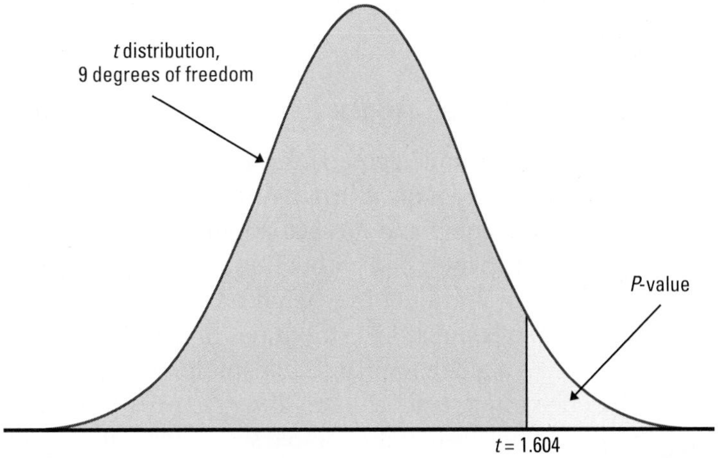

Step 4: **Interpretation** The experiment provides some evidence that calcium reduces blood pressure, but the evidence falls short of the traditional 5% and 1% levels. We would fail to reject H_0 at either of these significance levels.

We can estimate the difference in the mean decreases in blood pressure for the hypothetical calcium and placebo populations using a two-sample t interval. Example 13.4 shows how.

Example 13.4 *Calcium and blood pressure, again*

Two-sample t confidence interval

Meta-analysis Small samples have large margins of error. Large samples are expensive. Often we can find several studies of the same issue; if we could combine their results, we would have a large sample with a small margin of error. That is the idea of "meta-analysis." Of course, we can't just lump the studies together, because of differences in design and quality. Statisticians have more sophisticated ways of combining the results. Meta-analysis has been applied to issues ranging from the effect of secondhand smoke to whether coaching improves SAT scores.

For a 90% confidence interval, Table C shows that the $t(9)$ critical value is $t^* = 1.833$. We are 90% confident that the mean advantage of calcium over a placebo, $\mu_1 = \mu_2$, lies in the interval

$$(\bar{x}_1 - \bar{x}_2) \pm t^* \sqrt{\frac{s_1^2}{n_1} + \frac{s_2^2}{n_2}} = [5.000 - (-0.273)] \pm 1.833 \sqrt{\frac{8.743^2}{10} + \frac{5.901^2}{11}}$$
$$= 5.273 \pm 6.026$$
$$= (-0.753, 11.299)$$

Since the 90% confidence interval includes 0, we cannot reject H_0: $\mu_1 - \mu_2 = 0$ against the two-sided alternative at the $\alpha = 0.10$ level of significance.

Sample size strongly influences the P-value of a test. A result that fails to be significant at a specified level α in a small sample may be significant in a larger sample. With the rather small group sizes in Example 13.3, the difference in mean systolic blood pressures between the two groups, 5 points, was not significant. We suspect that larger groups might show a similar difference in mean blood pressure, which would indicate that calcium has a significant effect. (In fact, subsequent analysis of data from an experiment with more subjects resulted in a P-value of 0.008.)

Robustness Again

The two-sample t procedures are more robust than the one-sample t methods, particularly when the distributions are not symmetric. When the sizes of the two samples are equal and the two populations being compared have distributions with similar shapes, probability values from the t table are quite accurate for a broad range of distributions when the sample sizes are as small as $n_1 = n_2 = 5$.[7] When the two population distributions have different shapes, larger samples are needed.

CAUTION

As a guide to practice, adapt the guidelines given on page 655 for the use of one-sample t procedures to two-sample procedures by replacing "sample size" with the "sum of the sample sizes," $n_1 + n_2$ as long as n_1 and n_2 are both at least 5. These guidelines err on the side of safety, especially when the two samples are of equal size. **In planning a two-sample study, choose equal sample sizes if you can.** The two-sample t procedures are most robust against non-Normality in this case, and the conservative P-values (using Option 2) are most accurate.

Exercises

13.7 Bad inference In each of the following situations, explain what is wrong and why.

(a) A researcher wants to test $H_0: \bar{x}_1 = \bar{x}_2$ versus the two-sided alternative $H_a: \bar{x}_1 \neq \bar{x}_2$.

(b) A study recorded the scores of 20 children who were similar in age. The scores of the 10 boys in the study were compared with the scores of all 20 children using the two-sample methods of this section.

(c) A two-sample *t* statistic gave a *P*-value of 0.96. From this you can reject the null hypothesis with 95% confidence.

13.8 Web business You want to compare the daily sales for two different designs of Web pages for your Internet business. You assign the next 60 days to either Design A or Design B, 30 days to each.

(a) Describe how you would assign the days for Design A and Design B using the partial line of random digits provided below. Then use your plan to select the first three days for using Design A. Show your method clearly on your paper.

24005	52114	26224	39078

(b) Would you use a one-sided or two-sided significance test for this problem? Explain your choice. Then set up appropriate hypotheses.

(c) If you plan to use Table C to calculate the *P*-value, what are the degrees of freedom?

(d) The *t* statistic for comparing the mean sales is 2.06. Using Table C, what *P*-value would you report? What would you conclude?

13.9 Teaching reading, II Refer to Exercise 13.5 (page 786).

(a) Is there good evidence that the new activities improve the mean DRP score? Carry out a test and report your conclusions.

(b) Construct and interpret a 95% confidence interval for the difference in mean DRP scores.

13.10 Bone mineral loss by nursing mothers Refer to Exercise 13.6 (page 786).

(a) Do these data give good evidence that on the average nursing mothers lose more bone mineral? Carry out a test and report your conclusions.

(b) Construct and interpret a 95% confidence interval for the difference in mean bone mineral loss.

13.11 Paying for college College financial aid offices expect students to use summer earnings to help pay for college. But how large are these earnings? One college studied this question by asking a sample of students how much they earned. Omitting students who were not employed, there were 1296 responses. Here are the data in summary form:[8]

Group	n	$\bar{x}$	s
Males	675	\$1884.52	\$1368.37
Females	621	\$1360.39	\$1037.46

(a) The distribution of earnings is strongly skewed to the right. Nevertheless, use of t procedures is justified. Why?

(b) Give a 90% confidence interval for the difference between the mean summer earnings of male and female students.

(c) Once the sample size was decided, the sample was chosen by taking every 20th name from an alphabetical list of all undergraduates. Is it reasonable to consider the samples as SRSs chosen from the male and female undergraduate populations?

(d) What other information about the study would you request before accepting the results as describing all undergraduates?

13.12 Activity 13 analysis Locate your data from the paper airplane experiment (page 780).

(a) Carry out an appropriate test to determine whether one paper airplane model flies significantly farther on average than the other.

(b) Construct and interpret a 90% confidence interval for the difference in mean flight distance.

Software Approximation for the Degrees of Freedom

We noted earlier that the two-sample t statistic does not have a t distribution. In fact, the exact distribution changes as the unknown population standard deviations σ_1 and σ_2 change. However, an excellent approximation is available, thanks to statisticians Welch and Satterthwaite.

Approximate Distribution of the Two-Sample t Statistic

The distribution of the two-sample t statistic is close to the t distribution with degrees of freedom given by

$$\text{df} = \frac{\left(\frac{s_1^2}{n_1} + \frac{s_2^2}{n_2}\right)^2}{\frac{1}{n_1 - 1}\left(\frac{s_1^2}{n_1}\right)^2 + \frac{1}{n_2 - 1}\left(\frac{s_2^2}{n_2}\right)^2}$$

This is the approximation used by most statistical software. It is quite accurate when both sample sizes n_1 and n_2 are 5 or larger.

The t procedures remain exactly as before except that we use the t distribution with df given by the formula in the preceding box to give critical values and P-values.

Example 13.5 *Calcium and blood pressure, continued*

More accurate levels in t procedures

In the calcium experiment of Examples 13.2 to 13.4 the data gave

Group	Treatment	n	$\bar{x}$	s
1	Calcium	10	5.000	8.743
2	Placebo	11	−0.273	5.901

For improved accuracy, we can use critical points from the t distribution with df given by

$$\text{df} = \frac{\left(\frac{8.743^2}{10} + \frac{5.901^2}{11}\right)^2}{\frac{1}{9}\left(\frac{8.743^2}{10}\right)^2 + \frac{1}{10}\left(\frac{5.901^2}{11}\right)^2}$$

$$= \frac{116.848}{7.494} = 15.59$$

Notice that the degrees of freedom is not a whole number.

Figure 13.3 gives computer output for a significance test from Crunch It!, Fathom, and Minitab for the calcium experiment. Notice that the P-value of the one-sided test we performed in Example 13.3 (page 789) is now 0.0644, using 15.59 degrees of freedom.

Figure 13.3 *Computer output from CrunchIt!, Fathom, and Minitab for a two-sample* t *test in Example 13.5.*

CrunchIt!

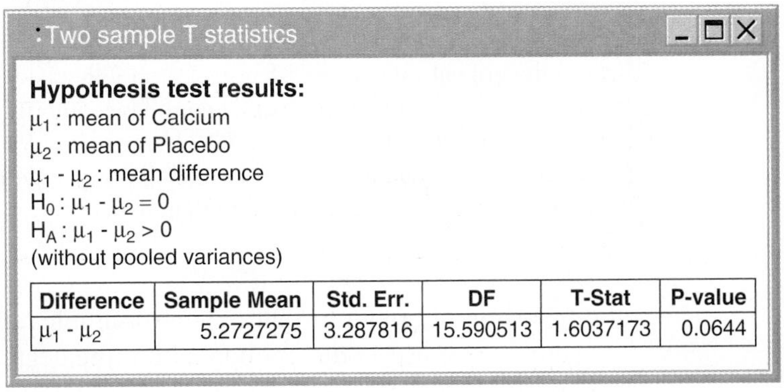
:Two sample T statistics

Hypothesis test results:

μ_1 : mean of Calcium
μ_2 : mean of Placebo
$\mu_1 - \mu_2$: mean difference
$H_0 : \mu_1 - \mu_2 = 0$
$H_A : \mu_1 - \mu_2 > 0$
(without pooled variances)

Difference	Sample Mean	Std. Err.	DF	T-Stat	P-value
$\mu_1 - \mu_2$	5.2727275	3.287816	15.590513	1.6037173	0.0644

(continued)

Figure 13.3 *(continued)*

Fathom

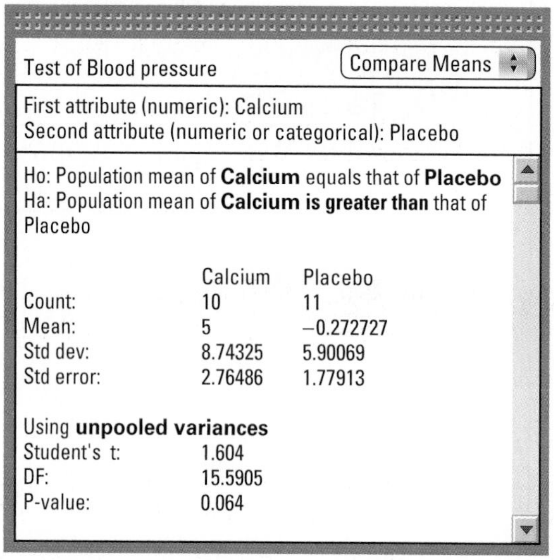
Test of Blood pressure
Compare Means
First attribute (numeric): Calcium
Second attribute (numeric or categorical): Placebo
Ho: Population mean of Calcium equals that of Placebo
Ha: Population mean of Calcium is greater than that of Placebo
Calcium Placebo
Count: 10 11
Mean: 5 −0.272727
Std dev: 8.74325 5.90069
Std error: 2.76486 1.77913
Using unpooled variances
Student's t: 1.604
DF: 15.5905
P-value: 0.064

Minitab

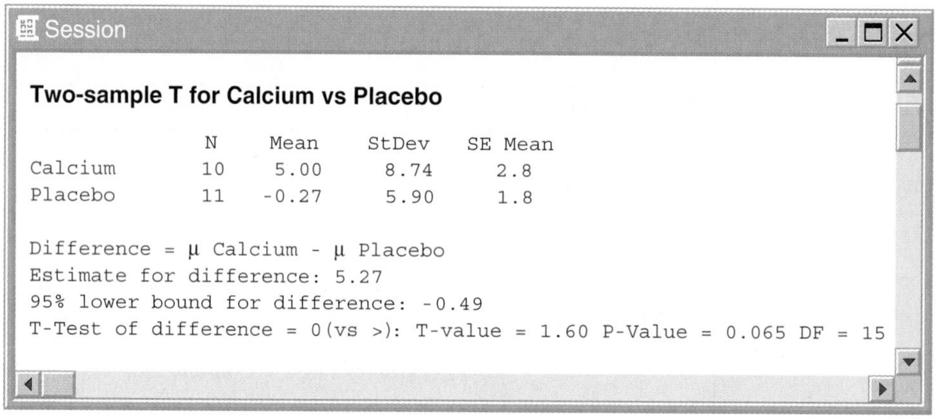
Session
Two-sample T for Calcium vs Placebo
N Mean StDev SE Mean
Calcium 10 5.00 8.74 2.8
Placebo 11 -0.27 5.90 1.8
Difference = μ Calcium - μ Placebo
Estimate for difference: 5.27
95% lower bound for difference: -0.49
T-Test of difference = 0 (vs >): T-value = 1.60 P-Value = 0.065 DF = 15

The conservative 90% confidence interval for $\mu_1 - \mu_2$ in Example 13.4 (page 790) used the critical value $t^* = 1.833$ based on 9 degrees of freedom. The resulting interval was $(-0.753, 11.299)$. A more exact confidence interval replaces $t^* = 1.833$ with the critical value for df $= 15.59$: $t^* = 1.749$. Using technology, we find that the resulting 90% confidence interval is $(-0.4767, 11.022)$. This confidence interval is a bit narrower (margin of error 5.749 rather than 6.026) than the conservative interval from Example 13.4.

As Example 13.5 illustrates, the two-sample t procedures are exactly as before except that we use a t distribution with more degrees of freedom. The number df from the box on page 792 is always at least as large as the smaller of $n_1 - 1$ and $n_2 - 1$. On the other hand, df is never larger than the sum $n_1 + n_2 - 2$ of the two individual degrees of freedom. The number of degrees of freedom is generally not a whole number. There is a t distribution for any positive degrees of freedom, even though Table C contains entries only for whole-number degrees of freedom.

Some software packages find df and then use the t distribution with the next smaller whole-number degrees of freedom. Others take care to use $t(\text{df})$ even when df is not a whole number. We recommend this approximation, but we do not suggest that you do the calculations by hand. With a graphing calculator or computer, the more accurate procedures are painless, as the following Technology Toolbox illustrates.

Technology Toolbox

Two-sample inference

Constructing confidence intervals and performing significance tests using two-sample t procedures on the TI-83/84/89 is very similar to the one-sample case. To illustrate, we will use the data on calcium supplements to lower blood pressure from Examples 13.2 to 13.5. The data are the decreases in systolic blood pressure after 12 weeks, in millimeters of mercury. The data for the 10 men in Group 1 (calcium) were

7	−4	18	17	−3	−5	1	10	11	−2

and for the 11 men in Group 2 (placebo),

−1	12	−1	−3	3	−5	5	2	−11	−1	−3

Tests of significance

- Enter the Group 1 (calcium) data into L_1/list1 and the Group 2 (placebo) data into L_2/list2.
- To perform the significance test, go to STAT/TESTS (Tests menu in the Statistics/List Editor APP on the TI-89) and choose 4:2-SampTTest.
- In the 2-SampTTest screen, specify "Data" and adjust your other settings as shown.

TI-83/84

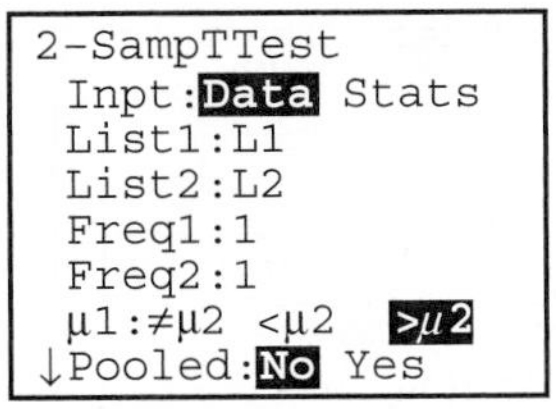

TI-89

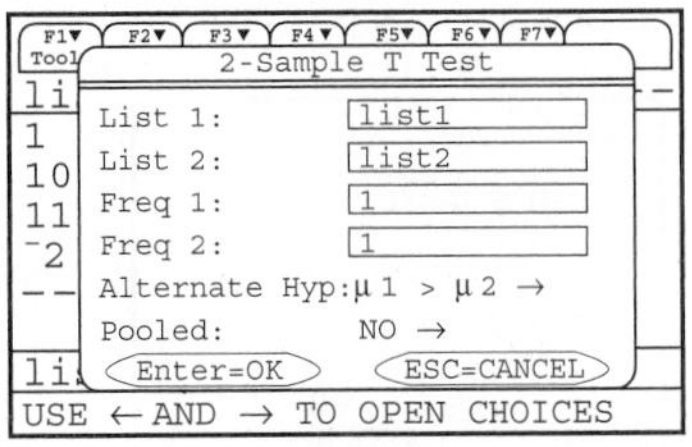

(continued)

Technology Toolbox

Two-sample inference *(continued)*

- Highlight "Calculate" and press ENTER. (The Pooled option will be discussed later.)

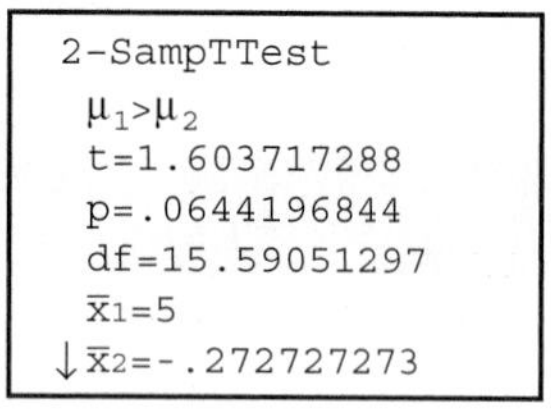

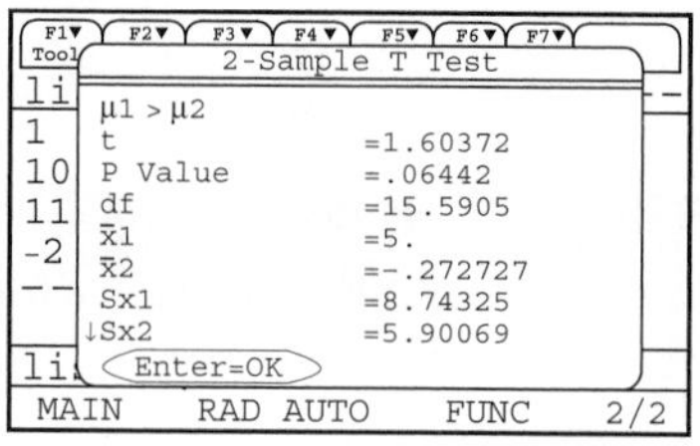

The results tell us that the two-sample t test statistic is $t = 1.6037$, and the P-value is $P = 0.0644$. There is enough evidence against H_0 to reject it at the 10% significance level but not at the 5% or 1% significance levels.

If you select "Draw" in the 2-SampTTest screen instead of "Calculate," the $t(k)$ distribution will be displayed, showing the test statistic $t = 1.6037$ and the upper 0.0644 critical area shaded.

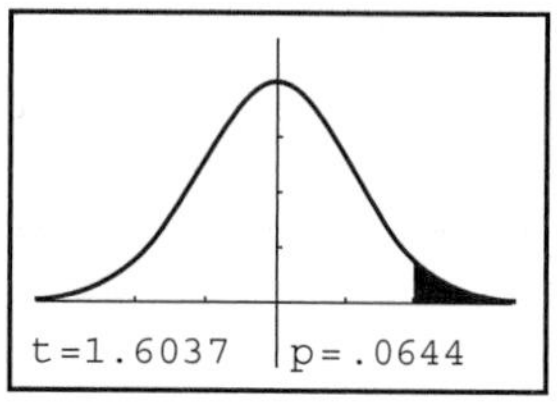

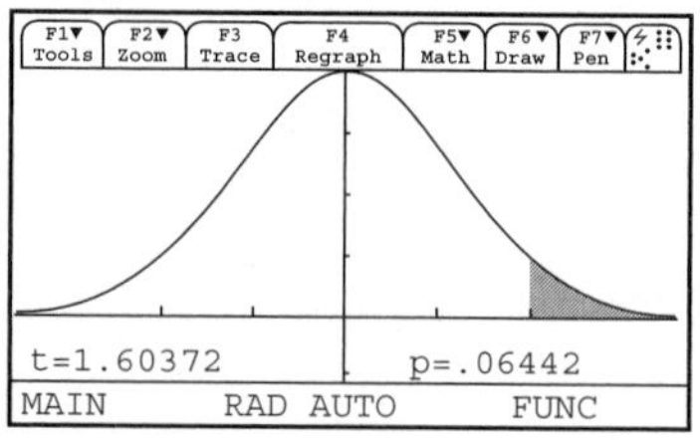

Confidence intervals

- With the data still stored in L_1/list1 and L_2/list2, select STAT/TESTS (Ints menu in the Stats/List Editor APP on the TI-89). Choose 2-SampTInt.
- In the 2-SampTInt screen, choose "Data" and adjust your settings as shown.

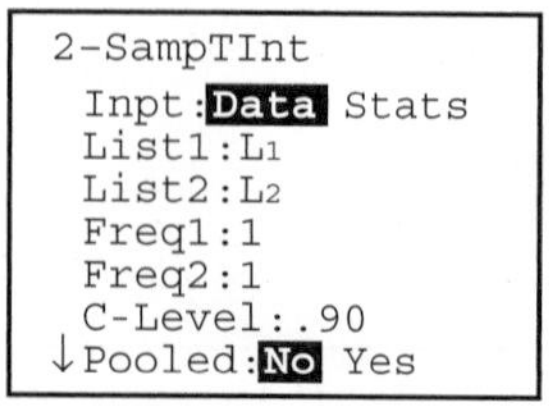

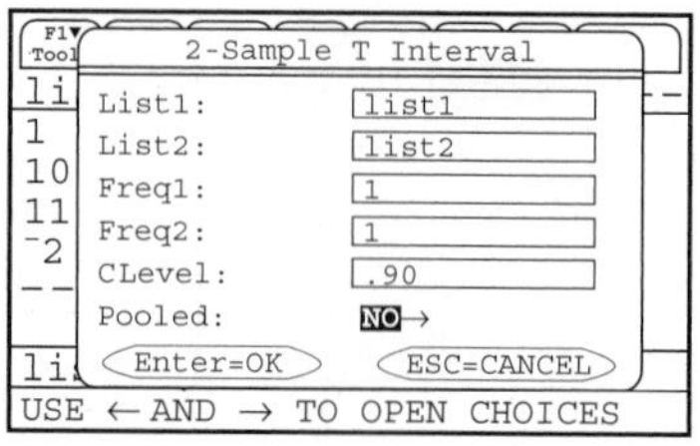

Technology Toolbox

Two-sample inference *(continued)*

- Highlight "Calculate" and press **ENTER**.

```
2-SampTInt
 (-.4767,11.022)
 df=15.59051297
 x̄1=5
 x̄2=-.272727273
 Sx1=8.74325137
↓Sx2=5.90069333
```

```
F1▼
Tool       2-Sample T Interval
li   CInt      ={⁻.477,11.02}
1    x̄1-x̄2     =5.27273
10   ME        =5.74941
11   df        =15.5905
⁻2   x̄1        =5.
--   x̄2        =⁻.272727
     Sx1       =8.74325
    ↓Sx2       =5.90069
li   Enter=OK
MAIN     RAD AUTO     FUNC     2/2
```

The 90% confidence interval for the true difference of the means $\mu_1 - \mu_2$ (or $\mu_{\text{CALCIUM}} - \mu_{\text{PLACEBO}}$, if you prefer) is $(-0.4767, 11.022)$.

If you are given the means $\bar{x}_1$ and $\bar{x}_2$ and sample standard deviations s_1 and s_2 instead of the original data, select the "Stats" option instead of "Data" in either the 2-SampTTest or the 2-SampTInt screen. Then provide the values requested.

Using the more accurate Welch-Satterthwaite df results in narrower confidence intervals and smaller *P*-values than the more conservative degrees of freedom we studied earlier. If you have technology available that uses the more accurate df, take advantage of it.

Example 13.6 *DDT poisoning*

Two-sample *t* with computer software

Poisoning by the pesticide DDT causes convulsions in humans and other mammals. Researchers seek to understand how the convulsions are caused. In a randomized comparative experiment, they compared 6 white rats poisoned with DDT with a control group of 6 unpoisoned rats. Electrical measurements of nerve activity are the main clue to the nature of DDT poisoning. When a nerve is stimulated, its electrical response shows a sharp spike followed by a much smaller second spike. The experiment found that the second spike is larger in rats fed DDT than in normal rats. This finding helped biologists understand how DDT poisoning works.[9]

The researchers measured the height (or amplitude) of the second spike as a percent of the first spike when a nerve in the rat's leg was stimulated. For the poisoned rats the results were

12.207	16.869	25.050	22.429	8.456	20.589

The control group data were

11.074	9.686	12.064	9.351	8.182	6.642

Figure 13.4 shows output from CrunchIt!, Fathom, and SAS statistical software for these data.[10] The difference in means for the two groups appears large, but in such small samples the sample mean is highly variable. A significance test can help confirm that we are seeing a real effect.

Figure 13.4 *Computer output from CrunchIt!, Fathom, and SAS for a two-sample* t *test using the more accurate fractional* df *value, for Example 13.6.*

CrunchIt!

Two sample T statistics

Hypothesis test results:
μ_1 : mean of DDT
μ_2 : mean of Control
$\mu_1 - \mu_2$: mean difference
$H_0 : \mu_1 - \mu_2 = 0$
$H_A : \mu_1 - \mu_2 \neq 0$
(without pooled variances)

Difference	Sample Mean	Std. Err.	DF	T-Stat	P-value
$\mu_1 - \mu_2$	8.100166	2.7080193	5.937621	2.9911776	0.0246

Fathom

Test of Spike — Compare Means

First attribute (numeric): DDT
Second attribute (numeric or categorical): Control

Ho: Population mean of **DDT** equals that of **Control**
Ha: Population mean of **DDT is not equal to** that of **Control**

	DDT	Control
Count:	6	6
Mean:	17.6	9.49983
Std dev:	6.34015	1.95006
Std error:	2.58835	0.796108

Using **unpooled variances**
Student's t: 2.991
DF: 5.93762
P-value: 0.025

```
                    TTEST PROCEDURE

Variable: SPIKE

GROUP     N     Mean           Std Dev        Std Error
-------------------------------------------------------
DDT       6     17.60000000    6.34014839     2.58835474

CONTROL   6     9.49983333     1.95005932     0.79610839

Variances       T        DF        Prob> |T|
-------------------------------------------------------
Unequal         2.9912   5.9       0.0247

Equal           2.9912   10.0      0.0135
```

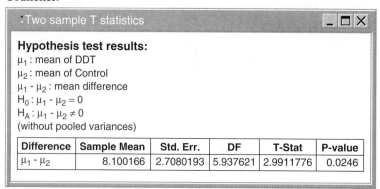

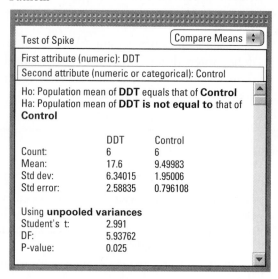

Step 1: Hypotheses We want to compare the mean height μ_{DDT} of the second-spike electrical response in the population of rats fed DDT with μ_{NORMAL}, the population mean second-spike height for normal rats. Because the researchers did not conjecture in advance that the size of the second spike would be higher in rats fed DDT, we use the two-sided alternative:

$$H_0: \mu_{\text{DDT}} = \mu_{\text{NORMAL}} \qquad \text{or, equivalently,} \qquad H_0: \mu_{\text{DDT}} - \mu_{\text{NORMAL}} = 0$$
$$H_a: \mu_{\text{DDT}} \neq \mu_{\text{NORMAL}} \qquad\qquad\qquad H_a: \mu_{\text{DDT}} - \mu_{\text{NORMAL}} \neq 0$$

Step 2: Conditions Since both population standard deviations are unknown, we should use a two-sample t test if the conditions are met.

The DDT data are much more spread out than the control data.

- **SRS** The researchers are willing to treat both samples as SRSs from their respective populations. Random assignment of the rats to the two treatments helps ensure that the only difference between the two groups is the treatment administered.
- **Normality** Normal probability plots (Figure 13.5) show no evidence of outliers or strong skewness. Both populations are plausibly Normal, as far as can be judged from 6 observations.

Figure 13.5 *Normal probability plots for the amplitude of the second spike as a percent of the first spike, for Example 13.6.*

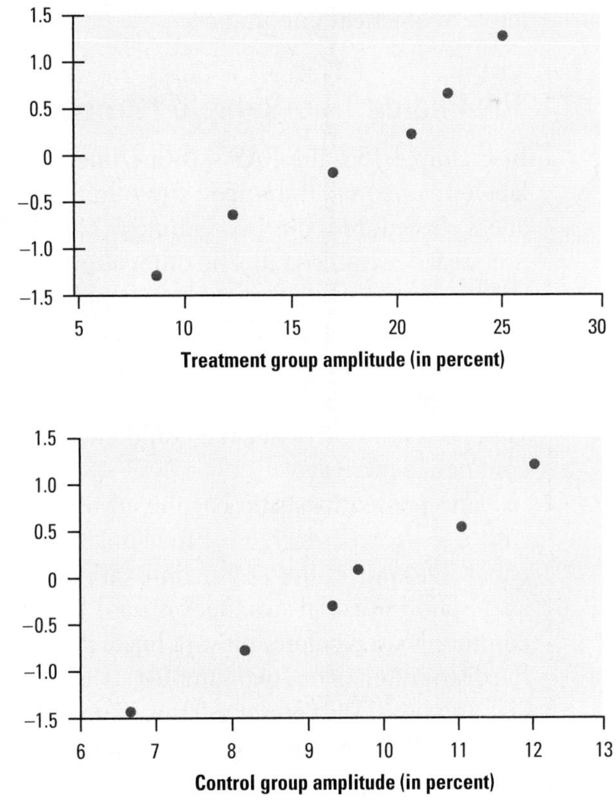

- **Independence** Due to the random assignment, the researchers can treat the two groups of rats as independent samples.

Step 3: Calculations

The SAS printout in Figure 13.4 reports the results of two t procedures: the general two-sample procedure ("unequal" variances) and a special procedure that assumes the two population variances are equal. We are interested in the first of these procedures.

- **Test statistic** The two-sample t statistic has the value $t = 2.9912$.
- **P-value** The degrees of freedom are df = 5.9, and the P-value from the $t(5.9)$ distribution is 0.0247.

Step 4: Interpretation The low P-value ($P = 0.0247$) provides strong evidence against H_0. We reject H_0 at the 5% significance level and conclude that the mean size of the secondary spike is larger in rats fed DDT.

Would the conservative test based on 5 degrees of freedom (both $n_1 - 1$ and $n_2 - 1$ are 5) have given a different result in Example 13.6? The statistic is exactly the same: $t = 2.9912$. The conservative P-value is $2P(T \geq 2.9912)$, where T has the $t(5)$ distribution. Table C shows that 2.9912 lies between the 0.02 and 0.01 upper critical values of the $t(5)$ distribution, so P for the two-sided test lies between 0.02 and 0.04. For practical purposes this is the same result as that given by the software. As Example 13.5 and Example 13.6 suggest, the difference between the t procedures using the conservative and the approximately correct distributions is rarely of practical importance.

The Pooled Two-Sample *t* Procedures (Don't Use Them!)

In Example 13.6 the SAS software offered a choice between two t tests. One is labeled "unequal" variances; the other, "equal" variances. The "unequal" variances procedure is our two-sample t. This test is valid whether or not the population variances are equal. The other choice is a special version of the two-sample t statistic that assumes that the two populations have the same variance. This procedure averages (the statistical term is "pools") the two sample variances to estimate the common population variance. The resulting statistic is called the pooled two-sample t statistic. It is equal to our t statistic if the two sample sizes are the same, but not otherwise. We could choose to use the pooled t for both tests and confidence intervals.

The pooled t statistic has the advantage that it has *exactly* the t distribution with $n_1 + n_2 - 2$ degrees of freedom if the two population variances really are equal. Of course, the population variances are often not equal. Moreover, the assumption of equal variances is hard to check from the data. The pooled t was commonly used before software made it easy to use the accurate approximation to the distribution of our two-sample t statistic. Now the pooled t is useful only in special situations. We cannot use the pooled t in Example 13.6, for example, because it is clear that the variance is much larger among rats fed DDT. **Our best advice for comparing two means is this: don't pool!**

Exercises

13.13 Is red wine better than white wine? Observational studies suggest that moderate use of alcohol reduces heart attacks, and that red wine may have special benefits. One reason may be that red wine contains polyphenols, substances that do good things to cholesterol in the blood and so may reduce the risk of heart attacks. In an experiment, healthy men were assigned at random to drink half a bottle of either red or white wine each day for two weeks. The level of polyphenols in their blood was measured before and after the two-week period. Here are the percent changes in level for the subjects in both groups:[11]

Red wine:	3.5	8.1	7.4	4.0	0.7	4.9	8.4	7.0	5.5
White wine:	3.1	0.5	−3.8	4.1	−0.6	2.7	1.9	−5.9	0.1

(a) Is there good evidence that red wine drinkers gain more polyphenols on average than white wine drinkers? Carry out an appropriate test using Table C to calculate the *P*-value.

(b) Use technology to perform the test in (a). Discuss any differences that you observe in the results.

(c) Does this study give reason to think that it is drinking red wine, rather than some lurking variable, that causes the increase in blood polyphenols? Justify your answer.

13.14 Red wine or white wine? Refer to the previous exercise.

(a) Construct and interpret a 95% confidence interval for the difference in mean polyphenol levels using Table C to find the critical value.

(b) Use technology to construct the 95% confidence interval described in (a). Discuss any differences that you observe in the results.

13.15 Competitive rowers, I What aspects of rowing technique distinguish between novice and skilled competitive rowers? Researchers compared two groups of female competitive rowers: a group of skilled rowers and a group of novices. The researchers measured many mechanical aspects of rowing style as the subjects rowed on a Stanford Rowing Ergometer. One important variable is the angular velocity of the knee, which describes the rate at which the knee joint opens as the legs push the body back on the sliding seat. The data show no outliers or strong skewness. Here is the SAS computer output:[12]

```
                    TTEST PROCEDURE

Variable: KNEE

GROUP          N         Mean       Std Dev      Std Error
----------------------------------------------------------
SKILLED       10     4.18283335   0.47905935   0.15149187

NOVICE         8     3.01000000   0.95894830   0.33903942

Variances      T         DF      Prob> |T|
----------------------------------------------------------
Unequal     3.1583      9.8      0.0104

Equal       3.3918     16.0      0.0037
```

(a) The researchers believed that the knee velocity would be higher for skilled rowers. Use the computer output to carry out an appropriate test of this belief. (Note that SAS provides two-sided *P*-values.) What do you conclude?

(b) Use technology to construct and interpret a 90% confidence interval for the mean difference between the knee velocities of skilled and novice female rowers.

(c) If you had used Table C to construct the confidence interval described in (b), how would the two results compare? Justify your answer *without* doing any calculations.

13.16 Competitive rowers, II The research in the previous exercise also wondered whether skilled and novice rowers differ in weight or other physical characteristics. Here is the SAS computer output for weight in kilograms:

```
                     TTEST PROCEDURE

Variable: WEIGHT

GROUP          N        Mean         Std Dev       Std Error
-------------------------------------------------------------
SKILLED       10     70.3700000   6.10034898    1.92909973

NOVICE         8     68.4500000   9.0399930     3.19612240

Variances      T         DF      Prob> |T|
-------------------------------------------------------------
Unequal                 11.8     0.6165

Equal       0.5376      16.0     0.5982
```

(a) Calculate the missing *t* statistic in the computer output. Show your work.

(b) Is there significant evidence of a difference in the mean weights of skilled and novice rowers? Justify your answer. (Note that SAS provides two-sided *P*-values.)

(c) If the *P*-value had been computed using the more conservative number of degrees of freedom, would it have been greater than, equal to, or less than 0.6165? Justify your answer *without* doing any calculations.

Section 13.1 Summary

The data in a **two-sample problem** should come from two independent SRSs, each drawn from a separate Normally distributed population.

Significance tests and confidence intervals for the difference between the means μ_1 and μ_2 of two Normal populations start from the difference $\bar{x}_1 - \bar{x}_2$ between the two sample means. Due to the central limit theorem, the resulting procedures are approximately correct for other population distributions when the sample sizes are large.

Draw independent SRSs of sizes n_1 and n_2 from two Normal populations with parameters μ_1, σ_1 and μ_2, σ_2. The **two-sample *z* statistic**

$$z = \frac{(\bar{x}_1 - \bar{x}_2) - (\mu_1 - \mu_2)}{\sqrt{\frac{\sigma_1^2}{n_1} + \frac{\sigma_2^2}{n_2}}}$$

has the standard Normal distribution.

Since we almost never know the population standard deviations in practice, we use the **two-sample *t* statistic**

$$t = \frac{(\bar{x}_1 - \bar{x}_2) - (\mu_1 - \mu_2)}{\sqrt{\dfrac{s_1^2}{n_1} + \dfrac{s_2^2}{n_2}}}$$

This test statistic does not have exactly a t distribution. However, good approximations are available with statistical software and calculators.

For conservative inference procedures to compare μ_1 and μ_2, use the two-sample t statistic with the $t(k)$ distribution. The number of degrees of freedom k is the smaller of $n_1 - 1$ and $n_2 - 1$.

For more accurate probability values, use the $t(k)$ distribution with degrees of freedom k estimated from the data. This is the usual procedure in statistical software.

The level C **confidence interval** for $\mu_1 - \mu_2$ given by

$$(\bar{x}_1 - \bar{x}_2) \pm t^* \sqrt{\frac{s_1^2}{n_1} + \frac{s_2^2}{n_2}}$$

has confidence level *at least* C if we use the more conservative number of degrees of freedom.

Significance tests for H_0: $\mu_1 - \mu_2$ are based on

$$t = \frac{(\bar{x}_1 - \bar{x}_2)}{\sqrt{\dfrac{s_1^2}{n_1} + \dfrac{s_2^2}{n_2}}}$$

These tests have a true P-value no higher than that calculated using the conservative df.

The guidelines for practical use of two-sample t procedures are similar to those for one-sample t procedures (just replace "sample size" with "sum of the sample sizes" in the box on page 788). Equal sample sizes are recommended.

Section 13.1 Exercises

In exercises that call for two-sample t *procedures, you may use either of the two approximations for the degrees of freedom that we have discussed, unless you are told otherwise. Be sure to state clearly which approximation you are using.*

13.17 Independent samples versus paired samples Deciding whether to perform a paired t test or a two-sample t test can be tricky. Your decision should be based on the design that produced the data. Which procedure would you choose in each of the following situations?[13]

(a) To test the wear characteristics of two tire brands, A and B, Brand A is mounted on 50 cars and Brand B on 50 other cars.

(b) To test the wear characteristics of two tire brands, A and B, one Brand A tire is mounted on one side of each car in the rear, while a Brand B tire is mounted on the other side. Which side gets which brand is determined by flipping a coin. The same procedure is used on the front.

(c) To test the effect of background music on productivity, factory workers are observed. For 1 month they had no background music. For another month they had background music.

(d) A random sample of 10 workers in Plant A is to be compared with a random sample of 10 workers in Plant B in terms of productivity.

(e) A new weight-reducing diet was tried on 10 women. The weight of each woman was measured before the diet and again after 10 weeks on the diet.

13.18 Treating scrapie in hamsters, I Scrapie is a degenerative disease of the nervous system. A study of the substance IDX as a treatment for scrapie used as subjects 20 infected hamsters. Ten, chosen at random, were injected with IDX. The other 10 were untreated. The researchers recorded how long each hamster lived. They reported, "Thus, although all infected control hamsters had died by 94 days after infection (mean ± SEM = 88.5 ± 1.9 days), IDX-treated hamsters lived up to 128 days (mean ± SEM = 116 ± 5.6 days)."[14]

(a) Fill in the values in this summary table:

Group	Treatment	n	$\bar{x}$	s
1	IDX	?	?	?
2	Untreated	?	?	?

(b) What degrees of freedom would you use in the conservative two-sample t procedures to compare the two treatments?

(c) Carefully describe the design of this study. Include a few sentences on how to carry out the random assignment.

13.19 Treating scrapie in hamsters, II The previous exercise contains the results of a study to determine whether IDX is an effective treatment of scrapie.

(a) Is there good evidence that hamsters treated with IDX live longer on the average? Justify your answer.

(b) Construct and interpret a 95% confidence interval for the mean amount by which IDX prolongs life.

Do birds learn to time their breeding? Blue titmice eat caterpillars. The birds would like lots of caterpillars around when they have young to feed, but they must breed much earlier. Have the birds learned from the past year's experience when they time their breeding the next year? Researchers randomly assigned 7 pairs of birds to have the natural caterpillar supply supplemented while feeding their young and another 6 pairs to serve as a control group relying on natural food supply only. The next year, they measured how many days after the caterpillar peak the birds produced their nestlings.[15] *Exercises 13.20 to 13.22 are based on this study.*

13.20 Did the randomization produce similar groups? First, compare the two groups of birds in the first year. The only difference should be the chance effect of the random assignment. The study report says: "In the experimental year, the degree of synchronization did not differ between food-supplemented and control females." For this comparison, the report gives $t = -1.05$.

(a) What type of t statistic (paired or two-sample) is this? Justify your answer.

(b) Explain how this t leads to the quoted conclusion.

13.21 Did the treatment have an effect? The investigators expected the control group to adjust their breeding date the next year, whereas the well-fed supplemented group had no reason to change. The report continues: "in the following year food-supplemented females were more out of synchrony with the caterpillar peak than the controls." Here are the data (days behind caterpillar peak):

Control:	4.6	2.3	7.7	6.0	4.6	−1.2	
Supplemented:	15.5	11.3	5.4	16.5	11.3	11.4	7.7

Carry out an appropriate test and show that it leads to the quoted conclusion.

13.22 Year-to-year comparison Rather than comparing the two groups in each year, we could compare the behavior of each group in the first and second years. The study report says: "Our main prediction was that females receiving additional food in the nestling period should not change laying date the next year, whereas controls, which (in our area) breed too late in their first year, were expected to advance their laying date in the second year."

Comparing days behind the caterpillar peak in Years 1 and 2 gave $t = 0.63$ for the control group and $t = -2.63$ for the supplemented group.

(a) What type of t statistic (paired or two-sample) are these? Justify your answer.

(b) What are the degrees of freedom for each t?

(c) Explain why these t-values do *not* agree with the prediction.

13.23 Students' self-concept Here is SAS output for a study of the self-concept of seventh-grade students. The variable SC is the score on the Piers-Harris Children's Self-Concept Scale. The analysis was done to see if male and female students differ in mean self-concept score.[16]

```
                          TTEST PROCEDURE

Variable: SC

SEX      N        Mean           Std Dev          Std Error
-----------------------------------------------------------
F       31    55.51612903     12.69611743     2.28029001

M       47    57.91489362     12.26488410     1.78901722

Variances        T             DF          Prob> |T|
-----------------------------------------------------------
Unequal      -0.8276          62.8         0.4110

Equal        -0.8336          76.0         0.4071
```

Write a sentence or two summarizing the comparison of females and males, as if you were preparing a report for publication.

13.24 The effect of logging How badly does logging damage tropical rain forests? One study compared forest plots in Borneo that had never been logged with similar plots nearby that had been logged 8 years earlier. The study found that the effects of logging were somewhat less severe than expected. Here are the data on the number of tree species in 12 unlogged plots and 9 logged plots:[17]

Unlogged:	22	18	22	20	15	21	13	13	19	13	19	15
Logged:	17	4	18	14	18	15	15	10	12			

(a) The study report says, "Loggers were unaware that the effects of logging would be assessed." Why is this important?

(b) The study report also explains why the plots can be considered to be randomly assigned. Why is this important?

(c) Does logging significantly reduce the mean number of species in a plot after 8 years? Give appropriate statistical evidence to support your conclusion.

(d) Construct and interpret a 90% confidence interval for the difference in mean number of species between unlogged and logged plots.

13.2 Comparing Two Proportions

Some women would like to have children but cannot do so for medical reasons. One option for these women is a procedure called in vitro fertilization, which involves injecting a fertilized egg into the woman's uterus. About 28% of women who undergo in vitro fertilization successfully get pregnant.[18] Can praying for these women help increase the pregnancy rate? Researchers designed an experiment to help answer this question.

Example 13.7 *Prayer and in vitro fertilization*
Comparing two proportions

A large group of women who were about to undergo in vitro fertilization served as subjects in an experiment. Each subject was randomly assigned to either a treatment group or a control group. Women in the treatment group were intentionally prayed for by several people (called intercessors) who did not know them, a process known as intercessory prayer. The praying continued for three weeks following in vitro fertilization. The intercessors did not pray for the women in the control group.

Here are the results: 44 of the 88 women (50%) in the treatment group got pregnant, compared to 21 out of 81 (26%) in the control group.[19] This difference seems large, but is it statistically significant?

In Example 13.7, a higher proportion of women who were prayed for got pregnant than of women who weren't prayed for. What does this difference in sample proportions tell us about the corresponding difference in population proportions? To answer this question, we need to know how to compare two proportions.

Two-Sample Problems: Proportions

We will use notation similar to that used in our study of two-sample t statistics. The groups we want to compare are Population 1 and Population 2. We have a separate SRS from each population or responses from two treatments in a randomized comparative experiment. A subscript shows which group a parameter or statistic describes. Here is our notation:

Population	Population proportion	Sample size	Sample proportion
1	p_1	n_1	$\hat{p}_1$
2	p_2	n_2	$\hat{p}_2$

We compare the populations by doing inference about the difference $p_1 - p_2$ between the population proportions. The statistic that estimates this difference is the difference between the two sample proportions, $\hat{p}_1 - \hat{p}_2$.

Example 13.8 *Does preschool help?*
Inference about $p_1 - p_2$

To study the long-term effects of preschool programs for poor children, the High/Scope Educational Research Foundation has followed two groups of Michigan children since early childhood. A control group of 61 children represents Population 1, poor children with no preschool. Another group of 62 from the same area and similar backgrounds attended preschool as 3- and 4-year-olds. This is a sample from Population 2, poor children who attend preschool. Thus, the sample sizes are $n_1 = 61$ and $n_2 = 62$.

One response variable of interest is the need for social services as adults. In the past ten years, 38 of the preschool sample and 49 of the control sample have needed social services (mainly welfare). The sample proportions are

$$\hat{p}_1 = \frac{49}{61} = 0.803 \quad \text{and} \quad \hat{p}_2 = \frac{38}{62} = 0.613$$

That is, about 80% of the control group uses social services, as opposed to about 61% of the preschool group.

To see if the study provides significant evidence that preschool reduces the later need for social services, we test the hypotheses

$$\begin{aligned} &H_0\colon p_1 = p_2 & \text{or, equivalently,} \quad & H_0\colon p_1 - p_2 = 0 \\ &H_a\colon p_1 > p_2 & & H_a\colon p_1 - p_2 > 0 \end{aligned}$$

where p_1 is the actual proportion of poor children who don't attend preschool and who need social services as adults, and p_2 is the corresponding population proportion for poor children who attend preschool.

To estimate how large the reduction is, we give a confidence interval for the difference. Both the test and the confidence interval start with the difference in sample proportions:

$$\hat{p}_1 - \hat{p}_2 = 0.803 - 0.613 = 0.190$$

The Sampling Distribution of $\hat{p}_1 - \hat{p}_2$

To use $\hat{p}_1 - \hat{p}_2$ for inference, we must know its sampling distribution. Both $\hat{p}_1$ and $\hat{p}_2$ are random variables. Their values would vary if we took repeated samples of the same size. The statistic $\hat{p}_1 - \hat{p}_2$ is the difference between these two random variables. In Chapter 7, we saw that if X and Y are any two random variables,

$$\mu_{X-Y} = \mu_X - \mu_Y$$

If X and Y are *independent* random variables,

$$\sigma^2_{X-Y} = \sigma^2_X + \sigma^2_Y$$

These results lead us to important facts about the sampling distribution of $\hat{p}_1 - \hat{p}_2$:

- **Center** The mean of $\hat{p}_1 - \hat{p}_2$ is

$$\mu_{\hat{p}_1 - \hat{p}_2} = \mu_{\hat{p}_1} - \mu_{\hat{p}_2}$$

From Chapter 9, we know that $\mu_{\hat{p}} = p$, so

$$\mu_{\hat{p}_1} - \mu_{\hat{p}_2} = p_1 - p_2$$

That is, the difference in sample proportions is an *unbiased estimator* of the difference in population proportions.

- **Spread** The variance of $\hat{p}_1 - \hat{p}_2$ is

$$\sigma^2_{\hat{p}_1 - \hat{p}_2} = \sigma^2_{\hat{p}_1} + \sigma^2_{\hat{p}_2}$$

provided that the sample proportions are *independent.* Note that the variances add. The standard deviations do not. Again from Chapter 9,

$$\sigma_{\hat{p}} = \sqrt{\frac{p(1-p)}{n}}$$

so

$$\sigma^2_{\hat{p}_1-\hat{p}_2} = \frac{p_1(1-p_1)}{n_1} + \frac{p_2(1-p_2)}{n_2}$$

Thus, the standard deviation of the difference in sample proportions is

$$\sigma_{\hat{p}_1-\hat{p}_2} = \sqrt{\frac{p_1(1-p_1)}{n_1} + \frac{p_2(1-p_2)}{n_2}}$$

- **Shape** When the samples are large, the distribution of $\hat{p}_1 - \hat{p}_2$ is approximately Normal. Recall from Chapter 9 that the sampling distribution of a sample proportion $\hat{p}$ will be approximately Normal if np and $n(1-p)$ are at least 10. The sampling distribution of $\hat{p}_1 - \hat{p}_2$ will be approximately Normal if the sampling distributions of both $\hat{p}_1$ and $\hat{p}_2$ are approximately Normal. That will happen if n_1p_1, $n_1(1-p_1)$, n_2p_2, and $n_2(1-p_2)$ are all at least 10. (*In fact, we'll be safe performing significance tests about* $p_1 - p_2$ *as long as all these values are at least* 5.)

Figure 13.6 displays the distribution of $\hat{p}_1 - \hat{p}_2$. The standard deviation of $\hat{p}_1 - \hat{p}_2$ involves the unknown parameters p_1 and p_2. Just as in Chapter 12, we must replace these by estimates in order to do inference. And just as in Chapter 12, we do this a bit differently for confidence intervals and for tests.

Figure 13.6 *Select independent SRSs from two populations having proportions of successes* p_1 *and* p_2*. The proportions of successes in the two samples are* $\hat{p}_1$ *and* $\hat{p}_2$*. When the samples are large, the sampling distribution of the difference* $\hat{p}_1 - \hat{p}_2$ *is approximately Normal.*

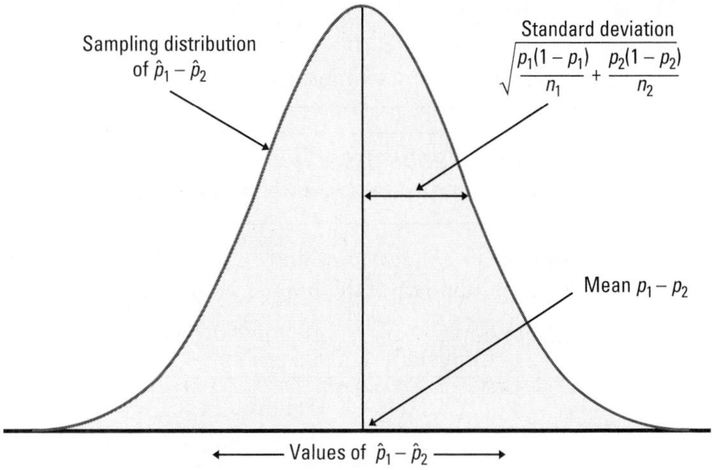

Confidence Intervals for $\hat{p}_1 - \hat{p}_2$

To obtain a confidence interval, replace the population proportions p_1 and p_2 in the expression for $\hat{p}_1 - \hat{p}_2$ by the sample proportions. The result is the ***standard error*** of the statistic $\hat{p}_1 - \hat{p}_2$:

$$\text{SE}_{\hat{p}_1-\hat{p}_2} = \sqrt{\frac{\hat{p}_1(1-\hat{p}_1)}{n_1} + \frac{\hat{p}_2(1-\hat{p}_2)}{n_2}}$$

The confidence interval again has the form

$$\text{estimate} \pm z^*\text{SE}_{\text{estimate}}$$

Confidence Interval for Comparing Two Proportions

Draw an SRS of size n_1 from a population having proportion p_1 of successes, and draw an independent SRS of size n_2 from another population having proportion p_2 of successes. When n_1 and n_2 are large, an approximate level C confidence interval for $\hat{p}_1 - \hat{p}_2$ is

$$(\hat{p}_1 - \hat{p}_2) \pm z^*\text{SE}_{\hat{p}_1-\hat{p}_2}$$

The **standard error** SE of $\hat{p}_1 - \hat{p}_2$ is

$$\text{SE}_{\hat{p}_1-\hat{p}_2} = \sqrt{\frac{\hat{p}_1(1-\hat{p}_1)}{n_1} + \frac{\hat{p}_2(1-\hat{p}_2)}{n_2}}$$

where z^* is the standard Normal critical value. This interval is called a **two-proportion z interval.**

Conditions: In practice, use this confidence interval when

- **SRS** The two samples can be viewed as SRSs from their respective populations or are the two groups in a randomized experiment.
- **Normality** The counts of "successes" and "failures" $n_1\hat{p}_1$, $n_1(1-\hat{p}_1)$, $n_2\hat{p}_2$, and $n_2(1-\hat{p}_2)$ are all at least 5. Notice that we use the sample statistics $\hat{p}_1$ and $\hat{p}_2$ in place of the unknown parameter values p_1 and p_2 when checking this condition.
- **Independence** The two samples are independent. When sampling without replacement, check that the two populations are at least 10 times as large as the corresponding samples.

Example 13.9 *How much does preschool help?*
Confidence interval for $p_1 - p_2$

Example 13.8 describes a study of the effect of preschool on later use of social services. Here is a summary of the important information:

Population	Population description	Sample size	Sample proportion
1	Control	$n_1 = 61$	$\hat{p}_1 = 49/61 = 0.803$
2	Preschool	$n_2 = 62$	$\hat{p}_2 = 38/62 = 0.613$

We completed Step 1 of the Inference Toolbox in Example 13.8.

Step 2: Conditions We should use a two-proportion z interval to calculate our estimate if the three important conditions are met.

- **SRS** Since we are not told how the two samples were selected, we must be cautious when drawing conclusions about the corresponding populations.
- **Normality** Look at the counts of successes and failures in the two samples.

$$n_1\hat{p}_1 = 49,\ n_1(1-\hat{p}_1) = 12,\ n_2\hat{p}_2 = 38,\ \text{and}\ n_2(1-\hat{p}_2) = 24$$

These are all at least 5, so the interval based on Normal calculations will be reasonably accurate.

- **Independence** We can be fairly confident that there are at least 610 poor children who did not attend preschool and at least 620 poor children who did in our populations of interest.

Step 3: Calculations The difference $p_1 - p_2$ measures the effect of preschool in reducing the proportion of people who later need social services. To compute a 95% confidence interval for $p_1 - p_2$, first find the standard error

$$\begin{aligned}
\text{SE}_{\hat{p}_1-\hat{p}_2} &= \sqrt{\frac{\hat{p}_1(1-\hat{p}_1)}{n_1} + \frac{\hat{p}_2(1-\hat{p}_2)}{n_2}} \\
&= \sqrt{\frac{(0.803)(0.197)}{61} + \frac{(0.613)(0.387)}{62}} \\
&= \sqrt{0.00642} = 0.0801
\end{aligned}$$

The 95% confidence interval is

$$\begin{aligned}
(\hat{p}_1 - \hat{p}_2) \pm z^*\text{SE}_{\hat{p}_1-\hat{p}_2} &= (0.803 - 0.613) \pm (1.960)(0.0801) \\
&= 0.190 \pm 0.157 \\
&= (0.033, 0.347)
\end{aligned}$$

Figure 13.7 displays output from CrunchIt!, Fathom, and Minitab for this example. As usual, you can understand the output even without knowledge of the program that produced it.

Figure 13.7 *CrunchIt!, Fathom, and Minitab output for Example 13.9.*

CrunchIt!

:One sample T statistics

95% Confidence interval results:

p_1 : proportion of successes for first population

p_2 : proportion of successes for second population

Difference	Count1	Total1	Count2	Total2	Sample Diff.	Std. Err.	L. Limit	U. Limit
p1 - p2	49	61	38	62	0.19037546	0.08010733	0.03336798	0.34738293

(continued)

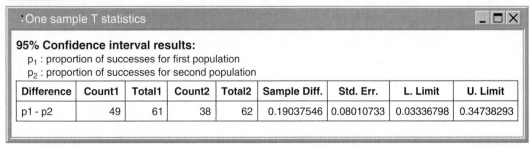

Fathom

```
From Summary Statistics                         Difference of Proportions
Attribute (categorical): unassigned
Attribute (categorical or grouping): unassigned

Interval estimate of the difference in proportions
49 out of 61, or 0.803279, Preschool are using social services later.
38 out of 62, or 0.612903, Control are using social services later.

Confidence level: 95.0%
Estimate:      0.190375 +/- 0.157007
Range:         0.033368 to 0.347383
```

Minitab

```
Session

Test and Confidence Interval for Two Proportions

Sample          X     N    Sample p
1              49    61    0.803279
2              38    62    0.612903

Estimate for p(1) - p(2):   0.190375
95% CI for p(1) - p(2):   (0.0333680, 0.347383)
Test for p(1) - p(2) = 0(vs not = 0): Z =2.32 P-Value = 0.020
```

Step 4: **Interpretation** We are 95% confident that the percent needing social services is somewhere between 3.3% and 34.7% lower among people who attended preschool. The confidence interval is wide because the sample sizes are a bit small for estimating the unknown proportions with precision.

The researchers in the study of Example 13.9 selected two separate samples from the two populations they wanted to compare. Many comparative studies start with just one sample, then divide it into two groups based on data gathered from the subjects. The two-proportion z procedures of this section are valid in such situations. This is an important fact about these methods.

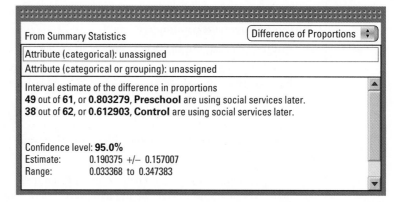

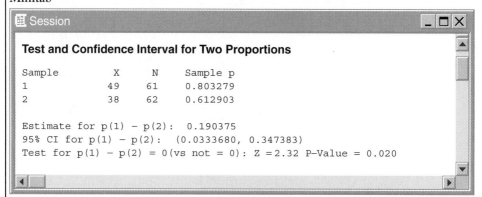

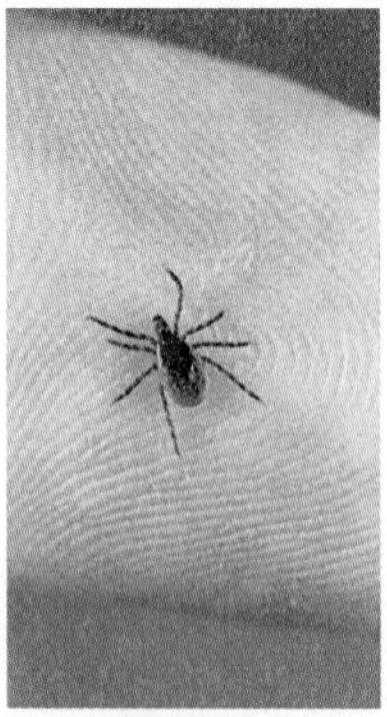

Exercises

13.25 Lyme disease Lyme disease is spread in the northeastern United States by infected ticks. The ticks are infected mainly by feeding on mice, so more mice result in more infected ticks. The mouse population in turn rises and falls with the abundance of acorns, their favored food. Experimenters studied two similar forest areas in a year when the acorn crop failed. They added hundreds of thousands of acorns to one area to imitate an abundant acorn crop, while leaving the other area untouched. The next spring, 54 of the 72 mice trapped in the first area were in breeding condition, versus 10 of the 17 mice trapped in the second area.[20] Construct and interpret a 90% confidence interval for the difference between the proportion of mice ready to breed in good acorn years and bad acorn years. Follow the Inference Toolbox.

13.26 Fear of crime among older blacks The elderly fear crime more than younger people, even though they are less likely to be victims of crime. One of the few studies that looked at older blacks recruited random samples of 56 black women and 63 black men over the age of 65 from Atlantic City, New Jersey. Of the women, 27 said they "felt vulnerable" to crime; 46 of the men said this.[21]

(a) What proportion of women in the sample feel vulnerable? Of men? (*Note*: Men are victims of crime more often than women, so we expect a higher proportion of men to feel vulnerable.)

(b) Construct and interpret a 95% confidence interval for the difference (men minus women).

13.27 Police radar and speeding Do drivers reduce excessive speed when they encounter police radar? Researchers studied the behavior of drivers on a rural interstate highway in Maryland where the speed limit was 55 miles per hour. They measured speed with an electronic device hidden in the pavement and, to eliminate large trucks, considered only vehicles less than 20 feet long. During some time periods, police radar was set up at the measurement location. Here are some of the data:[22]

	Number of vehicles	Number over 65 mph
No radar	12,931	5,690
Radar	3,285	1,051

(a) Construct and interpret a 95% confidence interval for the proportion of vehicles going faster than 65 miles per hour when no radar is present.

(b) Construct and interpret a 95% confidence interval for the effect of radar, as measured by the difference in proportions of vehicles going faster than 65 miles per hour with and without radar.

(c) The researchers chose a rural highway so that cars would be separated rather than in clusters, where some cars might slow because they see other cars slowing. Explain why such clusters might make inference invalid.

13.28 Information online A random digit dialing sample of 2092 adults found that 1318 used the Internet.[23] Of the users, 1041 said that they expect businesses to have Web sites that give product information; 294 of the 774 nonusers said this.

(a) Construct and interpret a 95% confidence interval for the proportion of all adults who use the Internet.

(b) Construct and interpret a 95% confidence interval to compare the proportions of users and nonusers who expect businesses to have Web sites.

Significance Tests for $p_1 - p_2$

An observed difference between two sample proportions can reflect a difference in the populations, or it may just be due to chance variation in random sampling. Significance tests help us decide if the effect we see in the samples is really there in the populations. The null hypothesis says that there is no difference between the two populations:

$$H_0: p_1 = p_2$$

The alternative hypothesis says what kind of difference we expect.

Example 13.10 ***Cholesterol and heart attacks***
Setting up a significance test

High levels of cholesterol in the blood are associated with higher risk of heart attacks. Will using a drug to lower blood cholesterol reduce heart attacks? The Helsinki Heart Study looked at this question. Middle-aged men were assigned at random to one of two treatments: 2051 men took the drug gemfibrozil to reduce their cholesterol levels, and a control group of 2030 men took a placebo. During the next five years, 56 men in the gemfibrozil group and 84 men in the placebo group had heart attacks.

The sample proportions who had heart attacks are

$$\hat{p}_1 = \frac{56}{2051} = 0.0273 \qquad \text{(gemfibrozil group)}$$

$$\hat{p}_2 = \frac{84}{2030} = 0.0414 \qquad \text{(placebo group)}$$

That is, about 4.1% of the men in the placebo group had heart attacks, against only about 2.7% of the men who took the drug. Is the apparent benefit of gemfibrozil statistically significant?

Step 1: Hypotheses We want to use this comparative randomized experiment to draw conclusions about p_1, the proportion of middle-aged men who would suffer heart attacks after taking gemfibrozil, and p_2, the proportion of middle-aged men who would suffer heart attacks if they only took a placebo. We hope to show that gemfibrozil reduces heart attacks, so we have a one-sided alternative:

$$H_0: p_1 = p_2 \qquad \text{or, equivalently,} \qquad H_0: p_1 - p_2 = 0$$
$$H_a: p_1 < p_2 \qquad\qquad\qquad\qquad\quad H_a: p_1 - p_2 < 0$$

We'll complete Steps 2 through 4 in the next example.

To do a test, standardize $\hat{p}_1 - \hat{p}_2$ to get a z statistic. If H_0 is true, all the observations in both samples really come from a single population of men of whom a single unknown proportion p will have a heart attack in a five-year period. So instead of estimating p_1 and p_2 separately, we combine the two samples and use the overall sample proportion to estimate the single population parameter p. Call this the ***combined sample proportion.*** It is

combined sample proportion

$$\hat{p}_c = \frac{\text{count of successes in both samples combined}}{\text{count of individuals in both samples combined}} = \frac{X_1 + X_2}{n_1 + n_2}$$

Use $\hat{p}_c$ in place of both $\hat{p}_1$ and $\hat{p}_2$ in the expression for the standard error SE of $\hat{p}_1 - \hat{p}_2$:

$$\text{SE}_{\hat{p}_1-\hat{p}_2} = \sqrt{\frac{\hat{p}_1(1-\hat{p}_1)}{n_1} + \frac{\hat{p}_2(1-\hat{p}_2)}{n_2}} = \sqrt{\frac{\hat{p}_c(1-\hat{p}_c)}{n_1} + \frac{\hat{p}_c(1-\hat{p}_c)}{n_2}}$$

$$= \sqrt{\hat{p}_c(1-\hat{p}_c)\left(\frac{1}{n_1} + \frac{1}{n_2}\right)}$$

This will yield a z statistic that has the standard Normal distribution when H_0 is true. Here is the test.

Significance Test for Comparing Two Proportions

To test the hypothesis

$$H_0: p_1 = p_2$$

first find the combined proportion $\hat{p}_C$ of successes in both samples combined. Then compute the z statistic

$$z = \frac{\hat{p}_1 - \hat{p}_2}{\sqrt{\hat{p}_c(1-\hat{p}_c)\left(\frac{1}{n_1} + \frac{1}{n_2}\right)}}$$

In terms of a variable Z having the standard Normal distribution, the P-value for a test of H_0 against

$H_a: p_1 > p_2$ is $P(Z \geq z)$

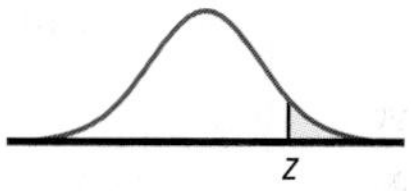

$H_a: p_1 < p_2$ is $P(Z \leq z)$

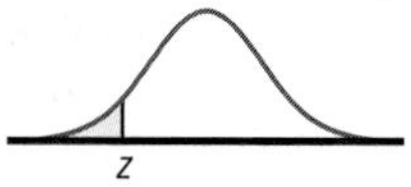

$H_a: p_1 \neq p_2$ is $2P(Z \geq |z|)$

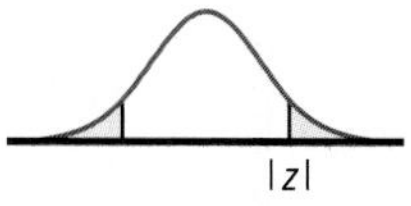

Conditions: Use these tests in practice when

- **SRS** The two samples can be viewed as SRSs from their respective populations or are the two groups in a randomized experiment.
- **Normality** The estimated counts of "successes" and "failures" $n_1\hat{p}_c$, $n_1(1-\hat{p}_c)$, $n_2\hat{p}_c$, and $n_2(1-\hat{p}_c)$ are all at least 10. (This is conservative. The **two-proportion z test** is still accurate as long as these values are at least 5.) Notice that we use the combined sample proportion $\hat{p}_c$ in place of the unknown parameter values p_1 and p_2 when checking this condition.
- **Independence** The samples are independent. When sampling without replacement, the two populations must be at least 10 times as large as the corresponding samples.

Example 13.11 *Cholesterol and heart attacks, continued*

Two-proportion z test for an experiment

Step 2: Conditions Since we are testing a claim about two population proportions, we should use a two-proportion z test if the conditions are satisfied.

- **SRS** Our samples are the two groups in a randomized comparative experiment. The random assignment should allow researchers to attribute any significant difference in the proportion of heart attacks between the two groups to gemfibrozil. We are viewing each group as a representative sample from the hypothetical population of middle-aged men who could receive these treatments. Since the subjects were not randomly selected from a larger population, however, this may limit the researchers' ability to generalize the findings of this study.
- **Normality** The combined proportion of heart attacks for the two groups in the Helsinki Heart Study is

$$\hat{p}_c = \frac{\text{count of heart attacks in both samples combined}}{\text{count of subjects in both samples combined}}$$

$$= \frac{56 + 84}{2051 + 2030}$$

$$= \frac{140}{4081} = 0.0343$$

Using this value, we find that

$$n_1\hat{p}_c = (2051)(0.0343) = 70.3 \qquad n_2\hat{p}_c = (2030)(0.0343) = 69.6$$

$$n_1(1 - \hat{p}_c) = (2051)(0.9657) = 1980.7 \qquad n_2(1 - \hat{p}_c) = (2030)(0.9657) = 1960.4$$

which are all 10 or larger.

- **Independence** Due to the random assignment, these two groups of men can be viewed as independent samples.

Step 3: Calculations

- **Test statistic** The z test statistic is

$$z = \frac{\hat{p}_1 - \hat{p}_2}{\sqrt{\hat{p}_c(1 - \hat{p}_c)\left(\frac{1}{n_1} + \frac{1}{n_2}\right)}}$$

$$= \frac{0.0273 - 0.0414}{\sqrt{(0.0343)(0.9657)\left(\frac{1}{2051} + \frac{1}{2030}\right)}}$$

$$= -\frac{0.0141}{0.005698} = -2.47$$

- ***P*-value** The one-sided P-value is the area under the standard Normal curve to the left of -2.47. Figure 13.7 shows this area. Table A gives $P = 0.0068$.

Figure 13.8 *The P-value for the one-sided test in Example 13.11.*

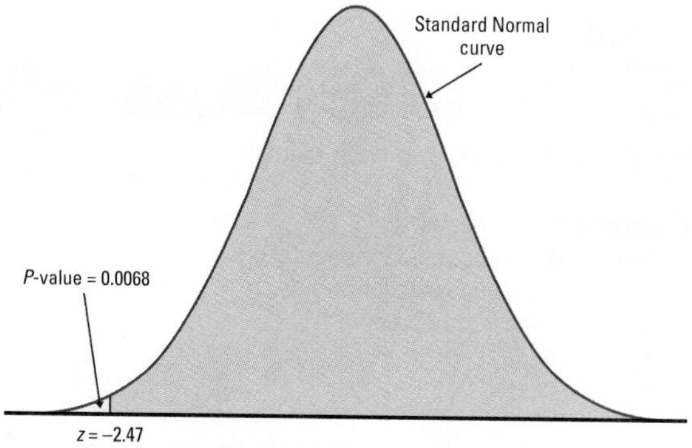

Step 4: Interpretation Since $P < 0.01$, the results are statistically significant at the $\alpha = 0.01$ level. There is strong evidence that gemfibrozil reduced the rate of heart attacks.

Example 13.12 *Don't drink the water!*

Two-proportion z test for an observational study

The movie *A Civil Action* tells the story of a major legal battle that took place in the small town of Woburn, Massachusetts. A town well that supplied water to East Woburn residents was contaminated by industrial chemicals. During the period that residents drank water from this well, a sample of 414 births showed 16 birth defects. On the west side of Woburn, a sample of 228 babies born during the same time period revealed 3 with birth defects. The plaintiffs suing the companies responsible for the contamination claimed that these data show that the rate of birth defects was significantly higher in East Woburn, where the contaminated well water was in use. How strong is the evidence supporting this claim? What should the judge for this case conclude?

The proportion of babies with birth defects in the East Woburn sample is

$$\hat{p}_1 = \frac{16}{414} = 0.0386$$

For the West Woburn sample, the corresponding proportion is

$$\hat{p}_2 = \frac{3}{228} = 0.0132$$

Is the difference, $\hat{p}_1 - \hat{p}_2 = 0.0386 - 0.0132 = 0.0254$, statistically significant?

Figure 13.9 (on the next page) shows computer output from Crunch It! and Minitab for a two-proportion z test.

Figure 13.9 *Crunch It! and Minitab output for a two-proportion z test for Example 13.12.*

CrunchIt!

:One sample T statistics

Two sample Proportion with summary

Hypothesis test results:
p_1 : proportion of successes for population 1
p_2 : proportion of successes for population 2
$p_1 - p_2$: difference in proportions
$H_0 : p_1 - p_2 = 0$
$H_A : p_1 - p_2 > 0$

Difference	Count1	Total1	Count2	Total2	Sample Diff.	Std. Err.	Z-Stat	P-value
$p_1 - p_2$	16	414	3	228	0.025489448	0.013976091	1.8237895	0.0341

Minitab

Session

Test and Confidence Interval for Two Proportions

```
Sample         X      N     Sample p
1             16    414     0.038647
2              3    228     0.013158

Estimate for p(1) - p(2):   0.0254894
95% lower bound for p(1) - p(2):   0.00556742
Test for p(1) - p(2) = 0 (vs > 0): Z = 1.82 P-Value = 0.034
```

Step 1: Hypotheses Let p_1 = the proportion of all East Woburn babies born with birth defects and p_2 = the proportion of all West Woburn babies born with birth defects. Our hypotheses are

$$H_0: p_1 = p_2 \qquad \text{or, equivalently,} \qquad H_0: p_1 - p_2 = 0$$
$$H_a: p_1 > p_2 \qquad\qquad\qquad\qquad H_a: p_1 - p_2 > 0$$

Step 2: Conditions We will use a significance test to compare the proportions of babies born with birth defects in East and West Woburn. Since we begin by assuming that $H_0: p_1 = p_2$ is true, we use

$$\hat{p}_c = \frac{X_1 + X_2}{n_1 + n_2} = \frac{16 + 3}{414 + 228} = 0.0296$$

as our combined estimate for the proportion of babies born with birth defects in all of Woburn.

We check the conditions:

- **SRS** We are willing to treat the two samples of babies as SRSs from the respective populations of interest.
- **Normality**

$$n_1\hat{p}_c = (414)(0.0296) = 12.25 \qquad n_2\hat{p}_c = (228)(0.0296) = 6.75$$
$$n_1(1 - \hat{p}_c) = (414)(0.9704) = 401.75 \qquad n_2(1 - \hat{p}_c) = (228)(0.9704) = 221.25$$

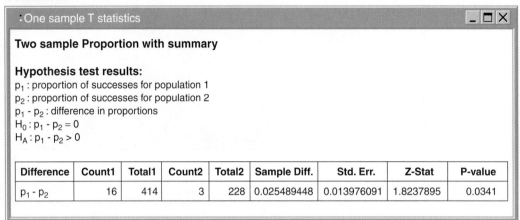

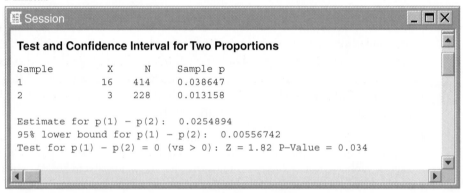

Since all four of these values are larger than 5, we are safe using a Normal approximation.

- **Independence** We must assume that both populations are at least 10 times as large as the samples of babies.

Step 3: **Calculations**

- **Test statistic** From the computer output, the test statistic is $z = 1.82$.
- ***P*-value** The computer output shows that the *P*-value is 0.0341.

Step 4: **Interpretation** The *P*-value, 0.0341, tells us that it is unlikely that we would obtain a difference in sample proportions as large as we did if the null hypothesis is true. Judges have generally adopted a 5% significance level as their standard for convincing evidence. More than likely, the judge in this case would conclude that the companies who contaminated the well water were responsible for the higher proportion of birth defects in East Woburn.

Exercises

13.29 Drug testing in schools In 2002 the Supreme Court ruled that schools could require random drug testing of students participating in competitive after-school activities such as athletics. Does drug testing reduce use of illegal drugs? A study compared two similar high schools in Oregon. Wahtonka High School tested athletes at random and Warrenton High School did not. In a confidential survey, 7 of 135 athletes at Wahtonka and 27 of 141 athletes at Warrenton said they were using drugs.[24] Regard these athletes as SRSs from the population of athletes at similar schools with and without drug testing.

Do the data give good reason to think that drug use among athletes is lower in schools that test for drugs? Carry out an appropriate test to help answer this question.

13.30 Preventing strokes Aspirin prevents blood from clotting and so helps prevent strokes. The Second European Stroke Prevention Study asked whether adding another anticlotting drug, named dipyridamole, would be more effective for patients who had already had a stroke. Here are the data on strokes and deaths during the two years of the study:[25]

	Number of patients	Number of strokes	Number of deaths
Aspirin alone	1649	206	182
Aspirin + dipyridamole	1650	157	185

(a) The study was a randomized comparative experiment. Outline the design of the study.

(b) Is there a significant difference in the proportion of strokes between these two treatments? Carry out an appropriate test to help answer this question.

(c) Construct and interpret a 95% confidence interval for the difference in the proportion of deaths for these two treatments.

(d) Describe a Type I and a Type II error in this setting. Which is more serious? Explain.

13.31 Access to computers A sample survey by Nielsen Media Research looked at computer access and use of the Internet. Whites were significantly more likely than blacks to own a home computer, but the black-white difference in computer access at work was not significant. The study team then looked separately at the households with annual incomes of at least \$40,000. The sample contained 1916 white and 131 black households. Here are the sample counts for these households:[26]

	Blacks	**Whites**
Own home computer	86	1173
Computer access at work	100	1132

Do higher-income blacks and whites differ significantly at the 5% level in the proportion who own home computers? Do they differ significantly in the proportion who have computer access at work? Give appropriate statistical evidence to justify your answers.

13.32 Prayer and pregnancy Return to Example 13.7 (page 807).

(a) Show that the results of the study are statistically significant.

(b) Does this study show that intercessory prayer causes an increase in pregnancy for women undergoing in vitro fertilization? Why or why not?

(c) Describe a Type I and a Type II error in this setting, and give the consequences of each. Which is more serious?

Section 13.2 Summary

We want to **compare the proportions p_1 and p_2 of successes in two populations.** The comparison is based on the difference $\hat{p}_1 - \hat{p}_2$ between the sample proportions of successes. When the sample sizes n_1 and n_2 are large enough, we can use z procedures because the sampling distribution of $\hat{p}_1 - \hat{p}_2$ is close to Normal.

An approximate level C **confidence interval** for $p_1 - p_2$ is

$$(\hat{p}_1 - \hat{p}_2) \pm z^* \mathrm{SE}_{\hat{p}_1 - \hat{p}_2}$$

where the **standard error** of $\hat{p}_1 - \hat{p}_2$ is

$$\mathrm{SE}_{\hat{p}_1 - \hat{p}_2} = \sqrt{\frac{\hat{p}_1(1-\hat{p}_1)}{n_1} + \frac{\hat{p}_2(1-\hat{p}_2)}{n_2}}$$

and z^* is a standard Normal critical value.

Significance tests of H_0: $p_1 = p_2$ use the **combined sample proportion**

$$\hat{p}_c = \frac{\text{count of successes in both samples combined}}{\text{count of individuals in both samples combined}}$$

and the **z statistic**

$$z = \frac{\hat{p}_1 - \hat{p}_2}{\sqrt{\hat{p}_c(1-\hat{p}_c)\left(\dfrac{1}{n_1} + \dfrac{1}{n_2}\right)}}$$

P-values can be determined using the standard Normal table.

Section 13.2 Exercises

13.33 What's wrong with this? For each of the following, explain what is wrong and why.

(a) A z statistic is used to test the null hypothesis that $\hat{p}_1 = \hat{p}_2$.

(b) A 95% confidence interval for the difference in two proportions includes errors due to nonresponse.

13.34 Reducing nonresponse? Telephone surveys often have high rates of nonresponse. When the call is handled by an answering machine, perhaps leaving a message on the machine will encourage people to respond when they are called again. Here are data from a study in which (at random) a message was or was not left when an answering machine picked up the first call from a survey:[27]

	Total households	Eventual contact	Completed survey
No message	100	58	33
Message	291	200	134

(a) Is there good evidence that leaving a message increases the proportion of households that are eventually contacted? Justify your answer.

(b) Is there good evidence that leaving a message increases the proportion who complete the survey? Justify your answer.

(c) If you find significant effects, look at their size. Do you think these effects are large enough to be important to survey takers?

13.35 Treating AIDS, I The drug AZT was the first drug that seemed effective in delaying the onset of AIDS. Evidence for AZT's effectiveness came from a large randomized comparative experiment. The subjects were 1300 volunteers who were infected with HIV, the virus that causes AIDS, but did not yet have AIDS. The study assigned 435 of the subjects at random to take 500 milligrams of AZT each day, and another 435 to take a placebo. (The others were assigned to a third treatment, a higher dose of AZT. We will compare only two groups.) At the end of the study, 38 of the placebo subjects and 17 of the AZT subjects had developed AIDS. We want to test the claim that taking AZT lowers the proportion of infected people who will develop AIDS in a given period of time.

(a) State hypotheses, and check that you can safely use the z procedures.

(b) How significant is the evidence that AZT is effective?

(c) The experiment was double-blind. Explain what this means.

(*Comment:* Medical experiments on treatments for AIDS and other fatal diseases raise hard ethical questions. Some people argue that because AIDS is always fatal, infected people should get any drug that has any hope of helping them. The counterargument is that we will then never find out which drugs really work. The placebo patients in this study were given AZT as soon as the results were in.)

13.36 Treating AIDS, II Refer to the previous exercise. Describe a Type I and a Type II error in this setting, and give the consequences of each. Which is more serious?

13.37 Golf club repairs The Ping Company makes custom-built golf clubs and competes in the $4 billion golf equipment industry. To improve its business process, Ping decided to study the time it took to repair golf clubs sent to the company by mail.[28] The company determined that 16% of orders were sent back to the customers in 5 days or less. Ping examined the processing of repair orders and made changes. Following the changes, 90% of orders were completed within 5 days. Assume that each of the estimated percents is based on a random sample of 200 orders.

(a) Construct and interpret a 95% confidence interval for the proportion of orders completed in 5 days or less before the changes.

(b) Do the same for the orders after the changes.

(c) Construct and interpret a 95% confidence interval for the improvement. Is this interval directly related to the intervals from parts (a) and (b)? Explain.

13.38 Steroids in high school A study by the National Athletic Trainers Association surveyed 1679 high school freshmen and 1366 high school seniors in Illinois. Results showed that 34 of the freshmen and 24 of the seniors had used anabolic steroids. Steroids, which are dangerous, are sometimes used to improve athletic performance.[29]

(a) In order to draw conclusions about all Illinois freshmen and seniors, how should the study samples be chosen?

(b) Construct and interpret a 95% confidence interval for the proportion of all high school freshmen in Illinois who have used steroids.

(c) Is there a significant difference between the proportions of freshmen and seniors who have used steroids? Justify your answer using your work from (b).

13.39 Children make choices Many new products introduced into the market are targeted toward children. The choice behavior of children with regard to new products is of particular interest to companies that design marketing strategies for these products. As part of one study, children in different age groups were compared on their ability to sort new products into the correct product category (milk or juice).[30] Here are some of the data:

Age group	n	Number who sorted correctly
4- to 5-year-olds	50	10
6- to 7-year-olds	53	28

Are these two age groups equally skilled at sorting? Use information from the CrunchIt! computer output below to support your answer.

Two sample Proportion with summary

95% confidence interval results:

p_1 : proportion of successes for population 1
p_2 : proportion of successes for population 2
$p_1 - p_2$: difference in proportions

Difference	Count1	Total1	Count2	Total2	Sample Diff.	Std. Err.	L. Limit	U. Limit
p1 - p2	10	50	28	53	−0.32830188	0.08889245	−0.5025279	−0.15407588

Hypothesis test results:
p_1 : proportion of successes for population 1
p_2 : proportion of successes for population 2
$p_1 - p_2$: difference in proportions
$H_0 : p_1 - p_2 = 0$
$H_A : p_1 - p_2 \neq 0$

Difference	Count1	Total1	Count2	Total2	Sample Diff.	Std. Err.	Z-Stat	P-value
$p_1 - p_2$	10	50	28	53	−0.32830188	0.09512769	−3.4511707	0.0006

13.40 In-line skaters A study of injuries to in-line skaters used data from the National Electronic Injury Surveillance System, which collects data from a random sample of hospital emergency rooms. In the six-month study period, 206 people came to the sample hospitals with injuries from in-line skating. We can think of these people as an SRS of all people injured while skating. Researchers were able to interview 161 of these people. Wrist injuries (mostly fractures) were the most common.[31]

(a) The interviews found that 53 people were wearing wrist guards and 6 of these had wrist injuries. Of the 108 who did not wear wrist guards, 45 had wrist injuries. What are the two sample proportions of wrist injuries?

(b) Use an appropriate statistical method to determine whether there is a significant difference between the two population proportions of wrist injuries. State carefully what populations your inference compares. (We would like to draw conclusions about all in-line skaters, but we have data only for injured skaters.)

(c) What was the percent of nonresponse among the original sample of 206 injured skaters? Explain why nonresponse may bias your conclusions.

(d) If the 45 individuals who did not respond had been interviewed, could your answer to (b) change? Justify your answer.

Technology Toolbox

Comparing proportions

The TI-83/84/89 can be used to compare proportions using significance tests and confidence intervals. Here, we use the information from the Helsinki Heart Study of Examples 13.10 and 13.11.

Significance tests
In the treatment (gemfibrozil) group of 2051 middle-aged men, 56 had heart attacks. In the control (placebo) group, 84 of the 2030 men had heart attacks. The hypotheses were

$$H_0: p_1 = p_2$$
$$H_a: p_1 < p_2$$

The alternative hypothesis says that gemfibrozil reduces heart attacks. To perform a test of the null hypothesis:

(continued)

Technology Toolbox

Comparing proportions *(continued)*

TI-83/84

- Press STAT, choose TESTS, then choose 6:2-PropZTest.

TI-89

- In the Statistics/List Editor, press 2nd F1 ([F6]) and choose 6:2-PropZTest.

- When the 2-PropZTest screen appears, enter the values $x_1 = 56$, $n_1 = 2051$, $x_2 = 84$, $n_2 = 2030$. Specify the alternative hypothesis $p_1 < p_2$.

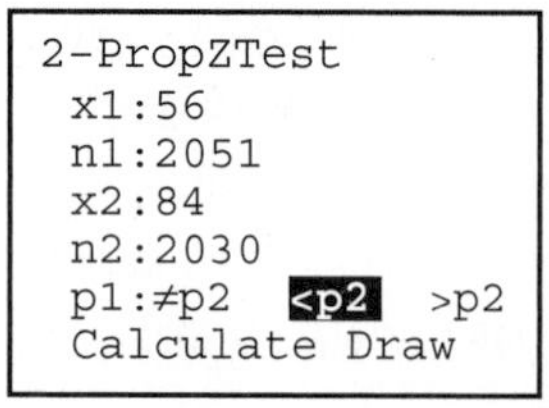

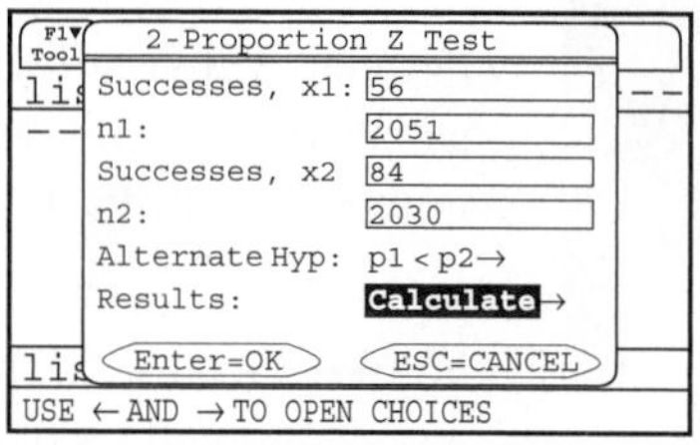

- If you select "Calculate" and press ENTER, you are told that the z statistic is $z = -2.47$ and the P-value is 0.0068, as shown here. Do you see the combined proportion of heart attacks? Does this agree with the value calculated in Example 13.11?

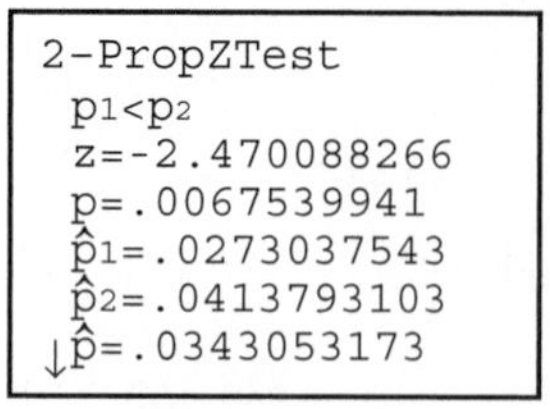

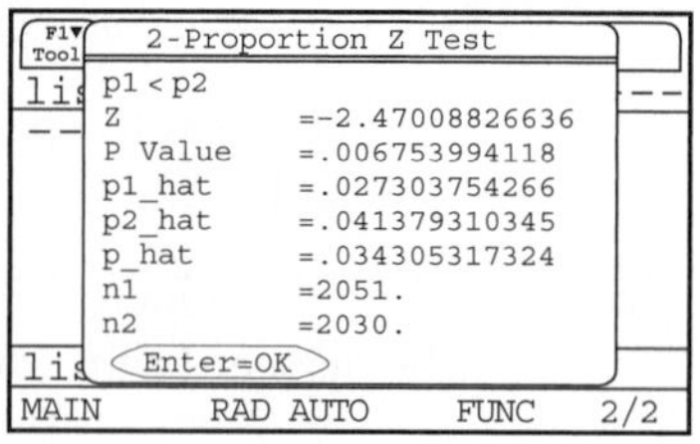

- If you select the "Draw" option, you will see the screen shown here. Compare these results with those in Example 13.11.

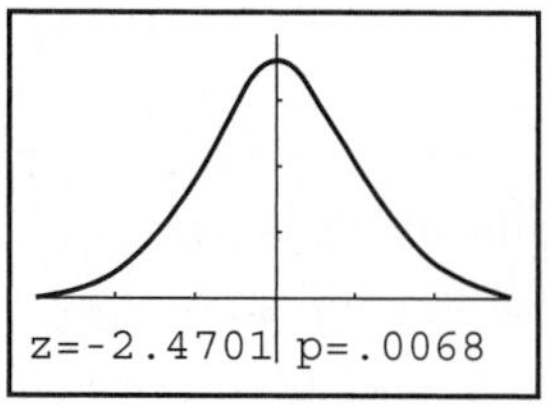

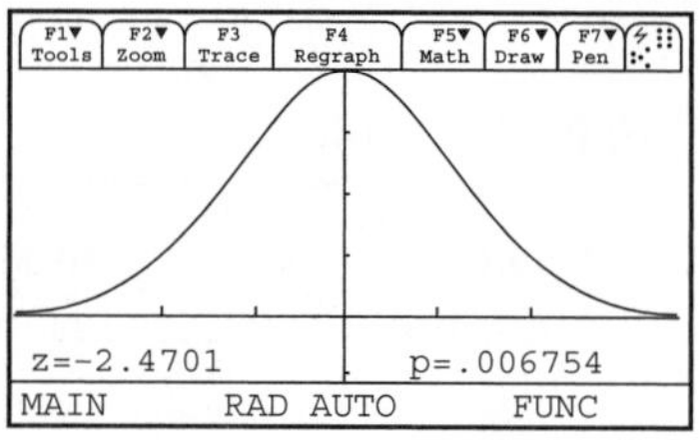

Confidence intervals

To use the TI-83/84/89 to construct a 95% confidence interval for the difference $p_1 - p_2$:

TI-83/84

- Press STAT, then choose TESTS and B:2-PropZInt.

TI-89

- In the Statistics/List Editor, press 2nd F2 ([F7]) and choose 6:2-PropZInt.

- When the 2-PropZInt screen appears, verify the values $x_1 = 56$, $n_1 = 2051$, $x_2 = 84$, $n_2 = 2030$, and specify the confidence level, 0.95.

Technology Toolbox

Comparing proportions *(continued)*

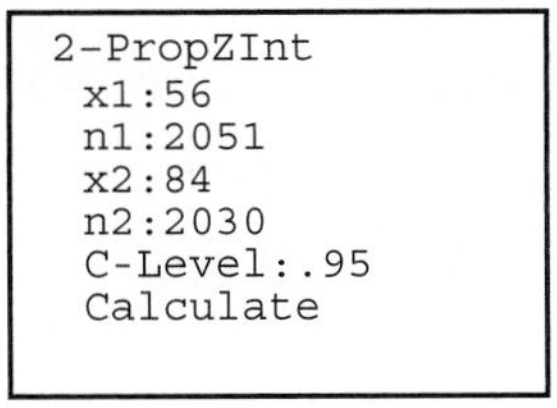

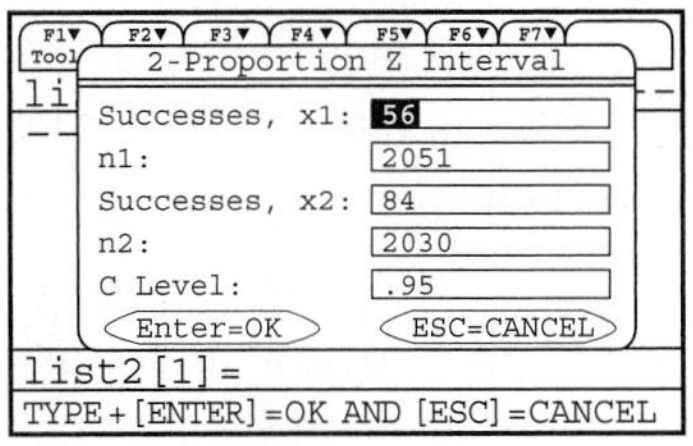

- Highlight "Calculate" and press ENTER. The 95% confidence interval for $p_1 - p_2$ is reported along with the two sample proportions and the two sample sizes, as shown here.

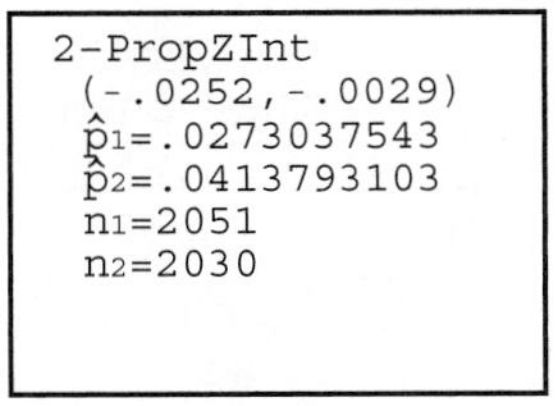

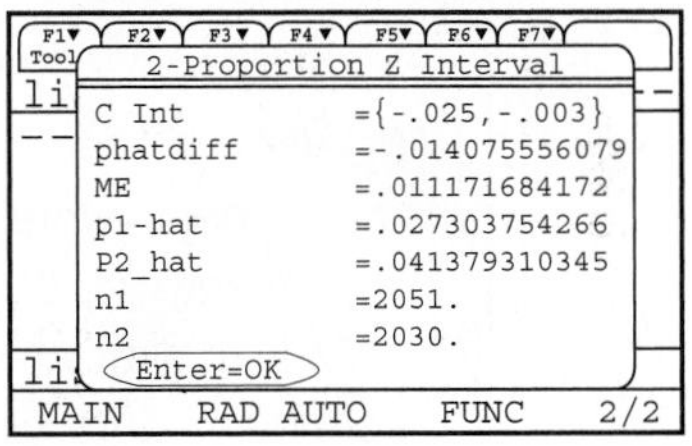

CASE CLOSED!

Fast-food frenzy

Let's return to the chapter-opening Case Study (page 779) about drive-thru service at fast-food restaurants. You are now ready to use what you have learned about comparing population parameters to make inferences about service time and accuracy in the drive-thru lane.

1. The 2001 *QSR* survey showed that McDonald's average drive-thru service time had increased for the second consecutive year to 170.85 seconds. Following that survey, many McDonald's restaurants implemented incentive programs that gave employees rewards for fast drive-thru times. That same year, Burger King instituted several measures designed to help employees more accurately fill drive-thru orders. For example, they tested see-though bags at some locations. Both of these chains would like to know if the programs they instituted had an effect. Give an appropriate pair of hypotheses for each chain to test. Be sure to define your parameters.

2. According to the 2004 *QSR* survey, McDonald's average service time for 596 drive-thru visits was 152.52 seconds with a standard deviation of 16.49 seconds. In 2001, McDonald's service time for 750 drive-thru visits had a standard deviation of 17.06 seconds. Estimate McDonald's improvement in average service time between 2001 and 2004 using a 95% confidence interval. Be sure to interpret your interval.

(continued)

3. In 2004, McDonald's ranked fourth in average service time behind Taco Bell. Researchers made 590 trips through Taco Bell drive-thrus. The average service time was 148.16 seconds, and the standard deviation was 18.71 seconds. Do these data provide evidence that drive-thru service times at Taco Bell were significantly faster than those at McDonald's in 2004? Give appropriate evidence to justify your answer.

4. In 2001, 730 of the 890 Burger King restaurants visited gave correct change and accurately filled the order requested. In 2002, 654 of the 742 Burger King restaurants visited gave correct change and accurately filled the order requested. Was there a significant improvement in accuracy between 2001 and 2002 following Burger King's corporate emphasis on order accuracy? Carry out the appropriate statistical procedure to address this question. Explain your results to Burger King's management so that they could be understood by someone who is not an expert in statistics.

5. When the four aspects of drive-thru quality are combined, Chick-fil-A earned the highest overall ranking in 2002. For example, Chick-fil-A had the lowest inaccuracy rate of any of the chains (196 visits; 14 inaccurate orders). McDonald's inaccuracy rate was 90 out of 750 orders. Construct and interpret a 95% confidence interval for the difference between the inaccuracy rate at Chick-fil-A and the rate at McDonald's. Explain the meaning of this interval.

Chapter Review

Summary

Problems involving comparison of two population parameters are among the most common in statistical practice. Such problems arise from randomized comparative experiments and from observational studies comparing two distinct populations. In either case, we get data from two independent samples. This chapter presents two-sample inference methods for comparing population means and proportions.

We use two-sample t procedures for confidence intervals and significance tests about the difference in population means. There are two options for the degrees of freedom: a more conservative value for use with Table C and a larger value used by technology. The conditions for the two-sample t test and interval are familiar: SRS, Normality, and independence.

Two-proportion z procedures require the same three conditions to be met. Use the sample proportions to verify the Normality condition for a confidence interval, and the combined proportion to verify the Normality condition for a significance test. Use these same values in Step 3 of the Inference Toolbox.

What You Should Have Learned

Here is a checklist of the major skills you should have acquired by studying this chapter.

A. Recognition

1. Determine whether a problem requires inference about comparing means or proportions.
2. Recognize from the design of a study whether one-sample t, paired t, or two-sample t procedures are needed.

B. Two-Sample t Procedures

1. Calculate and interpret a confidence interval for the difference between two means. Use the two-sample t statistic with conservative degrees of freedom if you do not have statistical software. Use the TI-83/84/89 or software if you have it.
2. Test the hypothesis that two populations have equal means against either a one-sided or a two-sided alternative. Use the two-sample t test with conservative degrees of freedom if you do not have statistical software. Use the TI-83/84/89 or software if you have it.
3. Recognize when the two-sample t procedures are appropriate in practice.

C. Comparing Two Proportions

1. Use the two-sample z procedure to give a confidence interval for the difference $p_1 - p_2$ between proportions in two populations based on independent SRSs from the populations.
2. Use a two-proportion z test to test the hypothesis H_0: $p_1 = p_2$ that proportions in two distinct populations are equal.
3. Check that you can safely use these z procedures in a particular setting.

Web Links

For more information about fast-food restaurants, visit www.qsrmagazine.com/drive-thru/

Chapter Review Exercises

13.41 Expensive ads Consumers who think a product's advertising is expensive often also think the product must be of high quality. Can other information undermine this effect? To find out, marketing researchers did an experiment. The subjects were 90 women from the clerical and administrative staff of a large organization. All subjects read an ad that described a fictional line of food products called "Five Chefs." The ad also described the major TV commercials that would soon be shown, an unusual expense for this type of product. The 45 women in the control group read nothing else. The 45 in the "undermine group" also read a news story headlined "No Link between Advertising Spending and New Product Quality."

All the subjects then rated the quality of Five Chefs products on a seven-point scale. The study report said, "The mean quality ratings were significantly lower in the undermine treatment ($\overline{x}_A = 4.56$) than in the control treatment ($\overline{x}_C = 5.05$; $t = 2.64$, $P < 0.01$)."[32]

(a) Is the matched pairs t test or the two-sample t test the right test in this setting? Why?

(b) What degrees of freedom would you use for the t statistic you chose in (a)?

(c) The distribution of individual responses is not Normal, because there is only a seven-point scale. Why is it nonetheless proper to use a t test?

13.42 Seat belt use The proportion of drivers who use seat belts depends on things like age (young people are more likely to go unbelted) and gender (women are more likely to buckle up). It also depends on local law. In New York City, police can stop a driver who is not belted. In Boston (as of late 2000), police can cite a driver for not wearing a seat belt only if the driver has been stopped for some other violation. Here are data from observing random samples of female Hispanic drivers in these two cities:[33]

City	Drivers	Belted
New York	220	183
Boston	117	68

(a) Is this an experiment or observational study? Why?

(b) Comparing local law suggests that a smaller proportion of drivers wear seat belts in Boston than in New York. Do the data give good evidence that this is true for female Hispanic drivers? Justify your answer.

13.43 Ethnicity and seat belt use Here are data from the study described in the previous exercise for Hispanic and white male drivers in Chicago:

Group	Drivers	Belted
Hispanic	539	286
White	292	164

Is there a significant difference between Hispanic and white drivers? How large is the difference? Use appropriate inference methods to answer both questions. Then write a few sentences summarizing your results.

13.44 Each day I am getting better in math A "subliminal" message is below our threshold of awareness but may nonetheless influence us. Can subliminal messages help students learn math? A group of students who had failed the mathematics part of the City University of New York Skills Assessment Test agreed to participate in a study to find out.

All received a daily subliminal message, flashed on a screen too rapidly to be consciously read. The treatment group of 10 students (chosen at random) was exposed to "Each day I am getting better in math." The control group of 8 students was exposed to a neutral message, "People are walking on the street." All students participated in a summer program designed to raise their math skills, and all took the assessment test again at the end of the program. Table 13.1 gives data on the subjects' scores before and after the program.

Table 13.1 *Mathematics skills scores before and after a subliminal message*

Treatment Group		Control Group	
Pretest	**Posttest**	**Pretest**	**Posttest**
18	24	18	29
18	25	24	29
21	33	20	24
18	29	18	26
18	33	24	38
20	36	22	27
23	34	15	22
23	36	19	31
21	34		
17	27		

Source: Data provided by Warren Page, New York City Technical College, from a study done by John Hudesman.

(a) Is there good evidence at the 5% level that the treatment brought about a greater improvement in math scores than the neutral message? Justify your answer.

(b) Construct and interpret a 90% confidence interval for the mean difference in gains between treatment and control.

(c) Carefully describe the design of this study.

13.45 Men versus women The National Assessment of Educational Progress (NAEP) Young Adult Literacy Assessment Survey interviewed a random sample of 1917 people 21 to 25 years old. The sample contained 840 men, of whom 775 were fully employed. There were 1077 women, and 680 of them were fully employed.[34]

(a) Use a 99% confidence interval to describe the difference between the proportions of young men and young women who are fully employed. Is the difference statistically significant at the 1% significance level? Justify your answer.

(b) The mean and standard deviation of scores on the NAEP's test of quantitative skills were $\bar{x}_1 = 272.40$ and $s_1 = 59.2$ for the men in the sample. For the women, the results were $\bar{x}_2 = 274.73$ and $s_2 = 57.5$. Is the difference between the mean scores for men and women significant at the 1% level? Give appropriate statistical evidence to justify your answer.

13.46 Independent samples versus paired samples Deciding whether to perform a matched pairs t test or a two-sample t test can be tricky.[35] Your decision should be based on the design that produced the data. Which procedure would you choose in each of parts (a)–(e)? Be sure to look on the next page.

(a) To compare the average weight gain of pigs fed two different rations, nine pairs of pigs were used. The pigs in each pair were littermates.

(b) To test the effects of a new fertilizer, 100 plots are treated with the new fertilizer, and 100 plots are treated with another fertilizer.

(c) A sample of college teachers is taken. We wish to compare the average salaries of male and female teachers.

(d) A new fertilizer is tested on 100 plots. Each plot is divided in half. Fertilizer A is applied to one half and B to the other.

(e) Consumers Union wants to compare two types of calculators. They get 100 volunteers and ask them to carry out a series of 50 routine calculations (such as figuring discounts, sales tax, totaling a bill). Each calculation is done on each type of calculator, and the time required for each calculation is recorded.

13.47 Comparing welfare programs A major study of alternative welfare programs randomly assigned women on welfare to one of two programs, called "WIN" and "Options." WIN was the existing program. The new Options program gave more incentives to work. An important question was how much more (on the average) women in Options earned than those in WIN. Here is Minitab output for earnings in dollars over a 3-year period:

```
TWOSAMPLE T FOR 'OPT' VS 'WIN'

            N      MEAN        STDEV        SE MEAN

OPT      1362      7638        289          7.8309

WIN      1395      6595        247          6.6132

95 PCT CI FOR MU OPT - MU WIN: (1022.90, 1063.10)
```

(a) Calculate and interpret a 99% confidence interval for the amount by which the mean earnings of Options participants exceeded the mean earnings of WIN subjects. (Minitab will give a 99% confidence interval if you instruct it to do so. Here we have only the basic output, which includes the 95% confidence interval.)

(b) The distribution of incomes is strongly skewed to the right but includes no extreme outliers because all the subjects were on welfare. What fact about these data allows us to use *t* procedures despite the strong skewness?

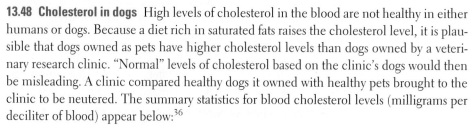

13.48 Cholesterol in dogs High levels of cholesterol in the blood are not healthy in either humans or dogs. Because a diet rich in saturated fats raises the cholesterol level, it is plausible that dogs owned as pets have higher cholesterol levels than dogs owned by a veterinary research clinic. "Normal" levels of cholesterol based on the clinic's dogs would then be misleading. A clinic compared healthy dogs it owned with healthy pets brought to the clinic to be neutered. The summary statistics for blood cholesterol levels (milligrams per deciliter of blood) appear below:[36]

Group	n	$\bar{x}$	s
Pets	26	193	68
Clinic	23	174	44

(a) Is there strong evidence that pets have higher mean cholesterol level than clinic dogs? Justify your answer.

(b) Calculate and interpret a 95% confidence interval for the difference in mean cholesterol levels between pets and clinic dogs.

(c) Construct and interpret a 95% confidence interval for the mean cholesterol level in pets.

(d) In (a), (b), and (c), assuming that the cholesterol measurements have no outliers and are not strongly skewed, what is the chief threat to the validity of the results of this study?

13.49 Air pollution and wheezing A study that evaluated the effects of a reduction in exposure to traffic-related air pollutants compared respiratory symptoms of 283 residents of an area with congested streets with 165 residents in a similar area where the congestion was removed because a bypass was constructed. The symptoms of the residents were evaluated at the beginning of the study and again a year after the bypass was completed. For the residents of the congested streets, 17 reported that symptoms of wheezing improved during the year, while 35 of the residents of the bypass streets reported improvement.

(a) Find the two sample proportions. Report the difference in the proportions and the standard error of the difference.

(b) What are the appropriate null and alternative hypotheses for examining the question of interest? Be sure to explain your choice of the alternative.

(c) Find the test statistic. Construct a sketch of the distribution of the test statistic under the assumption that the null hypothesis is true. Find the *P*-value and use your sketch to explain its meaning.

(d) Construct and interpret a 95% confidence interval to estimate the size of the effect of reducing congestion on wheezing. Summarize your ideas in a way that could be understood by someone who has very little experience with statistics.

(e) The study was done in the United Kingdom. To what extent do you think that the results can be generalized to other locations?

13.50 Child care workers The Current Population Survey (CPS) is the monthly government sample survey of 60,000 households that provides data on employment in the United States. A study of child care workers drew a sample from the CPS data tapes. We can consider this sample to be an SRS from the population of child care workers.[37]

(a) Out of 2455 child care workers in private households, 7% were black. Of 1191 nonhousehold child care workers, 14% were black. Construct and interpret a 99% confidence interval for the difference in the proportions of these groups of workers who are black. Is the difference statistically significant at the $\alpha = 0.01$ level? Justify your answer.

(b) The study also examined how many years of school child care workers had. For household workers, the mean and standard deviation were $\overline{x}_1 = 11.6$ years and $s_1 = 2.2$ years. For nonhousehold workers, $\overline{x}_2 = 12.2$ years and $s_2 = 2.1$ years. Construct and interpret a 99% confidence interval for the difference in mean years of education for the two groups. Is the difference significant at the $\alpha = 0.01$ level? Justify your answer.

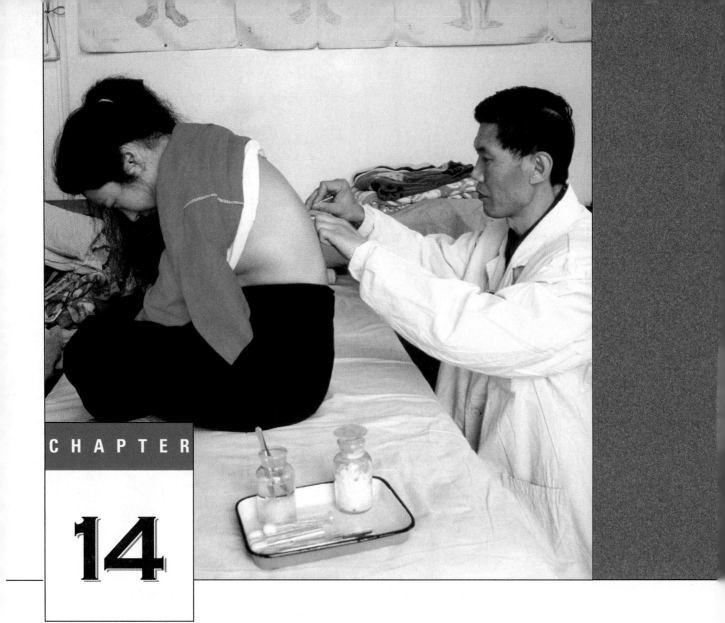

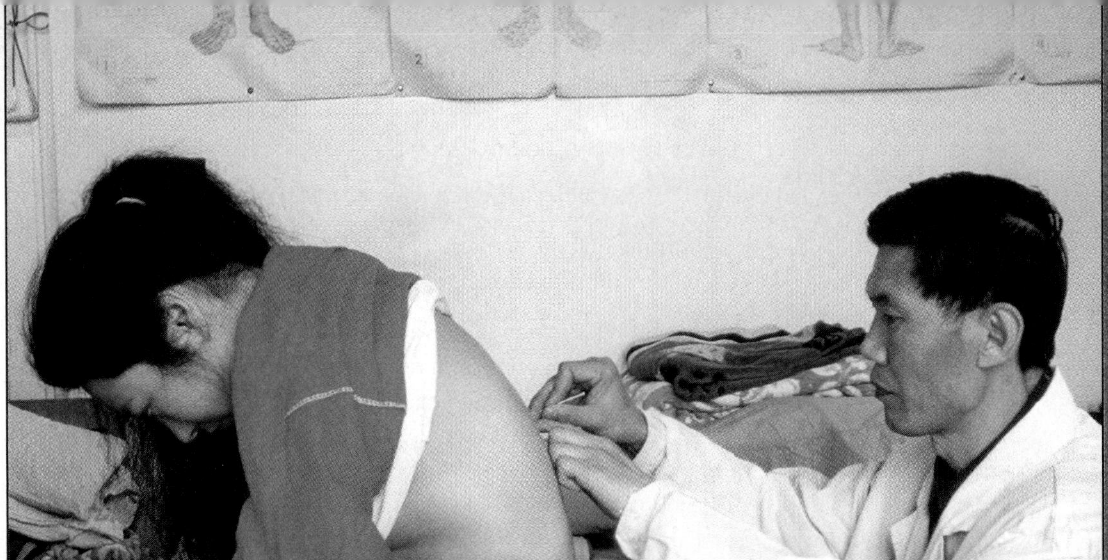

CHAPTER

14

Inference for Distributions of Categorical Variables: Chi-Square Procedures

CASE STUDY

Does acupuncture promote pregnancy?

A study reported in the medical journal *Fertility and Sterility* in 2002 sought to determine if the ancient Chinese art of acupuncture could help infertile women become pregnant.[1] One hundred sixty healthy women undergoing treatment with in vitro fertilization (IVF) or intracytoplasmic sperm injection (ICSI) were recruited for the study. The purpose of the study was to determine if acupuncture improves the clinical pregnancy rate after IVF or ICSI treatment.

Only patients with good embryo quality were accepted into the study. The ages of the subjects ranged from 21 to 43. The cause of the infertility was the same for both groups. Half of the subjects (80) were randomly assigned to an experimental (acupuncture) treatment group, and the remaining 80 were assigned to a control group.

The subjects in the treatment/acupuncture group received acupuncture treatment 25 minutes before embryo transfer and again 25 minutes after the transfer. Subjects in the control group were instructed to lie still for 25 minutes after the embryo transfer. Results are tabulated in the table below.

	Acupuncture group	Control group
Pregnant	34	21
Not pregnant	46	59
Total	**80**	**80**

When you finish this chapter, you will be able to apply an appropriate inference procedure to answer the question the researchers were investigating.

Activity 14A

"I didn't get enough reds!"

Materials needed: One 1.69-ounce bag of milk chocolate M&M's per student.

The M&M/Mars Company, headquartered in Hackettstown, New Jersey, makes milk chocolate candies. In 1995, they decided to replace the tan-colored M&M's with a new color. After conducting an extensive national preference survey, they decided to replace the tan M&M's with blue M&M's. The company has since changed the proportions of colors, and the company's Consumer Affairs Department announced:

> *On average, the new mix of colors of M&M's Milk Chocolate Candies will contain 13 percent of each of browns and reds, 14 percent yellows, 16 percent greens, 20 percent oranges and 24 percent blues. . . .While we mix the colors as thoroughly as possible, the above ratios may vary somewhat, especially in the smaller bags. This is because we combine the various colors in large quantities for the last production stage (printing). The bags are then filled on high-speed packaging machines by weight, not by count.*

The purpose of this Activity is to compare the color distribution of M&M's in your individual bag with the advertised distribution. We want to see if there is sufficient evidence to dispute the company's claim for their distribution. In order to use as random a sample as possible, it is best if the bags of M&M's are purchased at different stores and not obtained from one or a few sources of supply.

1. Open your bag and carefully count the number of M&M's of each color—brown, yellow, red, orange, green, and blue—as well as the total number of M&M's in the bag.
2. Fill in the counts, by color, and the total number of M&M's in your bag in the "Observed" row in a table like this:

	Brown	Yellow	Red	Orange	Green	Blue	Total
Observed, O							
Expected, E							
$(O - E)^2/E$							$X^2 =$

3. To obtain the expected counts, multiply the total number of M&M's in your bag by the company's stated percents (expressed in decimal form) for each of the colors.

4. For each color, perform this calculation:

$$\frac{(\text{observed} - \text{expected})^2}{\text{expected}} = \frac{(O - E)^2}{E}$$

and enter the result in the last row of the table. Then add up all of these calculated values, and name the sum X^2. Keep this number handy—you will use it later in the chapter.

5. If your sample reflects the distribution advertised by the M&M/Mars Company, there should be very little difference between the observed counts and the expected counts, and therefore, the calculated values making up the sum X^2 should be very small. Are the entries in the last row all about the same, or do any of the quantities stand out because they are "significantly" larger than the others? Did you get more of a particular color than you expected? Did you get fewer of a particular color than you expected?

6. Combine the counts obtained by all the students in your class to obtain a total count of M&M's of each color. Record the results in a table like this:

Color:	Brown	Yellow	Red	Orange	Green	Blue	Total
Observed:							

You will need these data in the exercises.

7. Record the total number of M&M's in each student's bag in your class. How did your bag compare with those of your classmates?

Introduction

In the previous chapter, we discussed inference procedures for comparing two population proportions. Sometimes we want to examine the distribution of proportions in a single population. The *chi-square test for goodness of fit* allows us to determine whether a specified population distribution seems valid. We can compare two or more population proportions using a *chi-square test for homogeneity of populations.* In doing so, we will organize our data in a two-way table. It is also possible to use the information provided in a two-way table to determine whether the distribution of one variable has been influenced by another variable. The *chi-square test of association/independence* helps us decide this issue.

Of course, we need to do a careful job of describing the data before we apply our methods of statistical inference. In Chapter 4, Section 2, we discussed data analysis methods for two-way tables. To prepare for Section 14.2, you may want to do a quick review of Section 4.2, beginning on page 292.

The methods of this chapter help us answer questions such as these:

- Are you more likely to have a motor vehicle collision when using a cell phone?
- Does background music influence wine purchases?
- How does the presence of an exclusive-territory clause in a franchisee's contract relate to the success of the business?

14.1 Test for Goodness of Fit

Suppose you open a 1.69-ounce bag of M&M's Milk Chocolate Candies and discover that out of 56 total M&M's in the bag, there are only 2 red M&M's. Knowing that 13% of all plain M&M's made by the M&M/Mars Company are red, and that in your sample of size 56, the proportion of red M&M's is only 2/56 = 0.036, you might feel that you didn't get your fair share of reds. You could use the z test described in Chapter 12 to test the hypotheses

$$H_0\colon p = 0.13$$
$$H_a\colon p < 0.13$$

where p is the proportion of red M&M's. You could then perform additional tests of significance for each of the remaining colors. But this would be inefficient. More important, it wouldn't tell us how likely it is that six sample proportions differ from the values stated by M&M/Mars Company as much as our sample does. There is a single test that can be applied to see if the observed sample distribution is significantly different in some way from the hypothesized population distribution. It is called the ***chi-square (χ^2) test for goodness of fit.***

chi-square (χ^2) test for goodness of fit

Example 14.1 *Auto accidents and cell phones*
Hypotheses for a goodness of fit test

Are you more likely to have a motor vehicle collision when using a cell phone? A study of 699 drivers who were using a cell phone when they were involved in a collision examined this question.[2] These drivers made 26,798 cell phone calls during a 14-month study period. Each of the 699 collisions was classified in various ways. Here are the counts for each day of the week:

Day:	Sun	Mon	Tue	Wed	Thu	Fri	Sat	**Total**
Number:	20	133	126	159	136	113	12	699

We have a total of 699 accidents involving drivers who were using a cell phone at the time of their accident. Let's explore the relationship between these accidents and the day of the week. Are the accidents equally likely to occur on any day of the week?

one-way table

cells

We can think of this table of counts as a ***one-way table*** with seven ***cells***, each with a count of the number of accidents that occurred on the particular day of the week. Our question is translated into the following hypotheses:

H_0: Motor vehicle accidents involving cell phone use are equally likely to occur on each of the seven days of the week.
H_a: The probabilities of a motor vehicle accident involving cell phone use vary from day to day (that is, they are not all the same).

We can also state the hypotheses in terms of population proportions:

H_0: $p_{\text{Sunday}} = p_{\text{Monday}} = \ldots = p_{\text{Saturday}} = \frac{1}{7}$

H_a: At least one of the proportions differs from the stated value.

The idea of the goodness of fit test is this: we compare the observed counts for our sample with the counts that would be expected if the number of motor vehicle accidents involving cell phones occurred uniformly throughout the week. The uniform distribution of accidents is the population. The more the observed counts differ from the expected counts, the more evidence we have to reject H_0 and conclude that the probabilities of an accident involving cell phone use are not the same for each day of the week.

expected count

In general, the ***expected count*** for any categorical variable is obtained by multiplying the proportion of the distribution for each category by the sample size.

Example 14.2 ***Accidents and cell phones***

Carrying out a χ^2 test for goodness of fit

Before proceeding with a significance test, it's always a good idea to plot the data. In this case, a bar graph allows you to compare the observed number of accidents by day with the expected number of accidents by day. The counts as well as the percents are shown in Table 14.1.

Table 14.1 ***The counts and percents of accidents involving cell phone use by day of week***

Day	Count	Percent	Cumulative percent
Sunday	20	2.86	2.86
Monday	133	19.0	21.86
Tuesday	126	18.0	39.86
Wednesday	159	22.7	62.56
Thursday	136	19.5	82.06
Friday	113	16.2	98.26
Saturday	12	1.72	99.98
Total	699	99.98	

Notice that the percents do not add to 100% because of roundoff error.

Next we calculate the expected counts for each day. Since there are 699 accidents, the expected number of accidents for each day is 699/7 = 99.86. Under the null hypothesis we expect one-seventh of the accidents to occur on each day. Now we are ready to picture the data. The graph is shown in Figure 14.1.

To determine whether the distribution of accidents is uniform, we need a way to measure how well the observed counts (O) fit the expected counts (E) under H_0. The procedure is to calculate the quantity

$$\frac{(\text{observed count} - \text{expected count})^2}{\text{expected count}} = \frac{(O - E)^2}{E}$$

for each day and then add up these terms. The sum is denoted X^2 and is called the ***chi-square statistic.***

Figure 14.1 *Bar graph for the data in Example 14.2.*

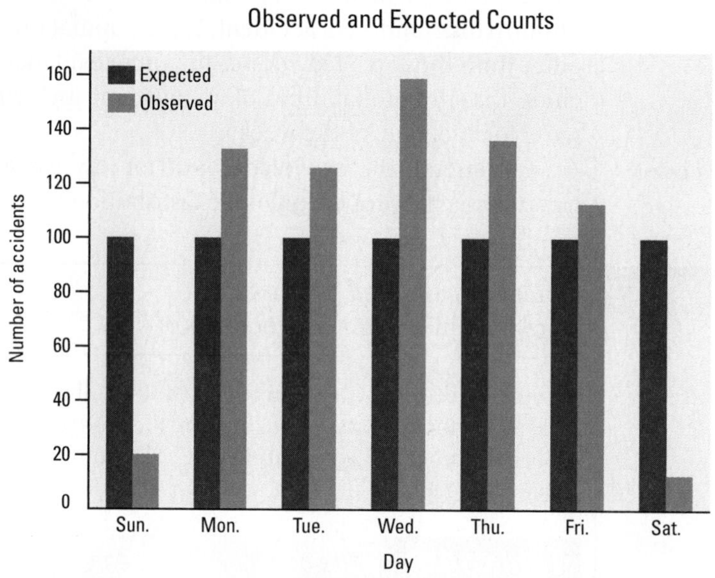

For Sunday

$$\frac{(O - E)^2}{E} = \frac{(20 - 99.86)^2}{99.86} = 63.86$$

and for Monday,

$$\frac{(O - E)^2}{E} = \frac{(133 - 99.86)^2}{99.86} = 11.0$$

and so forth. A summary of the calculations is shown in Table 14.2.

Table 14.2 *Calculating the chi-square statistic*

Day	Observed count, O	Expected count, E	$\frac{(O-E)^2}{E}$
Sunday	20	99.857	63.86
Monday	133	99.857	11.00
Tuesday	126	99.857	6.84
Wednesday	159	99.857	35.03
Thursday	136	99.857	13.08
Friday	113	99.857	1.73
Saturday	12	99.857	77.30
			$X^2 = 208.84$

The sum of the terms in the last column is

$$X^2 = \sum \frac{(O-E)^2}{E} = 208.84$$

The larger the differences between the observed and expected values, the larger X^2 will be, and the more evidence there will be against H_0.

The chi-square family of distribution curves is used to assess the evidence against H_0 represented in the value of X^2. The specific member of the family that is used is determined by the ***degrees of freedom.*** A χ^2 distribution with k degrees of freedom is written $\chi^2(k)$. Since we are working not only with counts but also with percents, six of the seven percents are free to vary, but the seventh is not, since all seven have to add to 100. In this case, we say that there are $7 - 1 = 6$ degrees of freedom. Remember that the degrees of freedom are 1 less than the number of cells in the one-way table of counts (not including the "Total" column). In the back of the book, Table D, Chi-Square Distribution Critical Values, shows a typical chi-square curve with the right-tail area shaded. The chi-square test statistic is a point on the horizontal axis, and the area to the right under the curve is the P-value of the test. This P-value is the probability of observing a value of X^2 at least as extreme as the one actually observed. The

degrees of freedom

larger the value of the chi-square statistic, the smaller the *P*-value and the more evidence you have against the null hypothesis H_0.

In Table D, for a *P*-value of 0.0005 and 6 degrees of freedom, we find that the critical value is 24.10. Since our $X^2 = 208.84$ is more extreme (larger) than the critical value, we say that the probability of observing a result as extreme as the one we actually observed, by chance alone, is less than 0.05%. There is sufficient evidence to reject H_0 and conclude that these types of accidents are not equally likely to occur on each of the seven days of the week.

The chi-square test applied to the hypothesis that a categorical variable has a specified distribution is called the ***test for goodness of fit***. The idea is that the test assesses whether the observed counts "fit" the hypothesized distribution. Here are the details.

The Chi-Square Test for Goodness of Fit

A **goodness of fit test** is used to help determine whether a population has a certain hypothesized distribution, expressed as proportions of individuals in the population falling into various outcome categories. Suppose that the hypothesized distribution has *k* outcome categories. To test the hypothesis

H_0: the actual population proportions are *equal to* the hypothesized proportions

first calculate the **chi-square test statistic**

$$X^2 = \sum \frac{(\text{observed count} - \text{expected count})^2}{\text{expected count}} = \sum \frac{(O - E)^2}{E}$$

Then X^2 has approximately a χ^2 distribution with $(k - 1)$ degrees of freedom. For a test of H_0 against the alternative hypothesis

H_a: at least two of the actual population proportions *differ from* their hypothesized proportions

the *P*-value is $P(\chi^2 \geq X^2)$.

Conditions: You may use this test with critical values from the chi-square distribution when all individual expected counts are at least 1 and no more than 20% of the expected counts are less than 5.

Notice the conditions for using the chi-square goodness of fit test. Look again at the difference terms $\frac{(O - E)^2}{E}$. We don't want to divide by zero, and since we're

CAUTION

working with *counts*, we require that *all expected counts* be greater than zero. **In checking conditions, remember that it's the expected counts that are critical, not the observed counts.** In Example 14.2, note that all of the expected cell counts are greater than 5.

Properties of the Chi-Square Distributions

Example 14.2 illustrated the mechanics of the chi-square goodness of fit test. Now we turn our attention to the ***chi-square distributions.***

The Chi-Square Distributions

The **chi-square distributions** are a family of distributions that take only positive values and are skewed to the right. A specific chi-square distribution is specified by one parameter, called the *degrees of freedom*.

Figure 14.2 shows the density curves for three members of the chi-square family of distributions. As the degrees of freedom increase, the density curves become less skewed and larger values become more likely. Table D in the back of the book gives critical values for chi-square distributions. To get *P*-values for a chi-square

Figure 14.2 *Density curves for the chi-square distributions with 1, 4, and 8 degrees of freedom. Chi-square distributions take only positive values.*

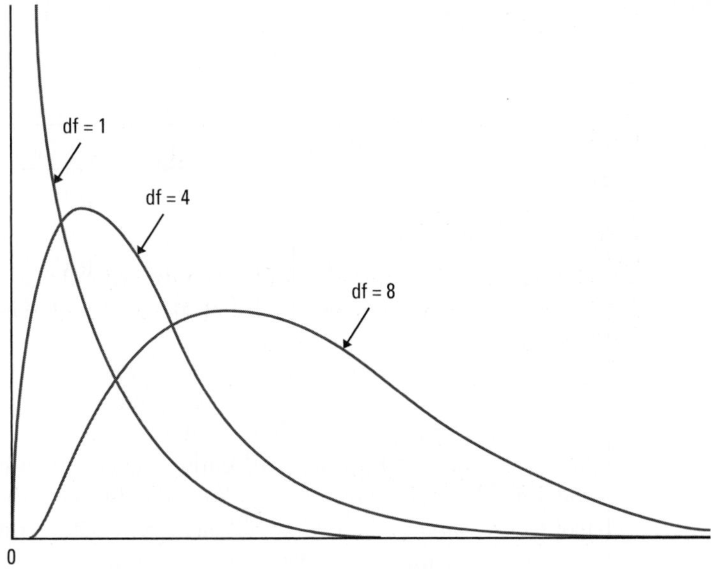

test, you can use Table D, computer software, or your TI-83/84/89. The details on using the calculator will be provided later in a Technology Toolbox.

The chi-square density curves have the following properties:

1. The total area under a chi-square curve is equal to 1.
2. Each chi-square curve (except when degrees of freedom = 1) begins at 0 on the horizontal axis, increases to a peak, and then approaches the horizontal axis asymptotically from above.
3. Each chi-square curve is skewed to the right. As the number of degrees of freedom increase, the curve becomes more and more symmetrical and looks more like a Normal curve.

One of the most common applications of the chi-square goodness of fit test is in the field of genetics. Scientists want to investigate the genetic characteristics of offspring that result from mating (also called "crossing") parents with known genetic makeups. Scientists use rules about dominant and recessive genes to predict the ratio of offspring that will fall in each possible genetic category. Then, the researchers mate the parents and classify the resulting offspring. The chi-square goodness of fit test helps the scientists assess the validity of their hypothesized ratios.

Example 14.3 *Red-eyed fruit flies*

Goodness of fit test

Biologists wish to mate two fruit flies having genetic makeup RrCc, indicating that each has one dominant gene (R) and one recessive gene (r) for eye color, along with one dominant (C) and one recessive (c) gene for wing type. Each offspring will receive one gene for each of the two traits from each parent. The following table, often called a Punnett square, shows the possible combinations of genes received by the offspring:

		Parent 2 passes on			
		RC	**Rc**	**rC**	**rc**
	RC	RRCC (x)	RRCc (x)	RrCC (x)	RrCc (x)
Parent 1 passes on	**Rc**	RRCc (x)	RRcc (y)	RrCc (x)	Rrcc (y)
	rC	RrCC (x)	RrCc (x)	rrCC (z)	rrCc (z)
	rc	RrCc (x)	Rrcc (y)	rrCc (z)	rrcc (w)

Any offspring receiving an R gene will have red eyes, and any offspring receiving a C gene will have straight wings. So based on this Punnett square, the biologists predict a ratio of 9 red-eyed, straight-winged (x):3 red-eyed, curly-winged (y):3 white-eyed, straight-winged (z):1 white-eyed, curly-winged (w) offspring. To test their hypothesis about the distribution of offspring, the biologists mate the fruit flies. Of 200 offspring, 99 had red eyes and straight wings, 42 had red eyes and curly wings, 49 had white eyes and straight wings, and 10 had white eyes and curly wings. Do these data differ significantly from what the biologists have predicted?

We return to the familiar structure of the Inference Toolbox to carry out the significance test.

Step 1: Hypotheses The biologists are interested in the proportion of offspring that fall into each genetic category for the population of all fruit flies that would result from crossing two parents with genetic makeup RrCc. Their hypotheses are

H_0: $p_{\text{red,straight}} = 0.5625, p_{\text{red,curly}} = 0.1875, p_{\text{white,straight}} = 0.1875, p_{\text{white,curly}} = 0.0625$

H_a: at least one of these proportions is incorrect

Step 2: Conditions We can use a chi-square goodness of fit test to measure the strength of the evidence against the hypothesized distribution, provided that the expected cell counts are large enough. Here are the expected counts:

Red-eyed, straight-winged:	(200)(0.5625) = 112.5
Red-eyed, curly-winged:	(200)(0.1875) = 37.5
White-eyed, straight-winged:	(200)(0.1875) = 37.5
White-eyed, curly-winged:	(200)(0.0625) = 12.5

Since all the expected counts are greater than 5, we can proceed with the test.

Step 3: Calculations

- The test statistic is

$$X^2 = \sum \frac{(O - E)^2}{E} = \frac{(99 - 112.5)^2}{112.5} + \frac{(42 - 37.5)^2}{37.5} + \frac{(49 - 37.5)^2}{37.5} + \frac{(10 - 12.5)^2}{12.5}$$

$$= 1.62 + 0.54 + 3.5267 + 0.5 = 6.187$$

- For df = 4 − 1 = 3, Table D shows that our test statistic falls between the critical values for a 0.15 and a 0.10 significance level. Technology produces the actual *P*-value of 0.1029.

Step 4: Interpretation The *P*-value of 0.1029 indicates that the probability of obtaining a sample of 200 fruit fly offspring in which the proportions differ from the hypothesized values by at least as much as the ones in our sample is over 10%, assuming that the null hypothesis is true. This is not sufficient evidence to reject the biologists' predicted distribution.

You can simplify the computations of Example 14.3 and other goodness of fit problems by using your calculator's list operations. The calculator also allows you to compute and visualize the *P*-value for a goodness of fit test.

Technology Toolbox

Goodness of fit tests

- Clear lists L_1/listl, L_2/list2, and L_3/list3.
- Enter the observed counts from Example 14.3 in L_1/listl. Calculate the expected counts separately and enter them in L_2/list2.

(continued)

Technology Toolbox

Goodness of fit tests *(continued)*

TI-83/84

- Define L_3 as $(L_1 - L_2)^2 / L_2$.

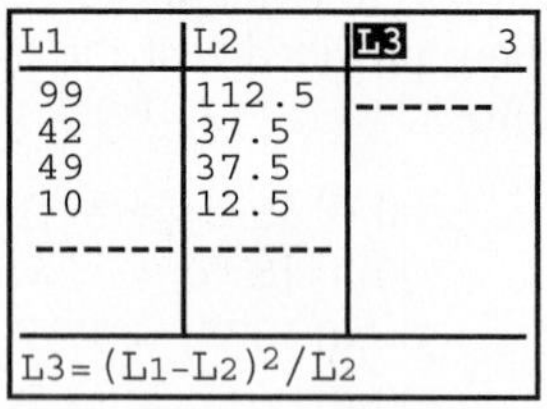

- Use the command `sum(L3)` to calculate the test statistic X^2. (`sum` is located in `MATH/LIST`.)

TI-89

- Define list3 as $[(\text{list1} - \text{list2})^2 / \text{list2}]$.

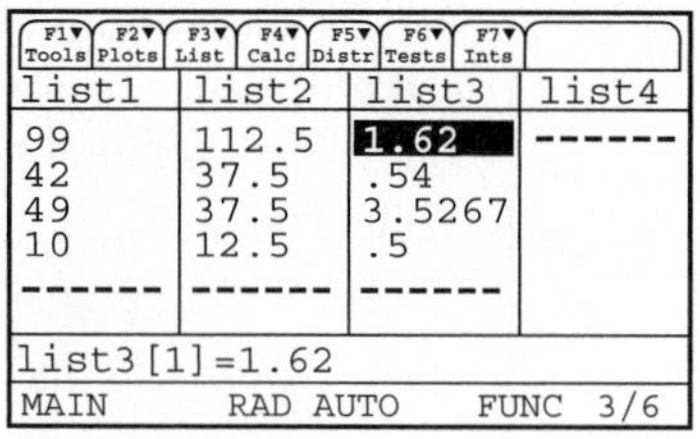

- Use the command `sum(list3)` to calculate X^2. (`sum` is located in the CATALOG.)

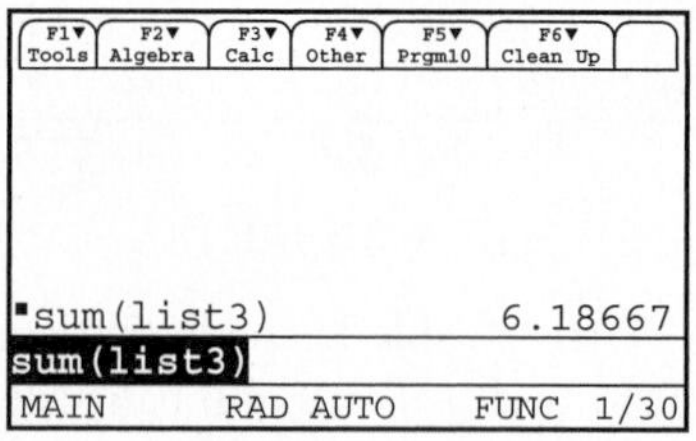

- Find the *P*-value using the χ^2`cdf` command (in the distributions, DISTR, menu on the TI-83/84 and in the CATALOG under Flash Apps on the TI-89).
- We ask for the area between $X^2 = 6.187$ and a very large number (1E99) and specify the degrees of freedom, as shown.

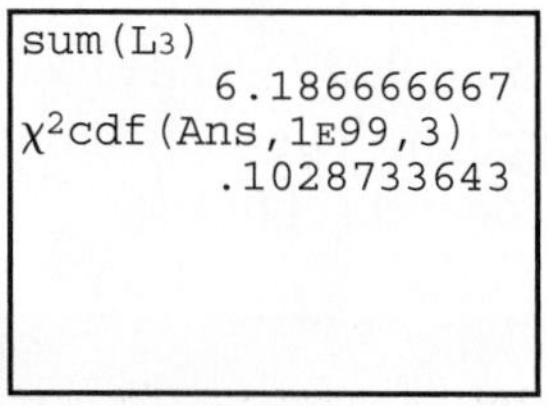

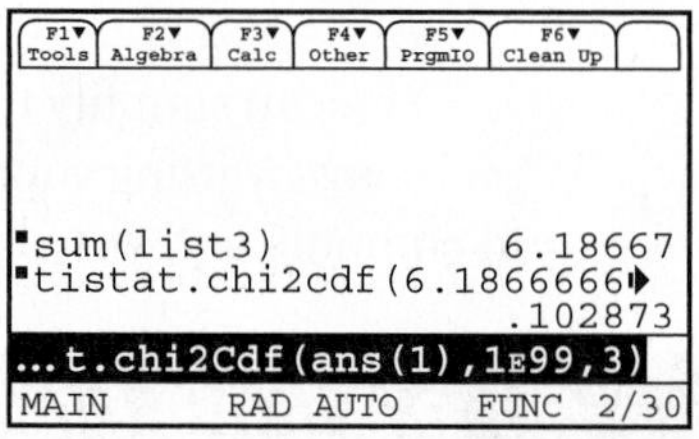

The *P*-value, 0.1029, indicates that $X^2 = 6.187$ is a possible result if H_0 is true, which does not provide strong enough evidence to reject H_0.

- To visualize this *P*-value, first set your viewing window as shown. Then enter the command `Shade`χ^2`(6.187,1E99,3)`. (This command is located in the DISTR/DRAW menu on the TI-83/84 and in the CATALOG under Flash Apps on the TI-89.) Note that the Shadeχ^2 command requires a left endpoint and a right endpoint, so use a sufficiently large right endpoint to achieve four-decimal-place accuracy.

Technology Toolbox

Goodness of fit tests *(continued)*

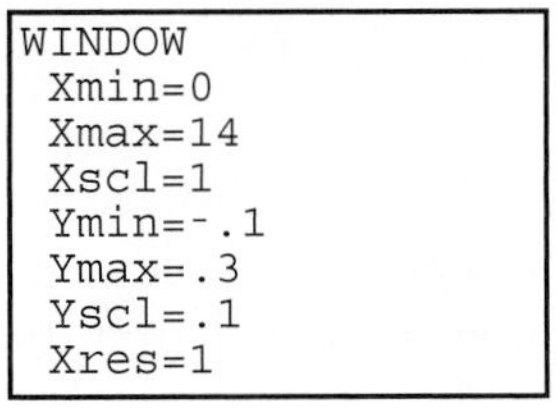

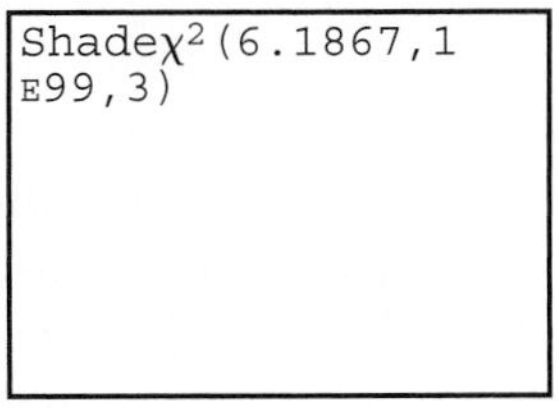

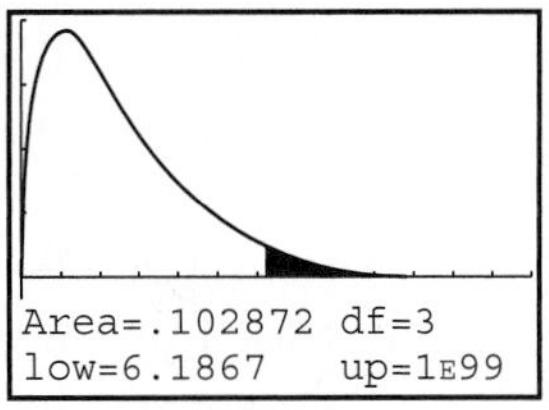

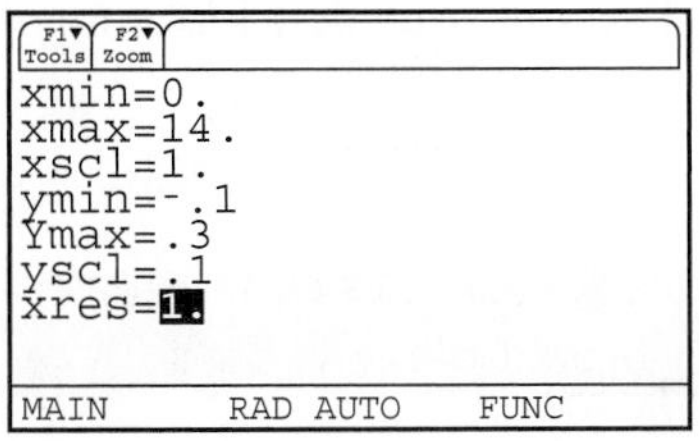

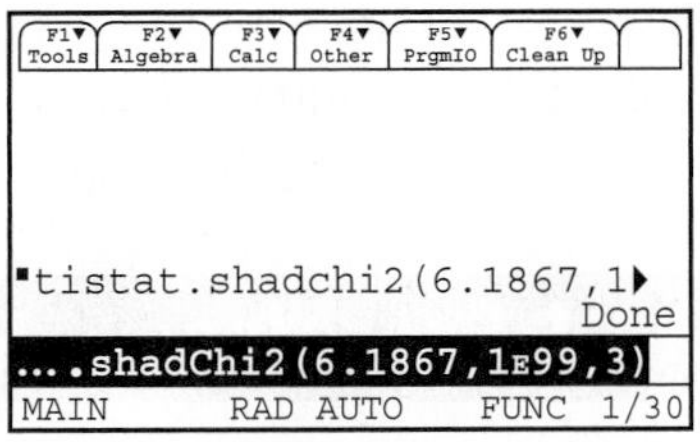

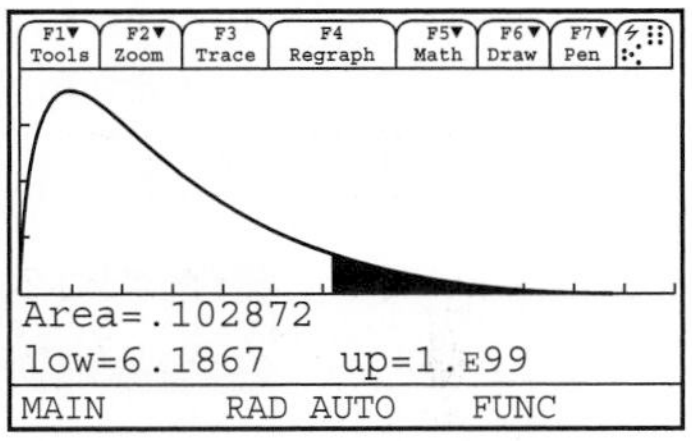

The shaded area, 0.1029, matches the P-value we found using χ^2cdf.

Follow-up Analysis

In the chi-square test for goodness of fit, we test the null hypothesis that a categorical variable has a specified distribution. If we find significance, we can conclude that our variable has a distribution different from the specified one. In this case, it's always a good idea to determine which categories of the variable provide the greatest differences between observed and expected counts. To do this, look at the individual terms $(O - E)^2/E$ that are added together to produce the test statistic X^2.

component

Consider again Example 14.3 (page 842). Clearly, the offspring contributing the largest amount to the X^2 statistic are the white-eyed, straight-winged flies (3.5267). This is known as the largest ***component*** of the chi-square statistic. In Example 14.2 (page 837) the largest components of X^2 are the ones for Saturday (77.30) and Sunday (63.86).

Section 14.1 Summary

The **chi-square (χ^2)** test for goodness of fit tests the null hypothesis that a categorical variable has a specified distribution.

The **expected count** for any variable category is obtained by multiplying the proportion of the distribution for each category times the sample size.

The **chi-square statistic** is $X^2 = \Sigma(O - E)^2/E$, where the sum is over k variable categories.

The chi-square test compares the value of the statistic X^2 with critical values from the **chi-square distribution** with $k - 1$ **degrees of freedom.**

For a test of

H_0: the population proportions equal the hypothesized values

against

H_a: at least one of the population proportions *differs from* its hypothesized value

the *P*-value is the area under the density curve to the right of X^2. Large values of X^2 are evidence against H_0.

The chi-square distribution is an approximation to the distribution of the statistic X^2. You can safely use this approximation when all expected counts are at least 1 and no more than 20% of all expected counts are less than 5. If the chi-square test finds a statistically significant *P*-value, do a follow-up analysis that compares the observed counts with the expected counts and that looks for the largest components of the chi-square statistic.

Section 14.1 Exercises

14.1 Finding *P*-values

(a) Find the *P*-value corresponding to $X^2 = 1.41$ for a chi-square distribution with 1 degree of freedom: (i) using Table D and (ii) with your graphing calculator.

(b) Find the area to the right of $X^2 = 19.62$ under the chi-square curve with 9 degrees of freedom: (i) using Table D and (ii) with your graphing calculator.

(c) Find the *P*-value corresponding to $X^2 = 7.04$ for a chi-square distribution with 6 degrees of freedom: (i) using Table D and (ii) with your graphing calculator.

14.2 M&M's activity Use the class M&M's data that you recorded in Steps 6 and 7 of Activity 14A to answer the following questions. Consider the entire count of M&M's in the class as one large sample from the production process.

(a) Do these data give you reason to doubt the color distribution of M&M's advertised by the M&M/Mars Company? Give appropriate statistical evidence to support your conclusion.

(b) For which color M&M does your sample proportion differ the most from the proportion claimed by the company? Use an appropriate statistical procedure to determine whether this difference is statistically significant.

(c) Can you use the data you collected in Step 7 to construct a confidence interval for the mean number of M&M's in the population of all 1.69-ounce bags produced by M&M/Mars Company? If so, do it. If not, explain why not.

14.3 Course grades Most students in a large college statistics course are taught by teaching assistants (TAs). One section is taught by the course supervisor, a full-time professor. The distribution of grades for the hundreds of students taught by TAs this semester was

Grade:	A	B	C	D/F
Probability:	0.32	0.41	0.20	0.07

The grades assigned by the professor to the 91 students in his section were

Grade:	A	B	C	D/F
Count:	22	38	20	11

(These data are real. We won't say when and where, but the professor was not any of the authors of this book.)

(a) What percents of students in the professor's section earned A, B, C, and D/F? In what ways does this distribution of grades differ from the TA distribution?

(b) Because the TA distribution is based on hundreds of students, we are willing to regard it as a fixed probability distribution. If the professor's grading follows this distribution, what are the expected counts of each grade in his section?

(c) Does the chi-square test for goodness of fit give good evidence that the professor's grade distribution differs from the TA distributions? Use the Inference Toolbox.

14.4 Saving birds from windows Many birds are injured or killed by flying into windows. It appears that birds see windows as open space. Can tilting windows down so that they reflect earth rather than sky reduce bird strikes? Suppose we place three windows at the edge of a woods: one vertical, one tilted 20 degrees, and one tilted 40 degrees. During the next four months, there are 53 bird strikes: 31 on the vertical window, 14 on the 20-degree window, and 8 on the 40-degree window.[3] If tilting had no effect, we would expect strikes on all three windows to have equal probability. Test this null hypothesis. What do you conclude?

14.5 Genetics: crossing tobacco plants Researchers want to cross two yellow-green tobacco plants with genetic makeup (Gg). Here is a Punnett square for this genetic experiment:

	G	g
G	GG	Gg
g	Gg	gg

This suggests that the expected ratio of green (GG) to yellow-green (Gg) to albino (gg) tobacco plants should be 1:2:1. When the researchers cross 84 pairs of yellow-green plants, the resulting offspring distribution is 22 green, 50 yellow-green, and 12 albino seedlings. Use a chi-square goodness of fit test to assess the validity of the researchers' genetic model.

14.6 Accidents and cell phones In Examples 14.1 (page 836) and 14.2 (page 837) we found that accidents involving cell phones are more likely to occur on weekdays. If we restrict attention to weekdays, are accidents more or less likely on certain weekdays than on other weekdays? Use the data for Monday through Friday in Example 14.1 to answer this question. Use the Inference Toolbox to carry out a significance test.

14.7 More candy The M&M/Mars Company reports the following distribution for other M&M varieties. For Peanut Chocolate Candies, the ratio is 23% of each of blue and orange, 15% of each of yellow and green, and 12% of each of brown and red. For Almond M&M's, the distribution is 20% of each of yellow, green, blue, and orange and 10% of each of brown and red. Buy a bag of one of these varieties of M&M's, perform a goodness of fit test of the company's reported distribution, and report your results. (Better still, obtain a larger sample by using multiple bags and do this problem as another class activity.)

14.8 What's your sign? The University of Chicago's General Social Survey (GSS) is the nation's most important social science sample survey. For reasons known only to social scientists, the GSS regularly asks its subjects their astrological sign. Here are the counts of responses in the most recent year this question was asked:[4]

Sign:	Aries	Taurus	Gemini	Cancer	Leo	Virgo
Count:	225	222	241	240	260	250
Sign:	Libra	Scorpio	Sagittarius	Capricorn	Aquarius	Pisces
Count:	243	214	200	216	224	244

If births were spread uniformly across the year, we would expect all 12 signs to be equally likely. Are they?

14.9 Is your random number generator working? In this exercise you will use your calculator to simulate sampling from the following uniform distribution:

X:	0	1	2	3	4	5	6	7	8	9
$P(X)$:	0.1	0.1	0.1	0.1	0.1	0.1	0.1	0.1	0.1	0.1

You will then perform a goodness of fit test to see if a randomly generated sample distribution comes from a population that is different from this distribution.

(a) State your null and alternative hypotheses for this test.

(b) Use the `randInt` function to randomly generate 200 digits from 0 to 9, and store these values in L_4/list4.

(c) Plot the data as a histogram with Window dimensions set as follows: $X[-.5, 9.5]_1$ and $Y[-5,30]_5$. (You may have to increase the vertical scale.) Then TRACE to see the frequencies of each digit. Record these frequencies (observed values) in L_1/list1.

(d) Determine the expected counts for a sample size of 200, and store them in L_2/list2.

(e) Complete a goodness of fit test. Report your chi-square statistic, the *P*-value, and your conclusion with regard to the null and alternative hypotheses.

14.10 Trix are for kids Trix cereal comes in five fruit flavors, and each flavor has a different shape. A curious student methodically sorted an entire box of the cereal and found the following distribution of flavors for the pieces of cereal in the box:

Flavor:	Grape	Lemon	Lime	Orange	Strawberry
Frequency:	530	470	420	610	585

Test the null hypothesis that the flavors are uniformly distributed versus the alternative that they are not.

14.2 Inference for Two-Way Tables

Activity 14B

Should marijuana be legalized for medical purposes?

The issue of whether to legalize marijuana is contentious. Proponents of legalization for medical uses state that cannabis (marijuana's other name) is valuable in a wide range of clinical applications, including pain relief—particularly of neuropathic pain (pain from nerve damage)—nausea, glaucoma, and movement disorders. They argue that physicians should be allowed to prescribe marijuana for patients for whom alternative treatments aren't nearly as effective. Opponents of legalization argue that the use of marijuana is addictive and that, once addicted to marijuana, users are more likely to move on to other, more debilitating drugs that are extremely hard to quit. This is one of many issues where the opinion of males may be different from the opinion of females. In this Activity, you will find out if there is a difference of opinion on the marijuana question between males and females in your class. In a follow-up exercise, you will determine if the differences you observed are statistically significant.

1. On an index card, write the word "Legalize" if you think that physicians should be legally able to prescribe marijuana for their patients. Write "Do not legalize" if you think the use of marijuana should not be legal for medical applications. (Feel free to substitute a different question for which you think the opinions of males and females are quite different.)
2. The teacher will collect the cards and report the results. Make a table like the one below, and record the results. Calculate the row and column totals.

	Male	Female	Total
Legalize			
Do not legalize			
Total			

3. Make a new table, and calculate the conditional percents. That is, calculate the percent of males who say, "Legalize," and the percent of males who say, "Do not legalize." Do the same for the females.
4. Construct a bar graph that compares the percents you calculated.
5. Write a statement to interpret what the bar graph tells you.

Keep these data handy. Once you learn how to perform an appropriate significance test in this section, you will be asked to perform it for these data to determine if the differences you observed are significant.

The two-sample z procedures of Chapter 13 allow us to compare the proportions of successes in two groups (either two populations or two treatment groups in an experiment). What if we want to compare more than two groups? We need a new statistical test. The new test starts by presenting the data as a two-way table. Two-way tables have more general uses than comparing the proportions of successes in several groups. As we saw in Section 2 of Chapter 4, they can be used to describe relationships between any two categorical variables. The same test that compares several proportions also tests whether the row and column variables are related in any two-way table. We will start with the problem of comparing several proportions.

Example 14.4 ***Does background music influence wine purchases?***
Conditional distributions

Market researchers know that background music can influence the mood and purchasing behavior of customers. One study in a supermarket in Northern Ireland compared three treatments: no music, French accordion music, and Italian string music. Under each condition, the researchers recorded the numbers of bottles of French, Italian, and other wine purchased.[5] Here is a table that summarizes the data:

	Music			
Wine	**None**	**French**	**Italian**	**Total**
French	30	39	30	**99**
Italian	11	1	19	**31**
Other	43	35	35	**113**
Total	84	75	84	**243**

The conditional distributions of types of wine sold for each kind of music being played and the marginal distribution of the types of wine sold are shown in the following table:

Column percents for wine and music

	Music			
Wine	**None**	**French**	**Italian**	**Total**
French	35.7	52.0	35.7	**40.7**
Italian	13.1	1.3	22.6	**12.8**
Other	51.2	46.7	41.7	**46.5**
Total	100.0	100.0	100.0	100.0

Figure 14.3 compares the distributions of types of wine sold for each of the three music conditions.

Figure 14.3 *Comparison of the percents of different types of wine sold for different music conditions.*

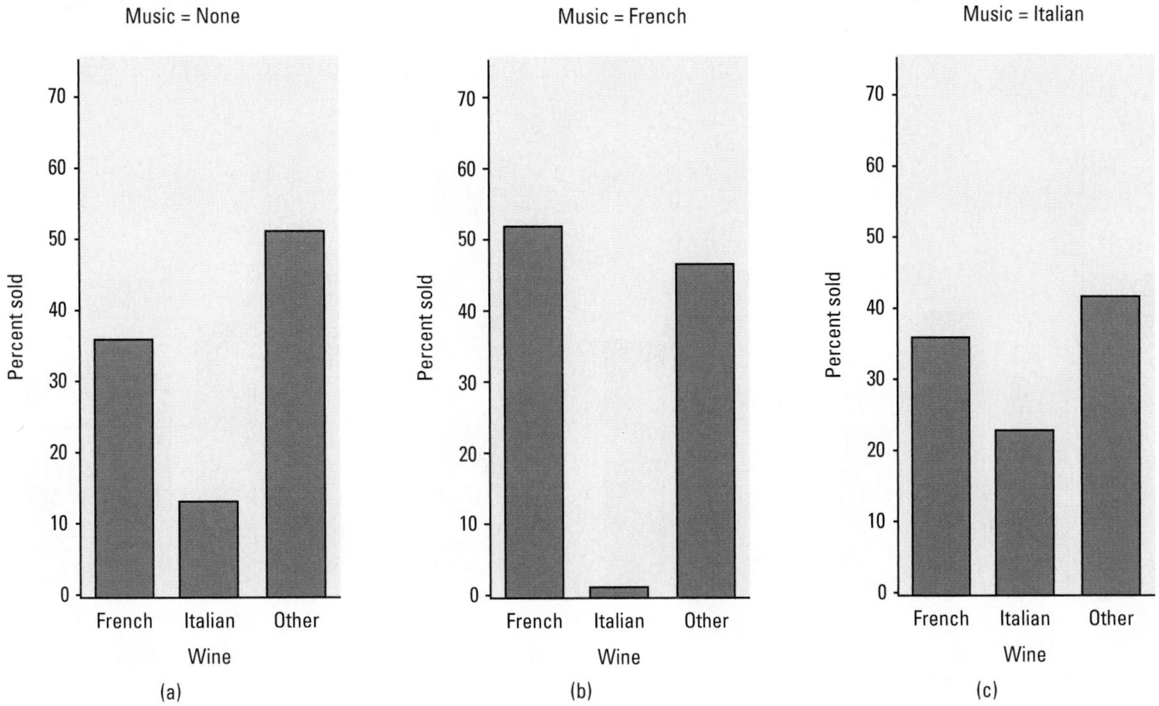

There appears to be an association between the music played and the type of wine that customers buy. Sales of Italian wine are very low when French music is playing but are higher when Italian music or no music is playing. French wine is popular in this market, selling well under all music conditions but notably better when French music is playing.

Another way to look at these data is to examine the row percents. These fix a type of wine and compare its sales when different types of music are playing. Figure 14.4 on the next page displays these results.

We see that more French wine is sold when French music is playing. Similarly for Italian wine. The negative effect of French music on sales of Italian wine is dramatic.

The Problem of Multiple Comparisons

First, we summarize the observed relation between the music being played and the type of wine purchased. The researchers expected that music would influence sales, so music type is the explanatory variable and the type of wine purchased is the response variable. In general, the clearest way to describe this kind of relationship is to compare the conditional distributions of the response variable for each value of the explanatory variable. So we will compare the column percents that give the conditional distribution of purchases for each type of music played.

Figure 14.4 *Comparison of the percents of different types of wine sold for different music conditions.*

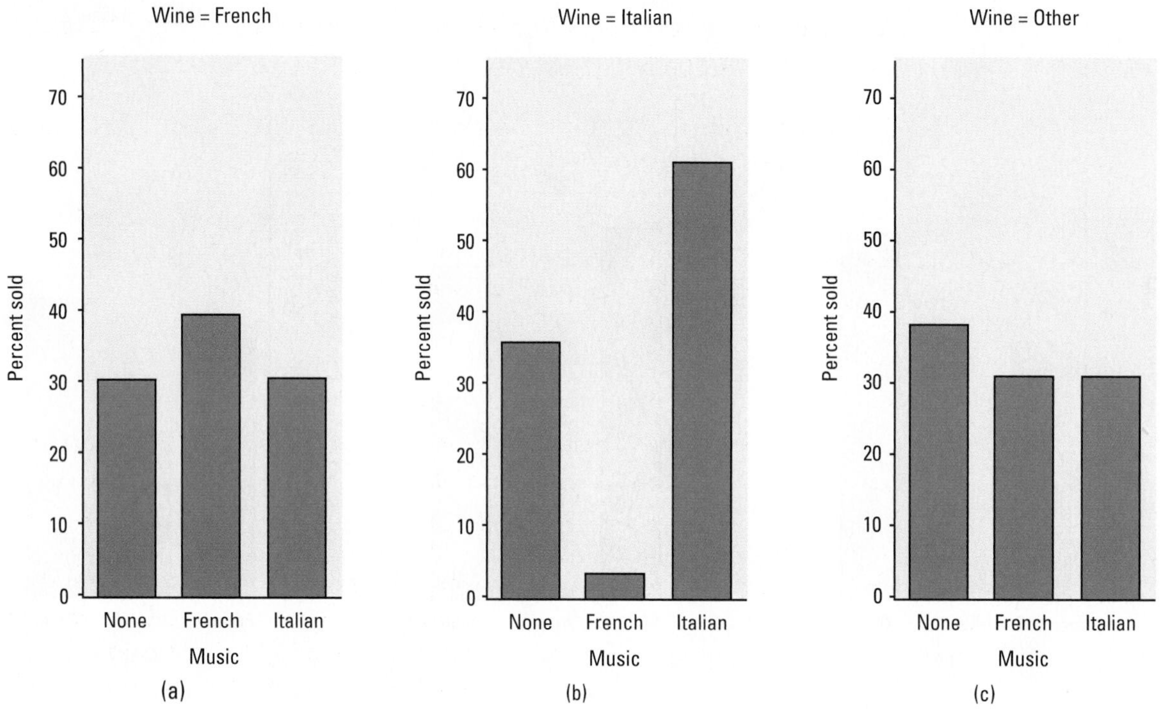

To compare the three population distributions in Example 14.4 (page 850), we might use chi-square goodness of fit procedures several times:

- Test H_0: the distribution of wine types for no music is the same as the distribution of wine types for French music.
- Test H_0: the distribution of wine types for no music is the same as the distribution of wine types for Italian music.
- Test H_0: the distribution of wine types for French music is the same as the distribution of wine types for Italian music.

The weakness of doing three tests is that we get *three* results, one for each test alone. That doesn't tell us how likely it is that three sample distributions are as different as these are if the corresponding population distributions are the same. It may be that the sample distributions are significantly different if we look at just two groups, but not significantly different if we look at two other groups. We can't safely compare many parameters by doing tests or confidence intervals for two parameters at a time.

The problem of how to do many comparisons at once with some overall measure of confidence in all our conclusions is common in statistics. This is the problem of *multiple comparisons*. Statistical methods for dealing with multiple comparisons usually have two parts:

1. An *overall* test to see if there is good evidence of any differences among the parameters that we want to compare.

2. A detailed *follow-up analysis* to decide which of the parameters differ and to estimate how large the differences are.

The overall test is one with which we are familiar—the chi-square test—but in this new setting it will be used for comparing several population proportions. The follow-up analysis can be quite elaborate.

Two-Way Tables

two-way table

$r \times c$ table

We've taken the first step in the overall test for comparing several proportions, which is to arrange the data in a ***two-way table*** that gives *counts* for both successes and failures. We call our table for Example 14.4 a 3×3 table because it has 3 rows and 3 columns (not counting the "Total" row and column). A table with r rows and c columns is an ***$r \times c$ table.*** The table shows the relationship between two categorical variables. The explanatory variable is the type of music. The response variable is the type of wine purchased. The two-way table gives the counts for all 9 combinations of values of these variables. Each of the 9 counts occupies a cell of the table.

Stating Hypotheses

We observe a clear relationship between music type and wine sales for the 243 bottles sold during the study. The chi-square test assesses whether this observed association is statistically significant, that is, too strong to occur often just by chance. The market researchers changed the background music and took three samples of wine sales in three distinct environments. Each column in the table represents one of these samples, and each row a type of wine. This is an example of *separate and independent random samples* from each of c populations. The c columns of the two-way table represent the populations. There is a single categorical response variable, wine type. The r rows of the table correspond to the values of the response variable.

The $r \times c$ table allows us to compare more than two populations, more than two categories of response, or both. In this setting, the null hypothesis becomes

H_0: The distribution of the response variable is the same in all c populations.

Because the response variable is categorical, its distribution just consists of the proportions of its r possible values or categories. The null hypothesis says that these population proportions are the same in all c populations.

Example 14.5 *The music and wine study*
Stating hypotheses

In the market research study of Example 14.4, we compare three populations:

Population 1: bottles of wine sold when no music is playing
Population 2: bottles of wine sold when French music is playing
Population 3: bottles of wine sold when Italian music is playing

We have three independent samples, of sizes 84, 75, and 84, with a separate sample being taken from each population. The null hypothesis for the chi-square test is

H_0: The proportions of each wine type sold are the same in all three populations.

The parameters of the model are the proportions of the three types of wine that would be sold in each of the three environments. There are three proportions (for French wine, Italian wine, and other wine) for each environment.

Computing Expected Cell Counts

The null hypothesis is that the distribution of wine selected is the same for all three populations of music types. The alternative is that the distributions of wine types are not all the same. Here is the formula for the ***expected cell counts*** under the null hypothesis.

Expected Cell Counts

$$\text{expected count} = \frac{\text{row total} \times \text{column total}}{n}$$

Here's the idea behind the formula for expected counts in a two-way table. If we have n independent trials and the probability of a success on each trial is p, we expect np successes. If we draw an SRS of n individuals from a population in which the proportion of successes is p, we expect np successes in the sample. Let's apply this idea to the music and wine study.

Example 14.6 *The music and wine study*
Calculating expected counts

The two-way table with row and column totals is

	Music			
Wine	**None**	**French**	**Italian**	**Total**
French	30	39	30	**99**
Italian	11	1	19	**31**
Other	43	35	35	**113**
Total	84	75	84	243

We will find the expected count for the cell in row 1 (French wine) and column 1 (no music). The proportion of no music among all 243 subjects is

$$\frac{\text{count of no music}}{\text{table total}} = \frac{\text{column 1 total}}{\text{table total}} = \frac{84}{243}$$

Think of this as p, the overall proportion of no music. If H_0 is true, we expect (except for random variation) this same proportion of no music in all three groups. So the expected count of no music among the 99 subjects who order French wine is

$$np = (99)\,\frac{84}{243} = 34.222$$

This expected count has the form announced in the box:

$$\frac{\text{row 1 total} \times \text{column 1 total}}{\text{table total}} = \frac{(99)(84)}{243} = 34.222$$

Calculate the expected counts for the remaining cells in the same way. Results are summarized in the following table.

Expected counts for music and wine

	Music			
Wine	**None**	**French**	**Italian**	**Total**
French	34.222	30.556	34.222	**99.000**
Italian	10.716	9.568	10.716	**31.000**
Other	39.062	34.877	39.062	**113.001**
Total	**84.000**	**75.001**	**84.000**	**243.001**

CAUTION

Note that although any count of bottles of wine sold must be a whole number, an expected count need not be.

We can check our work by adding the expected counts to obtain the row and column totals, as in the table. These should be the same as those in the table of observed counts, except for small roundoff errors, such as 113.001 rather than 113 for the total number of bottles of other wine sold.

Exercises

14.11 How to quit smoking, I It's hard for smokers to quit. Perhaps prescribing a drug to fight depression will work as well as the usual nicotine patch. Perhaps combining the patch and the drug will work better than either treatment alone. Here are data from a randomized, double-blind trial that compared four treatments.[6] A "success" means that the subject did not smoke for a year following the beginning of the study.

Treatment	Subjects	Successes
Nicotine patch	244	40
Drug	244	74
Patch plus drug	245	87
Placebo	160	25

(a) Summarize these data in a two-way table.

(b) Calculate the proportion of subjects who refrain from smoking in each of the four treatment groups.

(c) Make a graph to display the association. Describe what you see.

(d) Explain in words what the null hypothesis H_0: $p_1 = p_2 = p_3 = p_4$ says about subjects' smoking habits.

(e) Find the expected counts if H_0 is true, and display them in a two-way table similar to the table of observed counts.

(f) Compare the tables of observed and expected counts. Explain how the comparison expresses the same association you see in (b) and (c).

14.12 Why men and women play sports, I Do men and women participate in sports for the same reasons? One goal for sports participants is social comparison—the desire to win or to do better than other people. Another is mastery—the desire to improve one's skills or to try one's best. A study on why students participate in sports collected data from two independent random samples of 67 male and 67 female undergraduates at a large university.[7] Each student was classified into one of four categories based on his or her responses to a questionnaire about sports goals. The four categories were high social comparison-high mastery (HSC-HM), high social comparison-low mastery (HSC-LM), low social comparison-high mastery (LSC-HM), and low social comparison-low mastery (LSC-LM). One purpose of the study was to compare the goals of male and female students. Here are the data displayed in a two-way table:

Observed counts for sports goals

	Sex	
Goal	**Female**	**Male**
HSC-HM	14	31
HSC-LM	7	18
LSC-HM	21	5
LSC-LM	25	13

(a) This is an $r \times c$ table. What numbers do r and c stand for?

(b) Calculate the proportions of females falling into each of the four categories of sports goals. Then do the same for males.

(c) Make a bar graph to compare the distribution of sports goals for males and females.

(d) The null hypothesis says that the proportions of females falling into the four sports goal categories are the same as the proportions of males in those categories. The overall proportion of *students* in the HSC-HM category is $45/134 = 0.336$. Assuming H_0 is true, we expect that 33.6% of the females surveyed, $(0.336)(67) = 22.5$, will land in the HSC-HM category. Our expected count for the males in the HSC-HM category is also $(0.336)(67) = 22.5$. Find the rest of the expected counts and display them in a two-way table.

(e) Compare the observed counts with the expected counts. Are there any large deviations between them? Explain how the comparison expresses the same association you saw in (b) and (c).

14.13 Preventing strokes, I Aspirin prevents blood from clotting and so helps prevent strokes. The Second European Stroke Prevention Study asked whether adding another anticlotting drug named dipyridamole would be more effective for patients who had already had a stroke. Here are the data on strokes and deaths during the two years of the study:[8]

Treatment	Number of patients	Number of strokes	Number of deaths
Placebo	1649	250	202
Aspirin	1649	206	182
Dipyridamole	1654	211	188
Both	1650	157	185

(a) Make a two-way table of treatment by whether or not a patient had a stroke during the two-year study period.

(b) Make a graph to compare the rates of strokes for the four treatments. Which treatment appears most effective in preventing strokes?

(c) Explain in words what the null hypothesis H_0: $p_1 = p_2 = p_3 = p_4$ says about the incidence of strokes.

(d) Find the expected counts if H_0 is true, and display them in a two-way table similar to the table of expected counts.

14.14 Nonsmokers and education in France Smoking remains more common in much of Europe than in the United States. In the United States, there is a strong relationship between education and smoking: well-educated people are less likely to smoke. Does a similar relationship hold in France? Here is a two-way table of the level of education and smoking status (smoker, nonsmoker) of a sample of 459 French men aged 20 to 60 years.[9] The subjects are a random sample of men who visited a health center for a routine checkup. We are willing to consider them an SRS of men from their region of France.

Education	Smoker	Nonsmoker
Primary school	131	56
Secondary school	102	37
University	80	53

Carry out three significance tests of the three null hypotheses

$$H_0\colon p_{\text{primary}} = p_{\text{secondary}}$$

$$H_0\colon p_{\text{primary}} = p_{\text{university}}$$

$$H_0\colon p_{\text{secondary}} = p_{\text{university}}$$

against the two-sided alternatives. Give *P*-values for each test. These three *P*-values don't tell us how often the three proportions for the three education groups will be spread this far apart just by chance.

The Chi-Square Test for Homogeneity of Populations

In Example 14.4 (page 850) comparing the sample proportions of wine sold for different music types describes the differences among the three types of music. But the statistical test that tells us whether those differences are statistically significant doesn't use the sample proportions. It compares the observed and expected *counts*. The test statistic that makes the comparison is the ***chi-square statistic.***

Chi-Square Statistic

The **chi-square statistic** is a measure of how far the observed counts in a two-way table are from the expected counts. The formula for the statistic is

$$X^2 = \sum \frac{(\text{observed count} - \text{expected count})^2}{\text{expected count}}$$

The sum is over all $r \times c$ cells in the table.

This is an extension of the X^2 statistic for *one-way tables* of Section 14.1. There $r = 1$ and $c = 1$.

The X^2 Statistic and Its *P*-value

The expected counts in Example 14.6 (page 854) are all large, so we proceed with the chi-square test. Just as we did with the goodness of fit test, we compare the table of observed counts with the table of expected counts using the X^2 statistic.[10] We must calculate the term for each cell, then sum over all nine cells. For French wine with no music, the observed count is 30 bottles and the expected count is 34.222. The contribution to the X^2 statistic for this cell is therefore

$$\frac{(O - E)^2}{E} = \frac{(30 - 34.222)^2}{34.222} = 0.5209$$

The X^2 statistic is the sum of nine such terms:

$$\begin{aligned}
X^2 &= \sum \frac{(\text{observed} - \text{expected})^2}{\text{expected}} \\
&= \frac{(30 - 34.222)^2}{34.222} + \frac{(39 - 30.556)^2}{30.556} + \frac{(30 - 34.222)^2}{34.222} \\
&\quad + \frac{(11 - 10.716)^2}{10.716} + \frac{(1 - 9.568)^2}{9.568} + \frac{(19 - 10.716)^2}{10.716} \\
&\quad + \frac{(43 - 39.062)^2}{39.062} + \frac{(35 - 34.877)^2}{34.877} + \frac{(35 - 39.062)^2}{39.062} \\
&= 0.5209 + 2.3337 + 0.5209 + 0.0075 + 7.6724 \\
&\quad + 6.4038 + 0.3971 + 0.0004 + 0.4223 \\
&= 18.28
\end{aligned}$$

As in the test for goodness of fit, you should think of the chi-square statistic X^2 as a measure of the distance of the observed counts from the expected counts. Like any distance, it is always zero or positive, and it is zero only when the observed counts are exactly equal to the expected counts. Large values of X^2 are evidence against H_0 because they say that the observed counts are far from what we would expect if H_0 were true. Although the alternative hypothesis H_a is ***many-sided***, the chi-square test is one-sided because any violation of H_0 tends to produce a large value of X^2. Small values of X^2 are not evidence against H_0.

many-sided

The same chi-square procedure that we used to test goodness of fit allows us to compare the distribution of proportions in several populations, provided that we take *separate and independent random samples* from each population.

Chi-Square Test for Homogeneity of Populations

Select independent SRSs from each of c populations. Classify each individual in a sample according to a categorical response variable with r possible values. There are c different sets of proportions to be compared, one for each population.

The null hypothesis is that the distribution of the response variable is the same in all c populations. The alternative hypothesis says that these c distributions are not all the same.

If H_0 is true, the chi-square statistic X^2 has approximately a χ^2 distribution with $(r - 1)(c - 1)$ degrees of freedom (df).

The P-value for the chi-square test is the area to the right of X^2 under the chi-square density curve with $(r - 1)(c - 1)$ degrees of freedom.

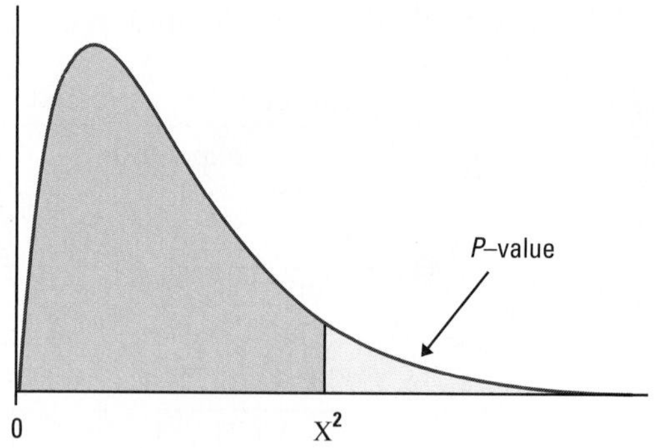

The chi-square test, like the z procedure for comparing two proportions, is an approximate method that becomes more accurate as the counts in the cells of the table get larger. Fortunately, the approximation is accurate for quite modest counts. Here is a practical guideline.[11]

Cell Counts Required for the Chi-Square Test

You can safely use the chi-square test with critical values from the chi-square distribution when no more than 20% of the expected counts are less than 5 and all individual expected counts are 1 or greater. In particular, all four expected counts in a 2×2 table should be 5 or greater.

We have examined the data in Example 14.4 (page 850) informally. Now we proceed to formal inference.

Example 14.7 *Are music and wine linked?*
The chi-square test

Step 1: Populations and parameters We want to use X^2 to compare the distribution of types of wine selected for each type of music. Our hypotheses are

H_0: The distributions of wine selected are the same in all three populations of music types.

H_a: The distributions of wine selected are not all the same.

Step 2: Conditions To use the chi-square test for homogeneity of populations:

- The data must come from independent SRSs from the populations of interest. We are willing to treat the subjects in the three groups as SRSs from their respective populations.
- All expected cell counts are greater than 1, and no more than 20% are less than 5. In Example 14.6, we saw that the smallest expected count was 9.568.

Step 3: Calculations

- The test statistic is

$$X^2 = \sum \frac{(O-E)^2}{E} = 0.5209 + 2.3337 + 0.5209 + 0.0075 + 7.6724$$
$$+ 6.4038 + 0.3971 + 0.0004 + 0.4223$$
$$= 18.28$$

Because there are $r = 3$ types of wine and $c = 3$ music conditions, the degrees of freedom for this statistic are

$$\text{df} = (r-1)(c-1) = (3-1)(3-1) = 4$$

Under the null hypothesis that music and wine sales are independent, the test statistic X^2 has a χ^2 distribution with df = 4.

df = 4

p	.0025	.001
χ^2	16.42	18.47

- To obtain the P-value, look at the df = 4 row in Table D. The calculated value $X^2 =$ 18.28 lies between the critical values for probabilities 0.0025 and 0.001. The P-value is therefore between 0.001 and 0.0025.

Step 4: Interpretation Because the expected cell counts are all large, the P-value from Table D will be quite accurate. There is strong evidence to reject H_0 ($X^2 = 18.28$, df = 4, $p < 0.0025$) and conclude that the type of music being played has a significant effect on wine sales.

Performing the Chi-Square Test with Technology

Calculating the expected counts and then the chi-square statistic by hand is a bit time-consuming. As usual, computer software saves time and always gets the arithmetic right. The TI-83/84 and TI-89, on a smaller scale, have also been programmed to conduct inference for two-way tables.

Technology Toolbox

Chi-square tests with Minitab

We enter the two-way table (the 9 counts) for the study of Example 14.4 (page 850) into the Minitab software package and request the chi-square test. The output appears in Figure 14.5. Most statistical software packages produce chi-square output similar to this.

Minitab repeats the two-way table of observed counts and puts the expected count for each cell below the observed count. It numbers the rows (1, 2, and 3) and the columns (C1, C2, and C3) and also puts in the row and column totals. Then the software calculates the chi-square statistic X^2. For these data, $X^2 = 18.28$. The statistic is a sum of 9 terms, one for each cell in the table. The chi-square display in the output shows the individual terms, as well as their sum.

The *P*-value is the probability that X^2 would take a value as large as, or larger than, 18.28 if H_0 were really true. The small *P*-value, 0.001, gives us good reason to conclude that there are differences among the effects of the three music treatments.

Figure 14.5 *Minitab output for the two-way table in the study of Example 14.4. The output gives the observed counts, the expected counts, and the value 18.279 for the chi-square statistic. The last line gives 0.001 as the probability of a value* greater than *18.279 if the null hypothesis is true. The* P*-value is calculated to be 0.0011.*

```
MTB >  ChiSquare C1  C2  C3.

Expected counts are printed below observed counts

           C1       C2       C3    Total
    1      30       39       30       99
        34.22    30.56    34.22

    2      11        1       19       31
        10.72     9.57    10.72

    3      43       35       35      113
        39.06    34.88    39.06

Total      84       75       84      243

ChiSq =  0.521 + 2.334 + 0.521 +
         0.008 + 7.672 + 6.404 +
         0.397 + 0.000 + 0.422 = 18.279
DF = 4, P-Value = 0.001
```

Technology Toolbox

Chi-square tests on the TI-83/84/89

To perform a chi-square test of the music and wine study on the TI-83/84/89, use a matrix, say matrix [A], to store the observed counts. Here are the keystrokes, along with several calculator screens for you to check your progress.

Enter the observed counts in the matrix [A].

TI-83/84

- Press **2nd** **x^{-1}** (MATRIX), arrow to EDIT, choose 1:[A].

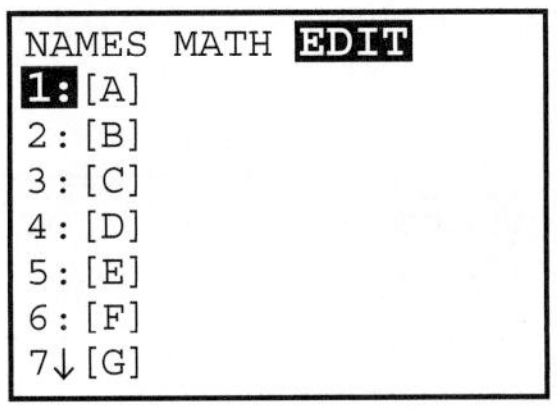

TI-89

- Press **APPS**, select 6:Data/Matrix Editor, and then 3:New. . . .
- Adjust your settings to match those shown.

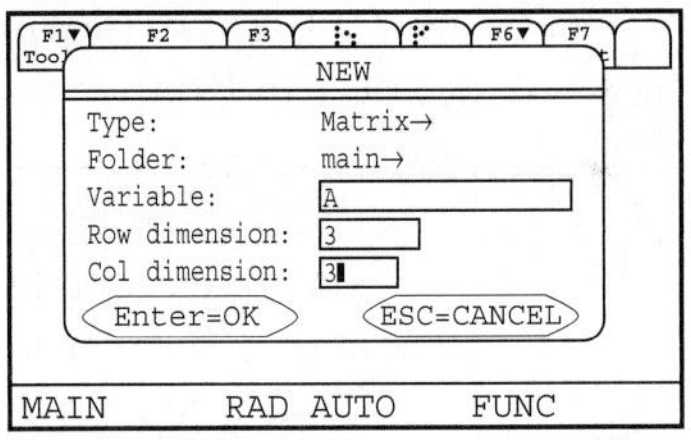

- Enter the observed counts from the two-way table in the matrix in the same locations.

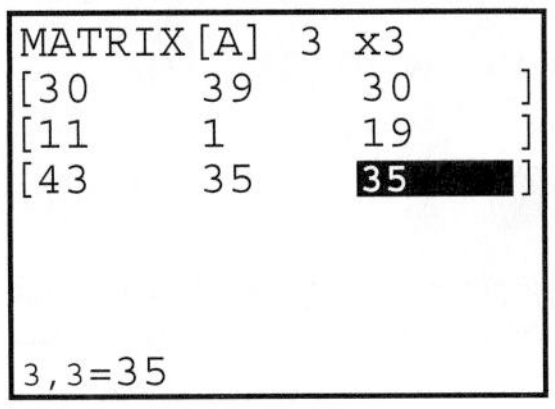

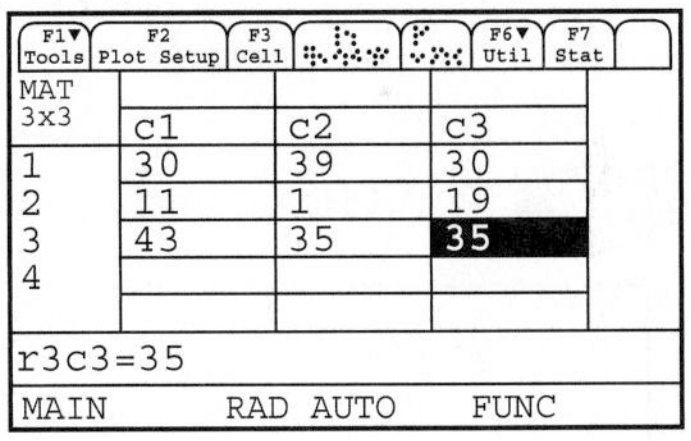

Specify the chi-square test, the matrix where the observed counts are found, and the matrix where the expected counts will be stored.

TI-83/84

- Press **STAT**, arrow to TESTS, and choose C: χ^2−Test.

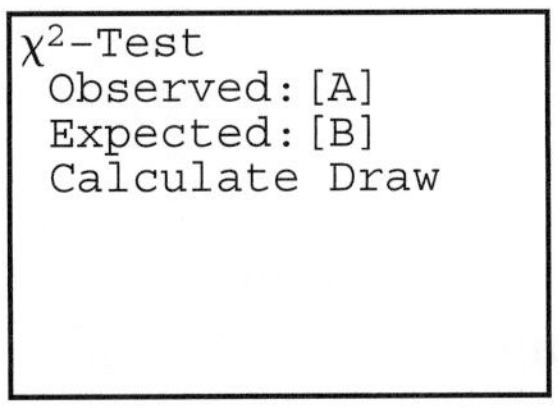

TI-89

- In the Statistics/List Editor, press **2nd** **F1** ([F6]) and choose 8:Chi2 2-way. . . .
- Adjust your settings as shown.

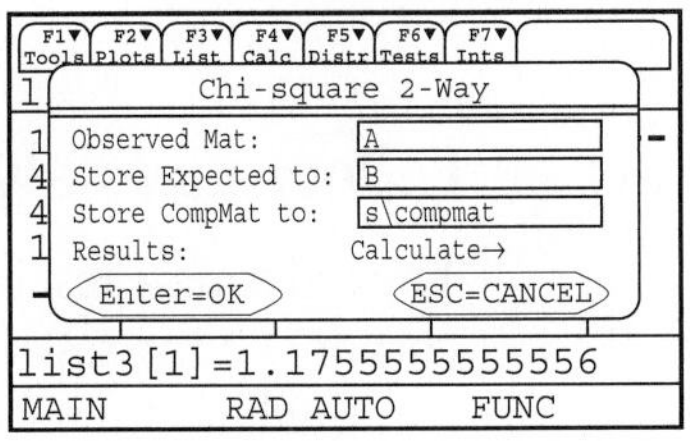

(continued)

Technology Toolbox

Chi-square tests on the TI-83/84/89 *(continued)*

- Choose "Calculate" or "Draw" to carry out the test. If you choose "Calculate," you should get these results:

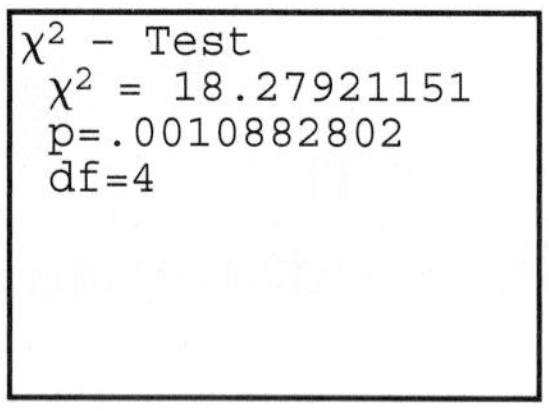

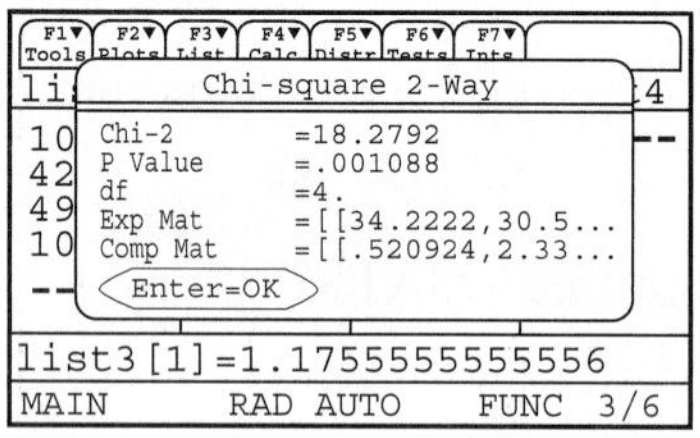

If you specify "Draw," the χ^2 curve with 4 degrees of freedom will be drawn, the critical area in the tail will be shaded, and the *P*-value will be displayed.

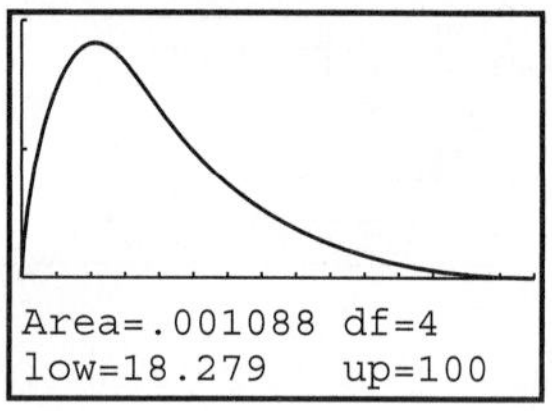

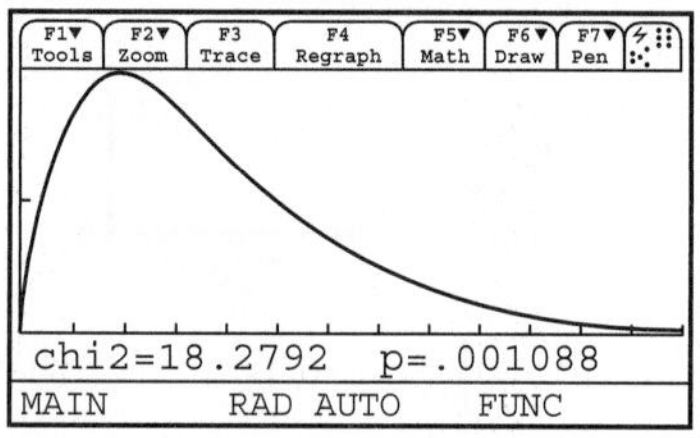

If you want to see the expected counts, simply ask for a display of the matrix [B].

- Press 2nd x^{-1} (MATRIX), and choose 2:[B].
- Press 2nd – (Var-LINK) and choose B.

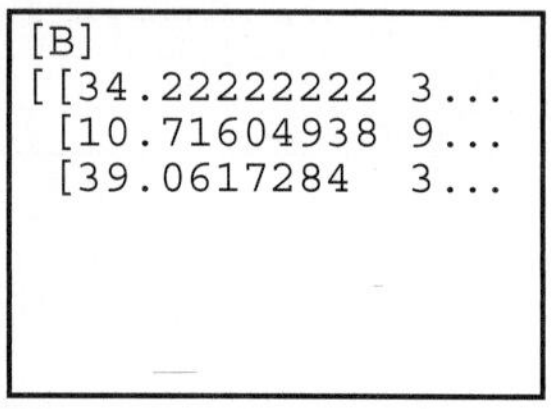

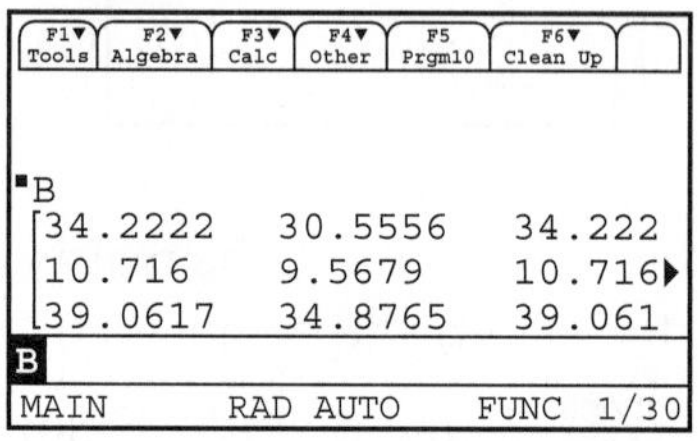

Verify that these calculator results agree with the Minitab results from the Technology Toolbox on page 862.

Follow-up Analysis

The chi-square test is the overall test for comparing any number of population proportions. If the test allows us to reject the null hypothesis that all the proportions are equal, we then want to do a follow-up analysis that examines the differences in detail. We won't describe how to do a formal follow-up analysis, but you should look at the data to see what specific effects they suggest.

Example 14.8 *A final look at the music and wine study*
Follow-up analysis

The study in Example 14.4 (page 850) found significant differences among the distributions of wines. We can see the specific differences in three ways.

First, the size and nature of the relationship between music and wine sales are described by row and column percents. These are displayed in Figures 14.3 and 14.4 on pages 851–852. These suggest major differences in wine selections for each music type. This is what the study hoped to find. Italian wine sells very poorly when French music is played. French wine sales are strong across the board, but particularly when French music is played.

Second, *compare the observed and expected counts* in Figure 14.5 on page 862.

components of X^2

Finally, Minitab prints under the table the 9 individual "distances" between the observed and expected counts that are added to get X^2. The arrangement of these ***components of*** $\mathbf{X^2}$ is the same as the 3×3 arrangement of the table. The largest components show which cells contribute the most to the overall distance X^2. Looking at the data, we see that just two of the nine terms that make up the chi-square sum contribute about 14 (almost 77%) of the total $X^2 = 18.28$. Comparing the observed and expected counts in these two cells, we see that sales of Italian wine are much below expectation when French music is playing and well above expectation when Italian music is playing. We are led to a specific conclusion: sales of Italian wine are strongly affected by Italian and French music. Figure 14.4(b) displays this effect.

All three ways of examining the data point to the same conclusion: the type of music being played has an effect on wine sales. This is an informal conclusion. More advanced methods provide tests and confidence intervals that make this follow-up analysis formal.

CAUTION

One last warning. The test confirms only that there is some relationship. The percents we have compared describe the nature of the relationship. **The chi-square test does not in itself tell us what population our conclusion describes.** If the study was done in one market on a Saturday, the results may apply only to Saturday shoppers at this market. The researchers may invoke their understanding of consumer behavior to argue that their findings apply more generally, but that is beyond the scope of the statistical analysis.

The Chi-Square Test and the *z* Test

We can use the chi-square test to compare any number of proportions. If we are comparing r proportions and make the columns of the table "success" and "failure," the counts form an $r \times 2$ table. P-values come from the chi-square distribution with $r - 1$ degrees of freedom. If $r = 2$, we are comparing just two proportions. We have two ways to do this: the z test from Section 13.2 and the chi-square test with 1 degree of freedom for a 2×2 table. These two tests always agree. In fact, the chi-square statistic X^2 is just the square of the z statistic, and the P-value for X^2 is exactly the same as the two-sided P-value for z. We recommend using the z test to compare two proportions because it gives you the choice of a one-sided test and is related to a confidence interval for $p_1 - p_2$. See Exercise 14.18.

Exercises

Securing the Net Professor Nong Ye, director of the Information and Systems Assurance Laboratory at Arizona State University, uses chi-square procedures to help make the World Wide Web safer and more secure from attackers (hackers). A hacker often carries out activities that are different from normal user activities in order to accomplish an attacking goal. A chi-square distance-monitoring procedure determines the difference between observed activities on computers and networks and the statistical profile of normal user activities.

14.15 How to quit smoking, II In Exercise 14.11 (page 855), you began to analyze data on the effectiveness of several treatments designed to help smokers quit.

(a) Starting from the table of expected counts, find the 8 components of the chi-square statistic and then the statistic X^2 itself.

(b) Use Table D to find the *P*-value for the test. Explain what it tells you.

(c) Which term contributes the most to X^2? Does this surprise you?

(d) What conclusion would you draw from this study?

(e) Perform the chi-square test on your calculator. Are your results the same?

14.16 How are schools doing? The nonprofit group Public Agenda conducted telephone interviews with 3 randomly selected groups of parents of high school children. There were 202 black parents, 202 Hispanic parents, and 201 white parents. One question asked was "Are the high schools in your state doing an excellent, good, fair, or poor job, or don't you know enough to say?" Here are the survey results:[12]

	Black parents	Hispanic parents	White parents
Excellent	12	34	22
Good	69	55	81
Fair	75	61	60
Poor	24	24	24
Don't know	22	28	14
Total	**202**	**202**	**201**

Write a brief analysis of these results. Include a graph or graphs, a test of significance, and your own discussion of the most important findings.

14.17 Why men and women play sports, II In Exercise 14.12 (page 856), you began to analyze data on the reasons why men and women play sports.

(a) Starting from the table of expected counts, find the 8 components of the chi-square statistic and then the statistic X^2 itself.

(b) Use Table D to find the *P*-value for the test. What decision would you make concerning H_0? Explain what this means in plain language.

(c) Which term(s) contribute most to the X^2 statistic? What specific relation between gender and sports goals do the term(s) point to?

(d) Figure 14.6 gives Minitab output for this test. Compare your work in parts (a) through (c) with the computer output.

Figure 14.6 *Minitab output for the study of gender and sports goals, for Exercise 14.17.*

```
Expected counts are printed below observed counts

             Female        Male        Total
HSC-HM         14           31          45
             22.50        22.50

HSC-LM          7           18          25
             12.50        12.50

LSC-HM         21            5          26
             13.00        13.00

LSC-LM         25           13          38
             19.00        19.00

 Total          67           67         134

 ChiSq   = 3.211  + 3.211  + 2.420  + 2.420  +
           4.923  + 4.923  + 1.895  + 1.895  = 24.898
 DF      = 3,  P-Value   = 0.000
```

14.18 Treating ulcers Gastric freezing was once a recommended treatment for ulcers in the upper intestine. Use of gastric freezing stopped after experiments showed it had no effect. One randomized comparative experiment found that 28 of the 82 gastric freezing patients improved, while 30 of the 78 patients in the placebo group improved.[13] We can test the hypothesis of "no difference" between the two groups in two ways: using the two-sample z statistic or using the chi-square statistic.

(a) State the null hypothesis with a two-sided alternative and carry out the z test. What is the P-value (from software or Table A)?

(b) Present the data in a 2×2 table. Use the chi-square test to test the hypotheses from (a). Verify that the X^2 statistic is the square of the z statistic. Use software or Table D to verify that the chi-square P-value agrees with the z result (up to the accuracy of the table if you do not use software).

(c) What do you conclude about the effectiveness of gastric freezing as a treatment for ulcers?

14.19 Preventing strokes, II Exercise 14.13 (page 857) asked whether adding a second anti-clotting drug to aspirin would be more effective for patients who had already had a stroke.

(a) Is there a significant difference among the four rates of strokes? Which components of the chi-square test statistic account for most of the total?

(b) The data report two response variables: whether the patient had a stroke and whether the patient died. Repeat your analysis for patient deaths. Is there evidence to conclude that the proportions of stroke patients who died during the two years of the study are different for the four treatment groups?

14.20 Preventing strokes, III This is a continuation of Exercises 14.13 (page 857) and 14.19. Consider only two of the treatment groups: aspirin only and aspirin + dipyridamole.

(a) Is there a significant difference in the proportion of strokes in the two groups? Give statistical evidence to support your decision.

(b) Is there a significant difference in death rates for the two groups?

(c) Is a chi-square procedure needed to answer (a) and (b)? Explain.

The Chi-Square Test of Association/Independence

Two-way tables can arise in several ways. The music and wine study in Example 14.4 (page 850) is an experiment that compared three music treatments using separate and independent samples. Each group is a sample from a separate population corresponding to a separate treatment. The study design fixes the size of each sample in advance, and the data record which of three outcomes (types of wine purchased) occurred for each category of the explanatory variable (types of music). The null hypothesis of "no difference among the treatments" takes the form of "equal proportions among the three populations." Example 14.9 illustrates a different setting for a two-way table.

Example 14.9 *Franchises that succeed*
Two-way table

Many popular businesses are franchises—think of McDonald's. The owner of a local franchise benefits from the brand recognition, national advertising, and detailed guidelines provided by the franchise chain. In return, he or she pays fees to the franchise firm and agrees to follow its policies. The relationship between the local entrepreneur and the franchise firm is spelled out in a detailed contract.

One clause that the contract may or may not contain is the entrepreneur's right to an exclusive territory. This means that the new outlet will be the only representative of the franchise in a specified territory and will not have to compete with other outlets of the same chain. How does the presence of an exclusive-territory clause in the contract relate to the survival of the business? A study designed to address this question collected data from a sample of 170 new franchise firms.[14]

Two categorical variables were measured for each firm. First, the firm was classified as successful or not based on whether or not it was still franchising as of a certain date. Second, the contract each firm offered to franchisees was classified according to whether or not there was an exclusive-territory clause. Here are the count data, arranged in a two-way table:

Observed numbers of firms

	Exclusive Territory		
Success	**Yes**	**No**	**Total**
Yes	108	15	123
No	34	13	47
Total	**142**	**28**	**170**

The two categorical variables in Example 14.9 are "Success," with values "Yes" and "No," and "Exclusive territory," with values "Yes" and "No." The objective in Example 14.9 is to compare franchises that have exclusive territories with those that do not. We view "Exclusive territory" as an explanatory variable and therefore we make it the column variable. The row variable is a categorical response variable, "Success."

The two-way table in Example 14.9 does not compare several populations. Instead, it arises by classifying observations from a *single* population in two ways: by exclusive territory and success. Both of these variables have two levels, so a statement of the null hypothesis is

H_0: There is no association between exclusive territory and success.

The setting of Example 14.9 is very different from a comparison of several proportions. Nevertheless, we can apply a chi-square test. One of the most useful properties of chi-square is that it tests the hypothesis "the row and column variables are not related to each other" whenever this hypothesis makes sense for a two-way table.

The Chi-Square Test of Association/Independence

Use the **chi-square test of association/independence** to test the null hypothesis

H_0: There is no association between two categorical variables.

when you have a two-way table from a single SRS, with each individual classified according to both of two categorical variables.

We start our analysis by computing descriptive statistics that summarize the observed relation between exclusive territory and success. We describe a relationship between categorical variables by comparing percents. The researchers suspected that exclusive territory helps explain a firm's success, so in this situation exclusive territory is the explanatory variable and success is the response variable. We should therefore compare the column percents that give the conditional distribution of success within each exclusive-territory category.

Example 14.10 ***Exclusive territory and success***
Looking at data numerically and graphically

To compare firms that have an exclusive territory, we examine column percents. For the "exclusive territory-yes" group, there are 108 successful firms out of a total of 142 firms. The column proportion for this cell is

$$\frac{108}{142} = 0.761$$

That is, 76.1% of the "exclusive territory-yes" franchises are successful. Similarly, 34 of the 142 firms in this group were not successful. The column proportion is

$$\frac{34}{142} = 0.239$$

or 23.9%. In all, we must calculate four percents. Here are the results, rounded for clarity:

Column percents for firms

	Exclusive Territory	
Success	**Yes**	**No**
Yes	76%	54%
No	24%	46%
Total	**100%**	**100%**

The "Total" row reminds us that 100% of each type of firm has been classified as successful or not. (The sums sometimes differ slightly from 100% because of roundoff error.) The bar graph in Figure 14.7 compares the percents of firms that are successful. In this Example, we are interested in the effect of having an exclusive territory on the success of the firm. The data reveal a clear relationship: 76% of firms with exclusive territories are successful, as opposed to only 54% of firms that were not offered exclusivity. The bar graph in Figure 14.7 helps us compare the distribution of success in the two exclusive-territory groups.

Figure 14.7 *Bar graph of the percents of firms that are successful, for Example 14.10.*

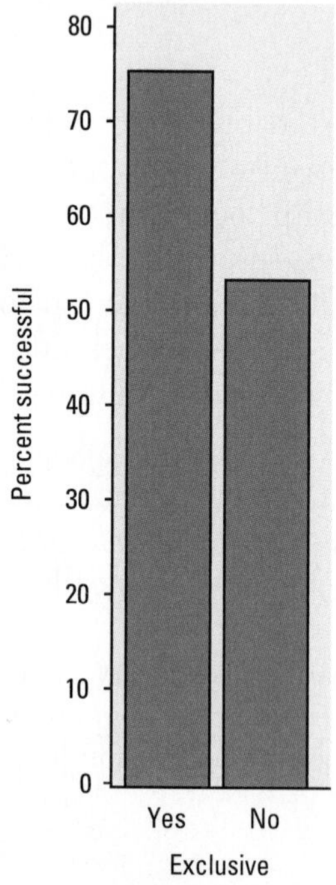

To look at the problem another way: among franchisees who had exclusive territories, the number of successful firms was about three times the number of unsuccessful firms. Among franchisees who did not have exclusive territories, only about half were successful.

The difference between the percents of successes between the two types of firms is quite large. A statistical test will tell us whether or not these differences can be plausibly attributed to chance. Specifically, if there is no association between success and having an exclusive territory, how likely is it that a sample would show a difference as large or larger than that displayed in Figure 14.7?

Computer packages can help you avoid tedious hand calculations. Here is output from one such package.

Example 14.11 *Successful franchises*
Using technology

Figure 14.8 shows the output from CrunchIt! for the data of Example 14.10. It includes some additional information needed for statistical inference. Check Figure 14.8 carefully. Be sure that you can identify the joint distribution, the marginal distributions, and the conditional distributions.

Figure 14.8 *CrunchIt! output for Example 14.11.*

	Yes	No	Total
Yes	108 87.8% 76.06% 63.53%	15 12.2% 53.57% 8.824%	123 100.00% 72.35% 72.35%
No	34 72.34% 23.94% 20%	13 27.66% 46.43% 7.647%	47 100.00% 27.65% 27.65%
Total	142 83.53% 100.00% 83.53%	28 16.47% 100.00% 16.47%	170 100.00% 100.00% 100.00%

Cell format: Count / Row percent / Column percent / Total percent

Test for independence of Success and Exclusive Territory:

Statistic	DF	Value	P-value
Chi-square	1	5.9111857	0.015

The *chi-square test of association/independence* assesses whether this observed association is statistically significant. That is, is the exclusive territory-success relationship in the sample sufficiently strong for us to conclude that it is due to a relationship between these two variables in the underlying population and not merely to chance? Note that the test asks only whether there is evidence of *some* relationship. To explore the direction or nature of the relationship we must examine the column or row percents. Note also that in using the chi-square test we are acting as if the subjects were a simple random sample from a single population of interest. If the franchises are a biased sample—for example, if unsuccessful franchisees are reluctant to participate in the study—then conclusions about the entire population of franchises are not justified.

Computing Expected Cell Counts

The null hypothesis is that there is no relationship between exclusive territory and success in the population. The alternative is that these two variables are related. If we assume that the null hypothesis is true, then success and exclusive territory are independent. We can find the expected cell counts using the multiplication rule for independent events (Chapter 6).

Example 14.12 *Successful franchises*
Calculating expected cell counts

What is the expected count for the cell corresponding to the event "successful" and "exclusive territory"? Under the null hypothesis that success and exclusive territory are independent,

$$P(\text{success and exclusive}) = P(\text{success}) \times P(\text{exclusive})$$
$$= \frac{123}{170} \times \frac{142}{170}$$

The expected count of "success and exclusive" can be found by multiplying this probability by the total number of franchises in the sample:

$$170\left(\frac{123}{170} \times \frac{142}{170}\right) = 102.74$$

Simple arithmetic shows that this is the same as calculating $(123 \times 142)/170$. In other words,

$$\text{expected count} = \frac{\text{row total} \times \text{column total}}{n}$$

where n is the sample size (table total).

Here is the completed table of expected counts:

Expected counts for success and exclusive territory

	Exclusive Territory		
Success	**Yes**	**No**	**Total**
Yes	102.74	20.26	**123.00**
No	39.26	7.74	**47.00**
Total	**142.00**	**28.00**	**170.00**

Performing the χ^2 Test

We are now ready to carry out the chi-square procedure for the success and exclusive-territory data, following the steps in the Inference Toolbox.

Example 14.13 *The franchise example*

Chi-square test for association/independence

Step 1: Hypotheses

H_0: Success and exclusive territory are independent.

H_a: Success and exclusive territory are dependent.

or, equivalently,

H_0: There is no association between success and exclusive territory.

H_a: There is an association between success and exclusive territory.

Step 2: Conditions To use the chi-square test of association/independence, we must check that all expected cell counts are at least 1 and that no more than 20% are less than 5. From the table in Example 14.12, we can see that these conditions are easily met.

Step 3: Calculations

- The test statistic is

$$X^2 = \sum \frac{(\text{observed} - \text{expected})^2}{\text{expected}}$$
$$= \frac{(108 - 102.74)^2}{102.74} + \frac{(15 - 20.26)^2}{20.26} + \frac{(34 - 39.26)^2}{39.26} + \frac{(13 - 7.74)^2}{7.74}$$
$$= 0.2693 + 1.3656 + 0.7047 + 3.5746$$
$$= 5.9112$$

- Because there are $r = 2$ categories of the row variable and $c = 2$ categories of the column variable, the degrees of freedom for this statistic are

$$\text{df} = (r - 1)(c - 1) = (2 - 1)(2 - 1) = 1$$

df = 1

p	.02	.01
χ^2	5.41	6.63

Under the null hypothesis that success and exclusive territory are independent, the test statistic X^2 has a χ^2 distribution with df = 1. To obtain the *P*-value, refer to the row in Table D corresponding to df = 1. The calculated value $X^2 = 5.91$ lies between upper critical points corresponding to probabilities 0.01 and 0.02. The *P*-value is therefore between 0.01 and 0.02.

Step 4: Interpretation There is sufficient evidence ($X^2 = 5.91$, df = 1, $0.01 < P < 0.02$) of an association between success and exclusive territory in the population of franchisees. The size and nature of this association are described by the table of percents in Figure 14.8 (page 871). Of course, this association does not show that exclusive territory *causes* success.

Conclusion

You can distinguish between the two types of chi-square tests for two-way tables by examining the design of the study. In the test of association/independence, there is a single sample from a single population. The individuals in the sample are classified according to two categorical variables. For the test of homogeneity of populations, there is a sample from each of two or more populations. Each individual is classified based on a single categorical variable. The precise statement of the hypotheses differs, depending on the sampling design.

Exercises

14.21 Extracurricular activities and grades, I North Carolina State University studied student performance in a course required by its chemical engineering major. One question of interest is the relationship between time spent in extracurricular activities and whether a student earned a C or better in the course. Here are the data for the 119 students who answered a question about extracurricular activities:[15]

	Extracurricular Activities (hours per week)		
Grade	**<2**	**2 to 12**	**>12**
C or better	11	68	3
D or F	9	23	5

(a) This is an $r \times c$ table. What are the numbers r and c?

(b) Find the proportion of successful students (C or better) in each of the three extracurricular activity groups. What kind of relationship between extracurricular activities and succeeding in the course do these proportions seem to show?

(c) Make a bar graph to compare the three proportions of successes.

(d) What null hypothesis will a chi-square procedure test in this setting?

(e) Find the expected counts if this hypothesis is true, and display them in a two-way table.

(f) Compare the observed counts with the expected counts. Are there large deviations between them? These deviations are another way of describing the relationship you described in (b).

14.22 Smoking by students and their parents, I How are the smoking habits of students related to their parents' smoking habits? Here are data from a survey of students in eight Arizona high schools:[16]

	Student smokes	**Student does not smoke**
Both parents smoke	400	1380
One parent smokes	416	1823
Neither parent smokes	188	1168

(a) This is an $r \times c$ table. What are the numbers r and c?

(b) Calculate the proportion of students who smoke in each of the three parent groups. Then describe in words the association between parent smoking and student smoking.

(c) Make a graph to display the association.

(d) Explain in words what the null hypothesis H_0 says about smoking.

(e) Find the expected counts if H_0 is true, and display them in a two-way table similar to the table of observed counts.

(f) Compare the tables of observed and expected counts. Explain how the comparison expresses the same association you see in (b) and (c).

14.23 Extracurricular activities and grades, II In Exercise 14.21, you began to analyze data on the relationship between time spent on extracurricular activities and success in a tough course. Figure 14.9 gives Minitab output (with some values deliberately omitted) for the two-way table in Exercise 14.21.

Figure 14.9 *Minitab output for Exercise 14.23.*

```
Chi-Square Test
Expected counts are printed below observed counts
                <2            2 to 12            >12        Total
A, B, C,        11              68                3           82
             13.78
D or F           9              23                5           37
                             28.29             2.49
Total           20              91                8          119
Chi-Sq      = 0.561  +                +       1.145  +
              1.244  +                +       2.538  =      6.926
DF =      ,  P-Value =
1 cells with expected counts less than 5.0
```

(a) Starting from the table of expected counts, find the 6 components of the chi-square statistic and then the statistic X^2 itself. Copy the computer output on your paper and fill in the five related missing values.

(b) Use Table D to find the *P*-value for this test. Then use your calculator to help you fill in the df and *P*-value entries on the computer output.

(c) Which term contributes the most to X^2? What specific relation between extracurricular activities and academic success does this term point to?

(d) Does the North Carolina State study convince you that spending more or less time on extracurricular activities causes changes in academic success? Explain your answer.

14.24 Smoking by students and their parents, II In Exercise 14.22, you began to analyze data on the relationship between smoking by parents and smoking by high school students.

(a) Use the Inference Toolbox to carry out the appropriate significance test.

(b) Which term(s) contribute the most to X^2? What specific relation between parent smoking and student smoking do the term(s) point to?

(c) Does the study convince you that parent smoking *causes* student smoking? Explain your answer.

14.25 Do you use cocaine? Sample surveys on sensitive issues can give different results depending on how the question is asked. A University of Wisconsin study randomly divided 2400 respondents into 3 groups. All participants were asked if they had ever used cocaine. One group of 800 was interviewed by phone; 21% said they had used cocaine. Another 800 people were asked the question in a one-on-one personal interview; 25% said "Yes." The remaining 800 were allowed to make an anonymous written response; 28% said "Yes."[17] Are there statistically significant differences among these proportions? Give appropriate statistical evidence to support your conclusion.

14.26 Opinions about the death penalty "Do you favor or oppose the death penalty for persons convicted of murder?" When the General Social Survey asked this question in its 2002 survey, the responses of people whose highest education was a bachelor's degree and of people with a graduate degree were as follows:[18]

	Favor	Oppose
Bachelor	135	71
Graduate	64	50

(a) Is there evidence that the proportions of all people at these levels of education who favor the death penalty differ? Find the two sample proportions, the z statistic, and its P-value.

(b) Is there evidence that the opinions of all people at these levels of education differ? Find the chi-square statistic X^2 and its P-value. If your work is correct, X^2 should be the same as z^2 and the two P-values should be identical. Are they?

Section 14.2 Summary

For **two-way tables,** we first compute percents or proportions that describe the relationship of interest. Then we turn to formal inference. Two different methods of generating data for two-way tables lead to the **chi-square test for homogeneity of populations** and the **chi-square test of association/independence**.

In the first design, independent SRSs are drawn from each of several populations, and each observation is classified according to a categorical variable of interest. The null hypothesis is that the distribution of this categorical variable is the same for all of the populations. We use the chi-square test for homogeneity of populations to test this hypothesis.

One common use of the chi-square test for homogeneity of populations is to compare several population proportions. The null hypothesis states that all of the population proportions are equal. The alternative hypothesis states that they are not all equal but allows any other relationship among the population proportions.

In the second design, a single SRS is drawn from a population, and observations are classified according to two categorical variables. The chi-square test of association/independence tests the null hypothesis that there is no relationship between the row variable and the column variable.

The **expected count** in any cell of a two-way table when H_0 is true is

$$\text{expected count} = \frac{\text{row total} \times \text{column total}}{n}$$

where n = sample size. The **chi-square statistic** is

$$X^2 = \sum \frac{(\text{observed count} - \text{expected count})^2}{\text{expected count}} = \sum \frac{(\text{O} - \text{E})^2}{\text{E}}$$

where the sum is over all $r \times c$ cells.

The chi-square test compares the value of the statistic X^2 with critical values from the **chi-square distribution** with $(r - 1)(c - 1)$ degrees of freedom. Large values of X^2 are evidence against H_0, so the *P*-value is the area under the chi-square density curve to the right of X^2.

The chi-square distribution is an approximation to the distribution of the statistic X^2. You can safely use this approximation when all expected cell counts are at least 1 and no more than 20% of all expected cell counts are less than 5.

Section 14.2 Exercises

14.27 Stress and heart attacks You read a newspaper article that describes a study of whether stress management can help reduce heart attacks. The 107 subjects all had reduced blood flow to the heart and so were at risk of a heart attack. They were assigned at random to three groups. The article goes on to say:

> One group took a four-month stress management program, another underwent a four-month exercise program, and the third received usual heart care from their personal physicians.
>
> In the next three years, only three of the 33 people in the stress management group suffered "cardiac events," defined as a fatal or non-fatal heart attack or a surgical procedure such as a bypass or angioplasty. In the same period, seven of the 34 people in the exercise group and 12 out of the 40 patients in usual care suffered such events.[19]

(a) Use the information in the news article to make a two-way table that describes the study results.

(b) What are the success rates of the three treatments in avoiding cardiac events?

(c) Find the expected cell counts under the null hypothesis that there is no difference among the treatments. Verify that the expected counts meet our guideline for use of the chi-square test.

(d) Is there a significant difference among the success rates for the three treatments? Give appropriate statistical evidence to support your answer.

14.28 Regulating guns The National Gun Policy Survey asked respondents, "Do you think there should be a law that would ban possession of handguns except for the police and other authorized persons?" Here are the responses, broken down by the respondent's level of education:[20]

Education	Yes	No
Less than high school	58	58
High school graduate	84	129
Some college	169	294
College graduate	98	135
Postgraduate degree	77	99

(a) How does the proportion of the sample who favor banning possession of handguns differ among people with different levels of education? Make a bar graph that compares the proportions, and briefly describe the relationship between education and opinion about a handgun ban.

(b) Show whether or not the sample provides good evidence that the proportion of the adult population who favor a ban on handguns changes with level of education.

14.29 Where do young adults live? A survey by the National Institutes of Health asked a random sample of young adults (aged 19 to 25 years), "Where do you live now? That is, where do you stay most often?" Here is the full two-way table (omitting a few who refused to answer and one who claimed to be homeless):[21]

	Female	Male
Parents' home	923	986
Another person's home	144	132
Own place	1294	1129
Group quarters	127	119

What are the most important differences between young men and women? Are their choices of living places significantly different?

14.30 Legalize marijuana? This is a continuation of Activity 14B.

(a) What population are you trying to draw conclusions about?

(b) An alert student in the back row suggests that you can't perform inference because your class is not a random sample. How do you respond to his concern?

(c) For this part, pretend that your class is an SRS. You want to perform inference on a two-way table of count data. Would this be a chi-square test of homogeneity of populations or a chi-square test of independence? Explain.

(d) Do the data that your class collected give you reason to believe that males and females from populations that your class represents have different views on whether physicians should be legally able to prescribe marijuana for their patients? Give appropriate statistical evidence to support your conclusion.

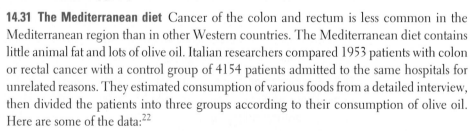

14.31 The Mediterranean diet Cancer of the colon and rectum is less common in the Mediterranean region than in other Western countries. The Mediterranean diet contains little animal fat and lots of olive oil. Italian researchers compared 1953 patients with colon or rectal cancer with a control group of 4154 patients admitted to the same hospitals for unrelated reasons. They estimated consumption of various foods from a detailed interview, then divided the patients into three groups according to their consumption of olive oil. Here are some of the data:[22]

	Olive Oil Consumption			
	Low	**Medium**	**High**	**Total**
Colon cancer	398	397	430	**1225**
Rectal cancer	250	241	237	**728**
Controls	1368	1377	1409	**4154**

(a) Is this study an experiment? Explain your answer.

(b) The investigators report that "less than 4% of cases or controls refused to participate." Why does this fact strengthen our confidence in the results?

(c) The researchers conjectured that high olive oil consumption would be more common among patients without cancer than in patients with colon cancer or rectal cancer. What do the data say?

14.32 Students and catalog shopping What is the most important reason that students buy from catalogs? The answer may differ for different groups of students. Here are results for samples of American and East Asian students at a large midwestern university:[23]

	American	East Asian
Save time	29	10
Easy	28	11
Low price	17	34
Live far from stores	11	4
No pressure to buy	10	3
Other reason	20	7
Total	115	69

Describe the most important differences in shopping habits between American and East Asian students. Is there a significant overall difference between the two distributions of responses?

14.33 Secondhand stores, I Shopping at secondhand stores is becoming more popular and has even attracted the attention of business schools. A study of customers' attitudes toward secondhand stores interviewed samples of shoppers at two secondhand stores of the same chain in two cities. The breakdown of the respondents by sex is as follows:[24]

	City 1	City 2
Men	38	68
Women	203	150
Total	241	218

Is there a significant difference between the proportions of women customers in the two cities?

(a) State the null hypothesis, find the sample proportions of women in both cities, do a two-sided z test, and give a P-value using software or Table A.

(b) Calculate the chi-square statistic X^2 and show that it is the square of the z statistic. Show that the P-value from Table D agrees (up to the accuracy of the table) with your result from (a).

(c) Give a 95% confidence interval for the difference between the proportions of women customers in the two cities.

14.34 Secondhand stores, II The study of shoppers in secondhand stores cited in the previous exercise also compared the income distributions of shoppers in the two stores. Here is a two-way table of counts:

Income	City 1	City 2
Under \$10,000	70	62
\$10,000 to \$19,999	52	63
\$20,000 to \$24,999	69	50
\$25,000 to \$34,999	22	19
\$35,000 or more	28	24

A statistical calculator gives the chi-square statistic for this table as $X^2 = 3.955$. Is there good evidence that customers at the two stores have different income distributions?

CASE CLOSED!

Does acupuncture promote pregnancy?

1. Refer to the data in the Case Study on page 833. Use a chi-square test to determine if there is a significant difference in pregnancy rate between the control group and the acupuncture group. Follow the Inference Toolbox. Make sure that you comment on whether this is a test of homogeneity of populations or a test of association/independence.
2. We can also use inference methods for comparing two population proportions to analyze the results of this study. Do so now. Then compare the results you got using the chi-square procedure with the results you got from comparing two population proportions.
3. Based on the method and results of this study, why can't we conclude that acupuncture has significant physiological effects on the reproductive system?

Chapter Review

Summary

This chapter develops several settings where variations of the chi-square test of significance are useful. In a **goodness of fit test,** the object is to determine if a population distribution is different from some specified distribution. The null hypothesis states that there is no difference between the distributions, while the alternative hypothesis states that there is some kind of difference. The chi-square test tells whether there is sufficient reason to reject the null hypothesis, but further analysis is needed to determine how and where the differences occur.

A goodness of fit test begins by finding the expected counts for each category, assuming the distribution is the same as the specified distribution. The chi-square statistic is a measure of how much the sample distribution diverges from the hypothesized distribution. For a given number of degrees of freedom, large values of the chi-square statistic provide evidence to reject the null hypothesis of no difference.

The **chi-square test for homogeneity** of populations is an overall test that tells us whether the data give good reason to reject the hypothesis that the distribution of a categorical variable is the same in several populations. It can be used when the data come from independent SRSs from the populations of interest. This procedure is also useful in testing the equality of proportions of successes in any number of populations. The alternative to this hypothesis is "many-sided" because it allows any relationship other than "all equal."

Two-way tables also arise when an SRS is taken from a single population, and each individual is classified according to two categorical variables. In this setting, use a **chi-square test of association/independence.** This procedure tests the null hypothesis that there is no relationship between the row variable and the column variable in a two-way table.

The chi-square test is actually an approximate test that becomes more accurate as the cell counts in the two-way table increase. Fortunately, chi-square *P*-values are quite accurate even for small counts. You should always accompany a chi-square test by data analysis to see what kind of relationship is present.

What You Should Have Learned

Here is a checklist of the major skills you should have acquired by studying this chapter.

A. Choose the Appropriate Chi-Square Procedure

1. For goodness of fit tests, use percents and bar graphs to compare hypothesized and actual distributions.
2. Distinguish between tests of homogeneity of populations and tests of association/independence.
3. Organize categorical data in a two-way table. Then use percents and bar graphs to describe the relationship between the categorical variables.

B. Perform Chi-Square Tests

1. Explain what null hypothesis is being tested.
2. Calculate expected counts.
3. Calculate the component of the chi-square statistic for any cell, as well as the overall statistic.
4. Give the degrees of freedom of a chi-square statistic.
5. Use the chi-square critical values in Table D to approximate the *P*-values of a chi-square test.

C. Interpret Chi-Square Tests

1. Locate expected cell counts, the chi-square statistic, and its *P*-value in output from computer software or a calculator.
2. If the test is significant, use percents, comparison of expected and observed counts, and the components of the chi-square statistic to see which deviations from the null hypothesis are most important.

Web Links

Information about all varieties of M&M/Mars Company candies can be found at us.mms.com/us/about/products/

The *New England Journal of Medicine* report that is the basis for Example 14.1 is discussed at www.findarticles.com/p/articles/mi_m4021/is_ISSN_0163-4089/ ai_55403024

An interesting paper, "Cell Phone Use and Traffic Accidents, Revisited," can be found at www.nsc.org/issues/idrive/cellfone.htm

For discussion of the music and wine connection of Example 14.4, see www.le.ac.uk/psychology/acn5/nature.html

Chapter Review Exercises

14.35 AP exam scores The Advanced Placement (AP) Statistics examination was first administered in May 1997. Students' papers are graded on a scale of 1 to 5, with 5 being the highest score. Over 7600 students took the exam in the first year, and the distribution of scores was as follows (not including exams that were scored late):

Score:	5	4	3	2	1
Percent:	15.3	22.0	24.8	19.8	18.1

A sample of students who took the exam had the following distribution of grades:

Score:	5	4	3	2	1
Frequency:	167	158	101	79	30

Calculate marginal percents and make a bar graph of the population scores and the sample scores, so that the two distributions can be compared visually. Then perform an appropriate test to determine if the distribution of scores for this particular sample is significantly different from the distribution of scores for all students who took the inaugural exam.

14.36 Effects of alcohol and nicotine on children Alcohol and nicotine consumption during pregnancy may harm children. Because drinking and smoking behaviors may be related, it is important to understand the nature of this relationship when assessing the possible effects on children. One study classified 452 mothers according to their alcohol intake prior to pregnancy recognition and their nicotine intake during pregnancy. The data are summarized in the following table:[25]

	Nicotine (milligrams/day)		
Alcohol (ounces/day)	**None**	**1–15**	**16 or more**
None	105	7	11
0.01–0.10	58	5	13
0.11–0.99	84	37	42
1.00 or more	57	16	17

Carry out a complete analysis of the association between alcohol and nicotine consumption. That is, describe the nature and strength of this association and assess its statistical significance. Include charts or figures to display the association.

14.37 College distribution At your local college 29% of the students are in their first year, 27% in their second, 25% in their third, and 19% in their fourth year. You have taken a survey of students, and when you classify them by year of study, you have 54, 66, 56, and

30 students in the first, second, third, and fourth years, respectively. Use a goodness of fit test to examine how well your sample reflects the population of your college.

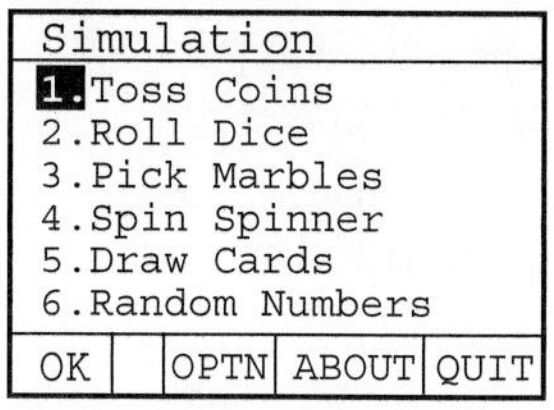

14.38 Prob Sim The Prob Sim APP for the TI-83/84/89 allows you to simulate tossing coins, rolling dice, picking marbles, spinning a spinner, drawing cards, and playing the lottery. If you have a TI-83/84/89, download this APP from your teacher.

- To run the APP, press the APPS key. On the TI-83/84 Plus, choose Prob Sim. On the TI-89, choose 1:FlashApps, then Prob Sim. Press ENTER. You should see the introductory screen shown in the margin. Press ENTER again to see the main menu.
- Choose 4. Spin Spinner. Spin the spinner a total of 200 times with 4 sets of 50 spins. Record the number of times that the spinner lands in each of the four numbered sections.
- Perform a significance test of the hypothesis that this program yields an equal proportion of 1s, 2s, 3s, and 4s. If you do not have the Prob Sim APP, use the following sample data: 51 1s, 39 2s, 53 3s, 57 4s.

14.39 Python eggs How is the hatching of water python eggs influenced by the temperature of the snake's nest? Researchers assigned newly laid eggs to one of three temperatures: hot, neutral, or cold. Hot duplicates the extra warmth provided by the mother python, and cold duplicates the absence of the mother. Here are the data on the number of eggs and the number that hatched:[26]

	Eggs	Hatched
Cold	27	16
Neutral	56	38
Hot	104	75

(a) Make a two-way table of temperature by outcome (hatched or not).

(b) Calculate the percent of eggs in each group that hatched. The researchers anticipated that eggs would not hatch in cold water. Do the data support that anticipation?

(c) Are there significant differences among the proportions of eggs that hatched in the three groups?

14.40 Cocaine addiction is hard to break Cocaine addicts need cocaine to feel any pleasure, so perhaps giving them an antidepressant drug will help. A 3-year study with 72 chronic cocaine users compared an antidepressant drug called desipramine with lithium and a placebo. (Lithium is a standard drug to treat cocaine addiction. A placebo is a dummy drug, used so that the effect of being in the study but not taking any drug can be seen.) One-third of the subjects, chosen at random, received each treatment.[27] Here are the results:

	Cocaine Relapse?	
Treatment	**Yes**	**No**
Desipramine	10	14
Lithium	18	6
Placebo	20	4

Compare the effectiveness of the three treatments in preventing relapse. Use percents and draw a bar graph. Write a brief summary of your conclusions.

14.41 Do pets increase survival? Psychological and social factors can influence the survival of patients with serious diseases. One study examined the relationship between survival of patients with coronary heart disease (CHD) and pet ownership. Each of 92 patients was classified as having a pet or not and by whether they survived for one year. Here are the data:[28]

	Pet Ownership	
Patient status	**No**	**Yes**
Alive	28	50
Dead	11	3

(a) Was this study an experiment? Why or why not?

(b) The researchers thought that having a pet might improve survival, so pet ownership is the explanatory variable. Compute appropriate percents to describe the data, and state your preliminary findings.

(c) Carry out an appropriate inference procedure to test the researchers' claim.

(d) What do you conclude? Do the data give convincing evidence that owning a pet is an effective treatment for increasing the survival of CHD patients?

(e) Did you use a χ^2 test or a z test in part (c)? Carry out the other test and compare the results. Which is the more appropriate test in this setting? Why?

14.42 Identity theft A study of identity theft looked at how well consumers protect themselves from this increasingly prevalent crime. The behaviors of 61 college students were compared with the behaviors of 59 nonstudents.[29] One of the questions was "When asked to create a password, I have used either my mother's maiden name, or my pet's name, or my birth date, or the last four digits of my social security number, or a series of consecutive numbers." For the students, 22 agreed with this statement while 30 of the nonstudents agreed.

(a) Display the data in a two-way table and perform the chi-square test. Summarize the results.

(b) Reanalyze the data using the methods for comparing two proportions that we studied in the previous chapter. Compare the results and verify that the chi-square statistic is the square of the z statistic.

(c) The students in this study were junior and senior college students from two sections of a course in Internet marketing at a large northeastern university. The nonstudents were a group of individuals who were recruited to attend commercial focus groups on the West Coast conducted by a lifestyle marketing organization. Discuss how the method of selecting the subjects in this study relates to the conclusions that can be drawn from it.

14.43 Sex sells In what ways do advertisers in magazines use sexual imagery to appeal to youth? One study classified each of 1509 full-page or larger ads as "not sexual" or "sexual," according to the amount and style of the dress of the male or female model in the ad. The ads were also classified according to the target readership of the magazine.[30] Here is the two-way table of counts:

	Magazine Readership			
Model dress	**Women**	**Men**	**General interest**	**Total**
Not sexual	351	514	248	1113
Sexual	225	105	66	396
Total	576	619	314	1509

(a) Summarize the data numerically and graphically.

(b) Perform the significance test that compares the model dress for the three categories of magazine readership. Summarize the results of your test and give your conclusion.

(c) All of the ads were taken from the March, July, and November issues of six magazines in one year. Discuss this fact from the viewpoint of the validity of the significance test and the interpretation of the results.

14.44 Student athletes and gambling A survey of student athletes that asked questions about gambling behavior classified students according to the National Collegiate Athletic Association (NCAA) division that their school belonged to.[31] For male student athletes, the percents who reported wagering on collegiate sports are given here, along with the numbers of respondents in each division.

Division	**I**	**II**	**III**
Percent:	17.2%	21.0%	24.4%
Number:	5619	2957	4089

(a) Use a significance test to compare the percents for the three NCAA divisions. Give details and a short summary of your conclusion.

(b) The percents in the table above are given in the NCAA report, but the numbers of male student athletes in each division who responded to the survey question are estimated based on other information in the report. To what extent do you think this has an effect on the results? (*Hint*: Rerun your analysis a few times, with slightly different numbers of students but the same percents.)

(c) Some student athletes may be reluctant to provide this kind of information, even in a survey where there is no possibility that they can be identified. Discuss how this fact may affect your conclusions.

(d) The chi-square test for this set of data assumes that the responses of the student athletes are independent. However, some of the students are at the same school and even on the same team. Discuss how you think this might affect the results.

CHAPTER

15

Inference for Regression

CASE STUDY

Three-pointers in college basketball

The three-point line was installed in college basketball for the 1986–1987 season and forever changed the nature of the game. The three-point line is a painted arc 19 feet, 9 inches from the center of the basket rim. It is so called because successful shots from beyond the line are awarded one point more than the two points awarded for "field goals" made from closer to the basket. Closer shots have always been higher-percentage shots, and that hasn't changed. But it can be argued that since the three-point line was introduced, there have been more upsets, more near upsets, more dramatic comebacks, and more parity among teams, hence making college basketball a more exciting game to watch.

A look back at the effect of this new feature 20 years later reveals some interesting results. You might expect, for example, that as the average number of three-point shots taken per game increases, the average number of three-point shots made also increases—a positive association. This appears to be borne out by the data:[1]

Season	Average number taken	Average number made	Percent made
1986–87	9.2	3.5	38.4
1987–88	10.4	4.0	38.3
1988–89	11.8	4.4	37.8
1989–90	12.8	4.7	36.8
1990–91	13.8	5.0	36.2
1991–92	14.0	5.0	35.6
1992–93	14.9	5.3	35.4
1993–94	16.5	5.7	34.5
1994–95	17.2	5.9	34.5
1995–96	17.1	5.9	34.3
1996–97	17.1	5.8	34.1
1997–98	17.4	6.0	34.4
1998–99	17.4	5.9	34.2
1999–00	17.7	6.1	34.4
2000–01	17.7	6.1	34.6
2001–02	18.3	6.3	34.6
2002–03	18.1	6.3	34.8
2003–04	18.3	6.3	34.6
2004–05	18.3	6.4	34.7

But now look at the percent of shots made versus the number of shots taken. This is clearly a negative association. This says that as more three-point shots are taken, the lower the percent of successful shots. Is this a contradiction? Is it possible that knowing the number of three-point shots taken can help us determine the percent of shots made? If so, how confident can we be in making our predictions? After you have studied this chapter, you will be better able to answer interesting questions like these.

Activity 15
Ideal proportions

Materials: Fabric tape measure; calculator for each team of students

The architect Vitruvius said that "if you open your legs so much as to decrease your height by 1/14 and spread and raise your arms til your middle fingers touch the level of the top of your head, you must know that the center of the outspread limbs will be in the navel and the space between the legs will be an equilateral triangle. . . . The length of a man's outspread arms is equal to his height."[2]

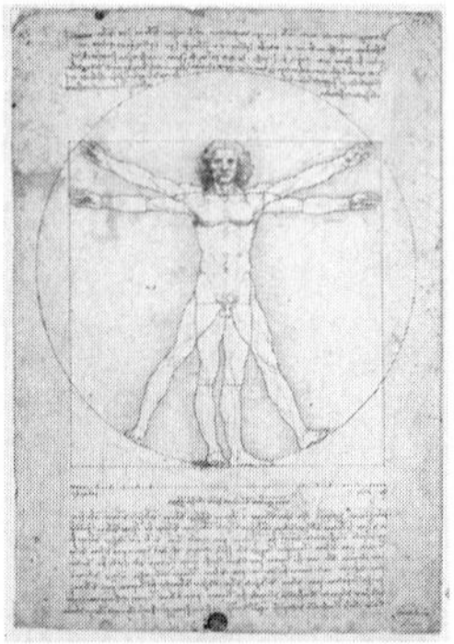

Leonardo da Vinci, the renowned painter, drew the illustration shown in the margin for a book on the works of Vitruvius. Da Vinci believed that the human body conformed to a set of geometric proportions as shown by the lines and circles in this drawing.

In this Activity, we want to determine if arm span can predict height. You will need a fabric measuring tape, and you should work in teams of three: the person to be measured and two people to hold the ends of the tape. You should collect at least 18 to 20 pairs of measurements. If your class has fewer students, recruit some volunteers from other classes. Remember: the more data, the better.

1. Take turns making and recording the following two measurements. First measure your arm span: the distance between the tips of the fingers when you stretch your arms out to the sides (the x-values). Then measure your height (the y-values). Unlike Vitruvius's man, who made an equilateral triangle with his legs, you should keep your heels together and stand tall. Combine your results with those of the other groups.
2. Make a scatterplot of the data. Clearly, the association should be positive. Is it? Would you describe the association as strong, moderate, or weak?
3. Use your calculator to perform least-squares regression and find the values of r and r^2. Plot the least-squares line on your scatterplot. Write a statement that interprets the meaning, in context, of the least-squares line and value of r^2 that you found.
4. Construct a residual plot to assess whether a line is an appropriate model for these data. Write a sentence that interprets your residual plot.

The Regression Model

When a scatterplot shows a linear relationship between a quantitative explanatory variable x and a quantitative response variable y, we can use the least-squares line fitted to the data to predict y for a given value of x. Now we want to do tests and confidence intervals in this setting.

Example 15.1 Crying and IQ
Modeling the data

Infants who cry easily may be more easily stimulated than others. This may be a sign of higher IQ. Child development researchers explored the relationship between the crying of infants four to ten days old and their later IQ test scores. A snap of a rubber band on the sole of the foot caused the infants to cry. The researchers recorded the crying and measured its intensity by the number of peaks in the most active 20 seconds. They later measured the children's IQ at age three years using the Stanford-Binet IQ test. Table 15.1 contains data on 38 infants. We will use the Data Analysis Toolbox to analyze these data.

Table 15.1 *Infants' crying and IQ scores*

Crying	IQ	Crying	IQ	Crying	IQ	Crying	IQ
10	87	20	90	17	94	12	94
12	97	16	100	19	103	12	103
9	103	23	103	13	104	14	106
16	106	27	108	18	109	10	109
18	109	15	112	18	112	23	113
15	114	21	114	16	118	9	119
12	119	12	120	19	120	16	124
20	132	15	133	22	135	31	135
16	136	17	141	30	155	22	157
33	159	13	162				

Source: Samuel Karelitz et al., "Relation of crying activity in early infancy to speech and intellectual development at age three years," *Child Development,* 35 (1964), pp. 769–777.

- **Data**

1. **Who?** We are told only that the individuals in the study were 38 infants who were studied when they were four to ten days old and again when they were three years old.
2. **What?** The explanatory variable is crying intensity and the response variable is children's IQ.
3. **Why?** Researchers wanted to determine if there is an association between crying activity in early infancy and IQ at age 3 years.
4. **When, where, how, and by whom?** The data come from an experiment described in 1964 in the journal *Child Development.*

- **Graphs** As always, we first examine the data. Figure 15.1, on the next page, is a scatterplot of the paired data. Plot the explanatory variable (count of crying peaks) horizontally and the response variable (IQ) vertically.

We look for the form, direction, and strength of the relationship as well as for outliers or other deviations. There is a moderate positive linear relationship, with no extreme outliers or potentially influential observations.

- **Numerical summaries** Because the scatterplot shows a roughly linear (straight-line) pattern, the *correlation* describes the direction and strength of the relationship. The correlation between crying and IQ is $r = 0.455$.

Figure 15.1 *Scatterplot, with regression line, of the IQ score of infants at age three years against the intensity of their crying soon after birth, for Example 15.1.*

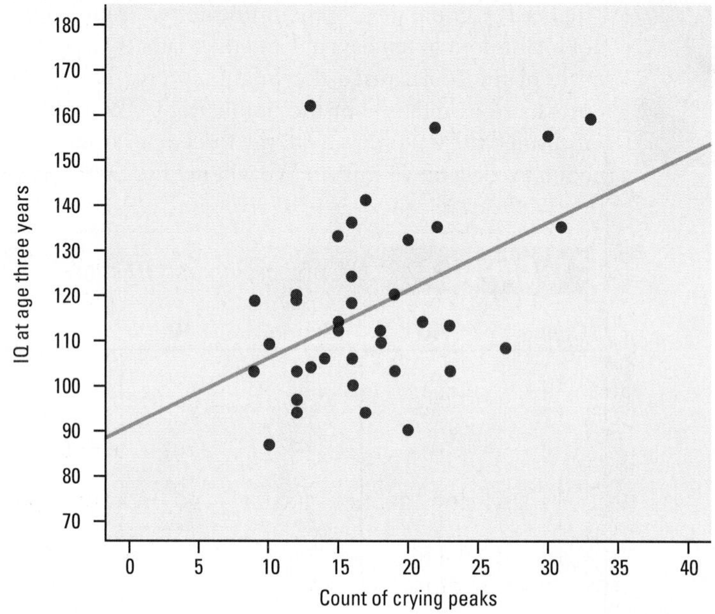

- **Model** We are interested in predicting the response from information given about the explanatory variable. So we find the least-squares regression line for predicting IQ from crying. This line lies as close as possible to the points (in the sense of least squares) in the vertical (y) direction. The equation of the least-squares regression line is

$$\begin{aligned}\hat{y} &= a + bx \\ &= 91.27 + 1.493x\end{aligned}$$

Here are the relevant TI-83/84 screens. The TI-89 results are similar.

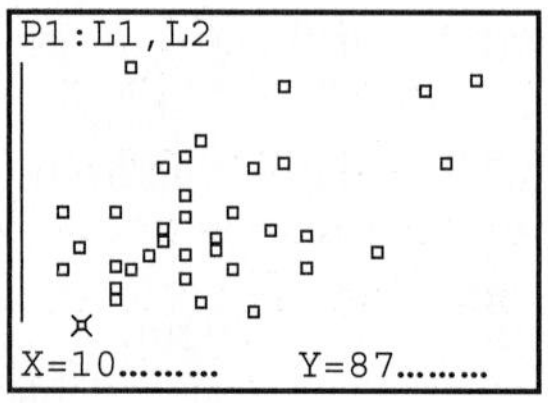


```
LinReg
 y=a+bx
 a=91.26829865
 b=1.492896598
 r²=.2069999897
 r=.4549725153
```

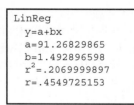

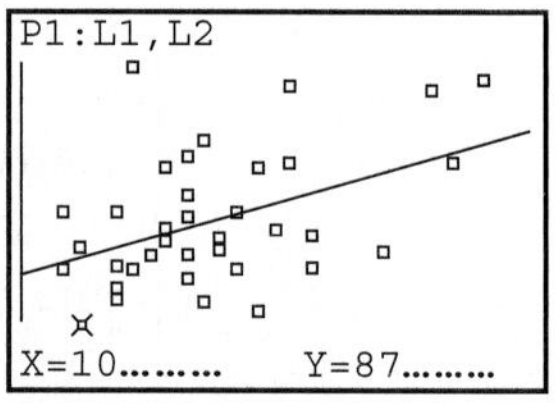

- **Interpretation** We use the notation $\hat{y}$ to remind ourselves that the regression line gives predictions of IQ. The predictions usually won't agree exactly with the actual values of the IQ measured several years later. Drawing the least-squares line on the scatterplot helps us see the overall pattern. Because $r^2 = 0.207$, only about 21% of the variation in IQ scores is explained by crying intensity. Prediction of IQ from crying intensity will not be very accurate. It is nonetheless impressive that behavior soon after birth can even partly predict IQ several years later.

Conditions for the Regression Model

The slope b and intercept a of the least-squares line are statistics. That is, we calculate them from the sample data. In the setting of Example 15.1, *these statistics would take somewhat different values if we repeated the study with different infants.* To do formal inference, we think of a and b as estimates of unknown parameters. The parameters appear in a mathematical model of the process that produces our data. Here are the required conditions for performing inference about the regression model.

Conditions for Regression Inference

We have n observations on an explanatory variable x and a response variable y. Our goal is to study or predict the behavior of y for given values of x.

- Repeated responses y are independent of each other.
- The mean response μ_y has a straight-line relationship with x:

$$\mu_y = \alpha + \beta x$$

  The slope β and intercept α are unknown parameters.
- The standard deviation of y (call it σ) is the same for all values of x. The value of σ is unknown.
- For any fixed value of x, the response y varies according to a Normal distribution.

true regression line

The heart of this model is that there is an "on the average" straight-line relationship between y and x. The ***true regression line*** $\mu_y = \alpha + \beta x$ says that the mean response μ_y moves along a straight line as the explanatory variable x changes. We can't observe the true regression line. The values of y that we do observe vary about their means according to a Normal distribution. If we hold x fixed and take many observations of y, the Normal pattern will eventually appear in a stemplot or histogram. In practice, we observe y for many different values of x, so that we see an overall linear pattern formed by points scattered about the true line. The standard deviation σ determines whether the points fall close to the true regression line (small σ) or are widely scattered (large σ).

Figure 15.2 on the next page shows the regression model in picture form. The line in the figure is the true regression line. The mean of the response y moves along this line as the explanatory variable x takes different values. The Normal curves show how y will vary when x is held fixed at different values. All of the curves have the same σ, so the variability of y is the same for all values of x. **You should check the conditions for inference when you do inference about regression.**

Checking the Regression Conditions

You can fit a least-squares line to any set of explanatory-response data when both variables are quantitative. If the scatterplot doesn't show a roughly linear pattern,

Figure 15.2 *The regression model. The line is the true regression line, which shows how the mean response μ_y changes as the explanatory variable x changes. For any fixed value of x, the observed response y varies according to a Normal distribution having mean μ_y.*

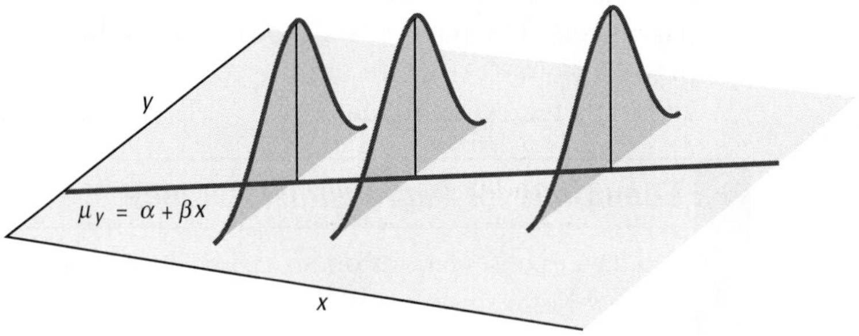

the fitted line may be almost useless. But it is still the line that fits the data best in the least-squares sense. To use regression inference, however, the data must satisfy the regression model conditions. Before we do inference, we must check these conditions one by one.

- **The observations are independent.** In particular, repeated observations on the same individual are not allowed. So we can't use ordinary regression to make inferences about the growth of a single child over time, for example.
- **The true relationship is linear.** We can't observe the true regression line, so we will almost never see a perfect straight-line relationship in our data. Look at the scatterplot to check that the overall pattern is roughly linear. A plot of the residuals against x magnifies any unusual pattern. Draw a horizontal line at zero on the residual plot to orient your eye. Because the sum of the residuals is always zero, zero is also the mean of the residuals.
- **The standard deviation of the response about the true line is the same everywhere.** Look at the scatterplot again. The scatter of the data points about the line should be roughly the same over the entire range of the data. A plot of the residuals against x, with a horizontal line at zero, makes this easier to check. It is quite common to find that as the response y gets larger, so does the scatter of the points about the fitted line. Rather than remaining fixed, the standard deviation σ, which measures the variability of responses, may be changing with x as the mean response changes with x. You cannot safely use our inference procedures when this happens.
- **The response varies Normally about the true regression line.** We can't observe the true regression line. We can observe the least-squares line and the residuals, which show the variation of the response about the fitted line. The residuals estimate the deviations of the response from the true regression line, so they should follow a Normal distribution. Make a histogram or stemplot of the residuals and check for clear skewness or other major departures from Normality. It turns out that inference for regression is not very sensitive to a

minor lack of Normality, especially when we have many observations. Do beware of influential observations, which move the regression line and can greatly affect the results of inference.

Fortunately, it is not hard to check for gross violations of the conditions for regression inference. Checking conditions uses the residuals. Most regression software will calculate and save the residuals for you.

Example 15.2 *Crying and IQ*

Checking conditions

Table 15.1 shows that the first infant studied had 10 crying peaks and a later IQ of 87. The predicted IQ for $x = 10$ is

$$\begin{aligned}\hat{y} &= 91.27 + 1.493x \\ &= 91.27 + 1.493(10) = 106.2\end{aligned}$$

The residual for this observation is

$$\begin{aligned}\text{residual} &= y - \hat{y} \\ &= 87 - 106.2 = -19.2\end{aligned}$$

That is, the observed IQ for this infant lies 19.2 points below the least-squares line on the scatterplot.

Repeat this calculation 37 more times, once for each subject. The 38 residuals are

−19.20	−31.13	−22.65	−15.18	−12.18	−15.15	−16.63	−6.18
−1.70	−22.60	−6.68	−6.17	−9.15	−23.58	−9.14	2.80
−9.14	−1.66	−6.14	−12.60	0.34	−8.62	2.85	14.30
9.82	10.82	0.37	8.85	10.87	19.34	10.89	−2.55
20.85	24.35	18.94	32.89	18.47	51.32		

If you haven't entered the crying and IQ data into your calculator, do that now as L_1/list1 and L_2/list2. Then on the TI-83/84, define list L_3 to be the observed minus the predicted values of y: $L_2 - Y_1(L_1) \rightarrow L_3$. On the TI-89, define list3 to be list2 $- Y_1$(list1). Verify that the 38 residuals are as shown and that the sum of the residuals is zero.

```
L1    L2    L3   3
10    87    ----
12    97
9     103
16    106
18    109
15    114
12    119
L3=L2-Y1(L1)
```

```
L1    L2    L3      3
10    87    -19.2
12    97    -12.18
9     103   -1.704
16    106   -9.155
18    109   -9.14
15    114   .33825
12    119   9.8169
L3(1)=-19.1972646...
```

```
1-Var Stats
 x̄=3.157895E-12
 Σx=1.2E-10
 Σx²=11023.3888
 Sx=17.26063229
 σx=17.03200455
↓n=38
```

Notice that the sum of the residuals is shown in the calculator screen as 1.2E−10, which is zero, up to round-off error. Another reason to use technology in doing regression is that round-off errors in hand calculation can accumulate and make the results inaccurate.

A residual plot appears in Figure 15.3 on the next page. The values of x are on the horizontal axis. The residuals are on the vertical axis, with a horizontal line at zero.

Figure 15.3 *Plot of the regression residuals for the IQ data against the explanatory variable, crying, for Example 15.2. The mean of the residuals is always 0.*

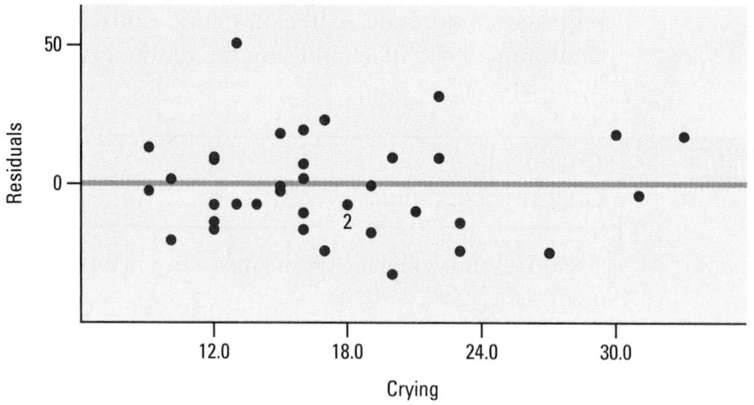

Now examine the distribution of the residuals for signs of strong non-Normality. Here is a stemplot of the residuals:

1	−3	1
4	−2	322
10	−1	965522
(11)	−0	99986666211
17	0	002289
11	1	0004889
4	2	04
2	3	2
1	4	
1	5	1

There is a slight right-skew, but we see no gross violation of the conditions. The residuals are approximately Normally distributed.

Exercises

15.1 An extinct beast, I See Exercise 3.13 (page 188). *Archaeopteryx* is an extinct beast having feathers like a bird but teeth and a long bony tail like a reptile. Here are the lengths in centimeters of the femur (a leg bone) and the humerus (a bone in the upper arm) for the five fossil specimens that preserve both bones:

Femur:	38	56	59	64	74
Humerus:	41	63	70	72	84

The strong linear relationship between the lengths of the two bones helped persuade scientists that all five specimens belong to the same species. Make a scatterplot with femur length as the explanatory variable. Use your calculator to obtain the correlation r and the equation of the least-squares regression line. Do you think that femur length will allow good prediction of humerus length?

15.2 Ideal proportions, I The students in Mr. Shenk's class measured their arm spans and heights (see Activity 15), entered their results into a Minitab worksheet, requested least-squares regression of height on arm span (both in inches), and obtained the following output:

Predictor	Coef	Stdev	t-ratio	p
Constant	11.547	5.600	2.06	0.056
armspan	0.84042	0.08091	10.39	0.000
s = 1.613	R-sq = 87.1%		R-sq(adj) = 86.3%	

A residual plot for the data is shown in Figure 15.4.

Figure 15.4 *Residual plot for Exercise 15.2.*

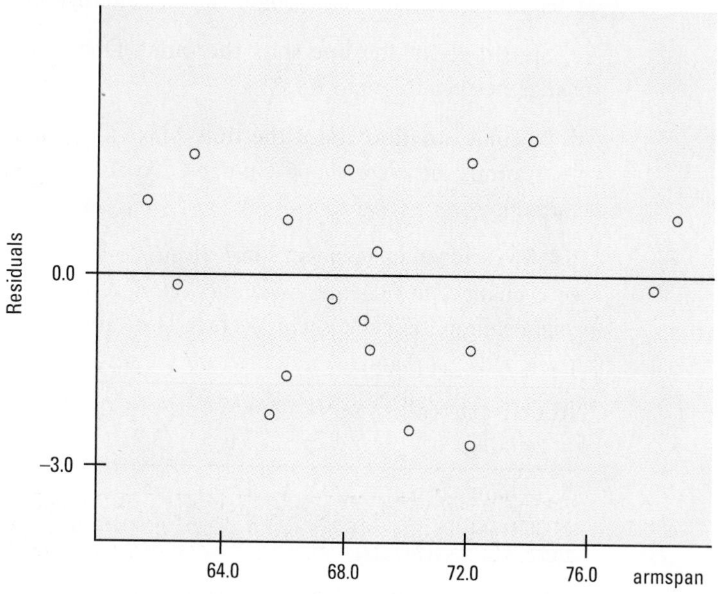

(a) Determine the equation of the least-squares regression line from the "Coef" column in the printout.

(b) In your opinion, is the least-squares line an appropriate model for the data? Would you be willing to predict a student's height if you knew that his arm span is 76 inches? Explain.

15.3 Sparrowhawk colonies Refer to Exercise 3.6 on page 180. One of nature's patterns connects the percent of adult birds in a colony that return from the previous year and the number of new adults that join the colony. Here are data for 13 colonies of sparrowhawks:

Percent return, x:	74	66	81	52	73	62	52	45	62	46	60	46	38
New adults, y:	5	6	8	11	12	15	16	17	18	18	19	20	20

This is an example of data that satisfy the conditions for regression inference well. Here are the residuals for the 13 colonies.

Percent return:	74	66	81	52	73	62	52
Residual:	−4.44	−5.87	0.69	−5.13	2.26	1.92	−0.13
Percent return:	45	62	46	60	46	38	
Residual:	−1.25	4.92	0.05	5.31	2.05	−0.38	

(a) **Independent observations.** Can we assume that the 13 observations are independent?

(b) **Linear relationship.** Make a residual plot. A plot of the residuals against the explanatory variable x magnifies the deviations from the least-squares line. Does the plot show any systematic deviation from a roughly linear pattern?

(c) **Spread about the line stays the same.** Does your plot in (b) show any systematic change in spread as x changes?

(d) **Normal variation about the line.** Make a histogram of the residuals. With only 13 observations, no clear shape emerges. Do strong skewness or outliers suggest lack of Normality?

15.4 Does fidgeting keep you slim? Example 3.9 (page 200) described an experiment to see if changes in fidgeting and other "nonexercise activity" (NEA) might help explain weight gain in individuals who overeat. Here are the data again:

NEA change (cal):	−94	−57	−29	135	143	151	245	355
Fat gain (kg):	4.2	3.0	3.7	2.7	3.2	3.6	2.4	1.3
NEA change (cal):	392	473	486	535	571	580	620	690
Fat gain (kg):	3.8	1.7	1.6	2.2	1.0	0.4	2.3	1.1

Answer questions (a) to (d) from the previous exercise for these data.

Estimating the Parameters

The first step in inference is to estimate the unknown parameters α, β, and σ. When the regression model describes our data and we calculate the least-squares line $\hat{y} = a + bx$, the slope b of the least-squares line is an unbiased estimator of the true slope β, and the intercept a of the least-squares line is an unbiased estimator of the true intercept α.

Example 15.3 *Crying and IQ*
Slope and intercept

The data in Figure 15.1 (page 890) fit the regression model of scatter about an invisible true regression line reasonably well. The least-squares line is $\hat{y} = 91.27 + 1.493x$. The slope is particularly important. A *slope is a rate of change*. The true slope β says how much higher average IQ is for children when the number of peaks in their crying measurement increases by 1. Because $b = 1.493$ estimates the unknown β, we estimate that, on the average, IQ is about 1.5 points higher for each additional crying peak.

We need the intercept $a = 91.27$ to draw the line, but it has no statistical meaning in this example. No child had fewer than 9 crying peaks, so we have no data near $x = 0$. We suspect that all normal children would cry when snapped with a rubber band, so that we will never observe $x = 0$.

The remaining parameter of the model is the standard deviation σ, which describes the variability of the response y about the true regression line. The least-squares line estimates the true regression line. So the residuals estimate how much y varies about the true line. The residuals are the vertical deviations of the data points from the least-squares line:

$$\text{residual} = \text{observed } y - \text{predicted } y$$
$$= y - \hat{y}$$

There are n residuals, one for each data point. Because σ is the standard deviation of responses about the true regression line, we estimate it by *a sample standard deviation of the residuals*. You encountered this error measure before (Chapter 3, page 218). We call this sample standard deviation a *standard error* to emphasize that it is estimated from data. The residuals from a least-squares line always have mean zero. That simplifies their standard error.

Standard Error about the Least-Squares Line

The standard error about the line is

$$s = \sqrt{\frac{\sum \text{residuals}^2}{n-2}}$$
$$= \sqrt{\frac{\sum (y - \hat{y})^2}{n-2}}$$

Use s to estimate the unknown σ in the regression model.

Because we use the standard error about the line so often in regression inference, we just call it s. Notice that s^2 is an average of the squared deviations of the data points from the line, so it qualifies as a variance. We average the squared deviations by dividing by $n - 2$, the number of data points less 2. It turns out that if we know $n - 2$ of the n residuals, the other two are determined. That is, $n - 2$ is the ***degrees of freedom*** of s. We first met the idea of degrees of freedom in the case of the ordinary sample standard deviation of n observations, which has $n - 1$ degrees of freedom. Now we are observing two variables rather than one, and the proper degrees of freedom is $n - 2$ rather than $n - 1$.

degrees of freedom

Calculating s begins with finding the predicted response for each x in your data set, then the residuals, and then s. In practice you will use technology that does this arithmetic instantly.

Example 15.4 ***Crying and IQ***
Calculating standard error

From Example 15.2, we know that the 38 residuals are

−19.20	−31.13	−22.65	−15.18	−12.18	−15.15	−16.63	−6.18
−1.70	−22.60	−6.68	−6.17	−9.15	−23.58	−9.14	2.80
−9.14	−1.66	−6.14	−12.60	0.34	−8.62	2.85	14.30
9.82	10.82	0.37	8.85	10.87	19.34	10.89	−2.55
20.85	24.35	18.94	32.89	18.47	51.32		

The variance about the line is

$$s^2 = \frac{1}{n-2}\sum \text{residual}^2$$

$$= \frac{1}{38-2}[(-19.20)^2 + (-31.13)^2 + \ldots + (51.32)^2]$$

$$= \frac{1}{36}(11{,}023.3) = 306.20$$

And so the standard error about the line is

$$s = \sqrt{306.20} = 17.50$$

Software gives 17.4987 to four decimal places, so the error resulting from rounding in this hand calculation is small.

Technology tip: Here's a quick way to calculate s. With the x-values in L_1/list1 and the y-values in L_2/list2, perform least-squares regression. The calculator creates or updates a list named RESID. Store the results in L_3/list 3. Then specify `1-Var Stats L3` and look at the value Σx^2. That's the sum of the squares of the residuals. Divide this number by $(n - 2)$ to get s^2. Take the square root to obtain s.

We will study several kinds of inference in the regression setting. The standard error s about the line is the key measure of the variability of the responses in regression. It is part of the standard error of all the statistics we will use for inference.

Regression for lawyers
Jury Verdict Research (www.juryverdictresearch.com) makes money from regression. The company maintains data on more than 193,000 jury verdicts in personal-injury lawsuits, recording more than 30 variables describing each case. Multiple regression using this mass of data allows Jury Verdict Research to predict how much a jury will award in a new case. Lawyers pay for these predictions and use them to negotiate settlements with insurance companies.

Confidence Intervals for the Regression Slope

The slope β of the true regression line is usually the most important parameter in a regression problem. The slope is the rate of change of the mean response as the explanatory variable increases. We often want to estimate β. The slope b of the least-squares line is an unbiased estimator of β. A confidence interval is more useful because it shows how accurate the estimate b is likely to be. The confidence interval for β has the familiar form

$$\text{estimate} \pm t^* \text{SE}_{\text{estimate}}$$

Because b is our estimate, the confidence interval becomes

$$b \pm t^* \text{SE}_b$$

Here are the details.

Confidence Interval for Regression Slope

A level C confidence interval for the slope β of the true regression line is

$$b \pm t^* \text{SE}_b$$

In this expression, the standard error of the least-squares slope b is

$$\text{SE}_b = \frac{s}{\sqrt{\sum (x - \bar{x})^2}}$$

and t^* is the critical value for the $t(n - 2)$ density curve with area C between $-t^*$ and t^*.

As advertised, the standard error of b is a multiple of s. Although we give the formula for this standard error, you should rarely have to calculate it by hand. Regression software gives the standard error SE_b along with b itself.

Example 15.5 *Crying and IQ*
Regression output

Figure 15.5 on the next page shows the basic output for the crying study from the regression command in the Minitab software package. Most statistical software provides similar output. (Minitab, like other software, produces more than this basic output. When you use software, just ignore the parts you don't need.)

The first line gives the equation of the least-squares regression line. The slope and intercept are rounded off there, so look in the "Coef" column of the table that follows for more accurate values. The intercept $a = 91.268$ appears in the "Constant" row. The slope $b = 1.4929$ appears in the "Crycount" row because we named the x variable "Crycount" when we entered the data.

The next column of output, headed "StDev," gives standard errors. In particular, $\text{SE}_b = 0.4870$. The standard error about the line, $s = 17.50$, appears below the table.

Figure 15.5 *Minitab regression output for the crying and IQ data, for Example 15.5.*

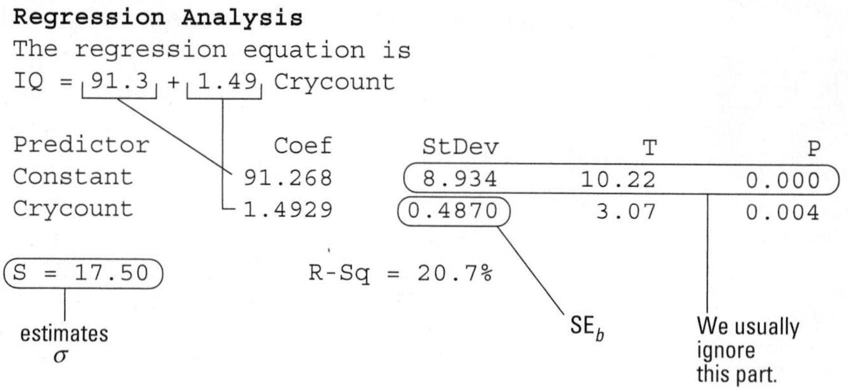

There are 38 data points, so the degrees of freedom are $n - 2 = 36$. For a 95% confidence interval for the true slope β, we will use the critical value $t^* = 2.042$ from the df = 30 row of Table C. This is the table degrees of freedom next smaller than 36. The interval is

$$\begin{aligned} b \pm t^*\text{SE}_b &= 1.4929 \pm (2.042)(0.4870) \\ &= 1.4929 \pm 0.9944 \\ &= 0.4985 \text{ to } 2.4873 \end{aligned}$$

Interpretation: We are 95% confident that mean IQ increases by between about 0.5 and 2.5 points for each additional peak in crying.

Exercises

15.5 An extinct beast, II Refer to Exercise 15.1 on page 894.

(a) Explain in words what the slope β of the true regression line says about *Archaeopteryx*.

(b) What is the estimate of β from the data? What is your estimate of the intercept α of the true regression line?

(c) Calculate the residuals for the five data points. Check that their sum is 0 (up to roundoff error). Use the residuals to estimate the standard deviation σ in the regression model. You have now estimated all three parameters in the model.

15.6 Competitive runners, I Exercise 3.81 (page 253) provided data on the speed of competitive runners (in feet per second) and the number of steps they took per second. Good runners take more steps per second as they speed up. Here are the data again:

Speed (ft/s):	15.86	16.88	17.50	18.62	19.97	21.06	22.11
Steps per second:	3.05	3.12	3.17	3.25	3.36	3.46	3.55

(a) Enter the data into your calculator, perform least-squares regression, and plot the scatterplot with the least-squares line. What is the strength of the association between speed and steps per second?

(b) Find the residuals for all 7 data points. Check that their sum is 0 (up to roundoff error).

(c) The model for regression inference has three parameters, α, β, and σ. Estimate these parameters from the data.

15.7 The Leaning Tower of Pisa The Leaning Tower of Pisa leans more as time passes. Here are measurements of the lean of the tower for the years 1975 to 1987.[3] The lean is the distance between where a point on the tower would be if the tower were straight and where the point actually is. The distances are recorded as tenths of a millimeter in excess of 2.9 meters. For example, the 1975 lean, which was 2.9642 meters, appears in the table as 642. We use only the last two digits of the year as our time variable.

Year:	75	76	77	78	79	80	81	82	83	84	85	86	87
Lean:	642	644	656	667	673	688	696	698	713	717	725	742	757

Here is part of the output from the Data Desk regression procedure with year as the explanatory variable and lean as the response variable:

Variable	Coefficient	s.e. of Coeff	t-ratio	prob
Constant	−61.1209	25.13	−2.43	0.0333
year	9.31868	0.3099	30.1	≤0.0001

(a) Plot the data. Briefly describe the direction, form, and strength of the relationship. The tower is tilting at a steady rate.

(b) The main purpose of the study is to estimate how fast the tower is tilting. What parameter in the regression model gives the rate at which the tilt is increasing, in tenths of a millimeter per year?

(c) We would like to determine a 95% confidence interval for the tilt rate. Unfortunately, we are not able to perform this inference. Why not?

15.8 Prey attract predators Here is one way in which nature regulates the size of animal populations: high population density attracts predators, who remove a higher proportion of the population than when the density of the prey is low. One study looked at kelp perch and their common predator, the kelp bass. The researcher set up four large circular pens on sandy ocean bottoms off the coast of southern California. He chose young perch at random from a large group and placed 10, 20, 40, and 60 perch in the four pens. Then he dropped the nets protecting the pens, allowing bass to swarm in, and counted the perch left after 2 hours. Here are data on the proportions of perch eaten in four repetitions of this setup:[4]

Number of perch	Proportion killed			
10	0.0	0.1	0.3	0.3
20	0.2	0.3	0.3	0.6
40	0.075	0.3	0.6	0.725
60	0.517	0.55	0.7	0.817

The explanatory variable is the number of perch (the prey) in a confined area. The response variable is the proportion of perch killed by bass (the predator) in 2 hours when the bass are allowed access to the perch. A scatterplot shows a linear relationship. Minitab output for regression is also shown.

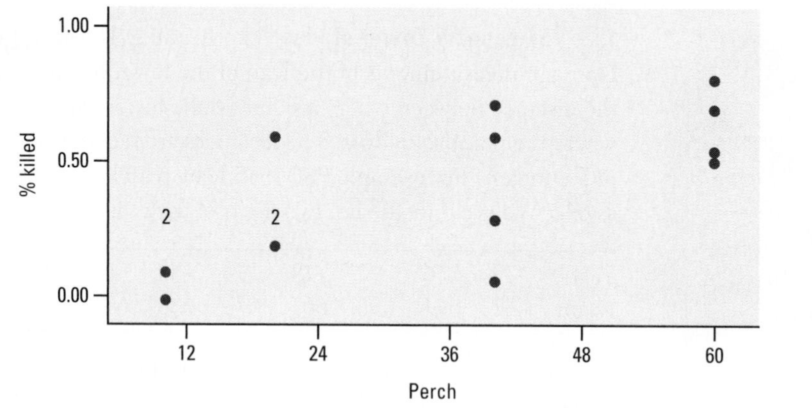

Predictor	Coef	Stdev	t-ratio	p
Constant	0.12049	0.09269	1.30	0.215
Perch	0.008569	0.002456	3.49	0.004

s = 0.1886 R-sq = 46.5% R-sq(adj) = 42.7%

(a) What is the equation of the least-squares line for predicting proportion killed from count of perch? What part of this equation shows that more perch does result in a higher proportion being killed by bass?

(b) What is the regression standard error s?

(c) Do the data support the principle that "more prey attract more predators, who drive down the number of prey"? Follow the Data Analysis Toolbox.

(d) Construct and interpret a 95% confidence interval for the slope of the true regression line.

15.9 Time at the table Does how long young children remain at the lunch table help predict how much they eat? Here are data on 20 toddlers observed over several months at a nursery school.[5] "Time" is the average number of minutes a child spent at the table when lunch was served. "Calories" is the average number of calories the child consumed during lunch, calculated from careful observation of what the child ate each day.

Time:	21.4	30.8	37.7	33.5	32.8	39.5	22.8	34.1	33.9	43.8
Calories:	472	498	465	456	423	437	508	431	479	454
Time:	42.4	43.1	29.2	31.3	28.6	32.9	30.6	35.1	33.0	43.7
Calories:	450	410	504	437	489	436	480	439	444	408

Make a scatterplot of the data and find the equation of the least-squares line for predicting calories consumed from time at the table. Describe briefly what the data show about the behavior of children. Then construct and interpret a 95% confidence interval for the slope of the true regression line.

Figure 15.6 *Excel output for the regression of proportion of perch killed by bass on count of perch in a pen, for Exercise 15.10.*

Microsoft Excel

	A	B	C	D	E	F	G
1	SUMMARY OUTPUT						
2							
3	*Regression Statistics*						
4	Multiple R	0.6021					
5	R Square	0.4652					
6	Adjusted R Square	0.4270					
7	Standard Error	0.1886					
8	Observations	16.0000					
9							
10		*Coefficients*	*Standard Error*	*t Stat*	*P-value*	*Lower 95%*	*Upper 95%*
11	Intercept	0.1205	0.0927	1.2999	0.2146	-0.0783	0.3193
12	Perch	0.0086	0.0025	3.4899	0.0036	0.0033	0.0138
13							

Sheet5 Sheet1 Sheet2 Sheet2

15.10 Prey attract predators: estimating the slope Refer to Exercise 15.8.

(a) Excel gives a 95% confidence interval for the slope of the population regression line. See Figure 15.6. What is it? Interpret this interval.

(b) Starting from the Minitab values of the least-squares slope b and its standard error, verify this confidence interval.

(c) Give a 90% confidence interval for the population slope. As usual, this interval is shorter than the 95% interval.

Testing the Hypothesis of No Linear Relationship

The most common hypothesis about the slope is

$$H_0: \beta = 0$$

A regression line with slope 0 is horizontal. That is, the mean of y does not change at all when x changes. So this H_0 says that there is *no true linear relationship* between x and y. Put another way, H_0 says that *straight-line dependence on x is of no value for predicting y.* Put yet another way, H_0 says that there is *no correlation* between x and y in the population from which we drew our data. You can use the test for zero slope to test the hypothesis of zero correlation between any two quantitative variables. That's a useful trick. **Do notice that testing correlation makes sense only if the observations are a random sample.** That is often not the case in regression settings, where researchers may fix in advance the values of x they want to study.

The test statistic is just the standardized version of the least-squares slope b. It is another t statistic. Here are the details.

Significance Tests for Regression Slope

To test the hypothesis H_0: $\beta = 0$, compute the **t statistic**

$$t = \frac{b}{\mathrm{SE}_b}$$

In terms of a random variable T having the $t(n-2)$ distribution, the P-value for a test of H_0 against

H_a: $\beta > 0$ is $P(T \geq t)$

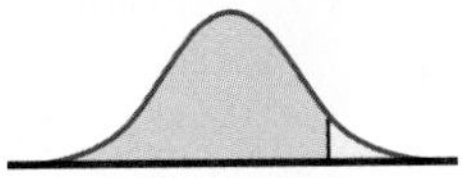

H_a: $\beta < 0$ is $P(T \leq t)$

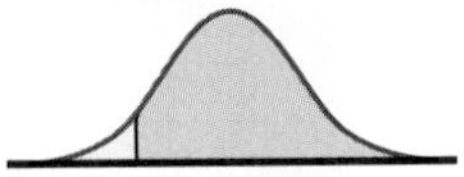

H_a: $\beta \neq 0$ is $2P(T \geq |t|)$

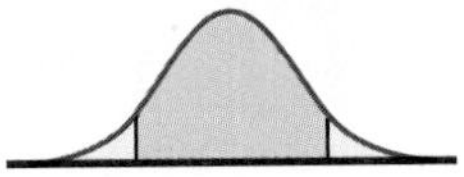

This test is also a test of the hypothesis that the correlation is 0 in the population.

Regression output from statistical software usually gives t and its two-sided P-value. For a one-sided test, divide the P-value in the output by 2.

Example 15.6 ***Crying and IQ***
Testing regression slope

The hypothesis H_0: $\beta = 0$ says that crying has no straight-line relationship with IQ. Figure 15.1 (page 890) shows that there is a relationship, so it is not surprising that the computer output in Figure 15.5 (page 900) gives $t = 3.07$ with two-sided P-value 0.004. There is very strong evidence that IQ is correlated with crying.

Example 15.7 ***Beer and blood alcohol***
Testing regression slope

TOOLBOX

Example 3.5 (page 177) looked at how well the number of beers a student drinks predicts his or her blood alcohol content (BAC). Sixteen student volunteers at Ohio State University drank a randomly assigned number of cans of beer. Thirty minutes later, a police officer measured their BAC. Here are the data again from Example 3.5:

Student:	1	2	3	4	5	6	7	8
Beers:	5	2	9	8	3	7	3	5
BAC:	0.10	0.03	0.19	0.12	0.04	0.095	0.07	0.06
Student:	9	10	11	12	13	14	15	16
Beers:	3	5	4	6	5	7	1	4
BAC:	0.02	0.05	0.07	0.10	0.085	0.09	0.01	0.05

The scatterplot in Figure 15.7 shows a clear linear relationship. Figure 15.8 gives part of the Minitab regression output. The blue line on the scatterplot is the least-squares line

$$\hat{y} = -0.0127 + 0.0180x$$

Figure 15.7 *Scatterplot of students' blood alcohol content against the number of cans of beer consumed, for Example 15.7.*

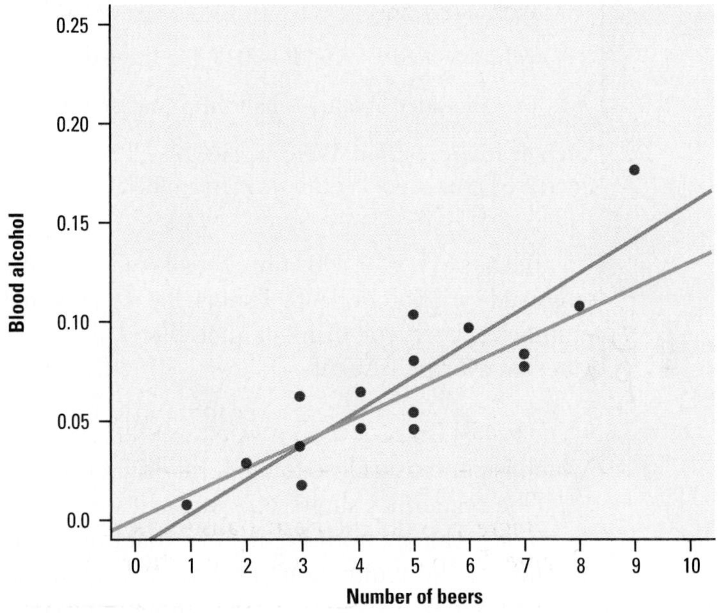

Figure 15.8 *Minitab output for the blood alcohol content data in Example 15.7.*

```
The regression equation is
BAC = - 0.0127 + 0.0180 Beers


Predictor        Coef       StDev          T          P
Constant     -0.01270     0.01264      -1.00      0.332
Beers        0.017964    0.002402       7.48      0.000

S = 0.02044     R-Sq = 80.0%
```

Now we're ready to perform formal inference. Once again, we are guided by the Inference Toolbox.

Step 1: Hypotheses We can test the hypothesis that the number of beers has no effect on BAC versus the one-sided alternative that more beers increases BAC. The hypotheses, in words and in symbols, are

$H_0: \beta = 0$ The number of beers has *no* effect on BAC.

$H_a: \beta > 0$ The number of beers has a positive effect on BAC.

Step 2: Conditions The scatterplot shows a straight-line association between number of beers and BAC. We will assume that the 16 responses (BAC levels) are independent, and that the standard deviation of BAC is the same for all values of x.

Step 3: Calculations

- The t test statistic is

$$t = \frac{b}{\text{SE}_b} = \frac{0.017964}{0.002402} \cong 7.48$$

  with two-sided P-value $P = 0.000$ to three decimal places.

- The one-sided P-value is half this value, so it is also close to 0.

Step 4: Interpretation We can reject H_0. There is strong evidence to conclude that an increased number of beers does increase BAC.

Because $r^2 = 0.800$, number of beers consumed accounts for 80% of the observed variation in BAC. That is, the data say that student opinion is wrong: the number of beers you drink predicts blood alcohol level quite well. Five beers produce an average BAC of

$$\hat{y} = -0.0127 + (0.0180)(5) = 0.077$$

which is perilously close to the legal driving limit of 0.08 in many states.

The scatterplot shows one unusual point: student number 3, who drank 9 beers. You can see from Figure 15.7 that this observation lies farthest from the fitted line in the y direction. That is, this point has the largest residual. Student number 3 may also be influential, though the point is not extreme in the x direction. To verify that our results are not too dependent on this one observation, do the regression again omitting student 3. The new regression line is the green line in Figure 15.7. Omitting student 3 decreases r^2 from 80% to 77%, and it changes the predicted BAC after 5 beers from 0.077 to 0.073. These small changes show that this observation is not very influential.

Technology Toolbox

Linear regression *t* test

The TI-83/84/89 can perform a linear regression t test. We will use the crying versus IQ data from Example 15.1 (page 889) to illustrate the process.

Enter the x-values (crying) into L_1/list1 and the y-values (IQ) into L_2/list2.

TI-83/84

- Press STAT/◄/E:LinRegTTest.

TI-89

- Press 2nd F1 ([F6]: Tests). Then select A:LinRegTTest.

Technology Toolbox

Linear regression test *(continued)*

```
EDIT CALC TESTS
0↑2-SampTInt...
A:1-PropZInt...
B:2-PropZInt...
C:χ²-Test...
D:2-SampFTest...
E:LinRegTTest...
F:ANOVA(
```

```
F1▼   F2▼   F3▼  F4▼  F5▼   F6▼   F7▼
Tools Plots List Calc Distr Tests Ints
list1  li  3↑2-SampZTest...
10     87  4:2-SampTTest...
12     97  5:1-PropZTest...
9      10  6:2-PropZTest...
16     10  7:Chi2 GOF...
18     10  8:Chi2 2-way...
15     11  9:2-SampFTest
           A↓LinRegTTest...
list1[1]=10
MAIN   RAD AUTO   FUNC   1/7
```

- In the LinRegTTest screen, specify L_1 for Xlist, L_2 for Ylist, and $\neq 0$ for the hypothesized slope β. You can leave the RegEQ space blank. Highlight the command "Calculate" and press ENTER.

```
LinRegTTest
  Xlist:L1
  Ylist:L2
  Freq:1
  β & ρ:≠0  <0  >0
  RegEQ:
  Calculate
```

- Specify list1 for X List, list2 for Y List, and 1 for Frequency. You have three choices for Alternate Hyp: we will choose $\beta \neq 0$. For results, choose "Calculate." (The other choice is "Draw.") Press ENTER.

```
Linear Regression T Test
X List:            list1
Y List:            list2
Freq:              1
Alternate Hyp:     β & ρ ≠ 0→
Store RegEqn to:   y1(x)→
Results:           Calculate→
 Enter=OK           ESC=CANCEL
USE   RAD AUTO   FUNC
```

The linear regression t test results take two screens to present.

```
LinRegTTest
  y=a+bx
  β≠0 and ρ≠0
  t=3.065489379
  p=.0041053001
  df=36
↓a=91.26829865
```

```
LinRegTTest
  y=a+bx
  β≠0 and ρ≠0
↑b=1.492896598
  s=17.49872122
  r²=.2069999897
  r=.4549725153
```

```
Linear Regression T Test
  y=a+bx
  β & ρ ≠0
  t             =3.06549
  p Value       =.004105
```

```
Linear Regression T Test
↑p Value        =.004105
  df            =36.
  a             =91.2683
  b             =1.4929
  s             =17.4987
  SE Slope      =.487001
  r²            =.207
  r             =.454973
 Enter=OK
USE   RAD AUTO   FUNC   3/7
```

The first screen reports that the t statistic is 3.07 with df $= 36$. The P-value is 0.004. Scrolling down, you find the intercept $a = 91.2683$ and slope $b = 1.4929$ of the least-squares line, as well as the correlation $r = 0.455$, the coefficient of determination $r^2 = 0.207$, and the standard error about the line $s = 17.50$.

Note that the TI-83/84/89 now have a provision for calculating a confidence interval for the slope. If your calculator has the most recent operating system, go to TESTS and select G:LinRegTInt. Usually, determining confidence intervals is provided by software.

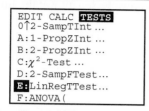

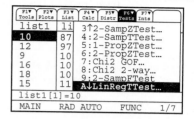

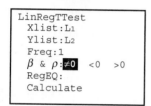

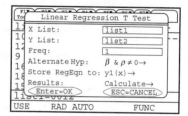

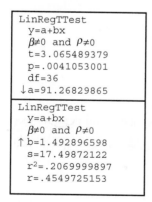

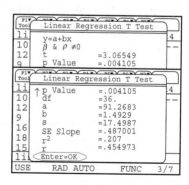

Exercises

15.11 An extinct beast, III Exercise 15.1 (page 894) presents data on the lengths of two bones in five fossil specimens of the extinct beast *Archaeopteryx*. Here is part of the output from the S-PLUS statistical software when we regress the length y of the humerus on the length x of the femur.

Coefficients:	Value	Std. Error	t value	Pr(>|t|)
(Intercept)	−3.6596	4.4590	−0.8207	0.4719
Femur	1.1969	0.0751		

(a) What is the equation of the least-squares regression line?

(b) We left out the t statistic for testing H_0: $\beta = 0$ and its P-value. Use the output to find t.

(c) How many degrees of freedom does t have? Use Table C to approximate the P-value of t for testing H_0: $\beta = 0$ against the one-sided alternative H_a: $\beta > 0$.

(d) Write a sentence to describe your conclusions about the slope of the true regression line.

(e) Calculate and interpret a 99% confidence interval for the true slope of the regression line.

15.12 Competitive runners, II Exercise 15.6 (page 900) presents data on the relationship between the speed of runners (x, in feet per second) and the number of steps y that they take in a second. Here is part of the Data Desk regression output for these data:

R squared = 99.8%
s = 0.0091 with 7 − 2 = 5 degrees of freedom

Variable	Coefficient	s.e. of Coeff	t-ratio	prob
Constant	1.76608	0.0307	57.6	<0.0001
Speed	0.080284	0.0016	49.7	<0.0001

(a) How can you tell from this output, even without the scatterplot, that there is a very strong straight-line relationship between running speed and steps per second?

(b) What parameter in the regression model gives the rate at which steps per second increase as running speed increases by 1 ft/s? Give a 99% confidence interval for this rate.

15.13 Are jet skis dangerous? Propelled by a stream of pressurized water, jet skis and other so-called wet bikes carry from one to three people and have become one of the most popular types of recreational vehicle sold today. But critics say that they're noisy, dangerous, and damaging to the environment. An article in the August 1997 issue of the *Journal of the American Medical Association* reported on a survey that tracked emergency room visits at randomly selected hospitals nationwide. Here are data on the number of jet skis in use, the number of accidents, and the number of fatalities for the years 1987–1996:[6]

Year	Number in use	Accidents	Fatalities
1987	92,756	376	5
1988	126,881	650	20
1989	178,510	844	20
1990	241,376	1162	28
1991	305,915	1513	26
1992	372,283	1650	34
1993	454,545	2236	35
1994	600,000	3002	56
1995	760,000	4028	68
1996	900,000	4010	55

(a) Make a scatterplot of number of accidents against number of jet skis in use. What does the scatterplot show about the relationship between these variables?

(b) Formulate null and alternative hypotheses about the slope of the true regression line. State a one-sided alternative hypothesis.

(c) What conditions are necessary in order to perform a linear regression significance test? Are these conditions satisfied in this situation?

(d) Perform a linear regression t test. Report the t statistic, the degrees of freedom, and the P-value. Write your conclusion in plain language.

(e) Calculate and interpret a 98% confidence interval for the true slope of the regression line.

15.14 Is wine good for your heart? There is some evidence that drinking moderate amounts of wine helps prevent heart attacks. Exercise 3.79 (pages 251–252) gives data on yearly wine consumption (liters of alcohol from drinking wine, per person) and yearly deaths from heart disease (deaths per 100,000 people) in 19 developed nations.

(a) Is there statistically significant evidence of a negative association between wine consumption and heart disease deaths? Carry out the appropriate test of significance and write a summary statement about your conclusions.

(b) Calculate and interpret a 95% confidence interval for the true regression slope.

15.15 Beavers and beetles Exercise 3.48 (pages 228–229) described a study that seemed to show that beavers benefit beetles. The researchers measured the number of stumps from trees cut by beavers and the number of clusters of beetle larvae. Here are the data again:

Stumps:	2	2	1	3	3	4	3	1	2	5	1	3
Beetle larvae:	10	30	12	24	36	40	43	11	27	56	18	40

Stumps:	2	1	2	2	1	1	4	1	2	1	4
Beetle larvae:	25	8	21	14	16	6	54	9	13	14	50

(a) Make a scatterplot that shows how the number of beaver-caused stumps influences the number of beetle larvae clusters. What does your plot show?

(b) Here is part of the Minitab regression output for these data:

Predictor	Coef	StDev	T	P
Constant	−1.286	2.853	−0.45	0.657
Stumps	11.894	1.136	10.47	0.000
S = 6.419		R-Sq = 83.9%		

Find the least-squares regression line and draw it on your plot. What percent of the observed variation in beetle larvae counts can be explained by straight-line dependence on beaver stump counts?

(c) Is there strong evidence that beaver stump counts help explain beetle larvae counts? Give appropriate statistical evidence to support your conclusion.

standardized residuals

15.16 Beaver and beetle residuals Software often calculates ***standardized residuals*** as well as the actual residuals from regression. Because standardized residuals have the standard z-score scale, it is easier to judge whether any are extreme. Here are the standardized residuals from the previous exercise, rounded to two decimal places:

−1.99	1.20	0.23	−1.67	0.26	−1.06	1.38	0.06	0.72	−0.40	1.21	0.90
0.40	−0.43	−0.24	−1.36	0.88	−0.75	1.30	−0.26	−1.51	0.55	0.62	

(a) Find the mean and standard deviation of the standardized residuals. Why do you expect values close to those you obtain?

(b) Make a stemplot of the standardized residuals. Are there any striking deviations from Normality? The most extreme residual is $z = -1.99$. Would this be surprisingly large if the 23 observations had a Normal distribution? Explain your answer.

(c) Plot the standardized residuals against the explanatory variable. Are there any suspicious patterns?

Summary

Least-squares regression fits a straight line to data in order to predict a response variable y from the explanatory variable x. **Inference about regression** requires that several conditions be met.

To use regression inference, data must satisfy the following conditions:

- The observations must be independent.
- The true relationship is linear.
- The standard deviation of the response about the true line is the same everywhere.
- The response varies Normally about the true regression line.

Verifying conditions uses the residuals.

The **regression model** says that there is a **true regression line** $\mu_y = \alpha + \beta x$ that describes how the mean response varies as x changes. The observed response y for any x has a Normal distribution with a mean given by the true regression line and with the same standard deviation σ for any value of x. The parameters of the regression model are the intercept α, the slope β, and the standard deviation σ.

The slope a and intercept b of the least-squares line estimate the slope α and intercept β of the true regression line. To estimate σ, use the **standard error about the line** s.

The standard error s has $n - 2$ degrees of freedom. All t procedures in regression inference have $n - 2$ degrees of freedom.

Confidence intervals for the slope of the true regression line have the form $b \pm t^* \text{SE}_b$. In practice, we use software to find the slope b of the least-squares line and its standard error SE_b.

To test the hypothesis that the true slope is zero, use the ***t* statistic** $t = b/\text{SE}_b$, also given by software. This null hypothesis says that straight-line dependence on x has no value for predicting y. It also says that the population correlation between x and y is zero.

CASE CLOSED!

Three-pointers in college basketball

Return to the chapter-opening Case Study (page 887). With what you have learned about inference in a regression setting, you are now ready to answer some questions about the relationship between the number of three-point shots taken in a college game and the percent of successes since the three-pointer was introduced in college play during the 1986–1987 season. The data (number of three-point shots taken as the explanatory variable, and percent of successes as the response variable) were entered into Minitab, and regression was requested. Here are the residual plot and a partial printout by the software.

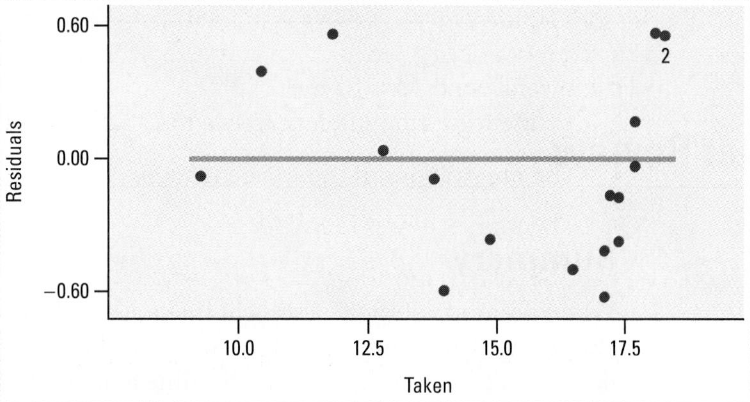

(continued)

Predictor	Coef	Stdev	t-ratio	p
Constant	42.8477	0.5536	77.40	0.000
Taken	−0.47620	0.03475	−13.70	0.000

s = 0.4224   R-sq = 91.7%   R-sq(adj) = 91.2%

1. Prior to performing formal inference, analyze the information provided, supplemented with any additional calculations or plots you wish to make. Then describe the relationship between the two variables.

2. What is the equation of the least-squares regression line? Is this a reasonably good model for these data? Justify your answer.

3. Minitab flags a particular data point, $(x, y) = (9.2, 38.4)$, and says that this point is "influential."

Unusual Observations

Obs.	Taken	Percent	Fit	Stdev.Fit	Residual	St.Resid
1	9.2	38.4000	38.4667	0.2453	−0.0667	−0.19 X

X denotes an obs. whose X value gives it large influence.

What does it mean to say that this point has large influence? How could a point that has such a small residual (−0.0667) have large influence?

4. Perform a test of significance on the slope of the true regression line. Follow the steps in the Inference Toolbox. What are your conclusions?

5. Construct a 95% confidence interval for the true slope. Interpret this confidence interval in context.

Chapter Review

Summary

When a scatterplot shows a straight-line relationship between an explanatory variable x and a response variable y, we often fit a least-squares regression line to describe the relationship and use this line to predict y from x. Statistical inference in the regression setting, however, requires more than just an overall linear pattern on a scatterplot.

The regression model says that there is a true straight-line relationship between x and the mean response μ_y. We can't observe this true regression line.

The responses y that we do observe vary if we take several observations at the same x. The regression model requires that for any fixed x, the responses have a Normal distribution. Moreover, the standard deviation σ of this distribution must be the same for all values of x.

The standard deviation σ describes how much variation there is in responses y when x is held fixed. Estimating σ is the key to inference about regression. Use the standard error s (roughly, the sample standard deviation of the residuals) to estimate σ. We can then do these types of inference:

- Give confidence intervals for the slope of the true regression line.
- Test the null hypothesis that the slope of the true regression line is zero. This hypothesis says that a straight-line relationship between x and y is of no value for predicting y. It is the same as saying that the correlation between x and y in the entire population is zero.

What You Should Have Learned

Here is a checklist of the major skills you should have acquired by studying this chapter.

A. Preliminaries

1. Make a scatterplot to show the relationship between an explanatory and a response variable.
2. Use a calculator or software to find the correlation and the least-squares regression line.

B. Recognition

1. Recognize the regression setting: a straight-line relationship between an explanatory variable x and a response variable y.
2. Recognize which type of inference you need in a particular regression setting.
3. Inspect the data to recognize situations in which inference isn't safe: a nonlinear relationship, influential observations, strongly skewed residuals in a small sample, or nonconstant variation of the data points about the regression line.

C. Doing Inference Using Software and Calculator Output

1. Explain in any specific regression setting the meaning of the slope β of the true regression line.
2. Understand computer output for regression. From the output, find the slope and intercept of the least-squares line, their standard errors, and the standard error about the line.
3. Use the output to carry out tests and calculate confidence intervals for β.

Web Links

A company that makes money from regression by maintaining jury verdict data is www.juryverdictresearch.com

Lots of college basketball data can be found by Googling "Official 2004 NCAA basketball." Look for a 300+ page document in pdf (portable document format).

Chapter Review Exercises

15.17 Does fast driving waste fuel? Exercise 3.9 (page 182) gives data on the fuel consumption of a small car at various speeds from 10 to 150 kilometers per hour. Is there evidence of straight-line dependence between speed and fuel use? Make a scatterplot and use it to explain the result of your test.

15.18 Sarah's growth Sarah's growth from age 3 years to 5 years was measured as follows:

Age (months):	36	48	51	54	57	60
Height (cm):	86	90	91	93	94	95

Explain why it would not be appropriate to perform inference on Sarah's data.

15.19 Coffee and deforestation Coffee is a leading export from several developing countries. When coffee prices are high, farmers often clear forest to plant more coffee trees. Here are five years' data on prices paid to coffee growers in Indonesia and the percent of forest area lost in a national park that lies in a coffee-producing region:[6]

Price (cents per pound)	29	40	54	55	71
Forest lost (percent)	0.49	1.59	1.69	1.82	3.10

After entering these data into your calculator, a regression analysis reports the following results:

$$\text{forest lost} = -1.0134 + 0.0552 \times \text{price}$$

$$r^2 = 0.91 \qquad P = 0.0125$$

(a) Explain in words what the slope β of the population regression line would tell us if we knew it. Based on the data, what are the estimates of β and the intercept α of the population regression line?
(b) What does $r^2 = 0.91$ add to the information given by the equation of the least-squares line? Interpret r^2.
(c) To what null and alternative hypotheses does the P-value refer? What does this P-value tell you?

(d) Calculate the residuals for the five data points. Check that their sum is 0 (up to round-off error). Use the residuals to estimate the standard deviation σ of percents of forest lost about the means given by the population regression line.
(e) Do you think that coffee price will allow good prediction of forest lost? Explain.

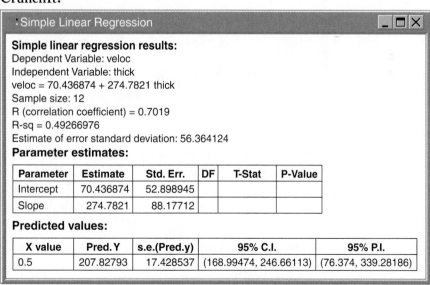

15.20 Casting aluminum In the casting of metal parts, molten metal flows through a "gate" into a die that shapes the part. The gate velocity (the speed at which metal is forced through the gate) plays a critical role in die casting. A firm that casts cylindrical aluminum pistons examined 12 types formed from the same alloy. What is the relationship between the cylinder wall thickness (inches) and the gate velocity (feet per second) chosen by the skilled workers who do the casting? If there is a clear pattern, it can be used to direct new workers or to automate the process. Here are the data:[8]

Thickness	Velocity	Thickness	Velocity	Thickness	Velocity
0.248	123.8	0.524	228.6	0.697	145.2
0.359	223.9	0.552	223.8	0.752	263.1
0.366	180.9	0.628	326.2	0.806	302.4
0.400	104.8	0.697	302.4	0.821	302.4

Figure 15.9 displays part of the CrunchIt! regression output. We left out t statistics and their P-values.

Figure 15.9 *CrunchIt! output for Exercise 15.20.*

CrunchIt!

Simple Linear Regression

Simple linear regression results:
Dependent Variable: veloc
Independent Variable: thick
veloc = 70.436874 + 274.7821 thick
Sample size: 12
R (correlation coefficient) = 0.7019
R-sq = 0.49266976
Estimate of error standard deviation: 56.364124

Parameter estimates:

Parameter	Estimate	Std. Err.	DF	T-Stat	P-Value
Intercept	70.436874	52.898945			
Slope	274.7821	88.17712			

Predicted values:

X value	Pred. Y	s.e.(Pred.y)	95% C.I.	95% P.I.
0.5	207.82793	17.428537	(168.99474, 246.66113)	(76.374, 339.28186)

(a) Make a scatterplot suitable for predicting gate velocity from thickness. Give the value of r^2 and the equation of the least-squares line. Draw the line on your plot.
(b) Based on the information given, test the hypothesis that there is no straight-line relationship between thickness and gate velocity. Give a test statistic, its approximate P-value using a table, and your conclusion.

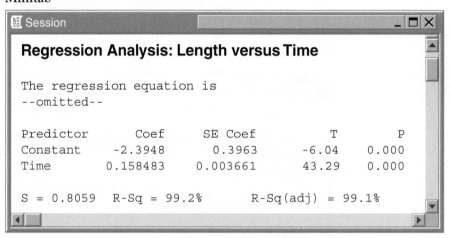

15.21 How fast do icicles grow? The rate at which an icicle grows depends on temperature, water flow, and wind. The data below are for an icicle grown in a cold chamber at −11°C with no wind and a water flow of 11.9 milligrams per second:[9]

Time (min):	10	20	30	40	50	60	70	80	90
Length (cm):	0.6	1.8	2.9	4.0	5.0	6.1	7.9	10.1	10.9
Time (min):	100	110	120	130	140	150	160	170	180
Length (cm):	12.7	14.4	16.6	18.1	19.9	21.0	23.4	24.7	27.8

We want to predict length from time. Figure 15.10 shows Minitab regression output for these data.

Figure 15.10 *Minitab regression analysis: length versus time for Exercise 15.21.*

Minitab

Session

Regression Analysis: Length versus Time

```
The regression equation is
--omitted--

Predictor        Coef     SE Coef        T       P
Constant      -2.3948      0.3963    -6.04   0.000
Time         0.158483    0.003661    43.29   0.000

S = 0.8059   R-Sq = 99.2%       R-Sq(adj) = 99.1%
```

(a) Make a scatterplot suitable for predicting length from time. The pattern is very linear. What is the coefficient of determination r^2? Time explains almost all of the change in length.
(b) Use the computer output to estimate the three parameters α, β, and σ.
(c) What is the equation of the least-squares regression line of length on time? Add this line to your plot.

15.22 Does fidgeting keep you slim? Exercise 15.4 (page 896) presented data showing that people who increased their nonexercise activity (NEA) when they were deliberately overfed gained less fat than other people. Use software to add formal inference to the data analysis for these data.
(a) Based on 16 subjects, the correlation between NEA increase and fat gain was $r = -0.7786$. Is this significant evidence that people with higher NEA increase gain less fat?
(b) The slope of the least-squares regression line was $b = -0.00344$, so that fat gain decreased by 0.00344 kilogram for each added calorie of NEA. Give a 90% confidence

interval for the slope of the population regression line. This rate of change is the most important parameter to be estimated.

15.23 Investing at home and overseas Investors ask about the relationship between returns on investments in the United States and investments overseas. Table 3.3 (page 232) gives the percent returns on U.S. and overseas common stocks over a 27-year period.
(a) Make a scatterplot suitable for predicting overseas returns from U.S. returns.
(b) Here is part of the output from the Minitab regression command:

```
Predictor      Coef        StDev        T        P
Constant       5.683       5.144      1.10    0.280
USreturn      0.6181      0.2369        *        *

S = 19.90     R-Sq = 21.4%
```

We have omitted the t statistic for β and its P-value. What is the value of t? What are its degrees of freedom? From Table C, how strong is the evidence for a linear relationship between U.S. and overseas returns?
(c) Is the regression prediction useful in practice? Use the r^2-value for this regression to help explain your finding.

15.24 Stock return residuals Exercise 15.23 presents a regression of overseas stock returns on U.S. stock returns based on 27 years' data. The residuals for this regression (in order by years across the rows) are

14.89	18.93	−11.44	−12.57	6.72	−17.77	16.99	22.96	−12.13
−3.05	−4.89	−20.87	4.17	−2.05	30.98	52.22	15.76	12.36
−14.78	−27.17	−12.04	−22.18	20.97	−0.29	−17.72	−13.80	−24.23

(a) Plot the residuals against x, the U.S. return. The plot suggests a mild violation of one of the regression conditions. Which one?
(b) Display the distribution of the residuals in a graph. In what way is the shape somewhat non-Normal? There is one possible outlier. Circle that point on the residual plot in (a). What year is this? This point is not very influential: redoing the regression without it does not greatly change the results. With 27 observations, we are willing to do regression inference for these data.

15.25 Weeds among the corn Lamb's-quarter is a common weed that interferes with the growth of corn. An agriculture researcher planted corn at the same rate in 16 small plots of ground, then weeded the plots by hand to allow a fixed number of lamb's-quarter plants to grow in each meter of corn row. No other weeds were allowed to grow. Here are the yields of corn (bushels per acre) in each of the plots:[10]

Weeds per meter	Corn yield	Weeds per meter	Corn yield	Weeds per meter	Corn yield	Weeds per meter	Corn yield
0	166.7	1	166.2	3	158.6	9	162.8
0	172.2	1	157.3	3	176.4	9	142.4
0	165.0	1	166.7	3	153.1	9	162.8
0	176.9	1	161.1	3	156.0	9	162.4

Use your calculator or software to analyze these data.

(a) Make a scatterplot and find the least-squares line. What percent of the observed variation in corn yield can be explained by a linear relationship between yield and weeds per meter?

(b) Is there good evidence that more weeds reduce corn yield? Give appropriate statistical evidence to support your answer.

(c) Explain from your findings in (a) and (b) why you would expect predictions based on this regression to be quite imprecise.

15.26 The professor swims Here are data on the time (in minutes) Professor Moore takes to swim 2000 yards and his pulse rate (beats per minute) after swimming:

Time:	34.12	35.72	34.72	34.05	34.13	35.72	36.17	35.57
Pulse:	152	124	140	152	146	128	136	144
Time:	35.37	35.57	35.43	36.05	34.85	34.70	34.75	33.93
Pulse:	148	144	136	124	148	144	140	156
Time:	34.60	34.00	34.35	35.62	35.68	35.28	35.97	
Pulse:	136	148	148	132	124	132	139	

A scatterplot shows a negative linear relationship: a faster time (fewer minutes) is associated with a higher heart rate. Here is part of the output from the regression function in an Excel spreadsheet:

	Coefficients	Standard Error	t Stat	P-value
Intercept	479.9341457	66.22779275	7.246718119	3.87075E-07
X Variable	-9.694903394	1.888664503	-5.1332057	4.37908E-05

Construct and interpret a 90% confidence interval for the slope of the true regression line. Explain what your result tells us about the relationship between the professor's swimming time and heart rate.

PART

IV Review Exercises

Communicate your thinking clearly in Exercises IV.1 to IV.10.

IV.1 Here is the description of the design of a study that compared subjects' responses to regular and fat-free potato chips:

During a given 2-week period, the participants received the same type of chip (regular or fat-free) each day. After the first 2-week period, there was a 1-week washout period in which no testing was performed. There was then another 2-week period in which the participants received the alternate type of potato chip (regular or fat-free) each day under the same protocol. The order in which the chips were presented was determined by random assignment.[1]

One response variable was the weight in grams of potato chips that a subject ate. We want to compare the amounts of regular and fat-free chips eaten.

(a) Forty-four women made up one set of subjects. Explain how to use the partial table of random digits below to do the random assignment required by the design. Then use your method to assign the first 5 women.

19223	95034	05756	28713
96409	12531	42544	82853

(b) State appropriate hypotheses for carrying out a significance test in this setting. What inference procedure should you use?

IV.2 Fruit flies show a daily cycle of rest and activity, but does the rest qualify as sleep? Researchers looking at brain activity and behavior finally concluded that fruit flies do sleep. A small part of the study used an infrared motion sensor to see if flies moved in response to vibrations. Here are results for low levels of vibration:[2]

	Response to Vibration?	
	No	Yes
Fly was walking	10	54
Fly was resting	28	4

Is there good reason to think that resting flies respond differently than flies that are walking? (That's a sign that the resting flies may actually be sleeping.) Give appropriate statistical evidence to justify your answer.

IV.3 The body's natural electrical field helps wounds heal. If diabetes changes this field, that might explain why people with diabetes heal slowly. A study of this idea compared normal mice and mice bred to spontaneously develop diabetes. The investigators attached sensors to the right hip and front feet of the mice and measured the difference in electrical potential (millivolts) between these locations. Here are the data:[3]

Diabetic mice					
14.70	13.60	7.40	1.05	10.55	16.40
10.00	22.60	15.20	19.60	17.25	18.40
9.80	11.70	14.85	14.45	18.25	10.15
10.85	10.30	10.45	8.55	8.85	19.20

Normal mice				
13.80	9.10	4.95	7.70	9.40
7.20	10.00	14.55	13.30	6.65
9.50	10.40	7.75	8.70	8.85
8.40	8.55	12.60		

(a) Make a stemplot suitable for comparing the two samples of potentials. Does it appear that potentials in the two groups differ in a systematic way? Justify your answer.

(b) Is there significant evidence of a difference in mean potentials between the two groups? Carry out an appropriate test at the $\alpha = 0.05$ significance level.

(c) Repeat your inference without the outlier in the diabetic group. Does the outlier affect your conclusion?

The National Assessment of Educational Progress (NAEP) includes a "long-term trend" study that tracks reading and mathematics skills over time in a way that allows comparisons between results from different years. Exercises IV.4 and IV.5 are based on information on 17-year-old students from the report on the latest long-term trend study, carried out in 2004.[4] The NAEP sample used a multistage design, but the overall effect is quite similar to an SRS of 17-year-olds who are still in school.

IV.4 In the 2004 sample of 2158 students, 1014 had at least one college graduate parent.

(a) Construct and interpret a 99% confidence interval for the proportion of all students in 2004 who had at least one college-educated parent.

(b) Notice that the sample excludes 17-year-olds who had dropped out of school. Given this information, would you expect your estimate in part (a) to be too high, too low, or approximately correct? Justify your answer.

IV.5 The 2004 NAEP sample contained 1122 female students and 1036 male students. The women had a mean mathematics score (on a scale of 0 to 500) of 305, with standard error 0.9. The male mean was 308, with standard error 1.0. Is there evidence that the mean mathematics scores of men and women differ in the population of all 17-year-old students? Give appropriate statistical evidence to justify your answer.

IV.6 After randomly assigning subjects to treatments in a randomized comparative experiment, we can compare the treatment groups to see how well the randomization worked. We hope to find no significant differences among the groups. A study of how to provide premature infants with a substance essential to their development assigned infants at random to receive one of four types of supplement, called PBM, NLCP, PL-LCP, and TG-LCP.[5]

(a) The subjects were 77 premature infants. Outline the design of the experiment if 20 are assigned to the PBM group and 19 to each of the other treatments.

(b) The random assignment resulted in 9 females in the TG-LCP group and 11 females in each of the other groups. Perform an appropriate test to see if there are significant gender differences among the groups. What do you find?

IV.7 Starting in the 1970s, medical technology allowed babies with very low birth weight (VLBW, less than 1500 grams, about 3.3 pounds) to survive without major handicaps. It was noticed that these children nonetheless had difficulties in school and as adults. A long study has followed 242 VLBW babies to age 20 years, along with a control group of 233 babies from the same population who had normal birth weight.[6]

(a) Is this an experiment or an observational study? Why?

(b) At age 20, 179 of the VLBW group and 193 of the control group had graduated from high school. Is the graduation rate among the VLBW group significantly lower than for the normal-birth-weight controls? Give appropriate statistical evidence to justify your answer.

(c) IQ scores were available for 113 men in the VLBW group. The mean IQ was 87.6, and the standard deviation was 15.1. The 106 men in the control group had mean IQ 94.7, with standard deviation 14.9. Construct and interpret a 90% confidence interval for the difference in mean IQ scores between VLBW and non-VLBW men.

IV.8 A study of the inheritance of speed and endurance in mice found a trade-off between these two characteristics, both of which help mice survive. To test endurance, mice were made to swim in a bucket with a weight attached to their tails. (The mice were rescued when exhausted.) Here are data on endurance in minutes for female and male mice:[7]

Group	n	Mean	Std. dev.
Female	162	11.4	26.09
Male	135	6.7	6.69

(a) The endurance data are skewed to the right. Why are t procedures reasonably accurate for these data?

(b) Calculate and interpret a 95% confidence interval for the mean endurance of female mice swimming.

(c) A 95% confidence interval for the mean endurance of male mice swimming is (5.56, 7.84). A 95% confidence interval for the difference in mean endurances of male and female mice swimming is (0.5, 8.9). Use the information in parts (b) and (c) to answer this question: do female mice have significantly higher endurance on the average than male mice? Explain your reasoning clearly.

IV.9 Intensive agriculture and burning of fossil fuels increase the amount of nitrogen deposited on the land. Too much nitrogen can reduce the variety of plants by favoring rapid growth of some species (think of putting fertilizer on your lawn to help grass choke out weeds). A study of 68 grassland sites in Britain measured nitrogen deposited (kilograms of nitrogen per hectare of land area per year) and also the "richness" of plant species (based on number of species and how abundant each species is). The authors reported a regression analysis as follows:[8]

$$\text{plant species richness} = 23.3 - 0.408 \times \text{nitrogen deposited}$$
$$r^2 = 0.55 \qquad P < 0.0001$$

(a)What does the slope $b = -0.408$ say about the effect of increased nitrogen deposits on species richness?

(b) What does $r^2 = 0.55$ add to the information given by the equation of the least-squares line?

(c) What null and alternative hypotheses do you think the *P*-value refers to? What does this *P*-value tell you?

IV.10 Many low- and middle-income families don't save enough for their future retirement. It would be to their advantage to contribute to an individual retirement account (IRA), which allows money to be invested for retirement without paying taxes on it now. Would more families contribute to an IRA if the money they invest were matched by their employer or other group? In an experiment on this question, the tax firm H&R Block offered to partly match IRA contributions of families with incomes below $40,000. The subjects were taxpayers visiting H&R Block offices in the St. Louis area. In all, 1681 married taxpayers were assigned at random to the control group (no match), 1780 to a 20% match, and 1831 to a 50% match. All were offered the chance to open an IRA.[9]

The study found that 49 married taxpayers in the control group, 240 in the 20% group, and 456 in the 50% group opened IRAs. Each taxpayer who opened an IRA decided how much to contribute. Here are the summaries of the contributions for the three groups:

Group	n	$\bar{x}$	s
Control	49	$1549	$1652
20% match	240	$1723	$1332
50% match	456	$1742	$1174

The amounts contributed exclude the matching amount.

(a) Do a significantly higher proportion of married taxpayers open an IRA when offered a 50% match than when offered a 20% match? Carry out an appropriate test to support your conclusion.

(b) Construct and interpret a 95% confidence interval for the mean contribution (among taxpayers who decide to contribute) due to offering a 20% match over no match.

(c) Recall H&R Block's original question: would more families contribute to an IRA if the money they invest were matched by their employer or other group? Use your result from either part (a) or part (b) to help answer this question.

(d) Can we generalize the results of this study to the population of all married U.S. taxpayers? Justify your answer.

Notes and Data Sources

Preliminary Chapter

1. Carlos Vallbona et al., "Response of pain to static magnetic fields in postpolio patients, a double blind pilot study," *Archives of Physical Medicine and Rehabilitation*, 78 (1997), pp. 1200–1203.
2. E. W. Campion, pp. 44–46. "Editorial: power lines, cancer, and fear," *New England Journal of Medicine*, 337, No. 1 (1997), The study report is M. S. Linet et al., "Residential exposure to magnetic fields and acute lymphoblastic leukemia in children," pp. 1–8 in the same issue. See also G. Taubes, "Magnetic field–cancer link: will it rest in peace?" *Science*, 277 (1997), p. 29.
3. See, for example, Martin Enserink, "The vanishing promises of hormone replacement," *Science*, 297 (2002), pp. 325–326; and Brian Vastag, "Hormone replacement therapy falls out of favor with expert committee," *Journal of the American Medical Association*, 287 (2002), pp. 1923–1924. A National Institutes of Health panel's comprehensive report is *International Position Paper on Women's Health and Menopause*, NIH Publication 02-3284, 2002.
4. J. E. Muscat et al., "Handheld cellular telephone use and risk of brain cancer," *Journal of the American Medical Association*, 284 (2000), pp. 3001–3007.
5. Seat belt data from the National Highway and Traffic Safety Administration, *NOPUS Survey*, 1998 and 2003.
6. Data on the Women's National Team are from the United States Soccer Federation Web site, www.ussoccer.com.
7. These data, from reports submitted by airlines to the Department of Transportation, appear in A. Barnett, "How numbers can trick you," *Technology Review*, October 1994, pp. 38–45.
8. Press release for the 2003 DuPont Automotive Color Survey, at www.automotive.dupont.com.
9. H. C. Sox, "Editorial: benefit and harm associated with screening for breast cancer," *New England Journal of Medicine*, 338, No. 16 (1998).
10. A. C. Nielsen, Jr., "Statistics in marketing," in *Making Statistics More Effective in Schools of Business*, University of Chicago, 1986.
11. Jeffrey G. Johnson et al., "Television viewing and aggressive behavior during adolescence and adulthood," *Science*, 295 (2002), pp. 2468–2471. The authors use statistical adjustments to control for the effects of a number of lurking variables. The association between TV viewing and aggression remains significant.
12. Contributed by Marigene Arnold of Kalamazoo College.

Chapter 1

1. Arbitron, *Radio Today, 2005 edition*, www.arbitron.com.
2. Arbitron, *Internet and Multimedia 2005*, www.arbitron.com.
3. United Nations data on literacy were found at www.earthtrends.wri.org.
4. U.S. Environmental Protection Agency, *Municipal Solid Waste in the United States: 2000 Facts and Figures*, document EPA530-R-02-001, 2002.
5. Based on experiments performed by G. T. Lloyd and E. H. Ramshaw of the CSIRO Division of Food Research, Victoria, Australia, 1982–1983.
6. Maribeth Cassidy Schmitt, "The effects of an elaborated directed reading activity on the metacomprehension skills of third graders," PhD thesis, Purdue University, 1987.
7. James T. Fleming, "The measurement of children's perception of difficulty in reading materials," *Research in the Teaching of English*, 1 (1967), pp. 136–156.
8. Wayne Nelson, "Theory and applications of hazard plotting for censored failure data," *Technometrics*, 14 (1972), pp. 945–966.
9. John K. Ford, "Diversification: how many stocks will suffice?" *American Association of Individual Investors Journal*, January 1990, pp. 14–16.
10. C. E. Brooks and N. Carruthers, *Handbook of Statistical Methods in Meteorology*, Her Majesty's Stationery Office, London, 1953.
11. C. B. Williams, *Style and Vocabulary: Numerological Studies*, Griffin, 1970.
12. *American Journal of Clinical Nutrition*, 62 (supplement) (1995): 178S–194S.
13. Monthly gasoline price index from the Consumer Price Index, from the Bureau of Labor Statistics, www.bls.gov, converted into dollars.
14. Debora L. Arsenau, "Comparison of diet management instruction for patients with non-insulin dependent diabetes mellitus: learning activity package vs. group instruction," MS thesis, Purdue University, 1993.
15. U.S. Department of Health and Human Services, National Center for Health Statistics.
16. National Center for Education Statistics, NEDRC Table Library, nces.ed.gov/das/library/index.asp.
17. These data were collected by students as a class project.
18. We thank Charles Cannon of Duke University for providing the data. The study report is C. H. Cannon, D. R. Peart, and M. Leighton, "Tree species diversity in commercially logged Bornean rainforest," *Science*, 281 (1998), pp. 1366–1367.
19. Recording Industry Association of America, *2003 Consumer Profile*.
20. Based on summaries in G. L. Cromwell et al., "A comparison of the nutritive value of opaque-2, floury-2, and normal corn for the chick," *Poultry Science*, 57 (1968), pp. 840–847.

21. Data from S. M. Stigler, "Do robust estimators work with real data?" *Annals of Statistics*, 5 (1977), pp. 1055–1078.
22. We thank Ethan J. Temeles of Amherst College for providing the data. His work is described in Ethan J. Temeles and W. John Kress, "Adaptation in a plant-hummingbird association," *Science*, 300 (2002), pp. 630–633.
23. Data for 1986 from David Brillinger, University of California, Berkeley. See David R. Brillinger, "Mapping aggregate birth data," in A. C. Singh and P. Whitridge (eds.), *Analysis of Data in Time*, Statistics Canada, 1990, pp.77–83. Boxplots similar to Figure 1.26 appear in David R. Brillinger, "Some examples of random process environmental data analysis," in P. K. Sen and C. R. Rao (eds.), *Handbook of Statistics*, Vol. 18, North-Holland, 2000.
24. Data from the most recent Annual Demographic Supplement can be found at www.bls.census.gov/cps/datamain.htm.
25. Data from the Annual Demographic Supplement for 2002.

Chapter 2

1. Stephen Jay Gould, "Entropic homogeneity isn't why no one hits .400 anymore," *Discover*, August 1986, pp. 60–66. Gould does not standardize but gives a speculative discussion instead.
2. Information on bone density in the reference populations was found on the Web at www.courses.washington.edu/bonephys/opbmd.html.
3. Summary information on PSAT scores of juniors was obtained from the College Board Web site at www.collegeboard.com.
4. Data from Gary Community School Corporation, courtesy of Celeste Foster, Department of Education, Purdue University.
5. Data from the *USA Today* online sports salaries data base at www.usatoday.com.
6. Detailed data appear in P. S. Levy et al., "Total serum cholesterol values for youths 12–17 years," *Vital and Health Statistics Series 11*, No. 150 (1975), U.S. National Center for Health Statistics.
7. The data for this example were provided by Nicholas Fisher.
8. Data found online at www.earthtrends.wri.org.
9. The data set was constructed based on information provided in P. D. Wood et al., "Plasma lipoprotein distributions in male and female runners," in P. Milvey (ed.), *The Marathon: Physiological, Medical, Epidemiological, and Psychological Studies*, New York Academy of Sciences, 1977.
10. *ACT High School Profile Report, HS Graduating Class 2004, 2005*, www.act.org.
11. National Center for Education Statistics, *Debt Burden: A Comparison of 1992–93 and 1999–2000 Bachelor's Degree Recipients a Year After Graduating*, 2005, nces.ed.gov.

Chapter 3

1. Data on mean SAT math scores and participation rates for the states was obtained from the research section of the College Board's Web site, www.collegeboard.com.
2. These are some of the data from the EESEE story "Blood Alcohol Content."
3. From a graph in Bernt-Erik Saether, Steiner Engen, and Erik Mattysen, "Demographic characteristics and population dynamical patterns of solitary birds," *Science*, 295 (2002), pp. 2070–2073.
4. Data provided by Darlene Gordon, Purdue University.
5. Based on T. N. Lam, "Estimating fuel consumption from engine size," *Journal of Transportation Engineering*, 111 (1985), pp. 339–357. The data for 10 to 50 km/h are measured; those for 60 and higher are calculated from a model given in the paper and are therefore smoothed.
6. A careful study of this phenomenon is W. S. Cleveland, P. Diaconis, and R. McGill, "Variables on scatterplots look more highly correlated when the scales are increased," *Science*, 216 (1982), pp. 1138–1141.
7. M. A. Houck et al., "Allometric scaling in the earliest fossil bird, *Archaeopteryx lithographica*," *Science*, 247 (1990), pp. 195–198. The authors conclude from a variety of evidence that all specimens represent the same species.
8. From a graph in Timothy G. O'Brien and Margaret F. Kinnaird, "Caffeine and conservation," *Science*, 300 (2003), p. 587.
9. *Consumer Reports*, June 1986, pp. 366–367.
10. Data for 1995, from the 1997 *Statistical Abstract of the United States.*
11. From a survey by the Wheat Industry Council reported in *USA Today*, October 20, 1983.
12. *T. Rowe Price Report*, winter 1997, p. 4.
13. W. L. Colville and D. P. McGill, "Effect of rate and method of planting on several plant characters and yield of irrigated corn," *Agronomy Journal*, 54 (1962), pp. 235–238.
14. Data from a plot in James A. Levine, Norman L. Eberhart, and Michael D. Jensen, "Role of nonexercise activity thermogenesis in resistance to fat gains in humans," *Science*, 283 (1999), pp. 212–214.
15. G. L. Kooyman et al., "Diving behavior and energetics during foraging cycles in king penguins," *Ecological Monographs*, 62 (1992), pp. 143–163.
16. W. M. Lewis and M. C. Grant, "Acid precipitation in the western United States," *Science*, 207 (1980), pp. 176–177.
17. Data from National Institute of Standards and Technology, *Engineering Statistics Handbook*, www.itl.nist.gov/div898/handbook. The analysis there does not comment on the bias of field measurements.
18. E. P. Hubble, "A relation between distance and radial velocity among extra-galactic nebulae," *Proceedings of the National Academy of Sciences*, 15 (1929), pp. 168–173.
19. D. A. Kurtz (ed.), *Trace Residue Analysis*, American Chemical Society Symposium Series No. 284, 1985, Appendix.

20. Based on a plot in G. D. Martinsen, E. M. Driebe, and T. G. Whitham, "Indirect interactions mediated by changing plant chemistry: beaver browsing benefits beetles," *Ecology*, 79 (1998), pp. 192–200.
21. T. Constable and E. McBean, "BOD/TOC correlations and their application to water quality evaluation," *Water, Air, and Soil Pollution*, 11 (1979), pp. 363–375.
22. From a presentation by Charles Knauf, Monroe County (New York) Environmental Health Laboratory.
23. Gary Smith, "Do statistics test scores regress toward the mean?" *Chance*, 10, No. 4 (1997), pp. 42–45.
24. From a graph in Tania Singer et al., "Empathy for pain involves the affective but not sensory components of pain," *Science*, 303 (2004), pp. 1157–1162. Data for other brain regions showed a stronger correlation and no outliers.
25. Gannett News Service article appearing in the *Lafayette (Ind.) Journal and Courier*, April 23, 1994.
26. This example is drawn from M. Goldstein, "Preliminary inspection of multivariate data," *The American Statistician*, 36 (1982), pp. 358–362.
27. David E. Bloom and David Canning, "The health and wealth of nations," *Science*, 287 (2000), pp. 1207–1208.
28. Data provided by Peter Cook, Department of Mathematics, Purdue University.
29. From a graph in Joaquim I. Goes et al., "Warming of the Eurasian landmass is making the Arabian Sea more productive," *Science*, 308 (2005), pp. 545–547.
30. The data are a random sample of 53 from the 1079 pairs recorded by K. Pearson and A. Lee, "On the laws of inheritance in man," *Biometrika*, November 1903, p. 408.
31. R. C. Nelson, C. M. Brooks, and N. L. Pike, "Biomechanical comparison of male and female distance runners," in P. Milvy (ed.), *The Marathon: Physiological, Medical, Epidemiological, and Psychological Studies*, New York Academy of Sciences, 1977, pp. 793–807.
32. "Dancing in step," *Economist*, March 22, 2001.
33. Michael Winerip, "On education: SAT essay test rewards length and ignores errors," *New York Times*, May 4, 2005.

Chapter 4

1. Sample Pennsylvania female rates provided by Life Quotes, Inc. in *USA Today*, December 20, 2004.
2. This activity was described in Elizabeth B. Applebaum, "A simulation to model exponential growth," *Mathematics Teacher*, 93, No.7 (October 2000), pp. 614–615.
3. G. A. Sacher and E. F. Staffelt, "Relation of gestation time to brain weight for placental mammals: implications for the theory of vertebrate growth," *American Naturalist*, 108 (1974), pp. 593–613. We found these data in F. L. Ramsey and D. W. Schafer, *The Statistical Sleuth: A Course in Methods of Data Analysis*, Duxbury, 1997.
4. There are several mathematical ways to show that log t fits into the power family at $p = 0$. Here's one. For powers $p \neq 0$, the indefinite integral $\int t^{p-1}dt$ is a multiple of t^p. When $p = 0$, $\int t^{-1}dt$ is log t.
5. Data from the World Bank's *1999 World Development Indicators*. Life expectancy is estimated for 1997, and GDP per capita (purchasing-power parity basis) is estimated for 1998.
6. From the Intel Web site, www.intel.com/technology/silicon/index.htm.
7. Gypsy moth data provided by Chuck Schwalbe, U.S. Department of Agriculture.
8. Joel Best, *Damned Lies and Statistics: Untangling Numbers from the Media, Politicians, and Activists*, University of California Press, 2001.
9. Stillman Drake, *Galileo at Work*, University of Chicago Press, 1978. We found these data in D. A. Dickey and J. T. Arnold, "Teaching statistics with data of historic significance," *Journal of Statistics Education*, 3 (1995), www.amstat.org/publications/jse/.
10. Planetary data from hyperphysics.phy-astr.gsu. edu/hbase/solar/soldata2.html.
11. Data originally from A. J. Clark, *Comparative Physiology of the Heart*, Macmillan, 1927, p. 84. Obtained from Frank R. Giordano and Maurice D. Weir, *A First Course in Mathematical Modeling*, Brooks/Cole, 1985, p. 56.
12. Jérôme Chave, Bernard Riéra, and Marc-A. Dubois, "Estimation of biomass in a neotropical forest of French Guiana: spatial and temporal variability," *Journal of Tropical Ecology*, 17 (2001), pp. 79–96.
13. These data were provided by Thomas Kuczek, Purdue University. Similar experiments are described by A. P. J. Trinci, "A kinetic study of the growth of *Aspergillus nidulans* and other fungi," *Journal of General Microbiology*, 57 (1969), pp. 11–24.
14. H. Lindberg, H. Roos, and P. Gardsell, "Prevalence of coxarthritis in former soccer players," *Acta Orthopedica Scandinavica*, 64 (1993), pp. 165–167.
15. S. V. Zagona (ed.), *Studies and Issues in Smoking Behavior*, University of Arizona Press, 1967, pp. 157–180.
16. R. Shine, T. R. L. Madsen, M. J. Elphick, and P. S. Harlow, "The influence of nest temperature and maternal brooding on hatchling phenotypes in water python," *Ecology*, 78 (1997), pp. 1713–1721.
17. F. D. Blau and M. A. Ferber, "Career plans and expectations of young women and men," *Journal of Human Resources*, 26 (1991), pp. 581–607.
18. Siem Oppe and Frank De Charro, "The effect of medical care by a helicopter trauma team on the probability of survival and the quality of life of hospitalized victims," *Accident Analysis and Prevention*, 33 (2001), pp. 129–138. The authors give the data in Example 4.15 as a "theoretical example" to illustrate the need for their more elaborate analysis of actual data using severity scores for each victim.
19. M. Radelet, "Racial characteristics and imposition of the death penalty," *American Sociological Review*, 46 (1981), pp. 918–927.
20. P. J. Bickel and J. W. O'Connell, "Is there a sex bias in graduate admissions?" *Science*, 187 (1975), pp. 398–404.

21. D. M. Barnes, "Breaking the cycle of addiction," *Science*, 241 (1988), pp. 1029–1030.
22. Lien-Ti Bei, "Consumers' purchase behavior toward recycled products: an acquisition-transaction utility theory perspective," MS thesis, Purdue University, 1993.
23. National Center for Education Statistics, *Digest of Education Statistics, 2003*, Table 249.
24. Laura L. Calderon et al., "Risk factors for obesity in Mexican-American girls: dietary factors, anthropometric factors, physical activity, and hours of television viewing," *Journal of the American Dietetic Association*, 96 (1996), pp. 1177–1179.
25. Saccharin appears on the National Institutes of Health list of suspected carcinogens but remains in wide use. In October 1997, an expert panel recommended by a 4 to 3 vote to keep it on the list, despite recommendations from other scientific groups that it be removed.
26. A detailed study of this correlation appears in E. M. Remolona, P. Kleinman, and D. Gruenstein, "Market returns and mutual fund flows," *Federal Reserve Bank of New York Economic Policy Review*, 3, No. 2 (1997), pp. 33–52.
27. M. S. Linet et al., "Residential exposure to magnetic fields and acute lymphoblastic leukemia in children," *New England Journal of Medicine*, 337 (1997), pp. 1–7.
28. *The Health Consequences of Smoking: 1983*, U.S. Health Service, 1983.
29. Gannett News Service article appearing in the *Lafayette (Ind.) Journal and Courier*, April 23, 1994.
30. D. E. Powers and D. A. Rock, *Effects of Coaching on SAT I: Reasoning Test Scores*, Educational Testing Service Research Report 98-6, College Entrance Examination Board, 1998.
31. National Center for Health Statistics, *Deaths: Preliminary Data for 2003*, National Vital Statistics Report, 53, No. 15, 2005.
32. Janice E. Williams et al., "Anger proneness predicts coronary heart disease risk," *Circulation*, 101 (2000), pp. 2034–2039.
33. S. Chatterjee and B. Price, *Regression Analysis by Example*, Wiley, 1977.
34. D. R. Appleton, J. M. French, and M. P. J. Vanderpump, "Ignoring a covariate: an example of Simpson's paradox," *The American Statistician*, 50 (1996), pp. 340–341.

Part I Review Exercises

1. From a graph in Craig Packer et al., "Ecological change, group territoriality, and population dynamics in Serengeti lions," *Science*, 307 (2005), pp. 390–393.
2. Mei-Hui Chen, "An exploratory comparison of American and Asian consumers' catalog patronage behavior," MS thesis, Purdue University, 1994.
3. Louie H. Yang, "Periodical cicadas as resource pulses in North American forests," *Science*, 306 (2004), pp. 1565–1567. The data are simulated Normal values that match the means and standard deviations reported in this article.

Chapter 5

1. *Early Human Development*, 76, No. 2 (February 2004), pp. 139–145.
2. *Journal of Hypertension*, December 2003.
3. Study conducted by cardiologists at Athens Medical School, Greece, and announced at a European cardiology conference in February 2004.
4. National Institute of Child Health and Human Development, Study of Early Child Care and Youth Development. The article appears in the July 2003 issue of *Child Development*. The quotations are from the summary on the NICHD Web site, www.nichd.nih.gov.
5. Reported by D. Horvitz in his contribution to "Pseudo-opinion polls: SLOP or useful data?" *Chance*, 8, No. 2 (1995), pp. 16–25.
6. Randall Rothenberger, "The trouble with mall interviewing," *New York Times*, August 16, 1989.
7. K. J. Mukamal et al., "Prior alcohol consumption and mortality following acute myocardial infarction," *Journal of the American Medical Association*, 285 (2001), pp. 1965–1970.
8. L. E. Moses and F. Mosteller, "Safety of anesthetics," in J. M. Tanur et al. (eds.), *Statistics: A Guide to the Unknown*, 3rd ed., Wadsworth, 1989, pp. 15–24.
9. *The ASCAP Survey and Your Royalties*, ASCAP, undated.
10. The most recent account of the design of the CPS is Bureau of Labor Statistics, *Design and Methodology*, Current Population Survey Technical Paper 63, March 2000 (available in print or online at www.bls.census.gov/cps/tp/tp63.htm). The account here omits many complications, such as the need to separately sample "group quarters" like college dormitories.
11. The nonresponse rate for the CPS comes from the online version of the Bureau of Labor Statistics *Handbook of Methods*, modified April 17, 2003, at www.bls.gov. The GSS reports its response rate on its Web site, www.norc.org/projects/gensoc.asp. The Pew study is described in Gregory Flemming and Kimberly Parker, "Race and reluctant respondents: possible consequences of non-response for pre-election surveys," Pew Research Center for the People and the Press, 1997, found at www.people-press.org.
12. George Bishop, "Americans' belief in God," *Public Opinion Quarterly*, 63 (1999), pp. 421–434.
13. For more detail on the limits of memory in surveys, see N. M. Bradburn, L. J. Rips, and S. K. Shevell, "Answering autobiographical questions: the impact of memory and inference on surveys," *Science*, 236 (1987), pp. 157–161.
14. Robert F. Belli et al., "Reducing vote overreporting in surveys: social desirability, memory failure, and source monitoring," *Public Opinion Quarterly*, 63 (1999), pp. 90–108.

15. Cynthia Crossen, "Margin of error: studies galore support products and positions, but are they reliable?" *Wall Street Journal*, November 14, 1991.
16. M. R. Kagay, "Poll on doubt of Holocaust is corrected," *New York Times*, July 8, 1994.
17. Giuliana Coccia, "An overview of non-response in Italian telephone surveys," *Proceedings of the 99th Session of the International Statistics Institute*, 1993, Book 3, pp. 271–272.
18. *New York Times*, August 21, 1989.
19. For a full description of the STAR program and its follow-up studies, go to www.heros-inc.org/star.htm.
20. Simplified from Arno J. Rethans, John L. Swasy, and Lawrence J. Marks, "Effects of television commercial repetition, receiver knowledge, and commercial length: a test of the two-factor model," *Journal of Marketing Research*, 23 (February 1986), pp. 50–61.
21. L. L. Miao, "Gastric freezing: an example of the evaluation of medical therapy by randomized clinical trials," in J. P. Bunker, B. A. Barnes, and F. Mosteller (eds.), *Costs, Risks, and Benefits of Surgery*, Oxford University Press, 1977, pp. 198–211.
22. Simplified from David L. Strayer, Frank A. Drews, and William A. Johnston, "Cell phone–induced failures of visual attention during simulated driving," *Journal of Experimental Psychology: Applied*, 9 (2003), pp. 23–32. See Example 5.23 for more detail.
23. Steering Committee of the Physicians' Health Study Research Group, "Final report on the aspirin component of the ongoing Physicians' Health Study," *New England Journal of Medicine*, 321 (1989), pp. 129–135.
24. Christopher Anderson, "Measuring what works in health care," *Science*, 263 (1994), pp. 1080–1082.
25. Joel Brockner et al., "Layoffs, equity theory, and work performance: further evidence of the impact of survivor guilt," *Academy of Management Journal*, 29 (1986), pp. 373–384.
26. John H. Kagel, Raymond C. Battalio, and C. G. Miles, "Marijuana and work performance: results from an experiment," *Journal of Human Resources*, 15 (1980), pp. 373–395.
27. Geetha Thiagarajan et al., "Antioxidant properties of green and black tea, and their potential ability to retard the progression of eye lens cataract," *Experimental Eye Research*, 73 (2001), pp. 393–401.
28. See note 26.
29. Martin Enserink, "Fickle mice highlight test problems," *Science*, 284 (1999), pp. 1599–1600. There is a full report of the study in the same issue.
30. Evan H. DeLucia et al., "Net primary production of a forest ecosystem with experimental CO2 enhancement," *Science*, 284 (1999), pp. 1177–1179. The investigators used the block design.
31. R. C. Shelton et al., "Effectiveness of St.-John's-wort in major depression," *Journal of the American Medical Association*, 285 (2001), pp. 1978–1986.
32. Based on the Electronic Encyclopedia of Statistical Examples and Exercises (EESEE) story "Surgery in a Blanket," found on the book's Web site www.whfreeman.com/tps3e.
33. Pierre J. Meunier et al., "The effects of strontium renelate on the risk of vertebral fracture in women with postmenopausal osteoporosis," *New England Journal of Medicine*, 350 (2004), pp. 459–468.
34. G. Kolata, "New study finds vitamins are not cancer preventers," *New York Times*, July 21, 1994. Look in the *Journal of the American Medical Association* of the same date for the details.
35. Warren McIsaac and Vivek Goel, "Is access to physician services in Ontario equitable?" Institute for Clinical Evaluative Sciences in Ontario, October 18, 1993.
36. Stephen A. Woodbury and Robert G. Spiegelman, "Bonuses to workers and employers to reduce unemployment: randomized trials in Illinois," *American Economic Review*, 77 (1987), pp. 513–530.
37. E. M. Peters et al., "Vitamin C supplementation reduces the incidence of postrace symptoms of upper-respiratory tract infection in ultramarathon runners," *American Journal of Clinical Nutrition*, 57 (1993), pp. 170–174.
38. This exercise is based on the EESEE story "Blinded Knee Doctors." The study was reported in W. E. Nelson, R. C. Henderson, L. C. Almekinders, R. A. DeMasi, and T. N. Taft, "An evaluation of pre- and postoperative nonsteroidal antiinflammatory drugs in patients undergoing knee arthroscopy," *Journal of Sports Medicine*, 21 (1994), pp. 510–516.
39. Sanjay K. Dhar, Claudia González-Vallejo, and Dilip Soman, "Modeling the effects of advertised price claims: tensile versus precise pricing," *Marketing Science*, 18 (1999), pp. 154–177.

Chapter 6

1. An informative and entertaining account of the origins of probability theory is Florence N. David, *Games, Gods and Gambling*, Charles Griffin, 1962.
2. From the EESEE story "Home-Field Advantage." The study is W. Hurley, "What sort of tournament should the World Series be?" *Chance*, 6, No. 2 (1993), pp. 31–33.
3. Information for the 1999–2000 academic year is from the 2003 *Statistical Abstract of the United States*, Table 286.
4. You can find a mathematical explanation of Benford's law in Ted Hill, "The first-digit phenomenon," *American Scientist*, 86 (1996), pp. 358–363; and Ted Hill, "The difficulty of faking data," *Chance*, 12, No. 3 (1999), pp. 27–31. Applications to fraud detection are discussed in the second paper by Hill and in Mark A. Nigrini, "I've got your number," *Journal of Accountancy*, May 1999, available online at www.aicpa.org/pubs/jofa/joaiss.htm.
5. Royal Statistical Society news release, "Royal Statistical Society concerned by issues raised in Sally Clark case," October 23, 2001, at www.rss.org.uk. For background see

an editorial and article in the *Economist*, January 22, 2004. The editorial is titled "The probability of injustice."

6. Peter Sheridan Dodds, Roby Muhamad, and Duncan J. Watts, "An experimental study of search in global social networks," *Science*, 301 (2003), pp. 827–829.

7. This is one of several tests discussed in Bernard M. Branson, "Rapid HIV testing: 2003 update," a presentation by the Centers for Disease Control and Prevention, found online at www.cdc.gov.

8. Corey Kilgannon, "When New York is on the end of the line," *New York Times*, November 7, 1999.

9. This and similar psychology experiments are reported by A. Tversky and D. Kahneman, "Extensional versus intuitive reasoning: the conjunction fallacy in probability judgement," *Psychological Review*, 90 (1983), pp. 293–315.

10. Table 6.1 closely follows the grade distributions for these three schools at the University of New Hampshire for the fall of 2000, found in a self-study document at www.unh.edu/academic-affairs/neasc/. The counts of grades mirror the proportions of UNH students in these schools. The table is simplified to refer to a university with only these three schools.

11. Information about Internet users comes from sample surveys carried out by the Pew Internet and American Life Project, found online at www.pewinternet.org. The music-downloading data were collected in 2003.

12. These probabilities came from studies by the sociologist Harry Edwards, reported in the *New York Times*, February 25, 1986.

13. See Note 11.

14. From the 2003 *Statistical Abstract of the United States*, Table 491.

15. Probabilities from trials with 2897 people known to be free of HIV antibodies and 673 people known to be infected are reported in J. Richard George, "Alternative specimen sources: methods for confirming positives," 1998 Conference on the Laboratory Science of HIV, found online at the Centers for Disease Control and Prevention, www.cdc.gov.

16. National Science Foundation, as reported in the 2000 *Statistical Abstract of the United States*.

17. Projections from the 2002 *Digest of Education Statistics*, found online at nces.ed.gov.

Chapter 7

1. We use $\bar{x}$ both for the random variable, which takes different values in repeated sampling, and for the numerical value of the random variable in a particular sample. Similarly, s and $\hat{p}$ stand both for random variables and for specific values. This notation is mathematically imprecise but statistically convenient.

2. In most applications X takes a finite number of possible values. The same ideas, implemented with more advanced mathematics, apply to random variables with an infinite but still countable collection of values. An example is a geometric random variable, considered in Section 8.2.

3. Data from the Census Bureau's 1998 American Housing Survey.

4. The probability 0.12 is based on a small study at one university: Michele L. Head, "Examining college students' ethical values," Consumer Science and Retailing honors project, Purdue University, 2003.

5. The mean of a continuous random variable X with density function $f(x)$ can be found by integration:

$$\mu_X = \int xf(x)dx$$

This integral is a kind of weighted average, analogous to the discrete-case mean

$$\mu_X = \Sigma xP(X = x)$$

The variance of a continuous random variable X is the average squared deviation of the values of X from their mean, found by the integral

$$\sigma_X^2 = \int (x - \mu)^2 f(x)dx$$

6. See A. Tversky and D. Kahneman, "Belief in the law of small numbers," *Psychological Bulletin*, 76 (1971), pp. 105–110, and other writings of these authors for a full account of our misperception of randomness.

7. Probabilities involving runs can be quite difficult to compute. That the probability of a run of three or more heads in 10 independent tosses of a fair coin is $(1/2) + (1/128) = 0.508$ can be found by clever counting, as can the other results given in the text. A general treatment using advanced methods appears in Section XIII.7 of William Feller, *An Introduction to Probability Theory and Its Applications*, vol. 1, 3rd ed., Wiley, 1968.

8. R. Vallone and A. Tversky, "The hot hand in basketball: on the misperception of random sequences," *Cognitive Psychology*, 17 (1985), pp. 295–314. A later series of articles that debate the independence question is A. Tversky and T. Gilovich, "The cold facts about the 'hot hand' in basketball," Chance, 2, No. 1 (1989), pp. 16–21; P. D. Larkey, R. A. Smith, and J. B. Kadane, "It's OK to believe in the 'hot hand,'" *Chance*, 2, No. 4 (1989), pp. 22–30; and A. Tversky and T. Gilovich, "The 'hot hand': statistical reality or cognitive illusion?" *Chance*, 2, No. 4 (1989), pp. 31–34.

9. The data on returns are from several sources, especially the *Fidelity Insight* newsletter, www.fidelityinsight.com.

Chapter 8

1. N. S. Don, B. E. McDonough, and C. A. Warren, "Psi testing of a controversial psychic under controlled conditions," *Journal of Parapsychology*, 56 (1992), pp. 87–96.

2. From a Gallup Poll taken in 2002, www.galluppoll.com.

3. The survey question is reported in Trish Hall, "Shop? Many say 'Only if I must,'" *New York Times*, November 28, 1990. In fact, 66% (1650 of 2500) in the sample said "Agree."

4. Office of Technology Assessment, *Scientific Validity of Polygraph Testing: A Research Review and Evaluation*, Government Printing Office, 1983.
5. John Schwartz, "Leisure pursuits of today's young men," *New York Times*, March 29, 2004. The source cited is comScore Media Matrix.

Part II Review Exercises

1. Charles S. Fuchs et al., "Alcohol consumption and mortality among women," *New England Journal of Medicine*, 332 (1995), pp. 1245–1250.
2. Based on a news item, "Bee off with you," *Economist*, November 2, 2002, p. 78.
3. Rita F. Redburg, "Vitamin E and cardiovascular health," *Journal of the American Medical Association*, 294 (2005), pp. 107–109.
4. Simplified from Sanjay K. Dhar, Claudia Gonzalez-Vallejo, and Dilip Soman, "Modeling the effects of advertised price claims: tensile versus precise pricing," *Marketing Science*, 18 (1999), pp. 154–177.

Chapter 9

1. Test results were reported in *PC World*, April 2005. We found them on the Web at www.pcworld.com.
2. In this book we discuss only the most widely used kind of statistical inference. This is sometimes called *frequentist* because it is based on answering the question "What would happen in many repetitions?" Another approach to inference, called *Bayesian*, can be used even for one-time situations. Bayesian inference is important but is conceptually complex and much less widely used in practice.
3. U.S. Census Bureau, *Income, Poverty, and Health Insurance Coverage in the United States: 2004*, Current Population Reports P60-229, 2005. The median income is of course lower, \$44,389.
4. Results from a poll taken October 2000 and reported at www.galluppoll.com.
5. "Pupils to judge Murphy's Law with toast test," *Sunday Telegraph* (UK), March 4, 2001.
6. Results from a poll taken January–April 2000 and reported at www.galluppoll.com.
7. Strictly speaking, the recipes we give for the standard deviations of $\bar{x}$ and $\hat{p}$ assume that an SRS of size n is drawn from an infinite population. If the population has finite size N, the standard deviations in the recipes are multiplied by $\sqrt{1-(n/N)}$. This "finite population correction" approaches 1 as N increases. When $n/N \le 0.1$, it is ≥ 0.948.
8. Henry Wechsler et al., *Binge Drinking on America's College Campuses*, Harvard School of Public Health, 2001.
9. John K. Ford, "A method for grading 1987 stock recommendations," *American Association of Individual Investors Journal*, March 1988, pp. 16–17.
10. This Activity is suggested in Richard L. Schaeffer, Ann Watkins, Mrudulla Gnanadeskian, and Jeffrey A. Witmer, *Activity-Based Statistics*, Springer, 1996.

Chapter 10

1. Francisco L. Rivera-Batiz, "Quantitative literacy and the likelihood of employment among young adults," *Journal of Human Resources*, 27 (1992), pp. 313–328.
2. This and similar results of Gallup polls are from the Gallup Organization Web site, www.galluppoll.com.
3. Data provided by Darlene Gordon, Purdue University.
4. Karl Pearson and A. Lee, "On the laws of inheritance in man," *Biometrika*, 2 (1902), p. 357. These data also appear in D. J. Hand et al., *A Handbook of Small Data Sets*, Chapman & Hall, 1994. This book offers more than 500 data sets that can be used in statistical exercises.
5. The values $\mu = 22$ and $\sigma = 50$ for the gains of uncoached students on the SAT Mathematics exam come from a study of 2733 students reported on the College Board Web site, www.collegeboard.com.
6. Data provided by Drina Iglesia, Department of Biological Sciences, Purdue University. The data are part of a larger study reported in D. D. S. Iglesia, E. J. Cragoe, Jr., and J. W. Vanable, "Electric field strength and epithelization in the newt (*Notophthalmus viridescens*)," *Journal of Experimental Zoology*, 274 (1996), pp. 56–62.
7. These data are from "Results report on the vitamin C pilot program," prepared by SUSTAIN (Sharing United States Technology to Aid in the Improvement of Nutrition) for the U.S. Agency for International Development. The report was used by the Committee on International Nutrition of the National Academy of Sciences/Institute of Medicine (NAS/IOM) to make recommendations on whether or not the vitamin C content of food commodities used in U.S. food aid programs should be increased. The program was directed by Peter Ranum and Françoise Chomé.
8. The vehicle is a 1997 Pontiac Transport Van owned by George P. McCabe.
9. R. A. Berner and G. P. Landis, "Gas bubbles in fossil amber as possible indicators of the major gas composition of ancient air," *Science*, 239 (1988), pp. 1406–1409. The 95% t confidence interval is 54.78 to 64.40. A bootstrap BCa interval is 55.03 to 62.63. So t is reasonably accurate despite the skew and the small sample.
10. These recommendations are based on extensive computer work. See, for example, Harry O. Posten, "The robustness of the one-sample t-test over the Pearson system," *Journal of Statistical Computation and Simulation*, 9 (1979), pp. 133–149; and E. S. Pearson and N. W. Please, "Relation between the shape of population distribution and the robustness of four simple test statistics," *Biometrika*, 62 (1975), pp. 223–241.
11. F. H. Rauscher et al., "Music training causes long-term enhancement of preschool children's spatial-temporal reasoning," *Neurological Research*, 19 (1997), pp. 2–8.
12. Shailja V. Nigdikar et al., "Consumption of red wine polyphenols reduces the susceptibility of low-density lipoproteins to oxidation in vivo," *American Journal of Clinical*

Nutrition, 68 (1998), pp. 258–265. (There were in fact only 30 subjects, some of whom received more than one treatment with a four-week period intervening.) The data are simulated from the Normal distribution with mean and standard deviation observed in this study.

13. Alan S. Banks et al., "Juvenile hallux abducto valgus association with metatarsus adductus," *Journal of the American Podiatric Medical Association*, 84 (1994), pp. 219–224.

14. David R. Pillow et al., "Confirmatory factor analysis examining attention deficit hyperactivity disorder symptoms and other childhood disruptive behaviors," *Journal of Abnormal Child Psychology*, 26 (1998), pp. 293–309.

15. Data from the College Alcohol Study Web site, www.hsph.harvard.edu/cas/.

16. From the "National Survey of Student Engagement, the college student report," 2003, available online at www.indiana.edu/~nssel/.

17. Linda Lyons, "Teens: Sex Can Wait," December 14, 2004, from the Gallup Poll Web site, www.galluppoll.com.

18. Data provided by Jude M. Werra & Associates, Brookfield, Wis.

19. "Poll: men, women at odds on sexual equality," Associated Press dispatch appearing in the *Lafayette (Ind.) Journal and Courier*, October 20, 1997.

20. Bryan E. Porter and Thomas D. Berry, "A nationwide survey of self-reported red light running: measuring prevalence, predictors, and perceived consequences," *Accident Analysis and Prevention*, 33 (2001), pp. 735–741.

21. Data from Guohua Li and Susan P. Baker, "Alcohol in fatally injured bicyclists," *Accident Analysis and Prevention*, 26 (1994), pp. 543–548.

22. Information about the survey can be found online at s6.library.arizona.edu/natcong/.

23. Based on information in "NCAA 2003 national study of collegiate sports wagering and associated health risks," which can be found on the NCAA Web site, www.ncaa.org.

24. John Paul McKinney and Kathleen G. McKinney, "Prayer in the lives of late adolescents," *Journal of Adolescence*, 22 (1999), pp. 279–290.

25. James A. Levine, Norma L. Eberhardt, and Michael D. Jensen, "Role of nonexercise activity thermogenesis in resistance to fat gain in humans," *Science*, 283 (1999), pp. 212–214. Data for this study are available from the *Science* Web site, www.sciencemag.org.

26. Jane E. Brody, "Alternative medicine makes inroads," *New York Times*, April 28, 1998.

27. Gary Edwards and Josephine Mazzuca, "Three quarters of Canadians support doctor-assisted suicide," Gallup Poll press release, March 24, 1999, at www.galluppoll.com.

28. From the August 2000 supplement to the Current Population Survey, from the Census Bureau Web site, www.census.gov.

29. From program 19, "Confidence Intervals," in the *Against All Odds* video series.

30. These data were collected in conjunction with a bone health study at Purdue University and were provided by Linda McCabe.

Part III Review Exercises

1. Based on interviews in 2000 and 2001 by the National Longitudinal Study of Adolescent Health. Found at the Web site of the Carolina Population Center, www.cpc.unc.edu.

2. Projections from the 2002 *Digest of Education Statistics*, found online at nces.ed.gov.

3. From the EESEE Case Study "Baby Hearing Screening: The Sound of Silence."

Chapter 11

1. Based on G. Salvendy et al. "Impact of personality and intelligence on job satisfaction of assembly line and bench work—an industrial study," *Applied Ergonomics*, 13 (1982), pp. 293–299.

2. Julie Ray, "Few Teens Clash with Friends," May 3, 2005, on the Gallup Poll Web site, www.galluppoll.com.

3. R. A. Fisher, "The arrangement of field experiments," *Journal of the Ministry of Agriculture of Great Britain*, 33 (1926), p. 504, quoted in Leonard J. Savage, "On rereading R. A. Fisher," *Annals of Statistics*, 4 (1976), p. 471. Fisher's work is described in a biography by his daughter: Joan Fisher Box, *R. A. Fisher: The Life of a Scientist*, Wiley, 1978.

4. Manisha Chandalia et al., "Beneficial effects of high dietary fiber intake in patients with type 2 diabetes mellitus," *New England Journal of Medicine*, 342 (2000), pp. 1392–1398.

5. Data provided by Diana Schellenberg, Purdue University School of Health Sciences.

6. D. L. Shankland et al., "The effect of 5-thio-D-glucose on insect development and its absorption by insects," *Journal of Insect Physiology*, 14 (1968), pp. 63–72.

7. For a discussion of statistical significance in the legal setting, see D. H. Kaye, "Is proof of statistical significance relevant?" *Washington Law Review*, 61 (1986), pp. 1333–1365. Kaye argues: "Presenting the *P*-value without characterizing the evidence by a significance test is a step in the right direction. Interval estimation, in turn, is an improvement over *P*-values."

8. The editorial was written by Phil Anderson. See *British Medical Journal*, 328 (2004), pp. 476–477. A letter to the editor on this topic by Doug Altman and J. Martin Bland appeared shortly after. See "Confidence intervals illuminate absence of evidence," *British Medical Journal*, 328 (2004), pp. 1016–1017.

9. A. Kamali et al., "Syndromic management of sexually-transmitted infections and behavior change interventions on transmission of HIV-1 in rural Uganda: a community randomised trial," *Lancet*, 361 (2003), pp. 645–652.

10. T. D. Sterling, "Publication decisions and their possible effects on inferences drawn from tests of significance—or

vice versa," *Journal of the American Statistical Association*, 54 (1959), pp. 30–34. Related comments appear in J. K. Skipper, A. L. Guenther, and G. Nass, "The sacredness of 0.05: a note concerning the uses of statistical levels of significance in social science," *American Sociologist*, 1 (1967), pp. 16–18.
11. Warren E. Leary, "Cell phones: questions but no answers," *New York Times*, October 26, 1999.
12. Information obtained from the Fornitek Canada Corporation *Wood Protection Bulletin*, July 2003.
13. The idea for this exercise was provided by Michael Legacy and Susan McGann.

Chapter 12

1. P. A. Mackowiak, S. S. Wasserman, and M. M. Levine, "A critical appraisal of 98.6 degrees F, the upper limit of the normal body temperature, and other legacies of Carl Reinhold August Wunderlich," *Journal of the American Medical Association*, 268 (1992), 1578–1580.
2. From the NIST Web site, www.nist.gov/srd/online.htm.
3. Data provided by Timothy Sturm.
4. Based on I. Cuellar, L. C. Harris, and R. Jasso, "An acculturation scale for Mexican American normal and clinical populations," *Hispanic Journal of Behavioral Sciences*, 2 (1980), pp. 199–217.
5. This exercise is based on events that are real. The data and details have been altered to protect the privacy of the individuals involved.
6. National Institute for Occupational Safety and Health, *Stress at Work*, 2000, available online at www.cdc.gov/niosh/stresswk.html. Results of this survey were reported in *Restaurant Business*, September 15, 1999, pp. 45–49.
7. Based on Raul de la Funete-Fernandez et al., "Expectation and dopamine release: mechanism of the placebo effect in Parkinson's disease," *Science*, 293 (2001), pp. 1164–1166.
8. Ajay Ghei, "An empirical analysis of psychological androgeny in the personality profile of the successful hotel manager," MS thesis, Purdue University, 1992.
9. Data provided by Chris Olsen, who found the information in scuba diving magazines.
10. "Euro coin accused of unfair flipping," *New Scientist*, January 4, 2002.

Chapter 13

1. This Case Study is based on the story "Drive-Thru Competition" in the Electronic Encyclopedia of Statistical Examples and Exercises (EESEE). The EESEE story idea originated from the article "Navigating the loop: the best drive-thru in America '04" by Howard Reill, *QSR*, October 2004, pp. 41–59. The article is available at the Web site www.*qsr*magazine.com/drive-thru.
2. C. P. Cannon et al., "Intensive versus moderate lipid lowering with statins after acute coronary syndromes," *New England Journal of Medicine*, 350 (2004), 1495–1504.
3. This study is reported in Roseann M. Lyle et al., "Blood pressure and metabolic effects of calcium supplementation in normotensive white and black men," *Journal of the American Medical Association*, 257 (1987), pp. 1772–1776. The data were provided by Dr. Lyle.
4. Adapted from Maribeth Cassidy Schmitt, "The effects of an elaborated directed reading activity on the metacomprehension skills of third graders," PhD dissertation, Purdue University, 1987.
5. M. Ann Laskey et al., "Bone changes after 3 mo of lactation: influence of calcium intake, breast-milk output, and vitamin D–receptor genotype," *American Journal of Clinical Nutrition*, 67 (1998), pp. 685–692.
6. Detailed information about the conservative *t* procedures can be found in Paul Leaverton and John W. Birch, "Small sample power curves for the two sample location problem," *Technometrics*, 11 (1969), pp. 299–307; in Henry Scheffé, "Practical solutions of the Beherns-Fisher problem," *Journal of the American Statistical Association*, 65 (1970), pp. 1501–1508; and in D. J. Best and J. C. W. Rayner, "Welch's approximate solution for the Beherns-Fisher problem," *Technometrics*, 29 (1987), pp. 205–210.
7. See the extensive simulation studies in Harry O. Posten, "The robustness of the two-sample *t*-test over the Pearson system," *Journal of Statistical Computation and Simulation*, 6 (1978), pp. 295–311, and in Harry O. Posten, H. Yeh, and Donald B. Owen, "Robustness of the two-sample *t*-test under violations of the homogeneity assumption," *Communications in Statistics*, 11 (1982), pp. 109–126.
8. Data for 1982, provided by Marvin Schlatter, Division of Financial Aid, Purdue University.
9. This Example is loosely based on D. L. Shankland, "Involvement of spinal cord and peripheral nerves in DDT-poisoning syndrome in albino rats," *Toxicology and Applied Pharmacology*, 6 (1964), pp. 197–213.
10. We did not use Minitab or DataDesk in Example 13.6 because these packages shortcut the two-sample *t* procedure. They calculate the degrees of freedom using the formula in the box on page 792 but then truncate to the next lower whole-number degrees of freedom to obtain the *P*-value. The result is slightly less accurate than the *P*-value from the *t*(df) distribution.
11. Shailija V. Nigdikar et al., "Consumption of red wine polyphenols reduces the susceptibility of low-density lipoproteins to oxidation in vivo," *American Journal of Clinical Nutrition*, 68 (1998), pp. 258–265.
12. Based on W. N. Nelson and C. J. Widule, "Kinematic analysis and efficiency estimate of intercollegiate female rowers," unpublished manuscript, 1983.
13. The idea for this exercise was provided by Robert Hayden.
14. F. Tagliavini et al., "Effectiveness of anthracycline against experimental prion disease in Syrian hamsters," *Science*, 276 (1997), pp. 1119–1121.
15. From a graph in Fabrizio Grieco, Arie J. van Noordwijk, and Marcel E. Visser, "Evidence for the effect of learning on

timing of reproduction in blue tits," *Science*, 296 (2002), pp. 136–138.

16. Data provided by Darlene Gordon, School of Education, Purdue University.

17. Data provided by Charles Cannon, Duke University. The study report is C. H. Cannon, D. R. Peart, and M. Leighton, "Tree species diversity in commercially logged Bornean rainforest," *Science*, 281 (1998), pp. 1366–1367.

18. "Assisted reproductive technology in the United States: 1997 results," generated from the American Society for Reproductive Medicine/Society for Assisted Reproductive Technology Registry, *Fertil Steril*, 74 (2000), pp. 641–653.

19. Kwang Y. Cha, Daniel P. Wirth, and Rogerio A. Lobo, "Does prayer influence the success of *in vitro* fertilization–embryo transfer?" *Journal of Reproductive Medicine*, 46 (2001), pp. 781–787.

20. Clive G. Jones et al., "Chain reactions linking acorns to gypsy moth outbreaks and Lyme disease risk," *Science*, 279 (1998), pp. 1023–1026.

21. Janice Joseph, "Fear of crime among black elderly," *Journal of Black Studies*, 27 (1997), pp. 698–717.

22. These are some of the data from the EESEE story "Radar Detectors and Speeding." The study is reported in N. Teed, K. L. Adrian, and R. Knoblouch, "The duration of speed reductions attributable to radar detectors," *Accident Analysis and Prevention*, 25 (1991), pp. 131–137.

23. Pew Internet Project, "Counting on the Internet," December 29, 2002, at www.pewinternet.org.

24. From an Associated Press dispatch appearing on December 30, 2002. The study report appeared in the *Journal of Adolescent Health.*

25. Martin Enserink, "Fraud and ethics charges hit stroke drug trial," *Science*, 274 (1996), pp. 2004–2005.

26. Donna L. Hoffman and Thomas P. Novak, "Bridging the racial divide on the Internet," *Science*, 280 (1998), pp. 390–391.

27. These are some of the data from the EESEE story "Leave Survey after the Beep." The study is reported in M. Xu, B. J. Bates, and J. C. Schweitzer, "The impact of messages on survey participation in answering machine households," *Public Opinion Quarterly*, 57 (1993), pp. 232–237.

28. Based on Robert T. Driescher, "A quality swing with Ping," *Quality Progress*, August 2001, pp. 37–41.

29. National Athletic Trainers Association, press release dated September 30, 1994.

30. Based on Deborah Roedder John and Ramnath Lakshmi-Ratan, "Age differences in children's choice behavior: the impact of available alternatives," *Journal of Marketing Research*, 29 (1992), pp. 216–226.

31. Modified from Richard A. Schieber et al., "Risk factors for injuries from in-line skating and the effectiveness of safety gear," *New England Journal of Medicine*, 335 (1996), Internet summary.

32. Based on Amna Kirmani and Peter Wright, "Money talks: perceived advertising expense and expected product quality," *Journal of Consumer Research*, 16 (1989), pp. 344–353.

33. JoAnn K. Wells, Allan F. Williams, and Charles M. Farmer, "Seat belt use among African Americans, Hispanics, and whites," *Accident Analysis and Prevention*, 34 (2002), pp. 523–529.

34. Francisco L. Rivera-Batiz, "Quantitative literacy and the likelihood of employment among young adults," *Journal of Human Resources*, 27 (1992), pp. 313–328.

35. The idea for this exercise was provided by Robert Hayden.

36. From V. D. Bass, W. E. Hoffmann, and J. L. Dorner, "Normal canine lipid profiles and effects of experimentally induced pancreatitis and hepatic necrosis on lipids," *American Journal of Veterinary Research*, 37 (1976), pp. 1355–1357.

37. David M. Blau, "The child care labor market," *Journal of Human Resources*, 27 (1992), pp. 9–39.

Chapter 14

1. W. E. Paulus et al., "Influence of acupuncture on the pregnancy rate in patients who undergo assisted reproductive therapy," *Fertility and Sterility*, 77, No. 4 (2002), pp. 721–724.

2. D. A. Redelmeier and R. J. Tibshirani, "Association between cellular-telephone calls and motor vehicle collisions," *New England Journal of Medicine*, 336 (1997), pp. 453–458.

3. Based on a news item in *Science*, 305 (2004), p. 1560. The study, by Daniel Klem, appeared in the *Wilson Journal.*

4. From the GSS data base at the University of Michigan, webapp.icpsr.umich.edu/GSS.

5. C. M. Ryan et al., "The effect of in-store music on consumer choice of wine," *Proceedings of the Nutrition Society*, 57 (1998), p. 1069A.

6. Douglas E. Jorenby et al., "A controlled trial of sustained-release bupropion, a nicotine patch, or both for smoking cessation," *New England Journal of Medicine*, 340 (1990), pp. 685–691.

7. Joan L. Duda, "The relationship between goal perspectives, persistence, and behavioral intensity among male and female recreational sports participants," *Leisure Sciences*, 10 (1988), pp. 95–106.

8. Martin Enserink, "Fraud and ethics charges hit stroke drug trial," *Science*, 274 (1996), pp. 2004–2005.

9. Karine Marangon et al., "Diet, antioxidant status, and smoking habits in French men," *American Journal of Clinical Nutrition*, 67 (1998), pp. 231–239.

10. An alternative formula that can be used for hand or calculator computations is

$$X^2 = \sum \frac{(\text{observed})^2}{\text{expected}} - n$$

11. There are many computer studies of the accuracy of chi-square critical values for X^2. For a brief discussion and

some references, see Section 3.2.5 of David S. Moore, "Tests of chi-squared type," in Ralph B. D'Agostino and Michael A. Stephens (eds.), *Goodness-of-Fit Techniques*, Marcel Dekker, 1986, pp. 63–95. If the expected cell counts are roughly equal, the chi-square approximation is adequate when the average expected counts are as small as 1 or 2. The guideline given in the text protects against unequal expected counts. For a survey of inference for smaller samples, see Alan Agresti, "A survey of exact inference for contingency tables," *Statistical Science*, 7 (1992), pp. 131–177.

12. Data compiled from a table of percents in "Americans view higher education as key to the American dream," press release by the National Center for Public Policy and Higher Education, www.highereducation.org, May 3, 2000.

13. Lillian Lin Miao, "Gastric freezing: an example of the evaluation of medical therapy by randomized clinical trials," in John P. Bunker, Benjamin A. Barnes, and Frederick Mosteller (eds.), *Costs, Risks, and Benefits of Surgery*, Oxford University Press, 1977, pp. 198–211.

14. P. Azoulay and S. Shane, "Entrepreneurs, contracts, and the failure of young firms," *Management Science*, 47 (2001), pp. 337–358.

15. Richard M. Felder et al., "Who gets it and who doesn't: a study of student performance in an introductory chemical engineering course," 1992 ASEE Annual Conference Proceedings, American Society for Engineering Education, Washington, D.C., 1992, pp. 1516–1519.

16. S. V. Zagona (ed.), *Studies and Issues in Smoking Behavior*, University of Arizona Press, 1967, pp. 157–180.

17. Modified from Felicity Barringer, "Measuring sexuality through polls can be shaky," *New York Times*, April 25, 1993.

18. From the GSS data archive at the Survey Documentation and Analysis site at the University of California, Berkeley, sda.berkeley.edu.

19. Brenda C. Coleman, "Study: heart attack risk cut 74% by stress management," Associated Press dispatch appearing in the *Lafayette (Ind.) Journal and Courier*, October 20, 1997.

20. Based closely on Susan B. Sorenson, "Regulating firearms as a consumer product," *Science*, 286 (1999), pp. 1481–1482. Because the results in the paper were "weighted to the U.S. population," we have changed some counts slightly for consistency.

21. The National Longitudinal Study of Adolescent Health interviewed a stratified random sample of 27,000 adolescents, then reinterviewed many of the subjects six years later, when most were aged 19 to 25. These data are from the Wave III reinterviews in 2000 and 2001, found at the Web site of the Carolina Population Center, www.cpc.unc.edu.

22. Claudia Braga et al., "Olive oil, other seasoning fats, and the risk of colorectal carcinoma," *Cancer*, 82 (1998), pp. 448–453.

23. Mei-Hui Chen, "An exploratory comparison of American and Asian consumers' catalog patronage behavior," MS thesis, Purdue University, 1994.

24. William D. Darley, "Store-choice behavior for preowned merchandise," *Journal of Business Research*, 27 (1993), pp. 17–31.

25. Ann P. Streissguth et al., "Intrauterine alcohol and nicotine exposure: attention and reaction time in 4-year-old children," *Developmental Psychology*, 20 (1984), pp. 533–541.

26. R. Shine, et al., "The influence of nest temperatures and maternal brooding on hatchling phenotypes in water pythons," *Ecology*, 78 (1997), pp. 1713–1721.

27. D. M. Barnes, "Breaking the cycle of addiction," *Science*, 241 (1988), pp. 1029–1030.

28. Erika Friedmann et al., "Animal companions and one-year survival of patients after discharge from a coronary care unit," *Public Health Reports*, 96 (1980), pp. 307–312.

29. George R. Milne, "How well do consumers protect themselves from identity theft?" *Journal of Consumer Affairs*, 37 (2003), pp. 388–402.

30. Tom Reichert, "The prevalence of sexual imagery in ads targeted to young adults," *Journal of Consumer Affairs*, 37 (2003), pp. 403–412.

31. Based on information in "NCAA 2003 national study of collegiate sports wagering and associated health risks," which can be found at the NCAA Web site, www.ncaa.org.

Chapter 15

1. Source: www.ncaa.org/library/records/basketball/m_basketball_records_book/2004/2004MbasketballRecords.pdf

2. azothgallery.com/vitruvian.html

3. G. Geri and B. Palla, "Considerazioni sulle più recenti osservazioni ottiche alla Torre Pendente di Pisa," *Estratto dal Bollettino della Società Italiana di Topografia e Fotogrammetria*, 2 (1988), pp. 121–135. Professor Julia Mortera of the University of Rome provided a translation.

4. Todd W. Anderson, "Predator responses, prey refuges, and density-dependent mortality of a marine fish," *Ecology*, 81 (2001), pp. 245–257.

5. Based on Marion E. Dunshee, "A study of factors affecting the amount and kind of food eaten by nursery school children," *Child Development*, 2 (1931), pp. 163–183. This article gives the means, standard deviations, and correlation for 37 children but does not give the actual data.

6. "Personal Watercraft-Related Injuries: A Growing Public Health Concern" by Branche, Christine M., et al. *Journal of the American Medical Association*, 1997; 278: 663–665.

7. From a graph in Timothy G. O'Brien and Margaret F. Kinnaird, "Caffeine and conservation," *Science*, 300 (2003), p. 587.

8. Peter H. Chen, Neftali Herrera, and Darren Christiansen, "Relationships between gate velocity and casting features among aluminum round castings," no date. Provided by Darren Christiansen.

9. N. Maeno et al., "Growth of icicles," *Journal of Glaciology*, 40 (1994), pp. 319–326.

10. Data provided by Samuel Phillips, Purdue University.

Part IV Review Exercises

1. Debra L. Miller et al., "Effect of fat-free potato chips with and without nutrition labels on fat and energy intakes," *American Journal of Clinical Nutrition*, 68 (1998), pp. 282–290.

2. Based on the online supplement to Paul J. Shaw et al., "Correlates of sleep and waking in *Drosophila melanogaster*," *Science*, 287 (2000), pp. 1834–1837.

3. Data provided by Corinne Lim, Purdue University, from a student project supervised by Professor Joseph Vanable.

4. Marianne Perle, Rebecca Moran, and Anthony D. Lutkus, *NAEP 2004 Trends in Academic Progress: Three Decades of Student Performance in Reading and Mathematics*, National Center for Education Statistics, 2005, at nces.ed.gov. The data given are approximate due to rounding in the study report.

5. Virgilio P. Carnielli et al., "Intestinal absorption of long-chain polyunsaturated fatty acids in preterm infants fed breast milk or formula," *American Journal of Clinical Nutrition*, 67 (1998), pp. 97–103.

6. Maureen Hack et al., "Outcomes in young adulthood for very-low-birth-weight infants," *New England Journal of Medicine*, 346 (2002), pp. 149–157. Exercise IV.7 is simplified, in that the measures reported in this paper have been statistically adjusted for "sociodemographic status."

7. Michael R. Dohm, Jack P. Hayes, and Theodore Garland, Jr., "Quantitative genetics of sprint running speed and swimming endurance in laboratory house mice (*Mus domesticus*)," *Evolution*, 50 (1996), pp. 1688–1701.

8. Carly J. Stevens et al., "Impact of nitrogen deposition on the species richness of grasslands," *Science*, 303 (2004), pp. 1876–1879.

9. Esther Duflo et al., "Savings incentives for low- and middle-income families: evidence from a field experiment with H&R Block," The Retirement Security Project, published online at www.retirementsecurityproject.org.

Photo Credits

Preliminary Chapter:
p.2 Markus Dlouhy/Das Fotoarchiv/Peter Arnold
p.5 © Tim Downs/Universal Press Syndicate
p.10 Used by permission of Susie MacNelly
p.14 © The New Yorker Collection, 1976, Dana Fradon. From cartoonbank.com
p.16 John Gress/Reuters/CORBIS
p.19 CORBIS
p.21 SuperStock
p.31 (top) CORBIS
p.31 (bottom) Lucy Nicholson/CORBIS
p.33 Brendan Byrne/Agefotostock

Part I:
p.35 Darrell Ingham/AllSport Concepts/Getty Images

Chapter 1:
p.36 Moshe Brakha/ABC-TV/The Kobal Collection
p.37 © 1998 Ziggy and Friends, Inc. Reprinted with permission of Universal Press Syndicate. All Rights Reserved.
p.38 Zits Partnership. King Features Syndicate
p.41 Michael A. Keller/CORBIS
p.48 Taxi/Getty Images
p.58 Zits Partnership. King Features Syndicate
p.65 Taxi/Getty Images
p.90 Randy Glasbergen
p.91 Victor Baldizon/Getty Images
p.102 Alvis Upitis/The Image Bank/Getty Images
p.103 Jeff Stahler, *The Cincinnati Post*

Chapter 2:
p.112 Charles Gupton/CORBIS
p.118 Getty Images
p.120 Charles Barsotti/cartoonbank.com
p.131 Spencer Grant/PhotoEdit
p.142 By permission of John L. Hart FLP and Creators Syndicate, Inc.
p.147 (top) CORBIS
p.147 (bottom) Mel Yates/Taxi/Getty Images
p.159 Knight-Ridder/Tribune Media Information Services. All Rights Reserved. Reprinted with permission.

Chapter 3:
p.168 Reuters/CORBIS
p.174 Stuart Westmorland/Getty Images
p.178 Dennis–King Features Syndicate
p.179 Daniel J. Cox/CORBIS
p.180 Mother Goose & Grimm–Grimmy, Inc. King Features Syndicate
p.189 Bill Ross/CORBIS
p.193 Neal Preston/CORBIS
p.204 R. Lynn/Photo Researchers
p.229 Carlyn Galati/Visuals Unlimited
p.252 Digital Vision/Getty Images
p.255 SuperStock/SuperStock/Picture Quest

Chapter 4:
p.256 Stephen Frink/CORBIS
p.262 Hebert PDN, Wearing-Wilde J, eds. Canada's Polar Life. www.polarlife.ca
p.265 Alamy
p.271 Don Mason/CORBIS
p.276 Hans Reinhard/zefa/CORBIS
p.282 Bettmann/CORBIS
p.289 Martin Harvey/CORBIS
p.298 Royalty-Free/CORBIS
p.300 Ashley/Cooper/PICIMPACT/CORBIS
p.302 © 2003 Dana Fradon from cartoonbank.com. All Rights Reserved.
p.306 Bill Keane, Inc. King Features Syndicate
p.312 Stahler/*Columbus Dispatch*/Dist. by Newspaper Enterprise Association, Inc.
p.318 Henryk Kaiser/eStock Photography/PictureQuest

Part II:
p.325 Brad Rickerby/Getty Images

Chapter 5:
p.326 William Brady/SuperStock
p.333 David Young-Wolff/Photoedit, Inc.
p.339 Ingram Publishing/SuperStock
p.341 (top) Cartoons by Landers, cartoons@landers.co.uk
p.341 (bottom) Andre Jenny/The Image Works
p.345 © 1995 Watterson. Universal Press Syndicate. Reprinted with permission. All Rights Reserved.
p.353 Chris Gardner/AP Wide World Photo
p.366 © The New Yorker Collection 2005 Mike Twohy/cartoonbank.com. All Rights Reserved.
p.366 (bottom) Ole Graf/Corbis
p.381 (top) Wade Payne/AP Photos
p.381 (bottom) Frank Franklin II/AP Wide World Photos

Part III:
p.387 Robert Glusic/Getty Images

Chapter 6:
p.388 Nick Ut/AP Wide World Photo
p.398 David Madison/The Image Bank/Getty Images
p.401 John Russell/AP Wide World
p.405 Paul J. Sutton/CORBIS
p.410 Dennis–North America Syndicate
p.411 Richard Levine/Alamy
p.412 SuperStock
p.424 (top) Courtesy of Nicole Yates
p.424 (bottom) Fabio Pili/Alamy
p.429 By permission of Dave Coverly and Creators Syndicate, Inc.
p.430 Greg Vote/Index Stock
p.434 © 2006 Dana Fradon from cartoonbank.com. All Rights Reserved.
p.450 Jim Craigmyle/CORBIS
p.455 Tom and Dee Ann McCarthy/CORBIS

Chapter 7:
p.462 Historic Golf Photos ©/The Ron Watts Collection
p.464 Image Source/Index Stock
p.480 Bob Daemmrich/The Image Works
p.484 Scott Adams/Dist. by United Feature Syndicate, Inc.
p.486 Jean-Loup Gautreau/Getty Images
p.489 Scott Adams/Dist. by United Feature Syndicate, Inc.
p.491 Russell Underwood/CORBIS
p.501 Nancy Sheehan/Index Stock
p.502 Adam Woolfitt/CORBIS

Chapter 8:

p.510 PhotoObjects.net/JupiterImages
p.515 © 1996 John McPherson. Reprinted with permission of Universal Press Syndicate. All Rights Reserved.
p.527 Erica Shires/zefa/CORBIS
p.536 Robert E. Klein/CORBIS
p.538 Universal/Photofest
p.549 (left) Courtesy of Dan Yates
p.549 (right) Courtesy of Dan Yates
p.557 Kayte Deioma/Photo Edit
p.558 Bob Daemmrich/The Image Works

Chapter 9:

p.560 Roland Weihrauch/dpa/CORBIS
p.577 Dorling Kindersley/Getty Images
p.581 FOXTROT © 2005 Bill Amend. Reprinted with permission of Universal Press Syndicate. All Rights Reserved.
p.588 Tony Arruza/CORBIS
p.596 Gandee Vasan/Getty Images
p.602 Jeff Greenberg/The Image Works
p.604 Frank Pedrick/Index Stock Imagery/PictureQuest
p.609 (top) Alan Hicks/Getty Images
p.609 (bottom) Kyle Krause/Inndex Stock Imagery/PictureQuest

Part IV:

p.613 Joseph Sohm/VisionsofAmerica.com/Getty Images

Chapter 10:

p.614 Stockbyte/Getty Images
p.616 Kurt Vinion/Getty Images
p.626 Jim Naughten/CORBIS
p.633 M. Thomsen/zefa/CORBIS
p.635 Phototake Inc./Alamy
p.638 Joe Raedle/Getty Images
p.641 Royalty-Free/CORBIS
p.650 David Sanger Photography/Alamy
p.657 Stock Connection Distribution/Alamy
p.668 © 1992 Dan Fradon from cartoonbank.com. All Rights Reserved.
p.672 Bettmann/CORBIS
p.676 Dave Bartruff/CORBIS
p.681 Gary Conner/Phototake

Chapter 11:

p.684 Francis Li/Alamy
p.689 CORBIS
p.694 Zits Partnership. King Features Syndicate
p.699 Comstock
p.706 W&D Mcintyre/Photo Researchers
p.710 Thomas Pflaum/The Image Works
p.720 Edward Bock/CORBIS
p.727 Zuma Press/CORBIS
p.731 CORBIS
p.733 HIRB/Index Stock
p.737 © The New Yorker Collection 1987 Dana Fradon from cartoon-bank.com. All Rights Reserved.
p.738 Gk Hart/Getty Images

Chapter 12:

p.740 Randy Faris/CORBIS
p.742 Getty Images
p.743 Zits Partnership/King Features Syndicate
p.747 Ramin/Talaie/CORBIS
p.761 Peter Bowater/Alamy
p.763 Shmuel Thaler/Index Stock Imager/PictureQuest
p.775 Sally Forth/King Features Syndicate
p.776 Simon Marcus/CORBIS
p.777 David Fleetham/Taxi/Getty Images

Chapter 13:

p.778 Tim Boyle/Getty Images
p.780 Digital Archive Japan/Alamy
p.785 Carolyn A. McKeone/Photo Researchers
p.803 Scott Adams/Dist. by United Feature Syndicate, Inc.
p.805 Hugh Clark/Frank Lane Picture Agency/CORBIS
p.806 Georg Gerster/Photo Researchers
p.812 Scott Camazine/Photo Researchers
p.822 By permission of Dave Coverly and Creators Syndicate, Inc.
p.823 David Young-Wolff/PhotoEdit, Inc.
p.828 Ted Horowitz/CORBIS
p.830 Arthur Tilley/i2i Images/PictureQuest

Chapter 14:

p.832 Peter Turnley/CORBIS
p.834 Richard Levine/Alamy
p.836 Richard Levine/Alamy
p.839 By permission of Dave Coverly and Creators Syndicate, Inc.
p.848 Lon C. Diehl/PhotoEdit
p.856 By permission of Dave Coverly and Creators Syndicate, Inc.
p.878 Foodcollection/Getty Images
p.884 Ted Horowitz/CORBIS

Chapter 15:

p.886 Jim McIsaac/Getty Images
p.888 Scala/Art Resource
p.889 Benelux Press/Index Stock Imagery/PictureQuest
p.894 Joe Tucciarone
p.901 Angelo Cavalli/CORBIS
p.902 SuperStock
p.908 Getty Images
p.915 IFA/eStock Photography/PictureQuest
p.916 AEF-David Sharrock/The Image Bank/Getty Images
p.917 Peter Beck/CORBIS

Solutions to Odd-Numbered Exercises

Preliminary Chapter

P1 Getting data from Jamie and her friends is convenient, but it does not provide a good snapshot of the opinions held by all young people. In short, Jamie and her friends are not a representative sample from the population of interest (young people).

P3 (a) An experiment. Animation is used for one group and a textbook for the other in order to compare the performance of the two different groups. (b) The company could conclude that they have solid evidence that computer animation was more successful than a textbook for these students. If the company wants to generalize this conclusion to a broader population, then they must assume that these groups are representative samples of the population of interest. The information provided does not indicate whether or not these students were randomly selected from some larger population. If they were, then inferences to that population are reasonable. Similarly, the information provided does not indicate whether or not randomization was used to partition the students into two groups. If randomization was used, then cause-and-effect conclusions can be drawn. Finally, we must assume that the tests adequately assess biology concepts.

P5 (a) In an observational study, people who drink alcohol (beer or wine) would be randomly selected and then variables that measure health characteristics would be collected and compared. In an experiment, the researchers would randomly assign the treatment (beer or wine) to the participants, who would be required to drink that type of alcohol. The variables that measure health characteristics would be collected and compared after a reasonable amount of time. (b) Wine drinkers tend to be wealthier, exercise more frequently, and have better eating habits than beer drinkers.

P7 (b) Consumers seem to prefer lighter colors (silver and white) to darker colors. Silver seems to be the most popular color. The percent of vehicles with other colors is 14.2%.

P9 (b) The dotplot is centered around 2 with a long right tail. Only two of the 34 differentials are negative, which indicates that the U.S. women's soccer team had a very good season.

P11 Who? The individuals are the AP Statistics students who completed a questionnaire on the first day of class. What? The categorical variables are gender (female or male), handedness (right or left), and favorite type of music (classical, gospel, rock, rap, country, R&B, top 40, oldies, etc.). The quantitative variables are height (in inches), amount of time the student is expecting to spend on homework (in minutes per week), and the total value of coins in a student's pocket (in cents). Why? The data were collected for the teacher to learn more about her/his students and to provide an interesting data set for the students to analyze. When, where, how, and by whom? A teacher collected these data on the first day of class at a high school using a questionnaire.

P13 Reality shows (yes or no indicating whether the student watches reality shows), Music (yes or no indicating whether the student watches music videos, concerts, or documentaries about musicians and singers), Time (the average amount of time, in minutes per day, spent watching television), Network (the average number of network programs—shows, movies, sporting events, etc.—watched per week on ABC, CBS, NBC, and FOX).

P15 The result of 14 out of 21 correct identifications is further away from what we expect (7) if the students are guessing, so it would provide more convincing evidence.

P17 Results will vary from student to student.

P19 (a) Available data from family interviews and police records were used. No treatments were imposed in order to observe various responses. (b) Parental involvement, profession of parents, educational priorities, amount of reading, type of child care, and participation in sports and other activities with peers are just a few other variables that may be related to the amount of TV watched. The effects of these other variables are mixed up with and cannot be separated from the effect of watching TV. This is known as confounding.

P21 Who? The individuals are motor vehicles produced in 2004. What? The categorical variables are make and model; vehicle type; transmission type. The quantitative variables are number of cylinders (integer count); city MPG (miles per gallon); highway MPG (miles per gallon). Why? The data were compiled to compare fuel economy. When, where, how, and by whom? A statement on www.fueleconomy.gov reveals that "The data included in the Department of Energy's Fuel Economy Guide are the result of vehicle testing done at the Environmental Protection Agency's National Vehicle and Fuel Emissions Laboratory in Ann Arbor, Michigan and by vehicle manufacturers themselves with oversight by EPA."

P23 There are 13 different types of cards in a standard deck of cards. The chance of getting exactly 3 of the same type, 1 of a second type, and 1 of a third type in a hand of 5 cards is 1/50. Thus, if we kept dealing 5 card hands over and over again, the proportion of hands that would contain three of a kind approaches 0.02.

P25 From Figure P.7, roughly 50 of the simulated differences were 2.5 or larger. Thus, the chance of getting a difference of 2.5 or larger if there is no difference in the magnets is approximately 0.005. Our conclusion would remain the same; we have evidence that the active magnets relieve pain in polio patients.

P27 The distribution of MPG is roughly symmetric with a center at about 27 miles per gallon. The range is 14, with the Chevrolet Malibu and the Honda Accord both getting the best mileage (34 MPG). The Mercedes-Benz E500 had the worst gas mileage (20 MPG).

Chapter 1

1.1 (a) 231.8 million tons.

1.3 (a) The stemplot does a better job because the dots in the dotplot are so spread out that it is difficult to identify the shape of the distribution. (b) Counts of the observations from the bottom up and the top down. By the use of parentheses in this column. (c) The sample with the concentration of 0.99.

1.5 The center of the distribution is 35, and there is approximately the same number of points to the left and right of the center. There are no gaps or outliers. The distribution is approximately symmetric.

1.7 (a) Roughly symmetric, maybe slightly skewed to the right. (b) About 15%. (c) The smallest return was between −70% and −60%, while the largest was between 100% and 110%. (d) 23%.

1.9 The distribution has a single peak at noon and falls off on either side of the peak in a symmetrical pattern. The range is from the 7th hour to the 17th hour of a day.

1.11 (b) The distribution is approximately symmetric with a single peak of about 55 years at the center. The range is 27 years. (c) The youngest was Teddy Roosevelt at 42; the oldest was Ronald Reagan at 69. (d) Clinton was 46 and not an outlier, so was not unusually young.

1.13 (a) $28; the center corresponds to the 50th percentile. (b) 9/50 = 0.18.

1.15 (b) The rate decreases from 1960 to 1970 and then levels off from 1970 to 2000. (c) Answers will vary but possible factors are the introduction of the contraceptive pill and legalization of abortion. (e) Total births decreased in the 1960s and 1970s, but there was a sharp rise in the 1980s followed by a small reduction in the 1990s. (f) Rate of births is not affected by a change in the population, but the total number of births is affected, assuming that the number of births per mother remains constant.

1.17 (b) The histogram is approximately symmetric with two unusually low observations at −44 and −2. Since these observations are strongly at odds with the general pattern, it is highly likely that they represent observational errors. (d) Newcomb's worst measurement errors occurred early in the observation process. As the observations progressed, they became more consistent.

1.19 Answers may vary but might include graduation rate, class sizes, faculty salaries, student-faculty ratio, quality of entering students (ACT/SAT scores, high school class rank, enrollment-to-admission ratio), and the percent of alumni who give to the school.

1.21 Unmet needs are greatest at private schools. We do not have information on all types of institutions and so a pie chart is incorrect (even if we did, it would be very difficult to detect the difference between the two private institutions and between the two public institutions in a pie chart).

1.23 (a) Alaska, 5.7%; Florida, 17.6%. (b) The distribution is roughly symmetric (perhaps slightly skewed left) and centered near 13%. Ignoring the outliers, the percents are spread from 8.5% to 15.6%.

1.25 (a) The histogram is skewed right with a center at 4 letters. The spread is from 1 to 15 letters. (b) There are more 2-, 3-, and 4-letter words in Shakespeare's plays and more very long words in *Popular Science* articles.

1.27 (a) Mean = 85. (b) 79.3; the mean is not resistant to outliers. (c) Stem-and-leaf plot.

1.29 $30 million. No.

1.31 Mean is $59,067; median is $43,318. A few very large salaries will pull the mean upward.

1.33 (b) Women's five-number summary: 101, 126, 138.5, 154, 200, $\bar{x}$ = 141.16. Men's five-number summary: 70, 98, 114.5, 143, 187, $\bar{x}$ = 121.25. (c) Women generally score higher than men. The men's scores are more spread out than the women's (even if we don't ignore the outlier). The shapes of the distributions are reasonably similar, with each displaying right skewness.

1.35 For example, take the data set 1, 2, 3, 4, 5, 6, 7, 8. In this case the median = 4.5, Q_1 = 2.5, Q_3 = 6.5, and *IQR* = 4. Changing any "extreme" value (that is, any value outside the interval between Q_1 and Q_3) will have no effect on the *IQR*. For example, if 8 is changed to 88, *IQR* will still equal 4.

1.37 (a) $Q_1 = 25$, $Q_3 = 45$. (b) $Q_3 + 1.5(IQR) = 45 + 1.5(20) = 75$. Bonds's 73 home runs is not an outlier.

1.39 (a) $\bar{x} = 5.4$. (b) $s = 0.6419$.

1.41 $\Sigma x = 11{,}200$, $\bar{x} = 1600$, $\Sigma(x-\bar{x}) = 0$, $\Sigma(x-\bar{x})^2 = 214{,}870$, $s^2 = 35{,}811.667$, $s = 189.240$.

1.43 (a) Any identical numbers between 0 and 10. (b) 0, 0, 10, 10. (c) Yes for (a) but only one for (b).

1.45 (a) $1,000, $1,000. (b) No. (c) No.

1.47 The distribution of scores on the statistics exam is roughly symmetric with a peak at 3. The distribution of AB calculus scores shows a very different pattern, with a peak at 1 and another slightly lower peak at 5. It's impossible to say which exam is easier.

1.49 (a) Use relative frequency histograms, since there are considerably more men than women. (b) Both histograms are skewed to the right, with the women's salaries generally lower than the men's. (c) Roundoff error.

1.51 (a) Most people will round their answers when asked to give an estimate like this. The values of 360 and 300 are suspicious. (b) The graphs and midpoints (men, 120; women, 175) suggest that women (claim to) study more than men.

1.53 (a) Group 1: $\bar{x} = 23.750$, $s = 5.065$. Group 2: $\bar{x} = 14.083$, $s = 4.981$. Group 3: $\bar{x} = 15.778$, $s = 5.761$ (all in counts of trees). (b) The means, along with stemplots, suggest that logging reduces the number of trees per plot and that recovery is slow. (c) Use of $\bar{x}$ and s should be acceptable, since there is only one outlier (2) in group 2 and the distributions show no extreme outliers or strong skewness (given the small sample sizes).

1.55 From 1994 to 1996, sales were similar for the two age groups. Between 1996 and 2003, sales increased for the over-35 group and decreased for the 15–34 group.

1.57 (a) Normal: 272, 337, 358, 400.5, 462. New: 318, 383.5, 406.5, 428.5, 477. Chicks fed new corn generally gain more weight than chicks fed normal corn. (b) Normal: $\bar{x} = 366.3$ g, $s = 50.805$ g. New: $\bar{x} = 402.95$ g, $s = 42.729$ g. About 36.6 g. (c) Normal: $\bar{x} = 12.9$ oz, $s = 1.792$ oz. New: $\bar{x} = 14.2$ oz, $s = 1.507$ oz.

1.59 Answers may vary but might include number of employees, value of company stock, total salaries.

1.61 (a) *H. bihai*: 46.34, 46.71, 47.12, 48.245, 50.26. *H. caribaea* red: 37.4, 38.07, 39.16, 41.69, 43.09. *H. caribaea* yellow: 34.57, 35.45, 36.11, 36.82, 38.13. The distributions have very different medians (*H. bihai* has the largest; *H. caribaea* yellow has the smallest); the distribution for *H. caribaea* yellow is skewed to the left; the other two distributions are skewed right. (b) *H. bihai*: $\bar{x} = 47.5975$ mm, $s = 1.2129$ mm. *H. caribaea* red: $\bar{x} = 39.7113$ mm, $s = 1.7988$ mm. *H. caribaea* yellow: $\bar{x} = 36.1800$ mm, $s = 0.9753$ mm. (c) *Bihai* and red appear right-skewed, although it is difficult to tell with such small samples. (d) *H. bihai*: $\bar{x} = 1.8739$ in., $s = 0.04775$ in. *H. caribaea* red: $\bar{x} = 1.5634$ in., $s = 0.07082$ in. *H. caribaea* yellow: $\bar{x} = 1.4244$ in., $s = 0.0384$ in.

1.63 (a) The stemplot is symmetric with no apparent outliers. (b) 50.7%. (c) 57.4%; 1956, 1964, 1972, 1984.

1.65 (a) Roughly symmetric; no. (b) Graphs will vary depending on class intervals chosen. (c) 8.42 min; about 9 min. (d) About the 28th percentile.

1.67 (a) −34.04%, −2.95%, 3.47%, 8.45%, 58.68%. (b) Roughly symmetric, unimodal with high and low outliers. (c) $1586.78; $659.57. (d) *IQR* = 11.401. Yes, it does appear that the software uses this criterion.

1.69 The time plot shows that women's times decreased quite rapidly from 1972 until the mid-1980s. Since that time, they have been fairly consistent at between 141 and 147 minutes.

Chapter 2

2.1 Eleanor: $z = (680 - 500)/100 = 1.8$. Gerald: $z = (27 - 18)/6 = 1.5$. Eleanor's score is higher.

2.3 (a) Judy's bone density score is about 1.5 standard deviations below the average score for all women her age. (b) $\sigma = 5.52$.

2.5 (b) $\bar{x} = 4.896$, $s = 0.976$; five-number summary: 2.7, 4.1, 4.8, 5.5, 7.1. The distribution is approximately symmetric with no outliers. It has a center at about 4.9% and a range of 4.4%. (c) 84th percentile; Illinois has one of the higher unemployment rates in the country. (d) Minnesota; $z = -0.61$. (e) $k = 1$: 35 values, 70% of all values in interval; $k = 2$: 47 values, 94% of all values in interval; $k = 3$: 50 values, 100% of all values in interval; $k = 4$: 50 values, 100% of all values in interval; $k = 5$: 50 values, 100% of all values in interval.

2.7 (a) National, 94.8th percentile; school, 68th percentile. (b) National, $z = 1.57$; school, $z = 0.62$. (c) The boys at Scott's school did well. Scott's score was relatively better compared to the national group than to his peers at school. Only 5.2% nationally scored 65 or higher, yet 32% scored 65 or higher at Scott's school. (d) Nationally, at least 89% of the scores are between 20 and 79.6, so at most 11% score a perfect 80. At Scott's school, at least 89% of the scores are between 30 and 80, so at most 11% score 29 or less.

2.9 Solutions will vary.

2.11 It has equal distances between the quartiles.

2.13 (a) The curve satisfies the two conditions of a density curve: curve is on or above horizontal axis, and the area under the curve is exactly 1. (b) 0.2. (c) 0.6. (d) 0.35. (e) The area between 0 and 0.2 is 0.35. The area between 0 and 0.4 is 0.6. Therefore, the "equal-areas point" must be between 0.2 and 0.4.

2.15 Women, $z = 2.96$; men, $z = 0.96$. 6 ft is quite tall for a woman but not at all extraordinary for a man.

2.17 Between 2004 and 2005, McCarty's salary increased by $50,000 (10%), while Damon's increased by $250,000 (3.125%). The *z*-score for McCarty decreased from −0.70 in 2004 to −0.80 in 2005, while the *z*-score for Damon increased from 0.71 in 2004 to 0.79 in 2005. Damon's salary percentile increased from the 87th in 2004 to the 93rd in 2005, while McCarty's decreased from the 20th in 2004 to the 14th in 2005.

2.19 (a) Erik is a relatively good runner compared to all those in the state but had a poor race that day by his own standards. (b) Erica is slow compared to the other swimmers at the state meet.

2.21 (b) 50%. (c) median = mean = 1, $Q_1 = 0.5$, $Q_3 = 1.5$. (d) 0.4.

2.23 Approximately 0.2 (for the tall one) and 0.5.

2.25 (a) 2.5%. (b) 64 to 74 inches. (c) 16%. (d) 84th percentile.

2.27 Answers will vary.

2.29 (a) 0.9978. (b) $1 - 0.9978 = 0.0022$. (c) $1 - 0.0485 = 0.9515$. (d) $0.9978 - 0.0485 = 0.9493$.

2.31 (a) $x > 0.4$ corresponds to $z > 0.75$. Table A gives $1 - 0.7734 = 0.2266$. (b) $0.40 < x < 0.50$ corresponds to $0.75 < z < 3.25$, so the proportion is $0.9994 - 0.7734 = 0.2260$. (c) $x > 0.4$ now corresponds to $z > -0.5$. Table A gives $1 - 0.3085 = 0.6915$. $0.40 < x < 0.50$ now corresponds to $-0.5 < z < 4.5$. Table A gives ≈ 1 for $z < 4.5$ and 0.3085 for $z < -0.5$, so the proportion is $1 - 0.3085 = 0.6915$.

2.33 (a) About 5.2%. (b) About 55%. (c) Approximately 279 days or longer.

2.35 (a) ±0.675. (b) Quartiles are 0.675 standard deviations above and below the mean; $266 \pm 0.675(16)$, or 255.2 days and 276.8 days.

2.37 The values at the low end and at the high end of the data set do not fit a Normal distribution. At the low end, the values are smaller than expected, and at the high end, the values are larger than expected from a Normal distribution.

2.39 (a) The distribution is moderately symmetric with two gaps and one high outlier. (b) Median = 15.75, mean = 15.59. Similar values, as you would expect in a symmetric distribution. (c) Yes. Mean ± 1 std. dev.: 68.2% of the data. Mean ± 2 std. dev.: 95.5% of the data. Mean ± 3 std. dev.: 100% of the data. (d) Except for one small shark and one large shark, the plot is fairly linear. It suggests that the distribution is approximately Normal. (e) Yes, all of the above observations point to approximate Normality of the data.

2.41 Answers will vary.
2.43 (a) 0.8997. (b) 0.6628. (c) 0.5625. (d) 0.2369.
2.45 (a) $x > 750$ corresponds to $z > 1.84$. Table A gives $1 - 0.9671 = 0.0329$, or approximately 3.3%. (b) $x > 750$ corresponds to $z > 2.26$. Table A gives $1 - 0.9881 = 0.0119$, or approximately 1.2%.
2.47 (a) At about ± 1.28. (b) 64.5 ± 3.2, or 61.3 inches and 67.7 inches.
2.49 The plot is nearly linear. Because heart rate is measured in whole numbers, there is a slight "step" appearance to the graph.
2.51 About 13.
2.53 A score of 27 corresponds to $z = 1.27$. From Table A, we can determine that 89.8% of scores in a standard Normal distribution would fall below this level; the actual fraction in this class was 1,052,490/1,171,460 = 0.8984. This is a good match, suggesting that the actual data could fit a Normal distribution.
2.55 Soldiers whose head circumference is outside the range 22.8 ± 1.81, approximately less than 21 inches or greater than 24.6 inches.
2.57 (a) $\bar{x}$ = \$17,776 is greater than M = \$15,532. $M - Q_1$ = \$5632 and $Q_3 - M$ = \$6968, so Q_3 is farther from the median than Q_1. A few students have very large student loans, pulling the mean and Q_3 away from the median. (b) Normal distribution Q_3: \$25,899.
2.59 (a) Approximately Normal with the exception of one outlier. (b) Approximately Normal data set. (c) Right-skewed data set with several outliers. (d) 3 mounds (at either end and in the middle) with a gap in the data toward the lower values.
2.61 Use window of `X[55,145]`$_{15}$ and `Y[-0.008, 0.028]`$_{.01}$.
(a) The command `shadeNorm(135,1E99,100,15)` produces an area of 0.009815.
(b) The command `shadeNorm(-1E99,75,100,15)` produces an area of 0.04779.
(c) `shadeNorm(70,130,100,15)` =0.9545. Alternatively, `1-2(shadeNorm(-1E99,70,100,15))` =0.9545.
2.63 invNorm (.05,22.8,1.1) = 20.99 and invNorm (.95,22.8,1.1) = 24.61.

Chapter 3

3.1 (a) Explanatory: time spent studying. Response: grade. (b) Explanatory: height. Response: weight. (c) Explanatory: inches of rain. Response: yield of corn. (d) Explore the relationship. (e) Explanatory: family's income. Response: years of education of eldest child.
3.3 Explanatory variable is water temperature and response variable is growth. Both are quantitative.
3.5 (a) Powerboat registrations. (b) Positive association. (c) Positive association. (d) Linear. (e) Strong relationship, can be used for prediction; about 48 manatees.
3.7 (a) A positive association would mean that students with higher IQs tend to have higher GPAs, and those with lower IQs generally have lower GPAs. The plot shows a positive association. (b) The relationship is positive, roughly linear, and moderately strong (except for 3 outliers). (c) The student with the lowest GPA (0.5) has an IQ of about 103.
3.9 (a) Speed is the explanatory variable. (b) The relationship is curved. High amounts of fuel were used for low and high values of speed, and low amounts of fuel were used for moderate speeds. This makes sense because the best fuel efficiency is obtained by driving at moderate speeds (60 km/h is about 37 mph). (c) Poor fuel efficiency (above-average fuel consumption) is found at both high and low speeds, and good fuel efficiency (below-average fuel consumption) is found at moderate speeds. (d) The relationship is very strong, with little scatter about the curve.
3.11 The calculator confirms the answers to Exercise 3.9.
3.13 (a) Plot shows strong, positive, linear relationship with no outliers; all five specimens could have come from the same species. (b) With x as the femur length and y as humerus length: $\bar{x} = 58.2$, $s_x = 13.2$, $\bar{y} = 66.0$, $s_y = 15.89$, $r = 0.994$. (c) Obviously, the same answer as (b).
3.15 (a) Lowest: about 107 calories (about 145 mg of sodium). Highest: about 195 calories, (about 510 mg of sodium). (b) Positive association; high-calorie hot dogs tend to be high in salt, and low-calorie hot dogs tend to have low sodium. (c) The lower-left point is an outlier. Ignoring this point, the relationship is moderately strong.
3.17 (b) $r = 0.2531$. (d) $r = 0.2531$. Although the points have been transformed, the distances between the corresponding points and the strength of the association have not changed.
3.19 $r = 1$.
3.21 (a) New York's median household income is about \$32,800, and the mean income per person is about \$27,500. (b) You would expect that the more money each person has, the more money each household would have, so you would expect the association to be positive. Household income will generally be higher because most households have one primary source of income and at least one other smaller source of income. (c) In the District of Columbia, where there are a few individuals earning a great deal of money. This causes the income distribution to become right-skewed, which would raise the mean per capita income above the median. (d) Alaska's median household income is about \$47,900. (e) Ignoring the outliers, the relationship is strong, positive, and moderately linear.
3.23 (a) Gender is a categorical variable. (b) r has a maximum value of 1. (c) Correlation has no units.
3.25 (b) There is a positive, linear relationship between the correct and guessed calories. The guessed calories for spaghetti with tomato sauce and the cream-filled snack cake do not appear to fit the overall pattern displayed for the other foods. (c) The correlation is $r = 0.825$. This agrees with the positive association observed in the plot. (d) The fact that the guesses are all higher than the true calorie count does not influence the correlation. The correlation r would not change if every guess were 100 calories higher. (e) The correlation without these two

foods is $r = 0.984$. The correlation is closer to 1 because the relationship is much stronger without these two foods.

3.27 The correlation is $r = 0.5$. The correlation is greatly lowered by the one outlier (10, 1). Outliers tend to have fairly strong effects on correlation; it is even stronger here because there are so few observations.

3.29 (a) For every extra week of age, a rat will increase its weight by an average of 40 grams. (b) The estimated weight of this rat at birth is 100 grams. (d) This equation would predict a weight of 4260 grams (about 9.4 lb) for a 2-year-old rat! Don't extrapolate!

3.31 (a) Slope = 0.0138 minute/meter. For every extra meter deeper the penguin dives, the dive lasts on average 0.0138 minutes (about 0.83 seconds) longer. (b) 5.45 minutes (5 minutes 27 seconds). (d) The intercept would mean that a dive of no depth would only last an average of 2.69 minutes; this obviously does not make any sense.

3.33 (b) Number killed = −41.43 + 0.125 (registrations). (c) About 48 manatees killed. (d) Yes, the measures seem to be succeeding, because for 3 of the last 4 years, their points are below the regression line, showing that fewer manatees were killed than expected. (e) Mean manatee deaths is 42. The prediction of 48 was too high.

3.35 (a) If y = blood alcohol content (BAC) and x = number of beers, $\hat{y} = -0.01270 + 0.017964x$. (b) Slope: for every extra beer consumed, the BAC will increase by an average of 0.017964. Intercept: if no beers are consumed, the BAC will be, on average, −0.01270 (obviously meaningless). (c) 0.0951. (d) 0.0049.

3.37 $b = rs_y/s_x = 0.8944(0.0441/2.1975) = 0{:}0179$; $a = \bar{y} - b\bar{x} = 0.07375 - (0.0179)(4.8125) = -0.0124$. $\hat{y} = -0.0124 + 0.017964x$, which is close to the equation in Exercise 3.35.

3.39 (a) There is a positive, linear association between the two variables. The larger the lab measurement, the more deviation from the line. (b) The points for the larger defects fall systematically below the line $y = x$, showing that the field measurements are too small compared to the lab results. (c) The slope would decrease and the intercept would increase. (d) The residual plot clearly shows the increased variability of field measurements with high values of lab measurements. The LSRL does not provide a great fit, especially for larger depths.

3.41 (b) METF = 201.162 + 24.026(MASSF). (c) Slope: for every extra kilogram of lean body mass, a female will increase her metabolic rate by an average of 24.026 calories. Intercept: when the lean body mass is zero, the metabolic rate in a woman is, on average, 201 calories (clearly meaningless). (d) The residual plot with MASSF on the horizontal axis shows no clear pattern, so the least-squares line is an adequate model for the data. (e) The residual plot with the predicted value on the horizontal axis looks exactly like the previous plot of the residuals versus lean body mass.

3.43 (b) The regression equation is $\hat{y} = -14.4 + 46.6x$. (c) The residuals for the extreme x-values ($x = 0.25$ and $x = 20.0$) are almost all positive; all those for the middle two x-values are negative. (d) $r^2 = 0.9997$; 99.97% of the variation in the response is explained by the least-squares regression with the amount of substance.

3.45 (a) $\hat{y} = 479.93 - 9.69x$. (b) Prediction is 147.4 beats per minute, 4.6 beats per minute less than the actual value. (c) With y = time and x = pulse, $\hat{y} = 43.10 - 0.0574x$. The prediction for time is 34.37 minutes, only 0.07 minutes (4.2 seconds) too high. (d) The results depend on which variable is viewed as the explanatory variable.

3.47 (a) $r^2 = 35.5\%$. (b) $\hat{y} = 6.083 + 1.707x$. (c) $\hat{y} = 9.07$. Regression line must pass through $(\bar{x}, \bar{y}) = (1.75, 9.07)$.

3.49 (a) On average, BOD rises by 1.507 mg/l for every 1 mg/l increase in TOC. (b) −55.43 mg/l; extrapolation.

3.51 (b) No, the pattern is not linear. (c) The sum is 0.01. The first two and the last four residuals are negative and those in the middle are positive, showing a curve in the residual plot.

3.53 (b) If x = age and y = height, $\hat{y} = 71.95 + 0.383x$. (c) When x = 40 months, $\hat{y}$ = 87.27 cm. When x = 60 months, $\hat{y}$ = 94.93 cm. (d) A change of 6 cm in 12 months is 0.5 cm/month. Sarah is growing at about 0.38 cm/month, more slowly than normal.

3.55 (a) Slope = 0.16; for every extra point earned on the midterm, the score on the final exam increases by an average of 0.16. Intercept = 30.2; if the score on the midterm was 0, the predicted score on the final would be 30.2. (b) $\hat{y} = 78.2$. (c) Only $r^2 = 36\%$ of the variability in the final score is accounted for by the regression, so Julie has good reason to think this is not a good estimate.

3.57 (a) U.S.: −26.4%, 5.1%, 18.2%, 30.4%, 37.6%. Overseas: −23.2%, 2.1%, 11.2%, 29.6%, 69.4%. (b) The quartiles and the median of the U.S. five-number summary are higher, but the maximum overseas return is higher. (c) Overseas stocks are more volatile. The boxplot is more widely spread.

3.59 (b) Let y = "guessed calories" and x = "actual calories". Using all points: $\hat{y} = 58.59 + 1.3036x$ (and $r^2 = 0.68$). Excluding spaghetti and snack cake: $\hat{y} = 43.88 + 1.14721x$ (and $r^2 = 0.97$). (c) The two removed points could be called influential because, when they are included, the regression line passes above every other point; after removing them, the new regression line passes through the "middle" of the remaining points.

3.61 (a) Point A is a horizontal outlier; that is, it has a much smaller x-value than the others. Point B is a vertical outlier; it has a higher y-value than the others. (b) The three regression formulas are $\hat{y} = 31.9 - 0.304x$ (the original data), $\hat{y} = 22.8 - 0.156x$ (with Point A), $\hat{y} = 32.3 - 0.293x$ (with Point B). Adding Point B has little impact. Point A is influential; it pulls the line down and changes the line's position relative to the original 13 data points.

3.63 Higher income can cause better health: higher income means more money to pay for medical care, drugs, and better nutrition, which in turn results in better health. Better health can cause higher income: if workers enjoy better health, they are better able to get and hold a job, which can increase their income.

3.65 (a) If y = population and x = year, $\hat{y} = 1166.93 - 0.5868x$. (b) The farm population decreased by about 0.59 million (590,000) people per year. The regression explains 97.7% of the variation in the population. (c) The predicted population for the year 2010 is −12,538,000; clearly impossible, as population must be greater than or equal to zero.

3.67 The correlation would be lower; individual stock performances will be more variable, weakening the relationship.

3.69 (a) Yes. The relationship is not very strong. (b) The mortality rate is extremely variable for those hospitals that treat few heart attacks. As the number of patients treated increases, the variability decreases and the mortality rate appears to decrease, giving the appearance of an exponentially decreasing pattern of points in the plot. The nonlinearity strengthens the conclusion that heart attack patients should avoid hospitals that treat few heart attacks.

3.71 Age is a lurking variable. We would expect both quantities, shoe size and reading level, to increase as the child ages.

3.73 (a) The scatterplot suggests a negative linear association, with correlation $r = -0.9179$. The regression line is $\hat{y} = 0.212 - 0.00561x$ (x = snow cover, y = wind stress); $r^2 = 0.843$; the linear relationship explains 84.3% of the variation in wind stress. We have good evidence that decreasing snow cover is strongly associated with increasing wind stress. (b) The graph shows 3 clusters of 7 points.

3.75 (b) The right-hand points lie below the left-hand points. (This means that the right-hand times are shorter, so the subject is right-handed.) (c) Right hand: $\hat{y} = 99.4 + 0.0283x$. Left hand: $\hat{y} = 172 + 0.262x$. The left-hand regression ($r^2 = 0.101$) is slightly better than the right-hand ($r^2 = 0.093$), but neither is very good.

3.77 The seriousness of the fire is a lurking variable: more serious fires need more attention. It would be more accurate to say that a large fire "causes" more firefighters to be sent, rather than vice versa.

3.79 (a) There is a negative association between alcohol consumption and heart disease. (b) If x = alcohol consumption (in l/yr) and y = heart disease death rate (per 100,000) (HDDR), $\hat{y} = 261 - 23.0x$. Slope: for every extra liter of alcohol consumed, the HDDR decreases on average by 23 per 100,000. Intercept: if no wine is consumed, the average HDDR is 261 per 100,000. (c) About 71% of the variation in death rate can be explained by the linear relationship. (d) 169. (e) No.

3.81 (a) Speed should go on the horizontal axis. (b) There is a very strong positive linear relationship; $r = 0.999$. (c) $\hat{y} = 1.76608 + 0.080284x$. (d) $r^2 = 0.998$, so 99.8% of the variation in steps per second is explained by the linear relationship. (e) No, the regression line would change because the role of x and y would be reversed. r^2 would stay the same.

3.83 (a) One possible measure is mean response: 106.2 spikes/second for pure tones, 176.6 spikes/second for monkey calls. (b) If x = pure tone response and y = monkey call response, $\hat{y} = 93.9 + 0.778x$. The third point has the largest residual. The first point is an outlier in the x direction. (c) The correlation drops only slightly (from 0.6386 to 0.6101) when the third point is removed; it drops more drastically (to 0.4793) when the first point is removed. (d) Without the first point, the line is $\hat{y} = 101 + 0.693x$; without the third point, it is $\hat{y} = 98.4 + 0.679x$.

3.85 Slope = 0.54; $\hat{y} = 33.67 + 0.54x$. Predict husband's height to be 69.85 inches.

Chapter 4

4.1 (a) Yes. (b) Let x = length and $y = \sqrt[3]{\text{weight}}$. The least-squares regression line is $\hat{y} = -0.0220 + 0.2466x$. The intercept of −0.0220 clearly has no practical interpretation in this situation, since weight and the cube root of weight must be positive. The slope of 0.2466 indicates that for every 1 cm increase in length, the cube root of weight will increase, on average, by 0.2466. (c) 694.5 g; compare with 689.9 g from Example 4.2. (d) The residuals are negative for lengths below 17 cm, positive for lengths between 18 cm and 27 cm, and have no clear pattern for lengths above 28 cm. (e) Nearly all (99.88%) of the variation in the cube root of the weight can be explained by the linear relationship with the length.

4.3 (a) The relationship is strong, negative, and curved, with no outliers. (b) Yes. (c) $\hat{P} = 0.3677 + 15.8994(1/V)$. $r^2 = 0.9958$ indicates almost a perfect fit. The residual plot shows a definite pattern, which should be of some concern, but the model still provides a good fit. (d) Let $y = 1/P$; the least-squares regression line is $\hat{y} = 0.1002 + 0.0398V$. $r^2 = 0.9997$, and the residual plot shows a random scatter. This transformation achieves linearity because $V = k/P$. (e) Model in part (c): 1.4277 atmospheres; model in part (d): 1.4343 atmospheres. They are the same to the nearest one-hundredth of an atmosphere.

4.5 (a) The relationship is strong, negative, and curved. (b) The ratios are all 0.717, so an exponential model is appropriate. (c) It achieves linearity. (d) If x = Depth and y = ln(Light Intensity), then $\hat{y} = 6.7891 - 0.3330x$. The intercept, 6.7891, provides an estimate for the average value of the natural log of the light intensity at the surface of the lake. The slope, −0.3330, indicates that the natural log of the light intensity decreases on average by 0.3330 for each one meter increase in depth. (e) The residual plot shows a fairly random scatter and relatively small residuals, so the linear model is appropriate. (f) If x = depth and y = light intensity, $\hat{y} = (e^{6.789})(e^{-0.333x})$, or $\hat{y} = 888(0.717)^x$. It is a satisfactory model. (g) At 22 m, the predicted light intensity would be 0.584 lumens. No, not surprised.

4.7 (a) If y = number of transistors and x = number of years since 1970, $y(1) = ab^1 = 2250$, and $y(4) = ab^4 = 9000$, so $a \approx 1417.41$ and $b \approx 1.5874$. $\hat{y} = (1417.41)(1.5874^x)$.

(b) $\ln \hat{y} = 7.257 + 0.462x$. (c) The slope for Moore's model (0.4621) is larger than the estimated slope in Example 4.6 (0.332), so the actual transistor counts have grown more slowly than Moore's law suggests.

4.9 (b) The transformed data appear to be linear from 1790 to about 1880, and then linear again but with a smaller slope. This linear trend indicates that an exponential model is still appropriate and the smaller slope reflects a slower growth rate. (c) The least-squares regression line for predicting y = Log(population) from x = time since 1790 is $\hat{y} = 1.329 + 0.0054x$. Transforming back to the original variables yields the exponential equation $\hat{y} = 21.348(1.0124)^x$. (d) The residual plot shows random scatter and $r^2 = 0.995$, so the exponential model provides an excellent fit. (e) Predicted population in 2010 is 323,531,803. The prediction is probably too low, because these estimates probably do not include homeless people and illegal immigrants.

4.11 (a) If x = body weight and y = life span, $\log \hat{y} = 0.7617 + 0.2182\log(x)$. (b) Residual plot shows random scatter, $r^2 = 0.7117$; 71.17% of the variation in the log of the lifespans is explained by the linear relationship with the log of the body weight. (c) $\hat{y} = 10^{0.7617}x^{0.2182} = 5.7770x^{0.2182}$. (d) Predicted average life span for a human weighing 65 kg would be 14.36 years! (e) Try plotting a graph of (weight$^{0.2}$, lifespan) or (weight, lifespan^5) to achieve linearity. The prediction depends on the model you use.

4.13 (a) As height increases, weight increases. Since weight is a 3-dimensional characteristic and height is 1-dimensional, weight should be proportional to the cube of the height. A model of the form Weight = a(Height)^b would be a good place to start. (b) Explanatory: height. Response: weight. (c) If x = height and y = weight, $\log \hat{y} = -1.391 + 2.003 \log x$. $r^2 = 0.9999$; almost all (99.99%) of the variation in log of weight is explained by the linear relationship with log of height. (d) The residual plot for the transformed data shows that the residuals are very close to zero with no discernible pattern. This model clearly fits the transformed data very well. (e) The inverse transformation gives the estimated power model $\hat{y} = 10^{-1.391}x^{2.003} = 0.0406x^{2.003}$. For a 5′10″ adult, the model predicts 201.4 lb. For a 7 foot adult, the model predicts 290.2 lbs.

4.15 (a) Relationship is curved, strong, and positive. (b) If x = time and y = distance, $\hat{y} = 0.99 + 490.416x^2$. (c) $r^2 = 0.9984$ and the residual plot shows random scatter and fairly small-sized residuals, so this looks like an appropriate model. (d) Yes. $\sqrt{\hat{y}} = 0.1046 + 22.0428x$. (e) $r^2 = 0.9986$ and the residual plot shows no pattern, which suggest a good model. (f) Using model from (b): 109.32 cm. Using model from (d): 109.51 cm. Answers will vary.

4.17 *Data*: Answer key questions. *Graphs*: A scatterplot of abundance y versus body mass x shows a curved relationship. Using the log transformation for both variables produces a moderately strong, negative, linear relationship. *Numerical summaries:* The correlation between log body mass and log abundance is -0.912. *Model:* The least-squares regression line for the transformed data is $\log \hat{y} = 1.9503 - 1.0481 \log x$, with an $r^2 = 0.8325$ and residual plot showing no obvious patterns. *Interpretation:* The inverse transformation gives the estimated power model $\hat{y} = 10^{1.9503}x^{-1.0481} = 89.1867x^{-1.0481}$, which provides a good fit for these data.

4.19 (b) The first phase is from 0 to 6 hours, when the mean colony size actually decreases. In the second phase, from 6 to 24 hours, the mean colony size increases exponentially. At 36 hours, mean growth is in the third phase, when growth is still increasing but at a slower rate than the previous phase. (c) If y = mean colony size and x = time, $\log \hat{y} = -0.5942 + 0.0851x$. When $x = 10$, $\log \hat{y} = 0.2568$, so $\hat{y} = 1.806$.

4.21 (a) Strength = c(weight)$^{2/3}$. (b) A graph of $y = x^{2/3}$ shows that strength does not increase linearly with body weight, as would be the case if a person 1 million times as heavy as an ant could lift 1 million times more than the ant. Rather, strength increases more slowly. For example, if weight is multiplied by 1000, strength will increase by a factor of $(1000)^{2/3} = 100$.

4.23 (a) 858 people. (b) 43 had arthritis. (c) 71 (8.3%) played elite soccer, 215 (25.1%) played non-elite soccer, and 572 (66.7%) did not play. (d) 14.1% of the elite soccer players, 4.2% of the non-elite soccer players, and 4.2% of those who did not play got arthritis. This suggests an association between playing elite soccer and developing arthritis later.

4.25 (a) 5375 students. (b) 18.7%. (c) Both parents smoke: 1780, 33.1%. One parent smokes: 2239, 41.7%. Neither parent smokes: 1356, 25.2%. (d) Student smokes, given both parents smoke: = 400/(400 + 1380) = 0.2247. Student doesn't smoke, given both parents smoke: = 1380/(400 + 1380) = 0.7753. Student smokes, given that one parent smokes: = 416/(416 + 1823) = 0.1858. Student doesn't smoke, given one parent smokes: = 1823/(416 + 1823) = 0.8142. Student smokes, given that neither parent smokes: = 188/(188 + 1168) = 0.1386. Student doesn't smoke, given that neither parent smokes: = 1168/(188 + 1168) = 0.8614. Students who smoke are more likely to come from families where one or more of their parents smoke.

4.27 (a) Female: accting., 30.2%; admin., 40.4%; econ., 2.2%; fin., 27.1%. Male: accting., 34.8%; admin., 24.8%; econ., 3.7%; fin., 36.6%. Biggest difference between men and women is in administration: a higher percent of women chose this major. A greater proportion of men chose other fields, especially finance. (b) 386 out of 722 responded. 46.5% did not respond.

4.29 (a) White defendant: 19 yes, 141 no. Black defendant: 17 yes, 149 no. (b) Overall death penalty: 11.9% of white defendants, 10.2% of black defendants. For white victims, 12.6% and 17.5%; for black victims, 0% and 5.8%. (c) The death penalty is more likely when the victim was white (14%) rather than black (5.4%). Because most convicted killers are of the same race as their victims, whites are more often sentenced to death.

Table A.

Combined —All People	Early Death Yes	No
Overweight	41	59
Not overweight	50	50

Table B.

Smokers	Early Death Yes	No
Overweight	10	0
Not overweight	40	20

Table C.

Nonsmokers	Early Death Yes	No
Overweight	31	59
Not overweight	10	30

4.31 Marital status: 4.1%, 93.9%, 1.5%, 0.5%. Job grade: 11.6%, 51.5%, 30.2%, 6.7%.

4.33 17.2%, 65.9%, 14.8%, 2.1%. They should add to 100%.

4.35 Married men would generally be older than single men, so they would have been in the work force longer, and therefore had more time to advance in their careers.

4.37 (a) 36.8%, 24.1%, 39.1%. About 39% of users think the recycled product is of lower quality. This would be a serious barrier to sales. (b) Buyers: higher (55.6%), same (19.4%), lower (25%). Nonbuyers: higher (29.9%), same (25.8%), lower (44.3%). We can't draw any conclusion about causation.

4.39 Examples will vary; one very simple possibility is shown above. The key is to be sure that there is a lower percent of overweight people among the smokers than among the nonsmokers.

4.41 No, rich nations have more TV sets than poor nations. Rich nations also have longer life expectancies because they have better nutrition, clean water, and better health care. There is a common response relationship between TV sets and length of life.

4.43 No. The "number of hours standing up at work" is a confounding variable.

4.45 It could be that children with lower intelligence watch many hours of television. It could be that children from lower socioeconomic level households, where parent(s) are less likely to limit television viewing and are unable to help their children with their schoolwork because the parents themselves lack education, watch more TV. The variables "number of hours watching television" and "grade point average" change in common response to "socioeconomic level."

4.47 The variable "knowledge gained as a result of taking the SAT previously" is a confounding variable.

4.49 Spending more time watching TV means that less time is spent on other activities. Answers will vary, but some possible lurking variables are: the amount of time parents spend at home, the amount of exercise, and the economy. For example, parents of heavy TV watchers may not spend as much time at home as other parents.

4.51 (a) Yes, it achieves linearity. (b) Let x = distance and y = intensity. The least-squares regression line for the transformed data is $\hat{y} = -0.0006 + 0.30(1/x^2)$. (c) $0.3/2.1^2 - 0.0005 = 0.0673$ candelas. (d) The predictions are very similar. Yes.

4.53 (a) 110,113; roundoff error. (b) Single: 23,331 (21.2%); Married: 62,834 (57.1%); Widowed: 11,288 (10.3%); Divorced: 12,660 (11.5%). (c) 15–24: Single: 80.6%; Married: 18.2%; Widowed: 0.1%; Divorced: 1.1%. 40–64: Single: 8.6%; Married 69.4%; Widowed: 5.0%; Divorced: 17.0%. Among the younger women, more than 4 out of 5 have not yet married, and those who are married have had little time to become widowed or divorced. Most of the older group is or has been married—only about 8.64% are still single. (d) Among single women, 46.9% are 15–24, 32.8% are 25–39, 17.2% are 40–64 and 3.1% are 65 or older.

4.55 The lurking variable is temperature or season. More flu cases occur in winter when less ice cream is sold, and less flu cases occur in the summer when more ice cream is sold. This is an example of common response.

4.57 *Data:* Answer key questions. *Graphs:* A scatterplot of the transformed data using the logarithm of count shows a strong, negative, linear relationship. *Numerical summaries:* $r = -0.994$ for the transformed data. *Model:* If x = time and y = log(count), $\log \hat{y} = 2.5941 - 0.0949x$. *Interpretation:* The residual plot shows no clear pattern and $r^2 = 98.8\%$, so the exponential decay model provides an excellent model for the number of surviving bacteria.

Chapter 5

5.1 The population is all local businesses. The sample is the 73 businesses that return the questionnaire (or, under certain circumstances, could be the 150 businesses selected). The nonresponse rate is 51.3%.

5.3 This is an experiment: a treatment is imposed. The explanatory variable is the teaching method (computer assisted or standard), and the response variable is the increase in reading ability based on the pre- and posttests.

5.5 Observational. The researcher did not attempt to change the amount that people drank. The explanatory variable is alcohol consumption. The response variable is survival after 4 years.

5.7 Only persons with a strong opinion on the subject (strong enough that they are willing to spend the time and at least 50 cents) will respond to this advertisement.

5.9 Labeling from 001 to 440, select 400, 077, 172, 417, 350, 131, 211, 273, 208, and 074.

5.11 Assign 01 to 30 to the students (in alphabetical order). The exact selection will depend on the starting line chosen in Table B; starting on line 123 gives 08-Ghosh, 15-Jones, 07-Fisher, and 27-Shaw. Assigning 0 to 9 to the faculty members gives (from line 109) 3-Gupta and 6-Moore. (We could also number faculty from 01 to 10, but this requires looking up two-digit numbers.)
5.13 (a) This is a stratified random sample. (b) Label each area code from 01 through 25; beginning at line 111, we choose 12 (559), 04 (209), 11 (805), 19 (562), 02 (707), 06 (925), 08 (650), 25 (619), 17 (626), and 14 (661).
5.15 (a) Households without telephones or with unlisted numbers. Such households would likely be made up of poor individuals (who cannot afford a phone), those who choose not to have phones, and those who do not wish to have their phone number published. (b) Those with unlisted numbers would be included in the sampling frame when a random digit dialer is used.
5.17 The first wording would pull respondents toward a tax cut because the second wording mentions several popular alternative uses for tax money.
5.19 (a) There were 14,484 responses. (Note that we have no guarantee that these came from 14,484 distinct people; some may have voted more than once.) (b) This voluntary response sample collects only the opinions of those who visit this site and feel strongly enough to respond.
5.21 (a) The adults in the country. (b) Members of the local business and professional women's clubs. (c) All households in the United States.
5.23 Number the bottles across the rows from 01 to 25, then select 12–B0986, 04–A1101, and 11–A2220. (If numbering is done down columns instead, the sample will be A1117, B1102, and A1098.)
5.25 One could use the labels already assigned to the blocks, but that would mean skipping a lot of four-digit combinations that do not correspond to any block. An alternative would be to drop the second digit and use labels 100, 101, 102, . . . , 105; 200, . . . , 211; 300, . . . , 325. But by far the simplest approach is to assign labels 01, . . . , 44 (in numerical order by the four-digit numbers already assigned), enter the table at line 125, and select 21 (3002), 37 (3018), 18 (2011), 44 (3025), and 23 (3004).
5.27 It is not an SRS, because some samples of size 250 have no chance of being selected (e.g., a sample containing 250 women).
5.29 A sample from a smaller subgroup gives less information about the population. "Men" constituted only about one-third of our sample, so we know less about that group than we know about all adults.
5.31 Answers will vary. One possible approach: Obtain a list of schools, stratified by size or location (rural, suburban, urban). Choose SRSs (not necessarily all the same size) of schools from each stratum. Then choose SRSs (again, not necessarily the same size) of students from the selected schools.
5.33 Experimental units: pine seedlings. Factor: light intensity. Treatments: full light, 25% light, and 5% light. Response variable: dry weight at the end of the study.
5.35 The units are the individuals who were called. One factor is what information is offered. Treatments are (1) giving name, (2) identifying university, (3) both of these. Second factor is offering to send a copy of the results. The treatments are (1) offering or (2) not offering. The response is whether the interview was completed.
5.37 (a) Chest pain. (b) Placebo effect. (c) Control.
5.39 (a) In a serious case, when the patient has little chance of surviving, a doctor might choose not to recommend surgery; it might be seen as an unnecessary measure, bringing expense and a hospital stay with little benefit to the patient.
(b)

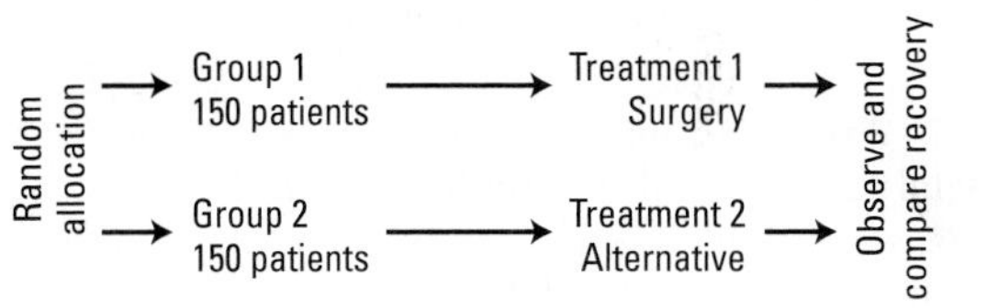

5.41 (a) Diagram below. (b) Assigning the students numbers from 001 to 120 and using line 123 from Table B, the first four subjects are 102, 063, 035, and 090.

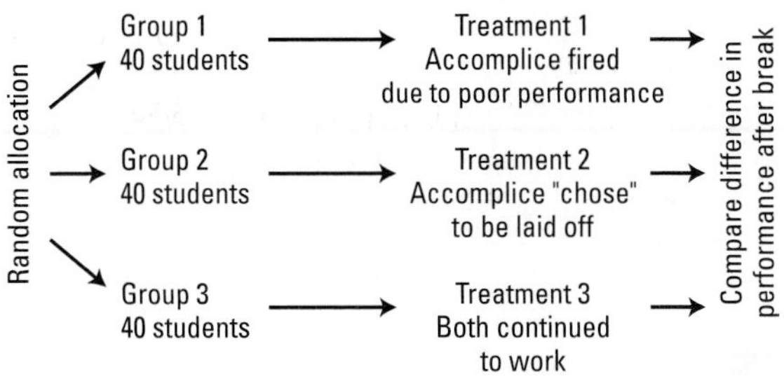

5.43 The second design is an experiment—a treatment is imposed on the subjects. The first is an observational study; the explanatory variable (exercise) may be confounded by the types of men in each group. In spite of the researcher's attempt to match "similar" men from each group, those in the exercise group could be somehow different from men in the nonexercise group.
5.45 Because the experimenter knew which subjects had learned the meditation techniques, he or she may have had some expectations about the outcome of the experiment: if the experimenter believed that meditation was beneficial, he or she may have subconsciously rated that group as being less anxious.

5.47 (a) Ordered by increasing weight, the five blocks are (1) Williams-22, Deng-24, Hernandez-25, and Moses-25; (2) Santiago-27, Kendall-28, Mann-28, and Smith-29; (3) Brunk-30, Obrach-30, Rodriguez-30, and Loren-32; (4) Jackson-33, Stall-33, Brown-34, and Cruz-34; (5) Birnbaum-35, Tran-35, Nevesky-39, and Wilansky-42. (b) The exact randomization will vary with the starting line in Table B. Different methods are possible; perhaps the simplest is to number from 1 to 4 within each block, then assign the members of Block 1 to a weight-loss treatment, then assign Block 2, etc. For example, starting on line 133, we assign 4-Moses to Treatment A, 1-Williams to B, and 3-Hernandez to C (so that 2-Deng gets Treatment D), then carry on for Block 2, etc. (either continuing on the same line or starting over somewhere else).

5.49 (a) Assign 10 subjects to Group 1 (the 70° group) and the other 10 to Group 2 (the 90° group). Record the number of correct insertions in each group. (b) All subjects will perform the task twice; once in each temperature condition. Randomly choose which temperature each subject works in first, either by flipping a coin or by placing 10 subjects in Group 1 (70°, then 90°) and the other 10 in Group 2.

5.51 (a) "Randomized" means that patients were randomly assigned either Saint-John's-wort or the placebo. "Placebo controlled" means that we will compare the results for the group using Saint-John's-wort with the results for the group that received the placebo. "Double-blind" means that neither the subjects nor the researchers interacting with them (including those who measured depression levels) know who is receiving which treatment.
(b)

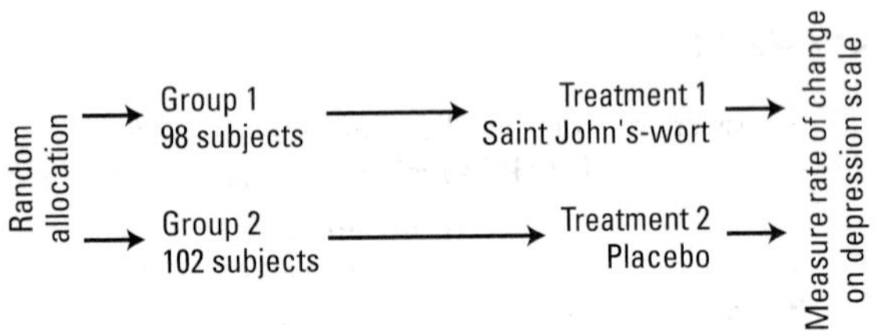

5.53 (a) Experimental subjects: patients. Factor: temperature. Response variable: presense of infection.
(b)

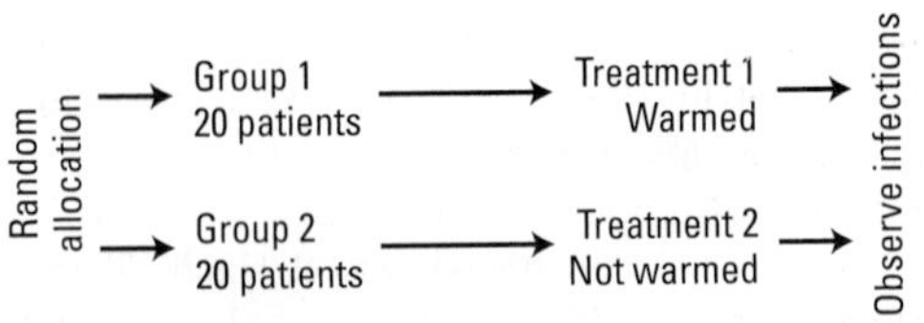

(c) Assign each subject a number, 01 to 40, by alphabetical order. Starting at line 121 in Table B, the first 20 different subjects are 29-Ng, 07-Cordoba, 34-Sugiwara, 22-Kaplan, 10-Devlin, 25-Lucero, 13-Garcia, 38-Ullmann, 15-Green, 05-Cansico, 09-Decker, 08-Curzakis, 27-McNeill, 23-Kim, 30-Quinones, 28-Morse, 18-Howard, 03-Afifi, 01-Abbott, 36-Travers. These subjects will be assigned to Treatment Group 1; the remaining subjects go into Group 2. (d) We want the treatment groups to be as alike as possible. If the same operating team was used to operate on "warmed" and "unwarmed" patients, then the effect of the "warming" on the occurrence of infection might be confounded with the effect of the surgical team (e.g., how skillful the team was in performing the necessary preventive measures). (e) Double-blindness. We would prefer a double-blind experiment here to ensure that the patients would not be treated differently with regard to preventing and monitoring infections due to prior knowledge of how they were treated during surgery.

5.55 (a) Randomly assign 20 men to each of two groups. Record each subject's blood pressure, then apply the treatments: a calcium supplement for Group 1 and a placebo for Group 2. After sufficient time has passed, measure blood pressure again and observe any change. (b) Number from 01 to 40 down the columns. Group 1 is 18-Howard, 20-Imrani, 26-Maldonado, 35-Tompkins, 39-Willis, 16-Guillen, 04-Bikalis, 21-James, 19-Hruska, 37-Tullock, 29-O'Brian, 07-Cranston, 34-Solomon, 22-Kaplan, 10-Durr, 25-Liang, 13-Fratianna, 38-Underwood, 15-Green, and 05-Chen. (c) Block on race (make two groups, one of black men and one of white men) and then apply the design in (a) to each block. Observe the change in blood pressure for each block.

5.57 Responding to a placebo does not imply that the complaint was not "real"—38% of the placebo group in the gastric freezing experiment improved, and those patients really had ulcers. The placebo effect is a psychological response, but it may make an actual physical improvement in the patient's health.

5.59 Three possible treatments are (1) fine, (2) jail time, and (3) attending counseling classes. The response variable would be the rate at which people in the three groups are rearrested.

5.61 (a) Explanatory variable: type of cookie (Toll House or dark chocolate). Response variable: cookie preference. (b) The population is all cookies manufactured of each type. We can assume that the cookies used were not systematically different from the population and so, hopefully, a representative sample was used, but it was not an SRS, because not all sets of cookies in the population had an equal chance of being used.
(c)

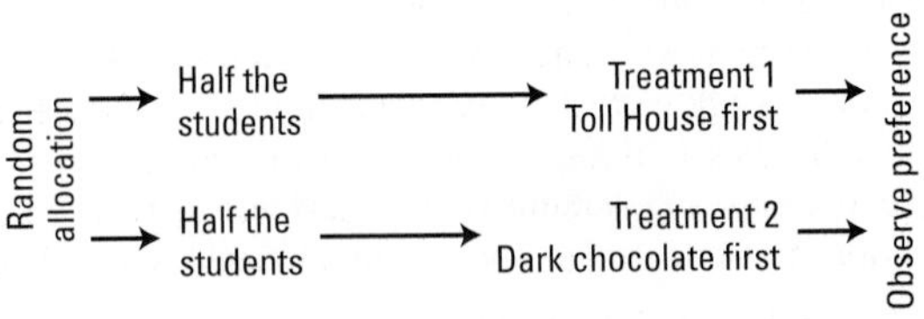

(d) Variability was controlled by the students being randomly assigned to the two groups, the cookies being tasted in different orders, the cookies being as similar as possible and given to the student in the same way. (e) Half the class was assigned to each method of tasting the cookies, which provides sufficient student-to-student variability in both groups to show that any difference is due to the cookies and not the variability. (f) The experiment was blind (the students didn't know which cookie they were eating—it has nothing to do with their blindfolds!) but not double-blind (the person handing them the cookie knew—which cookie the student was being given). (g) Answers will vary. Could they tell the difference at all? You could add two more treatments: Toll House followed by another Toll House and a dark chocolate followed by another dark chocolate. They may strongly prefer one cookie when, in fact, they are both the same! You might want to block by gender, or if you wanted to reach a conclusion about the whole school, you could take a stratified sample and block by grade.

5.63 (a) A matched pairs design (two halves of the same board would have similar properties). (b) A sample survey (with a stratified sample: smokers and nonsmokers). (c) A block design (blocked by gender).

5.65 Each player will complete the experiment twice, once with oxygen during the rest period, and once without oxygen. The order for each player will be determined by the flip of a coin.

5.67 (a)

		Factor B: Administration method		
		Injection	Skin patch	IV drip
Factor A: Dosage	5 mg	Treatment 1	Treatment 2	Treatment 3
	10 mg	Treatment 4	Treatment 5	Treatment 6

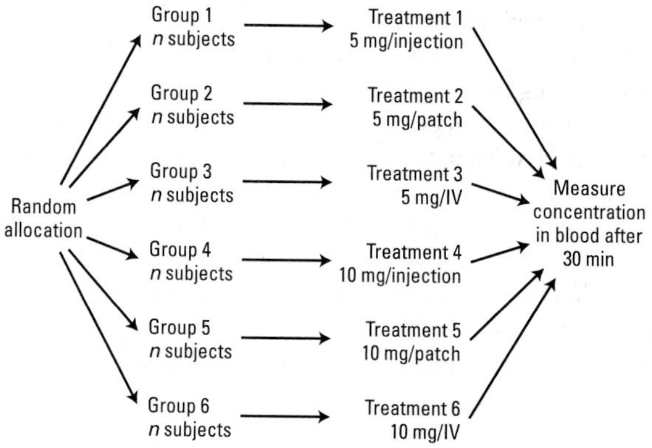

(b) Large samples give more information; in particular, with large samples we reduce variability in the observed mean concentrations so that we can have more confidence that the differences we might observe are due to the treatment applied rather than random fluctuation. (c) Use a block design. Separate men and women and randomly allocate the 6 treatments to each gender.

5.69 As described, there are two factors: zip code (three levels: none, 5-digit, 9-digit) and the day on which the letter is mailed (three levels: Monday, Thursday, or Saturday) for a total of 9 treatments. To control confounding variables, aside from mailing all letters to the same address, all letters should be the same size and either printed in the same handwriting or typed. The design should also specify how many letters will be in each treatment group. Also, the letters should be sent randomly over many weeks.

5.71 (a)

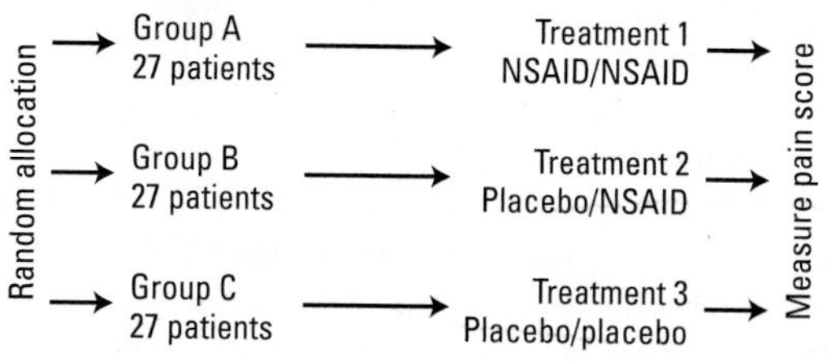

The two extra patients can be randomly assigned to two of the three groups
(b) No one involved in administering treatments or assessing their effectiveness or the patients themselves knew which subjects were in which group. (c) If the pain score for Group A were significantly lower than that for Group C, it suggests that the NSAID was successful in reducing pain. However, because the pain score for Group A was not significantly lower than that for Group B, this suggests that it was the application of the NSAID *before* the surgery that made the drug effective.

Chapter 6

6.1 Answers will vary but here are some possibilities: (a) Flip the coin twice. Let HH ↔ failure, and let the other three outcomes, HT, TH, TT ↔ success. (b) Let 1, 2, 3 ↔ success, and let 4 ↔ failure. If 5 or 6 comes up, ignore it and roll again. (c) Peel off two consecutive digits from the table; let 00 through 74 ↔ success, and let 75 through 99 ↔ failure. (d) Let diamonds, spades, clubs ↔ success, and let hearts ↔ failure.

6.3 (a) Obtain an alphabetical list of the student body, and assign consecutive numbers to the students on the list. Use a random process (table or random digit generator) to select 10 students from this list. (b) Let the two-digit groups 00 to 83 represent a "Yes" to the question of whether or not to abolish evening exams and the groups 84 to 99 represent a "No." (c) Starting at line 129 in Table B ("Yes" in boldface) and moving across rows:
Repetition 1: **36, 75**, 95, 89, 84, **68, 28, 82, 29, 13** (7 "Yes")
Repetition 2: **18, 63**, 85, **43, 03, 00, 79, 50**, 87, **27** (8 "Yes")
Repetition 3: **69, 05, 16, 48, 17**, 87, **17, 40**, 95, **17** (8 "Yes")
Repetition 4: 84, **53, 40, 64**, 89, 87, **20, 19, 72, 45** (7 "Yes")
Repetition 5: **05, 00, 71, 66, 32, 81, 19, 41, 48, 73** (10 "Yes")
(Theoretically, we should achieve 10 "Yes" results approximately 17.5% of the time.)

6.5 The choice of digits in these simulations may of course vary from that made here. In (a) to (c), a single digit simulates the response; for (d), two digits simulate the response of a single voter. **(a)** Odd digits ↔ voter would vote Democratic; even digits ↔ voter would vote Republican. **(b)** 0, 1, 2, 3, 4, 5 ↔ Democratic; 6, 7, 8, 9 ↔ Republican. **(c)** 0, 1, 2, 3 ↔ Democratic; 4, 5, 6, 7 ↔ Republican; 8, 9 ↔ Undecided. **(d)** 00, 01, . . . , 52 ↔ Democratic; 53, 54, . . . , 99 ↔ Republican

6.7 Let 1 ↔ girl and 0 ↔ boy. The command `randInt(0,1,4)` produces a 0 or 1 with equal likelihood in groups of 4. Continue to press ENTER. In 50 repetitions, we got at least one girl 47 times, and no girls 3 times. Our simulation produced a girl 94% of the time, versus a probability of 0.938 obtained in Example 6.6.

6.9 Let 00 to 14 ↔ breaking a racquet, and let 15 to 99 ↔ not breaking a racquet. Starting with line 141 in the random digit table, we peel two digits off at a time and record the results: 96 76 73 59 64 23 82 29 60 12. In the first repetition, Brian played 10 matches until he broke a racquet. Additional repetitions produced these results: 3 matches, 11 matches, 6 matches, 37 matches, 5 matches, 3 matches, 4 matches, 11 matches, and 1 match. The average for these 10 repetitions is 9.1. We will learn later that the expected number of matches until a break is about 6.67. More repetitions should improve our estimate.

6.11 Let integers 1 to 25 ↔ passenger who fails to appear and 26 to 100 ↔ passenger who shows up. The command `randInt(1,100,9)` represents one van. Continue to press ENTER. In 50 repetitions, our simulation produced 12 buses with 8 people and 3 buses with 9 people, so 15 buses had more than 7 people, suggesting a probability of 0.3 that the bus will be overbooked. This compares with a theoretical probability of 0.3003.

6.13 **(a)** Read two random digits at a time from Table B. Let 01 to 13 represent a heart, let 14 to 52 represent another suit, and ignore the other two-digit numbers. **(b)** You should beat Slim about 44% of the time; no, it is not a fair game.

6.15 The command `randInt(1,365,23)→L1: SortA (L1)` randomly selects 23 (numbers) birthdays and assigns them to L_1. Then it sorts the day in increasing order. Scroll through the list to see duplicate birthdays. Repeat many times. For a large number of repetitions, there should be duplicate birthdays about half the time. To simulate 41 people, change 23 to 41 in the command and repeat many times. We assume that there are 365 days for birthdays, and that all days are equally likely to be a birthday.

6.17 **(a)** One digit simulates System A's response: 0 to 8 shut down the reactor, and 9 fails to shut it down. **(b)** One digit simulates System B's response: 0 to 7 shut down the reactor, and 8 and 9 fail. **(c)** A pair of consecutive digits simulates the response of both systems, the first giving A's response as in (a), and the second B's response as in (b). If a single digit were used to simulate both systems, the reactions of A and B would be dependent—for example, if A fails, then B must also fail. **(d)** The true probability that the reactor will shut down is 1 − (0.2)(0.1) = 0.98.

6.19 **(b)** In our simulation, Shaq hit 52% of his shots. **(c)** The longest sequence of misses in our run was 6, and the longest sequence of hits was 9. Of course, results will vary.

6.21 Long trials of this experiment often approach 40% heads. One theory attributes this surprising result to a "bottle-cap effect" due to an unequal rim on the penny. We don't know. But a teaching assistant claims to have spent a profitable evening at a party betting on spinning coins after learning of the effect.

6.23 There are 21 0s among the first 200 digits; the proportion is 0.105.

6.25 Answers will vary.

6.27 The study looked at regular season games, which included games against poorer teams, and it is reasonable to believe that the 63% figure is inflated because of these weaker opponents. In the World Series, the two teams will (presumably) be nearly the best, and home game wins will not be so easy.

6.29 **(a)** S = {germinates, fails to grow}. **(b)** If measured in weeks, for example, S = {0, 1, 2, . . .}. **(c)** S = {A, B, C, D, F}. **(d)** Using Y for "yes (shot made)" and N for "no (shot missed)," S = {YYYY, NNNN, YYYN, NNNY, YYNY, NNYN, YNYY, NYNN, NYYY, YNNN, YYNN, NNYY, YNYN, NYNY, YNNY, NYYN}. (There are 16 items in the sample space.) **(e)** S = {0, 1, 2, 3, 4}.

6.31 S = {all numbers between ___ and ___}. The numbers in the blanks may vary. Table 1.9 has values from 86 to 195 cal; the range of values should include at least those numbers. Some students may play it safe and say S = {all numbers greater than 0}.

6.33 **(a)** $10 \times 10 \times 10 \times 10 = 10{,}000$. **(b)** $10 \times 9 \times 8 \times 7 = 5040$. **(c)** There are 10,000 four-digit tags, 1000 three-digit tags, 100 two-digit tags, and 10 one-digit tags, for a total of 11,110 license tags.

6.35 **(a)** Sum of 2: (1, 1); sum of 3: (1, 2) (2, 1); sum of 4: (1, 3) (2, 2) (3, 1); sum of 2: (1, 4) (2, 3) (3, 2) (4, 1); sum of 2: (1, 5) (2, 4) (3, 3) (4, 2) (5, 1); sum of 2: (1, 6) (2, 5) (3, 4) (4, 3) (5, 2) (6, 1); sum of 2: (2, 6) (3, 5) (4, 4) (5, 3) (6, 2); sum of 2: (3, 6) (4, 5) (5, 4) (6, 3); sum of 2: (4, 6) (5, 5) (6, 4); sum of 2: (5, 6) (6, 5); sum of 2: (6, 6). **(b)** 18. **(c)** There are 4 ways to get a sum of 5 and 5 ways to get a sum of 8. **(d)** Answers will vary.

6.37 **(a)** P(type AB) = 0.04. The sum of all possible outcomes is 1. **(b)** P(type O or B) = 0.49 + 0.20 = 0.69.

6.39 P(either cardiovascular disease or cancer) = 0.45 + 0.22 = 0.67; P(other cause) = 1 − 0.67 = 0.33.

6.41 **(a)** The sum is 1, as we expect since all possible outcomes are listed. **(b)** 1 − 0.41 = 0.59. **(c)** 0.41 + 0.23 = 0.64.

6.43 **(a)** Since all 4 faces have the same shape and the same area, it is reasonable to assume that the probability of a face being down is the same as for any other face. Since the sum of the probabilities must be 1, the probability of each should be 0.25. **(b)** Outcomes (1,1) (1,2) (1,3) (1,4) (2,1)

(2,2) (2,3) (2,4) (3,1) (3,2) (3,3) (3,4) (4,1) (4,2) (4,3) (4,4). The probability of any pair is 1/16 = 0.0625. P(sum = 5) = P(1,4) + P(2,3) + P(3,2) + P(4,1) = (0.0625) (4) = 0.25.

6.45 Fight one big battle because his probability of winning is 0.6, which is higher than $0.8^3 = 0.512$.

6.47 No, it is unlikely that these events are independent. In particular, it is reasonable to expect that college graduates are less likely to be laborers or operators.

6.49 An individual light remains lit for 3 years with probability 1 − 0.02; the whole string remains lit with probability $(1 - 0.02)^{20} = (0.98)^{20} = 0.6676$.

6.51 **(a)** P(one call does not reach a person) = 0.8. Thus, P(none of the 5 calls reaches a person) = $(0.8)^5$ = 0.32768. **(b)** P(one call to NYC does not reach a person) = 0.92. Thus, P(none of the 5 calls to NYC reach a person) = $(0.92)^5$ = 0.6591.

6.53 **(a)** S = {right, left}. **(b)** S = {all numbers between 150 and 200 cm}. (Choices of upper and lower limits will vary.) **(c)** S = {all numbers greater than or equal to 0}, or S = {0, 0.01, 0.02, 0.03, . . .}. **(d)** S = {all numbers between 0 and 1440}. (There are 1440 minutes in one day, so this is the *largest* upper limit we could choose; many students will likely give a smaller upper limit.)

6.55 **(a)** Legitimate. **(b)** Not legitimate: the total is more than 1. **(c)** Legitimate.

6.57 **(a)** The sum of all 8 probabilities equals 1 and all probabilities satisfy $0 \le P \le 1$. **(b)** P(A) = 0.000 + 0.003 + 0.060 + 0.062 = 0.125. **(c)** The chosen person is not white. $P(B^c) = 1 - P(B) = 1 - (0.060 + 0.691) = 1 - 0.751 = 0.249$. **(d)** $P(A^c \cap B) = 0.691$.

6.59 **(a)** P(undergraduate and score ≥ 600) = (0.40)(0.50) = 0.20. P(graduate and score ≥ 600) = (0.60)(0.70) = 0.42. **(b)** P(score ≥ 600) = P(UG and score ≥ 600) + P(G and score ≥ 600) = 0.20 + 0.42 = 0.62.

6.61 **(a)** P(under 65) = 0.321 + 0.124 = 0.445. P(65 or older) = 1 − 0.445 = 0.555. **(b)** P(tests done) = 0.321 + 0.365 = 0.686. P(tests not done) = 1 − 0.686 = 0.314. **(c)** P(A and B) = 0.365; P(A)P(B) = (0.555)(0.686) = 0.3807. A and B are not independent; tests were done less frequently on older patients than if these events were independent.

6.63 Look at the first five rolls in each sequence. All have one G and four R's, so those probabilities are the same. In the first sequence, you win regardless of the sixth roll; for the second, you win if the sixth roll is G, and for the third sequence, you win if it is R. The respective probabilities are 0.00823, 0.00549, and 0.00274.

6.65 **(b)** P(neither admits Zack) = 1 − P(P or S) = 1 − (0.4 + 0.5 − 0.2) = 0.3. **(c)** P(S and not P) = P(S) − P(P and S) = 0.3.

6.67 **(a)** {A and B}: household is both prosperous and educated; P(A and B) = 0.082 (given). **(b)** {A and B^c}: household is prosperous but not educated; P(A and B^c) = P(A) − P(A and B) = 0.056. **(c)** {A^c and B}: household is not prosperous but is educated; P(A^c and B) = P(B) − P(A and B) = 0.179. **(d)** {A^c and B^c}: household is neither prosperous nor educated; P(A^c and B^c) = 0.683 (so that the probabilities add to 1).

6.69 **(a)** 15% drink only cola. **(b)** 20% drink none of these.

6.71 **(a)** P(the vehicle is a car) = A^c; $P(A^c) = 1 - P(A) = 1 - 0.69 = 0.31$. **(b)** P(the vehicle is an imported car) = P(A^c and B) = P(B) − P(A and B) = 0.22 − 0.14 = 0.08. **(c)** $P(A^c \mid B) = P(A^c \cap B)/P(B) = 0.08/0.22 = 0.364$. **(d)** The events A^c and B are not independent, if they were, $P(A^c \mid B)$ would be the same as $P(A^c)$.

6.73 $P(premium \mid \$20) = P(premium \cap \$20)/P(\$20)$ = 0.15/0.445 = 0.337. About 34%.

6.75 If F = {dollar falls} and R = {renegotiation demanded}, then P(F and R) = P(F) P(R | F) = (0.4)(0.8) = 0.32.

6.77 First, concentrate on (say) spades. The probability that the first card dealt is one of those five cards (A♠, K♠, Q♠, J♠, or 10♠) is 5/52. The conditional probability that the second is one of those cards, given that the first was, is 4/51. Continuing like this, we get 3/50, 2/49, and finally 1/48; the product of these five probabilities gives P(royal flush in spades) = 0.00000038477. Multiplying by four (there are four suits) gives P(royal flush) = 0.000001539.

6.79 $P(A \mid B) = P(A \cap B)/P(B) = 0.082/0.261 = 0.3142$. If A and B were independent, $P(A \mid B)$ would equal P(A) and also $P(A \cap B)$ would equal P(A)P(B).

6.81 Let I = the event that the operation results in infection, F = the event that the operation fails. We seek $P(I^c \cap F^c)$. We are given that P(I) = 0.03, P(F) = 0.14, and P(I and F) = 0.01. Then $P(I^c \cap F^c)$ = 1 − P(I or F) = 1 − (0.03 + 0.14 + 0.01) = 0.82.

6.83 **(a)** P(antibody test pos.) = 0.1428. **(b)** P(antibody test pos.) = 0.9487. **(c)** A positive result does not always indicate that the antibody is present. How common a factor is in the population can impact the test probabilities.

6.85 **(a)** P(switch bad) = 0.1, P(switch OK) = 1 − P(switch bad) = 0.9. **(b)** Of the 9999 remaining switches, 999 are bad. P(second bad | first bad) = 0.09991. **(c)** Of the 9999 remaining switches, 1000 are bad. P(second bad | first good) = 0.10001.

6.87 Let W be the event "the person is a woman" and P be "the person earned a professional degree." **(a)** P(W) = 1119/1944 = 0.5756. **(b)** P(W | P) = 39/83 = 0.4699. **(c)** W and P are not independent; if they were, the two probabilities in (a) and (b), P(W) and P(W | P), would be equal.

6.89 Let M be the event "the person is a man" and B be "the person earned a bachelor's degree." **(a)** P(M) = 825/1944 = 0.4244. **(b)** P(B | M) = 559/825 = 0.6776. **(c)** $P(M \cap B) = P(M)P(B \mid M)$ = (0.4244)(0.6776) = 0.2876. This agrees with the directly computed probability: P(M and B) = 559/1944 = 0.2876.

6.91 Percent of A students involved in an accident is $P(C \mid A) = P(C \cap A)/P(A) = 0.01/0.10 = 0.10$.

6.93 P(correct) = P(knows answer) + P(doesn't know but guesses correctly) = 0.75 + (0.25)(0.20) = 0.8.

6.95 P(knows the answer | gives the correct answer)= 0.75/0.80 = 0.9375.

6.97 (a) A single run: spin the 1–10 spinner twice; see if the larger of the two numbers is larger than 5. The player wins if either number is 6, 7, 8, 9, or 10. (b) If using the random digit table, let 0 represent 10, and let the digits 1–9 represent themselves. (c) `randInt(1,10,2)`. (d) In our simulation of 20 repetitions, we observed 13 wins, for a 65% win rate. Using the methods of the next chapter, it can be shown that there is a 75% probability of winning this game.
6.99 (a) Since Carla makes 80% of her free throws, let a single digit represent a free throw, and let 0–7 ↔ "hit" and 8, 9 ↔ "miss." (b) We instructed the calculator to simulate a free throw and store the result in L_1. Then we instructed the calculator to see if the attempt was a hit (1) or a miss (0) and to record that fact in L_2. Continue to press ENTER ENTER until there are 20 simulated free throws. Scroll through L_2 and determine the longest string of 1s (consecutive baskets). In our first set of 20 repetitions, we observed 8 consecutive baskets. Additional sets of 20 repetitions produced 11, 9, 9, and 8. The average longest run was 9 consecutive baskets in 20 attempts. The five-number summary is Min. = 8, Q_1 = 8, Med. = 9, Q_3 = 10, and Max. = 11. There was surprisingly little variation in our five groups of 20 repetitions.
6.101 (a) $P(\text{type AB}) = 1 - (0.45 + 0.40 + 0.11) = 0.04$. (b) $P(\text{type B or type O}) = 0.11 + 0.45 = 0.56$. (c) Assuming that the blood types for husband and wife are independent, $P(\text{type B and type A}) = (0.11)(0.40) = 0.044$. (d) $P(\text{type B and type A}) + P(\text{type A and type B}) = (0.11)(0.40) + (0.40)(0.11) = 0.088$. (e) $P(\text{husband type O or wife type O}) = P(\text{husband type O}) + P(\text{wife type O}) + P(\text{husband and wife both type O}) = 0.45 + 0.45 + (0.45)^2 = 1.1025$.
6.103 (a) To find $P(A \text{ or } C)$, we would need to know $P(A \text{ and } C)$. (b) To find $P(A \text{ and } C)$, we would need to know $P(A \text{ or } C)$.
6.105 Let H = adult belongs to health club, and let G = adult goes to club at least twice a week. $P(G \text{ and } H) = P(H)P(G \mid H) = (0.1)(0.4) = 0.04$.
6.107 Let R_1 be Taster 1's rating and R_2 be Taster 2's rating. $P(R_2 > 3 \mid R_1 = 3) = P(R_2 > 3 \cap R_1 = 3/P = (R_1 = 3)$. $P(R_1 = 3) = 0.01 + 0.05 + 0.25 + 0.05 + 0.01 = 0.37$, and $P(R_2 > 3 \cap R_1 = 3) = 0.05 + 0.01 = 0.06$. $P(R_2 > 3 \mid R_1 = 3) = 0.06/0.37 = 0.1622$.

Chapter 7

7.1 (a) 1/3. (b) and (c) Answers vary.
7.3 (a) "At least one nonword error" is the event $X > 0$. $P(X > 0) = 1 - P(X = 0) = 0.9$. (b) $\{X \le 2\}$ is "no more than two nonword errors," or "fewer than three nonword errors." $P(X \le 2) = (X = 0) + P(X = 1) + P(X = 2) = 0.1 + 0.2 + 0.3 = 0.6$. $P(X < 2) = P(X = 0) + P(X = 1) = 0.1 + 0.2 = 0.3$.
7.5 (a) $P(X \ge 5) = 0.868$. (b) $\{X > 5\}$ = the event that the unit has more than five rooms. $P(X > 5) = P(X = 6) + P(X = 7) + \ldots + P(X = 10) = 0.658$. (c) A discrete random variable has a countable number of values, each of which has a distinct probability; $P(X = x)$. $P(X \ge 5)$ and $P(X > 5)$ are different because the first event contains the value X = 5 and the second does not.
7.7 (a) 0.49. (b) 0.49. (c) 0.73. (d) $P(0.27 < X < 1.27) = P(0.27 < X < 1) = 0.73$. (e) $P(0.1 \le X \le 0.2 \text{ or } 0.8 \le X \le 0.9) = 0.2$. (f) $P(\text{not } [0.3 \le X \le 0.8]) = 1 - 0.5 = 0.5$. (g) $P(X = 0.5) = 0$.
7.9 (a) $P(\hat{p} \ge 0.45) = P\left(Z \ge \frac{0.45 - 0.4}{0.024}\right) = P(Z \ge 2.083) = 0.0186$. (b) $P(\hat{p} < 0.35) = P(Z < -2.086) = 0.0186$. (c) $P(0.35 < \hat{p} < 0.45) = P(-2.086 < Z < 2.086) = 0.9628$.
7.11 (a) The 36 possible pairs of up-faces are (1, 1) (1, 2) (1, 3) (1, 4) (1, 5) (1, 6) (2, 1) (2, 2) (2, 3) (2, 4) (2, 5) (2, 6) (3, 1) (3, 2) (3, 3) (3, 4) (3, 5) (3, 6) (4, 1) (4, 2) (4, 3) (4, 4) (4, 5) (4, 6) (5, 1) (5, 2) (5, 3) (5, 4) (5, 5) (5, 6) (6, 1) (6, 2) (6, 3) (6, 4) (6, 5) (6, 6) (b) Each pair must have probability 1/36. (c) Let X = sum of up-faces. Then $P(X = 2) = 1/36$, $P(X = 3) = 2/36$, $P(X = 4) = 3/36$, $P(X = 5) = 4/36$, $P(X = 6) = 5/36$, $P(X = 7) = 6/36$, $P(X = 8) = 5/36$, $P(X = 9) = 4/36$, $P(X = 10) = 2/36$, $P(X = 11) = 2/36$, $P(X = 12) = 1/36$. (d) $P(7 \text{ or } 11) = 6/36 + 2/36 = 8/36$, or 2/9. (e) P (any sum other than 7) $= 1 - P(7) = 1 - 6/36 = 30/36 = 5/6$.
7.13 (a) $\{Y > 1\}$ or $\{Y \ge 2\}$. $P(Y > 1) = 1 - P(Y = 1) = 0.75$. (b) $P(2 < Y \le 4) = P(Y = 3) + P(Y = 4) = 0.32$. (c) $P(Y \ne 2) = 1 - P(Y = 2) = 1 - 0.42 = 0.58$.
7.15 (a) All probabilities are between 0 and 1; the probabilities add to 1. (b) 75.2%. (c) $P(X \ge 6) = 1 - 0.010 - 0.007 = 0.983$. (d) $P(X > 6) = 1 - 0.010 - 0.007 - 0.007 = 0.976$. (e) Either $X \ge 9$ or $X > 8$. The probability is $0.068 + 0.070 + 0.041 + 0.752 = 0.931$.
7.17 (a) The height should be $^1/_2$. since the area under the curve must be 1. (b) $P(Y \le 1) = {}^1/_2$. (c) $P(0.5 < Y < 1.3) = 0.4$. (d) $P(Y \ge 0.8) = 0.6$.
7.19 The resulting histogram should approximately resemble the triangular density curve of Figure 7.8, with any deviations or irregularities depending upon the specific random numbers generated.
7.21 In this case, we will simulate 500 observations from the N(0.15, 0.0092) distribution. The required TI-83 commands are as follows:

```
ClrList L₁
randNorm(0.15,0.0092,500) →L₁
sortA(L₁)
```

Scrolling through the 500 simulated observations, we can determine the relative frequency of observations that are at least 0.16 by using the complement rule. For a sample simulation, there are 435 observations less than 0.16; thus, the desired relative frequency is $1 - 435/500 = 65/500 = 0.13$. The actual probability that $P \ge 0.16$ is 0.1385. 500 observations yield a reasonably close approximation.

7.23 $\mu_x = \Sigma x_i p_i = 0\left(\frac{1}{8}\right) + 1\left(\frac{3}{8}\right) + 2\left(\frac{3}{8}\right) + 3\left(\frac{1}{8}\right) = 1.5$

$\sigma_x^2 = \Sigma (x_i - \mu_x)^2 p_i = (0 - 1.5)^2\left(\frac{1}{8}\right) + (1 - 1.5)^2\left(\frac{3}{8}\right)$

$+ (2 - 1.5)^2\left(\frac{3}{8}\right) + (3 - 1.5)^2\left(\frac{1}{8}\right) + (0 - 1.5)^2\left(\frac{1}{8}\right)$

$= 0.75$, so $\sigma_x = 0.866$ girls.

7.25 Owner-occupied units: $\mu = (1)(0.003) + (2)(0.002) + \ldots + (10)(0.035) = 6.284$.
Renter-occupied units: $\mu = (1)(0.008) + (2)(0.027) + \ldots + (10)(0.003) = 4.187$.
The larger value of μ for owner-occupied units reflects the fact that the owner distribution was symmetric rather than skewed to the right, as was the case with the renter distribution.

7.27 **(a)** The payoff is either \$0, $P(X = 0) = 0.75$, or \$3, $P(X = 3) = 0.25$. **(b)** For each \$1 bet, $\mu_X = (\$0)(0.75) + (\$3)(0.25) = \$0.75$. **(c)** The casino makes 25 cents for every dollar bet (in the long run).

7.29 Household size: $\mu =$
$(1)(0.25)+(2)(0.32)+(3)(0.17)+(4)(0.15)+(5)(0.07)+(6)(0.03)+(7)(0.01) = 2.6$

Family size: $\mu =$
$(1)(0)+(2)(0.42)+(3)(0.23)+(4)(0.21)+(5)(0.09)+(6)(0.03)+(7)(0.02) = 3.14$

Household size:
$\sigma^2 = (1 - 2.6)^2(0.25) + (2 - 2.6)^2(0.32) + \ldots + (7 - 2.6)^2(0.01) = 2.02$

and
$\sigma = \sqrt{2.02} = 1.421$

Family size:
$\sigma^2 = (1 - 3.14)^2(0) + (2 - 3.14)^2(0.42) + \ldots + (7 - 3.14)^2(0.02) = 1.5604$

and
$\sigma = \sqrt{1.5604} = 1.249$

The family distribution has a slightly larger mean than the household distribution, reflecting the fact that the family distribution assigns more "weight" (probability) to the values 3 and 4. The two standard deviations are roughly equivalent; the household standard deviation may be larger because of the fact that it assigns a nonzero probability to the value 1.

7.31 The graph for $X_{max} = 10$ for the first 10 values of $\bar{x}$, whereas the graph for $X_{max} = 100$ gets closer and closer to $\mu = 64.5$ as $\bar{x}$ increases. This illustrates that the larger the sample size (represented by the integers 1, 2, 3, . . . , in L_1), the closer the sample means x get to the population mean $\mu = 64.5$. (In other words, this exercise illustrates the law of large numbers in a graphical manner.)

7.33 $P(L = 1) = 1/512$, $P(L = 2) = 88/512$, $P(L = 3) = 185/512$, $P(L = 4) = 127/512$, $P(L = 5) = 63/512$, $P(L = 6) = 28/512$, $P(L = 7) = 12/512$, $P(L = 8) = 5/512$, $P(L = 9) = 2/512$, $P(L = 10) = 1/512$. P(you win) = P(run of 1 or 2), so the expected outcome is $\mu = (\$2)(0.1738) + (-\$1)(0.8262) = -\$0.4785$. On the average, you will lose about 48 cents each time you play. (Simulated results should be close to this exact result; how close depends on how many trials are used.)

7.35 **(a)** Expected result of a single die is 3.5. The green mean of the applet does not agree with the expected sum. As the number tossed increases, the mean fluctuates less and stabilizes close to the expected sum. This is called the Law of Large Numbers. **(b)** The expected result for two dice is 7. Again, the mean fluctuates and then stabilizes close to the expected sum. **(c)** Expected sum for 3 dice is 10.5; for 4 dice, 14; for 5 dice, 17.5. The greatest number of dice possible for this applet is 10, with an expected value of 35. The expected value is 3.5 × the number of dice.

7.37 **(a)** When $a + bX = 5$, $P(a + bX) = 0.2$; when $a + bX = 8$, $P(a + bX) = 0.5$; $a + bX = 17$, $P(a + bX) = 0.3$. **(b)** $\mu_{a+bx} = (5)(0.2) + (8)(0.5) + (17)(0.3) = 10.1$. $\sigma^2_{a+bx} = (5 - 10.1)^2(0.2) + (8 - 10.1)^2(0.5) + (17 - 10.1)^2(0.3) = 21.69$. **(c)** $\mu_x = 2.7$ and $\sigma_X = 1.5524$, so $\mu_{a+bx} = a + b\,\mu_x = 2 + (3)(2.7) = 10.1$ and $\sigma^2_{a+bx} = b^2\sigma_x^2$, so $\sigma_{a+bx} = b\,\sigma_x = (3)(1.5524) = 4.657$. Answers in (b) and (c) are the same. **(d)** $\sigma^2_{a+bX} = b^2\sigma_x^2 = 3^2(2.41) = 21.69$, the same as part (b). **(e)** Using the rules is much simpler.

7.39 **(a)** Dependent: since the cards are being drawn from the deck without replacement, the nature of the third card (and thus the value of Y) will depend upon the nature of the first two cards that were drawn (which determine the value of X). **(b)** Independent: X relates to the outcome of the first roll, Y to the outcome of the second roll, and individual dice rolls are independent.

7.41 **(a)** 11 + 20 = 31 seconds. **(b)** No, the mean time for the entire operation is not changed by the decrease in the standard deviation. **(c)** The standard deviation for the total time is $\sqrt{2^2 + 4^2} \cong 4.4721$.

7.43 **(a)** $\mu_X = (0)(0.03) + (1)(0.16) + (2)(0.30) + (3)(0.23) + (4)(0.17) + (5)(0.11) = 2.68$. $\sigma_X^2 = (0 - 2.68)^2(0.03) + (1 - 2.68)^2(0.16) + (2 - 2.68)^2(0.30) + (3 - 2.68)^2(0.23) + (4 - 2.68)^2(0.17) + (5 - 2.68)^2(0.11) = 1.7176$, and $\sigma = \sqrt{1.7176} = 1.3106$. **(b)** Answers will vary.

7.45 **(a)** Randomly selected students would presumably be unrelated. **(b)** $\mu_{f-m} = \mu_f - \mu_m = 120 - 105 = 15$. $\sigma^2_{f-m} = \sigma_f^2 + \sigma_m^2 = 28^2 + 35^2 = 2009$, so $\sigma_{f-m} = 44.82$. **(c)** Knowing only the mean and standard deviation, we cannot find that probability (unless we assume that the distribution is Normal). Many different distributions can have the same mean and standard deviation.

7.47 **(a)** $\mu_x = 550°C$; $\sigma_x^2 = 32.5$, so $\sigma_x = 5.701°C$. **(b)** Mean: 0°C. Standard deviation: 5.701°C. **(c)** $\mu_y = \frac{9}{5}\mu_x + 32 = 1022°F$ and $\sigma_x = \sqrt{\left(\frac{9}{5}\right)^2 \sigma_x^2} = \left(\frac{9}{5}\right)\sigma_x = 10.26°F$.

7.49 (**a**) Yes, this is always true; it does not depend on independence. (**b**) No, it is not reasonable to believe that X and Y are independent.

7.51 (**a**) $\sigma_Y^2 = (0 - 0.7)^2(0.4) + (1 - 0.7)^2(0.5) + (2 - 0.7)^2(0.1) = 0.41$, so $\sigma_Y = 0.640$. (**b**) $\sigma_{X+Y}^2 = \sigma_X^2 + \sigma_Y^2 = 0.890 + 0.41 = 1.3$, so $\sigma_{X+Y} = 1.14$. (**c**) $\sigma_{350X+400Y}^2 = \sigma_{350X}^2 + \sigma_{400Y}^2 = (350^2)(\sigma_X^2) + (400^2)(\sigma_Y^2) = (350^2)(0.890) + (400^2)(0.41) = 174{,}625$, so $\sigma_{350X+400Y} = 417.882$.

7.53 Let V = vault, P = parallel bars, B = balance beam, and F = floor exercise; $\mu_{V+P+B+F} = \mu_V + \mu_P + \mu_B + \mu_F = 9.314 + 9.553 + 9.461 + 9.543 = 37.871$. $\sigma_{V+P+B+F}^2 = \sigma_V^2 + \sigma_P^2 + \sigma_B^2 + \sigma_F^2 = 0.216^2 + 0.122^2 + 0.203^2 + 0.099^2 = 0.11255$, so $\sigma_{V+P+B+F} = 0.3355$. The distribution of Carly Patterson's total scores will be $N(37.871, 0.3355)$. The probability that she will beat the score of 38.211 is $P(X > 38.211) = P(Z > 1.01) = 0.155$.

7.55 The missing probability is 0.99058 (so that the sum is 1). This gives mean earnings $\mu_X = \$303.3525$.

7.57 $\sigma_X^2 = 94{,}236{,}826.64$, so that $\sigma_X = \$9707.57$

7.59 $X - Y$ is $\approx N(0, 0.4243)$, so $P(|X - Y| \geq 0.8) = P(|Z| \geq 1.8856) = 1 - P(|Z| \leq 1.8856) = 0.0593$ (table value: 0.0588).

7.61 (**a**) $\mu_{Y-X} = \mu_Y - \mu_X = 2.001 - 2.000 = 0.001$ g. $\sigma^2_{Y-X} = \sigma^2_Y + \sigma^2_X = 0.002^2 + 0.001^2 = 0.000005$, so $\sigma^2_{Y-X} = 0.002236$ g. (**b**) $\mu_Z = \frac{1}{2}\mu_x = \frac{1}{2}\mu_y = 2.0005$ g

$\sigma_Z^2 = \frac{1}{4}\ \sigma_x = \frac{1}{4}\ \sigma_y = 0.00000125$, so $\sigma_Z = 0.001118$ g

Z is slightly more variable than Y, since $\sigma_Z > \sigma_Y$.

7.63 (**a**) A single random digit simulates each toss, with (say) odd = heads and even = tails. The first round is two digits, with two odds a win; if you don't win, look at two more digits, again with two odds a win. The calculator command `randInt(0,1,2)` provides two digits, either 0 (tail) or 1 (head). (**b**) In 50 plays, we got 25 wins for an esimate of \$0. (**c**) To win a dollar, you can win on the first round by getting two heads or win on the second round by not getting two heads on the first round and then getting two heads on the second round. So the probability of

winning is $\frac{1}{4} + \left(\frac{3}{4}\right)\left(\frac{1}{4}\right) = \frac{7}{16}$. The expected value is

$(\$1)\left(\frac{7}{16}\right) + (-\$1)\left(\frac{9}{16}\right) = -\frac{2}{16} = \$0.125$

Chapter 8

8.1 Not binomial: there is no fixed n (i.e., there is no definite upper limit on the number of defects).

8.3 Yes: (1) "Success" means reaching a live person; "failure" is any other outcome. (2) We have a fixed number of observations (n = 15). (3) It is reasonable to believe that each call is independent of the others. (4) Each randomly dialed number has chance $p = 0.2$ of reaching a live person.

8.5 Not binomial: because the student receives instruction after incorrect answers, her probability of success is likely to increase.

8.7 $P(X = k) = \binom{n}{k}(p)^k(1 - p)^{nk}$, so $P(X = 3) = \binom{5}{3}(0.25)^3(0.75)^2 = 10(0.25)^3(0.75)^2 = 0.0879$.

8.9 Let X = number of children with blood type O. X is $B(5, 0.25)$. $P(X = k) = \binom{n}{k}(p)^k(1 - p)^{n-k}$. $P(X \geq 1) = 1 - P(X = 0) = 1 - \binom{5}{0}(0.25)^0(0.75)^5 = 1 - (0.75)^5 = 1 - 0.237 = 0.763$.

8.11 (**a**) $P(X = 3) = \binom{15}{3}(0.3)^3(0.7)^{12} = 0.17004$. (**b**) $P(X = 0) = \binom{15}{0}(0.3)^0(0.7)^{15} = 0.00475$.

8.13 (**a**) $P(X = 2)$ = `binompdf(5,0.25,2)` = 0.2637. (**b**) The binomial probabilities for X = 0, . . . , 5 are 0.2373, 0.3955, 0.2637, 0.0879, 0.0146, 0.0010. (**c**) The cumulative probabilities for X = 0, . . . , 5 are 0.2373, 0.6328, 0.8965, 0.9844, 0.9990, 1. Compared with Corinne's cdf histogram, the bars in this histogram get taller sooner. Both peak at 1 on the extreme right.

8.15 (**a**) Let X = the number of correct answers. X is $B(10, 0.25)$. The probability of at least one correct answer is $PX \geq 1) = 1 - P(X = 0) = 1-$`binompdf(10,0.25,0)` $= 1 - 0.056 = 0.944$. (**b**) Let X = the number of correct answers. $P(X \geq 1) = 1 - P(X = 0)$. $P(X = 0)$ is the probability of getting every question wrong. Note that this is not a binomial, as each question has a different probability of success. The probability of getting the first question wrong is 2/3, the second question wrong 3/4, and the third question wrong 4/5. The probability of getting all the questions wrong is the product of these, or 0.4. So $P(X \geq 1) = 1 - P(X = 0) = 1 - 0.4 = 0.6$.

8.17 (**a**) $P(X = 11)$ = `binompdf(20,0.8,11)` = 0.0074. (**b**) $P(X = 20)$ = `binompdf(20,0.8,20)` = 0.8^{20} = 0.0115. (**c**) $P(X \leq 19) = 1 - P(X = 20) = 1 - 0.0015 = 0.9985$.

8.19 (**a**) $np = 1500$, $n(1-p) = 1000$; both values are greater than or equal to 10. (**b**) Let X = the number of people in the sample who find shopping frustrating. X is $B(2500, 0.6)$. Then $P(X \geq 1520) = 1 - P(X \leq 1519) = 1 -$ `binomcdf(2500,0.6,1519)` = 1−0.7868609113 = 0.2131390887, which rounds to 0.2131. The probability correct to six decimal places is 0.213139. (**c**) $P(X \leq 1468)$ = `binomcdf(2500,0.6,1468)` =0.0994. Using the Normal approximation to the binomial, $P(X \leq 1468) = 0.0957$, a difference of 0.0037.

8.21 (**a**) $\mu = np = (20)(0.8) = 16$. (**b**) $\sigma = \sqrt{(20)(0.8)(0.2)} = \sqrt{29.44} = 1.789$. (**c**) If $p = 0.9$, $\sigma = 1.342$. If $p = 0.99$, $\sigma = 0.445$. As the probability of success gets closer to 1, the standard deviation decreases.

8.23 (**a**) X = the number of people in the sample of 400 adult Richmonders who approve of the president's reaction. X is approximately binomial because the sample size is small compared to the population size (all adult Richmonders), and as a result, the individual responses may be considered independent and the probability of success (approval) remains essentially the same from trial to trial. $n = 400$ and $p = 0.92$. (**b**) $P(X \leq 358)$ = `binomcdf(400,0.92,358)` = 0.0441.

(c) $\mu = (400)(0.92) = 368$, $\sigma = \sigma = \sqrt{(400)(0.92)(0.08)} = \sqrt{29.44} = 5.426$ (d) $P(X \le 358) \approx P\left(Z \le \frac{358 - 368}{5.426}\right) = Z \le -1.84 \approx 0.0327$. Yes, even though p is close to 1, both np and $n(1 - p)$ are greater than 10.

8.25 The command `cumSum(L2)→L3` calculates and stores the values of $P(X \le x)$ for $x = 0, 1, 2, \ldots, 12$. The entries in L_3 and the entries in L_4 defined by `binomcdf(12,0.75,L1)` → L_4 are identical.

8.27 (a) If S is the number of the 3 eggs chosen by Sara that are contaminated, S is a binomial with $p = 0.25$; S is $B(3, 0.25)$. (b) Using the calculator and letting $0 \leftrightarrow$ contaminated egg and 1, 2, or $3 \leftrightarrow$ good egg, simulate choosing 3 eggs by `RandInt(0,3,3)`. Repeating this 50 times leads to 30 occasions when at least one of the eggs is contaminated; $P(S > 0) = 0.6$. (c) $P(S > 0) = 1 - P(S = 0) = 1 - (0.75)^3 = 0.5781$. The value obtained by simulation is close.

8.29 If X is the count of 0s among n random digits and has a binomial distribution with $p = 0.1$, $B(n, 0.1)$. (a) When $n = 40$, $P(X = 4) =$ `binompdf(40,0.1,4)` $= 0.206$. (b) When $n = 5$, $P(X \ge 1) = 1 - P(X = 0) = 1 - (0.9)^5 = 1 - 0.590 = 0.410$.

8.31 (a)A binomial distribution is *not* an appropriate choice for field goals made by the NFL player, because given the different situations the kicker faces, his probability of success is likely to change from one attempt to another. (b) It would be reasonable to use a binomial distribution for free throws made by the NBA player because we have $n = 150$ attempts, presumably independent (or at least approximately so), with chance of success $p = 0.8$ each time because of similar external conditions each time.

8.33 (a) Answers will vary. (b) $n = 40$, $p = 0.1$. (c) As the number of lines used increases, the mean gets closer to 4, the standard deviation becomes closer to 1.8974 and the shape will still be skewed to the right because we are simulating a binomial distribution with $n = 40$ and $p = 0.1$. (d) As n increases, $np \ge 10$ and $n(1-p) \ge 10$, and so the distribution meets the requirements for the approximation to the Normal distribution with a mean of np and a standard deviation of $\sqrt{np(1-p)}$.

8.35 X, the number of correctly answered questions, is binomial with $n = 10$ and $p = 0.2$. X is $B(10, 0.2)$. (a) $P(X = 0) = 0.107$. (b) $P(X = 1) = 0.268$. (c) $P(X = 2) = 0.302$. (d) $P(X \le 3) = 0.879$. (e) $P(X > 3) = 1 - P(X \le 3) = 0.121$. (f) $P(X = 10) \approx 0$!! (g) $P(X = x)$ for $x = 0, 1, 2, 3, \ldots, 7 = 0.107, 0.268, 0.302, 0.201, 0.088, 0.026, 0.006, 0.001$; $P(X = 8)$, $P(X = 9)$, and $P(X = 10)$ are all less than 0.001.

8.37 In this case, $n = 20$ and the probability that a randomly selected basketball player graduates is $p = 0.8$. We will estimate $P(X \le 11)$ by simulating 30 observations of X = number graduated and computing the relative frequency of observations that are 11 or smaller. The sequence of TI-83 commands is: `randBin(1,0.8,20)` → `L1 : sum(L1)` → `L2(1)`, where 1's represent players who graduated. Press **ENTER** sufficient times to obtain 30 numbers. The actual value of $P(X \le 11)$ is `binomcdf(20,0.8,11)` $= 0.01$.

8.39 (a) X, the number of auction site visitors, is binomial with $n = 12$ and $p = 0.5$. X is $B(12, 0.5)$. (b) $P(X = 8) =$ `binompdf(12,0.5,8)` $= 0.1208$; $P(X \ge 8) = 1 - P(X \le 7) = 1 -$ `binomcdf(12,0.5,7)` $= 1 - 0.806 = 0.194$.

8.41 (a) Geometric setting: success = tail, failure = head; trial = flip of a coin; p = ½. (b) Not a geometric setting. You are not counting the number of trials before the first success is obtained. (c) Geometric setting: success = jack, failure = any other card; trial = drawing of a card; $p = 4/52 = 0.0769$ (trials are independent because the card is replaced each time). (d) Geometric setting: success = match all 6 numbers, failure = do not match all 6 numbers; trial = drawing on a particular day; $p = \frac{1}{\binom{44}{6}} \cong 0.000142$. (e) Not a geometric setting. The probability of a success is not constant, because you are drawing without replacement. Also, you are interested in getting 3 successes rather than just the first success.

8.43 (a) This is a geometric setting because the trials (tests) on successive drives are independent, $p = 0.3$ on each trial, and X is the number of trials required to achieve the first success. X = number of drives tested in order to find the first defective one. Success = defective drive. (b) $P(X = 5) = (1 - 0.03)^{5-1}(0.03) = 0.0266$. (c) pdf; $P(X = x)$ for p = 0.03 and X = 1, 2, 3, and 4 = 0.03, 0.0291, 0.0282, and 0.0274.

8.45 (a) X = number of flips required in order to get the first head. X is a geometric random variable with $p = 0.5$. (b) pdf; $P(X = x)$ for $p = 0.5$ and X = 1, 2, 3, 4, and 5 = 0.5, 0.25, 0.125, 0.0625, and 0.03125. (c) cdf; $P(X \le x)$ for $p = 0.5$ and X = 1, 2, 3, 4, and 5 = 0.5, 0.75, 0.875, 0.9375, and 0.96875.

8.47 (b) $P(X > 10) = (1 - 1/6)^{10} = (5/6)^{10} = 0.1615$. (c) Using the calculator, `geometcdf(1/6,25)` = 0.989517404 and `geometcdf(1/6,26)` = 0.9912645033, so the smallest positive integer k for which $P(X \le k) > 0.99$ is 26.

8.49 (a) The shots are independent, and the probability of success is the same for each shot. A "success" is a missed shot, so $p = 0.2$. (b) The first success (miss) is the sixth shot, so $P(X = 6) = (1-p)^{n-1}(p) = (0.8)^5(0.2) = 0.0655$. (c) $P(X \le 6) = 1 - P(X > 6) = 1 - (1-p)^6 = 1 - (0.8)^6 = 0.738$. Using a calculator, $P(X \le 6) =$ `geometcdf(0.2,6)` $= 0.738$.

8.51 (a) Geometric; X = number of marbles you must draw to find the first red marble. We chose geometric because the number of trials (draws) is the variable quantity.

(b) $p = \frac{20}{35} = \frac{4}{7}$ $\quad P(X = 2) = \left(1 - \frac{4}{7}\right)^{2-1}\left(\frac{4}{7}\right) = 0.2449$

$$P(X \le 2) = P(X = 1) + P(X = 2) = \frac{4}{7} + \left(\frac{3}{7}\right)\left(\frac{4}{7}\right) = 0.8163$$

$$P(X > 2) = \left(1 - \frac{4}{7}\right)^2 = 0.1837$$

(c) Using TI-83 commands: `seq(X,X,1,20)→L1`, `geompdf(4/7,L1)→L2`, and `geomcdf(4/7,L1)→L3` [or `cumSum(L2)→L3`].

8.53 (a) Success = getting a correct answer. X = number of questions Carla must answer in order to get the first correct answer. $p = 0.2$ (all choices equally likely to be selected). **(b)** $P(X = 5) = (1-0.2)^{5-1}(0.2) = 0.082$. **(c)** $P(X > 4) = (1-0.2)^4 = 0.4096$. **(d)** $P(X = 1) = 0.2$; $P(X = 2) = 0.16$; $P(X = 3) = 0.128$; $P(X= 4) = 0.1024$; $P(X = 5) = 0.082$; . . . **(e)** $\mu_X = 1/0.2 = 5$.

8.55 **(a)** Letting G = girl and B = boy, the outcomes are {G, BG, BBG, BBBG, BBBB}. Success = having a girl. **(b)** X can have values of 0, 1, 2, 3, and 4. Using the multiplication rule for independent events, $P(X = 0) = 1/2$; $P(X = 1) = (1/2)(1/2) = 1/4$; $P(X = 2) = (1/2)(1/2)(1/2) = 1/8$; $P(X = 3) = (1/2)(1/2)(1/2)(1/2) = 1/16$; and $P(X = 4) = (1/2)(1/2)(1/2)(1/2) = 1/16$. Note that $\sum P(X) = 1$. **(c)** Y = number of children produced until first girl is born. Then Y is a geometric variable for Y = 1 to 4 but not for values greater than 4, because the couple stops having children after 4. Note that BBBB is not included in the event Y = 4; $P(Y = 1) = 1/2$; $P(Y = 2) = 1/4$; $P(Y = 3) = 1/8$; $P(Y = 4) = 1/16$. Note that this is not a valid distribution, since $\sum P(Y) < 1$. **(d)** If T is the number of children in the family, $P(T = 1) = 0.5$, $P(T = 2) = 0.25$, $P(T = 3) = 0.125$, $P(T = 4) = 0.125$, so $\mu_T = 1(0.5) + 2(0.25) + 3(0.125) + 4(0.125) = 1.875$. **(e)** $P(T > 1.875) = 1 - P(T = 1) = 0.5$. **(f)** P(having a girl) = 1 − P(not having a girl) = 1 − P(BBBB) = 1 − 0.125 = 0.9375.

8.57 Find the mean of 25 randomly generated observations of X, the number of children. We can create a suitable string of random digits (say of length 100) by using the command `randInt(0,9,100)→L1`. Let the digits 0 to 4 represent a boy and 5 to 9 represent a girl. Scroll down the list and count until you get a "girl number" or "4 boy numbers in a row," whichever comes first. The number you have counted is X, the number in the family. Continue until you have 25 values of X. $\bar{x} = \sum X/n = 45/25 = 1.8$. Compare with the known mean $\mu = 1.875$.

8.59 **(a)** No, it is not reasonable to assume that the opinions of a husband and wife are independent. **(b)** No, trials are not independent, and the probability is not the same for each fraternity member. Notice also that the population is only 3 times as large as the sample; it should be at least 10 times as large.

8.61 **(a)** Symmetric; the shape depends on the value of the probability of success. **(c)** 0.0078125.

8.63 **(a)** $P(X = 10) = 0.165$. **(b)** $P(10 < X < 15) = 0.271$ or $P(10 \le X \le 15) = 0.442$, depending on your interpretation of "between" being exclusive or inclusive. **(c)** $P(X > 15) = 1 - P(X \le 15) = 0.002$. **(d)** $P(X < 8) = P(X \le 7) = 0.224$.

8.65 **(a)** X, the number of positive tests, has a binomial distribution with parameters $n = 1000$ and $p = 0.004$. **(b)** $\mu = np = 4$ positive tests. **(c)** To use the Normal approximation, we need np and $n(1 - p)$ both bigger than 10, and as we saw in (b), $np = 4$.

8.67 **(a)** X is geometric with $p = 0.325$. $P(X = 1) = 0.325$.
(b)
$$\begin{aligned} P(X \le 3) &= 1 - P(X > 3) \\ &= 1 - P(1 - P)^3 \\ &= 1 - 0.675^3 \\ &= 0.6925. \end{aligned}$$
(c) $P(X > 4) = (1 - 0.325)^4 = 0.208$. **(d)** $\mu = 1/p = 1/0.325 = 3.08$, just over 3 at-bats. **(e)** Use the commands `seq(X,X,1,10)→L1`, `geo pdf(0.325,L1)→L2`, and `geomcdf(0.325,L1)→L3`.

Chapter 9

9.1 **(a)** $\mu = 2.5003$ is a parameter (related to the population of all the ball bearings in the container) and $\bar{x} = 2.5009$ is a statistic (related to the sample of 100 ball bearings). **(b)** $\hat{p} = 7.2\%$ is a statistic (related to the sample of members of the work force who were unemployed).

9.3 **(a)** Since the proportion of times the toast will land butter-side down is 0.5, the result of 20 coin flips will simulate the outcomes of 20 pieces of falling toast (landing butter-side up or butter-side down). **(b)** Answers will vary. **(c)** Answers will vary; however, it is more likely that the center of this distribution will be close to 0.5, and it is more likely that the shape will be close to Normal. **(d)** Answers will vary. **(e)** By combining the results from many students, he can get a more accurate estimate of the value of p since the value of $\hat{p}$ approaches p as the sample size increases.

9.5 **(a)** The scores will vary depending on the starting row. Note that the smallest possible mean is 61.75 (from the sample 58, 62, 62, 65) and the largest is 77.25 (from 73, 74, 80, 82). **(b)** and **(c)** Answers will vary.

9.7 **(b)** $\mu = 141.847$ days. **(c)** Means will vary with samples. **(d)** It would be unlikely (though not impossible) for all five values to fall on the same side of μ. **(e)** The mean of the (theoretical) sampling distribution would be μ. **(f)** Answers will vary.

9.9 **(a)** Since the smallest number of total tax returns (i.e., the smallest population) is still more than 100 times the sample size, the variability will be (approximately) the same for all states. **(b)** Yes, it will change—the sample taken from Wyoming will be about the same size, but the sample from, e.g., California will be considerably larger, and therefore the variability will decrease.

9.11 $\hat{p} = 40.2\%$ and $\hat{p} = 31.7\%$ are both statistics (related to the sample of black high school students).

9.13 **(a)** If we choose many samples, the average of the $\bar{x}$-values from these samples will be close to μ. **(b)** The larger sample will give more information, and therefore more precise results. The variability in the distribution of the sample average decreases as the sample size increases.

9.15 **(a)** $\mu = 3.5$, $\sigma = 1.708$. **(b)** This is equivalent to rolling a pair of fair, six-sided dice. **(c)** SRS of size 2:
1,1; $\bar{x} = 1$
2,1; 1,2; $\bar{x} = 1.5$
3,1; 2,2; 1,3; $\bar{x} = 2$
4,1; 2,3; 3,2; 1,4; $\bar{x} = 2.5$
5,1; 2,4; 3,3; 4,2; 1,5; $\bar{x} = 3$
6,1; 2,5; 3,4; 4,3; 5,2; 1,6; $\bar{x} = 3.5$
6,2; 3,5; 4,4; 5,3; 2,6; $\bar{x} = 4$
6,3; 4,5; 5,4; 3,6; $\bar{x} = 4.5$
6,4; 5,5; 4,6; $\bar{x} = 5$
6,5; 5,6; $\bar{x} = 5.5$
6,6; $\bar{x} = 6$

(d) The center is identical to that of the population distribution, the shape is symmetrical and bell-shaped, and the spread is smaller than that of the population distribution.

9.17 Assuming that the poll's sample size was less than 870,000 (10% of the population of New Jersey), the variability would be practically the same for either population. (The sample size for this poll would have been considerably less than 870,000.)

9.19 **(a)** $\mu_{\hat{p}} = p = 0.7$; $\sigma_{\hat{p}} = \sqrt{\dfrac{p(1-p)}{n}} = \sqrt{\dfrac{(0.7)(0.3)}{1012}} =$ 0.0144. **(b)** The population (all U.S. adults) is clearly at least 10 times as large as the sample (the 1012 surveyed adults). **(c)** $np = (1012)(0.7) = 708.4$; $n(1-p) = (1012)(0.3) = 303.6$; both are at least 10. **(d)** $P(\hat{p} \leq 0.67) = P(Z \leq -2.08) = 0.0188$. This is a fairly unusual result if 70% of the population actually drink the cereal milk. **(e)** To halve the standard deviation of the sample proportion, multiply the sample size by 4; we would need to sample (1012)(4) = 4048 adults. **(f)** It would probably be higher, since teenagers (and children in general) have a greater tendency to drink the cereal milk.

9.21 For $n = 300$: $\sigma_{\hat{p}} = 0.02828$ and $P(0.37 \leq \hat{p} \leq 0.43) = 0.7108$. For $n = 1200$: $\sigma_{\hat{p}} = 0.01414$ and $P(0.37 \leq \hat{p} \leq 0.43) = 0.9660$. For $n = 4800$: $\sigma_{\hat{p}} = 0.00707$ and $P(0.37 \leq \hat{p} \leq 0.43) \approx 1$. Larger sample sizes give sample proportions that are more likely to be close to the true proportion.

9.23 **(a)** 0.86 (86%). **(b)** We can use the Normal approximation (Rule of Thumb 2 is just satisfied: $n(1-p) = 10$). $\sigma_{\hat{p}} = 0.03$, and $P(\hat{p} < 0.86) = P(Z < -1.33) = 0.0918$. (*Note:* The exact binominial probability is 0.1239.) **(c)** Even when the claim is correct, in about 10% of samples of this size, we can expect to observe 86% or fewer orders shipped on time, due to sampling variability alone.

9.25 **(a)** $\mu_{\hat{p}} = p = 0.15$; $\sigma_{\hat{p}} = \sqrt{\dfrac{p(1-p)}{n}} =$

$$\sqrt{\frac{(0.15)(0.85)}{1540}} = 0.0091.$$

(b) The population (U.S. adults) is considerably larger than 10 times the sample size (1540). **(c)** $np = 231$, $n(1-p) = 1309$; both are at least 10. **(d)** $P(0.13 \leq \hat{p} \leq 0.17) = P(-2.20 \leq Z \leq 2.20) = 0.9722$. **(e)** To reduce the standard deviation by a third, we need a sample nine times as large; about 13,860.

9.27 **(a)** $\hat{p} = 0.62$. **(b)** $\mu_{\hat{p}} = p = 0.67$, $\sigma_{\hat{p}} = \sqrt{\dfrac{p(1-p)}{n}} =$

$$\sqrt{\frac{(0.67)(0.33)}{100}} = 0.047.\ P(\hat{p} \leq 0.62) =$$

$$P\left(Z \leq \frac{0.62-0.67}{0.047}\right) = P(Z \leq -1.06) = 0.1446.$$

(c) Getting a sample proportion at or below 0.62 is not an unlikely event. The sample results are lower than the national percentage, but the sample was so small that such a difference could arise by chance even if the true campus proportion is 0.67.

9.29 **(a)** $\mu_{\hat{p}} = 0.75, \sigma_{\hat{p}} = \sqrt{\dfrac{(0.75)(0.25)}{100}} = 0.0433.$

$P(\hat{p} \leq 0.7) = P(Z \leq -1.15) = 0.1251$. **(b)** $\mu_{\hat{p}} = 0.75$,

$\sigma_{\hat{p}} = \sqrt{\dfrac{(0.75)(0.25)}{250}} = 0.0274$. $P(\hat{p} \leq 0.7) = P(Z \leq -1.83) = 0.0336$.

(c) To reduce the standard deviation by a fourth, we need a sample sixteen times as large; $n = 1600$. **(d)** The answer is the same for Laura.

9.31 **(a)** $\mu_{\bar{x}} = \mu = -3.5\%$, $\sigma_{\bar{x}} = \dfrac{\sigma}{\sqrt{n}} = \dfrac{26\%}{\sqrt{5}} = 11.628\%.$

(b) $P(X \geq 5\%) = P\left(Z \geq \dfrac{5-(-3.5)}{26}\right) = 0.3707.$

(c) $P(\bar{x} \geq 5\%) = P\left(Z \geq \dfrac{5-(-3.5)}{11.628}\right) = 0.2327.$

(d) $P(\bar{x} < 0\%) = P\left(Z < \dfrac{0-(-3.5)}{11.628}\right) = 0.6179.$

Approximately 62% of all five-stock portfolios lost money.

9.33 **(a)** $\sigma_{\bar{x}} = \dfrac{\sigma}{\sqrt{n}} = \dfrac{10}{\sqrt{3}} = 5.77$ milligrams. **(b)** Solve

$\dfrac{\sigma}{\sqrt{n}} = 3$; $n = 11.1$, so $n = 12$. The average of several measurements is more likely to be closer to the true mean than a single measurement.

9.35 $\bar{x}$ has approximately a N(1.6, 0.0849) distribution; the probability $P(\bar{x} > 2)$ is $P(Z > 4.71)$, which is essentially 0.

9.37 **(a)** No, this probability cannot be calculated, because we do not know the distribution of the weights. **(b)** If total weight = 4000 pounds and there are 20 passengers, $\bar{x} = 200$ pounds. Using the central limit theorem, $\bar{x}$ has approximately a N(190, 7.826) distribution. $P(\bar{x} > 200) = P(Z > 1.28) = 0.1003$.

9.39 **(a)** $X \sim N(125, 10)$ distribution. $P(X > 140) = P(Z > 1.5) = 0.668$. **(b)** $\bar{x} \sim N(125, 5)$ distribution. $P(\bar{x} > 140) = P(Z > 3) = 0.0013$. **(c)** No, the population distribution is Normally distributed so there is no need to use the central limit theorem.

9.41 **(a)** $X \sim N(298, 3)$. $P(X < 295) = P(Z < -1) = 0.1587$. **(b)** $\bar{x} \sim N(298, 1.225)$. $P(\bar{x} < 295) = P(Z < -2.45) = 0.0071$.

9.43 **(a)** $\bar{x} \sim N(2.2, 0.1941)$. **(b)** $P(\bar{x} < 2) = P(Z < -1.03) = 0.1515$. **(c)** $P(X < 100) = P(\bar{x} < 1.923) = P(Z < -1.43) = 0.0764$.

9.45 The mean loss from fire, by definition, is the long-term average of many observations of the random variable X = fire loss. The behavior of X is much less predictable if only a small number of observations are made. If only 12 policies were sold, then the company would have no protection against the large expense that would be incurred if at least 1 of the 12 policyholders happened to lose his or her home. If thousands of policies were sold, then the average fire loss for these policies would be far more likely to be close to μ, and the company's profit would not be endangered by the few large fire-loss payments that it would have to make.

9.47 **(a)** $p = 68\% = 0.68$ is a parameter; $\hat{p} = 73\% = 0.73$ is a statistic. **(b)** $\mu_{\hat{p}} = p = 0.68$, $\sigma_{\hat{p}} = \sqrt{\frac{p(1-p)}{n}} = \sqrt{\frac{(0.68)(0.32)}{150}} = 0.0381$. **(c)** $P(\hat{p} \geq 0.73) = P(Z \geq 1.31) = 0.0951$. There is about a 10% chance that an observation greater than or equal to the observed value of 0.73 will be seen. The random digit device could be working properly.

9.49 **(a)** $X \sim N(100, 15)$. $P(X \geq 105) = P(Z \geq 0.33) = 0.3707$. **(b)** $\mu_{\bar{x}} = \mu = 100$; $\sigma_{\bar{x}} = 15/\sqrt{60} = 1.936$. **(c)** $P(\bar{x} \geq 105) = P(Z \geq 2.58) = 0.0049$. **(d)** The answer to (a) could be quite different; the answer to (b) would be the same (it does not depend on Normality at all). The answer we gave for (c) would still be fairly reliable because of the central limit theorem.

9.51 **(a)** $\mu_{\bar{x}} = \mu = 0.5$; $\sigma_{\bar{x}} = 0.7/\sqrt{50} = 0.099$. **(b)** $P(\bar{x} > 0.6) = P(Z > 1.01) = 0.1562$.

9.53 **(a)** $\hat{p}$ has an approximately Normal distribution with center $\mu_{\hat{p}} = p = 0.2$ and $\sigma_{\hat{p}} = \sqrt{\frac{p(1-p)}{n}} = \sqrt{\frac{(0.2)(0.8)}{1555}} = 0.0101$. **(b)** $P(\hat{p} \leq 0.1929) = P(Z \leq -0.70) = 0.2420$.

9.55 The mean NOX level for 25 cars has a $N(0.2, 0.01)$ distribution, and $P(Z > 2.33) = 0.01$ if Z is $N(0, 1)$, so $L = 0.2 + 2.33 \times 0.01 = 0.2233$ g/mi.

9.57 **(a)** $\mu_{\hat{p}} = 0.5$, and $\sigma_{\hat{p}} = \sqrt{\frac{p(1-p)}{n}} = \sqrt{\frac{(0.5)(0.5)}{14{,}941}} = 0.0041$. **(b)** $P(0.49 < \hat{p} < 0.51) = P(-2.44 < Z < 2.44) = 0.9854$.

Chapter 10

10.1 **(a)** The sampling distribution of $\bar{x}$ is approximately Normal with mean $\mu_{\bar{x}} = 280$ and standard deviation $\sigma_{\bar{x}} = \frac{\sigma}{\sqrt{n}} = \frac{60}{\sqrt{840}} \approx 2.1$. **(b)** Mean = 280. One standard deviation from mean: 277.9 and 282.1. Two standard deviations from mean: 275.8 and 284.2. Three standard deviations from mean: 273.7 and 286.3. **(c)** Two standard deviations; $m \approx 4.2$. **(d)** The confidence intervals drawn will vary. **(e)** About 95% (by the 68–95–99.7 rule).

10.3 No. The student is misinterpreting "95% confidence." This is a statement about the mean score for all young men, not about individual scores.

10.5 **(a)** (48%, 54%). **(b)** 51% is a statistic from one sample. A different sample may give us a totally different answer. When taking samples from a population, not every sample of adults will give the same result, leading to *sampling variability*. A margin of error allows for this variability. **(c)** "95% confidence" means that this interval was found using a procedure that produces correct results (i.e., includes the true population proportion) 95% of the time. **(d)** Survey errors such as undercoverage or nonresponse, depending on how the Gallup Poll was taken, could affect the accuracy.

10.7 Search Table A for 0.0125 (half of the 2.5% that is *not* included in a 97.5% confidence interval). This area corresponds to $z^* = 2.24$.

10.9 **(a)** μ = the mean IQ score for all seventh-grade girls in the school district. The low score (72) is an outlier, but there are no other deviations from Normality. In addition, the central limit theorem tells us that the sampling distribution of $\bar{x}$ will be approximately Normal since $n = 31$. We are told to treat these 31 girls as an SRS. The 31 measurements in the sample should be independent if there are at least $10(31) = 310$ seventh-grade girls in this school district. $\bar{x} = 105.84$; the 99% confidence interval for μ is $105.84 \pm 2.576(15/\sqrt{31}) = 105.84 \pm 6.94 = (98.90, 112.78)$. With 99% confidence, we estimate the mean IQ score for all seventh-grade girls in the school district to be between 98.90 and 112.78 IQ points. **(b)** Unless the class was representative of the whole school district, we would not have been able to generalize our conclusions to the population of interest.

10.11 **(a)** We want to estimate μ = the mean length of time the general managers have spent with their current hotel chain. The sample of managers is an SRS (stated in question) and the sample size is large enough ($n = 114$) to use the central limit theorem to assume Normality for the sampling distribution. The managers' lengths of employment are independent, and so the conditions for a confidence interval for a mean are met. 99% C.I. for $\mu = 11.78 \pm 2.576(3.2/\sqrt{114}) = 11.78 \pm 0.77 = (11.01, 12.55)$. With 99% confidence, we estimate that the mean length of time the general managers have spent with their current hotel chain is between 11.01 and 12.55 years. **(b)** 46 out of 160 did not reply. This nonresponse could have affected the results of our confidence interval considerably, especially if those who didn't respond differed in a systematic way from those who did.

10.13 For 80 video monitors: margin of error = 7.9, which is half of 15.8, the margin of error for $n = 20$.

10.15 **(a)** (18.90, 25.10). With 95% confidence we estimate the mean gain for all Math SAT second-try scores to be between 18.90 and 25.10 points. **(b)** The 90% confidence interval = (19.4, 24.6), and the 99% confidence interval = (17.93, 26.07). **(c)** The confidence interval widens with increasing confidence level.

10.17 (a) The computations are correct. (b) Since the numbers are based on voluntary response, rather than an SRS, the methods of this section cannot be used; the interval does not apply to the whole population.

10.19 (a) We would want the sample to be an SRS of the population under study and the observations to be independently sampled. Because the sample size is only 25, the population should be approximately Normally distributed. (b) 95% C.I. for μ: (64, 88). We are 95% confident that the population mean is between 64 and 88, within a range of 12 on either side of the sample mean. (c) In repeated samples, using this method, 95% of the samples will capture μ.

10.21 No, a confidence interval describes a parameter, not an individual measurement.

10.23 (a) We can be 99% confident that between 63% and 69% of all adults favor such an amendment. (b) The survey excludes people without telephones, a large percent of whom would be poor, so this group would be underrepresented. Also, Alaska and Hawaii are not included in the sample.

10.25 $z^*(\sigma/\sqrt{n}) \leq 1$, $1.645(8/\sqrt{n}) \leq 1$, so $n \geq (1.645)^2(8)^2 = 173.19$. Choose $n = 174$.

10.27 (a) SEM $= 9.3/\sqrt{27} = 1.790$. (b) Since $s/\sqrt{3} = 0.01$, $s = 0.0173$.

10.29 (a) 0.228. (b), (c), and (d):

df	$P(t > 2)$	Absolute difference
2	0.0917	0.0689
10	0.0367	0.0139
30	0.0273	0.0045
50	0.0255	0.0027
100	0.0241	0.0013

(e) As the degrees of freedom increases, the area to the right of 2 under the $t(k)$ distribution gets closer to the area under the standard Normal curve to the right of 2.

10.31 (a) Methods of displaying will vary. The distribution is slightly left-skewed with mean $\bar{x} = 18.48$ and standard deviation $s = 3.12$. (b) Yes. The sample size is not large enough to use the central limit theorem for Normality. However, there are no outliers or severe skewness in the sample data that suggest the population distribution isn't Normal. (c) $\bar{x} = 18.48$, $s = 3.12$, and $n = 20$, so the standard error is 0.6977. $t^* = 2.093$, so the margin of error is 1.46. (d) (17.02, 19.94). With 95% confidence we estimate the mean gas mileage for all these cars to be between 17.02 and 19.94 mpg. (e) If these vehicles were representative of these similar vehicles and were being driven under the exact same circumstances, (i.e., from the same population), then the interval would apply, as long as the condition of Normality is met for the interval to be valid at all.

10.33 (a) Methods of displaying will vary. The distribution is roughly symmetric with mean $\bar{x} = 3.62$ and standard deviation $s = 3.055$. (b) Using df $= 30$, $t^* = 2.042$, and the interval is (2.548, 4.688). The TI calculator gives (2.552, 4.684) using df $= 33$. With 95% confidence we estimate that the mean change in reasoning score after 6 months of piano lessons for all preschool children is between 2.548 and 4.688 points. (c) No. We don't know that students were assigned at random to the groups in this study. Also, some improvement could come with increased age.

10.35 (a) Taking d = number of disruptive behaviors on moon days − number on other days, we want to estimate μ_d = the mean difference for dementia patients. We don't know how the sample was selected. If these patients aren't representative of the population of interest, we won't be able to generalize our results. The sample size is too small ($n = 15$) for the central limit theorem to apply, so we examine the sample data. The distribution is roughly symmetric with no outliers, which gives us no reason to doubt the Normality of the population of differences. We assume that these 15 difference measurements are independent. $\bar{x}_d = 2.43$, $s_d = 1.46$, $n = 15$, t^* (for df $= 14$) $= 2.145$: $2.43 \pm 2.145(1.46/\sqrt{15}) = (1.62, 3.24)$. On average, the patients have between 1.62 and 3.24 more episodes of aggressive behavior on moon days than on other days. (b) No, this is an observational study; there could be any number of reasons for increased aggressive behavior.

10.37 (a) Use df $= 9$: $t^* = 2.262$. (b) Use df $= 19$: $t^* = 2.861$. (c) Use df $= 6$: $t^* = 1.440$.

10.39 Let μ = the mean HAV angle in the indicated population of patients. The t procedure may be used (despite the outlier at 50) because n is large (close to 40), the patients were randomly selected and (it is assumed) independent. For these data, $\bar{x} = 25.42$, $s = 7.475$. Using df $= 30$, $t^* = 2.042$. The 95% C.I. is $25.42 \pm 2.042\left(\frac{7.475}{\sqrt{38}}\right) = (22.944, 27.896)$. With 95% confidence we estimate that the mean HAV angle for all patients in the indicated population is between 22.9 and 27.9 degrees.

10.41 (a) The data are slightly left-skewed, with no outliers; the t test should be reliable. (b) (0.212, 2.69). We are 90% confident that the mean increase in listening score after attending the summer institute is between 0.212 and 2.69 points. (c) No, they may have improved for a number of other reasons, for instance, by studying every night or living with families who spoke only Spanish. There was no control group for comparison, either.

10.43 Using df $= 100$, 90% C.I. $= (2.12, 2.32)$; 95% C.I. $= (2.098, 2.342)$. TI calculator gives 90% C.I. $= (2.12, 2.32)$ and 95% C.I. $= (2.099, 2.341)$ using df $= 281$. The width of the interval increases as the confidence level increases.

10.45 (a) Population is the 175 residents of Tonya's dorm; p is the proportion of all the residents who like the food. (b) $\hat{p} = 0.28$. (c) No, the population is not large enough relative to the sample.

10.47 (a) Population is all adult heterosexuals; p is the proportion of all adult heterosexuals who have had both a blood transfusion and a sexual partner from a high risk of AIDS group. (b) $\hat{p} = 0.002$. (c) No, there are only 5 or 6 "successes" in the sample.
10.49 (a) The population is all college undergraduates, and the parameter p is the proportion of them who are abstainers. (b) Example 10.15 states that the sample is an SRS. The population (all graduates) is at least 10 times the sample size of 10,904. $n\hat{p}$ and $n(1-\hat{p})$ are both at least 10, so conditions for the confidence interval are satisfied. (c) 99% C.I. for p: (0.183, 0.203). (d) With 99% confidence, we estimate that between 11.0% and 24.7% of all college undergraduates can be classified as nondrinkers.
10.51 (a) $\hat{p} = \frac{15}{84} = 0.179$.

$$\text{Standard error} = \sqrt{\frac{(0.179)(0.821)}{84}} = 0.042.$$

(b) (0.110, 0.247). With 90% confidence, we estimate that between 11.0% and 24.7% of all applicants lie about having a degree.
10.53 (a) See margins of error in table below. (b) With $n = 500$, the new margins of error are less than half their former size (in fact, they have decreased by a factor of $1/\sqrt{5} \approx 0.447$).
10.55 (a) (0.61, 0.67). With 95% confidence we estimate

p:	0.1	0.2	0.3	0.4	0.5
(a) m:	0.0588	0.0784	0.0898	0.0960	0.0980
(b) m:	0.0263	0.0351	0.0402	0.0429	0.0438

p:	0.6	0.7	0.8	0.9
(a) m:	0.0960	0.0898	0.0784	0.0588
(b) m:	0.0429	0.0402	0.0351	0.0263

that between 61% and 67% of all teens aged 13 to 17 have TVs in their rooms. (b) Not all samples will be the same, so there is some variability from sample to sample. The margin of error accounts for this variability. (c) The results may have been taken from a "phone-in" survey, so only those who were interested enough and had access to a phone responded.
10.57 (a) (0.1682, 0.2205). We are 95% confident that between 16.8% and 22.1% of all drivers would say that they had run at least one red light. (b) More than 171 respondents have run red lights. We would not expect very many people to claim they *have* run red lights when they have not, but some people will deny running red lights when they have.
10.59 (a) We do not know that the examined records came from an SRS, so we must be cautious. Both $n\hat{p}$ and $n(1-\hat{p})$ are at least 10. There are more than (10)(1711) fatally injured bicyclists in the United States. A 99% C.I. for p, the proportion of all fatally injured bicyclists aged 15 or older who tested positive for alcohol is

$$0.317 \pm 2.576\sqrt{\frac{(0.317)(0.683)}{1711}} = (0.288, 0.346).$$

With 99% confidence, we estimate that between 28.8% and 34.6% of all fatally injured bicyclists aged 15 or older would test positive for alcohol. (b) No, we do not know, for example, what percentage of cyclists who were not involved in fatal accidents had alcohol in their systems.
10.61 (a) (0.301, 0.354). (b) The nonresponse rate (3.6%) is quite small, which suggests that the results should be reliable. If we had information for the few congregations that failed to respond, our conclusions would probably not change very much. (c) Speakers and listeners probably perceive sermon length differently (just as, say, students and teachers have different perceptions of the length of a class period).
10.63 No, the data are not based on an SRS, and thus the z procedures are not reliable in this case. In particular, a voluntary response sample is typically biased.
10.65 (a) The distribution is slightly skewed to the right. (b) (223.973, 224.031). With 95% confidence, we estimate that the mean critical dimension for all auto engine crankshafts is between 223.973 and 224.031 mm. (c) In repeated samples of this size, 95% of the intervals obtained will contain the true parameter. (d) $n \geq 35$.
10.67 (a) (109.93, 119.87). With 99% confidence we estimate that the mean seated systolic blood pressure of all healthy white males is between 109.93 and 119.87 mm Hg. (b) We assume that the 27 members of the placebo group can be viewed as an SRS of the population, and that the distribution of seated systolic blood pressure in this population is Normal, or at least not too non-Normal. Since the sample size is somewhat large, the procedure should be valid as long as the data show no outliers and no strong skewness. We must assume that these 27 measurements are independent.
10.69 (a) (0.415, 0.465). With 95% confidence, we estimate that between 41.5% and 46.5% of all adults would use alternative medicine. (b) The news report should note the estimate and the margin of error (2.5%). A brief, nontechnical explanation of "95% confidence" might also be included.
10.71 (a) The distribution is roughly symmetric with three low and five high outliers. (b) It is stated that the data are from a random sample, and as the sample size is large ($n = 50$), we can use the central limit theorem to show that the sampling distribution of $\bar{x}$ is approximately Normal. These should represent 50 independent measurements since there are at least 10(50) = 500 users of commercial Internet services. (c) (19.08, 22.72) with df = 40; TI calculator gives (19.09, 22.71) with df = 49. With 90% confidence, we estimate the mean cost for all users of commercial Internet service to be between \$19.08 and \$22.72.
10.73 (a) (0.390, 0.450). With 99% confidence, we estimate that between 39% and 45% of all adults attended church or synagogue within the last 7 days. (b) $n \geq 16{,}590$. The use of $p^* = 0.5$ is reasonable because our confidence interval shows that the actual p is in the range 0.3 to 0.7.

Chapter 11

11.1 (a) μ = the mean score for all older students at this college. (b) N(115, 6). (c) Assuming H_0 is true, observing a value like 118.6 would not be surprising, but 125.7 is less likely and therefore provides more evidence against H_0. (d) Yes, the sample size is not large enough ($n = 25$) to use the central limit theorem for Normality. (e) No, the older students at this college may not be representative of older students at other colleges in the USA.

11.3 (a) H_0: $\mu = 115$; H_a: $\mu > 115$. (b) H_0: $\mu = 12$; H_a: $\mu < 12$.

11.5 (a) p = proportion of calls involving life-threatening injuries for which the paramedics arrived within 8 minutes. H_0: $p = 0.78$; H_a: $p > 0.78$. (b) μ = mean percent of local household food expenditures used for restaurant meals. H_0: $\mu = 30$; H_a: $\mu \neq 30$.

11.7 (a) Because the workers were chosen without replacement, randomly sampled from the assembly workers, and then randomly assigned to each group, the requirements of SRS and independence are met. The question states that the differences in job satisfaction follow a Normal distribution. (b) Yes, because the sample size ($n = 18$) is too small for the central limit theorem to apply.

11.9 (a) $z = 0.6$ and P-value $= 0.2743$. (b) $z = 1.78$ and P-value $= 0.0375$. (c) If μ really was 115, the probability of getting a sample mean of 118.6 or something more extreme by chance is 0.2743; that of getting a sample mean of 125.8 or something more exterme by chance is 0.0375—much more unlikely.

11.11 (a) $\bar{x} = 398$. (b) It is Normal because the population distribution is Normal. We must assume independence, and the three chosen weeks can be considered as a representative sample of all the weeks after the price is reduced. (c) $z = 2.31$ and P-value $= 0.0104$. (d) It is significant at $\alpha = 0.05$ but not at $\alpha = 0.01$. This is pretty convincing evidence against H_0.

11.13 Significance at the 1% level means that the P-value for the test is less than 0.01. So it must also be less than 0.05. A result that is significant at the 5% level, by contrast, may or may not be significant at the 1% level.

11.15 (a) Reject H_0 if $z > 1.645$. (b) Reject H_0 if $|z| > 1.96$. (c) For tests at a fixed significance level (α), we reject H_0 when we observe values of our statistic that are so extreme (far from the mean or other "center" of the sampling distribution) that they would rarely occur when H_0 is true. (Specifically, they occur with probability no greater than α.) For a two-sided alternative, we split the rejection region—this set of extreme values—into two pieces. With a one-sided alternative, all the extreme values are in one piece, which is twice as large (in area) as either of the two pieces used for the two-sided test.

11.17 The command `rand(100)`$\rightarrow L_1$ generates 100 random numbers in the interval (0, 1) and stores them in list L_1. The answers will vary but an example gave $\bar{x} = 0.4851$, $z = -0.516$, P-value $= 0.603$. There is no evidence to suggest that the mean of the random numbers generated is different from 0.5.

11.19 (a) (1) Take a random sample of several apartments and measure the area of each of them. (2) H_0: $\mu = 1250$ vs. H_a: $\mu < 1250$. (b) (1) Take a random sample of service calls over the year and find out how long the response time was on each call. (2) H_0: $\mu = 1.8$ vs. H_a: $\mu \neq 1.8$. (c) (1) Take a random sample of students from his school and find the proportion of lefties. (2) H_0: $p = 0.12$ vs. H_a: $p \neq 0.12$.

11.21 (a) For a two-sided alternative, z is statistically significant at $\alpha = 0.005$ if $|z| > 2.81$. (b) For a one-sided alternative (on the positive side), z is statistically significant at $\alpha = 0.005$ if $z > 2.575$.

11.23 (a) If the population mean really is 15, there's about an 8% chance of getting a sample mean as far from or even farther from 15 as we did in this sample. (b) Fail to reject H_0 at $\alpha = 0.05$. (c) The probability that you are wrong is either 0 or 1, depending on the true value of μ.

11.25 SRS is necessary to ensure generalizability; Normality is required to perform calculations using z; independence is required for the standard deviation of the sampling distribution of $\bar{x}$ to be accurate.

11.27 We test H_0: $\mu = 5$ vs. H_a: $\mu < 5$, where μ = the mean dissolved oxygen content in the stream. It is stated in the question that the samples are randomly chosen from locations without replacement, which also satisfies independence. The sample size is large ($n = 45$), so the central limit theorem can be used for Normality. The test statistic is

$$z = \frac{4.62 - 5}{0.92/\sqrt{45}} = -2.77, \text{ and the } P\text{-value is } P(Z < -2.77)$$

$= 0.0028$. Because the P-value is small, there is very strong evidence that the mean dissolved oxygen content in the stream is less than 5 mg.

11.29 (a) The data are slightly skewed to the right but there are no outliers. The mean of 11.516 is slightly larger than the median of 11.501 and the sample standard deviation of 0.095 is smaller than the population standard deviation of 0.2. 65% of the sample observations are between $\bar{x} \pm 1s$, and 95% of the observations are between $\bar{x} \pm 2s$. The sample data could have come from a Normal population. (b) H_0: $\mu = 11.5$ vs. H_a: $\mu \neq 11.5$, where μ = the mean target value of tablet hardness. The sample was randomly selected (stated) without replacement, and the question of Normality is discussed in part (a). The conditions for a one-sample, two-sided, z test of means are met. $z = 0.367$. P-value $= 0.7114$. Because the P-value is fairly large, we do not reject H_0. This is reasonable variation when the null hypothesis is true, so there is little evidence to suggest that the mean hardness of the tablets is not meeting the target value.

11.31 (a) Yes: P-value $= 0.06$ indicates that the results observed are not significant at the 5% level, so the 95% confidence interval will include 10. (b) No: Because P-value < 0.1, we can reject H_0: $\mu = 10$ at the 10% level. The 90%

confidence interval would include only those values a for which we could *not* reject H_0: $\mu = a$ at the 10% level.
11.33 (a) (99.86, 108.41). With 90% confidence, we estimate that the mean reading of all radon detectors exposed to 105 picocuries of radon is between 99.86 and 108.41 picocuries. (b) Because 105 falls in this 90% confidence interval, we cannot reject H_0: $\mu = 105$ in favor of H_a: $\mu \neq 105$.
11.35 (a) Sample may not be representative, as the women have taken themselves to the clinic. Normality should be OK due to the large sample size ($n = 160$). We would have to assume independent measurements. (b) H_0: $\mu = 9.5$ vs. H_a: $\mu \neq 9.5$. $z = 2.21$, so the P-value = 0.0272. Reject H_0 at $\alpha = 0.05$ and conclude that the mean calcium level in healthy, pregnant Guatemalan women likely differs from 9.5 grams per deciliter. (c) (9.508, 9.632). With 95% confidence, we estimate that the mean blood calcium of all healthy pregnant women in Guatemala at their first visit is between 9.508 and 9.632 grams per deciliter.
11.37 (a) Yes, 30 is in the 95% confidence interval, because P-value = 0.09 means that we would not reject H_0 at $\alpha = 0.05$. (b) No, 30 is not in the 90% confidence interval, because we would reject H_0 at $\alpha = 0.10$.
11.39 Yes. H_0: $\mu = 450$ vs. H_a: $\mu > 450$; $z = 2.46$; P-value = 0.0069.
11.41 (a) Yes. (b) Not significant for $\bar{x} \leq 0.5$ but significant for $\bar{x} \geq 0.6$. (c) Significant for $\bar{x} \geq 0.8$.
11.43 (a) $z = 1.64$; not significant at 5% level (P-value = 0.0505). (b) $z = 1.65$; significant at 5% level (P-value = 0.0495).
11.45 $n = 100$: (496.24, 547.76). $n = 1000$: (513.85, 530.15). $n = 10{,}000$: (519.42, 524.58).
11.47 (a) No, in a sample of size 500, we expect to see about 5 people who do better than random guessing, with a significance level of 0.01. These four might have ESP, or they may simply be among the "lucky" ones we expect to see. (b) The researcher should repeat the procedure on these four to see if they again perform well.
11.49 (a) H_0: $p = 0.75$, H_a: $p > 0.75$. (b) Type I: if the manager decided that they were responding to more than 75% of the calls within 8 minutes when in fact they were responding to less than 75% within that time. Type II: if the manager decided that they were responding to 75% or less of the calls within 8 minutes, when in fact they were responding to more than 75% in that time. (c) Type I: may be satisfied with response time and see no need to improve time, when they should. Type II: try and make them even faster when there would be no need. (d) Type I. (e) Students can give either answer with an appropriate defense.
11.51 (a) Type I: you decide the mean sales for the new catalog will be more than \$40 when in fact it is not. (b) Type II: you decide that the mean sales for the new catalog will be \$40 (or less) when it turns out to be more than \$40. (c) Type I. (d) $\alpha = 0.01$, $\beta = 0.379$. (e) Critical value of $\bar{x}$.
11.53 (a) H_0: $\mu = \$85{,}000$ vs. H_a: $\mu > \$85{,}000$, where μ = the mean income of those living near the restaurant. (b) Type I error: concluding that the local mean income exceeds \$85,000 when in fact it does not (and, therefore, opening your restaurant in a locale that will not support it). Type II error: concluding that the local mean income does not exceed \$85,000 when in fact it does (and, therefore, deciding not to open your restaurant in a locale that would support it). (c) Type I error is more serious. If you opened your restaurant in an inappropriate area, then you would sustain a financial loss before you recognized the mistake. If you failed to open your restaurant in an appropriate area, then you would miss out on an opportunity to earn a profit there, but you would not necessarily lose money (e.g., if you chose another appropriate location in its place). (d) $\alpha = 0.01$ would be most appropriate because it would minimize your probability of committing a Type I error. (e) When $\mu = 87{,}000$, the probability of committing a Type II error is 0.3078, or about 31%.
11.55 The power of this study is far lower than what is generally desired; for example, it is well below the "80% standard" mentioned in the text. Twenty percent power for the specified effect means that, if the effect is present, we will detect it only 20% of the time. With such a small chance of detecting an important difference, the study should probably not be run (unless the sample size is increased to give sufficiently high power).
11.57 (a) For both p_0 and p_1, each probability is between 0 and 1 and the sum of the probabilities for each distribution is 1. (b) P(Type I error) = $P(X = 0 \text{ or } X = 1$ when the distribution is $p_0) = 0.2$. (c) P(Type II error) = $P(X > 1$ when the distribution is $p_1) = 0.5$.
11.59 A larger sample gives more information and therefore gives a better chance of detecting a given alternative; that is, larger samples give more power.
11.61 (a) H_0: $\mu = 100$; H_a: $\mu > 100$. (b) Type I: assumes that the weight is too heavy when in reality it is OK. Captain Ben decides not to fly, even though it is safe to do so, which inconveniences the passengers. Type II: assumes that the weight is OK when in reality it is too heavy. Captain Ben takes off in unsafe conditions, and risks having to make an emergency landing. (c) 0.10. (d) Sample is a large proportion of population. Is population Normally distributed?
11.63 Finding something to be "statistically significant" is not really useful unless the significance level is sufficiently small. While there is some freedom to decide what "sufficiently small" means, $\alpha = 0.5$ would lead your team to incorrectly reject H_0 half the time, so it is clearly a bad choice. (This approach would be essentially equivalent to flipping a coin to make your decision!)
11.65 (a) H_0: $\mu = \$72{,}500$; H_a: $\mu > \$72{,}500$. (b) H_0: $p = 0.75$; H_a: $p < 0.75$. (c) H_0: $\mu = 20$ seconds; H_a: $\mu < 20$ seconds.
11.67 Because the rejection region for a one-sided test is all in one tail, a level α one-sided significance test rejects a hypothesis H_0: $\mu = \mu_0$ exactly when the value μ_0 falls outside a level $(1-\alpha)/2$ confidence interval for μ. (a) Yes,15 is

in the 99% confidence interval because P-value $= 0.02$ means that we would not reject H_0 at $\alpha = 0.005$. **(b)** No, 15 is not in the 95% confidence interval because we would reject H_0 at $\alpha = 0.025$.

11.69 For testing H_0: $\mu = 150$ min vs. H_a: $\mu < 150$ min, the test statistic is $z = -3.28$ and the P-value is 0.0005. This is very strong evidence that students study less than an average of 2.5 hours per night.

11.71 **(a)** Margin of error decreases. **(b)** The P-value decreases. **(c)** The power increases.

11.73 **(a)** H_0: $\mu = 300$ vs. H_a: $\mu < 300$. **(b)** Type I error: concluding that the company's claim is exaggerated ($\mu < 300$) when in fact it is legitimate ($\mu = 300$). Type II error: concluding that the company's claim is legitimate when in fact it is invalid. Type II error would be more serious in this case, because allowing the company to continue the "false advertising" of its chairs' strength could lead to injuries, lawsuits, and other serious consequences. **(c)** All values at or below 295.495 pounds would cause us to reject H_0. **(d)** 0.022. **(e)** Increase the sample size or increase the level of significance.

Chapter 12

12.1 **(a)** 2.015. **(b)** 2.518.

12.3 **(a)** 14. **(b)** 1.82 is between 1.761 ($p = 0.05$) and 2.145 ($p = 0.025$). **(c)** The P-value is between 0.025 and 0.05. **(d)** $t = 1.82$ is significant at $\alpha = 0.05$ but not at $\alpha = 0.01$.

12.5 **(a)** H_0: $\mu = 1200$ vs. H_a: $\mu < 1200$, where μ = mean daily calcium intake in mg. The t procedure is justified because the sample size is large ($n = 38$) and thus the distribution of $\bar{x}$ will be approximately Normal by the central limit theorem. The value of the t statistic is -3.95. With df $= 37$, we have P-value $= 0.00017$. Because the P-value is less than the chosen $\alpha = 0.05$, we reject H_0 and conclude that the daily mean intake is significantly less than the RDA recommendation. **(b)** Without the two high outliers, $t = -6.73$ and the P-value ≈ 0

12.7 H_0: $\mu = 16$ vs. H_a: $\mu \neq 16$, $t = -5.8$ with df $= 15$, P-value < 0.0001.

12.9 **(a)** Randomly assign 12 (or 13) into a group that will use the right-hand knob first; the rest should use the left-hand knob first. Alternatively, for each student, randomly select which knob he or she should use first. **(b)** H_0: $\mu = 0$ (no difference). If $\mu = \mu_{R-L}$, H_a: $\mu < 0$. $t = -2.9037$, P-value $= 0.0039$. **(c)** A Type I error is thinking there is a difference when in fact there isn't, which could lead to creating two different instruments when it is unnecessary. A Type II error would be assuming there is no difference when in fact there is. This could lead to a slower performance time for left-handed workers, leading to unnecessary discrimination. **(d)** Power $= 0.29$ (very low!).

12.11 Let $\mu = \mu_{\text{Vanguard} - \text{managed fund}}$. Then H_0: $\mu = 0$ vs. H_a: $\mu > 0$. $t = 1.19$, $0.10 < P\text{-value} < 0.15$. This is not convincing evidence that the index fund does better.

12.13 **(a)** Yes. $H_0: \mu = 0$ vs. H_a: $\mu > 0$. $t = 43.5$; the P-value is basically 0. **(b)** (\$312.14, \$351.86). **(c)** The sample size is very large, and we are told that we have an SRS. This means that outliers are the only potential snag, and there are none. **(d)** Make the offer to an SRS of 200 customers, and choose another SRS of 200 as a control group. Compare the mean increase for the two groups.

12.15 **(a)** 9. **(b)** Between 0.05 and 0.025.

12.17 **(a)** For a t distribution with df $= 4$, the P-value is 0.0704—not significant at the 5% level. **(b)** For a t distribution with df $= 9$, the P-value is 0.0368, which is significant at the 5% level. A larger sample size means that there is less variability in the sample mean. The "tails" of the $t(4)$ distribution are slightly "heavier," which is why the P-value is larger.

12.19 **(a)** Type I error: concluding that there is a mean difference in yield of 0.5 lb/plant when in fact there is none. Type II error: concluding that no mean difference in yield of 0.5 lb/plant exists when in fact one does. Type II error is more serious. **(b)** The power is 0.5287. (Reject H_0 if $t > 1.833$, i.e., if $\bar{x} > 0.4811$.) **(c)** The power is 0.9034. (Reject H_0 if $t > 1.711$, i.e., if $\bar{x} > 0.2840$.) **(d)** Any of the following: increasing the significance level, decreasing σ, or moving the alternative farther away from μ_0.

12.21 **(a)** Letting $\mu = \mu_{\text{Haiti} - \text{factory}}$, test H_0: $\mu = 0$ vs. H_a: $\mu < 0$. $t = -4.96$, P-value < 0.0005. **(b)** $(-7.54, -3.12)$. With 95% confidence, we estimate the mean loss in vitamin C content over the 50-month period to be between 3.12 and 7.54 mg/100 g. **(c)** For H_0: $\mu = 40$ vs. H_a: $\mu \neq 40$, $t = 3.092$, and P-value is between 0.0025 and 0.005.

12.23 **(a)** No, np_0 and $n(1-p_0)$ are less than 10 (they both equal 5). **(b)** No, the expected number of failures is less than 10: $n(1-p_0) = 2$. **(c)** Yes; we have an SRS, the population is more than 10 times as large as the sample, and $np_0 = n(1-p_0) = 10$.

12.25 **(a)** H_0: $p = 0.533$ vs. H_a: $p > 0.533$. $z = 1.67$, P-value $= 0.0475$. **(b)** Type I error: concluding that Shaq has improved his free-throwing when in fact he has not. Type II error: concluding that Shaq has not improved his free-throwing when in fact he has. **(c)** 0.21. **(d)** Type I: 0.05. Type II: 0.79.

12.27 **(a)** Type I: deciding that the proportion differs from the national proportion when in fact it doesn't. This may lead to the restaurant manager investigating the reason for the difference, which could waste time and money. Type II: deciding that the proportion is the same as the national proportion when in fact it isn't. This may lead the manager to conclude that no action is needed, which may result in disgruntled employees. **(b)** Power $= 0.04$. **(c)** $n = 200$, power $= 0.102$. **(d)** 1% sig. level, power $= 0.062$. 10% significance level, power $= 0.299$.

12.29 H_0: $p = 1/3$ vs. H_a: $p > 1/3$. $z = 2.72$, P-value $= 0.0033$.

12.31 **(a)** The margin of error equals z times the standard error; for 95% confidence, we would have $z = 1.96$. **(b)** H_0 should refer to p (the population proportion), not $\hat{p}$ (the sample proportion). **(c)** The Normal distribution (and a z

test statistic) should be used for significance tests involving proportions.

12.33 **(a)** There is borderline evidence. H_0: $\mu = 0\%$ vs. H_a: $\mu \neq 0\%$. $t = 2.0028$, P-value $= 0.0522$. **(b)** Even if we had rejected H_0, this would only mean that the *average* change was nonzero; this does not guarantee that each *individual* store changed.

12.35 **(a)** H_0: $p = 0.5$ vs. H_a: $p > 0.5$. $z = 1.697$, P-value $=$ 0.0446; reject H_0 at the 5% level. **(b)** (0.5071, 0.7329). **(c)** The coffee should be presented in random order; some should get the instant coffee first, and others the fresh-brewed first.

12.37 **(a)** The distribution looks reasonably symmetric; other than the one high outlier, it appears to be nearly Normal. $\bar{x} =$ 15.59 ft, and $s = 2.550$ ft. **(b)** (14.81, 16.36). Since 20 ft does not fall in (or even near) this interval, we reject this claim. **(c)** We need to know what population we are examining: Were these all full-grown sharks? Were they all male? (i.e., is μ the mean adult male shark length? Or something else?)

Chapter 13

13.1 **(a)** We will obtain our data using counts so this is a problem about comparing proportions. **(b)** It is an observational study comparing random samples selected from two independent populations.

13.3 **(a)** Two samples. The two types of text are used by two independent groups of children. **(b)** Matched pairs. The two types of text are both used by each child.

13.5 **(a)** $H_0: \mu_t = \mu_c$ vs. $H_a: \mu_t > \mu_c$; where μ_t and μ_c are the mean improvement due to treatment and control, respectively. **(b)** The treatment group is slightly left-skewed with a greater mean and smaller standard deviation ($\bar{x}$= 51.48, $s =$ 11.01) than the control group ($\bar{x}$= 41.52, $s = 17.15$). **(c)** Randomization was not really possible, because existing classes were used; the researcher could not shuffle the students.

13.7 **(a)** Hypotheses should involve μ_1 and μ_2 (population means) rather than $\bar{x}_1$ and $\bar{x}_2$ (sample means). **(b)** The samples are not independent; we would need to compare the 10 boys to the 10 girls. **(c)** We need the P-value to be small (e.g., less than 0.05) to reject H_0. A large P-value like this gives no reason to doubt H_0.

13.9 **(a)** H_0: $\mu_t = \mu_c$ vs. H_a: $\mu_t > \mu_c$. $t = 2.311$, $0.01 <$ P-value < 0.02 with df $= 20$ (TI calculator gives P-value $=$ 0.0132 with df $= 37.86$). **(b)** (0.97, 18.94) with df $= 20$; (1.233, 18.68) on TI calculator with df $= 37.86$. With 95% confidence, we estimate the mean improvement in reading ability using the new reading activities compared to not using them over an 8-week period to be between 1.23 and 18.68 points.

13.11 **(a)** Because the sample sizes are so large, the t procedures are robust against non-Normality in the populations. **(b)** (\$412.68, \$635.58) using df $= 100$; (\$413.54, \$634.72) using df $= 620$; (\$413.62, \$634.64) using df $= 1249.21$. **(c)** The sample is not really random, but there is no reason to expect that the method used should introduce any bias into the sample. **(d)** Students without employment were excluded, so the survey results can only (possibly) extend to employed undergraduates. Knowing the number of unreturned questionnaires would also be useful.

13.13 **(a)** Test H_0: $\mu_{RED} = \mu_{WHITE}$ vs. H_a: $\mu_{RED} > \mu_{WHITE}$. $t = 3.81$ with df $= 8$ gives a P-value of between 0.0025 and 0.005. **(b)** With the same test statistic, but with df $= 14.97$, the P-value is now 0.00085. The exact degrees of freedom give a smaller and less conservative P-value. **(c)** This study appears to have been a well-designed experiment, so it does provide evidence of causation. However, the sample size is fairly small, so assessing whether the sample is from a Normal population is not easy.

13.15 **(a)** $H_0: \mu_{skilled} = \mu_{novice}$ vs. $H_a: \mu_s > \mu_n$. $t = 3.1583$, P-value $= 0.0052$. **(b)** (0.4982, 1.8475). With 90% confidence, we estimate that skilled female rowers have a mean angular knee velocity of between 0.498 and 1.847 units higher than that of novice female rowers. (c) It would be wider.

13.17 **(a)** Two-sample t test. **(b)** Paired t test. **(c)** Paired t test. **(d)** Two-sample t test. **(e)** Paired t test.

13.19 **(a)** $t = -4.65$. With either df $= 9$ or df $= 11.04$, we have a significant result (P-value < 0.001 or P-value $<$ 0.0005, respectively), so there is strong evidence that IDX prolongs life. **(b)** If using df $= 9$, the interval is (14.12, 40.88). With 95% confidence we estimate that IDX hamsters live, on average, between 14.12 and 40.88 days longer than untreated hamsters. If using df $= 11.05$, the interval is (14.49, 40.51).

13.21 Test H_0: $\mu_C = \mu_S$ vs. H_a: $\mu_C \neq \mu_S$. $t = -3.74$ and the P-value is between 0.01 and 0.02 (df $= 5$) or 0.0033 (df $=$ 10.95), agreeing with the stated conclusion.

13.23 Answers will vary, but as an example: The difference between average female (55.5) and male (57.9) self-concept scores was so small that it can be attributed to chance variation in the samples ($t = -0.83$, df $= 62.8$, P-value $=$ 0.4110). In other words, based on this sample, we have no evidence that mean self-concept scores differ by gender.

13.25 $\hat{p}_1 = 54/72 = 0.75$ and $\hat{p}_2 = 10/17 = 0.5882$. Then SE $= 0.1298$, so the 90% confidence interval is $(0.75 - 0.5882) \pm (1.645)(0.1298) = (-0.0518, 0.3753)$. With 90% confidence, we estimate that the proportion of mice ready to breed in the good acorn years is between 5.2% lower and 37.5% higher than in the bad years. These methods can be used because the populations of mice are certainly more than 10 times as large as the samples, and the counts of successes and failures are at least 5 in both samples.

13.27 **(a)** (0.4315, 0.4486). With 95% confidence, we estimate that the proportion of cars that go faster than 65 mph when no radar is present is between 43.15% and 44.86%. **(b)** (0.102, 0.138). With 95% confidence, we estimate that the proportion of cars going over 65 mph is between 10.2% and

13.8% higher when no radar is present compared to when radar is present. **(c)** In a cluster of cars, where one driver's behavior might affect the others, we do not have independence, one of the important properties of a random sample.
13.29 H_0: $p_1 = p_2$ vs. H_a: $p_1 < p_2$. $z = -3.53$, P-value = 0.0002.
13.31 PC access at home: H_0: $p_B = p_W$ vs. H_a: $p_B \neq p_W$. $z = 1.01$ and P-value = 0.3124. With the same hypotheses for the proportions with PC access at work, $z = 3.90$ and P-value < 0.0004, so we have very strong evidence of a difference in the proportions who have computer access at work.
13.33 **(a)** H_0 should refer to p_1 and p_2 **(b)** Confidence intervals account only for sampling error.
13.35 **(a)** H_0: $p_1 = p_2$ vs. H_a: $p_1 > p_2$, where p_1 is the proportion of all HIV patients taking a placebo who develop AIDS and p_2 is the proportion of all HIV patients taking AZT who develop AIDS. The populations are much larger than the samples, $n_1\hat{p}_c$,$n_1(1-\hat{p}_c)$, $n_2\hat{p}_c$,$n_2(1-\hat{p}_c)$ are all at least 5. **(b)** $\hat{p}_1 = 38/435 = 0.0874$, $\hat{p}_2 = 17/435 = 0.0391$ and $\hat{p}_c = 0.0632$. $z = 2.926$, giving a P-value of 0.0017; a significantly smaller proportion of patients who took AZT developed AIDS than those taking a placebo. **(c)** Neither the subjects nor the researchers who had contact with them knew which subjects were getting which drug; if anyone had known, they might have affected the outcome by letting their expectations or biases affect the results.
13.37 **(a)** The proportion of orders completed in 5 days or less before the changes was (0.1092, 0.2108). **(b)** After the changes, the 95% confidence interval is (0.8584, 0.9416). **(c)** 95% confidence interval for $p_{after} - p_{before}$ is (0.6743, 0.8057), or about a 67.4% to 80.6% improvement. No, it is not directly related; simply to subtract the upper limits and the lower limits from (a) and (b) would not give the interval in (c).
13.39 H_0: $p_1 = p_2$ vs. H_a: $p_1 \neq p_2$. From the output, $z = -3.45$ and P-value = 0.0006, showing a significant difference in the proportion of children in the two age groups who sorted products correctly. A 95% confidence interval for $p_1 - p_2$ is $(-0.5025279, -0.15407588)$. With 95% confidence we estimate that between 15.4% and 50.3% more 6- to 7-year-olds can sort new products into the correct product category than 4- to 5-year-olds.
13.41 **(a)** This is a two-sample t test; the two groups of women are (presumably) independent. **(b)** df = 44. **(c)** The sample sizes are large enough that non-Normality has little effect on the reliability of the procedure.
13.43 H_0: $p_H = p_W$ vs. H_a: $p_H \neq p_W$; $z = -0.86$, P-value = 0.3898; 95% CI: $(-0.1018, 0.0398)$. No significant difference.
13.45 **(a)** (0.2465, 0.3359). Yes, because the confidence interval doesn't contain 0. **(b)** H_0: $\mu_M = \mu_W$ vs. H_a: $\mu_M \neq \mu_W$; $t = -0.8658$, P-value close to 0.4. No.
13.47 **(a)** (1016.56, 1069.44). **(b)** Skewness will have little effect because the sample sizes are very large.
13.49 **(a)** $\hat{p}_C = 17/283 = 0.0601$ and $\hat{p}_B = 35/165 = 0.2121$. $\hat{p}_C - \hat{p}_B = 0.1520$. $SE_{diff} = 0.03482$. **(b)** H_0: $p_s = p_B$ vs. H_a: $p_C < p_B$. The alternative reflects the reasonable expectation that reducing pollution might decrease wheezing. **(c)** $z = 4.85$. The P-value is very small ($P < 0.0001$). **(d)** $(-0.2203, -0.0838)$. The percent reporting improvement was between 8% and 22% higher for bypass residents. **(e)** There may be geographic factors (e.g., weather) or cultural factors (e.g., diet) that limit how much we can generalize the conclusions.

Chapter 14

14.1 **(a)** (i) 0.20 < P-value < 0.25. (ii) P-value = 0.235. **(b)** (i) 0.02 < P-value < 0.025. (ii) P-value = 0.0204. **(c)** (i) P-value > 0.25. (ii) P-value = 0.3172.
14.3 **(a)** "A": 24.2%, "B": 41.8%, "C": 22.0%, "D/F": 12.1%. Fewer A's and more D/F's than the TA sections. **(b)** "A": 29.12, "B": 37.31, "C": 18.20, "D/F": 6.37. **(c)** H_0: $p_1 = 0.32$, $p_1 = 0.41$, $p_1 = 0.20$, $p_1 = 0.07$ vs. H_a: at least one of these proportions is different. All the expected counts are greater than 5, so the condition for X^2 is satisfied. $X^2 = 5.297$ (df = 3), so the P-value = 0.1513; there is not enough evidence to conclude that the professor's grade distribution was different from the TA grade distribution.
14.5 H_0: The genetic model is valid (the different colors occur in the stated ratio of 1:2:1). H_a: The genetic model is not valid. Expected counts: 21GG, 42Gg, 21gg. $X^2 = 5.43$, df = 2, P-value = 0.0662.
14.7 Answers will vary.
14.9 **(a)** H_0: $p_0 = p_1 = \ldots = p_9 = 0.1$ vs. H_a: at least one of the p_i's is not equal to 0.1. **(b)** and **(c)** Answers will vary. Using `randInt(0,9,200)` →L_4, we obtained these counts for digits 0 to 9: 19, 17, 23, 22, 19, 20, 25, 12, 27, 16. **(c)** Answers will vary: for example, 19, 17, 23, 22, 19, 20, 25, 12, 27, 16. **(d)** Expected counts should all be 20. **(e)** For example, $X^2 = 8.9$, df = 9, P-value = 0.447. There is no evidence that the sample data were generated from a distribution that is different from the uniform distribution.
14.11 **(a)** See part (e). **(b)** $\hat{p}_N = \frac{40}{244} = 0.1639$ $\hat{p}_D = \frac{74}{244} = 0.3033$ $\hat{p}_{N+D} = \frac{87}{245} = 0.3551$, $\hat{p}_P = \frac{25}{160} = 0.15625$ **(d)** The success rate (proportion of those who quit) is the same for all four treatments. **(e)** Observed counts with Expected counts in parentheses: Patch: Successes 40 (61.75), Failures 204 (182.25). Drug: Successes 74 (61.75), Failures 170 (182.25). Patch + Drug: Successes 87 (62), Failures 158 (183). Placebo: Successes 25 (40.49), Failures 135 (119.51). **(f)** A higher percent than expected of the "patch plus drug" and "drug" subjects successfully quit, and a lower percent than expected of the other two groups quit. This reflects the fact that the "patch plus drug" and "drug" success rates were considerably higher than the success rates for the other two groups, as seen in (b) and (c).
14.13 **(a)** and **(d)** Observed counts with Expected counts in parentheses: Placebo: strokes 250 (205.81), no strokes 1399 (1443.19). Aspirin: strokes 206 (205.81), no strokes 1443 (1443.19). Dipyridamole: strokes 211 (206.44), no strokes

1443 (1447.56). Both: strokes 157 (205.94), no strokes 1493 (1444.06). **(b)** Taking both medications seems to be the most effective in preventing strokes. **(c)** The effectiveness of the treatment (the proportion of patients who have strokes) is the same for all four treatments.

14.15 **(a)** Reading row by row, the X^2 components are 7.662, 2.596, 2.430, 0.823, 10.076, 3.414, 5.928, and 2.008. $X^2 = 34.937$ (df = 3). **(b)** According to Table D, P-value < 0.0005. A P-value of this size indicates that it is extremely unlikely that such a result occurred due to chance; it represents very strong evidence against H_0. **(c)** 10.076, corresponding to the "patch plus drug/success" category, is the largest contributor. This is not surprising because, according to Exercise 14.11(e), the "patch plus drug" group contains a higher than expected number of successful quitters. **(d)** Treatment is strongly associated with success; specifically, the drug, or the patch together with the drug, seem to be most effective. **(e)** The X^2-value and conclusion are the same; the P-value is given more accurately, as 0.00000013.

14.17 **(a)** Reading row by row, the X^2 components are 3.211, 3.211, 2.420, 2.420, 4.923, 4.923, 1.895, and 1.895. $X^2 = 24.898$ (df=3). **(b)** From Table D, P-value < 0.0005. A P-value of this size indicates that it is extremely unlikely that such a result occurred due to chance; it represents very strong evidence against H_0. **(c)** The terms corresponding to HSC-HM and LSC-HM (for both sexes) provide the largest contributions to X^2. This reflects the fact that males are more likely to have "winning" (social comparison) as a goal, while females are more concerned with "mastery." **(d)** The terms and results are identical. The P-value of 0.000 in the Minitab output reflects the fact that the true P-value in part (b) was actually considerably smaller than 0.0005.

14.19 **(a)** $X^2 = 24.243$ with df=3, a very significant result ($P < 0.0005$). The largest contributions to the X^2 statistic come from Stroke/Placebo. Subjects taking a placebo had many more strokes than expected, while those taking both drugs had fewer strokes. **(b)** $X^2 = 1.418$ with df=3; P-value = 0.701. The combination of both drugs is effective at decreasing the risk of stroke, but no drug treatment had a significant impact on death rate.

14.21 **(a)** $r = 2$, $c = 3$. **(b)** 55.0%, 74.7%, and 37.5%. Some (but not too much) time spent in extracurricular activities seems to be beneficial. **(d)** H_0: There is no association between amount of time spent on extracurricular activities and grades earned in the course. H_a: There is an association. **(e)** C or better: 13.78, 62.71, 5.51; D or F: 6.22, 28.29, 2.49. **(f)** The first and last columns in the observed data have lower numbers in the "passing" row (and higher numbers in the "failing" row) than we would have expected, while the middle column has this reversed—more passed than we would have expected if the proportions were all equal.

14.23 **(a)** Missing entries in table of expected counts (row by row): 62.71, 5.51, 6.22. Missing entries in components of X^2: 0.447, 0.991. **(b)** df = 2, P-value is between 0.025 and 0.05. Rejecting H_0 means that we conclude that there is a relationship between hours spent in extracurricular activities and performance in the course. **(c)** The highest contribution comes from column 3, row 2 (">12 hours of extracurricular activities, D or F in the course"). Too much time spent on these activities seems to hurt academic performance. **(d)** No, this study demonstrates association, not causation. Certain types of students may tend to spend a moderate amount of time in extracurricular activities and also work hard in their classes; one does not necessarily cause the other.

14.25 H_0: all proportions are equal vs. H_a: at least one proportion is different. All observed counts are greater than 5, so we may proceed with a χ^2 analysis. X^2=10.619 with df = 2; P-value = 0.0049. There is good evidence that contact method makes a difference in response.

14.27 **(a)** Cardiac event: 3, 7, 12. No cardiac event: 30, 27, 28. **(b)** Success rates (% of No's): 90.91%, 79.41%, 70%. **(c)** Expected cell counts, cardiac event: 6.785, 6.991, 8.224. No cardiac event: 26.215, 27.009, 31.776. All expected cell counts exceed 5, so the chi-square test can be used. **(d)** $X^2 = 4.84$ with df=2; P-value = 0.0889. Though the success rate for the stress management group is slightly higher than for the other two groups, there does not appear to be a significant difference among the success rates.

14.29 H_0: There is no difference between the proportions of living places of males and females vs. H_a: At least one male/female proportion is different. All expected counts > 5. $X^2 = 11.038$ (df = 3) and P-value = 0.0115. The choices of living spaces are significantly different for males and females.

14.31 **(a)** This is not an experiment; no treatment was assigned to the subjects. **(b)** A high nonresponse rate might mean that our attempt to get a random sample was thwarted because of those who did not participate; this nonresponse rate is extraordinarily low. **(c)** H_0: There is no association between olive oil consumption and cancer. H_a: There is an association between olive oil consumption and cancer. $X^2 = 1.552$; df = 4, P-value = 0.8174.

14.33 **(a)** H_0: $p_1 = p_2$, $\hat{p}_1 = 203/241 = 0.8423$, and $\hat{p}_2 = 150/218 = 0.6881$. $z = 3.9159$, P-value = 0.00009. **(b)** $X^2 = 15.334$, which equals z^2. With df = 1, Table D tells us that P-value < 0.0005; a statistical calculator gives $P = 0.00009$. **(c)** (0.0774, 0.2311).

14.35 The observed frequencies, their marginal percents, and the expected values for the sample are score "5": 167, 31.2, 81.855; score "4": 158, 29.5, 117.7; score "3": 101, 18.9, 132.68; score "2": 79, 14.8, 105.93, score "1": 30, 5.6, 96.835. H_0: The distribution of scores in this sample is the same as the distribution of scores for all students who took this inaugural exam. H_a: The distribution of scores in this sample is different from the national results. All expected counts are greater than 5. $X^2 = 162.9$, df = 4. The P-value is ≈ 0.0000. The distribution of AP Statistics exam scores in this sample is different from the national distribution.

14.37 H_0: The survey results match the college population. H_a: The survey results do not match the college population. All expected counts are greater than 5. $X^2 = 5.016$, *P*-value $= 0.1706$.

14.39 (a) Cold: Hatched 16, Not hatched 11. Neutral: Hatched 38, Not hatched 18. Hot: Hatched 75, Not hatched 29. (b) Cold 59.3%; Neutral 67.9%; Hot 72.1%. The percent hatching increases with temperature; the cold water did not prevent hatching, but made it less likely. (c) All expected counts are greater than 5. A test of H_0: $p_C = p_N = p_H$ vs. H_a: at least one p_i is different yields $X^2 = 1.703$ with df $= 2$ and *P*-value $= 0.427$. There is sufficient evidence that the proportion of eggs that hatched are not all the same for the three groups.

14.41 (a) No, no treatment was imposed. (b) Among those who did not own a pet, 71.8% survived, while 94.3% of pet owners survived. Overall, 84.8% of the patients survived. It appears that you are more likely to survive CHD if you own a pet! (c) H_0: There is no relationship between patient status and pet ownership. H_a: There is a relationship between patient status and pet ownership. $X^2 = 8.851$, df $= 1$, *P*-value $= 0.003$. (d) Provided there are no confounding variables, there is strong evidence to suggest that owning a pet improves survival. (e) We used a X^2 test. In a z test, we would test H_0: $p_1 = p_2$ vs. H_a: $p_1 < p_2$. For this test, $z = -2.975$ and *P*-value $= 0.0015$. The *P*-value is half that obtained in (c). The z test enables us to use a more specific one-tailed test but if we were just interested in whether pet ownership made a *difference* to survival rate (a two-tailed test) and not just *improved* survival rates (a one-tailed test), then it wouldn't matter which test we used.

14.43 (a) Percent of sexual ads: Women 39.06%, Men 16.96%, General 21.02%. (b) $X^2 = 80.9$, df $= 2$, and the *P*-value is almost 0. (c) The sample is not an SRS: a set of magazines was chosen, and then all ads in three issues of those magazines were examined. It is not clear how this sampling approach might invalidate our conclusions, but it does make them suspect. The students may discuss the ads as a group, which may affect their opinion in one way or another.

Chapter 15

15.1 The correlation is $r = 0.994$, and the least-squares linear regression equation is $\hat{y} = -3.660 + 1.1969x$, where $y =$ humerus length and $x =$ femur length. The scatterplot shows a strong, positive, linear relationship, all of which suggests that femur length will be a good prediction of humerus length.

15.3 (a) The observations are independent because they come from 13 unrelated colonies. (b) The scatterplot of the residuals against the percent returning shows no obvious pattern. (c) The spread may be slightly wider in the middle, but not markedly so. (d) A stemplot (equivalent to a histogram) shows no clear deviation from Normality.

15.5 (a) β represents how much we can expect the humerus length to increase when femur length increases by 1 cm. (b) The estimate of β is $b = 1.1969$, and the estimate of α is $a = -3.660$. (c) The residuals are -0.8226, -0.3668, 3.0425, -0.9420, and -0.9110; the sum is -0.0001; $s = \sqrt{\dfrac{\Sigma(\text{resid}^2)}{n-2}} = \sqrt{\dfrac{11.79}{3}} = 1.982$.

15.7 (a) The plot shows a strong, positive, linear relationship. (b) β (the slope) is this rate; the estimate in this model is listed as the coefficient of "year": $b = 9.31868$. (c) The observations are not independent, consequently we can't perform inference.

15.9 The regression equation is $\hat{y} = 560.65 - 3.0771x$, where $y =$ calories and $x =$ time; this and the plot show that, generally, the longer a child remains at the table, the fewer calories he or she will consume. The conditions for regression inference are satisfied. 95% confidence interval is $(-4.8625, -1.2917)$. With 95% confidence, we estimate that for every extra minute a child sits at the table, he or she will consume an average of between 1.29 and 4.86 calories less during lunch.

15.11 (a) $\hat{y} = -3.660 + 1.1969x$, where $y =$ humerus length and $x =$ femur length. (b) $t = b/SE = 15.9374$. (c) df $= 3$; since $t > 12.92$, we know that *P*-value < 0.0005. (d) There is very strong evidence that $\beta > 0$, that is, that the line is useful for predicting the length of the humerus given the length of the femur. (e) $(0.7579, 1.6359)$. We are 99% confident that for every extra centimeter in femur length, the length of the humerus will increase between 0.7582 and 1.6356 cm.

15.13 (a) The horizontal axis is "Jet skis in use," and the vertical axis is "Accidents." There is a strong explanatory-response relationship between the number of jet skis in use (explanatory) and the number of accidents (response). (b) H_0: $\beta = 0$ (there is no association between number of jet skis in use and number of accidents). H_a: $\beta > 0$ (there is a positive association between number of jet skis in use and number of accidents). (c) The conditions are satisfied except for having independent observations. We will proceed with caution. (d) `LinRegTTest` With the earlier caveat, there is sufficient evidence to reject reports that $t = 21.079$ with df $= 8$. The *P*-value is 0.000. There is sufficient evidence to reject H_0 and conclude that there is a positive association between year and number of accidents. As the number of jet skis in use increases, the number of accidents increases. (e) $(0.00417, 0.00549)$. With 98% confidence, we estimate that for every extra thousand jet skis in use, the number of accidents increases by a mean of between 4.17 and 5.49 per year.

15.15 (a) Stumps (the explanatory variable) should be on the horizontal axis; the plot shows a positive linear association. (b) $\hat{y} = -1.286 + 11.894x$, where $y =$ number of beetle larvae and $x =$ stump counts. $r^2 = 83.9\%$, so regression on stump counts explains 83.9% of the variation in the number of beetle larvae. (c) H_0: $\beta = 0$ vs. H_a: $\beta \neq 0$, and the test statistic is $t = 10.47$ (df = 21). The output shows *P*-value = 0.000, so we have strong evidence that beaver stump counts help explain beetle larvae counts.

15.17 Regression of fuel consumption on speed gives $b = -0.01466$, $SE_b = 0.02334$, and $t = -0.63$. With df = 13, we see that *P*-value = 0.541, so we have no evidence to suggest a straight-line relationship. While the relationship between these two variables is very strong, it is definitely not linear. See solution to Exercise to 3.9.

15.19 (a) The value of β would tell us the mean percent of forest lost for every extra cent per pound increase in the price of coffee. The estimate of β is $b = 0.0552$ and the estimate for α is $a = -1.0134$. (b) That $r^2 = 0.91$ says that 91% of the total observed variation among the 5 years' data of the forest area lost is accounted for by the straight-line relationship with prices paid to coffee growers. This says that the straight-line relationship described by the least-squares line is very strong. (c) The *P*-value given is for a two-sided alternative: H_0: $\beta = 0$ vs. H_a: $\beta \neq 0$. The small *P*-value says there is very strong evidence that the population slope is positive, that is, that percent of forest loss does increase as coffee price increases. (d) $s = 0.327$. (e) Everything suggests a good model: should be a good predictor within range of given values of coffee price.

15.21 (a) $r^2 = 99.2\%$. (b) The estimates of α, β, and σ are $a = -2.3948$ cm, $b = 0.158483$ cm/min, and $s = 0.8059$ cm. (c) The regression equation is $\hat{y} = -2.39 + 0.158x$, where y = length and x = time.

15.23 (a) U.S. returns (the explanatory variable) should be on the horizontal axis. There is a positive association between investment returns in the United States and investments overseas, although it is not very strong. (b) $t = b/SE = 2.609$, df = 25; $0.01 < P\text{-value} < 0.02$. Thus, we have fairly strong evidence for a linear relationship, that is, that the slope is nonzero. (c) $r^2 = 21.4\%$. The linear relationship between the U.S. and overseas returns accounts for only about 21.4% of the variation in overseas returns—not a good indication that the regression will be useful in practice.

15.25 (a) $\hat{y} = 166.5 - 1.099x$, where y = corn yield (bushels/acre) and x = weeds per meter. $r^2 = 20.9\%$, so the linear relationship explains about 20.9%, of the variation in yield. (b) The t statistic for testing H_0: $\beta = 0$ vs. H_a: $\beta < 0$ is $t = -1.92$; with df = 14, the *P*-value is 0.0375. There is sufficient evidence to conclude that weeds influence corn yields. (c) The small number of observations for each value of the explanatory variable (weeds/meter), the large variability in those observations, and the small value of r^2 will make prediction with this model somewhat imprecise.

Index

S

T

U

V

W

X

Y

Z

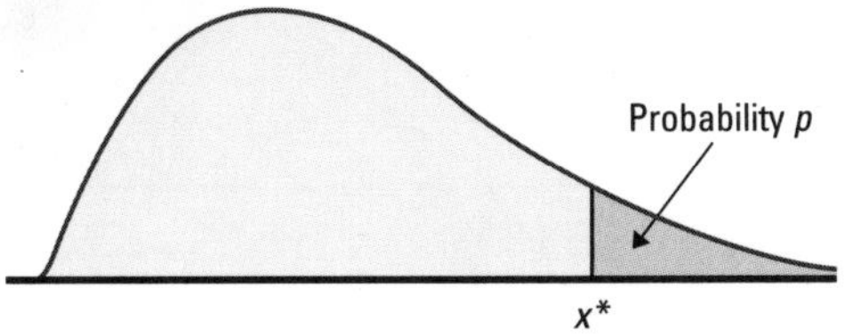

Table entry for p is the critical value x^* with probability p lying to its right.

Table D Chi-square distribution critical values

						p						
df	.25	.20	.15	.10	.05	.025	.02	.01	.005	.0025	.001	.0005
1	1.32	1.64	2.07	2.71	3.84	5.02	5.41	6.63	7.88	9.14	10.83	12.12
2	2.77	3.22	3.79	4.61	5.99	7.38	7.82	9.21	10.60	11.98	13.82	15.20
3	4.11	4.64	5.32	6.25	7.81	9.35	9.84	11.34	12.84	14.32	16.27	17.73
4	5.39	5.99	6.74	7.78	9.49	11.14	11.67	13.28	14.86	16.42	18.47	20.00
5	6.63	7.29	8.12	9.24	11.07	12.83	13.39	15.09	16.75	18.39	20.51	22.11
6	7.84	8.56	9.45	10.64	12.59	14.45	15.03	16.81	18.55	20.25	22.46	24.10
7	9.04	9.80	10.75	12.02	14.07	16.01	16.62	18.48	20.28	22.04	24.32	26.02
8	10.22	11.03	12.03	13.36	15.51	17.53	18.17	20.09	21.95	23.77	26.12	27.87
9	11.39	12.24	13.29	14.68	16.92	19.02	19.68	21.67	23.59	25.46	27.88	29.67
10	12.55	13.44	14.53	15.99	18.31	20.48	21.16	23.21	25.19	27.11	29.59	31.42
11	13.70	14.63	15.77	17.28	19.68	21.92	22.62	24.72	26.76	28.73	31.26	33.14
12	14.85	15.81	16.99	18.55	21.03	23.34	24.05	26.22	28.30	30.32	32.91	34.82
13	15.98	16.98	18.20	19.81	22.36	24.74	25.47	27.69	29.82	31.88	34.53	36.48
14	17.12	18.15	19.41	21.06	23.68	26.12	26.87	29.14	31.32	33.43	36.12	38.11
15	18.25	19.31	20.60	22.31	25.00	27.49	28.26	30.58	32.80	34.95	37.70	39.72
16	19.37	20.47	21.79	23.54	26.30	28.85	29.63	32.00	34.27	36.46	39.25	41.31
17	20.49	21.61	22.98	24.77	27.59	30.19	31.00	33.41	35.72	37.95	40.79	42.88
18	21.60	22.76	24.16	25.99	28.87	31.53	32.35	34.81	37.16	39.42	42.31	44.43
19	22.72	23.90	25.33	27.20	30.14	32.85	33.69	36.19	38.58	40.88	43.82	45.97
20	23.83	25.04	26.50	28.41	31.41	34.17	35.02	37.57	40.00	42.34	45.31	47.50
21	24.93	26.17	27.66	29.62	32.67	35.48	36.34	38.93	41.40	43.78	46.80	49.01
22	26.04	27.30	28.82	30.81	33.92	36.78	37.66	40.29	42.80	45.20	48.27	50.51
23	27.14	28.43	29.98	32.01	35.17	38.08	38.97	41.64	44.18	46.62	49.73	52.00
24	28.24	29.55	31.13	33.20	36.42	39.36	40.27	42.98	45.56	48.03	51.18	53.48
25	29.34	30.68	32.28	34.38	37.65	40.65	41.57	44.31	46.93	49.44	52.62	54.95
26	30.43	31.79	33.43	35.56	38.89	41.92	42.86	45.64	48.29	50.83	54.05	56.41
27	31.53	32.91	34.57	36.74	40.11	43.19	44.14	46.96	49.64	52.22	55.48	57.86
28	32.62	34.03	35.71	37.92	41.34	44.46	45.42	48.28	50.99	53.59	56.89	59.30
29	33.71	35.14	36.85	39.09	42.56	45.72	46.69	49.59	52.34	54.97	58.30	60.73
30	34.80	36.25	37.99	40.26	43.77	46.98	47.96	50.89	53.67	56.33	59.70	62.16
40	45.62	47.27	49.24	51.81	55.76	59.34	60.44	63.69	66.77	69.70	73.40	76.09
50	56.33	58.16	60.35	63.17	67.50	71.42	72.61	76.15	79.49	82.66	86.66	89.56
60	66.98	68.97	71.34	74.40	79.08	83.30	84.58	88.38	91.95	95.34	99.61	102.7
80	88.13	90.41	93.11	96.58	101.9	106.6	108.1	112.3	116.3	120.1	124.8	128.3
100	109.1	111.7	114.7	118.5	124.3	129.6	131.1	135.8	140.2	144.3	149.4	153.2

Table B Random digits

Line								
101	19223	95034	05756	28713	96409	12531	42544	82853
102	73676	47150	99400	01927	27754	42648	82425	36290
103	45467	71709	77558	00095	32863	29485	82226	90056
104	52711	38889	93074	60227	40011	85848	48767	52573
105	95592	94007	69971	91481	60779	53791	17297	59335
106	68417	35013	15529	72765	85089	57067	50211	47487
107	82739	57890	20807	47511	81676	55300	94383	14893
108	60940	72024	17868	24943	61790	90656	87964	18883
109	36009	19365	15412	39638	85453	46816	83485	41979
110	38448	48789	18338	24697	39364	42006	76688	08708
111	81486	69487	60513	09297	00412	71238	27649	39950
112	59636	88804	04634	71197	19352	73089	84898	45785
113	62568	70206	40325	03699	71080	22553	11486	11776
114	45149	32992	75730	66280	03819	56202	02938	70915
115	61041	77684	94322	24709	73698	14526	31893	32592
116	14459	26056	31424	80371	65103	62253	50490	61181
117	38167	98532	62183	70632	23417	26185	41448	75532
118	73190	32533	04470	29669	84407	90785	65956	86382
119	95857	07118	87664	92099	58806	66979	98624	84826
120	35476	55972	39421	65850	04266	35435	43742	11937
121	71487	09984	29077	14863	61683	47052	62224	51025
122	13873	81598	95052	90908	73592	75186	87136	95761
123	54580	81507	27102	56027	55892	33063	41842	81868
124	71035	09001	43367	49497	72719	96758	27611	91596
125	96746	12149	37823	71868	18442	35119	62103	39244
126	96927	19931	36809	74192	77567	88741	48409	41903
127	43909	99477	25330	64359	40085	16925	85117	36071
128	15689	14227	06565	14374	13352	49367	81982	87209
129	36759	58984	68288	22913	18638	54303	00795	08727
130	69051	64817	87174	09517	84534	06489	87201	97245
131	05007	16632	81194	14873	04197	85576	45195	96565
132	68732	55259	84292	08796	43165	93739	31685	97150
133	45740	41807	65561	33302	07051	93623	18132	09547
134	27816	78416	18329	21337	35213	37741	04312	68508
135	66925	55658	39100	78458	11206	19876	87151	31260
136	08421	44753	77377	28744	75592	08563	79140	92454
137	53645	66812	61421	47836	12609	15373	98481	14592
138	66831	68908	40772	21558	47781	33586	79177	06928
139	55588	99404	70708	41098	43563	56934	48394	51719
140	12975	13258	13048	45144	72321	81940	00360	02428
141	96767	35964	23822	96012	94591	65194	50842	53372
142	72829	50232	97892	63408	77919	44575	24870	04178
143	88565	42628	17797	49376	61762	16953	88604	12724
144	62964	88145	83083	69453	46109	59505	69680	00900
145	19687	12633	57857	95806	09931	02150	43163	58636
146	37609	59057	66967	83401	60705	02384	90597	93600
147	54973	86278	88737	74351	47500	84552	19909	67181
148	00694	05977	19664	65441	20903	62371	22725	53340
149	71546	05233	53946	68743	72460	27601	45403	88692
150	07511	88915	41267	16853	84569	79367	32337	03316